Visualization of Shock Wave Phenomena

Kazuyoshi Takayama

Visualization of Shock Wave Phenomena

Springer

Kazuyoshi Takayama
Sendai, Japan

ISBN 978-3-030-19450-5 ISBN 978-3-030-19451-2 (eBook)
https://doi.org/10.1007/978-3-030-19451-2

This Springer imprint is published by the registered company Springer Nature Switzerland AG
The registered company address is: Gewerbestrasse 11, 6330 Cham, Switzerland

To Chieko Takayama

Preface

This book presents mostly sequential double exposure holographic interferometric pictures of shock wave phenomena which cover from basic topics of the shock wave phenomena to their applications. If watching these sequential images, readers will readily grasp physics of shock wave motions sequentially without explicit description in words; for instance, comprehensive understanding of shock tube flows and the effect of boundary layers on the shock tube flows are well displayed through the sequential display of interferograms.

Chapter 1 describes introductory remarks. The importance of flow visualization to the shock wave phenomena in gases and liquids is emphasized.

Chapter 2 describes in detail the shock wave transitions from Mach to regular reflections and vice versa over wedges and various objects. The transitions over not only straight wedges but also cones, concave and convex double wedges, and curved concave and convex walls are discussed in detail. The design of a diaphragm-less shock tube is presented.

In Chap. 3, the diffraction from a backward facing 90° step and related topics are discussed. Three-dimensional shock wave diffraction from open ends of shock tubes is uniquely visualized by diffuse holography in detail.

In Chap. 4, the shock wave interactions with cylindrical objects are discussed. The direct measurement of unsteady drag force over a sphere is a highlight of this chapter. Chapter 5 describes the shock wave focusing on gas in horizontal and vertical annular coaxial shock tubes.

In Chap. 6, shock wave mitigations over various wall conditions are examined. Chapter 7 discusses the shock wave interaction with foreign gas interfaces. The three-dimensional formation of vortices over the interfaces was conducted in order to verify that the Richtmyer–Meshkov instability is inherently three-dimensional.

In Chap. 8, explosions in gases generated by the ignition of PbN_6 and AgN_3 pellets are described. The initiation of three-dimensional detonation waves in stoichiometric hydrogen and oxygen mixture is discussed.

In Chap. 9, as a continuation of Chap. 8, the generation of underwater shock waves is presented. Their interactions with a single air bubble, bubble cloud and water/silicone oil interfaces and their focusing from a truncated ellipsoidal cavity are presented.

In Chap. 10, as an extension of the study of underwater shock wave focusing, their applications to medicine such as an extracorporeal shock wave lithotripsy and ESWL and other applications of underwater shock wave research to medicine are presented.

Chapter 11 presents miscellaneous topics.

The photographs included in this book were taken under the collaborations with the author's students and colleagues. Many of them were not published in the open literatures. If any photographs presented in this book will be in benefit to readers, the author's efforts will have been rewarded.

Sendai, Japan
October 2018

Kazuyoshi Takayama

Acknowledgements

The author would like to acknowledge the late Professor M. Honda of the Institute of High-Speed Mechanics, Tohoku University for his guidance for the author to study high-speed gas dynamics and also to acknowledge the late Professor I. I. Glass of the Institute for Aerospace Studies, University of Toronto for encouraging the author to devote to the shock wave research. The author would like to express his heartfelt appreciations to them. A special appreciation is presented to Professor O. Igra of the Ben Gurion University of the Negev for his proofreading the manuscript.

The author expresses his thanks to Dr. O. Onodera and Mr. S. Hayasaka for their devotions to the author's experiments. The author also thank Professor T. Saito of the Muroran Institute of Technology, Dr. H. Yamamoto of Tohoku University Hospital, and Dr. K. Ohtani of the Institute of Fluid Science, Tohoku University for their assistance to the author's shock wave research. Lastly, the author acknowledges with thanks collaborations with the students and the friends who collaborated with him.

Contents

apparently transparent media, such as commercial glass plates and acrylic plates, and even in water and air exposed to a convection.

Fringes obtained in identical arrangement during the double exposures are named infinite fringes. In two-dimensional cases such as shock tube flows, the fringes are expressed with infinite width and correspond to equal-density contours. This is named infinite fringe interferometry.

When shifting and rotate the center of *RB* on FH in Fig. 1.1 either at the first or the second exposure and keeping *OB* unchanged, equal interval parallel fringes appear in reconstructed images. The orientation and the interval of fringes are determined by the degree of rotation and deviation from the initial state. In the holographic arrangement, *RB* was collimated with a 150 mm o.d. convex lens and superimposed with *OB* on FH. Figure 1.2a shows the 150 mm diameter lens, movable lens ML as illustrated in Fig. 1.1, mounted on a stage. As illustrated in Fig. 1.2b, the lens is rotated by θ and is swung by ε at either the first or second exposures. When adjusting ε and θ, the interval and orientation of fringes are arbitrarily selected. Changes of phase angles during the double exposures are expressed by the deviation of fringes from their regular intervals. This is named as a finite fringe interferogram. It was not easy to measure the deformation of fringe distributions. However, today, the deviation of fringe distribution from its initial position are relatively easily determined by using a computer-assisted image processing system (Houwing 2005).

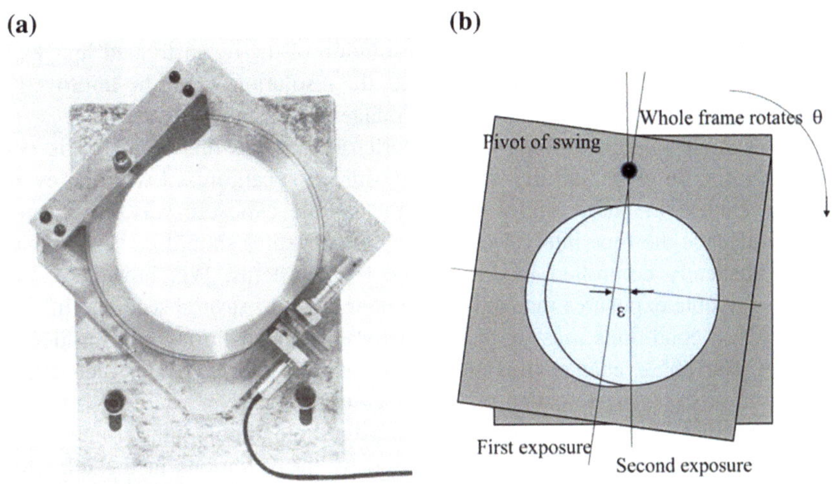

Fig. 1.2 Movable lens mount: **a** a movable lens mount; **b** an illustration of movable mount Takayama (1983)

In order to collimate *OB*, we used paraboloidal schlieren mirrors of 300 mm diameter and 3000 mm focal length and 500 mm diameter and 5000 mm focal length, The surface finish of the used mirrors was a quarter to one tenth of the light wavelength. In order to visualized larger fields of view, a pair of paraboloidal mirrors having 1000 mm dia. and 8000 mm focal length was introduced. The single mirror and its steel base weighing about 300 kg, we mounted them on a steel box the bottom of which had 1.5 mm diameter holes evenly distributed. We can make the box and the mirror move freely and align precisely.

The source laser beam is collimated with a plano-concave lens of short focal length to a slightly larger diameter *OB* and only the central part of the source light passes the test section.

The difference of path lengths between *OB* and *RB* is adjusted to be less than the coherent length of the source laser. The *OB* and *RB* illuminated simultaneously a holographic film @laced on a film holder, FH, at the approximately 20°. In order to satisfy the condition for linear transmittance of holographic recording materials the ratio of *OB* and *RB* intensities ranges between 2:1 and 3:1. Holographic films, AGFA 10E75 100 mm × 125 mm sheet films, was placed on a film holder FH made of thick aluminum plate and coated black, on the surface of which 2 mm dia. holes were uniformly distributed. Films were held flat, when slightly reduced pressures were applied through the perforation Takayama (1983).

In general, when a hologram was illuminated by a coherent beam, the reconstructed images are three-dimensional virtual ones. Although reconstructed images are three-dimensional and can be readily recognized with naked eyes but are hardly recorded on films. The three-dimensional images were recorded from various view angles. Although, as shown later, spatial resolution of the reconstructed images is not as sharp as that of the holograms' image, the resolution would be improved if properly processed by a computer assisted image processing systems.

In the double exposure holographic interferometry, the first exposure is performed prior to an event and the second exposure is synchronized with the event. Thus the change in phase angles during the double exposure was stored on a holo-film. Hence the time-interval of the double exposures should be short, which was automatically controlled by the source laser systems. We, however, often performed double exposures manually for time interval of several seconds. In such cases, the test conditions must be kept exactly unchanged throughout the double exposure. The phase change thus stored on the film can be reconstructed later through the process, the so-called process, reconstruction.

In flow visualizations by using double exposure holographic interferometry, the change in phase angles is uniquely expressed by the variations in the refractive indices. Hence, holographic interferometry is a method, unlike the Mach Zehnder interferometry, to quantitatively measure density fields with minimum restrictions on the optical arrangement. In double exposure holographic interferometry, inhomogeneity inherited from optical components, such as test section windows, media under study, may perturb the contrast of the background but will hardly distort fringe distributions. Hence, double exposure holographic interferometry is useful in visualizing shock waves in slightly inhomogeneous media such as liquids,

1.2 Double Exposure Holographic Interferometry Applied to Shock Wave Research

A schematic diagram of holographic interferometric system applied to shock tube flows at the Shock Wave Research Center (SWRC) of the Institute of Fluid Science, Tohoku University is illustrated in Fig. 1.1, Takayama (1983). Light sources were Q-switched ruby lasers (Apollo Lasers Ltd. double pulse holographic ruby lasers, pulse width of 25 ns, 695.4 nm wavelength and 10 and 2 J/pulse at TEM_{00} mode and a single pulse ruby laser, pulse width of 25 ns, 695.4 nm wavelength and 2 J/pulse at TEM_{00} mode). The beam was divided, by using a beam splitter *BS*, the 60% of the source laser beam was directed into object beam *OB* and 40% into reference *RB*.

The optical path of the *OB* is identical to the observation beam of the conventional optical arrangements, while the *RB* is unique to holographic interferometric arrangement, takes the identical light path length as the *OB* and independent of the events preserves the phase angle of the light source. Then when the *OB* and *RB* are superimposed on holographic films, we can identify the variation of phase angles caused by the event.

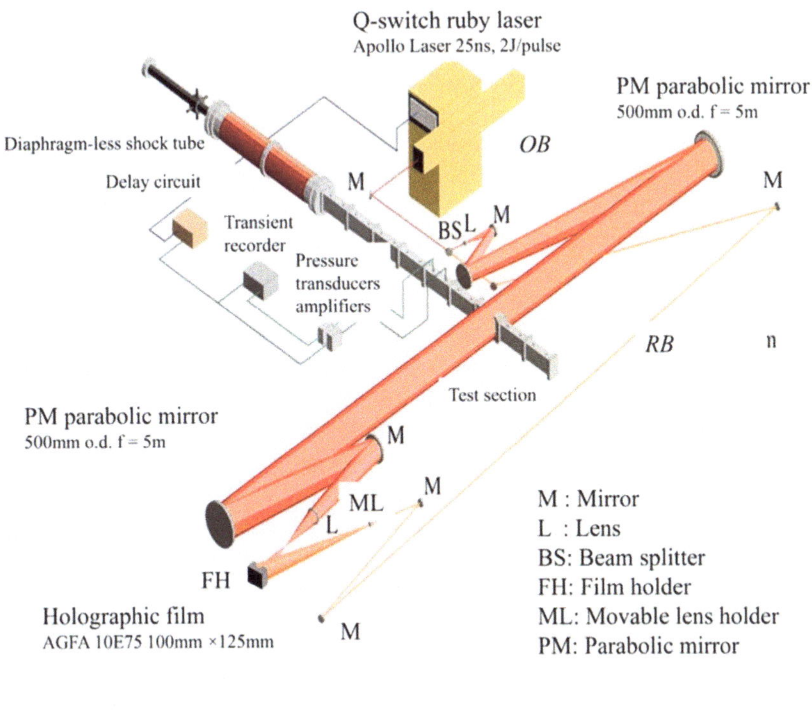

Fig. 1.1 An optical arrangement of holographic interferometry applied to shock tube research at the Shock Wave Research Center of the Institute of Fluid Science, Tohoku University

Chapter 1
Holographic Visualization of Shock Wave Phenomena

1.1 Introduction

Gabor (1948), for the first time, presented the concept holography. In 1971, the Nobel Prize in physics was awarded on him for his invention of holography, which opened a new era in the flow visualization. This was a long time before the advent of the lasers. Encouraged by the development of lasers, Light beams are characterized by their amplitude and phase angle. Leith and Upatnieks (1962) developed off-axis holography in which object and reference beams, *OB* and *RB*, illuminate on a holographic film with an off-axial direction and became a basis of modern holographic interferometry.

Shadowgraph and schlieren methods are traditional ways for visualizing compressible flows. Their principle is based on recording, on film, amplitude variations induced by the density changes associated with the flows, whereas holography uniquely records changes in phase angles of light beams induced by changes in considered flows.

Wortberg (1974) and Russell et al. (1974), for the first time, reported the usefulness in visualizing shock tube flows and their results were reported wave at the 9th International Shock Tube Symposium held at Stanford University. Russell et al. (1974) visualized the flow developed in a Ludwig tube flow. The vortex formation from the trailing edge of a wing-section was visualized by Mandella and Bershader (1986).

In 1975, equipped with a ruby laser which was one of the first generation products in Japan, we struggled to apply it as light source for visualizing shock tube flows. In 1980, we eventually purchased a double pulse holographic ruby laser and started to intensively visualize shock tube flows and underwater shock waves by holographic interferometry. Its first outcome was for the first time reported in Takayama (1983).

K. Takayama, *Visualization of Shock Wave Phenomena*,
https://doi.org/10.1007/978-3-030-19451-2_1

1.3 Analytical Background of Holographic Interferometry

To apply holographic interferometry to shock tube experiments, the light source should be a Q-switched laser having a high coherency. In 1980, Q-switched ruby lasers having a pulse duration of 25 ns and wavelength of 694.3 nm were used. The source laser beam is split into object beam *OB* and reference beam *RB,* to be called U_{ob} and U_{re}, respectively. Let subscripts 1 and 2 refer to the first and second exposures, respectively. ω is the frequency of the ruby laser, where the light speed $c = 2\pi\omega\lambda$ and λ is the wave length of 694.3 nm. Hence *OB* and *RB* in the first and second exposures are expressed as

$$U_{k,ob}(x, y) = a_{k,ob}(x, y)\exp\{i\omega t + i\phi_{k,ob}(x, y)\},$$

where k = 1 for first exposure and 2 for second exposure,

$$U_{k,re}(x, y) = a_{k,re}(x, y)\exp\{i\omega t + i\phi_{k,re}(x, y)\} \tag{1.1}$$

where $a_{k,ob}(x, y)$ and $a_{k,re}(x, y)$ are the amplitude distributions of the *OB* and *RB* for the first and second exposure and $\phi_{k,ob}(x, y)$ and $\phi_{k,re}(x, y)$ are their phase distributions, respectively.

The amplitude I_1 of light waves recorded on the holofilm at the first exposure is expressed as $U_{1ob} + U_{1re}$.

$$\begin{aligned} I_1 &= |U_{1ob} + U_{1re}|^2 \\ &= (U_{1ob} + U_{1re})\left(\underline{U_{1ob} + U_{1re}}\right), \end{aligned} \tag{1.2}$$

where $\underline{U_{1ob} + U_{1re}}$ is complex conjugate of $U_{1ob} + U_{1re}$.

In the second exposure, the *OB* which is represented as $U_{2ob}(x, y)$ carries not only the amplitude variation but also the phase variation, whereas $a_{re}(x, y)$ and $\phi_{re}(x, y)$ are spatially uniform and constant.

In constructing finite fringe interferograms, we inserted a collimating lens in the cutoff side of the *RB* path and displaced its center during the double exposures. By displacing the collimating lens and rotating the angle of the axis of the collimating lens, we controlled the phase angle Δϕ between the first and second exposures. The interval and orientation of finite fringes were readily adjusted.

In constructing infinite fringe interferograms, we fix the collimating lens unchanged during the double exposure, which means having identical Δϕ during the first and second exposures. Then the fringe interval becomes infinitely wide. Hence, infinite fringe interferometry was achieved.

The total intensity of illuminating light beams on the film which transmits only the amplitude of the light is given;

$$I(x, y) = (U_{1ob} + U_{1re})\overline{(U_{1ob} + U_{1re})}(U_{2ob} + U_{2re})\overline{(U_{2ob} + U_{2re})} \quad (1.3)$$

Although the relationship between the recorded light intensity I(x, y) and the transmittance of the film Ta is, in general, nonlinear, for some types of recording materials and light intensity, if the ratio of *OB* to *RB* is sufficiently large, a linear relationship is valid between Ta and I(x, y) (Gabor 1949). In order to achieve this condition, the ratio of *OB* to *RB* is, in practice, empirically chosen to be approximately 2:1–3:1. Therefore, the amplitude transmittance of the film Ta is given by, $Ta = k_0 + k_1 I(x, y)$ where k_0 and k_1 are constant.

If a hologram is illuminated with *RB,* $U_{re} = a_{re} \exp(i\omega\, t + i\phi_{re})$, the wave field emerging on the hologram with an amplitude transmittance Ta, is

$$TaU_{re} = \{k_0 + k_1 I(x, y)\}U_{re}.$$

The amplitude distribution T(x, y) of the reconstructed image has a physically meaningful term, which can be rewritten as,

$$T_3 = k_1 a_{re}^2 a_{2ob} \exp(i\phi_{1ob} + i\omega t) + \exp(i\phi_{2ob} + i\omega t) \quad (1.4)$$

This term represents the reconstructed phase change recorded during the double exposures. The relationship between phase changes and the fringe intensities of the reconstructed double exposure holographic interferogram is derived from Eq. (1.4) to read,

$$\begin{aligned} I_{reconstruct} &= T_3^2 \\ &= (k_1 a_{re})^2 a_{ob}^2 \{1 + \cos(\phi_1 - \phi_2)\} \end{aligned} \quad (1.5)$$

Equation (1.5) means that the fringes this reconstructed in interferograms correspond to the difference in the phase angle during the double exposures. In case $\phi_1 - \phi_2 = 2\pi N$, where the fringe number N is integer, $I_{reconstruct}$ is a maximum showing dark interference fringes. As changes in phase angles indicate the variations in the refractive indices along the light path; when the flow is two-dimensional flow, the fringe distribution just correspond to the variation of refractive indices. The following relationship is valid,

$$\phi_1 - \phi_2 = 2\pi L(n_1 - n_2)/\lambda, \quad (1.6)$$

where L and λ are the light path length and the light wave length. $n_1 - n_2$ stands for changes in refractive indices.

In gases, the refractive index n is related to the density ρ as follows.

$$n - 1 = K\rho, \quad (1.7)$$

where K is the Gladstone-Dale constant, which varies with λ. So far as the experiments are performed by using ruby laser, K is constant.

In liquids, the refractive index and density is related via Lawrence-Lawrentz relations

$$\left(n^2 - 1\right)/\left(n^2 + 2\right) = C\rho, \tag{1.8}$$

where C is constant. In gases, the refractive index n is written as n = 1 + ε and ε ≪ 1, and hence Eq. (1.8) reduces to Eq. (1.9).

$$\Delta\rho = N\lambda/KL \tag{1.9}$$

We can readily determine the density distribution by counting dark fringes which correspond to phase angle of 2 Nπ. By counting dark fringes, we can readily identify the density distribution. Brightest fringes correspond to the phase (2 N + 1) π. Likewise, by measuring the grey level of fringes between neighboring dark and bright fringes, enable one to correctly estimate the phase angle of corresponding grey level and hence to interpolate density between dark and bright fringes.

Figure 1.3 demonstrates, for example, an infinite fringe interferogram of a shock wave of Ms = 1.50 in air reflected from a 25° wedge. The b;ack fringes correspond to equal density contours indicting phase angle of 2 Nπ likewise the white fringes correspond to phase angle of (2 N + 1) π. Knowing fringe numbers, the density contours are readily determined.

The eight-digit number allocated to this interferogram shows a ID number, for example, #92090404 indicates that this interferogram was taken in 1997, 4th of September and the results of No. 4th experiment.

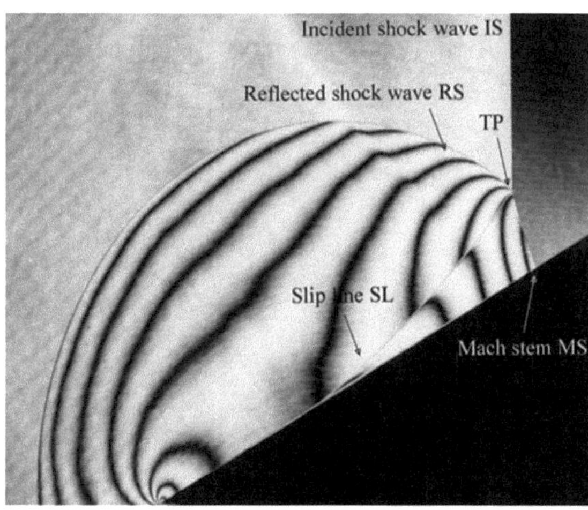

Fig. 1.3 #92090404, single Mach reflection SMR over a 25° wedge for Ms = 1.50 in air in 60 mm × 150 mm shock tube

References

Gabor, D. (1948). A new microscopic principle. *Nature, 161,* 777–778.

Gabor, D. (1949). Microscopy by reconstructed wave fronts. *Proceedings of the Royal Society of London, 197,* 454–487.

Houwing, A. F. P., Takayama, K., Jiang, Z., Sun, M., Yada, K., & Mitobe, H. (2005). Interferometric measurement of density in nonstationary shock wave reflection flow and comparison with CFD. *Shock Waves, 14,* 11–19.

Leith, E. N., & Upatnieks, J. (1962). Reconstructed wave fronts and communication theory. *Journal of the Optical Society of America A, 52,* 1123–1130.

Mandella, M., & Bershader, D. (1986) Quantitative study of shock-generated compressible vortex flows. In D. Bershader & R. Hanson (Eds.), *Proceeding 15th International Symposium on Shock Waves and Shock Tubes Shock Waves and Shock Tubes* (pp. 471–477) Berkeley.

Russell, D. A., Buonadonna, V. R., Jones, T. G. (1974). Double expansion nozzle for shock tunnel and Ludwieg Tube. In D. Bershader & W. Griffith (Eds.), Recent developments in shock tube research. In *Proceeding of 9th International Shock Tube Symposium* (pp. 238–249). Stanford.

Takayama, K. (1983). Application of holographic interferometry to shock wave research. In *Proceeding SPIE 298 International Symposium of Industrial Application of Holographic Interferometry* (pp. 174–181).

Wortberg, G. (1974). A holographic interferometer for gas dynamic measurement. In D. Bershader & W. Griffith (Eds.), *Proceedings of the International Symposium on Shock Waves Recent Developments in Shock Tube Research* (pp. 267–276) Stanford.

Chapter 2
Shock Waves in Gases

2.1 Shock Wave Reflections Over Straight Wedges

2.1.1 Introduction

When a shock wave is reflected from a steep wedge, the reflected patter of the incident shock wave, or in short IS, forms a V shaped wave pattern. This pattern is similar to a sound wave reflection from a plane wall and hence is named as a "regular reflection", and in short RR. A head-on-collision of a shock wave with a plane wall is an extreme case of the RR.

When an IS encounters a shallow wedge, it will not reflect from the wedge surface but intersects with the reflected shock wave above the wall surface. A third shock wave which is normal to the wall surface merges with their intersection point resulting in a Y-shaped shock wave pattern. Such a three shock confluence formed a triple point or in short TP and accompanied a slip line. Today, it is known that this pattern is one of typical representations of gas dynamic non-linearity. The reflection pattern is called "Mach reflection", or in short MR. A shock wave propagating perpendicularly along a straight wall is an extreme case of the MR.

If one recalls the history of the survey of shock wave reflections, Ernst Mach was one of pioneers who felt a puzzling feature of MR (Reichenbach 1983). He watched the interaction of two spherical shock waves created by simultaneous spark discharges on a soot covered glass plate. Eventually a V-shaped soot free region was discovered on the glass plate and people named this region as a Mach-V. The Mach-V was created by travelling vortices developed from the triple point, TP and indirectly proved the presence of the Mr. Teopler visualized, for the first time, this unusual shock wave reflection patter which was MR but he failed to claim the pattern as a MR (Krehl and van der Geest 1991).

Although the structure of the TP is not yet fully understood, the third shock wave emanating from the TP to the wedge surface is called a "Mach stem" or a Mach shock, MS. The transition of reflected shock waves between RR and MR is

K. Takayama, *Visualization of Shock Wave Phenomena*,
https://doi.org/10.1007/978-3-030-19451-2_2

governed by the wedge angle, the shock Mach number Ms, and the specific heats ratio γ of the test gas.

An MS is traditionally assumed to be straight and perpendicular to the wedge surface. Figure 1.4 shows a weak shock wave of Ms = 1.158 in atmospheric air reflected from a 7.0° wedge. It is observed that in a case of weak shock wave reflection from shallower wedges, the MS is slightly curved and SL is not observable. The point at which IS, RS and MS intersect is not a triple point but a point on the trajectory of the glancing incidence angle, which will be discussed later. Based on his private preference offered terminology, Birkhoff (1960) named the departure of MS's shape from that of traditional Mach reflection patterns as the von Neumann paradox. Today it is understood that this terminology never means a paradoxical phenomenon but exhibits a feature of weak shock waves, in which the isentropic feature of the state behind weak shock wave reflections are exaggerated. The reflection pattern is shown in grey color as seen in Fig. 2.1. Although we see a dark fringe but miss a straight slip line, SL. This Mach reflection pattern is named as von Neumann Mach reflection or in short vNMR.

Ben-Dor (1979) defined the domain and boundary of the family of Mach reflections in shock tube flows. The pattern shown in Fig. 1.3 is, according to the definition, named as a single Mach reflection SMR. The SMR has a distinct triple point TP and SL emanating from the TP. The reflected shock wave RS is curved, which indicates that the flow behind it is locally subsonic. With increasing the wedge angle, θ_w, the shape of the reflected shock, RS tends to be straightened indicating that the flow behind it is transonic. Then a merger of compression waves behind the straight RS intersects with a curved RS forming a kink point. Then the SMR terminates to become a transitional Mach reflection, TMR. With further increase of the θ_w, the transonic flow region became supersonic, the coalescence of trains of compression waves becomes a shock wave which is named as a secondary shock wave. Then, the kink point turns into a secondary triple point having three shock confluence and a slip line, SL. The second Mach reflection characterizes a

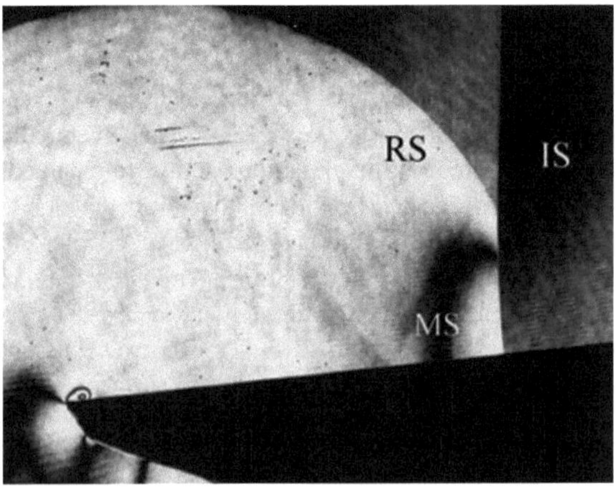

Fig. 2.1 Typical vNMR in a 60 mm × 150 mm shock tube, #87092210, for Ms = 1.158 in atmospheric air at 300 K, wedge angle of θ_w = 7.0°. Notice that no slip line, SL is observable

reflection pattern as a double Mach reflection, DMR. These transitions of reflected shock wave patterns exist only in unsteady flows.

2.1.2 *Shock Polar*

von Neumann studied systematically, for the first time, reflection of steady oblique shock waves from solid wall. Propagation and reflection of shock waves over wedges were visualized in shock tube experiments. Shock tube flows are essentially non-stationary phenomena but it can be reduced to steady flows by replacing the flow framework to a shock fixed coordinate. Flow conditions in front of and behind a given shock wave are at least locally uniform. The states behind these shock waves are determined by solving the Rankine-Hugoniot relations, R-H relations. Using the R-H relations, von Neumann derived algebraic equations deriving a transition from RR to MR. His analysis showed that a solution for either MR or RR could be determined, given the following set of parameters, ξ, θ_w, and γ. Here, ξ is the inverse strength in the pressure ratio across the shock wave;

$$\xi = (\gamma + 1)/\left(2\gamma Ms^2 - \gamma + 1\right) \tag{2.1}$$

For a given specific heats ratio γ, the reflected shock transition between MR and RR occurs with continuous change in the parameters ξ, θ_w. Figure 2.1 shows the definition of regions and angles in the frame of the triple point fixed coordinate for Ms = 1.497 # 92090404.

Regions (1) and (2) correspond to the ones in front of the incident shock IS, and the one across it. Regions (3) and (4) designate the one behind the the RS and that behind the MS, respectively. These regions are uniform in the vicinity of the triple point TP. The values of particle velocities' values in the region (3) and (4), obtained while deflected in crossing at the RS and MS, respectively, differ from each other

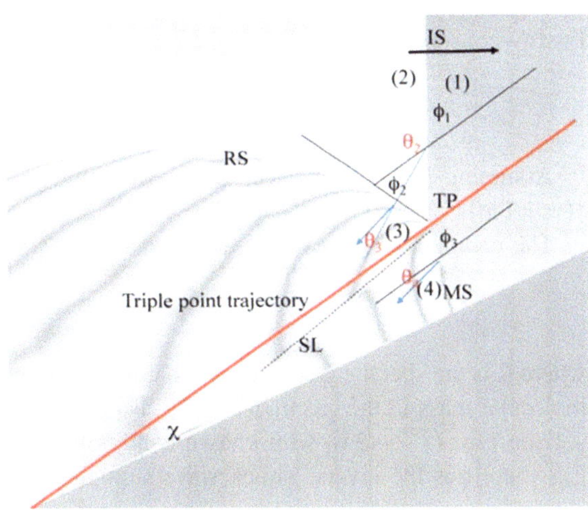

Fig. 2.2 Definition of regions and angles: #92090404, for Ms = 1.497 in atmospheric air at 300.1 K, wedge angle $\theta_w = 25°$

but both have the same direction. This means that the line separating the regions (3) and (4) is a contact discontinuity or a slip line, SL; across it, the pressures are constant, but the densities change discontinuously.

In coordinates fixed to the TP, the conservation of mass, and energy and the equation of motion for the individual shock waves are reduced to those of steady oblique shock waves. Tangential velocity components along the oblique shock should be identically given;

$$U_i \cos\phi_\iota = U_k \cos\left(\phi_j - \theta_k\right) \tag{2.2}$$

where subscripts i, j, and k denote the individual flow regions shown in Fig. 2.2 and are selected: for IS, i = j = 1 and k = 2; for RS, i = j = 2 and k = 3; and for MS, i = k = 1 and j = 4. Angle ϕ is defined, in SMR in Fig. 1.5 as the incident angle of the IS with respect to the trajectory angle of the TP. That is, $\phi_1 = 90° - \theta_w - \chi$, where χ is the trajectory angle of the TP.

This relation is derived based on the assumption that the MS is straight at least in the vicinity of the TP and perpendicular to the wedge surface. For RR, ϕ reduces to the incident angle with respect to the wedge surface,

$$\phi_1 = 90° - \theta_w.$$

Other equations are:

Mass conservation;

$$\rho_I U_i \sin\phi_I = \rho_k U_k \sin\left(\phi_j - \theta_k\right). \tag{2.3}$$

Equation of motion;

$$p_1 + \rho_\iota U_i^2 \sin^2\phi_I = p_k + \rho_k U_k^2 \sin^2\left(\phi_j - \theta_k\right). \tag{2.4}$$

Energy conservation;

$$2h_i + U_i^2 \sin^2\phi_i = 2h_k + U_k^2 \sin^2\left(\phi_j - \theta_k\right). \tag{2.5}$$

Assuming ideal and perfect gases, we define the specific enthalpy as: $h = \gamma p / \{(\gamma - 1)\rho\}$.

The equation of state for ideal gas is

$$p = \rho RT, \tag{2.6}$$

where R is the specific gas constant, that is the universal gas constant divided by the molecular mass of the gas under study. The value of θ_k is, therefore, determined by solving Eqs. (2.3)–(2.6) with known conditions in region (i, j), we have $\theta_k = F(\phi_j, \xi)$, where ξ is the inverse shock wave strength across regions (j) and (k),

$$F(\phi, \xi) = \tan^{-1}\left[(\xi - 1)\left\{(\mu - 1)M^2 - \mu - \xi\right\}^{1/2} / \left\{\left(1 - \gamma M^2 - \xi\right)(\mu + \xi)^{1/2}\right\}\right], \tag{2.7}$$

where $\mu = (\gamma - 1)/(\gamma + 1)$,

$$M = Ms/\sin\phi \quad \text{for } k = 1, 3$$
$$= \left\{\left(2/(\gamma - 1) + Ms^2\right)\left\{(\mu\xi + 1)^{1/2}/\left(\mu\xi + \xi^2\right)^{1/2} - 1\right\}\right. \quad \text{for } k = 2.$$

A family of solutions for a given $M_1 = Ms/\sin\phi_1$ and γ is expressed in terms of a pressure, p, and flow deflection angle, θ. A solution curve expressed in a p, θ—plane is known as a shock polar diagram. Shock polar analysis is a useful graphical explanation of wave interactions providing a physical insight into various, not necessarily simple, interactions of shock waves. Kawamura and Saito (1956), for the first time, adopted the shock polar diagram in analyzing shock reflections over wedges. The solution curve, I, for given initial conditions in region (1) is named as an I-polar. Another solution curve which starts from the condition in region (2) on the I-polar and is superimposed on the I-polar is named as a R-polar. The R-polar determines all the possible solutions R for a given ϕ_1 p_2, θ_2.

Depending upon the boundary conditions, all the possible solutions of regions (3) and (4) are described on the R-polar. For determining the state behind the reflected shock wave, boundary conditions have to be specified. In MR, the boundary conditions in the vicinity of TP are that flow vectors in the regions (3) and (4) are parallel to each other and the pressures are identical locally across SL.

That is,

$$p_3 = p_4 \text{ and } \theta_2 \pm \theta_3 = \theta_4. \tag{2.8}$$

In RR, the boundary condition is that the streamline in the region (3) is parallel to the wedge surface:

$$\theta_2 + \theta_3 = 0. \tag{2.9}$$

The second intersection point of the R-polar with the I-polar satisfies the boundary conditions given by Eq. (2.9) (Takayama and Sasaki 1983).

Figure 2.3 shows a shock polar for Ms = 1.50 in air. The ordinate designates pressure p normalized by the initial pressure p_1 and the abscissa designates angle θ in degree. I-polar is drawn for wedge angles ranging from 30° to 60° at every 5° and R-polar is also drawn for individual wedge angles for Ms = 1.50. Intersecting points in the first quadrant of individual I-polar and R-polar provide solutions of pressures, p, and the flow deflection angles, θ, in the vicinity of the TP corresponding to regions (3) and (4).

The reflected shock wave transition between MR and RR occurs theoretically either when the R-polar touches the p-axis or when the R-polar intersects both the I-polar and the p-axis at the same point. The former is called the detachment

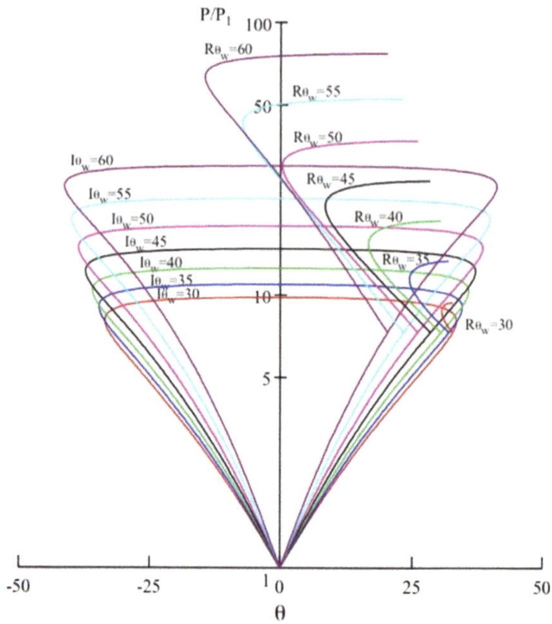

Fig. 2.3 Shock polar for a shock wave of Ms = 1.5 reflecting from wedges of 30° to 60° (Numata et al. 2009)

criterion and latter renamed as the von Neumann criterion (Ben-Dor 1979; von Neumann 1963; Courant and Friedrichs 1948). The detachment criterion satisfies the boundary condition of Eq. (2.8), when $\theta_3 = \theta_{max}$, where θ_{max} is a maximum flow deflection angle, and $\theta_4 = \theta_2 - \theta_3 = 0$. This relationship indicates that the slip line emanating from TP is parallel to the wall.

The von Neumann criterion also satisfies the above-mentioned flow deflection, so that, as will be seen later, the slip line is again parallel to the wedge surface. The wedge angle at which the transition takes place is called the critical transition angle, θ_{crit}. Hence, knowing θ_{crit} and ϕ, the triple point trajectory angle χ is readily obtained if we assume straight and perpendicular MS to the wedge surface.

Figure 2.4 shows experimental results Takayama and Sasaki (1983) for θ_{crit} obtained for concave and convex walls of various radii and initial angles against inverse shock strength ξ. In The experiments were performed in a 30 mm × 40 mm conventional shock tube and previously obtained results (Smith 1948) were also added. The detachment and von Neumann criteria are also presented. The ordinate denotes θ_{crit} in degree and the abscissa denotes the inverse shock strength ξ.

von Neumann concluded that the criterion for the reflected shock transition depended on the strength of incident shock waves and the detachment criterion was valid for weak shock waves and the von Neumann criterion was applicable to strong shock waves. Von Neumann defined weak and strong shock waves based on the shock polar diagrams (Courant and Friedrichs 1948). For strong shock waves, it is possible that either MR or RR exists in the region between the detachment and the von Neumann criteria. In the shock wave reflection over wedges in shock tubes, the transition consistently follows the detachment criterion, however, as will be

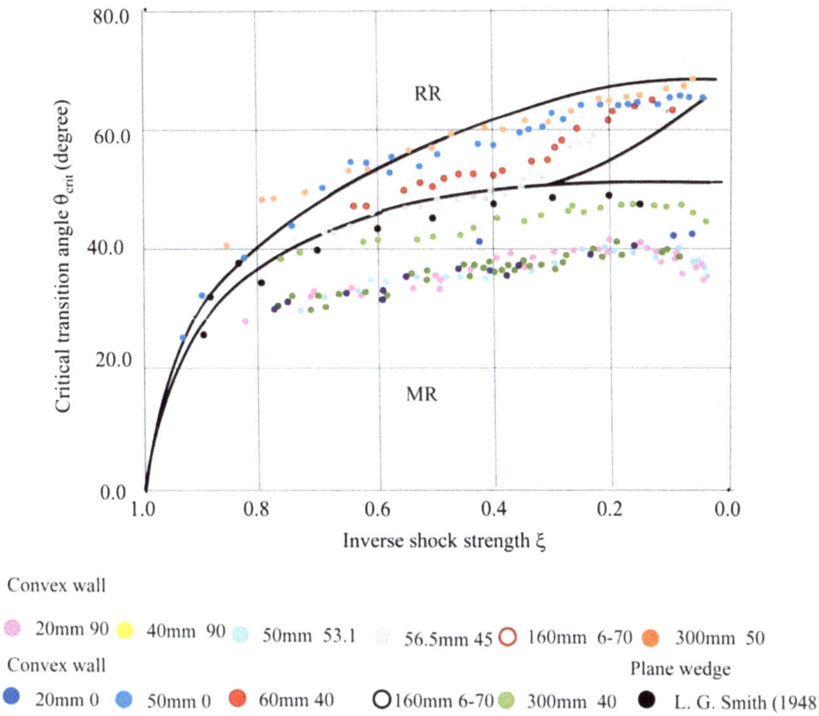

Fig. 2.4 The dependence of the critical transition angle θ_{crit} against the inverse shock strength ξ Kawamura and Saito (1956)

discussed later, along curved walls, that is, the variable wall angle with shock wave propagations, the flows are truly non-stationary; in such cases, either the transition follows the von Neumann criterion or the detachment criterion (Ben-Dor 1979; Takayama and Sasaki 1983; Courant and Friedrichs 1948).

However, the transition in stationary flows occurs at either the detachment criterion or the von Neumann criterion. Hornung et al. reported that, for a given strong shock, depending upon the operational condition of the wind tunnel or the setup of the wedge models, either MR or RR occurs in the region between the detachment and von Neumann criteria. People believe that the wind tunnels create stationary flows but it is not always true. The starting flow in wind tunnels or insertions of models into the wind tunnel flows always create the transitional flows which disturbs the stationary flows.

Another definition is that IS is weak, when the flow behind IS in MR is subsonic. Then, the I- and the R-polars intersect on the lower branch of the R-polar in Fig. 1.7. It is strong when the flow behind RS is supersonic, which occurs when the I- and the R-polars intersect on the upper branch of the R-polar. This implies that the weak and the strong boundary is at the sonic condition. However, these two definitions differ only slightly Ben-Dor (1979).

2.1.3 *Wedges*

2.1.3.1 Shock Wave Reflection Over Straight Wedges in a 40 mm × 80 mm Shock Tube

The transition of reflected shock waves between RR and MR is one of fundamental topics of shock wave research. In the late 1970, for shock wave reflection experiments, a 40 mm × 80 mm conventional shock tube was used by rupturing Mylar diaphragms in a double diaphragm system. At that time, the repeatability in terms of the scatter of shock Mach number ΔMs was ±2% for Ms ranging from 1.1 to 4.0 in air. Figure 2.5 shows the test section of the 40 mm × 80 mm conventional shock tube made of a single piece of carbon steel. The design and manufacturing were made in house. The parts of the shock tube were manufactured in the machine shop of the Institute of High Speed Mechanics of Tohoku University. In order to avoid thermal distortions of metal pieces, welding was not used at all. The wedge was placed on a movable stage. The wedge angle was controlled precisely ranging from 0° to 60°. As seen in Fig. 2.5, by adjusting the height of vertical shaft which sustained the height of movable stage with micrometer from outside the shock tube (Takayama and Sasaki 1983). At first, shock tube flows were visualized using conventional shadowgraph and schlieren methods. Its light source was a Q-switched ruby laser, the first product of ruby laser made in Japan. In 1980 a Q-switched ruby laser, manufactured in Apollo Laser Co. Ltd, was introduced. Then the previous optical arrangement of shadowgraph was combined with a double exposure holographic interferometry. Since then all the shock tube flow visualizations were conducted by double exposure holographic interferometry.

Figure 2.6a–j show shock wave reflections for Ms = 2.00 in air at 550 hPa and 295.6 K, at wedge angle ranging from θ_w = 0.0° and 3.0°. At small wedge angles the IS is reflected very slightly by passing along the wedge. Then this wedge angle is named as the so-called glancing incidence angle, χ_{glance}. The χ_{glance} is defined as

Fig. 2.5 Test section of a 40 mm × 80 mm shock tube (Takayama and Sasaki 1983)

following: the corner signal is transmitted at the particle velocity u and toward the IS at the local sound speed a. Therefore, the χ_{glance} is written as;

$$\tan\chi_{glance\cdot} = \left\{a^2 - (Ms - u)^2\right\}^{1/2} / Ms, \tag{2.10}$$

where a is the local sound speed normalized by the sound speed in front of the shock wave a_o and u is the local particle velocity u normalized by a_0. In Fig. 2.6a, the RS intersects with the MS and lies on the MS is not observable, the trajectory of glancing incidence angle and intersect and form a triple point agrees with of $\chi_{glance} = 27.6°$ for Ms = 2.0. In Fig. 2.6b, the trajectory angle of the TP is about 24.0° for $\theta_w = 3.0°$. SL is not visible in Fig. 2.5b, so that the pattern of the reflected shock wave is von Neumann Mach reflection, vNMR. Figure 2.6a–j shows shock wave reflections for Ms = 3.00 in nitrogen at in nitrogen at 200 hPa and 295.6 K. Figure 2.5c–f shows SMR, whereas in Fig. 2.6g, h, the RS has a kink point and hence the reflection patterns are TMR. In Fig. 2.6i, with increasing the wedge angle, the kink point transits to a secondary triple point and hence the reflection pattern becomes a DMR. In Fig. 2.6j, at $\theta_w = 60.0°$, the pattern of the regular reflection is straight at the reflection point, which eventually indicates the flow behind the RS is supersonic. Hence, the reflection pattern is defined as a "supersonic regular reflection", or in short, SPRR.

Figure 2.7 shows direct shadowgraphs of DMR over $\theta_w = 45.0°$ and 50.0° wedges for Ms = 4.90 in air at 15 hPa and their enlargements. The triple points emitted luminosity. When performing for stronger shock wave at reduced initial pressures, the entire test fields were filled with intense luminosity and hence the film was totally over exposed. In performing shock tube experiments as shown Fig. 2.7, the films were covered with neutral density filters and the incident shock wave Mach number was Ms = 4.90; this shock wave was a lowest Ms we succeeded to visualize.

2.1.3.2 Shock Wave Reflection Over Straight Wedges in a 60 mm × 150 mm Diaphragm-Less Shock Tube

A 60 mm × 150 mm conventional shock tube made of standard high tensile strength carbon steel was designed and manufactured in a machine shop of the Institute. To minimize the deformation of steel plates, without using heavy machine works, the shock tube and its test section were carefully manufactured. Although it took time but their tolerance was 10 μm.

A diaphragm-less operation system was developed (Yang 1995). Figure 2.8a shows the diaphragm section. The high pressure chamber and the low pressure channel had a co-axial structure and separated by a rubber membrane as seen in Fig. 2.8b. It was bulged by pressurized helium gas from behind in an auxiliary high pressure chamber and tightly sealed the driver gas and test gas. The bulged

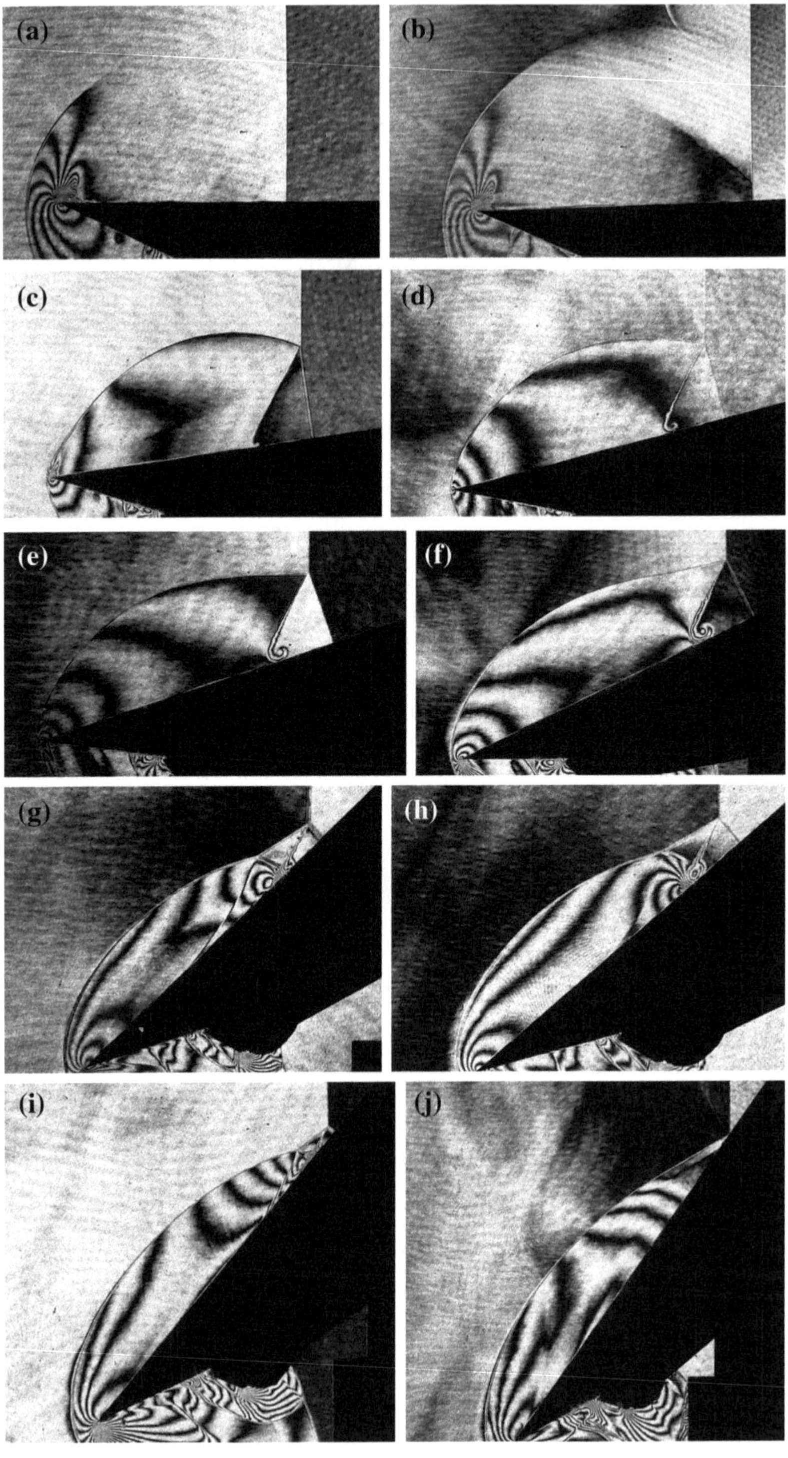
(a)
(b)
(c)
(d)
(e)
(f)
(g)
(h)
(i)
(j)

◂**Fig. 2.6** Evolution of shock wave reflection for Ms = 3.0 in nitrogen at 200 hPa and 295.6 K: **a** #81092917, Ms = 2.00 at 550 hPa and 295.6 K, $\theta_w = 0.0°$; **b** #81092920, Ms = 2.00 at 550 hPa and 295.6 K, $\theta_w = 3.0°$; **c** #81100219, $\theta_w = 11.0°$; **d** #81100218, $\theta_w = 18.0°$; **e** #81100218, $\theta_w = 18.0°$; **f** #81100225, $\theta_w = 23.0°$; **g** #81100305, $\theta_w = 40.0°$; **h** #81100304, $\theta_w = 45.0°$; **i** #81100505, $\theta_w = 50.0°$; **j** #81100307, $\theta_w = 60.0°$ (Takayama and Sasaki 1983)

membrane was supported by a grid B. The high pressure helium was separated by a Mylar diaphragm. Upon rupturing it, the rubber membrane quickly receded and the driver gas drove a shock wave into the test section as shown in Fig. 2.8c. The rubber membrane was sustained by a grid A, when it receded.

With such a simple arrangement, shock waves were generated ranging from Ms = 1.1 up to 5.0 in air. The scatter in shock wave Mach number ΔMs was ±0.2%. The merit of diaphragm-less operation system is not only its higher degree of repeatability and but also no diaphragm fragments are generated. The diaphragm-less shock tube was operational continuously without exposing the shock tube to ambient air. As ambient air contaminated the shock tube wall so that the degree purity of the test gas was maintained high if the test gas was a foreign gas other than air.

In conventional shock tubes, diaphragm fragments are always shattered, so far as Mylar diaphragms were used. If a cellophane diaphragm is used, its fragments spread out every where inside shock tubes. Soon after the shock tube run, the inside the test section was cleaned.

Using this shock tube, the evolution of moderately strong shock waves ranging from Ms = 2.585 to 2.654 in air at 100 hPa and 289 K was shown for the wedge angle from 2° to 60°. Selective interferograms are shown in Fig. 2.9a–z. In Fig. 2.9a, in the case of the wedge angle of 2°, a three- shock confluence, that is TP is observed but a slip line, SL is not visible. The three shock wave confluence lies on a trajectory of glancing incidence angle χ_{glance}, which indicates the outer boundary of the corner signal to reach. The MS is not straight but is curved in the vicinity of the TP. This pattern shows a typical vNMR.

The measured $\chi_{glance} = \theta_w$ in Fig. 2.9a is about 23° and $\theta_w = 2°$, whereas the predicted χ_{glance} in Eq. (2.10) being 26° for Ms = 2.61, hence the predicted value of $\chi_{glance} = \theta_w$ of Fig. 2.9a is 21°. The MS is slightly curved but intersects perpendicularly with the wedge surface as seen in Fig. 2.8a, b.

A bow shock wave generated at the leading edge of the wedge merged smoothly with a curved RS emanating from the TP. This implies that the flows in the entire region is subsonic. In the regions (3) and (4) in the vicinity of the TP, particle velocities are discontinuous but their flow directions and pressures are identical across the SL but the densities are discontinuous. However, in Fig. 2.9a. the density difference across the regions (3) and (4), is so small not sufficient for creating a visible SL. With increasing θ_w in Fig. 2.9b–d, the SL became visible and stagnated on the wedge surface. For shallower wedge angles, in Fig. 2.9b–d, the SL had a smooth laminar structure, their tail stagnated on the wedge surface. With increasing θ_w, the SL was broadened and eventually had a turbulent structure. Pressures in the regions (3) and (4) are identical in the vicinity of TP but, are higher on the wedge

(a)

(b)

(c)

(d)

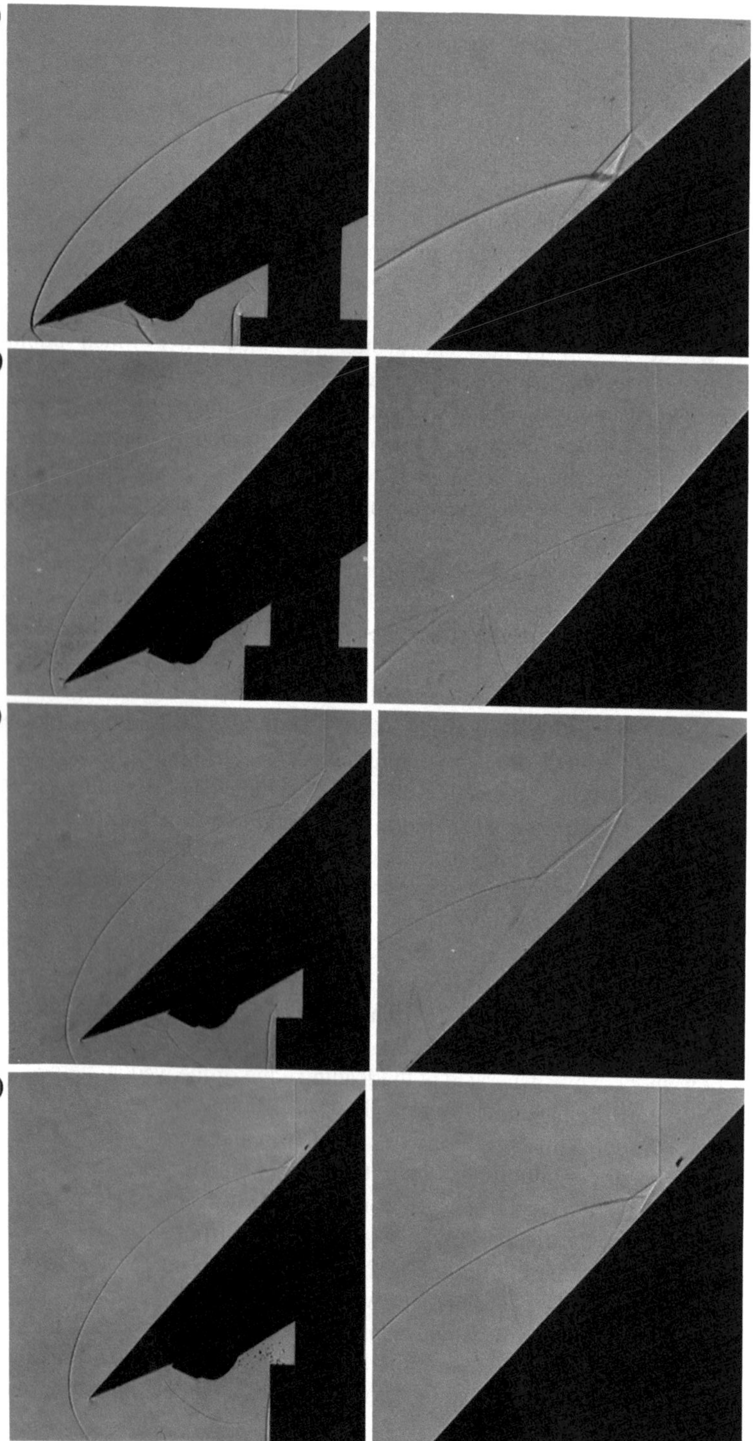

◂**Fig. 2.7** Generation of a luminous spot at a triple point: **a** #80091005, Ms = 4.90 in air at 15 hPa, 295.7 K, $\theta_w = 45°$ (Ben-Dor et al. 1983); enlargement of (**a**); **b** #80091006, Ms = 4.90, $\theta_w = 50°$; enlargement of (**b**); **e** #80090926, Ms = 4.10 in air at 25 hPa and 297.0 K, $\theta_w = 45°$; enlargement of (**c**); **d** #80090916, Ms = 3.10 in air at 40 hPa and 299.5 K, $\theta_w = 50°$; enlargement of (**d**)

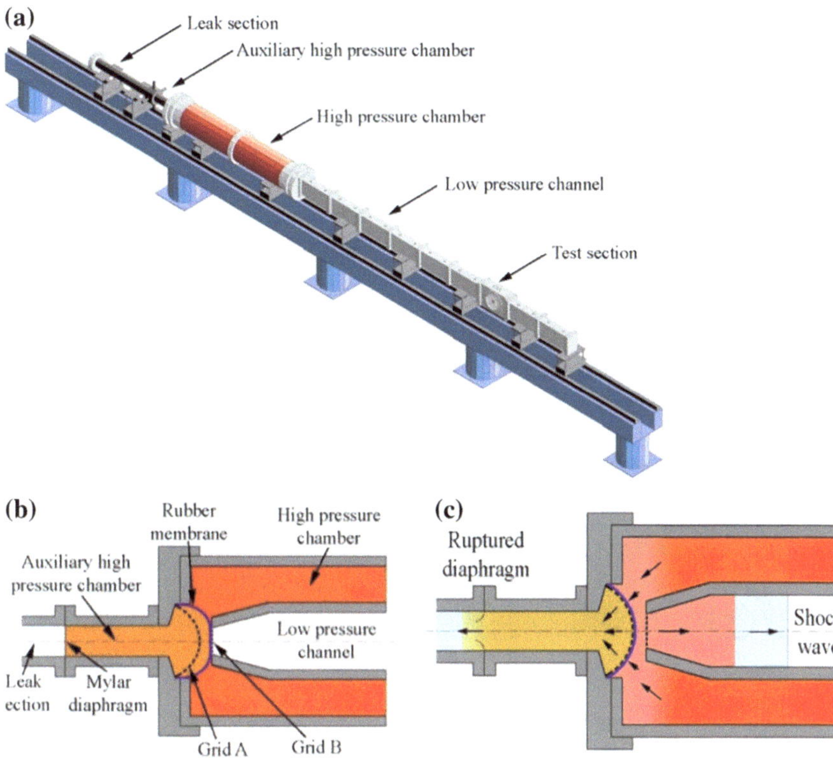

Fig. 2.8 A 60 mm 150 mm shock tube: **a** 60 mm × 150 mm diaphragm-less shock tube; **b** setup for operation; and **c** initiate to drive

surface in the region (3) than that in the region (4). Hence, in Fig. 2.9e, f, the SL rolled up forward as seen in Fig. 2.9g, h.

With increasing θ_w, the RS emanating from the TP was straightened in the vicinity of TP as seen in Fig. 2.9g, h. The straight RS implies that the flow behind the BS is transonic or supersonic. On shallow wedges, the straight RS and the curved RS intersect. This implies that the subsonic flow region and transonic flow region co-exist side by side and hence a kink point is formed and compression waves generated along the curved RS are going to coalesce if the wedge angle further increased.

The trend of dark fringe distributions indicates the assemblage of compression waves toward the kink point and then the TMR is formed. However, as observed in Fig. 2.9k–m, the TMR is not formed suddenly but takes a slightly long transient

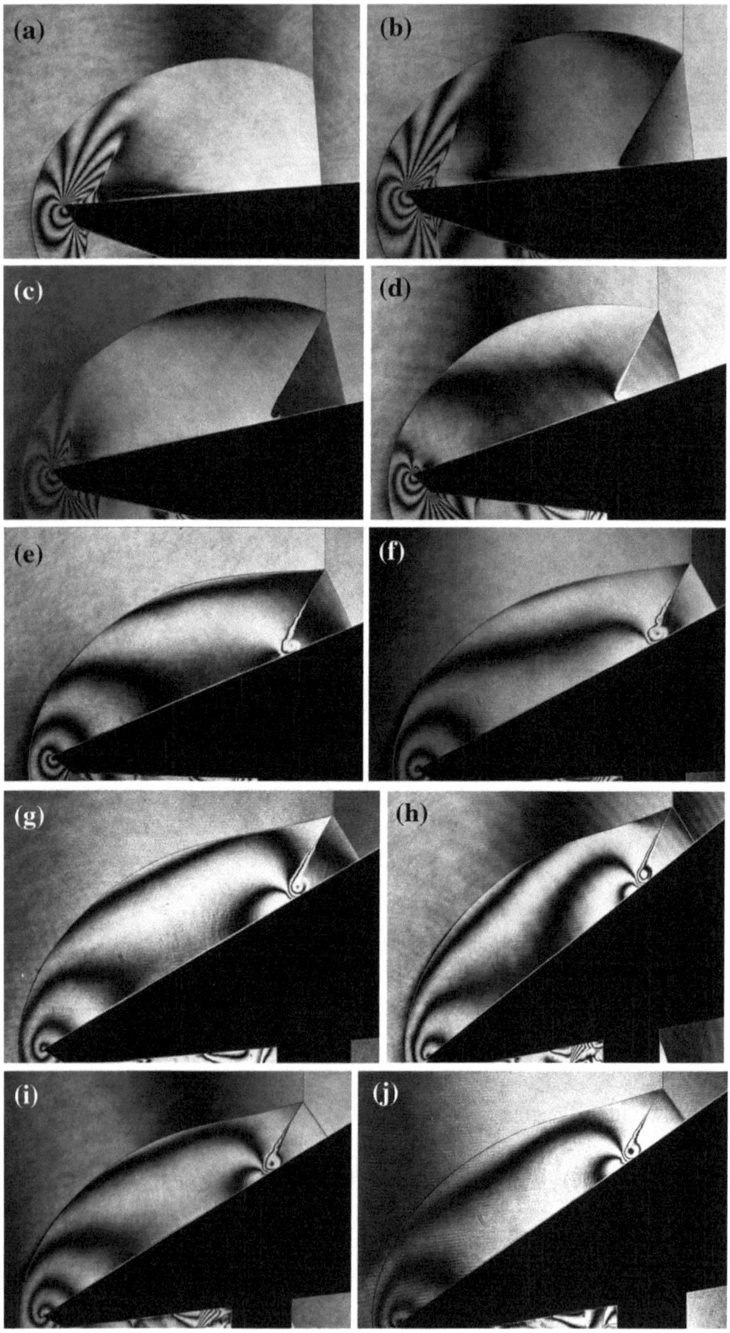
(a)
(b)
(c)
(d)
(e)
(f)
(g)
(h)
(i)
(j)

◂**Fig. 2.9** Sequential interferograms of shock wave reflection over movable wedge for Ms = 2.6 at 100 hPa, 298 K in air: **a** #83110516, Ms = 2.612, $\theta_w = 2°$, vNMR; **b** #83110515, Ms = 2.642, $\theta_w = 5°$, SMR; **c** #83110514, Ms = 2.653, $\theta_w = 11°$, SMR; **d** # 83110513, Ms = 2.632, $\theta_w = 17°$, SMR; **e** #83110512, Ms = 2.615, $\theta_w = 25°$, SMR; **f** #83110511, Ms = 2.606, $\theta_w = 29°$, reflected shock shows a sign of TRM; **g**#83110510, Ms = 2.602, $\theta_w = 31°$, SMR with a sign of TRM; **h** #83110508, Ms = 2.631, $\theta_w = 33°$, SMR with a sign of TMR; **i** #83110509, Ms = 2.631, $\theta_w = 34°$, SMR with a sign of TMR; **j** # 83110506, Ms = 2.605, $\theta_w = 35°$, SMR with a sign of TMR; **k** #83110505, Ms = 2.611, $\theta_w = 37°$, a kink point is initiated; **l** #83110504, Ms = 2.648, $\theta_w = 39°$, formation of TMR; **m** #83110503 Ms = 2.589, $\theta_w = 41°$, TRM; **n** #83110502, Ms = 2.613, $\theta_w = 43°$, termination of TMR; **o** #83110507, Ms = 2.599, $\theta_w = 46°$, transition from TMR to DMR; **p** #83110423, Ms = 2.611, $\theta_w = 47°$, DMR; **q** enlargement of (**p**), DMR; **r** #83110421, Ms = 2.624, $\theta_w = 48°$, DMR; **s** #83110422, Ms = 2.654, $\theta_w = 49.5°$, DMR; **t** enlargement of (s); **u** #83110420, Ms = 2.625, $\theta_w = 51°$ DMR; **v** enlargement of (**t**), the second TP moves toward the first TP; **w** #83110424, Ms = 2.640. $\theta_w = 52°$ DMR is going to terminate; **x** enlargement of (**w**), the second TP merges to the first TP, the termination of DMR; **y** #83110417, Ms = 2.628, $\theta_w = 54°$, SPRR; **z** #83110415, Ms = 2.585, $\theta_w = 60°$, SPRR

process. With a further increase in θ_w, the kink point becomes a triple point, TP and compression waves coalesce into a secondary shock wave. Eventually from the resulting TP, the three-shock confluence and a secondary slip line appear. It should be noticed that the initiation of double Mach reflection, DMR also takes a transient process. Figure 2.8l–m indicate the formation of TMR. In Fig. 2.9n, it is not decisively identified that a faint change in fringe patterns indicates the initiation of a SL in the region behind the curved RS. This also indicates a transition from a TMR to a DMR.

Ben-Dor (1979) analytically obtained, for the first time, the domain and boundary among a SMR, a TMR and a DMR. It is experimentally found that the SMR does not transit into TMR instantaneously at a predicted $\theta_{crit.}$ Similarly, the evolution from a TMR to a DMR also takes a transient process. The difference between the prediction and experimental findings is most probably based on the fact that in analytical predictions all the individual flow regions are assumed to be uniform while in reality these regions are non-uniform because of the presence of viscosity and the finite size of experimental facilities.

Figure 2.9p–x shows the transition from a DMR to a RR. At $\theta_w = \theta_{crit}$, the TP terminates and a supersonic region appears behind the straight RR. This RR pattern is defined as a “supersonic regular reflection”, SPRR as shown in Fig. 2.9y. The curved RR has a subsonic flow behind the reflection point and is defined as a “subsonic regular reflection”, or in short SbRR.

As θ_w approaches to the θ_{crit}, the length of the MS emanating from the TP of the DMR is shortened to the length barely recognizable. For example, Fig. 2.9s, u, w shows DMR at $\theta_w = 49.5°$, 50°, and 51°, respectively and their enlarged images are presented in Fig. 2.8t, v, x. The DMR pattern in Fig. 2.9x is barely observable. Even such a localized TP accompanies a short SL which turns into a vortex on the wedge surface. As the local Reynolds number in terms of the characteristic length of the MS has an order of magnitude of a few hundreds, the flow at the TP is subjected to viscous effects, which indicates that the transition from MR to RR

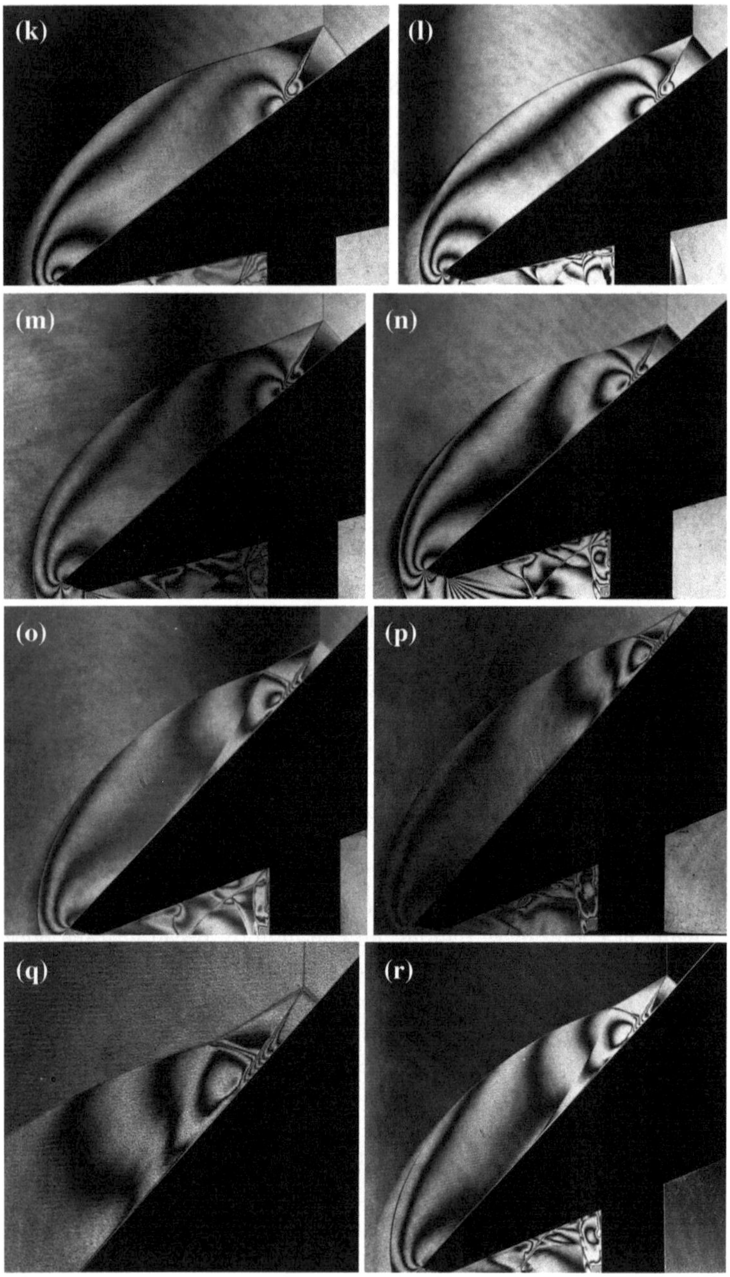

Fig. 2.9 (continued)

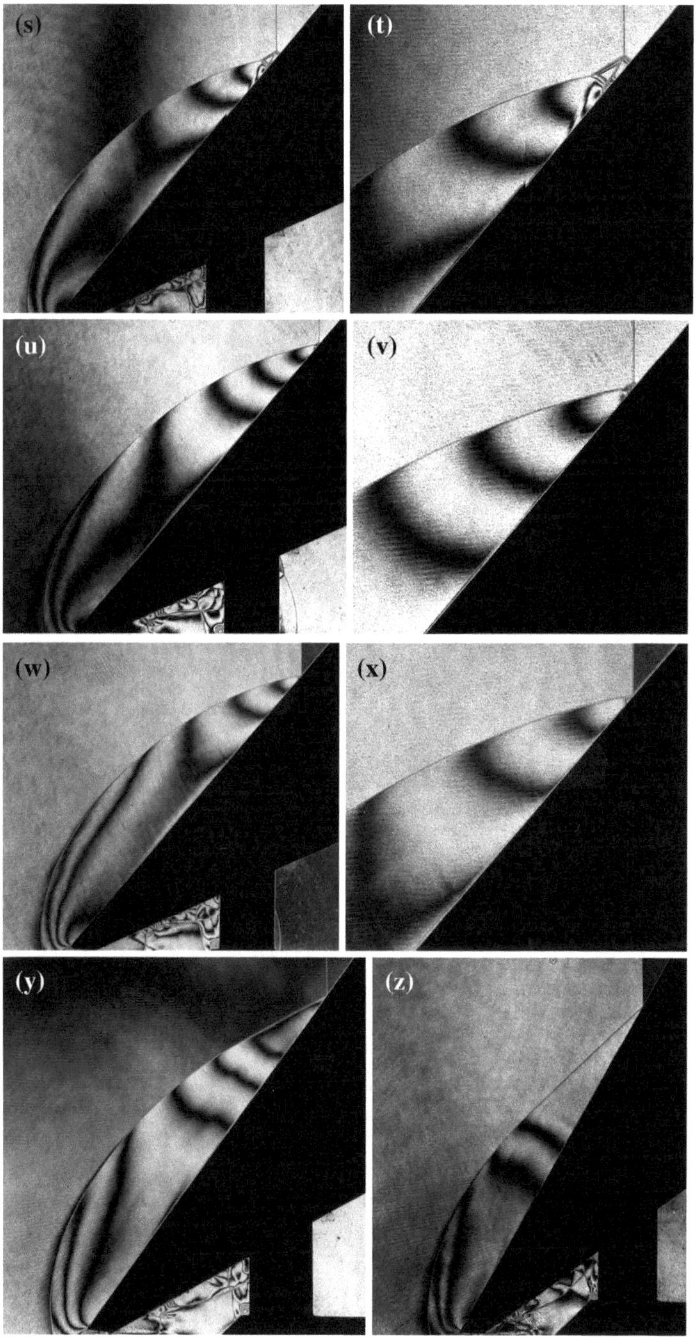

Fig. 2.9 (continued)

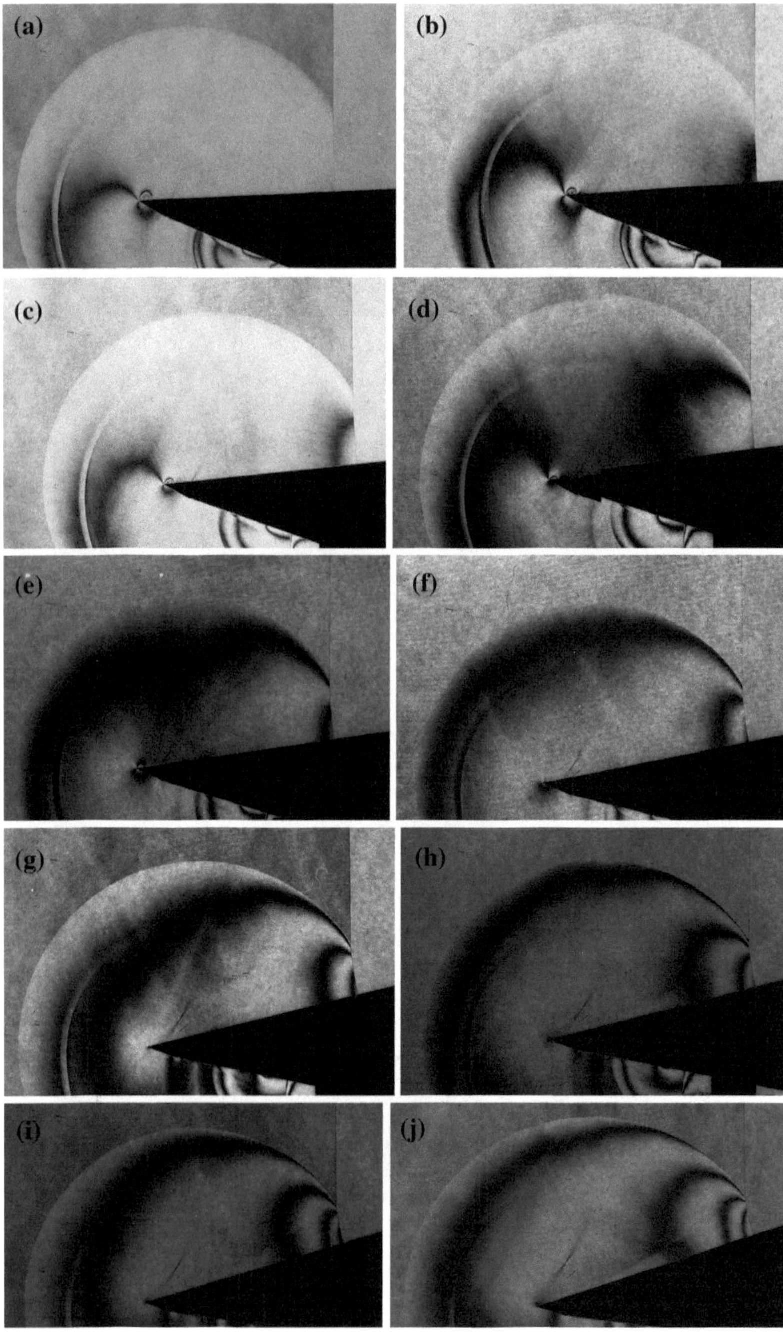
(a)
(b)
(c)
(d)
(e)
(f)
(g)
(h)
(i)
(j)

◂**Fig. 2.10** Selected images used to edit animated display of shock wave reflection over a movable wedge installed in a 60 mm × 150 mm diaphragm-less shock tube for Ms = 1.20 in air at 1013 hPa, 301 K: **a** #94100402, $\theta_w = 2.0°$; **b** #94100404, $\theta_w = 4.5°$; **c** #94100406, $\theta_w = 5.9°$; **d** #94100409, $\theta_w = 7.4°$; **e** #94100411, $\theta_w = 8.9°$; **f** #94100414, $\theta_w = 11.1°$; **g** #94100417, $\theta_w = 13.4°$; **h** #94100419, $\theta_w = 14.8°$; **i** #94100421, $\theta_w = 16.3°$; **j** #94100423, $\theta_w = 17.8°$; **k** #94100501, $\theta_w = 19.3°$; **l** #94100503, $\theta_w = 20.8°$; **m** #94100505, $\theta_w = 22.2°$; **n** #94100507, $\theta_w = 23.7°$; **o** #94100509, $\theta_w = 25.2°$; **p** #94100511, $\theta_w = 26.7°$; **q** #94100512, $\theta_w = 27.4°$; **r** #94100514,$\theta_w = 28.9°$; **s** #94100516, $\theta_w = 30.4°$; **t** #94100518, $\theta_w = 31.8°$; **u** #94100520, $\theta_w = 33.2°$; **v** #94100522, $\theta_w = 34.6°$; **w** #94100524, $\theta_w = 36.1°$; **x** #94100526, $\theta_w = 37.6°$; **y** #94100528, $\theta_w = 39.1°$; **z** #94100530, $\theta_w = 40.6°$; **A** #94100532, $\theta_w = 42.1°$; **B** #94100534, $\theta_w = 43.5°$; **C** #94100535, $\theta_w = 44.2°$; **D** #94100540, $\theta_w = 48.0°$; **E** #94100542, $\theta_w = 49.4°$; **F** #94100544, $\theta_w = 50.9°$; **G** #94100546, $\theta_w = 52.4°$; **H** #94100544, $\theta_w = 54.5°$

would be affected by the initial pressure. However, the effect of initial pressure on the transition of reflected shock waves from wedges were hardly detectable in conventional shock tubes.

When experiments are conducted in a shock tube having a higher degree of reproducibility, the viscous effect on the shock wave transitions would be resolved more clearly. It would be more straightforward to conduct numerical analysts for reproducing interferometric images.

2.1.3.3 Animated Display of Shock Wave Reflection from a Movable Wedge

A wedge was placed on a movable stand installed in the 60 mm × 150 mm shock tube. The movable stand had a similar structure to that shown in Fig. 2.5. The wedge angle was variable from 0° to 60°. The reflected shock waves were visualized by double exposure holographic interferometry for Ms = 1.20 in air at the atmospheric pressure and 301 K. The reconstructed holograms were edited in the form of animated display, in which the wedge surface was positioned horizontally on the bottom of the animated display and the reflection point was positioned at the center of the individual frames. Therefore, the IS was at first vertical to the center of horizontal wedge and started continuously tilting. The MS was also tilting and the TP was moving. The animated display only allowed such a unique evolution of shock wave reflection patterns. Figure 2.10 shows selective images taken from the interferometric images.

In the animation, the density variation around the TP and the growth of the MS are observed through the variation of fringes. When the IS is vertical to the wedge surface, the TP doesn't exist. However, when the MS appears over a shallow wedge surface. With the increase in the wedge angle θ_W. the MS was slightly curved, the TP lies on the trajectory angle of the glancing incidence angle. The MS, however, is not uniformly curved but has a minimal radius of curvature in a position slightly below the TP. With further increasing the wedge angle θ_W, the point of the minimal radius of curvature is gradually approaching toward the TP. At the same time, fringes appear behind the MS and increase their number. Fringes tend to merge

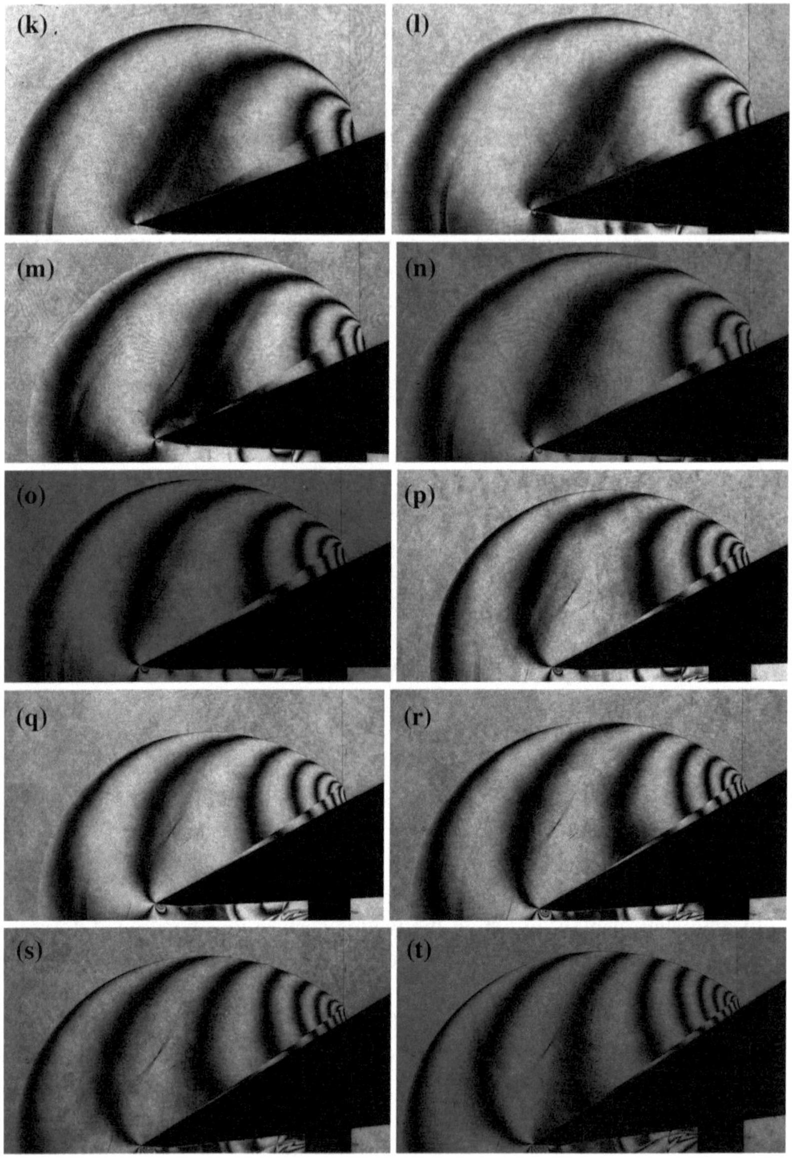

Fig. 2.10 (continued)

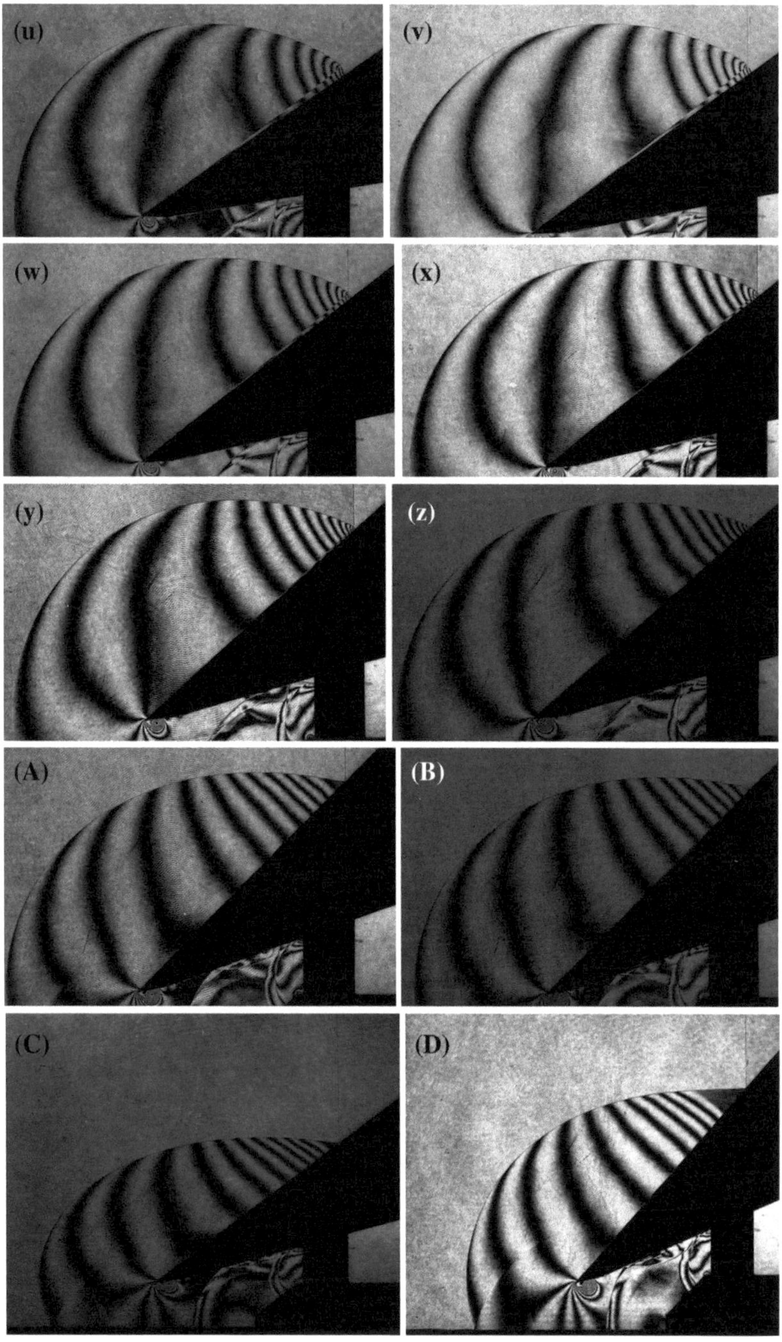

Fig. 2.10 (continued)

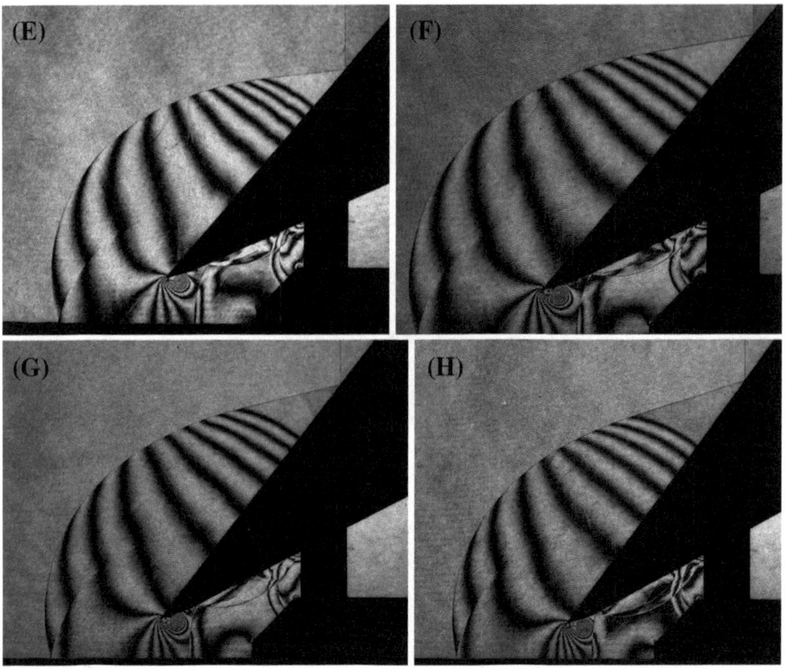

Fig. 2.10 (continued)

toward apparent TP. The fringes merged with the SL, was very vaguely visible in Fig. 2.10h, and was slightly clearly observed in Fig. 2.10i. The transition to a MR took place in Fig. 2.10y at $\theta_w = 39.1°$.

Figure 2.11 shows sequential images of very weak shock wave reflected from a wedge mounted on a movable stand installed in the 60 mm × 150 mm diaphragm-less shock tube. In Fig. 2.11a–h, the reflection patters are the vNMR so that the SL is missing. In Fig. 2.11i, the transition to the SbRR takes place.

2.1.3.4 Weak Reflected Shock Wave in a 150 mm × 60 mm Shock Tube

The density variation $\Delta\rho$ behind the shock wave in the double exposure holographic interferogram is related to the fringe number N in Eq. (1.9) as $N = \Delta\rho L/K\lambda$. This means that in an optical arrangement having a longer *OB* path, the sensitivity becomes higher. Then the 60 mm × 150 mm shock tube was turned 90° sideway forming a 150 mm × 60 mm cross sectional shock tube. Its optical sensitivity was as high as 2.5 times than the former shock tube. A 150 mm wide wedge model with a 60° edge angle was sandwiched between 150 mm diameter and 20 mm thick acrylic plates and installed into the test section of the shock tube.

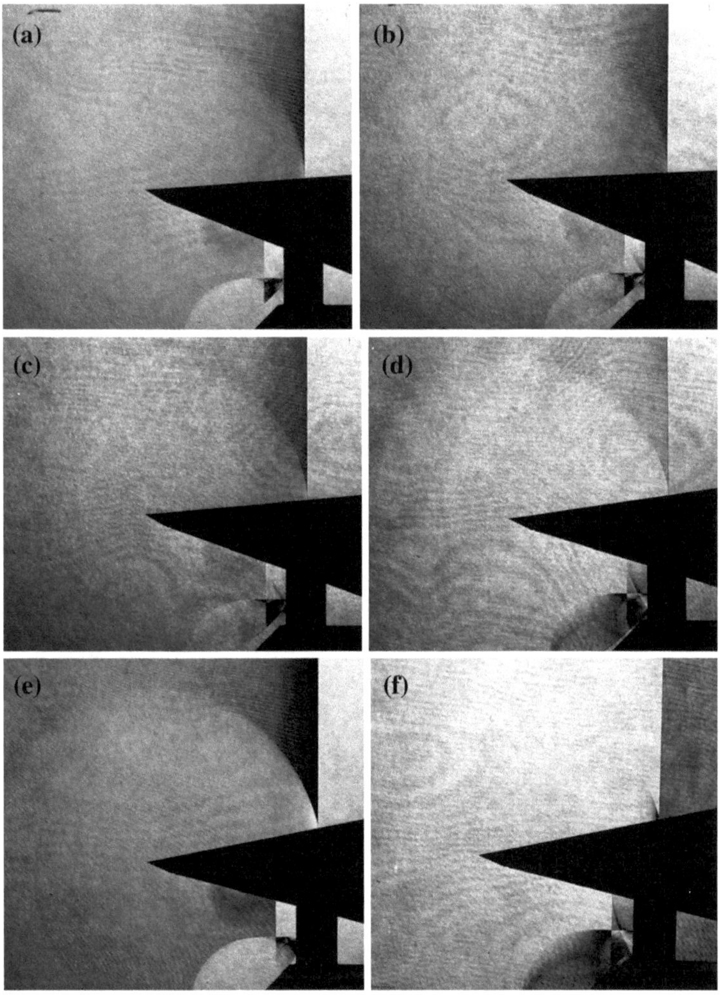

Fig. 2.11 Shock wave reflection over wedges mounted on a movable stage in a 60 mm × 150 mm diaphragm-less shock tube for Ms = 1.032 in atmospheric air, 299.5 K: **a** #94061102, $\theta_w = 3.0°$; **b** #94061101, $\theta_w = 5.0°$; **c** #94061103, $\theta_w = 6.0°$; **d** #94061105, $\theta_w = 7.5°$; **e** #94061107, $\theta_w = 10.0°$; **f** #94061108, $\theta_w = 12.0°$; **g** #94061110, $\theta_w = 14.5°$; **h** #94061111, $\theta_w = 16.0°$; **i** #94061114, $\theta_w = 17.5°$; **j** #94061115, $\theta_w = 18.5°$; **k** #94061117, $\theta_w = 20.0°$; **l** #94061120, $\theta_w = 21.5°$

Wedge angles were adjusted by rotating the observation windows and the whole wedge. Figure 2.12a–r shows sequential results for Ms = 1.20 in air at 800 hPa, taken at 80 μs after the IS passed the leading edge of the wedge. In Fig. 2.12a at $\theta_w = 2.0°$, no fringe is observed behind the MS, whereas a dark fringe appears behind the RS and the IS. In Fig. 2.12b at $\theta_w = 4.0°$ and in Fig. 2.12c at $\theta_w = 6.0°$, a dark broadened fringe appears behind the MS. In Fig. 2.12b, c, the MS is curved not uniformly but its radius of curvature varies. Two dark fringes intersect with the

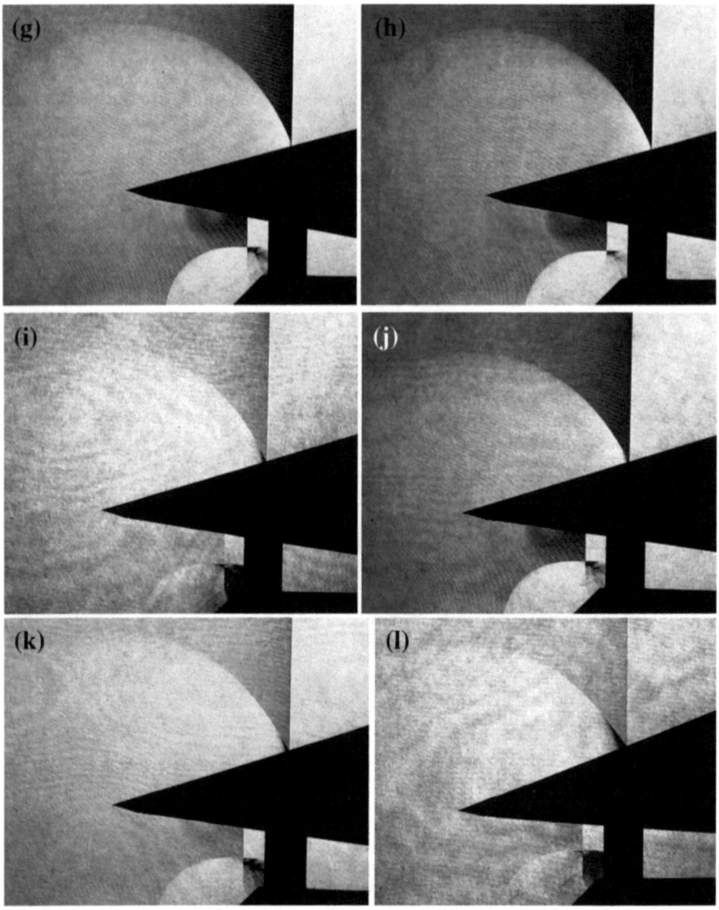

Fig. 2.11 (continued)

MS at which the radius of curvature becomes a minimum. In shallower wedge angles, the IS, RS, and MS merge at a point forming a three-shock confluence, or TP. However, the SL is not emanating from the TP as no density discontinuity exists behind the MS. Dark fringes running downward along the RS intersect with the MS, whereas other dark fringes starting from the wedge surface upward eventually intersect with the MS. These downward shifting fringes and upward shifting fringes intersect with the MS forming just like dividing stream lines. Such a fringe distribution is a typical pattern of a vNMR. Eventually, the radius of curvature along the MS becomes minimal at the point at which the fringe distribution looks like a dividing stream line. Then with the increase in the wedge angle, the fringes gradually converge into the TP and the MS becomes straight and is shortened. The shock wave reflection pattern eventually becomes a SMR. In Fig. 2.12s at $\theta_w = 34°$, the SMR transits to a RR.

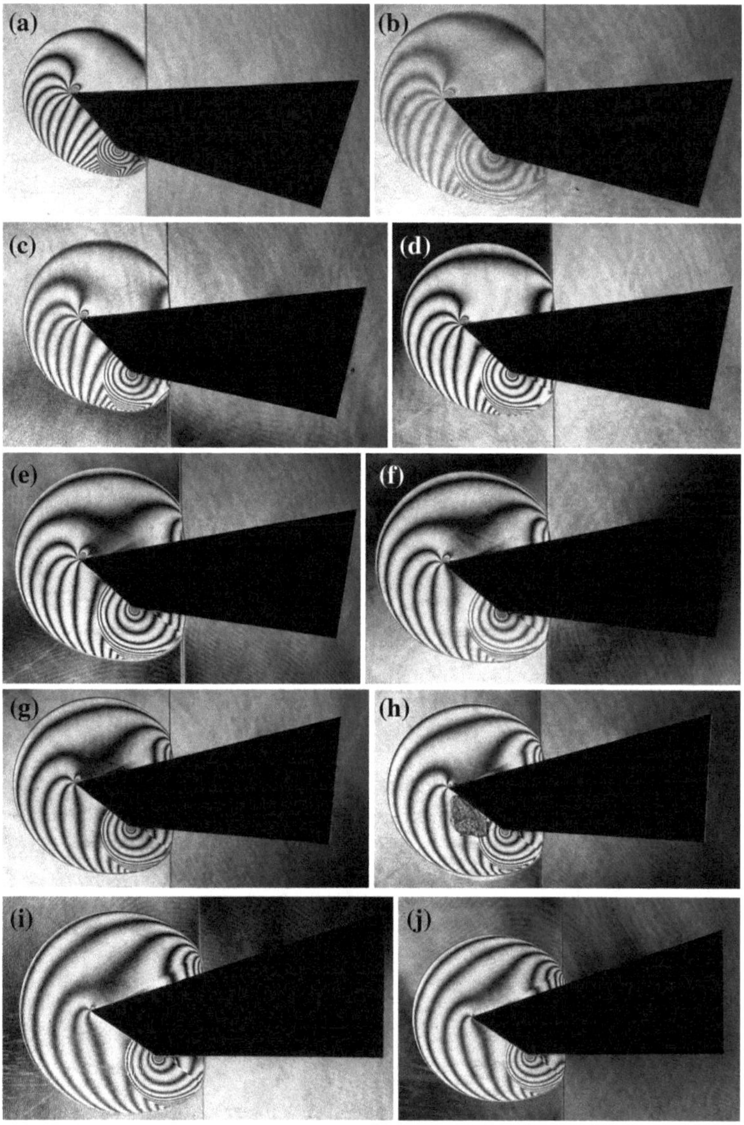

Fig. 2.12 Evolution of reflected shock wave over a 150 mm wide wedge for Ms = 1.20 in air at 800 hPa, 290.4 K in a 150 mm × 60 mm shock tube at 80 μs from the leading edge: **a** #88031004, Ms = 1.204, θ = 2°; **b** #88031006, Ms = 1.203, θ = 4°; **c** #88031101, Ms = 1.200, θ = 6°; **d** #88031103, Ms = 1.206, θ = 8°; **e** #88031105, Ms = 1.213, θ = 10°; **f** #88031107, Ms = 1.207, θ = 12°; **g** #88031111, Ms = 1.201, θ = 14°; **h** #88031113, Ms = 1.206, θ = 16°; **i** #88031115, Ms = 1.198, θ = 18°; **j** #88031103, Ms = 1.206, θ = 20°; **k** #88031118, Ms = 1.207, θ = 21°; **l** #88031119, Ms = 1.211, θ = 22°; **m** #88031123, Ms = 1.180, θ = 26°; **n** #88031125, Ms = 1.186, θ = 27°; **o** #88031126, Ms = 1.187, θ = 28°; **p** #88031129, Ms = 1.198, θ = 31°; **q** #88031130, Ms = 1.197, θ = 32°; **r** #88031132, Ms = 1.180, θ = 34°; **s** #88031133, Ms = 1.172, θ = 35°

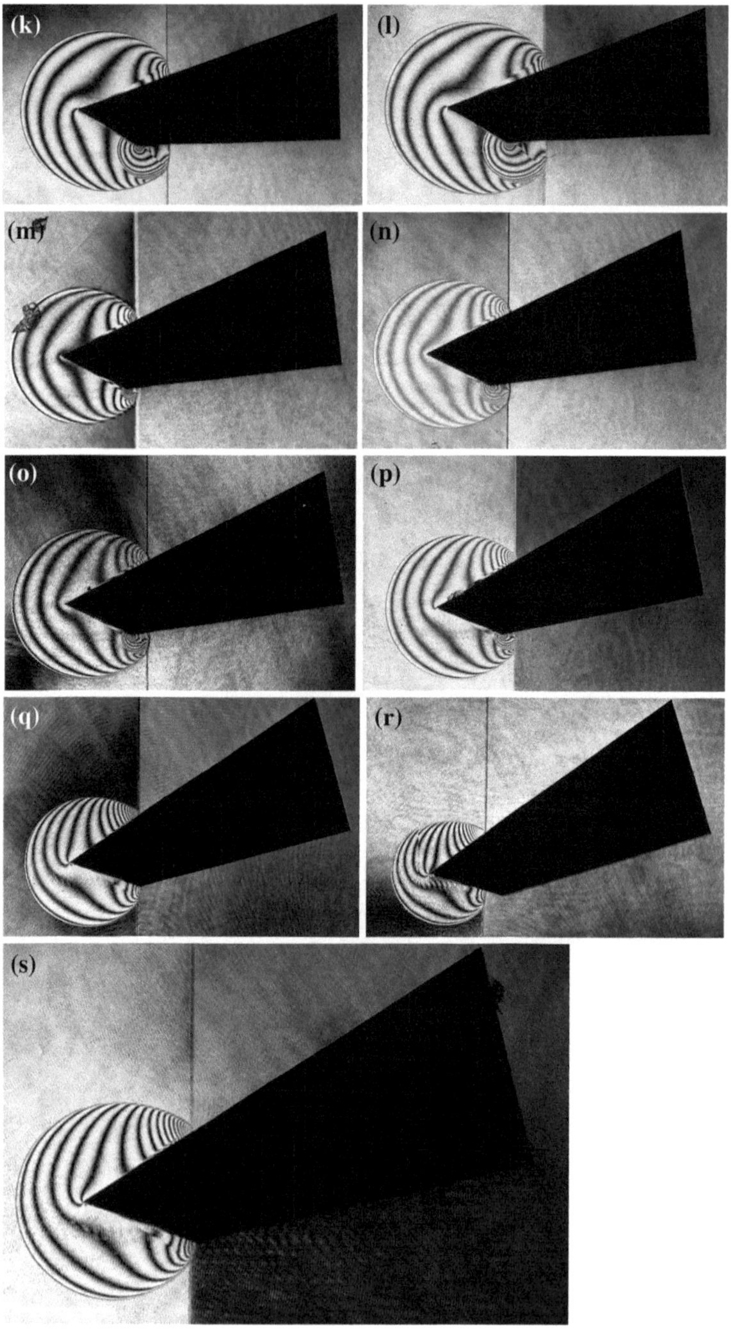

Fig. 2.12 (continued)

Figure 2.13 summarizes the variation of the TP trajectory for the shock wave of Ms = 1.20 reflected from wedges of variable angles as shown in Figs. 2.10 and 2.12. The ordinate denotes the triple point trajectory angle χ in degree and the abscissa denotes the wedge angle θ_w in degree. The glancing incidence angle of the shock wave of Ms = 1.20 in air is 25.47°; and red circles denotes the TP as shown in Fig. 2.10 and blue circles denotes the TP as shown in Fig. 2.12. At smaller θ_w, the TPs lie on the line of $\chi_{glance} - \theta_{w.}$ Such TPs have the curved MS and intersected perpendicularly to the wedge surface. The radius of curvature is not constant but has a minimal radius along the MS. The point of the minimal radius of curvature along the MS can be experimentally estimated. In Fig. 2.13, the grey circles denote the minimal radii of curvature or the kink points shown in Fig. 2.10 and faint pink circles denote these points shown in Fig. 2.12. The grey circles and faint pink circles are distributed below the line of $\chi_{glance} - \theta_w$ and typically show the pattern of vNMR. In increasing θ_w, the MS becomes straight and the reflected shock wave pattern becomes SMR. A further increase in the wedge angles causes the transition from the SMR to the RR.

In the case of very shallow edges and for weak shock waves, the points of the merger of an IS with a RS always lie on the trajectory of $\chi_{glance} - \theta_w$, which means that pressure variations $\Delta p = \varepsilon$ are so small that the isentropic condition of Δs $(\Delta p)^3 = \varepsilon^3$ is always satisfied. On the contrary, the shock-shock is defined by Whitham (1959) as a discontinuous front carrying the boundary condition that the MS is perpendicular to the wedge surface. Then the minimum radius of curvature or a kink point is formed in the area at which the isentropic condition is satisfied and the glancing incident condition is fulfilled. Figure 2.14 show reflection of shock wave for Ms = 1.05 in air over a movable wedge ranging from 2° to 30°. It should be noticed that the reflection patters are vNMR for the wedge angle up to the critical

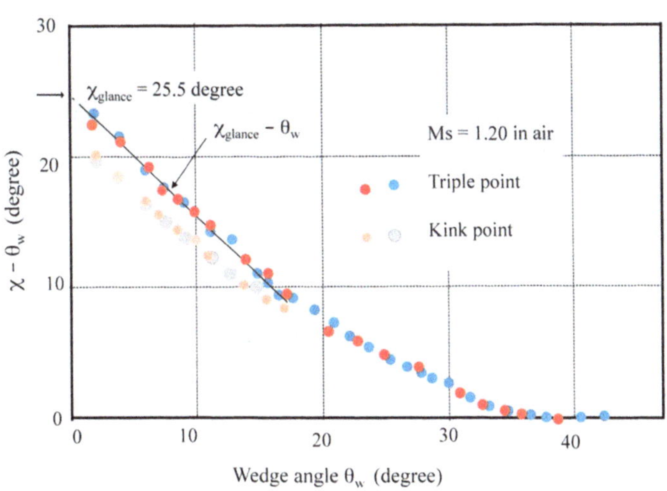

Fig. 2.13 Summary of variations of the triple point trajectory angle χ versus the wedge angle θ_w shown in Figs. 2.10 and 2.12

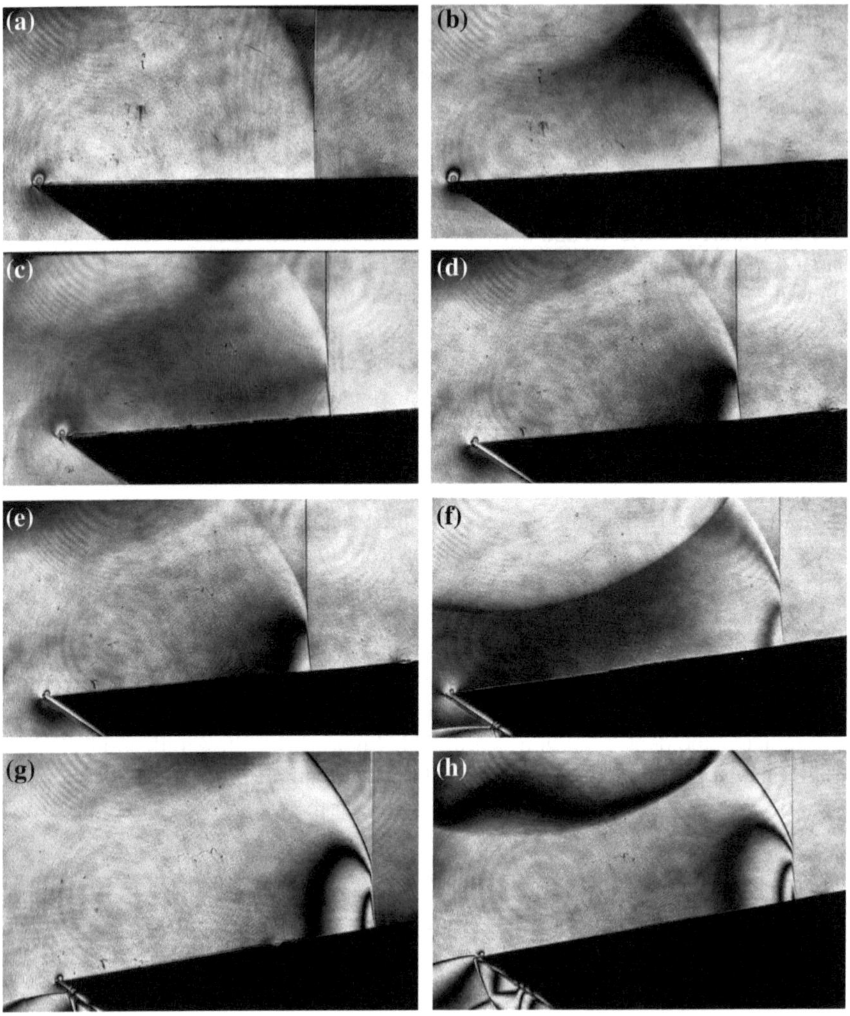

Fig. 2.14 Evolution of reflection of weak shock waves in a 150 mm × 60 mm shock tube for Ms = 1.05 in air at 1013 hPa, 290 K: **a** #88011108, Ms = 1.056, $\theta_w = 2.0°$; **b** #88011104, Ms = 1.051, $\theta_w = 3.0°$; **c** #88011102, Ms = 1.037, $\theta_w = 4.0°$; **d** #88010905, Ms = 1.050, $\theta_w = 5.0°$; **e** #88010903, Ms = 1.053, $\theta_w = 6.0°$; **f** #88010812, Ms = 1.049, $\theta_w = 8.0°$; **g** #88010810, Ms = 1.049, $\theta_w = 9.0°$; **h** #88010807, Ms = 1.061, $\theta_w = 10.0°$; **i** enlargement of (**h**); **j** #88010801, Ms = 1.058, $\theta_w = 12.0°$: **k** #88010716, Ms = 1.049, $\theta_w = 13.0°$; **l** #8801071, Ms = 1.058, $\theta_w = 14.0°$; **m** #88010711, Ms = 1.060, θw = 15.0°; **n** #88010707, Ms = 1.051, θw = 16.0°; **o** #88010704, Ms = 1.051, θw = 17.0°; **p** #87113005, Ms = 1.075, $\theta_w = 18.0°$, in a shock tube L = 60 mm; **q** #88010601, Ms = 1.072, $\theta_w = 20.0°$; **r** #87120106, Ms = 1.070, $\theta_w = 21.0°$ in a shock tube L = 60 mm; **s** #87120110, Ms = 1.073, $\theta_w = 23.0°$ in a shock tube L = 60 mm; **t** #87120118, Ms = 1.068, $\theta_w = 27.0°$ in a shock tube L = 60 mm; **u** #87120201, Ms = 1.071, $\theta_w = 29.0°$ in a shock tube L = 60 mm; **v** #87120203, Ms = 1.062, $\theta_w = 30.0°$ in a shock tube L = 60 mm

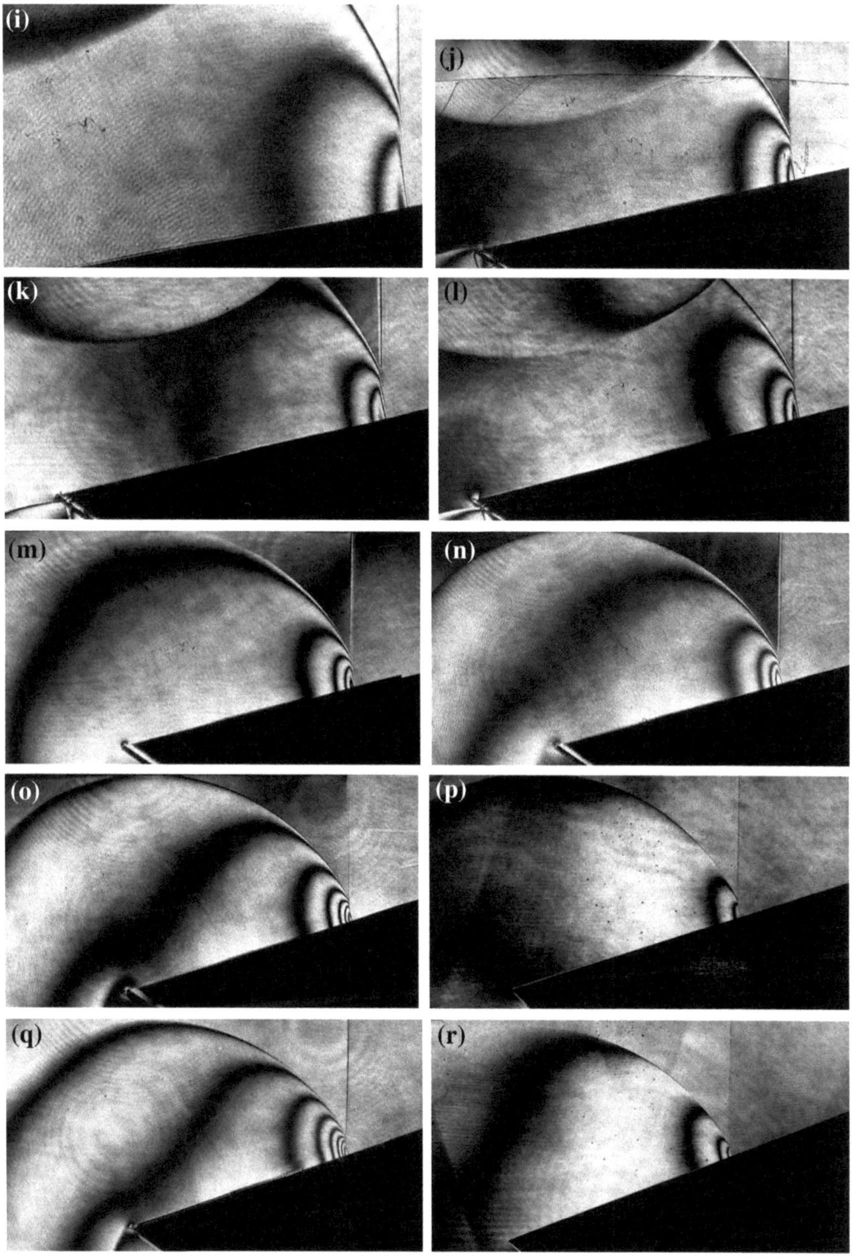

Fig. 2.14 (continued)

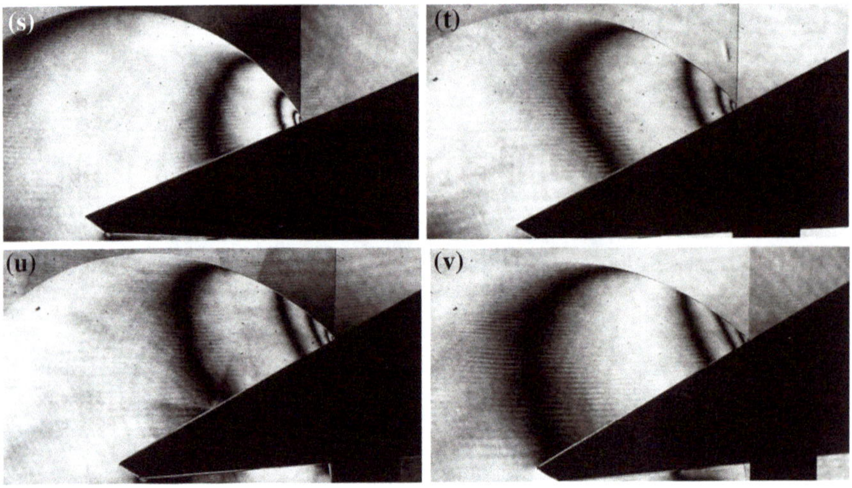

Fig. 2.14 (continued)

transition at about 29°. It is also noticed that the minimum radius of curvature point consistently appears just below the TP.

2.1.3.5 Shock Wave Reflection from Wedges in Dusty Gas

Figure 2.15 shows experimental arrangement composed of a 50 mm diameter conventional shock tube that was connected to a 30 mm × 40 mm shock tube test section. Almost spherically shaped fly ash were used as dust particles of about 5 μm in diameter; they were supplied from a dust feeder carried by an air flow placed just at the diaphragm section. At the end of the shock tube, a 150 mm diameter and 1.5 m long dump tank was connected to which a vacuum pump and a dust filter were attached Suguyama et al. (1986). The dust particle so far feeding was continuously circulated and recovered. The dust particles were distributed relatively uniformly over the test section at loading ratio of 0.02. The loading ratio was

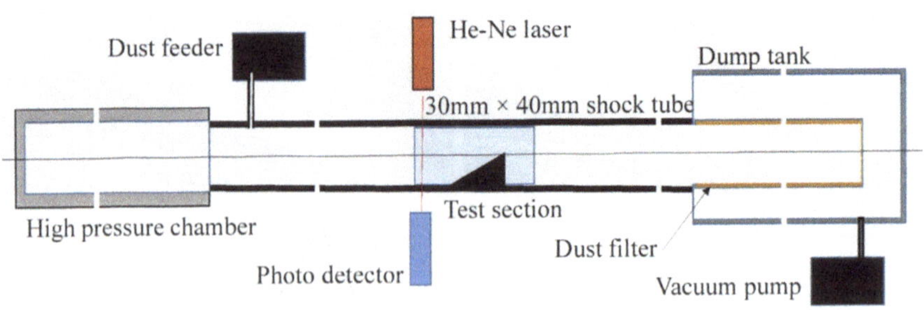

Fig. 2.15 Schematic diagram of dusty gas shock tube experiment (Suguyama et al. 1986)

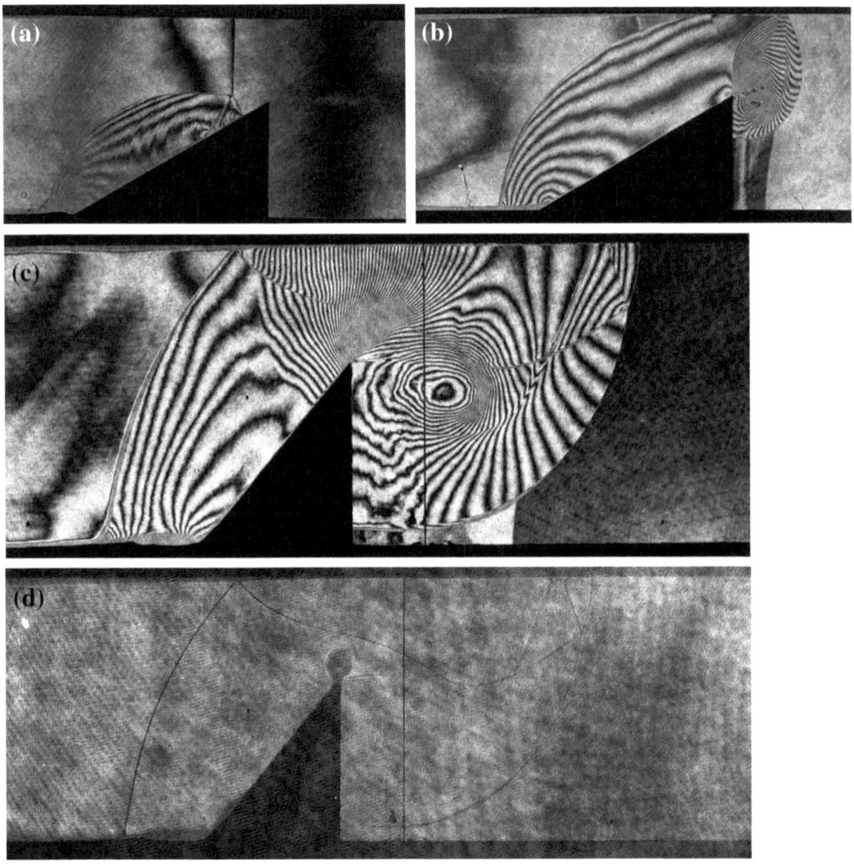

Fig. 2.16 Shock wave reflection over wedges in dusty gas for Ms = 2.05 in air at 1013 hPa, 289.5 K: **a** #87042305, Ms = 2.022, $\theta_w = 30°$; **b** #87042210, Ms = 2.044, $\theta_w = 30°$; **c** #87042215, Ms = 2.053, $\theta_w = 45°$; **d** unreconstructed hologram of (**c**)

defined as a ratio of particle mass to the mass of tested air. With this arrangement, the experiment was conducted for Ms ranging from 1.5 to 2.20 and corresponding Reynolds number ranging from 2.6×10^5 to 6.4×10^5 based on the characteristic length of the shock tube. Several different wedges in the shock tube test section. Unlike in the previously reported cases, in the present experiments, it was impossible to determine θ_{crit} as there was no movable stand for the tested wedges. In the present experiments, only shock wave reflection patterns were observed.

Figure 2.16 shows shock wave reflection patterns for Ms = 2.01. The reflection pattern seen in Fig. 2.16a is the SMR almost identical with the SMR shown in Fig. 2.9e. The boundary layer developing along the shock tube upper wall looks thicker whereas the sedimentation of dust particles on the bottom wall is observed; a layer of dust particle is blown off from the edge. In Fig. 2.16b, the transmitted shock wave was diffraction at the corner of the wedge and a vortex is formed.

The wedge was not tightly touched the observation windows and gap intervals of probably 0.05 mm existed. Then the IS leaked through this gap. The grey pattern observed at the rear corner of the wedge seen in Fig. 2.16b was the leakage of the IS. It would be a meaningful to clarify the effect of the dust loading ratio on the prevailing value of θ_{crit} of a shock wave reflection from wedges. The present experiment is a preliminary test for designing a dusty gas shock tube in which reproducible runs would produce reliable dusty shock tube flows. However, it would be appropriate to supply reliable interferometric image to support a numerical scheme for predicting the effect of dust loading ratio on θ_{crit}. Figure 2.16c, d shows the diffraction of the SMR over a 45° wedge, an interferogram and its unreconstructed image. The flow pattern is similar to shock wave diffraction over a backward facing step but it is impossible to identify the effect of dust particles on the shock wave reflection. In Fig. 2.16d, in an unreconstructed hologram, an elongated grey region just behind the bifurcated reflected shock wave at the leading edge and a circular grey region at the tip of the wedge. In such regions, the local stream lines sharply curve so that the spontaneously induced centrifugal forces would eject dust particles away from the regions. Then dust free regions are formed. However, it is a puzzle that the dust free region appears to be grey. While passing through the dusty gas shock tube flows, the collimated OB is scattered with uniformly distributed dust particles. Hence resulting *OB* were scattered due to the so-called Mie scattering Merzkirch (1974), whereas the *OB* passing the dust free regions was not affected by the light scattering. Therefore, the *OB* passing through the dust free region appears to look greyer than that through the dusty air. Then in Fig. 2.16d, the difference of light intensity is clearly distinguished. However, in the double exposures, the light scattering never contributed to the variation of the phase angle so that the dust free region did create no difference in fringe distributions.

Figure 2.17 shows the case for Ms = 1.75 and reflection patterns are similar to Fig. 2.9. In Fig. 2.17a, other than fringe distributions, as grey noises, sedimentations of dust particles on the bottom wall and on the wedge surface as well are identified. In Fig. 2.17a, b, as background noises, shadows of dust particles blown off from the corner of the wedge are observed. Figure 2.17c, d shows an interferogram and its unreconstructed image, respectively. Although the relationship between the dusty gas loading ratio and its refractive index is not known, dark fringe distribution behind the reflected shock wave in Fig. 2.17c and the grey noise distribution in Fig. 2.17d agreed with each other. The sedimentation of dust particles on the shock tube bottom wall and the wedge surface is faintly visible. The reflected shock wave is diffracted at the corner of the wedge. The transmitted shock wave forms a vortex at the corner as seen in Fig. 2.17c. Figure 2.17d shows the unreconstructed hologram of Fig. 2.17c. In Fig. 2.17d, dust free region is observed inside the region in which fringes are densely concentrated seen in Fig. 2.17c.

Figure 2.18a, b shows evolution of weak shock wave reflection of Ms = 1.40 from a wedge $\theta_w = 20°$ in dusty gas. The reflection patterns are almost identical with the reflection from a shallow wedge in a pure gas. Figure 2.18c shows a MR from a wedge of $\theta_w = 35°$. Figure 2.18d shows a RR from a wedge of $\theta_w = 45°$. So far observed, no significant difference is not observed in fringe distributions

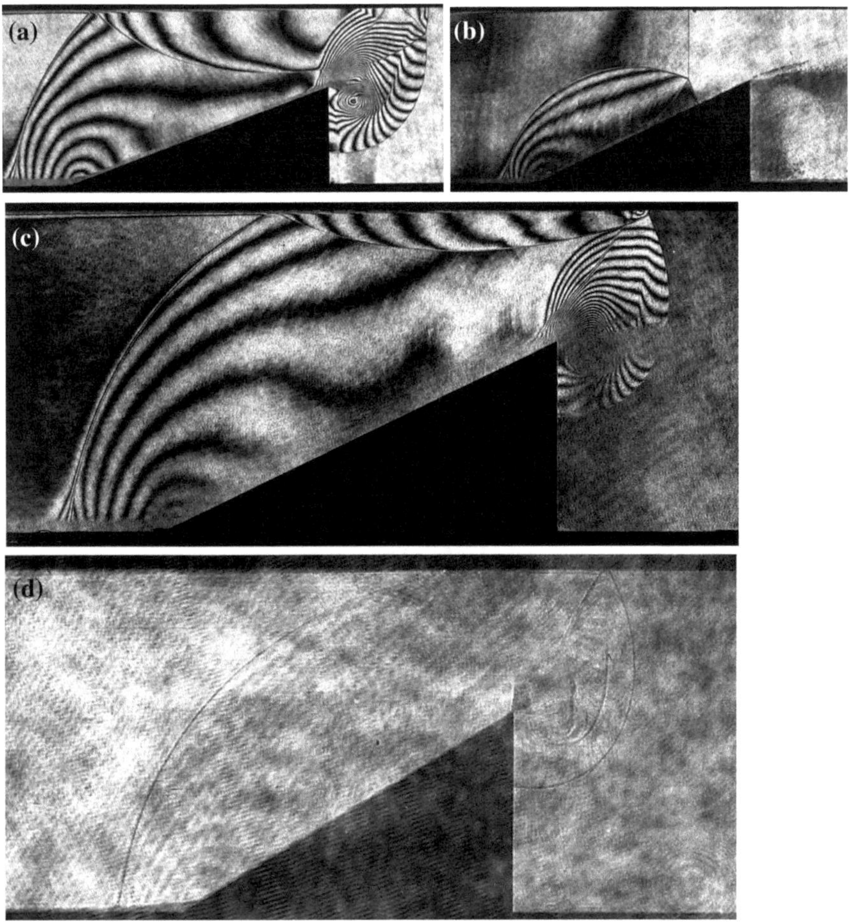

Fig. 2.17 Shock wave reflection over wedges in dusty gas for Ms = 1.75 in air at 1013 hPa, 289.5 K: **a** #87042108, Ms = 1.755, $\theta_w = 20°$; **b** #87042111, Ms = 1.718, $\theta_w = 25°$; **b** #87042305, Ms = 2.022, $\theta_w = 30°$; **c** #87042110, Ms = 1.718, $\theta_w = 25°$; **d** enlargement of (**c**)

between the present dusty gas shock tube flow and a pure air shock tube flow. These images would be useful to validate numerical schemes for reproducing dusty gas shock tube flows.

2.1.3.6 Delayed Transition in a 100 mm × 180 mm Shock Tube

In 1996, Professor Dewey organized the Mach Reflection Symposium for the second time in Victoria and reported a puzzling result of shock wave reflections over a wedge whose angle was slightly larger than θ_{crit}. in argon. Then the reflection pattern should be MR. However, Professor Dewey showed that at the leading edge of the

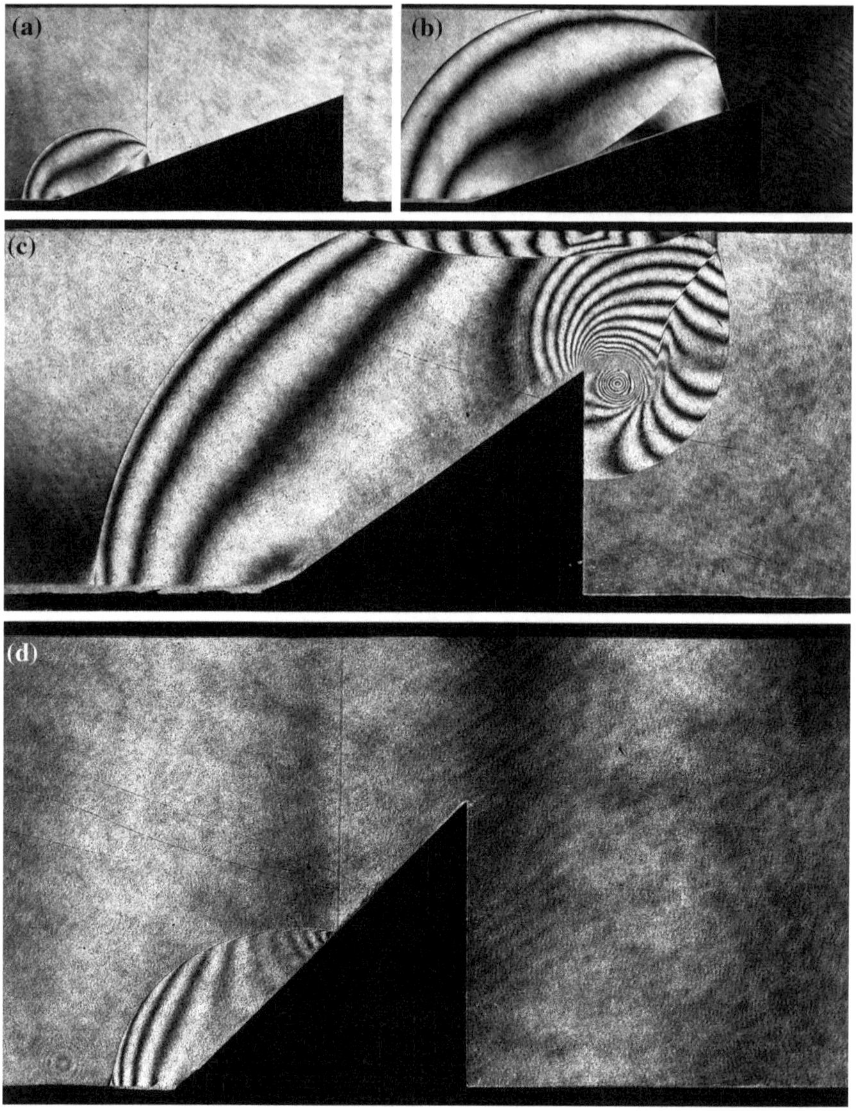

Fig. 2.18 Shock wave reflection over wedges in dusty gas for Ms = 1.44 in atmospheric air at 290.2 K: **a** #87042102, Ms = 1.432, $\theta_w = 20°$; **b** #87042103, Ms = 1.440, $\theta_w = 20°$; **c** #87042122, Ms = 1.432, $\theta_w = 35°$; **d** #87042204, Ms = 1.412, $\theta_w = 45°$

wedge, a RR appeared and it transited to a MR in the distance away from the leading edge. The audience were astonished at his lecture and expressed their own comments. No one at that time understood why delayed transitions occurred. Later Henderson proved numerically that it was the presence of the boundary layer developing along the wedge surface that caused the delayed transition (Henderson et al. 1997). In 1994, Professor Glass donated his 100 mm × 180 mm

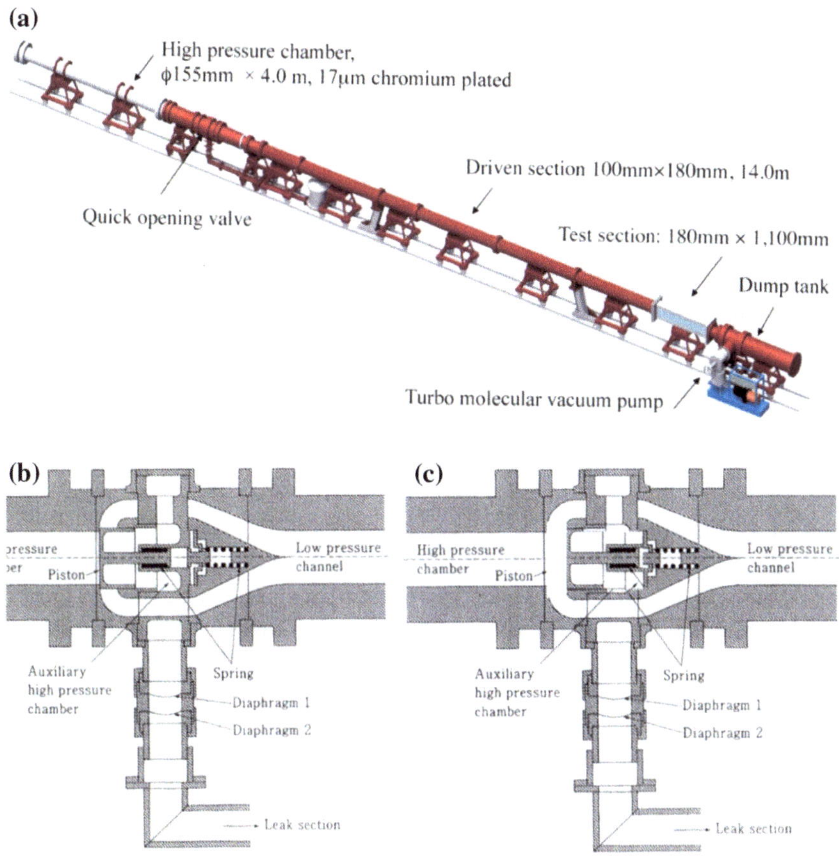

Fig. 2.19 Refurbished diaphragm-less shock tube in the SWRC of the IFS (Itabashi 1998): **a** 100 mm × 180 mm shock tube; **b** quick opening valve closed; **c** the piston was quickly stored inside the tear-drop shaped container

Hypervelocity Shock Tube of University of Toronto Institute of Aerospace Science (UTIAS) to SWRC of the Institute of Fluid Science (IFS) Tohoku University. When performing experiments in a small shock tube, results obtained in the shock tube consistently deviated from analytical predictions because of the size effect of the shock tubes. In 1994, when the UTIAS Shock Tube was delivered to SWRC, it was refurbished as shown in Fig. 2.19a. The driver chamber was manufactured at the Nippon Steel Co. Ltd., in Muroran; a single piece of high tensile strength steel of 4 m in length, 155 mm in inner diameter, 355 mm in outer diameter. The inner surface of the driver chamber was plated with chromium of 17 μm in thickness and its tolerance of the diameter was 3 μm. The test section was replaced with a stainless steel test section of the field of view of a 180 mm × 1,100 mm. Its length was wide enough to visualize shock tube flows with a 1000 mm diameter collimated *OB*. Glass (1975) stated in his book that a shock tube was a test tube of modern aerodynamics. Indeed, the shock tube contributed immensely to the development of high speed gas-dynamics and supported the atmospheric reentry technology. However, this shock tube was not repeatable. The UTIAS shock tube adopted a diaphragm-less operational system. A high-pressure driver gas was separated by an aluminum piston from the downstream side as shown in Fig. 2.19b. The piston was supported by high-pressure helium from behind. A Mylar diaphragm was ruptured by using a double diaphragm system. Then the piston receded very quickly and neatly fit into a tear-drop shaped container as shown in Fig. 2.19c. At the same time, a shock wave moved around the container's outer boundary. A higher degree of reproducibility was achieved by this system but the planar shock formation distance became slightly longer. The piston moved very quickly from its initial position to its recovery position at 75 mm distance. It was relatively easy to accelerate the piston but it was hard to safely attenuate the piston. The piston should be repeatedly operated at least for about 300 times. Hence, a spring and a damper mechanism were used inside the container. The spring and damper were regularly maintained and after about 300 runs the parts were repaired.

Figure 2.20 shows characteristic diagram of Ms against the pressure ratio in air and nitrogen. The ordinate denotes shock wave Mach number Ms and the abscissa denotes the pressure ratio of a driver gas pressure to a test gas pressure, p_4/p_1. In a simple shock tube theory, the ratio of the driver gas pressure p_4 and the test gas pressure p_1 is given by

$$p_4/p_1 = p_4/p_2 \cdot p_2/p_1, \tag{2.11}$$

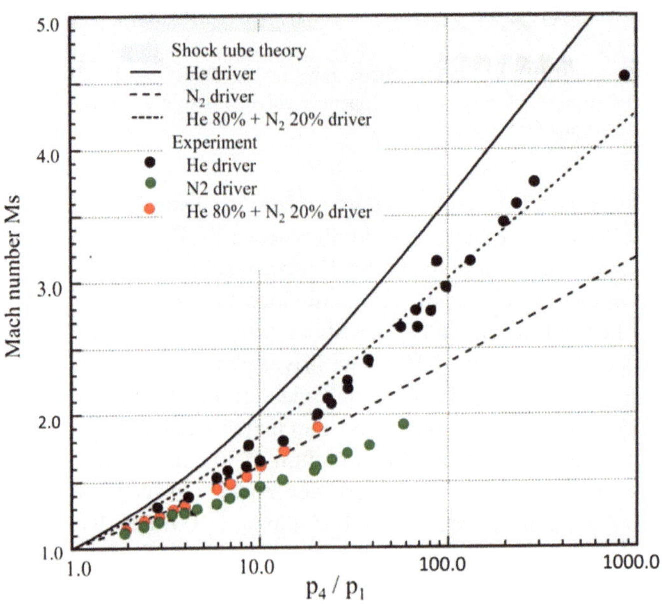

Fig. 2.20 Characteristics of the present refurbished shock tube (Itabashi 1998)

where $p_2/p_1 = (2\gamma_1 Ms^2 - \gamma_1 + 1)/(\gamma_1 + 1)$, $p_4/p_2 = \{1 - (\gamma_4 - 1)a_{41}(Ms - 1/Ms)/(\gamma_1 + 1)\}^{-m}$, $a_{41} = a_4/a_1$, and $m = 2\ \gamma_4/(\gamma_4 - 1)$. a_4 and a_1 are sound speeds of the driver and test gases, respectively. γ_4 and γ_1 are the specific heats ratio of the diver and test gases, respectively.

Experimental results were compared with a simple shock tube theory given by Eq. (2.11) (Gaydon and Hurle 1963) for high-pressure helium and nitrogen driver and a mixture of 80% helium and 20% nitrogen in volume ratio. Experimental results and the individual predictions lie neatly on lines. But in conventional diaphragm rupturing systems, experimental results scatter relatively widely. Helium drivers were contaminated with 80% helium and 20% nitrogen in volume and had a fair agreement between the prediction and experiments. Fine broken lines denote the driver of gas mixtures. The scatter of Ms obtained under the identical initial condition ΔMs is ±0.3% for Ms ranging from 1.5 to 5.0 in air. When the shock tube experiments are conducted continuously in a day, the ΔMs even decreases to ±0.1%. Conventional rupturing diaphragm system could achieve the reproducibility ΔMs = ±1.0% at best.

Professor Dewy's lecture inspired Professor Henderson, who encouraged numerical analysts to simulate the delayed transition in argon (Henderson et al. 1997). To confirm their numerical results, experiments were conducted using the diaphragm-less shock tube (Henderson et al. 2001). Figure 2.21 shows delayed transition over wedges installed in the 100 mm × 180 mm shock tube for wedge angles of 34.6°, 44.0°, 50.5° and 52.0° and at initial pressures ranging from 37 to 282 hPa in argon, for the Reynolds numbers, Re, ranging from 3×10^5 to 5×10^6. where the Re was defined referring the hydraulic radius of the shock tube and the shock tube flows. For double exposure holographic interferometric observation, the first exposure was performed when the IS reached close to the leading edge and the second exposure was performed as seen in Fig. 2.21a at Δt = 120 μs after the first exposure. At the second exposure, the IS reached at the center of the wedge surface. Images of reflected shock waves observed during the double exposures are superimposed. Even in these interferograms, the density contours so far recorded are normalized by the density of the air ahead of the IS. Hence the fringes are more densely distributed at enhancing the initial pressure.

Figure 2.21a, b shows the SMR over a 34.6° wedge and at the initial pressure of $p_0 = 144$ hPa. Figure 2.21c, d shows the SMR over a 38.6° wedge and at the initial pressure of $p_0 = 288$ hPa. In Fig. 2.21a–d, the RS emanating from the TP looks straight in the vicinity of TP but a kink point is not clearly identified. In Fig. 2.9, the evolution of shock wave reflection patterns for Ms = 2.60 in air was observed sequentially over the wedges of θ_w which varied from 0° to θ_w much larger than θ_{crit}. The reflected shock wave patterns are supposed to be self-similar. However, on a hologram seen in Fig. 2.21, the two shock wave patterns are not necessarily self-similar. The reflected shock wave patterns varied slightly with the distance away from the leading edge. This implies that the boundary layer developing along the wedge surface varied the local boundary conditions. Figure 2.21e, f shows the TMR over a 44.0° wedge and at $p_0 = 144$ hPa.

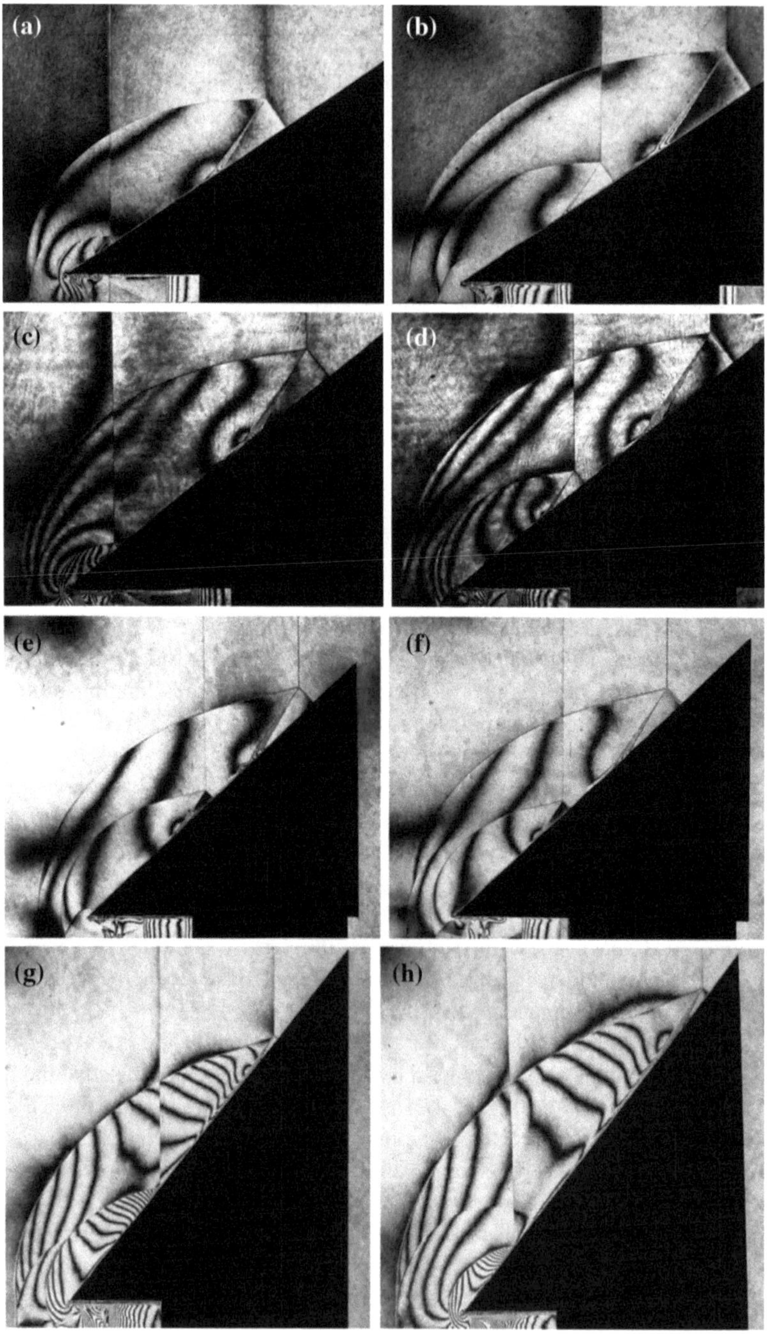
(a)
(b)
(c)
(d)
(e)
(f)
(g)
(h)

◂**Fig. 2.21** Delayed transition over wedges in 100 mm × 180 mm shock tube in argon: **a** #96091003, interval of double exposures of Δt = 120 μs, for Ms = 2.326 at 144 hPa and 295.9 K, wedge angle of θ_w = 34.6°; **b** #96090905, 80 μs, for Ms = 2.333 at 144 hPa and 296.4 K, θ_w = 34.6°; **c** #98112508 120 μs for Ms = 2.320 at 288 hPa and 291.6 K, θ_w = 38.6°; **d** #98112509, 80 μs, for Ms = 2.328 at 288 hPa and 291.5 K, θ_w = 38.6°; **e** #97040502, 60 μs, Ms = 2.315 at 144 hPa, 290.3 K, θ_w = 44.0°; **f** #97041501, 60 μs, for Ms = 2.334 at 144 hPa and 289.0 K, θ_w = 44.0°; **g** #97040409, 80 μs, Ms = 2.338 at 288 hPa and 291.3 K, θ_w = 52°; **h** #97040404, 80 μs, Ms = 2.329 at 288 hPa and 290.9 K, θ_w = 52°; **i** #97031007, 80 μs, Ms = 2.330 at 72 hPa and 289.4 K, θ_w = 52.0°; **j** #97031016, 60 μs, Ms = 2.329 at 72 hPa and 289.0 K, θ_w = 52.0°: **k** #97031109, 30 μs, Ms = 2.327 at 38 hPa and 291.3 K, θ_w = 52.0°; **l** #97031110, 30 μs, Ms = 2.331 at 38 hPa and 291.2 K, θ_w = 52.0°; **m** enlargement of (**k**)

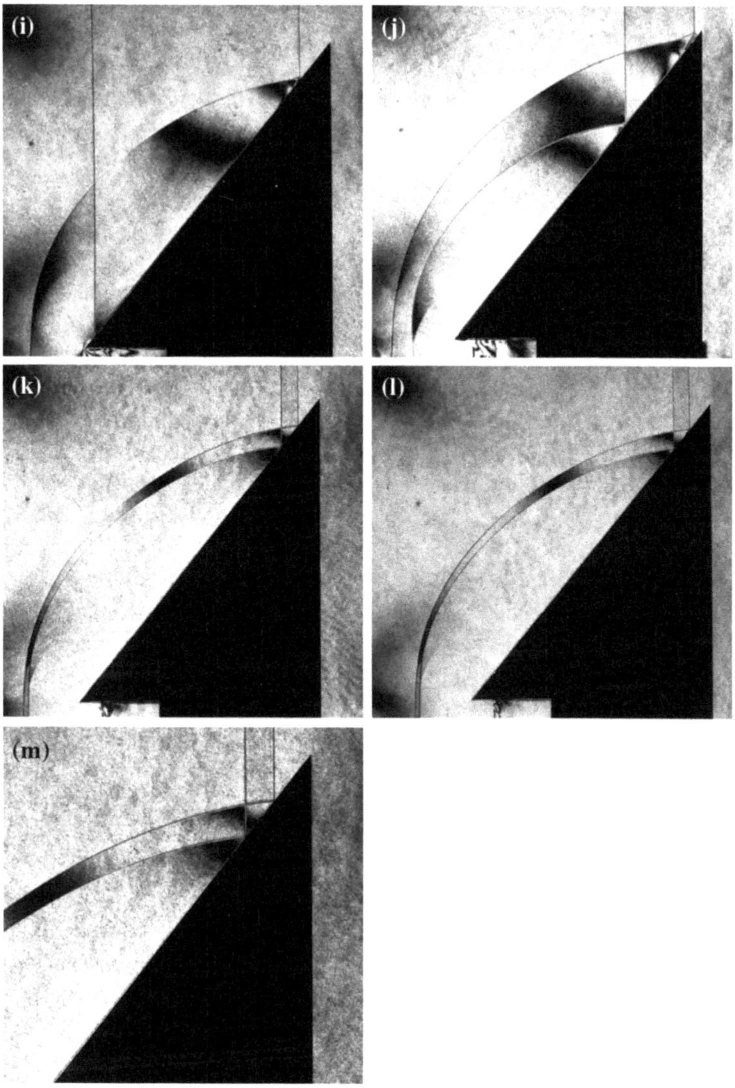

Fig. 2.21 (continued)

Figure 2.21g, h shows the DMR over a 52.0° wedge and at p_0 = 288 hPa. The two reflected shock wave patterns on a 52.0° wedge and at p_0 = 72 hPa seen in Fig. 2.21i shows a small DMR near the trailing edge of the wedge and RR close to the leading edge. In Fig. 2.21j, the reflection patterns show both small DMR near the trailing edge. When keeping the wedge angle identical and reducing the initial pressure to 38 hPa, the reflected shock wave patterns seen in Fig. 2.21k, l are supersonic RR or in short SuRR near the trailing edge of the wedge and later transits to the DMR. The initiation of MR was retarded. We now confirmed that the MR appeared delayed, the distance at which MR appeared was elongated, as the initial pressure decreased.

In shock tube flows, the IS are followed by the boundary layers, which satisfy the isothermal and non slip condition on the wedge surface. In shock fixed co-ordinates, the boundary layer displacement thickness modified the apparent contour of the shock tube wall and hence the ISs lean forward at their foot on the wall. In the neighborhood of the leading edge, apparently due to the effect of the boundary layer displacement thickness, the wedge angle θ_w exceeds θ_{crit}, and hence the reflected shock wave would be RR. When the IS moves away from the leading edge, the flow on the wedge surface satisfies the real boundary condition of $\theta_w <$ $_{crit}$. Then this information is transmitted to the reflection point through the boundary layer so that the reflection pattern reverts to MR. It is the boundary layer displacement thickness, δ, which causes delay in the RR transition. In the laminar boundary layer, the δ is inversely proportional to the Reynolds number, Re = ρuL/μ, where ρ, μ, u, and L are the density, viscosity, particle speed behind the IS, and hydraulic radius of the shock tube, respectively. The thickness effect of the boundary layer was enhanced if the initial pressure is reduced and a shock tube geometry is smaller. The interferometric images collected in a 40 mm × 80 mm shock tube as shown in Fig. 2.5 are slightly deviated from those collected in a 100 mm × 180 mm shock tube.

Figure 2.22 summarize results of visualizations (Itabashi 1998). In Fig. 2.22a, co-ordinates of triple points along a 52° wedge are presented at the initial pressures p_0 of 37, 60, 95, 141 and 282 hPa. The ordinate shows the height y of triple point in mm and the abscissa shows the distance x along the wedge in mm. Hence the intersections of estimated triple point trajectories at y = 0 mm give $x_{int,}$ the point at which the delayed transition starts. Figure 2.22b summarized the relationship between x_{int} and Re at 52° wedge which is close to θ_{crit}. The ordinate shows x_{int} in mm and the abscissa shows the Re. The x_{int} is inversely proportional to Re. It is shown that in a shock wave reflection from a wedge angle relatively closer to θ_{crit} and at very low initial pressure, the RR might have been dominant and MR wouldn't be observable even in a shock tube of, for example, a characteristic length of 100 mm. It is often reported that shock tube experiments conducted under the identical Ms and wedge angle never produce any consistent θ_{crit}. Results obtained in small shock tubes and in large shock tubes scatter widely. As seen in Fig. 2.22b, the x_{int} varies over 100 mm, which means it was tested with a shorter wedge installed in a small shock tube, and the reflection pattern is always RR, whereas with a long wedge installed in a large shock tube, it can be MR. Results of the Euler

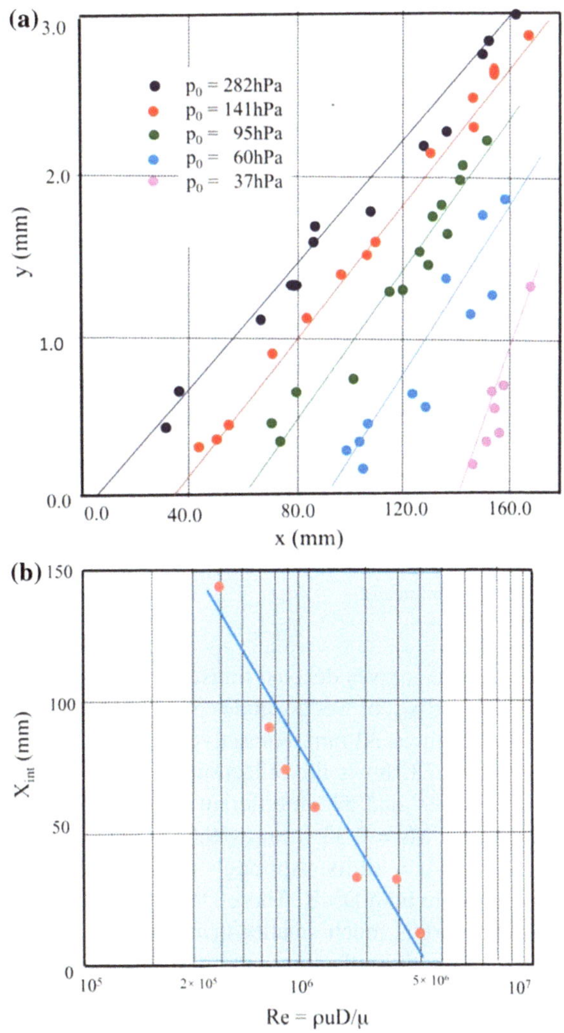

Fig. 2.22 A summary of delayed transitions in argon for Ms = 2.33. Parts of the data appeared Fig. 2.20. Dependence of delayed transition on the Reynold number, Re: **a** triple point position along 52° wedge at various initial pressures; **b** the x_{int} versus the Re; **c** triple point positions along wedges at 144 hPa

solvers disagree with experiments because inviscid solutions agree with shock tube flows of infinitely large Reynolds number. Hence to have a better agreement between numerical solutions and interferograms, for example, as shown in Fig. 2.22, Navier-Stokes solvers with a fine mesh zoning should be compared with experiments of appropriate Re. The effect of the boundary layer on the delayed transition appears not necessarily over wedge angles closer to the θ_{crit}. Figure 2.22c shows triple point trajectories of wedge angles 34.6°, 44.0°, 50.5°, and 52.0° at p_0 = 144 hPa in the x-y plane: the ordinate shows triple point height, y in mm and the abscissa shows the distance, x from the leading edge in mm. The numerical results are cited from Henderson et al. (1997). As the simulations agree well with experiments, we extraporated the trajectory lines. The intersection of the trajectories

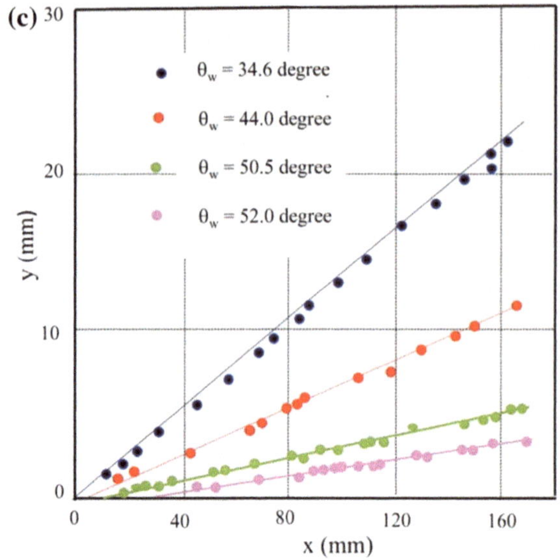

Fig. 2.22 (continued)

with the x-axis gives delayed transition distance, $x_{int.}$ For individual wedge angles, we estimated x_{int} of wedge angles 34.6°, 44.0°, 50.5°, and 52.0° to be 0 mm, about 5, 15, and about 30 mm, respectively.

Figure 2.23 shows the reflection of shock wave of Ms = 2.33 in nitrogen over wedges of 49° and 52° at different initial pressures. Figure 2.23a shows reflection patterns for Ms = 2.33 from a 49.0° wedge at 144 hPa and the double exposure interval is Δt = 80 μs. Although the reflection patterns at the first and second exposures are both DMR. These DMR patterns are not self-similar but DMR at the first exposure is much smaller than one at the second exposure. In Fig. 2.23b, at Δt = 120 μs, the reflection pattern at the first exposure is a RR near the leading edge, whereas that at the second exposure is a DMR near the rear edge. Figure 2.23b clearly shows the delayed transition. In Fig. 2.23c–f reducing the initial pressure to 72 hPa, the difference of the reflection patterns during the double exposures is observed. In Fig. 2.23g, h, the initial pressure is p_0 = 29 hPa. In Fig. 2.23i, j, p_0 = 15 hPa, the reflection pattern observed at the first exposure is RR, whereas the reflection pattern observed at the second exposure is DMR.

Figure 2.24 summarize results of visualizations: **a** triple point trajectories over a 49° wedge at 14.1, 28.2, 70.5 and 141.0 hPa displayed in the x, y-plane; **b** triple point trajectories over 34.6° wedge, 38.6° wedge, 44.0° wedge, and 49.0° wedge displayed in the x-y plane and comparison with numerical simulation offered by Professor Hatanaka of Muroran Institute of Technology. Similarly to argon, the transition distance x_{int} increases with the reduction of the initial pressures.

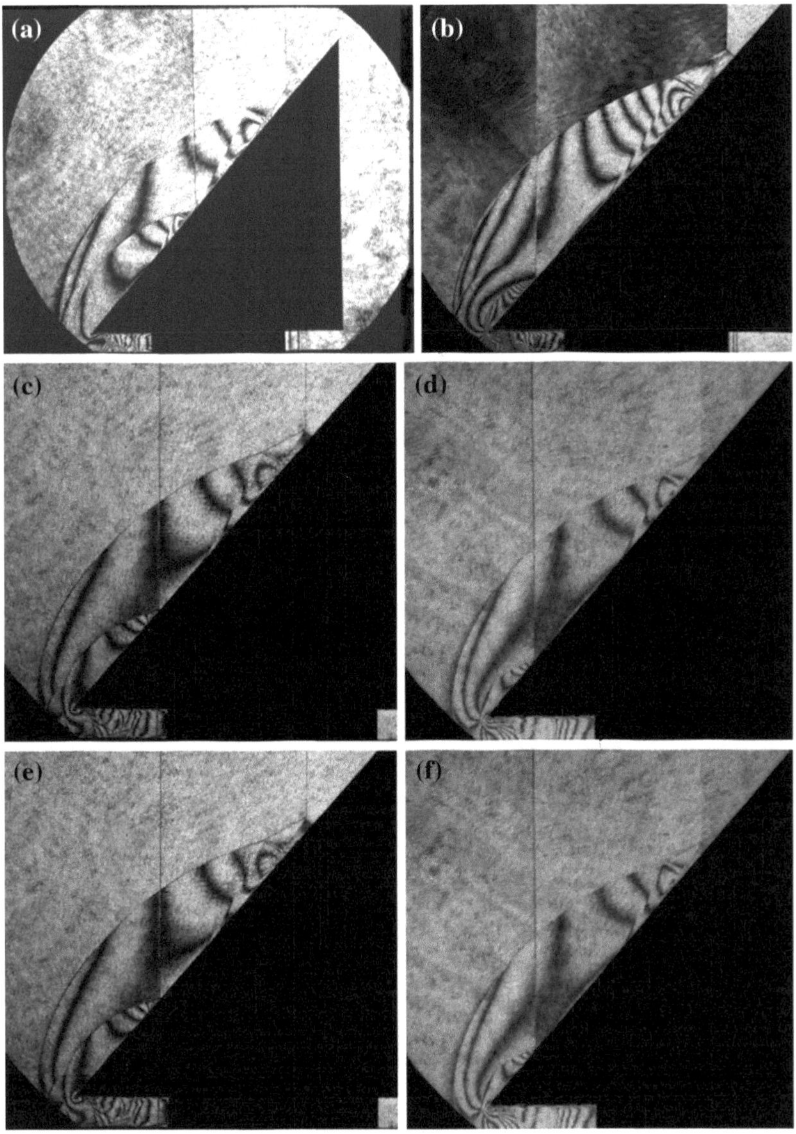

Fig. 2.23 Delayed transition over wedges in 100 mm × 180 mm shock tube in nitrogen: **a** #98090102, $\Delta t = 80\ \mu s$, Ms = 2.328 at 144 hPa and 295.0 K, $\theta_w = 49.0°$; **b** #98090104, 120 μs, Ms = 2.326 at 144 hPa and 295.0 K, $\theta_w = 49.0°$; **c** #98091701, 80 μs, Ms = 2.324 at 72 hPa and 295.2 K, $\theta_w = 49.0°$; **d** #98091702, 80 μs, Ms = 2.330, at 72 hPa and 296.2 K, $\theta_w = 49.0°$; **e** #98091701, 80 μs, Ms = 2.324 72 hPa 22.2C $\theta_w = 49.0°$; **f** #98091702, 80 μs, Ms = 2.330 at 72 hPa and 290.2 K, $\theta_w = 49.0°$; **g** #98091823, 60 μs, Ms = 2.333 at 29 hPa and 301.2 K, $\theta_w = 49.0°$; **h** #98091825, 60 μs, Ms = 2.327 at 28.2 hPa and 296.3 K, $\theta_w = 49.0°$; **i** #98092413, 40 μs, Ms = 2.327 at 15 hPa and 296.6 K, $\theta_w = 49.0°$; **j** #98092408, 60 μs, Ms = 2.327 at 15 hPa and 290.5 K, $\theta_w = 49.0°$ (Itabashi 1998; Henderson et al. 2001)

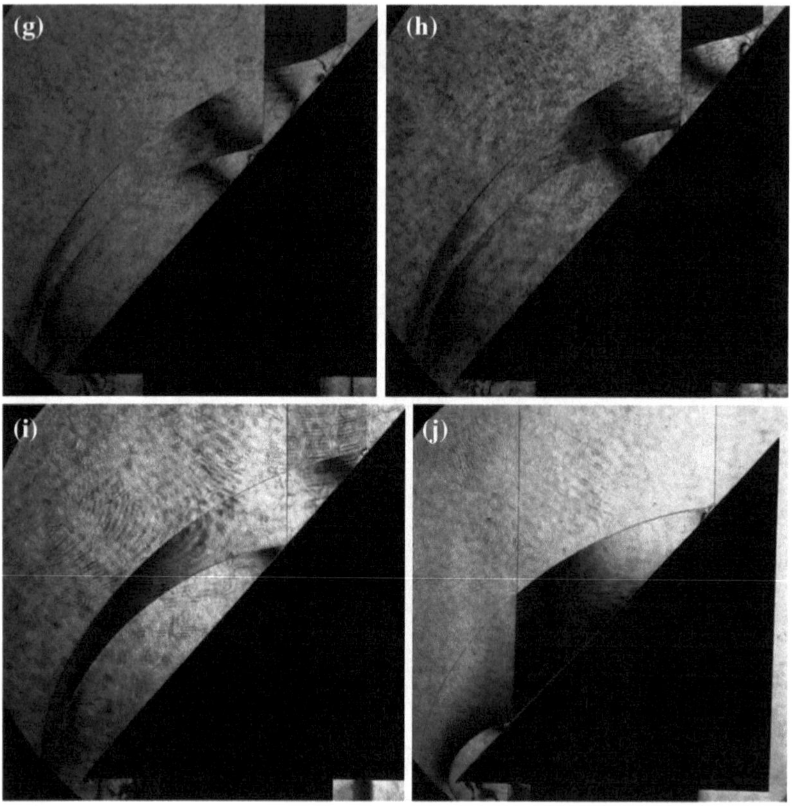

Fig. 2.23 (continued)

2.1.3.7 Evolution of Shock Wave Reflection Over Wedges in CO_2

Figure 2.25 shows the evolution of reflected shock waves over wedges for Ms = 2.0 in CO_2 at 400 hPa and 296 K in the 40 mm × 80 mm shock tube. Figure 2.25a shows a $\theta_w = 0°$ wedge and the predicted glancing incidence angle θ_{glance} is 27.0°, whereas the measured θ_{glance} is about 28°. Although in Fig. 2.25b the pattern of reflected shock wave is a SMR, the SL is only vaguely visible, whereas in Fig. 2.25c, the SL is clearly visible. However, a point from which the SL starts is not identical to the point at which the IS and the RS intersect. Figure 2.25d shows a SMR. Figure 2.25e–h shows earlier stages of TMR and its transition to DMR. Figure 2.25i shows a DMR at $\theta_w = 48°$. Figure 2.25j is a SuRR at $\theta_w = 51°$. Argon is a monatomic gas and has three translational degrees of freedom only. Nitrogen and air are diatomic gases and have three translational degrees of freedom, two rotational and one vibrational degrees of freedom, while CO_2 has, in addition to these degrees of freedom, two more vibrational degrees of freedom which are readily excited even at energy levels of modestly strong shock

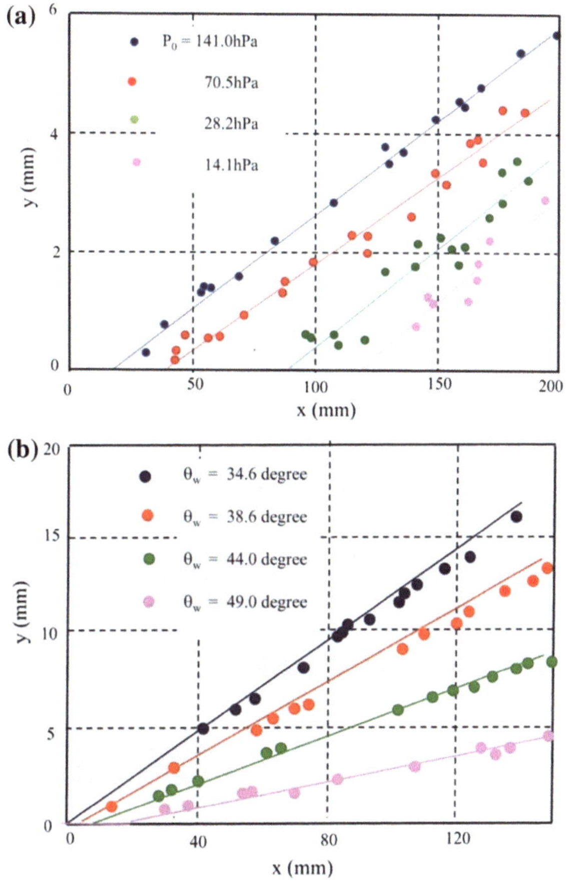

Fig. 2.24 A summary of delayed transitions for Ms = 2.33 in nitrogen. Parts of data appeared Fig. 2.17 indicating the dependence of delayed transition on the Reynold number, Re: **a** triple point position along a 49° wedge at various initial pressures; **b** triple point positions along 34.6°, 38.6°, 44.0° and 49.0° wedges at 144 hPa (Kosugi 2000)

waves. Then vibrational relaxation appears at a short distance from the frozen shock waves. It should be noticed that CO_2 is a polyatomic gas exhibiting readily the so-called real gas effect. The vibrational excitation takes place even at moderate temperature behind shock wave of Ms = 2.0. The incident shock waves seen in Fig. 2.25 shows a broadened structure that exhibits vibrational relaxation. This appears not only at the IS but also along the MS and the RS.

In an ideal gas, the density ratio ρ_{22}/ρ_1 across a strong oblique shock wave can be written as

$$\begin{aligned}\rho_{22}/\rho_1 &= (\gamma+1)\text{Ms}^2\sin^2\phi/\{(\gamma-1)\text{Ms}^2\sin^2\phi+2\},\\ &= \tan\phi/\tan(\phi-\theta)\end{aligned} \tag{2.12}$$

where γ is the ratio of specific heats of the gas under study, θ and ϕ are the deflection angle of flow across the shock wave and the inclination angle of the oblique shock wave, respectively. Then Mssinϕ is the normal component of the oblique shock

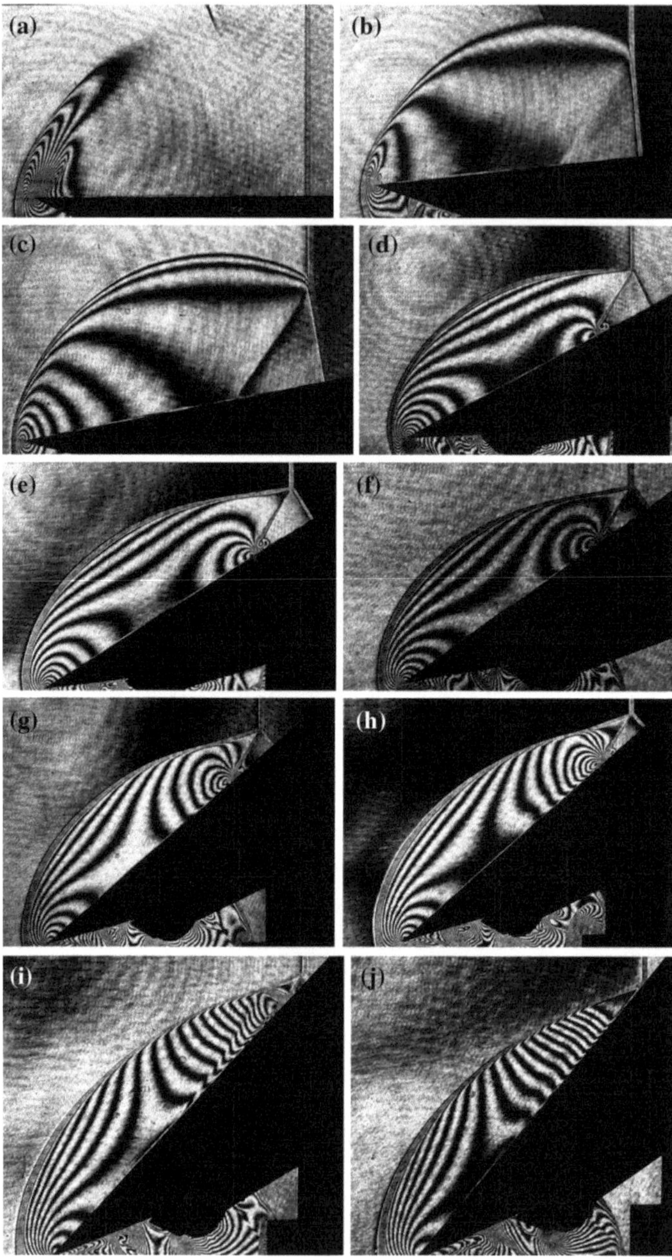

Fig. 2.25 Evolution of shock wave reflection over wedges in CO_2 at Ms = 2.0 at 400 hPa and 296 K: **a** #81100728, Ms = 1.991, $\theta_w = 0.0°$; **b** #81100726, Ms = 2.001, $\theta_w = 4.0°$; **c** #81100725, Ms = 1.955, $\theta_w = 10.0°$; **d** #81100710, Ms = 2.010, $\theta_w = 27°$; **e** #81100723, Ms = 2.001, $\theta_w = 32°$; **f** #81100722, Ms = 1.946, $\theta_w = 36°$; **g** #81100717, 130 μs Ms = 1.982, $\theta_w = 40°$ TMR; **h** #81100716, Ms = 1.991, $\theta_w = 44°$ TMR; **i** #81100713, Ms = 2.010, $\theta_w = 48°$ DMR; **j** #81100711, Ms = 1.973, $\theta_w = 51°$ SPRR

wave Mach number. For Mssin$\phi \gg 1$, ρ_2/ρ_1 is approximately $(\gamma + 1)/(\gamma - 1)$. This indicates that polytropic gases, whose ratio of specific heats is close to unity, are more compressible than monatomic or diatomic gases. Hence, at the identical oblique shock angle, $\phi - \theta$ is smaller in polytropic gases. Figure 2.26 shows a DMR over a $\theta_w = 35°$ wedge for Ms = 4.397 in CO_2 at 50 hPa. We define trajectory angles of the first and second triple points as χ_1 and χ_2. In the case of a stronger shock wave reflection in CO_2, the second triple point, is hence located closer to the wedge surface, that is, $\chi_2 < \chi_1$. The grey irregularly shaped zone attached to the RS is the remnant of bifurcated RS. The foot of the upward moving RS interacts with the side wall boundary layer and eventually bifurcates. This is a similar phenomenon which is observed when a reflected shock wave interacts with the boundary layer at the end wall of shock tubes. This phenomenon is discussed later.

Similarly to Fig. 2.25, the density distribution behind IS as observed in Fig. 2.27 shows vibrational relaxation distance. Figure 2.27a, b appeared to be a SMR over $\theta_w = 5.0°$ and 10.0° wedges, respectively, however, the RS is straightened in the vicinity of the TP. The reflection patterns look the stage of initiation of a TMR. The reflection pattern is a TMR in Fig. 2.27c and it is a DMR in Fig. 2.27d. In Fig. 2.27e–g, the reflection patterns are DMR. Figure 2.27g shows an enlargement of the DMR over a $\theta_w = 36°$ wedge in Fig. 2.27f. The secondary shock wave of the DMR interacts with a SL produces a vortex that is reversely rolling toward the MS as seen in Fig. 2.27g. In Fig. 2.27i, j, with increasing θ_w, the point of the secondary shock wave approaches to the first TP and the resulting MS is gradually shortened and at $\theta_w = \theta_{crit}$, DMR terminates and eventually transits to SPRR. An irregularly shaped grey shadow attached along the RS is the projection of the foot of the bifurcated RS. The structures of the MS, the RS, and the secondary shock wave accompanied vibrational relaxation zones. The SL turned into a vortex which moved downward. The MS tends to move downward. This trend is more enhanced when Ms increases and also in the case of SF_6 as shown later. As explained in Fig. 2.26, the secondary triple point trajectory angle χ_2 is smaller than χ_1.

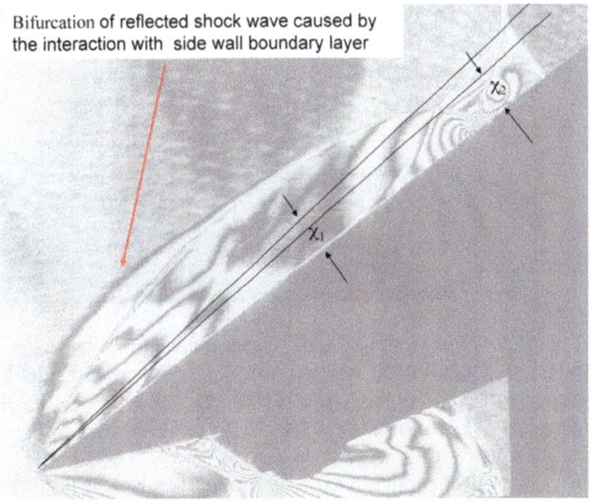

Fig. 2.26 Overeview of a shock wave reflection from a $\theta_w = 35°$ wedge for Ms = 4.397 in CO_2 at 50 hPa, #81100518

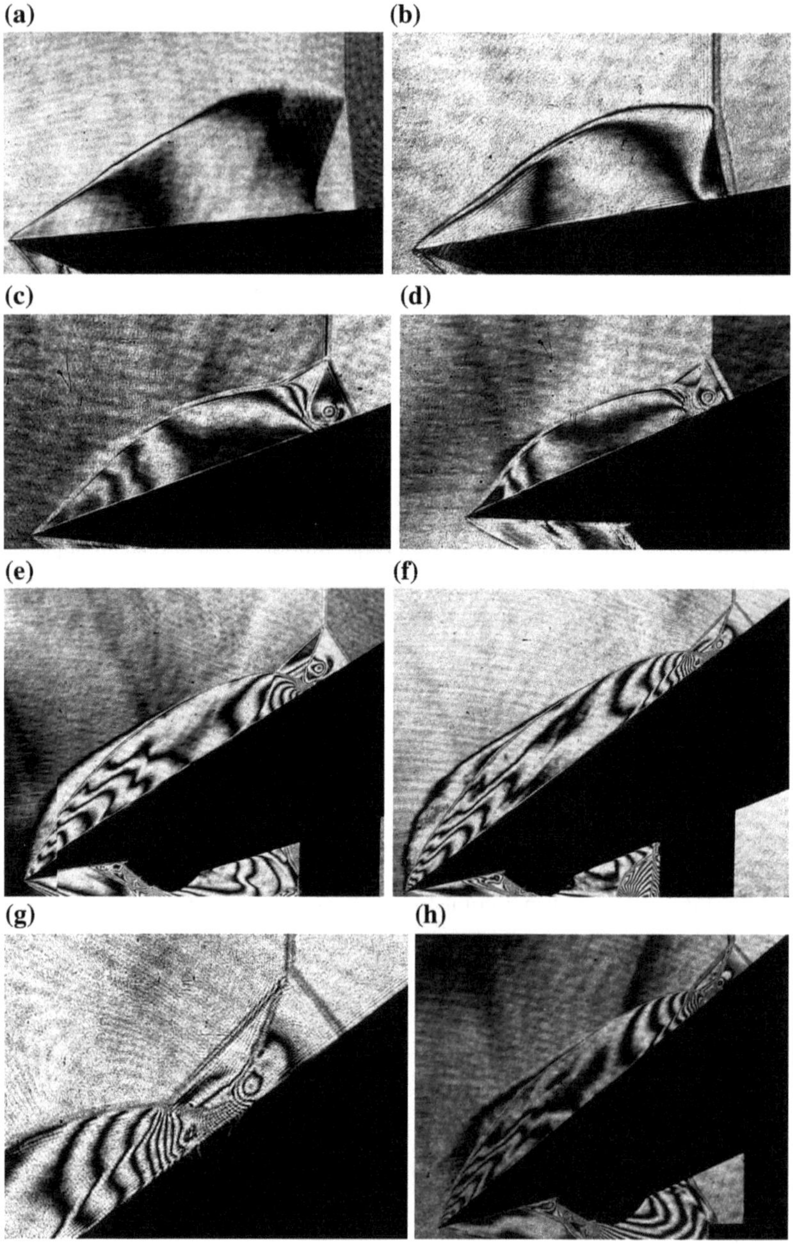

Fig. 2.27 Evolution of shock wave reflection over wedges in CO_2 for Ms = 4.3 at 50 hPa, 294.4 K: **a** #81100527, Ms = 4.307, $\theta_w = 5°$; **b** #81100526, Ms = 4.224, $\theta_w = 10°$; **c** #81100523, Ms = 4.352, $\theta_w = 21°$; **d** #81100522, Ms = 4.266, $\theta_w = 25°$; **e** #81100519, Ms = 4.396, $\theta_w = 32°$; **f** #81100517, Ms = 4.397, $\theta_w = 36°$; **g** enlargement of (**f**); **h** #81100516, Ms = 4.397, $\theta_w = 40°$; **i** #81100513, Ms = 4.352, $\theta_w = 42°$; **j** #81100512, Ms = 4.397, $\theta_w = 44°$; **k** #81100509, Ms = 4.309, $\theta_w = 50°$; **l** #81100511, Ms = 4.307, $\theta_w = 55°$

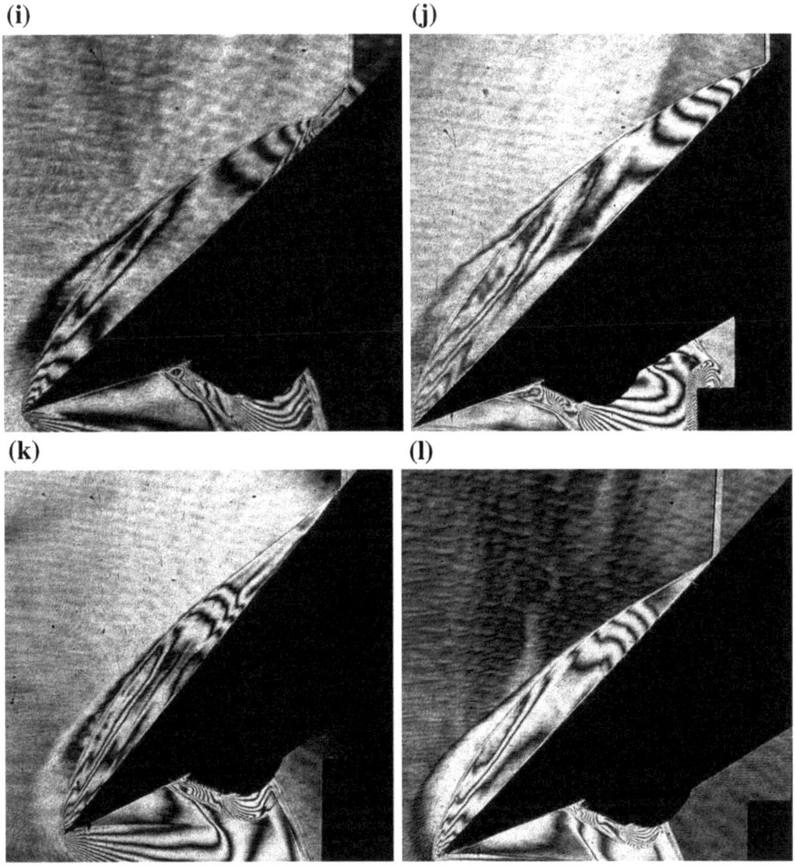

Fig. 2.27 (continued)

2.1.3.8 Evolution of Shock Wave Reflection Over Wedges in SF_6

SF_6 is frequently used for various industries such as electrical insulation etc. Regarding the basic study of isotope separation technology, SF_6 became often a replacement of radio active UF_6. The value of γ of SF_6 being 1.08, its gas dynamic behavior is unique. A shock tube experiment by using SF_6 was conducted. According to the industrial safety regulation, this gas belongs to a category of harmless gases. But after shock tube experiments, the laboratory air was contaminated with F_2 probably in a few parts per billion. Later when introducing a diaphragm-less shock tube, the shock tube was not exposed to open air and hence the contamination issue of F_2 was solved. Figure 2.28 shows direct shadowgraphs of evolution of the shock wave reflection in SF_6 in the 40 mm × 80 mm conventional shock tube. The reflection patterns are SMR seen in Fig. 2.28a–d, a TMR seen in Fig. 2.28e, f, and DMR seen in Fig. 2.28g, h.

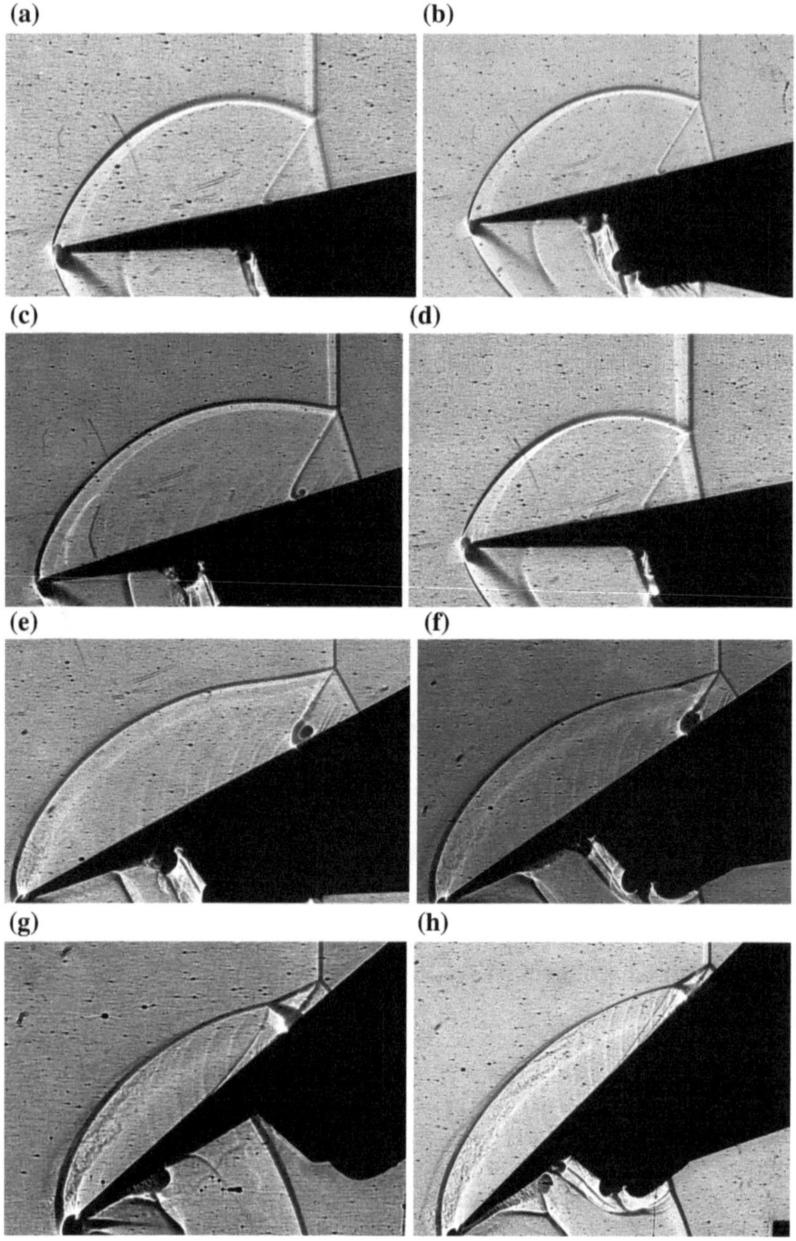

Fig. 2.28 Evolution of reflected shock wave for Ms = 2.0 in SF_6 at 240 hPa, 292.2 K: **a** #82020330, Ms = 1.940, SMR, $\theta_w = 10°$; **b** 82020329, Ms = 1.896, SMR, $\theta_w = 11°$; **c** 82020328, 880 μs, Ms = 1.903, SMR, $\theta_w = 13°$; **d** #82020324, Ms = 1.961, SMR, $\theta_w = 16°$; **e** #82020326, Ms = 1.916, TMR, $\theta_w = 27°$; **f** #82020325, Ms = 1.949, TMR, $\theta_w = 32°$; **g** #82020322, Ms = 1.985, DMR, $\theta_w = 39°$; **h** #82020321, Ms = 1.923, DMR, $\theta_w = 43°$

Figure 2.29 summarizes the triple point trajectories shown in Fig. 2.28 in SF_6 for Ms = 2.0 and shown in Fig. 2.27 in CO_2 for Ms = 4.3. The ordinate denotes the trajectory angle of the triple point $\chi - \theta_w$ in degree and the abscissa denotes the wedge angle θ_w in degree. Blue filled circles denote first triple point of CO_2 χ_1. Grey filled circles denotes the second tripe point of CO_2 χ_2. Red filled circles denote triple point of SF_6 χ_1. The glancing incidence angle χ_{glance} in SF_6 for Ms = 2.0 is about 27.0°, whereas χ_{glance} in CO_2 for Ms = 4.3 is about 22.0°. In CO_2, χ_1 is always larger than χ_2, whereas in SF_6 for Ms = 2.0, χ_1 is smaller than χ_2, the same as in diatomic and monatomic gases. However, in SF_6 for Ms = 2.0 as seen in Fig. 2.28, χ_1 is consistently larger than χ_2.

Figure 2.30a–q shows shock wave reflections from a wedge of variable wedge angles for Ms = 7.50 in SF_6 conducted in the 60 mm × 150 mm diaphragm-less shock tube. In order to generated strong shock wave in SF_6, high-pressure helium drove SF_6. A jaggedly shaped interface observed on the left of Fig. 2.30a is a contact discontinuity which drove a shock wave in SF_6. The helium/SF_6 interface is unstable due to the interfacial instability and then the interface generated obliquely running disturbances. Figure 2.30a shows the flow in the slug length. In Fig. 2.30a, b, an interferogram and its enlargement are shown. Figure 2.30c, d shows a unreconstructed hologram of Fig. 2.30a and its enlargement, respectively. It is a general trend of shock tube flows that the interface is accelerated and the shock wave is decelerated due to the presence of the boundary layer behind the shock wave developing along the sidewall. Hence, the slug length, high-pressure zone behind the shock wave is shortened with propagation.

Figure 2.30o illustrates the interfacial disturbances and the deformation of the shock wave. Figure 2.30c, d shows the un-reconstructed hologram of Fig. 2.30 and its enlargement, which is equivalent to a direct shadowgraph. The shock wave is significantly deformed in Fig. 2.30q so that the three-dimensionally distorted shock wave hardly creates a distinctly shaped first triple point. On wedges of large wedge angles, the reflected patterns are DMR and their secondary triple point trajectory

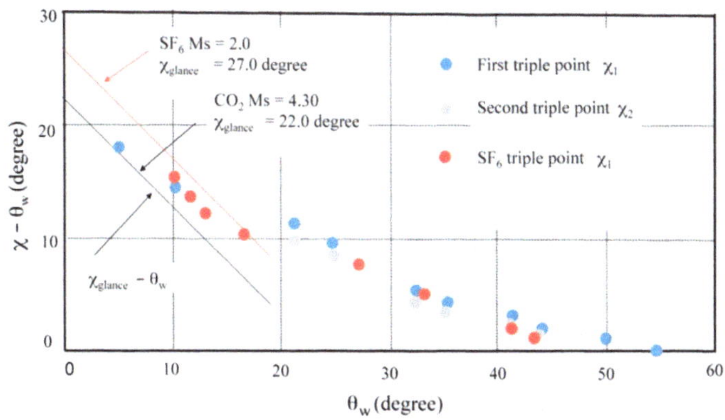

Fig. 2.29 Triple point trajectory angles χ against wedge angles θ_w. A summary of Figs. 2.26 and 2.27

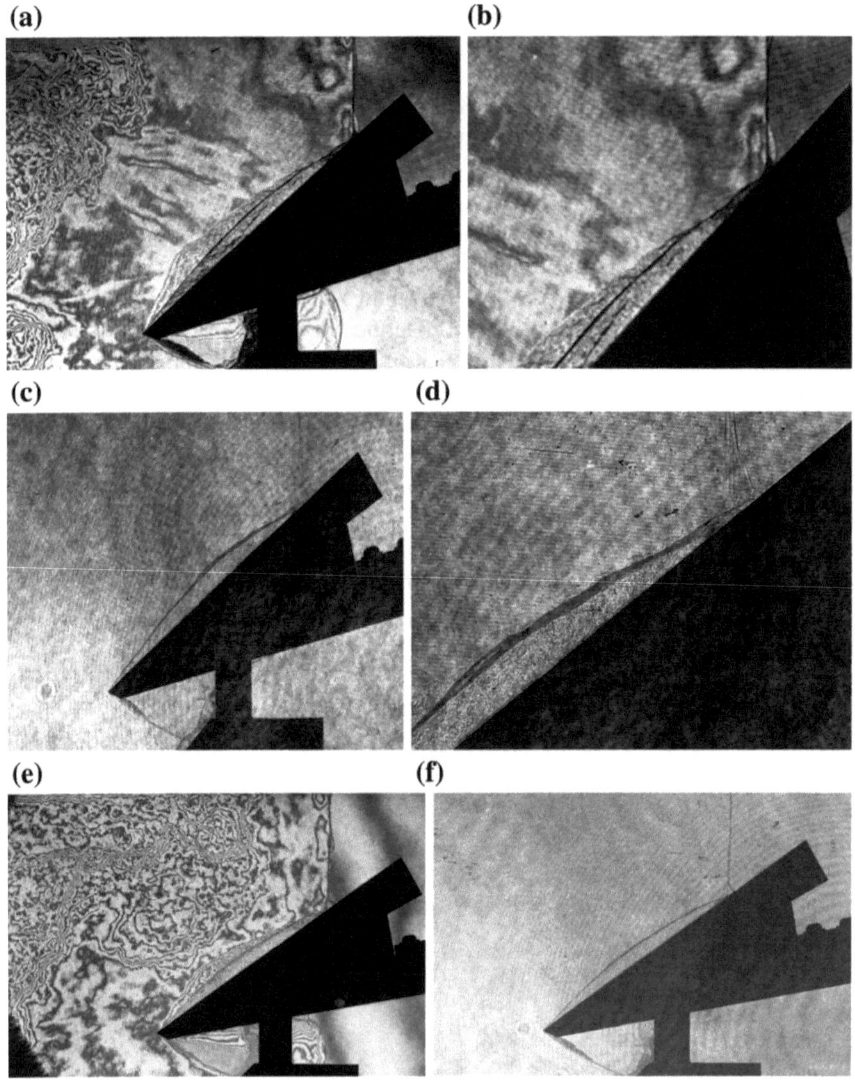

Fig. 2.30 Evolution of shock wave reflection over wedges for Ms = 7.50 in SF6 at 15 hPa and 295.5 K: **a** #91060405, Ms = 7.52, $\theta_w = 31°$; **b** enlargement of (**a**); **c** #91060405 hologram; **d** enlargement of (**c**); **e** #91060302, Ms = 7.42, $\theta_w = 32°$; **f** #91060302 hologram; **g** #91060406, Ms = 7.47, $\theta_w = 35°$; **h** #91060305, Ms = 7.62, $\theta_w = 36°$; **i** #91060404, Ms = 7.59, $\theta_w = 37°$; **j** #91060402, Ms = 7.59, $\theta_w = 37°$; **k** #91060403, Ms = 7.60, $\theta_w = 38°$; **l** #91060407, Ms = 7.47, $\theta_w = 38°$; **m** #91060407 hologram; **n** enlargement of (**m**); **o** #91060401 Ms = 7.52, $\theta_w = 41°$; **p** enlargement of (**o**); **q** #91053001, Ms = 7.36, $\theta_w = 45°$

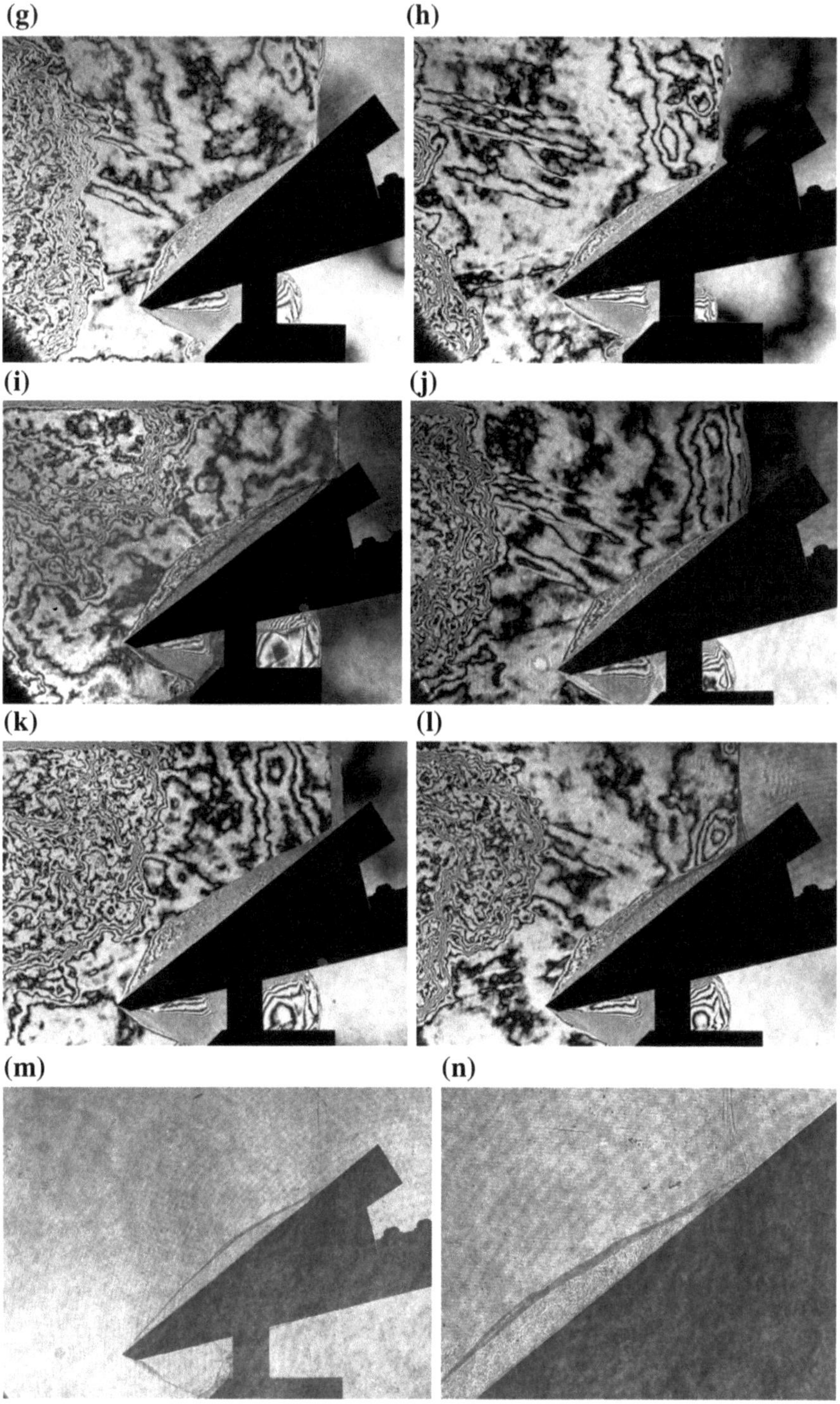

Fig. 2.30 (continued)

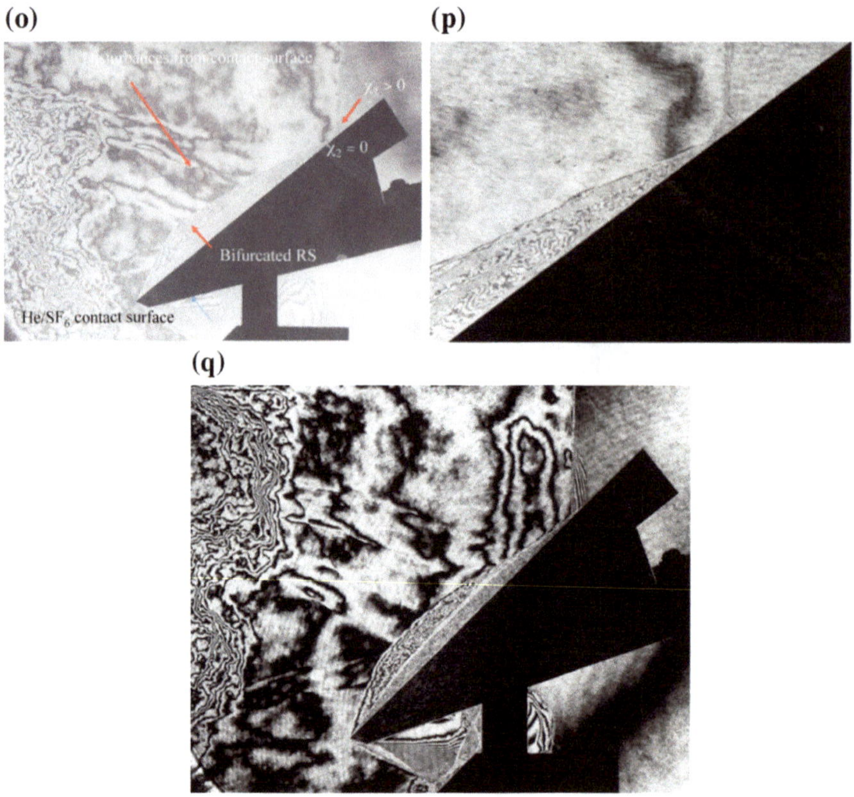

Fig. 2.30 (continued)

angle became smaller than the first triple point trajectory angle, that is, $\chi_1 > \chi_2$. With increasing the wedge angle, χ_2 tends to approach the wedge surface. In Fig. 2.30i, χ_2 is nearly lying on the wedge surface but will never stay on the wedge.

2.1.3.9 Interaction of Reflected Shock Waves Over 90° Intersecting Wedges

To simulate shock wave propagation in a rectangular supersonic air intake, three-dimensional observations of shock wave interactions from two perpendicularly intersecting wedges were conducted (Muguro 1998). Figure 2.31a illustrates the interaction of two reflected shock waves from two perpendicularly intersecting wedges. Two wedges of inclination angles 30°, 43.5°, 45° and 55° intersected perpendicularly were installed in the 60 mm × 150 mm diaphragm-less shock tube as shown in Fig. 2.31b. A reflected shock wave pattern is MR over a shallow wedge and RR over a steep wedge. Possible combinations of interaction pattern are MR-MR, MR-RR, and RR-RR. The visualization was conducted by diffuse

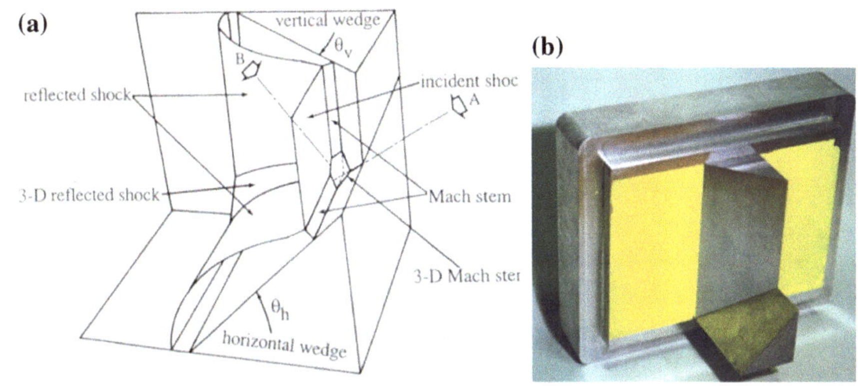

Fig. 2.31 Intersection of shock waves reflected from two intersecting wedges: **a** illustration of two shock wave interacting perpendicularly; **b** two wedge combination installed in the 60 mm × 150 mm diaphragm-less shock tube test section

holographic interferometry. The wedges and the test section were coated with a yellow fluorescent paint, which has higher degree of reflectivity to the ruby laser light (Meguro 1998). It should be noticed that as seen in Fig. 2.31a θ_h stands for the wedge angle of a vertical wedge, and θ_v that of a horizontal wedge.

A Q-switch ruby laser (Apollo Laser Co. Ltd.) has a pulse width of 25 ns and 1 J/pulse and its double pulse interval was variable from 1 μs to 1 ms. In order to visualize the shape of shock waves, the double pulse interval was selected to be 1 μs. The laser irradiation was synchronized with the shock wave motion with an appropriate delay time. Hence, the variation of phase angles occurred during the double pulse interval of 1 μs was recorded on the holographic film. As the shock wave of Ms = 2.0 moved at about 0.7 mm during 1 μs, then shock waves and their interaction over the intersecting wedges are recorded as three-dimensionally distributed thin lines. A collimated *OB* illuminated the coated wedges and was so far reflected from the coated wall. The diffused *OB* carrying holographic information irradiated a holographic film. The reconstructed images were recorded using a conventional reflex camera or a digital camera. Figure 2.32 summarize reconstructed images of interaction of the reflected shock waves from intersecting wedges. Experiments were conducted for Ms = 1.2, 1.5, 2.5 and 2.8 in air. Shock waves are described as three-dimensionally distributed thin lines having thickness of 0.5 mm to 1.0 mm, which are thin enough to explained not only the difference RR and MR but also fine reflected shock wave patterns.

Figure 2.33a shows a diagram of domain and boundary of reflection patterns for Ms from 3.0 to 1.2. The ordinate denotes the vertical wedge angle in degree and the abscissa denotes the horizontal wedge angle in degree. White circles denote MR-MR interaction accompanying 3-D MS. Black filled circles denote RR-RR. Red colored regions denote the MR-MR region, blue colored regions denote the RR-RR region, orange colored regions denote RR-MR or MR-RR region. Green colored regions denote the region outside the critical transition angles, MR-RR.

(a) (b) (c) (d) (e) (f)

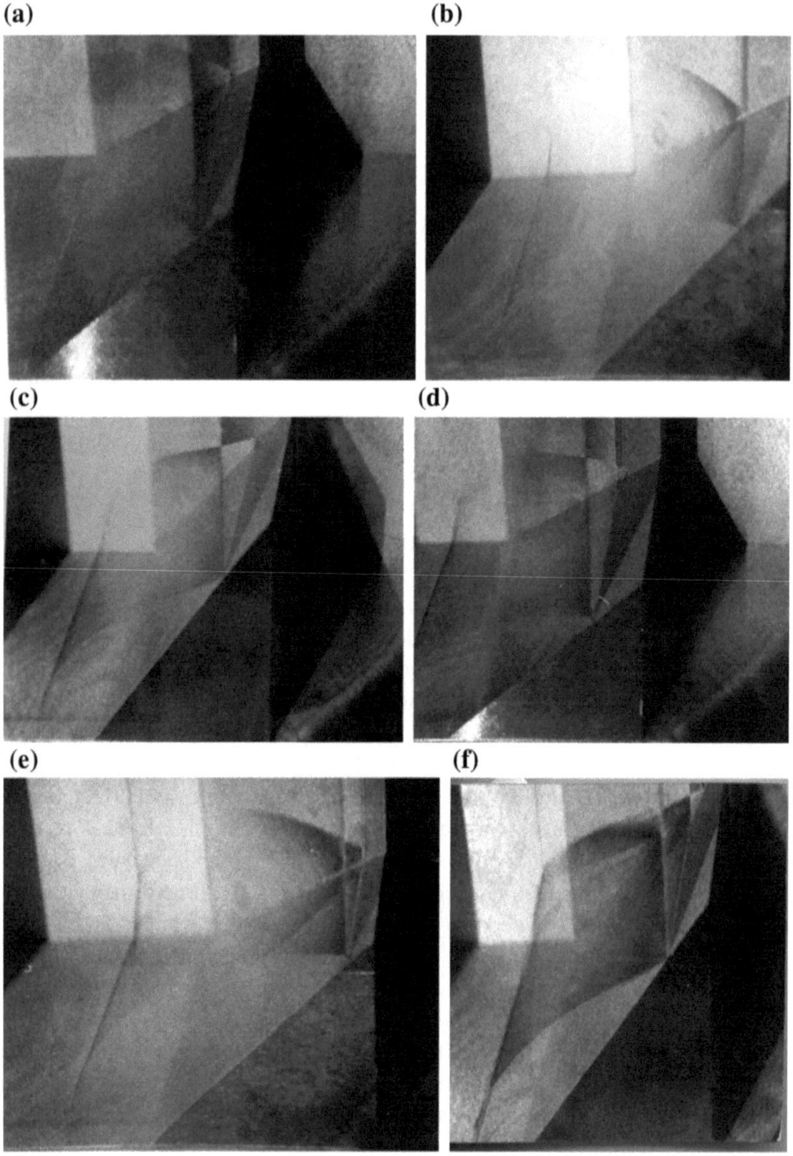

Fig. 2.32 Reconstruction of holograms: **a** #95111510, $\theta_h = 45°$, $\theta_v = 30°$, Ms = 1.201; **b** #95111201, $\theta_h = 45°$, $\theta_v = 45°$ Ms = 1.207; **c** #95111603, $\theta_h = 55°$, $\theta_v = 30°$, Ms = 1.50; **d** #95111508, $\theta_h = 43.5$, $\theta_v = 45°$ Ms = 1.500; **e** #95111406, $\theta_h = 45°$, $\theta_v = 45°$ Ms = 1.505; **f** #95111606, $\theta_h = 55°$, $\theta_v = 30°$, Ms = 2.00; **g** #95111507, $\theta_h = 43.5$, $\theta_v = 30°$ Ms = 2.000; **h** #95111503, $\theta_h = 43.5$, $\theta_v = 30°$, Ms = 2.501; **i** #95111310, $\theta_h = 45$ $\theta_v = 45°$, Ms = 2.501; **j** #95111607, $\theta_h = 55$, $\theta_v = 30°$, Ms = 2.50; **k** #95111314, $\theta_h = 45$, $\theta_v = 45°$, Ms = 2.818; **l** #95111502, $\theta_h = 43.5$, $\theta_v = 30°$, Ms = 2.845

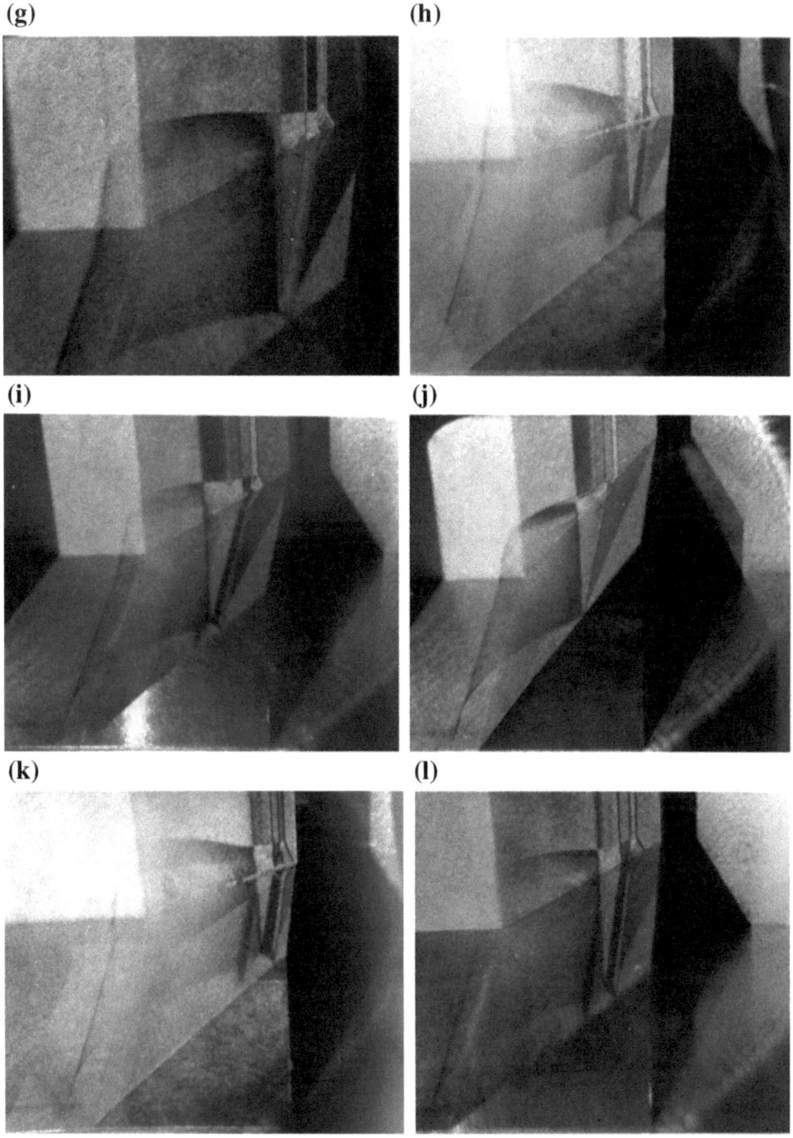

Fig. 2.32 (continued)

Figure 2.32b shows a result of numerical simulation by solving the Euler equations. The ordinate and abscissa denote vertical and horizontal wedge angles, respectively, see for details Meguro et al. (1997). The MS which is formed due to the intersection of two Mach reflections over wedges of different wedge angles, for example, over 30° and 45° wedges, has an asymmetric three-dimensional structure. Then this Mach stem can be define as a three-dimensional MS. However, the reflection

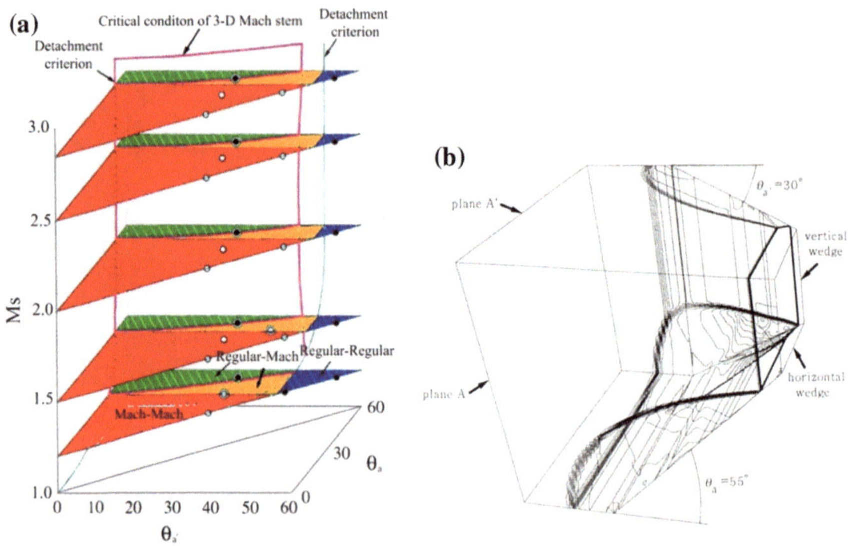

Fig. 2.33 Shock wave reflection from two intersecting wedges: **a** the domain and boundary of reflection patterns in air; **b** a numerical simulation of wedge angle 30° and 55° for Ms = 2.5 in air

patterns of RR-MR can not be identified in the present diffuse holographic observation. In conclusion, an appropriate numerical simulation may reproduce reflection patterns in wide ranges of vertical and horizontal wedge angles.

2.1.3.10 Shock Wave Reflection from a Skewed Wedge

This is a continuation of the previous subsection. If a shock wave is reflected from a three-dimensional skewed wedge whose its inclination angle varies continuously from 30° to 60°, the pattern of resulting shock wave from a 30° wedge angle will be a MR and that from a 60° wedge angle will be a RR. At a certain wedge angle, the transition from MR to RR will take place. Will this transition angle be the same as the θ_{crot} predicted by the detachment criterion reflected shock wave. The skewed wedge was installed in the 100 mm × 180 mm diaphragm-less shock tube, (Numata 2009). The shape is defined in x, y, z co-ordinates as $z = \tan\{\pi(1 + X)/6\}$, where x, y, z are normalized by the shock tube width W. This shape is so simple that it is easy to manufacture it in house. In Fig. 2.34a, the skewed wedge is placed in a 180 mm × 1100 mm test section (Numata 2009).

In order to apply the double exposure diffuse holographic interferometry of 1 µs pulse interval, the test section and the skewed wedge were coated with yellowish fluorescent spray. The collimated *OB* illuminated the test area obliquely and the reflected *OB* was collected on a 100 mm × 125 mm holo-film. Figure 2.34b shows the skewed wedge presented in x, y, z co-ordinates. The wedge is normalized by the width of the shock tube of 100 mm.

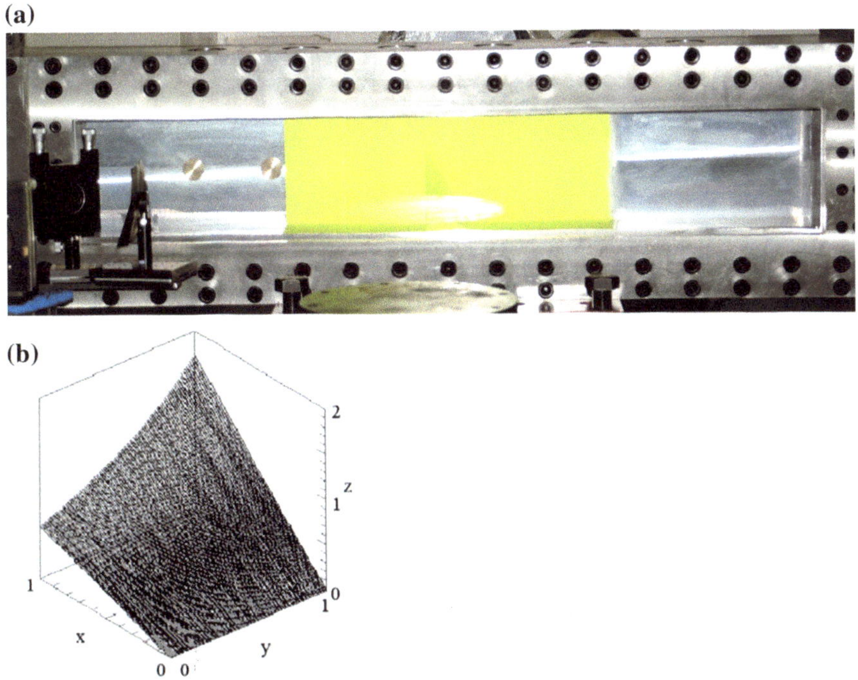

Fig. 2.34 A skewed wedge of variable inclination angles from 30° to 60° installed in a 180 mm × 1100 mm view field: **a** a skewed wedge; **b** normalized co-ordinate (Numata et al. 2009)

Figure 2.35 are reconstructed images for Ms = 1.50, 2.00, and 2.50. On the frontal side, the wedge angle was 30°, then the reflection patterns are SMR for Ms = 1.50 and 2.0 and TMR for Ms = 2.50, whereas the rear side, the wedge angle is 60° and the reflection pattern is RR. For example, when the I-polar for Ms = 1.50 is reflected from a wedge, the pressure behind its reflected shock wave readily obtained by drawing the shock polar. As seen in Fig. 2.3, the intersection point of the I-polar and the R-polar of a specified wedge angle gives the pressure behind the reflected shock wave. Then the pressure becomes higher with increase in the wedge angle. Therefore, the streamline behind the skewed wedge is not always straight along the generatrix line but move toward the shallower wedge angle. Therefore, although the transition would occur at a certain critical transition angle θ_{crit}, it is not sure whether or not this transition angle would be the same as the θ_{crit} predicted by the detachment criterion over a plane wedge.

To estimate the transition angle over the skewed wedge, the wedge surface was uniformly coated by the carbone soot. When the shock wave passed the soot covered wedge surface, the soots were removed by vortices accompanied by the slip line. Figure 2.36a shows the sooted surface exposed to a shock wave of Ms = 2.50. The soot covered surface shows the area the vortices did not scratch. A board oblique dark line shows that the transition of the reflected shock wave over

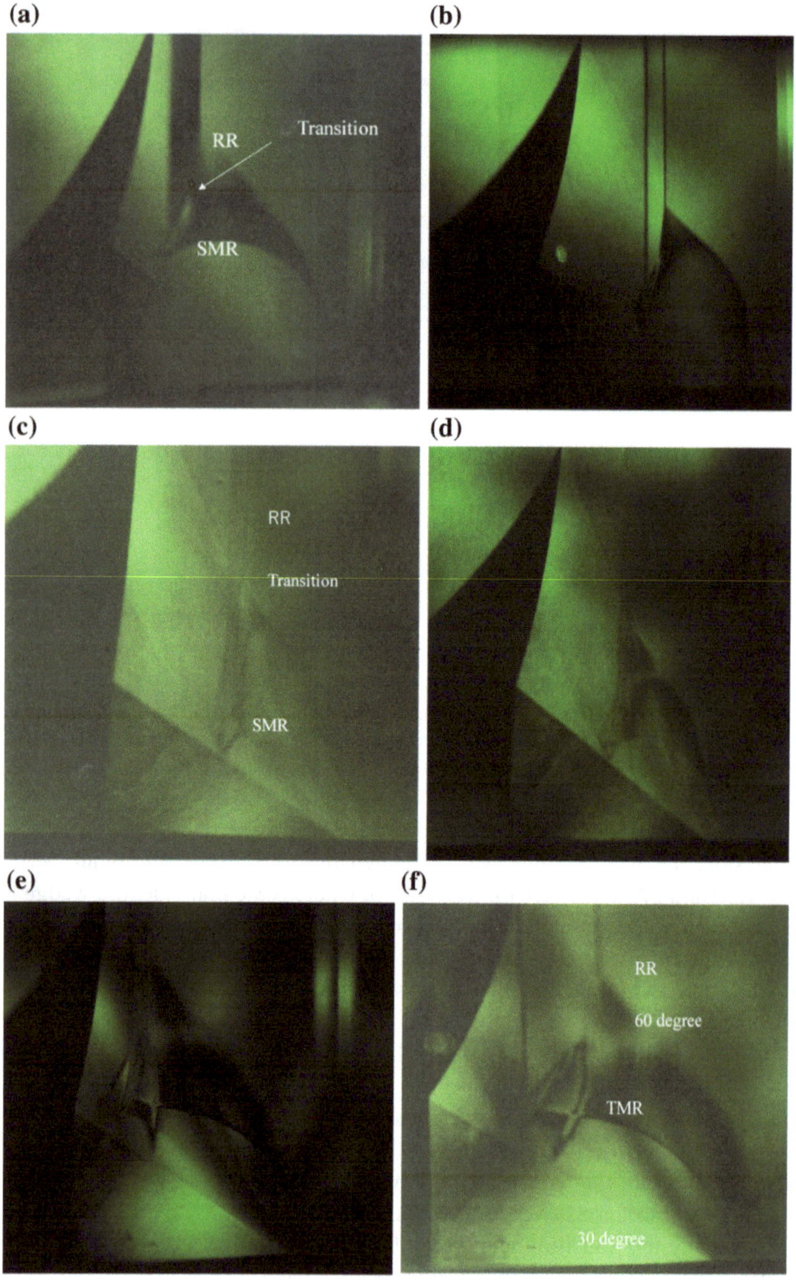

Fig. 2.35 Reconstruction of diffuse holographic interferograms: **a** #06050306, Ms = 1.55; **b** #06060101, Ms = 1.55; **c** #06051514, Ms = 2.0; **d** #06051523, Ms ~ 2.0; **e** #06050143, Ms = 2.55; **f** #06051855. Ms = 2.55 (Numata et al. 2009)

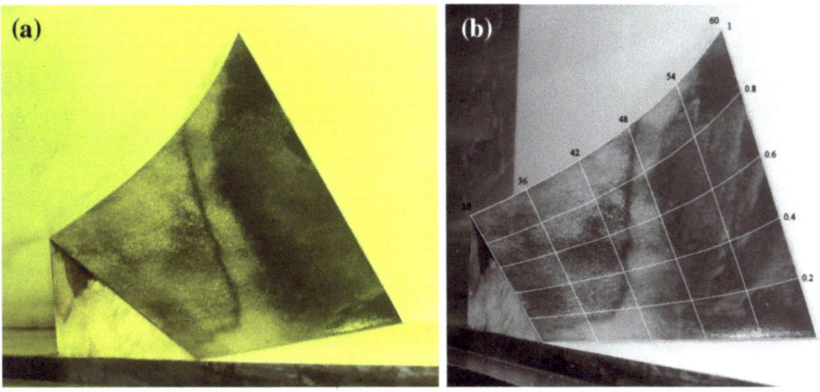

Fig. 2.36 Soot pattern created on the wedge surface for Ms = 2.5: **a** Soot pattern; **b** co-ordinate

the skewed wedge cannot be simply two-dimensional. In Fig. 2.36b, the ordinates denotes wedge angles in degree and the abscissa denotes the normalized distance along the 60° wedge surface. A black line obliquely remaining over the soot free area terminated at about 50° and indicated that the transition from TMR to RR occurred at this angle.

2.1.4 Evolution of Shock Wave Reflection from Roughened Wedges

Shock wave reflections over roughened wedges have been intensively investigated in the past. Saw tooth surface roughness was used often adopted to evaluated the effect of surface roughness on shock wave attenuations (Takayama et al. 1981). The critical transition angle θ_{crit}, in general, decreases as the degree of surface roughness increases. If the roughness is comparable to the boundary layer displacement thickness δ, it will apparently increase wedge angles. Therefore, the transition from MR to RR occurs in relatively smaller wedge angle. In the case of a very coarse surface roughness well over the boundary layer displacement thickness δ, the IS interacts separately with the roughness and creates many wavelets which promote the transition to RR. Wedges with 90° saw tooth roughnesses k = 0.1, 0.2, 0.8 and 2.0 mm are mounted on a movable stand and installed in the 40 mm × 80 mm conventional shock tube. Sequentially visualizations of shock wave reflections from these roughened wedges are carried out.

2.1.4.1 Wedges of a 0.1 mm Saw Tooth Surface Roughness

Figure 2.37 shows evolution of shock wave reflection over 0.1 mm roughness for Ms = 1.40 in air. A MR on wedges of saw tooth roughnesses, vortices are generated behind RS and MS and attenuated the transmitted shock wave. In Fig. 2.37d

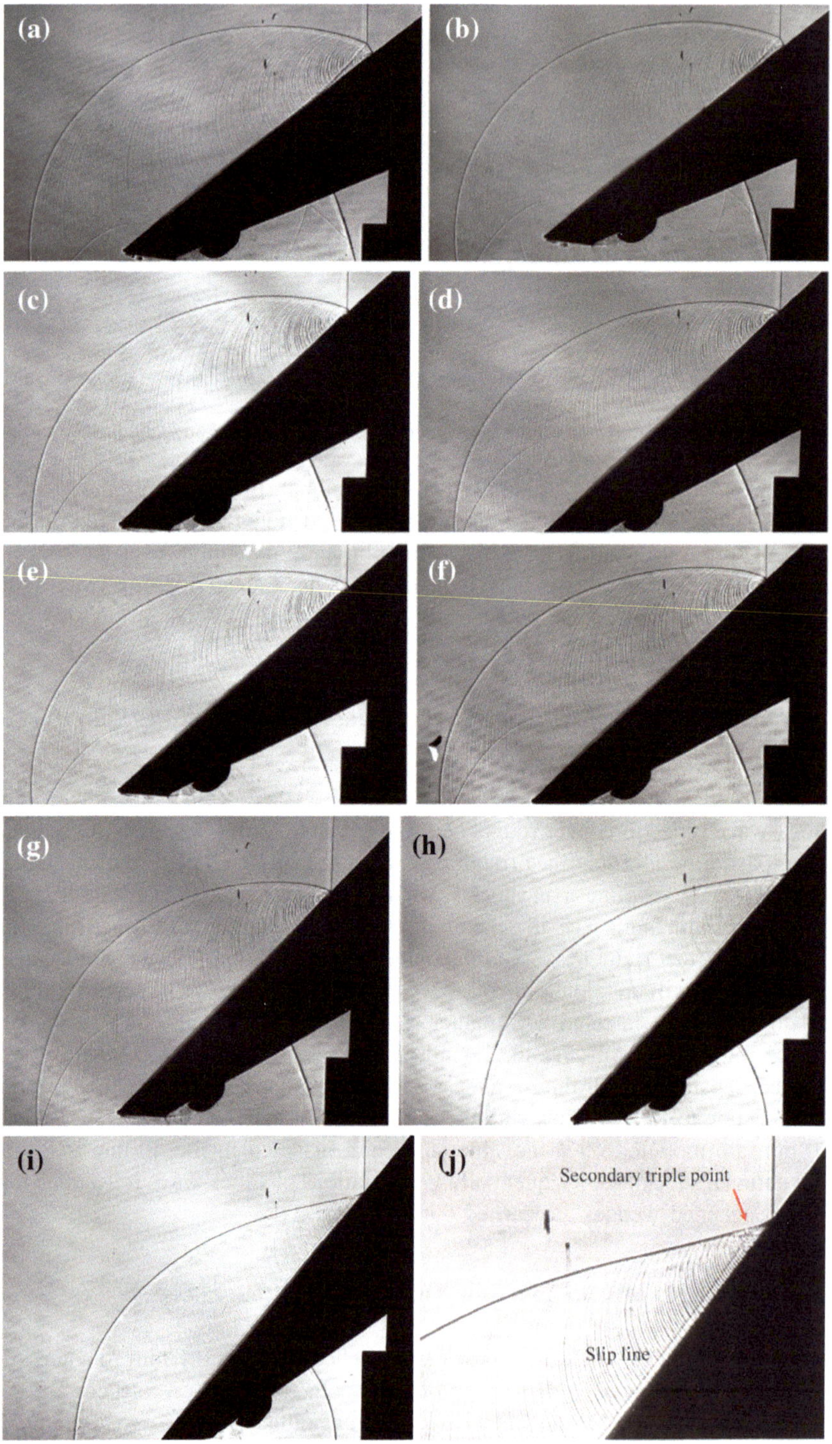
(a)
(b)
(c)
(d)
(e)
(f)
(g)
(h)
(i)
(j)
Secondary triple point
Slip line

◀**Fig. 2.37** Shock reflection over roughened wedges with a saw tooth roughness k = 0.1 mm for Ms = 1.46 in air at 500 hPa and 282.6 K: **a** #80050922, Ms = 1.463, $\theta_w = 38°$; **b** #80050921, Ms = 1.460, $\theta_w = 40°$; **c** #80050930, Ms = 1.465, $\theta_w = 41°$; **d** #80050927, Ms = 1.460, $\theta_w = 42.5°$; **e** #80050929, Ms = 1.466, $\theta_w = 43.0°$; **f** #80050928, Ms = 1.458, $\theta_w = 44°$; **g** #80050927, Ms = 1.460, $\theta_w = 45°$; **h** #80050924, Ms = 1.466, $\theta_w = 47°$; **i** #80050923, Ms = 1.463, $\theta_w = 49.5°$; **j** enlargement of (**i**)

the slip line is directed away from the wedge surface. This reflection pattern is a so-called inverse Mach reflection, or in short IvMR (Courant and Friedrichs 1948). This will be discussed in the Chap. 3 The IvMR is a final shape of a MR and transits to a RR, as seen in Fig. 2.37g. Figure 2.37h then shows a RR which accompanies a secondary triple point. The evolution of such a transition uniquely occurs not only in a fine surface roughness but also over a coarse surface roughness.

2.1.4.2 Wedges of a 0.2 mm Saw Tooth Surface Roughness

Figure 2.38 shows evolution of shock wave reflection from wedges of a 0.2 mm saw tooth roughness for weak shock waves of Ms = 1.09 in air. In Fig. 2.38a, the wedge angle was 0° and circular co-axially shaped shock waves were reflected from each saw tooth and the MS was slightly leaned forward. Courant and Friedrichs (1948) defined this reflection pattern, as already seen in Fig. 2.38b, as an inverse Mach, vNMR. This pattern will soon transit to a RR with increasing in the wedge angle. When the TP reached on the wedge surface as seen in Fig. 2.38c, d, it is succeeded by the secondary TP on the RS as seen in Fig. 2.38f.

Figure 2.39a–j shows the evolution of reflected shock wave from 0.2 mm saw tooth roughness for Ms = 2.0 in air at p_o = 500 hPa, 293.6 K. In Fig. 2.39a, the SL was faintly observed emanating from the TP. The MS was curved and leaned forward. The boundary layer developing along the wedge surface was obscured in the region between the foot of the MS and a point at which the SL touched the wedge surface. With the increase in wedge angles as seen in Fig. 2.39f, g, the reflection patterns transited from SMR to a TMR and its kink point merged to the TP. Figure 2.39g shows that the envelop of wavelets and a SL emanating from the TP coalesce into another SL. If the wedge angle became slightly larger than the angle shown in Fig. 2.39g of $\theta_w = 36.0°$, the TP would touch the wedge surface and eventually the RR would appear. However, the TP would not be terminated but changed to another TP on the RS. In short, the SL was maintained. As seen Fig. 2.39j, the RR accompanied a secondary TP on its RS, which resembled the pattern of a DMR. Figure 2.39k–p shows single exposure interferograms emphasizing the shock wave reflections over steeper wedges for Ms = 1.83 in air at 400 hPa, 286.5 K. Figure 2.39k shows a reflected shock pattern immediately after the transaction to a RR over an identical wedge angle of 46.0° as seen in Fig. 2.39j. Figure 2.39l shows an enlargement of the reflection pattern over a wedge $\theta_w = 47.0°$.

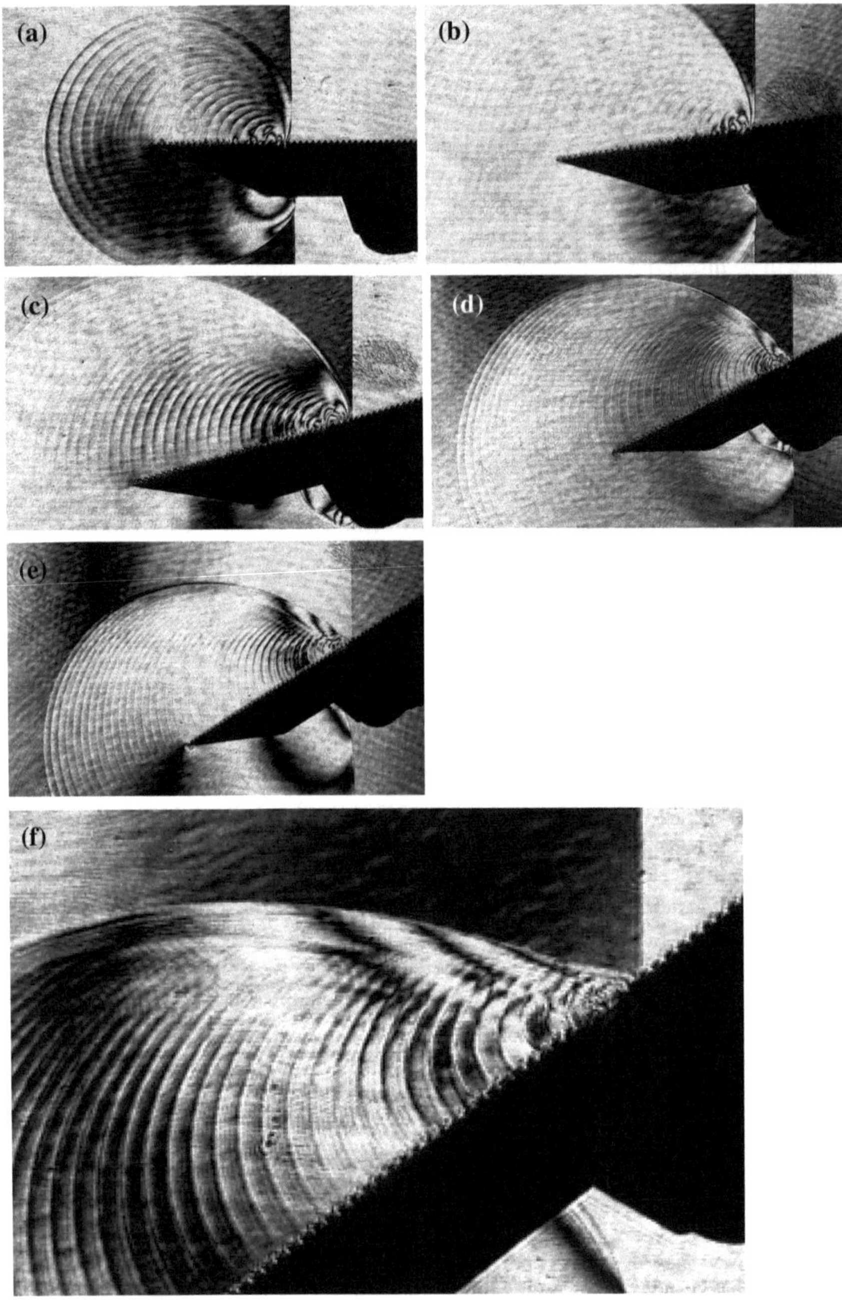

Fig. 2.38 Evolution of shock wave reflection from roughened wedges with saw tooth roughness k = 0.2 mm for Ms = 1.09: in air at 600 hPa and 293.6 K: **a** # 81101404, Ms = 1.090, $\theta_w = 0.0°$; **b** #81101407, Ms = 1.090, $\theta_w = 5°$; **c** #81101408, Ms = 1.090, $\theta_w = 17°$; **d** #81101412, Ms = 1.086, $\theta_w = 24°$; **e** #81101409, Ms = 1.093, $\theta_w = 24°$; **f** #81101411, Ms = 1.087, $\theta_w = 39°$

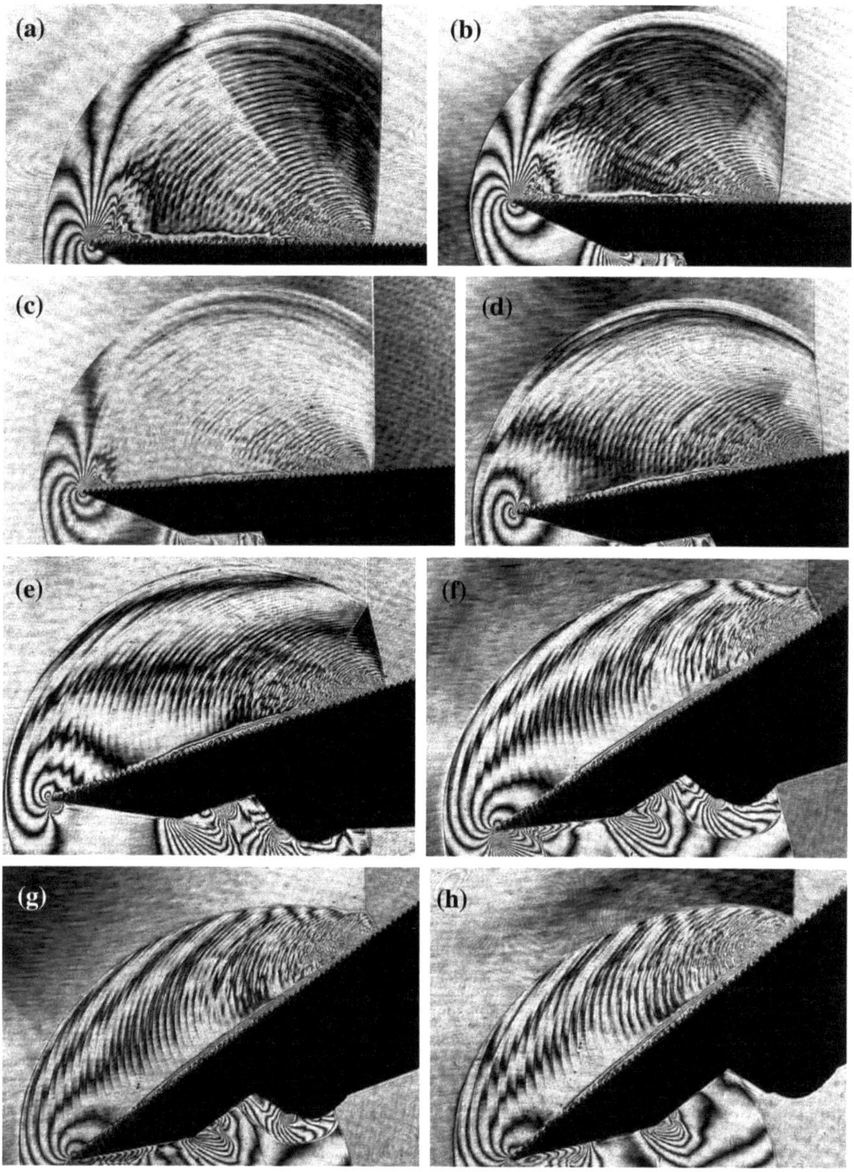

Fig. 2.39 Evolution of shock wave reflection over roughened wedges k = 0.2 mm **a–j** for Ms = 2.0 in air at 500 hPa and 293.6 K; **k–p** for Ms = 1.83 at 400 hPa, 286.5 K single exposure: **a** #81101423, Ms = 1.970, $\theta_w = 1.0°$; **b** #81101422, Ms = 2.004, $\theta_w = 2.0°$; **c** #81101421, Ms = 2.004, $\theta_w = 4.0°$; **d** #81101420, Ms = 1.993, $\theta_w = 10.0°$; **e** #81101419, Ms = 1.943, $\theta_w = 17.0°$; **f** 81101416, Ms = 2.016, $\theta_w = 31.0°$; **g** #81101418, Ms = 2.016, $\theta_w = 36.0°$; **h** #81101415, Ms = 1.981, $\theta_w = 41°$; **i** #81101414, Ms = 1.993, $\theta_w = 42.0°$; **j** #84012401, Ms = 1.862, $\theta_w = 46.0°$; **k** #84012322, Ms = 1.862, $\theta_w = 46.0°$ **l** #84012321, Ms = 1.862, $\theta_w = 47.0°$; **m** 84012320 Ms = 1.844, $\theta_w = 59.0°$; **n** 84012319, Ms = 1.884, $\theta_w = 59.0°$; **o** #84012318, Ms = 1.826, $\theta_w = 56.0°$; **p** #84012316, Ms = 1.871, $\theta_w = 59.0°$

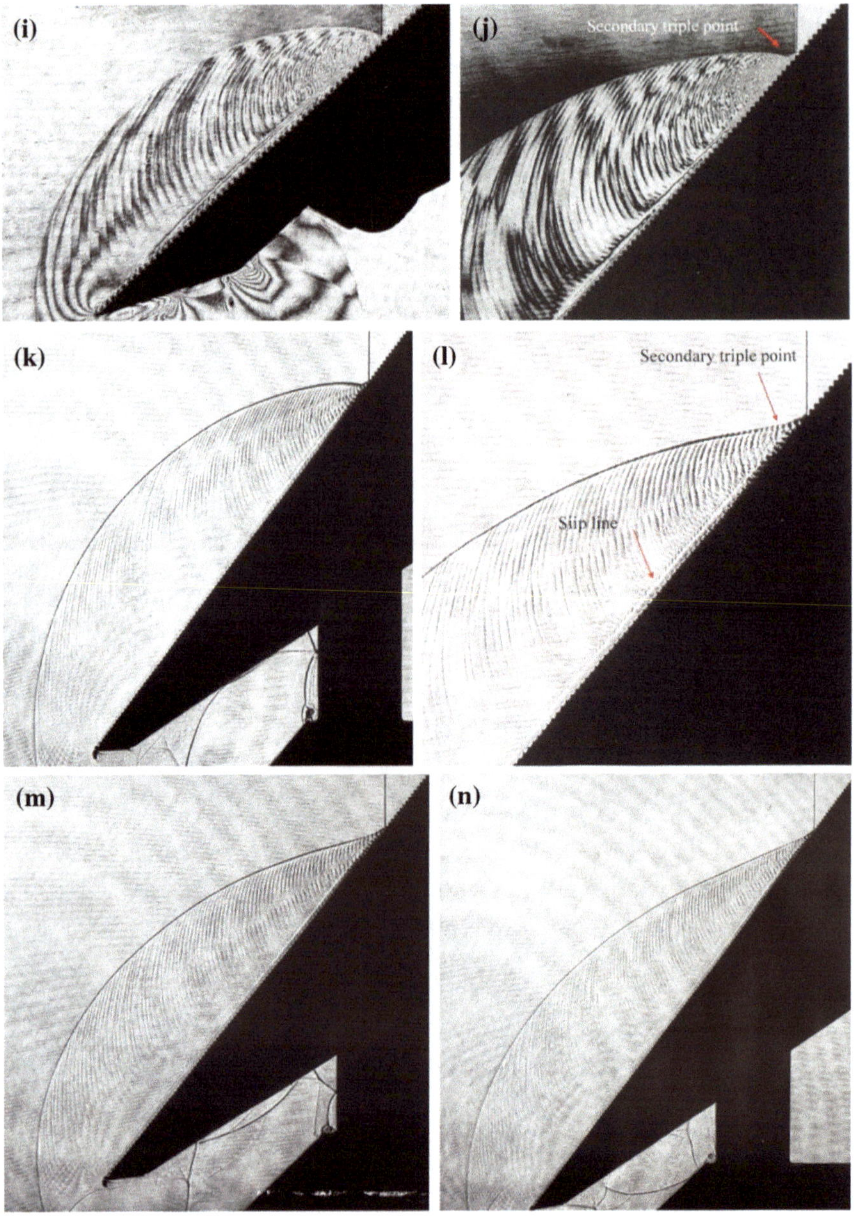

Fig. 2.39 (continued)

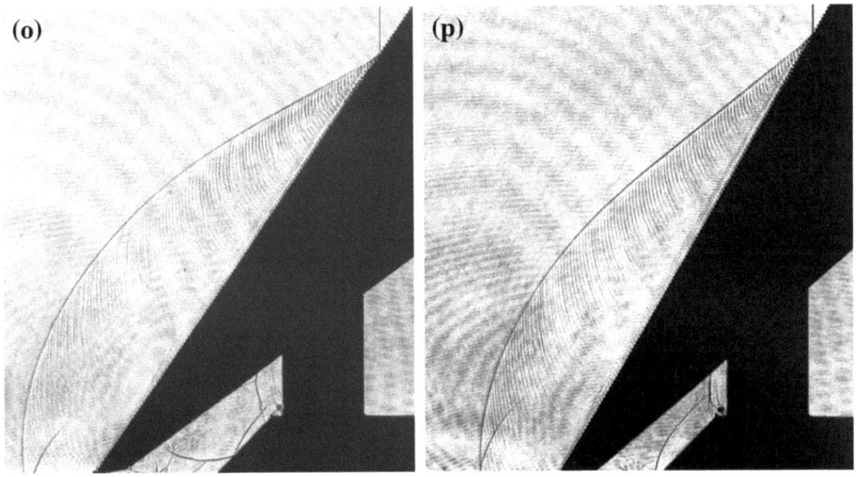

Fig. 2.39 (continued)

2.1.4.3 Wedges of a 0.8 mm Saw Tooth Surface Roughness

Figure 2.40 shows single exposure interferograms of evolution of reflected shock waves from a wedge having 0.8 mm saw tooth roughness for Ms = 1.46 in air at 500 hPa, 291.0 K. The pattern of the reflected shock wave is a SMR. Vortices were generated at the corners of the saw tooth roughness. The growth and dissipation of the vortices along the roughened wedge surface and their merger into a train of wavelets are well observed in Fig. 2.39c–g. The train of the wavelets becomes a SL emanating from the TP and is parallel to the wedge surface. Courant and Friedrichs (1948) defined this reflection pattern as a stationary Mach reflection, or in short, StMR. The shock tube flows are quasi-stationary and hence the StMR is held uniquely only for short time when the flow conditions fulfilled the presence of the StMR. Figure 2.40j shows a RR over a wedge of $\theta_w = 40.5°$. A secondary TP is seen from which the SL is emanating. The upstream flow information of the secondary TP are transmitted up to this point. Figure 2.40k shows a RR over a wedge of $\theta_w = 41.0°$. Figure 2.40l is its enlargement. The IS interacted independently with individual corners of the saw tooth.

2.1.4.4 Wedges of a 2.0 mm Saw Tooth Surface Roughness

Figure 2.41 shows later stages of single exposure interferograms of shock wave reflections from a coarse saw tooth wedge, k = 2.0 mm, for Ms = 1.47 in air at 500 hPa and 282.6 K. The saw tooth roughness is so coarse that the boundary layer displacement thickness would hardly contribute to the reflected shock wave transition. Due to interaction of IS with such a coarse roughness, the initiation, growth, and dissipation of vortices are better resolved. Analogously to the roughness of

(a) (b) (c) (d) (e) (f) (g) (h)

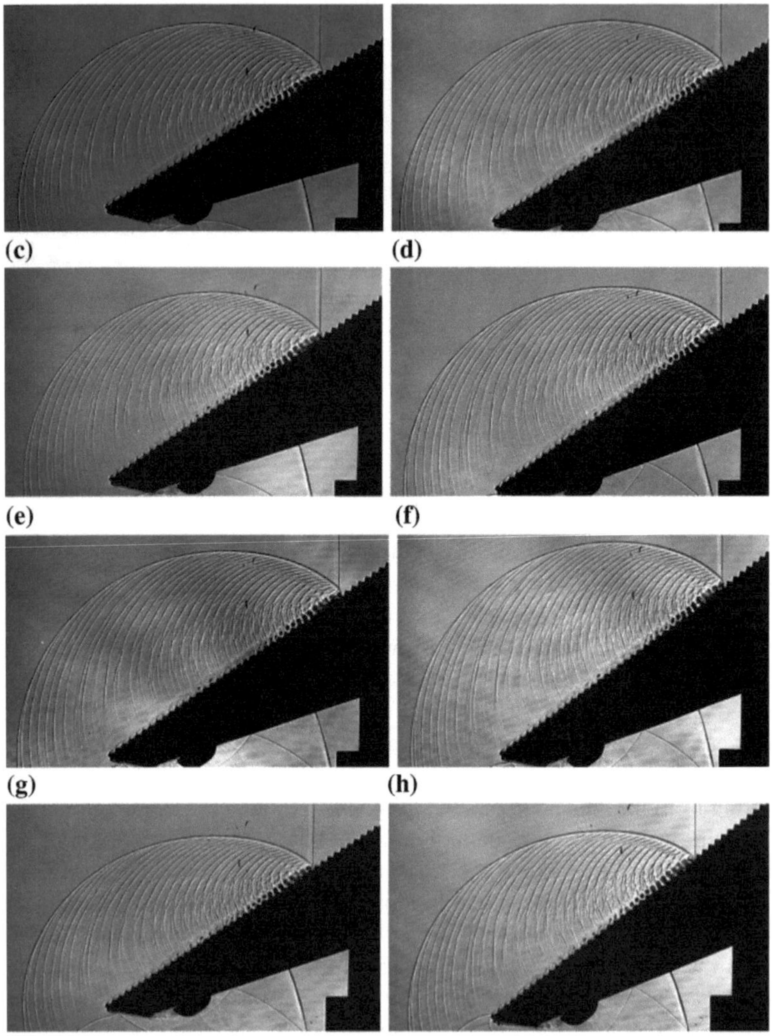

Fig. 2.40 Shock reflection over roughened wedges k = 0.8 mm for Ms = 1.45 in air at 500 hPa, 291.0 K: **a** #80050912, Ms = 1.456, $\theta_w = 32°$; **b** #80050916, Ms = 1.470, $\theta_w = 33°$; **c** #80050911, Ms = 1.456, $\theta_w = 35°$; **d** #80050913, Ms = 1.466, $\theta_w = 35.5°$; **e** #80050915, Ms = 1.473, $\theta_w = 36°$; **f** #80050914, Ms = 1.473, $\theta_w = 36°$; **g** #80050910, Ms = 1.472, $\theta_w = 36°$; **h** #80050909, Ms = 1.466, $\theta_w = 37°$; **i** #80050917, Ms = 1.472, $\theta_w = 38°$; **j** #80050907, Ms = 1.463, $\theta_w = 40.5°$; **k** #80050906, Ms = 1.466, $\theta_w = 41°$; **l** enlargement of (**k**)

0.8 mm as seen in Fig. 2.40, even in Fig. 2.41a, b, the envelop of wavelets which were created by the vortices formed a SL and merged with the TP. The SL looked parallel or was directed to the roughened wall surface. Then as already discussed in the case of k = 0.8 mm, the reflection patterns seen in Fig. 2.41a, b are StMR. With

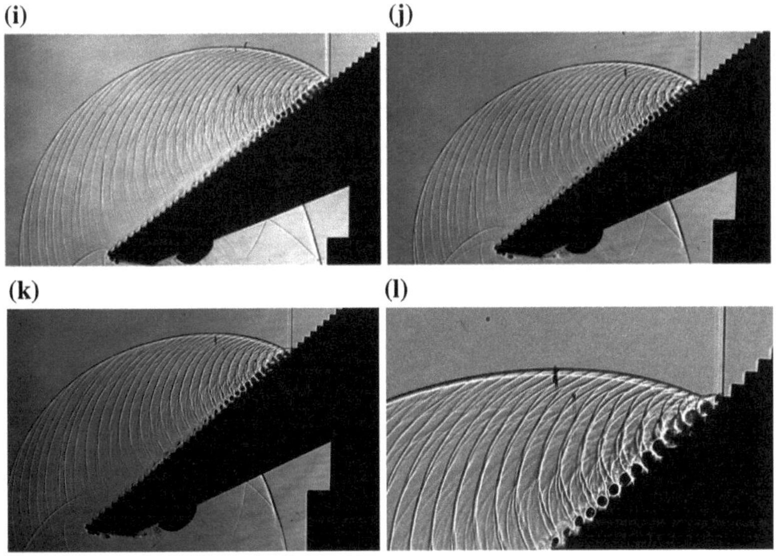

Fig. 2.40 (continued)

a slight increase in the wedge angle, the resulting reflection pattern would be an IvMR and the transition to a RR would occur. In Fig. 2.41c, a kink point which corresponds to a secondary TP is observable on the reflected shock wave. All the upstream information can reach up to the secondary TP and in the downstream region the IS interacts independently with the individual corner of the saw tooth. Figure 2.41j shows a RR and Fig. 2.41k is its enlargement. The IS interacts only locally with the individual saw tooth corners.

In order to determine the critical transition angles θ_{crit}, experiments were performed sequentially by changing in wedge angled and shock wave Mach numbers Ms. The reflection patterns were visualized by using double exposure and single exposure interferograms. Selective results of the visualizations were presented form Figs. 2.37, 2.38, 2.39, 2.40 and 2.41. The results obtained for various Ms and Re values are: Ms = 1.04, Re = 0.8×10^5, $\theta_{crit} = 27.5°$; Ms = 1.12, Re = 2.50×10^5, $\theta_{crit} = 32.5°$; Ms = 1.21, Re = 3.40×10^5, $\theta_{crit} = 40.5°$; Ms = 1.44, Re = 2.70×10^5, $\theta_{crit} = 43.7°$; Ms = 1.86, Re = 0.58×10^5, $\theta_{crit} = 45.8°$; and Ms = 3.80, Re = 0.32×10^5, $\theta_{crit} = 44.9°$.

Figure 2.42 summarized these results. The ordinate denote the θ_{crit} in degree and the abscissa denotes the inverse shock strength, the inverse pressure ratio $\xi = (\gamma + 1)/(2\gamma Ms^2 - \gamma + 1)$. Black filled circles denote θ_{crit} for smooth wedge (Smith 1948). Yellow, green, dark blue, and red filled circles denote the present results of k = 0.1, 0.2, 0.8, and 2.0 mm respectively (Takayama et al. 1981).

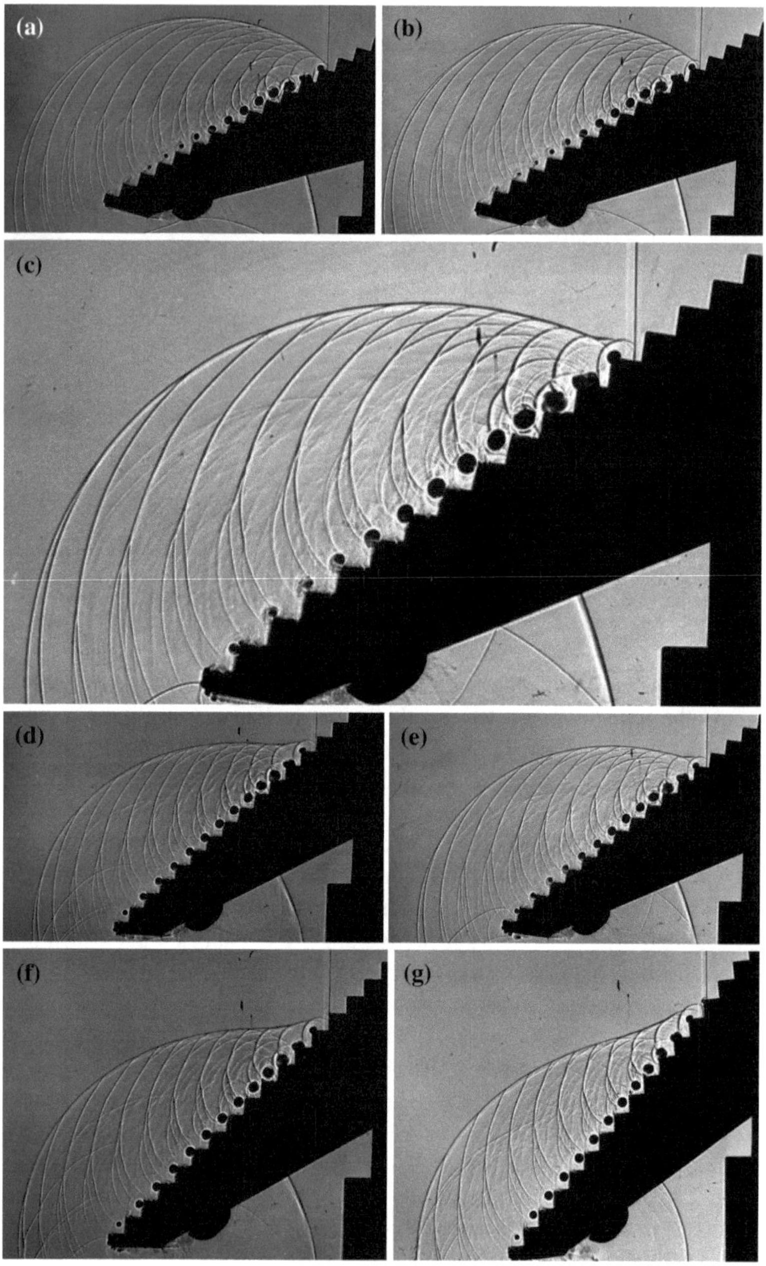

Fig. 2.41 Shock reflection over roughened wedges with saw tooth roughness k = 2.0 mm for Ms = 1.47 in air at 500 hPa and 282.6 K: **a** #80050811, Ms = 1.476, $\theta_w = 31°$; **b** #80050812, Ms = 1.470, $\theta_w = 33°$; **c** #80050839, Ms = 1.469, $\theta_w = 34°$; **d** #80050807, Ms = 1.456, $\theta_w = 37°$; **e** #80050808, Ms = 1.467, $\theta_w = 40°$; **f** #80050805, Ms = 1.473, $\theta_w = 45°$; **g** 80050802, Ms = 1.463, $\theta_w = 46°$; **h** #80050803, Ms = 1.468, $\theta_w = 47°$; **i** #80050804, Ms = 1.470, $\theta_w = 48°$; **j** #80050801, Ms = 1.456, $\theta_w = 51°$; **k** enlargement of (j)

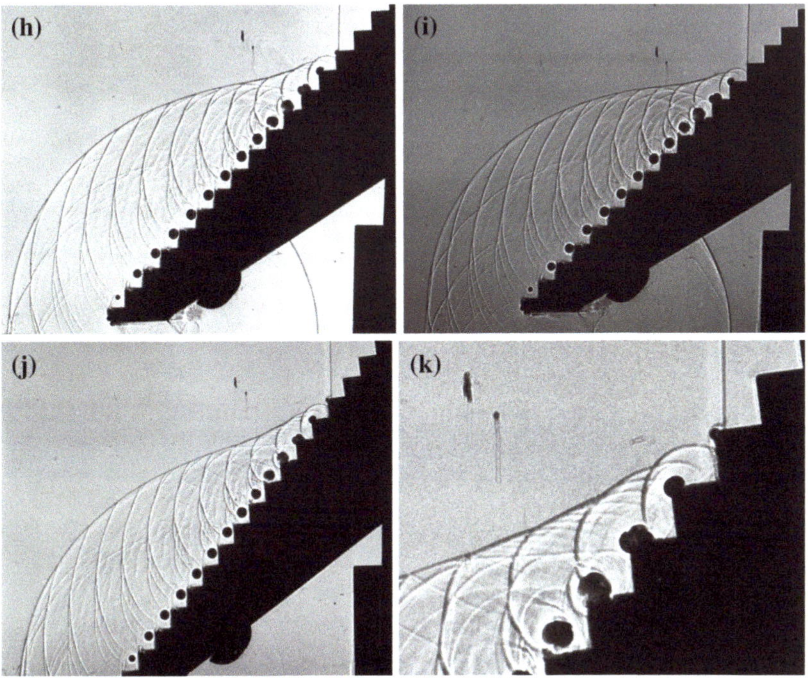

Fig. 2.41 (continued)

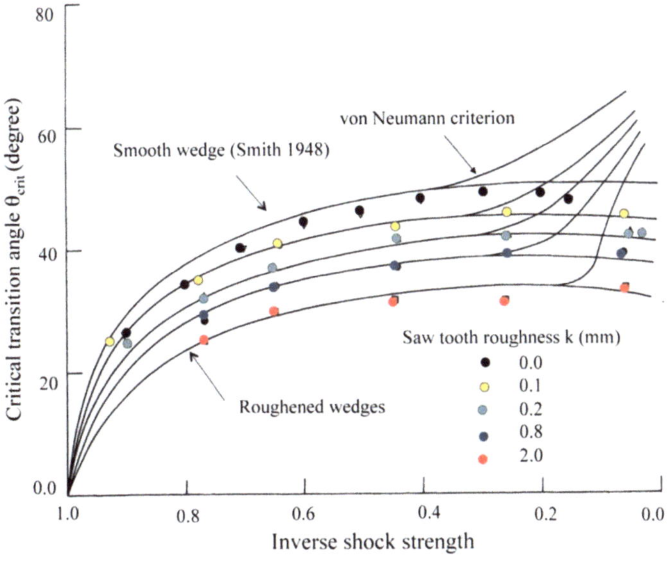

Fig. 2.42 Summary of experiments (Takayama et al. 1981)

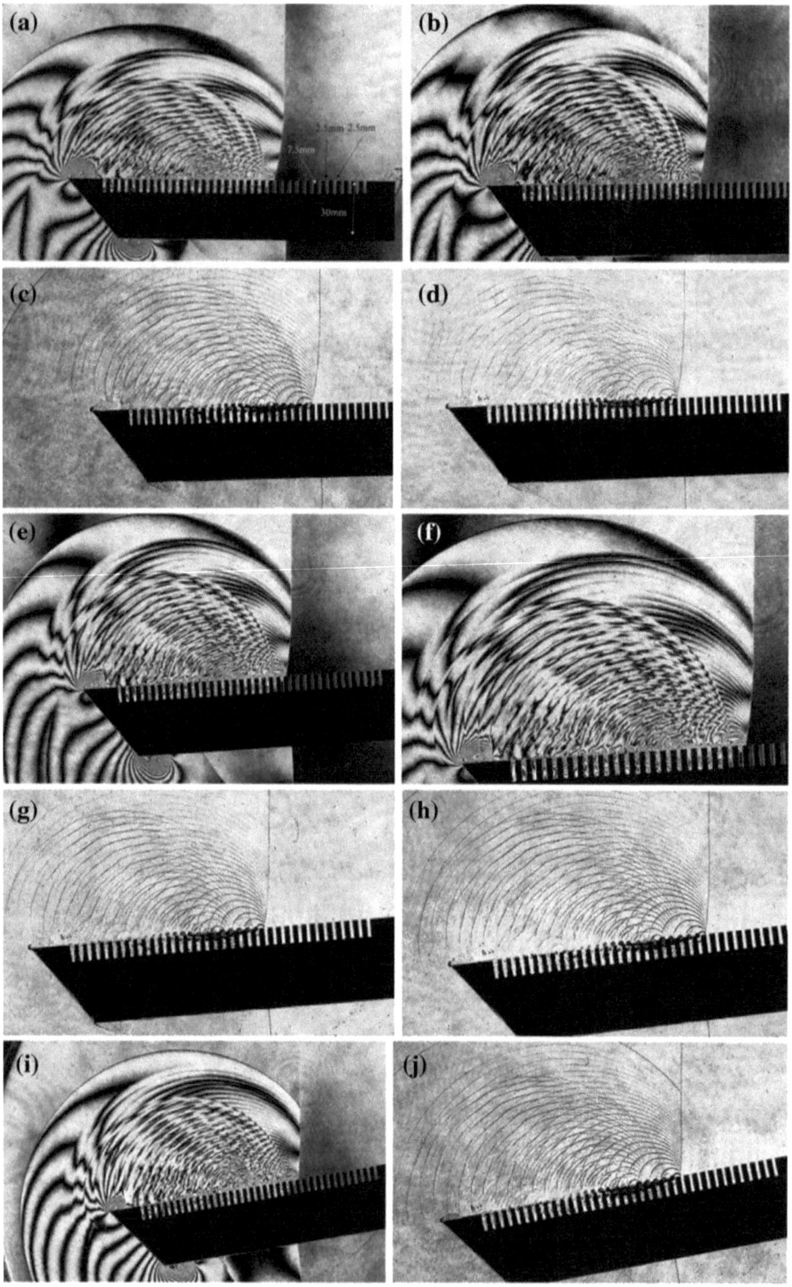
(a)
2.5mm 2.5mm
7.5mm
30mm
(b)
(c)
(d)
(e)
(f)
(g)
(h)
(i)
(j)

◀**Fig. 2.43** Evolution of reflected shock waves from a slotted wedge for Ms = 1.36 in air at atmospheric air at 295 K: **a** #86060612, Ms = 1.360, θ_w = 1.0°; **b** #86060611, Ms = 1.362, θ_w = 2.0; **c** #86060612, Ms = 1.368, θ_w = 3.0°, single exposure; **d** #86060609, Ms = 1.358, θ_w = 4.0°, single exposure; **e** #86060608, Ms = 1.358, θ_w = 5.0; **f** enlargement of (**e**); **g** #86060607, Ms = 1.367, θ_w = 6.0° single exposure; **h** #86060605, Ms = 1.370, θ_w = 8.0°, single exposure; **i** #86060603, Ms = 1.368, θ_w = 10.0°; **j** #86060601, Ms = 1.358, θ_w = 12.0° single exposure; **k** #86060517, Ms = 1.365, θ_w = 14.0°, single exposure; **l** #86060516, Ms = 1.352, θ_w = 15.0°; **m** #86060511, Ms = 1.365, θ_w = 20.0; **n** #86060510, Ms = 1.365, θ_w = 21.0°, single exposure; **o** #86060508, Ms = 1.360, θ_w = 23.0, single exposure; **p** #86060506, Ms = 1.359, θ_w = 25.0°; **q** #86060509, Ms = 1.365, θ_w = 28.0° single exposure; **r** #86053002, Ms = 1.359, θ_w = 30.0; **s** #86053004, Ms = 1.363, θ_w = 32.0; **t** #86060402, Ms = 1.349, θ_w = 34.0° single exposure; **u** #86060404, Ms = 1.344, θ_w = 36.0 single exposure; **v** #86060406, Ms = 1.346, θ_w = 38.0° single exposure; **w** #86060407, Ms = 1.346, θ_w = 39.0° single exposure

2.1.5 Evolution of Shock Wave Reflection from Slotted Wedges or Perforated Wedges

When a shock wave propagates along a roughened wedge, the critical transition angle decreases as seen in Fig. 2.42. When shock waves are reflected from slotted wedges or perforated wedges, what will happen?

2.1.5.1 Slotted Wedges

A slotted wedge was installed in the test section of the 60 mm × 150 mm diaphragm-less shock tube. The wedge was sandwiched between two acrylic disks of 200 mm in diameter and 20 mm in thickness and adjusted its angle by rotating the disks. On the tested slot is 1.5 mm wide and 7.0 mm deep and a gap between neighboring slots is 1.5 mm. 36 slits are distributed along the wedge model surface with a perforation ratio ε = 0.4.

Figure 2.43 shows evolution of shock wave reflected from the slotted wedge surface for Ms = 1.36 in atmospheric air at 295 K. The IS is diffracted at each slot opening, transmitted along the slot wall and reflected from its bottom. In the meantime, expansion waves are created from each slot corner coalescing into an envelop of expansion wave (Onodera and Takayama 1990). In the case of saw tooth roughened surface, vortices are created, it dissipates energy, and thereby attenuates the IS.

As seen in Figs. 2.37, 2.38, 2.39, 2.40 and 2.41, wavelets reflected from individual saw tooth roughness formed an envelop of a SL which interacted with the TP and eventually formed the IvMR. In the final reflection pattern, the SL emanating from the TP is directed away from the wedge surface, which is a typical wave pattern observed in an IvMR. When it transits to RR, the resulting RS has a secondary TP. However, over the slotted wedges, the flow is deflected at the individual slots and their reflections are expansion waves. As seen in Fig. 2.43r–t, a slip line emanating from the TP is directed toward the wedge surface. Therefore, the resulting reflection pattern is a direct Mach reflection or in short, DiMR. Hence with

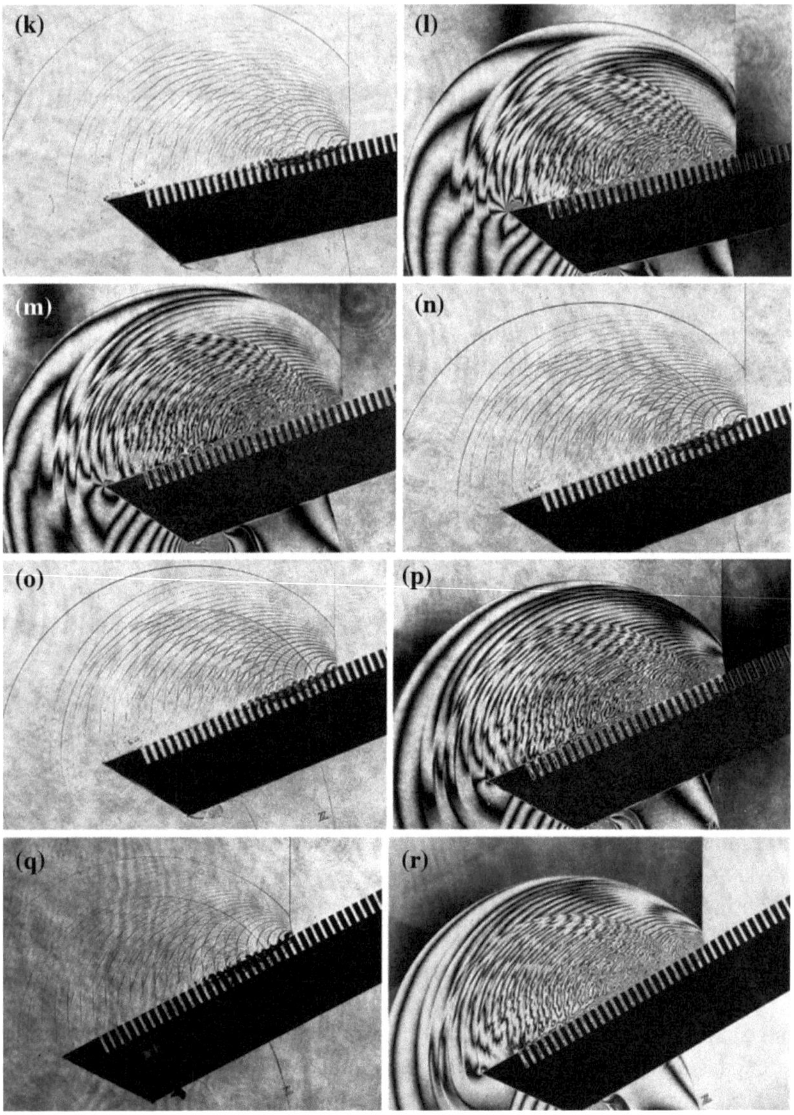

Fig. 2.43 (continued)

increase in wedge angles, MR transits to RR, without causing any complicated reflection patterns, as seen in Fig. 2.42v, w.

Figure 2.42w shows an enlargement of a RR at $\theta_w = 39°$. The transmitted shock wave propagates along the individual slots and reflected. Reflected waves from the bottom of slots released in the area behind the RS formed an envelop of compression wavelets.

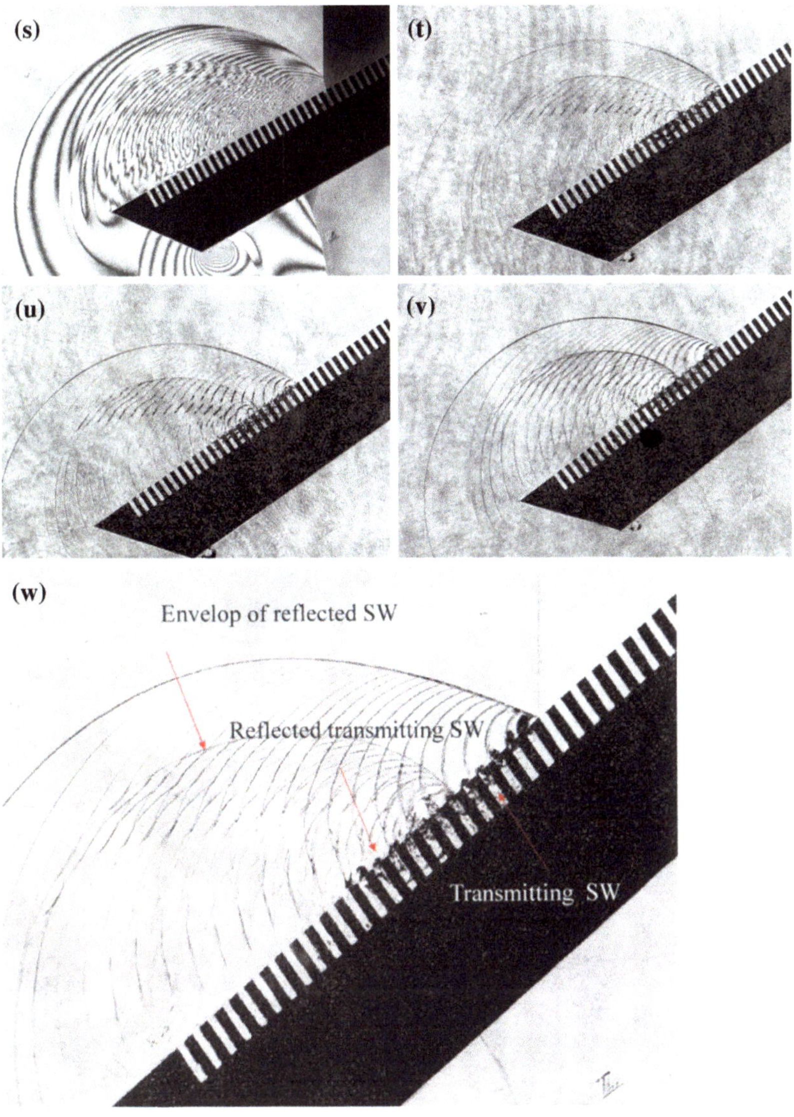

Fig. 2.43 (continued)

To assure whether or not the transition over slotted wedges occurs similarly to that seen in a case of the smooth and solid wedges, we performed the transition was checked for stronger shock waves and for wedge angles close to the θ_{crit}. In Fig. 2.43, the reflection pattern is DMR for Ms = 3.0 and the wedge angle was increased step by step over up to and over θ_{crit}. In Fig. 2.44a. a DMR pattern is seen and with increasing in the wedge angle, although the reflection pattern is maintained, it becomes smaller and eventually transits straightforwardly to RR.

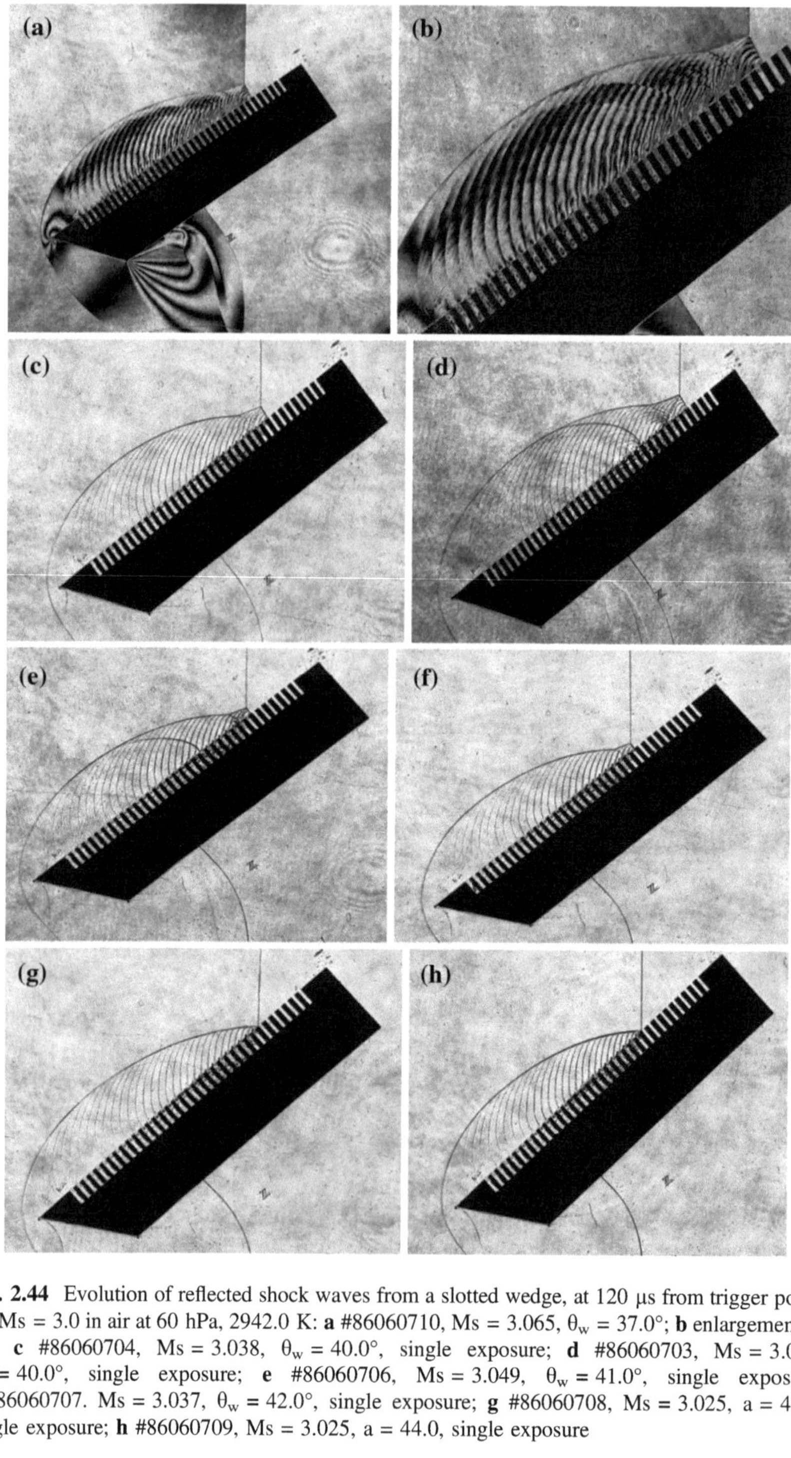

Fig. 2.44 Evolution of reflected shock waves from a slotted wedge, at 120 μs from trigger point, for Ms = 3.0 in air at 60 hPa, 2942.0 K: **a** #86060710, Ms = 3.065, θ_w = 37.0°; **b** enlargement of (**a**); **c** #86060704, Ms = 3.038, θ_w = 40.0°, single exposure; **d** #86060703, Ms = 3.016, θ_w = 40.0°, single exposure; **e** #86060706, Ms = 3.049, θ_w = 41.0°, single exposure; **f** #86060707. Ms = 3.037, θ_w = 42.0°, single exposure; **g** #86060708, Ms = 3.025, a = 43.0, single exposure; **h** #86060709, Ms = 3.025, a = 44.0, single exposure

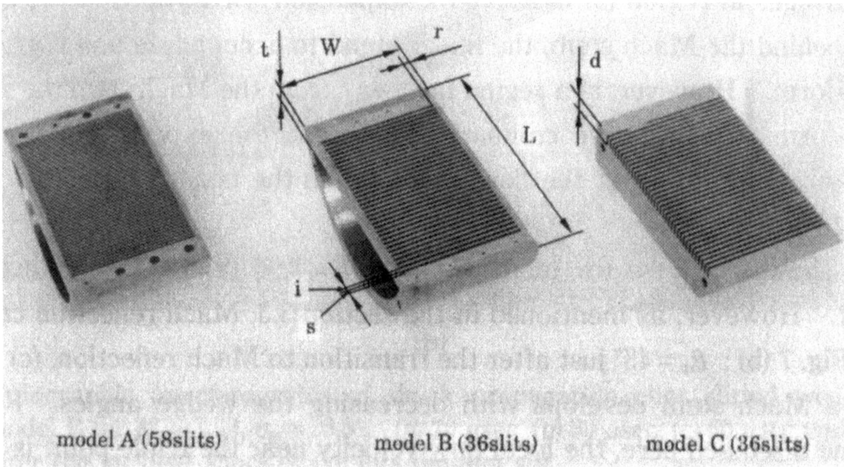

Fig. 2.45 Perforated and slotted wall Models (Onodera and Takayama 1990)

2.1.5.2 Perforated Wedges

Figure 2.45a shows slotted wall Models: Model C had the following dimensions: L = 110 mm, d = 7 mm, s = 1.5 mm, i = 1.5 mm and perforation ratio $\varepsilon = 0.4$. Model A hag s = 0.5 mm, i = 1.0 mm, $\varepsilon = 0.34$; and Model B had $\varepsilon = 0.4$, s = 1.5 mm, i = 1.5 mm, L = 100 mm, t = 5 mm, r = 3.0 mm.

In the early 1970s, many shock wave researchers investigated shock wave propagation over perforated walls. In the proceedings of the 8th International Symposium on Shock Tubes, papers were presented on this topic. At that time, researchers tried to Model the mass flow through the wedge perforations. For correctly measuring the mass flow through the perforations, Model A and Model B were constructed. Using these Models, it was possible to monitor mass flows leaked from the perforations. Hence it was possible to correctly estimate the deflection velocity, which would be a boundary condition in drawing shock polar.

Shock wave reflection from perforated wedges or slotted wedges is one of basic research topics of shock wave dynamics. Many reports were presented on the shock wave propagation over perforated walls (for example Szumowski 1972). A perforated plate with perforation ratio of 0.40 was manufacture in our machine shop, was sandwiched between 200 mm diameter and 20 mm thick PMMA plates, was installed in the test section of the 60 mm × 150 mm diaphragm-less shock tube. The wedge angles were arbitrarily adjusted by rotating the PMMA plates.

Figure 2.46 shows sequential observation of the shock wave reflection for Ms = 1.17 in air at 930 hPa, 290.8 K over the Model B as shown Fig. 2.45. The perforated wedges were supported by a half-cur cylinder which was fixed on the side wall. With this arrangement, it was observed that an oblique shock wave was driven by the gas flow which leaked out through the perforation behind the IS. The evolution of the inclination angle of the oblique shock wave indicates the attenuation of the transmitting shock wave along the perforated wall.

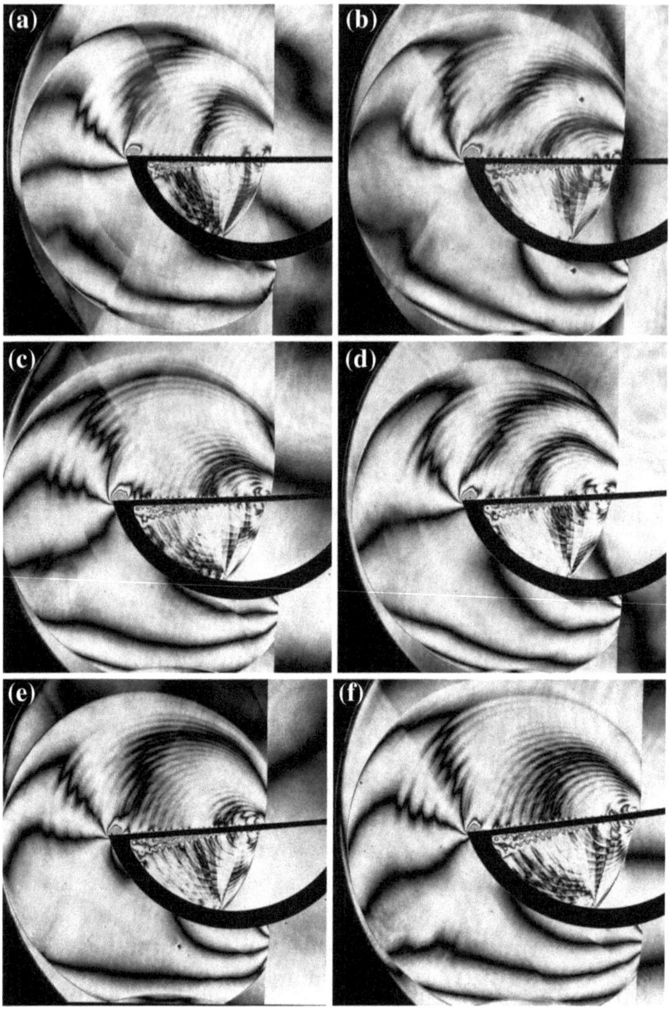

Fig. 2.46 Evolution of reflected shock waves over a perforated wall of Model B shown in Fig. 2.44 for Ms = 1.17 in air at 930 hPa, 290.8 K: **a** #86030320, Ms = 1.152, $\theta_w = 0.0°$; **b** #86030321, Ms = 1.169, $\theta_w = 1.0°$; **c** #86030318, Ms = 1.170, $\theta_w = 3.0°$; **d** #86030317, Ms = 1.170, $\theta_w = 4.0°$; **e** #86030316, Ms = 1.171, $\theta_w = 6.0°$; **f** #86030315, Ms = 1.169, $\theta_w = 9.0°$; **g** #86030314, Ms = 1.163, $\theta_w = 13.0°$; **h** #86030313, Ms = 1.163, $\theta_w = 17.0°$; **i** #86030312 Ms = 1.169, $\theta_w = 21.0°$; **j** #86030310, Ms = 1.172, $\theta_w = 27.0°$; **k** #86030307, Ms = 1.174, $\theta_w = 28.0°$; **l** #86030308, Ms = 1.170, $\theta_w = 29.0°$; **m** #86030307, Ms = 1.170, $\theta_w = 30.0°$; **n** #86030305, Ms = 1.165, $\theta_w = 32.0°$; **o** #86030304, Ms = 1.172, $\theta_w = 33.0°$; **p** #86030301, Ms = 1.173, $\theta_w = 34.0°$; **q** #86030302, Ms = 1.173, $\theta_w = 35.0°$; **r** #86030301, Ms = 1.162, $\theta_w = 36.0°$

Figure 2.46a shows the shock wave propagation along the perforated wedge at $\theta_w = 0.0°$, hence the triple point lies on the trajectory of the glancing incidence angle. Figure 2.46b–k shows the shock wave propagation along the perforated

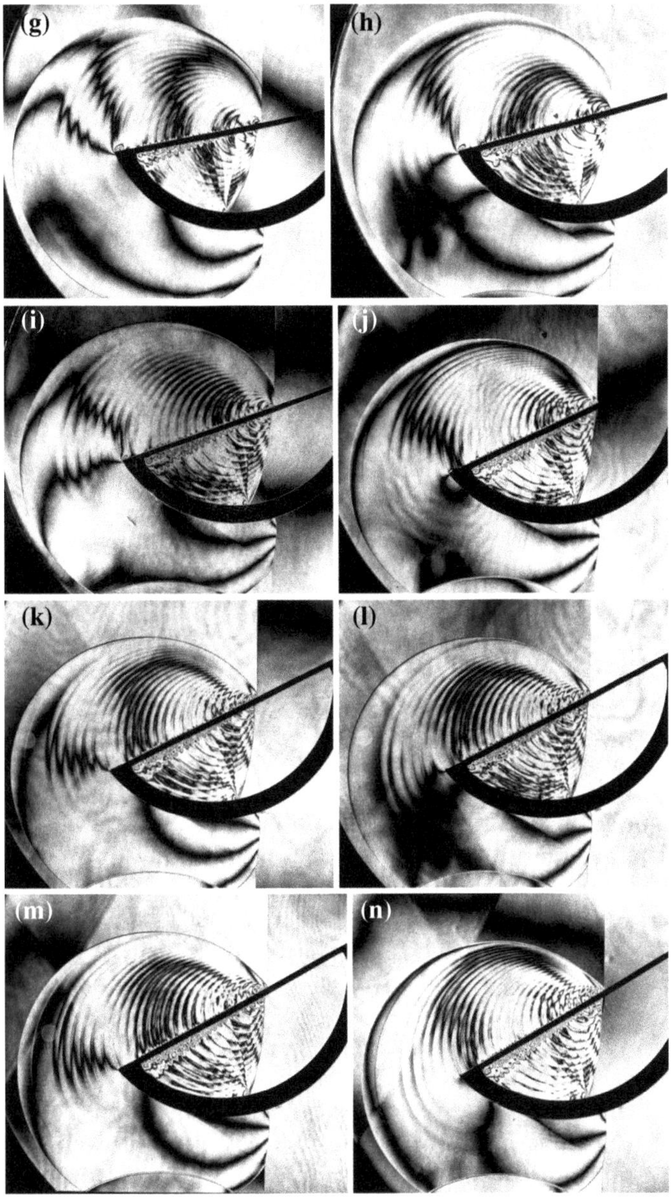

Fig. 2.46 (continued)

wedge at from $\theta_w = 1.0°$ to and 28.0°. The patterns of all the reflected shock waves show the MR but the triple points don't show any slip lines. Therefore, the reflection patterns are vNMR. In Fig. 2.46m at the wedge angle of $\theta_w = 30.0°$, the reflected shock wave misses a MS. Then the transition to a RR occurred indicating that the θ_{crit} would be about 30.0°.

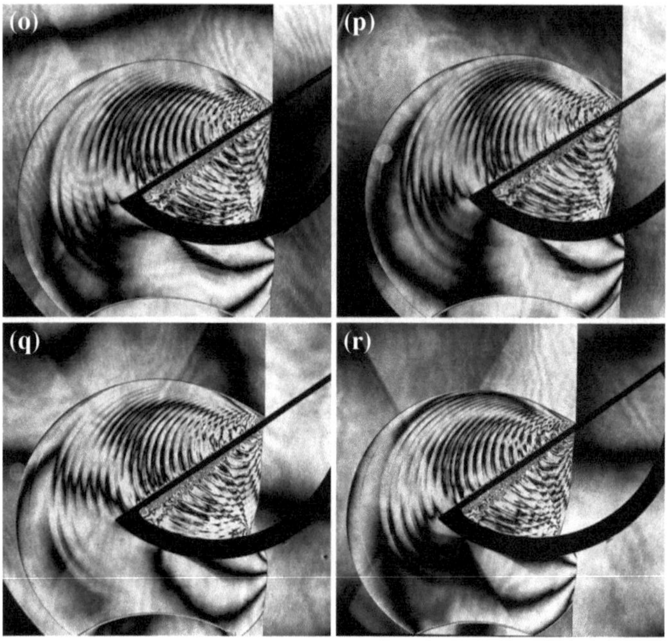

Fig. 2.46 (continued)

Figure 2.42 summarized the dependence of the critical transition angles θ_{crit} on the inverse shock strength, that is, the inverse of the pressure ratio ξ over wedges with the saw tooth surface roughness. In the case of the saw tooth roughness k = 0.8 mm, the critical transition angle θ_{crit} was about 30.0° for inverse shock strength of $\xi - 0.7$ which corresponds to Ms = 1.17. This result agreed with the result of the observation shown in Fig. 2.46.

Figure 2.47 shows the evolution of reflected shock waves over the perforated wedge of Model B for Ms = 1.56 in air or at the inverse shock strength of $\xi = 0.375$ at 755 hPa, 301 K. The wedge angle is increased from $\theta_w = 0°$ to $\theta_w = 90.0°$ by every 10.0°. Figure 2.47a shows the reflection from a wedge of the glancing incidence. In Fig. 2.47b–j, the reflection pattern is a SMR but the SL emanating from the TP is just visible, whereas in Fig. 2.46 the reflection pattern is a vNMR. The transition occurs at an angle slightly larger than $\theta_w = 40.0°$. The critical transition angle for Ms = 1.56; $\xi = 0.375$ is comparable to the value of θ_{crit} of the wedge with saw tooth roughness k = 0.2 mm as shown in Fig. 2.42. Figure 2.47j shows a symmetrical shock wave reflection from $\theta_w = 90.0°$.

Figure 2.48 shows the evolution of a shock wave reflection over a perforated wedge of the Model B for Ms = 2.90 or at the inverse shock strength of $\xi = 0.104$ in air at 40 hPa, 297.0 K. The reflection pattern is a SMR in Fig. 2.48a–c. However, with the increase in the wedge angle, it is a TMR in Fig. 2.48d–f and the transition to a RR takes place. As seen in Fig. 2.45, the Model B has wide flat edge

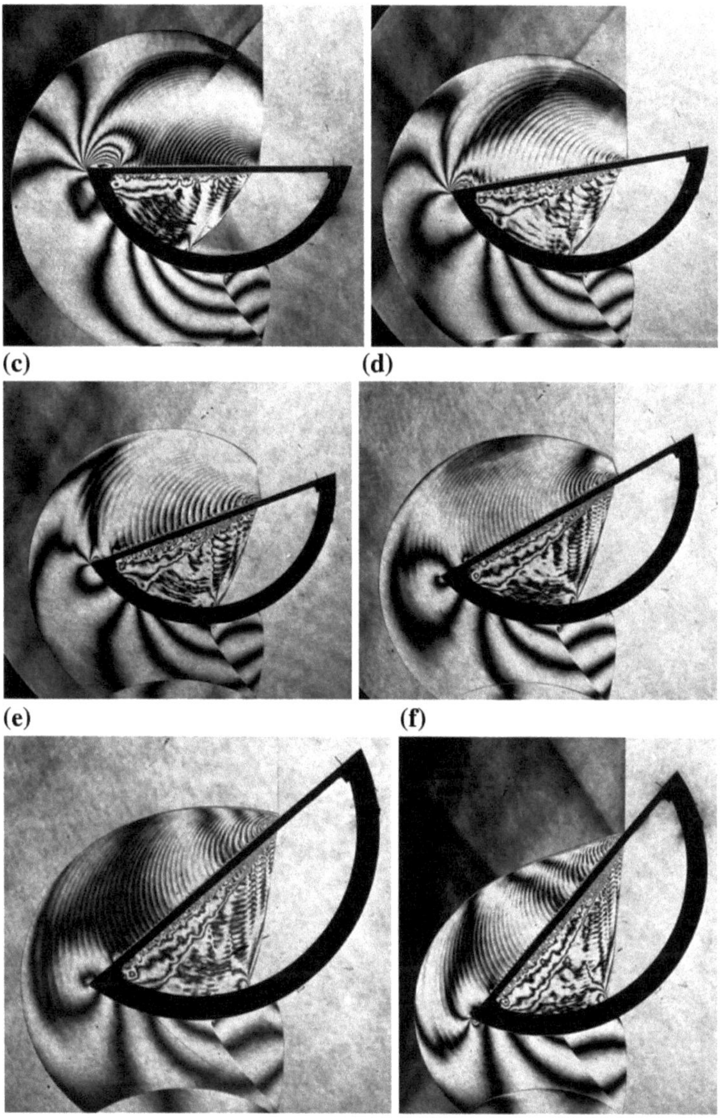

Fig. 2.47 Evolution of reflected shock waves over a perforated wall for Ms = 1.56 in air at 755 hPa, 301 K: **a** #94072801, Ms = 1.560, $\theta_w = 0°$; **b** #94072702, Ms = 1.558, $\theta_w = 10°$; **c** #94072703, Ms = 1.558, $\theta_w = 20°$; **d** #94072804, Ms = 1.560, $\theta_w = 30°$; **e** #94072805, Ms = 1.558, $\theta_w = 40°$; **f** #94072806, Ms = 1.780, $\theta_w = 50°$; **g** #94072807, Ms = 1.780, $\theta_w = 60°$; **h** #94072808, Ms = 1.558, $\theta_w = 70°$: **i** #94072809, Ms = 1.558, $\theta_w = 80°$; **j** #94072810, Ms = 1.556, $\theta_w = 90°$

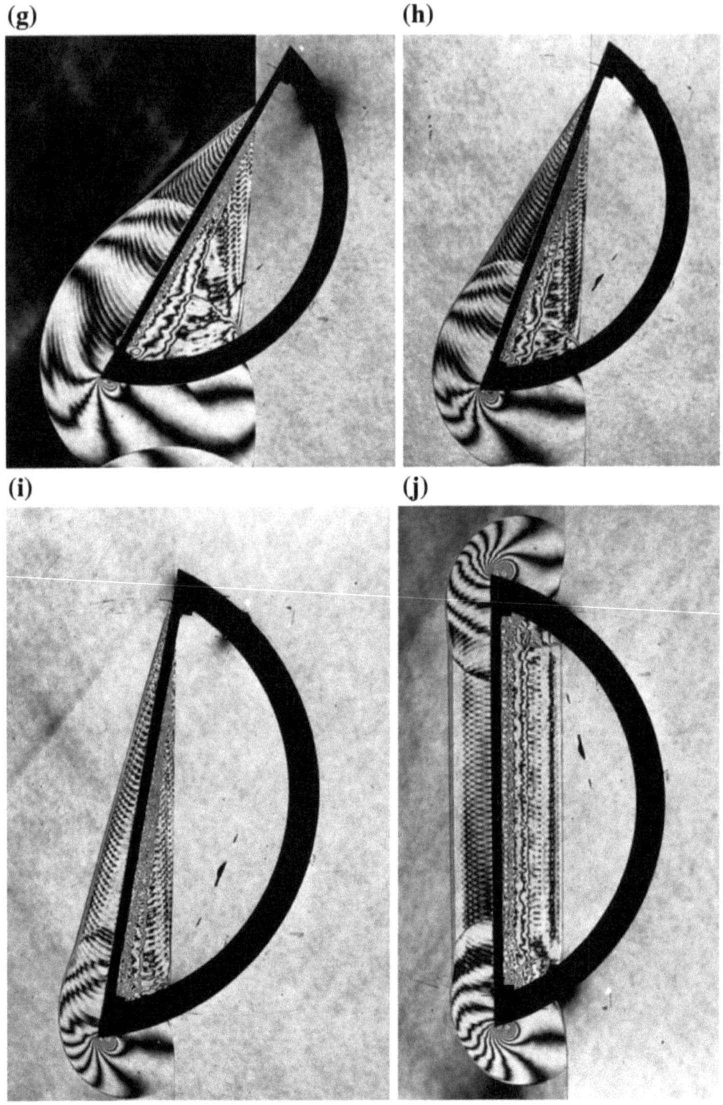

Fig. 2.47 (continued)

parts of r = 3 mm which support the perforation. It is noticed that the shock wave is reflected, at the same time, from the 54 mm wide perforated wedge and from 6 mm wide flat wedge. Then the θ_{crit} occurring over the Model B wedge would be slightly larger than the θ_{crit} occurring over a 60 mm wide perforated wedge.

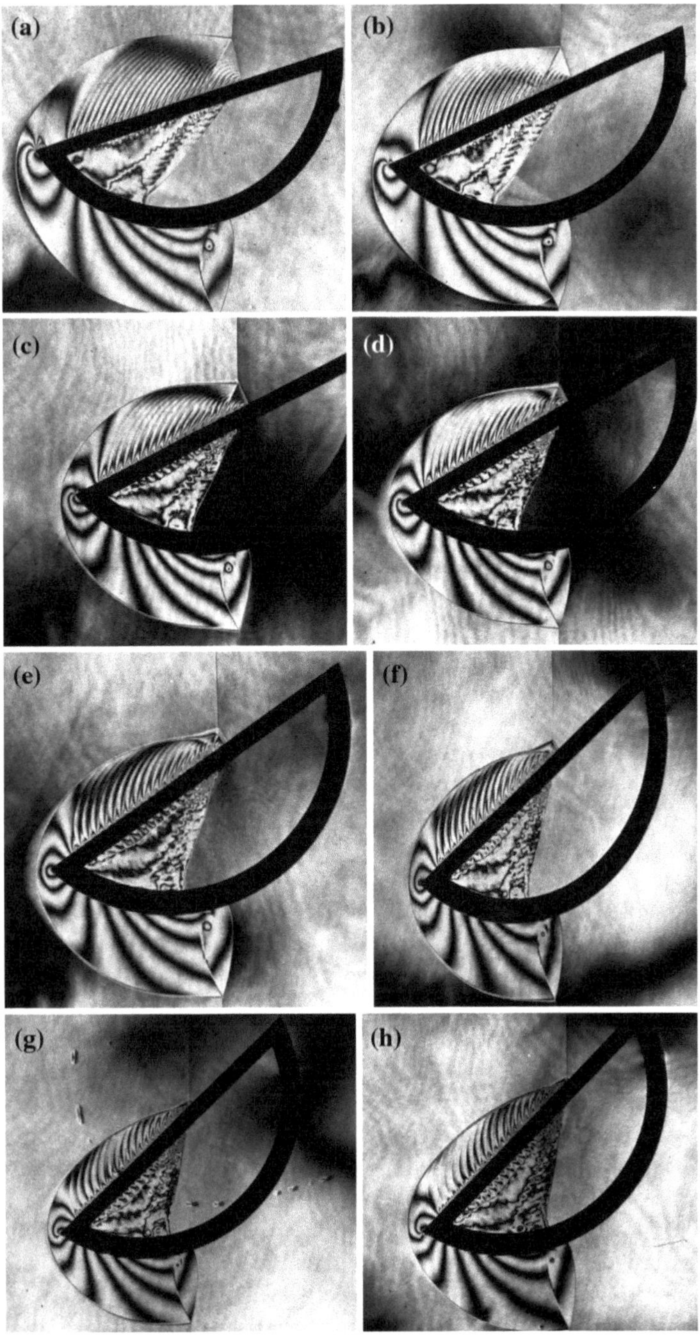

Fig. 2.48 Evolution of reflected shock waves over a perforated wall for Ms = 2.90 in air at 40 hPa, 297.0 K: **a** #86032506, Ms = 2.902, θ_w = 21.0°; **b** #86032505, Ms = 2.902, θ_w = 26.5°; **c** #86012319, Ms = 2.948, θ_w = 28.0°; **d** #86012318, Ms = 2.931, θ_w = 31.0°; **e** #86012315, Ms = 2.971, θ_w = 38.0°; **f** #86012301, Ms = 2.967, θ_w = 40.0°; **g** #86012306, Ms = 2.920, θ_w = 45.0°; **h** #86012308, Ms = 2.988, θ_w = 47.0°; **i** #86012310, Ms = 2.961, θ_w = 52.0°

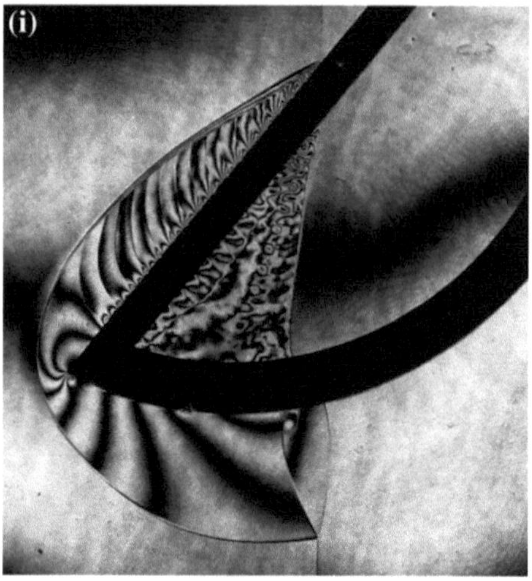

Fig. 2.48 (continued)

2.1.6 Evolution of Shock Reflection from Liquid Wedges

2.1.6.1 Water Wedges

In experiments of shock wave reflections from wedges, shock wave researchers always argue the effect of a surface roughness of wedges on the critical transition angles. It was, therefore, a motivation of shock wave reflection from water wedges. A water wedge is formed by filling water in a tilted low pressure channel of a shock tube seen in Fig. 2.49. The shock wave reflection from the water wedge is visualized in double exposure holographic interferometry (Miyoshi 1987).

A 240 mm diameter and 30 mm wide circular stainless steel test section was constructed for investigating shock wave reflection from water wedges. The test section was and connected to a 30 mm × 40 mm shock tube. The entire shock tube and the circular test section was rotated at an angle ranging from 0° up to 60° keeping the test section at the center of rotation. Water was filled and the whole shock tube was rotated at a required angle θ_w. The water wedge surface was kept perfectly flat. However, due to the surface tension in water, the water touched on the side window glasses at the wet angle forming a meniscus. The locally curved water surface working as a lens and broadened the shadow of the water surface. The pressure in the test section was reduced to the pressure slightly higher than water vapor pressure and then filled water slowly, and thereafter recovered to the desired test pressure level. Then the meniscus effect was minimized.

As the sound speed in water is much faster than that of the IS, the forward running wave is observed, in the wide range of Ms, ahead of the IS. However, the component of the IS along the water wedge surface is $u_s/\cos\theta_w$ where u_s is shock wave speed. Then, if $u_s/\cos\theta_w = a_{water}$ is valid where a_{water} is the sound speed in

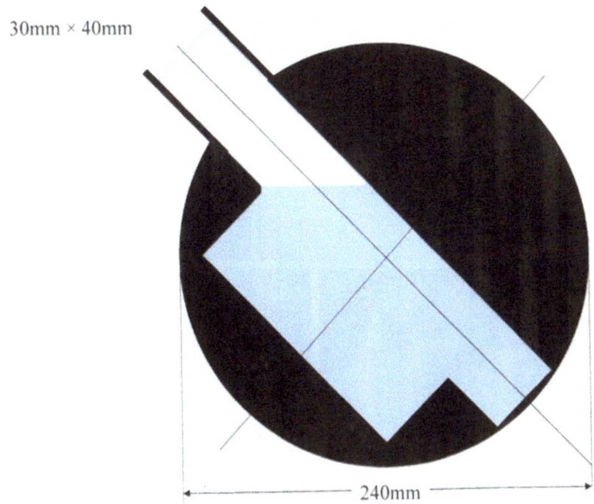

Fig. 2.49 Test section of water wedge (Miyoshi 1987)

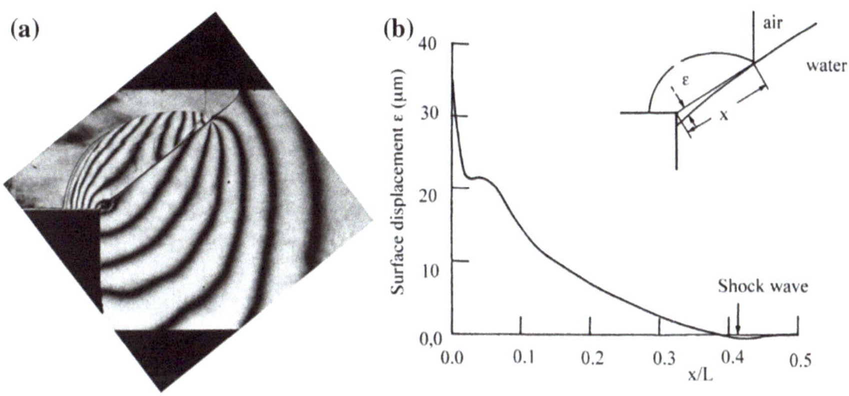

Fig. 2.50 Shock wave reflection from a water wedge for Ms = 1.666, in air at 650 hPa 294.5 K, $\theta_w = 37.5°$: **a** #87051303; **b** deformation of water surface corresponding to (**a**) (Miyoshi 1987)

water, the precursory wave merges with the shock wave in air. Figure 2.50a shows the reflection of a shock wave of Ms = 1.67 from a 37.5° water wedge. Figure 2.50b shows a result from a numerical simulation of this experiment in which a in house code based on the TVD scheme is used (Itoh 1986). The deformation of the water wedge surface due to the shock wave loading was simulated and was maximal at the foot of the shock wave on the water wedge. Although the contact point is a singular spot, the degree of the maximal deformation predicted numerically is presumably unphysical. The deformation at the contact point would be at most about 20 μm.

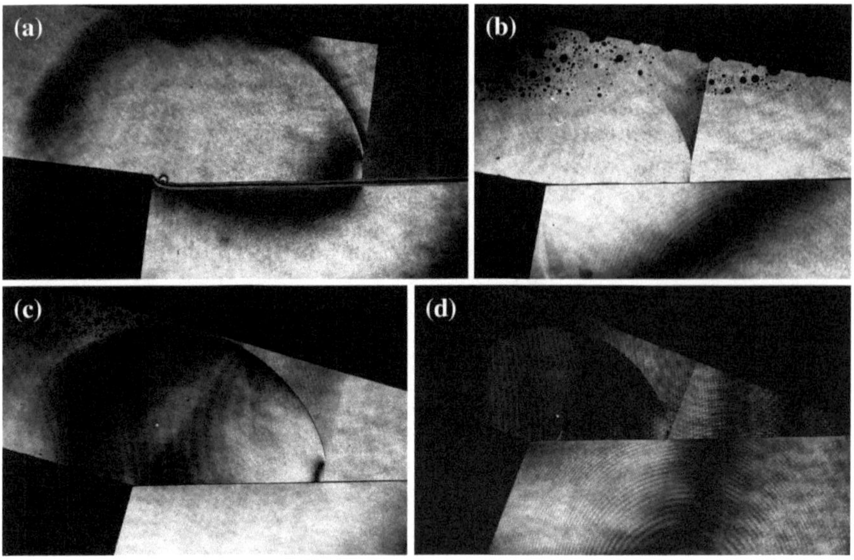

Fig. 2.51 Evolution of weak reflected shock wave over water wedges for Ms = 1.1 in air at 1013 hPa, 287.2 K: **a** #87041601, Ms = 1.148, θ_w = 8°; **b** #87041602, Ms = 1.148, θ_w = 12°; **c** #87041603, Ms = 1.088, θ_w = 17°; **d** #87041604, Ms = 1.056, θ_w = 22°

Figure 2.51 shows evolution of weak shock wave reflections over relatively shallow water wedge angles ranging from θ_w = 8° to θ_w = 22°. Fringes are generated in water due to the change in phase angles during the double exposures and are converted to density variations by using the formulation given in the Chap. 1. Among quantitative optical flow visualization methods, the double exposure holographic interferometry can record only the change in phase angles during the double exposure and is insensitive to inhomogeneity of the medium or slow moving convective flows in the medium. Then the double exposure holographic interferometry is very suited for underwater shock wave experiments.

The test gas was ambient air and the test water was degassed distilled water. After shock tube runs, tested water was shattered and spread inside the shock tube. Then the water wedge experiments were not necessarily pleasant ones because it was a time wasting work to clean the inside the shock tube and to check the valve ports and pressure transducers.

Figure 2.52 shows the evolution of a weak shock wave reflection over moderately tilted and steeply tilted water wedges of θ_w = 30° to θ_w = 46°. The transition occurred at an angle between θ_w = 30° as seen in Fig. 2.52c and θ_w = 36° as seen in Fig. 2.52c, d.

Figure 2.53a–h shows the evolution of reflected shock waves over water wedges from θ_w = 30° to θ_w = 50° for Ms = 1.77 in air at 650 hPa, 294.5 K. Disturbances created by the IS propagate along the water surface at the speed of Ms/cos $\theta_{w.}$ Then for shallow wedge angles in Fig. 2.53a, b, the disturbances propagate at subsonic speed to the the sound speed in water. Then a sonic wave appears ahead of the IS.

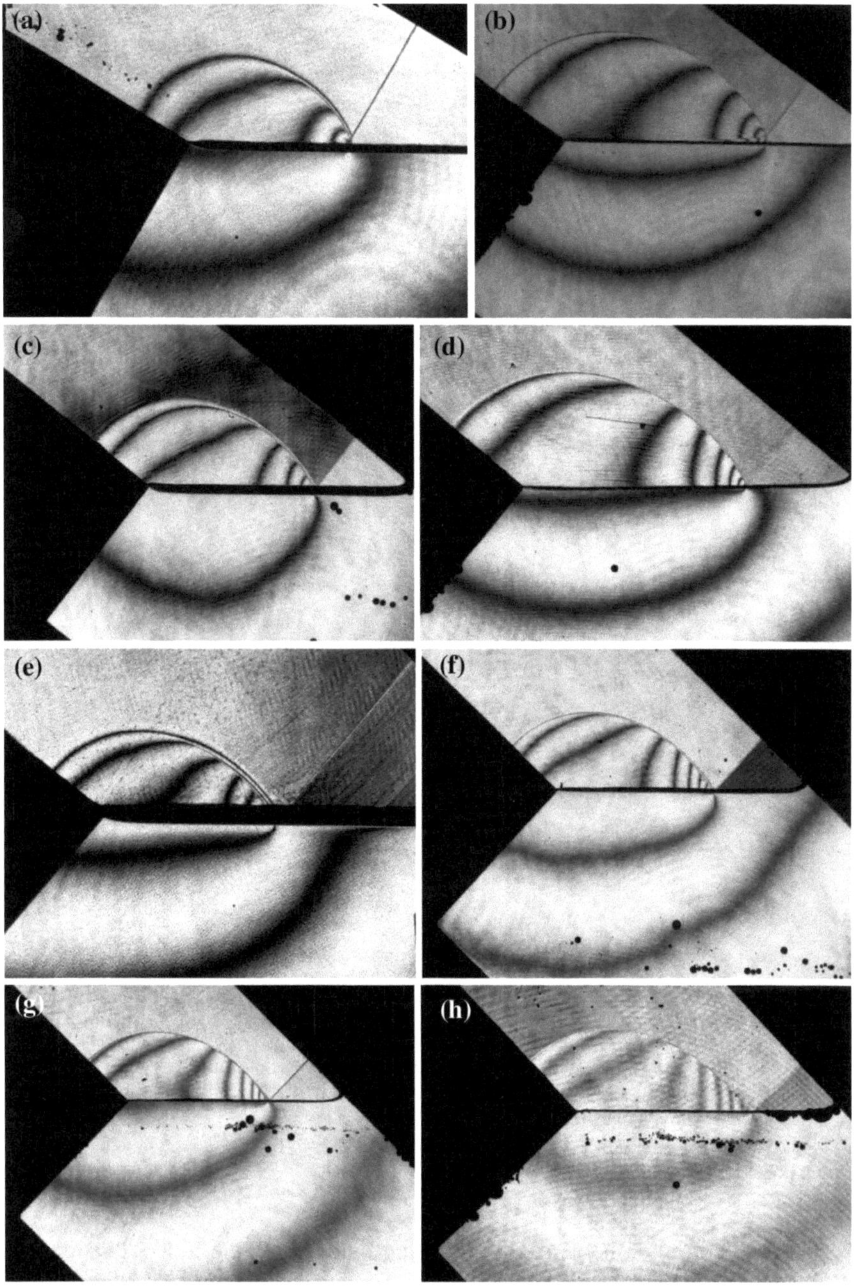

Fig. 2.52 Evolution of reflected shock wave over water wedge for Ms = 1.25 in atmospheric air at 298.0 K: **a** #87070804, Ms = 1.270, $\theta_w = 30°$; **b** #87061522 1 Ms = 1.256, $\theta_w = 37.5°$; **c** #87061606, Ms = 1.243, $\theta_w = 37.5°$; **d** #87061523, Ms = 1.261, $\theta_w = 40.0°$; **e** #87070711, Ms = 1.277, $\theta_w = 40.0°$; **f** #87061601, Ms = 1.214, $\theta_w = 43.0°$; **g** #87061602, Ms = 1.214, $\theta_w = 44.0°$; **h** #87061604, Ms = 1.267, $\theta_w = 46.0°$

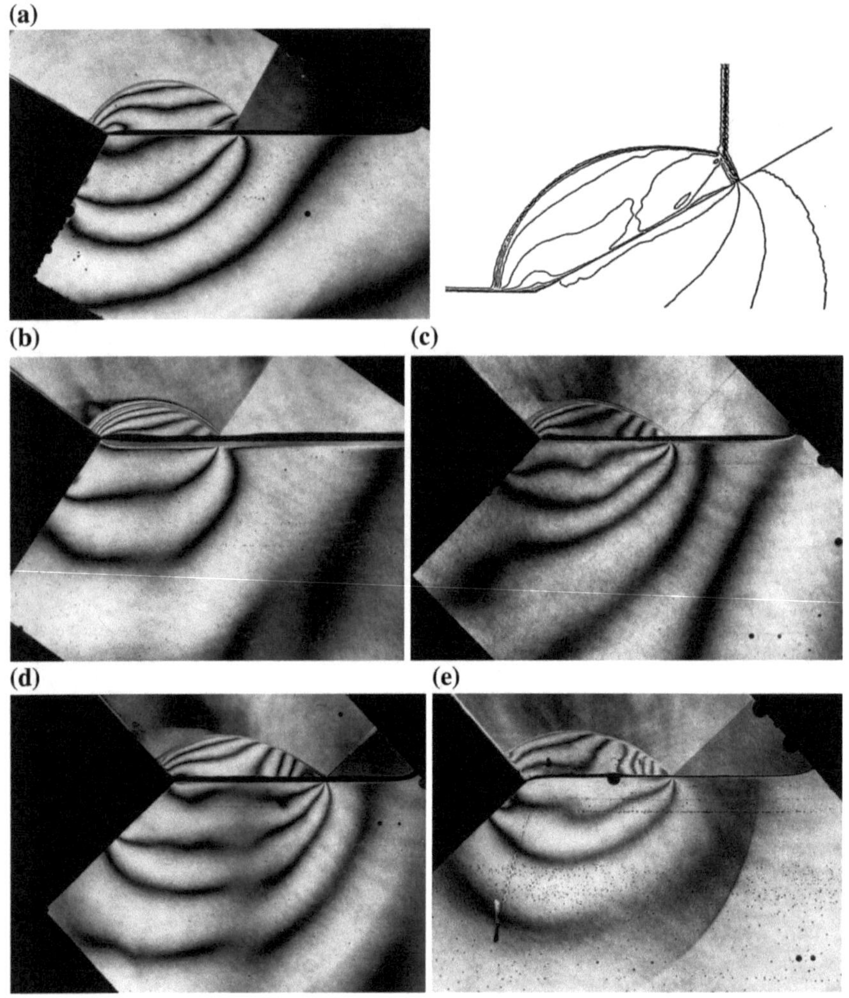

Fig. 2.53 Evolution of reflected shock wave over water wedge for Ms = 1.77 in air at 650 hPa, 294.5 K: **a** #87050705, Ms = 1.790, $\theta_w = 30°$ and numerical simulation; **b** #87050706, Ms = 1.775, $\theta_w = 35°$; **c** #87050708, Ms = 1.783, $\theta_w = 45°$; **d** #87050709, Ms = 1.778, $\theta_w = 47°$; e #87050802 343 ms Ms = 1.778, $\theta_w = 48°$; f #87051203, Ms = 1.665, $\theta_w = 49°$; **g** #87051204, Ms = 1.791, $\theta_w = 50°$; **h** #87050805, Ms = 1.787, $\theta_w = 52°$ and numerical simulation, (Miyoshi 1987)

A density distribution is obtained by solving numerically the Euler equations and displayed as a numerical fringe distribution. The numerical density is adjusted to be the same as an experimental one as shown in Fig. 2.53a. The numerical density distribution agreed well with the experimental one. Figure 2.53c shows a RR. At the foot of the IS, fringes are concentration which indicate a local pressure

(f) (g)

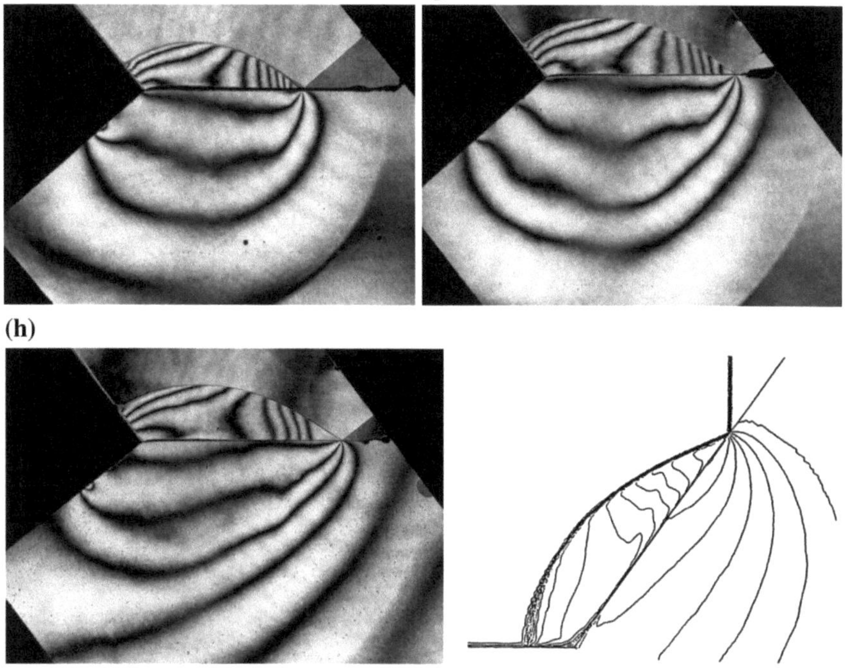

Fig. 2.53 (continued)

enhancement. Increasing θ_w, precursory wavelets form a train of compression waves and fringes are concentrated at the foot of the IS.

In Fig. 2.53d–g, a discontinuous shadow propagates at sonic speed in water. The discontinuous shadow is expected to merge with the foot of the IS, when When $\cos\theta_w = u_s/a_{water}$ where u_s and a_{water} are the shock speed in air and the sound speed in water. In Fig. 2.53h an experiment and its numerical simulation are compared for Ms = 1.787 and the wedge angle of $\theta_w = 52°$. The reflection pattern is a RR. The fringe patterns in the simulation and the interferogram agree well. The fringes in water are concentrating at the foot of the IS.

Figure 2.54a shows the evolutions of reflected shock wave propagations for Ms = 2.20 over water wedge of $\theta_w = 0.0°$. The SL is faintly visible. The IS is perfectly perpendicular to the water surface. The reflection pattern is a vNMR. In Fig. 2.54b, the reflection pattern is a SMR. In Fig. 2.54c, the fringes in water concentrate at the foot of the IS. As the shape of the RS are straight in the vicinity of the TP, which indicates that the reflection pattern is TMR. Figure 2.54d is its enlargement and the reflection pattern is a RR. With increase in the θ_w, in Fig. 2.54e, f, the reflection patterns are SPRR. It should be noticed that precursory waves are observed ahead of the foot of the IS.

(a) (b) (c) (d) (e) (f)

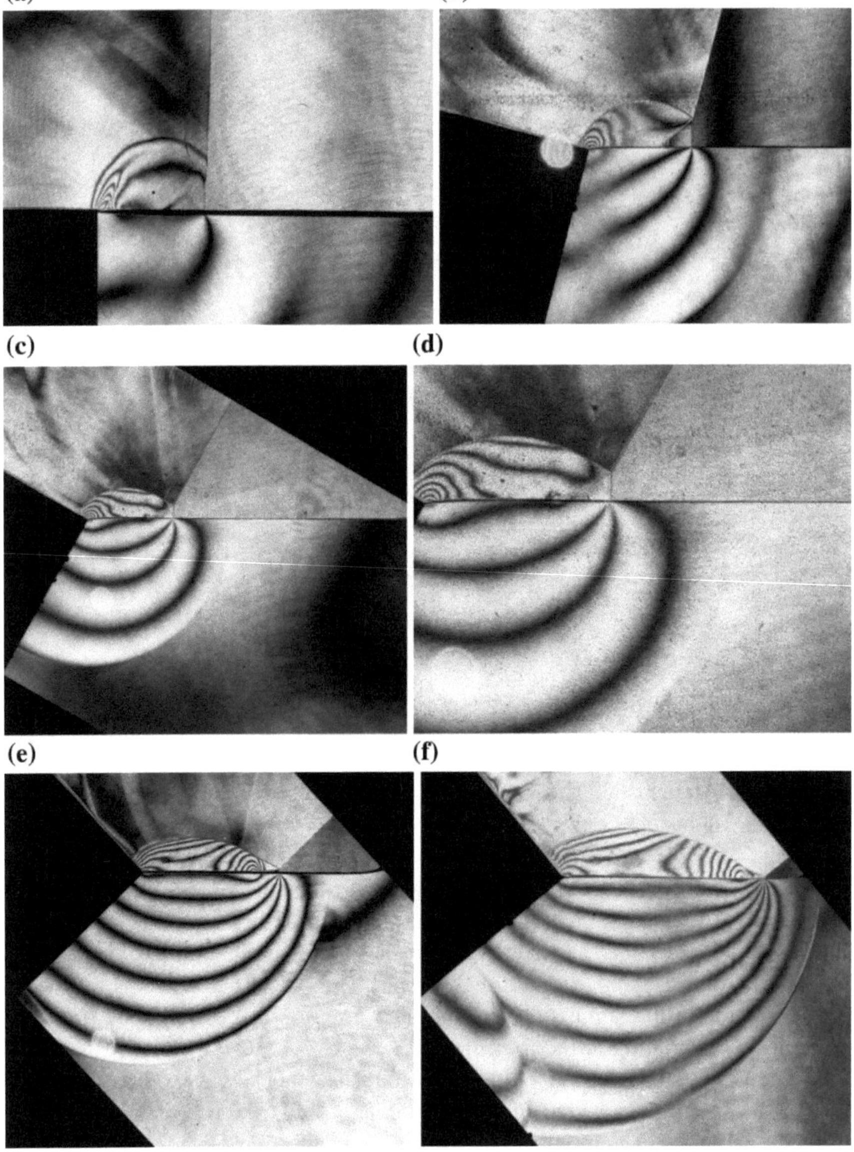

Fig. 2.54 Reflected shock wave over water wedge for Ms = 2.25 in air at 710 hPa, 294.7 K: **a** #85061901, Ms = 2.214, θ_w = 0,0°; **b** #85061705, Ms = 2.211, 14.3°, SMR; **c** #85061704, Ms = 2.217, θ_w = 29.8°, SMR; **d** enlargement of **c**; **e** #85061701, Ms = 2.217, θ_w = 46.2°, SuRR; **f** #85061407, Ms = 2.308, θ_w = 50.0°, SPRR

(a) (b)

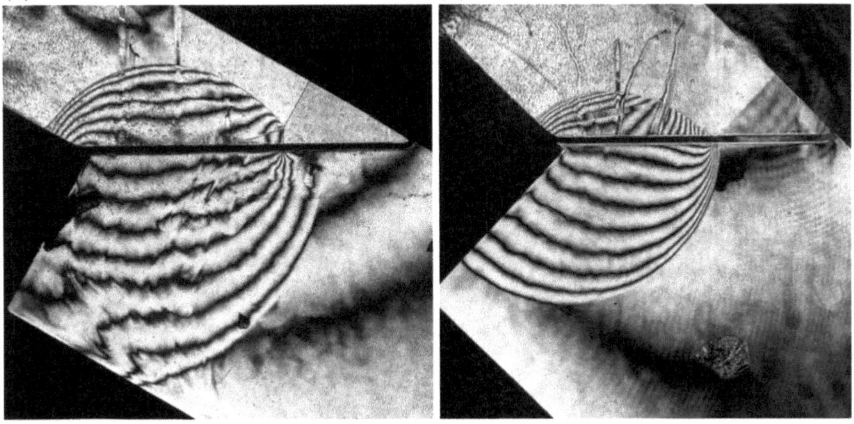

Fig. 2.55 Reflected shock wave from Aflud E10 wedge: **a** #87051810, for Ms = 1.262 in air/Aflut E10 mixture at atmospheric pressure, 289.2 K, θ_w = 35.0°; **b** #87051907, Ms = 1.450 in air/Aflut E10 mixture at atmospheric pressure, 294.0 K, θ_w = 44.0°

2.1.6.2 Retrograde Liquid Wedges

There are numerous combinations of gases and safe and non-toxic liquids for forming a gas/liquid interface. A retrograde liquid Aflud E10 has a retrograde medium having a peculiar character of creating condensation shock waves when the medium is exposed to expansion waves. The liquid has a high vapor pressure, stable, and non-toxic. Just for reference, a series of experiments was conducted by filling the Aflut E10 in the test section. Two images are presented here just for reference in air/Aflut E10 mixture at atmospheric pressure. In Fig. 2.55a, the reflection pattern is a SMR from the liquids wedge of θ_w = 35.0° for Ms = 1.26 in air-AfludE10 mixture and a precursory wave propagates in front of the MS. In Fig. 2.55b, the reflection pattern is a SPRR from the liquids wedge of θ_w = 44.0° for Ms = 1.45 in air-AfludE10 mixture.

2.1.7 Evolution of Shock Wave Reflection from Cones

2.1.7.1 Cones

Shock wave reflections from a wedge and a cone are self-similar phenomena. However, the reflections from cones are unique because the reflection pattern is solely affected by the cone angle and the distance from the leading edge to the foot of the shock wave. Then the cone size never affects the reflection process until it becomes the shock tube width. Figure 2.56 shows reflected shock wave

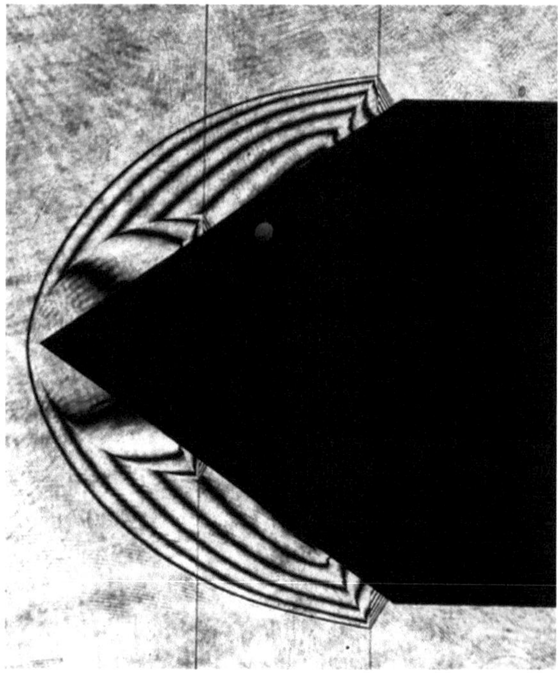

Fig. 2.56 Double exposure interferogram of shock wave reflection from a cone of apex angle is 69.2°: #98120304, Ms = 2.330, 144 hPa 292.9 K, in Argon (Kosugi 2000)

propagating over an apex angle 69.2° cone of 60 mm base diameter for Ms = 2.33 in argon. The cone was made of brass and supported rigidly from its base and was installed in the 100 mm × 180 mm diaphragm-less shock tube. This shock tube can accommodate large cones. The visualization was performed by the double exposure interferometry at 60 μs double pulse time interval. Hence two images of shock wave reflections were recorded in one hologram. When testing shock wave reflection over wedges installed in the 100 mm × 180 mm shock tube, shock wave motions along the wedge surface were affected by the presence of a boundary layer developing along the wedge surface as discussed, for example, in Fig. 2.21. The evolved flows at the foot of the IS is governed by the boundary layer displacement thickness depending on the distance from the leading edge and the shock tube sidewalls. The flows over cones are also affected by the boundary layer developing along the cone surface.

In Fig. 2.56, the superimposed reflected shock waves looked slightly different. The shock wave reflection over a cone is not necessarily self-similar.

It was decided to experimentally determine the θ_{crit} over cones. However, it was impossible to conduct experiments by continuously varying the cone apex angle. Therefore, a series of experiments were carried out in 1975 by using 50 pieces of cone models having 25 mm base diameter and apex angles ranged from 7.5° to 150° at every 2.5° step (Takayama and Sekiguchi 1977). At first, the cone models tested in a 40 mm × 80 mm conventional shock tube. In 1989, to refine the previous experiment, 88 pieces of new cone models made of brass having 25 mm base

Fig. 2.57 Cone models (Yang et al. 1996)

(a) (b) (c)

(d) (e)

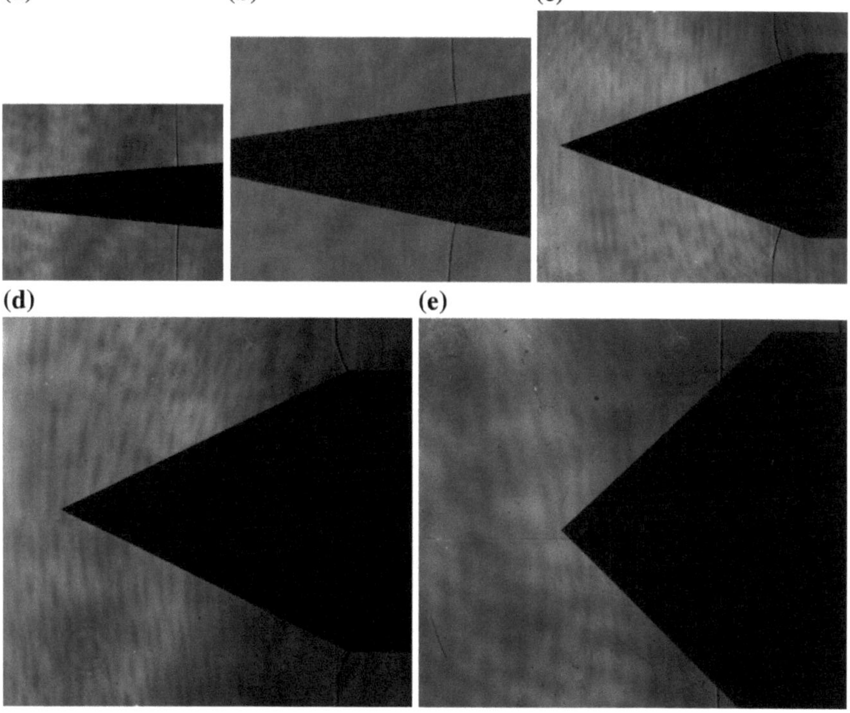

Fig. 2.58 Shock wave reflection from a cone at Ms = 1.20 in air, α apex angle of cone: **a** #75120407, α = 10°; **b** #75120406, α = 20°; **c** #75120404, α = 40°; **d** #75120401 α = 55°; **e** #75120103, α = 87.5°

diameters and their apex angles ranging from 10° to 120° at every 2° steps were manufactured. Figure 2.57 shows the cone models used in 1989 experiments.

Four models were installed in a 60 mm × 150 mm conventional shock tube's test section and the reflection patterns were observed simultaneously. Figure 2.58 shows selected results: In axial symmetric flows, the images of IS and MS were overlapped but it was still possible to identify the position of the TP. The triple point trajectory angle χ is defined by $\tan\chi = h/L$ where h is the height of the SL and L is the distance between the cone apex and the foot of MS. The extrapolation of the angle of the trajectory abled to estimate θ_{crit} as the angle at which χ became zero.

Figure 2.59 summarizes the 1975 experiments. The triple point positions were plotted against the semi-apex angle of cones θ_c (Takayama and Sekiguchi 1977). Although time consuming and laborious εξπεριμεντσ, τηε ϖισυαλιζατιονσ ωασ περφορμεδ βψ υσινγ διρεχτ σηαδοωγραπησ. Τηε ορδινατε δενοτεσ τηε trajectory angle $\chi - \theta_c$ in degree and the abscissa denotes half apex angle of cones θ_c in degree. Filled circles denote the measured results for Ms ranging from 1.04 to 3.09. The results of Bryson and Gross (1961) agreed well with the present experiments.

In January 1989, experiments by using the cones shown in Fig. 2.57 were carried out by installing four cone models in the 60 mm × 150 mm diaphragm-less shock tube. Experiments of were repeated four times. Figure 2.60 shows results for Ms = 1.20. Figure 2.60a shows four cones of half apex angles θ_c from 39° to 42°. The reflection patterns are SMR. Figure 2.60b shows four cones of θ_c = 51°–54°. The reflection patterns are also SMR, whereas in Fig. 2.60c θ_c is between 47.5° and 49°, the reflection patterns are RR.

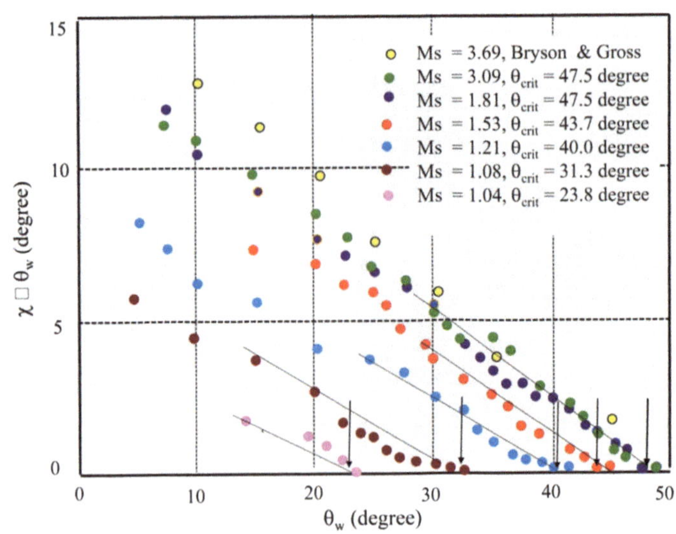

Fig. 2.59 Triple point trajectories, summary of Fig. 2.57

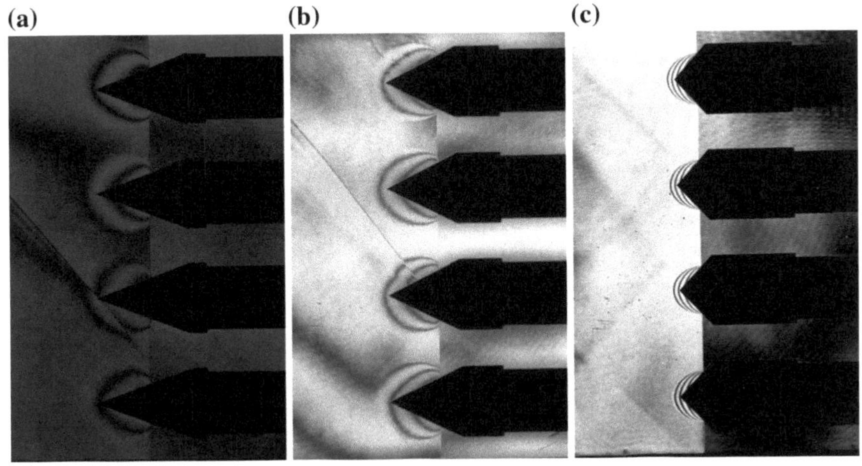

Fig. 2.60 Shock wave reflection from cones at Ms = 1.20 in atmospheric air at 291.8 K: **a** #89011304, θ_c = 19.5–21°; **b** #89011703, θ_c = 25.5–27°; **c** #89012414, θ_c = 47.5–49°

Figure 2.61a–d shows shock wave reflection for Ms = 1.17 in air from cone angles θ_c between 12° and 15°, θ_c between 21° and 24°, θ_c between 25° and 28°, θ_c between 35° and 38°, respectively. Their reflection patterns are SMR. In Fig. 2.61e, cone angles are θ_c between 43° and 46°, causing a reflection pattern RR. The transition occurred at about θ_c 40°. A question arises whether or not delayed transition may occur in the shock wave reflection from cones, similar to that witnessed in shock wave reflection from flat wedges.

In order to reply this question, cone experiments were repeated in the 100 mm × 180 mm diaphragm-less shock tube in 1995. Figure 2.62a–c shows sequential observation of shock wave reflections for Ms ≈ 1.4 in air and a cone angle of 10°. The reflection pattern is a SMR. Figure 2.62d–f, shows sequential images of shock waves obtained for Ms = 1.4 in air and a cone angle of θ_c = 15°. The reflected shock wave pattern is a SMR.

2.1.7.2 Evolution of Shock Wave Reflection from Cones Connected to Hollow Cylinders

A cone of apex angle 10° was connected to the rear part of a hollow cylinder. In order to investigate the effect of the shape of leading edge, a cylinder-cone combination having 8 mm diameter hollow cylinder, 1 mm wall thickness, and 40 mm in length was installed in the test section of the 100 mm × 180 mm diaphragm-less shock tube. In Fig. 2.63, the evolution of transmission and reflection of a shock wave for Ms = 1.40 was shown. The central part of the IS was transmitted through the hollow cylinder. The reflected shock wave is, therefore, weak and propagates along the 10° inclined conical wall. The reflected pattern is a SMR. Figure 2.63f

(a) (b) (c)

(d) (e) (f)

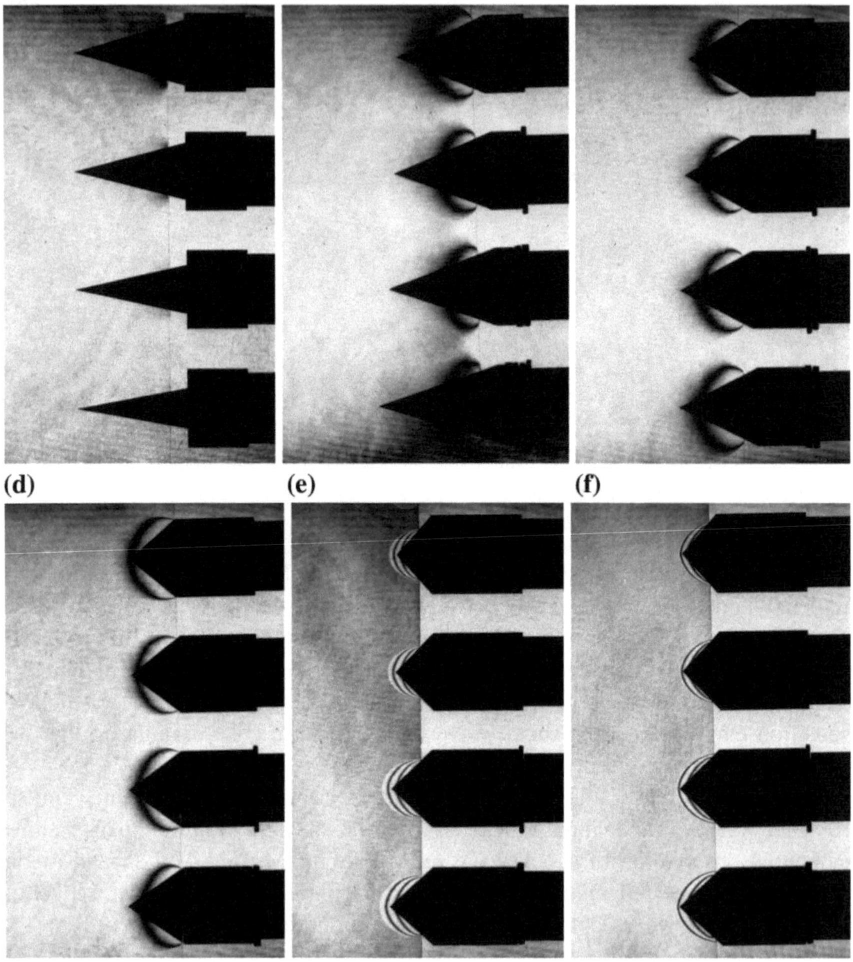

Fig. 2.61 Shock wave reflection from cones at Ms = 1.17 in atmospheric air at 290.0 K: **a** #94110503, Ms = 1.166, θ_c= 12°–15°; **b** #94110507, Ms = 1.177, θ_c = 21°–24°; **c** #94110601, Ms = 1.168, θ_c = 25°–28°; **d** #94110605, Ms = 1.168, θ_c = 35°–38°; **e** #94110609, Ms = 1.168, θ_c = 43°–46°; **f** #94110610, Ms = 1.168, θ_c = 46°–49°

was taken at the elapsed time 1060 μs and the transmitted shock wave was reflected from the bottom of the hollow cylinder and already released from the opening of the hollow cylinder. The resulting jet flow was ejected. Figure 2.64 shows evolution of reflected shock wave interaction with a 20° cone and 13 mm diameter hollow cylinder combination for Ms = 1.40. The resulted reflection pattern is a SMR.

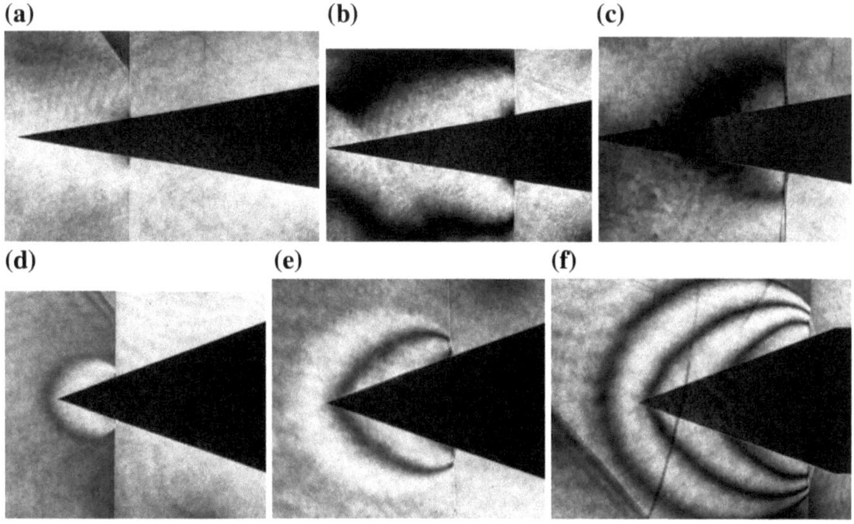

Fig. 2.62 Reflected shock wave pattern from a cone in air at 700 hPa, 290.3 K: **a** #95040601, 880 μs from trigger point, Ms = 1.389, $\theta_c = 10°$; **b** #95040602, 850 μs, Ms = 1.408, $\theta_c = 10°$; **c** #95040501, 960 μs, Ms = 1.404, $\theta_c = 10°$; **d** #95041102, 800 μs, Ms = 1.384, $\theta_c = 15°$; **e** #95041103, 900 μs, Ms = 1.411, $\theta_c = 15°$; **f** #95041105, 1000 μs, Ms = 1.400, $\theta_c = 15°$

2.1.7.3 Heat Transfer from Cones

Saito et al. (2004) distributed 16 pieces of 1.5 mm wide, 58 μm thick, and 90 mm long platinum thin film heat transfer gauges evenly along the circumference of a 40 mm diameter cylinder and measured time variations of heat flux distributions over the cylinder installed in the 100 mm × 180 mm diaphragm-less shock tube.

In order to investigate the transition of a reflected shock wave over a cone, platinum heat transfer gauges were distributed along the cone surface. Figure 2.65a shows a cone model along which 8 pieces of 1.0 mm wide, 58 μm thick, and 5.0 mm long platinum heat transfer gauges were distributed. Figure 2.65b shows the positions of the gauges and the geometry of the cone model; its base diameter is 50 mm and its apex angle α is 86°. The cone was made of ceramics, Photoveel (Sumikin Ceramics and Quartz Co. Ltd. Japan). The cone model was installed in the test section of the 100 mm × 180 mm diaphragm-less shock tube. Experiments were conducted for Ms = 2.38 in air and Re = 3 × 10^4 per characteristic length of 10 mm.

Figure 2.66 shows sequential interferograms. As the *OB* path lengths are so short that no dark fringes are observed and at 4 mm from the leading edge as shown in Fig. 2.66a, the reflection pattern is hardly identified but away from the leading edge at 11.6 mm. 15.3 and 23.5 mm in Fig. 2.66b–d, respectively, the reflection pattern is a TMR.

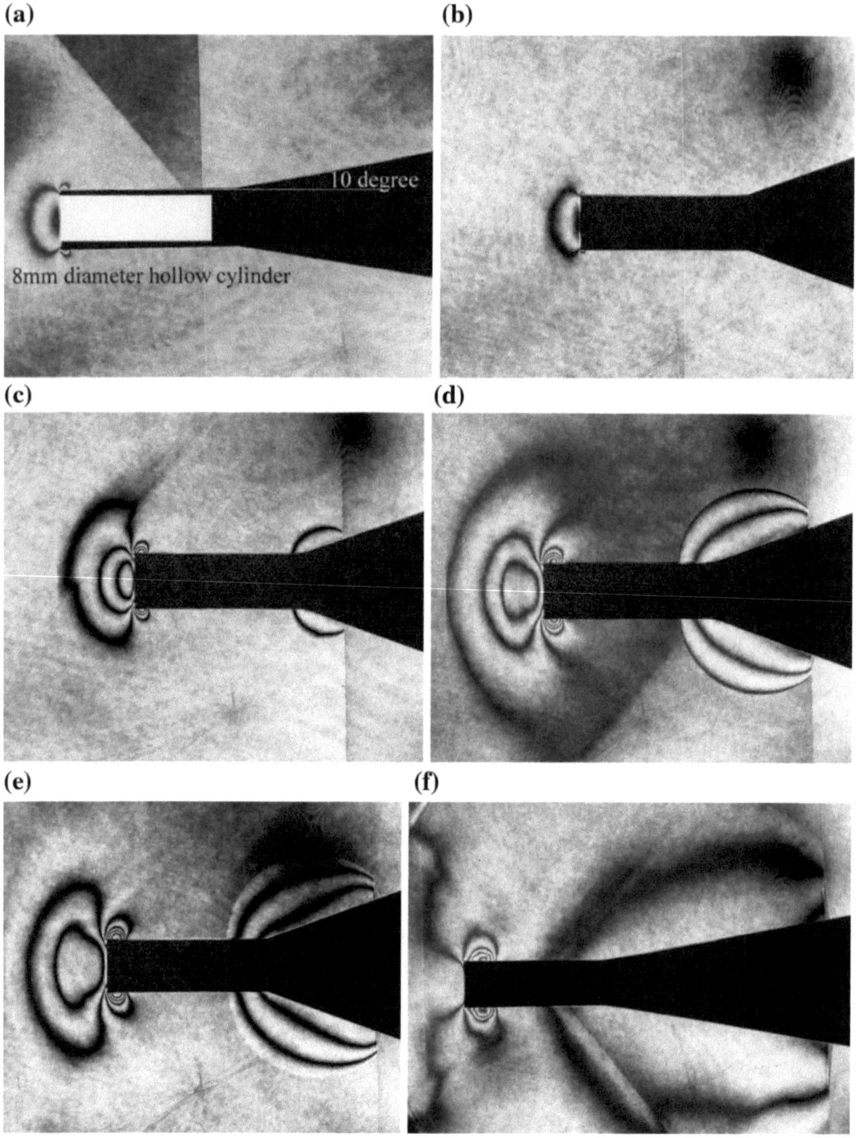

Fig. 2.63 Evolution of shock wave reflection from a 10° cone connected to a 8 mm inner diameter hollow cylinder for Ms = 1.40 in air at 700 hPa, 294.5 K: #95041404, 70 μs from trigger point, Ms = 1.380; **b** #95041409, 160 μs, Ms = 1.406; **c** #95041405, 200 μs, Ms = 1.409; **d** #95041406, 230 μs, Ms = 1.414; **e** #95041407, 250 μs Ms = 1.401; **f** #95041402, 1060 μs, Ms = 1.408

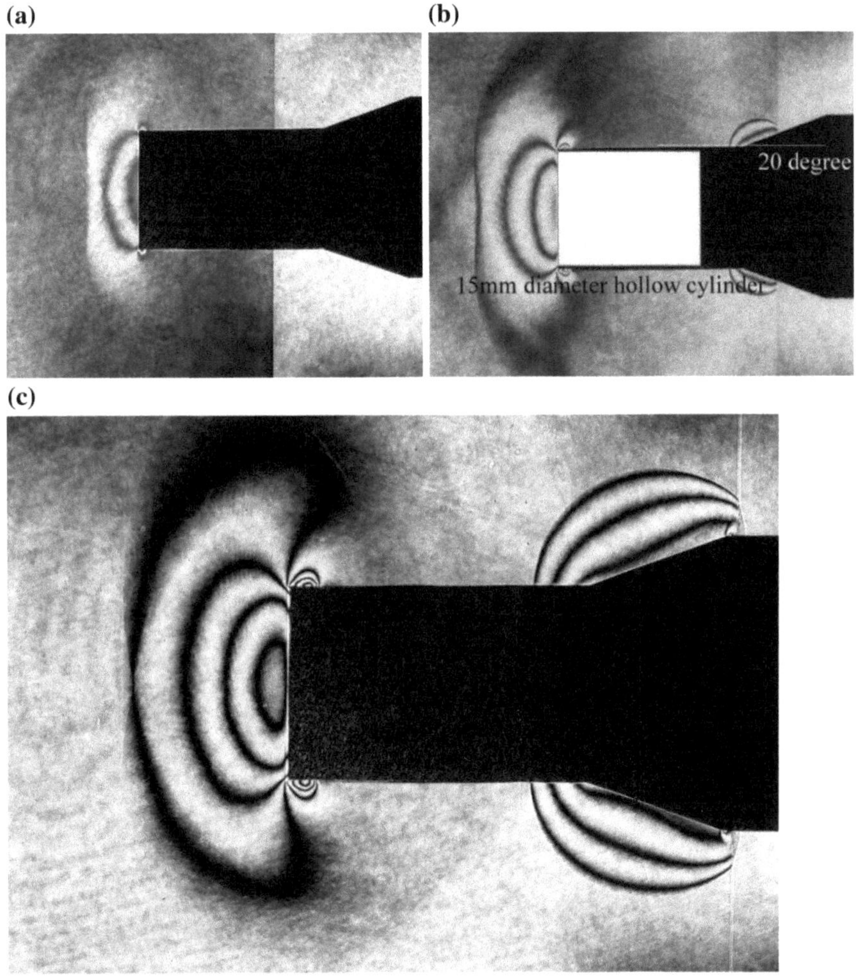

Fig. 2.64 Evolution of shock wave reflection from a 20° cone connected to a 13 mm hollow inner diameter hollow cylinder for Ms = 1.40 in air at 700 hPa, 294.5 K; **a** #95040703, 60 μs, Ms = 1.404; **b** #95041003, 100 μs, Ms = 1.400; **c** #95040705, 180 μs, Ms = 1.416

Figure 2.67a–f shows time variations of heat flux measured by platinum thin film heat transfer gauges distributed on the cone surface as shown in Fig. 2.65b: **a** 6.8 mm; **b** 10.4 mm; **c** 13.7 mm; **d** 17.0 mm; **e** 20.5 mm; **f** 23.6 mm. Dotted lines denote numerical results using in house code solving a Navier-Stokes equation. The ordinate denotes heat flux in MW/m^2 and the abscissa denotes the elapsed time in μs.

The scheme employed the following assumptions: $\gamma = 1.4$, the Sutherland formula for viscosity and heat conduction coefficients; Non-slip and isothermal

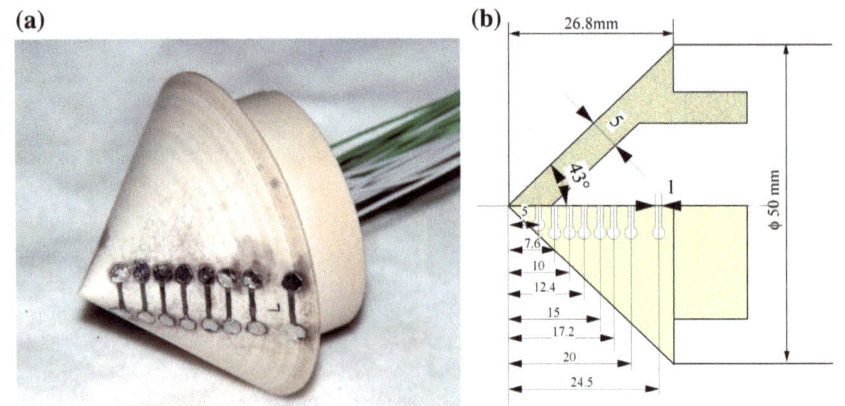

Fig. 2.65 Cone model: **a** the attachment of platinum heat transfer gauge; **b** cross section (Kuribayashi et al. 2007)

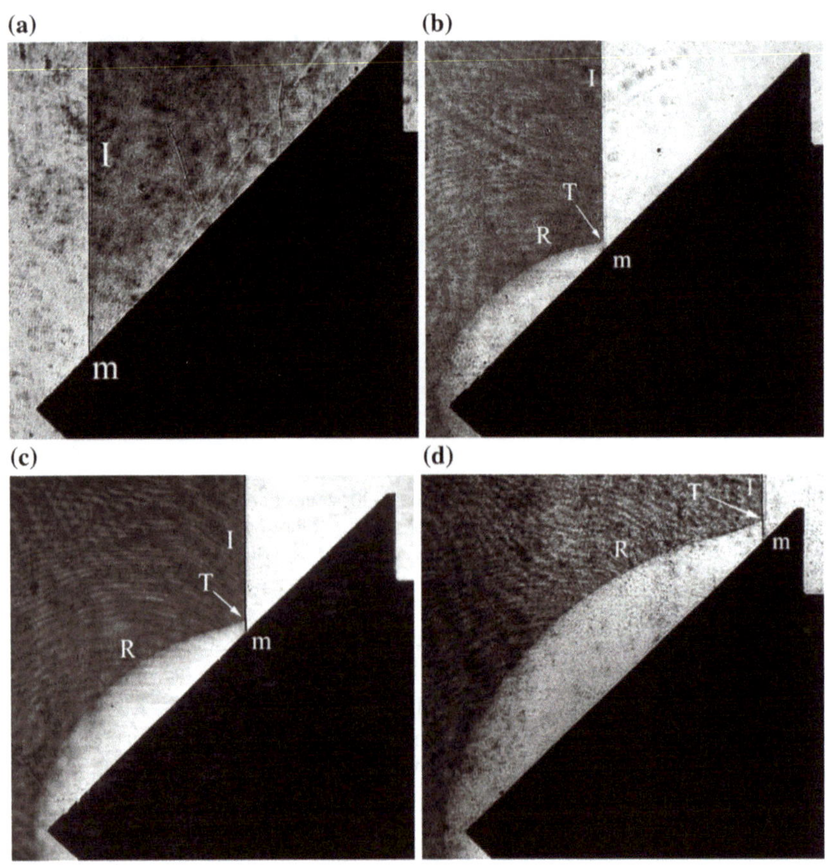

Fig. 2.66 Evolution of shock wave transition over the cone: location of the shock waves at: **a** 4.0 mm; **b** 11.6 mm; **c** 15.3 mm; **d** 23.5 mm

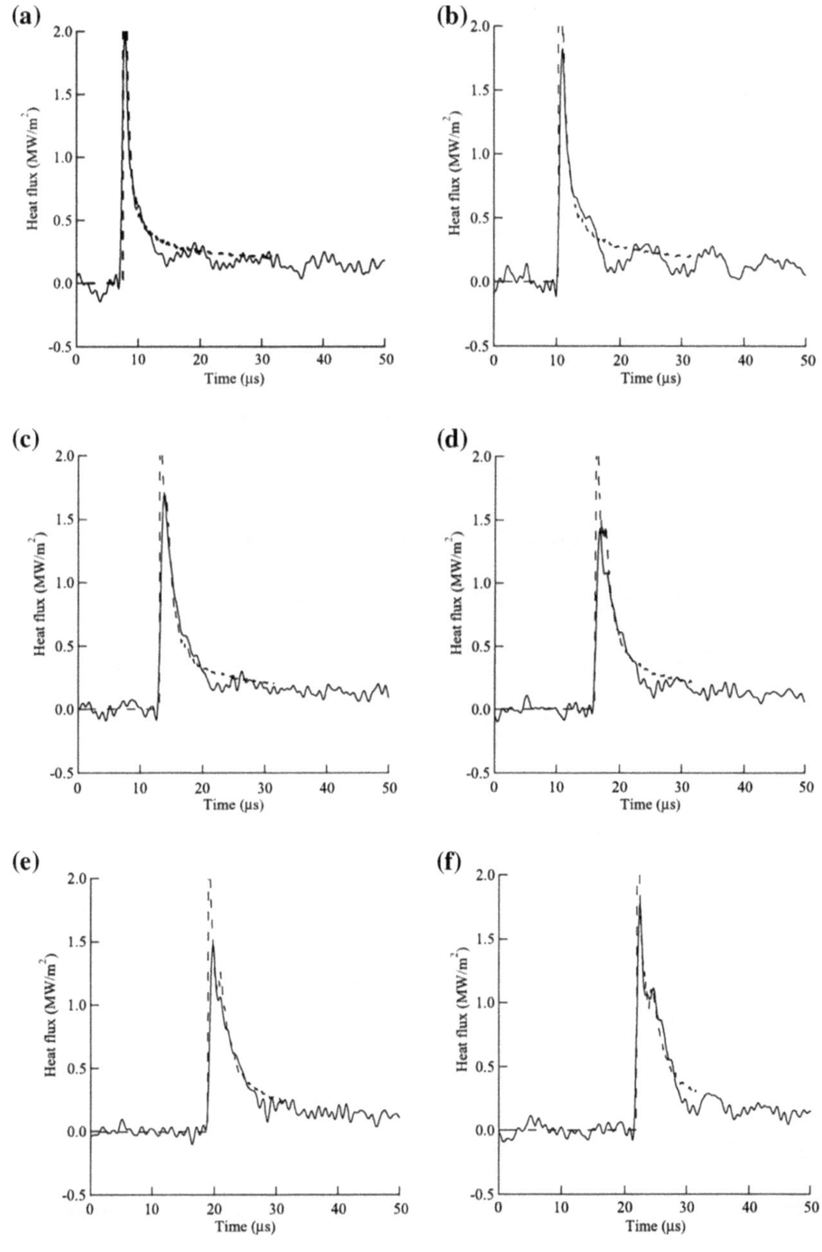

Fig. 2.67 Time variation of heat flux obtained by platinum thin film heat flux gauges shown in Fig. 2.59. Solid lines denote measured results and dotted lines denote numerical simulation: **a** 6.8 mm; **b** 10.4 mm; **c** 13.7 mm; **d** 17.0 mm; **e** 20.5 mm; **f** 23.6 mm (Saito et al. 2004)

boundary condition; Imposing quadrilateral unstructured grids. The finest grid size was a 1.5 μm around the cone surfaces. Numerical results agreed well with experimental findings.

It is clear from Fig. 2.67a that upon the arrival of shock wave, the heat flux reaches a maximum value and thereafter it decreases monotonously to appropriate steady value. Such behavior is typical to a RR reflection pattern. On the contrary, as seen in Fig. 2.67b–f, the heat flux changes its monotonous decrease in the middle. In Fig. 2.67e, f, this trend becomes eminent and the reflection pattern is SMR accompanying SL behind MS, which indicates the appearance of the delayed transition.

2.1.7.4 Effect of Initial Angle on Transition

To assure the delayed transitions over cones, 50 mm diameter cones were installed in the test section of the 100 mm × 180 mm diaphragm-less shock tube. Figure 2.68 shows a TMR over a cone whose apex angle is 34.6°. As the spatial resolution of high speed imaging was not necessarily high, it was decided to conduct double exposures at interval of 120 μs and compare directly two images recorded in one interferogram. In conclusion, the two resulting reflection never showed self-similar patterns. In Fig. 2.68e, $\theta_c = 44.0°$ the reflection pattern is a RR near the apex and transits to a TMR close to the end. In Fig. 2.68f, at $\theta_c = 49°$, the reflection pattern is a RR near the apex and transits to a DMR close to the end. In Fig. 2.68g, at $\theta_c = 50.2°$, the reflection pattern is a RR all the way along the cone surface.

The delayed transition does occur along cone surface. The ratio of the cross section of the outer boundary of the boundary layer displacement thickness to the cross section of the cones is significantly large in the vicinity of the apex but in the case of the wedge the ratio is not large in the vicinity of the leading edge. The departure of the reflected shock wave from a self-similar pattern occurs much significantly over the cones than that occurs over the wedges.

Figure 2.69 summarizes the triple point trajectories on cone surfaces having apex angle of 34.6°, 38.6°, 44.0°, 49.0°, and 50.2° at the initial pressure of 144 hPa. The ordinate denotes the TP height in mm and the abscissa the distance from the apex in mm. Dark blue, red, green, pink, and black circles denote the apex angles of 34.6°, 38.6°, 44.0°, 49.0°, and 50.2, respectively. The solid lines denote numerical results obtained by solving the Navier-Stokes Equation with fine mesh zoning (Kosugi 2000). The numerical results agree well with the experimental findings. The intersection of the solid lines with x-axis gives the length at which the transition occurs, which is the delayed transition distance as defined x_{int}. With increasing θ_c, x_{int} increases. At the cone apex angle of $\theta_c = 50.2°$ which exceeds the critical transition angle, a RR appears. Filled circles and open circles denote trajectories of upper and lower sides of the individual cone surfaces, respectively.

Figure 2.70 shows the relationship between the delayed transition distance and initial pressure for Ms = 2.327 in air. The ordinate denotes the delayed transition

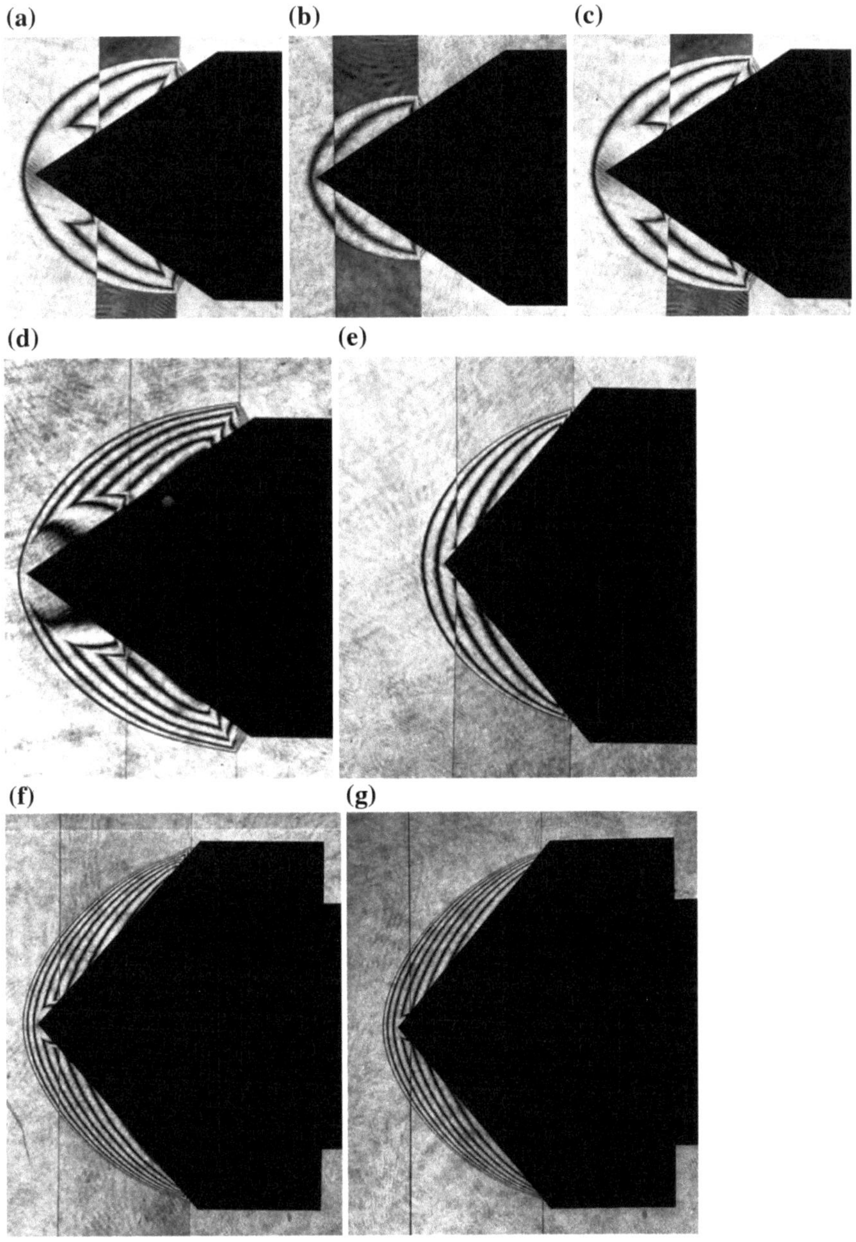

Fig. 2.68 Delayed transition of reflected shock wave over cones for Ms = 2.33 in air at 144 hPa, 293.9 K; **a** #98120103, in Argon $\theta_c = 34.6°$; **b** #98120304, Ms = 2.330 in Argon, $\theta_c = 34.6°$; **c** #98120103, Ms = 2.330 in Argon, $\theta_c = 34.6°$; **d** #98120304, Ms = 2.330, $\theta_c = 34.6°$; **e** #98120114, Ms = 2.326, $\theta_c = 44.0°$; **f** #98121102, Ms = 2.324, $\theta_c = 49.0°$; **g** #98121006, Ms = 2.327, $\theta_c = 50.2°$

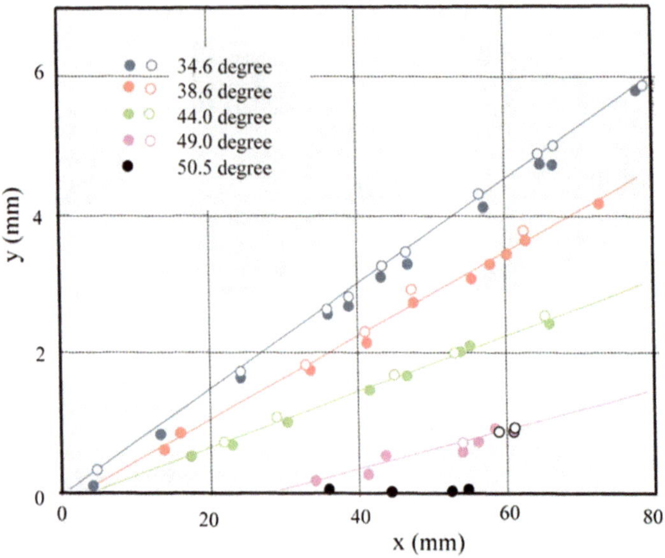

Fig. 2.69 Triple point position for Ms = 2.33 at 144 hPa

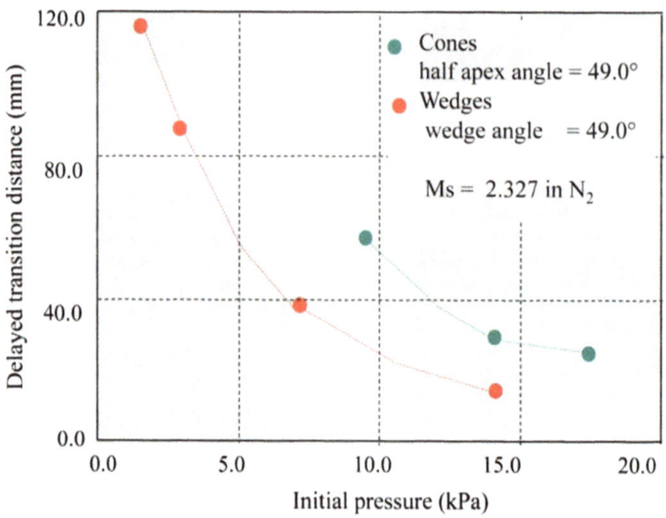

Fig. 2.70 Delayed transition distance x_{int} against initial pressure for Ms = 2.33 (Kitade 2001)

distance x_{int} in mm and the abscissa denotes the initial pressure in kPa. Green filled circles denote cone half apex angle of $\theta_c = 49°$ and red filled circles denote wedge of $\theta_w = 49°$. The delayed transition distance x_{int} increases reversely proportional to the initial pressures and this trend is more significant in cones.

Shock wave researchers often stated that shock tube experiments were not reliable, because the critical transition angles obtained by the experiments conducted under identical initial condition widely varied depending on the shock tube geometry. Figure 2.61 indicates that even if experiments are conducted for identical Ms and Re but in different shock tube geometries, although a MR is observed in a large shock tube, in a small shock tube only a RR is observed.

2.1.8 Tilted Cones

The shock wave reflection from a wedge is a two-dimensional phenomenon and that from a cone is also axial symmetric phenomenon. The shock wave reflection over a skewed wedge is a three-dimensional phenomenon. Similarly to a skewed wedge, the shock wave reflection over a tilted cone is a three-dimensional phenomenon. It is a challenge to quantitatively reconstruct three-dimensional shock wave phenomena by collections of two-dimensional images viewed from various view angles.

In order to quantitatively observe the three-dimensional interaction of a shock wave with a tilted cone interacting with a planar shock wave. Figure 2.71 shows a setup. A cone having the apex angle of 70° and the base diameter of 30 mm was tilted by 20° to the opening of a 230 mm diameter shock tube. The stand-off distance between the shock tube opening to the tip of the cone was 19 mm. The shock wave of Ms = 1.20 is diffracted at the edge of 230 mm diameter shock tube.

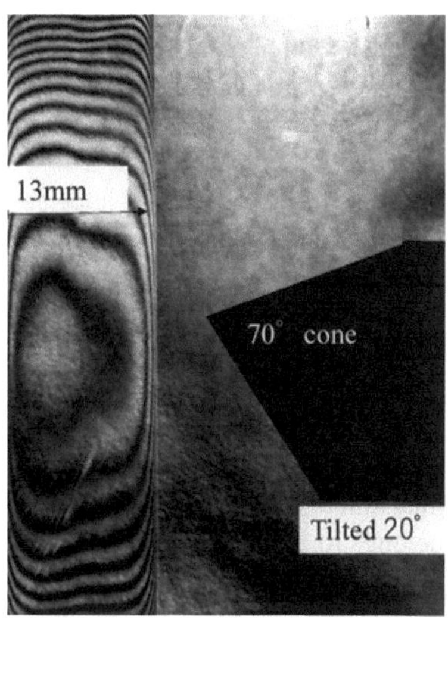

Fig. 2.71 Initial setup. The stand-off distance from the opening to the tip of tilted cone is 19 mm

The transmitted shock wave was planar when impacted the tilted cone. In Fig. 2.71, the planar shock wave reached 13 mm from the opening. Due to the diffraction at the edge of the shock tube opening, the edge part of the transmitted shock wave attenuated. The central part of the transmitted shock wave was unaffected by the diffraction so that Ms was unchanged when the shock wave interacted with the tilted cone. If the image of a three-dimensional interaction of the shock wave and the tilted cone is reconstructed, how can the initiation and the termination of RR and MR be restructures? (Fig. 2.72).

Unlike previous symmetric cone experiments, for example, as seen in Fig. 2.68, the present cone is tilted by 20° and rotated around the shock tube axis. Then the inclination angle of the cone is varied continuously from 10° to 45°.

2.2 Shock Wave Reflection Over Double Wedges

2.2.1 Concave Double Wedges

When a wedge of angle θ_2 is placed on a wedge of angle θ_1 in a staggered arrangement, such a combination of wedges would form, $\theta_1 < \theta_2$, a concave double wedge. On the contrary, when a wedge of angle θ_2 is removed from a wedge of angle θ_1 in a staggered arrangement, the resulting shape of the wedge would form, $\theta_1 > \theta_2$. a convex double wedge. In this case, if the first wedge angle is smaller than the critical transition angle, $\theta_1 1 < \theta_{crit}$, the MR would appear at first, whose MS is reflected from the corner of the second wedge. Depending on the value of $\theta_1 2$, the MS's trajectory angle χ tends to leave from the second wedge surface, $\chi > \theta_{1w}$. Courant and Friedrichs (1948) defined this reflection pattern as a direct Mach reflection, or in short, DiMR, which is locally self-similar. In the case of $\chi = \theta_{1w}$, the TP propagates parallel along the second wedge surface and the slip line SL is also parallel to the wedge surface. Therefore, the reflection pattern is locally unvaried with elapsed time. This pattern is defined as stationary Mach reflection, or in short, StMR. The StMR appears universally in steady flows, for example, in wind tunnel flows, but the presence of StMR is very unique in unsteady shock tube flows.

The SL of the MR on the second wedge surface shown in Fig. 2.73 looks locally parallel to the second wedge surface and would be, within the limitation of the length of the second wedge, parallel up to the end of the second wedge. This image does not warrant a StMR holding. If the second wedge is very long, the TP may either leave away from the wedge surface or approach toward it. When $\chi < \theta_{1w}$, the TP moves toward the second wedge surface. Courant and Friedrichs (1948) defined this reflection pattern as an inverse Mach reflection, or in short, IvMR, which belongs to a truly unsteady flow. As soon as the MS terminates and the TP touches the second wedge surface. Then, the TP was reflected from the wedge surface and formed another TP. Anyway the IvMR transited to the SPRR. However, the transition to the SPRR cannot proceed straightforwardly. The SPRR accompanies a

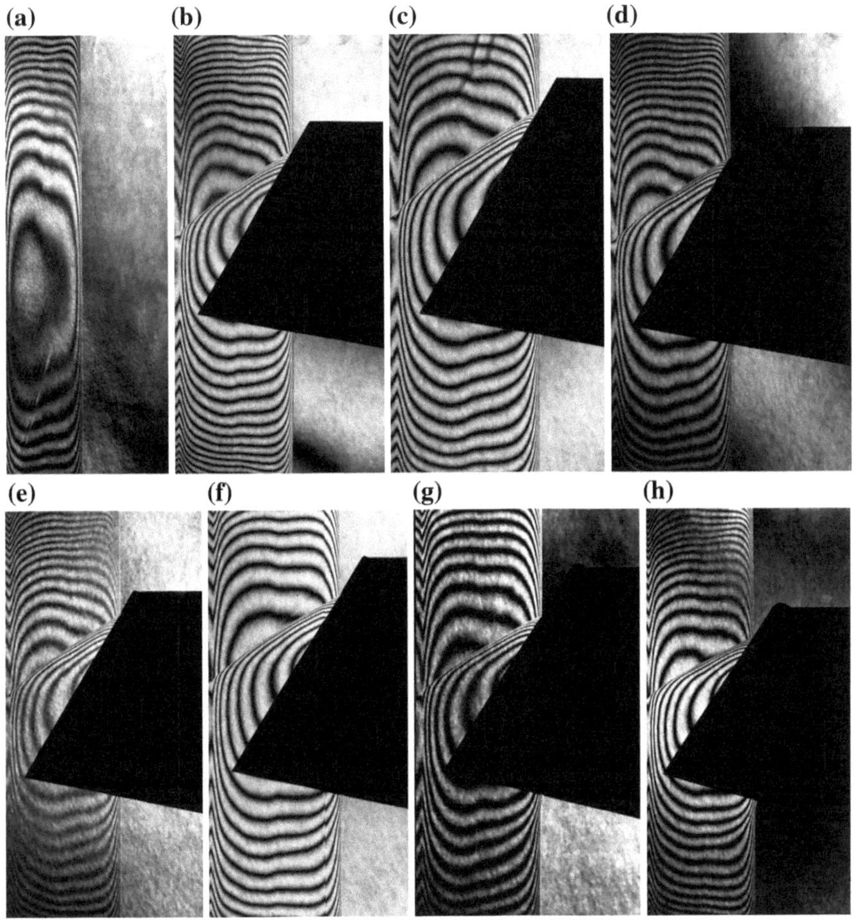

Fig. 2.72 Shock wave released from a 230 mm diameter shock tube reflected from a tilted cone for Ms = 1.20 in air at 1013 hPa. 291.8 K: **a** #92041501; **b** 92041103, Ms = 1.196, α = 4°; **c** 92041104, Ms = 1.199, α = 8°; **d** #92041323, Ms = 1.199, α = 10°; **e** #92041325, Ms = 1.198, α = 18°; **f** #92041105, Ms = 1.216, α = 20°; **g** #92041301, Ms = 1.189, α = 30°; **h** #92041312, Ms = 1.195, α = 38°; **i** #92041302, Ms = 1.191, α = 40°; **j** #92041313, Ms = 1.195, α = 44°; **k** #92041303, Ms = 1.186, α = 50°; **l** 92041316, Ms = 1.197, α = 58°; **m** #92041304, Ms = 1.1962 α = 60°; **n** #92041317, Ms = 1.197, α = 64°; **o** #92041320, Ms = 1.193, α = 78°; **p** #92041322, Ms = 1.194, α = 88°; **q** #92041307, Ms = 1.153, α = 100°

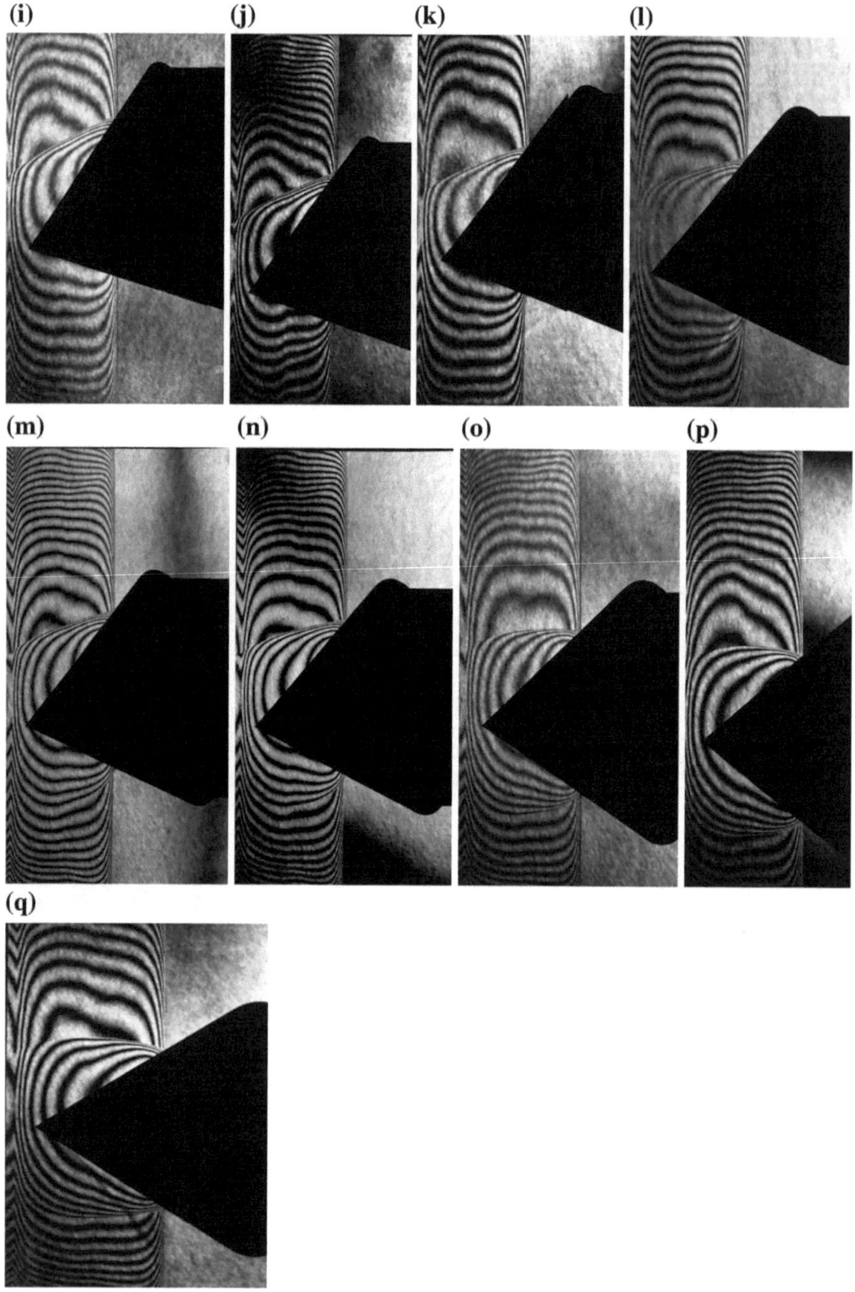

Fig. 2.72 (continued)

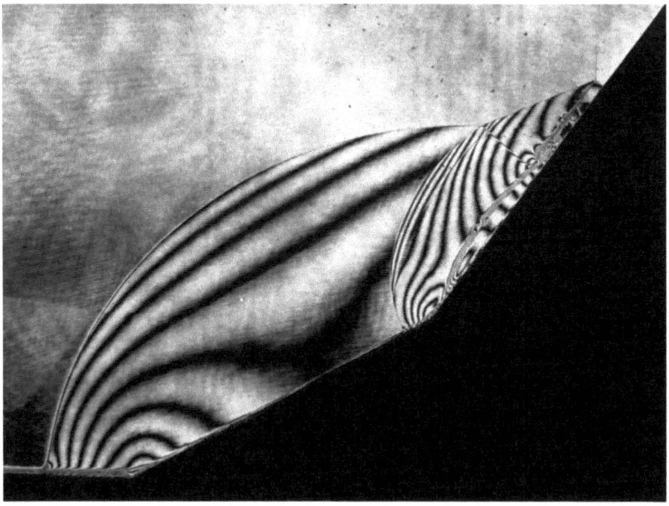

Fig. 2.73 Apparent StMR over a 30°/55° double wedge: #87120719, for Ms = 1.927 in air at 500 hPa 293.0 K, the first wedge length of 50 mm

distinct secondary TP on its RS. This transition proceeded in part analogously to the transition from a TMR to a DMR. The secondary TP succeeded the previous TP.

Figure 2.74 shows sequential interferograms of the reflection of shock waves for Ms = 1.50 over $\theta_1 = 15°$ and $\theta_2 = 35°$ double wedge. The allocated time to each image indicate the time of the second exposure from the triggering point. Figure 2.74b is enlargement of Fig. 2.74a. The MS interacted locally with the second wedge surface. In Fig. 2.74c, e, g, two SL interact and eventually form a DiMR as seen in Fig. 2.74h.

The shock wave transition over a concave double wedge is governed not only by the combination of wedge angles but also by first wedge length L. The height of the MS is approximately proportional to the first wedge length L. Hence the evolution of the reflection pattern on the second wedge surface is affected by L. The L can be a length scale.

Figure 2.75 compare the variation of the final reflection pattern for Ms = 1.60–1.80, wedge angles 30°/55° and 27°/52°, the first wedge lengths are L = 7, 12, 25, and 52 mm. Figure 2.75a, b shows final reflection patterns of SPRR in L = 7 mm. Figure 2.75c, d shows SPRR for Ms = 1.90 in L = 12 mm, and SPRR in L = 25 mm, respectively. Figure 2.75f shows an IvMR for Ms = 1.67 in L = 52 mm. Figure 2.75g shows StMR for Ms = 1.62 in L = 52 mm. The result indicates that the final reflection pattern is affected by the first wedge length. In Fig. 2.75e, g, the SL looks almost parallel to the wedge surface. Soon it turns into a two-dimensional array of vortices developing with elapsing time.

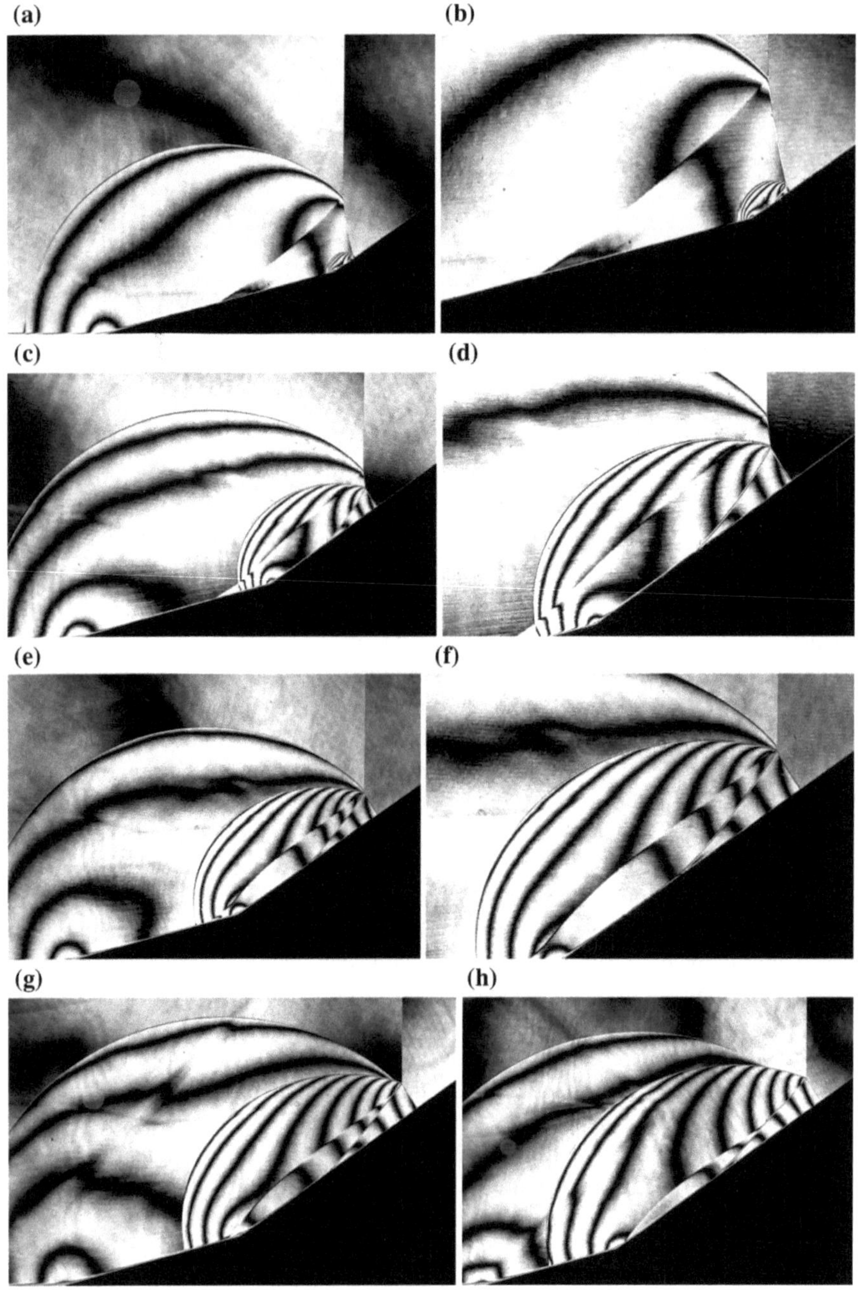

Fig. 2.74 DiMR of reflected shock wave over a θ_w = 15°/35° double wedge for Ms = 1.50 in air at 670 hPa, 290 K: **a** #86013107, 570 μs delay time from trigger point, Ms = 1.514; **b** enlargement of **a**; **c** #86013101, 620 μs, Ms = 1.506; **d** enlargement of **c**; **e** #86013104, 640 μs, Ms = 1.518; **f** enlargement of (**e**); **g** #86013106, 660 μs, Ms = 1.514; **h** #86013105, 720 μs, Ms = 1.514, DiMR

(a) (b)

(c) (d)

(e) (f)

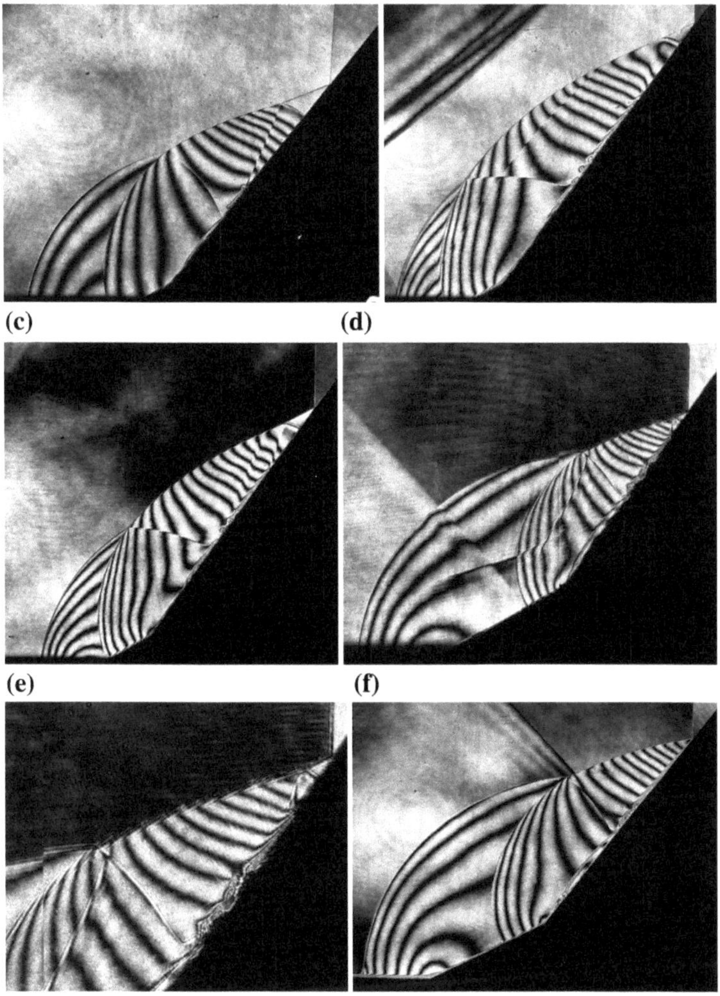

Fig. 2.75 Effect of the first wedge length on the transition for Ms = 1.5–1.9: **a** #87120715, Ms = 1.521 at 800 hPa, 293.0 K, θ_1 = 30°/55°, L = 7 mm, SPRR; **b** #87120709, Ms = 1.896 at 500 hPa, 293.0 K, θ_1 = 30°/55°, L = 7 mm, SPRR; **c** #87120708, Ms = 1.896 at 500 hPa 293.0 K, θ_1 = 30°/55°, L = 12 mm, SPRR; **d** #87120412, Ms = 1.676 at 620 hPa 293.0 K, θ_1 = 30°/55°, L = 25 mm, SPRR; **e** enlargement of (**d**); **f** #87120803, Ms = 1.639 at 680 hPa 294.1 K, θ_1 = 27°/°52, L = 52 mm, IvMR; **g** #87120801, Ms = 1.621 at 680 hPa 294.1 K, θ_1 = 27°/°52, L = 52 mm, StMR; **h** enlargement of (**g**), StMR

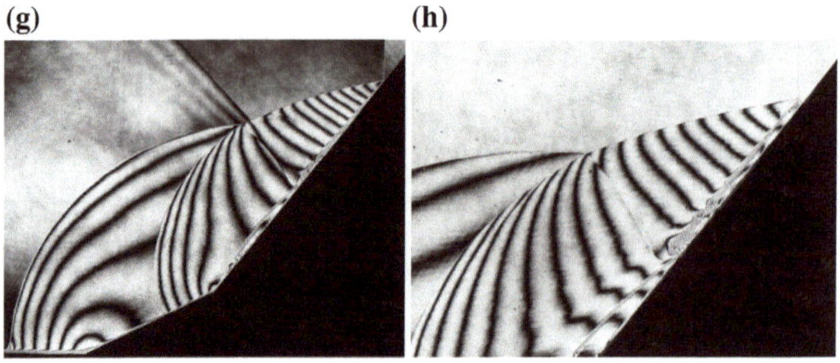

Fig. 2.75 (continued)

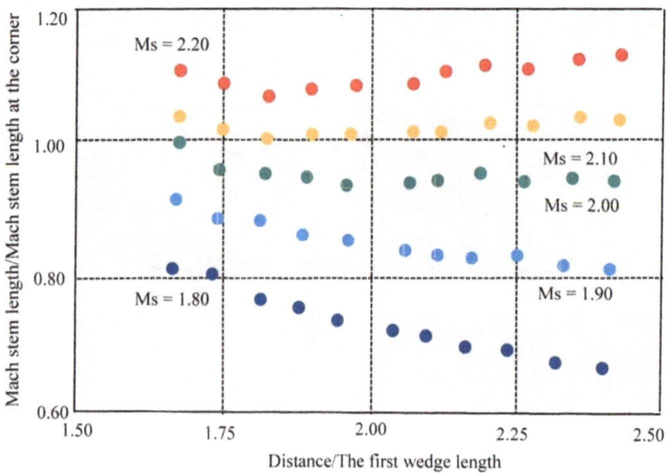

Fig. 2.76 Effect of Ms on the evolution of reflection pattern (Komuro 1990)

Figure 2.76 shows the evolution of the MS along the second wedge surface, this is a summary of numerical simulation. Komuro (1990) numerically simulated the transition of reflected shock wave along the double wedge of 27°/55° by solving the Euler Equations. The ordinate denotes the height of MS normalized by that defined at the corner of the double wedge. The abscissa denotes the distance along the second wedge surface normalized by the first wedge length. For Ms = 2.20, the height of the MS increases with it propagation, which means that the reflection pattern is a DiMR. For Ms = 2.10, the height of MS decreases, is temporarily holding as StMR, but eventually tends starts to gradually increase, which means that the reflection transits to DiMR. For Ms = 2.00, the height of MS remains constant, that is, the StMR is maintained. For Ms = 1.90 and 1.80, the height of the MS decreases monotonously, that is, the IvMR is maintained for a while and then transits to SPRR. It is known that the transition over the second wedge surface is influenced strongly depending on the first wedge length and the boundary layer

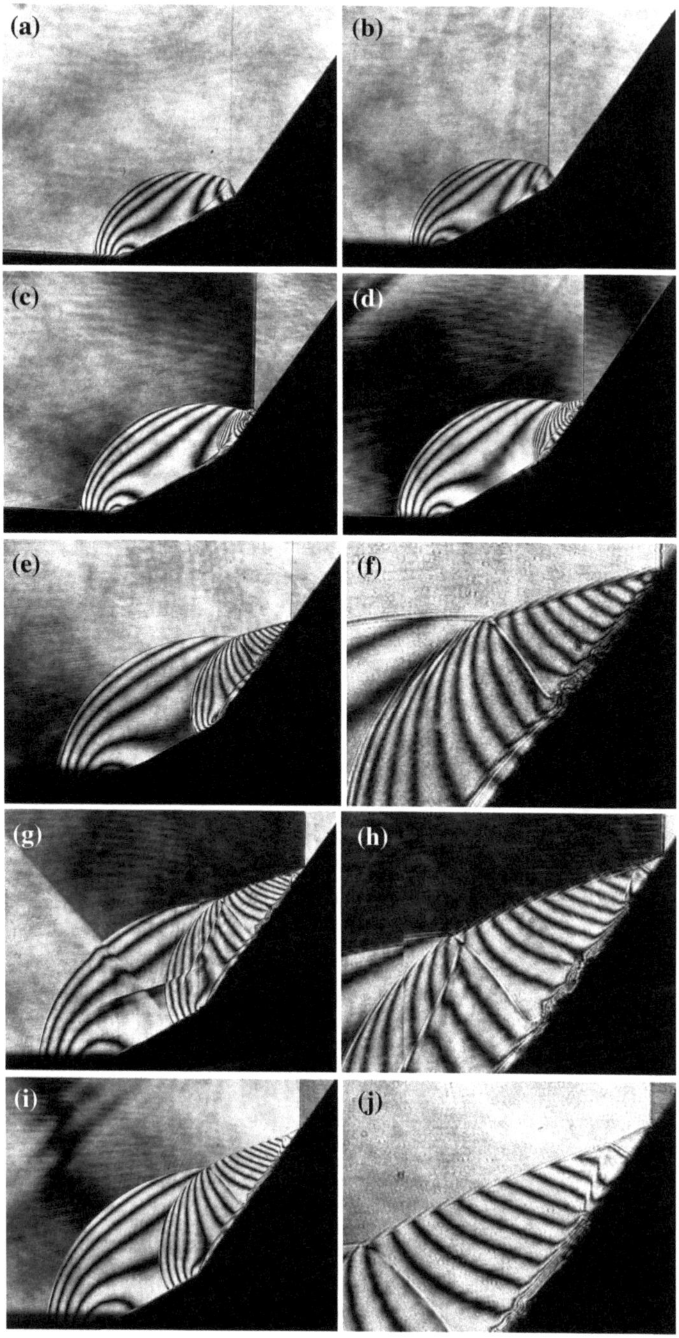

Fig. 2.77 Evolution of an IvMR over a $\theta_w = 30°/55°$ double wedge for Ms = 1.60 in air at 620 hPa, 293 K, the length of the first wedge L = 25 mm: **a** #87120404, Ms = 1.631; **b** #87120409, Ms = 1.669; **c** #87120407, Ms = 1.700; **d** #87120410, Ms = 1.676; **e** #87120411, Ms = 1.680; **f** enlargement of (**e**); **g** #87120412, Ms = 1.676; **h** enlargement of (**g**); **i** #87120413, Ms = 1.673; **j** enlargement of (**i**)

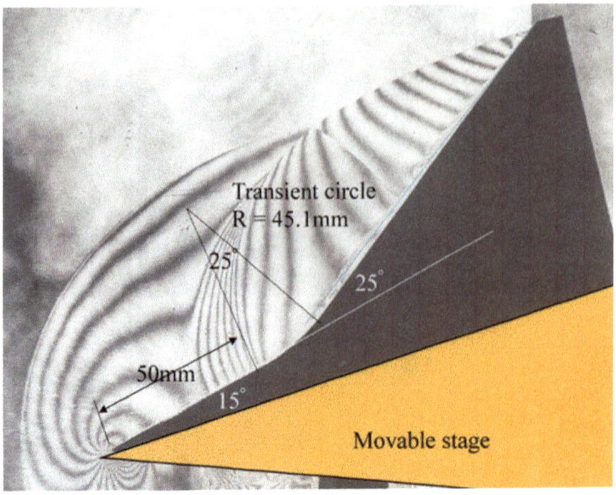

Fig. 2.78 A concave double wedge of 15°/25° installed on a movable stage #88022308, Ms = 1.764

developing on the wedges. Therefore, probably the Euler solvers shown in Fig. 2.76 may not precisely reproduce the transition over double wedges. Anyway the StMR may be present only in a special combination of parameters. Presumably this reflection pattern will not exist consistently in shock tube flows as the concept of StMR violates the unsteadiness of shock tube flows.

Figure 2.77 shows the evolution of an IvMR over a θ_w = 30°/55° double wedge for Ms = 1.60–1.80 in air at 600–650 hPa, 293 K and the first wedge length is L = 25 mm. As the second wedge angle θ_2 is larger than θ_{crit}, the reflection pattern on the second wedge is an IvMR and the transition to a SPRR would take a long distance. Figure 2.77 shows this sequence and is comparable to the numerical prediction shown in dark blue circles presented in Fig. 2.76.

The feature of the stationary Mach reflection, StMR, is that its slip line is consistently parallel to the second wedge surface. This means that the trajectory angle θ of its triple point should be θ = 0. In the (p, θ)-plane of a shock polar diagram, the solution exists only on the axis of θ = 0. In short, the solution of StMR is given as the intersection of R-polar with I-polar. This solution is valid only in steady flows. In Fig. 2.77, solutions suggested the existence of StMR for a specific Mach number and in a special combination of wedge angles. Then to experimentally verify the presence of StMR, a concave double wedge should be mounted on a movable stage. Figure 2.78 shows the concave double wedge mounted on a movable stage: the first wedge angle θ_1 is 15° and the second wedge angle θ_2 = 25°. The two wedges are connected smoothly with a transient circular arc of a radius of 45.1 mm.

Figure 2.79a–f shows the evolution of shock wave reflection from a 15°/25° double wedge mounted on a movable stage. As shown in Fig. 2.79c at θ_2 = 48° the reflection pattern is a StMR but in Fig. 2.79d at θ_2 = 53° the reflection pattern is an IvMR. Unlike the previous double wedge cases, the arc shaped transitional area created a band of wavelets continuously reflected via the curved area to the straight wedge of θ_2. However, the reflection pattern was not strongly affected by the circular arc.

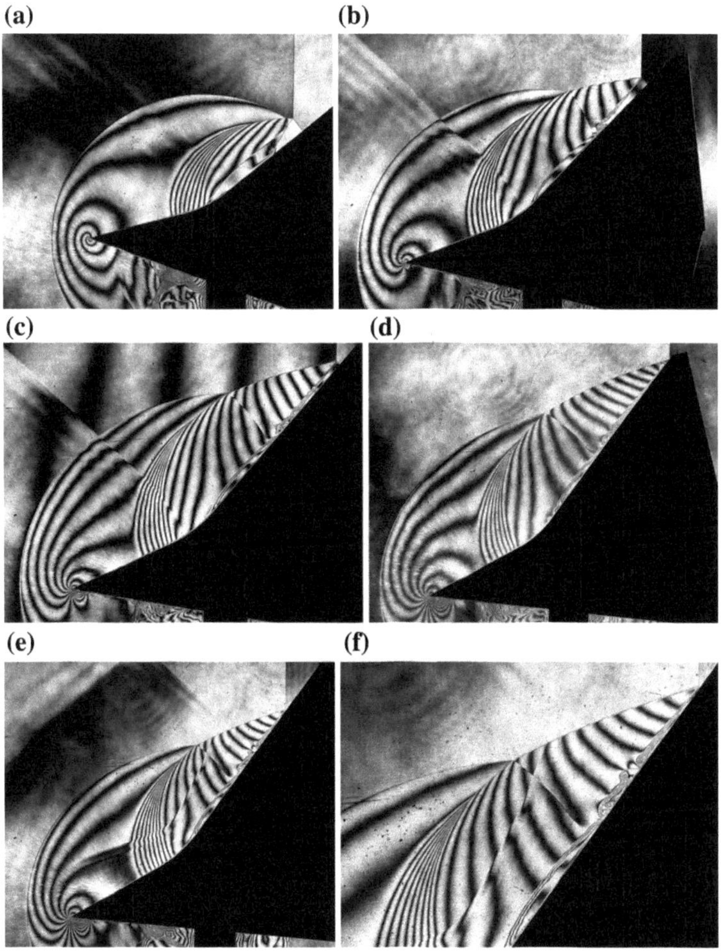

Fig. 2.79 Evolution of shock waves for Ms = 1.76 in air at 500 hPa, 288.8 K mounted on a 15°/25° movable wedge: **a** #88022402, 50 μs from the trigger point, Ms = 1.762, $\theta_2 = 37°$; **b** #88022406. 75 μs, Ms = 1.749, $\theta_2 = 48°$; **c** #88022407, 25 μs, Ms = 1.751, $\theta_2 = 48°$; **d** #88022408, 70 μs, Ms = 1.764, $\theta_2 = 53°$; **e** #88022409, 50 μs, Ms = 1.773, $\theta_2 = 55°$; **f** enlargement of (**e**)

Figure 2.80 shows the evolution of shock wave reflection from a double wedge mounted on a movable stage for Ms = 2.35 in air. In Fig. 2.80f, at $\theta_2 = 55°$ the reflection pattern is a StMR. The SL looked parallel to the second wedge surface. In the meantime, the SL became unstable and vortices started to appear along the SL.

For collecting images of a StMR, movable wedge experiments were conducted for Ms = 2.50 by varying the wedge angles from the later stage of a DiMR to earlier stage of an IvMR. Figure 2.81a shows results for $\theta_1 = 43.5°$, $\theta_2 - \theta_1 = 15°$ for Ms = 2.521 and the final reflection pattern was DiMR. In Fig. 2.81b, for

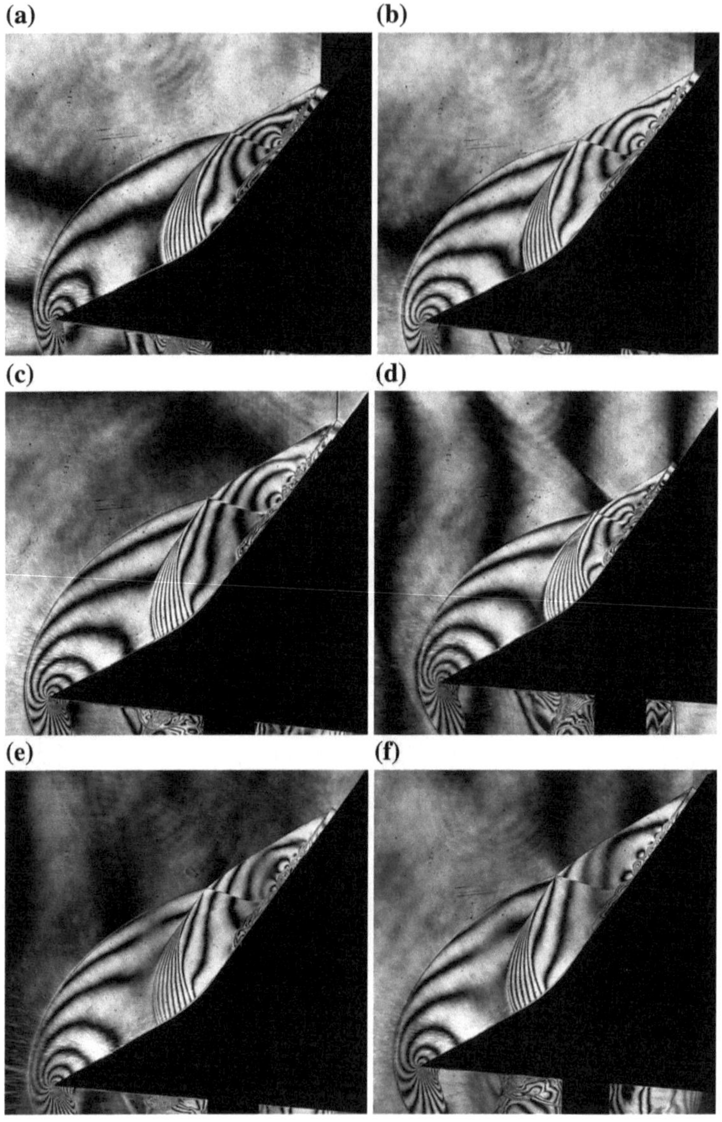

Fig. 2.80 Evolution of shock wave for Ms = 2.35 in air at 200 hPa, 288.8 K mounted on a 15°/25° movable wedge: **a** #88022504, 200 μs from the trigger point, Ms = 2.365, θ_2 = 50°; **b** #88022507, 200 μs, Ms = 2.357, θ_2 = 50°; **c** #88022508, 200 μs, Ms = 2.357, θ_2 = 50°; **d** #88022511, 185 μs, Ms = 2.349, θ_2 = 55°; **e** #88022509, 200 μs, Ms = 2.357, θ_2 = 55°; **f** #88022510, 185 μs, Ms = 2.412, θ_2 = 55°

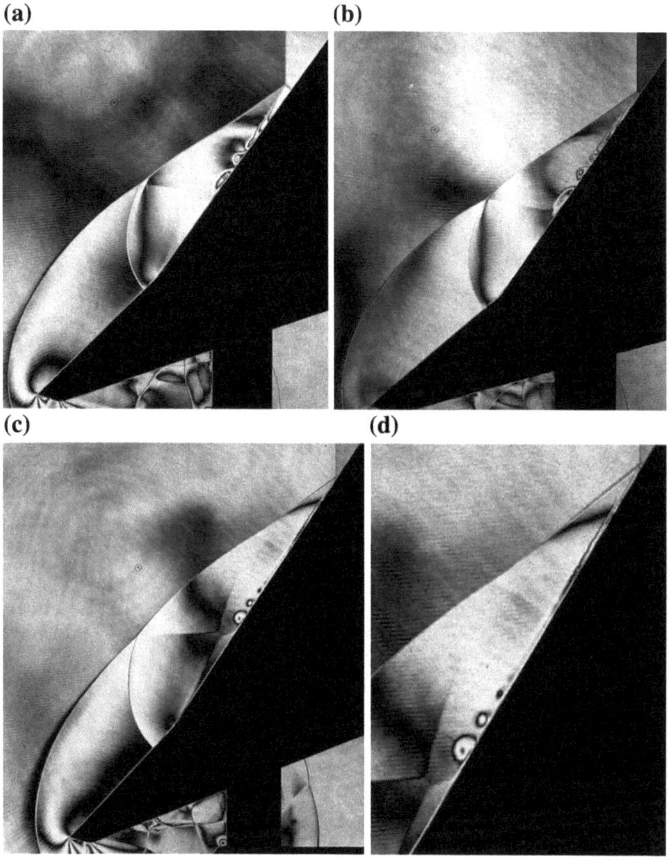

Fig. 2.81 IvMR at critical transition angles for Ms = 2.50 in air at 45 hPa, 284 K: **a** #84041906, delay time from the trigger point 40 μs, Ms = 2.521, $\theta_1 = 43.5°$, $\theta_2 - \theta_1 = 10°$, DiMR; **b** #84041909, 40 μs, Ms = 2.462, $\theta_1 = 42.0°$, $\theta_2 - \theta_1 = 15°$ StMR; **c** #84041904, 40 μs, Ms = 2.496, $\theta_1 = 42.5°$, $\theta_2 - \theta_1 = 10°$ StMR; **d** enlargement of (**c**); **e** #84041908, 40 μs, Ms = 2.521, $\theta_1 = 41.5°$, $\theta_2 - \theta_1 = 15°$ StMR; **f** enlargement of (**e**); **g** #84041913, 40 μs, Ms = 2.530, $\theta_1 = 40.0°$, $\theta_2 - \theta_1 = 20°$ StMR; **h** enlargement of (**g**); **i** #84041911, 40 μs, Ms = 2.379, $\theta_1 = 47.5°$, $\theta_2 - \theta_1 = 15°$ IvMR; **j** #84042003, 40 μs, Ms = 2.500, $\theta_1 = 47.5°$, $\theta_2 - \theta_1 = 10°$ IvMR

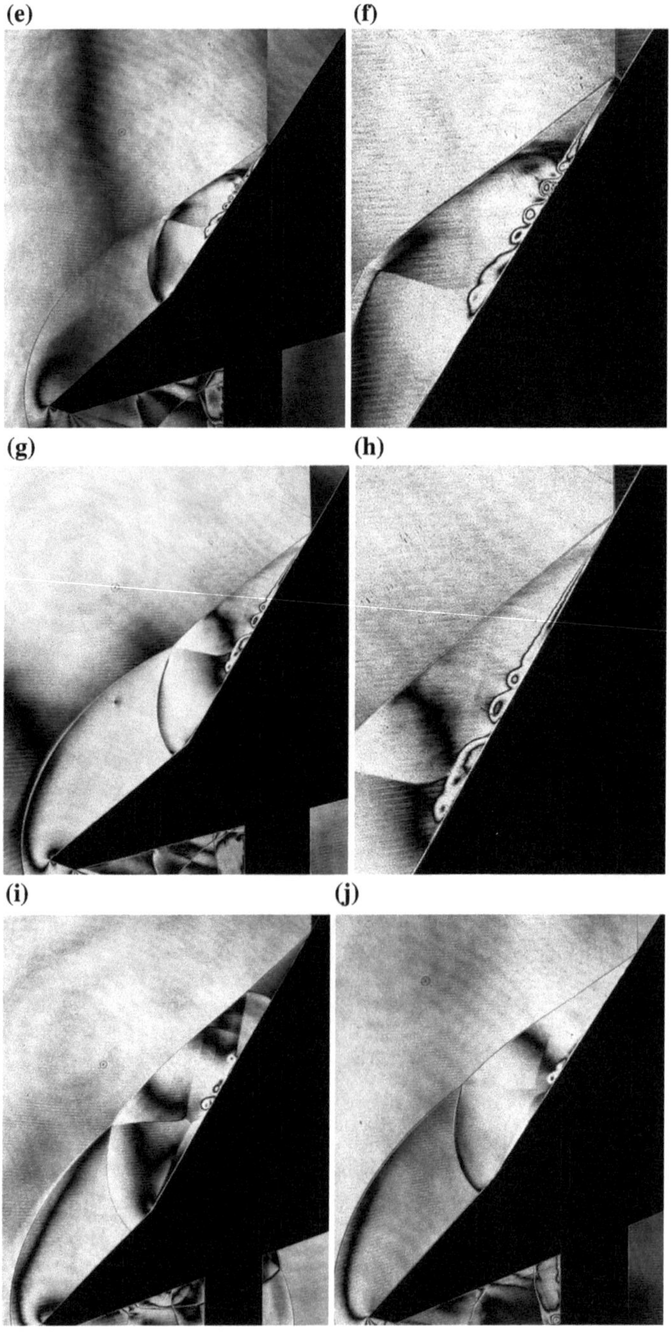

Fig. 2.81 (continued)

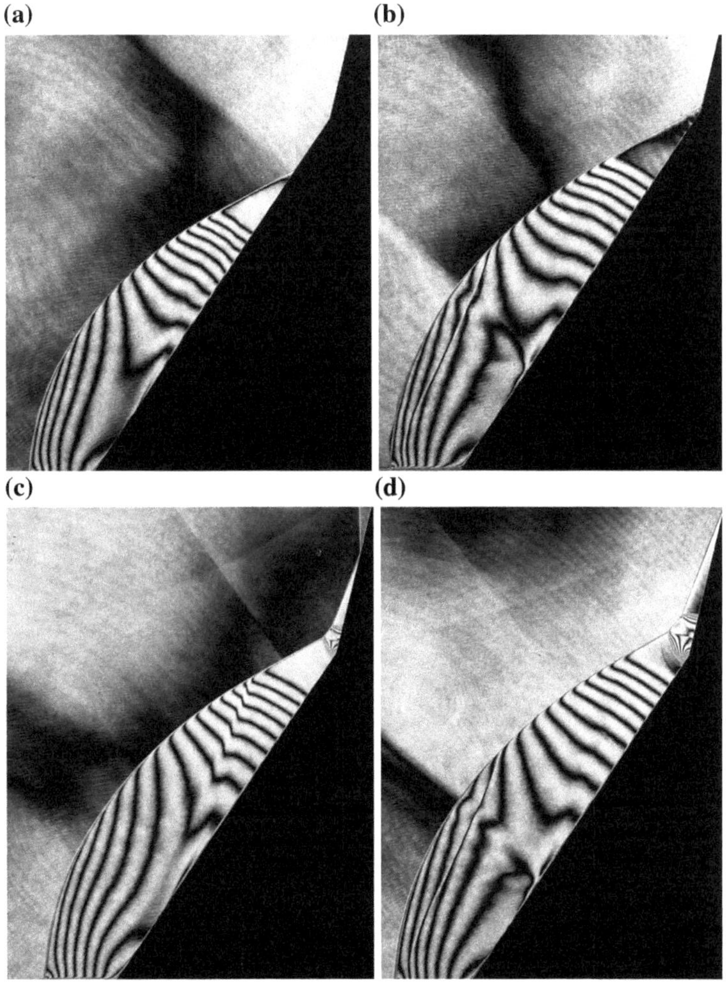

Fig. 2.82 Evolution of reflected shock wave IvMR for Ms = 2.0 over a 55°/75° concave wedge in air at 500 hPa, 298 K: **a** #86091704, 105 μs delay time from the trigger point, Ms = 1.928; **b** #86091701, 120 μs, Ms = 1.913; **c** #86091702, 130 μs, Ms = 1.922; **d** #86091703, 135 μs, Ms = 1.920

Ms = 2.462, $\theta_1 = 42.0°$, $\theta_2 - \theta_1 = 15°$, then MS is shortened but parallel to the wedge surface. Then, the reflection pattern is StMR. In Fig. 2.81c, for Ms = 2.496, $\theta_1 = 42.5°$, $\theta_2 - \theta_1 = 10°$, then the reflection pattern is StMR. Figure 2.81c is its enlargement. Ms is very shortened but looks parallel to the wedge surface. Figure 2.81e, g shows StMR. Figure 2.81f, h are their enlargements. With a slight increase in the wedge angle, the reflection pattern changed to IvMR.

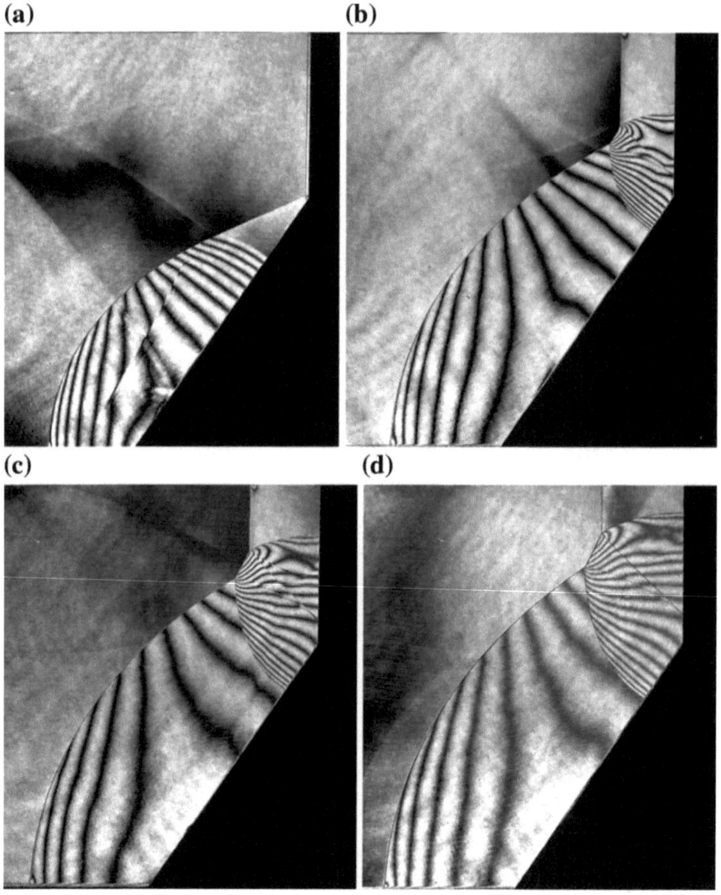

Fig. 2.83 The shock wave reflection from a 55°/90° concave double wedge for Ms = 1.470 in air at 700 hPa, 289.5 K. Experiments were conducted in 60 mm × 150 mm shock tube: **a** #92012101, delay time 610 μs, Ms = 1.468; **b** #92012005, delay time 640 μs, Ms = 1.468; **c** #92012007, 680 μs, Ms = 1.468; **d** #92012003, 700 μs, Ms = 1.470

One of the most trivial cases is the reflection from concave double wedges in which $\theta_1 > \theta_{crit}$. Figure 2.82a–d, shows the evolution of reflected shock waves for Ms = 2.0 in air from a 55°/75° concave double wedge. The reflection patterns are supersonic Mach reflections, SPRR. In the absence of triple points, the shock wave reflection patterns and their interactions are simple and straightforward.

Figure 2.83a–d shows the evolution of shock wave reflection for Ms = 1.47 from the combination of a 55°/90° wedge. The reflected shock wave pattern is SPRR already on the first wedge. At the same time, the IS is reflected from the vertical wall. Eventually, the transmitted shock wave and the reflected shock wave interacted resulting in a simple and straightforward pattern.

2.2.2 Convex Double Wedges

When the first wedge angle θ_1 is larger than that of the second wedge θ_2, it is called a convex double wedge. Then, three patterns of shock wave reflection are possible. The first case is a DiMR appearing on the first wedge. Its Mach stem is defined as MS_1. At the convex corner, it starts diffracting. At the same time, the first DiMR propagates independently on the second wedge surface. Second case is that a RR appears on the first wedge and interacts with the second wedge. MR appears on the second wedge. The third case is a trivial one. The RR is maintained on the second wedge surface, too.

Figure 2.84 shows a 60°/30° convex double wedge. A shock wave of Ms = 1.50 interacted with the wedge and eventually a MR appears. A train of expansion waves propagates reversely toward the upstream. The MR so far created at the corner continuously interacted with the reflected shock wave of the initial RR. At the corner of the convex wall, a vortex is generated.

Figure 2.85 shows the evolution of shock wave of Ms = 1.48 reflected from a 35°/15° convex wedge in air at 660 hPa and 300 K. The first wedge angle being $\theta_1 < \theta_{crit}$, a MR, appears in Fig. 2.85a. Interacting with the second wedge, the MR changes into a second MR. The SL of the first MR interacts with a vortex at the corner and its tail vanishes as seen in Fig. 2.85b. In Fig. 2.85c, d, the SL produced on the second wedge terminates with a vortex generated at the corner.

Figure 2.86 shows the evolution of shock wave of Ms = 1.50 reflected from a 60°/30° convex wedge.

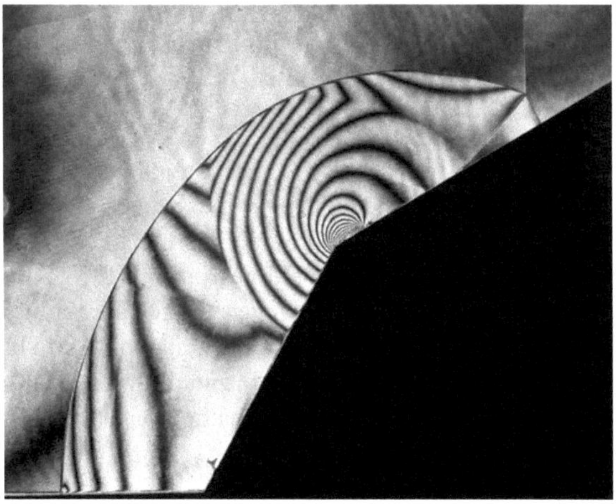

Fig. 2.84 Reflection of a MR over a 60°/30° convex double wedge: #86012709 630 μs, Ms = 1.498 in air at 670 hPa, 288 K

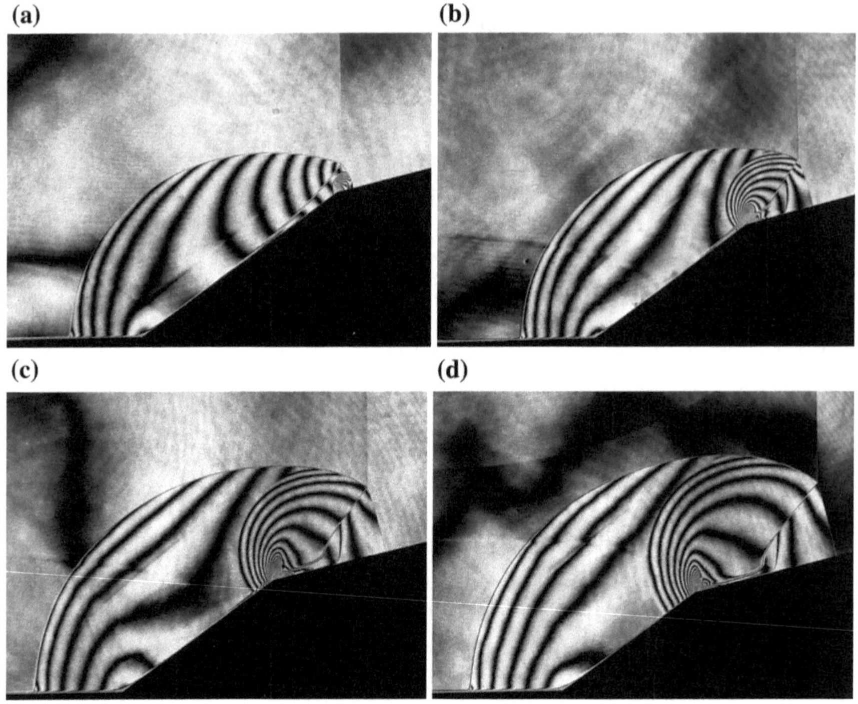

Fig. 2.85 Shock wave reflection from a 35°/15° convex wedge for Ms = 1.48 in air at 660 hPa and 300 K: **a** #86012803, 570 μs from the trigger point, Ms = 1.483; **b** #86012804, 600 μs, Ms = 1.483; **c** #86012805, 630 μs, Ms = 1.485; **d** 86012806, 640 μs, Ms = 1.489

2.3 Evolution of Shock Wave Reflection from Curved Walls

When a planar shock wave propagates over a continuously curved concave and convex walls, the main feature of the reflected shock wave transition is, in principle, similar to convex and concave double wedges. The transition from a RR to a MR occurs over a convex wedge and the transition from a MR to a RR occurs over a concave wedge (Ben-Dor et al. 1980). However, in the case of curved wedges, the critical transition angle varies significantly depending on its radius of curvature (Takayama and Sasaki 1983). The previous results of visualization regarding the dependence of the radii of curvature on the transition of reflected shock waves are summarized.

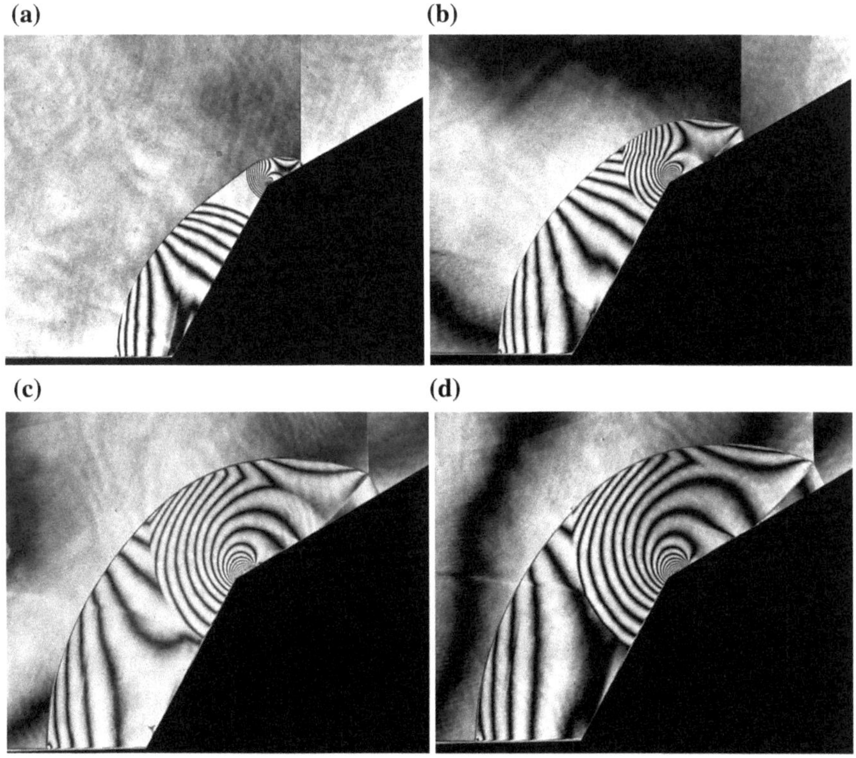

Fig. 2.86 Shock wave reflection from a 60°/30° convex wedge for Ms = 1.50 in air at 670 hPa and 288 K; **a** #86012706 570 μs, Ms = 1.496; **b** #86012707 590 μs, Ms = 1.500; **c** #86012709, 630 μs, Ms = 1.498; **d** #86012710, 650 μs, Ms = 1.490

2.3.1 Concave Walls

Unlike a concave double wedge, in a concave wall, its wall angle varies continuously from 0° to 90°. Figure 2.87 shows the sequences of reflected shock wave patterns. When the IS propagates along a shallow wall angle, the corner signal propagates along the trajectory of the glancing incidence angle as seen in Fig. 2.87a, b. Then the IS intersects with the curved MS. The MS being normal to the concave wall surface, the density behind it increases. However, at earlier stage of the reflection, although the TP is formed but not distinctly and the SL is missing. This reflection pattern is an vNMR but the trajectory angle being $\chi > 0$ so that the reflection type is DiMR as seen in Fig. 2.87c, d. With further increase in the wall angle, the TP tends to approaches toward the wall, that is $\chi < 0$. The reflection pattern becomes an IvMR as seen in Fig. 2.87e. Figure 2.87f is an enlargement of Fig. 2.87e. Figure 2.87g–j shows the termination of the IvMR and its transition to SPRR. Here the θ_{crit} is defined as a wall angle at which the IvMR transits to a

(a) (b) (c) (d) (e) (f) (g) (h)

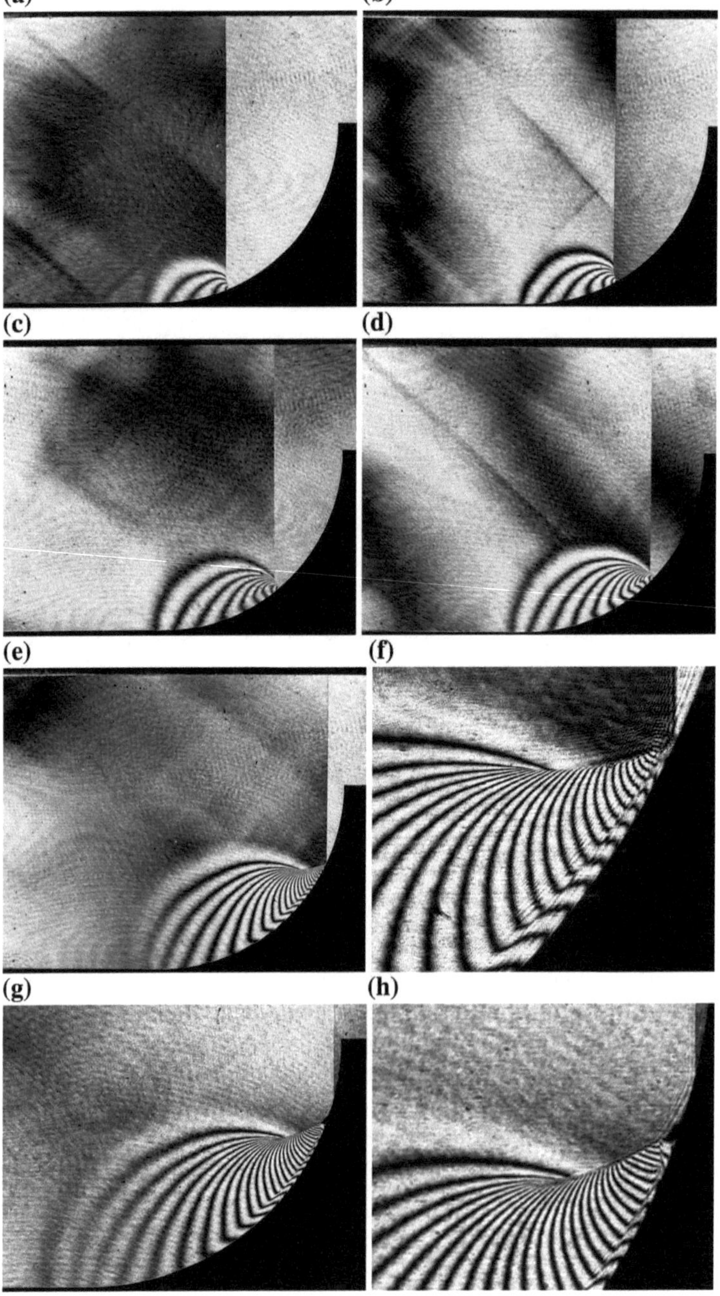

Fig. 2.87 Evolution of reflected shock wave from a 100 mm diameter circular wall for Ms = 1.53 in air at 800 hPa, 292 K conducted in the 40 mm × 80 mm conventional shock tube: **a** #81101503, 132 μs, Ms = 1.530; **b** #81101505, 142 μs, Ms = 1.530; **c** 81101507, 152 μs, Ms = 1.530; **d** #81101508, 162 μs, Ms = 1.530; **e** #81101509, us Ms = 1.530; **f** enlargement of (**e**) #81101505, 162 μs, Ms = 1.530; **g** #81101611, Ms = 1.530; **h** enlargement of (**g**); **i** #81101606, 177 μs, Ms = 1.575; **j** enlargement of (**i**)

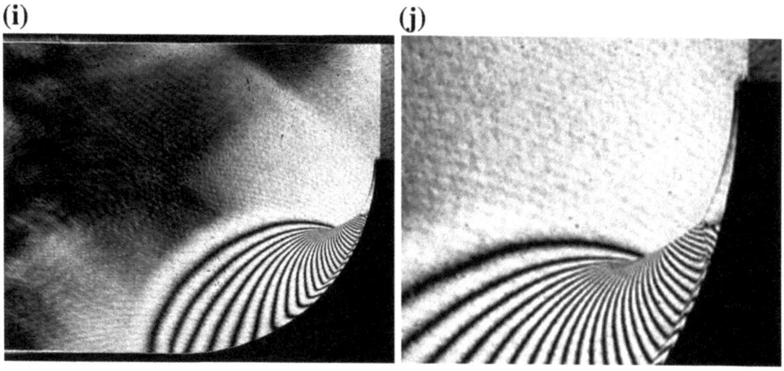

Fig. 2.87 (continued)

SPRR. As discussed later, the θ_{crit} is larger than the prediction of the Neumann criterion. In Fig. 2.87d, e, the slip lines look almost parallel or coaxial to the curved wall, but this type of reflection is analogous to StMR in the case of concave double wedges. In the present concave wall. StMR appears only transitionally and immediately transits to IvMR. As will discussed later, θ_{cri} on concave walls varied depending on the diameter of circular concave walls.

Figure 2.88 shows the evolution of shock wave reflection from a 100 mm diameter circular concave wall for Ms = 1.42 in air. Experiments were conducted in the 60 mm × 150 mm diaphragm-less shock tube having a higher degree of reproducibility. Figure 2.88a–h shows the gradual increase in fringes at earlier stage and their formation to a distinct TP. Along the concave wall, fringes increased in number with increasing the wall angle. Figure 2.88i–l shows the sequential formation of the distinct triple point. The type of reflection is a DiMR. With the further increase in the wall angle, the curved SL becomes gradually parallel to the circular wall. Such coaxial SL is maintained only for short time. The type of reflection pattern in the wedges is defined as a StMR. Along the concave walls, a similar pattern to StMR appears only temporarily as seen in Fig. 2.88k, l. When the TP touched the curved wall and the MS terminated but the entropy increase across the SL was maintained by the second TP. Therefore, at the transition from an IvMR to a SPRR, the TP carried by the IvMR became the second TP behind the SPRR. The structure of the secondary shock wave is very similar to that appearing in a DMR.

Figure 2.89 shows the initiation of an IvMR and its transition into a SPRR. Experiments were conducted in the 40 mm × 80 mm conventional shock tube. Ms scattered slightly widely but through these images the evolution of reflected shock waves was well observed. In Fig. 2.89a, b, fringe numbers increased gradually with the increase in the wall. In Fig. 2.89c–f, the formation of the TP and the transition from a StMR to an IvMR are observed. Figure 2.89g–i shows the termination of the IvMR. As soon as the TP of the IvMR reaches the wall and the MS terminate, the transition to a SPRR takes place. The value of this critical transition angle θ_{crit} is, as discussed later, larger than that predicted value by von Neumann criterion applied to straight wedges.

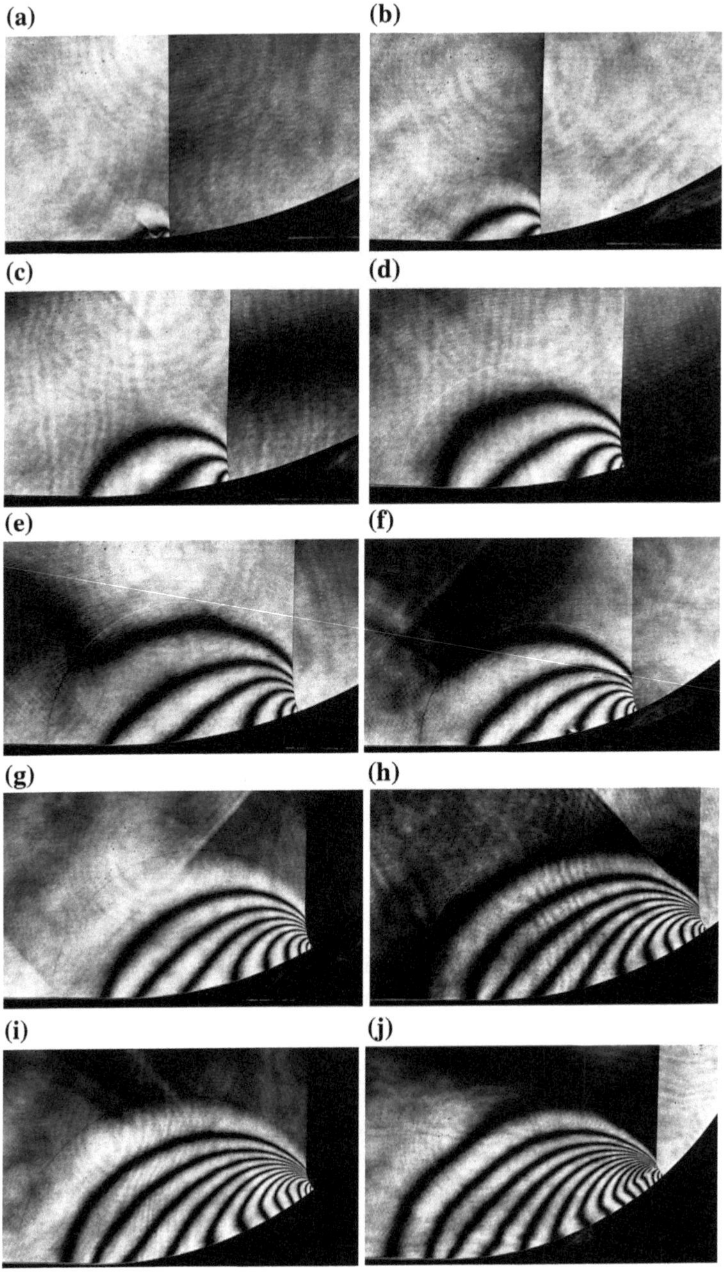

Fig. 2.88 Evolution of reflected shock wave from a 100 mm diameter circular wall for Ms = 1.42 in air at 800 hPa, 294.3 K: **a** #87111902, 240 μs delay time from the trigger point, Ms = 1.431; **b** #87111903, 260 μs, Ms = 1.432; **c** #87111904, 280 μs, Ms = 1.422; **d** #87111905, 300 μs, Ms = 1.422; **e** #87111906, 320 μs, Ms = 1.418; **f** #87111907, 340 μs, Ms = 1.410; **g** #87111908, 360 μs, Ms = 1.414; **h** #87111909, 380 μs, Ms = 1.428; **i** #87111910, 400 μs, Ms = 1.418; **j** #87112007, 415 μs, Ms = 1.422; **k** #87111912 440 μs, Ms = 1.426; **l** #87112001, 460 μs, Ms = 1.430

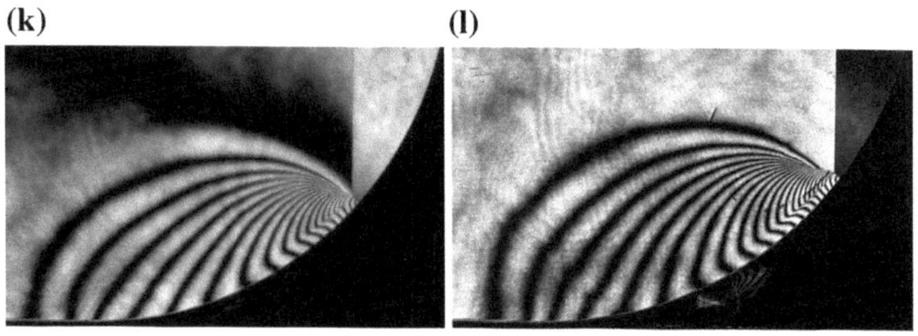

Fig. 2.88 (continued)

In propagating along a circular concave wall, the transition of a reflected shock wave from a MR to a RR occurs at the θ_{crit} at a given Ms. Its value, however, varies depending not only on Ms but also the Reynolds number, Re, that simply reflects the initial pressure and the diameter of the concave wall.

It was already observed that on the wedge having a slightly smaller wedge angle than the θ_{crit}, a RR appeared at the leading edge and the transition to a MR was delayed and that the delayed transition distance varied significantly depending on the initial pressure. Hence it is speculated that the initial pressure would affect the transition over not only concave walls but also convex walls. The θ_{crit} was obtained in the transition from an IvMR to a RR over a 100 mm diameter circular concave wall for Ms = 2.33 in air.

Figure 2.90 summarizes the experimental results: The height of MS against the concave wall angle. The ordinate denotes the height of Mach stem in mm and the abscissa shows concave wall angle in degrees. Blue and red filled circles denote initial pressures of 14.1 and 1.41 kPa, respectively. The angle at which MS vanishes is θ_{crit}. As shown in arrows on the abscissa, the detachment criterion suggests $\theta_{crit} = 51.5°$ and the mechanical equilibrium criterion offers $\theta_{crit} = 63.5°$, whereas experimentally deduced values are: $\theta_{crit} = 71.0°$ at 1.41 kPa and $\theta_{crit} = 73.0°$ at 14.1 Pa. θ_{crit} varies depending on the initial pressure, in other words, on the Reynolds number, Re. In the previous experiments, researchers paid attention mostly to Ms but were less aware of the effects of the initial pressures or Re.

2.3.2 *Convex Walls*

Upon the shock wave reflection from a double convex wedge whose wall angle is $\theta_w > \theta_{crit}$, the initial reflection pattern is a RR. With the decrease in the wall angle θ_w, the RR transits to a MR. This is a reverse reflection sequence from the one observed in the concave wall as discussed previously (Takayama and Sasaki 1983).

Bryson and Gross, by using shadowgraph and the schlieren methods, visualized the reflection and diffraction of shock waves over cylinders, and spheres. Shock wave reflection over the 80 mm diameter cylinder was visualized in the

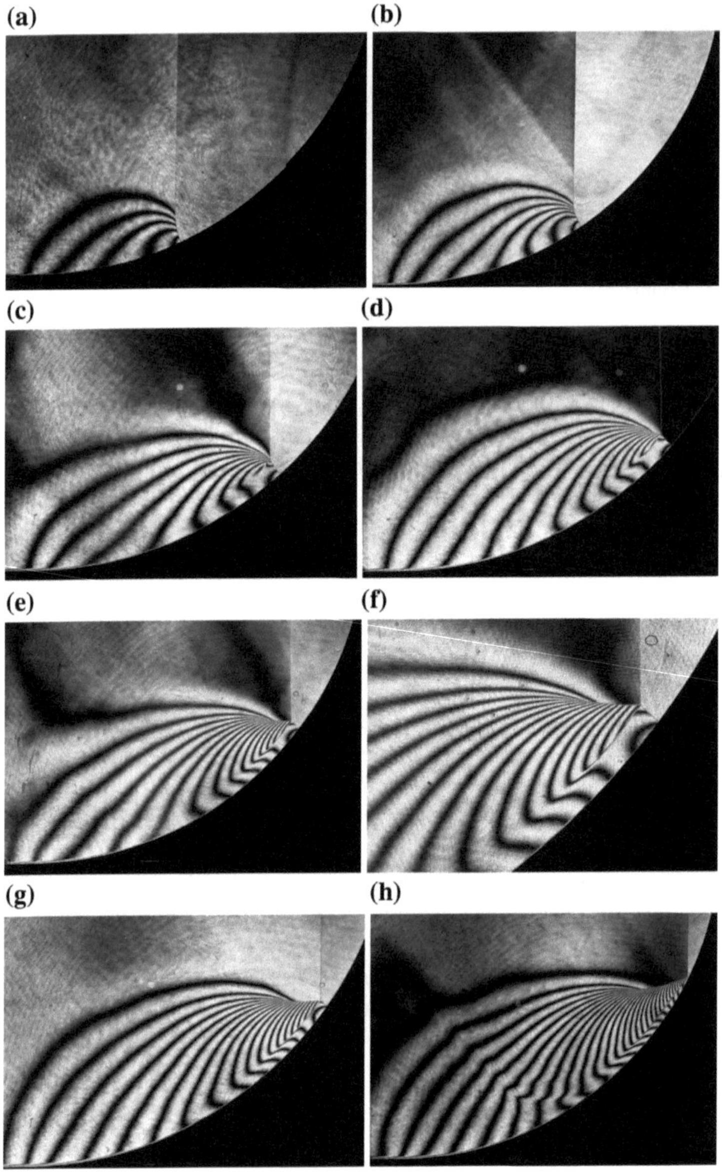

Fig. 2.89 Evolution of reflected shock wave over a 100 mm diameter concave wall for Ms = 1.60 in air at 400 hPa, 291 K: **a** #83101914, delay time from the trigger point 50 μs; **b** #83101912, 90 μs; **c** #83101911, 110 μs; **d** #83101909, 130 μs; **e** #83101902,130 μs; **f** enlargement of (**e**); **g** #83101908, 160 μs; **h** #83101904, 180 μs; **i** #83101903, 210 μs; **j** enlargement of (**i**); **k** 83101905, 210 μs; l enlargement of (**k**); **m** #83101907, 220 μs; **n** enlargement of (**m**)

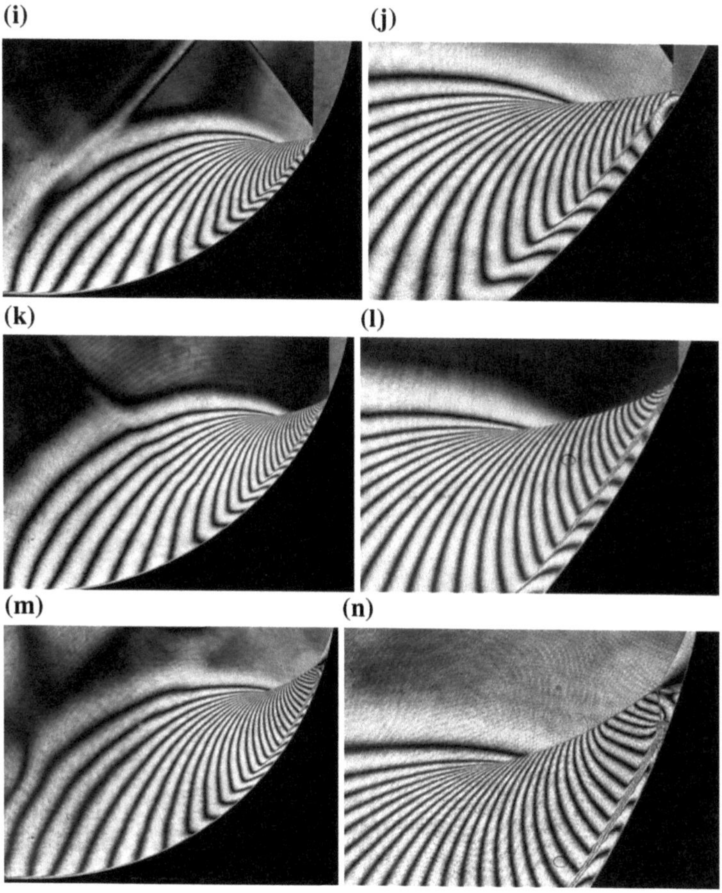

Fig. 2.89 (continued)

40 mm × 80 mm conventional shock tube for Ms = 1.035 in air Takayama and Sasaki (1983). In the present experiment, a shock wave being weak and the *OB* path length being only 40 mm, to enhance the sensitivity of interferograms at such low pressures, the initial air pressures were increased up to 1610 hPa.

Obtained interferograms are shown in Fig. 2.91. Along the frontal side of the cylinder, a RR appears as seen in Fig. 2.91a. Its enlargement is shown in Fig. 2.91b. With propagation, the RR transits to SMR as seen in Fig. 2.91c, and its enlargement in Fig. 2.91d.

Figure 2.92a–f shows reflections of stronger shock waves for Ms = 2.5. The reflection pattern shows SPRR. Hence immediately after the transition, the RR has a kink point and the reflection pattern is TMR in Fig. 2.92c.

Figure 2.93 shows direct shadowgraphs of the evolution of shock wave for Ms = 3.15 reflected from a convex wall accommodated in a 40 mm × 80 mm

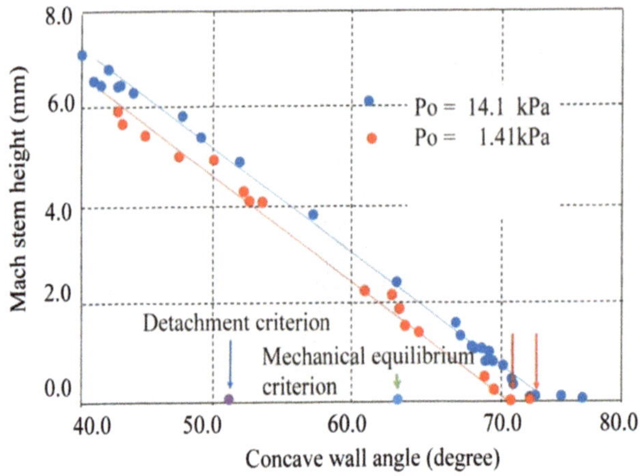

Fig. 2.90 The effect of initial pressures on θ_{crit} at Ms = 2.33 in air. The IvMR on 100 mm diameter concave wall installed in a 100 mm × 180 mm diaphragm-less shock tube (Kitade 2001)

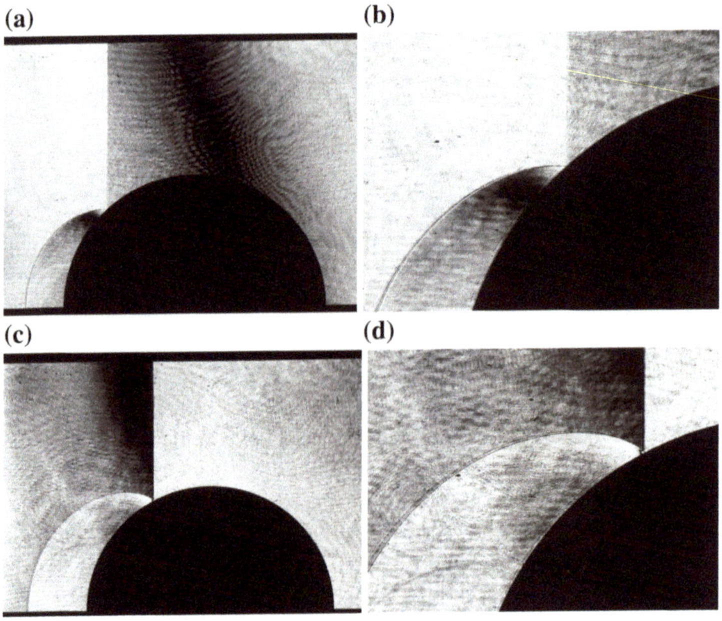

Fig. 2.91 Reflection of shock wave over a 80 mm diameter cylinder for Ms = 1.035 in air at 1610 hPa, 293.8 K: **a** #81101933, delay time from the trigger point 120 μs, RR; **b** enlargement of (**a**); **c** #81101937, 140 μs, SMR; **d** enlargement of (**c**)

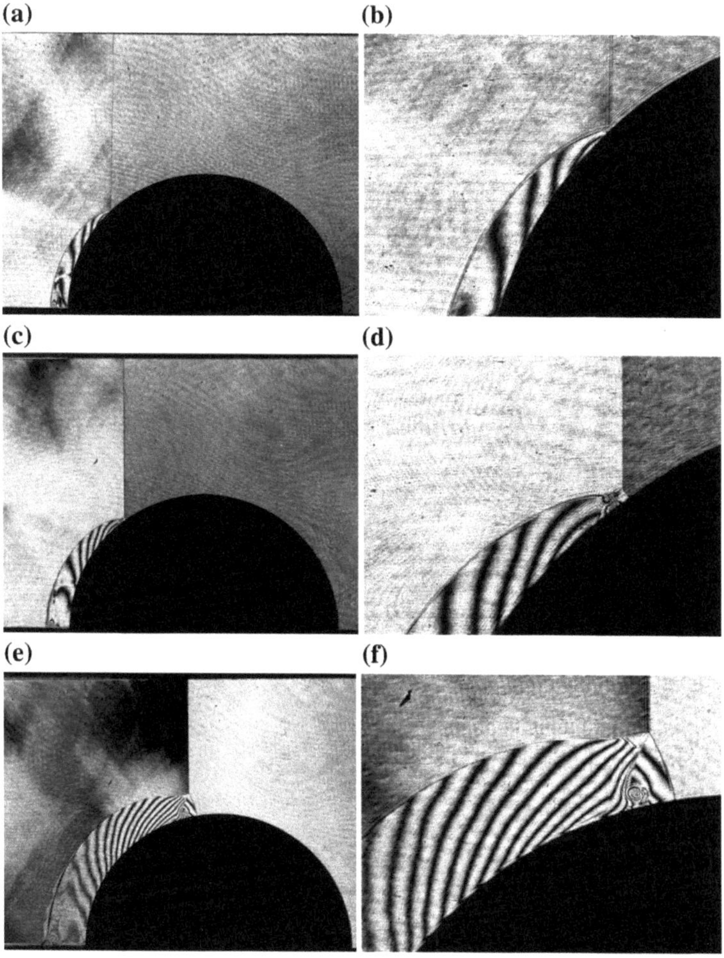

Fig. 2.92 Evolution of shock wave reflection over a 80 mm diameter cylinder for 2.50 in air at 300 hPa, 291.7 K: **a** #81102213, Ms = 2.50, RR; **b** enlargement of (**a**); **c** #81102206, Ms = 2.50, TMR; **d** enlargement of (**c**); **e** #81102208, Ms = 2.55 TMR, **f** enlargement of (**e**)

conventional shock tube. The convex wall has 300 mm diameter is mounted on a movable stage at the initial inclination angle of $\theta_w = 50°$. In Fig. 2.93a the reflection pattern at the leading edge is a RR. In Fig. 2.93b–d, reflection patterns are DMR.

Figure 2.94 shows shock wave reflection from a 100 mm diameter circular cylinder for Ms = 2.326 in air at 144 hPa, 290.2 K in a 100 mm × 180 mm diaphragm-less shock tube. Visualizations are conducted using double pulse interferogram at double pulse interval of 200 μs. The reflection pattern is a RR, with propagation, becomes a TMR, and on further propagation it turns into a SMR.

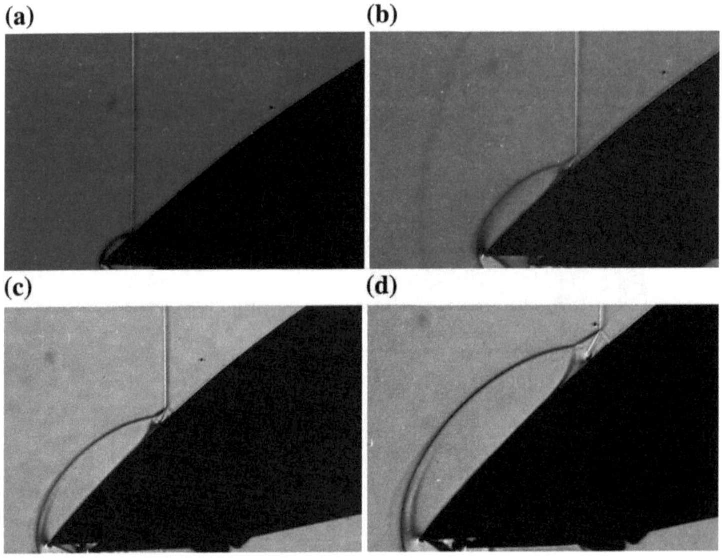

Fig. 2.93 Evolution of reflected shock wave over a 300 mm diameter convex wall mounted on a movable stage for Ms = 3.150 in air at 40 hPa, 295.1 K, at $\theta_w = 50°$, visualization was conducted by direct shadowgraph; **a** #80072529, 14 μs, Ms = 3.135, RR; **b** #80072520, 20 μs, Ms = 3.175, DMR; **c** #80072510, 24 μs; ms = 3.158, DMR; **d** #80072513, 30 μs. Ms = 3.071, DMR

In Fig. 2.95, triple point positions deduced from individual experiments are plotted. The ordinates denote the height of the MS in mm and the abscissa denotes the wall angles in degree. The blue and red filled circles denote the initial pressure 14.1 and 1.41 hPa, respectively. A filled grey circle shows θ_{crit} of detachment criterion $\theta_{crit} = 51.0°$ whereas $\theta_{crit} = 43.0°$ at $p_0 = 14.1$ hPa and $\theta_{crit} = 38.0°$ at $p_0 = 1.41$ hPa.

It is found that the initial pressure significantly affects $\theta_{crit.}$ The higher the initial pressure is, (or in other words, the higher the Reynolds number becomes) the closer the θ_{crit} approaches to the detachment criterion. For reducing the initial pressures, the larger the departure from the detachment criterion becomes.

Figure 2.96 summarizes the series of present observations. Images observed through a slit was recorded by using a streak mode of Ima-Con high speed camera (John Hadland Photonics Model 675).

Figure 2.96a shows a streak photograph and a sketch of a shock wave motion in atmospheric air. The shock wave of Ms = 1.3 was reflecting over a 100 mm diameter convex cylinder. A 0.5 mm wide slit was attached on the edge of the cylinder. The image of the shock wave viewed through the slit was visualized by direct shadowgraph and recorded in the streak mode of the Ima Con high speed camera. In order to achieve a better resolution of the streak image the slit image was rotated by 49.5° relative to the direction of the shock wave propagation. The image rotator was designed and manufactured in house. The picture in Fig. 2.96a shows

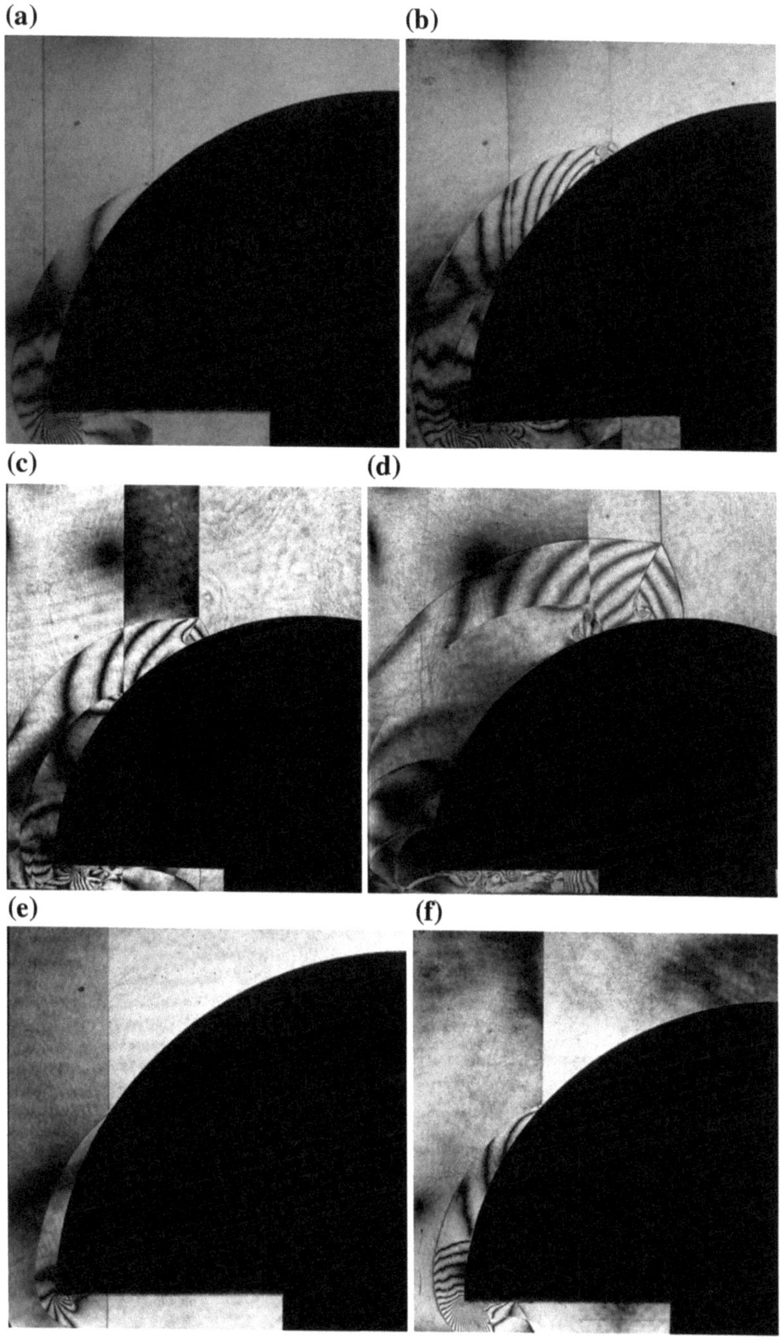

Fig. 2.94 Shock wave reflection from a 100 mm diameter circular cylinder in a 100 mm × 180 mm diaphragm-less shock tube for Ms = 2.33 in air in Fig. 2.83a–f at 144 hPa, 290.2 K, in Fig. 2.83e, f at 28.2 hPa, 291.5 K: a #98122103, Ms = 2.335; **b** #98122109, Ms = 2.326; **c** #98122104, Ms = 2.371; **d** #98122104, Ms = 2.371 144 hPa; **e** #99011909, Ms = 2.333; f #99011203, Ms = 2.330

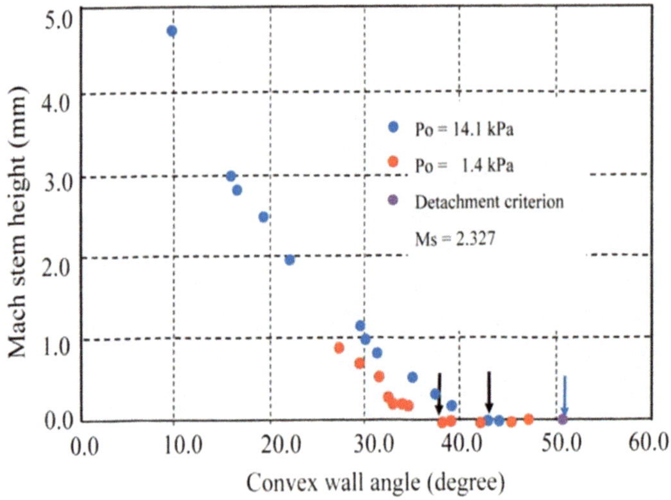

Fig. 2.95 Effect of initial pressures on θ_{crit} at Ms = 2.327 in air, 100 mm diameter convex wall in a 100 mm × 180 mm shock tube (Kitade 2001)

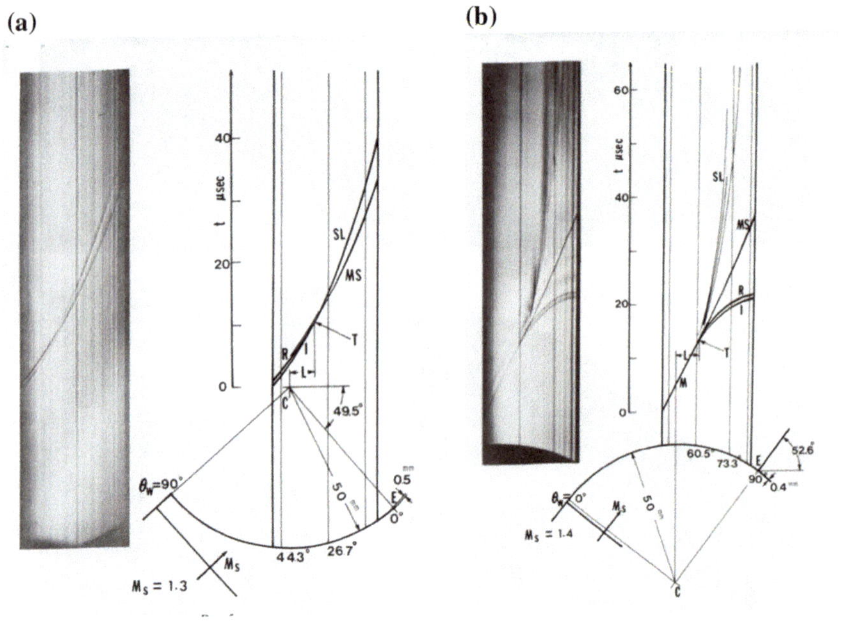

Fig. 2.96 Streak photographs applied to convex and concave walls: **a** convex wall; **b** concave wall

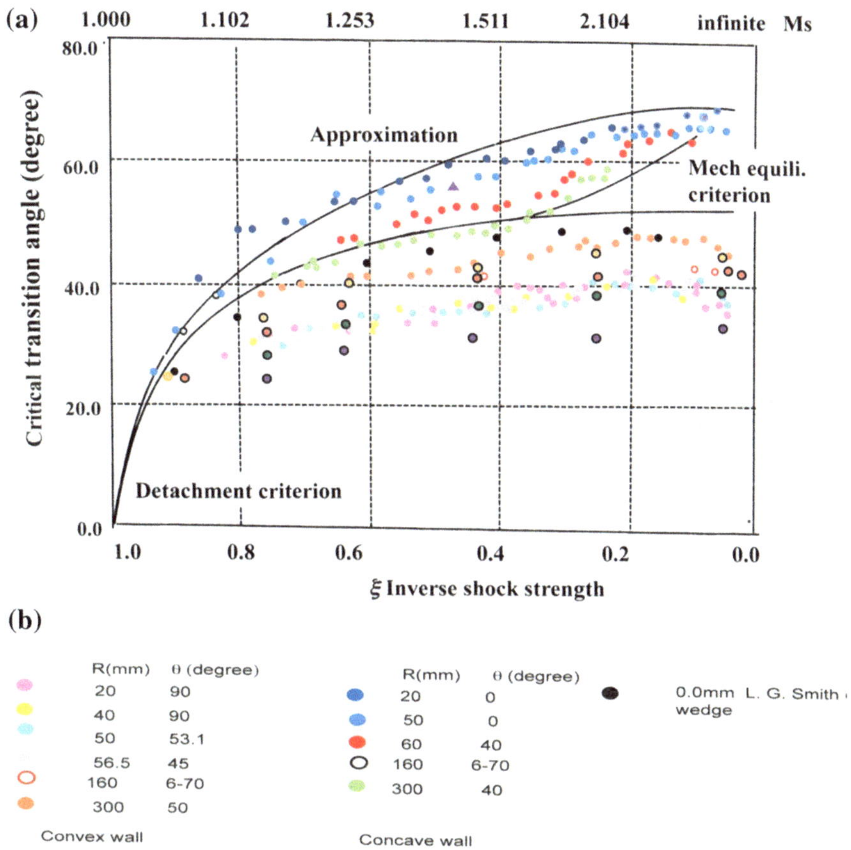

Fig. 2.97 Summary of the critical transition angle θ_{crit} versus inverse shock strength ξ: **a** summary; **b** symbols (Takayama et al. 2016)

the rotated streak image. In the streak picture, a RR in the frontal side the cylinder shows trajectories of the incident shock I and that of the reflected shock wave R. At the transition point T, the trajectories of Mach stem MS and a slip line SL appear. Interpolating the gradient change of these trajectories, the critical transition angle θ_{crit} is readily estimated.

In Fig. 2.96b, a streak photograph of the transition of a shock wave of Ms = 1.4 over a 100 mm concave wall is shown. A 0.4 mm wide slit was attached on the edge of a 100 mm diameter concave cylinder. The streak image was rotated by 52.6° in order to have the highest resolution of the images as seen in the streak picture. Along a shallow concave wall, the reflection pattern is a MR accompanying a MS. Then image through the slit was a Mach stem M. After the transition, a SPRR appears and which accompanies the incident shock wave I and a reflected shock R and later the secondary MS and a slip line SL. Then the interpolating these trajectories, their intersection provides the critical transition angle at T.

Takayama and Sasaki (1983) collected values of θ_{crit} by repeating streak recording over concave and convex walls with various radii of curvature. In addition to this, updated recent results are summarized in Fig. 2.97 (Takayama et al. 2016).

Figure 2.97a summarizes the values of θ_{crit} over concave and convex walls against the inverse strength of the shock wave ξ. The ordinate denotes the critical transition angle θ_{crit} in degree and the abscissa denotes the inverse strength of the shock wave ξ. Figure 2.79b shows the nomenclature. Data points were estimated by the above-mentioned streak recording in air. Pink. yellow, light blue, grey filled circles, red open circles, red filled circles denotes convex walls having combinations of radius and initial angle of 20 mm—90°, 40 mm—90°, 50 mm—53.1°, 160 mm—variable from 6 to 70°, and 300 mm—50°, respectively. Dark blue, blue, red filled circles, black open circles, grey filled circles indicate concave walls having combinations of radius and initial angle of 20 mm—0°, 50 mm—0°, 60 mm—40°, 160 mm—variable from 6 to 70°, and 300 mm—40°, respectively. Curve A designates the detachment criterion and curve B the von Neumann criterion. Black filled circles denote values of the critical transition angles over straight wedges measured by Smith (1948). A laminar boundary layer develops along the shock tube side wall and its displacement thickness δ grows in proportion to the square root of the inverse of the Reynolds number Re defined as Re = $(ux/\nu)^{1/2}$ where ν, u, and x is a particle speed, kinematic viscosity, and a distance from the I, respectively. Hence, the boundary layer effect appears in proportion to $x^{1/2}$.

On convex walls of a smaller radius, values of θ_{crit} significantly differ from from the detachment criterion, while on concave walls of a smaller radius, the θ_{crit} departs from the von Neumann criterion. However, for convex and concave walls of larger radii, values of θ_{crit} tend to agree with the detachment and von Neumann criteria, respectively. This implies that the larger the radii is, the less pronouncing the contribution of boundary layers to the transitions were.

References

Ben-Dor, G. (1979). *Shock wave reflection phenomena*. New York: Springer.

Ben-Dor, G., Takayama, K., & Kawa'uchi, T. (1980). The transition from regular to Mach reflection and from Mach to regular reflection in truly nonstationary flows. *Journal of Fluid Mechanics, 100,* 147–160.

Birkhoff, G. (1960). *Hydrodynamics. A study in logic, fact, and similitude*. Princeton: Princeton University Press.

Bryson, A. E., & Gross, R. W. F. (1961). Diffraction of strong shocks by cone, cylinder, and spheres. *Journal of Fluid Mechanics, 10,* 1–16.

Courant, R., & Friedrichs, K. O. (1948). *Supersonic flows and shock waves*. NY: Wiley Inter-Science.

Gaydon, A. G., & Hurle, I. R. (1963). *The shock tube high-temperature chemical physics*. London: Chapmam and Hall Ltd.

Glass, I. I. (1975). *Shock wave and man*. Toronto: Toronto University Press.

Henderson, L. F., Crutchfield, W. Y., & Virgona, R. J. (1997). The effect of heat conductivity and viscosity of argon on shock wave diffraction over rigid ramp. *Journal of Fluid Mechanics, 331,* 1–49.

Henderson, L. F., Takayama, K., Crutchfield, W. Y., & Itabashi, S. (2001). The persistence of regular reflection during strong shock diffraction over rigid ramps. *Journal of Fluid Mechanics, 431,* 273–296.

Hornung, H. G., & Kychakoff, G. (1978). Transition from regular to Mach reflection of shock waves in relaxing gases. In: A. B. Hertzberg & D. Russell (Eds.), *Proceeding of 11th International Symposium on Shock Tubes and Waves, Shock Tube and Shock Wave Research* (pp. 296–302). Seattle.

Itabashi, S. (1998). *Effects of viscosity and heat transfer on reflected shock wave transition over wedges* (Master thesis). Faculty of Engineering, Graduate School of Tohoku University.

Itoh, K. (1986). *Study of transonic flow in a shock tube* (Master thesis). Graduate School of Engineering, Faculty of Engineering Tohoku University.

Kawamura, R., & Saito, H. (1956). Reflection of shock waves. *Journal Physics Society of Japan, 11,* 584–592.

Kitade, M. (2001). *Experimental and numerical study of effect of viscosity on reflected shock wave transition* (Master thesis). Graduate School of Engineering, Faculty of Engineering Tohoku University.

Komuro, R. (1990). *Study of shock wave reflection from concave double wedges* (Master thesis). Graduate School of Engineering, Faculty of Engineering Tohoku University.

Kosugi, T. (2000). *Experimental study of delayed transition of reflected shock wave over various bodies* (Master thesis). Graduate School of Engineering, Faculty of Engineering, Tohoku University.

Krehl, P. O. K., & van der Geest, M. (1991). The discovery of the Mach reflection effect and its demonstration in an auditorium. *Shock Waves, 1,* 3–15.

Kuribayashi, T., Ohtani, K., Takayama, K., Menezes, V., Sun, M., & Saito, T. (2007). Heat flux measurement over a cone in a shock tube flow. *Shock Wave, 16,* 275–285.

Merzkirch, W. (1974). *Flow visualization.* New York: Academic Press.

Miyoshi, H. (1987). *Study of reflection and propagation of shock wave over water surface* (Master thesis). Graduate School of Engineering, Faculty of Engineering Tohoku University.

Meguro, T., Takayama, K., & Onodera, O. (1997). Three-dimensional shock wave reflection over a corner of two intersecting wedges. *Shock Waves, 7,* 107–121.

Muguro, T. (1998). *Study of three-dimensional reflection of shock waves* (Ph.D. thesis). Graduate School of Engineering, Faculty of Engineering, Tohoku University.

Numata, D. (2009). *Experimental study of hypervelocity impact phenomena at low temperature in a ballistic range* (Ph.D. thesis). Graduate School of Engineering, Faculty of Engineering Tohoku University.

Numata, D., Ohtani, K., & Takayama, K. (2009). Diffuse holographic interferometric observation of shock wave reflection from a skewed wedge. *Shock Waves, 19,* 103–112.

Onodera, H., & Takayama, K. (1990). Shock wave propagation over slitted wedge. *Reports Institute of Fluid Science, Tohoku University, 1,* 45–66.

Reichenbach, H. (1983). Contribution of Ernst mach to fluid mechanics. *Annual Review Fluid Mechanics, 15,* 1–28.

Saito, T., Menezes, V., Kuribayashi, T., Sun, M., Jagadeesh, G., & Takayama, K. (2004). Unsteady convective surface heat transfer measurements on cylinder for CFD code validation. *Shock Waves, 13,* 327–337.

Smith, L. G. (1948). *Photographic investigation of the reflection of plane shocks in air.* Off Sci Res Dev OSRD Rep 6271 Washington DC USA.

Suguyama, H., Takayama, K., Shirota, R., & Doi, H. (1986). An experimental study on shock waves propagating through a dusty gas in a horizontal channel. In: D. Bershader & R. Hanson (Eds.), *Proceedings of 15th International Symposium on Shock Waves and Shock Tubes, Shock Waves and Shock Tubes* (pp. 667–673). Berkeley.

Szumowski, A. P. (1972). Attenuation of a shock wave along a perforated tube. In: L. J. Stollery, A. G. Gaydon & P. R. Owen (Eds.), *Proceedings of 8th International Shock Tube Symposium Shock Tube Research* (pp. 14/1–14/14). London.

Takayama, K., Abe, A., & Chernyshoff, M. (2016). Scale effects on the transition of reflected shock waves. In: K. Kontis (Ed.) *Proceedings of 22nd ISSI Glasgow Shock Wave Interactions* (pp. 1–29).

Takayama, K., Gotoh, J., & Ben-Dor, G. (1981). Influence of surface roughness on the shock transition in quasi stationary and truly non-stationary flows. In C. E. Treanor & J. G. Hall (Eds.), *Proceedings of 13th International Symposium on Shock Tubes and Waves Shock Tubes and Waves* (pp. 326–334). Niagara Falls.

Takayama, K., & Sasaki, M. (1983). Effects of radius of curvature and initial angle on shock wave transition over a concave or convex wall. *Reports of Institute High Speed Mechanics, Tohoku University,* (Vol. 6, pp. 238–308)

Takayama, K., & Sekiguchi, H. (1977). An experiment on shock diffraction by cones. *Reports of Institute High Speed Mechanics, Tohoku University, 36*, 53–74.

von Neumann, J. (1963). *Collected works 6*, pp. 238–308.

Whitham, G. B. (1959). A new approach to problems of shock dynamics. Part II three dimensional problems. *Journal of Fluid Mechanics, 5*, 359–378.

Yamanaka, T. (1972). *An investigation of secondary injection of thrust vector control (in Japanese). NAL TR-286T.* Chofu, Japan.

Yang, J.-M. (1995). *Experimental and analytical study of behavior of weak shock waves* (Doctoral thesis). Graduate School of Tohoku University, Faculty of Engineering.

Yang, J. M., Sasoh, A., & Takayama, K. (1996). Reflection of a shock wave over a cone. *Shock Waves, 6,* 267–273.

Chapter 3
Shock Wave Diffraction

3.1 Shock Wave Diffraction at a Backward Facing Step

In the 18th ISSW, a poster session regarding diffraction of shock wave of Ms = 1.50 in air at a 90° sharp corner was organized. Results of 16 numerical simulations and 3 visualizations were presented (Takayama and Inoue 1991). This poster session was a bench mark test demonstrating the state–of-art of the shock wave diffraction study in the 1990.

The Shock Wave Research Center in the Institute of Fluid Science, Tohoku University submitted a results of interferometric flow visualization to this poster session. Figure 3.1 show sequential double exposure holographic interferograms of the shock wave diffraction at a 90° backward facing step for Ms = 1.45 in air. To submit the best result to this poster session, it was decided to turn the 60 mm × 150 mm conventional shock tube over sideway in the similar way as shown in Figs. 2.12 and 2.13. Then, the experiment was conducted in the 150 mm × 60 mm shock tube with the elongated *OB* path length increasing the fringe number 2.5 time more than that in the previous 60 mm path length.

The curved fringes that appear to be centered at the corner point represent expansion waves created behind the diffracting shock wave. The IS diffracted at the corner constitutes a transmitted shock wave. It propagated at the same time toward the reverse direction at the sonic speed. Figure 3.1a–h show the evolution of the transmitted shock wave and their enlargements. The transmitting shock wave and the IS intersected smoothly and continuously.

A vortex generated at the corner developed with elapsed time. At the same time, the pressure was decreasing across the expansion wave so that the flow was accelerated along the vortex and eventually the flow speed became supersonic. Although the Prandtl-Meyer expansion fan was originally defined in steady supersonic flows, it appears in the diffracting flows at a corner, as well. The Prandtl-Meyer expansion fan terminated forming a normal shock wave as seen in

K. Takayama, *Visualization of Shock Wave Phenomena*,
https://doi.org/10.1007/978-3-030-19451-2_3

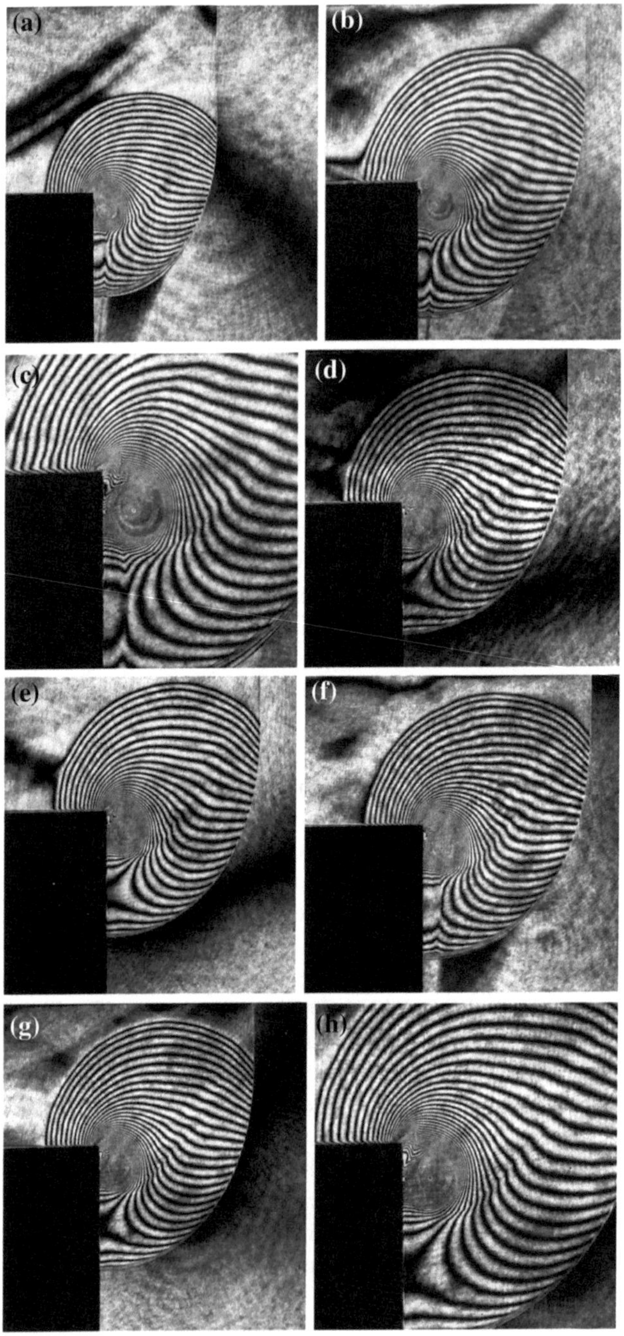

Fig. 3.1 Shock wave diffraction at 90° corner for Ms = 1.45 in air at 800 hPa, 283.5 K, *OB* path length is 150 mm: **a** #91071204, Ms = 1.460. This was submitted to the poster session; **b** #91071205, Ms = 1.416; **c** enlargement of (**b**); **d** #91071504, Ms = 1.460; **e** #91071503, Ms = 1.455; **f** #91071209, Ms = 1.454; **g** #91071501, Ms = 1.438; **h** enlargement of (**g**)

Fig. 3.1b, c. The edges of the shock wave terminated merging with shear layers associated with the vortex flow.

The local supersonic flow developed along the outer boundary of the vortex. Then between the supersonic flow and the outer boundary of the vortex, a shear layer developed. It became a marker demonstrated the sensitivity of the numerical codes to the present poster session. In the series of interferograms shown in Fig. 3.1, the Reynolds numbers defined in the shock tube experiments were ranging from 5×10^4 to 5×10^5. Structures of shear layers presented poster session papers (Takayama and Inoue 1991) were different depending on the Reynolds number adopted in the individual simulations. In the present experiment, the shear layer developed smoothly with elapsing time and no sign of instability was observed along the shear layer.

Figure 3.2a–m show the evolution of a shock wave diffraction over a 90° backward facing step for Ms = 1.40 in atmospheric air at 283.5 K. The experiments were conducted in the diaphragm-less 60 mm × 150 mm shock tube. The time allocated to each figure shows the sequential time from the second exposure. The port at which the pressure transducer was installed for triggering the light source was positioned at 200 mm upstream of the test section. Therefore, the time allocated to individual images was needed in order to edit these images in animated display (Babinsky et al. 1995). At the corner, the IS is diffracting and the vortex is growing with elapsing time. The expansion wave propagates reversely as shown in Fig. 3.2j at speed of u–a, where a and u are the sound speed in the shocked air and the particle velocity. The flow behind the IS is subsonic, $u < a$.

The transmitted shock wave intersects with the incident shock wave smoothly at earlier stages as seen in Fig. 3.2a–d. Later, the slip line generated by the vortex merges with the intersection point of the two shock waves, as seen in Fig. 3.2j. Due to such a wave diffraction pattern, the transmitted shock wave never evolves into a simple spherical shape but will take temporally a slightly non-spherical shape and eventually approach a spherical shape.

However, in the case of a very weak incident shock wave, the diffracted shock wave directly transforms into a spherical shock wave as seen in Fig. 3.3. Upon leaving a shock tube opening, a shock wave will take a transitional shape for a short time interval and eventually evolve into a spherical shock wave.

The shape of such a transitional shock wave depends significantly on the shape of the shock tube opening. Therefore, in the case of complex opening, for example, triangular or star shaped openings, vortices so far released have three-dimensional structures so that it is not easy to visualize the sequence of three-dimensional transition to the final spherical shock waves. Noises induced by high-speed jets are also one of the topics related to the sequence of transition to spherical shock wave or shock wave diffraction over a backward facing step. The earlier transition of complexly shaped shock waves to final spherical shock wave is sequentially visualized.

Figure 3.3 show a diffraction of very weak shock wave for Ms = 1.04 in the atmospheric air at 294.1 K. The experiment was carried out in the 150 mm × 60 mm diaphragm-less shock tube and the *OB* path was elongated. Only very small vortex

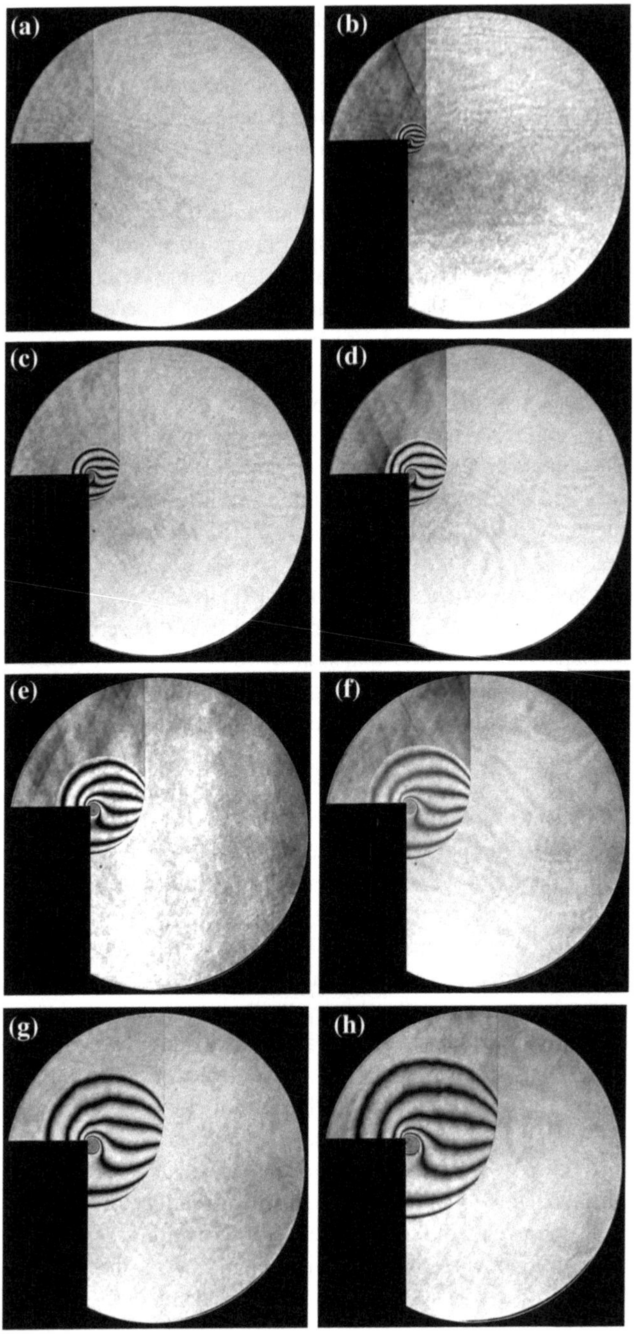

Fig. 3.2 Shock wave diffraction at 90° corner, Ms = 1.416, in air, at atmospheric pressure, 283.5 K: **a** #94032805, 11 μs; **b** #94032912, 49 μs; **c** #94032919, 79 μs; **d** #94032928, 97 μs; **e** #94040503,133 μs; **f** #94040507,163 μs; **g** #94040514, 205 μs; **h** #94040605, 241 μs; **i** #94040610, 265 μs; **j** #94040701, 295 μs; **k** #94042395, 325 μs; **l** #94042492, 385 μs; **m** #94040405, 445 μs; **n** #94040407, 515 μs (Babinski et al. 1995)

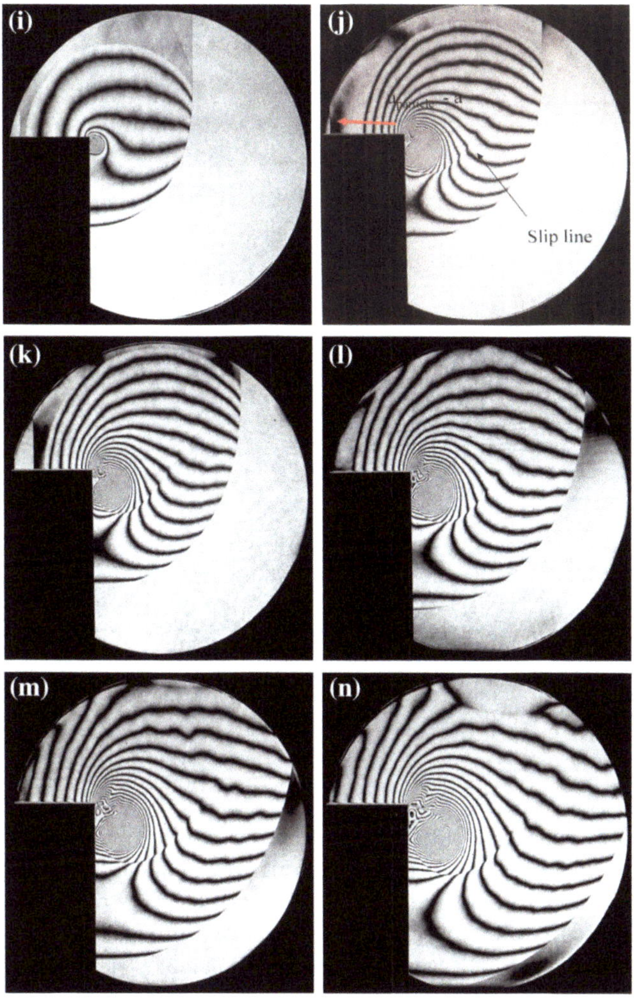

Fig. 3.2 (continued)

was formed at the corner, as the particle velocity was about 22 m/s. Figure 3.4 show the diffraction of very weak shock waves for Ms = 1.02 in the 60 mm × 150 mm shock tube but the *OB* path length was only 60 mm. Although the black fringes are broadened and only one fringe shift was observed, a sign of corner vortex was observable in Fig. 3.4b. The time interval of the double exposures was 200 μs so that two images of shock waves were superimposed in one interferogram.

Figure 3.5 shows shock wave diffractions observed in a 150 mm × 60 mm diaphragm-less shock tube for Ms = 1.50. The distance allocated to individual images is the position of the transmitted shock wave from the corner. The evolution of the vortex and the expansion fan are observed. The grey zones seen in Fig. 3.5a–c

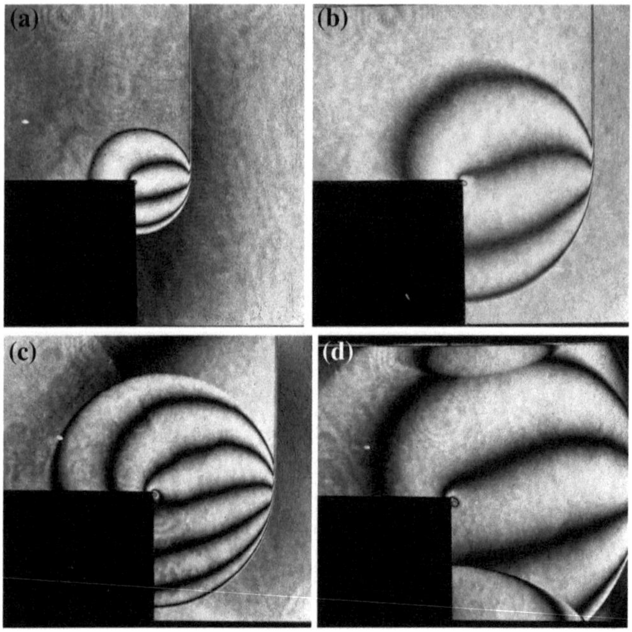

Fig. 3.3 Shock wave propagation at 90° corner of 150 mm × 60 mm shock tube for Ms = 1.04 in the atmospheric air at 294.1 K: **a** #91100904, 270 μs from trigger point, Ms = 1.032; **b** #91100903, 310 μs, Ms = 1.031; **c** #91101101, 310 μs, Ms = 1.047; **d** #91101103, 350 μs, Ms = 1.049

are the area in which the collimated light beams concentrate so densely that the fringes are no longer resolved. With finer fringe distributions, the sequential vortex growth and its interaction with the expansion fan are clearly observed. It would be an interesting work to reproduce the fine structures of fringe distributions seen in Fig. 3.5 by applying an appropriate numerical scheme.

Figure 3.6 show the evolution of a diffracting shock wave over a 90° corner for Ms = 2.20 in air at 250 hPa, 294.0 K. The experiment was conducted in the 60 mm × 150 mm diaphragm-less shock tube. The local flow Mach number behind the incident shock wave is supersonic so that u > a and then the fringes diverging from the corner are remnants of the Prandtl Meyer expansion fan in a steady supersonic flow over a corner.

3.2 Shock Wave Released from Openings

Shock waves released from openings of a circular cross sectional shock tube and a rectangular cross sectional shock tube are presented.

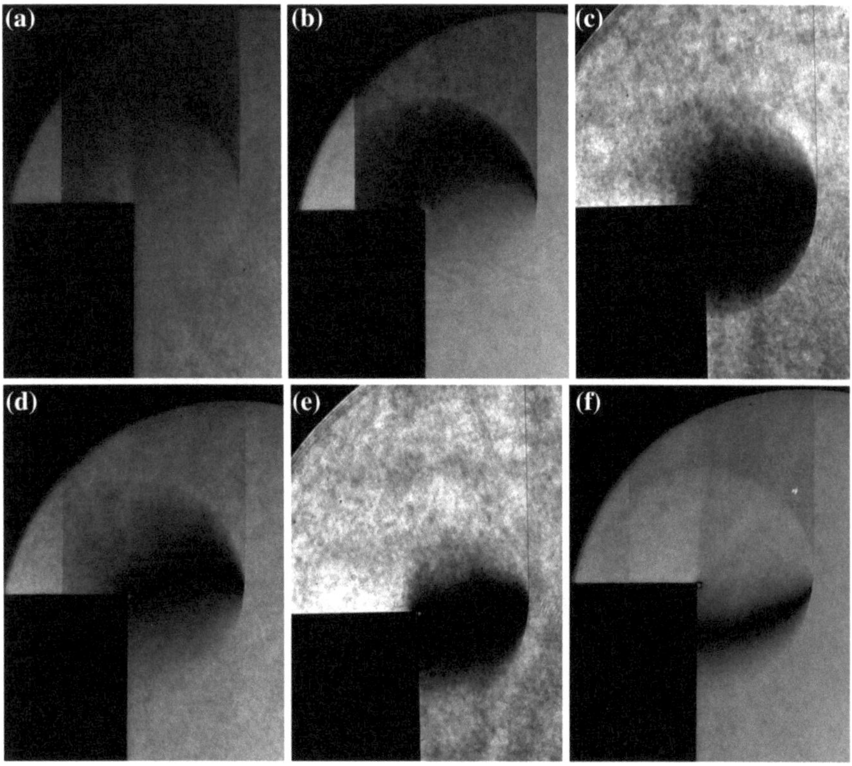

Fig. 3.4 Diffraction of a very weak shock wave for Ms = 1.015–1.036 in the atmospheric air at 293.0 K at about 100 μs from the corner: **a** #94101101, Ms = 1.015; **b** #94101102, Ms = 1.020; **c** #94102502, Ms = 1.020; **d** #94101103, Ms = 1.024; **e** #94102509, Ms =1.030; **f** #94101105, Ms = 1.036

3.2.1 Circular Opening

Figure 3.7 show direct shadowgraphs of the transformation of a planar shock wave to a spherical one for Ms = 1.25 in air at 500 hPa. The shock waves generated in a 25 mm diameter conventional shock tube are released into a 250 mm diameter and 250 mm wide cylindrical chamber. Figure 3.7a, b show shock waves diffracting from the opening. Then its resulting shape consists of a planar shock wave intersecting continuously with a curved shock wave. In Fig. 3.7c, a vortex loop was released from the opening. Figure 3.7d was its enlargement and the boundary layer released from the opening. The perturbation in the boundary layer released from the opening rolled up forming a vortex.

The planar shock wave intersects smoothly and continuously with the diffracting curved shock wave. When the planar part of the intersecting shock wave converged

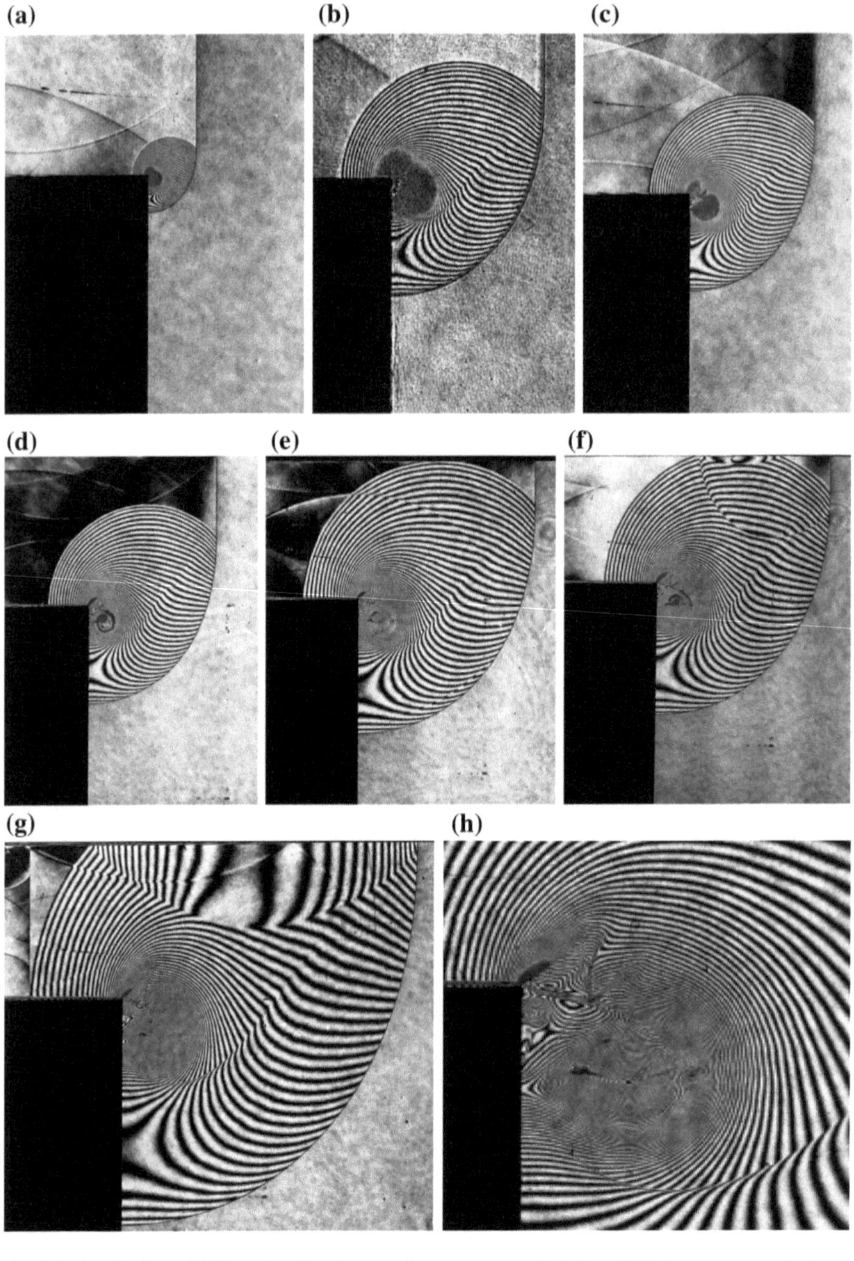

Fig. 3.5 Evolution of diffracting shock wave from a corner for Ms = 1.50 in the atmospheric air at 288.5 K: **a** #94020303, Ms = 1.503, the position of transmitted shock wave at 10 mm from the corner; **b** enlargement of (**a**); **c** #94020301, Ms = 1.502, 26 mm; **d** #94020304, Ms = 1.503, 51.5 mm; **e** #94020203, Ms = 1.486, 78 mm; **f** #94020202, Ms = 1.486, 82 mm; **g** #94020204, Ms = 1.503, 115 mm; **h** enlargement of (**g**)

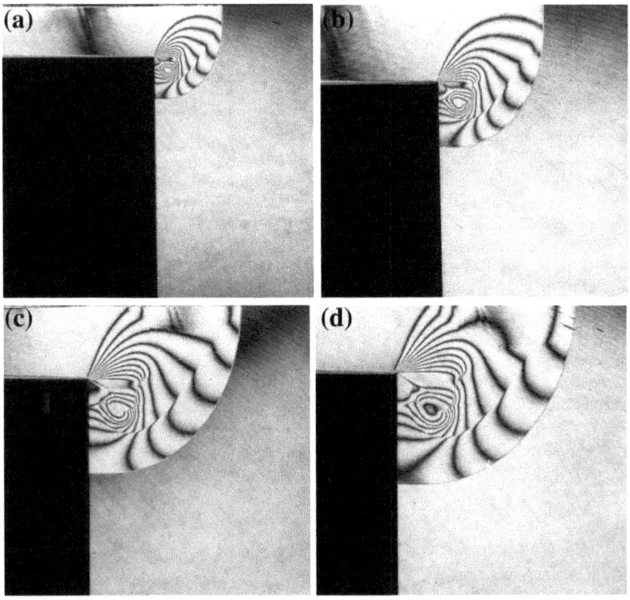

Fig. 3.6 Evolution of diffracting shock wave over a 90° corner for Ms = 2.20 in air at 250 hPa, 294.0 K: **a** #90101306, 210 μs, Ms = 2.172; **b** #90101309, 260 μs, Ms = 2.184; **c** #90101305, 220 μs, Ms = 2.200; **d** #90101307, 240 μs, Ms = 2.234

at the center, the shape of intersecting shock wave not necessarily take simply a spherical shape as seen in Fig. 3.7c. The radius of curvature at its center of convergence became a minimum for a short period of time. In short, the resulting shape looked like a peach. Such a singular shape was observable in Fig. 3.7e, f.

This trend is observable not only during the shock wave diffraction from the circular cross sectional shock tube but also during the shock wave diffraction from the opening of rectangular cross sectional shock tube. Abe (1989) reproduced numerically the transformation of diffracting shock wave shape (Abe 1989).

Figure 3.8 show the evolution of a shock wave released from a 25 mm diameter conventional shock tube for Ms = 1.55 in atmospheric air at 291 K. The transformation of a diffracting shock wave from a planar shape to the spherical shape was sequentially visualized. In Fig. 3.8a–g, the intersection of a planar shock wave with a diffracting curved shock wave eventually formed almost a spherical shock wave. When the two intersecting shock waves converged at the center, the resulting shape showed a peach shape. In Fig. 3.8h–j, such a singular shape was maintained only for a short period of time. The shape of a vortex loop released from the opening formed like a horizontally stretched symbol "∞". In Fig. 3.8k, l, when the vortex loop shape became a thin horizontal loop, the transmitted shock wave became a complete spherical shape.

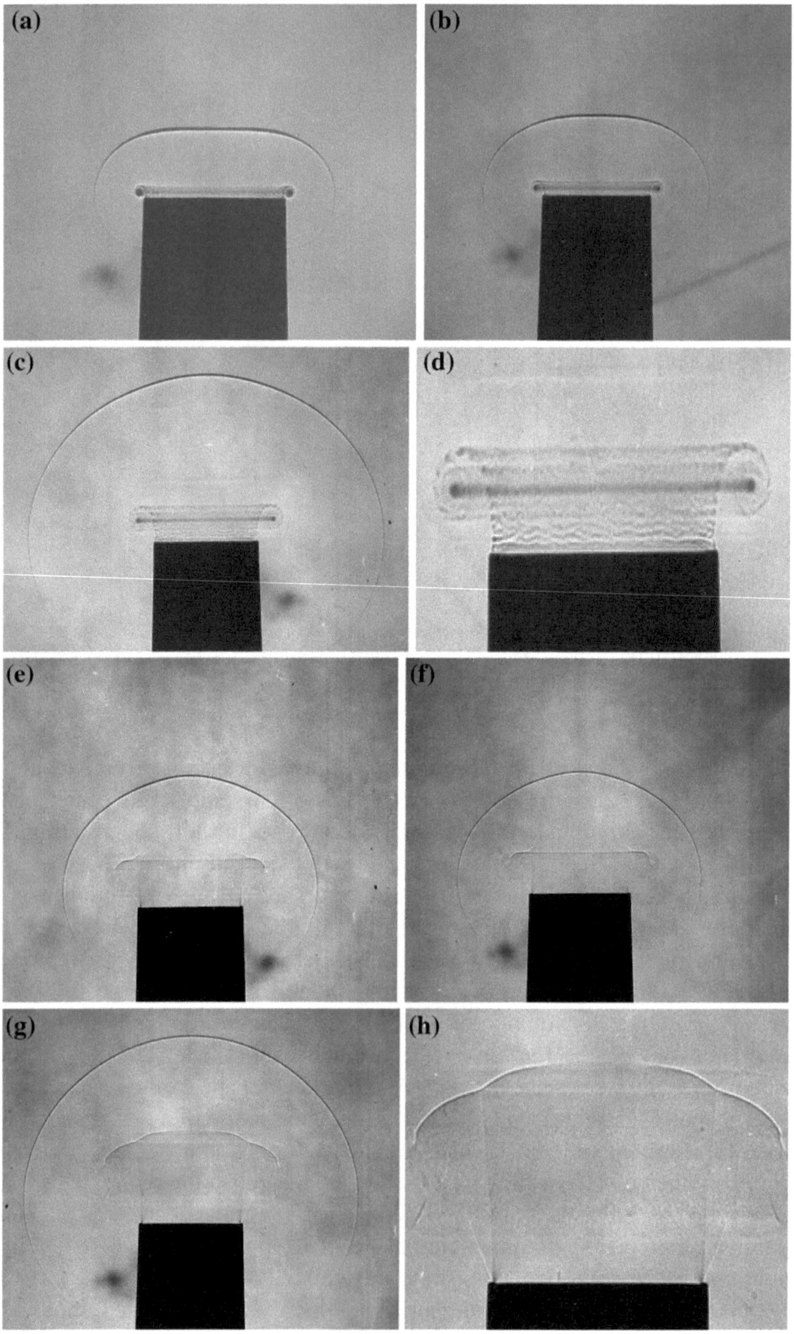

Fig. 3.7 Evolution of a shock wave release from a 25 mm diameter tube (**a–d**) for Ms = 1.250 in air at 500 hPa, (**e–h**) for Ms = 1.895 in air at 300 hPa, direct shadowgraph: **a** #77021401; **b** #77021403; **c** #77021405; **d** enlargement of **c**; **e** #77021501; **f** #77021612; **g** #77021505; **h** enlargement of (**g**)

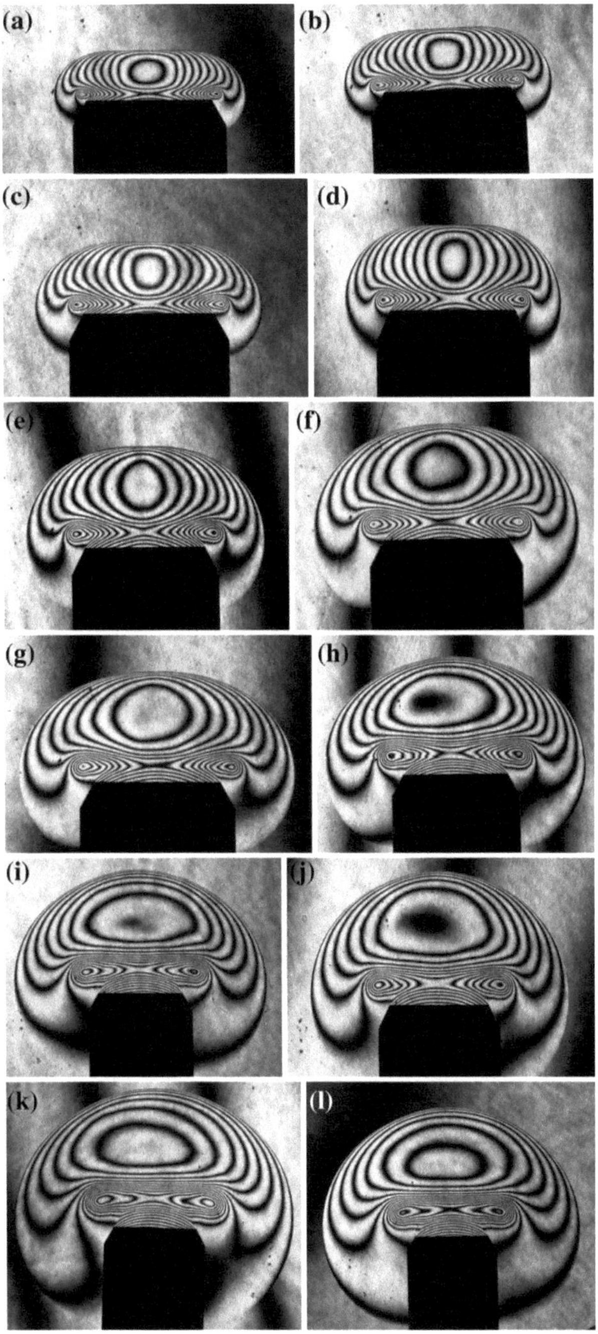

Fig. 3.8 Evolution of shock wave diffracting from a 25 mm diameter shock tube for Ms = 1.55 in atmospheric air at 291 K: **a** #89050804, 20 μs from exit, Ms = 1.570; **b** #89050805, 25 μs, Ms = 1.566; **c** #89050807, 35 μs, Ms = 1.578; **d** #89050810, 40 μs, Ms = 1.562; **e** #89050808, 45 μs, Ms = 1.564; **f** #89050811, 50 μs, Ms = 1.576; **g** #89050813, 55 μs, Ms = 1.555; **h** #89050814, 60 μs, Ms = 1.535; **i** #89050816, 70 μs, Ms = 1.579; **j** #89050819, 80 μs, Ms = 1.529; **k** #89050821, 90 μs, Ms = 1.557; **l** #89050823, 100 μs, Ms = 1.536

3.2.2 Two-Dimensional Opening

Figure 3.9 show the evolution of a shock wave diffracting from a 20 mm × 60 mm and 50 mm long two-dimensional duct installed in the test section of the 60 mm 150 mm conventional shock tube for Ms = 1.4 in atmospheric air at 296.8 K. The time allocated to each figure caption indicates the time from the trigger point to the time instant at which the second exposure was conducted. Figure 3.9a shows the experimental setup: the shock wave diffraction at a 90° corner. The transition took a similar procedure as seen in Figs. 3.7 and 3.8.

Figure 3.9a, b show the diffraction of a planar shock wave at the opening. Figure 3.9c–r show the transition of a distorted cylindrical shape to completely cylindrical shock waves. Then in Fig. 3.9k–m, the cylindrical shock wave was reflected from the side walls.

Figure 3.10 show the later stage of shock wave discharged from a 25 mm diameter shock tube for Ms = 1.39 and 1.55 in atmospheric air. A close shape of a contact surface drove a diverging spherical shock wave. Behind the spherical shock wave, a relatively uniform flow region and the core flow region are clearly distinguished. The fringe pattern forms an image resembling the face of the movie character ET.

3.2.3 Interaction of a Diffracting Shock Wave with Droplets in a Line

Figure 3.11 show the interaction of 0.75 mm diameter water droplets with a spherical shock wave diffracting from a 25 mm diameter shock tube for Ms = 1.50 in atmospheric air. The shock wave was released into a 250 mm diameter and 250 mm wide test chamber and the droplets were falling in a line at 70 mm stand-off distance from the shock tube. The first exposure was carried out before the introduction of droplets into shock tube and the second exposure was synchronized with the droplets introduction and their interaction with the IS. The time allocated to individual interferograms the elapsed time from the trigger point. The droplets just impinged by the diffracting shock wave were not deformed nor much displaced. At this stage, these droplets were almost unaffected by the diffracting shock wave.

In Fig. 3.11, the interferograms and their unreconstructed holograms are compared to observe the procedure of droplet shattering. Although the image resolution of the unreconstructed holograms is poorer than single exposure holograms, the image of the unreconstructed hologram works well for interpreting the procedure of the droplet shattering.

In Fig. 3.11a, b, the droplets in a line still keep their initial distribution. In Fig. 3.10 c, d, the droplets located in the central part of the line are deformed, whereas those outside the central part of the line remain undisturbed. The deformed droplets are shattered by the exposure to the shear flows and jet flow inside the central core. Figure 3.11e, f show a single exposure interferogram and its enlargement,

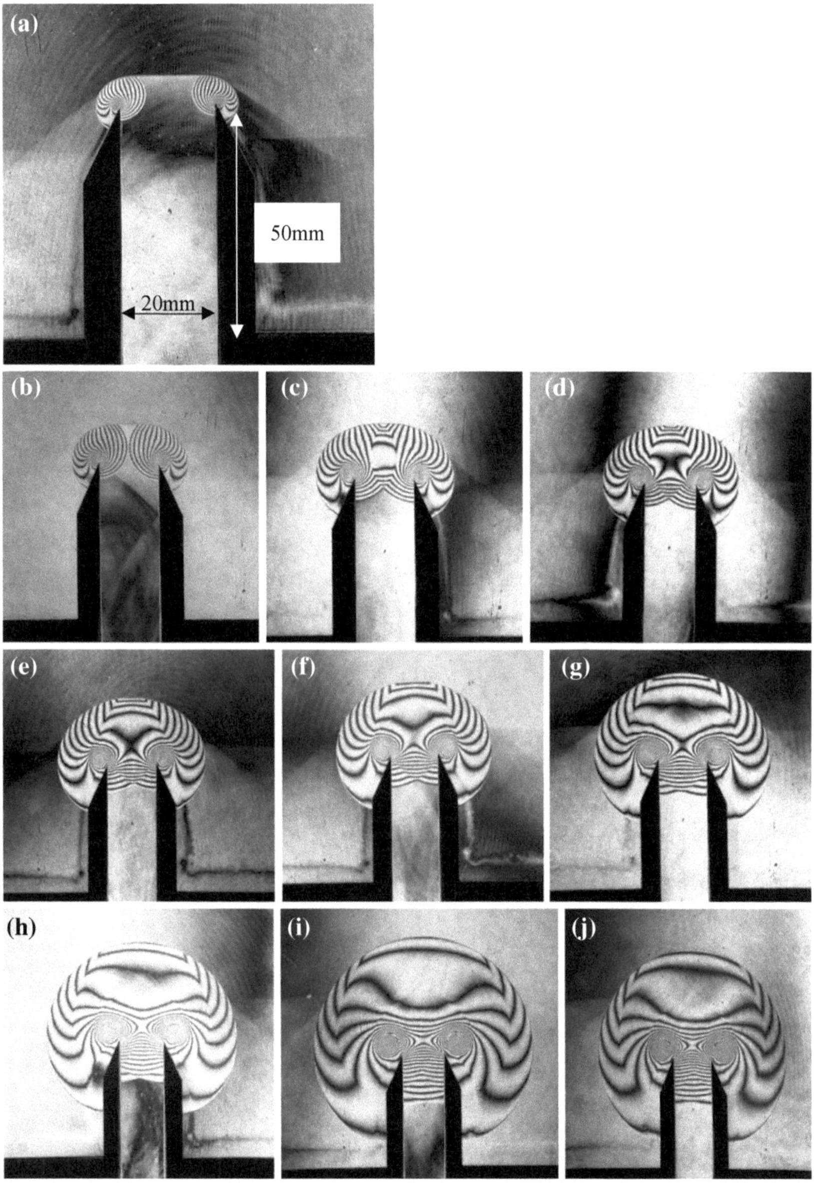

Fig. 3.9 Evolution of shock waves diffracted from a two-dimensional duct for Ms = 1.4 in atmospheric air at 296.8 K: **a** #90100402, 50 μs from trigger point, Ms = 1.414; **b** #90100403, 60 μs from trigger point, Ms = 1.414; **c** #90100404, 70 μs, Ms = 1.414; **d** #90100501, 80 μs, Ms = 1.409; **e** #90100502, 90 μs, Ms = 1.409; **f** #90100503, 100 μs, Ms = 1.406; **g** #90100504, 110 μs, Ms = 1.407; **h** #90100506, 130 μs, Ms = 1.406; **i** #90100508, 150 μs, Ms = 1.404; **j** #90100509, 160 μs, Ms = 1.391; **k** #90100511, 180 μs, Ms = 1.405; **l** #90100803, 230 μs, Ms = 1.410; **m** #90100805, 250 μs, Ms = 1.413; **n** #90100807, 270 μs, Ms = 1.417; **o** #90100808, 300 μs, Ms = 1.418; **p** #90100809, 330 μs, Ms = 1.417; **q** #90100810, 360 μs, Ms = 1.402; **r** #90100813, 420 μs, Ms = 1.406

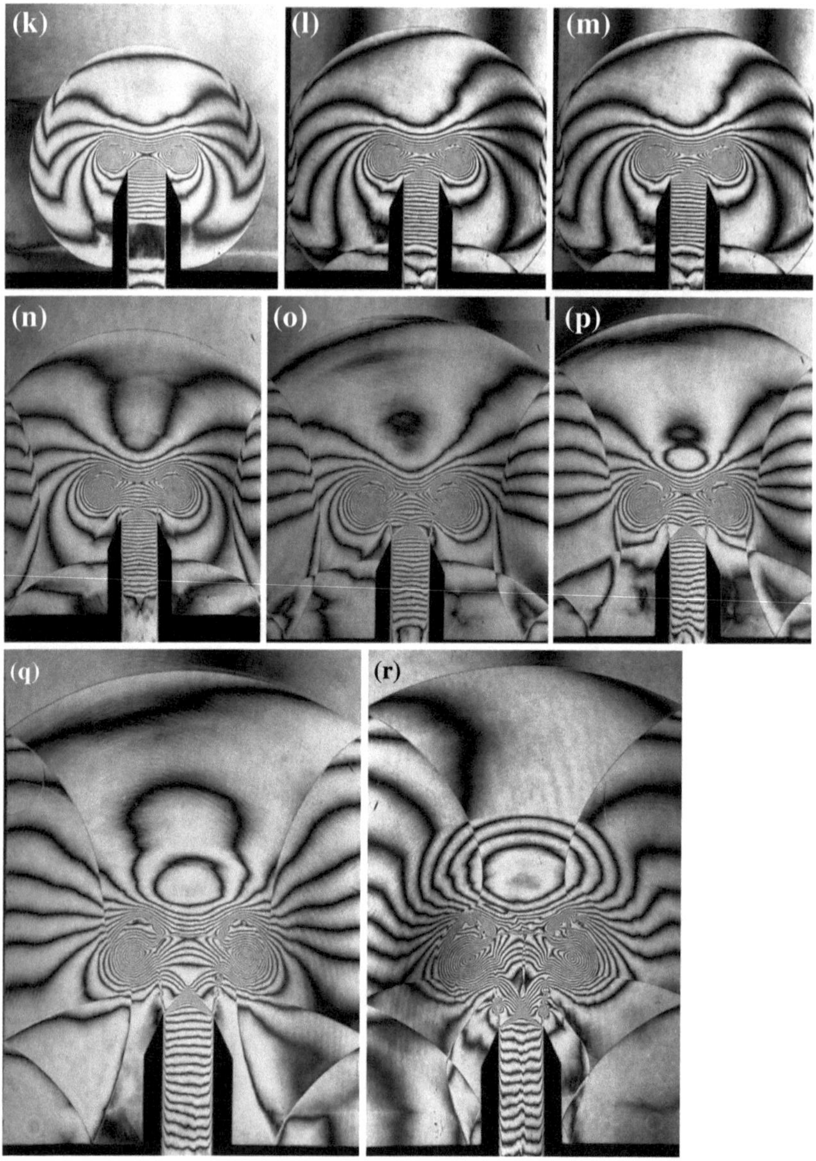

Fig. 3.9 (continued)

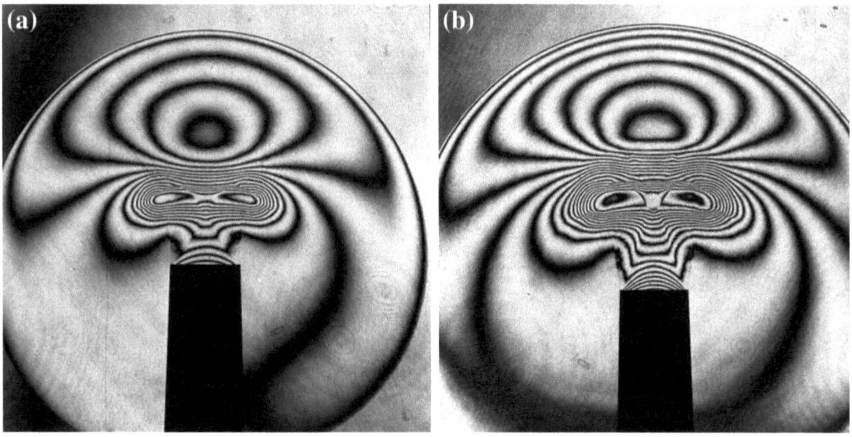

Fig. 3.10 Interferometric observation of shock waves discharged from a 25 mm o.d. tube at the later stage in air 1013 hPa, 289.9 K: **a** #83031003, Ms = 1.387 at about 200 μs from the trigger point; **b** #83031007, Ms = 1.548, at 195 μs from the trigger point

respectively. The particle flow is uniform and its Mach number is M = 0.56 behind the IS of Ms = 1.56. The diffracting flow has a central core which is continuously accelerated and the flow outside the central core flow remains subsonic is accelerated and becomes supersonic. The droplets located in the central core are deformed non-uniformly as seen in Fig. 3.11c, d. Figure 3.11e is a single exposure interferogram. Figure 3.11f is its enlargement. The flow in the core region is accelerated and becomes supersonic. It should be noticed, however, that the bow shock waves appear ahead of the droplets located in the central core. The droplets located outside the central core are in the subsonic flow and stay unchanged. Figure 3.11g, h are taken nearly the same time delay. However, in unreconstructed hologram in Fig. 3.11h, the bow shock is not observed. Then the difference in flow structures affected distinctly droplet deformations. In Fig. 3.11i, j, the droplets located in the central core flow shattered significantly shattered with elapsing time. It should be noticed, however, that the droplets located outside the core region are deformedly modestly.

In Fig. 3.11m, n, the Mach number of the IS is 1.78 and the particle flow Mach number M is 0.71. In Fig. 3.11a–l this hologram was taken at relatively earlier time and the particle flows are relatively slow but in Fig. 3.11m the hologram was taken later time. Figure 3.11n is an unreconstructed hologram and shows remarkable deformation of the droplets. The droplets showed catastrophic disintegration in the central core flow region and stretched in the transient region between the central core flow and the region behind the transmitted shock wave. The shadows of shattering droplets are clearly observed in unreconstructed holograms.

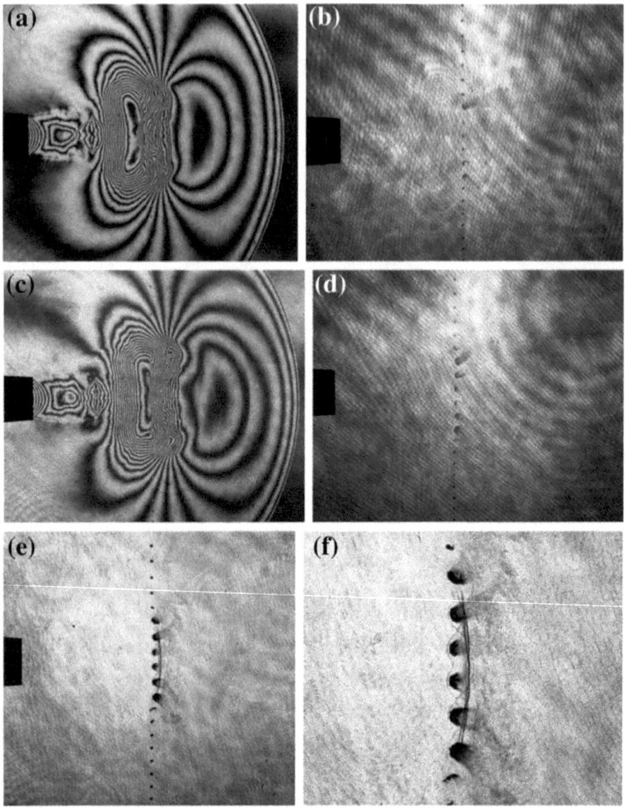

Fig. 3.11 Droplets in a line interacting with a shock wave diffracting from a 25 mm diameter shock tube for Ms = 1.50 in atmospheric air at 289.0 K, comparison with unreconstructed hologram. Water droplets were vertically falling from the upper wall by an ultrasonic oscillation: **a** #83031608, 1.95 ms, Ms = 1.512; **b** unreconstructed hologram of (**a**); **c** #83031604, 1.96 ms, Ms = 1.516; **d** unreconstructed hologram of (**c**); **e** #83031511, 2.0 ms. Ms = 1.513, single exposure; **f** enlargement of (**e**); **g** #83031509, 2.0 ms, Ms = 1.529; **h** unreconstructed hologram of (**g**); **i** #83031506. 2.1 ms Ms = 1.509; **j** unreconstructed hologram of (**i**); **k** #83031607, 1.95 ms, Ms = 1.586#; **l** unreconstructed hologram of (**k**); **m** #83031602 1.97 ms Ms = 1.783; **n** unreconstructed hologram of (**m**)

3.2.4 Interaction with a Helium Plume

Figure 3.12 show the evolution of a spherical shock wave exposed to a helium plume (sound speed a_{helium} = 1060 m/s). A shock wave was released from a 25 mm diameter shock tube opening into the 250 mm diameter and 250 mm wide chamber. A spherical shock wave interacted with helium plume. The test condition was for Ms = 2.2 (u_s = 760 m/s) in air at 500 hPa, 290 K. The helium plumes were diffusively released from a slightly pressurized helium plenum chamber through a piece of filter paper contained in an orifice placed on the upper side the chamber.

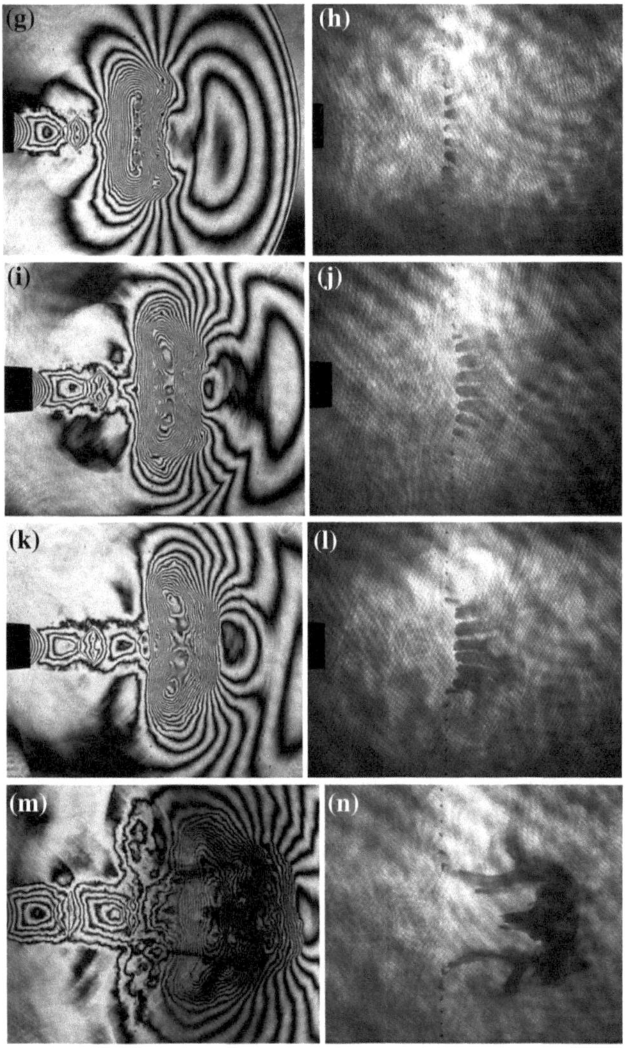

Fig. 3.11 (continued)

The helium plume was released with an appropriate synchronization with the shock wave motion. Figure 3.12 show sequences of shock waves interacting with the helium plumes using direct shadowgraphs. When the spherical shock wave moved into the area of dense helium concentration, the shock wave disappeared in the region at which the shock speed became slower than that in air-helium mixture.

Figure 3.13 show the spherical shock wave for Ms = 3.45 (u_s = 1190/s) interacting with the helium plume. In Fig. 3.13b, the spherical shock wave is visible inside the helium plume as $u_s > a_{helium}$. As the spherical shock wave quickly attenuates, it vanishes as observed in Fig. 3.13e, f.

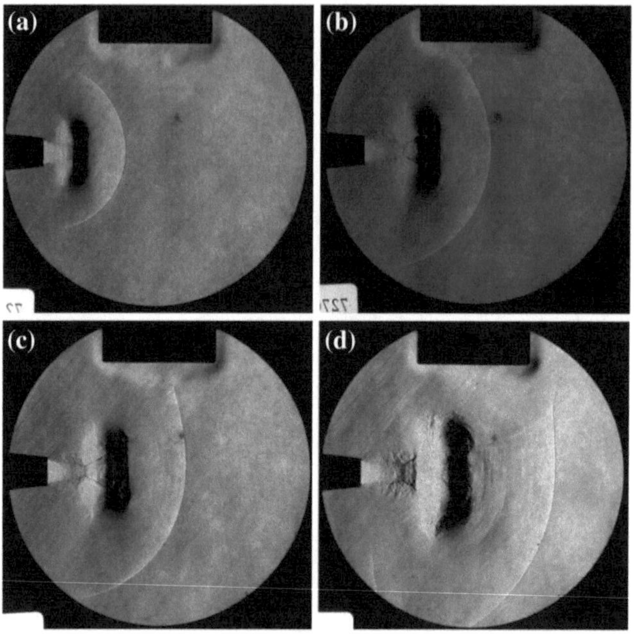

Fig. 3.12 Spherical shock wave interaction with a helium plume for Ms = 2.20 in air at 500 hPa, 290 K: **a** #76072705; **b** #76072706; **c** #76072709; **d** #76072710

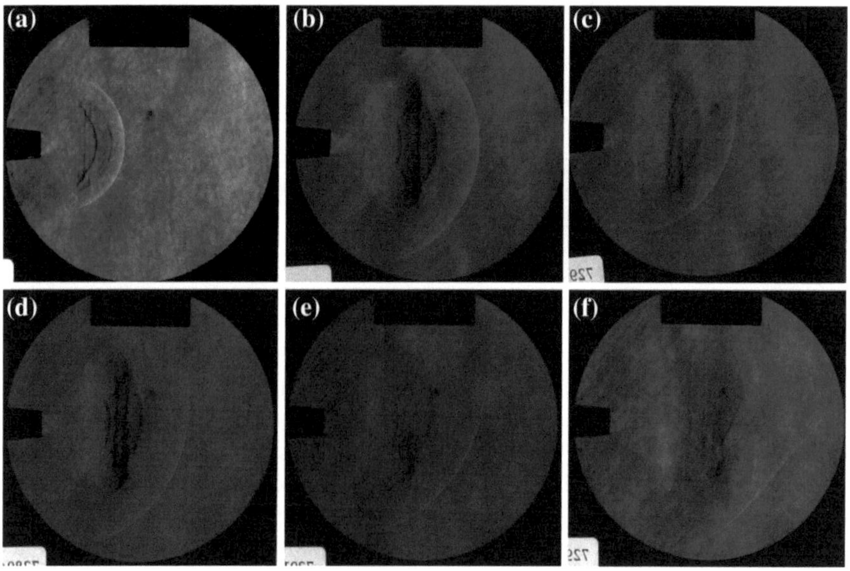

Fig. 3.13 Spherical shock wave interaction with a helium plume for Ms = 3.48 in air at 100 hPa: **a** #76072901, Ms = 3.40; **b** #76072920; **c** #76072912; **d** #76072804, Ms = 3.45; **e** #76072917, Ms = 3.48; **f** #76072804, Ms = 3.45

3.3 Square Opening

3.3.1 Diffuse Holographic Observation

When a shock wave was released from a 40 mm × 40 mm shock tube into an open test section consisting of a 250 mm diameter and a 250 mm wide cylindrical space, the shape of the diffracting shock wave had not a two-dimensional but three-dimensional shape. In order to visualize the three-dimensional diffracting shock wave, diffuse holographic interferometry was applied. The event in the test section was illuminated by a diffusely scattered *OB*. It was formed by transmitting a collimated *OB* through a thin smoked glass plate. The diffusely scattered *OB* carried the three-dimensional information of diffracting shock waves and the test section. The hologram was later reconstructed to visualize three-dimensional shock wave diffraction but it should be noticed that the spatial resolution of images was not as sharply recorded as that of image holograms. In Fig. 3.14, the reconstructed images of transmitted shock waves of Ms = 1.50 viewed from view angles β of about 20° and 30° were presented. If a hologram was reconstructed by recording the resulting images with a video camera from various view angles, the three-dimensional images of the shock wave and flow fields would be readily obtained. Figure 3.14a–f show sequential reconstructed images of shock waves of Ms = 1.50 from the square opening viewed from the corner. Fringe distributions look variable depending on

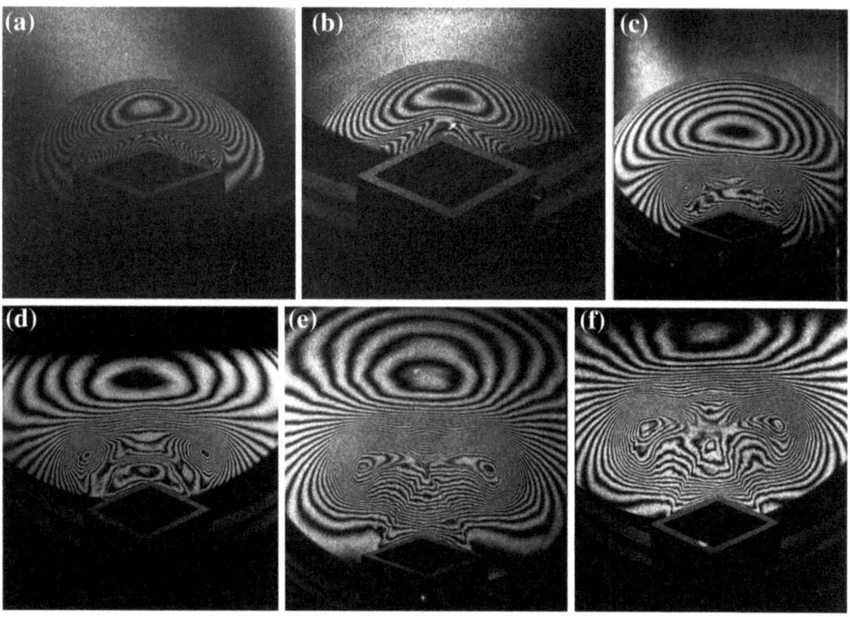

Fig. 3.14 Reconstruction of Ms = 1.50 shock waves released from a 40 mm × 40 mm square opening: **a** #90040106, reconstruction view angle β = 20.1°; **b** #90040103, β = 25.2°; **c** #90040105, β = 19.5°; **d** #90040101, β = 28.1°; **e** #90040109, β = 15.9°; **f** #90040108, β = 25.2°

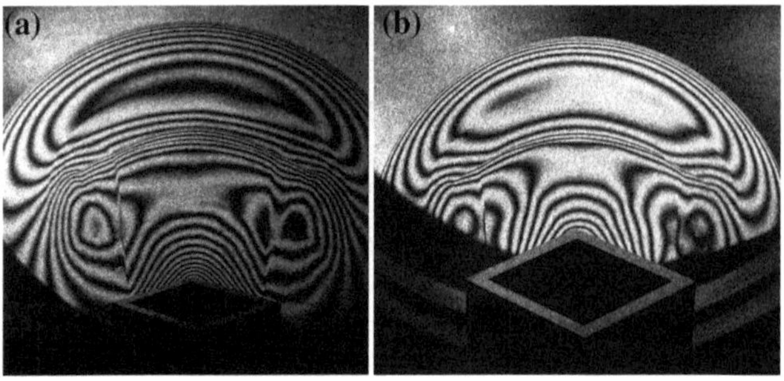

Fig. 3.15 Reconstruction of Ms = 2.95 shock waves released from a 40 mm × 40 mm square opening: **a** #90040116, reconstruction view angle β = 15.9°; **b** #90040114, reconstruction view angle β = 28.1°

the different view angles β. Figure 3.15a and b show reconstructed images of Ms = 2.95. Due to reduced test pressure, the fringe numbers are less but their distributions are similar to those in Fig. 3.14.

3.3.2 *Two-Dimensional Observation Square Opening*

Figure 3.16 show the evolution of Ms = 1.50 shock wave diffracting from a 40 mm × 40 mm shock tube opening. The transition of the planar shock wave to the spherical shock wave is viewed from the corner direction. The vortices created at the edge of the square opening look different if they were viewed from various directions. Figure 3.16e shows a spherical shock wave. Figure 3.16f shows an unreconstructed hologram taken at almost identical time instant of Fig. 3.16e.

Figure 3.17 show the evolution of shock waves for Ms = 1.50 diffracted from a 40 mm shock tube. The transition of the planar shock wave to the spherical shock wave observed from the side is sequentially visualized. Figure 3.17a shows the image of the direct shadowgraph. The image observed from the side is almost identical with the image in the two-dimensional observation.

Figure 3.18a, c, e show images observed from the side. Figure 3.18b, d, f show images observed from the corner direction. The visualizations were conducted for Ms = 1.50, 2.0 and 2.50 at almost identical time instants. The patterns of the vortices are very different from each others depending on the observation from the corner direction and the side.

3.4 Diffraction of Shock Waves from Opening

Holographic interferometry was useful in visualizing three-dimensional shock wave diffraction from a 40 mm × 40 mm square opening. Three-dimensional shock wave diffractions from similar openings are visualized from the axial direction by using diffuse holography.

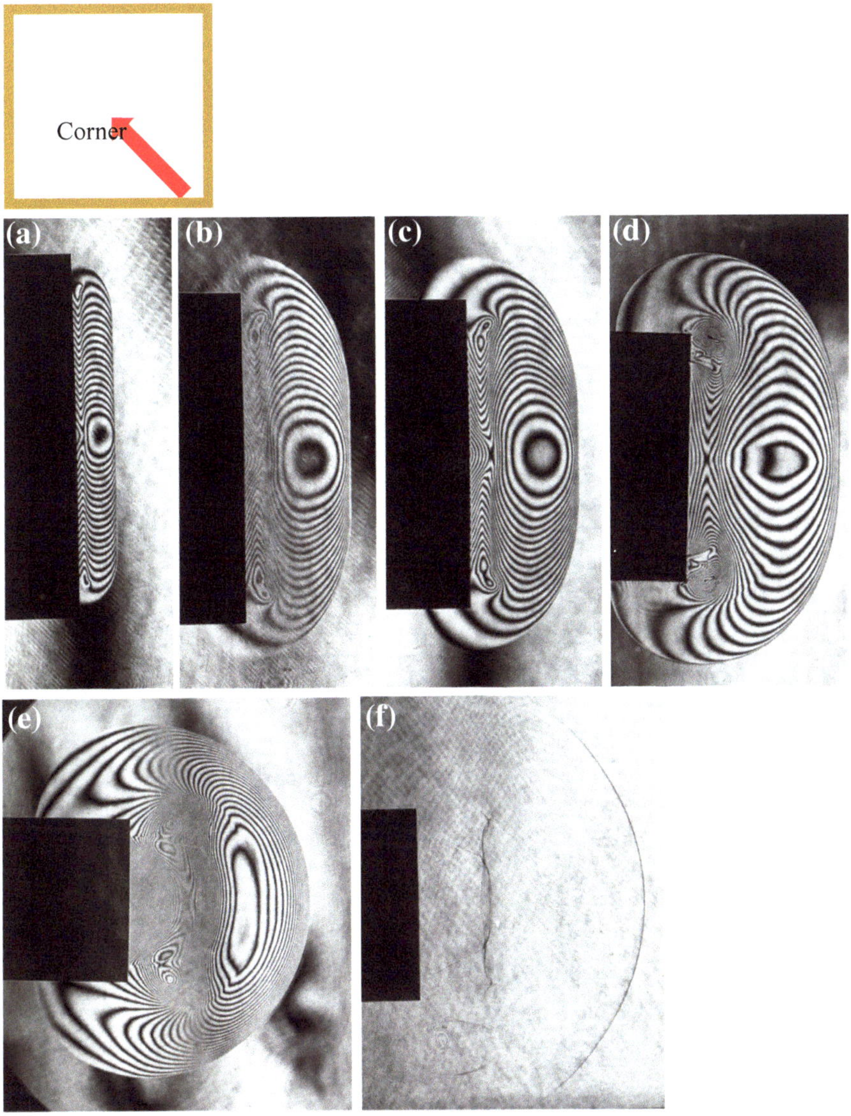

Fig. 3.16 Evolution of shock wave diffracted from an opening of a 40 mm × 40 mm shock tube, viewed from the corner direction for Ms = 1.50 in atmospheric air at 291 K: **a** #90040413; **b** #90040906; **c** #90040412; **d** #90040308; **e** #90041001; **f** #90040415, single exposure

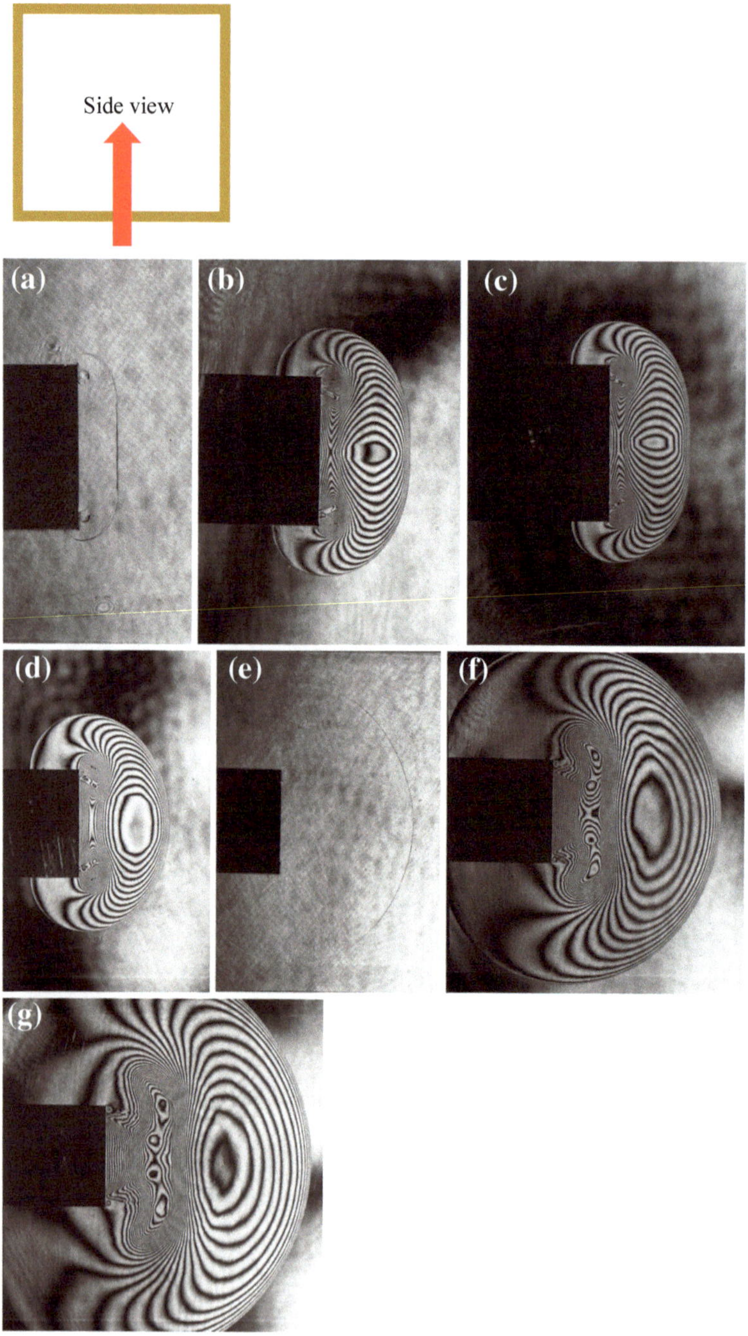

Fig. 3.17 Evolution of shock wave diffracting from a 40 mm × 40 mm shock tube, observation from the side for Ms = 1.50 in atmospheric air at 291 K: **a** #90040313, single exposure; **b** #9000307; **c** #9000308; **d** #9000309; **e** #9000315, single exposure; **f** #9000306; **g** #9000304

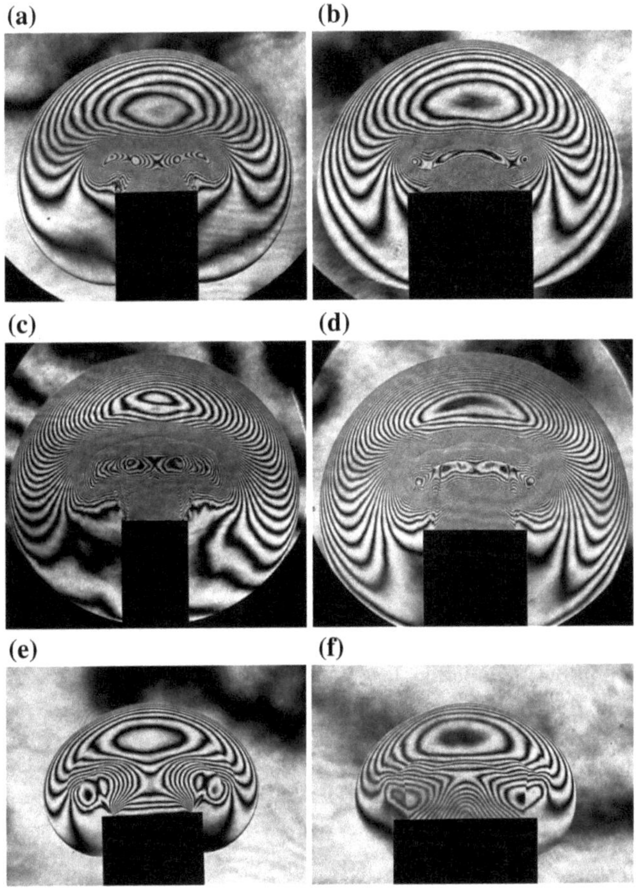

Fig. 3.18 Shock waves diffracted from a 40 mm × 40 mm opening, observations from the corner direction and the side: **a** #90041625, observation from side for Ms = 1.50 at 1013 hPa; **b** #90040911, observation from corner for Ms = 1.50 at 1013 hPa; **c** #90041620, observation from side for Ms = 2.0 at 1013 hPa; **d** #90041006, observation from corner for Ms = 2.0 at 1013 hPa; **e** #90041601, observation from side for Ms = 2.50 at 200 hPa; **f** #90041014, observation from corner for Ms = 2.50 at 200 hPa

3.4.1 Square Opening

Figure 3.19a shows a 40 mm × 40 mm shock tube and a flange. In order to visualize diffracting shock wave and a release of vortex loop from the opening, the shock tube opening and the flange were coated with the yellow fluorescent paint, which had a higher degree of reflectivity to the wavelength of 6943 nm of ruby laser beam. When a collimated object beam *OB* illuminated slightly obliquely the area coated with the fluorescent paint, the reflected *OB* would carry the holographic information of the diffracting shock wave and the resulting vortex loop and were

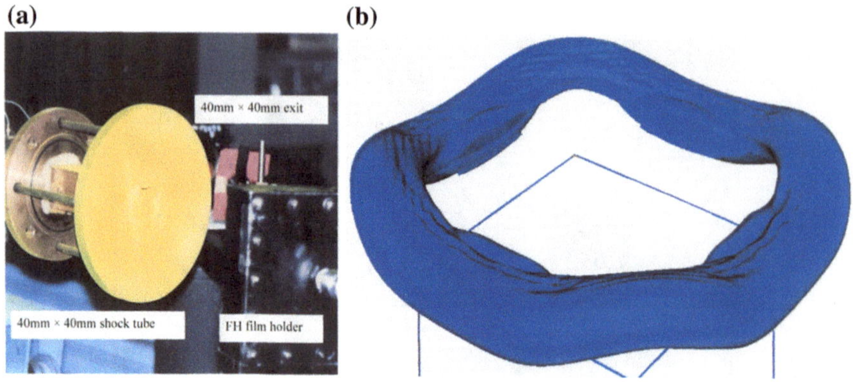

Fig. 3.19 Shock wave diffracted from a square opening: **a** experimental setup of observation with diffuse holographic observation for Ms −1.29; **b** Numerical simulation of a vortex loop for Ms = 1.29 in air (Onodera et al. 1997)

recorded on holographic films (Onodera et al. 1997). Figure 3.19b shows a numerical vortex loop obtained by solving numerically the Euler Equations for Ms = 1.29 in atmospheric air.

Figure 3.20 show the evolution of a transmitted shock wave and the formation of a vortex loop. Illuminating the test area with collimated *OB*, the reflected *OB* carried the holographic information contained in the diffracting shock wave and the vortex loop was recorded on a holographic film. The holographic film was reconstructed slightly oblique direction. Hence the shock waves and vortex loops were the imaged observed from slightly oblique direction. In Fig. 3.20a, the shape of the transmitted shock wave changes from a square shape to a circular shape. If the speed of vortex moving to the axial direction is defined as axial velocity, as observed in Fig. 3.19b, the axial velocity is different between the 90° corner and the side. The axial vortex velocity leaves from the corner faster than that from the side. In Fig. 3.20f, the transmitted shock wave already left far away from the opening and was out of the field of view. The vortex loop approaches to a ring shape.

3.4.2 *Triangular Opening*

Figure 3.21 show the combination of a square opening and a 60° triangular opening. It was already found that the motion of the transmitted shock wave and the growth of the vortex loop were affected by the corner angle. The transformation of the transmitted shock wave's shape and the growth of the vortex loop took a complex sequence. In Fig. 3.21a, at earlier stage, 20 μs from the moment at which the shock wave was released, no clear vortex loop was observed at the 30° corner. In Fig. 3.21b, a faint ring shape was just observed at the sharp corner. The findings imply that at the earlier stage, the vortex moved straight to the axial direction was

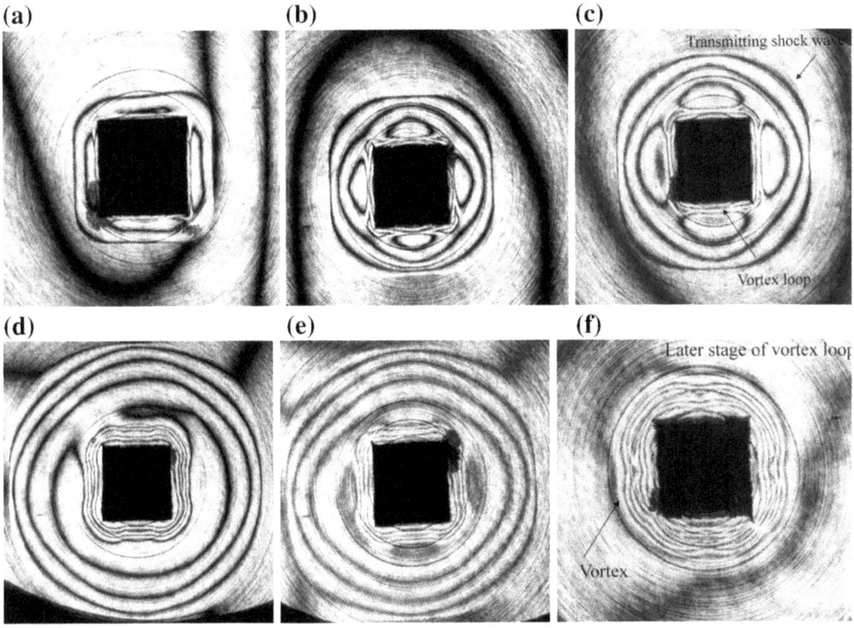

Fig. 3.20 Evolution of diffracting shock wave and vortex formation from a 40 mm × 40 mm square opening viewed from the end wall at Ms = 1.29 in atmospheric air at 298.0 K: **a** #96072411, 20 μs after the departure of the shock wave from the opening; **b** #96072308, 100 μs; **c** #96072307, 120 μs; **d** #96072306, 140 μs; **e** #96072304, 180 μs; **f** #96072401, 240 μs

the fastest but hardly spread at first toward the side direction. This trend was already observed during the shock wave diffraction at 90° corner seen in Fig. 3.19b.

In this triangular opening, the transmitted shock waves from two sharp corners interacted with each other. Figure 3.21c shows that the two shock waves' interaction produced a reflection pattern of a RR. In Fig. 3.21d, with elapsed time the reflection pattern became a SMR. At long time later, the shock waves from triangular openings and one from a square opening merged into one spherical shock wave as seen in Fig. 3.11f.

3.4.3 Semi-circular Opening

Figure 3.22 show an opening shape comprising the combination of a concave semicircle shape with a square shape. The edge of the semicircle opening touched the flange at zero angle. The shock wave diffracted from the semicircular opening generated a semicircular shaped transmitted shock wave. In Fig. 3.22a, the

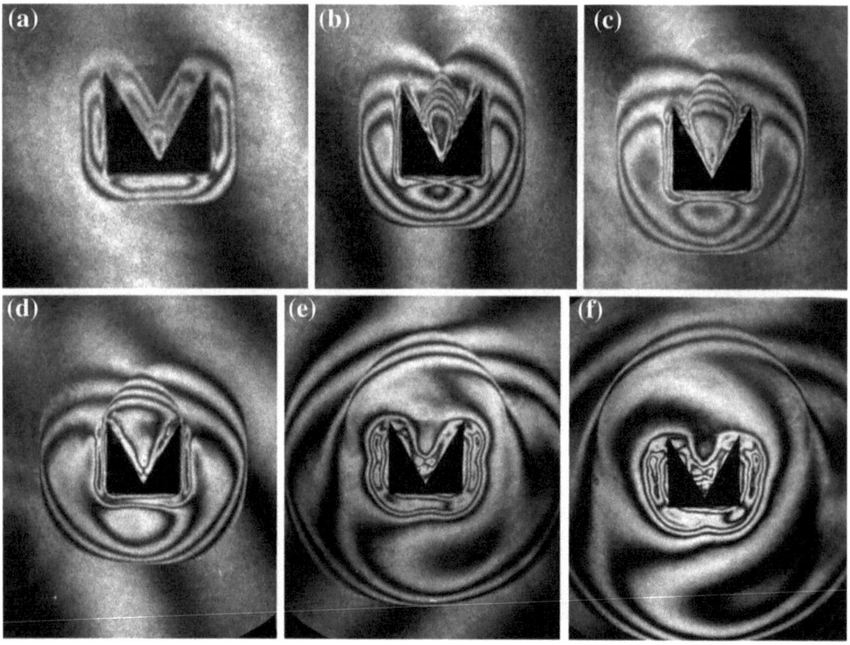

Fig. 3.21 Evolution of diffracting shock wave and vortex formation from a 40 mm × 40 mm square opening with a 60° insertion. Diffuse holographic observation from the end wall for Ms = 1.29 in atmospheric air at 298.0 K: **a** #970605, 20 μs from the release of the shock wave from the opening; **b** 40 μs; **c** 60 μs; **d** 80 μs; **e** #97052304, 100 μs; **f** #97060507, 120 μs

visualization was conducted at 20 μs from the shock wave transmission at the opening. From the two edges, the two circular transmitted shock wave were released and interacted with each other producing a SMR. However, any vortex loops were not observed. In Fig. 3.22b, the visualization was conducted at 40 μs from the starting of the shock wave diffraction. The vortex loops were observed along the semicircular edge and also the square edge. In Fig. 3.21e, the transmitting shock waves from the semicircular opening and also from the square opening merged. In the Chap. 5, the wave interaction pattern created from the focusing a weak shock wave from a shallow reflector in shock tube experiments was discussed. The pattern seen in Fig. 3.21e is similar to that of shock wave focusing from a shallow reflector. At the edges at which the semicircular opening touched with square opening at zero angle, the vortex loop hardly appeared throughout the observation period of time. This would indicate that the vortex loops did move straight forward probably at relatively high-speed. It would be speculated that shock waves diffract from a narrow opening whose contact angle approaches is almost zero, the resulting vortex will be released at very high-speed. If a shock wave is released from a lip shaped opening, the vortices will be released straight to the axial direction at very high-speed.

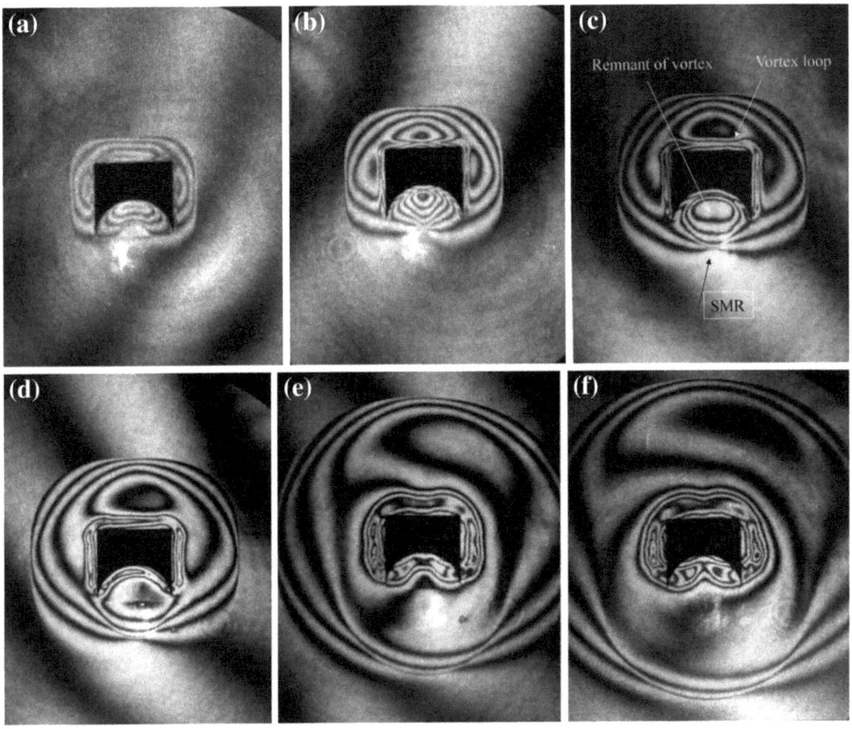

Fig. 3.22 Evolution of diffracting shock wave and vortex formation from a 40 mm × 40 mm square opening with a 60° insertion at Ms = 1.29. Diffuse holographic observation from the end wall at Ms = 1.29 in atmospheric air at 298.0 K: **a** #97052605, 20 μs after the release of the shock wave from the opening; **b** #97052603, 40 μs; **c** #97052601, 60 μs; **d** #97052607, 80 μs; **e** #97052602, 100 μs; **f** #97052609, 120 μs

3.5 Evolution of a Vortex Loop

Holographic interferometric visualization was applied intensively to observe the evolution of vortex loops released from a 50 mm inner diameter ×90 mm outer diameter shock tube made of aluminum alloy. Figure 3.23a shows the shock tube, which was positioned co-axially between field of view of two schlieren mirrors having diameter 1000 mm and focal length of 8 m. In order to specially visualize shock wave motions released from the shock tube's opening, it was supported by thin pillars and was movable to any position between the two mirroes (Kainuma 2004).

The high-pressure chamber and the low pressure channel were 2.2 and 4.5 m long, respectively. A double diaphragm section was placed between these sections. Figure 3.23b showed a sketch of the shock tube. Mylar films of various thicknesses were inserted between two sections as marked in Fig. 3.23b. In the intermediate driver section, the filled gas pressure was about one half of the driver gas pressure.

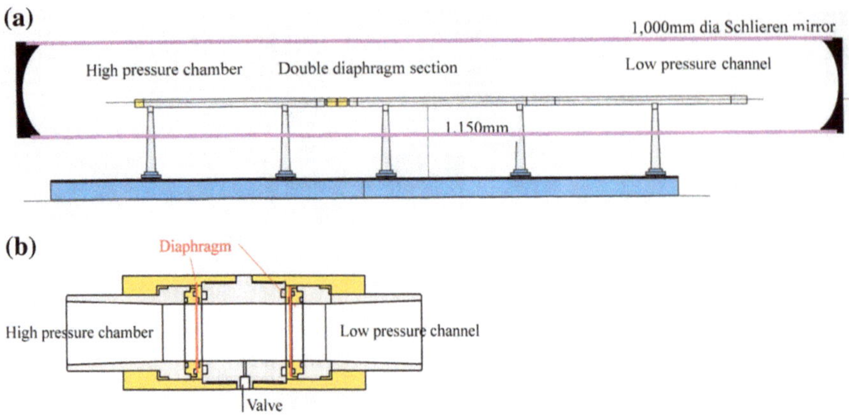

Fig. 3.23 A shock tube specially designed to study vortex motion: **a** Shock tube; **b** structure of double diaphragm section (Kainuma 2004)

Upon the sudden release of the pressure in the intermediate driver section, the diaphragms ruptured abruptly producing shock waves of a higher degree of reproducibility.

Figure 3.24a, b show sequential observations of the shock wave diffraction over a 25 mm diameter orifice for Ms = 1.24 in atmospheric air. The vortex loop has a uniform ring shape and propagates at a constant speed. The transmitted shock wave evolves, in due course, into a spherical shock wave and the jet released from the orifice has a laminar structure. Figure 3.24c–f show sequential observations of shock wave diffraction over a four-star shaped orifice of geometry of 6.7 mm 13.4 mm. The vortex loop released from a sharp corner propagates faster than that released from the orifice having flatter edges. Hence the vortex loop is three-dimensionally distorted. A similar trend is observable in the case of an eight-star shaped orifice shown in Fig. 3.24 g, h. Affected by the motion of these vortex loops, in Fig. 3.24h the jet at later time shows already a slightly turbulent structure. The series of experiments were conducted during #040518–#040520.

Figure 3.25 shows the effect of the orifice shape on the development of the vortex loop. The ordinate denotes the dimension-less distance normalized by the orifice diameter a, b in mm for Ms = 1.24. The abscissa denotes elapsed time in ms. Red and blue filled circles denote blunt and sharp edged orifice, respectively. The vortex loops develop faster from the smaller the orifice diameters. Edge shapes would not significantly affect the development of vortex loops.

In Fig. 3.26 the time variation of vortex loops released from a 50 mm diameter orifice observed from side and the observation from the radial direction are sequentially compared for Ms = 1.24. Figure 3.26a, b shows the vortex loop at elapsed time of 2 ms. The transmitted shock wave moves at 0.4 m/ms and the vortex loop moves at about 0.1 m/ms. In Fig. 3.26a, b, the transmitted shock wave is located at about 0.4 m from the orifice and the vortex loop at about 0.1 m from the orifice. The vortex loop released from the opening of the 50 mm inner diameter

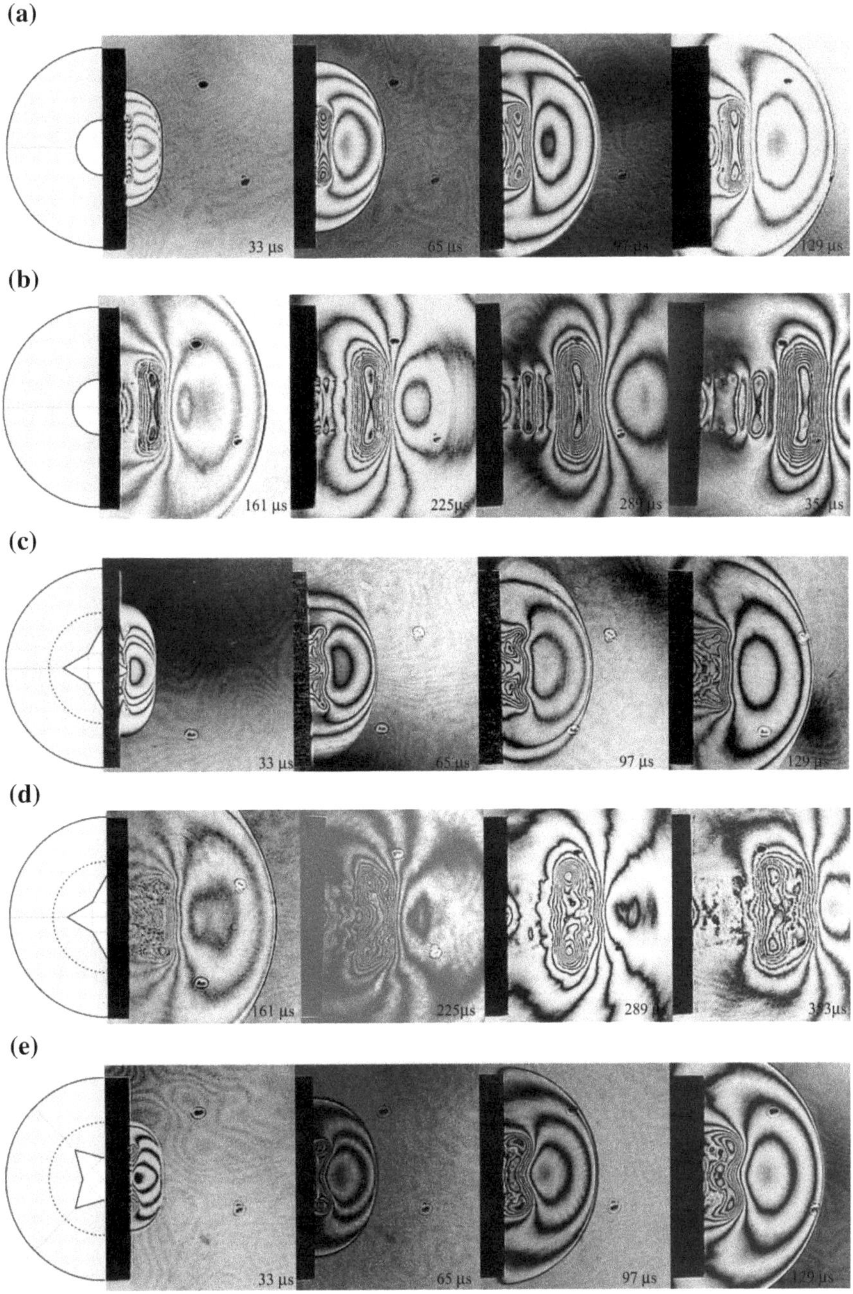

Fig. 3.24 Evolution of diffracting shock waves passing through orifices for Ms = 1.24 in atmospheric air at 290 K: **a**, **b** ϕ50 mm circular opening; **c**, **d** four-star shape; **e**, **f** four-star shape; **g**, **h** eight-star shape

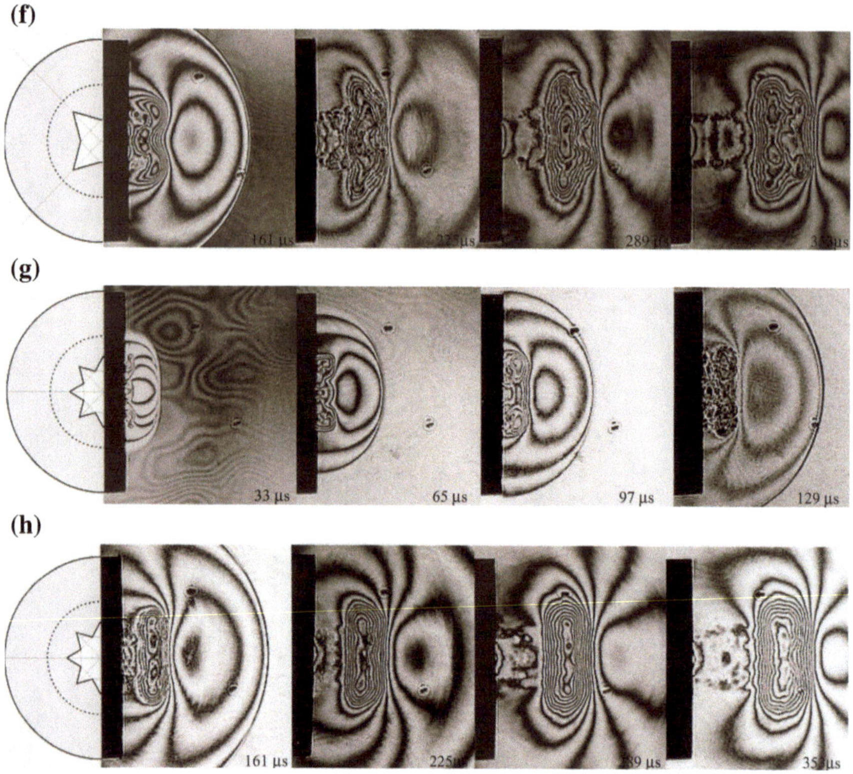

Fig. 3.24 (continued)

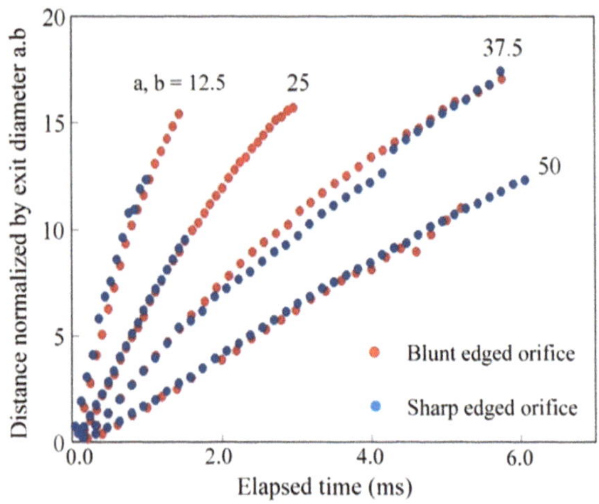

Fig. 3.25 Evolution of vortex loops, dependence of orifice shape

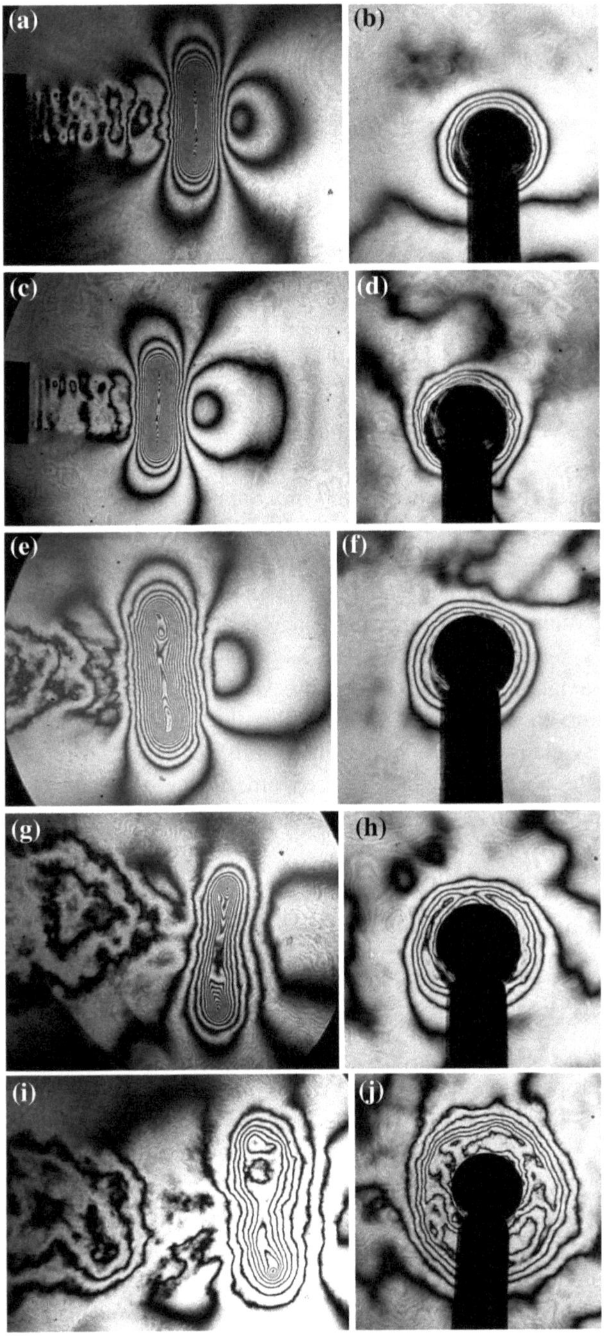

Fig. 3.26 Deformation of vortices released from a 50 mm diameter orifice viewed from side and axial direction for Ms = 1.24 in atmospheric air: **a** #04052008, 2 ms departure from orifice; **b** #04060106, 2 ms; **c** #04052007, 3 ms; **d** #04060104, 3 ms; **e** #04052001, 6 ms; **f** #04060107, 6 ms; **g** #04052509, 9 ms; **h** #04060108, 9 ms; **i** #04052509, 10 ms; **j** #04060110, 10 ms (Kainuma 2004)

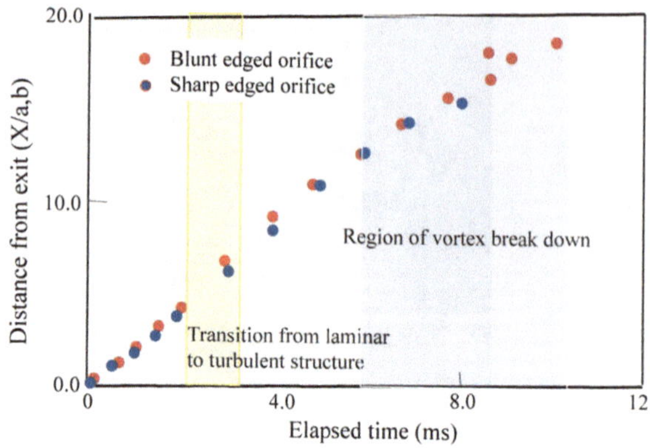

Fig. 3.27 Evolution of vortex breakdown, dependence of orifice edge

shock tube was visualized using double exposure holographic interferometer. The shadow in Fig. 3.26b shows the 90 mm outer diameter shock tube placed in a co-axial position between a pair of 1000 mm diameter schlieren mirrors. The vortex loop shows initially co-axial fringe distribution which indicates a laminar structure. With elapsed time of 6 ms, the fringes on becomes vortex loop is positioned at about 0.4 m from the orifice and the transition to turbulent structure started in Fig. 3.26e, f. With elapsing time, the fringe distribution became slightly irregular which shows a sign of the vortex break down seen in Fig. 3.26h. The fringes show no longer co-axial and the vortex loop breaks down observed in Fig. 3.26i, j.

Summarizing these images, in Fig. 3.27, the position of the vortex loop is plotted against elapsed time in ms. The ordinate denotes the non-dimensional distance and the abscissa denotes elapsed time in ms. Red and blue filled circles denote the blunt edged orifice and the sharp edged orifice, respectively. At about 3 ms, the transition of vortex loop structures from laminar structure to turbulent one was observed. At elapsed time from 6 ms to 10 ms, the vortex breakdown occurs. The edge shapes hardly affected the movement of the vortex loops.

3.6 Shock Wave Propagation Along 90° Bends

Shock wave propagation along a 90° bend is one of the elementary topics of shock wave research. In 1975, when starting shock tube experiments, the first research project was a visualization of shock wave propagation through a 90° bend using direct shadowgraph.

At first a 30 mm × 40 mm shock tube was constructed in house. The material of this shock tube comprised of a commercial extruded square cross sectional brass tubes of 5 mm wall thickness having a smooth surface. This shock tube was connected to the circular test section in which a 90° bend was installed as shown in Fig. 3.28. Shock waves were generated still by rupturing the Mylar membranes. The shock wave generation by rupturing diaphragm shattered small fragments which were shattered along the low pressure channel so that the cleaning the low

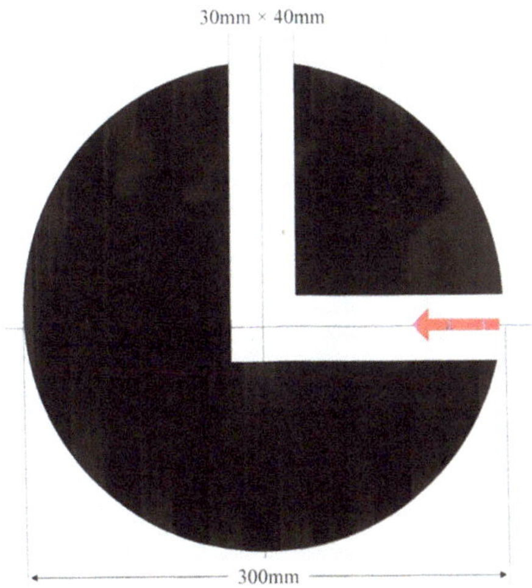

Fig. 3.28 An illustration of 90° bend section connected to a 30 mm × 40 mm conventional shock tube (Honda et al. 1977)

pressure channel became a routine work. At that time, the scatter of the shock wave Mach number ΔMs was at best ±2%. Hence, no one believed that shock tubes were reliable experimental tools (Heilig 1969).

Figure 3.29 show series of first direct shadowgraphs of shock wave of Ms = 1.3 propagating along a 90° bend. At that time polaroid films were used for recording material. The shock wave diffraction at the inside corner and the transmitted shock wave reflection from the outside wall were observed. The shock wave diffraction at the inside corner generated a triple point as shown in Fig. 3.29b.

Figure 3.30 show, similar to a converging and diverging nozzle flow, a bend comprised of the combination of a 160 mm diameter circular outer wall and a 90° inside corner. This shape was expected to conduct smooth shock wave transmission and to suppress the vortex formation at the inside corner.

Figure 3.31 show a series of direct shadowgraphs of the shock wave transmission along a bend having a smoothly converging and diverging throat. In order to make the IS turn 90° smoothly without causing any boundary layer separations at the inner corner, a small pocket was attached at the inner corner. Figure 3.31a, b show the transmission of the incident shock wave for Ms = 1.45 in air along a 90° bend having an outer wall radius of 160 mm and an inner wall radius of 20 mm. Figure 3.31c–l show the curved wall of the same geometry having a small pocket at the inner wall.

In Fig. 3.31c, the shock wave was diffracted at the inner corner. In Fig. 3.31d, a shocked air flew into the pocket. In Fig. 3.31e–g, a reflected shock wave was bifurcated and the boundary layer growth was suppressed along the inner wall. Anyway in Fig. 3.31j the transmitted shock wave became planar at the distance away from the inner corner.

Figure 3.32 show the evolution of shock wave propagating along a 90° bend in the 150 mm × 60 mm shock tube for Ms = 1.26 in atmospheric air at 292 K. The inner and outer walls were covered with a sponge sheet. Such a wall would suppress the boundary layer developing along the wall and the reflection of wavelets

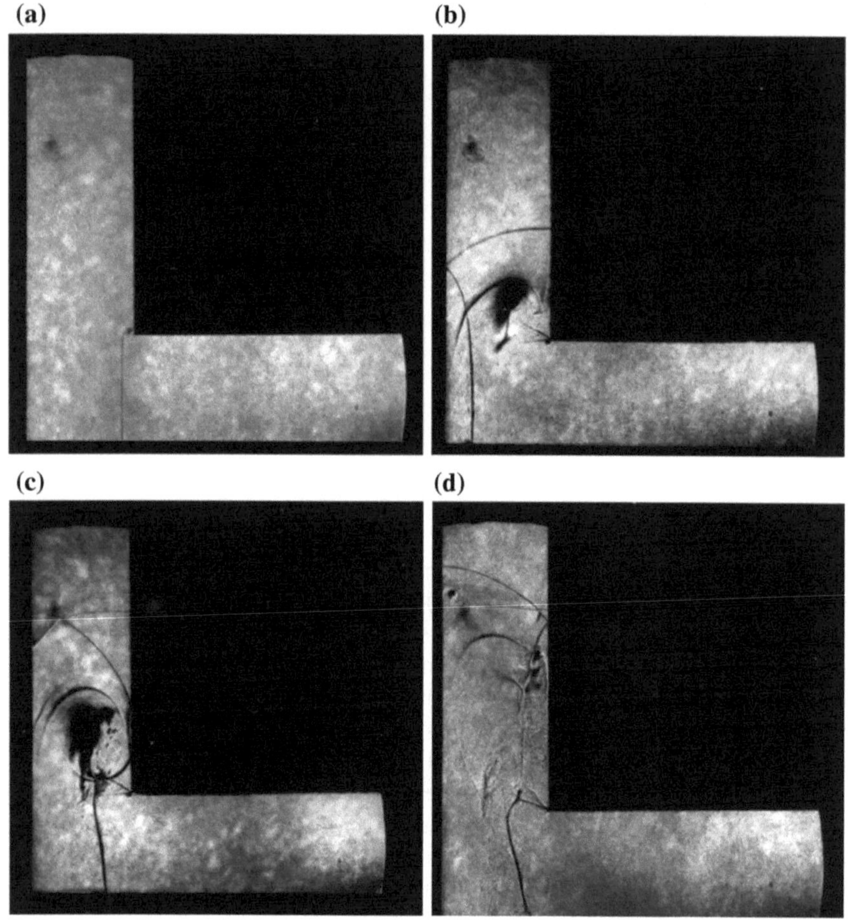

Fig. 3.29 Shock wave transmission along a 90° bend for Ms = 1.30 in air at 400 hPa, 292 K; **a** #75032401; **b** #75032402; **c** #75032403; **d** #75032404 (Honda et al. 1977)

from the wall surface. Then the transmission of shock waves is different from that over a metal wall. Comparing the images shown in Fig. 3.32 with those shown in Fig. 3.29, the vortex formation and the shock wave reflection seem to be very different. Wavelets seen in Fig. 3.29 were no longer observed.

Figure 3.33 summarizes the previous result of a sharp 90° bend seen in Fig. 3.29 and another result of the curved bend having the outer wall radius of 100 mm and the inner wall radius of 60 mm. The ordinate denotes Ms of the transmitted shock waves and the abscissa denotes Ms of the incident shock waves. Blue and red filled circles denote the results of sharp 90° bend and the curved bend, respectively. The effect of the incident shock wave mitigation through the bend is maximal in a sharp 90° bend. The larger the radius of curvature is, the degree of the incident mitigation is reduced.

Figure 3.34 show sequential observations of shock waves for Ms = 2.30 in air propagating along a 90° bend having an outer wall radius 160 mm and an inner

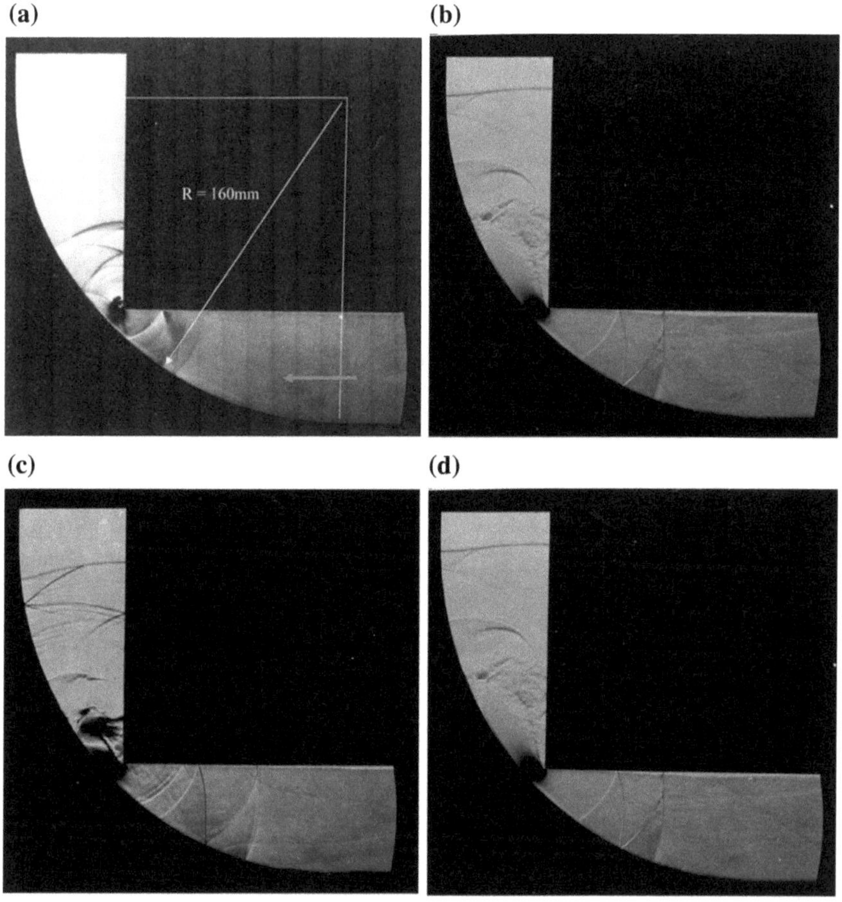

Fig. 3.30 Shock wave transmission along a 120 mm × 0 mm bend having a throat: **a** #76030201, Ms = 1.106, 1013 hPa; **b** #76030203 Ms = 1.313, 666 hPa; **c** #76030205, Ms = 1.318, 400 hPa; **d** #76030211, Ms = 1.822, 133 hPa (Honda et al. 1977)

wall radius of 120 mm. Along the inner wall, the shock wave is gradually diffracted and hence a vortex formation along the inner wall is suppressed. At the same time, the transmitted shock wave was gradually reflected from the outer wall in a similar way to the reflection from the concave wall. The shock wave reflection pattern is a SMR as seen in Fig. 3.34c, d. However, the incident shock wave is, unlike that over a concave wall, normal to the outer wall and its triple point interacts with the diffracting shock wave. The resulting reflection pattern was always a DiMR. The triple point migrated between the outer and inner walls as shown in Fig. 3.34e, f and the resulting transmitted shock wave became a planar relatively quickly as the curvature became larger.

Figure 3.35 show the evolution of a shock wave propagation along a curved bend having the outer wall radius of 120 mm and the inner wall radius of 80 mm

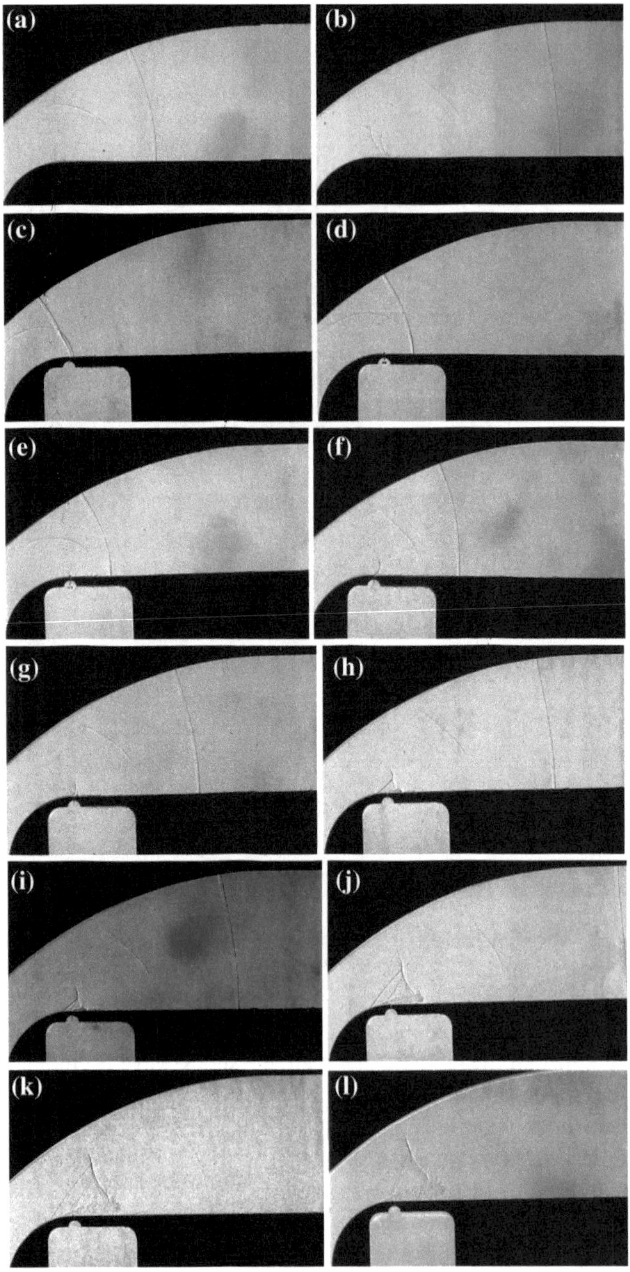

Fig. 3.31 Shock wave transmission along 90° corner having a small pocket at the inner wall for Ms = 1.45 in air at 800 hPa: **a** #77051322, Ms = 1.452; **b** #77051324, Ms = 1.446; **c** #77051922; **d** #77051903; **e** #77051902; **f** #77051913; **g** #77051914; **h** #77051912; **i** #77051911; **j** #77051997; **k** #77051911; **l** #77051909

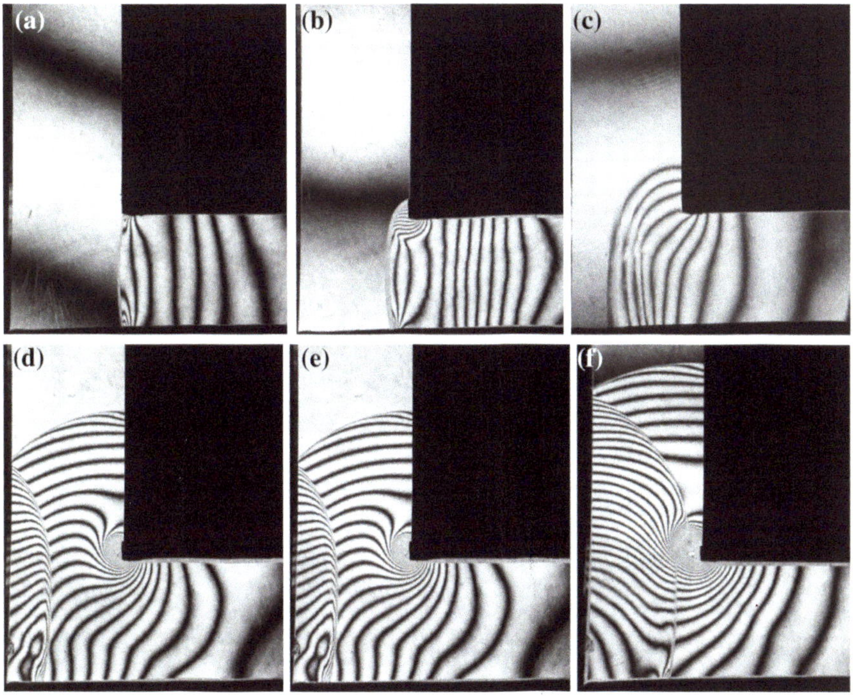

Fig. 3.32 Shock wave transmission along 90° bend covered by a sponge sheet. *OB* path length was150 mm for Ms = 1.14 − 1.26 in atmospheric air at 292 K: **a** #90091807, Ms = 1.140; **b** #90091814, Ms = 1.269; **c** #90091808, Ms = 1.137; **d** #90091816, Ms = 1.269; **e** #90091816, Ms = 1.269; **f** #90091817, Ms = 1.273

Fig. 3.33 Mitigation of the incident shock wave, the dependence of shapes of 90° bends (Honda et al. 1977)

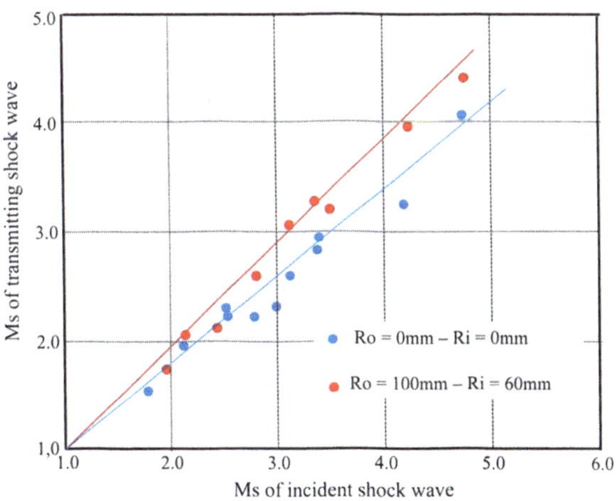

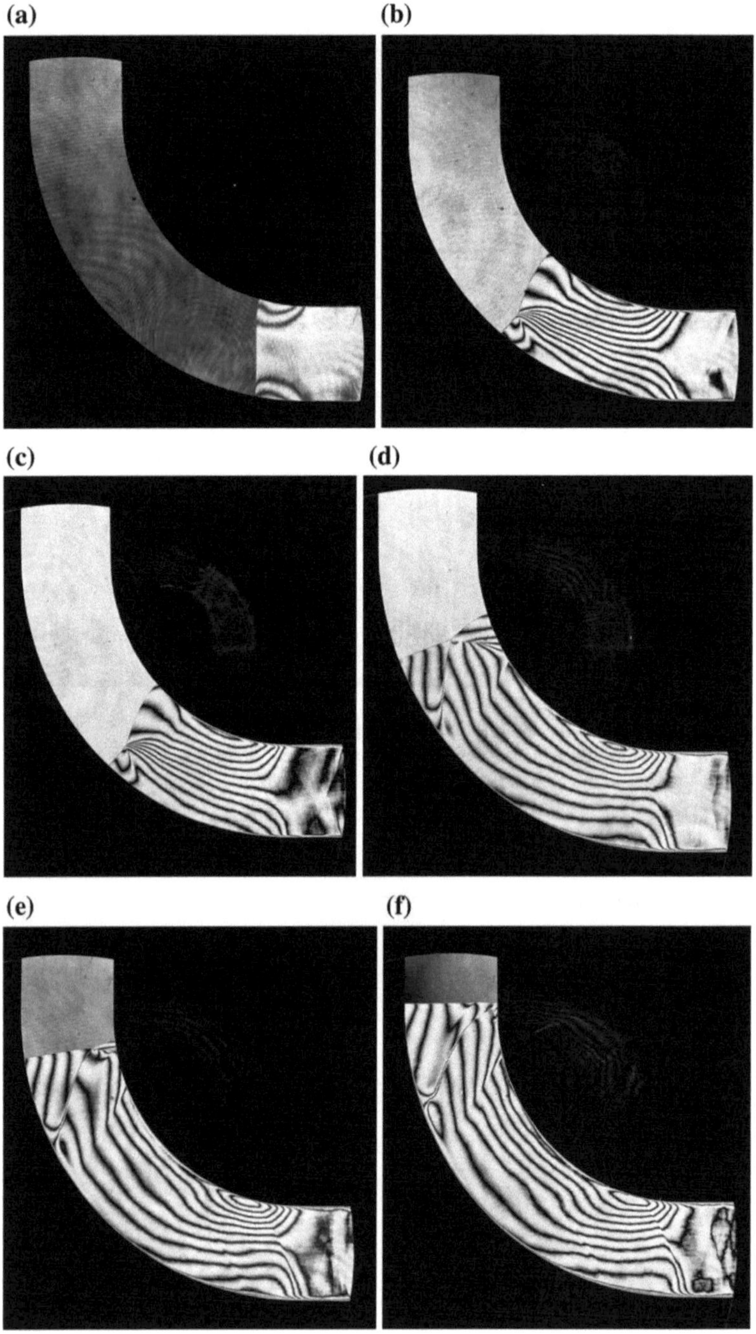

Fig. 3.34 Shock wave propagating along a R = 120 mm × R = 80 mm bend for Ms = 2.30 in air at 810 hPa, 297.5 K; **a** #85082109 60 μs from trigger point, Ms = 2.378; **b** #85082111 120 μs, Ms = 2.118; **c** #85082112 150 μs, Ms = 2.118; **d** #85082115 210 μs, Ms = 2.344; **e** #85082116 240 μs, Ms = 2.363; **f** #85082117 270 μs, Ms = 2.345

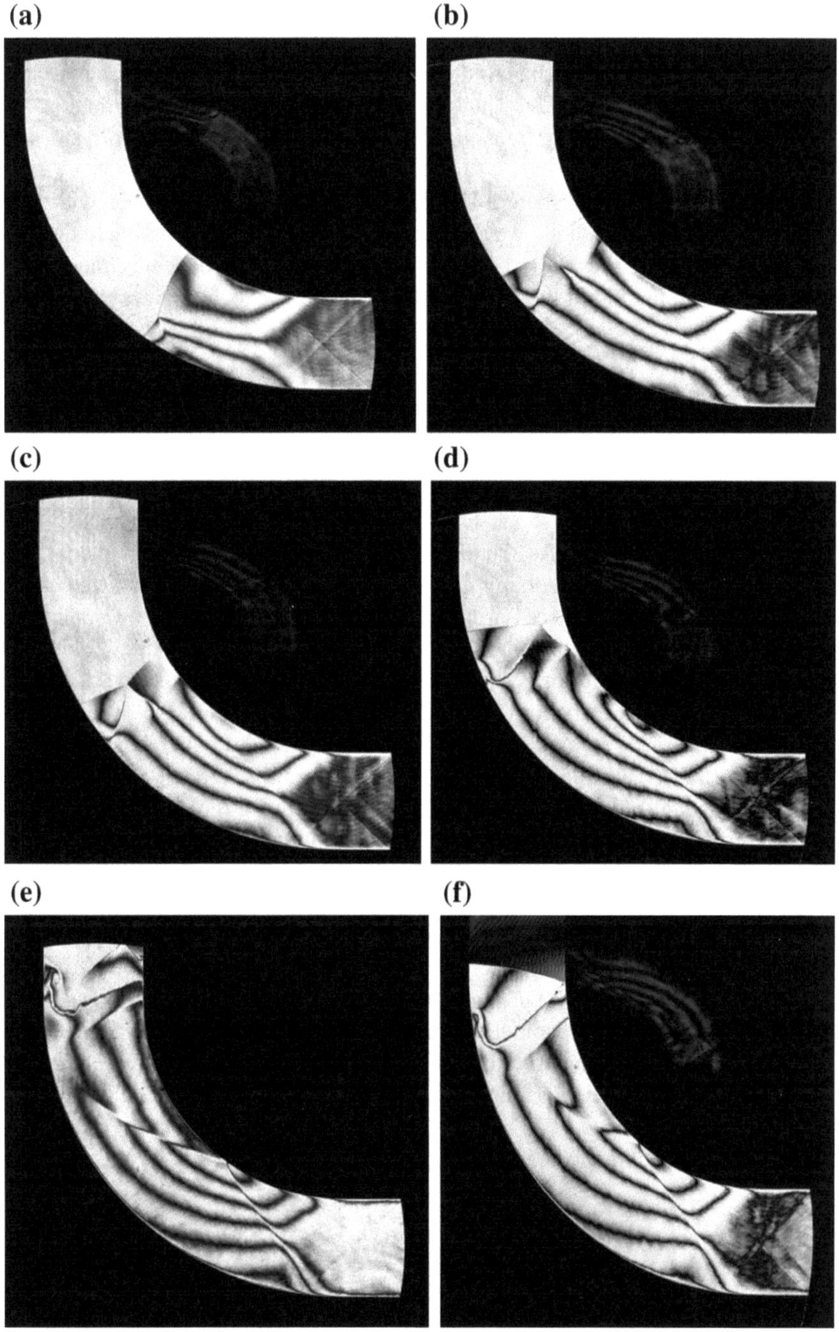

Fig. 3.35 Shock wave propagating along a 120 mm × 80 mm bend for Ms = 4.0 in CO_2 at 405 hPa and 297.9 K: **a** #85082201, 42 μs from trigger point, Ms = 3.959; **b** #85082203, 460 μs, Ms = 3.955; **c** 85082202, 440 μs, Ms = 4.023; **d** #85082204, 480 μs, Ms = 3.979; **e** #85082206, 520 μs, Ms = 4.028, **f** #85082205, 500 μs, Ms = 4.005

for Ms = 4.0 in CO_2 at 405 hPa and 297.9 K. Along the outer wall, the reflection pattern was identical with the shock wave reflection over a 120 mm radius wall. The ratio of the outer wall radius to the shock tube wall width is 3. Along the inner wall, the diffracting shock wave changes its pattern and the vortex formation does not occur. Hence the resulting transmitted shock wave was very similar to shown in Fig. 3.34.

In Fig. 3.35e, f, the transmitted shock wave quickly became planar with propagation. The transmitted shock wave recover planar shape in this gradual bend. It is concluded that if a shock tube needed to have a 90° curved bend connected by various reasons, the ratio of the radius of the bend to the shock tube width should be over 3.

3.7 Aspheric Lens

Rectangular cross sectional shock tubes and their test sections are suited for the optical flow visualization. As circular cross sectional shock tubes have simple structures, they are used for many experiments. However, the circular cross sectional shock tubes are not necessarily the best configuration for flow visualization. Yamanaka (1972) proposed to quantitatively visualize the effect side jets injected on the thrust of a conical nozzle by introducing an aspheric lens shaped diverging nozzle Takayama and Onodera (1983). At that time, the Nikon Co. Ltd manufactured this nozzle made of high quality optical glass. Based on the idea presented by Yamanaka (1972), Takayama (1983) at first measured the refractive index of acryl to ruby laser beam of wavelength 6943 nm, then designed the aspheric shape, and manufactured in house a 100 mm long cylinder having the aspheric cross section out of commercially available acryl. Trial and error improved the skill of manufacturing and polishing test pieces. Eventually variously sized test sections having were manufactured and applied to shock wave research.

Figure 3.36 illustrates a ray tracing through an aspheric lens shaped cross section. The collimated incident *OB* transmits the aspheric lens shaped cross section in parallel and comes out in parallel. With this alignment, circular shock tube flows can be quantitatively visualized by illumination with a collimated light beams. Tracing individual light rays passing through the aspheric lens shape, the wall shape is analytically determined.

Let **X**, and $\mathbf{X_n}$ be position vectors of the circular inner wall and its normal vector. Then,

$$\mathbf{X} = \mathbf{X}(x = r_0\cos\theta, y = r_0\sin\theta)$$
$$\mathbf{X_n} = \mathbf{X_n}(x_n = \cos\theta, y_n = \sin\theta),$$

where r_0 is the radius of the inner wall and angle θ is defined as seen in Fig. 3.36. For a parallel incident light ray to the x-axis, the unit vector of the ray is represented

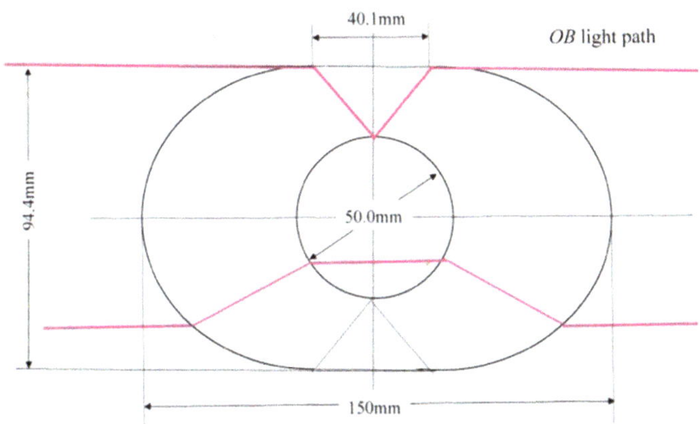

Fig. 3.36 A 50 mm diameter cross sectional aspheric lens and ray tracing (Takayama and Onodera 1983)

by r′(1, 0) and that of the reflected light ray inside the wall is r′(l, m), where l and m are direction cosines. At the wall surface, the Snell's law $\sin\theta_{11} = n\sin\theta_2$ is applied, where n is the refractive index of the wall material, in the present case, it is PMMA, to air at wave length of ruby laser.

A following relationship is valid for the collimated light ray which refracts at a cylinder surface made of acryl,

$$\begin{aligned}\cos\theta &= \mathbf{X_n}r' \\ &= \cos\theta\end{aligned} \tag{3.1}$$

For a light ray emanating from the lens/air interface,

$$\begin{aligned}l\cos\theta + m\sin\theta &= \mathbf{X_n}r' \\ &= \cos\theta_2, \text{where } l^2 + m^2 = 1\end{aligned} \tag{3.2}$$

l and m are then obtained by solving the Eqs. (3.1) and (3.2) and, therefore, the aspheric lens shape **X**(x, y) is readily determined. Let's define a local wall thickness s. Then, the wall thickness is determined as light rays entering the aspheric lens come out simultaneously. Equation (3.3) summarizes the wall shape where s is normalized with the maximum wall thickness s_0 at y = 0 and θ = 0.

$$\begin{aligned}S/S_0 &= (n-1)/(n-l) = (n\cos\theta_2 + \cos\theta_1)/\{(n+1)\cos\theta_2\} \\ \mathbf{X}'(x,y) &= \mathbf{X}(X,Y) + sr'(l,m)\end{aligned} \tag{3.3}$$

The aspheric lens shape is readily determined for the measured value of refractive index of acryl: n = 1.4915 ± 0.0005 to the wavelength of a ruby laser, $\lambda = 694.3$ nm.

A 50 mm diameter and 100 mm long cylindrical aspheric lens was manufactured in house. Today probably a 3-D printer will be able to produce this shape without serious technical difficulty. Polishing the surface may require the skill and patience of craftsmen. If the given diameter s_0 is 50 mm, the thinnest wall thickness, in this case, is 27.2 mm. This wall thickness is thick enough to minimize the potential distortion during machining. As seen in Fig. 3.36, there is a triangular zone in which the collimated *OB* cannot pass. The length of this zone is 40.1 mm long along the circumference of the aspheric lens and occupies a wide space enough to drill a port of accommodating pressure transducers or other gauges.

To calibrate the aspheric lens test section, a 1 mm × 1 mm test chart of 50 mm in width and 100 mm in length was inserted into the test section. The test chart width fitted the diameter of the test section. The first exposure was carried out in the absence of the test chart and the second exposure was conducted by insertion of the test chart to the aspheric lens test section. Figure 3.37a shows a double exposure interferogram of the test chart in the aspheric lens test section. The 50 mm wide test chart was uniformly broadened in horizontal direction which was 94.4 mm but any distortions were observed. The test chart was not distorted to the lateral direction. The edge of the aspheric was blurred due to inaccuracy of machining. Hence, the maximum width so far observable was 48.5 mm. If the field of view was expressed in terms of the view angle, it was approximately 75°. The contrast of the

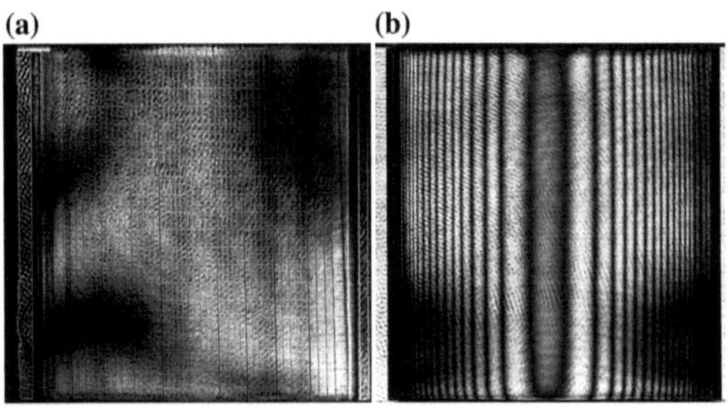

Fig. 3.37 Calibration of aspheric lens test section in air: **a** Test chart #81110401; **b** #81112002 the first exposure at 670 hPa, 287 K and second exposure at 1013 hPa, 287 K (Takayama and Onodera 1983)

background was slightly non-uniform. Such a non-uniformity was attributable to inhomogeneity of acryl.

Figure 3.37b shows the accuracy of the aspheric lens. Firstly, air at 670 hPa and at the room temperature was filled in the test section and the first exposure was conducted. Then the atmospheric air at the room temperature was filled in the test section, the second exposure was conducted. Two exposures recorded the change in the phase angles corresponding to the density increase of about 150% in the test section.

Figure 3.37b shows the resulting fringe distribution indicating uniform density distribution in the test section. The fringe intervals corresponding to the path length distribution for the collimated *OB* to cross a circular cross section.

Grey patterns observable in the background were created due to the inhomogeneity of the acrylic. Although the contrast of the image in Fig. 3.37b was slightly non-uniform, the fringes were distributing straight and seemed to be locally broadened but this was an optical illusion because of disturbed non-uniformity in the background.

3.7.1 Shock Wave Formation Driven by a Moving Piston

The aspheric test section was connected to a 50 mm diameter shock tube. Then a 50 mm diameter and 100 mm long nylon piston was inserted at the entrance of the aspheric lens test section. The initial condition in front of the piston was in air at 10 hPa and 284.3 K. Figure 3.38 illustrates the experimental setup. Upon the impingement of a shock wave on the rear surface of the piston, it impulsively moved into the aspheric lens test section. The piston speed at the entry of the test section was 69 m/s. If the shock Mach number Ms possibly driven by the piston can be written as $Ms = 1 + \varepsilon$, where $\varepsilon \ll 1$, and $\varepsilon = (\gamma + 1)u/4a$ where u and a are the

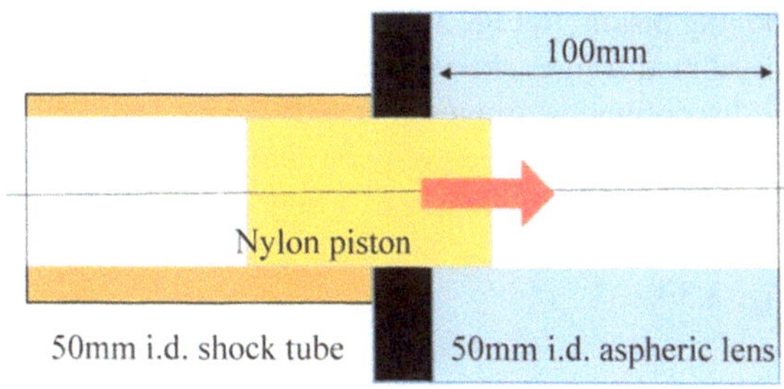

Fig. 3.38 A nylon piston sliding into aspheric lens test section

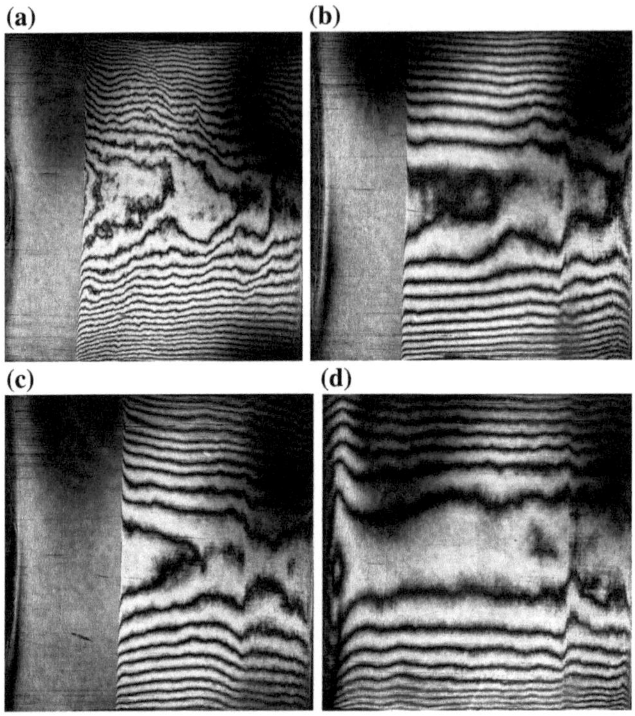

Fig. 3.39 Formation of weak shock wave driven by a nylon piston at apparent Ms = 1.12 in air at 10 hPa, 294.3 K: **a** #92060101, 3100 μs from trigger point; **b** #92052903, 3450 μs; **c** #92052904, 3450 μs; **d** #92052907 3900 μs

particle velocity u = 69 m/s and the sound speed in air a = 345 m/s, respectively. Eventually the estimated Ms is 1.12.

Figure 3.39 show sequential observations of weak shock wave driven by the moving piston. In Fig. 3.39a, the grey uniform area shows the piston and irregularly distributed fringes show compression wavelets driven by the piston. In Fig. 3.39b–d, discontinuous fringes indicate the formation of a weak shock wave of about Ms = 1.1. It should be noticed that a planar shock was formed already at about 20 mm ahead of the piston. Fringes were distributed almost horizontally and in parallel indicating that the density was distributed almost uniformly as shown in Fig. 3.39.

3.7.2 *Circular Cross Sectional Shock Tube Passing a 90° Elbow*

Visualization of a shock wave propagating along a 50 mm diameter circular cross sectional 90° bend is one of the important research topics of shock wave dynamics. Today a 3-D printer will be able to produce an aspheric lens shaped circular cross sectional transparent 90° bend. However, in the early 1980, a 50 mm diameter circular cross sectional 90° bend was made of brass in order to investigate the character of transmitted shock wave after passing the circular cross sectional 90° bend. A 50 mm diameter conventional shock tube was connected to the 90° bend. Figure 3.40 shows the 90° bend section made of brass and a 100 mm long aspheric lens shaped test section connected at the distance 160 mm and 910 mm from the center of the shock tube.

Through two-dimensional 90° bends, the diffracting shock waves interacted with reflected shock waves created series of vortices. Therefore, it will take a longer distance for the distortion of the two-dimensional wave to recovers to a planar shape.

Figure 3.41a–f show sequential visualization of a transmitted shock wave propagating downstream of the 90° bend for Ms = 1.6 in atmospheric air at 289 K. The shock wave diffraction at the inside corner would create a corner vortex only locally at the central area. The shock wave reflection from the outside wall varies only locally depending on the local inclination angle. The reflected shock wave is no longer two-dimensional but curved three-dimensional. It is easy to imagine such the three-dimensional wave pattern and their interaction with the three-dimensional 90° bend. The transmitted shock wave became gradually flat with elapsing time but waves were migrating along the shock tube. Figure 3.40d is a single exposure interferogram indicating the transmitted shock wave is flattened. Three-dimensionally distorted reflected shock wave and the transmitted shock wave are clearly observed. In the case of the circular cross sectional 90° bend, the wave interacted with three-dimensionally curved wall and the distorted wave fronts were quickly suppressed with elapsing time. The position of aspheric test section was

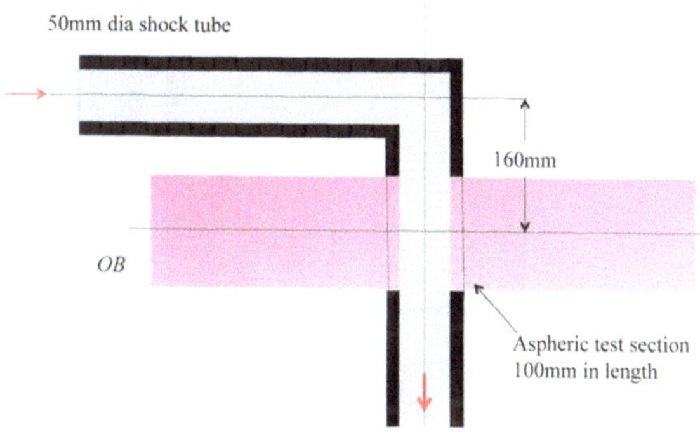

Fig. 3.40 Circular cross sectional shock tube connected to 90°

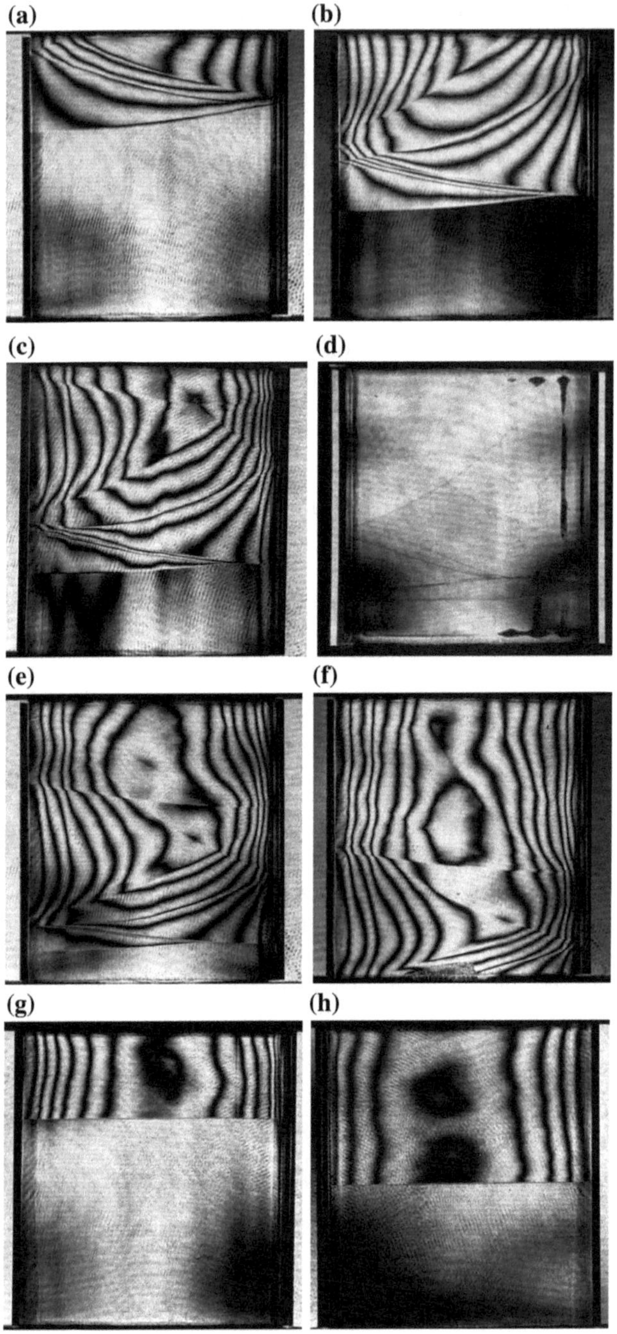

Fig. 3.41 Transmitted shock wave at 160 mm from a 90° bend for Ms = 1.6 in atmospheric air at 289 K: **a** #81110606, Ms = 1.583; **b** #81110604, Ms = 1.589; **c** #81110602, Ms = 1.593; **d** #82091001, Ms = 1.550; **e** #81110605, Ms = 1.583: **f** #81110704, Ms = 1.576: **g** #81111207, Ms = 1.274 at 1013 hPa, 289.3 K, at 910 mm from 90° bend; **h** #81111213, Ms = 1.165 at 1016 hPa, 289.3 K at 910 mm from 90° bend

moved to the 910 mm distance downstream from the center of the circular tube. In Fig. 3.41g, h, at the characteristic distance of 18.2, the transmitted shock wave was well flattened. The parallel fringe patterns seen in Fig. 3.41g, h indicates a uniform density distribution across the shock tube. The transmitted shock wave is recovered to almost a planar shape. In conclusion, circular cross sectional 90° bends can recover diffracting shock waves to planar shapes in shorter distance. Recent advanced super-computations will reproduce the shock wave transmission through a circular cross sectional 90° bend. At that time, Fig. 3.41will provide a useful comparison with numerical result.

3.7.3 Co-axial Shock Wave Diffraction at Area Change

A 100 mm diameter and 100 mm long aspheric lens shaped test section was connected to a 50 mm diameter shock tube. Figure 3.42 shows an experimental arrangement for visualizing a shock wave passing through the circular cross sectional area expansion. This is, in short, a shock wave diffraction from a circular cross sectional duct.

Figure 3.43 show a sequential observation of a shock wave diffraction at a 50 mm/100 mm area expansion for Ms = 1.50 in atmospheric air and at 287 K. The transition of a planar shock wave to a spherical shock wave and the deformation of the vortex loop are sequentially observed. A planar shock wave is diffracted at the corner of area expansion and the shape of diffracting shock wave changes from a planar front to circular shape. However, the ratio of area expansion is only 2.0 so that the circular expanding shock front is reflected from a 100 mm diameter wall.

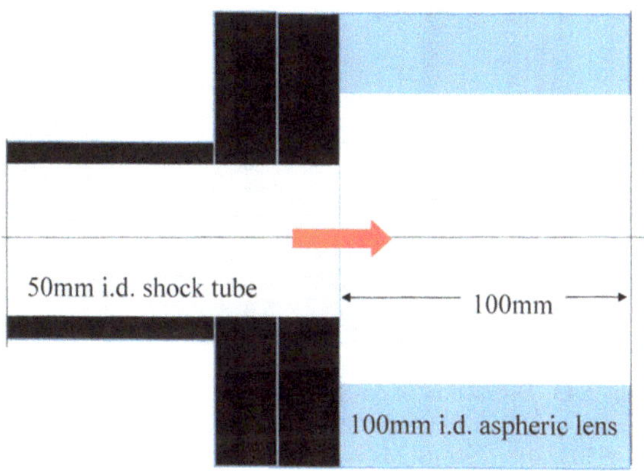

Fig. 3.42 Experimental setup of 50 mm/100 mm area change

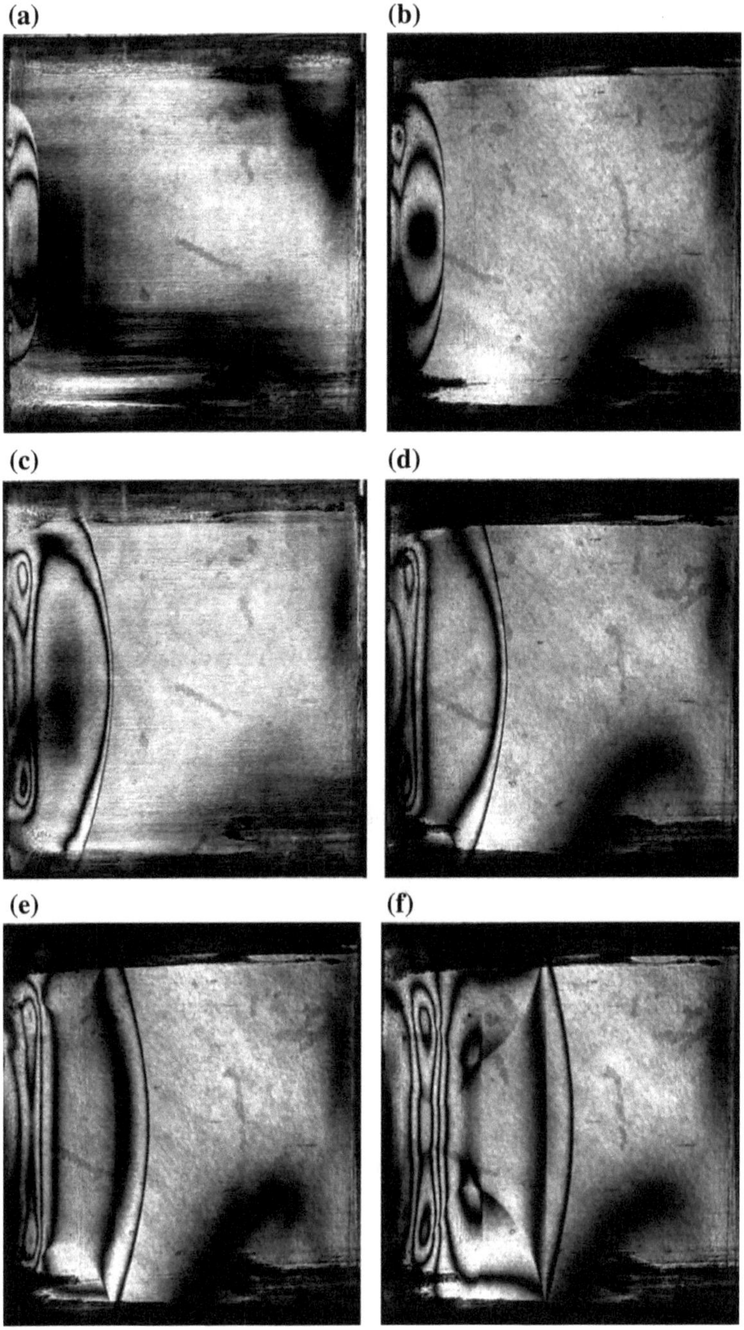

Fig. 3.43 Diffraction of co-axial shock wave diffraction at 50 mm/100 mm diameter area change for Ms = 1.50 in atmospheric air at 287 K: **a** #95030313, 220 μs from trigger point, Ms = 1.482; **b** #95032202, 230 μs Ms = 1.500; **c** #95030312, 240 μs, Ms = 1.508; **d** #95032204, 270 μs, Ms = 1.500; **e** #95032205, 290 μs, Ms = 1.500; **f** #95032206, 310 μs, Ms = 1.500; **g** #95030309, 300 μs, Ms = 1.508; **h** #95030308, 320 μs, Ms = 1.504; **i** #95030307, 340 μs, Ms = 1.501; **j** #95030314, 360 μs, Ms = 1.508

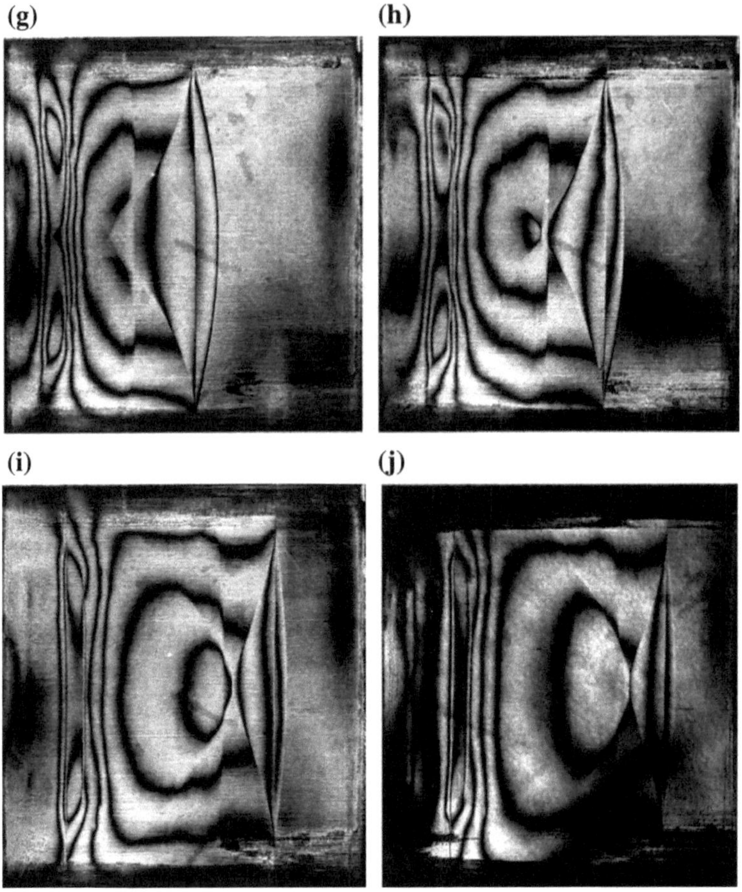

Fig. 3.43 (continued)

References

Abe, A. (1989). *Study of diffraction of shock wave released from the open end of a shock tube* (Ph.D. thesis). Graduate School of Engineering, Faculty of Engineering Tohoku University.

Babinsky, H., Yang, J. M., & Takayama, K. (1995). Animated visualization of shock wave flow field for dynamic comparison between experiment and numerical prediction. In *Proceedings of SPIE* (Vol. 2410, pp. 101–1).

Heilig, W. (1969). Diffraction of shock waves by a cylinder. *Physics Fluids, 12,* 154–157.

Honda, M., Takayama, K., & Onodera, O. (1977). Shock wave propagation over 90° bends. *Reports of the Institute of High Speed Mechanics, Tohoku University, 35,* 74–81.

Kainuma, M. (2004). *Study of diffracting shock waves and vortex released from released from a circular shock tube* (Ph.D. Thesis). Graduate School of Engineering, Faculty of Engineering Tohoku University.

Onodera, O., Jiang, Z. L., & Takayama, K. (1997). Holographic interferometric observation of shock waves discharged from the open end of square tubes. In A. P. F. Houwing, & A. Paul (Eds.), *Proceedings of 21st ISSW* (Vol. 2, pp. 1139–1444). The Great Keppel Island, Australia.

Takayama, K. (1983). Application of holographic interferometry to shock wave research. In *International Symposium of Industrial Application of Holographic Interferometry, Proceedings of SPIE* (Vol. 298, pp. 174–181).

Takayama, K., & Inoue, O. (1991). Shock wave diffraction over a 90° sharp corner—poster presented at 18th ISSW. *Shock Wave, 1,* 301–312.

Takayama, K., & Onodera, O. (1983). Shock wave propagation past circular cross sectional 90° bends. In D. Archer, & B. E. Milton (Eds.), *Shock Tubes and Waves, Proceedings of 14th International Shock Tubes and Waves, Sydney* (pp. 207–212).

Yamanaka. T. (1972). An investigation of secondary injection of thrust vector control (in Japanese). NAL TR-286T, Chofu, Japan.

Chapter 4
Shock Wave Interaction with Bodies of Various Shapes

Shock wave interactions with cylinders and other bodies are one of the fundamental topics of shock-dynamics. In this chapter results of flow visualizations over these body are presented.
Counting fringe orders, the density distribution over the bodies can be determined. Then the density distributions can be converted to the pressure distributions. However, the quantitative density distributions demonstrate slip lines and hence useful to validate numerical schemes.

The procedure of estimating density distribution from interferograms are conducted as follows: Interferometric fringe number N is determined integrating the density along the *OB* path,

$$\mathrm{N}\lambda/\mathrm{K} = \int_0^L (\rho - \rho_0)\mathrm{dz} \tag{4.1}$$

where L is the light path length, that is, the shock tube width and z is the distance of light path length. The λ is wave length of ruby laser 694.3 nm. The ρ and the ρ_0 are densities behind and in front of the shock wave. If the density profile in the boundary layer is known, the path integral is rewritten as following,

$$\begin{aligned}\mathrm{N}\lambda/\mathrm{K} = & \int_{\text{upper boundary layer}} (\rho - \rho_0)\mathrm{dz} + \int_{\text{core flow}} (\rho - \rho_0)\mathrm{dz} \\ & + \int_{\text{lower boundary layer}} (\rho - \rho_0)\mathrm{dz}\end{aligned} \tag{4.2}$$

Assume that the shock tube wall is at room temperature and, for the sake of simplicity, the Karmann-Pohlhausen velocity profile in the boundary layer, Eq. (4.2) can be rewritten as follows,

K. Takayama, *Visualization of Shock Wave Phenomena*,
https://doi.org/10.1007/978-3-030-19451-2_4

$$N\lambda/K = L(\rho_2 - \rho_W) + 2\delta\left(\rho_W \int_0^1 f(\eta)d\eta - \rho_2\right)dz \tag{4.3}$$

where ρ_w is the density on the wall and $f(\eta)$ is the Karmann Pohlhausen velocity profile across the boundary layer, where $\eta = z/\delta$, $A = -(\gamma - 1/2)M^2\ \rho_w/\rho_2$, and $B = -\{(\gamma - 1/2)M^2 + 1 - (T_w/T_2))\rho_w/\rho_2$.

$$f(\eta) = 1/\left\{A(2\eta - 2\eta^3 + \eta^4)^2\square + B(2\eta - 2\eta^3 + \eta^4) + 1\right\}^{\square} \tag{4.4}$$

T_w is the wall temperature and T_2 is the uniform flow temperature behind the shock wave. M is the uniform flow Mach number behind the shock wave. For Ms = 1.7 and a wall temperature T_w = 300 K, M = 0.77, ρ_w/ρ_2 = 1.458, and Tw/T2 = ρ_w/ρ_2, respectively. A = 0.1729 and B = 0.4327 and eventually $\int_0^1 f(\eta)d\eta = 0.836$. Therefore, the fringe number N is given by

$$N = KL/\lambda(\rho_2 - \rho_0)(1 + 0.836\delta/L). \tag{4.5}$$

The density increment $\Delta\rho$ corresponding to one fringe shift is given by

$$\Delta\rho/\rho_0 = \lambda/(KL\rho_0)/(1 + 0.836\delta/L) \tag{4.6}$$

For L = 60 mm, λ = 694.3 nm and the Gladstone-Dale constant K of this wavelength, $\Delta\rho/\rho_0$ is equivalent to approximately 4.6% of the ambient density. The ratio of the density increment taking the boundary layer displacement thickness into consideration, that corresponding to one fringe shift is written as

$$\Delta\rho/\rho_0 = 1 + 0.88\delta/L. \tag{4.7}$$

This implies that the contribution of the boundary layer displacement thickness δ to the total fringe shift is 0.88δ/L. For laminar boundary layer, according to the boundary layer theory (Schlichting 1960), the boundary layer displacement thickness is given by

$$\delta/L = 7.812(x/Re)^{1/2} \tag{4.8}$$

where x is the distance measured from the incident shock wave.

δ/L for the present shock tube flow is approximately 0.025 at x = 1 m. The effect of the boundary layer on fringe number decreases by 2%. The correction of fringe number is so small that it may be neglected if the experiments are not conducted at extremely low pressures.

The density distribution over cylinders can be determined simply by counting the fringe numbers. The resulting density distribution is convertible to a pressure distribution in the neighborhood of the cylinder surface, if the isentropic relation is

applied at a given value of stagnation density. From known pressure distributions over the cylinders, the pressure coefficient is readily determined and then the drag and lift forces are obtained by counting fringes on interferograms.

4.1 Circular Cylinder

4.1.1 Cylinder in Air

A 40 mm diameter circular cylinder was put between two 15 mm thick acrylic plates which were glued in the frame of the test section of the 60 mm × 150 mm conventional shock tube. Visualizations were conducted for a wide range of Ms and initial pressure using double exposure holographic interferometry. Acrylic plates are slightly inhomogeneous but would not distorted fringe distributions but only affected the image contrast.

Figure 4.1 shows the installation of a 20 mm diameter cylinder in the test section windows.

Figure 4.2 show sequential observations of a shock wave for Ms = 1.50 over a 40 mm diameter cylinder. The reflected shock wave pattern is RR when the shock wave impinged at the frontal stagnation area of the cylinder. The reflection pattern became SMR, when the shock wave approached to the equator of the cylinder. The transmitting shock waves propagated along the upper and lower sides of the cylinder toward the rear stagnation point. At the same time, the triple points TP are formed emanating curved Mach stems MS and slip lines SL.

Figure 4.3 show sequential observation of a shock wave for Ms = 2.6 in air propagating along a 40 mm diameter cylinder installed in the 60 mm × 150 mm conventional shock tube. At frontal surface, the reflection pattern was initially a RR.

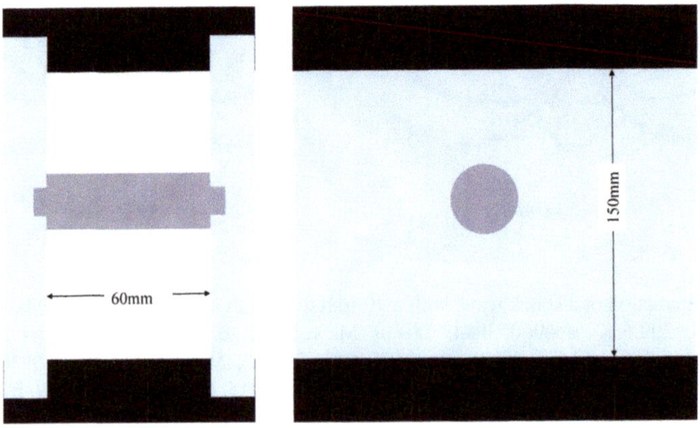

Fig. 4.1 Installation of a cylinder in the test section

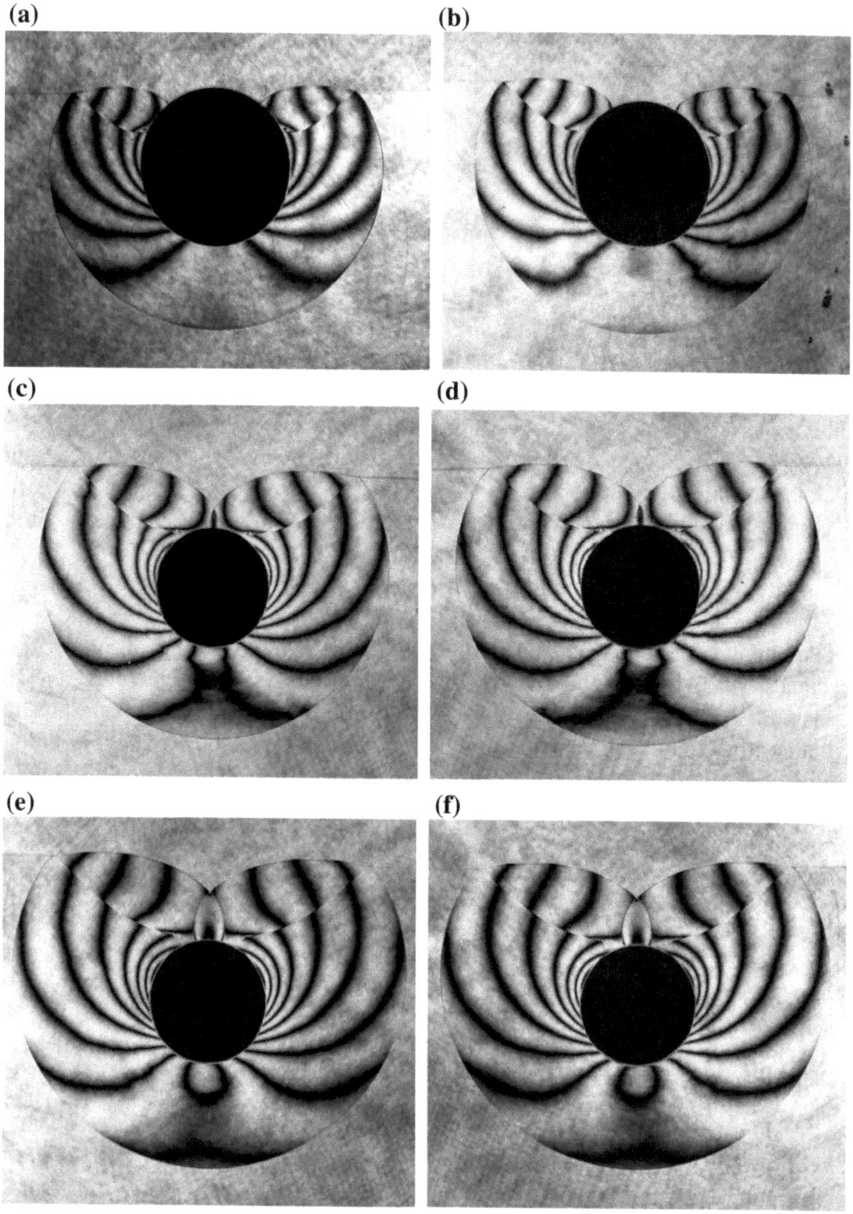

Fig. 4.2 Interaction of a shock wave with a 40 mm diameter circular cylinder for Ms = 1.50 in air at 275 hPa, 299.6 K: **a** #96053044, 180 μs Ms = 1.499; **b** #96053008, 190 μs Ms = 1.499; **c** #96053011, 200 μs Ms = 1.501; **d** #96053012, 220 μs, Ms = 1.497; **e** #96053013, 230 μs Ms = 1.503; **f** #96053014, 240 μs Ms = 1.500; **g** #96053015, 250 μs Ms = 1.499; **h** #96053016, 260 μs Ms = 1.501

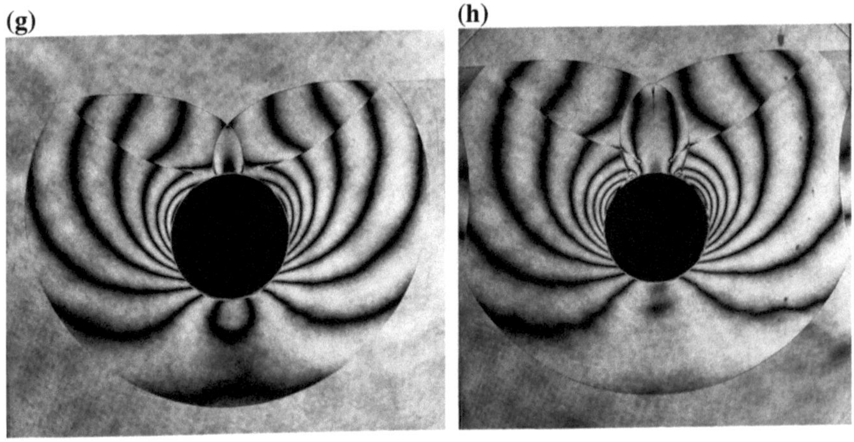

Fig. 4.2 (continued)

At the equator seen in Fig. 4.3a the SMR appear accompanying TP and curved MS seen in Fig. 4.3b. The Mach stems propagated along the upper and lower wall were reflected at the rear stagnation point and the reflected MS moved to the reverse direction. The reflected MS interacted with boundary layer developing along the cylinder surface. This process of the interaction is similar to the reflected shock interaction with the side wall boundary layer at the shock tube side wall (Mark 1956). Another reflected shock wave/boundary layer interaction was observed. In Fig. 4.3e an arrow indicates the bifurcation of RS interacting with the boundary layer developing along the shock tube sidewall. Figure 4.3g, h show the patterns of reflected MS interacting with boundary layer. The MS interacts with a SL emanating from the TP.

Figure 4.4 show sequential observations of interaction of shock wave of Ms = 1.70 with a 20 mm diameter cylinder in air at 900 hPa. The experiment was performed in the 60 mm × 150 mm conventional shock tube.

4.1.2 Cylinder in CO_2

Figure 4.5 show the evolution of shock wave interactions with a 20 mm diameter cylinder in CO_2 in a conventional 60 mm × 150 mm shock tube. A 20 mm cylinder was sandwiched between two acrylic plates, which were once used to measure pressures and then had holes accommodating the pressure transducers. In Fig. 4.5c, the black circular shadows were remnants of the smoothly plugged holes accommodating the pressure transducers.

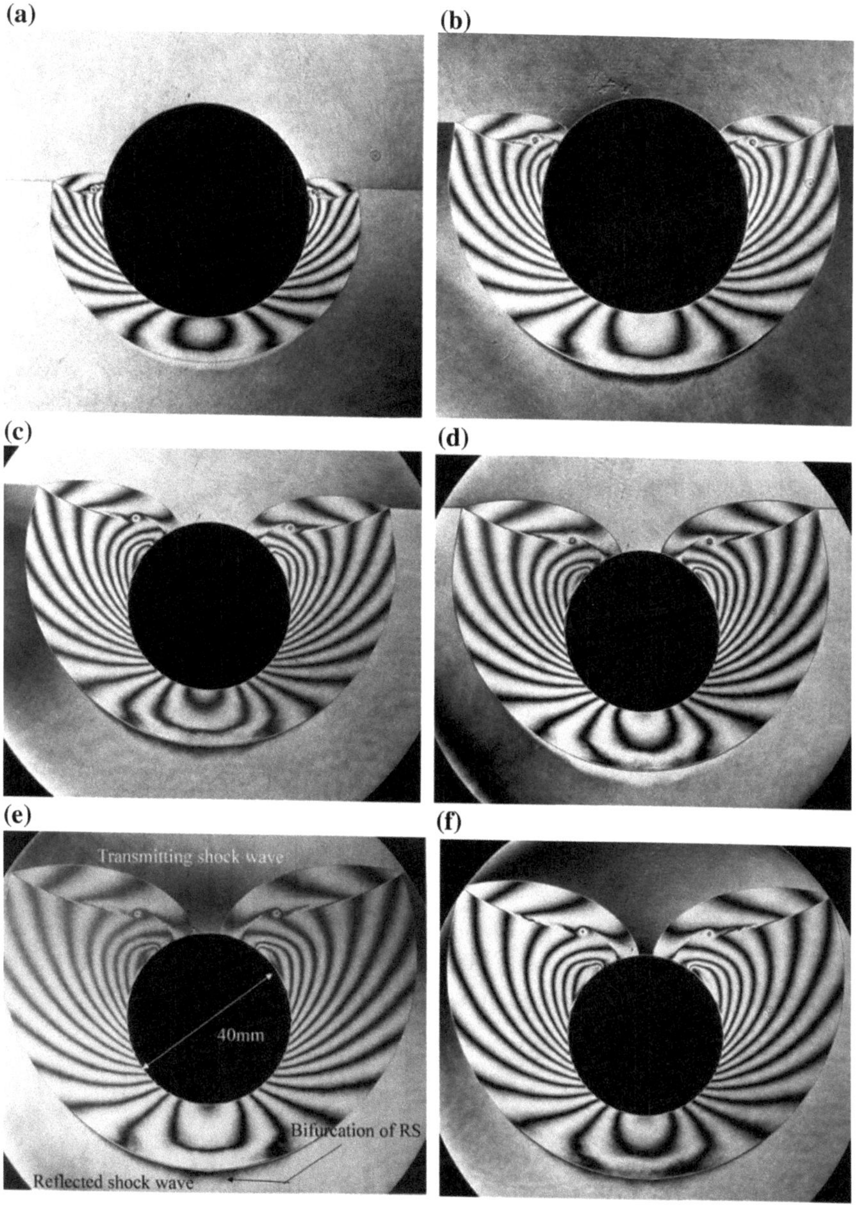

Fig. 4.3 Shock interaction with a 40 mm diameter cylinder for Ms = 2.60 in air at 100 hPa, 291.1 K: **a** #83110803, Ms = 2.591; **b** #83110901, Ms = 2.591; **c** #83110904, Ms = 2.617; **d** #83110802, Ms = 2.591; **e** #83050902, Ms = 2.620; **f** #83110906, Ms = 2.607; **g** #83050903, Ms = 2.589; **h** #83050901, Ms = 2.589; **i** enlargement of (**h**); **j** #83050904, Ms = 2.579

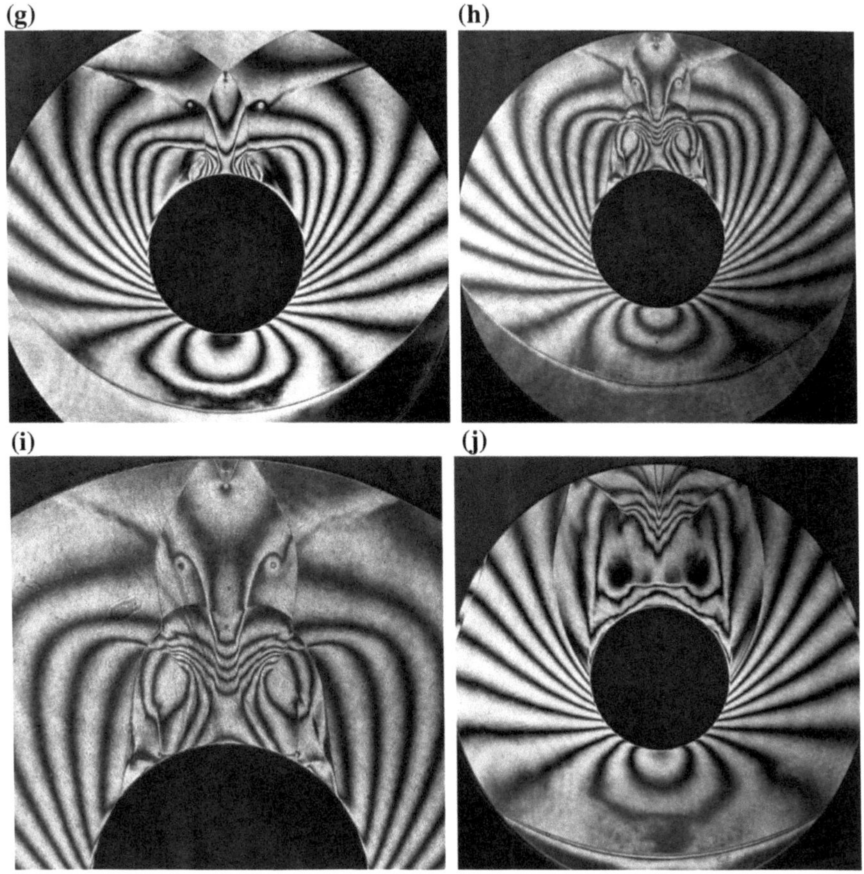

Fig. 4.3 (continued)

The polyatomic gases are favorable media from the point of view of interferometric visualization. The Gladstone-Dale constant in CO_2 and air are 0.00045 and 0.00027. respectively and the sensitivity of the fringes are proportional to the Gladstone-Dale constant. Therefore, the sensitivity of CO_2 is 1.7 times higher than that of air. This effect is much pronounced in $SF_{6.}$

The specific heats ratio, γ, in monoatomic gases is 1.667, that in air is 1.4, that in CO_2 is 1.29, and that in SF_6 is 1.08. The reflected shock wave will bifurcate due to the interacting with the sidewall boundary layer. With the value of γ approaching to unity, the range of Ms at which the bifurcation occurs becomes wider and the effect of such an interaction is more pronounced. The interaction between the reflected MS and the boundary layer along the cylinder surface is enhanced with the decrease in the value of γ (Mark 1956).

In monatomic gas, on the contrary, the effect of the reflected shock wave/ boundary layer is minimized.

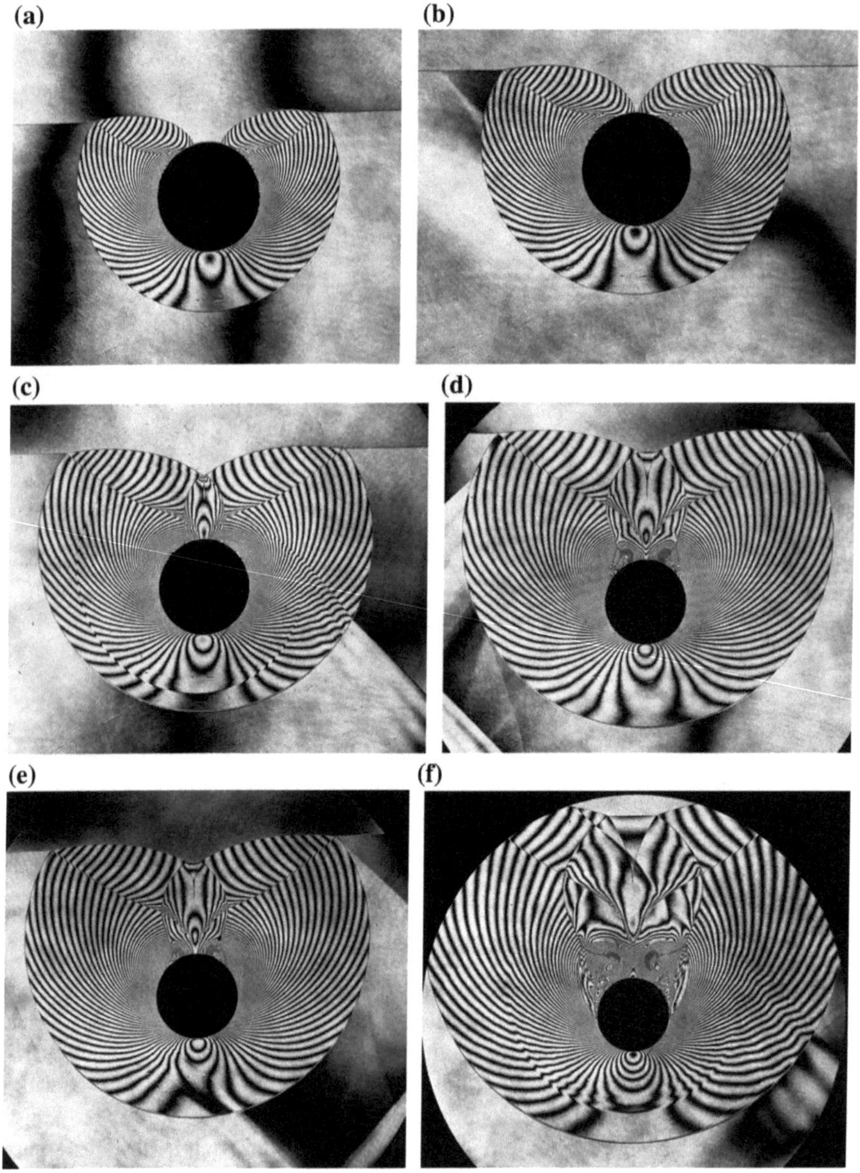

Fig. 4.4 Evolution of shock wave interaction with a 20 mm diameter cylinder for Ms—1.70, Re = 0.3 × 10^5 in air at 900 hPa, 2891.1 K: **a** #88012703, Ms = 1.673; **b** #88012705, Ms = 1.680; **c** #88012706, Ms = 1.678; **d** #88012707, Ms = 1.719; **e** #88012801, Ms = 1.723; **f** #88012708; Ms = 1.687, 900 hPa; **g** 88012802 623 μs Ms = 1.701; **h** enlargement of (**g**); **i** #88012715, Ms = 1.703; **j** #88012709, Ms = 1.713; **k** 88012803, Ms = 1.723; **l** #88012804; Ms = 1.681; **m** #88012710, Ms = 1.737; **n** #88012714, Ms = 1.675; **o** #88012809, Ms = 1.728; **p** #88012807. Ms = 1.672

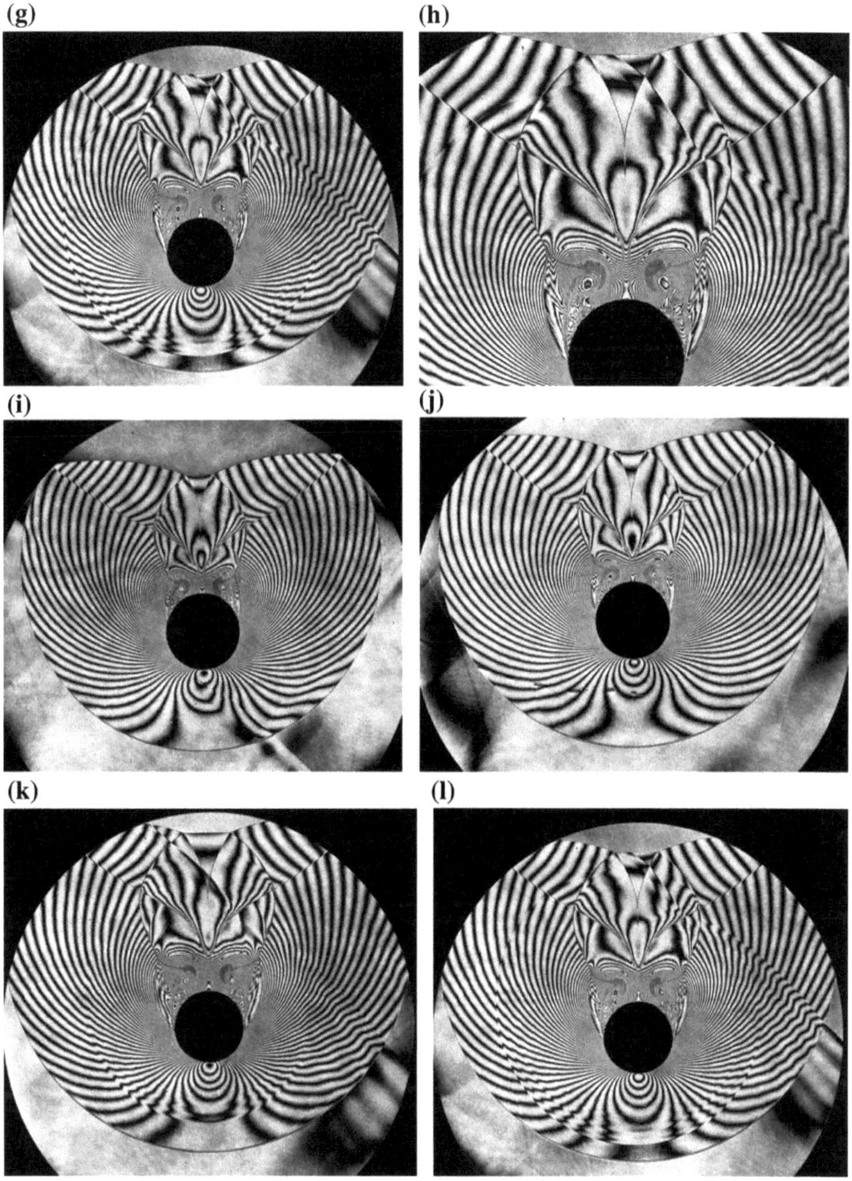

Fig. 4.4 (continued)

Figure 4.6 show the evolution of shock wave interaction with a 20 mm cylinder for Ms = 2.20 in CO_2 at 300 hPa, 289.1 K. Experiments were performed in a 60 mm × 150 mm conventional shock tube. Fringes in a densely populated area are hardly resolved but the density distribution precisely. In Fig. 4.6a–f, nearly

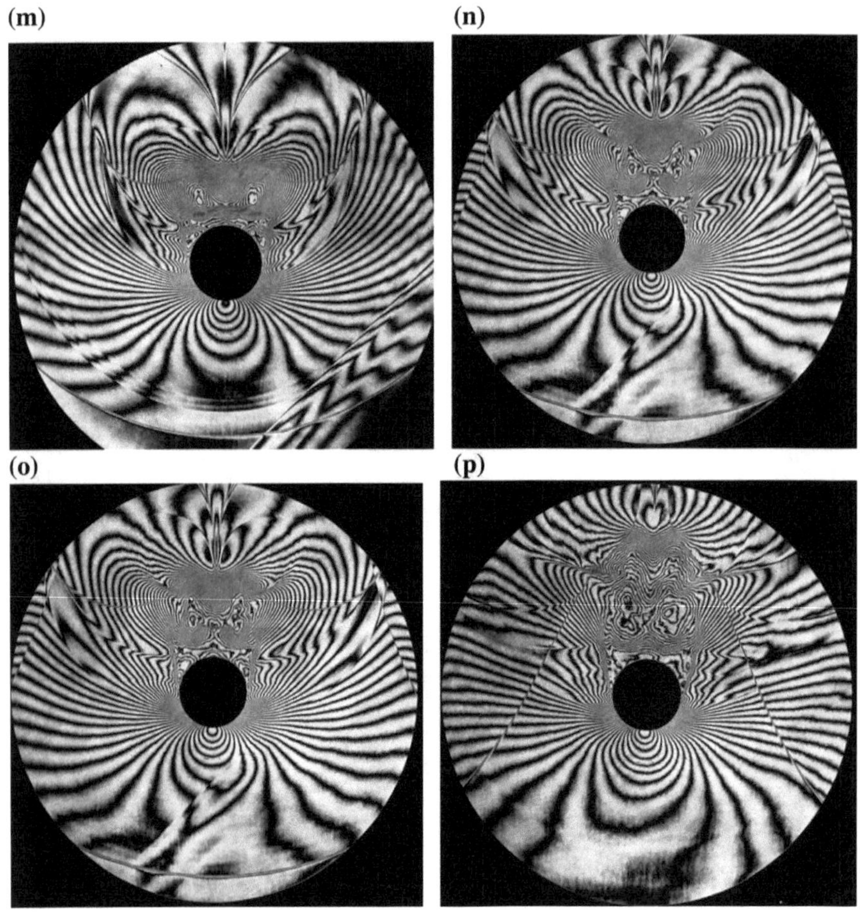

Fig. 4.4 (continued)

steady supersonic shock tube flows were built up so that the shock stand-off distance over the 20 mm diameter cylinder was maintained for a while.

4.1.3 *Cylinder in SF_6*

Figure 4.7 show sequential observations of shock wave/cylinder interaction in SF_6 at 100 hPa. The RS interacted with the sidewall boundary layer and caused a wide bifurcation region which were missing in Fig. 4.6.

Figure 4.8 show later stage of shock wave interaction with a 20 mm diameter cylinder for Ms = 4.20 in SF_6 at 20 hPa, 290.2 K. The RS along the frontal surface of the cylinder interacted with the boundary layer developing along the shock tube

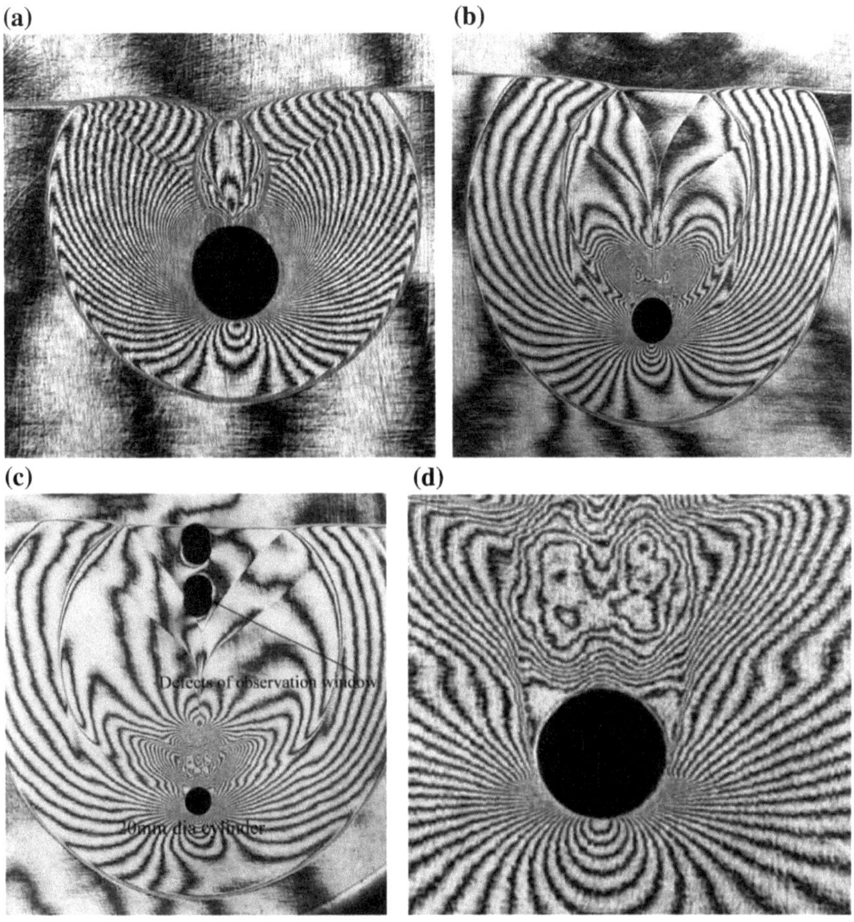

Fig. 4.5 Interaction of shock wave with a 20 mm diameter cylinder for Ms = 1.66 in CO_2 at 500 hPa, 290.2 K: **a** #89030204, Ms = 1.67; **b** #89030209, Ms = 1.66; **c** #89030211, Ms = 1.66 l; **d** enlargement of (**c**)

sidewall. The boundary layer was significantly bifurcated, which resulted in irregularly shaped fringes distributed at the frontal side of the cylinder.

4.1.4 Cylinder in Dusty Gas

In Fig. 2.14, the shock wave interaction with a 10 mm diameter cylinder installed in the test section of a dusty gas shock tube flow was investigated. The experiments were conducted in a 30 mm × 40 mm conventional shock tube which was already explained in Fig. 2.14. The dust particles under study were fly ashes of about 5 μm

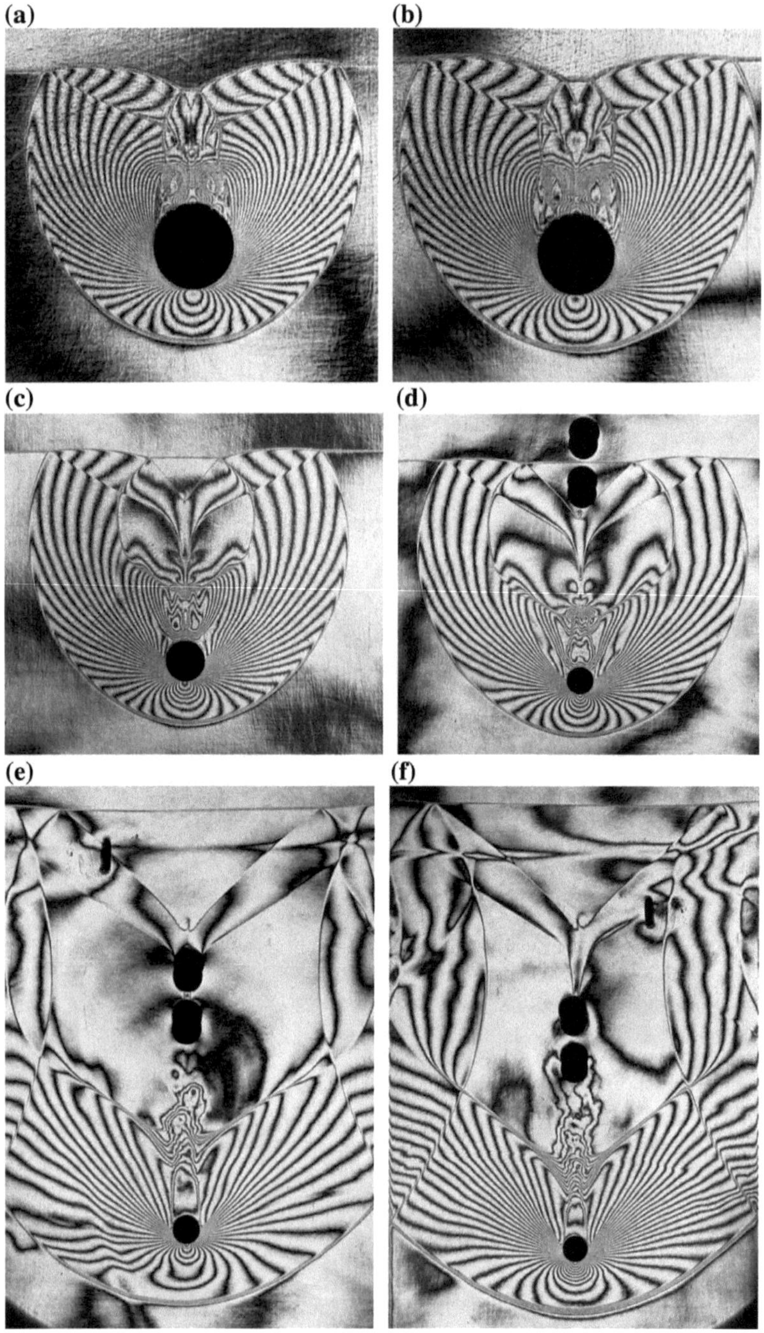

Fig. 4.6 Evolution of shock wave interaction with a 20 mm diameter cylinder for Ms = 2.20 in CO_2 at 300 hPa, 289.1 K: **a** #89030313, Ms = 2.21; **b** #89030312, Ms = 2.24; **c** #89030311, Ms = 2.24; **d** #89030403, Ms = 2.20; **e** #89030406, Ms = 2.28; **f** #89030404

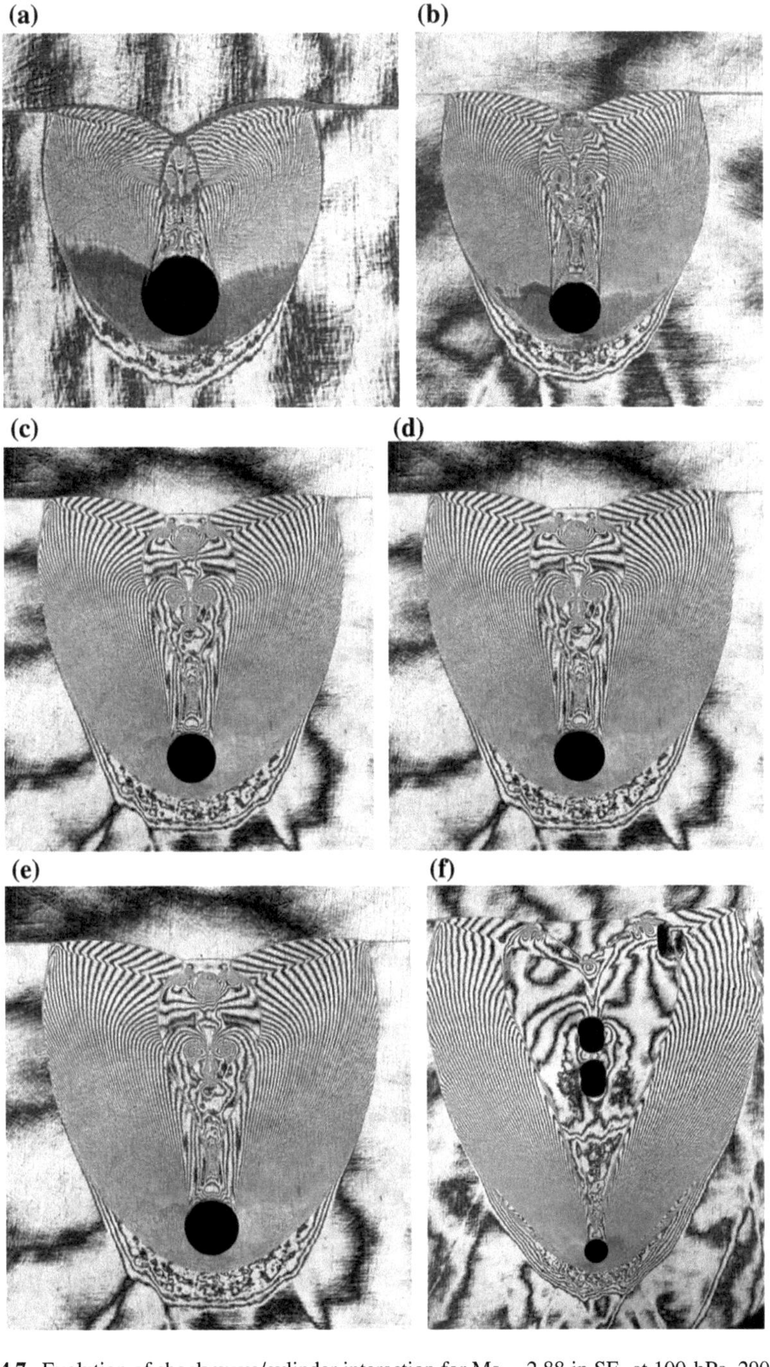

Fig. 4.7 Evolution of shock wave/cylinder interaction for Ms = 2.88 in SF_6 at 100 hPa, 290.2 K: **a** #89030606, 140 μs from trigger point, Ms = 2.86; **b** #89030604, 180 μs, Ms = 2.88; **c** #890306032301, 200 μs, Ms = 2.89; **d** #89030602, 220 μs, Ms = 2.24; **e** #89030705, 300 μs, Ms = 2.82; **f** #89030709, 300 μs, Ms = 2.88; **g** #89030711, 400 μs, Ms = 2.83; **h** #89030713, 450 μs, Ms = 2.81

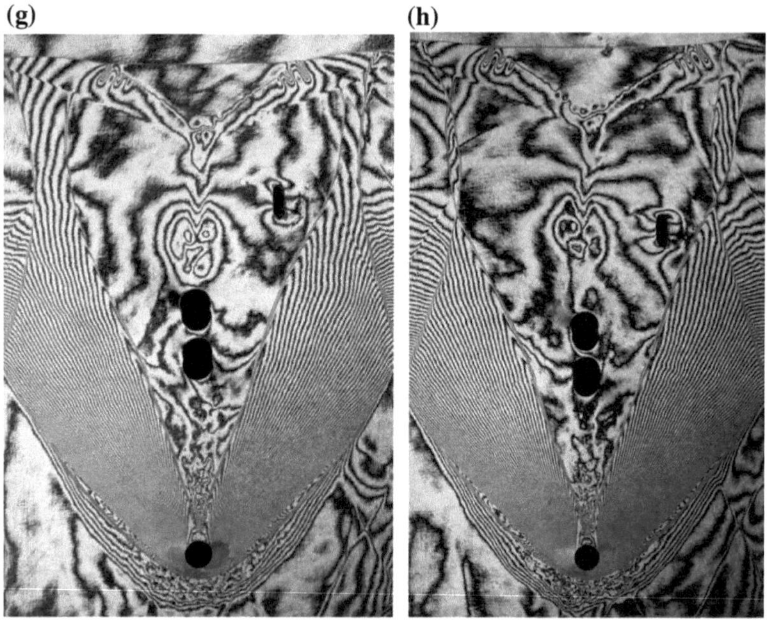

Fig. 4.7 (continued)

diameter. The dust loading ratio was defined as a ratio of mass of dust particle to the mass of air volume under study and was approximately 0.02 (Sugiyama et al. 1988). Figure 2.14 explained the dust circulation system: the dust particles were supplied from the dust hopper and recovered at the dump tank via a filter separating the dust particles from the air. The shock wave Mach number Ms ranges from 1.3 to 2.15 in air and the corresponding Reynolds number referred to the cylinder diameter ranges from 6.5×10^4 to 1.6×10^5, respectively.

Figure 4.9 show shock wave reflections over the cylinder for Ms = 1.3 in dusty air. The procedure of exposures were as follows: The room light was off and the holographic film was placed on the film holder. The first exposure was conducted before the dust circulation started. The film was covered with a thick cloth. The dust particles circulation was started and the room light was on. The dust concentration was continuously monitored. Several minutes later when the dust concentration reached to a given value, the dust circulation was stopped and the room light was off. Then second exposure was conducted. Meantime, the whole system was kept motionless except the dust circulation systems.

The present optical arrangement is similar to the shadowgraph and hence unreconstructed holograms in Fig. 4.9 are equivalent to direct shadow pictures. Figure 4.9a shows an interferogram of a dusty gas shock tube flow. But its fringe distribution never shows any deviations from that of dust free flows.

Grey regions around the cylinder surface on the unreconstructed holograms seen in Fig,4.9b show the dust free region in which the dust particles were removed due

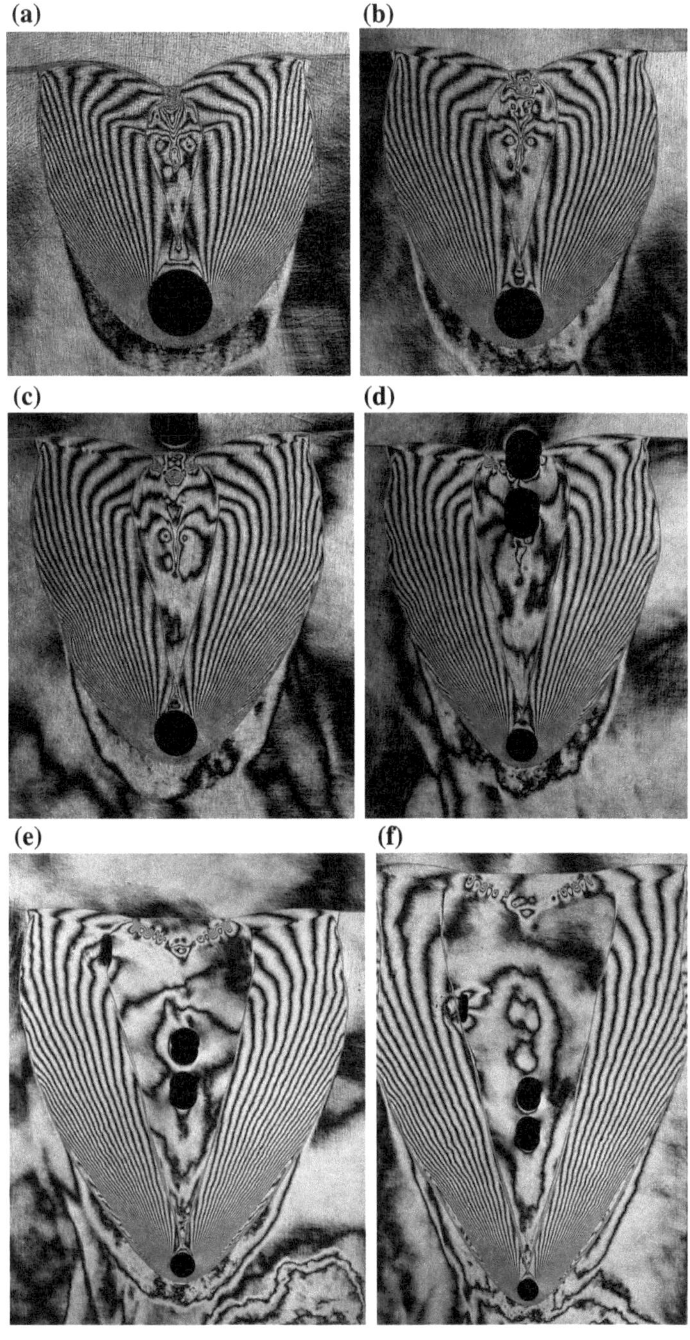

Fig. 4.8 Evolution of shock wave/cylinder interaction at later stage for Ms = 4.20 in SF_6 at 20 hPa, 290.2 K; **a** #89030718, 340 μs from trigger point, Ms = 4.18; **b** #89030717, 370 μs, Ms = 4.20; **c** #89030716, 400 μs, Ms = 4.22; **d** #89030715, 430 μs, Ms = 4.22; **e** #89030804, 580 μs, Ms = 4.05; **f** #89030805, 630 μs, Ms = 4.18

to centrifugal forces. At the second exposure, the *OB* carried all the holographic information which included the light absorption by the dust particles and the light scattering with the dust particles. Although the loading ratio was 0.02, the presence

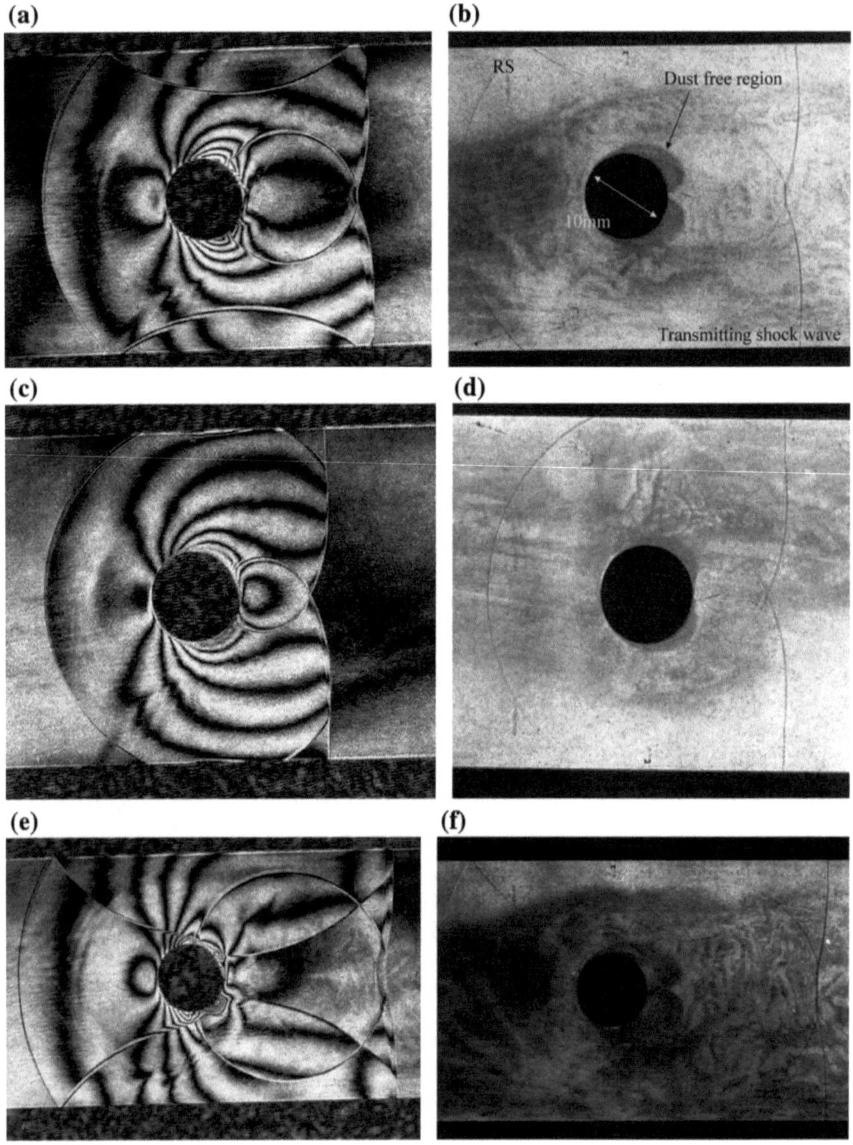

Fig. 4.9 Shock wave/cylinder interaction in a dusty gas at 1013 hPa, 186 K: **a** #86111207, 340 μs time from the trigger point, Ms = 1.320; **b** unreconstructed hologram of (**a**); **c** #86111202, Ms = 1.304; **d** unreconstructed hologram of (**c**); **e** #86111208, 350 μs, Ms = 1.320; **f** unreconstructed hologram of (**e**); **g** #86111504 400 μs Ms = 2.174; **h** unreconstructed and hologram of (**g**); **i** #86111507 360 μs Ms = 2.200; **j** unreconstructed hologram of (**i**)

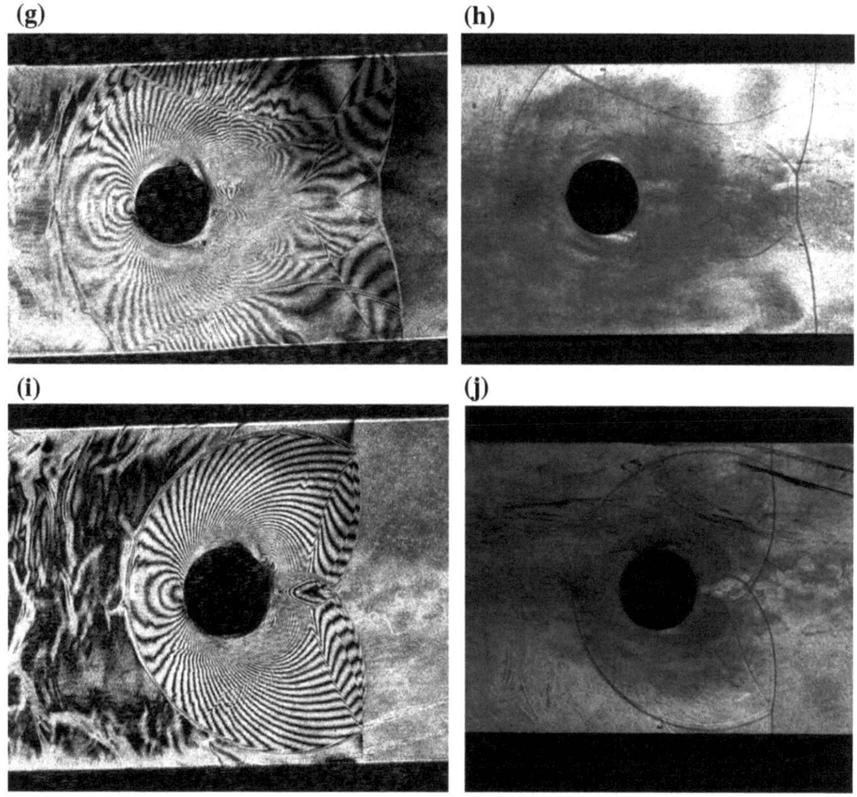

Fig. 4.9 (continued)

of the uniformly distributed dust particles scattered the laser beam. Due to the light scattering and diffraction from the particles, the background contrast looked brighter. This is the so-called the Mie scattering, which occurs when the particles having larger sizes than the light wave length scatter the source light. Then in the region where the dust particles were completely absent, the contrast of background was darker than the region in which the dust particles are uniformly distributed. Molecules in our sky scattered the sun light so that the sky is blue due to the so-called Rayleigh scattering. On the moon surface there are no air no dust particles in the sky and the sun light was not scattered. Then the sky looks dark.

The dust free region is created by vortex motion which was created due to the interaction of the reflected MS with the boundary layer developing along the cylinder surface. The flow in the vortex induced centrifugal forces strong enough to eject the dust particles. As the shock tube size is small, the proportion of vortex is large if compared with a large shock tube, for example, as shown in Fig. 4.2. Izumi (1988) numerically simulated the presence of the dust free region using the TVD scheme. The dust free region existed in nearly the same area as dark pattern appeared.

4.1.5 Rotating Cylinder

When a shock wave propagated along a rotating cylinder, the transition form a RR to SMR would be different along the cylinder surface depending on the direction of rotation. Just to observe this effect, a hollow cylinder of radius R made of aluminum alloy was rotated in the 60 mm × 150 mm diaphragm-less shock tube. Using an inverter the rotation speed ω was about 2000 rad/s to counter-clockwise rotation and the angular velocity Rω was about 50 m/s. Figure 4.10a shows the test section coated with the fluorescent paint on the shock tube wall and the cylinder surface. Diffuse holographic observation was conducted. In Fig. 4.10b, a result of double exposure diffuse holographic observation is shown. A shock wave of Ms = 1.19 propagated along the cylinder from left to right hand side. In Fig. 4.10b along the upper side, the flow direction was counter clockwise. The transition from RR to MR was retarded, whereas along the lower side the transition was promoted. The positions of triple points on the upper and lower sides were clearly observed and

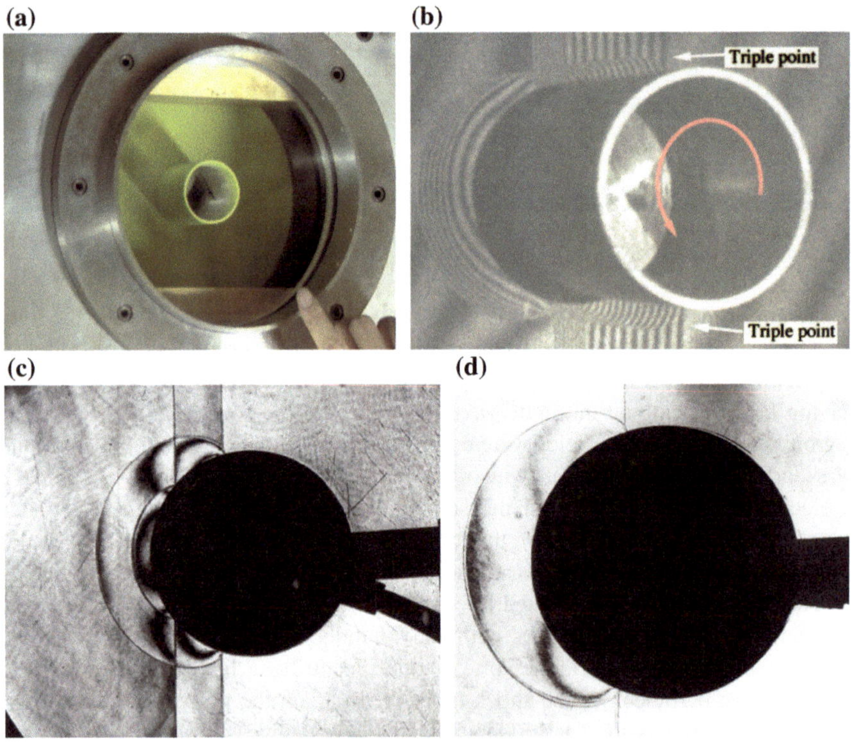

Fig. 4.10 Shock wave reflection from a counter-clockwise rotating cylinder 25,000 rpm of 50 mm diameter for Ms = 1.19 in air: **a** #99061502 l; **b** #99061503; **c**, **d** 70 mm cylinder (Yada 2001)

then the height of the MS was short on the upper side but was long on the lower side. The image of the reflected shock wave was projected on the sidewall (Sun et al 2001). Figure 4.10c, d show a double exposure interferogram conducted later stage (Yada 2001). The test piece was supported out side the test section. The visualization was conducted by the standard image hologram. The shapes of Mach stems on the upper and lower sides are slightly different indicating the difference of counter flows. This was a preparatory experiment and in the future the experimental system would be refined to double the rotation speed.

4.1.6 Partially Perforated Cylinders

Figure 4.11 show sequential observations of shock wave diffractions over a 100 mm diameter hollow cylinder made of brass. The cylinder had 10 mm wall thickness and was sandwiched between two acrylic observation plates of the test section of the 60 mm × 150 mm conventional shock tube for Ms = 1.165 in atmospheric air. As shown in Fig. 4.11a, along the upper side of the cylinder, fine slots of 1.0 mm wide and 1.5 mm interval were distributed and the lower part was a 10 mm thick brass tube connected to the slots. The slots were supported by 2.5 mm wide rims on the bottom. Then the ratio of opening was 40%. The structure was so delicate that it survived only for 50 runs.

Figure 4.11a shows the fine slot structure and the early stage of its interaction with the IS. The reflection pattern on the slotted surface was a subsonic regular reflection or in short SbRR, whereas it was a simple Mach reflection SMR on the solid surface. On roughened or perforated wedge surfaces, θ_{crit} is smaller than the solid surface. In Fig. 4.11b, c, the effect of slotted wall was distinguished comparing the reflection pattern between those along the upper and lower surfaces. The trajectory of the TP position along the slotted wall is lower than that along the solid surface.

As the test piece was not rigidly sandwiched between the observation windows so that the IS or the transmitting shock wave leaked through the narrow gap forming faint shadows of the shock wave leakage inside the cylinder.

The shock waves transmitting through the slots were reflected from the 2.5 mm wide rims at the bottom of slots. From the sequential observations, the critical transition angle θ_{crit} cn be readily estimated. The value of the resulting θ_{crit} is much smaller than that over solid cylinders.

Figure 4.12 show the evolution of shock wave for Ms = 1.52 propagating along the slotted cylinder. The transmitting shock wave through the slotted wall propagated obliquely inside the cylinder. The convexly shaped shock wave propagated toward the convex wall and its reflection formed the shock wave focusing as seen in Fig. 4.12j.

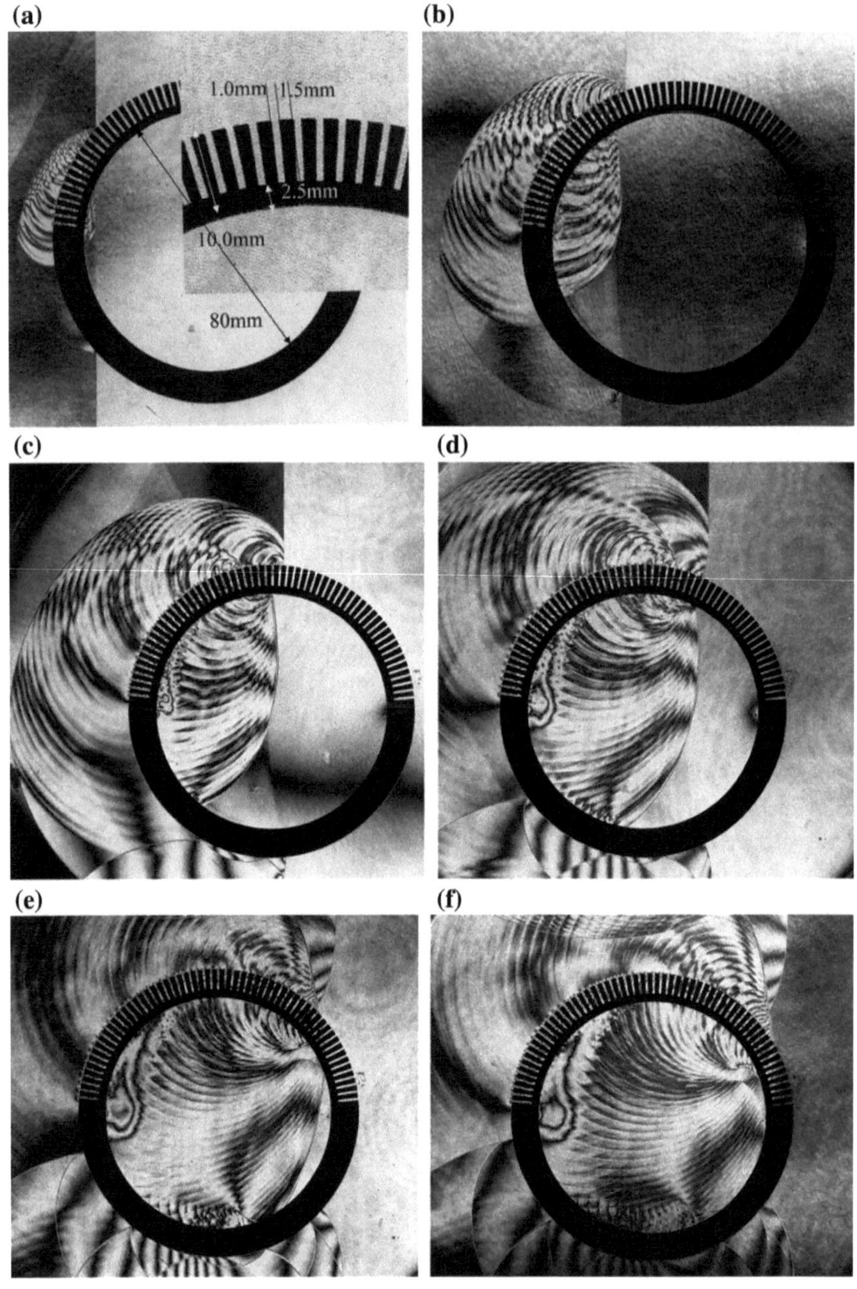

Fig. 4.11 Shock wave interaction with a perforated cylinder for Ms = 1.17 in atmospheric air, at 288 K: **a** #86121908, 275 μs from trigger time, Ms = 1.165; **b** #86121904, 375 μs, Ms = 1.171; **c** #86121911, 450 μs, Ms = 1.170; **d** #86121912, 475 μs, Ms = 1.170; **e** #86121914, 525 μs, Ms = 1.171; **f** #86121915, 550 μs, Ms = 1.161; **g** #86121916, 575 μs, Ms = 1.174; **h** #86121917 600 μs, Ms = 1.166

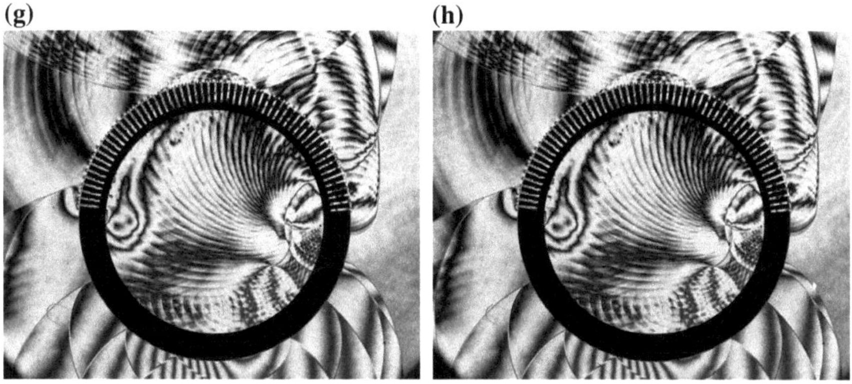

Fig. 4.11 (continued)

4.1.7 Tilted Cylinders

The shock wave interaction with a cylinder was one of the basic research topics of the shock wave research. The critical transition angle θ_{crit} of reflected shock waves over cylinders were experimentally determined. It would be a natural course of questions to know how the value of θ_{crit} over tilted cylinders would be different from that from the two-dimensional cylinders. Tilted cylinders were installed in the test section of the 60 mm × 150 mm conventional shock tube.

4.1.7.1 30° Tilted Cylinder

Figure 4.13a–f show sequential observation of reflection and diffraction of shock waves over a 30 mm diameter cylinders tilted 30°. The observations were conducted by the double exposure holographic interferometry for Ms ranging from 1.7 to 3.05 in air. All the reflection patterns are three-dimensional and show a SMR along the frontal side. If one can visualize the reflection patter along the sidewall and the other side of the tilted cylinder, the MS would reflect and causing complex wave interactions. Probably at this inclination angle of cylinder, the SMR will prevail. It will be interesting to reproduce the shock wave reflection over the tilted cylinders in the similar physical significance to reproduced the shock wave reflection over a tilted cone as shown in Sect. 2.1.7.5.

4.1.7.2 45° Tilted Cylinder

Figure 4.14a–c show sequential observations of the shock wave reflection and diffraction along a 45° tilted cylinder for Ms = 1.26 in air. Figure 4.14d–f show the sequential observation of the shock wave reflection and diffraction along a 45° tilted

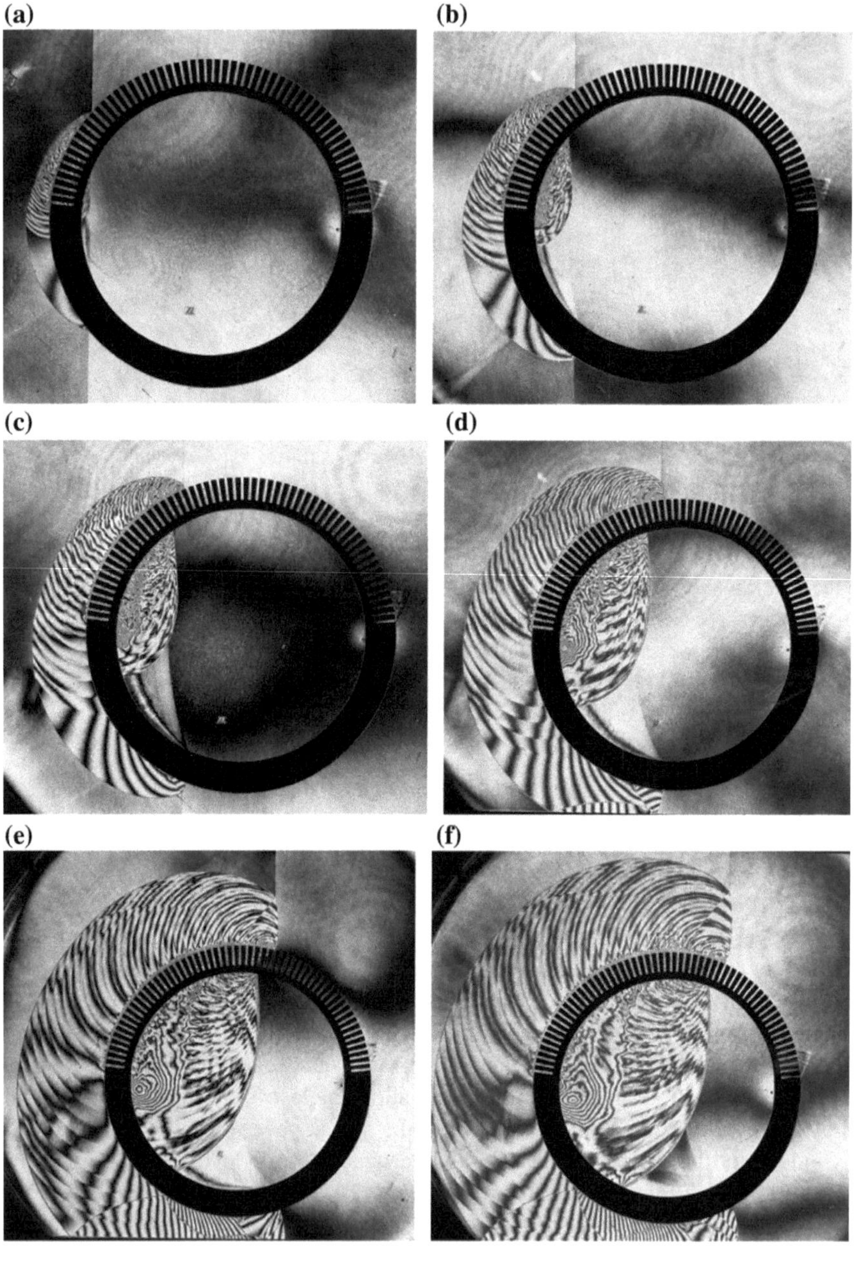

Fig. 4.12 Shock wave interaction with a perforated cylinder at Ms = 1.52 in air at 866 hPa, 289.0 K: **a** #86121918, 135 μs from the trigger point, Ms = 1.512; **b** #86121919, 155 μs, Ms = 1.512; **c** #86121920, 175 μs, Ms = 1.528; **d** #86121904,195 μs, Ms = 1.517; **e** #86121923, 235 μs, Ms = 1.522; **f** #86121924, 255 μs, Ms = 1.517; **g** #86121925, 275 μs, Ms = 1.517; **h** #86121926, 295 μs, Ms = 1.512; **i**#86121927, 315 μs, Ms = 1.513; **j** #86121930, 375 μs, Ms = 1.514

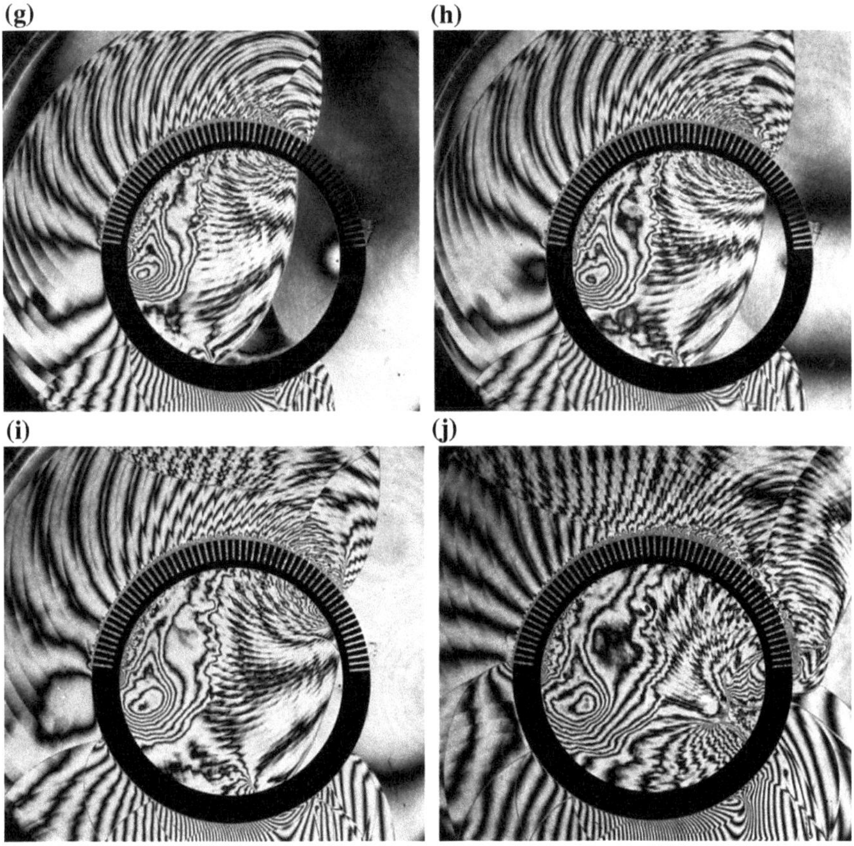

Fig. 4.12 (continued)

cylinder for Ms = 1.70. In Fig. 4.14a, the reflection pattern is a supersonic regular reflection SuRR, whereas in Fig. 4.14b–f, it is SMR. When the shock wave passed the top corner, the reflection pattern is SMR in a similar manner to the case of a convex double wedge. In Fig. 4.14a, a second discontinuous line visible closer to the tilted cylinder is a reflection of the incident shock wave from the side walls.

4.1.7.3 60° Tilted Cylinder

Figure 4.15 show sequential observation of the reflection and diffraction of shock waves over 60° tilted cylinder. The reflection patterns are SuRR all over the cylinder surface. In Fig. 4.15d, when the shock wave passed the corner of the flat top of the tilted cylinder, the reflection pattern became SMR. This trend is similar to

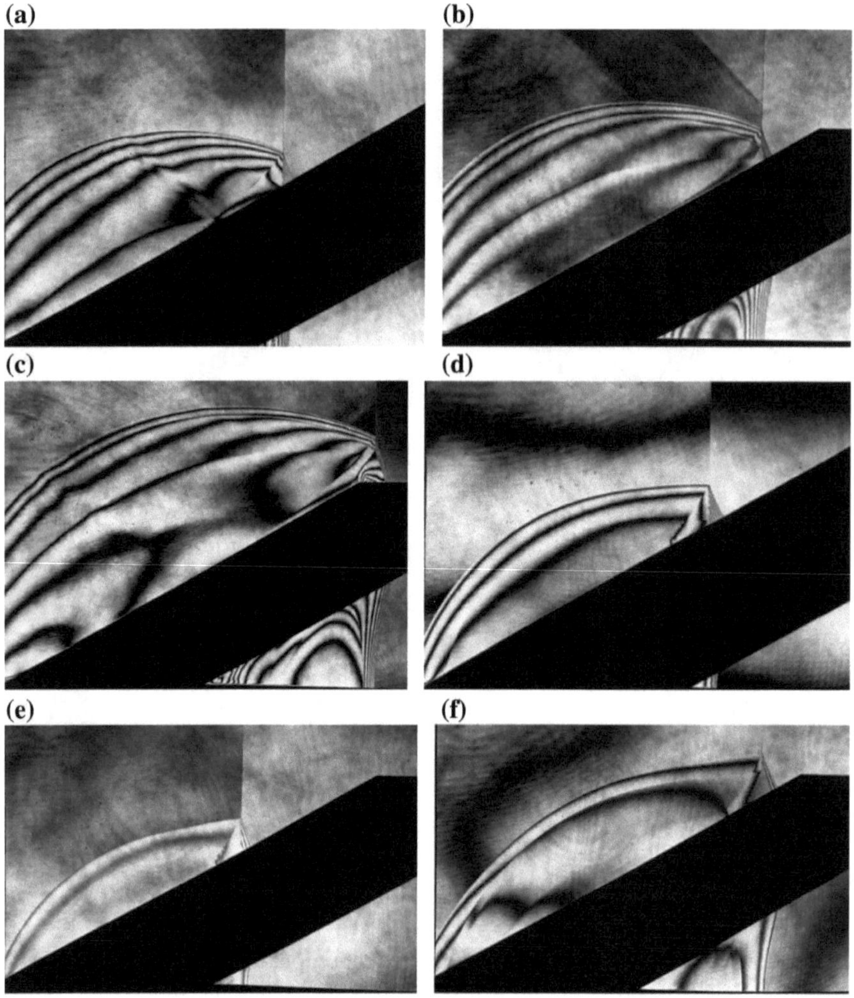

Fig. 4.13 Shock wave diffraction over a 30° tilted cylinder: **a** #87121501, Ms = 1.709, 500 hPa 292.7 K; **b** #87121502, Ms = 1.712, 500 hPa 292.7 K; **c** #87121503, Ms = 1.716, 500 hPa 291.8 K; **d** #87121511, Ms = 2.568, 130 hPa, 294.3 K; **e** #87121513, Ms = 3.067, 50 hPa, 294 K; **f** #87121512, Ms = 3.053, 50 hPa, 294 K

the shock wave propagation over a convex wall. Other than this effect, the reflected shock waves will transit to the SMR when the local inclination angle approaches to the critical transition angle θ_{crit} somewhere in the sidewall. However, there is no method to quantitatively visualize the transition of the reflected shock wave over tilted cylinder.

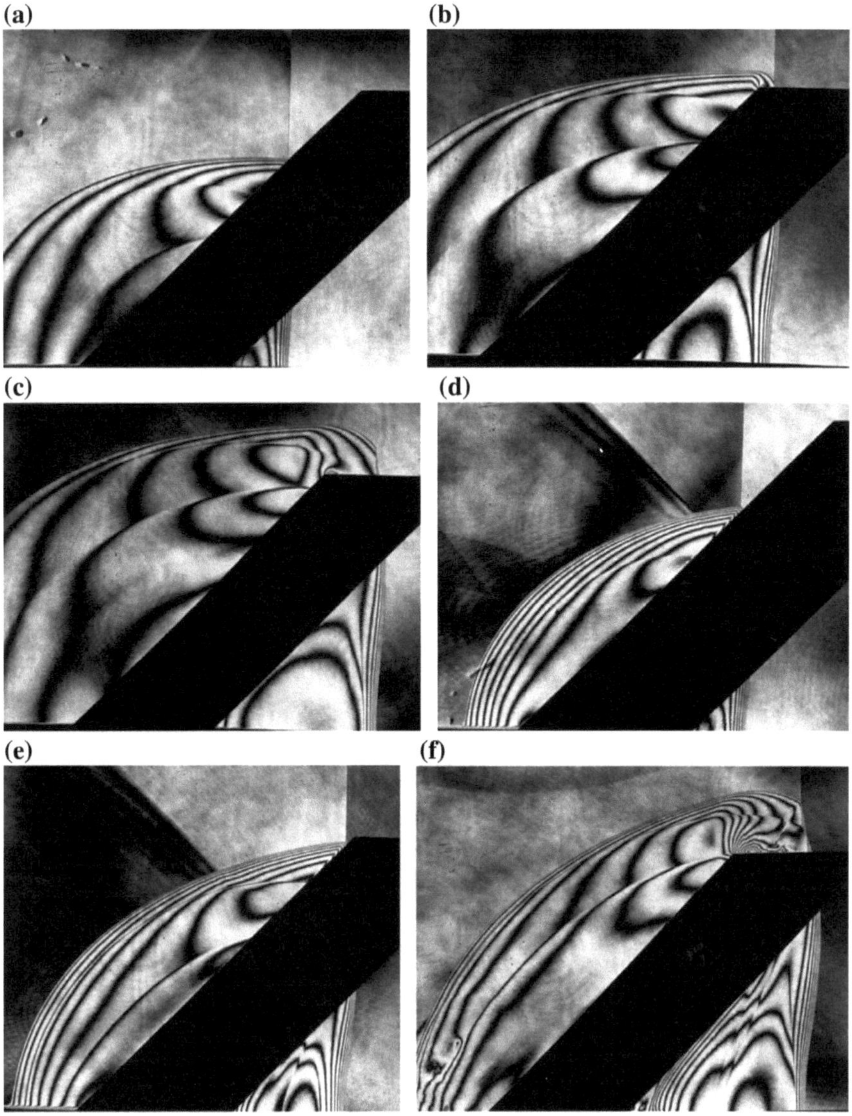

Fig. 4.14 Shock wave reflection and diffraction from a 45° tilted cylinder: **a** #87121204, Ms = 1.258, at 1013 hPa, 293.1 K; **b** #87121206, Ms = 1.259, 1013 hPa, 293.6 K; **c** #87121207, Ms = 1.263 at 1013 hPa, 293.6 K; **d** #87121211, Ms = 1.721 at 500 hPa, 293.0 K; **e** #87121212, Ms = 1.706, at 500 hPa, 291.0 K; **f** #87121213, Ms = 1.711 at 500 hPa, 292.6 K

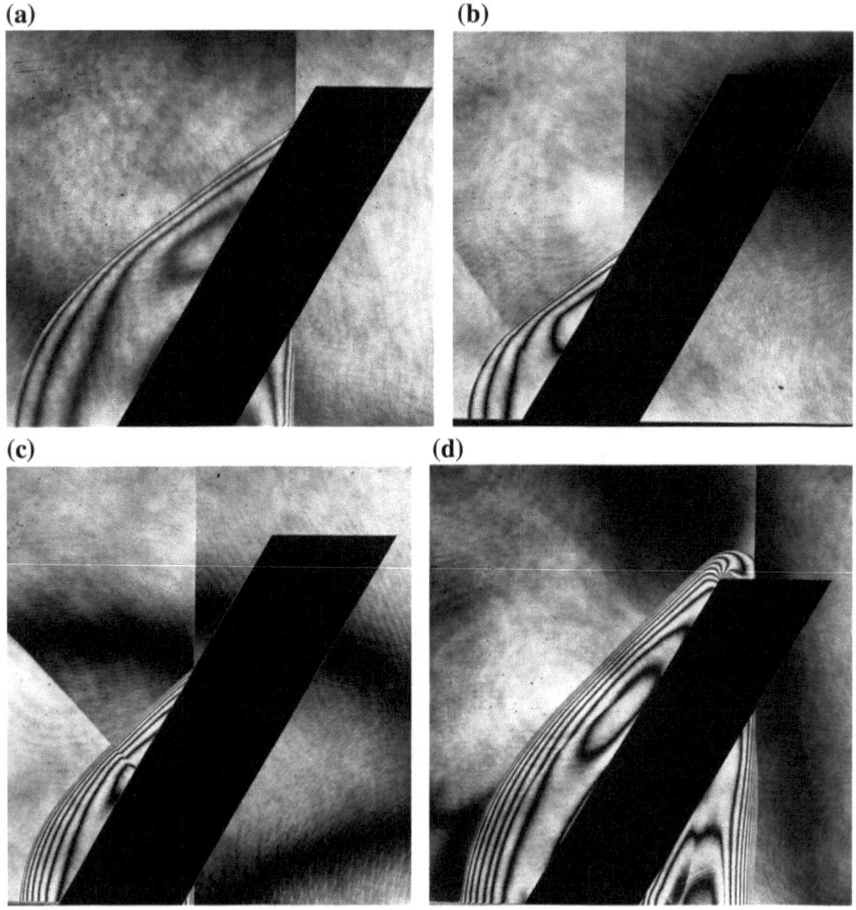

Fig. 4.15 Shock wave reflection and diffraction from a 60° tilted cylinder: **a** #87121109, Ms = 1.260 at 1013 hPa 294.4 K; **b** #87121110, Ms = 1.580 at 1013 hPa, 294.4 K; **c** #87121108, Ms = 1.713 at 500 hPa 294.5 K; **d** #87121106, Ms = 1.722 at 500 hPa 294.5 K

4.1.7.4 Truncated Vertical Cylinder

This is a very trivial experiment. The shock wave reflection and transmission over a truncated 30 mm diameter cylinder standing in a 60 mm × 150 mm shock tube was visualized. Figure 4.16 show the sequence of shock wave transitions for Ms—1.70. The reflection pattern is RR at the frontal surface but the pattern of the shock wave at the side and the top of cylinder is SMR. Unlike the shock interaction with a two-dimensional cylinder, the three-dimensional diffraction at the edge of the upright cylinder shows a complex pattern.

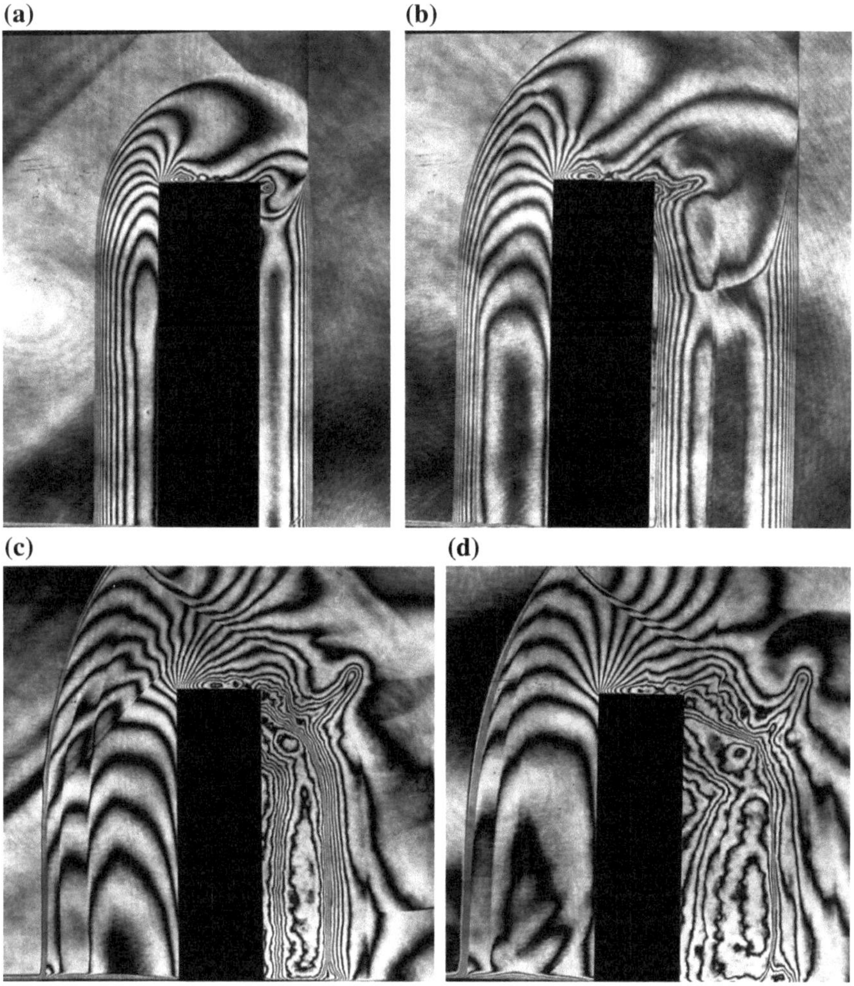

Fig. 4.16 Shock wave reflection and diffraction from truncated cylinder for Ms = 1.70 in air at 500 hPa, 295.1 K: **a** #87121001, Ms = 1.690; **b** #87121002, Ms = 1.696; **c** #87121004, Ms = 1.707; **d** #87121005, Ms = 1.878

4.1.8 Diffuse Holographic Observation Over a 60° Tilted Cylinder

This is a summary of the previous experiments. A 60° tilted cylinder was placed in the 60 mm × 150 mm conventional shock tube. The cylinder and the entire test section of the shock tube was coated with a fluorescent paint in the same way as the diffuse holographic observation. Then synchronizing the shock wave motion, the *OB* illuminated the shock tube test section from a slightly oblique direction. The *OB*

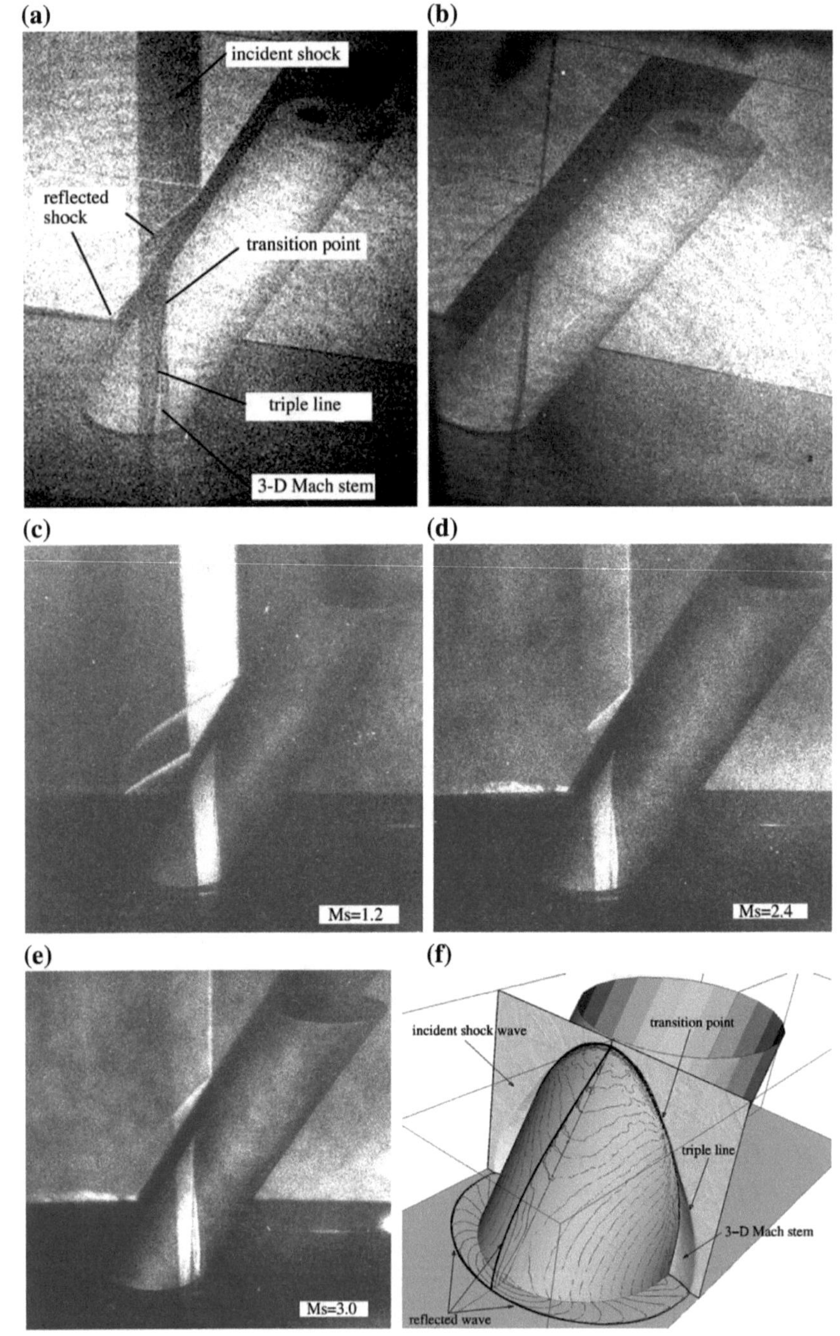

Fig. 4.17 Shock wave reflection over a tilted cylinder: **a** explanation of 60° tilted cylinder for Ms = 1.50; **b** holographic display of Ms = 1.50 different view angle; **c** holographic display of Ms = 1.20 different view angle; **d** holographic display of Ms = 2.4; **e** holographic display of Ms = 3.0; **f** determination of critical transition angle

was diffused passing a smoked glass plate. Then, the diffused *OB* reflected from the coated wall carried holographic information of the event and was recorded on a holo-film. Figure 4.17a is a reconstructed image which clearly shows the reflected shock wave from the cylinder for Ms = 1.5. In Fig. 4.17b, the reconstruction angle is normal to the view field (Timofeev et al. 1997).

A SuRR is observed on the frontal surface. The spot at which dark grey shadows discontinuously change the contract indicates the triple point and is clearly identified in Figs. 4.13, 4.14, 4.15 and 4.16. From these images, the θ_{crit} can be determined relatively accurately. Figure 4.17b–d show the three-dimensional images for Ms = 1.2, 2.4 and 3.0. The positions of the triple points on the side wall were determined. Timofeev et al. (1997) numerically identified the transition point of the shock wave reflection in Fig. 4.17f. Figure 4.17d shows that the transition from RR to SMR occurred at the angle similar to the transition over an elliptic cylinder. The boundary layer separations neither on the bottom wall nor on the truncated flat top are observed as seen in Figs. 4.13, 4.14, 4.15 and 4.16.

Figure 4.18 summarizes the θ_{crit} over the tilted cylinder against the inverse strength of shock wave ξ. The ordinate designates the θ_{crit} in degree and the abscissa shows inverse shock strength ξ. Red filled circles show the θ_{crit} over 60° tilted cylinder. A solid line shows the detachment criterion (Courant and Friedrichs 1948). Open circles show experimental results of wedges (Smith 1948) and black filled circles show the results of water wedges as discussed in Sect. 2.6.1. Filled triangles show the results of cones (Yang 1995). The transition over the tilted cylinder is 3-D phenomena but the results of θ_{crit} agree reasonably well with results over wedges, in particular, at stronger shock waves. For weak shock waves, the present results depart from the wedge and cone experiments.

This is a preliminary experiments of tilted 60° and should be compared with transition over concave wall experiments and to refer numerical results of solving Navier-Stokes Equations.

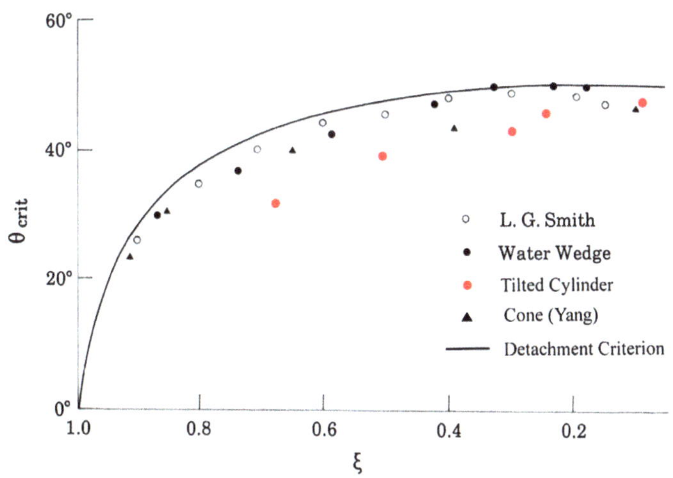

Fig. 4.18 Critical transition angle θ_{crit} versus inverse shock strength ξ

4.2 Unsteady Drag Force on a Sphere

Shock wave reflections over spheres are one of the fundamental experimental topics of shock wave dynamics. In the past there were efforts for measuring drag forces of shock wave laden spheres. The aim of this experiment is the direct measurement of a drag force of a shock laden 80 mm diameter sphere (Tanno et al. 2004).

Figure 4.19a shows a vertical conventional shock tube having about 7 m in total height. The shock tube is comprised of a 1.8 m long and 250 mm diameter high pressure chamber made of stainless steel and a 3 m long and 300 mm × 300 mm cross sectional low pressure chamber, a 600 mm long and 300 mm × 300 mm cross sectional test section made of stainless steel, and a 1.5 m long and 1.0 m diameter dump tank. A double diaphragm system is used. The reproducibility of the IS was poor. A 80 mm diameter sphere made of aluminum alloy was suspended vertically and placed in a test section positioned just before the dump tank. Figure 4.19b, c show the view of the shock tube and a 80 mm diameter sphere model suspended in the test section. In order to reliably measure the drag force, an accelerometer (Endevco piezoelectric accelerometer 2250A-10, 80 kHz) was installed in the sphere model.

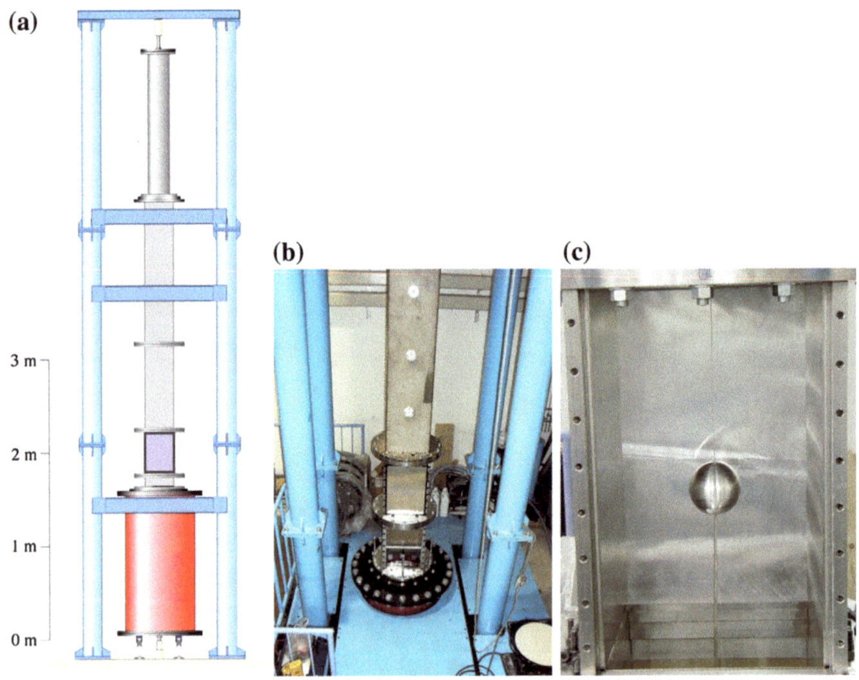

Fig. 4.19 Unsteady drag force of a 80 mm sphere: **a** facility; **b** vertical section; **c** 80 mm suspended sphere model

Heilig (1969) distributed pressure transducers over a 100 mm diameter cylinder installed in a shock tube and measured the drag force of a 100 mm diameter cylinder. He measured, for the first time, directly the time variation of pressure distributions over the cylinder exposed to the shock wave for Ms = 1.25 in air. He then reported that unsteady drag forces on a cylinder showed temporarily a peak value and monotonously reduced to the value of the drag force in the steady flow behind the shock wave. It was a question whether or not the unsteady drag force of a sphere might change in a similar fashion as that of cylinders. This was one of the motivations of the present experiment.

A 80 mm diameter sphere was suspended by a thin wire from just below the position of the double-diaphragm section along the center of the shock tube. When the shock impacted the sphere, the stress wave propagated along the wire and reflected back from the supporting point. The accelerometer installed in the sphere measured the acceleration of the shock laden sphere. The measurement would be terminated when the reflected stress wave from the support arrived at the sphere. However, the wire was so long that when the reflected stress wave arrived at the sphere, the sphere was already impinged by the reflected shock wave from the sidewall, which was the end of unsteady drag force measurement.

The output signal of the accelerometer was transmitted to the recorder through a cable from a hole on the bottom of the sphere. The uniform flow condition behind the IS was maintained for over 600 μs. This duration of time was long enough to measure the entire sequence of the drag forces. Applying the convolution function of the accelerometer to the output signal, the time variation of unsteady drag force was obtained. The convolution function was determined by measuring the frequency response of the accelerometer when it was impacted by a hummer.

Figure 4.20a–u show sequential schlieren pictures recorded by a high-speed video camera Shimadze SH 100 at 10^6 frame/s. The schlieren pictures were compared with interferometric images. The arrival of wavelets observed in Fig. 4.20e–n created pressure perturbations which were also detected by the accelerometer. In Fig. 4.20i–l, when the MS of the transmitting shock wave focused at the rear stagnation area, pressures detected by the accelerometer became maximal. In the two-dimensional shock wave cylinder interaction, the pressure at the rear stagnation point was just enhanced, when the MS merged. However, in the case of the sphere, it was focusing that the reflected MS merged at the rear stagnation point. Therefore, the pressure created by shock wave focusing was so high that the peak pressure exceeded the drag force. Hence, the drag force became negative. In Fig. 4.20s–u, the arrival of reflected shock waves from the sidewall terminated the presence of negative drag force. The measured drag force was compared with a result of the Navier-Stokes solver.

In Fig. 4.21, the measured drag force and numerical results were compared (Tanno et al. 2004). The ordinate denotes the drag forces in N and the abscissa denotes the elapsed time in μs. Green lines show output signal directly recorded by the accelerometer. Applying the deconvolution function to the measured force date, the unsteady drag force was deduced. The variation of the processed drag force was shown in a red line. The measured drag forces were compared with the numerical

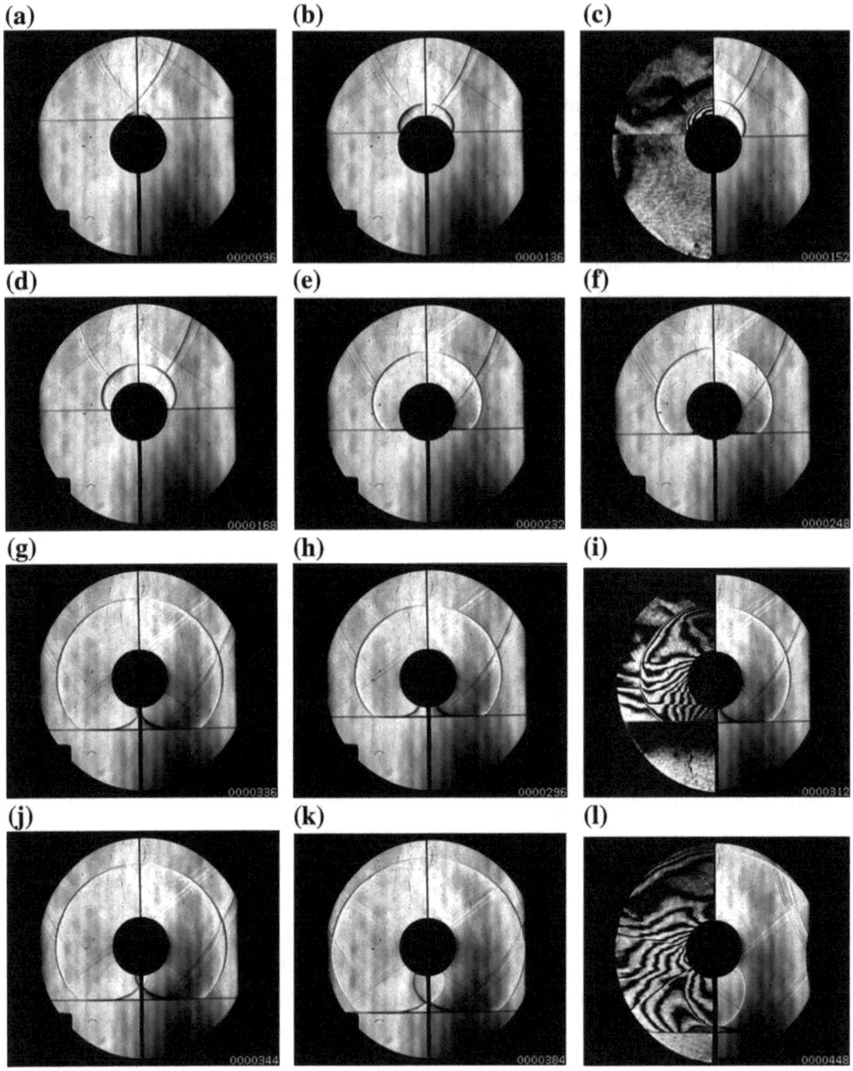

Fig. 4.20 Sequential observation of shock wave interaction with a 80 mm sphere for Ms = 1.25, comparison with interferograms

simulation solving the Navier Stokes Equations up to the elapsed time of 700 μs. The measured results agreed well with the result of the numerical simulation. The drag force had a maximum value then decreased to the value of the steady flow dag force. As seen in Fig. 4.20s–u, the focusing of MS at the rear stagnation point induced a peak pressure, and eventually the resulting drag force became negative. The negative drag force was maintained for about 150 μs. This is a unique feature of the sphere exposed by shock waves.

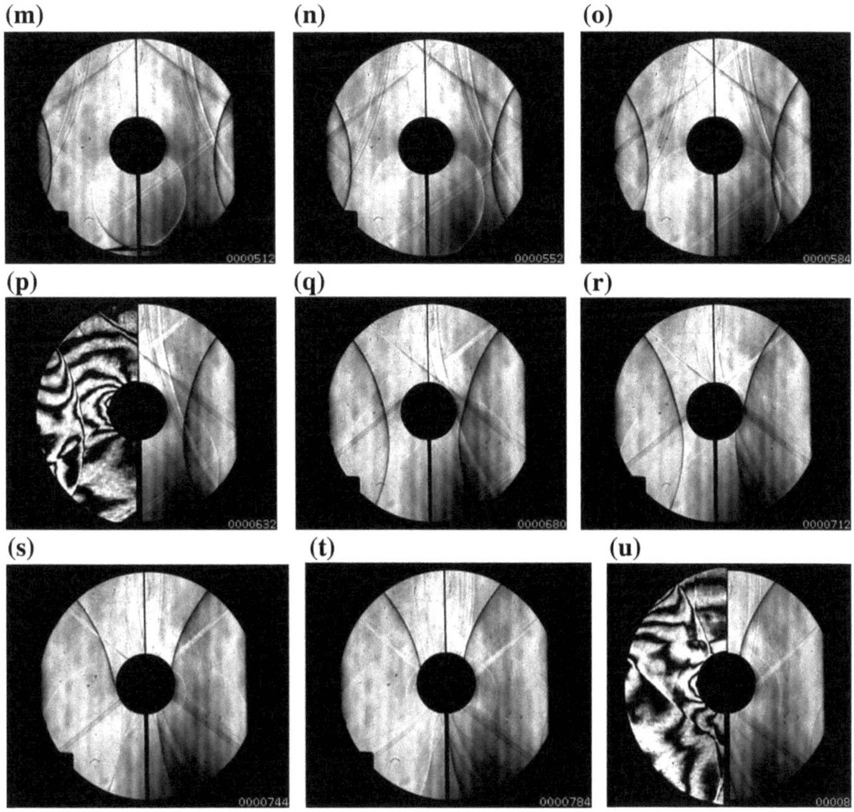

Fig. 4.20 (continued)

However, the presence of negative drag forces of shock laden spheres does not necessarily universally occur. Do dust particles of diameter about 5 μm as discussed in Sect. 4.1.4 have a negative drag forces when exposed to shock wave? Based on the comparison shown in Fig. 4.21, Sun et al. (2005) solved the Navier Stokes Equations and reproduced shock wave interaction for wide range of diameters of spheres ranging from 8 μm to 8 mm, in terms of the Reynolds number, Re ranging from 49 to 4.9×10^4 and Knudsen number, Kn ranging 9.4×10^{-2} to 9.4×10^{-6}.

The variation of so-defined drag coefficients, that is, the drag force normalized by condition behind the incident shock wave flow, was presented in Fig. 4.22. The ordinate denote the so-defined drag coefficient $C_D = 2f/\rho u^2 A$, where f, ρ, u, A are the drag force, the density and the particle velocity behind the shock wave, and the cross sectional area of the sphere under study, respectively. In conclusion, drag forces are consistently positive for spheres of 8 μm and 80 μm diameter whose Kn are still small enough to consider the continuum medium. The drag coefficient

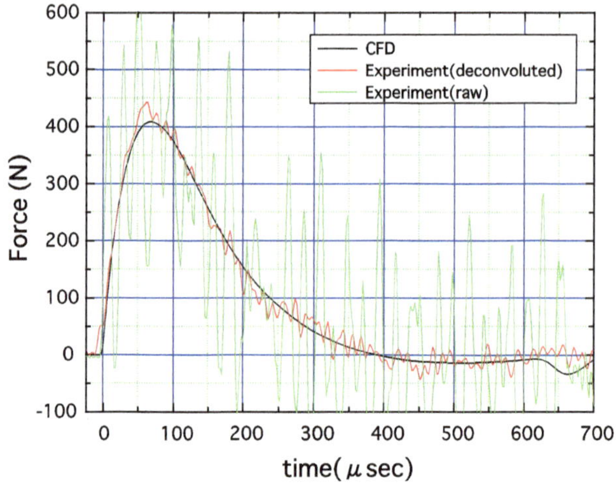

Fig. 4.21 Unsteady drag force over a 80 mm diameter sphere

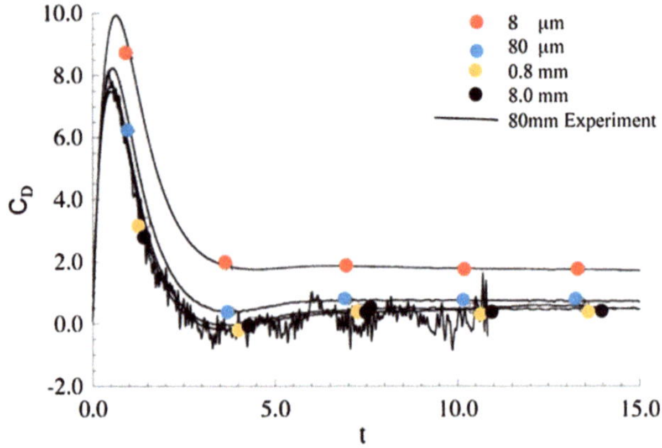

Fig. 4.22 Effect of time dependent drag force on diameter of sphere (Sun et al. 2005)

curves of 0.8 and 8 mm diameter spheres show nearly identical with the distribution in which negative pressure regions appear. It is noticed that in the shock wave interaction with a 10 mm diameter cylinder in the dusty gas shock tube flow, rarefied gas dynamic effects are negligible.

4.3 Shock Stand-off Distance Over a Free Flight Sphere

In supersonic steady flows, a bow shock wave appears in front of a blunt body. When the flow speed approaches to the sonic speed, a bow shock would appear in front of the blunt body. From an engineering point of view, wind tunnels can not technically generate a steady sonic flow. Hence, it is impossible to produce a bow shock wave in front of a blunt body in the sonic flows. However, shock tubes, in principle, can be a tool to produce transonic flows.

For testing such flows, a 10 mm diameter bearing ball was placed in the test section of the 60 mm × 150 mm conventional shock tube in order to produce a wide range of transonic flows. The flow Mach number behind an incident shock wave of Ms = 2.350 in air has a local flow Mach number is 1.10. Figure 4.23a, b show the interaction of the sphere with this flow at a delay time of 110 and 500 μs measured from the moment when the second exposure was synchronized with the shock wave, respectively. When the shock tube flow is used as a replacement of a transonic wind tunnel flow, the steady flow will be established at later time when the interaction of the incident shock wave with the model was suppressed. The conversion of shock tube flows in the 60 mm × 150 mm shock tube to a transonic flow can not be straightforward. This is so due to boundary layers developing along the shock tube walls as seen in Fig. 4.23b, and flow unsteadiness prevailing over the entire flow field. The shock tubes can never be a reliable transonic flow simulator (Kikuchi et al 2016).

(a) (b)

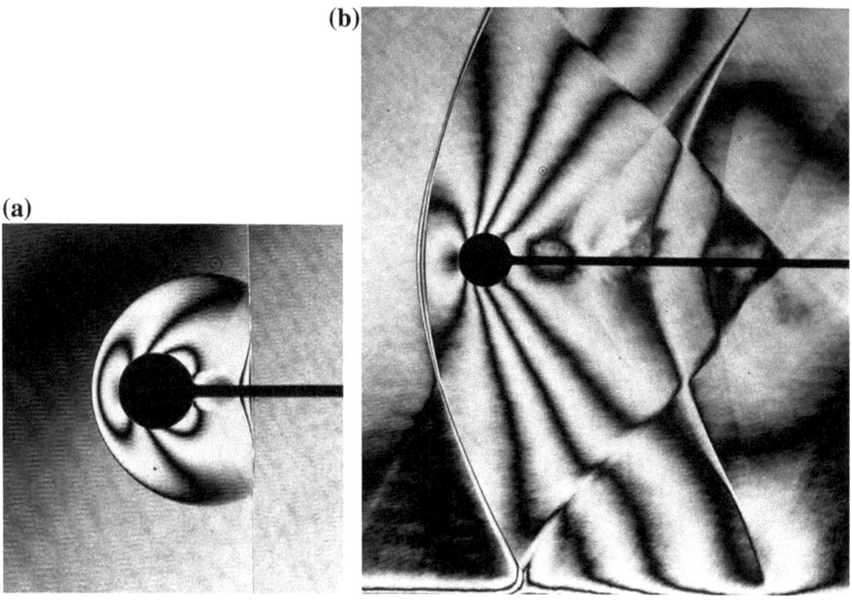

Fig. 4.23 Bow shock waves in transonic shock tube flows Ms = 2.350 in air at 150 hPa, 290.2 K, post-shock flow Mach number M = 1.10: **a** #83112215 110 μs from trigger point Ms = 2.350; **b** #83112220 500 μs

It was decided to use a ballistic range to accurately project spheres at transonic speed range. Figure 4.24a–c show a launching facility, a 40 mm diameter sphere and a 50 mm diameter sabot which can split into four pieces, and the arrangement to connect the launcher to a test chamber, respectively. A 40 mm diameter nylon sphere was contained in a 50 mm diameter polycarbonate sabot. The combination of the sabot and the sphere was launched into a ballistic range. Figure 4.24c explains the experimental arrangement. The sphere and sabot flew into the sabot remover through which sabot split into four pieces and sphere was separated from the sabot. Then sphere passed through the arrayed orifice plates removing blast wave. Eventually the sphere alone flew into the observation section and was visualized with 600 mm diameter interferometry.

Figure 4.25 shows 40 mm diameter spheres at the free flight speed ranging Ms = 0.986 to Ms = 1.104. In subsonic free flight, for example, in Fig. 4.25a–d, an apparent bow shock appears in front of the sphere. The waves observed were

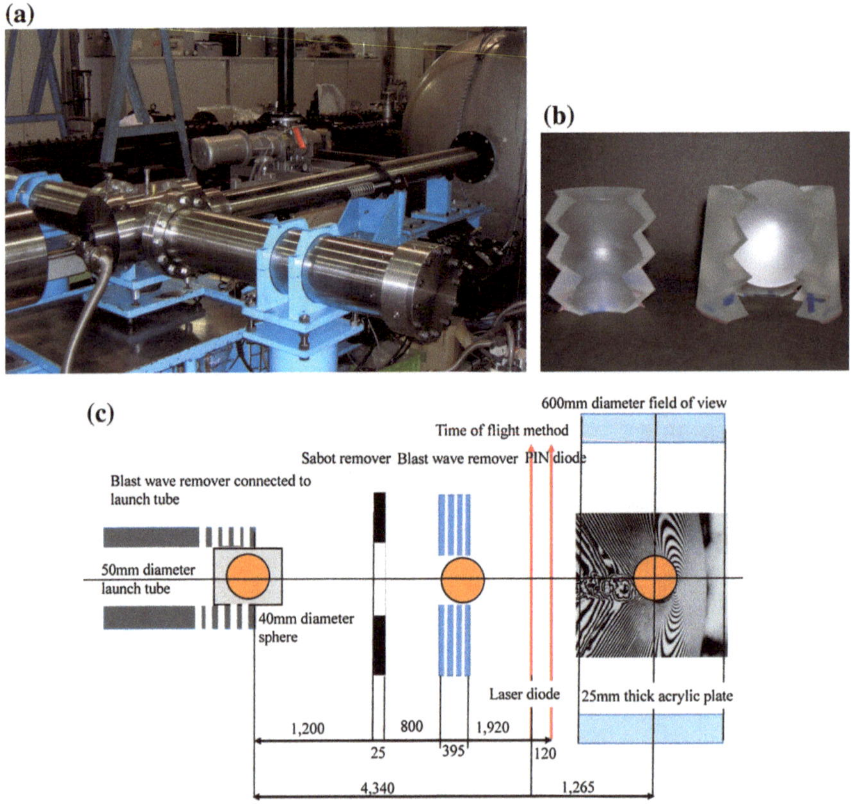

Fig. 4.24 Experimental setup of observing shock wave detachment distance from 40 mm sphere: **a** 10 mm bore gas gun; **b** a 40 mm nylon sphere and 50 mm policarbonate sabot; **c** experimental arrangement (Kikuchi et al. 2016)

subsonic, Ms < 1.0, which are not necessarily shock waves but must be a train of compression waves propagating at sonic speed. If a steady high subsonic wind tunnel flow, say Ms = 0.99, were operational, a bow shock wave might have been

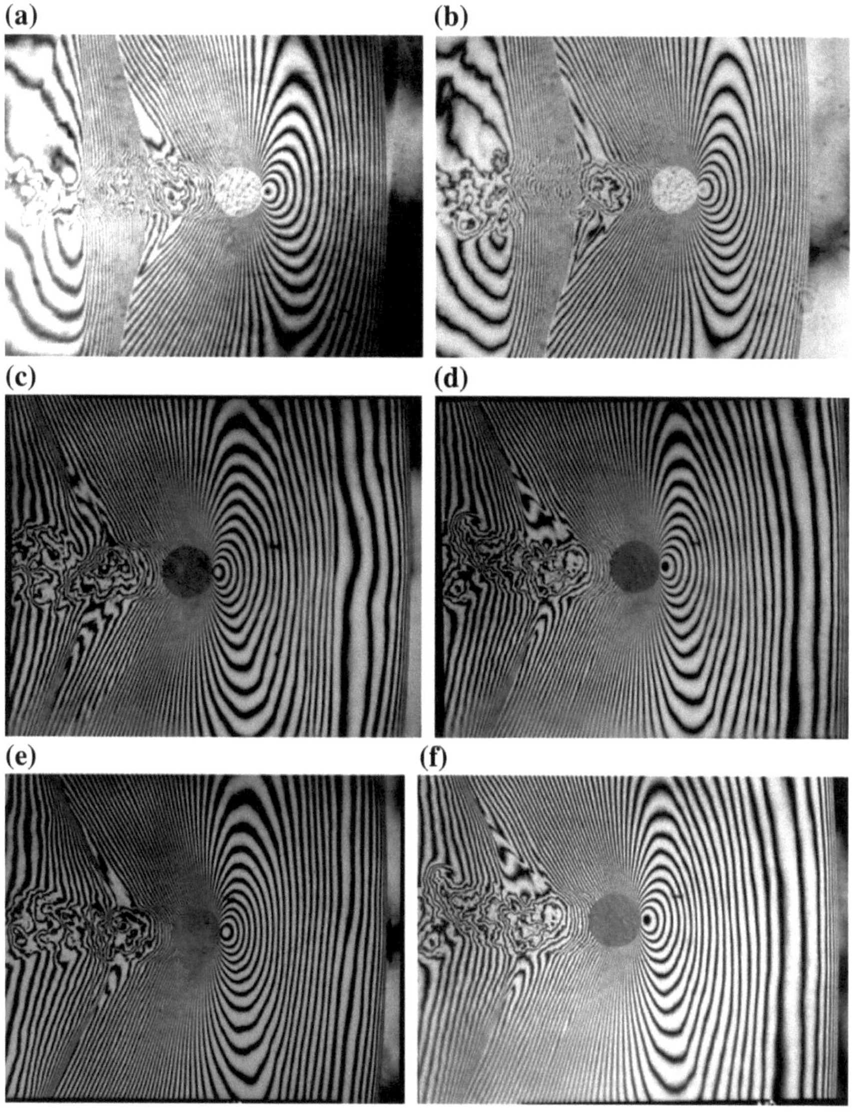

Fig. 4.25 Detached shock waves in front of a 40 mm diameter sphere: **a** Ms = 0.986, Re = 0.901 × 10^5; **b** Ms = 0.993, Re = 0.906 × 10^5; **c** Ms = 0.997, Re = 0.917 × 10^5; **d** Ms = 0.998, Re = 0.920 × 10^5; **e** Ms = 1.003, Re = 0.924 × 10^5; **f** Ms = 1.011; **g** Ms = 1.027, Re = 0.950 × 10^5; **h** Ms = 1.043, Re = 0.961 × 10^5; **i** Ms = 1.056, Re = 0.973 × 10^5; **j** Ms = 1.062, Re = 0.977 × 10^5; **k** Ms = 1.067, Re = 0.983 × 10^5; **l** Ms = 1.070, Re = 0.985 × 10^5; **m** Ms = 1.084; **n** Ms = 1.104

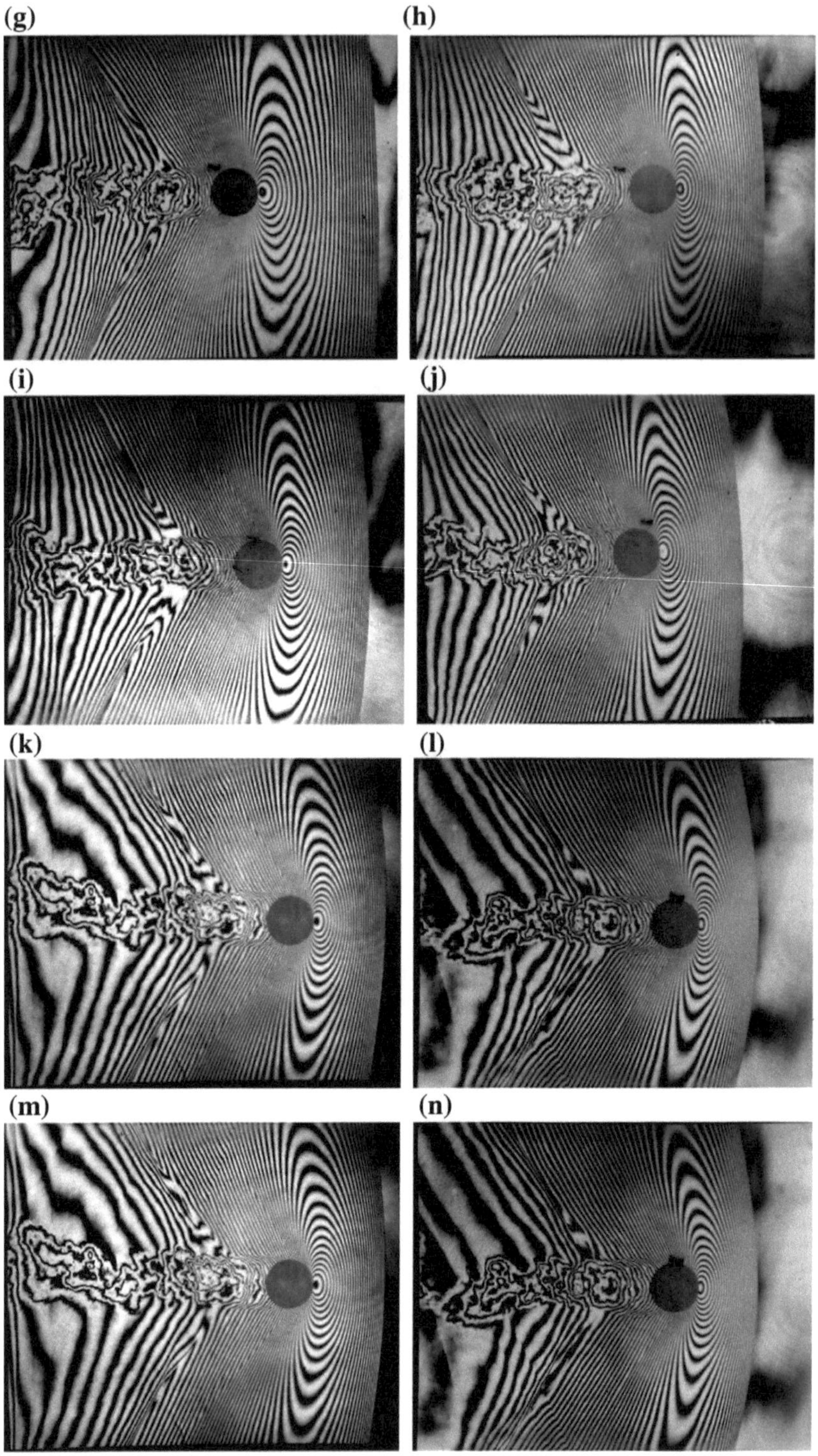

Fig. 4.25 (continued)

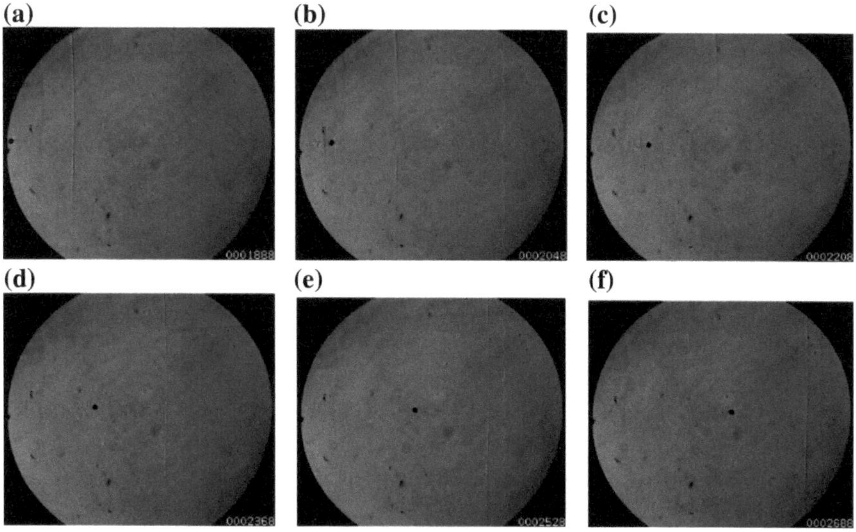

Fig. 4.26 Selective images of a 10 mm sphere in free flight for Ms = 0.949 at entry: **a** 60 μs; **b** 65 μs; **c** 70 μs; **d** 75 μs; **e** 80 μs; **f** 85 μs (Kikuchi et al. 2016)

observable in front of a blunt body. If a wind tunnel was operational at speed of Ms = 1 + ε, 0 < ε ≪ 0.1, the bow shock would be detached reasonably far away from a spherical model. However, no wind tunnel is operational at the flow Mach number unity. Such a wind tunnel is only imaginary. Such a critical condition never be stable, even if it may appear. In shock tube flows, it is nearly a useless effort to create a local flow Mach number of unity behind the incident shock waves.

Figure 4.26 show sequential direct shadow pictures of a 10 mm diameter bearing ball flying across a 500 mm view field. Images were recorded with the high speed video camera, Shimadzu SH100 at framing rate of 10^6 frame/s. The entry speed at the left hand side is Ms = 0.949, which attenuated to Ms = 0.939 at the right hand side. A detached wave was observed ahead of the sphere.

Figure 4.27 summarizes the trajectories of the shock wave and the sphere shown in Fig. 4.26. The ordinate denotes the flight distance in mm. The abscissa denotes the elapsed time in ms. Red filled circles show the position of the shock wave and green filled circles show the position of the sphere. Blue filled circles show the resulting shock stand-off distance δ/d. During a fight distance of 500 mm, the dimension-less stand-off δ/d increases from about 11 to 15, while the apparent Ms of the sphere changes from 0.949 to 0.939, and the shock wave propagates at the sonic speed. The shock stand-off distance δ/d increases with elapsing time. A bow shock wave is physically a train of compression wave propagating at sonic speed.

In Fig. 4.28, results of previous visualizations are summarized. The ordinate denotes the dimensionless stand-off distance δ/d and the abscissa denotes the spheres' Ms in free flight. Red filled circles show interferometric images of the 40 mm diameter spheres and blue filled circles show the 10 mm diameter spheres

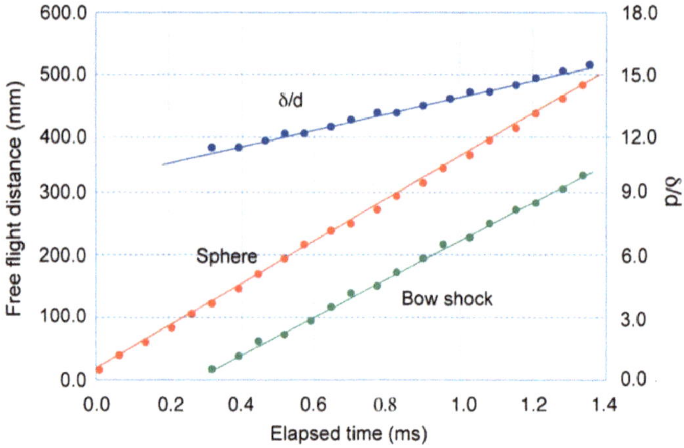

Fig. 4.27 Time variation of the detachment distance and the trajectories of the detached shock wave and the sphere (Kikuchi et al. 2016)

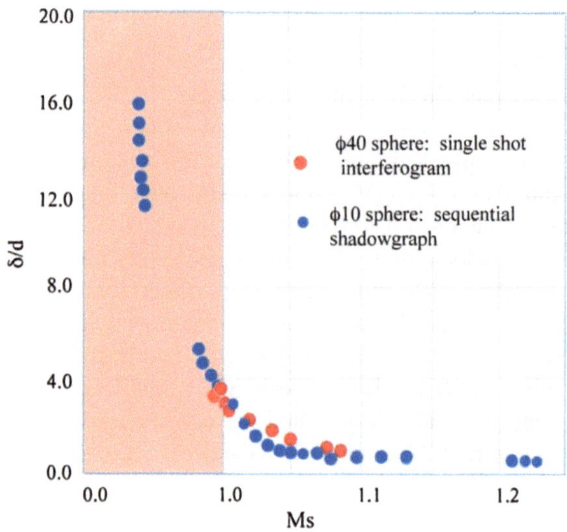

Fig. 4.28 Shock stand-off distance versus Ms (Kikuchi et al. 2016)

recorded by a high-speed video camera. It is noticed that bow shock waves appeared even in subsonic flows, Ms < 1. 0 The trend of data collected with 40 and 10 mm spheres agreed well with each other. Surprisingly the bow shock waves continuously crossed the boundary of Ms = 1.0 as if a bow shock exists in front of subsonic moving spheres.

Professor Ben-Dor once told the author his experience in a battle front. While listening to booms induced by a gun shell, soldiers immediately judged whether or not it would land close to them or a distance away. The sonic booms were composed of trains of compression waves driven by gun shells propagating at high subsonic speed. In short, booms were detached at reasonably long distance from the gun shells. Figure 4.28 demonstrates that in unsteady flows sonic waves exist in front of subsonic moving blunt bodies.

A train of compression waves and a weak shock wave are both visualized as a discontinuous sharp line and it is hard to distinguish between weak shock waves and a train of compression waves.

4.4 Elliptic Cylinders

Elliptical cylinders of the major radius of 40 mm and the minor radii of 40 mm, 20 mm, and 10 mm, respectively, were installed in the 60 mm × 150 mm conventional shock tube in a similar manner as shown in Fig. 4.1. The 40 mm cylinder was equivalent to an elliptical cylinder having the ratio of the major radius to the minor radius of 1:1. Therefore, in the series of the present experiments, the shock wave interaction with ellipse of their aspect ratios from 4:1. 2:1, 4:3, 1:1 were visualized for Ms = 1.30 in air and the Reynolds number referred to the major diameter $Re = 5.0 \times 10^5$.

4.4.1 4:3 Elliptic Cylinders

Figure 4.29a–d show the evolution of shock wave reflections from the 4:3 elliptical cylinders for Ms = 1.30 in atmospheric air. The reflection patterns are similar to those shown in Fig. 4.2. Figure 4.30a–g show the identical case for Ms = 2.60 at 120 hPa in air. In Fig. 4.30a, the reflection pattern is RR at the frontal side of the ellipse. In Fig. 4.30b it becomes SMR at the equator. It is noticed that since the flow behind the shock wave for Ms = 2.60 is supersonic, the reflected shock wave departs gradually from the frontal side of the ellipse with elapsed time and reaches a consistent shock stand-off distance δ. This trend is clearly observable in Fig. 4.30f, g. Slightly irregular pattern visible along the reflected shock wave shows the bifurcation of the reflected shock wave which is created due to its interaction with the boundary layer developing along the shock tube side wall. The bifurcation of the reflected shock wave will be discussed in the Sect. 4.6.2 (Mark 1956). In Fig. 4.30e–g, the SL

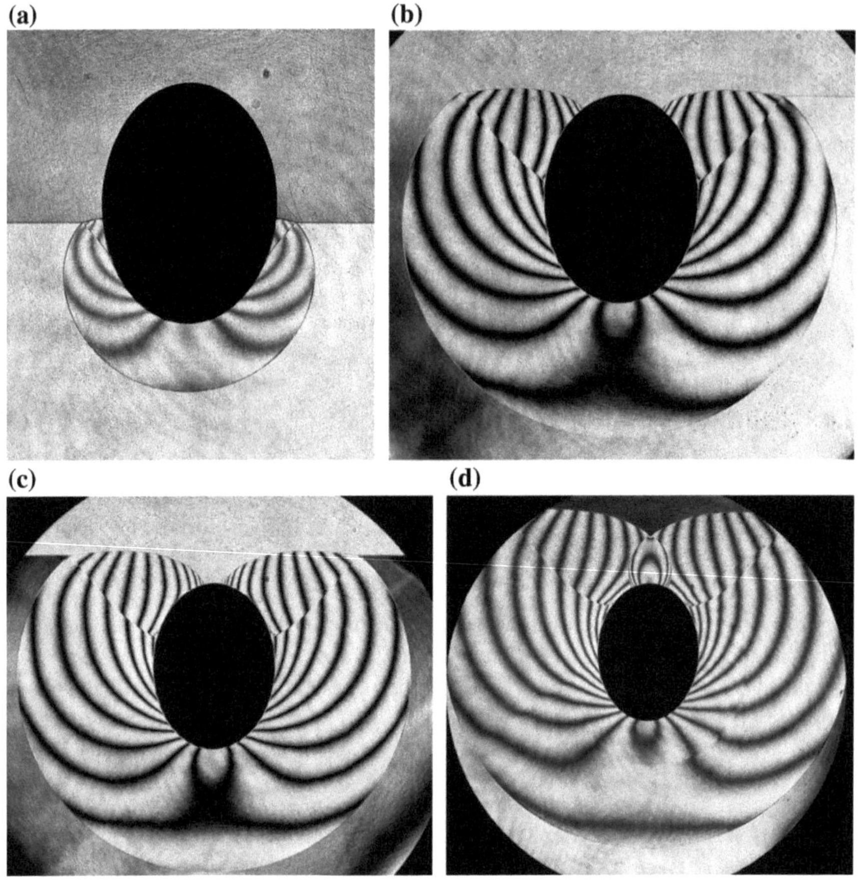

Fig. 4.29 The evolution of shock wave interaction with a 4:3 elliptical cylinder interaction for Ms = 1.30 in atmospheric air at 298.7 K attack angle $\alpha = 0°$: **a** #85082816, Ms = 1.294; **b** # 85082820, Ms = 1.302; **c** #85082822, Ms = 1.306; **d** #85082902, Ms = 1.298

emanating from the TP intersects with the ellipsoid surface and its interaction with the reflected MS forms complex wave interactions. The fringe pattern reminds one of deformation of a barking tiger face.

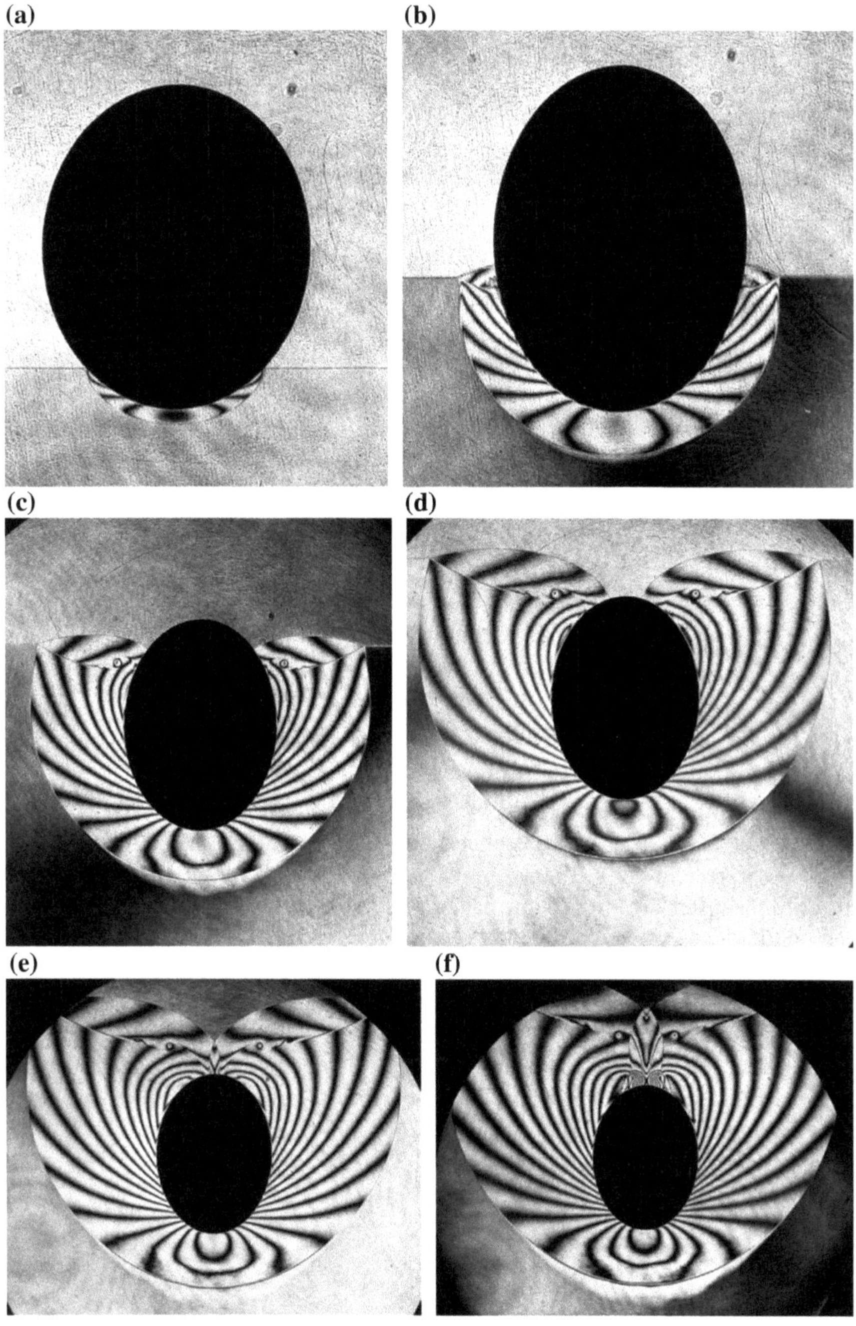

Fig. 4.30 The evolution of the shock wave interaction with a/4:3 elliptical cylinder interaction for Ms = 2.60 in air at 120 hPa, 298.7 K, attack angle α = 0°: **a** #85082922, Ms = 2.605; **b** #85083007, Ms = 2.593; **c** 85083006, Ms = 2.613; **d** #85083012, Ms = 2.578; **e** #85083013, Ms = 2.583; **f** #85083014, Ms = 2.584; **g** enlargement of (**f**)

(g)

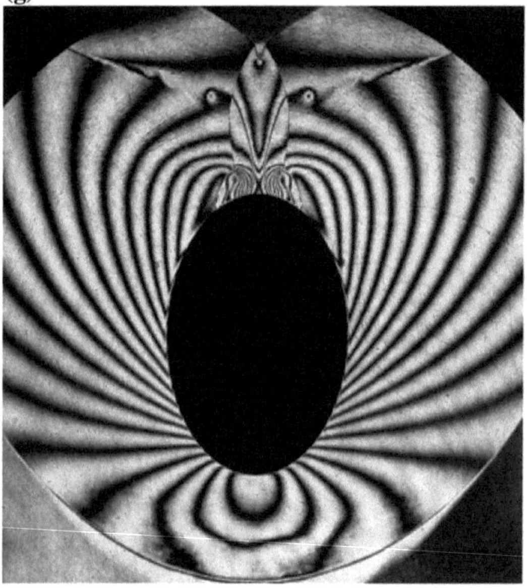

Fig. 4.30 (continued)

4.4.2 2:1 Elliptic Cylinders

Figure 4.31a–g show the evolution of shock wave reflections from the 2:1 elliptical cylinder for Ms = 1.30 in atmospheric air. The reflection patterns are similar to those observable in Fig. 4.29. Figure 4.32a–f show the identical case for Ms = 2.60 at 120 hPa in air. The reflection patterns are similar to the patterns observed in Fig. 4.30.

Figures 4.33 and 4.34 show the evolution of the shock wave reflection from a 2:1 elliptic cylinder for Ms = 1.70 in air with attack angles of $\alpha = 5°$ and $10°$, respectively. The interaction of the reflected MS with the boundary layer developing along the ellipse is enhanced with elapsing time and with increase in the attack angle. Figure 4.33c–d show development of vortices at the rear side of the ellipse. Figure 4.33d–i show the vortices shedding from the rear side.

A similar trend is observed in Fig. 4.33 for Ms = 1.70 and $\alpha = 5°$; it is also observed in Fig. 4.34 for Ms = 1.70 and $\alpha = 10°$.

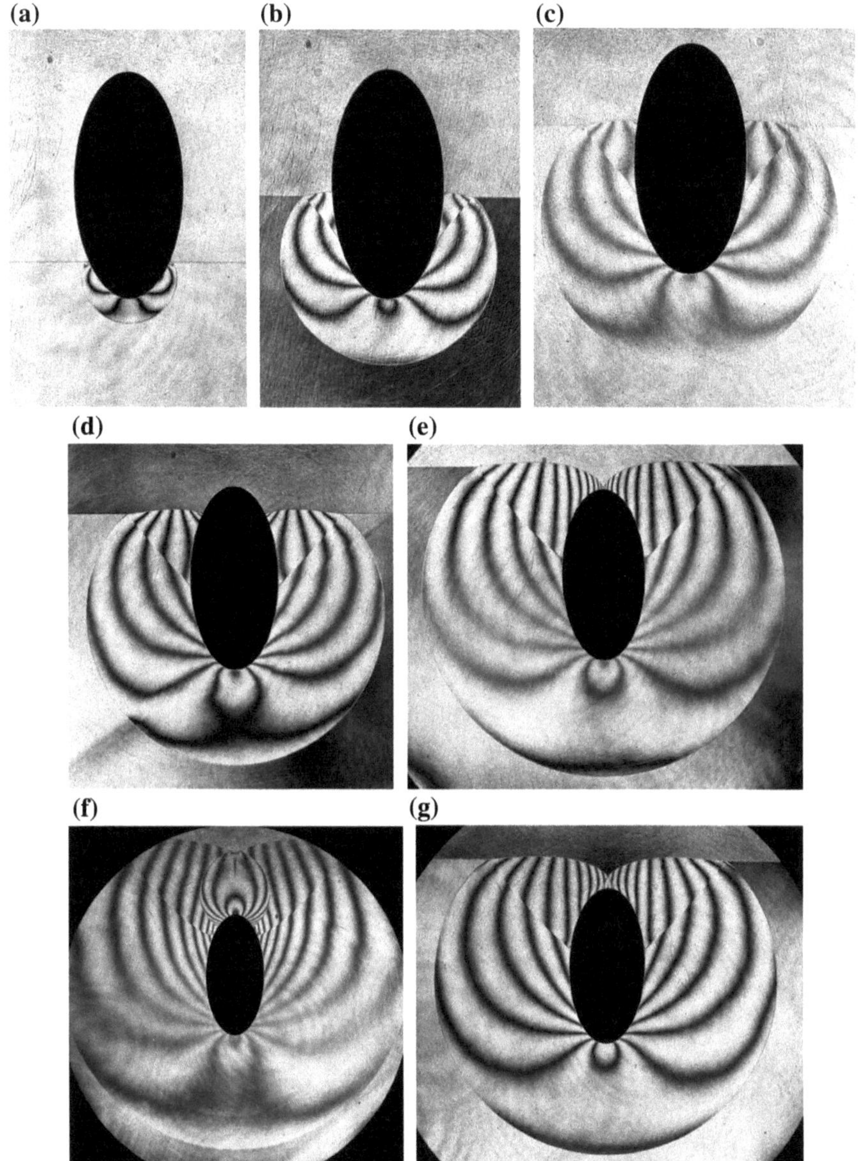

Fig. 4.31 The evolution of the shock wave interaction with a 1:2 elliptical cylinder for Ms = 1.30 in atmospheric air at 298.9 K, attack angle α = 0°: **a** #85083017, Ms = 1.300; **b** #85083020, Ms = 1.295; **c** #85083104, Ms = 1.298; **d** #85083106, Ms = 1.302; **d** #85083106, Ms = 1.302; **e** #85083107, Ms = 1.304; **f** #85083109, Ms = 1.298; **g** #85083108, Ms = 1.295

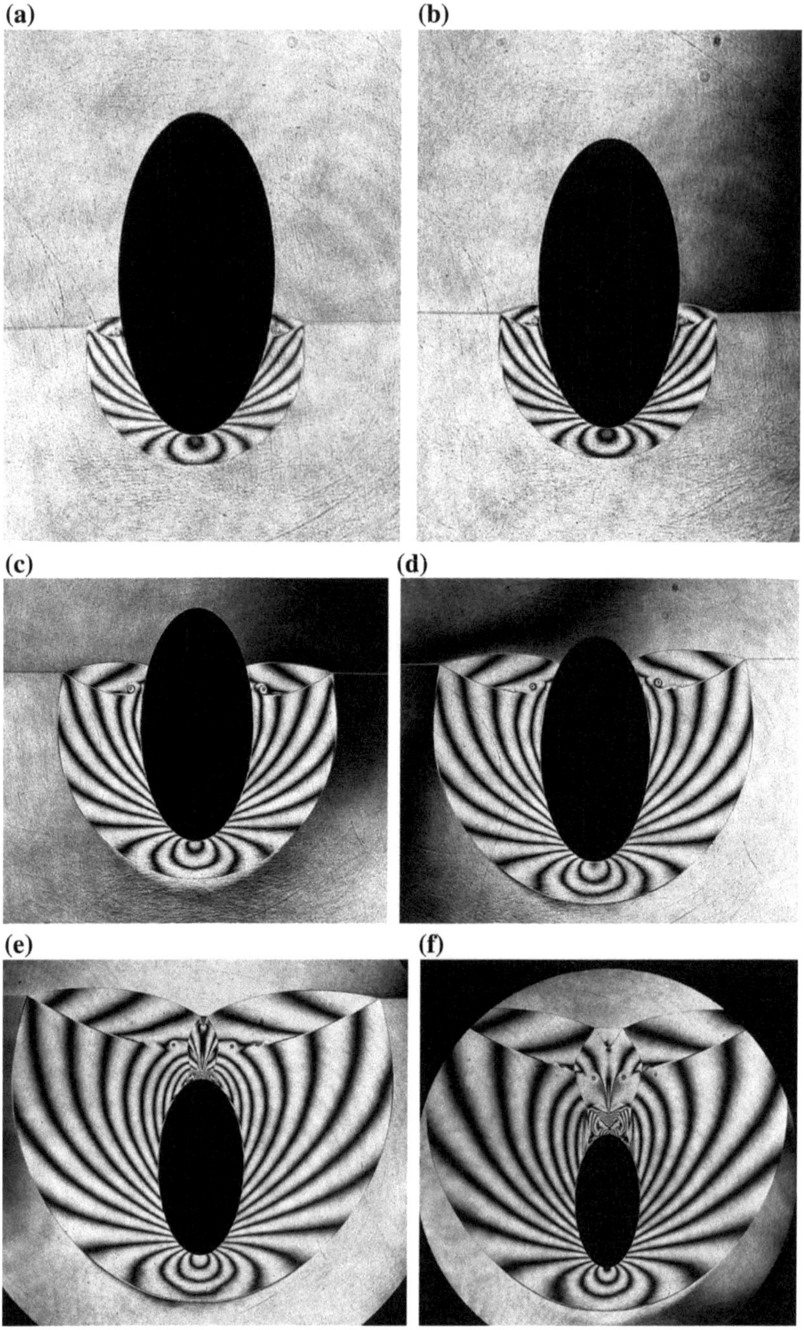

Fig. 4.32 Shock wave interaction with a 2:1 elliptical cylinder interaction for Ms = 2.60 in air at 120 hPa, 298.8 K, attack angle $\alpha = 0°$: **a** #85090215, Ms = 2.589; **b** #85090207, Ms = 2.611; **c** #85090206, Ms = 2.574; **d** #85090204, Ms = 2.589; **e** #85090216, Ms = 2.571; **f** #85090205, Ms = 2.649

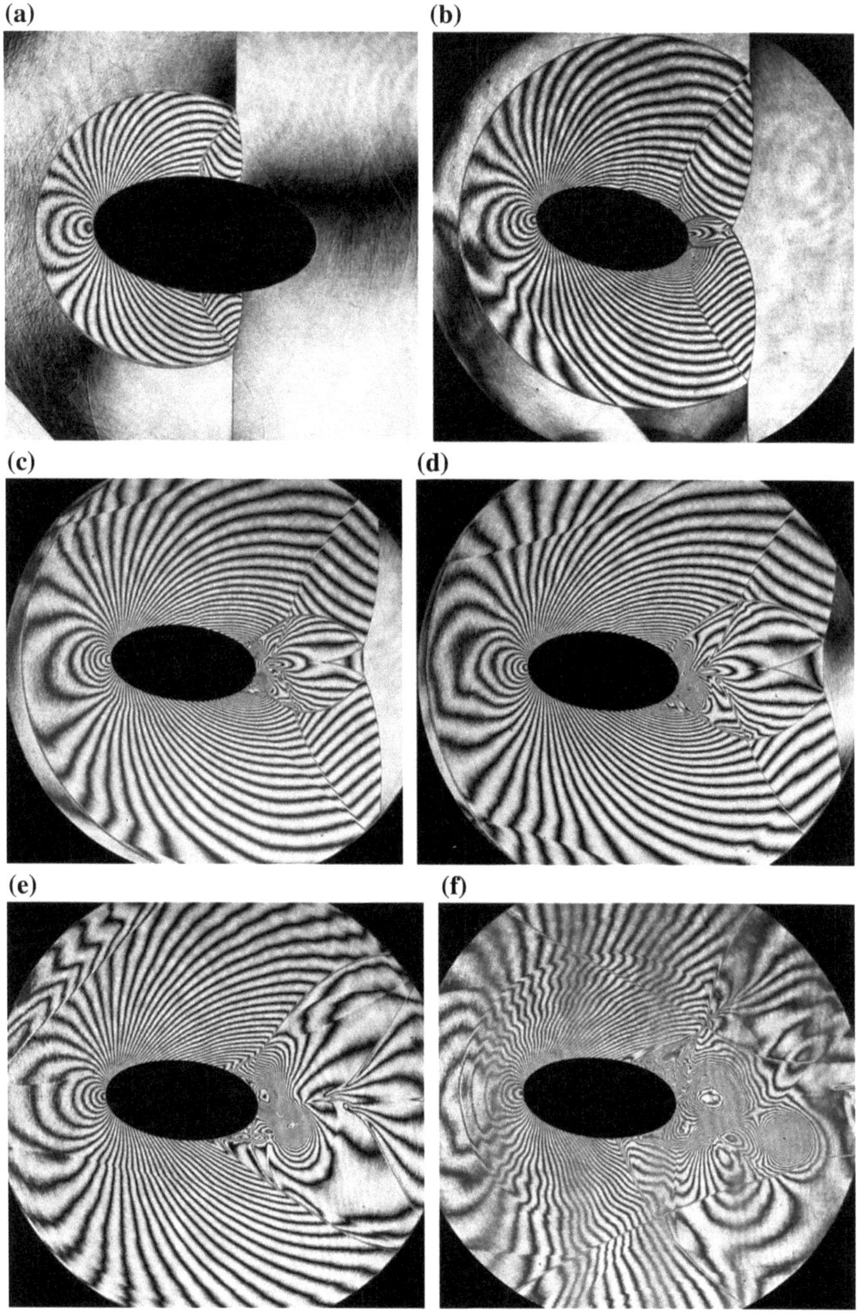

Fig. 4.33 Evolution of shock wave interaction with a 2:1 elliptical cylinder interaction for Ms = 1.70 in air at 900 hPa, 288.1 K, attack angle α =5°: **a** #87012204 1.5 ms, Ms = 1.677; **b** #87012215 150 μs Ms = 1.687; **c** #87012207 160 μs Ms = 1.686; **d** #87012202 200 μs Ms = 1.701; **e** #87012209 200 μs Ms = 1.701; **f** #87012212 350 μs Ms = 1.601; **g** #87012306 250 μs Ms = 1.682; **h** #87012213 400 μs Ms = 1.683; **i** enlargement of (**h**)

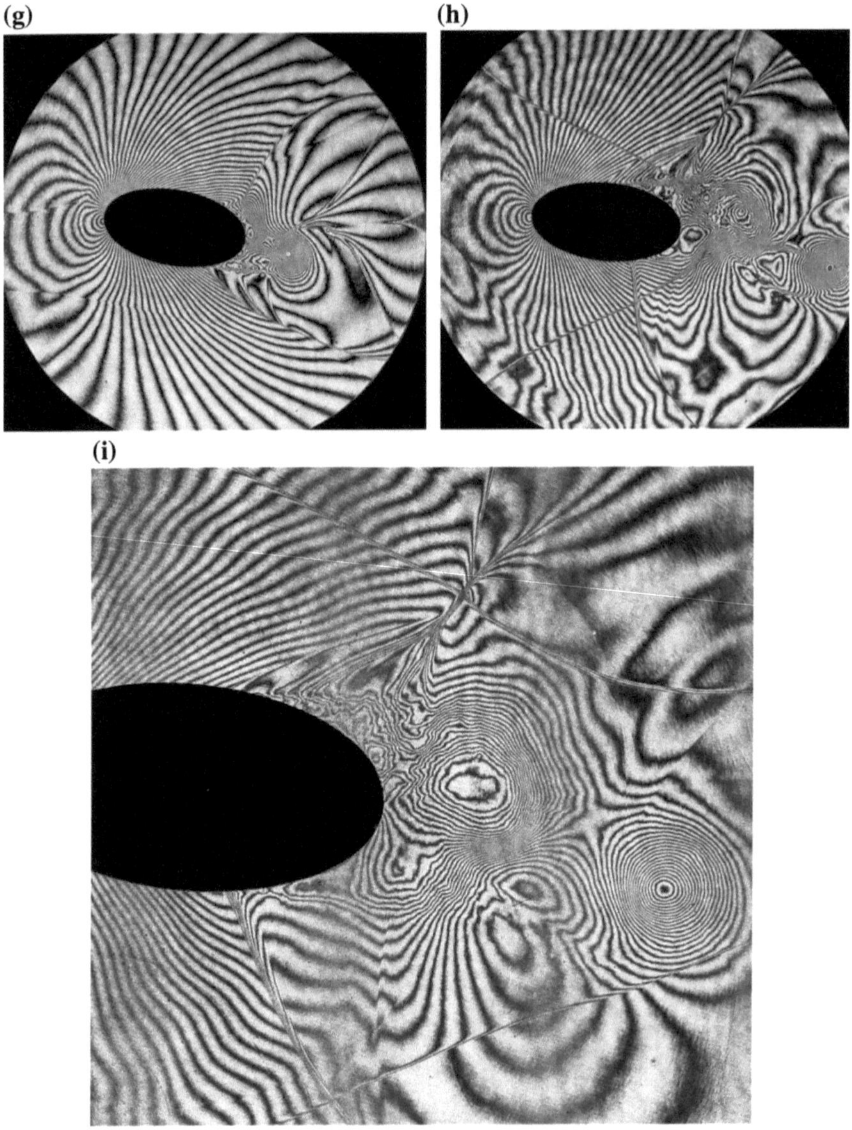

Fig. 4.33 (continued)

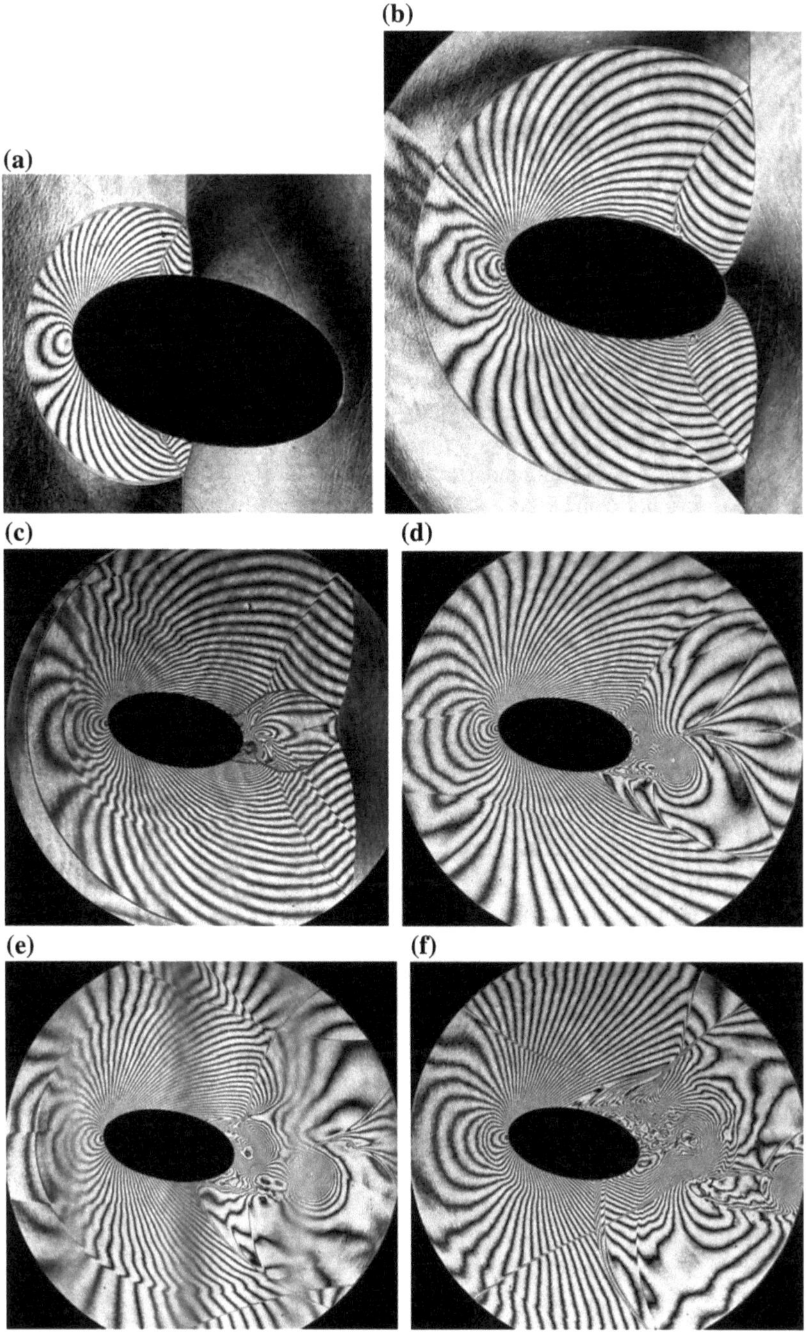

Fig. 4.34 Evolution of shock wave interaction with a2:1 elliptical cylinder interaction at Ms = 1.70 in air at 900 hPa, 288.1 K, attack angle α = 10°: **a** #87012302, Ms = 1.702; **b** #87012305, Ms = 1.671 8701; **c** #87012301, Ms = 1.679; **d** #87012306, Ms = 1.682; **e** #87012307 300 μs Ms = 1.693; **f** #87012602 400 μs Ms = 1.691

4.4.3 4:1 Elliptic Cylinders

Figures 4.35 and 4.36 show the evolution of the shock wave interaction with a 4:1 elliptic cylinder at attack angle $\alpha = 0°$ for Ms = 1.7 and 2.6 respectively.

The interferograms have precise spatial distributions over the elliptic cylinders enough to resolve the density distributions on elliptical cylinder surface as observed, for example in Fig. 4.29 and Fig. 4.31. The time variation of density distributions over the elliptical cylinders can be experimentally determined out of the sequential interferograms. If assuming isothermal condition on the cylinder surface, the pressure p on the cylinder can be estimated out of the density distribution on the cylinder surfaces by assuming $p/\rho^{\gamma} = p_0/\rho_0^{\gamma}$ where p_0 is ambient pressure, ρ_0 is ambient density. Then integrating the pressure profile along the cylinder surface, the drag force on the cylinders can be estimated. Itoh (1986) summarized the time variation of the drag coefficient C_D over 40 mm major diameter and 30 mm, 20 mm and 10mm minor diameter elliptical cylinders for Ms = 1.30 in atmospheric air. Figure 4.37 summarizes the results. The ordinate denotes C_D and the abscissa denotes dimensionless time tUs/D where t, Us, D denote elapsed time in μs, the shock speed m/s, and the major diameter of the cylinder. Red, black, blue, and green filled circles denote the experimental results of the aspect ratio of the cylinders 1.0, 4:3, 2:1 and 4:1, respectively. Just for reference, numerical simulation based on TVD finite difference scheme solving the Navier Stokes solver (Itoh 1986). Fair agreements between the experiments and the simulation are obtained. It should be noticed that the C_D can be maximal and monotonously decreases to the steady flow value. The unsteady drag force over a shock laden sphere was experimentally measured by Tanno et al. (2004) and discussed in the Sect. 2.4.

Figure 4.37 summarized the time variation in the drag coefficients for Ms = 1.30 of a 40 mm diameter cylinder, and 4:3, 2:1, 4:1 elliptic cylinders, all having a major diameter of 40 mm. The ordinate denotes drag force normalized by initial condition, C_D, and the abscissa denotes dimension-less time tUs/D. Black, blue, green, and red filled circles denote 4:3, 2:1, and 4:2 elliptic cylinders, respectively. Solid lines denote numerical simulation using TVD scheme (Itoh 1986). Fair agreement is observed between the simulation and experimental findings.

Figures 4.38 and 4.39 show the evolution of shock wave interaction with a 4:1 elliptic cylinder for Ms = 1.70 with attack angles of $\alpha = 10°$ and $\alpha = 45°$, respectively. In Fig. 4.38, the evolution of interaction of the reflected MS with the boundary layer is observed. At an attack angle of $\alpha = 10°$, a vortex is formed at the rear edge of the 4:1 slender shaped ellipse and it is shading with the elapsed time as shown in Fig. 4.38g–h, creating a lift force. On the contrary at attack angle $\alpha = 45°$ in Fig. 4.39, the boundary layer separation occurs at the leading edge. The flow pattern is typical to stall.

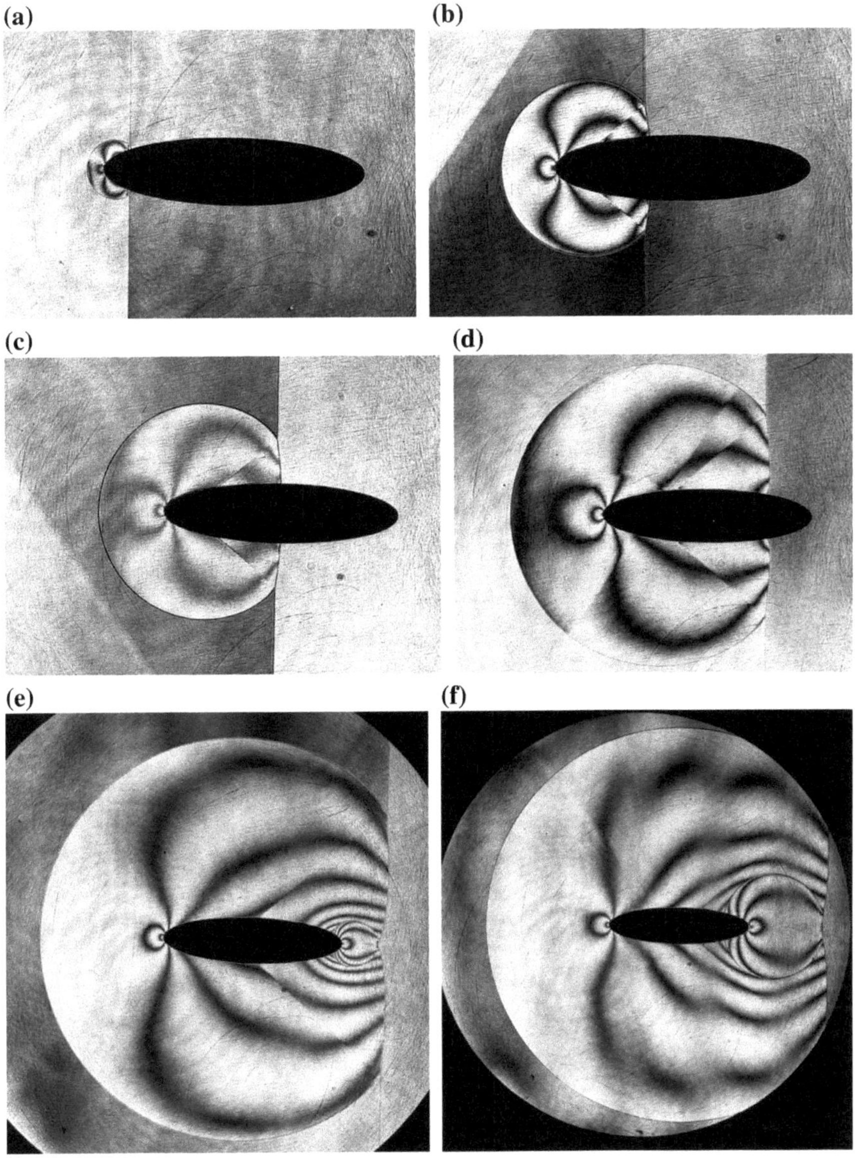

Fig. 4.35 The evolution of the shock wave interaction with a 4:1 elliptical cylinder interaction at Ms = 1.70 in air at 900 hPa, 288.1 K, attack angle α = 0°: **a** #85090310, Ms = 1.305; **b** #85090312, Ms = 1.310; **c** #85090314, Ms = 1.300; **d** #85090315, Ms = 1.308; **e** #85090317, Ms = 1.300; **f** #85090319, Ms = 1.288

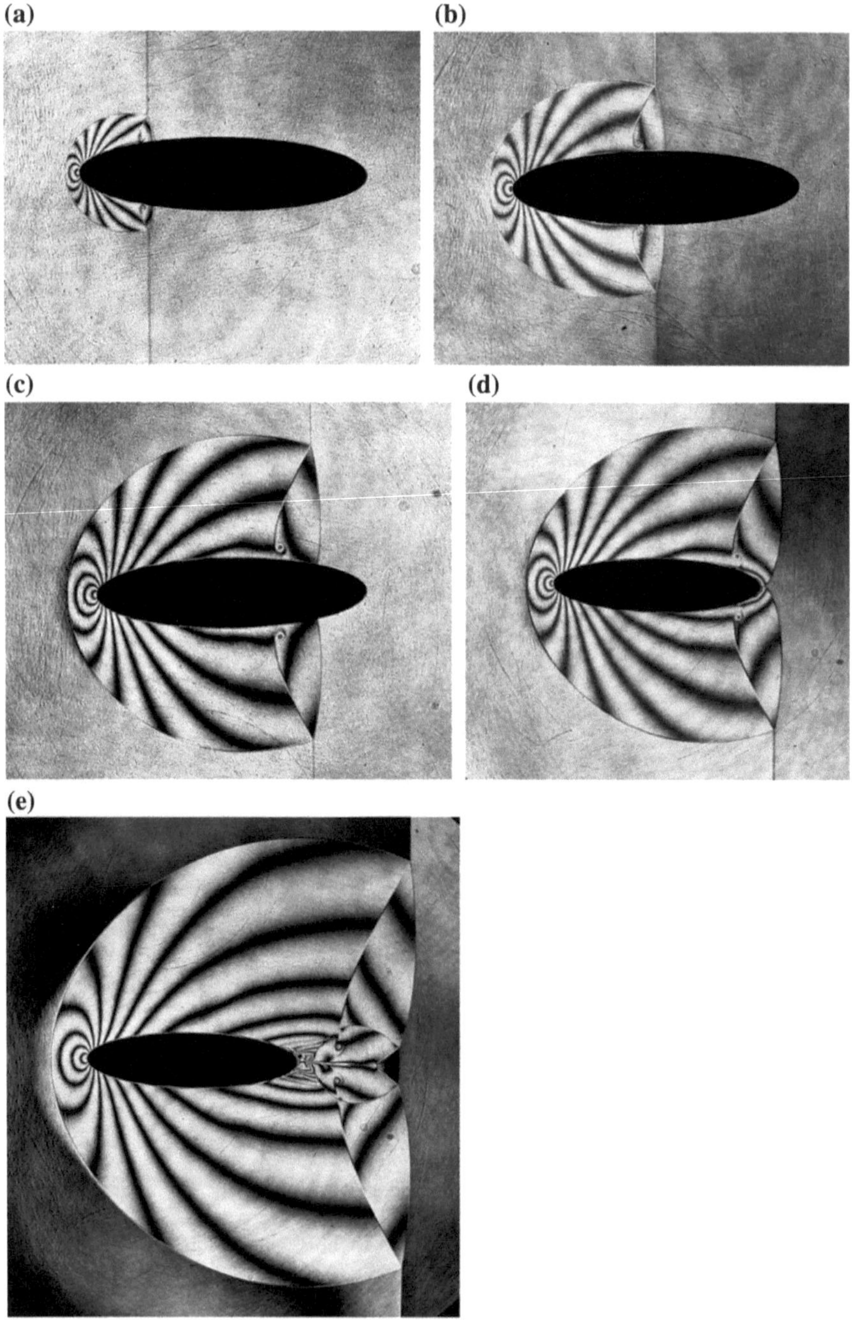

Fig. 4.36 The evolution of the shock wave interaction with a 4:1 elliptical cylinder for Ms = 2.60 in air at 120 hPa, 283.7 K, attack angle α = 0°: **a** #85090219, Ms = 2.571; **b** #85090220, Ms = 2.597; **c** #85090218, Ms = 2.601; **d** #85090222, Ms = 2.594; **e** #85090301, Ms = 2.588; **f** enlargement of (**e**)

(f)

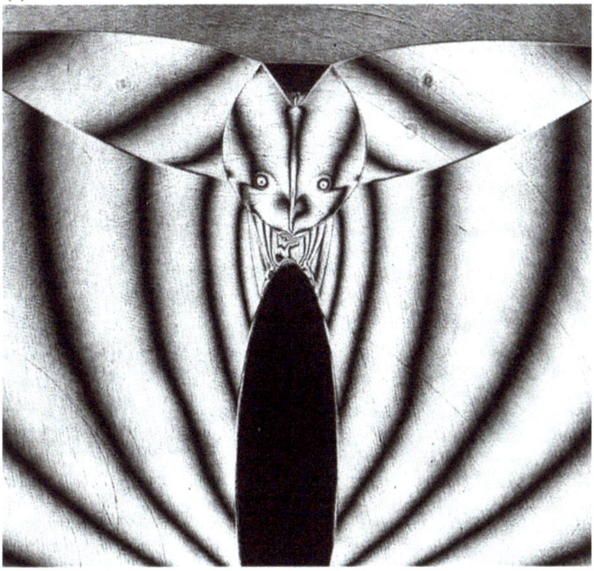

Fig. 4.36 (continued)

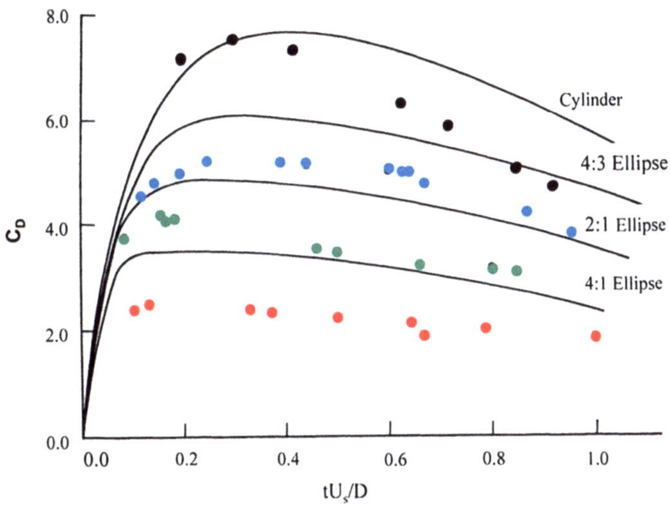

Fig. 4.37 Time variation of drag forces over a cylinder and ellipses at $\alpha = 0°$ for Ms = 1.30 in air (Itoh 1986)

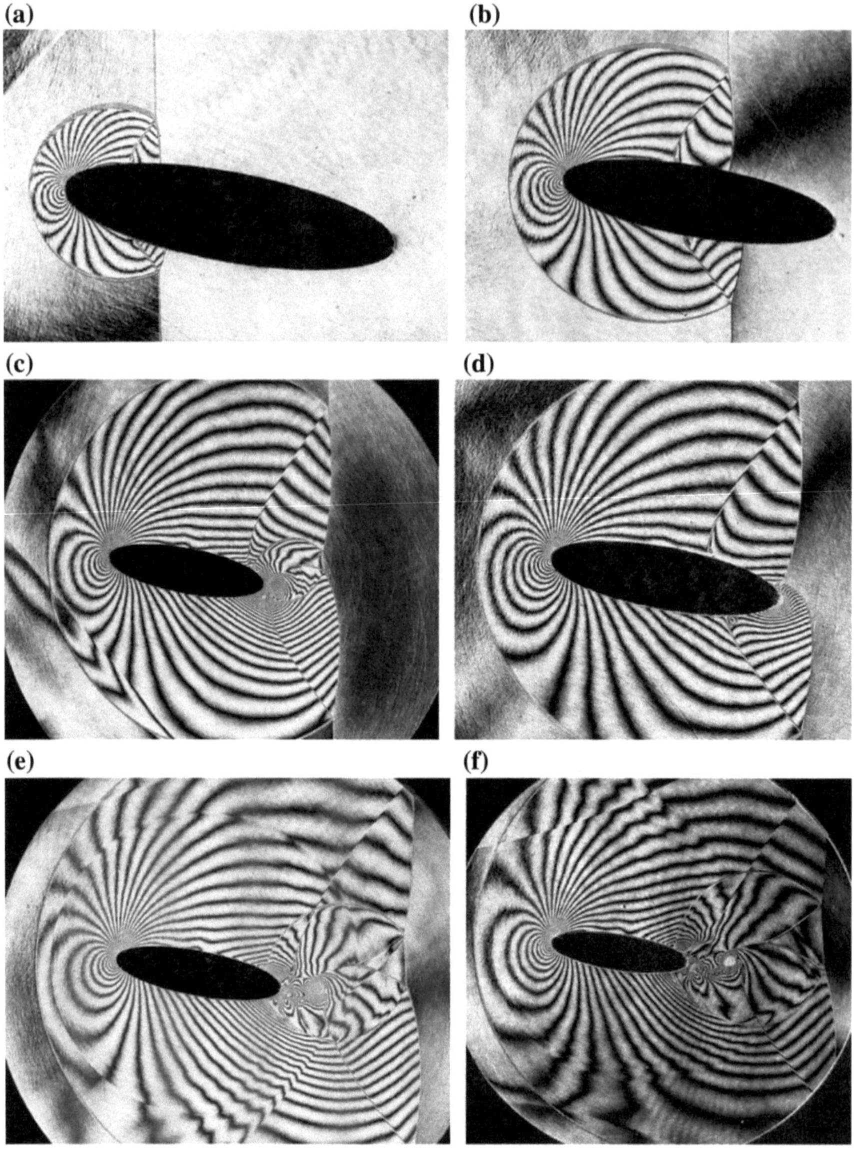

Fig. 4.38 Evolution of shock waves interaction with a 4:1 elliptical cylinder for Ms = 1.70 in air at 900 hPa, 283.7 K, attack angle α = 10°: **a** #87012802, Ms = 1.700; **b** #87012803, Ms = 1.704; **c** #87012801, Ms = 1.721; **d** #87012805, Ms = 1.713; **e** #87012808, Ms = 1.707; **f** #87012701, Ms = 1.661; **g** #87012703, Ms = 1.680; **h** #87012704, Ms = 1.685; **i** enlargement of (**h**)

(g)

(h)

(i)

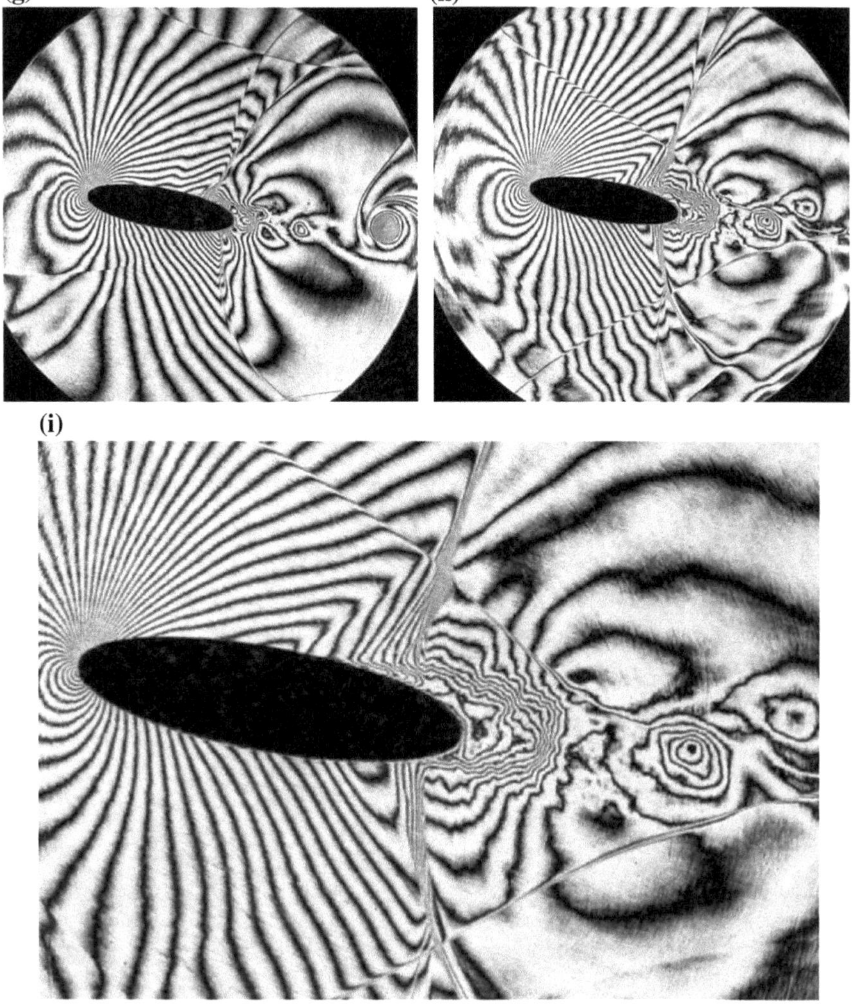

Fig. 4.38 (continued)

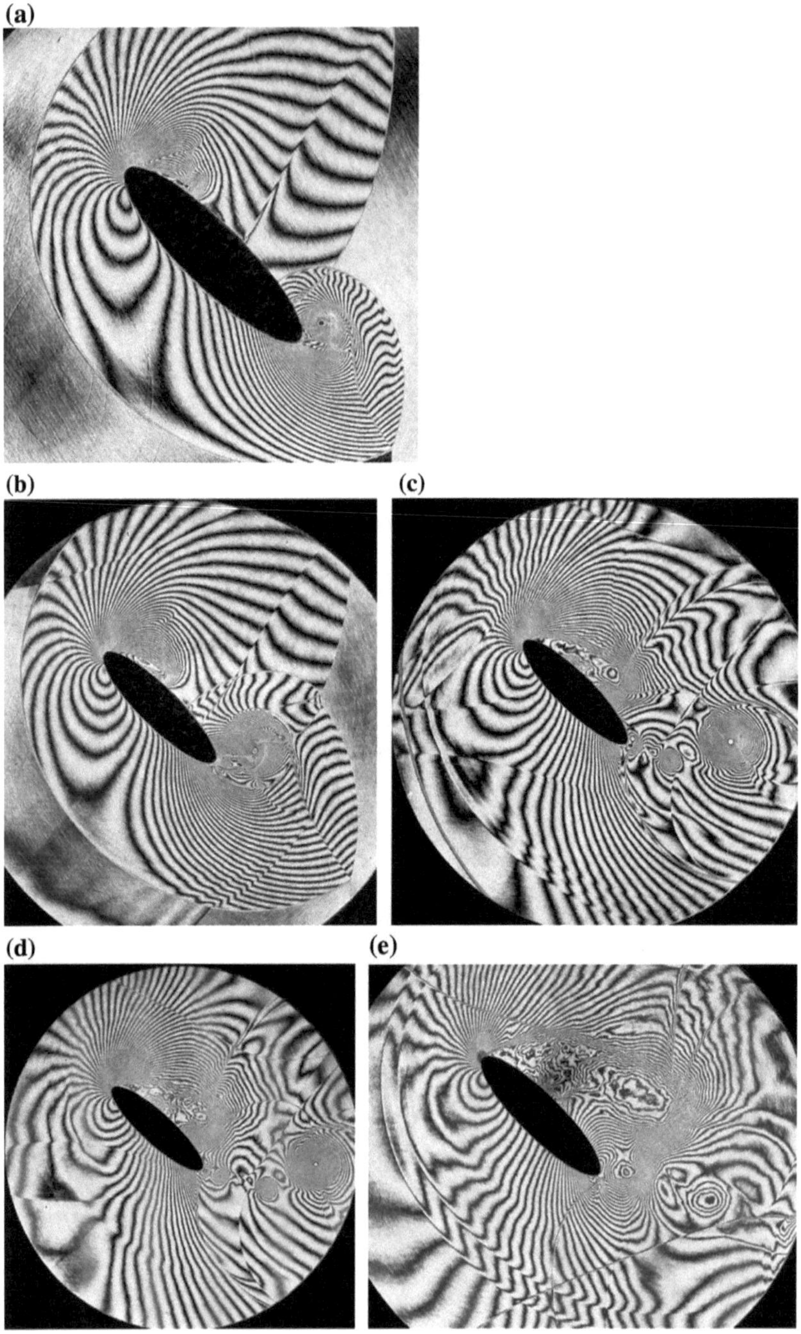

Fig. 4.39 The evolution of the shock wave interaction with a 4:1 elliptical cylinder for Ms = 1.70 in air at 900 hPa, 290.7 K, attack angle $\alpha = 45°$: **a** #87012904, Ms = 1.674; **b** #87012908, Ms = 1.690; **c** #87012910, Ms = 1.697; **d** #87012901, Ms = 1.675; **e** #87012902, Ms = 1.689

4.4.4 Rectangular Plates

4.4.4.1 Rectangular Plate with Attack Angle $\alpha = 0°$

Figure 4.40a–j show the evolution of shock wave interaction for Ms = 1.40 in atmospheric air with a 10 mm thick and 40 mm wide rectangular plate of aspect ratio of 4:1 installed in the 60 mm $\times$150 mm conventional shock tube at attack angle $\alpha = 0°$. The resulting Reynolds number referred to the test condition is about Re = 5.0 $\times 10^5$. The transmitting shock wave is diffracted at the rear corners and the reflected expansion wave propagates to the reverse direction. Figure 4.41a–i show

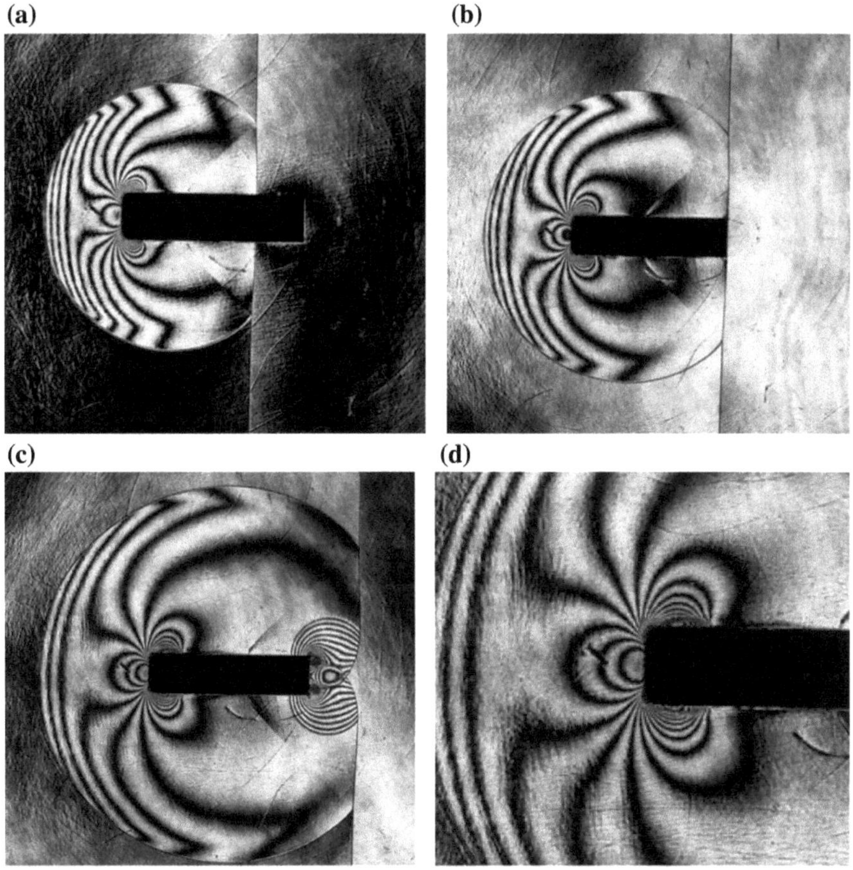

Fig. 4.40 The evolution of the shock wave interaction with a 4:1 rectangular plate for Ms = 1.40 in atmospheric air at 290.0 K and attack angle $\alpha = 0°$: **a** #88021524, Ms = 1.387; **b** #88021525, Ms = 1.392; **c** #88021522, Ms = 1.400; **d** enlargement of (**c**); **e** #88021516, Ms = 1.398; **f** 88021517, Ms = 1.401; **g** #88021518, Ms = 1.398; **h** #88021519, Ms = 1.395; **i** #88021520, Ms = 1.403; **j** 88021521, Ms = 1.405

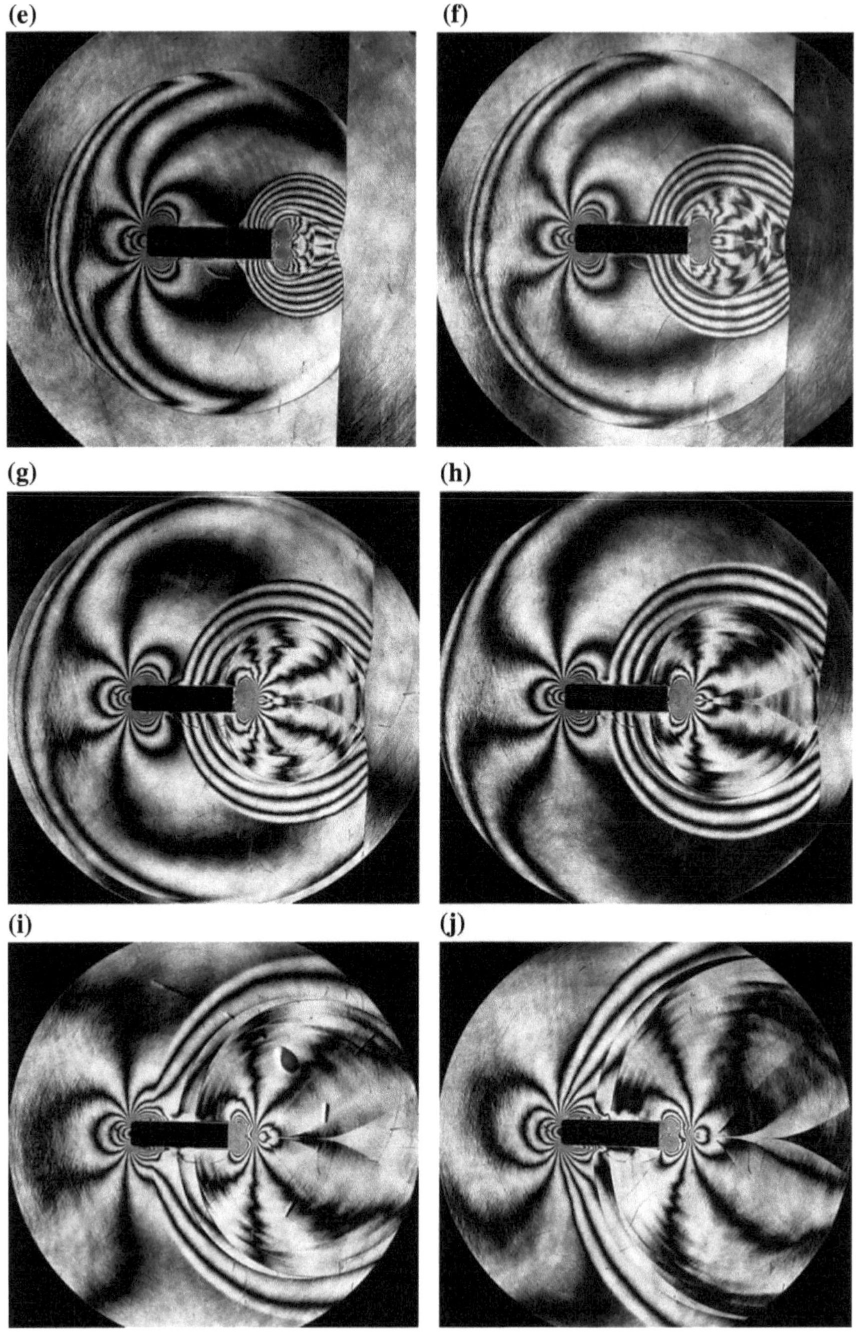

Fig. 4.40 (continued)

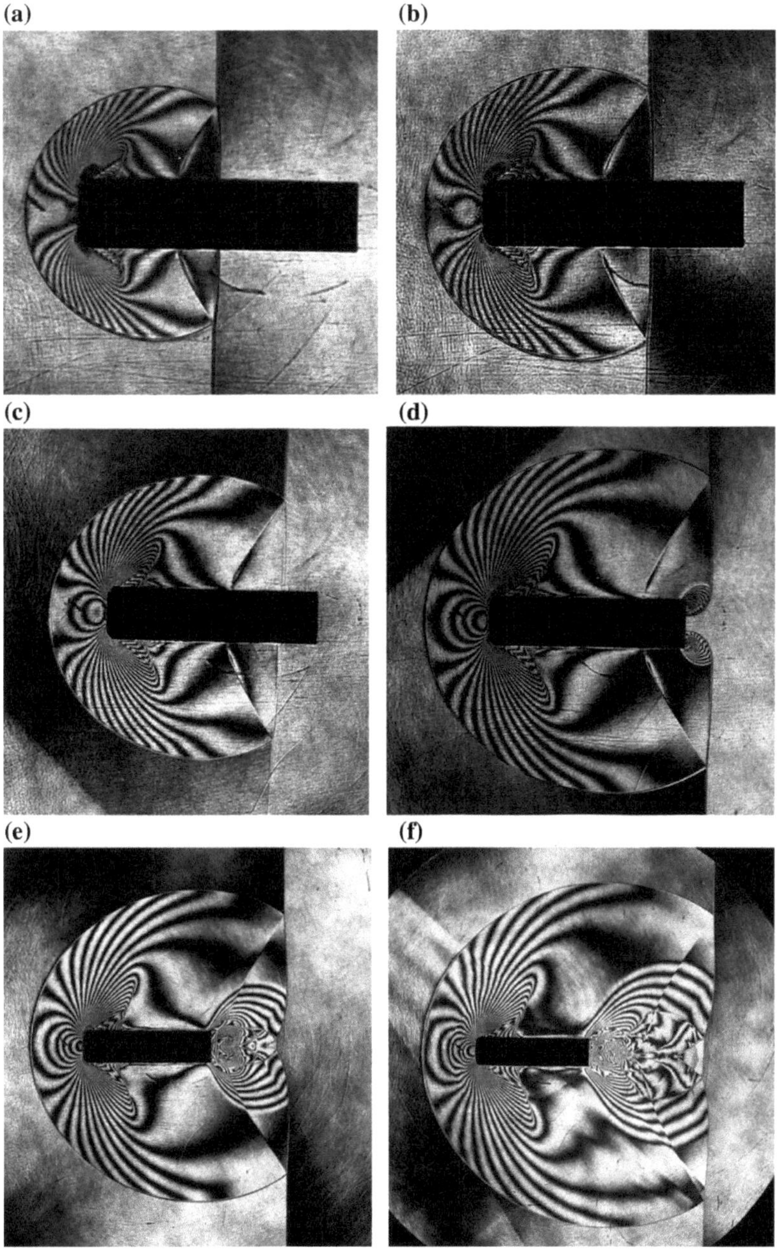

Fig. 4.41 The evolution of the shock wave interaction with a 4:1 rectangular plate for Ms = 2.20 in air at 300 hPa, 291.2.0 K, attack angle α = 0°: **a** #88021503, Ms = 2.168; **b** #88021504, Ms = 2.150; **c** #88021505, Ms = 2.141; **d** #88021506, Ms = 2.188; **e** #88021508, Ms = 2.210; **f** #88021510, Ms = 2.157; **g** #88021512, Ms = 2.118; **h** #88021513, Ms = 2.185; **i** enlargement of (**h**)

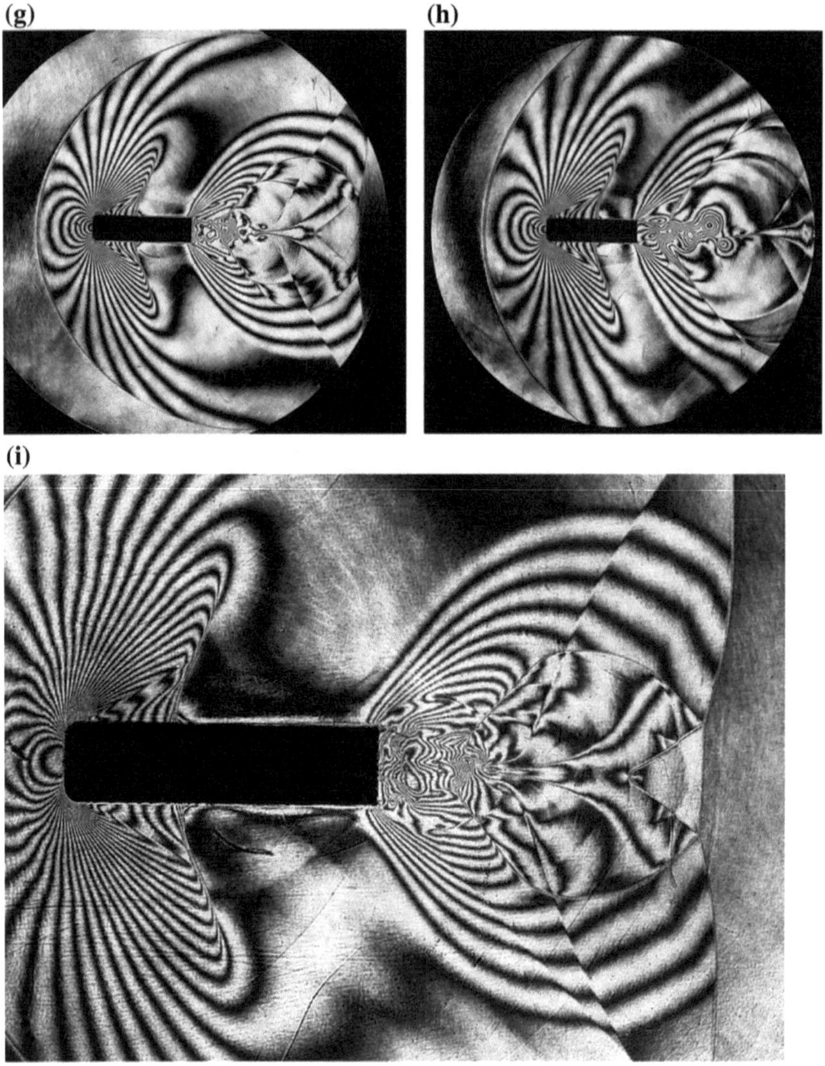

Fig. 4.41 (continued)

the evolution of the shock wave interaction of the rectangular plate of aspect ratio 4:1 for Ms =2.20. Impinging on the rectangular plate, the IS is diffracted at the frontal corner forming separation bubbles. The transmitting shock wave is diffracted again at the rear corners.

4.4.4.2 Attack Angle α = 5°

Figure 4.42a–f show reflected shock wave interactions for Ms = 1.70 in atmospheric air with the 4:1 rectangular plate at attack angle α = 5°. Upon the shock wave impingement, the IS was diffracted at the upper corner of the frontal side forming a separation bubble which developed with elapsed time. The transmitting shock wave was diffracted at the corner of the rear side of the rectangular plate.

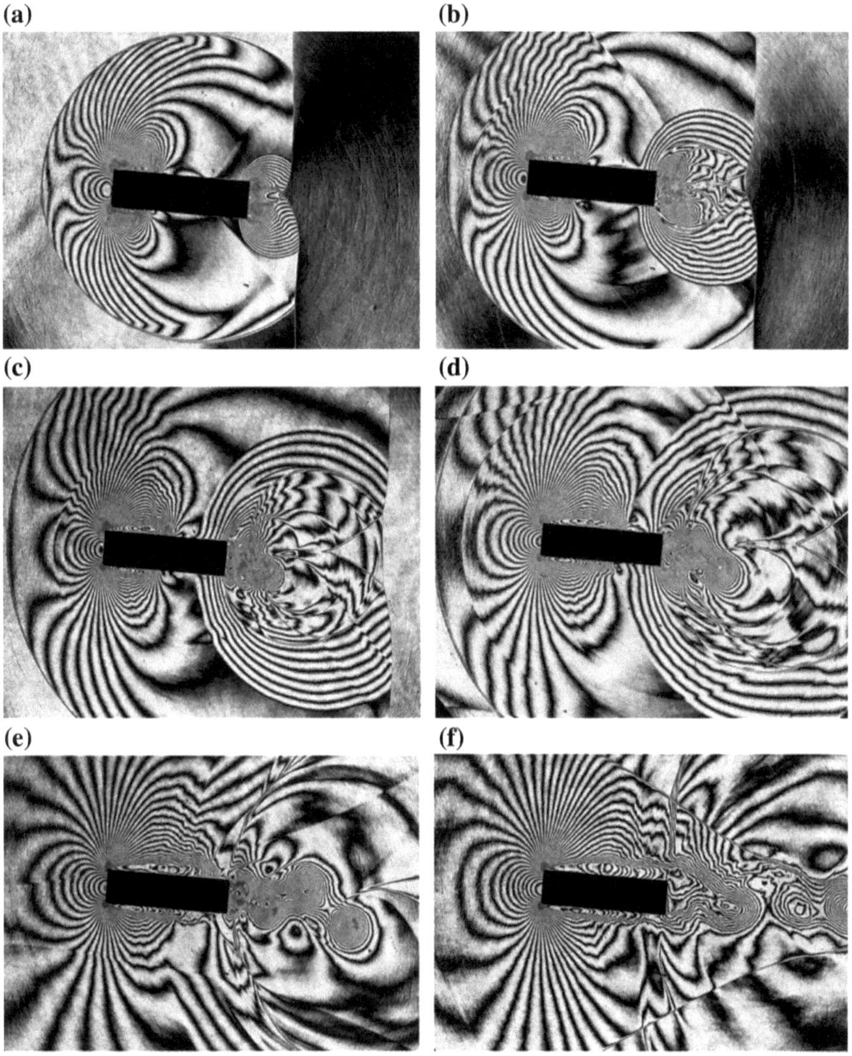

Fig. 4.42 The evolution of the shock wave interaction with a 4:1 rectangular plate for Ms = 1.70 in atmospheric air at 290.0 K, attack angle α = 5°: **a** #87012006, Ms = 1.714; **b** #87012002, Ms = 1.700; **c** #87012010, Ms = 1.684; **d** #87012001, Ms = 1.696; **e** #87012007, Ms = 1.722; **f** #87012009. Ms = 1.685

4.4.4.3 Attack Angle α = 10°

Figure 4.43a–d show reflected shock wave interaction for Ms = 1.70 in atmospheric air with 4:1 rectangular plate at attack angle α = 10°. Upon the shock wave impingement, the IS was diffracted more significantly at the upper corner of the frontal side forming a separation bubble which developed with elapsing time. The train of vortices was intermittently released from the corners of the rear side.

4.4.4.4 Attack Angle α = 45°

Figure 4.44a–f show reflected shock wave interaction for Ms = 1.70 in atmospheric air with 4:1 rectangular plate at α = 45°. In Fig. 4.44a the IS was reflected from the lower corner and was diffracted at the upper corner of the frontal side.

(a) (b) (c) (d)

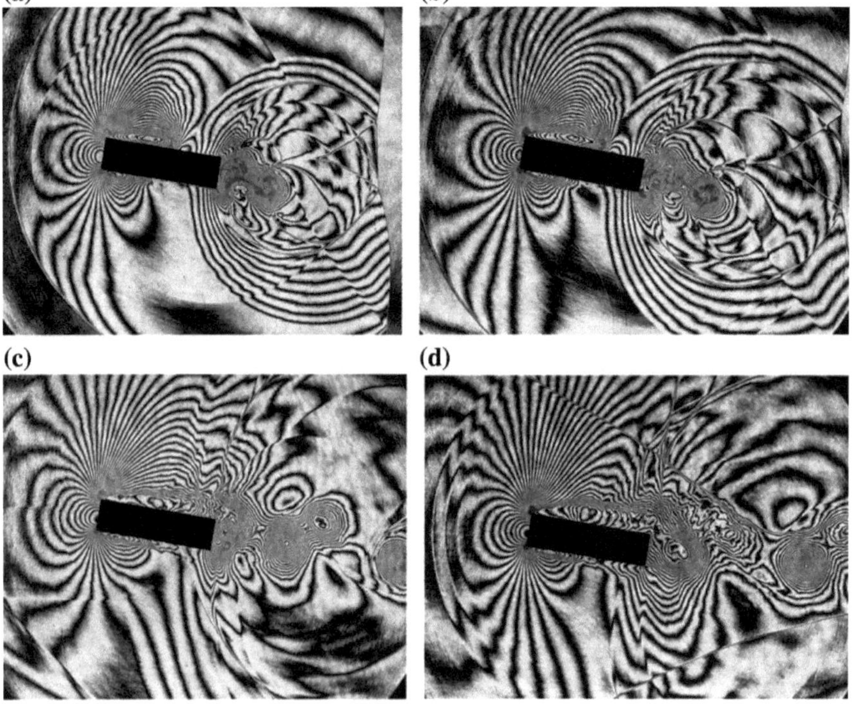

Fig. 4.43 Evolution of shock wave interaction with a 4:1 rectangular plate for Ms = 1.70 in atmospheric air at 290.0 K, attack angle α = 10°: **a** #87012103, Ms = 1.689; **b** #87012011, Ms = 1.663; **c** #87012107, Ms = 1.696; **d** #87012013, Ms = 1.658

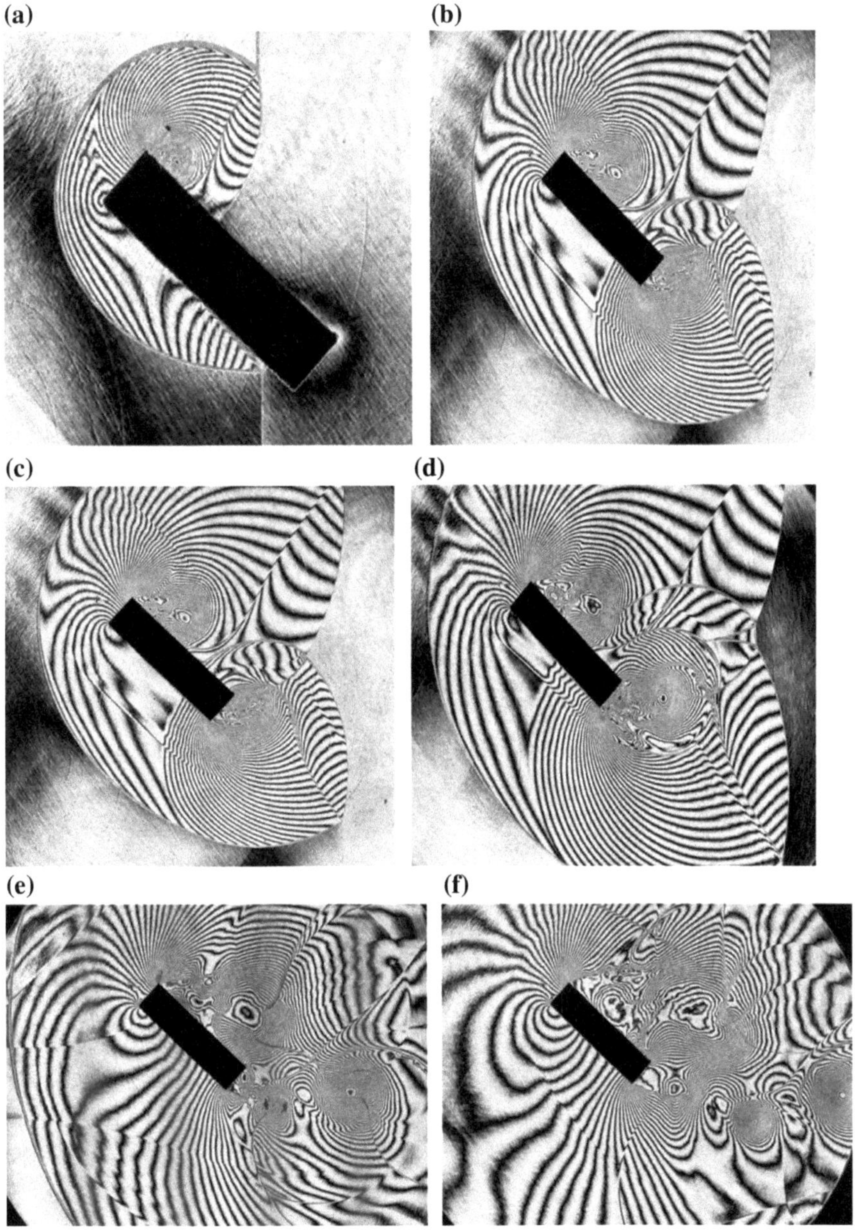

Fig. 4.44 The evolution of the shock wave interaction with a 4:1 rectangular plate for Ms = 1.70 in air at 900 hPa, 284.6 K, attack angle α = 45°: **a** #87013009, Ms = 1.673; **b** #87013008, Ms = 1.684; **c** #87013003, Ms = 1.686; **d** #87013006, Ms = 1.706; **e** #87013005, Ms = 1.687; **f** #87013010, Ms = 1.679

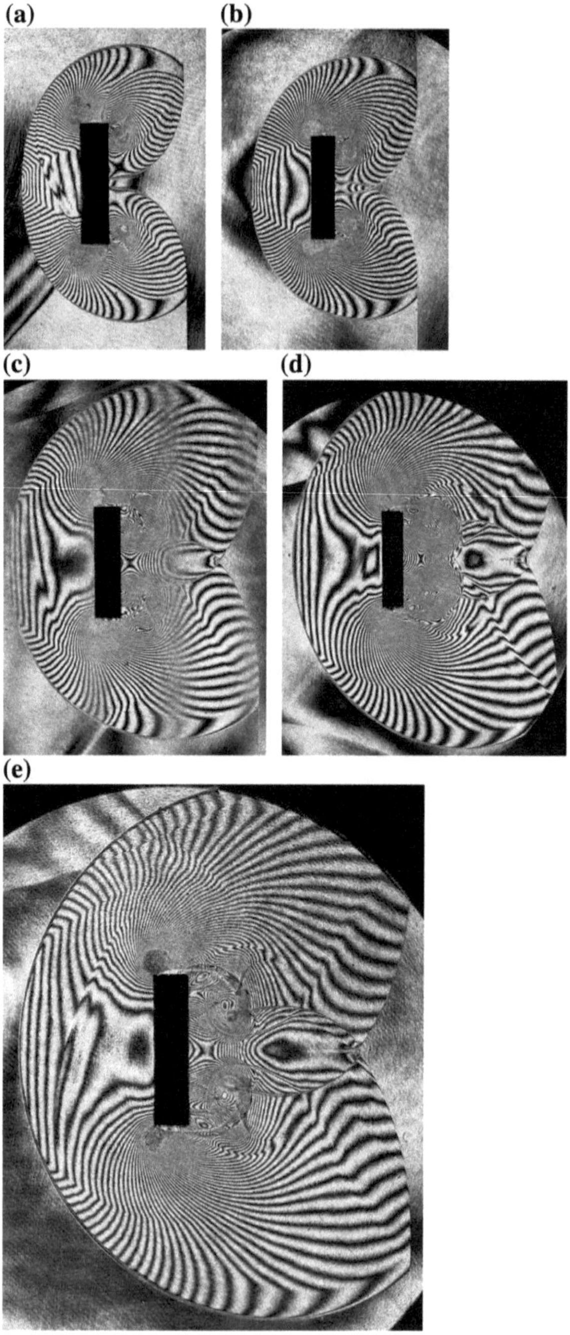

Fig. 4.45 Interaction with a 4:1 rectangular plate for Ms = 1.70 in air at 900 hPa, 284.6 K and attack angle α = 90°: **a** #87020601, Ms = 1.719; **b** #87020902, Ms = 1.718; **c** #87020603, Ms = 1.719; **d** #87020702, Ms = 1.703; **e** #87020901, Ms = 1.718

4.4.4.5 Attack Angle α = 90°, Head-on Collision

Figure 4.45a–e show head reflection of shock wave for Ms = 1.70 from the 4:1 rectangular plate at α = 90°.

4.4.5 *NACA 0012 Airfoil*

An NACA0012 airfoil having a 60 mm chord length is sandwiched between two circular acrylic plates and installed in the test section of the 60 mm × 150 mm conventional shock tube. However, shock tubes are not necessarily useful tools for generating transonic flows ranging the flow Mach number from 0.95 to 1.05. However, moderately strong transonic flows are relatively easily reproduced in shock tubes. As the blockage ratio of the airfoil was about 0.2 and wavelets were readily suppressed, the resulting transonic flow was established after 1.3 ms transient period of time and was maintained for about 1.5 ms.

For a Ms = 1.74 shock wave in air at 700 hPa, 293 K, the local flow Mach number M is 0.8 and the Reynolds number Re is 5×10^5. To adjust an attack angle, the entire observation windows were rotated to a specified attack angle (Itoh 1986). Figure 4.46 show the sequence of establishing the transonic flow. The IS of Ms = 1.80 impinged on the airfoil as seen in Fig. 4.46a–c. The reflected Mach stems propagated reversely and the reflected waves from the upper and lower walls passing along the airfoil surface were suppressed with elapsing time as seen in Fig. 4.46d.

Figure 4.47 show an NACA0012 airfoil placed in the transonic flow of M = 0.8 and Re = 5×10^5 while changing its attack angle from 0° to 7.0°.

Figure 4.48c, d show a finite fringe double exposure interferogram and an infinite fringe double exposure interferogram, respectively. The finite fringe interferogram and the infinite fringe interferogram have their inherited merits and demerits when analyzing their fringes. Therefore, it would be very appropriate for image analysis, if the two types of interferograms can be combined in one interferogram. Figure 4.48a is a triple exposure interferogram. The first exposure was conducted under no flow condition and the second exposure was conducted with a *RB* the collimating lens as shown in Fig. 1.2 was rotated and shifted appropriately. Then the third exposure was conducted synchronizing the motion of the incident shock wave. Eventually the triple exposure interferogram was obtained as shown in Fig. 4.48a.

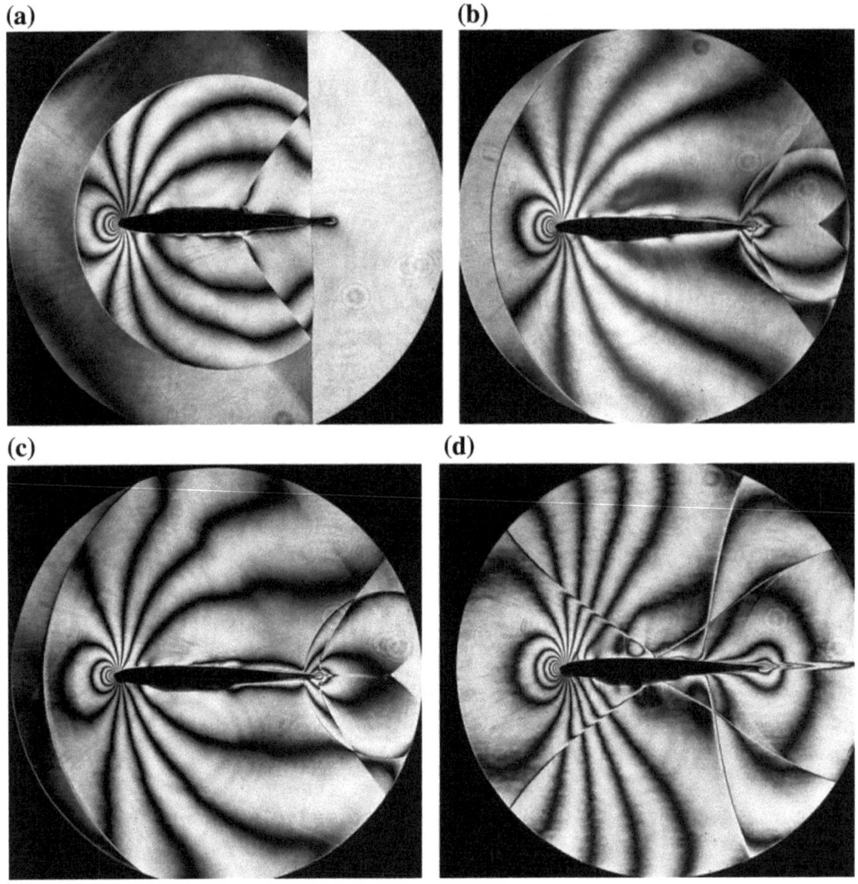

Fig. 4.46 Formation of transonic flows over an NACA0012 airfoil for Ms = 1.80, M = 0.85 in air at 500 hPa, 287.5 K: **a** #83020916 340 μs Ms = 1.798; **b** #83020912, 360 μs Ms = 1.799; **c** #83020913, 370 μs Ms = 1.798; **d** #83020920, 800 μs Ms = 1.804

Figure 4.49 show supersonic flows over an NACA 0012 airfoil at different attack angles ranging from $\alpha = 0.5°$ to 6.5° for Ms = 2.38 and local flow Mach number M = 1.03 in air. Comparing airfoils in subsonic flow to that observed in supersonic flows, it is relatively easy to analyze. Measuring fringe distributions recorded in interferograms, the density contours over the airfoil are readily deduced. Hence assuming that the isothermal wall condition prevails, the density contours are readily converted to an appropriate pressure distribution along the airfoil surface.

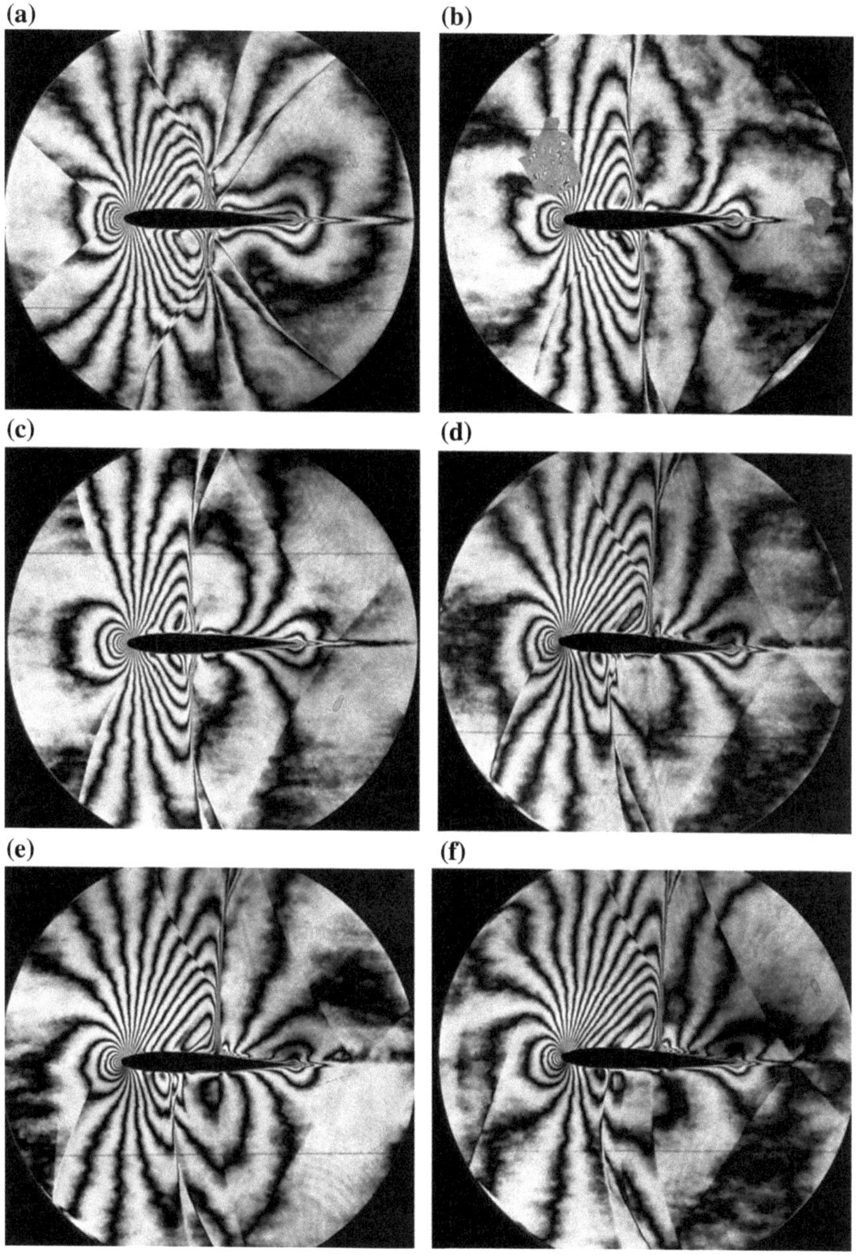

Fig. 4.47 Shock wave interaction with an NACA 0012 airfoil for Ms = 1.74, local transonic flow M = 0.80, Re ~5 × 10^5 at 700 hPa, 293 K: **a** #83040603, 1.3 ms time delay from trigger point and attack angle $\alpha = 0.0°$; **b** #83040610, $\alpha = 0.5°$; **c** #83040607, $\alpha = 1.0°$; **d** #83040703, $\alpha = 1.25°$; **e** #83040705, $\alpha = 1.5°$; **f** #83040708, $\alpha = 1.75°$; **g** #83040802, $\alpha = 2.0°$; **h** #83040805, $\alpha = 2.5°$; **i** #83040808, $\alpha = 3.0°$; **j** #83040815, $\alpha = 3.5°$; **k** #83040813, $\alpha = 4.0°$; **l** #83040815, $\alpha = 4.5°$; **m** #83040817, $\alpha = 5.0°$; **n** #83040809, $\alpha = 5.5°$; **o** #83041101, $\alpha = 6.5°$; **p** #83040818, $\alpha = 6.9°$; **q** #83040707, $\alpha = 7.0°$; **r** enlargement of (**q**)

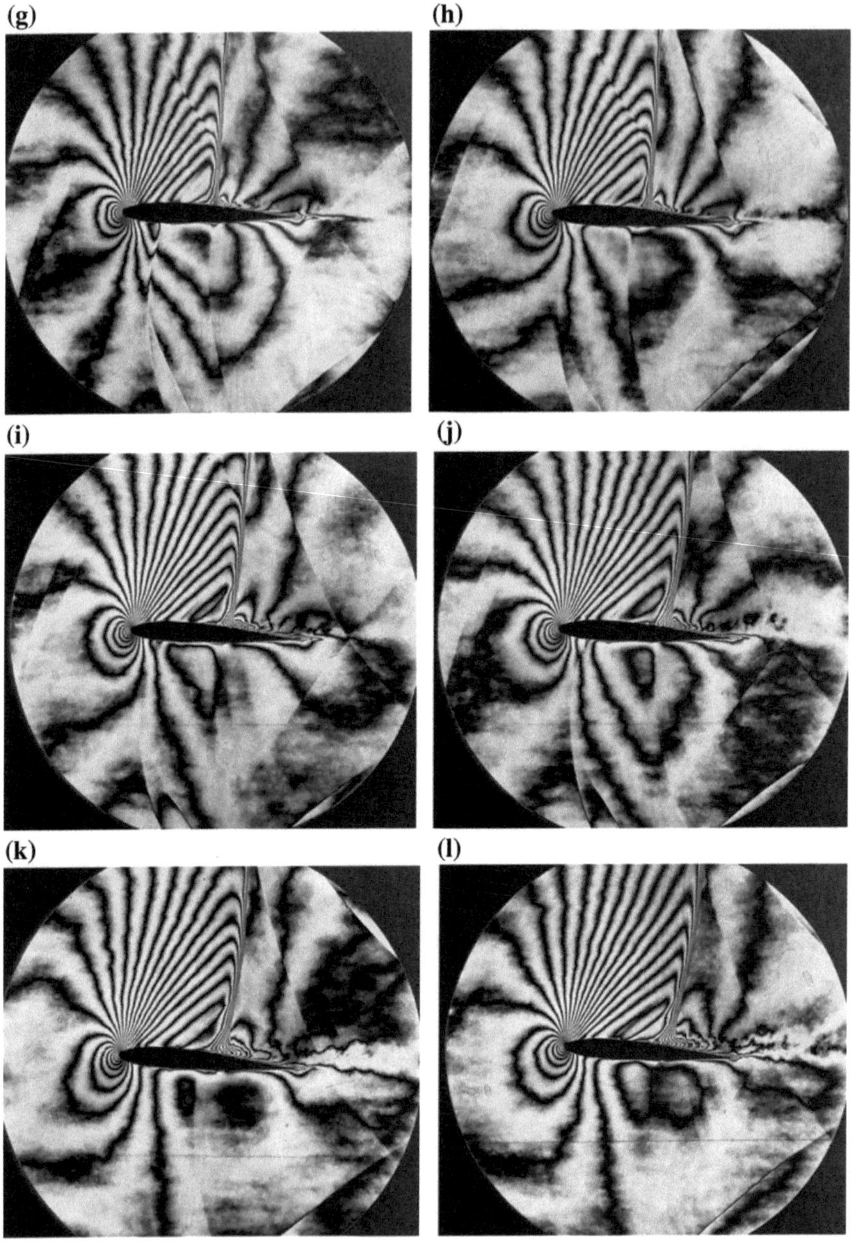

Fig. 4.47 (continued)

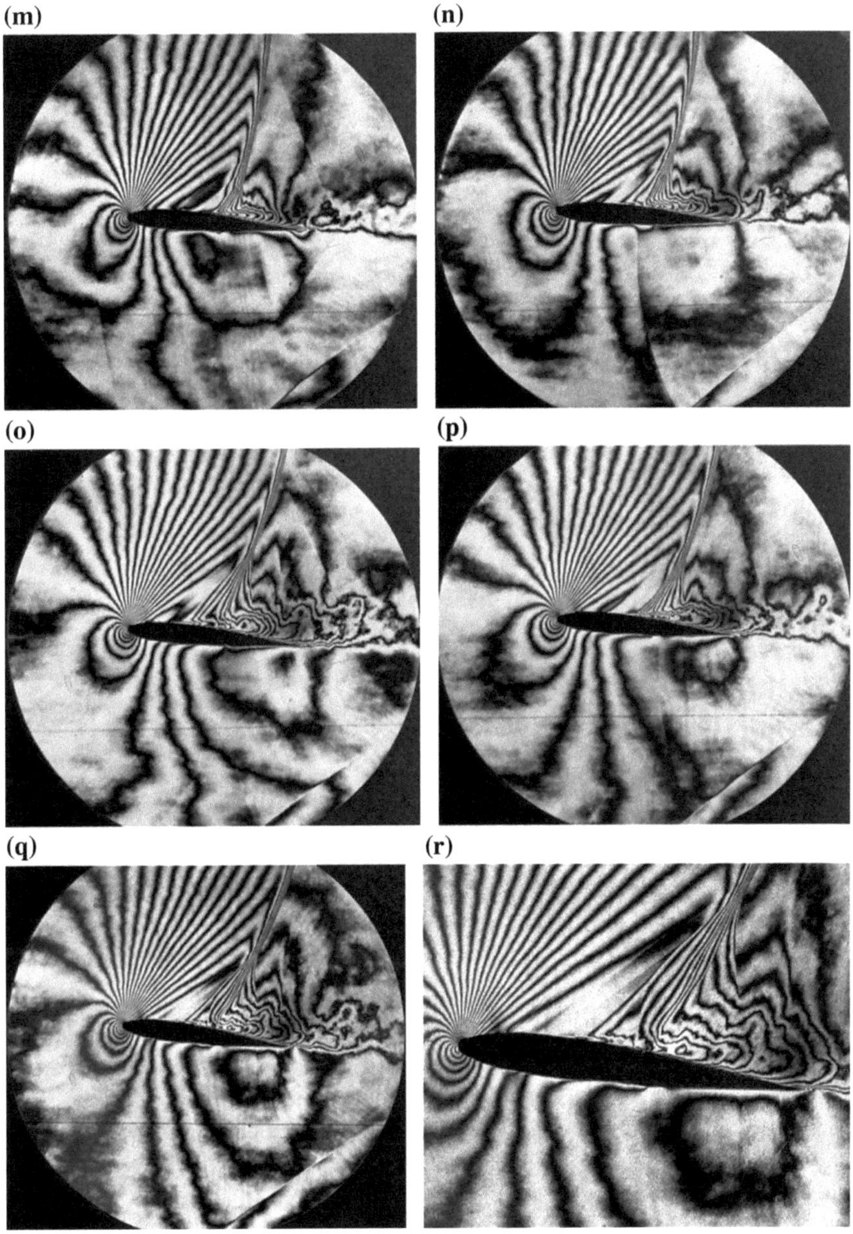

Fig. 4.47 (continued)

(a) (b) (c) (d)

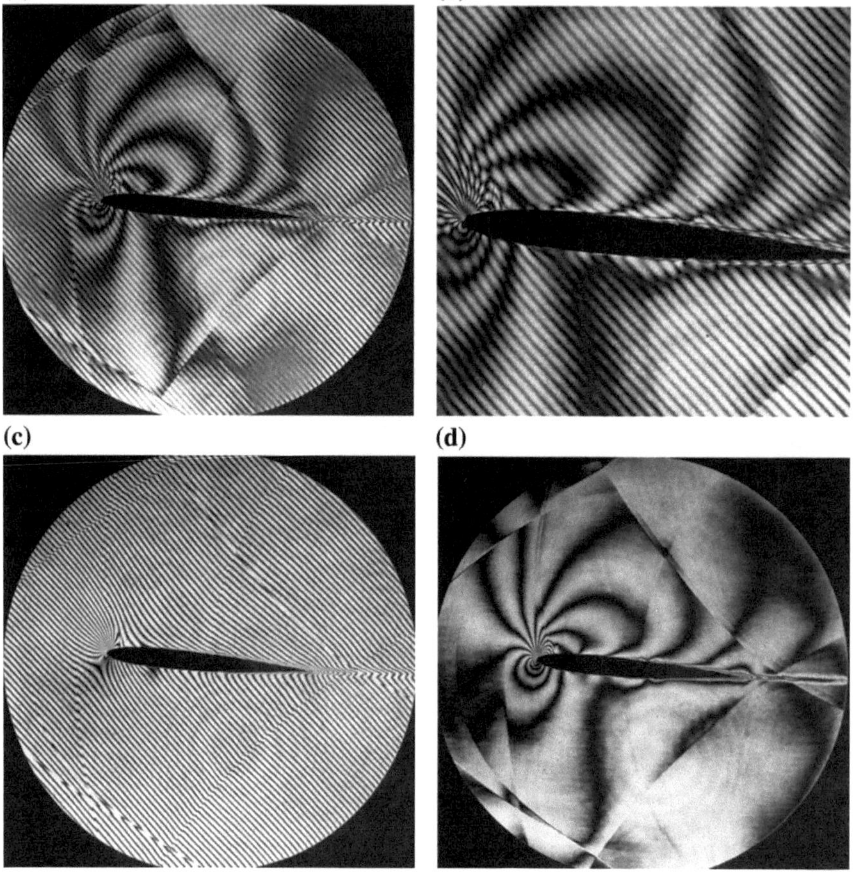

Fig. 4.48 Finite and infinite fringe double exposure interferograms and triple exposure interferogram: **a** triple exposure interferogram #83021710, Ms = 1.483; **b** enlargement of (**a**); **c** finite fringe #83021711, Ms = 1.483; **d** infinite fringe #83021712, Ms = 1.484

Therefore, the pressure coefficient of the airfoil is experimentally determined; that is, the lift coefficient and drag coefficients are determined. In Fig. 4.49, the shock stand-off distance from the airfoil was unvaried during the experiments. The black circle was the plug filling the hole of the pressure transducer. The surface of the plug was so smooth that it never disturbed the flows.

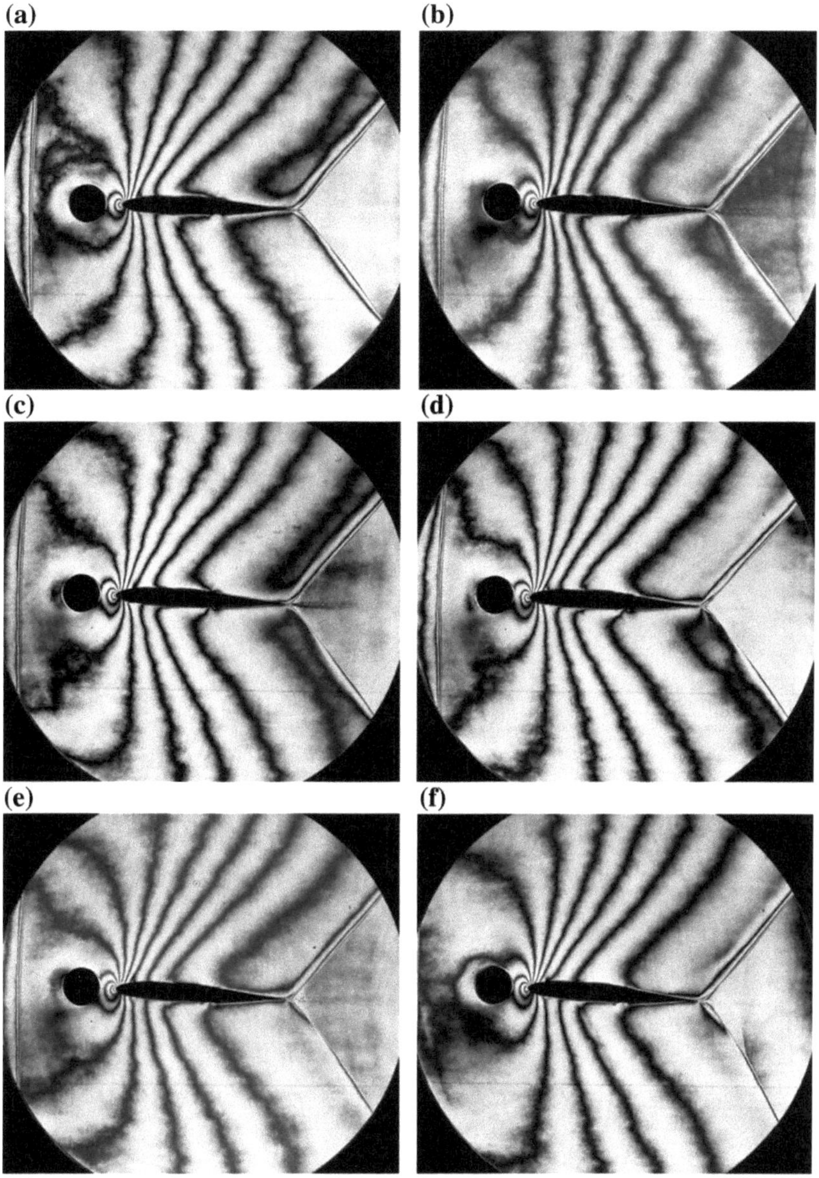

Fig. 4.49 Transonic flow of M = 1.15, Re = 6 × 10^4 over an NACA 0012 airfoil at variable attack angles α for Ms = 2.38 at 1.5 ms time delay from trigger point in air at 150 hPa, 293: **a** #83111512, Ms = 2.382, α = 0.5°; **b** #83111511, Ms = 2.375, α = 1.0°; **c** #83111509, Ms = 2.375, α = 1.5°; **d** #83111508, Ms = 2.367, α = 2.0°; **e** #83111507, Ms = 2.359, α = 2.5°; **f** #83111506, Ms = 2.355, α = 3.0°; **g** #83111505, Ms = 2.378, α = 3.5°; **h** #83111415, Ms = 2.384, α = 4°; **i** #83111503, Ms = 2.409, α = 4.5°; **j** #83111418, Ms = 2.368, α = 5°; **k** # 83111419, Ms = 2.346, α = 5.5°; **l** # 83111501, Ms = 2.333, α = 6.0°; **m** #83111502, Ms = 2.348, α = 6.5°

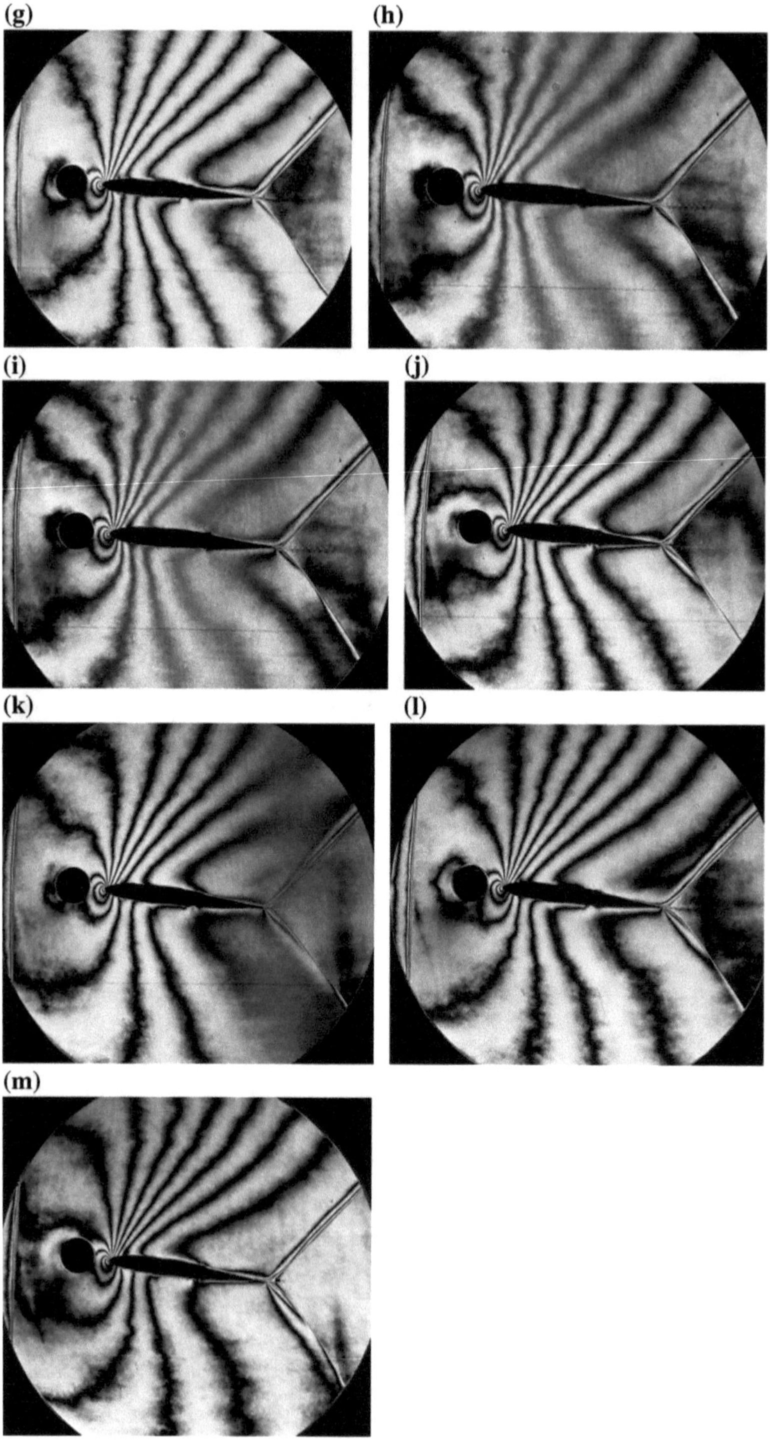

Fig. 4.49 (continued)

4.5 Nozzle Flows

4.5.1 Diverging Nozzle

In Fig. 4.50, a diverging nozzle of apex angle 25° was installed in the 60 mm 150 mm conventional shock tube (Saito et al. 2000) A shock wave of Ms = 2.40 at 150 hPa, 292.1 K impinged at the entry wall is shown in Fig. 4.50a. The shock wave was diffracted at the corner and the interaction took place. With the elapsed time, so-called nozzle starting process occurred. In Fig. 4.50i, a uniformly expanding flow region appeared. Usually the shape of diverging section is designed to promote the uniformly expanding flow region.

4.5.2 Converging and Diverging Nozzle

A diverging nozzle as shown in Fig. 4.50 is the simplest way for establishing a steady supersonic in a shock tube. In order to obtain a more uniform flow region by suppressing the unsteady starting process, converging diverging nozzle having a throat was used. From practical point of view, the nozzle starting is a benign topic for numerical code validations (Saito et al. 2000). However, it was not a simple problem. It was found that the Euler solvers worked only at early stage of the shock wave propagation inside the nozzle. The Navier-Stokes solvers which assumed the existence of a laminar boundary layer worked well only up to early stage of nozzle starting. To reproduce the recorded fringe distributions, it is required to establish a reliable turbulent model.

Figures 4.50 and 4.51 showed experimental results conducted repeatedly for identical Mach numbers and Reynolds numbers and identical initial conditions. The comparison between Figs. 4.50 and 4.51 revealed that the nozzle starting processes are significantly different depending on the throat configuration. The nozzle throat had a straight shape as shown in Fig. 4.50 but is rounded as shown in Fig. 4.47.

Figure 4.52 show very early stage in flow establishment inside a converging and diverging nozzle starting. At very early stage, a laminar boundary layer prevails and therefore, a viscous simulation is relatively benign. When the flow built-up completely, a fully turbulent boundary layer developed. For simulating the flow at this stage, the selection of a suitable turbulent models becomes an important issue. At this time, these images will be useful in validating the numerical code.

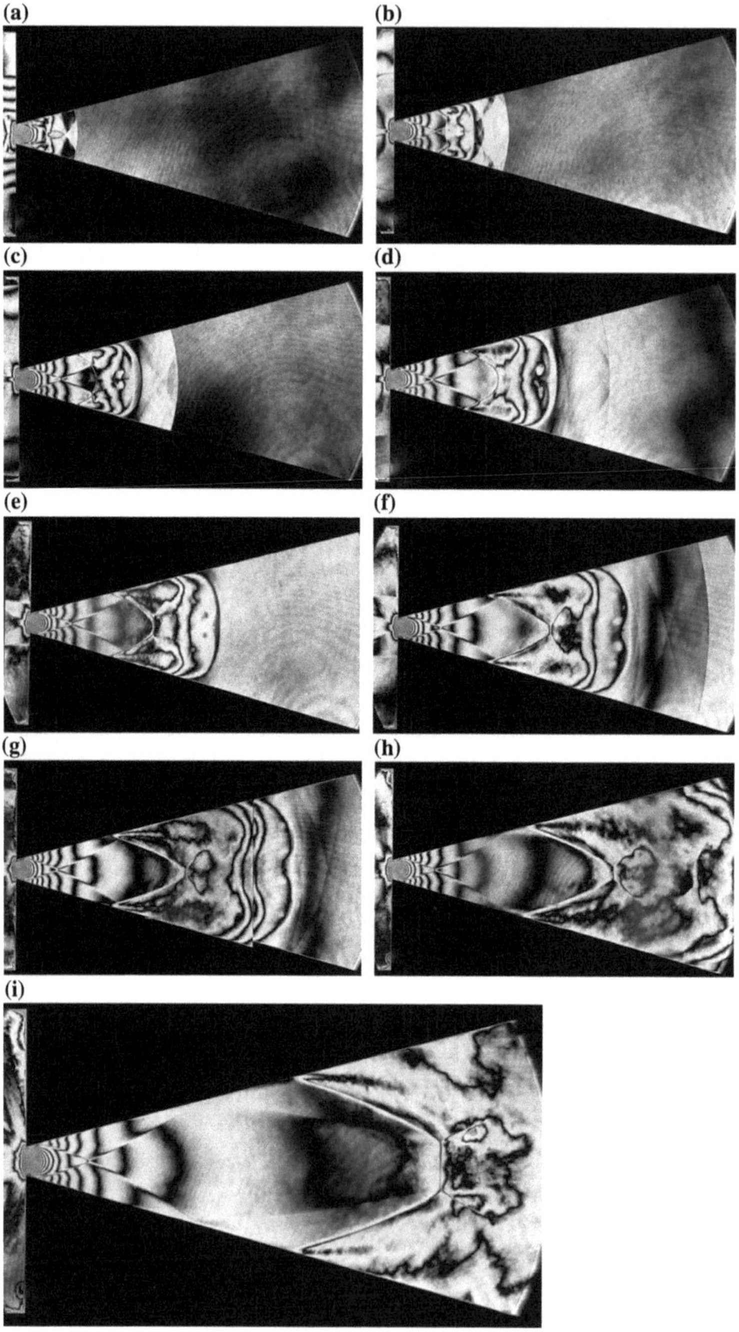

Fig. 4.50 Shock wave propagation in a diverging nozzle in air for Ms = 2.40 at 150 hPa, 292.1 K: **a** #83090804, Ms = 2.430, 298.4 K; **b** #83090805, Ms = 2.430; **c** #83090806, Ms = 2.430; **d** #83090807, Ms = 2.448; **e** 83090809, Ms = 2.391; **f** #83090810, Ms = 2.473: **g** #83090811, Ms = 2.368; **h** #83090812, Ms = 2.408; **i** #83090814, Ms = 2.415

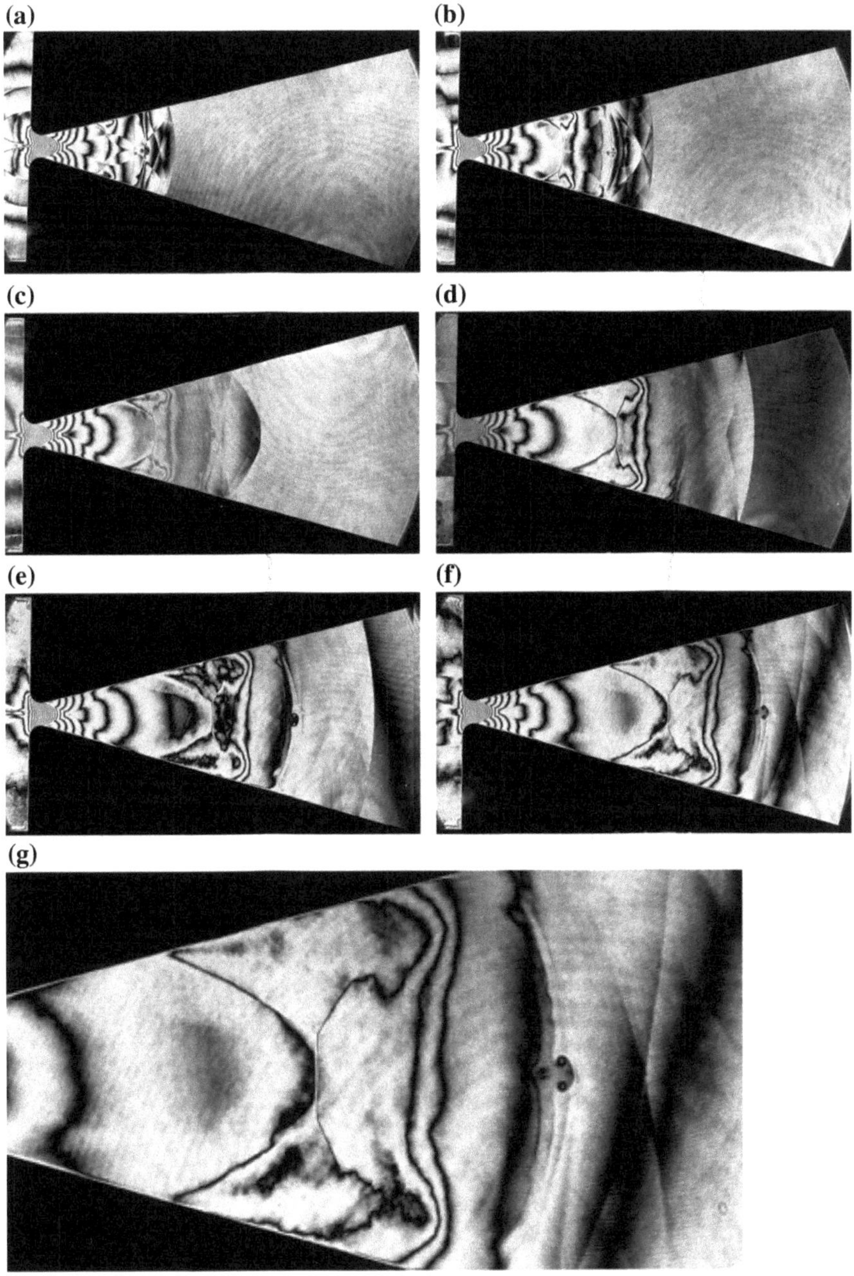

Fig. 4.51 Shock wave propagation inside a converging and diverging nozzle for Ms = 2.45 in air at 150 hPa, 292.1 K: **a** #83082902, 20 μs from starting, Ms = 2.437; **b** #83082905, 50 μs, Ms = 2.455; **c** #83082906, 100 μs, Ms = 2.455; **d** #83082907, 150 μs, Ms = 2.447; **e** #83082909, 200 μs, Ms = 2.422; **f** #83083001, 250 μs, Ms = 2.437; **g** enlargement of (**f**)

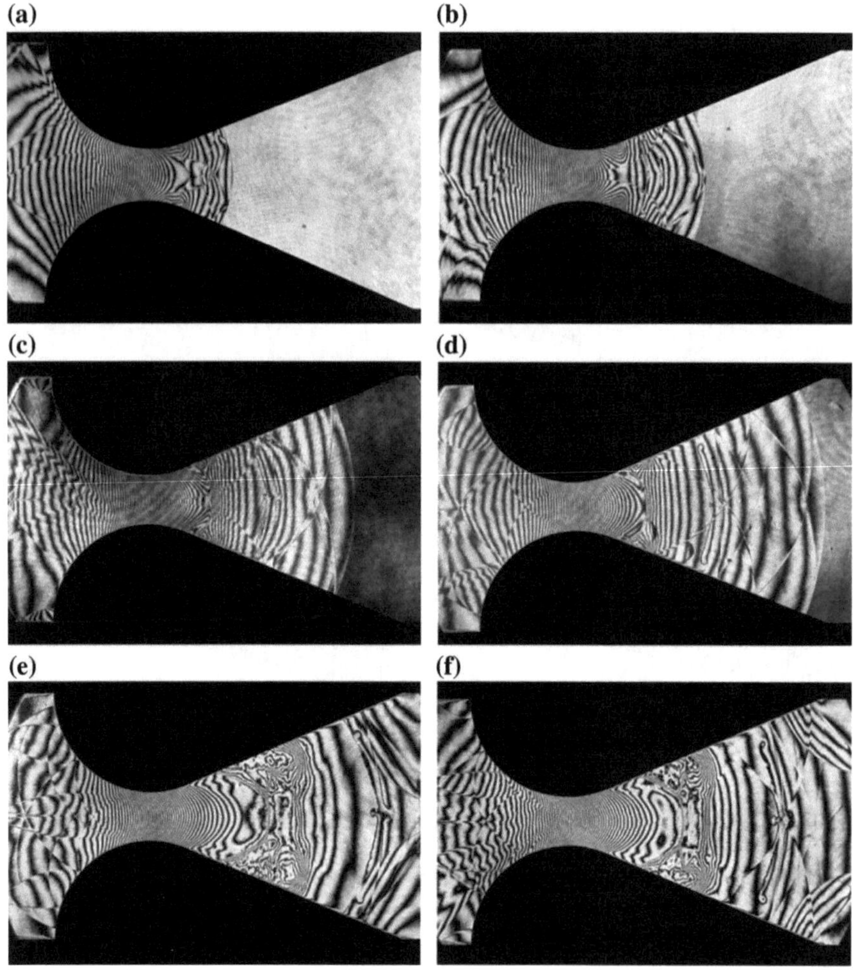

Fig. 4.52 Shock wave propagation in a de Laval nozzle for Ms = 1.50 in atmospheric air at 294.4 K: **a** #85101103 600 μs, Ms = 1.500, atm; **b** #85101104, 650 μs, Ms = 1.496; **c** #85101105, 750 μs, Ms = 1.496; **d** #85101106, 850 μs, Ms = 1.460; **e** #85101107, 950 μs, Ms = 1.524; **f** #85101108, 1050 μs, Ms = 1.487

4.6 Boundary Layers

Some shock wave researchers were tempted to trust analytical models rather than to believe experimental results. They sometimes dared to say "Experiments had errors and uncertainty, whereas theory had a clear background" Analytical models have a clear background. The clear that analytical model have is sometime akin to fiction. Experimentalists working in gas-dynamics are struggling with the flow non-uniformity and unsteadiness or turbulence. The presence of wall boundary layers is a problem to overcome. Nevertheless, in shock tube experiments, the boundary layer plays an important role (Schlichting 1960).

4.6.1 Boundary Layer in Shock Tube Flows

Figure 4.53a, b show shock tube flows and boundary layers developing along the shock tube side walls of the 60 mm × 150 mm conventional shock tube. Shock waves are propagating from the left to the right. The particle flow is moving from the left to the right. Figure 4.53a–d are displayed upside down. 3.2 mm plastic beads distributed on the wall disturbed the boundary layer development. Figure 4.53c, d show the wavelets created by the beads and their interactions with boundary layers. This was a transient flow and hence the density was uniformly distributed and hence the boundary layer profile could be estimated. However, it should be noticed that these fringe distribution was not truly two-dimensional.

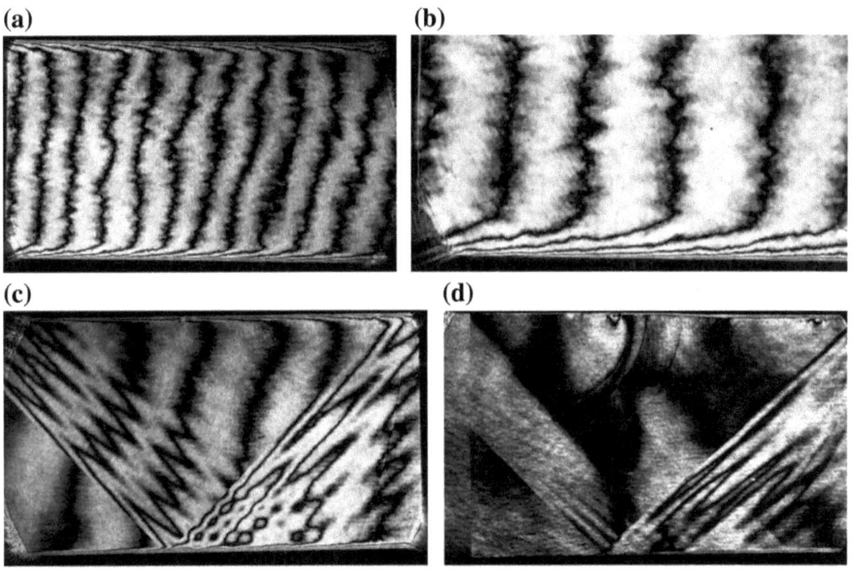

Fig. 4.53 Boundary layers developing behind incident shock wave, 3.2 mm diameter plastic beads are distributed on the upper floors: **a** #90020606, 2.0 ms from the trigger point, Ms = 1.514, in atmospheric air at 289.9 K; **b** enlargement of (**a**); **c** #90020604, 1.0 ms from the trigger point, Ms = 1.498 in atmospheric air at 289.4 K; **d** #90020904, 170 μs from the trigger point, Ms = 1.814 in air at 700 hPa 290.2 K

4.6.2 Reflected Shock Wave/Boundary Layer Interaction

A reflected shock wave reflected from a shock tube end wall interacts with the boundary layer developed behind the incident shock wave. In the advent of the hypersonic flow experiments, a throat and a diverging nozzle were connected to the end of the shock tube and the high enthalpy stagnant condition behind the reflected shock wave was used as the reservoir for the hypersonic flows. Such a shock tube facility is called shock tunnel. In the past, studying high-pressure and high temperature generated behind the reflected shock waves was the hot topics in shock tube technology. The operation of shock tubes was so tuned as to elongate the stagnation condition behind the reflected shock waves near the end wall. This operational method was named as Tailoring (Gaydon and Hurle 1963). This funny name was taken from the profession of making suits. Then in the 1960s, the topic attracted shock wave researchers. It was reported that, depending on the condition, the reflected shock waves interact significantly with the sidewall boundary layer and then bifurcate.

Mark (1956) proposed an analytical model for the bifurcation criterion. When the stagnation pressure in the boundary layer is higher than that prevailing the reflected shock wave, then the reflected shock wave propagates straight in upstream direction. However, if the stagnation pressure in the boundary layer is lower than the pressure ahead of the reflected shock wave, the boundary layer separates and the separation bubble develops with evolving time, Therefore, the foot of the reflected shock wave bifurcates and eventually forms an oblique shock wave.

The degree of the bifurcation is variable, depending on the value of the specific heats ratio of the working gas. The bifurcation becomes violent with the value of γ approaches unity. As in monatomic gas $\gamma = 1.667$, it occurs only in a limited region of Ms and the degree of the bifurcation degree becomes modest. The effect of γ on the bifurcation was experimentally investigated in Honda et al. (1975). Figure 4.54 are sequential observation of reflected shock waves interacting with the boundary layer for Ms = 2.5 in air at 200 hPa 290.4 K. The series of experiments was conducted in a 40 mm × 80 mm conventional shock tube. Bifurcation was not observed in this experiment of Ms = 2.50 as seen in Fig. 4.54.

Figure 4.55 shows the enlargement of a reflected shock wave shown in Fig. 4.56a for Ms = 3.60 in CO_2 at 50 hPa, 290.3 K. As seen, the foot of the reflected shock wave bifurcated. An oblique shock wave is formed starting from the boundary layer separation point. The boundary layer separation generated a separation bubble. The bifurcated shock wave and the reflected shock wave intersected forming a triple point. The third shock wave and a faintly visible slip line emerge from this triple point. Sequential observation shown in Fig. 4.56 indicates that the triple point is moving toward the center of the shock tube, which implies that the triple point and its three-shock confluence forms an inverse MR. Then, the oblique shock wave is equivalent to the IS, the reflected shock wave is equivalence to the MS and the third shock wave is equivalent to the RS.

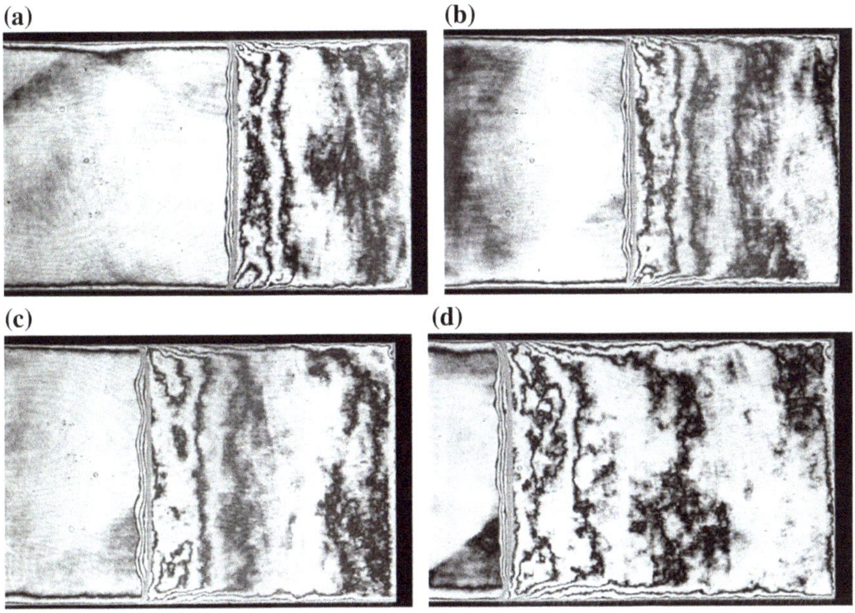

Fig. 4.54 Interaction of the reflected shock waves with the boundary layers for Ms = 2.5 in air at 200 hPa 290.4 K, $\gamma = 1.4$: **a** #81102601 for Ms = 2.561; **b** #81102602, Ms = 2.523; **c** #81102603, Ms = 2.506; **d** #81102606, Ms = 2.506

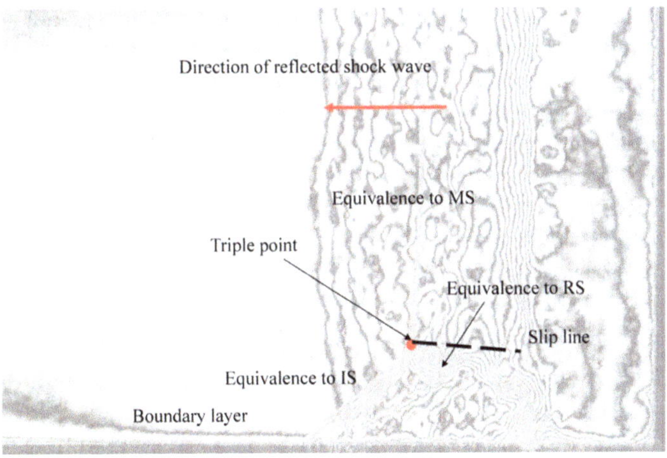

Fig. 4.55 The bifurcation of the reflected shock wave, enlargement of Fig. 4.52a #81102710, for Ms = 3.60 in CO_2 at 50 hPa 290.3 K

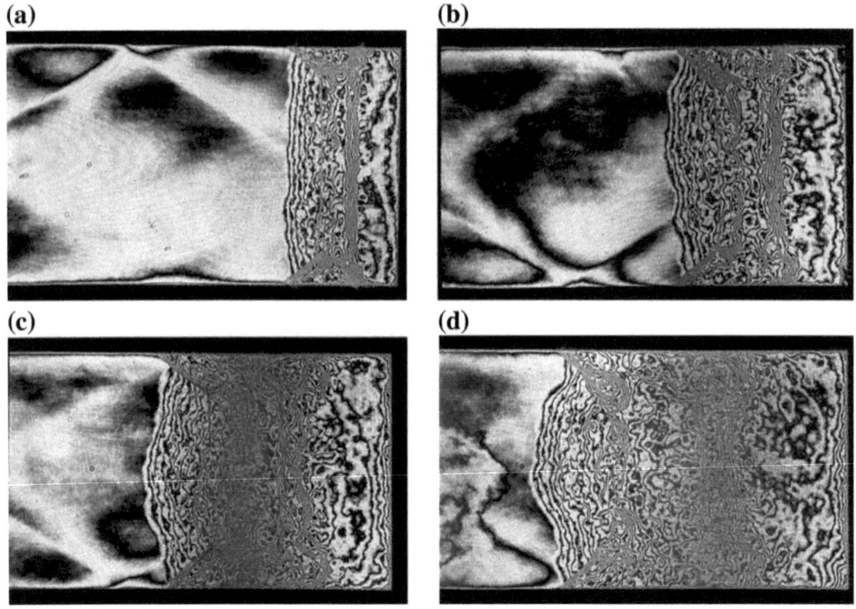

Fig. 4.56 Reflected shock wave/boundary layer interaction for Ms = 3.60 in CO_2 at 50 hPa 290.3 K, $\gamma = 1.29$: **a** #81102710, Ms = 3.647; **b** #81102711, Ms = 3.647; **c** #81102713, Ms = 3.774; **d** #81102712, Ms = 3.647

Some of the airbags installed in cars have a structure similar to a shock tube. When such airbags are activated, shock waves propagate in the tube, in which high pressure argon over 20 MPa is filled. Recently it was cleared why argon is used. Argon is a monatomic gas $\gamma = 1.667$. It minimizes the bifurcation and the pressure behind the reflected shock wave does not decrease.

In airbags that have similar structures as shock tubes; its operation is initiated by the explosion of explosive material. The explosion transmitted shock waves through the argon toward the end wall. When the end wall ruptured the shock gas flew into the inflator. At that time their bifurcation was suppressed at the end wall then the shocked argon and the resulting gases which were created by the chemical reaction of the explosive material flew into the inflator. At this stage, the role of argon was important.

Figure 4.56 show sequential observation of the reflected shock wave interaction with the side wall boundary layer. The bifurcation developed with the elapsed time. The triple points moved toward the center of the shock tube as seen in Fig. 4.56d.

Figure 4.57 shows an enlargement of Fig. 4.58a #81102718 for Ms = 5.20 in CO_2 at 10 hPa. At the reduced initial pressure, the image contrast seen in Fig. 4.57 is clearer than that seen in Fig. 4.55.

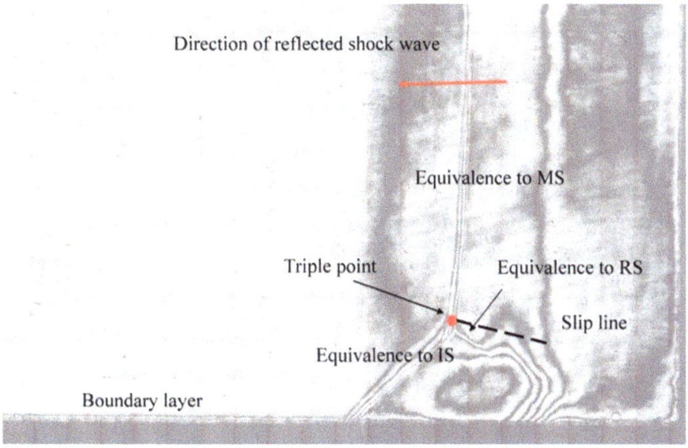

Fig. 4.57 Bifurcation of a reflected shock wave, enlargement of Fig. 4.54a, #81102718, for Ms = 5.20 in CO_2 at 10 hPa 290.3 K

Interpolating the positions of the triple points along the upper and lower walls from Figs. 4.56 and 4.58d, indicates that the trajectories intersected at the center of the shock tube. This means that the reflection patterns is the IvMR. The image data was missing for speculating how this reflection pattern transited to a RR.

As the sidewall boundary layers develop independently of the shock tube size. Therefore, in small shock tubes, the triple points would merger at a short distance from the end wall. Then, the significant pressure reduction would be induced. The effect of the reflected shock interaction with the sidewall boundary layer is governed by the shape of the shock tube cross section. In circular cross sectional shock tubes, the merger of triple point trajectories would be equivalent to focusing.

4.7 Pseudo Shock Waves in a Duct

When supersonic flows in a straight duct decelerate to subsonic flows, a train of shock waves appear, which successively interact with sidewall boundary layer. These shock waves are called pseudo shock waves. This phenomenon appears often in duct flows and pneumatic machineries. Often the pseudo shock wave induced oscillation in pipelines or generated noises. Professor Sugiyama of Muroran Institute of Technology visualized pseudo shock waves using a conventional schlieren method (Sugiyama et al 1987). We were once invited to visualize pseudo shock wave generated in his facility with double exposure holographic

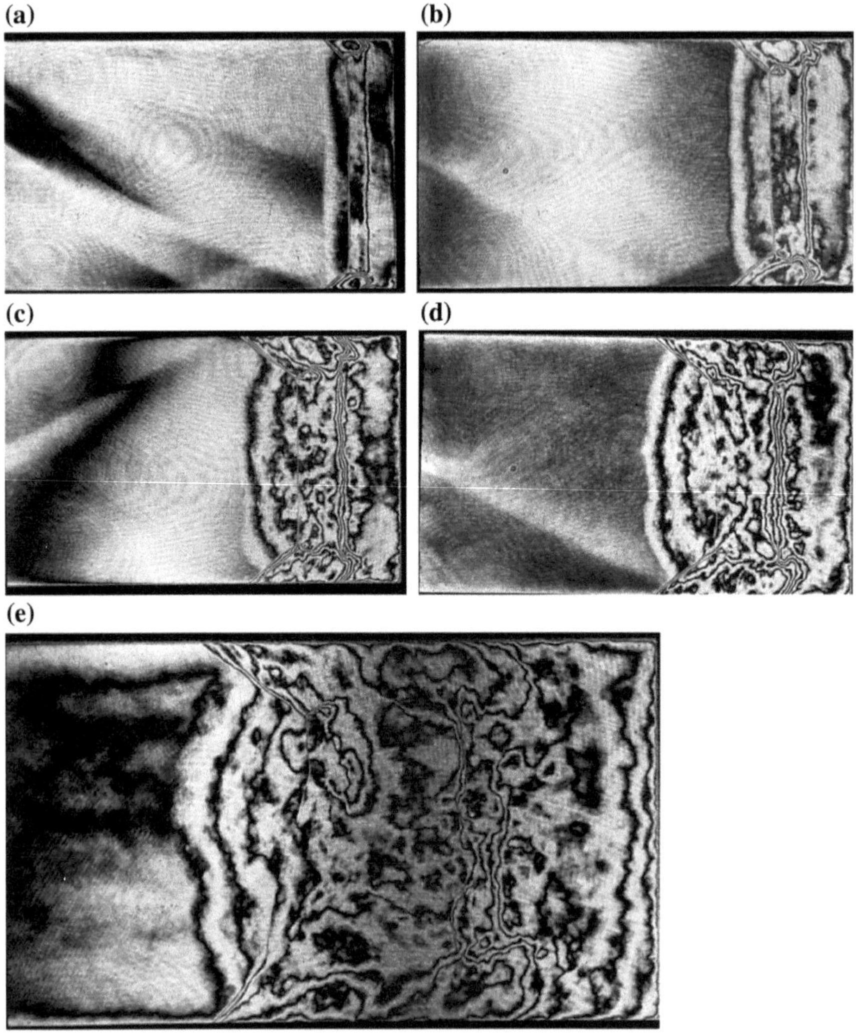

Fig. 4.58 Reflected shock wave/boundary layer interaction for Ms = 5.20 in CO_2 at 10 hPa 290.3 K, γ = 1.29: **a** #81102718, 100 μs from trigger point, Ms = 5.225; **b** #81102717, 150 μs, Ms = 5.043; **c** #81102801, 120 μs Ms = 5.120; **d** #81102716, 200 μs, Ms = 5.102; **e** #81102715, 250 μs, Ms = 5.237

interferometry. The facility was a 50 mm × 50 mm straight duct, the length to diameter ratio was L/D = 20.6–23.6 where the L was duct length, D = 50 mm and the flow Mach number ranged from 1.72 to 1.88. Figure 4.59 show sequential interferograms. It is noticed that the duct has a square cross section and its flow field

(a)

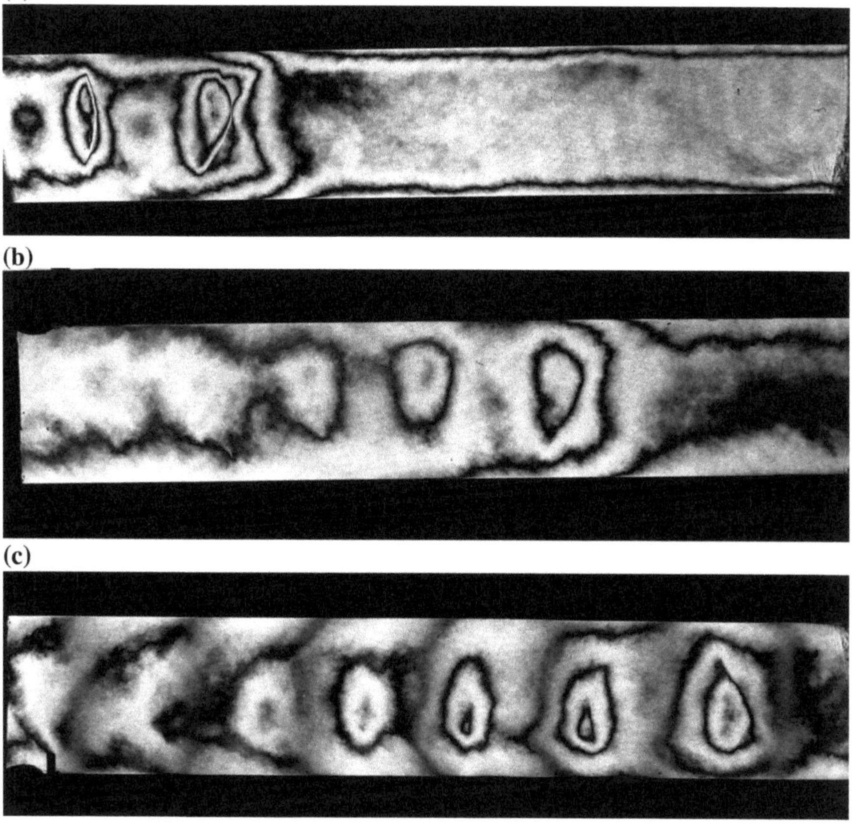

(b)

(c)

Fig. 4.59 Pseudo shock wave generated in a duct: **a** #87101304; **b** #87101202; **c** #87101302 (Sugiyama et al. 1987)

is not necessarily two-dimensional. The boundary layer developed along the square wall and hence the pseudo shock wave looked slightly blurred. The reflected shock wave bifurcated, when the stagnation pressures in main flows are higher than the stagnation pressures in the boundary layer. This condition is fulfilled in in the region where the pseudo shock waves bifurcate (Inoue et al. 1995).

References

Courant, R., & Friedrichs, K. O. (1948). *Supersonic flows and shock waves*. New York, NY: Wiley InterScience.

Gaydon, A. G., & Hurle, I. R. (1963). *The shock tube high-temperature chemical physics*. London: Chapman and Hall Ltd.

Heilig, W. (1969). Diffraction of shock waves by a cylinder. *Physics Fluids, 12*, 154–157.

Honda, M., Takayama, K., Onodera, O., & Kohama, Y. (1975). Motion of reflected shock wave in shock tube. In G. Kamimoto (Ed.), *Modern Developments in Shock Tube Research, Proceedings 10th International Shock Tube Symposium*, Kyoto (pp. 320–327).

Inoue, O., Imuta, G., Milton, B. E., & Takayama, K. (1995). Computational study of shock wave focusing in a log-spiral duct. *Shock Waves, 5,* 183–188.

Itoh, K. (1986). *Study of transonic flow in a shock tube* (Master thesis). Graduate School of Engineering, Faculty of Engineering Tohoku University.

Izumi, M. (1988). *Study of particle-gas two phase shock tube flows* (Master thesis). Graduate School of Engineering, Faculty of Engineering Tohoku University.

Kikuchi, T., Takayama, K., Igra, D., & Falcovitz, J. (2016). Shock stand-off distance over spheres in unsteady flows. In G. Ben-Dor, O. Sadat, & O. Igra (Eds.), *Proceedings of 30th ISSW*, Tel Aviv (Vol. 1, pp. 275–278).

Mark, H. (1956). *The interaction of a reflected shock wave with the boundary layer in a shock tube*. NACA TM 1418.

Saito, T., Timofeev, E. V., Sun, M., & Takayama, K. (2000). Numerical and experimental study of 2-D nozzle starting processes. In G. J. Ball, R. Hillier, & G. T. Roberts (Eds.), *Proceedings of 2nd ISSW*, London (Vol. 2, pp. 1071–1076).

Schlichting, H. (1960). *Boundary layer theory*. New York, NY: McGraw Hill Book Company Ltd.

Smith, L. G. (1948). *Photographic investigation of the reflection of plane shocks in air* (Office of Scientific Research and Development OSRD Report 6271), Washington DC, USA.

Sugiyama, H., Takeda, H., Zhang, J., & Abe, F. (1987). Multiple shock wave and turbulent boundary layer interaction in a rectangular duct. In G. Groenig (Ed.), *Shock Tube and Waves, Proceedings of 16th International Symposium on Shock Tubes and Waves*, Aachen (pp. 185–191).

Sugiyama, H., Doi, H., Nagumi, H., & Takayama, K. (1988). Experimental study of high-speed gas particle unsteady flow past blunt bodies. *Proceedings of the16th International Symposium on Space Technology and Science*, Sapporo (pp. 781–786).

Sun, M., Saito, T., Takayama, K., & Tanno, H. (2005). Unsteady drag on a sphere by shock wave loading. *Shock Waves, 14,* 3–9.

Sun, M., Yada, K., Ojima, H., Ogawa, H, & Takayama, K. (2001). Study of shock wave interaction with a rotating cylinder. In K. Takayama (Ed.), *Proceedings of SPIE 24th International Congress of High Speed Photography and Photonics*, Sendai (pp. 682–687).

Tanno, H., Komuro, T., Sato, K., Itoh, K., Ueda, S., Takayama, K., & Ojima, H. (2004). Unsteady drag force measurement in shock tube. In Z. L. Jiang (Ed.), *Proceedings of 24th ISSW*, Beijing (Vol. 1, pp. 371–376).

Timofeev, E. V., Takayama, K., Voinovich, P. V., Sislian, J., & Saito, T. (1997). Numerical and experimental study of three-dimensional unsteady shock wave interaction with an oblique cylinder. In A. P. F. Houwing, & A. Paul (Eds.), *Proceedings of 21st ISSW*, The Great Keppel Island (Vol. 2, pp. 1487–1492).

Yada, K. (2001). *Unsteady transition of reflected shock wave over bodies in shock tube flows* (Ph.D. thesis). Graduate School of Engineering, Faculty of Engineering Tohoku University.

Yang, J.-M. (1995). *Experimental and analytical study of behavior of weak shock waves* (Ph.D. thesis). Graduate School of Tohoku University, Faculty of Engineering.

Chapter 5
Shock Wave Focusing in Gases

5.1 Introduction

The two-dimensional shock wave focusing is divided into two patterns: the reflection from concave walls and; the convergence of curved incident shock waves, which is called implosion. This is a reverse process of an explosion.

Three-dimensional shock wave focusing is also defined: convergence of a reflected planar shock wave from a concave wall. However, the implosion of a spherical shock wave is, from engineering point of view, difficult to conduct. In 1989 a workshop on shock wave focusing was organized inviting well known researchers in those days. Ten specialists presented their current works on the topics Takayama (1990).

5.2 Two-Dimensional Focusing

5.2.1 Circular Wall

Figure 5.1a–m shows the focusing of reflected shock waves from a 60 mm diameter circular wall installed in the 40 mm × 80 mm conventional shock tube for Ms = 1.25 in atmospheric air at 297.0 K. The visualization was conducted in 1980 by a direct shadowgraph. The focusing process strongly depends on the wall shapes, Ms and γ.

The reflection pattern follows all the types of shock wave reflection: at first it is a DiMR including a vNMR and a SMR; it transits to a StMR; and lastly it becomes a IvMR Courant and Friedrichs (1948). With the increase in the wall angle θ_w, the IvMR terminates to a supersonic regular reflection, a SPRR which accompanies a secondary TP as seen in Fig. 5.1a. Figure 5.1d is the enlargement of Fig. 5.1c. The secondary triple point TP, the curved secondary MS and the SL are reflected from the curved wall. In Fig. 5.1, the shock wave reflected from the edge of the concave

K. Takayama, *Visualization of Shock Wave Phenomena*,
https://doi.org/10.1007/978-3-030-19451-2_5

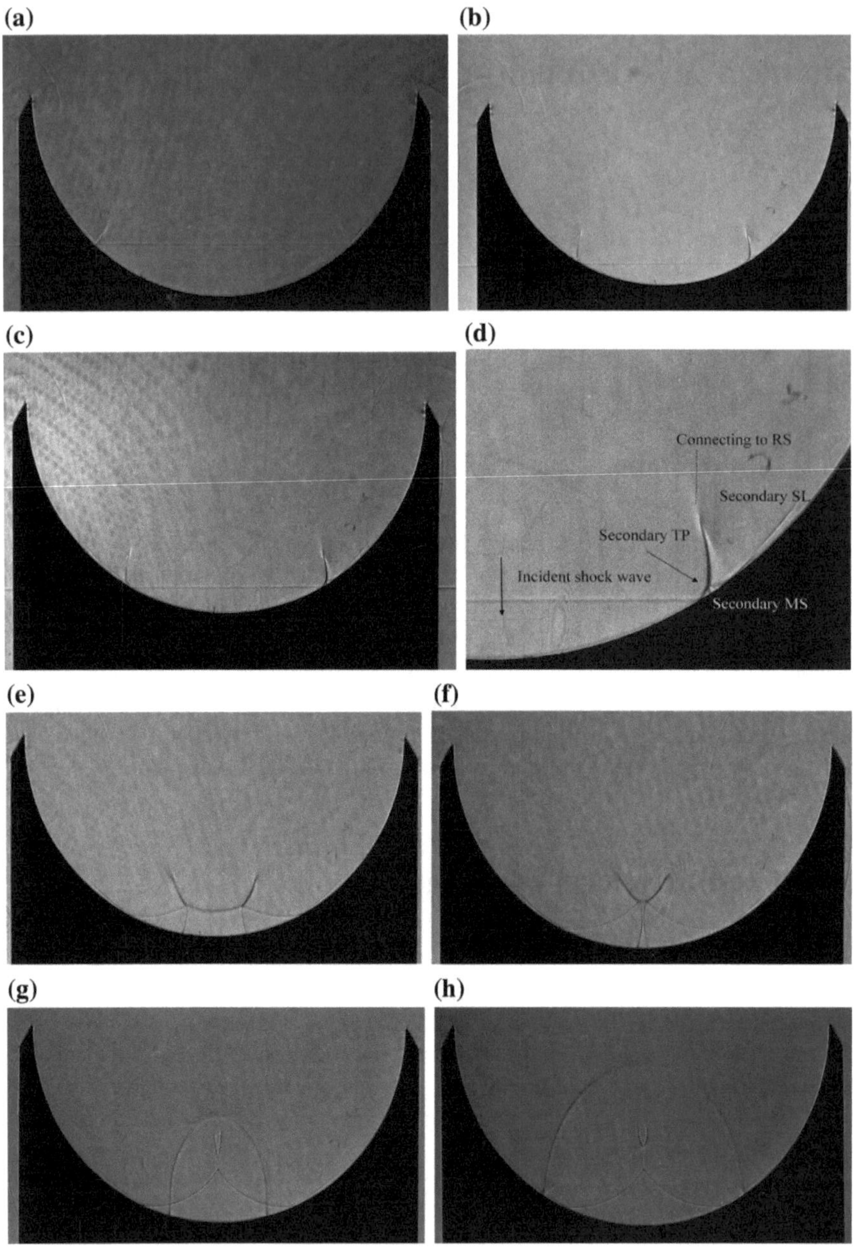

Fig. 5.1 Sequential observation of the focusing from a 60 mm diameter circular wall for Ms = 1.25 in atmospheric air at 297.0 K, direct shadowgraph: **a** #80082908, 100 μs from trigger point; **b** #80082909, 100 μs, Ms = 1.220; **c** #80082911, 110 μs, Ms = 1.219; **d** enlargement of (**c**); **e** #80082912, 130 μs, Ms = 1.219; **f** #80082906, 140 μs, Ms = 1.218; **g** #80082905, 160 μs, Ms = 1.218; **h** #80082904, 180 μs, Ms = 1.218; **i** enlargement of (**h**); **j** #80082901, 200 μs, Ms = 1.218; **k** enlargement of (**j**); **l** #80082902, 220 μs, Ms = 1.218; **m** #80082903, 240 μs, Ms = 1.218

(i) (j) (k) (l) (m)

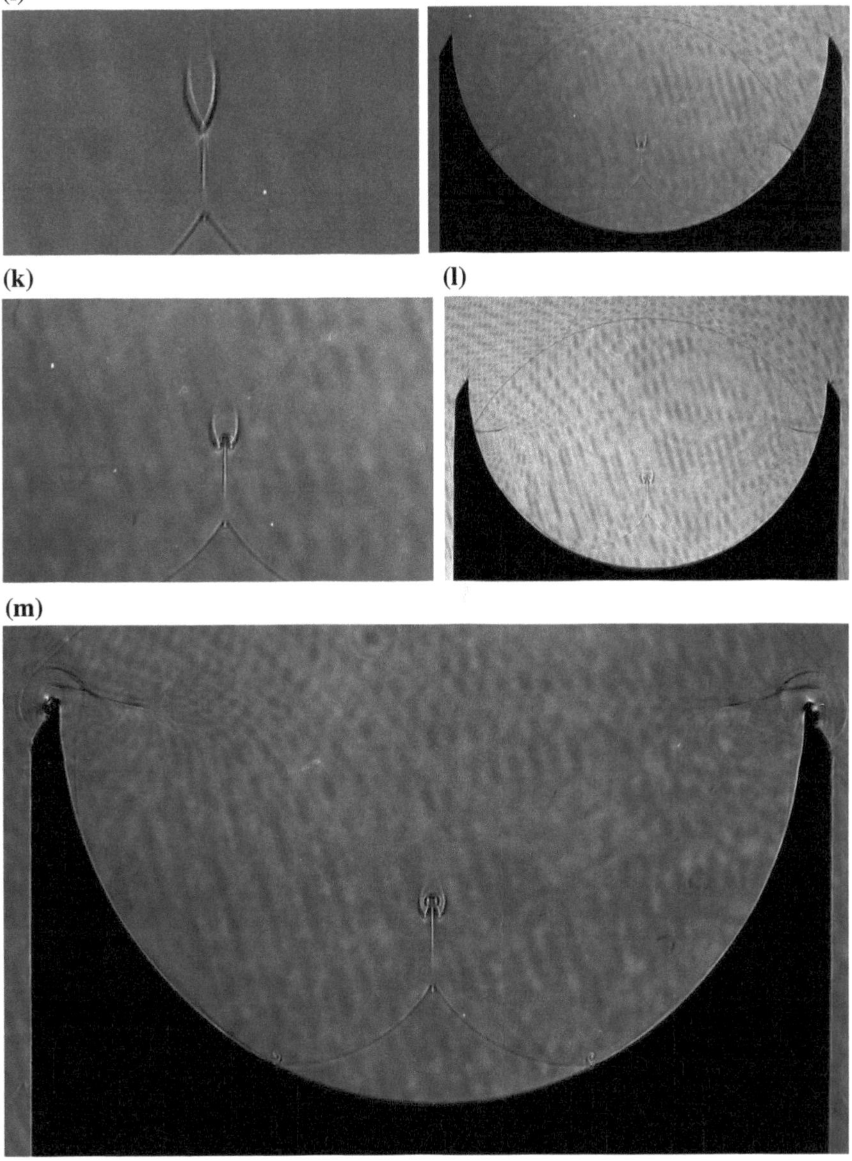

Fig. 5.1 (continued)

wall is so weakened that it is invisible. In the interferograms, the reflected shock wave is visible. In Fig. 5.1e, f, the secondary TP merged at the center and became stagnant. The vortices emanating from the second TP merged and at the same time the dissipation started which smeared out the energy accumulation. In Fig. 5.1i–m,

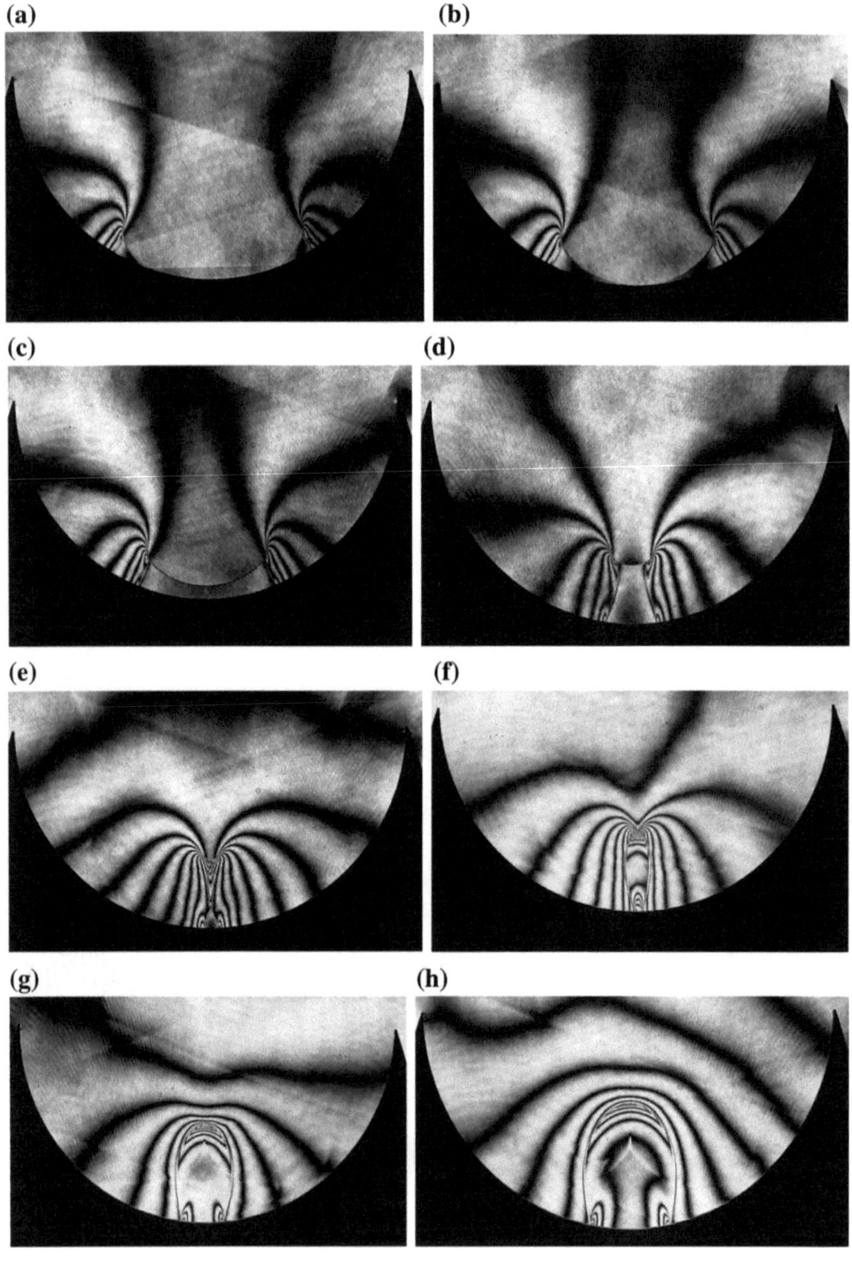

Fig. 5.2 Shock wave focusing from a 120 mm diameter circular wall for Ms = 1.07 in atmospheric air at 287.0 K: **a** #86030516, 370 μs from the trigger point, Ms = 1.072; **b** #86030515, 380 μs, Ms = 1.074; **c** #86030514, 300 μs, Ms = 1.077; **d** #86030510, 430 μs, Ms = 1.070; **e** #86030511, 440 μs, Ms = 1.073; **f** #86030602, 450 μs, Ms = 1.071; **g** #86030604, 470 μs, Ms = 1.072; **h** #86030605, 480 μs, Ms = 1.072; **i** #86030609 550 μs, Ms = 1.075; **j** #86030611, 650 μs, Ms = 1.068

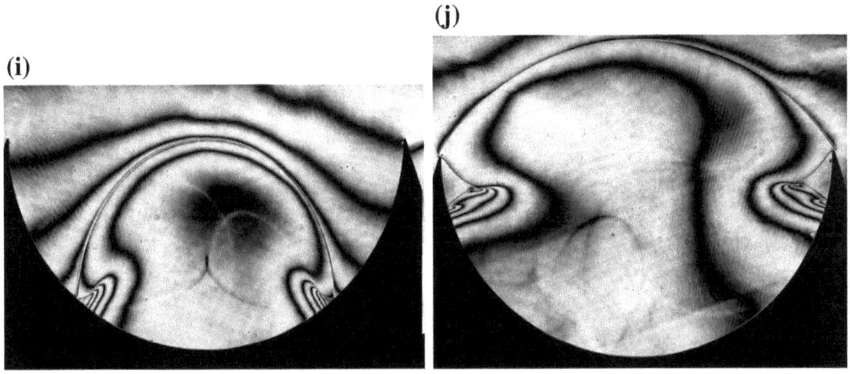

Fig. 5.2 (continued)

the contrast of slip lines emanating from the TP became fain showing the dissipation of density jump across the SL. The time attached to individual figures indicates the delay time when the second exposure was conducted. The trigger point was the position the pressure transducer was installed at some distance before the test section.

Figure 5.2 shows the evolution of shock wave focusing from a 120 mm diameter circular reflector installed in the 60 mm × 150 mm conventional shock tube for Ms = 1.07 in atmospheric air. Figure 5.2a–c shows a SPRR and its reflection. Figure 5.2d–f shows the intersection of the secondary triple points and their merger with the secondary Mach stems. The reflection pattern leaning steeply forward is SuRR in Fig. 5.2j. In Fig. 5.2h, i, the remains of the slip lines SL is visible. The density jump across the SL gradually disappears. The pressures and the density focused at the area at which fringes concentrated but the temperature was not necessarily high.

Figure 5.3 shows sequentially shock wave focusing for Ms = 1.47 in air. Experiments were conducted in a 60 mm × 150 mm conventional shock tube. Figure 5.3a shows a reflection of a SPRR from the circular wall. The two triple points just reflected at the center and the curved secondary Mach stems are moving outward in Fig. 5.3b. The merger of the secondary TP and the secondary Mach stems created the intersection of the slip lines. The waves are moving outward as seen in Fig. 5.3c–f. The slip lines merged and formed the vortices accumulation at a localized spot as seen in Fig. 5.3c–f. The accumulation of vortices is disconnected from the wave motion and is stagnant in a localized area as seen in Fig. 5.3g–j. The initial sharply accumulated fringe distribution became loosed and the fringe number decreased with the elapsed time as seen in Fig. 5.3g–j. During these wave interactions, the pressure just fluctuates and nearly constant and the fringe number also unvaried. Hence the temperature did not increase high.

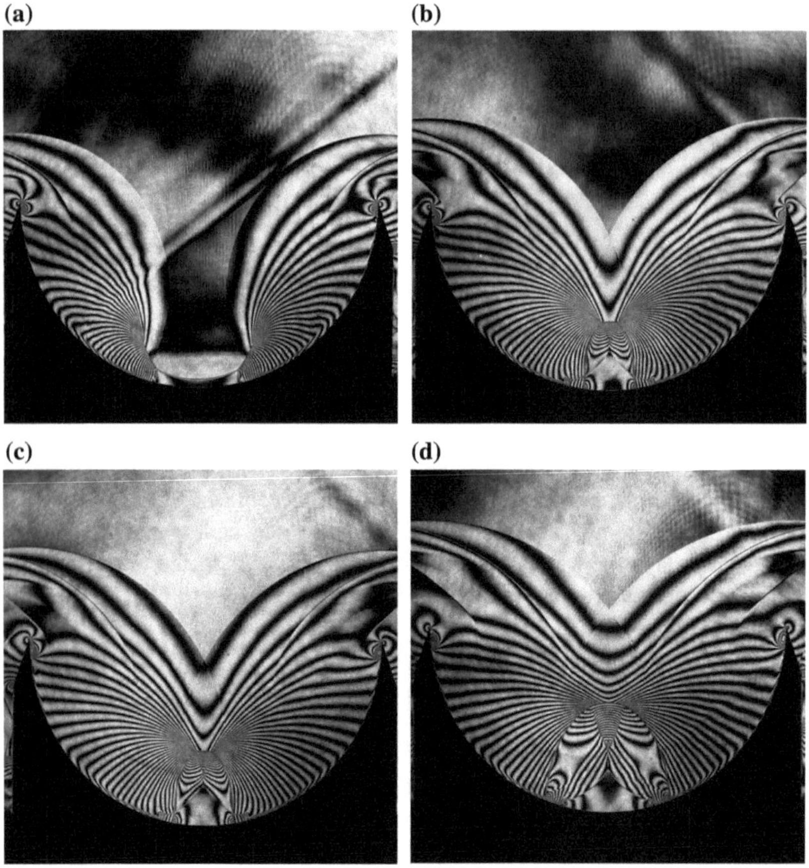

Fig. 5.3 Evolution of shock wave focusing from a 120 mm diameter circular wall in stalled in the 60 mm × 150 mm conventional shock tube for Ms = 1.46 in air at 800 hPa, 289.5 K: **a** #84042628, 120 s from the trigger point, Ms = 1.468; **b** #84042624, 70 s, Ms = 1.480; **c** #84042623 80 μs, Ms = 1.471; **d** #84042620, 100 s, Ms = 1.465; **e** #84042619, 110 s, Ms = 1.454; **f** #84042618, 130 s, Ms = 1.468; **g** #84042615, 160 s, Ms = 1.468; **h** #84050209, 200us, Ms = 1.475; **i** #8405210, 144 s, Ms = 1.468; **j** #84050211, Ms = 1.481

Figure 5.4 shows the evolution of the reflected shock wave focusing from a 120 mm circular wall installed in the 60 mm × 150 mm diaphragm-less shock tube for Ms = 2.02 in air at 450 hPa, 290 K. The pattern of focusing is similar to that shown in Fig. 5.3.

Figure 5.5 show the reflection and focusing from the 120 mm diameter reflector for Ms = 3.0 in air at 60 hPa, 291 K. Figure 5.5a shows a transitional Mach reflection, TMR. With increasing the wall angle, the reflection pattern transits to a

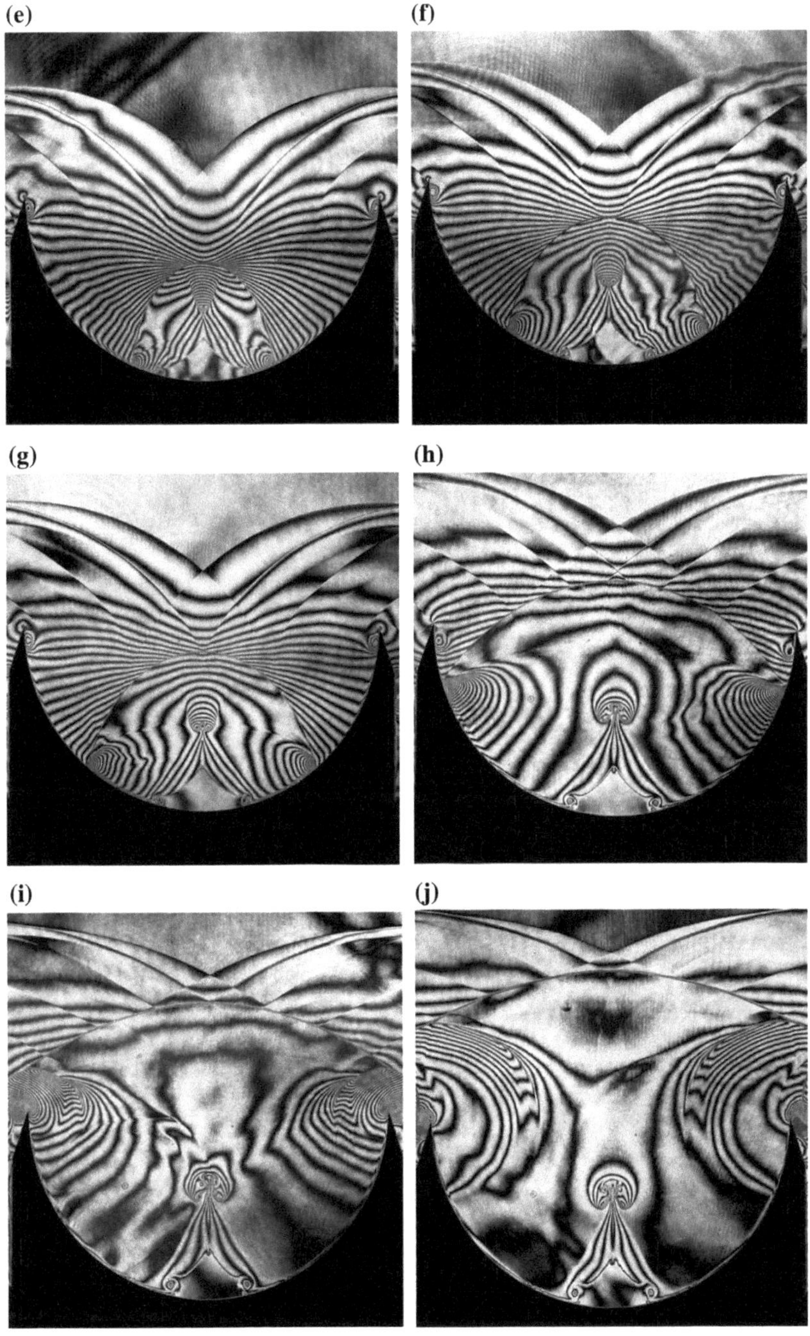

Fig. 5.3 (continued)

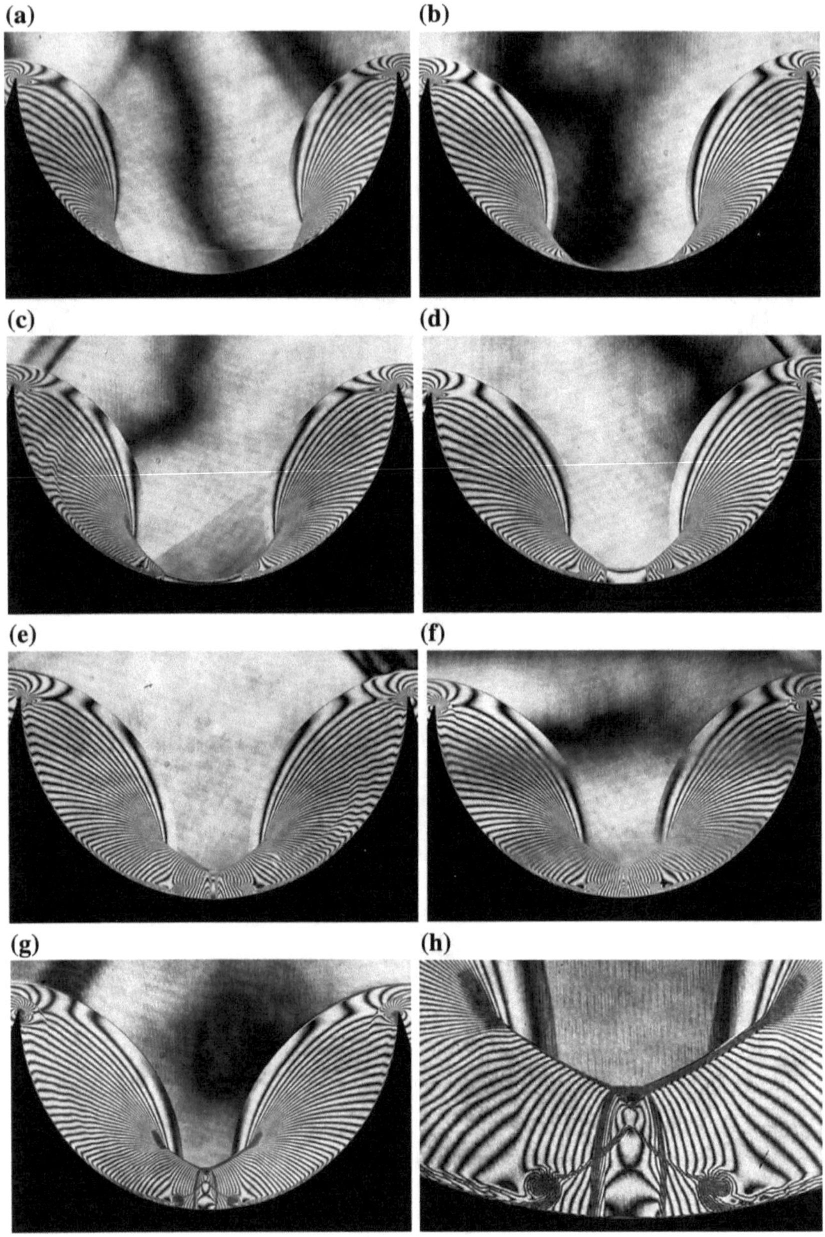

Fig. 5.4 Shock wave focusing from a 120 mm diameter circular wall for Ms = 2.02 in air at 450 hPa, 290 K: **a** #86030615, 265 μs, Ms = 2.033; **b** #86030618 280 μs, Ms = 2.030; **c** #86030701 300 μs, Ms = 2.020; **d** #86030617 275 μs, Ms = 2.030; **c** #86030704 315 μs, Ms = 2.010; **f** #86030619 290 μs, Ms = 2.015; **g** #86030707 330 μs, Ms = 2.002; **h** enlargement of (**g**); **i** #86030706 325 μs, Ms = 2.025; **j** #86030708 335 μs, Ms = 2.030; **k** #86030711 350 μs, Ms = 2.019; **l** #86030713 360 μs, Ms = 2.011; **m** #86030716 395 μs, Ms = 2.035; **n** #86030718, 430 μs, Ms = 2.026; **o** enlargement of (**n**)

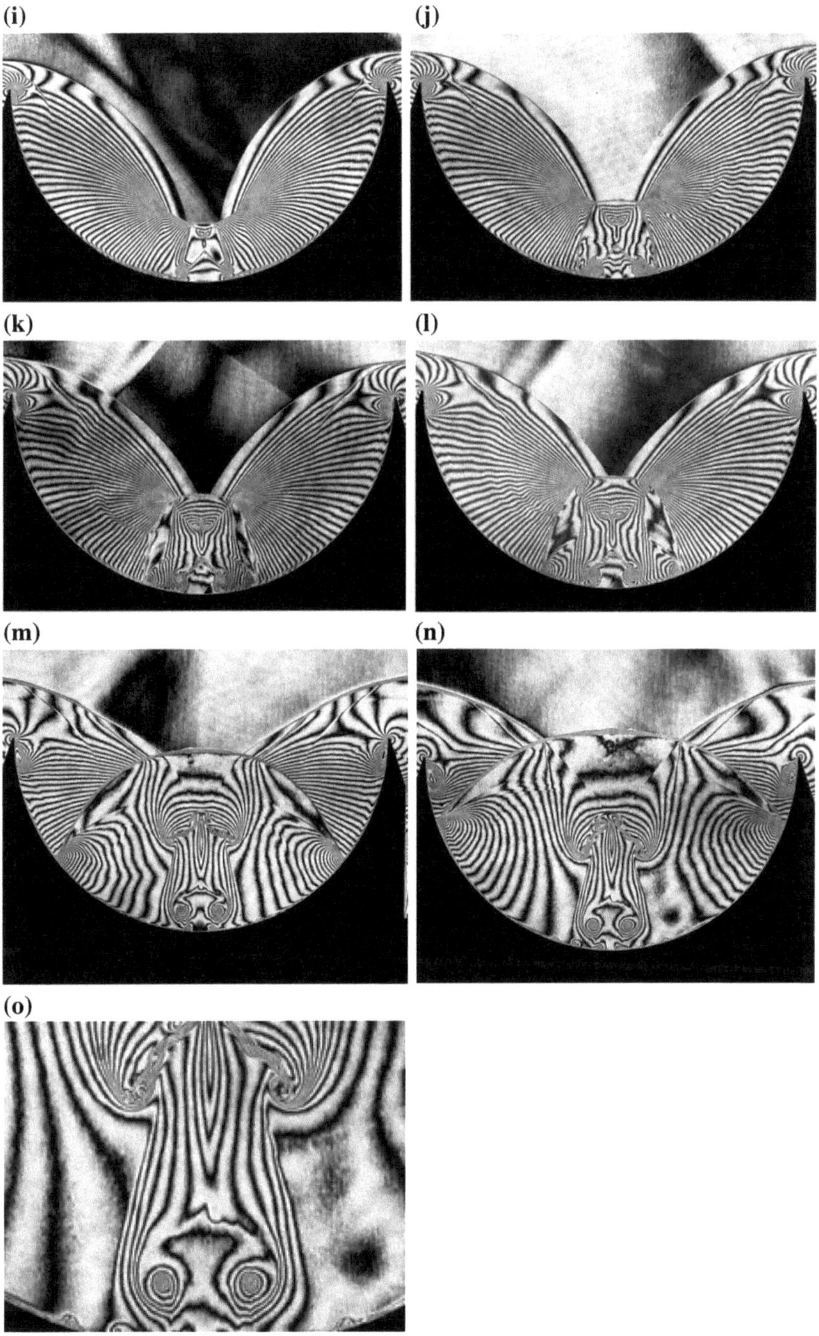

Fig. 5.4 (continued)

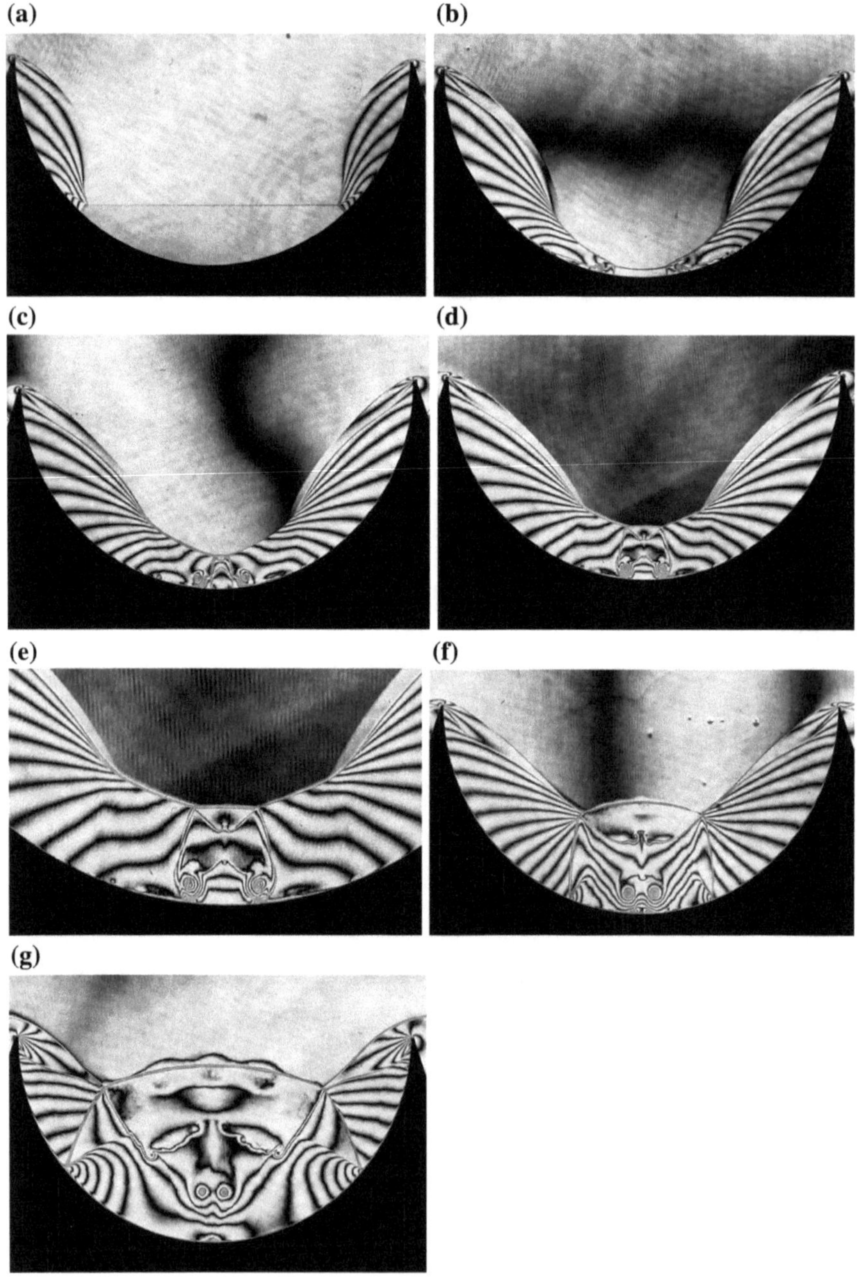

Fig. 5.5 Shock wave focusing from a 120 mm diameter circular wall for Ms = 3.0 in air at 60 hPa, 291 K: **a** #86031303, 100 μs from the trigger point, Ms = 2.966; **b** #86031305, 120 μs, Ms = 2.954; **c** #86031407, 105 μs, Ms = 3.015; **d** #86031408, 110 μs, Ms = 3.054; **e** enlargement of (**d**); **f** #86031001, 50 μs, Ms = 3.013; **g** #86031505, 180 μs, Ms = 3.027

DMR and successively becomes an IvMR. Figure 5.5b shows a reflection of the resulting SPRR. The secondary Mach stems moving from the both sides move toward the center and eventually intersect with each other at the center. A symmetrical interaction of slip lines is observed in Fig. 5.5c, d as the initial pressures is reduced, the pattern observed so far formed coincidentally an exotic human face as seen in Fig. 5.5d. In Fig. 5.5g the bifurcated reflected shock wave induced by its interaction with the sidewall boundary layer, the figure that looks like a crown attached to the human face.

5.2.2 *Closed Circle*

A 30 mm × 300 mm diameter test section was connected the 30 mm × 40 mm conventional shock tube, which became a closed circular test section. Unlike previous truncated cylinders, this closed circular test section had no singular geometry. Figure 5.6 shows the evolution of shock wave propagating and reflecting inside the closed circular test section for Ms = 1.5 in atmospheric air at 290 K. In Fig. 5.6a, b, the incident shock waves is diffracting at the entrance corner and generating a pair of twin vortices. The transmitting shock waves continuously change their reflection patterns as discussed in the Sect. 5.1.1. The reflected patterns eventually became IvMR as seen in Fig. 5.6c. The secondary Mach stems interacted as is successively observable in Fig. 5.6d–f. Meantime, the vortices generated at the entrance corners developed with increasing time and started interacting with the reflected waves as seen in Fig. 5.6g–l.

The early stage of focusing is similar to that seen from concave walls but at later stage the interactions with the vortices generated at the entrance corner appeared (Sun 2005).

5.2.3 *Effects of Entrance Angles on Focusing*

Focusing of shock waves from a concave wall is affected by the initial angle of the curved walls. Therefore, Fig. 5.7 shows a reflector having initial angles of θ = 75°, 45°, 30°, and 15° installed in the 40 mm × 80 mm conventional shock tube.

5.2.3.1 Wall Angle 75°

Figure 5.8 shows the evolution of a weak shock wave reflected from a circular reflector having a wall angle 75° and a radius of 154.6 mm for Ms = 1.13 in atmospheric air at 295.0 K. The initial reflection pattern is SPRR. Then the two SPRR propagating along the upper and the lower walls intact with each other and the final reflection pattern becomes a vNMR as seen in Fig. 5.8e–f.

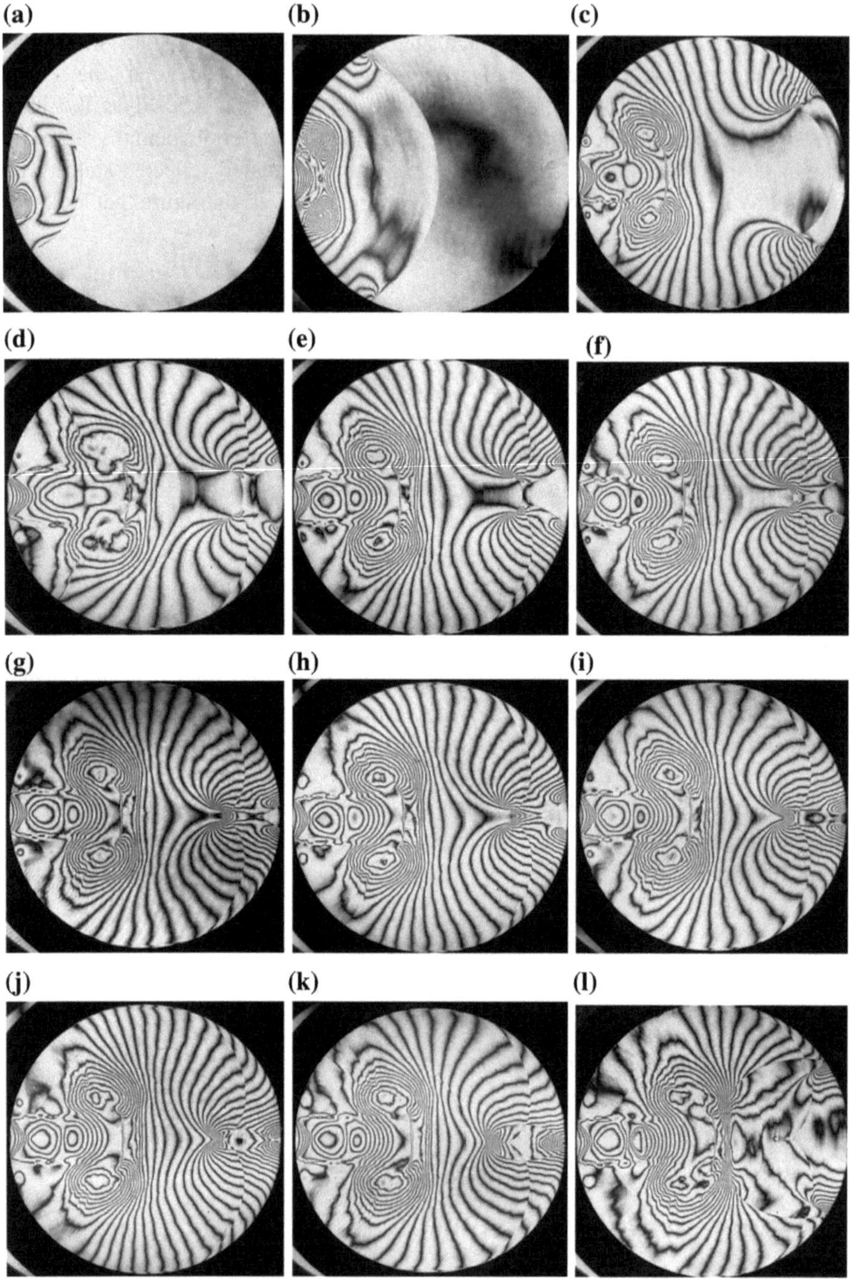

Fig. 5.6 Shock wave focusing from a 300 mm diameter circular wall for Ms = 1.5 in atmospheric air at 290 K (Sun et al. 2005): **a** #95100401; **b** #95100507; **c** #95100403; **d** #95100408; **e** #95100405; **f** #95100504; **g** #95100404; **h** #95100509; **i** #95100510; **j** #95100505; **k** #95100406; **l** #95100506

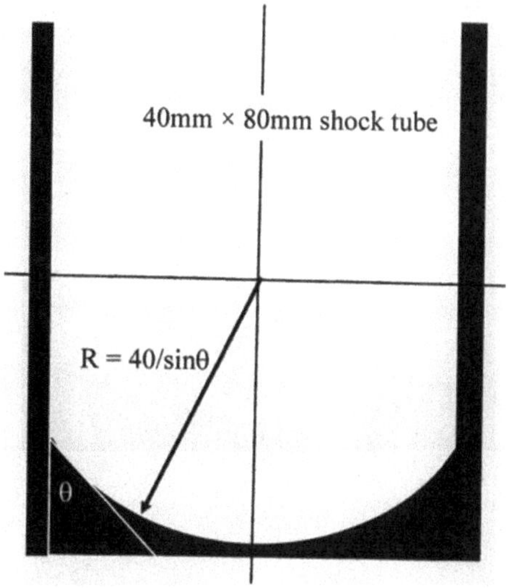

Fig. 5.7 A reflector of radius R = 40/sinθ mm installed in a 40 mm × 80 mm shock tube, where θ is the wall angle

5.2.3.2 Wall Angle 45°

Figure 5.9 shows the evolution of the reflected shock wave focusing from concave wall angle of 45° for Ms = 1.13 in atmospheric air at 295.0 K. The IvMR transits to SuRR which accompany a secondary TP. Then, the secondary triple points contribute to the final reflected shock wave pattern. The vortices remain at the end wall, see in Fig. 5.9d–f. The effect of the initial wall angle 45° on the reflection pattern is similar to that observed in the case where the initial wall angle was 75°.

5.2.3.3 Wall Angle 30°

Figure 5.10 shows the evolution of the reflected shock wave focusing from the concave wall set angle at 30° for Ms = 1.13 in atmospheric air at 295.0 K. The reflection pattern is almost identical to the one observed in the case where the wall angle was 45°. The reflection pattern transits from an IvMR to a SPRR.

5.2.3.4 Wall Angle 15°

The evolution of the reflected shock wave focusing from a concave wall angle set at 15° for Ms = 1.07 in atmospheric air at 295.0 K. The inverse Mach reflections, IvMR, merge at the center as seen in Fig. 5.11a, b. The shock waves reflected from the corner are now so weakened that the waves are not observable as seen in Fig. 5.11e–g.

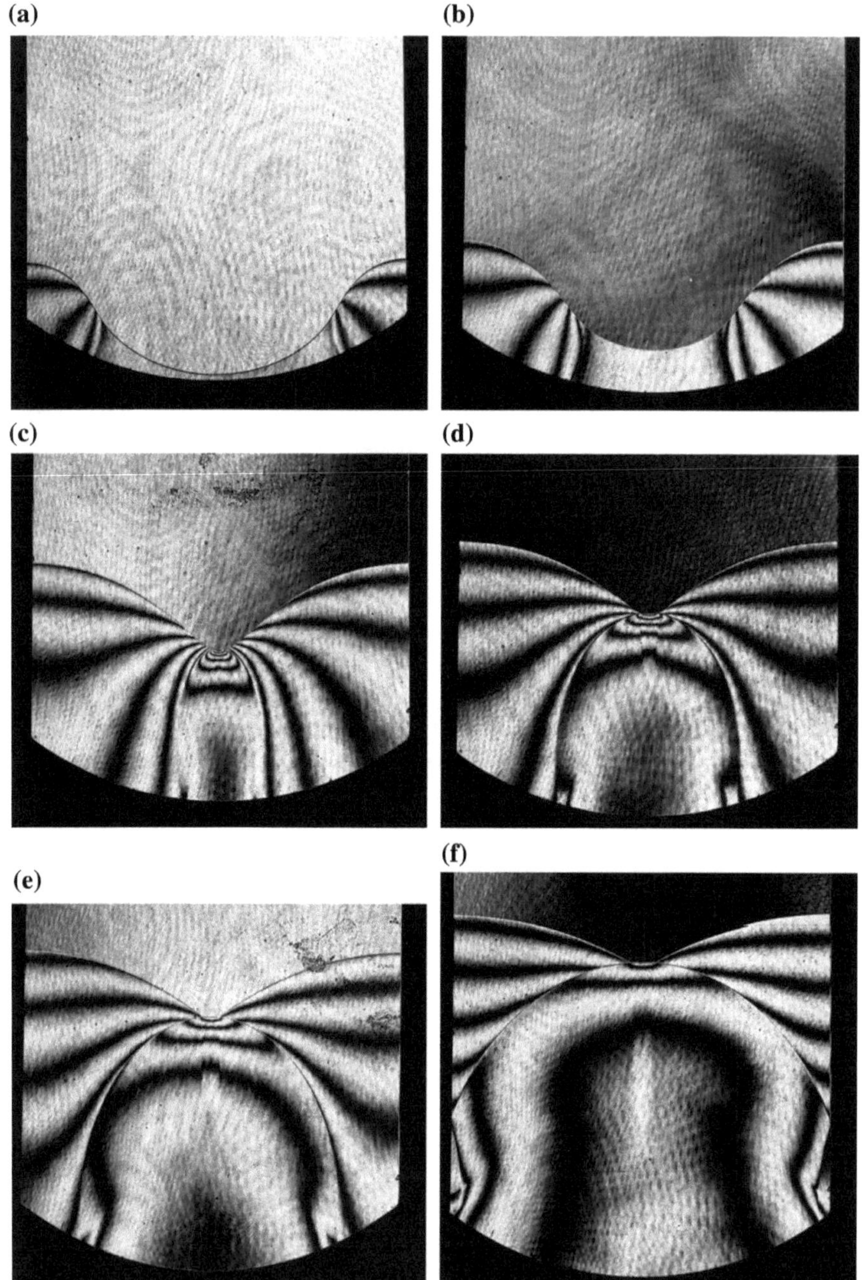

Fig. 5.8 The evolution of the reflected shock wave focusing from concave wall angle of 75° for Ms = 1.13 in atmospheric air at 295.0 K: **a** #81061904, 50 μs from trigger point. Ms = 1.131; **b** #81061906, Ms = 1.136; **c** #81061909, 100 μs, Ms = 1.136; **d** #81061910, 110 μs, Ms = 1.148; **e** #81061911, 130 μs, Ms = 1.148; **f** #81061913, 200 μs, Ms = 1.137

5.2.3.5 Entrance Angle 40°

Circular reflectors with various wall angles of 40°, 50°, 60°, and 70° were placed in the 60 mm × 150 mm conventional shock tube for visualizing the evolution of shock wave focusing for Ms = 1.30 (Fig. 5.12).

Figure 5.13a shows the wall angle of $\theta = 40°$ and Ms = 1.30. The radius R of the circular wall is given by $R = 30/\sin\theta$ in mm. The initial reflection pattern was a SMR which eventually transited to an IvMR. Figure 5.13a shows the resulting regular reflection SuRR. A pair of secondary triple points and curved Mach stems merged between Fig. 5.13b, c. Figure 5.13d–f shows the situation at later stages.

5.2.3.6 Entrance Angle 50°

Figure 5.14 shows the evolution of the shock wave reflection from the circular wall when it is set at a wall angle of 50°. The reflection pattern is SuRR. In Fig. 5.14c, d, the triple points merged at the center and reflected. Meantime, the high-pressure area coalesced into a shock wave. This is a focusing sequence from a shallow reflector. In Fig. 5.14c, d, the flattened part of the reflected shock wave is gradually enlarged and the resulting reflection pattern is SMR.

5.2.3.7 Entrance Angle 60°

Figure 5.15 shows the reflection pattern of SuRR. The entire sequences of focusing are similar to that shown in Fig. 5.14.

5.2.3.8 Entrance Angle 70°

The reflection pattern is SuRR. The sequence of focusing is very similar to that observed in Fig. 5.15. The general trend is that the reflection pattern approaches to the reflection from a slightly perturbed plane wall (Fig. 5.16).

5.3 Shock Wave Reflection from Convex and Concave Walls

A combination of convex and concave reflectors as illustrated in Fig. 5.17 is installed in the 60 mm × 150 mm conventional shock tube. Sequence of focusing was observed for the varying radius R. For the given depth H of the reflectors, R is determined by $2R = (L^2 + H^2)/(4H)$, where L is the width of test section L = 150 mm.

(a) (b) (c) (d) (e) (f)

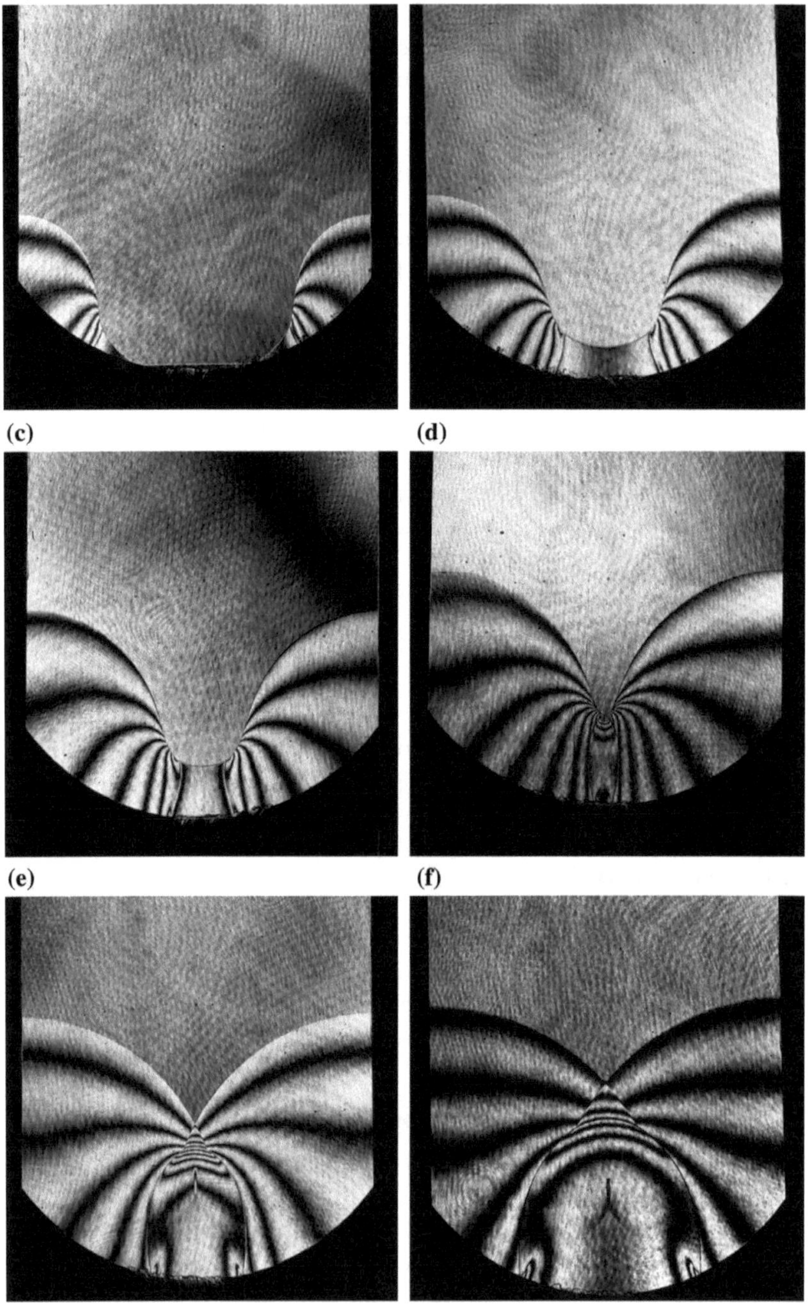

◀**Fig. 5.9** Evolution of reflected shock wave focusing from concave wall angle of 45° for Ms = 1.13 in atmospheric air at 295.0 K: **a** 81061924, 50 μs from trigger point, Ms = 1.135; **b** #81061923 70 μs, Ms = 1.140; **c** #81061922, Ms = 1.147; **d** #81061921, 110 μs, Ms = 1.139; **e** #81061920 130 μs, Ms = 1.133; **f** #81061919, 150 μs, Ms = 1.137; **g** #81061918, Ms = 1.109; **h** #81061917, Ms = 1.137

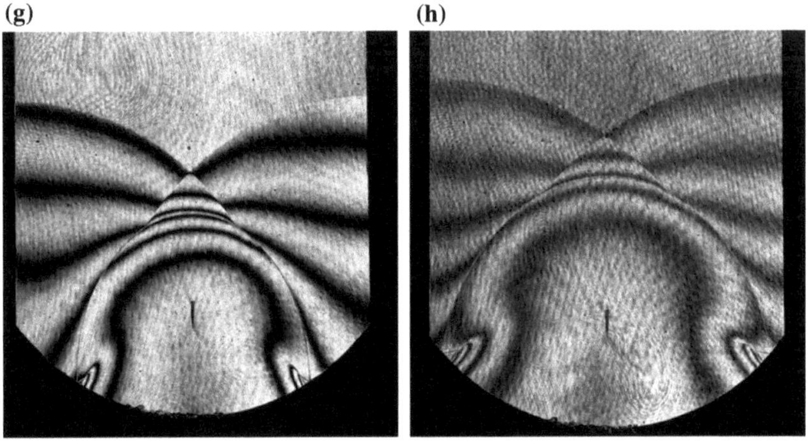

Fig. 5.9 (continued)

5.3.1 75 mm Depth

For a given depth of 75 mm, the radius R is 37.5 mm. Figure 5.18 shows the evolution of the reflection and focusing for Ms = 1.40, in atmospheric air at 297 K. The circular entry shape decisively affected the process of the propagation and focusing of the transmitting shock wave. Figure 5.18a–h showed the sequence of focusing. The waves reflected from the upper and lower walls affected the sequence of focusing. The focusing of slip lines observed in Fig. 5.18g, h is very similar to the pattern as seen in Fig. 5.3. Mean time during the shock vortex interaction, wave patterns with exotic shapes appeared. The comparison between the present experimental findings and an appropriate numerical simulation would be a challenging task for nfor code validation.

5.3.2 Depth 57 mm

Figure 5.19 shows a reflection from a convex and concave reflector of H = 57 mm and R = 56.5 mm for Ms = 1.44. The reflection pattern is, at first, a RR and along a concave section, transits to a SMR in Fig. 5.19a and eventually becomes a SuRR in Fig. 5.19b. Then repeating the wave interactions, the focusing is terminated in Fig. 5.19d.

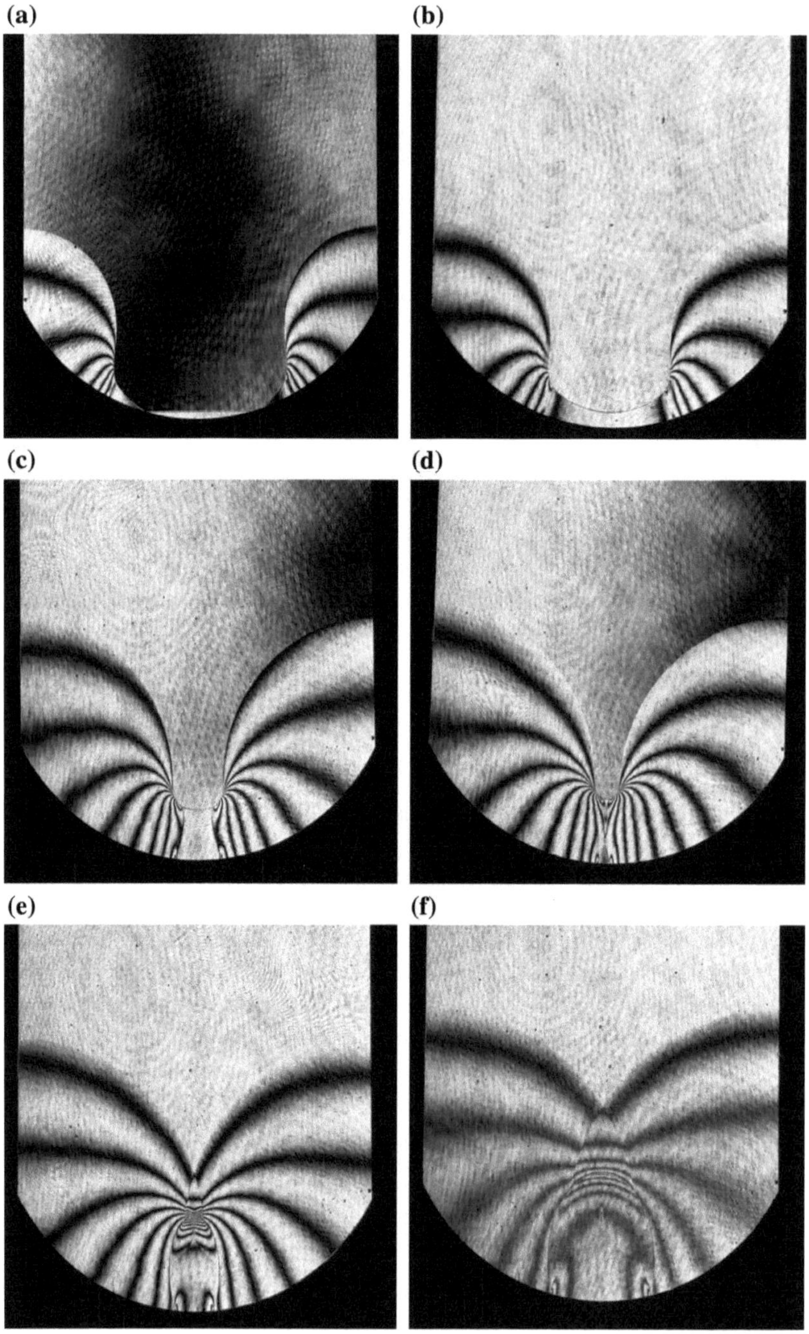

◄**Fig. 5.10** The evolution of the reflected shock wave focusing from a concave wall angle of 30° for Ms = 1.13 in atmospheric air at 295.0 K: **a** #81062002, 50 μs after the trigger point, Ms = 1.139; **b** #81062004, 70 μs, Ms = 1.136; **c** #81062005, 90 μs, Ms = 1.149; **d** #81062009, 97 μs, Ms = 1.136; **e** #81062006, 110 μs, Ms = 1.150; **f** #81062007, 130 μs, Ms = 1.138; **g** #81062009, 140 μs, Ms = 1.136; **h** #81062009, 150 μs, Ms = 1.136

(g) (h)

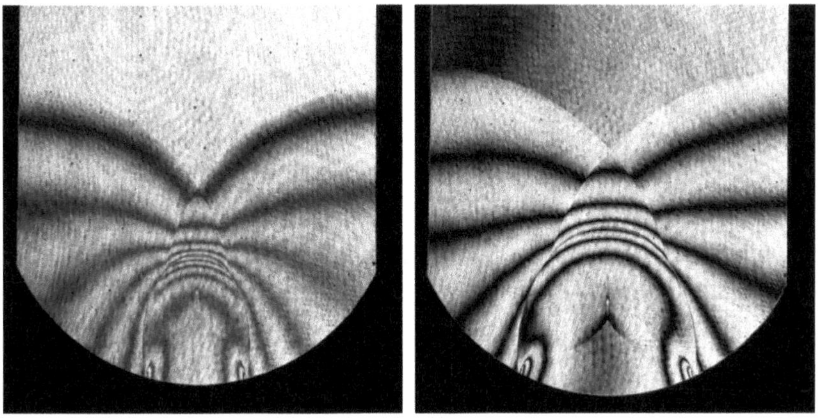

Fig. 5.10 (continued)

5.3.3 Depth 42 mm

Figure 5.20 shows the evolution of the shock wave reflection from a combined reflector having R = 72.2 mm and H = 42 mm for Ms = 1.43 in atmospheric air at 297.6 K. In Fig. 5.20a, the reflection pattern is SuRR accompanying a secondary shock wave. The sequence of shock wave reflections from a shallow reflector is similar to that observed in the case of shallow entrance angles. The reflection pattern shown in Fig. 5.20f is similar to that shown in Fig. 5.19f.

5.3.4 Depth 31 mm

Figure 5.21 shows the shock wave reflection from a shallow reflector of R = 94.6 mm and H = 31 mm for Ms = 1.42 in atmospheric air at 293.0 K. The reflection and focusing from the shallow reflector shown in Fig. 5.21 is similar to the focusing of IvMR and similar to the reflection from the shallow reflectors seen in Figs. 5.8 and 5.25.

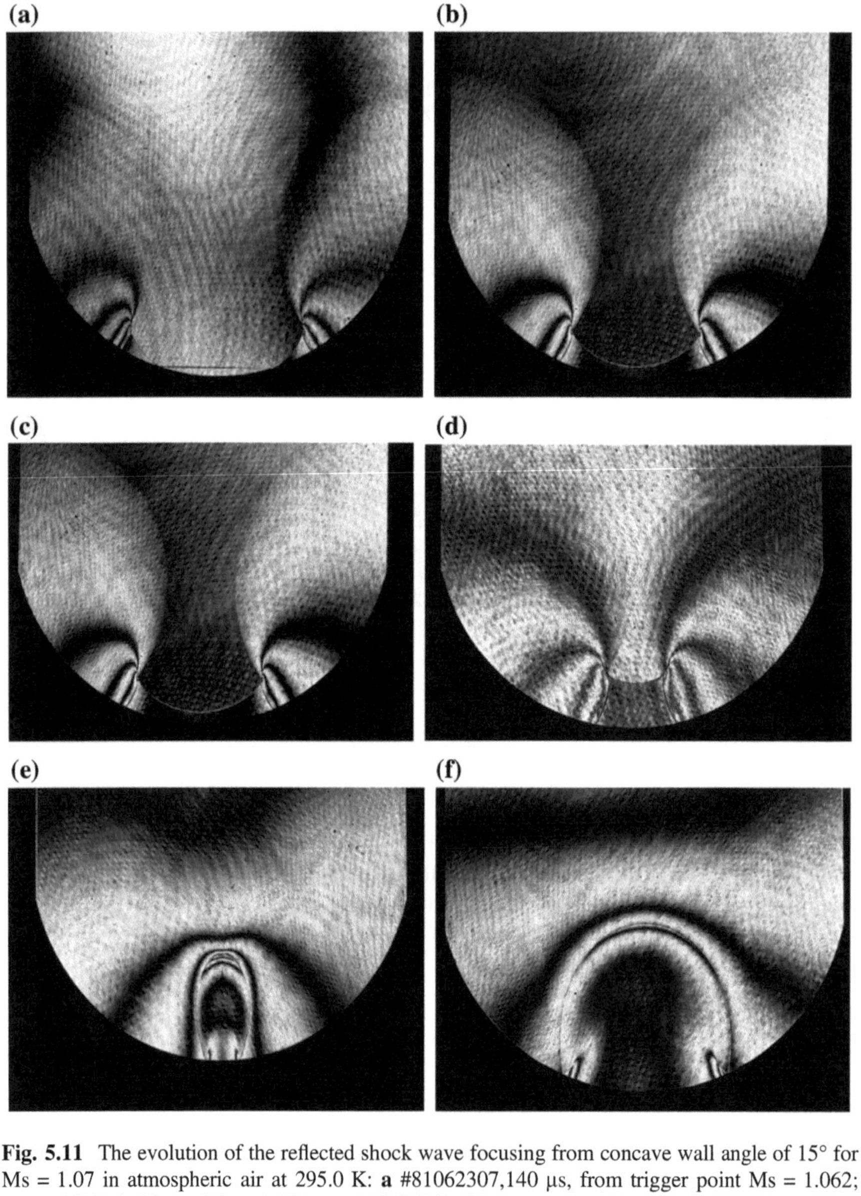

Fig. 5.11 The evolution of the reflected shock wave focusing from concave wall angle of 15° for Ms = 1.07 in atmospheric air at 295.0 K: **a** #81062307,140 μs, from trigger point Ms = 1.062; **b** #81062306, 80 μs, Ms = 1.081; **c** #81062305 100 μs, Ms = 1.069; **d** #81062304, 120 μs, Ms = 1.060; **e** #81062303, 150 μs, Ms = 1.060; **f** #81062309, 180 μs, Ms = 1.063; **g** #81062301, 200 μs, Ms = 1.048

(g)

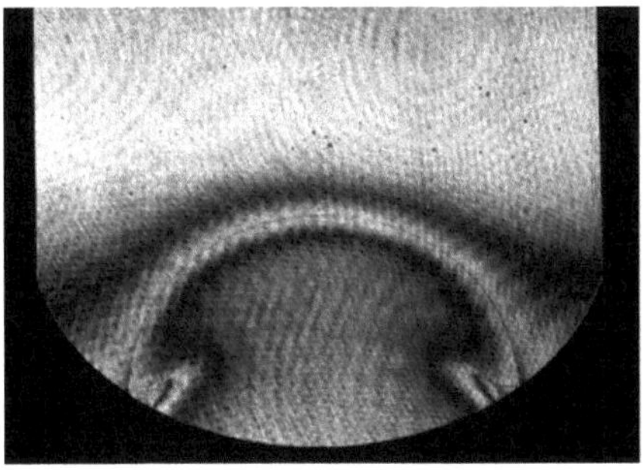

Fig. 5.11 (continued)

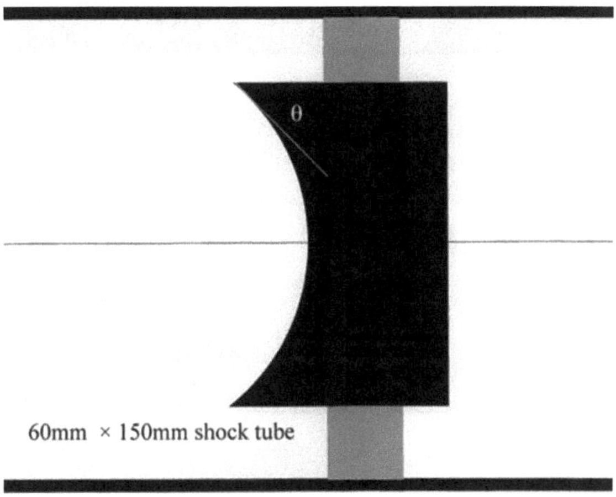

Fig. 5.12 Circular reflector of wall angle θ installed in a 60 mm × 150 mm conventional shock tube

5.4 Focusing from a Logarithmic Spiral Shaped Area Convergence

Shock waves are focused while passing along a duct having area convergence or reflected from concave walls. However, such a shock wave focusing is always associated with wave interactions. If a planar shock wave could be focused on a localized area while propagating along a specially shock tube having a special shape, it would have been wonderful. Milton et al. (1975) proposed a logarithmic spiral shaped passage in which a planar shock wave can be focused to a spot and thereby generating high pressures and temperatures. The theory was based on Whitham's ray shock theory (Whitham 1959).

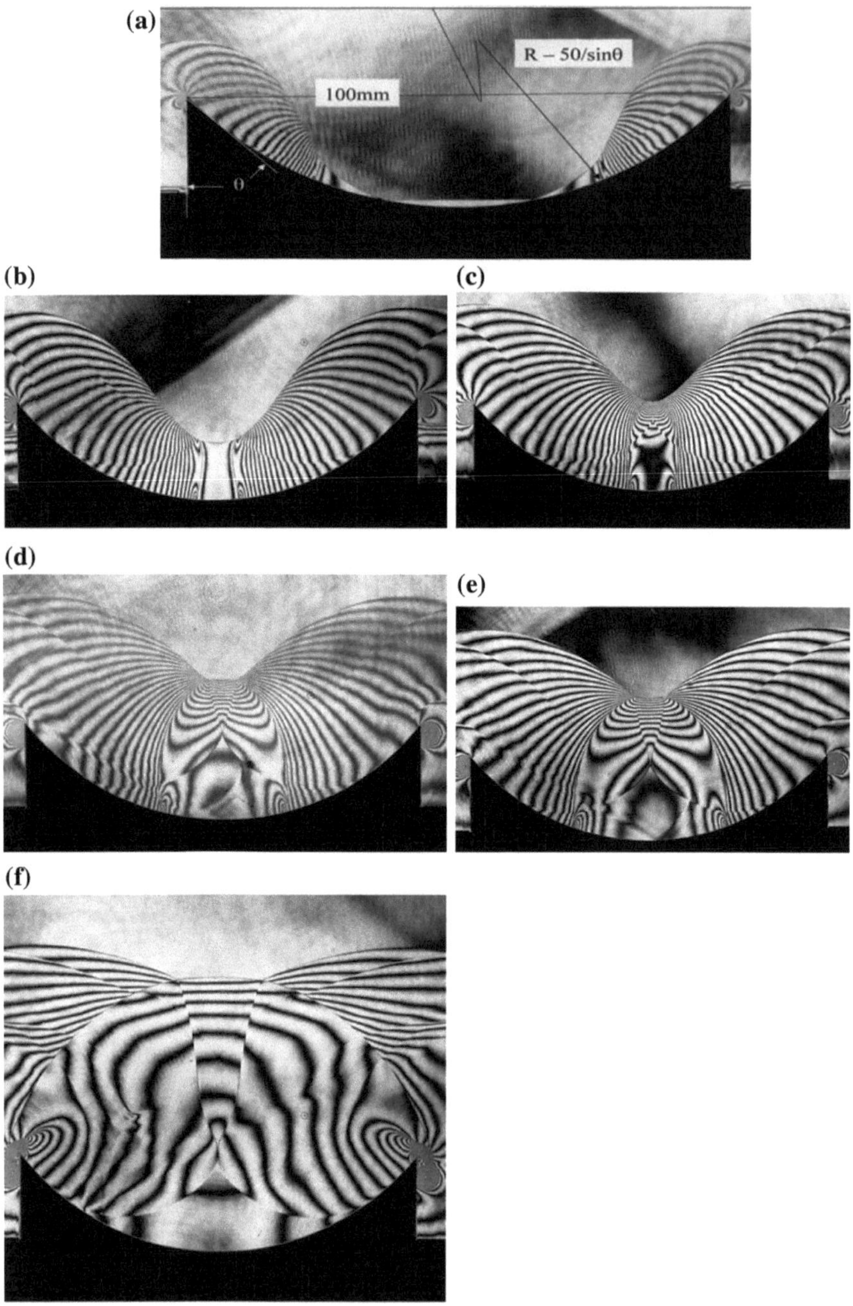

Fig. 5.13 Effects of the wall angle of 40° for Ms = 1.30, in atmospheric air at 286.4 K: **a** #84042515, 120 μs from trigger point, Ms = 1.304; **b** #84042521, 170 μs, Ms = 1.306; **c** #84042513, 240 μs, Ms = 1.304; **d** #84042519, 320 μs, Ms = 1.315; **e** #84042513, 240 μs, Ms = 1.304; **f** #84042519, 320 μs, Ms = 1.315

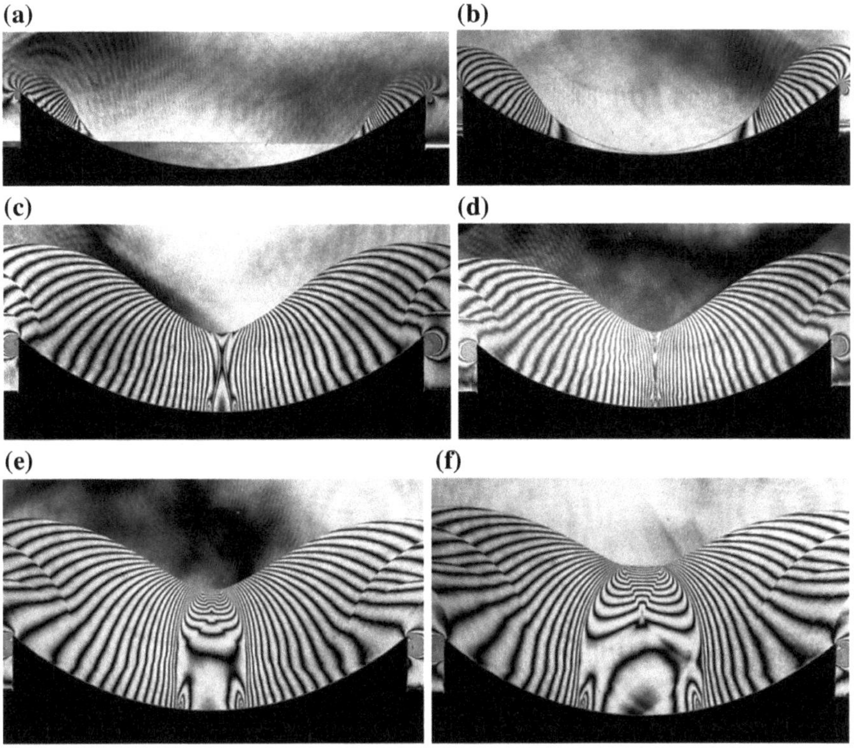

Fig. 5.14 Effects of the wall angle of 50° for Ms = 1.30 in atmospheric air at 286.4 K; **a**#84042509, 110 μs from trigger point, Ms = 1.306; **b** #84042507, 130 μs, Ms = 1.301; **c** #84042504, 160 μs, Ms = 1.306; **d** #84042501, 200 μs, Ms = 1.306; **e** #84042506, 220 μs, Ms 1.308; **f** #84042508, 240 μs, Ms 1.311

Such a configuration is a logarithmic spiral defined in the (r, θ)-plane as:

$$r = R\exp\{(\chi - \theta)/\tan\chi\}$$

where R and χ are given for specifying the Ms and the value of γ (Milton et al. 1975) and indicate the radius of the starting point of log-spiral shape from the origin and the initial angle between the sidewall and the line from the origin to the starting point as shown in Fig. 5.22, respectively.

The logarithmic spiral shape is determined for specified values of Ms and the specific heats ratio of the test gas of γ and hence a given R and χ.

For specified values of Ms and χ, log-spiral shaped models having 129 mm long and 30 mm wide were manufactured and installed in the 60 mm × 150 mm conventional shock tube. The models had 30 mm long straight entry and 99 mm long log-spiral section. The Mach number, Ms, ranged from Ms = 1.4 to 2.7 in air (Milton 1989). Figure 5.23 shows the evolution of a shock wave focusing in a

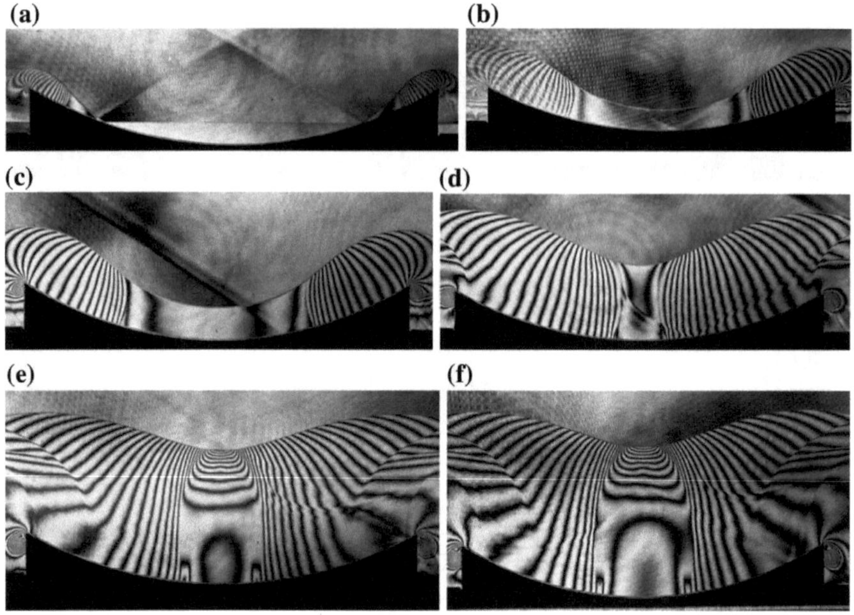

Fig. 5.15 Effect of wall angle of 60° for Ms = 1.30 in atmospheric air at 286.4 K: **a** #84042406, 400 μs from the trigger point, Ms = 1.303; **b** #84042405, 400 μs, Ms = 1.311; **c** #84042407, 450 μs, Ms = 1.303; **d** #84042409, 490 μs, Ms = 1.308; **e** #84042411, 530 μs, Ms = 1.311; **f** #84042412, 550 μs, Ms = 1.308

logarithmic spiral shaped passage manufactured for specially for Ms = 1.40 and their enlargements for Ms = 1.40 in atmospheric air at 294.8 K.

Figure 5.23a shows the early stage of focusing. The IS's foot is always perpendicular to the log-spiral wall so that at the earlier stage, the shock wave in the vicinity of the foot of the IS on the log-spiral is curved smoothly. This trend was observed previously when discussing the IS propagation along a concave wall, for example see in Fig. 5.3. The shock wave smoothly converges keeping the triple point position initially at its glancing incidence angle. The foot of the is always perpendicular to the log-spiral wall and the central part of the transmitted shock wave is perpendicular to the central plane of the shock tube.

In Fig. 5.23a, even though the fringe number increased, the transmitting shock wave still curved continuously connecting smoothly with its foot. This trend became enhanced as seen in Fig. 5.23b. When the fringe concentration was maximal along the smoothly curved transmitted shock wave, a distinct triple points appear as seen in Fig. 5.23c. The reflection pattern quickly turned into an IvMR and merged at the center. The resulting focusing was a merger of vortices shown in Fig, 5.23d. Anyway, the merit of the log-spiral shape was effective in focusing of a planar shock wave to a point while minimizing the generation of undesirable wave interactions. At a later stage of focusing, the reflection pattern was similar to the one observed in the case of shock wave focusing from a concave reflector, the

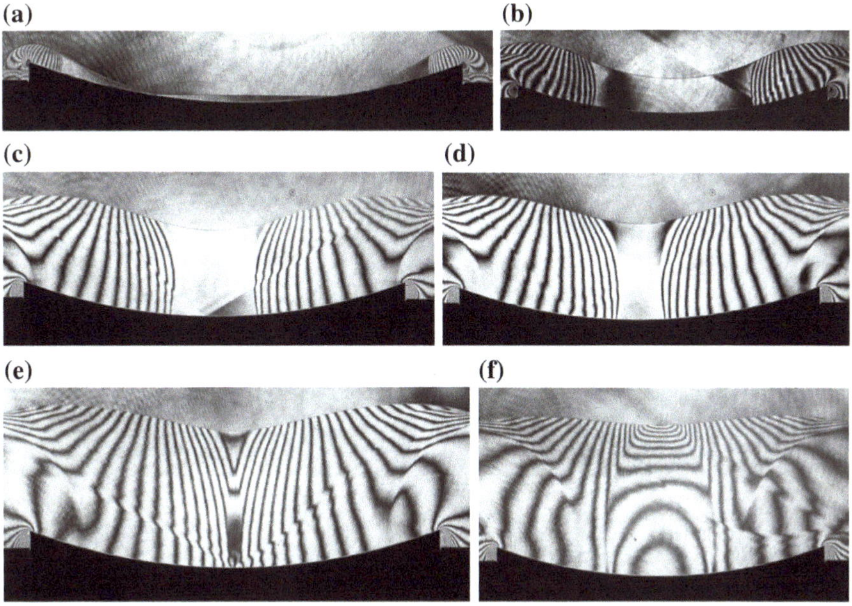

Fig. 5.16 Effects of wall angle of 70° for Ms = 1.30, in atmospheric air at 286.4 K: **a** #84042005, 120 μs, Ms = 1.314; **b** #84042009, 160 μs, Ms = 1.320; **c** #84042013, 200 μs, Ms = 1.320; **d** #84042014, 220 μs, Ms = 1.311; **e** #84042015, 240 μs, Ms = 1.311; **f** #84042017, 280 μs, Ms = 1.313

Fig. 5.17 A concave and convex wall

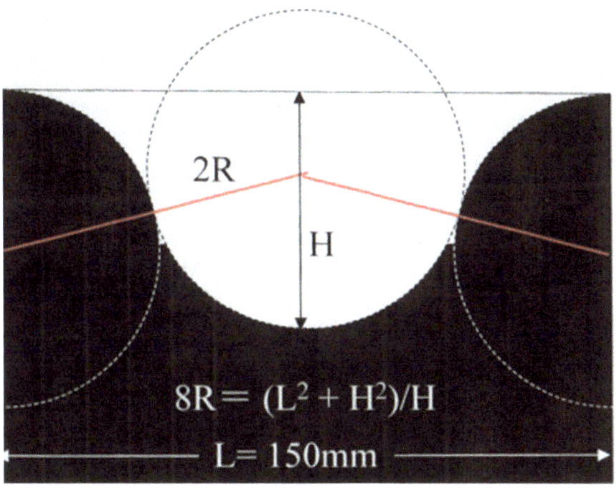

accumulation of vortices played an important role during a later stage of focusing. In Fig. 5.23d, the dense accumulation of fringes turned into vortices. After the vortex interaction, the flow fields become more or less similar to the case of shock wave reflections from circular reflectors. When the vortices converged at the end of the log-spiral reflector as seen in Fig. 5.23d and in its enlargement in Fig. 5.23h, the

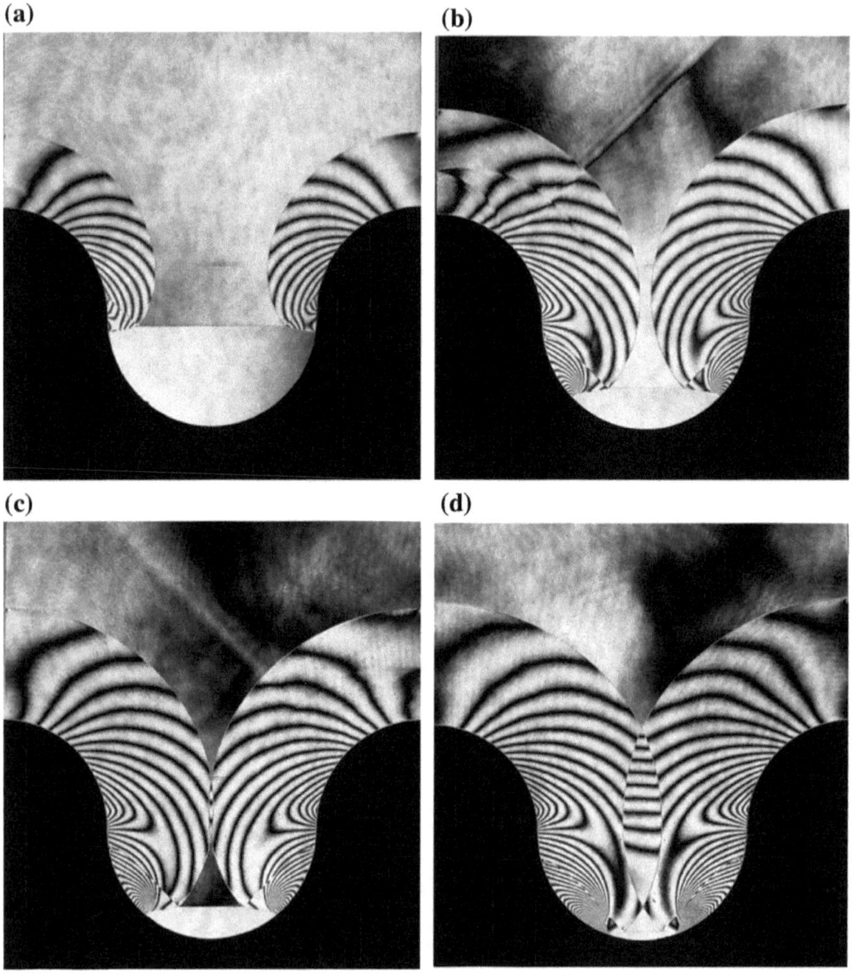

Fig. 5.18 Effect of convex and concave walls on focusing, 75 mm depth for Ms = 1.40, in atmospheric air at 297 K: **a** #86091816, Ms = 1.427; **b** #86091810, Ms = 1.435; **c** #86091803, Ms = 1.411; **d** #86091806, Ms = 1.416; **e** #86091803, Ms = 1.411; **f** #86091805, 310 μs, Ms = 1.414; **g** #86091811, Ms = 1.433; **h** #86091801, Ms = 1.407

temperature would be enhanced as to emit luminosity from the spot where the vortex focused. In holographic interferometry, the luminous emissions are not observable because the sensitivity of holographic films could not detect the luminosity but Milton et al. (1975) used direct shadowgraph and observed the luminosity.

Figure 5.24 shows the evolution of a shock wave focusing from a log-spiral passage for Ms = 1.480 in air at 900 hPa, 292.5 K. Each image is compared with its enlarged one. This log-spiral shape gradually curved the IS and eventually focused it at the tip of the log-spiral.

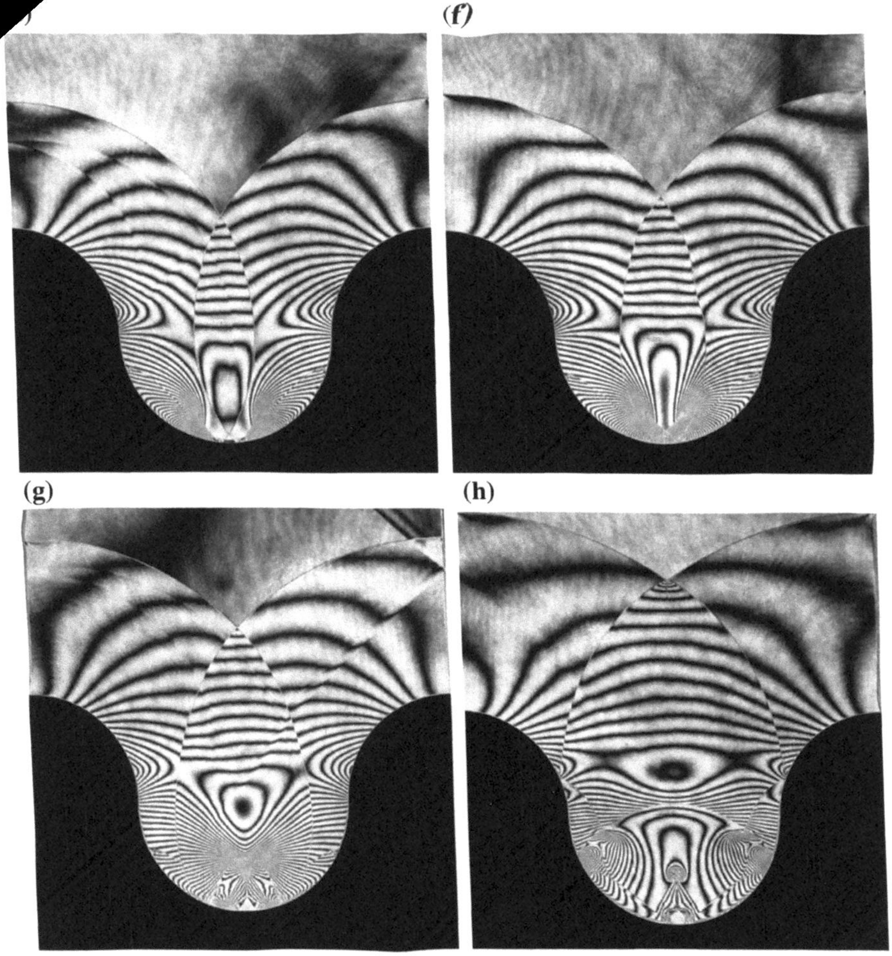

Fig. 5.18 (continued)

Figure 5.25 shows the evolution of shock wave focusing from a log-spiral passage for Ms = 2.70 in air at 110 hPa, 287.0 K. Figure 5.25a, c, e, f offer comparisons between recorded observations taken at a different test time with appropriate numerical simulations (Inoue et al. 1995). Figure 5.25g shows a magnified view of Fig. 5.25f. Figure 5.25d, e shows the fringe accumulation, and convergence of the vortices at its the tip of the log-spiral passage. The convergence resulted in temperature enhancement. Milton reported (Milton et al. 1975) spontaneous luminous emission due to the generation of high temperature at the tip.

As seen in Fig. 5.25, the numerical simulation agreed well with the experimental findings. Using very fine adoptive meshes, the present numerical simulation succeeded in reproducing the shock wave focusing along the log-spiral passage.

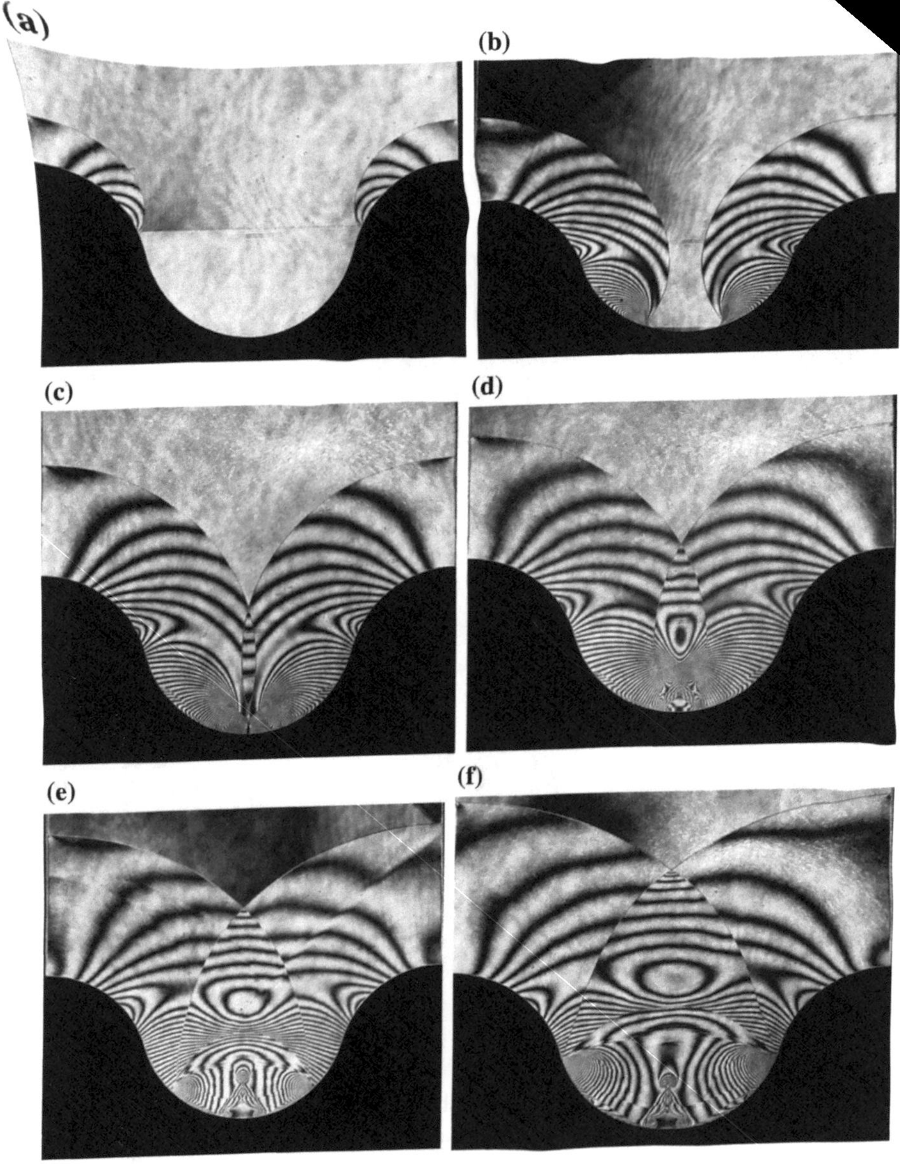

Fig. 5.19 Evolution of shock wave reflection from a convex and concave wall of 57 mm depth for Ms = 1.43 in atmospheric air at 293.0 K: **a** #86093006, 340 μs from trigger point, Ms = 1.439; **b** #86093002, Ms = 1.432; **c** #86092908, 385 μs, Ms = 1.432; **d** #86093001, 395 μs, Ms = 1.415; **e** #86093008, 415 μs, Ms = 1.438; **f** #86092905, 420 μs, Ms = 1.439

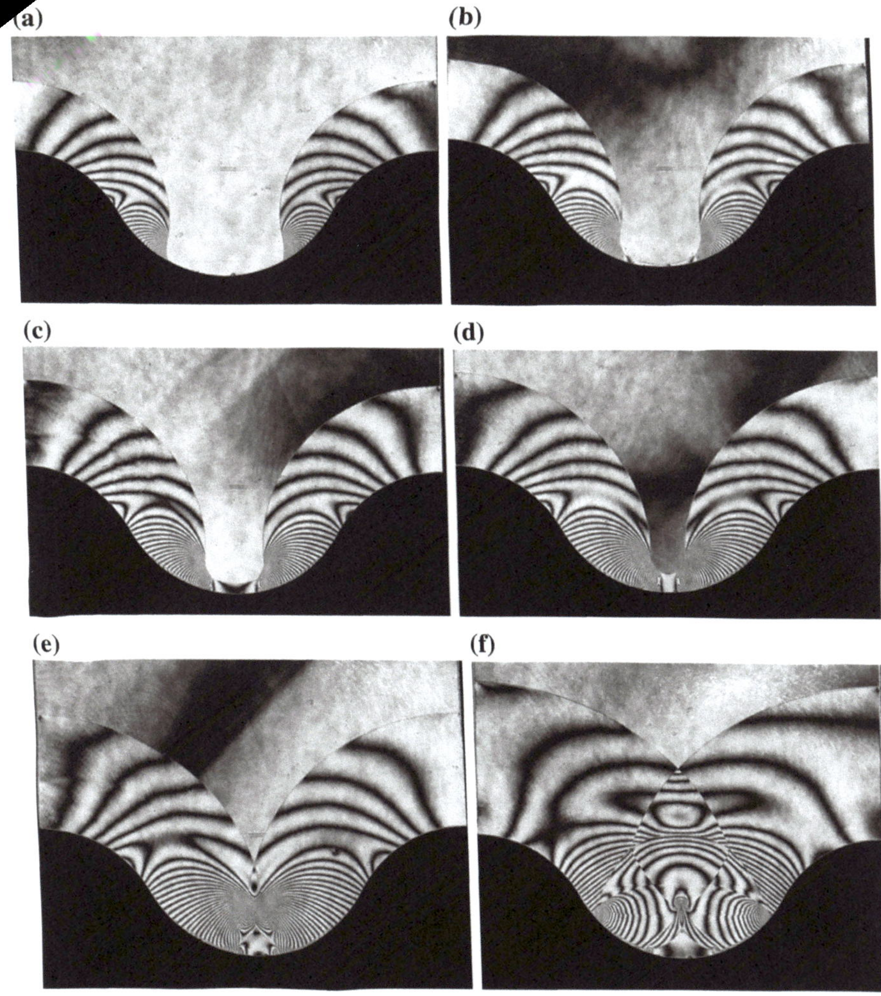

Fig. 5.20 The evolution of shock wave reflection from a convex and concave wall reflector of R = 42 mm for Ms = 1.43 in atmospheric air at 297.6 K: **a** #86100105, Ms = 1.425; **b** #86100107, Ms = 1.428; **c** #86100108, Ms = 1.430; **d** #86100109, Ms = 1.431; **e** #86100111, Ms = 1.429; **f** #86100116, Ms = 1.430

5.5 Shock Wave Focusing from Area Contraction

Shock waves are strengthened while propagating along ducts having area contraction. In a broad sense, this is a kind of shock wave focusing. When shock waves propagate along a duct having a V-shaped area contraction, the pressure enhancement varies depending on the angle of the area contraction. Figure 5.26 illustrates a V-shaped area contraction installed in the 60 mm × 150 mm

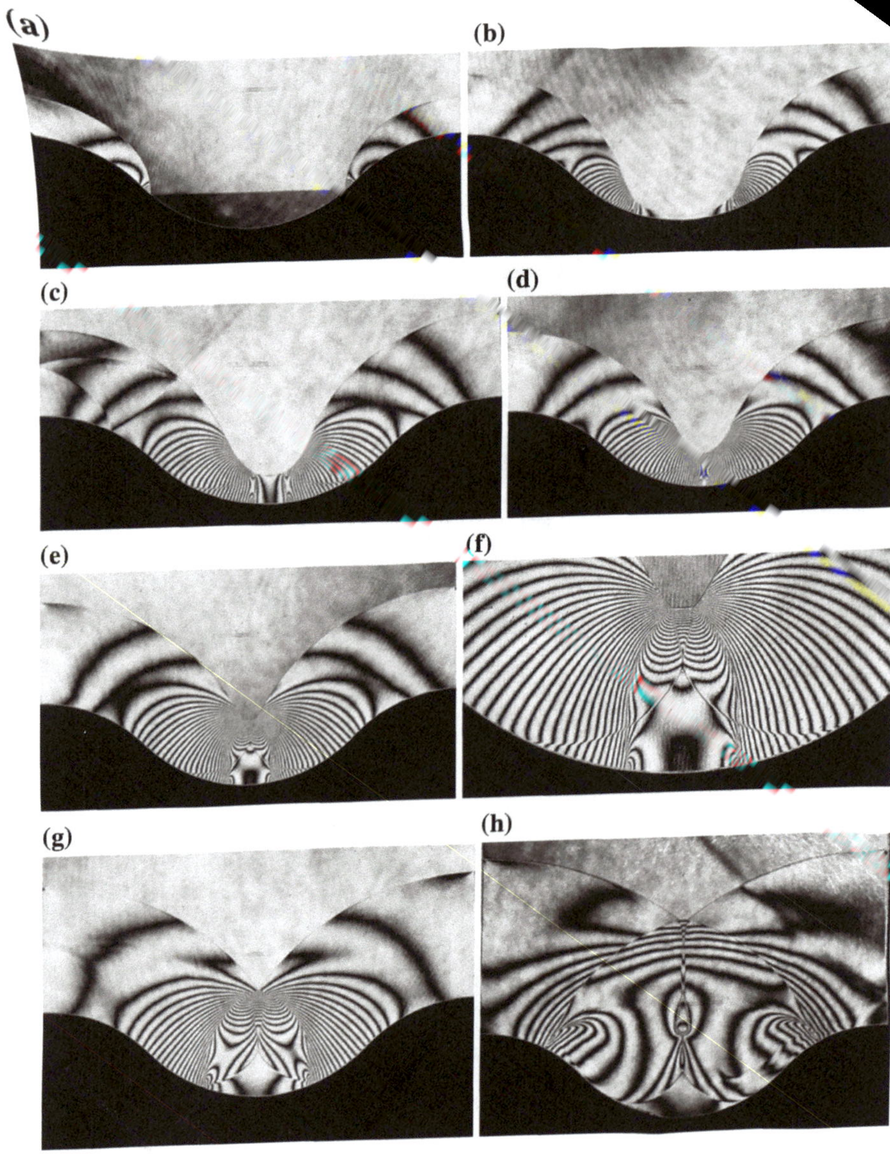

Fig. 5.21 The evolution of the shock wave reflection from a reflector of R = 94.6 mm and H = 31 mm for Ms = 1.42 in atmospheric air at 293.0 K: **a** #86092103, Ms = 1.421; **b** #86092105, Ms = 1.438; **c** #86092108, Ms = 1.430; **d** #86092109, Ms = 1.435; **e** #86092111, 410 μs, Ms = 1.432; **f** enlargement of (**e**); **g** #86092112, Ms = 1.423; **h** #86092116, Ms = 1.443

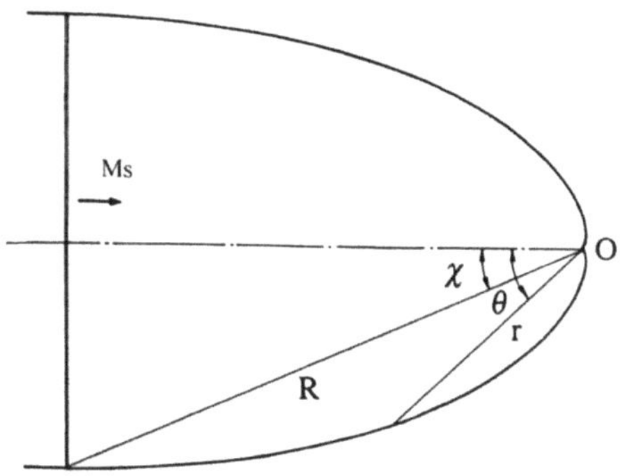

Fig. 5.22 Logarithmic spiral shape

conventional shock tube. When the duct inclination angle of θ_w is smaller than the critical transition angle θ_{crit}, the reflection pattern is always a SMR. Hence the triple points of reflected shock waves along the upper and lower walls intersect and eventually merge at the end of the area contraction. Meantime, resulting reflected shock waves RS and slip lines SL repeatedly interact and move toward the end of the V-shaped duct.

5.5.1 *V-Shaped Area Contraction*

In August 1985, the B747 of JAL Flight 123 clashed against a mountain and over 500 passengers were killed. The B747's pressure bulkhead was ruptured while cruising, resulting in blown-off of the vertical rudder. The B747 lost control and crashed. The cabin pressure was higher than the pressure at the cruising altitude. The pressure bulkhead worked as a diaphragm in a shock tube sustaining this pressure difference. Therefore, its rupture generated a shock wave, which was not a planar shock wave but propagated into the space inside the vertical tail. As this structure had an area contraction, the pressure there increased high enough to blow-off the vertical rudder.

For checking the dependence of area reduction on the resulting pressure amplification, Shock wave strengths were amplified propagating in a duct of area contractions. V-shaped area contractions of 30°, 60°, 90°, 120° were installed in the 60 mm × 150 mm conventional shock tube and tested for various shock strengths. In Fig. 5.27, the SMR is reflected from these area contractions.

5.5.1.1 Angle 30°

Figure 5.27 shows the evolution of the shock wave reflection and focusing along 30° area contraction for for Ms = 1.30 in atmospheric air at 289.3 K. Multiple reflections of the triple points and vortices enhanced pressures stepwise are visible in Fig. 5.27f (Figs. 5.28, 5.29).

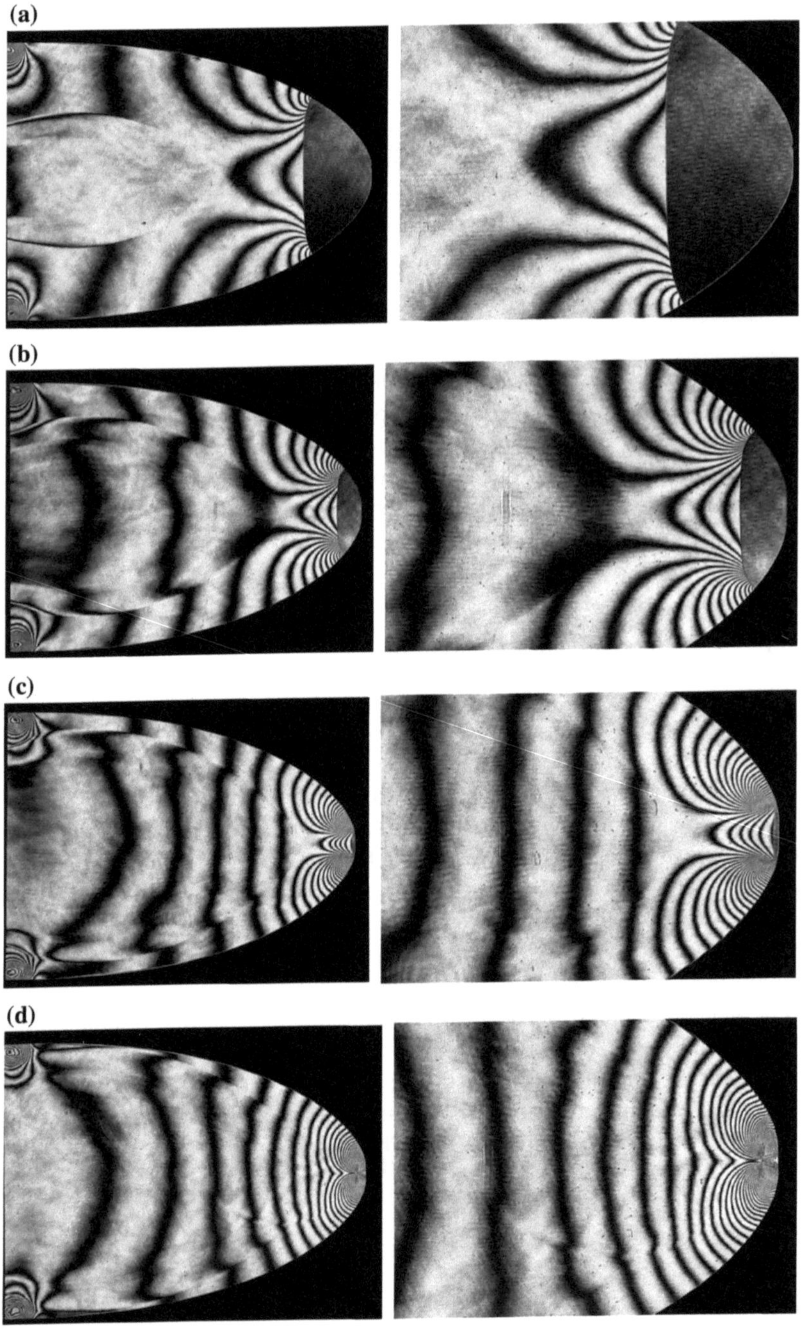

Fig. 5.23 The evolution of shock wave focusing from a log-spiral passage for Ms = 1.40 in atmospheric air at 294.8 K, comparison with enlarged images: **a** #86100909, 370 μs from the trigger point, Ms = 1.396; **b** #86100803, 390 μs, Ms = 1.396; **c** #86100804, 400 μs, Ms = 1.398; **d** #86100805, Ms = 1.383; **e** #86100808, 407 μs, Ms = 1.389; **f** #86100806, 410 μs, Ms = 1.398; **g** #86100809, 415 μs, Ms = 1.390; **h** enlargement of (**d**)

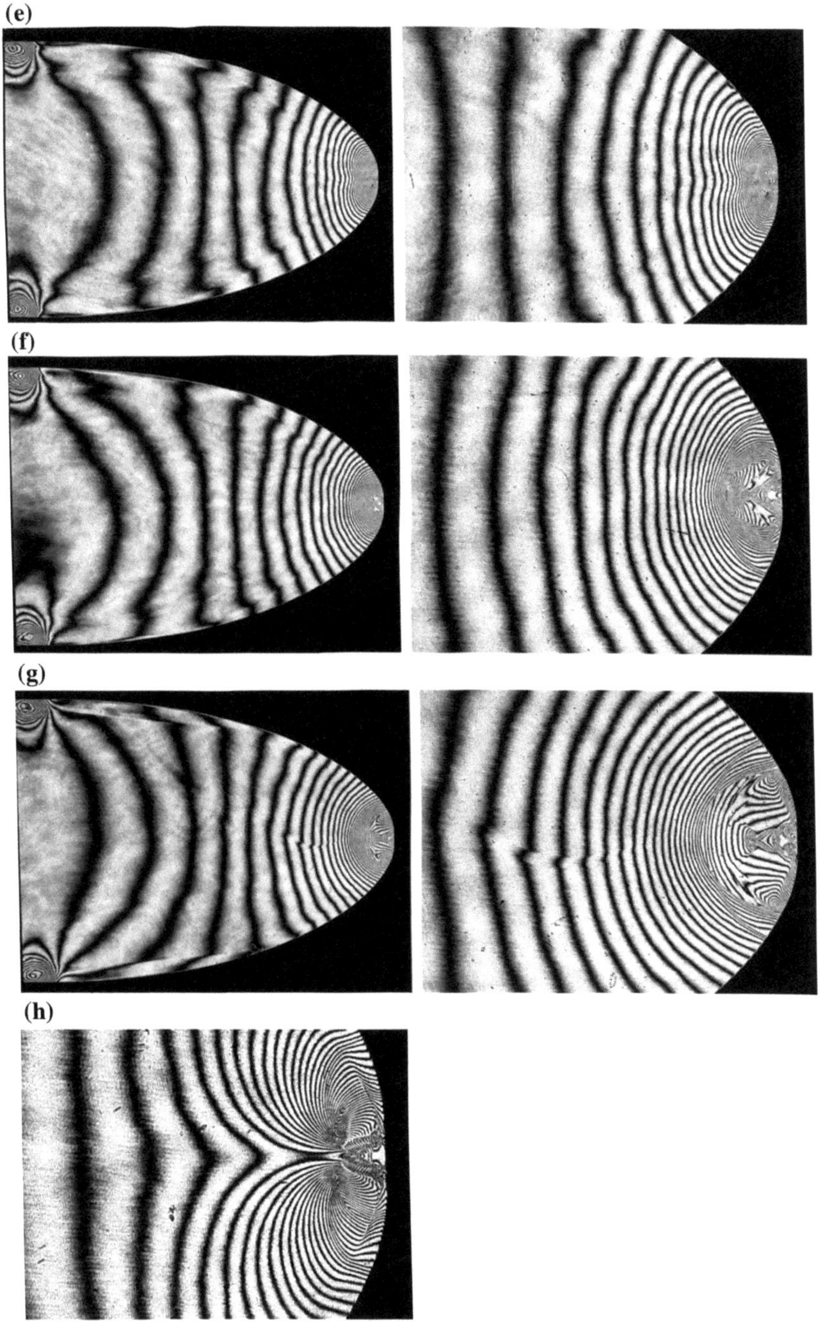

Fig. 5.23 (continued)

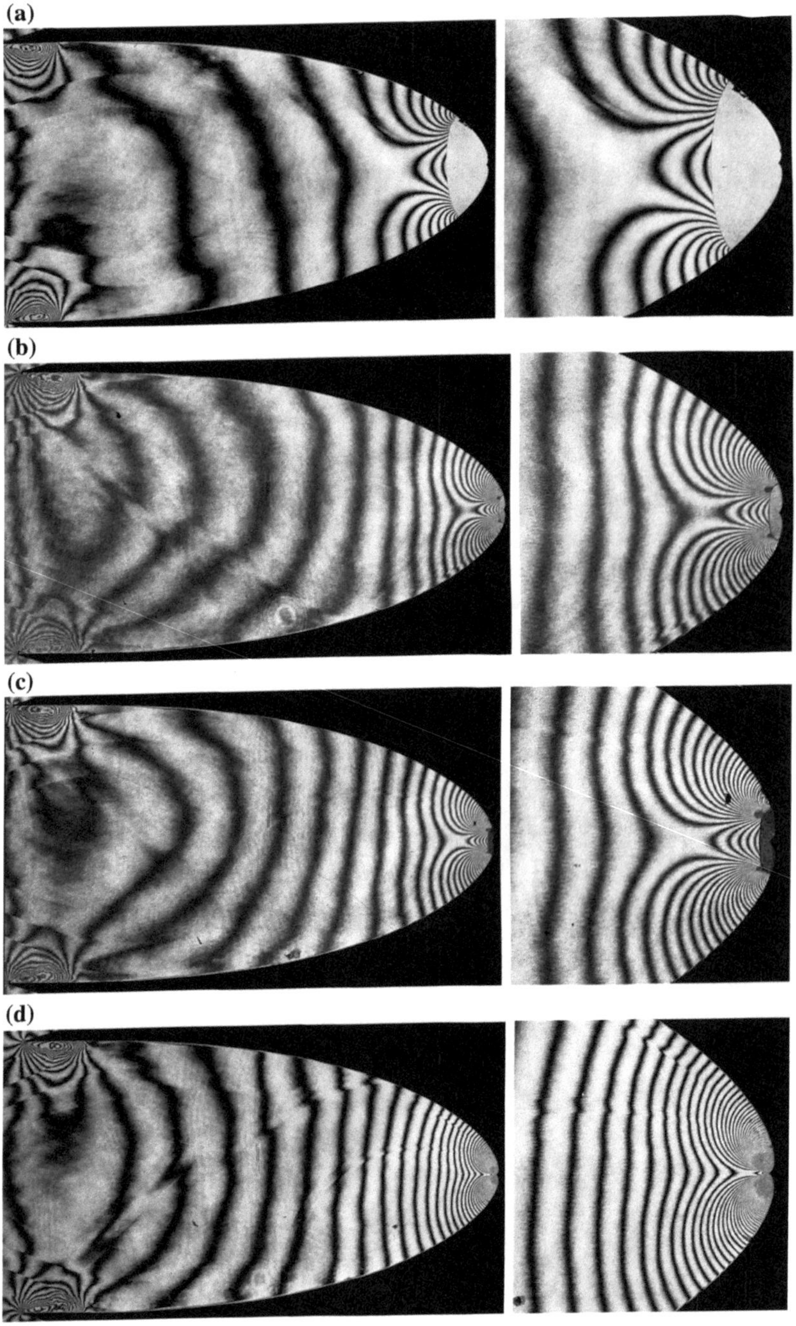

Fig. 5.24 The evolution of the shock wave focusing from a log-spiral passage for Ms = 1.480 in air at 900 hPa, 287.1 K: **a** #86112713, 495 μs from trigger point, Ms = 1.479; **b** #86112707, 510 μs, Ms = 1.480; **c** #86112712, 500 μs, Ms = 1.476; **d** #86122711, 505 μs, Ms = 1475; **e** #86122708, 515 μs, Ms = 1.480; f #86122709, 517 μs, Ms = 1.276

(e)

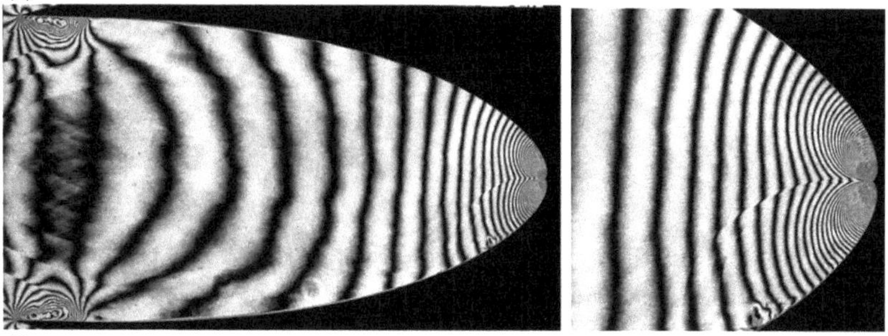

(f)

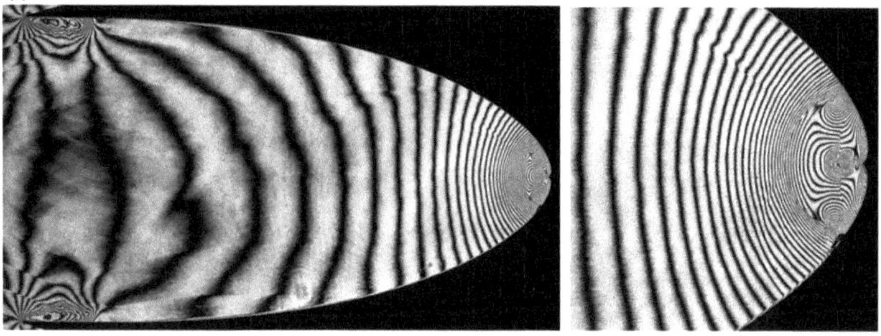

Fig. 5.24 (continued)

5.5.1.2 Angle 60°

Figure 5.30 shows the evolution of the shock wave focusing from 60° area contraction for Ms = 1.27 in atmospheric air at 290.3 K. Propagating toward the tip, the transmitting shock waves interacted repeatedly along the upper and lower walls and enhanced the pressure discontinuously behind the interacting shock wave. Although the procedure of shock wave interaction is simple, it contains complex shock dynamic effects. Figure 5.31a–i show the same V-shape angle but higher Ms = 2.17. The intersection of the TP from the upper and lower walls accompany SL. The convergence of the SL and their interactions are observed in Fig. 5.31i.

5.5.1.3 Angle 90°

Figure 5.32 show shock wave focusing from a 90° area contraction placed in a 60, 150 mm shock tube for Ms = 1.27 in atmospheric air at 287.4 K. The reflection patterns are SbRR and hence the pressure enhancement is caused simply by repeated reflection of oblique shock waves.

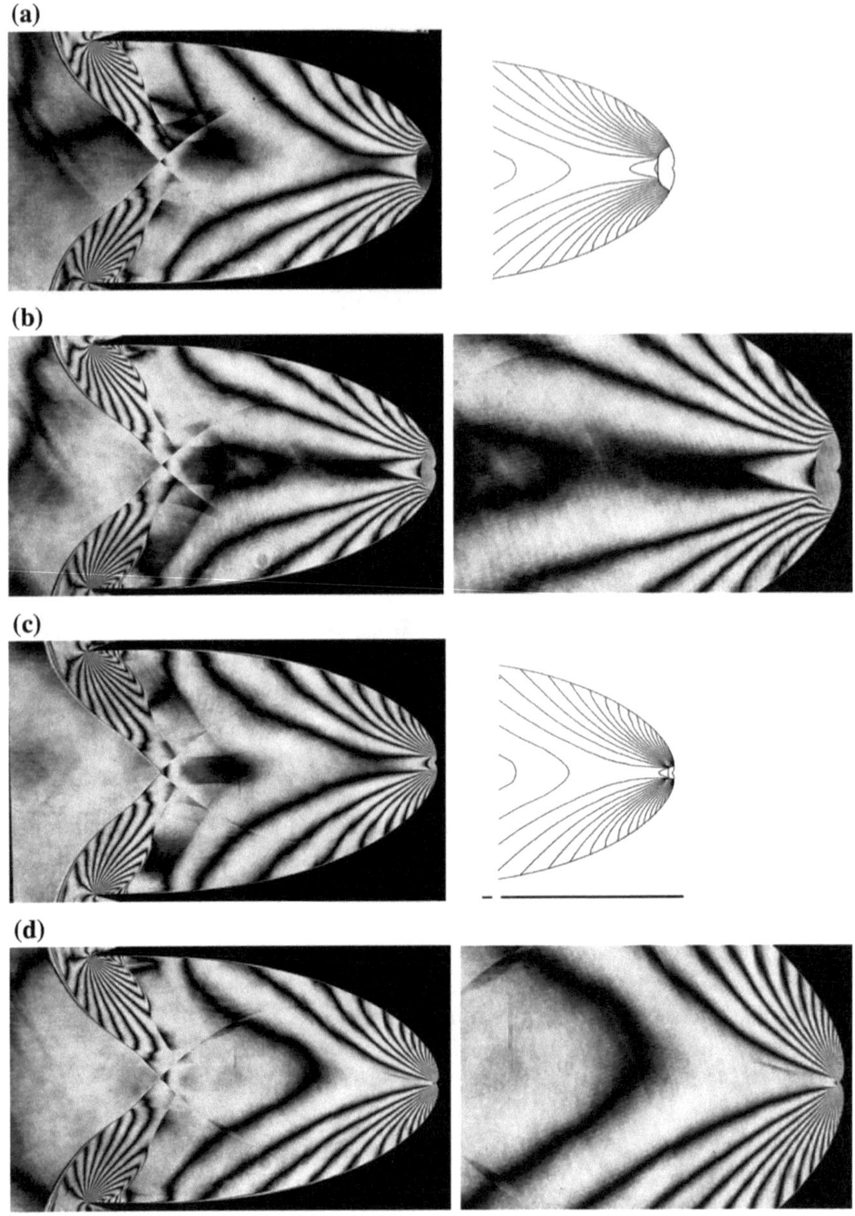

Fig. 5.25 Evolution of shock wave focusing from a log-spiral passage for Ms = 2.70 in air at 110 hPa, 287.0 K: **a** #86112812, 510 μs from trigger point, Ms = 2.722; **b** #86112808, 500 μs, Ms = 2.739; **c** #86112811, 507 μs, Ms = 2.722; **d** #86112809, 505 μs, Ms = 2.698; **e** #86112813, 510 μs, Ms = 2.723; **f** #86112814, 510 μs, Ms = 2.723; **g** enlargement of (**f**)

(e) (f) (g)

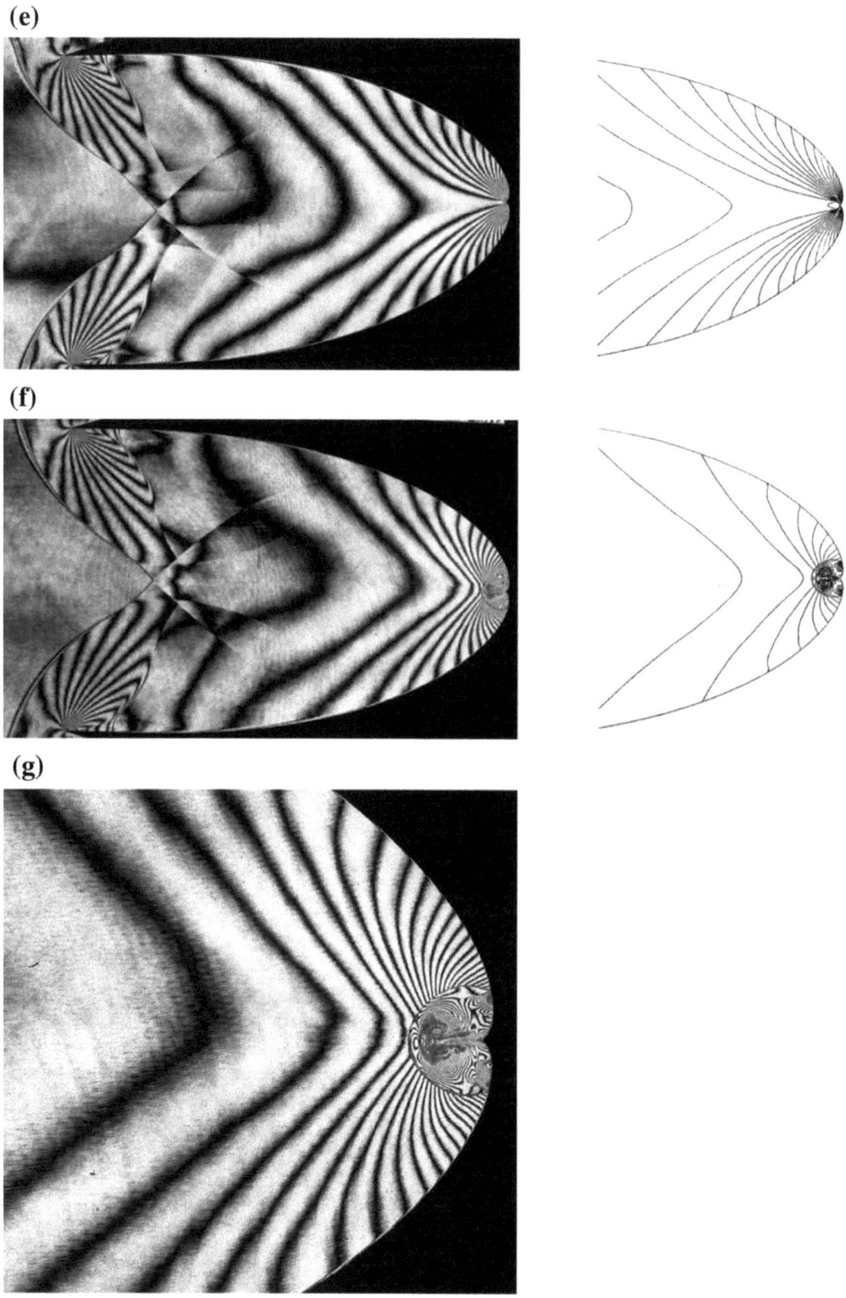

Fig. 5.25 (continued)

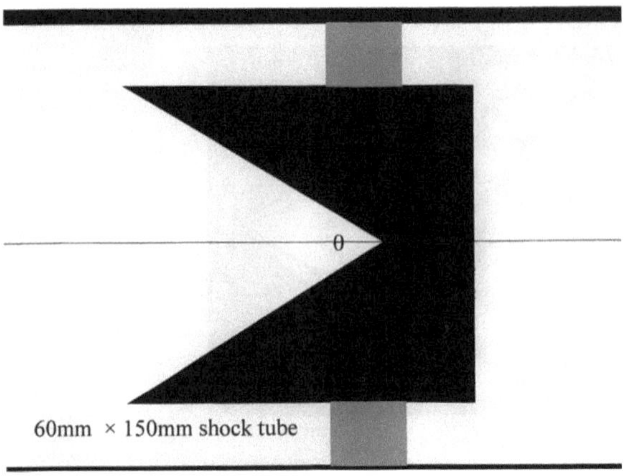

Fig. 5.26 V-shaped duct, θ = 30°, 60°, 90°, 120°

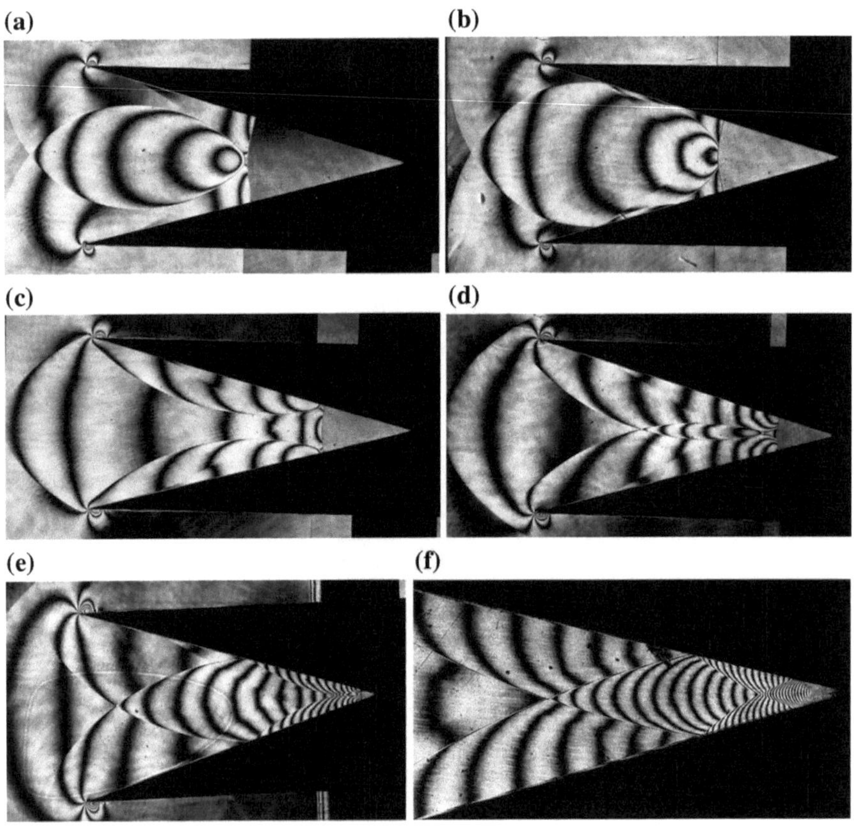

Fig. 5.27 The evolution of the shock wave focusing along 30° shaped area contraction for Ms = 1.30 in atmospheric air at 289.3 K: **a** #88020220, 260 μs from trigger point, Ms = 1.301: **b** #88020218, 280 μs, Ms = 1.299; **c** #88020216, 310 μs, Ms = 1.301; **d** #88020224, 330 μs, Ms = 1.303; **e** #88020227, 360 μs, Ms = 1.299; **f** #88020228, 370 μs, Ms = 1.301

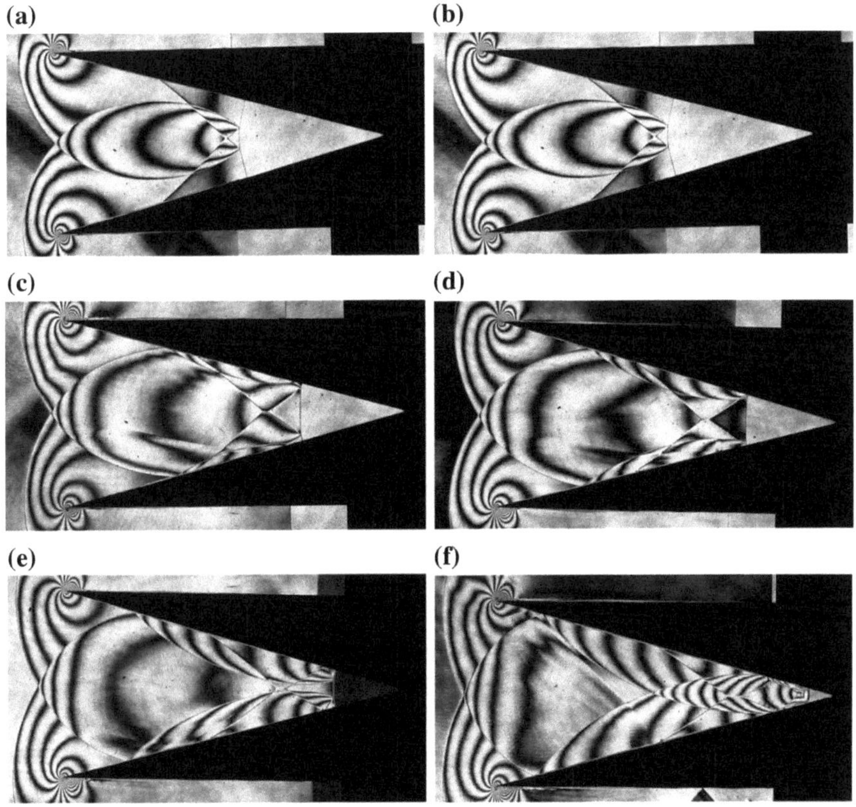

Fig. 5.28 The evolution of the shock wave focusing along 30° shaped area contraction for Ms = 1.60 in air at 600 hPa, 289.5 K: **a** #88020301, 80 μs, Ms = 1.685; **b** #88020303, 90 μs, Ms = 1.682; **c** #88020305, 110 μs, Ms = 1.686; **d** #88020307, 120 μs, Ms = 1.685; **e** #88020308, 130 μs, Ms = 1.674; **f** #88020309, 140 μs, Ms = 1.607

5.5.1.4 Angle 120°

In Fig. 5.33, a SPRR appears on the wedge surface. Their reflection pattern created a stepwise pressure enhancement on the wall surface.

5.6 Conical Area Convergence

Figure 5.34 shows a 10° conical area converging section made of acryl. In order to quantitatively visualize the shock wave propagation through the test section, it has an aspheric lens shape which enables to quantitatively visualized the flows in a circular cross sectional duct. It had 50 mm in diameter it and connected to the 50 mm diameter shock tube.

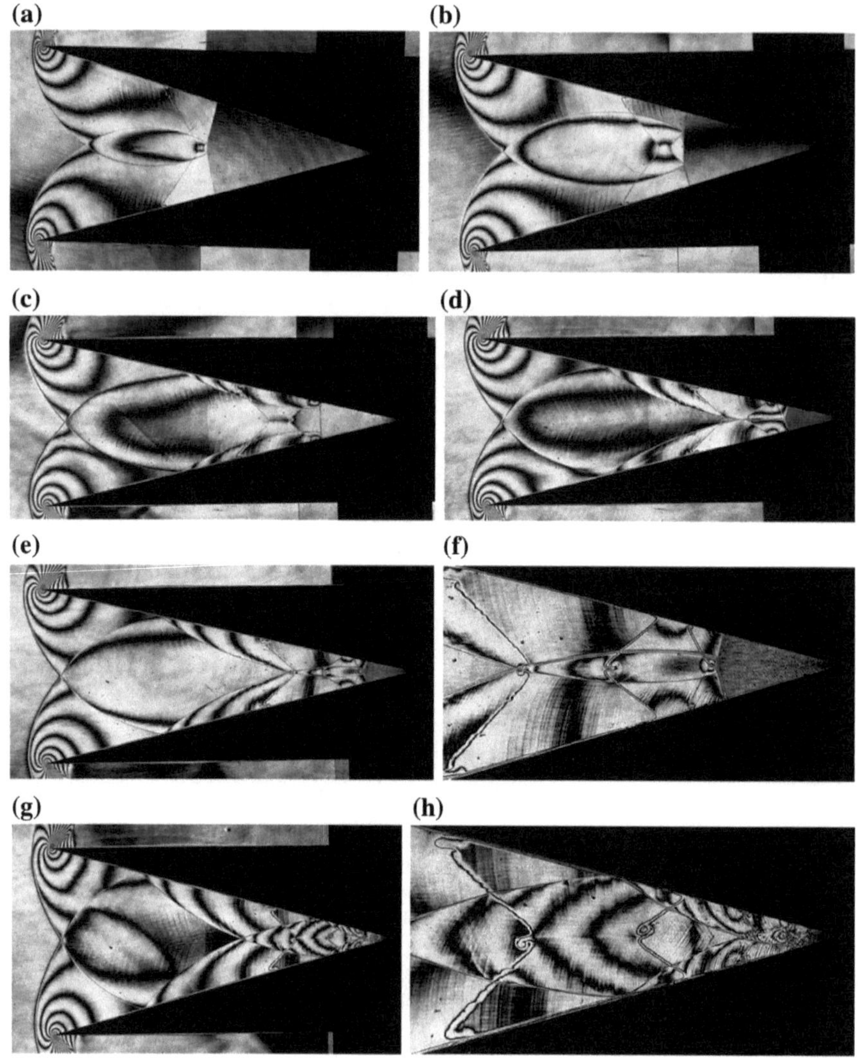

Fig. 5.29 The evolution of shock wave focusing from 30° area contraction for Ms = 2.10 in air at 300 hPa, 291.5 K: **a** #88020318, 290 μs from trigger point, Ms = 2.169; **b** #88020320, 310 μs, Ms = 2.142; **c** #88020321, 320 μs, Ms = 2.180; **d** #88020322 350 μs, Ms = 2.154; **c** #88020323, 340 μs, Ms = 2.157; **f** enlargement of (**e**); **g** #88020324 350 μs, Ms = 2.159; **h** enlargement of (**g**)

The shock wave propagation in a 10° (half apex angle) conical test section was visualized using double exposure holographic interferometry and diffuse holography as well. The pattern of the transmitting shock wave is, in principle, identical with the shock wave propagation from a shallow V shaped duct. Therefore, the reflection pattern of the shock wave will be a SMR. However, unlike two-dimensional shock tube flows, neither the reflected shock wave, RS, nor the

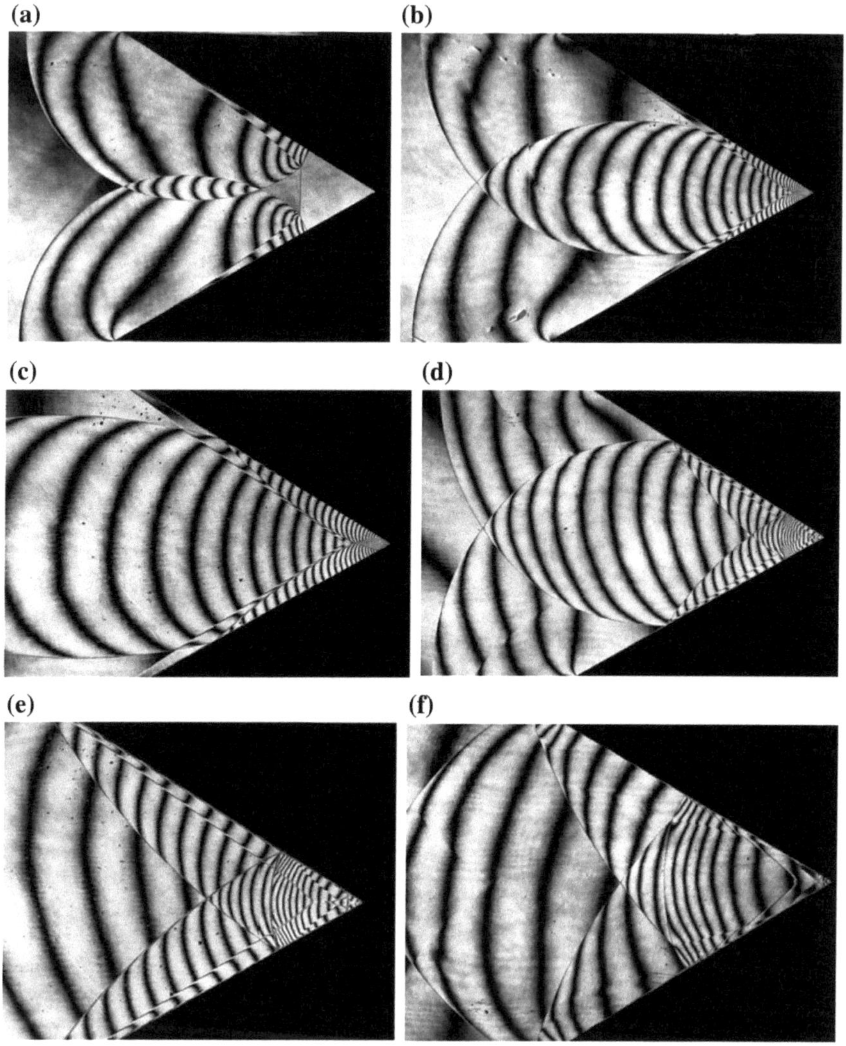

Fig. 5.30 Evolution of shock wave focusing from 60° area contraction for Ms = 1.27 in atmospheric air at 290.3 K: **a** #88020202, 330 μs from trigger point, Ms = 1.274; **b** #88020206, 380 μs, Ms = 1.274; **c** enlargement of (**b**); **d** #88020209, 410 μs, Ms = 1.274; **e** enlargement of (**d**); **f** #88020214, 500 μs, Ms = 1.266; **g** enlargement of (**f**)

slip line, SL are distinctly visible. At the shock front, the loop formed by the intersection of the triple point, TP, and the Mach stem, MS, is the distinctly observed. Hence the trajectories of the TP are recognized correctly as seen in Fig. 5.35a, b. Figure 5.35b is a single exposure interferogram Milton (1989). Figure 5.35c, d shows the second stages of the reflection.

Figure 5.36 summarizes triple point trajectories. The triple points intersect and converge, in other words, focus. Then the shock wave started diverging.

(g)

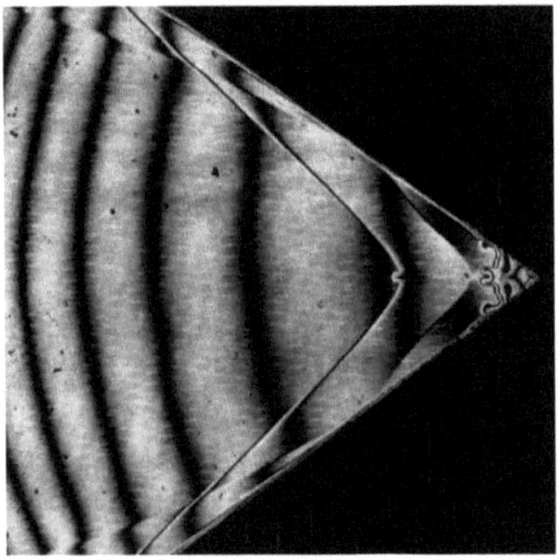

Fig. 5.30 (continued)

Figure 5.35c, d correspond to the stage of diverging shock waves. But the patterns so far observed was not clear. Figure 5.37 shows shock wave convergence in a 10°–20° combined duct for Ms = 1.46 in air.

Figure 5.38 shows the shock wave propagation along a duct having 10° and 20° combined area convergence. Figure 5.38a shows a single exposure interferogram for Ms = 1.46 in air at 750 hPa 289.7 K. Figure 5.38b Summarizes the evolution of triple point trajectories for Ms = 1.46.

5.7 Circular Co-axial Annular Shock Wave Focusing

5.7.1 Horizontal Annular Co-axial Shock Tube

The convergence of a cylindrical or spherical shock wave into a spot is named an implosion, while the divergence of a shock wave from a point source is called explosion. Guderley (1942) studied analytically the implosion of shock wave and derived a self-similar solution. Perry and Kantrowitz (1951) experimentally investigated the convergence of annular coaxial shock waves. In this experiment, the annular co-axial planar shock wave smoothly diffracted over an axisymmetric tear drop shaped inner core placed at the end of horizontal shock tube. They visualized the converging cylindrical shock waves using shadowgraph system and discussed the stability of converging shock waves.

Regarding the convergence of annular shock waves, experiments were carried out, for example, by Wu et al. (1978), Hoshizawa (1987), Neemeh and Less (1990), Watanabe (1993), and Apazidis et al. (2011). Knystautus and Lee (1971) investigated the convergence of detonation waves in an annular detonation tube, which

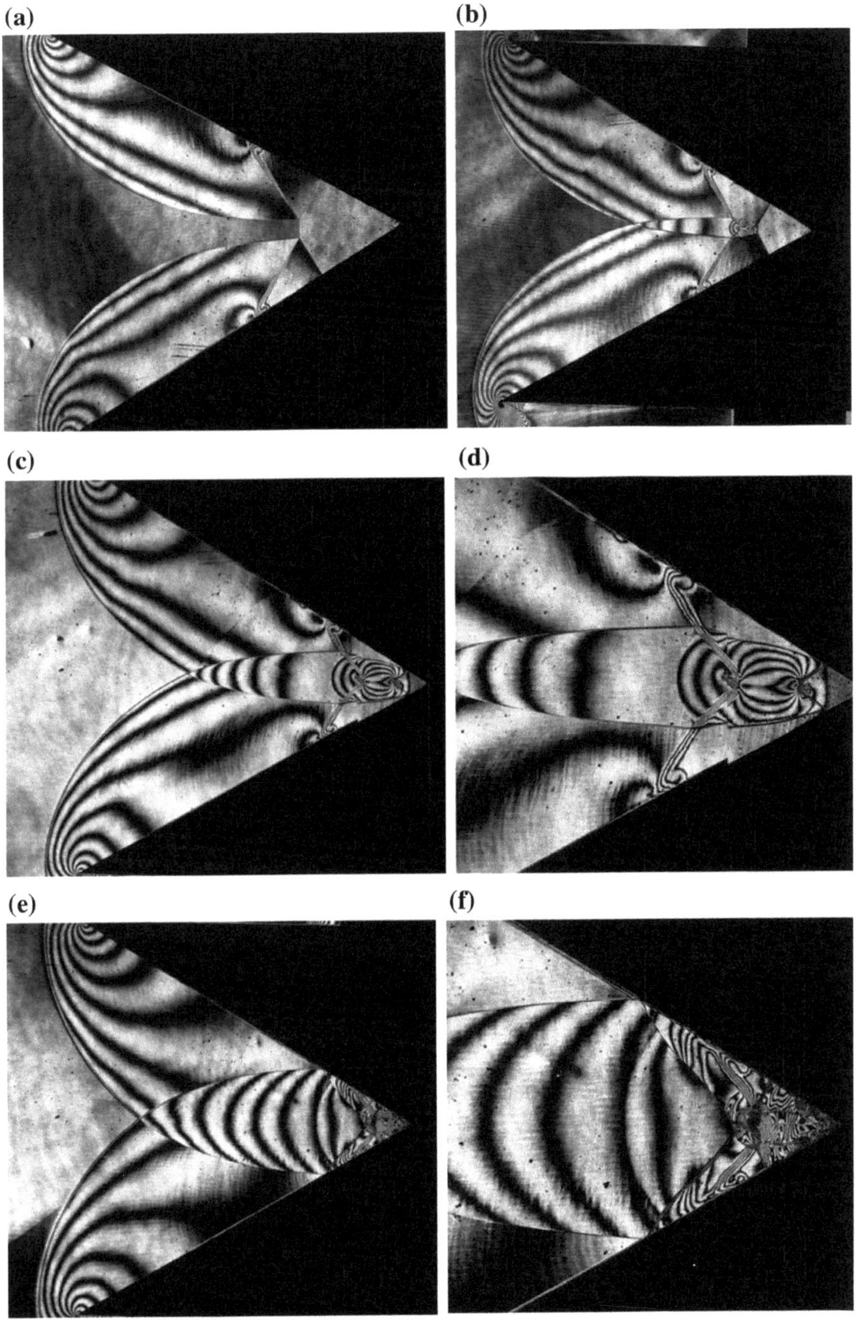

Fig. 5.31 The evolution of shock wave focusing from 60° area contraction for Ms = 2.17 in air at 300 hPa, 288.6 K: **a** #88020403. 340 μs from trigger point, Ms = 2.169; **b** #88020404, 350 μs, Ms = 2.152; **c** #88020405, 360 μs, Ms = 2.152; **d** enlargement of (**c**); **e** #88020406, 370 μs, Ms = 2.140; **f** enlargement of (**e**); **g** #88020409, 370 μs, Ms = 2.152; **h** enlargement of (**g**); **i** #88020411, 400 μs, Ms = 2.129

(g)

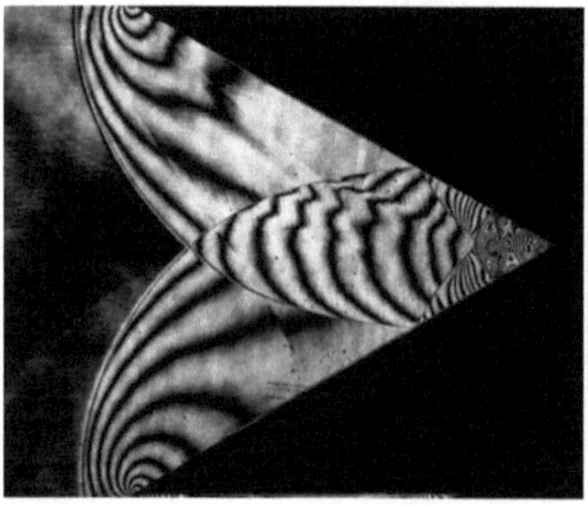

(h)

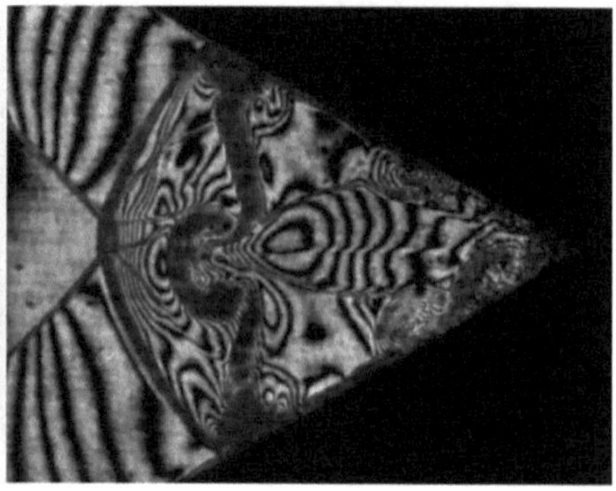

(i)

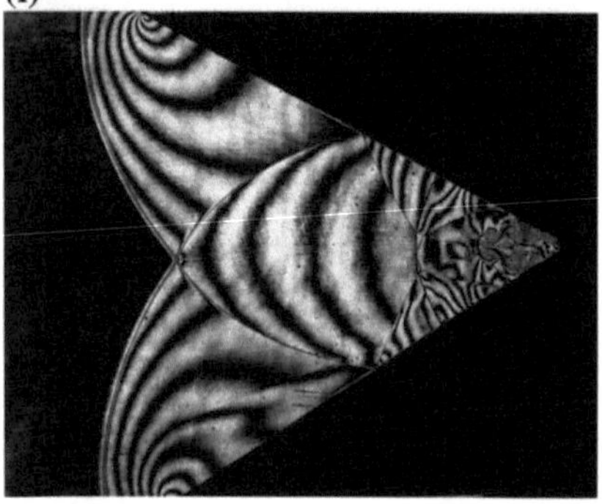

Fig. 5.31 (continued)

had the similar structure to that of Perry and Kantrowitz (1951). They observed the spiral shaped soot patterns over the plane where the detonation wave converged. Using a similar facility reported by Knystautus and Lee (1971), Fujiwara et al. (1979) observed the luminous emission at the center of converging detonation waves. Terao (1973) invesigated the converging detonation wave using a very large 800 mm diameter detonation chamber and also observed converging detonation waves propagating at a velocity exceeding the Chapman-Jouget detonation velocity. He witnessed the generation of a very high pressure at the spot of the center of convergence.

Figure 5.38 illustrates a 500 mm long, a 210 mm inner diameter, and a 230 mm outer diameters annular co-axial test section Hoshizawa (1987). It was connected to a 50 mm diameter conventional shock tube was connected via a 45° conical section to the test section. The inner cylinder was supported by two pairs of two-18 mm diameter. cylindrical struts located at the S_1 and S_2, as seen in Fig. 5.38. The struts diffracted the transmitting shock wave and created wakes in the flow behind the transmitted shock wave. Therefore, in order to suppress these flow disturbances, the test section was 500 mm long. Two pressure transducers were installed at the position of P_1 and P_2.

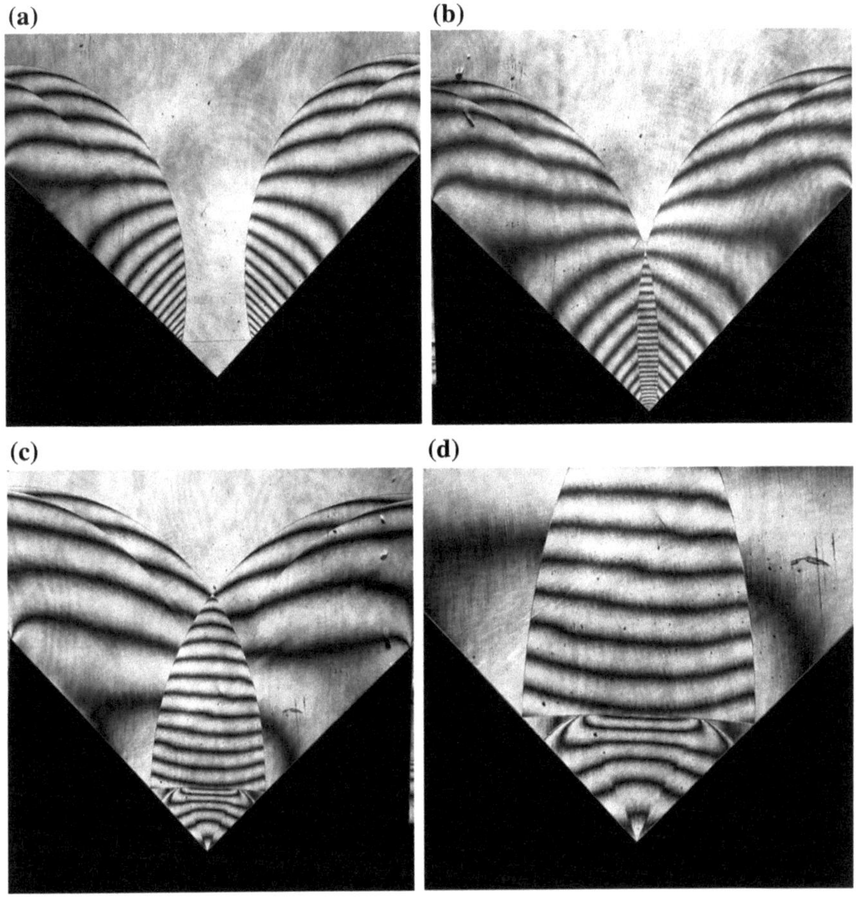

Fig. 5.32 The evolution of shock wave convergence from 90° reflector for Ms = 1.27 in atmospheric air at 287.4 K: **a** #88020114, 370 μs from trigger point, Ms = 1.282; **b** #88020114, 370 μs, Ms = 1.282; **c** #88020115 420 μs, Ms = 1.275; **d** enlargement of (**c**)

The interval between the inner and outer tubes was 10.0 ± 0.02 mm. The blockage ratio of the struts to the annular shock tube cross section was 0.12. At the end of the test section, an annular co-axial 90° bend was connected, its inner and outer radii of which were 2.5 and 12.5 mm, respectively. The result of the previous experiment with the two-dimensional bends Honda et al. (1977) were used for having the optimal shape of the present bend. A 130 mm diameter glass plate was placed on the outside end wall and a 130 mm diameter aluminum plated glass mirror was placed on the inside end wall. In the experiments, the ring shaped shock wave turning 90° and the end of the straight test section eventually became an annular co-axial converging shock wave in a similar manner as described in Perry and Kantrowitz (1951). This arrangement smoothly re-directed a ring shaped shock wave to a co-axial converging shock wave and suppressed the flow non-uniformity the struts created. However, the passage through the co-axial bend diffracted and

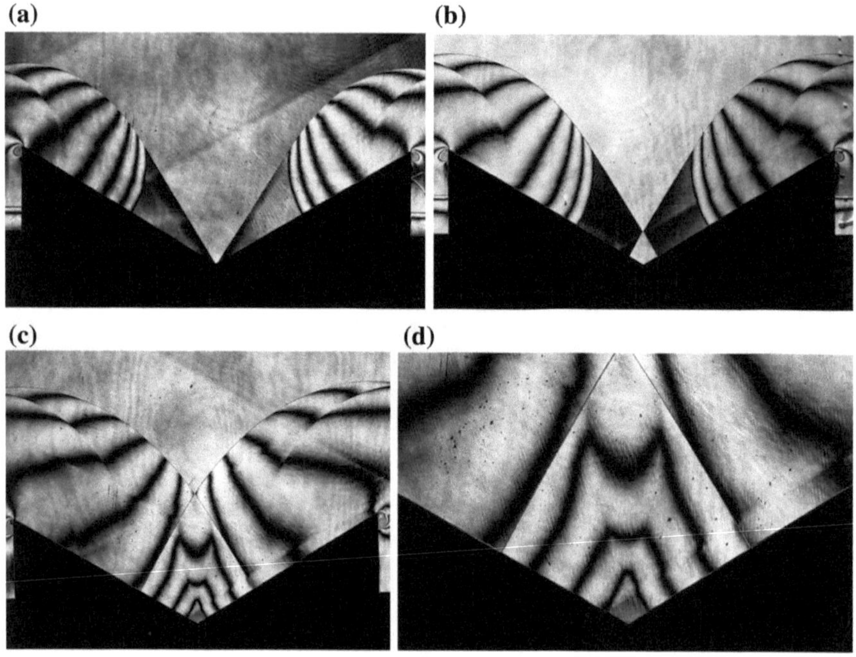

Fig. 5.33 Evolution of shock wave convergence from 120° area contraction for Ms = 1.16 in atmospheric air at 287.0 K: **a** #88020106, 460 μs from trigger point, Ms = 1.153; **b** #88020107, 470 μs, Ms = 1.162; **c** #88020109 520 μs, Ms = 1.164; **d** enlargement of (**c**)

Fig. 5.34 A 10° (half apex angle) conical test section

reflected the transmitting shock wave induced wavelets, which propagated behind the converging shock wave. The test gas was air at the initial pressure ranging from 50 hPa to 100 kPa. Visualization was conducted by double exposure Twyman-Green interferometry Takayama (1983).

Perry and Kantrowitz (1951) concluded, from results of their visualization, that the convergence of weak annular co-axial cylindrical shock waves was stable. However, in our double exposure holographic interferometric observations, it was not always achievable. The initial disturbances created by the reflection of shock

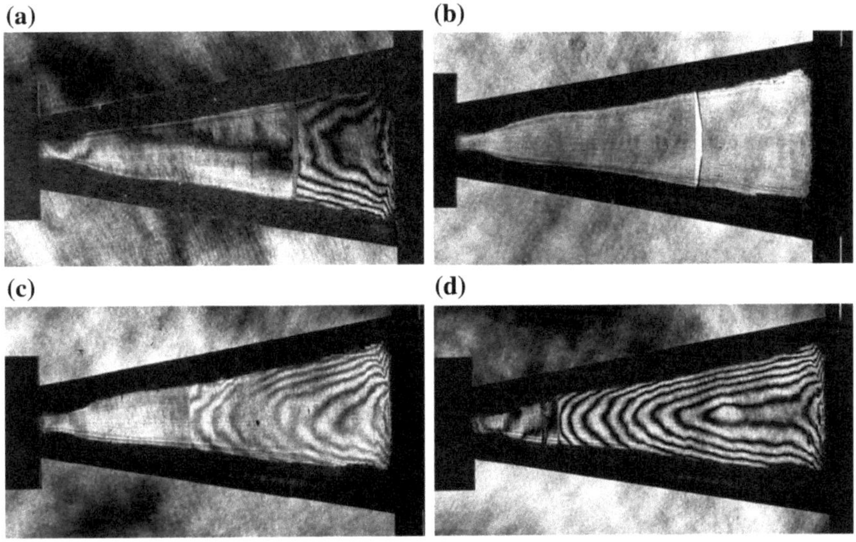

Fig. 5.35 Shock wave propagation along a 10° area convergence for Ms = 1.74 in air at 500 hPa, 288 K: **a** #84122701, 60 μs from trigger point. Notice the IS propagating from the right to the left; **b** #84122704, 62 μs, single exposure; **c** #84122605, 120 μs. Ms = 1.742; **d** #84122610, 150 μs, Ms = 1.742

Fig. 5.36 Triple point trajectories in the 10° area convergence for Ms = 1.73 (Milton 1989)

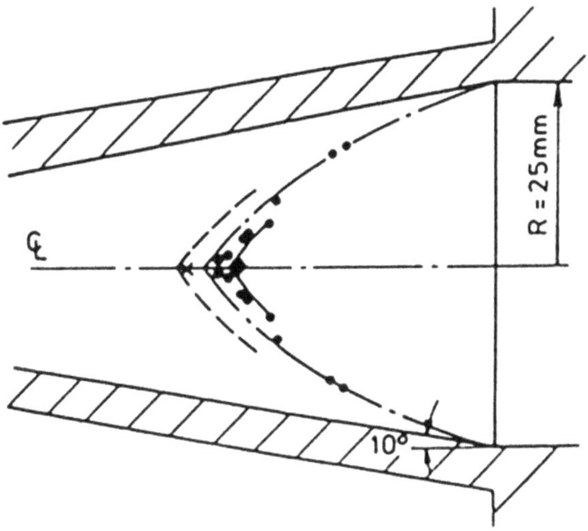

wave reflection from the struts and its connection with wakes hardly disappeared from the flow. These flow perturbations affected the annular shock wave and prevented its focusing. The presence of the four struts resulted in flow instability that prevent proper converging of the annular shock wave.

Figure 5.39 shows the convergence process of annular co-axial shock wave convergence for Ms = 1.38 in CO_2 at 400 hPa and 295.0 K. As seen in Fig. 5.39a,

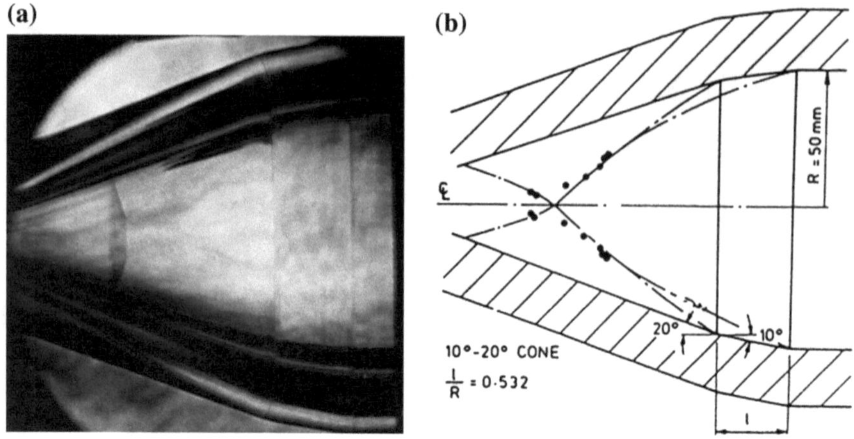

Fig. 5.37 Shock wave propagation along a 10°–20° combined duct and 50 mm diameter area convergence: **a** #84121228, 560 μs from trigger point, for Ms = 1. 456 in air at 750 hPa 289.7 K, single exposure; **b** summary of triple point trajectory for Ms = 1.46 (Milton 1989)

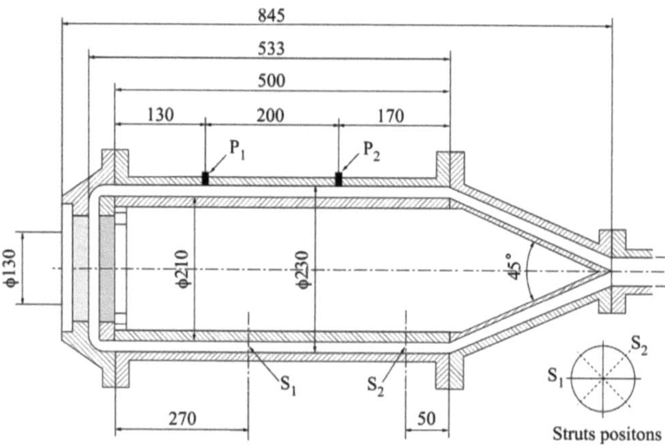

Fig. 5.38 The test section

the shape of the incident shock wave looks nearly cylindrical but shows a sign of mode-four perturbations. Figure 5.39b–d shows the development of the mode four perturbations and the deformation of the circular shock wave toward a square shape although not yet reachable in these interferograns. The four circular fringes attached to the converging shock wave increase their numbers with elapsed time representing the amplification of the initial disturbances. This created four pairs of triple points at the final stage of convergence. In Fig. 5.39e, the transmitted shock wave converged at the center but as the fringes are so densely populated so it is impossible to resolve them. When the transmitted shock wave converged and turned to the cylindrical diverging shock wave. Figure 5.39f shows the cylindrical

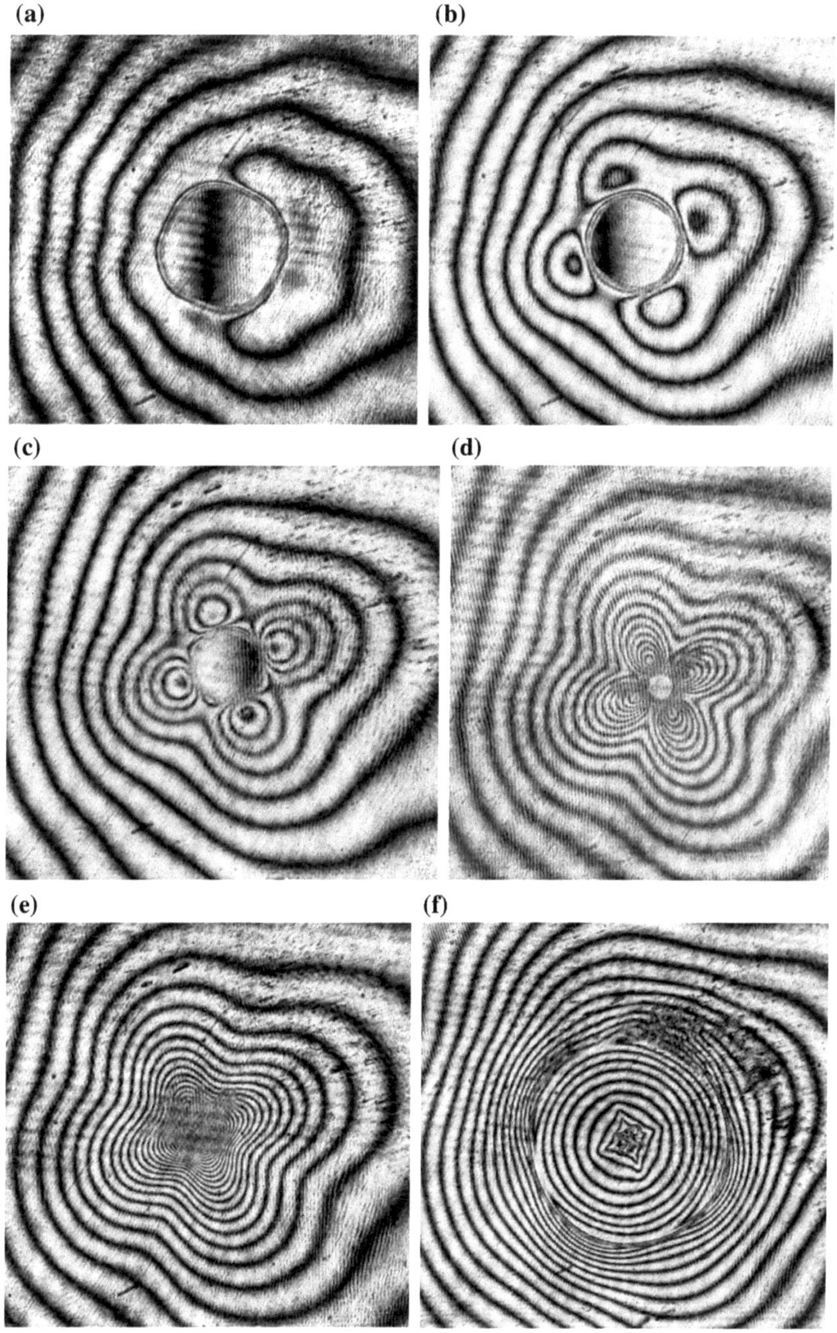

Fig. 5.39 The evolution of shock wave convergence in the test section shown in Fig. 5.37 for Ms = 1,38 in CO_2 at 400 hPa. 295.0 K: **a** #82051207, 280 μs from trigger point, Ms = 1.373; **b** #82051208, 330 μs, Ms = 1.373; **c** #82051205, 283 μs, Ms = 1.373; **d** #82051204, 285 μs, Ms = 1.385; **e** #82051210, 290 μs, Ms = 1.386; **f** #82051211, 310 μs, Ms = 1.383; **g** #82051213, 325 μs, Ms = 1.386

(g)

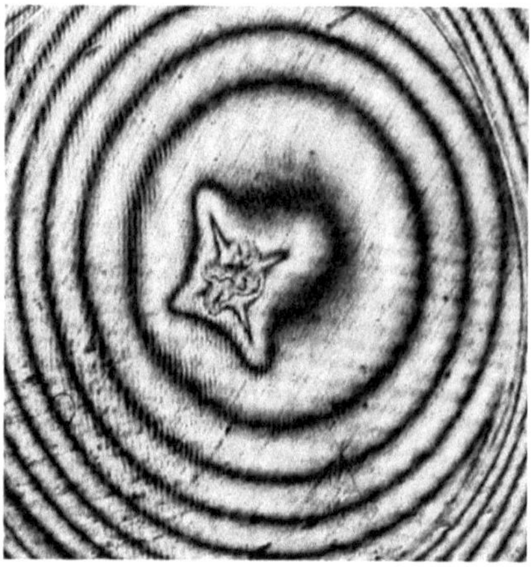

Fig. 5.39 (continued)

diverging shock wave. Behind the diverging shock wave the remnants of the vortices are visible at the center. Figure 5.40g shows the enlarged image. In Fig. 5.39g, the cross shaped remnants of four pairs of vortices are visible far behind the diverging shock wave.

Figures 5.40 show the evolution of a converging shock wave for Ms = 2.00 in air at 100 hPa and 298.5 K. In Fig. 5.40a, the shape of the shock wave was slightly deformed affected by mode-four disturbances. In Fig. 5.40b, d, the shock wave reached a square shape being composed of a SMR having four pairs of triple points at the four corners. Figure 5.40c is a magnified image of Fig. 5.40b. Four pairs of triple points consisting of three shock confluence and a slip line are clearly observable in the magnified photo. Figure 5.40e was taken a few μs after taking Fig. 5.40d, when the shock wave was converged and the shock wave just started diverging. The diverging shock wave was always stable. Figure 5.40g shows four pairs of vortices at the center behind the cylindrical diverging shock wave which were generated by the triple point interaction. Each pair of vortices had the same pattern as seen at the shock wave focusing from a concave circular reflector, for example, as seen in Fig. 5.3g.

Guderley (1942) derived a self-similar solution for imploding shock waves in ideal gas; he suggested the following form,

$$r/r_o = (1 - t/t_o)^n$$

where r and t are averaged radial distance and the elapsed time, respectively. r_o and t_o are reference radius and time and n is a self-similar exponent. Streak recordings of shock wave focusing were conducted for incident shock wave for Ms = 1.10, 1.50, and 2.10 in air. Figure 5.41a shows the results obtained for Ms = 1.5 in air. A 1.0 mm wide slit was attached to, crossing the observation window glass. The

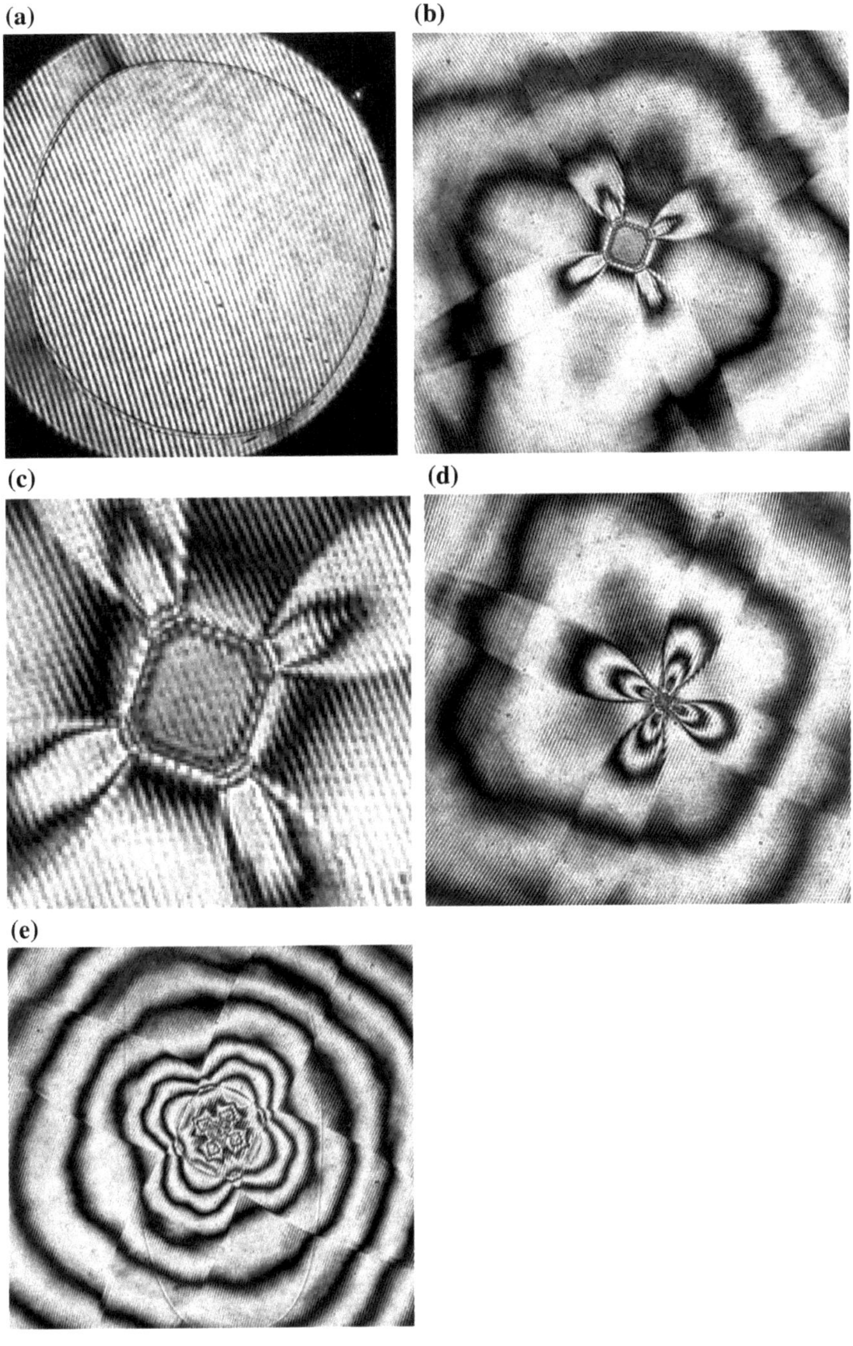

Fig. 5.40 The evolution of the shock wave convergence for Ms = 2.00 in air at 100 hPa, 298.5 K; **a** #89060903, 42 μs from trigger time, Ms = 2.007; **b** #89060718, 43 μs, Ms = 2.043; **c** enlargement of (**c**); **d** #89060716, 45 μs, Ms = 2.043; **e** #89060708, 46 μs, Ms = 2.050

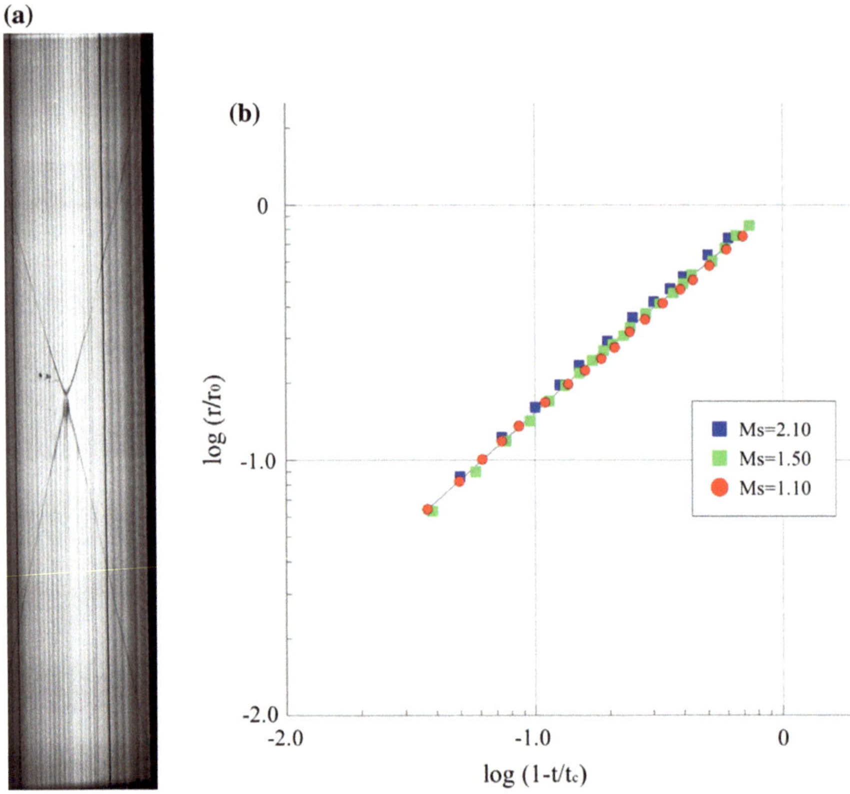

Fig. 5.41 A streak photograph of shock wave convergence: **a** #82061414, for Ms = 1.5. **b** Summary of streak recording of time variation of triple point trajectory data were collected from the tests #82061112 to #82061622. The self-similar exponent n varies from 0.828 to 0.833 (Hoshizawa 1987)

light source was an Argon-ion laser equipped with a mechanical shutter having the opening and closing time of about 100 μs. Direct shadowgraph images of shock waves viewed through the slit were recorded by an ImaCon high speed camera operated in streak mode (John Hadland Ima Con 675). Recording speed was 1 mm/μs. The time is running from the bottom upward. When the shock wave converges, the shock wave diameter decreases quickly toward the point of convergence. The slightly dark shadows gradually generated behind the accelerating converging shock waves would indicate the formation of vortices and their accumulation; this can be observed in Fig. 5.40. Immediately after the implosion, the shock wave turned to a diverging wave. Measurements of the radius-time variations of the converging cylindrical shock waves are shown in the *x*-*t* plane presented in Fig. 5. 41a; both the radii and the elapsed time were plotted in a logarithmic plane.

Figure 5.41b summarizes the results of the streak recordings conducted from tests #82061112 to #82061622. The ordinate denotes the non-dimensional radius $\log_e(r/r_o)$ and the abscissa denotes the non-dimensional time $\log_e(1 - t/t_o)$. The time variations of the radius are displayed for Ms = 1.1, 1.5. and 2.1 in the $x{-}t$ plane. The self-similar exponents were deduced from this display; n is 0.828 for Ms = 1.1, n is 0.830 for Ms = 1.5, and n is 0.833 for Ms = 2.1, whereas Guderley's (1942) analytically derived n = 0.835 for $\gamma = 1.4$. Terao (1973), based on his detonation wave focusing chamber, obtained n = 0.82. The present results Hoshizawa (1987) agree well with previous results.

In reality the converging cylindrical shock waves were deformed by the exposure to the perturbation from behind as seen in Fig. 5.40. Hence the shape would be deformed from a circular shape. From the sequential interferograms, deviations of the local radius Δr from the average radius R could be readily obtained. Figure 5.42 summarizes the normalize Δr/R during the shock wave convergence. The ordinate denotes the normalized Δr/R and the abscissa denotes the averaged radius R in mm for Ms = 1.1, 1.5 and 2.1. As seen in Figs. 5.39, 5.40, Δr/R increases with decreasing in R. The trend is nearly independent of the Ms.

5.7.2 *Vertical Annular Co-axial Shock Tube*

The mode-four instability shown in Fig. 5.42 is a direct result from the presence of four struts which were supporting the horizontal inner core as shown in Fig. 5.38. In order to construct an annular shock tube without using the struts, Watanabe (1993) constructed a vertical annular co-axial shock tube composed from a 230 mm inner diameter and a 210 mm outer diameter tube as seen in Fig. 5.43. The shock tube had a massive and rigid structure sustaining its vertical position by itself. The shock tube had a high pressure driver section and a low pressure channel; a piston

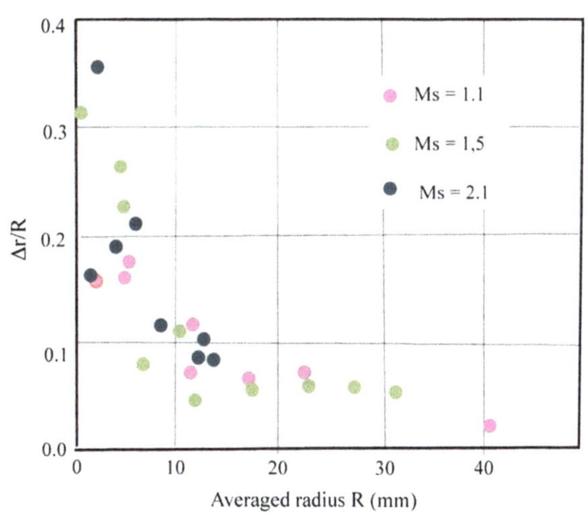

Fig. 5.42 Dimensionless deformation of converging shock waves in air Δr/R and averaged radius R (Hoshizawa 1987)

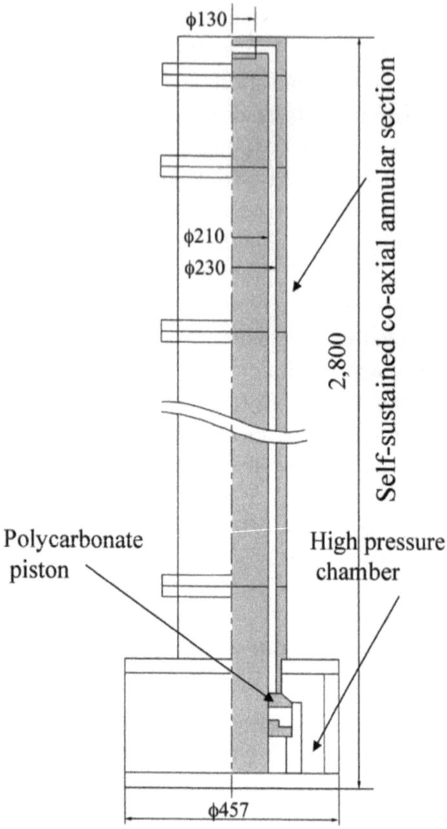

Fig. 5.43 The first generation diaphragm-less vertical annular co-axial shock tube (Watanabe 1993)

driving mechanism was used for separating between the two sections of the shock tube.

The high pressure helium was filled into the driver chamber, from a reservoir that was attached outside the main structure. A light weigh ring shaped polycarbonate piston separated the low-pressure channel from the high pressure driver, see in Fig. 5.43. Its quick movement served as replacement of a diaphragm rupturing mechanism. The light weight piston was backed up with a high-pressure auxiliary helium from behind. Its quick movement served as effectively released the high pressure helium into the low pressure channel eventually forming an annular co-axial shaped shock wave. Therefore, this diaphragm-less shock tube produced annular co-axial shock waves associated with a minimal degree of disturbances.

This shock tube system had no inherent mode number and even produced mode number zero, that is m = 0. The only possibly created inherited disturbances in this shock tube would be those associated with the wavelets induced by the shock wave diffracting at the corner and reflecting from the 90° bend.

To control mode numbers, disturbances created by the insertions of 2, 3, 4, 6, 8, 12 and 24 pieces of 10 mm long and 10 mm diameter cylinders were examined.

The disturbances generated by small cylinders positioned in equal circumference at the 20 mm distance before the 90° corner resulted in evenly deformed converging cylindrical shock wave.

Watanabe (1993) used his vertical shock tube shown in Fig. 5.43 and observed converging shock waves exposed to the mode four perturbations. He measured the deviation of local radius Δr from the averaged diameter R and as discussed in Fig. 5.42 obtained the growth of perturbed wave shape for Ms = 2.0 in air with decreasing the averaged radius at 12.0, 5.7, and 3.0 mm. Figure 5.3 summarized the results for Ms = 2.0 in air. The ordinate denotes $\Delta r/R$ and the abscissa denotes the angles in circumference of the circular shock wave from 0 to 2π in radian.

Figure 5.44a shows the distribution of perturbed shock wave at R = 12 mm. The four coherent distribution of maximal peaks shows the converging shock is perturbed by the mode-four disturbances. In Fig. 5.43b, the mode-four disturbance was amplified. However, the mode-four disturbance never reached to a catastrophic amplitude; it was suppressed by introduction of the formation of triple points. Previously very densely populated fringe distribution was never maintained continuously but formed a triple point and, in short, a SMR was formed. In Fig. 5.44c, the increase in $\Delta r/R$ resulted in the formation of the triple points. The plateaus in Fig. 5.44c indicated the appearance of Mach stems.

Figure 5.45 shows results of pressure measurements along the test section shown in Fig. 2.43. Pressure transducers Kistler model 603B were distributed along the test section at the radii of 0, 15 mm, 30, and 45 mm. The experiments were conducted in the same initial condition for Ms = 1.50 as shown Fig. 5.44. The ordinate denoted the dimension-less pressure, p/p_o, where p_o is atmospheric pressure. The abscissa denotes the elapsed time in μs. Open circles show the measured pressures. Solid lines indicate the results of a numerical simulation solving the Navier-Stokes Equations (Watanabe 1993). Numerical results agree reasonably well with measured pressures. The time variation of pressure at the center (R = 0 mm) indicated that when the shock wave converged the pressure exponentially reached maximal value. The pressure jump across the converging shock waves measured at R = 15 mm, 30 mm and 45 mm increased linearly toward the center. However, the pressure increased exponentially toward the center.

Immediately after convergence, the converged shock wave turned into a diverging shock wave the. The peak pressure started to decrease. Then the diverging shock wave was reflected from the 90° bend and its reflected shock propagated in the reverse direction and converged again at the center. The pressure spikes at the center following the first maximal pressure were caused by these waves.

5.7.2.1 Mode 0

Figure 5.46 shows the evolution of a converging shock wave of m = 0 for Ms = 1.50 in atmospheric air at 289.0 K. The shock waves were not perturbed and converged toward the enter as seen in Fig. 5.46a–c. It was noticed that while

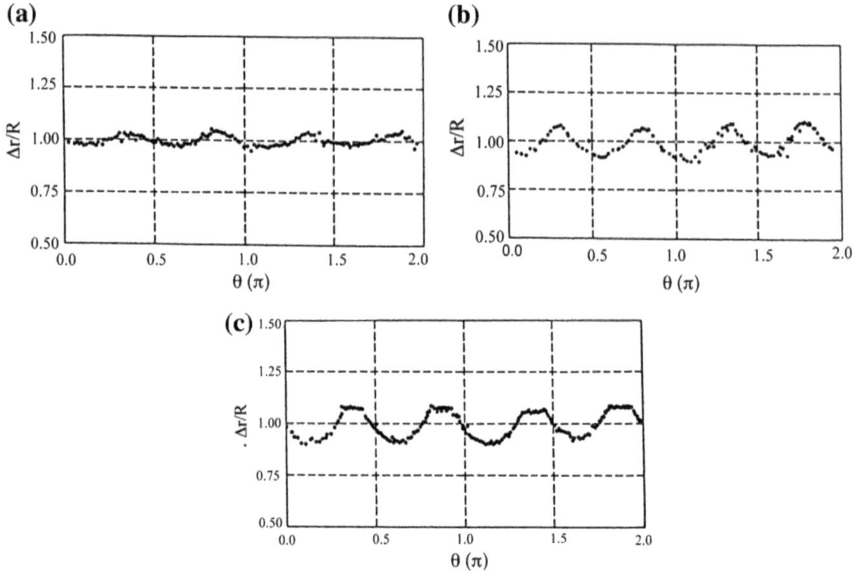

Fig. 5.44 The formation of mach reflections during the shock wave convergence for Ms = 2.0 in air, refer to Fig. 5.39: **a** R = 12.0 mm; **b** R = 5.7 mm; **c** R = 3.0 mm (Watanabe 1993)

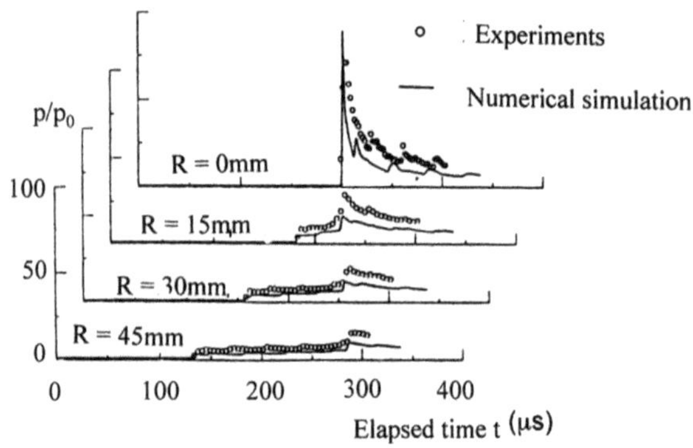

Fig. 5.45 Time variation of pressures measured at R = 0, 15, 30, and 45 mm for Ms = 1.50 in atmospheric air at 289.0 K referred from Fig. 5.43 (Watanabe 1993)

converging, concentric fringes appeared and gradually increased in their number increases. The test section was designed similarly to that in the horizontal shock tube and its interval was exactly set 10.0 mm ± 0.01 mm. However, the outside wall of the the test section was not rigidly fixed to the main shock tube structure. When the converging shock wave was focused in the test section and the high pressure was maintained for a short time, say for 100 μs, the interval of the test section would be widened very slightly, probably the width of the laser beam wave length. The first exposure was conducted at long time before the arrival of the shock wave at the test section and the second exposure was conducted when the shock

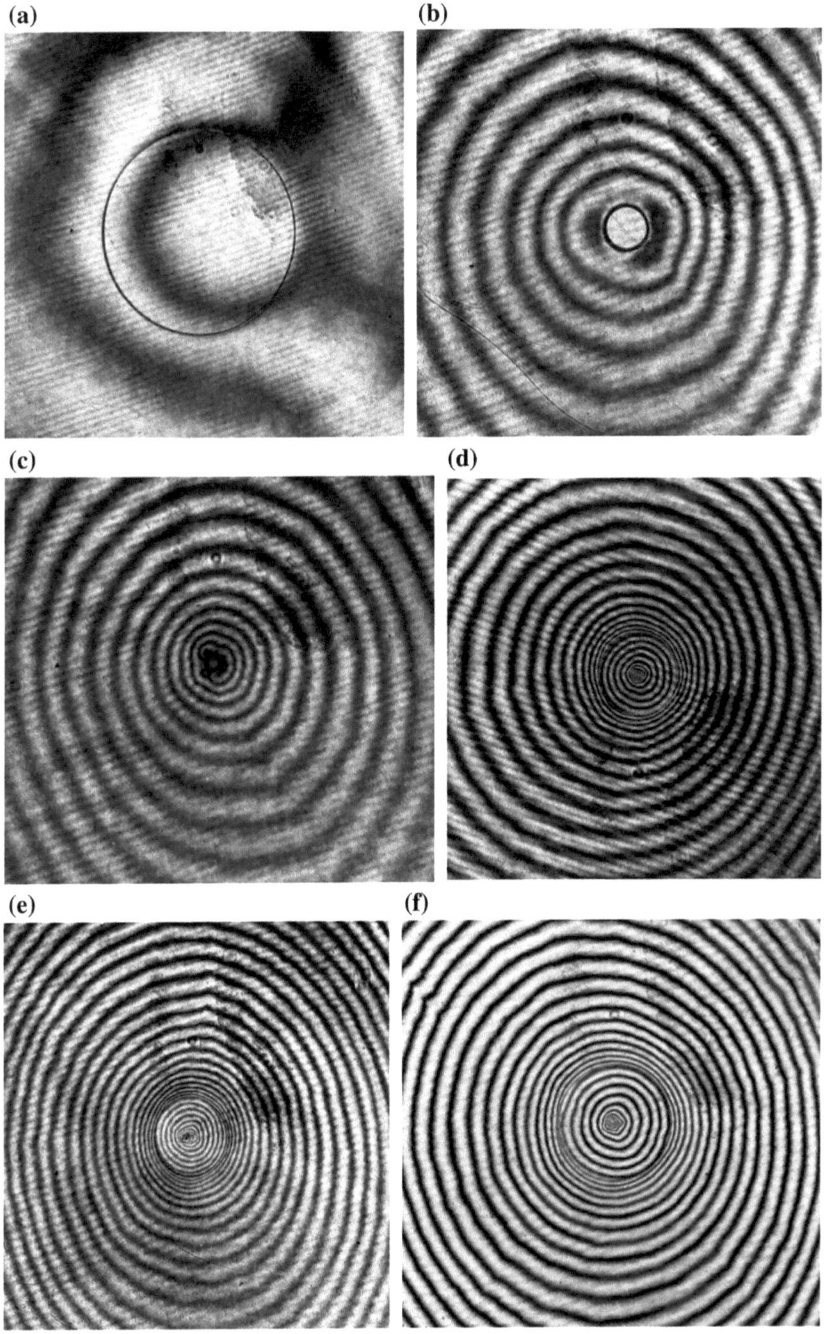

Fig. 5.46 The evolution of an annular co-axial shock wave in a vertical shock tube for Ms = 1.50 in atmospheric air at 289.0 K, m = 0: **a** #92112908 Ms = 1.53; **b** #92112902, Ms = 1.52; **c** #92112906, Ms = 1.54; **d** #92112802, Ms = 1.57; **e** #92112903, Ms = 1.54; **f** #92112805, Ms = 1.53; **g** enlargement of (**f**)

(g)

Fig. 5.46 (continued)

wave converged inside the test section. Then the change in the phase angle created the undesired co-axial fringes.

This effect would be a drawback of the present vertical shock tube. In order to overcome this drawback the outer wall should be made of massive metal frame and should be supported by an independent mechanism of the vertical channel. Anyway since the time variation in the fringe number was measurable, in the future experiments, the fringes thus created due to the out wall deformation would be deleted by introducing computer assisted image processing systems.

In Fig. 5.46d, e, the converging shock wave was supposed to converge in the absence of without creating any initial disturbances. So far during the observation until very last stage of the convergence, the fringe patterns looked symmetrical. However, Fig. 5.46f showed that the fringes just behind the diverging shock wave showed asymmetry. Figure 5.46g shows its enlargement. Although a symmetric fringe distribution was observed, the convergence of the shock wave having the mode zero may not be an easy task to achieve.

5.7.2.2 Mode 2

Figure 5.47 shows the convergence of the shock wave having mode two (m = 2) disturbance for Ms = 1.50 in atmospheric air at 290.0 K. Figure 5.47e shows the enlargement of Fig. 5.47d.

(a) (b) (c) (d) (e)

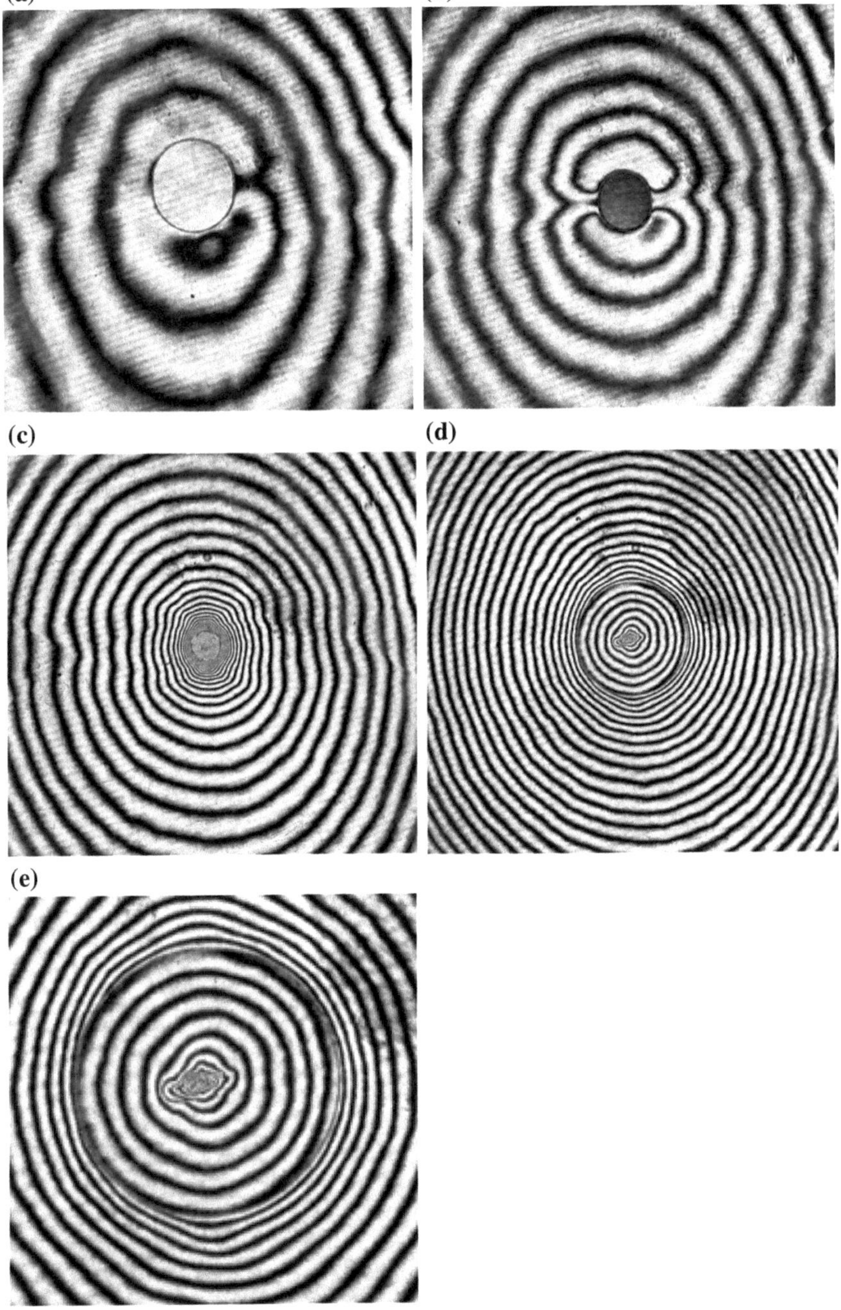

Fig. 5.47 The evolution of converging shock waves having m = 2 for Ms = 1.50 in atmospheric air at 290.0 K: **a** #92112917 Ms = 1.53; **b** #92112914 Ms = 1.54; **c** #92112912 Ms = 1.55; **d** #92112909, Ms = 1.55; **e** enlargement of (**d**)

5.7.2.3 Mode 3

Figure 5.48 shows the evolution of converging shock wave having m = 3 for Ms = 1.59 in atmospheric air 290.0 K.

5.7.2.4 Mode 4

Figure 5.49 shows the evolution of converging shock wave having m = 4 for Ms = 1.50 in atmospheric air 290.0 K. Figure 5.49e shows the enlargement of Fig. 5.49d. Behind the diverging shock wave, the four pairs of vortices were observed.

5.7.2.5 Mode 6

Figure 5.50 shows the evolution of converging shock wave having m = 6 for Ms = 1.50 in atmospheric air 290.0 K. Figure 5.50e shows the enlargement of Fig. 5.50d. Behind the diverging shock wave, the irregularly shaped remnant of vortices were observed.

5.7.2.6 Mode 8

Figure 5.51 shows the evolution of converging shock wave having m = 8 for Ms = 1.50 in atmospheric air. It should be noticed that the speed of shock front perturbations differs depending on the mode number. The results will be summarized in Fig. 5.54.

5.7.2.7 Mode 12

Figure 5.52 shows the evolution of the converging shock waves having m = 12 for Ms = 1.50 in atmospheric air at 290.0 K.

5.7.2.8 Mode 24

Figure 5.53 shows the evolution of the converging shock waves having m = 24 for Ms = 1.50 in atmospheric air at 290.0 K.

Figure 5.54 summarizes the results from the interferometric observations shown in Figs. 5.46, 5.47, 5.48, 5.49, 5.50, 5.51, 5.52 and 5.53. Variations of the deviated radius of the converging shock waves, Δr from the averaged radius R are measured and normalized by the R for the mode numbers m = 2, 4, 8, 12, and 24. The ordinate denotes Δr/R and the abscissa denotes the radius R in mm. Red filled

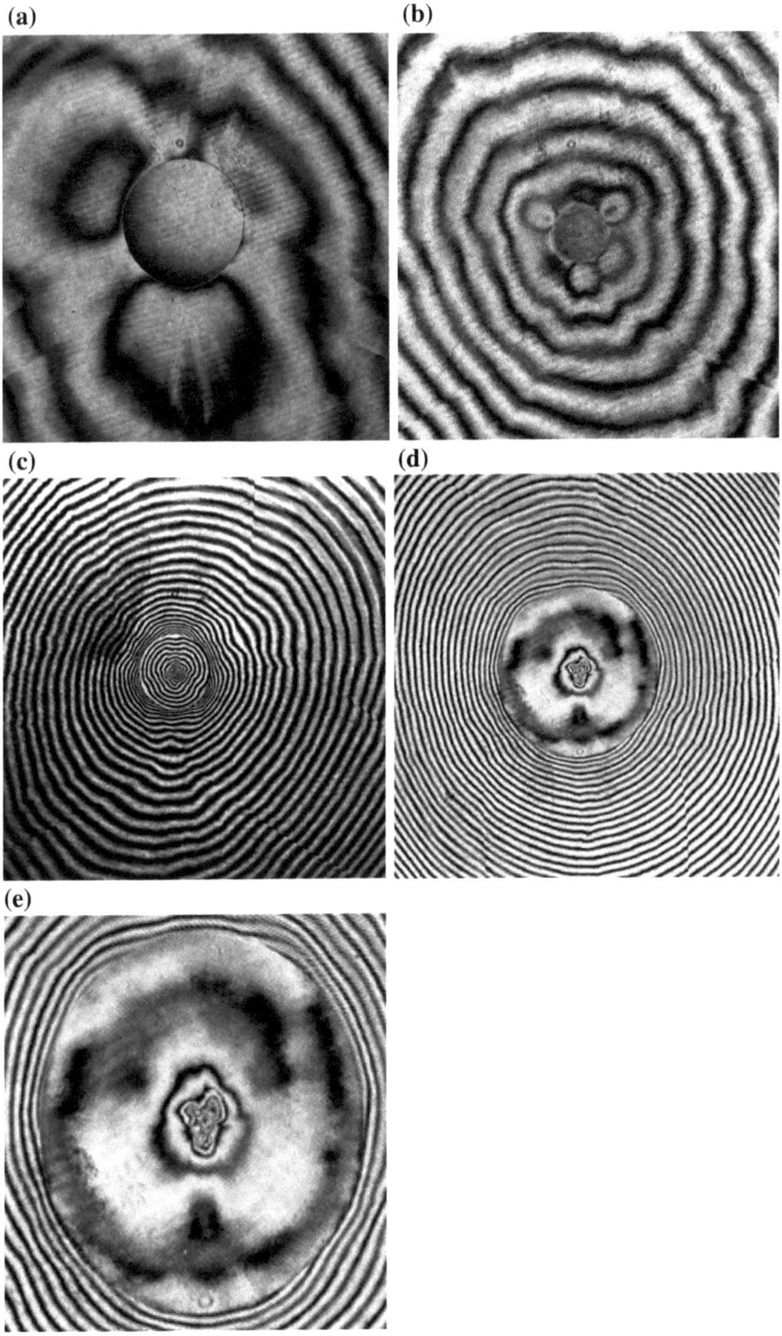

Fig. 5.48 The evolution of converging shock waves having m = 3 for Ms = 1.59 in atmospheric air 290.0 K: **a** #92120107, Ms = 1.51; **b** #92120212, Ms = 1.51; **c** #92120109, Ms = 1.53; **d** #92120118, Ms = 1.54; **e** enlargement of (**d**)

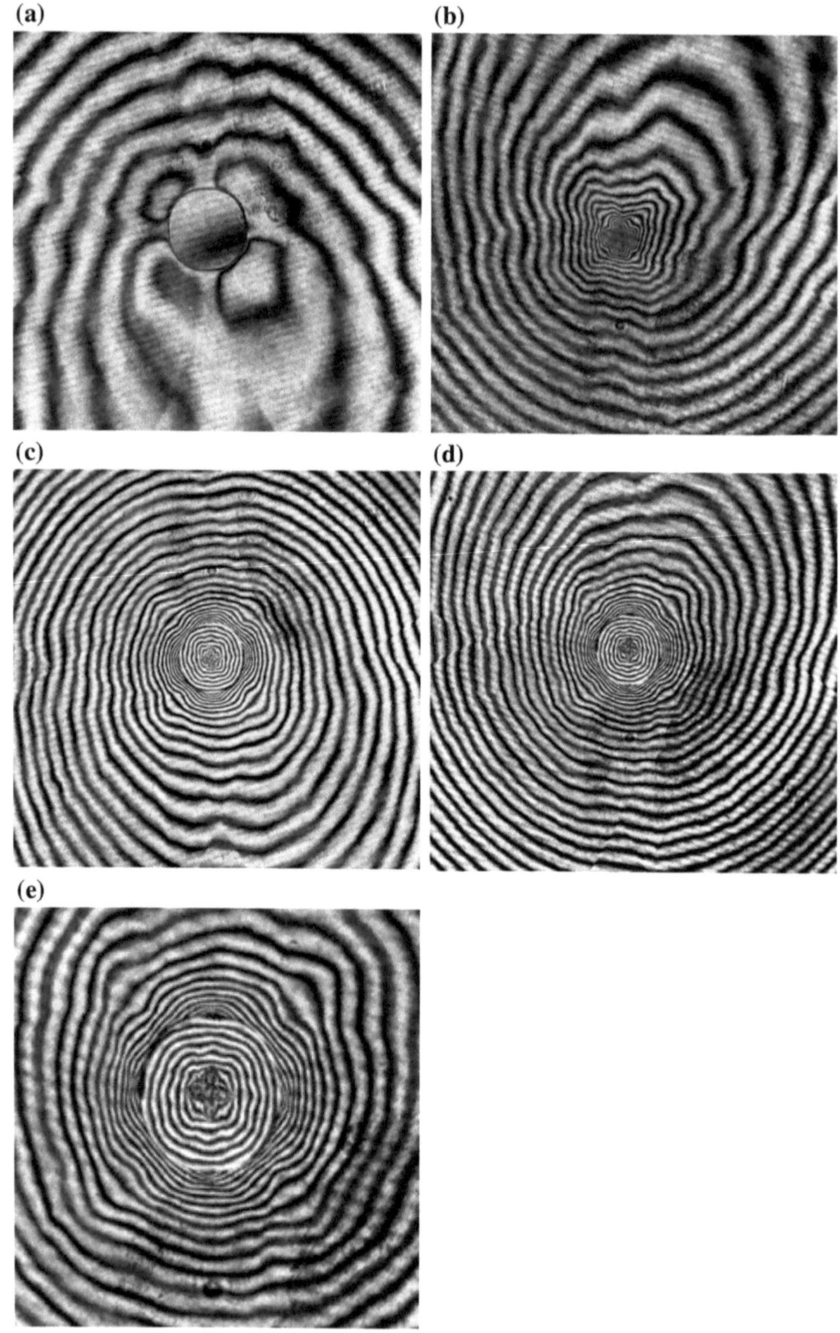

Fig. 5.49 The evolution of converging shock waves having m = 4 for Ms = 1.50 m = 4, in atmospheric air at 290.0 K: **a** #92112925 Ms = 1.55; **b** #92112924 Ms = 1.53; **c** #92112922 Ms = 1.55; **d** #92112923 Ms = 1.55; **e** enlargement of (**d**)

(a) (b) (c) (d) (e)

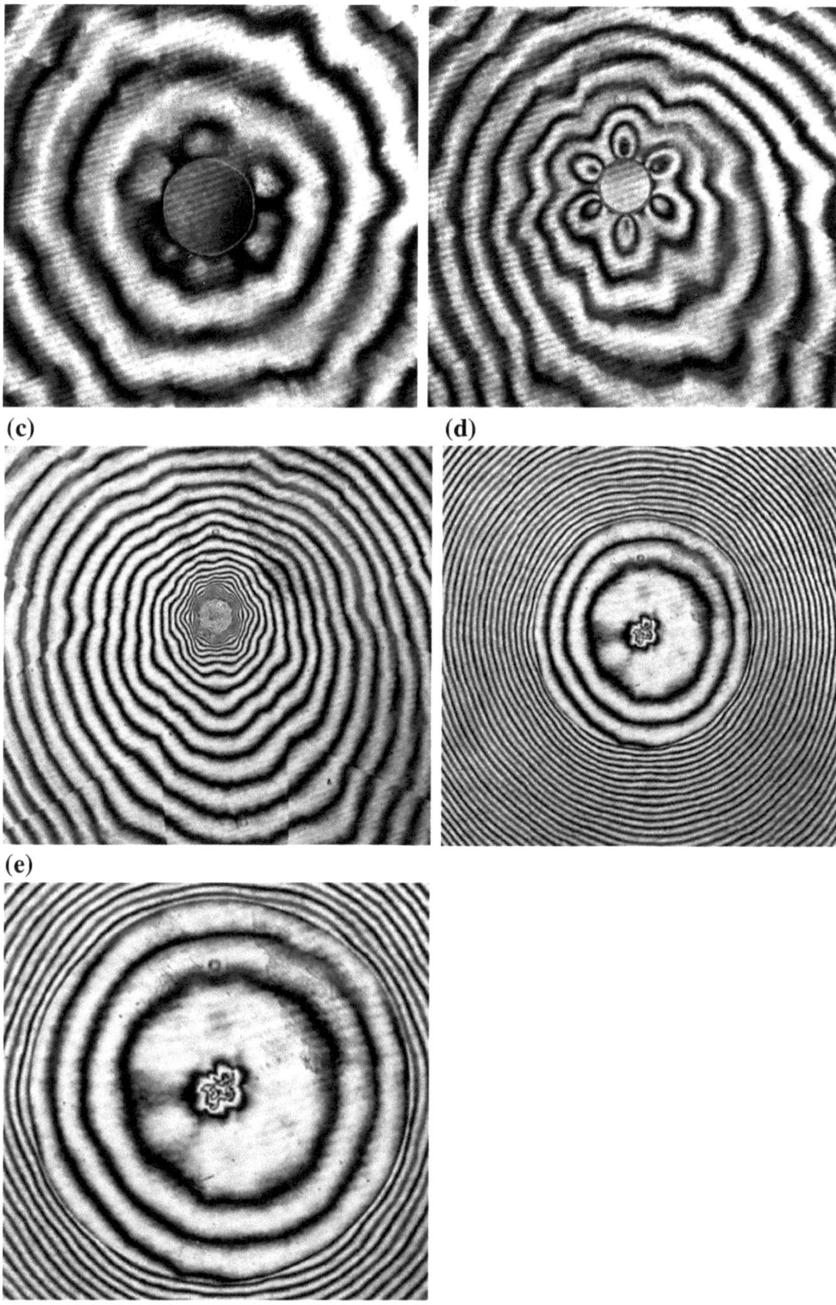

Fig. 5.50 The evolution of the converging shock waves having m = 6, for Ms = 1.50 in atmospheric air at 290.0 K: **a** #92113028, Ms = 1.52; **b** #92113027, Ms = 1.56; **c** #92113019, Ms = 1.54; **d** #92113023, Ms = 1.57; **e** enlargement of (**d**)

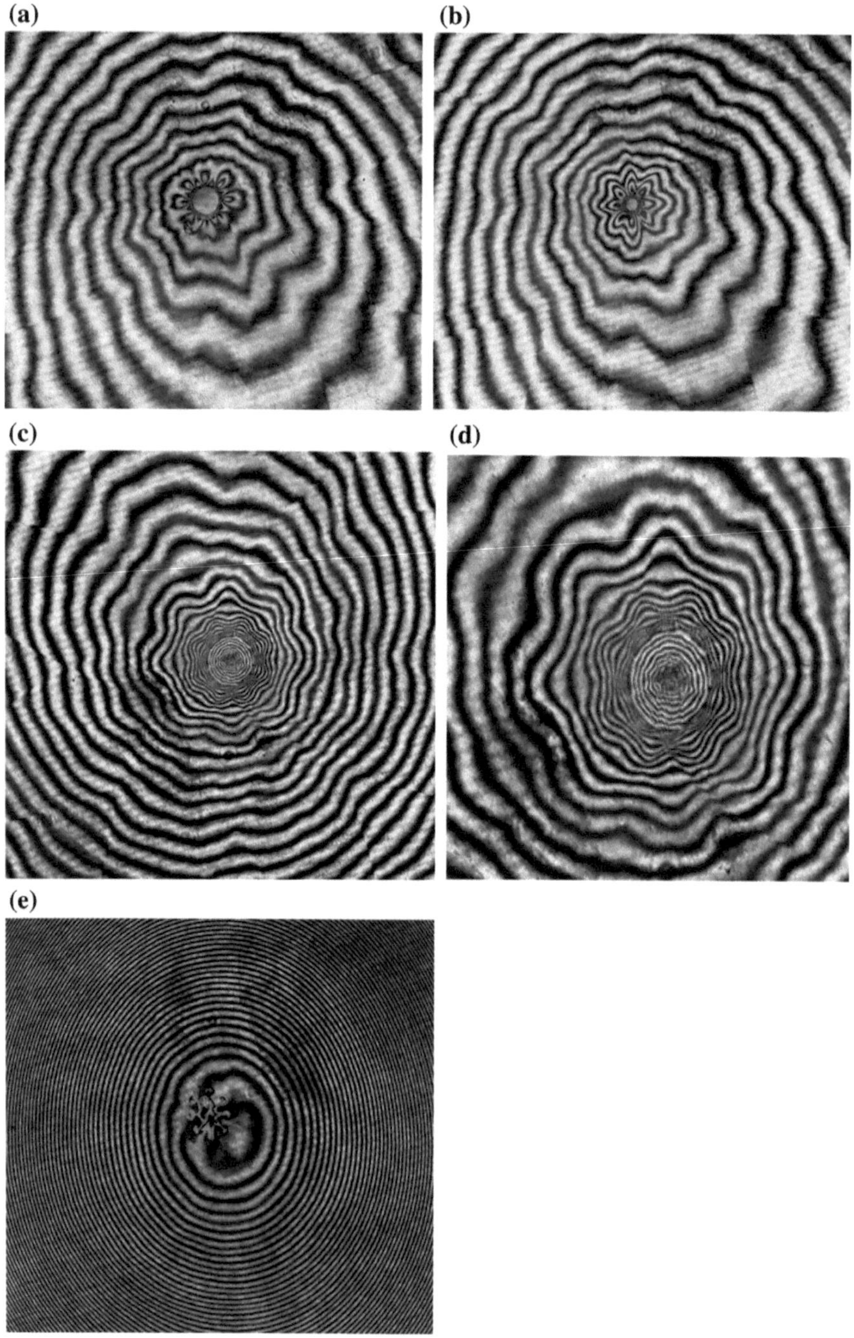

Fig. 5.51 The evolution of the converging shock waves having m = 8 for Ms = 1.50 in atmospheric air at 290.0 K: **a** #92112935, Ms = 1.54; **b** #92112932, Ms = 1.53; **c** #92112930. Ms = 1.52; **d** enlargement of (**c**); **e** #92112929, Ms = 1.67

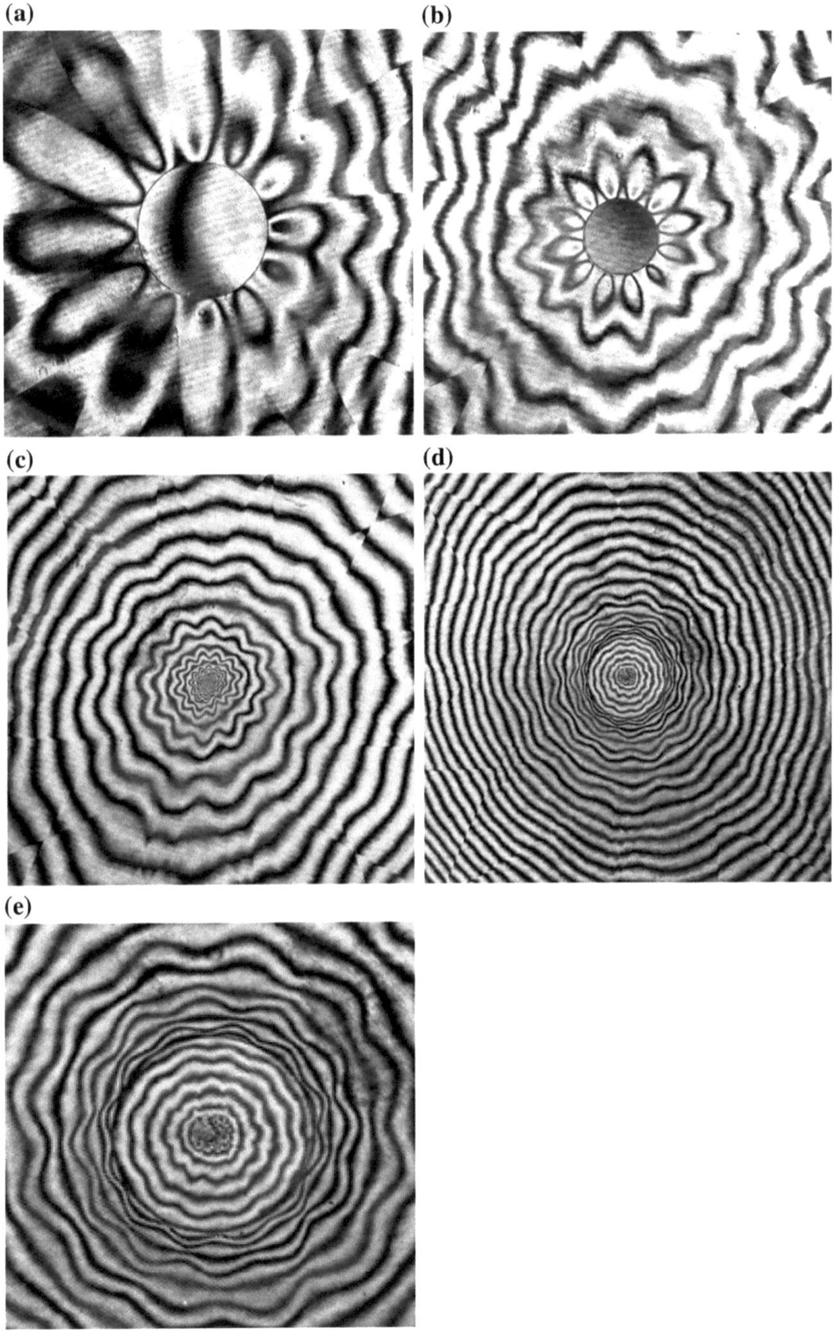

Fig. 5.52 The evolution of the converging shock waves m = 12 for Ms = 1.50 in atmospheric air at 290.0 K: **a** #92113010, Ms = 1.52; **b** #92113009, Ms = 1.60; **c** #92113016, Ms = 1.53; **d** #92113015, Ms = 1.53; **e** enlargement of (**d**)

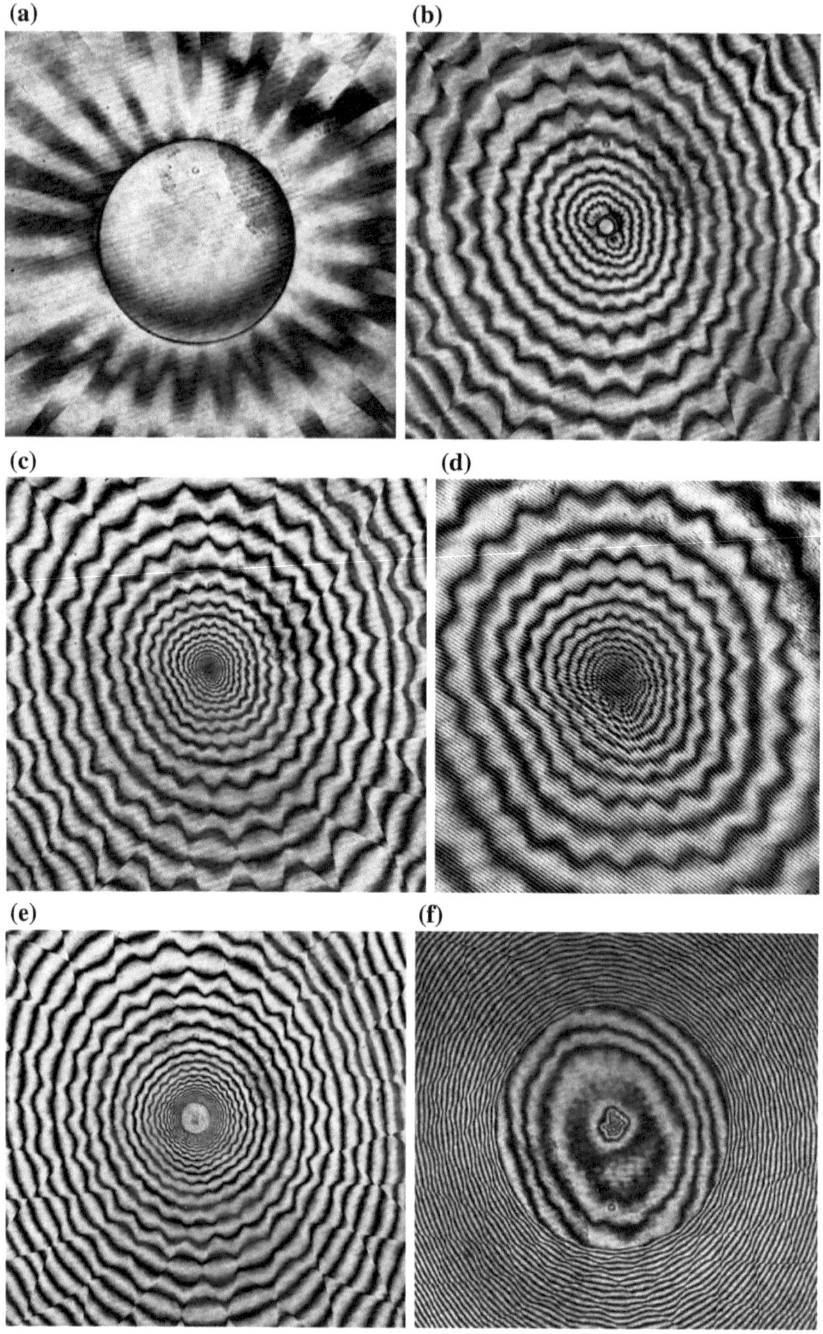

Fig. 5.53 Evolution of converging shock waves having m = 24 for Ms = 1.50 m = 24, in air at 1013 hPa, 290.0 K: **a** #92112937 Ms = 1.52; **b** #92113007 Ms = 1.53; **c** #92113003 Ms = 1.53; **d** enlargement of (**c**); **e** #92113004 Ms = 1.52; **f** #92113005 Ms = 1.53

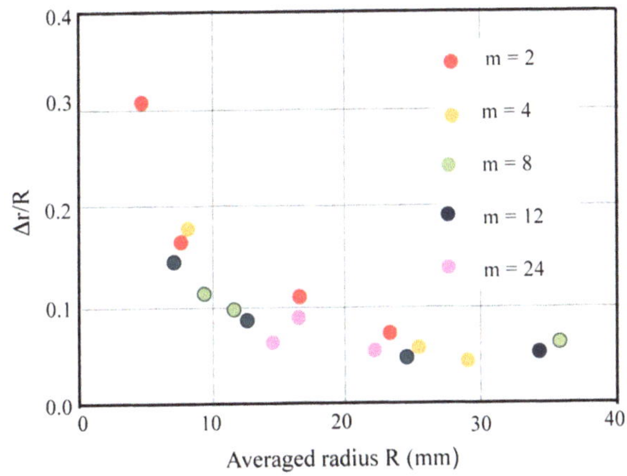

Fig. 5.54 Growth of converging shock wave deformation along radius for Ms = 1.50 (Watanabe 1993)

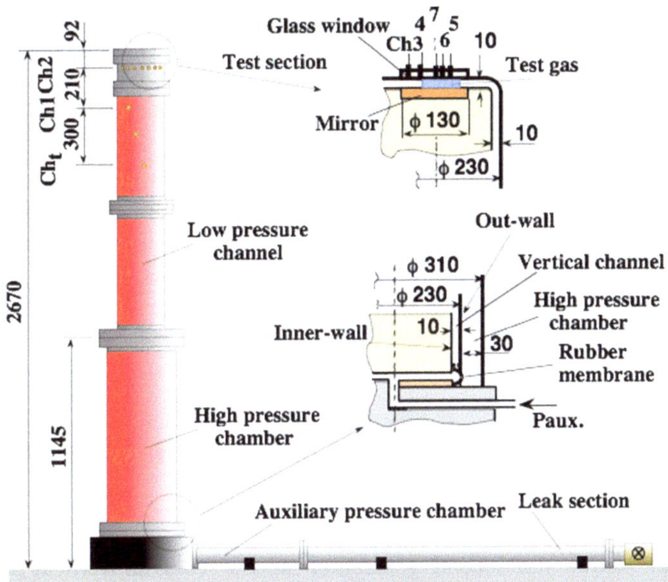

Fig. 5.55 A second generation vertical shock tube in which a rubber membrane serves as a moving diaphragm (Hosseini et al. 1997)

circles refer to m = 2, pink filled circles refer to m = 4, green filled circles refer to m = 8, dark blue filled circles refer to m = 12, and pale pink filled circles refer to m = 24. In general, the smaller the mode number is, the larger Δr/R becomes while approaching toward the center of convergence. In the case of m = 24, the remnant of the vortices at the center as seen Fig. 5.53e, is smaller than that seen in the case

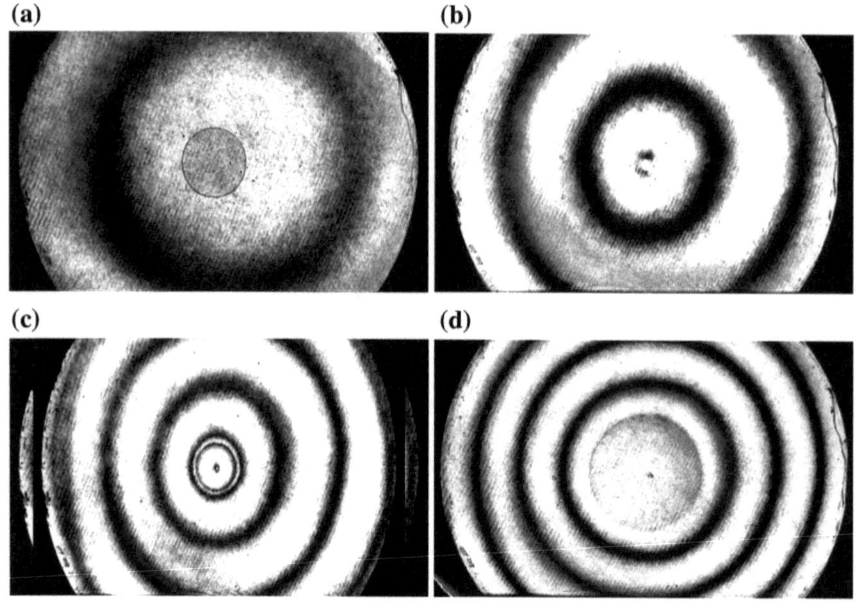

Fig. 5.56 the evolution of the converging shock wave having m = 0 for Ms = 1.50 m = 0 in atmospheric air: **a** 322 μs from trigger time; **b** 332 μs; **c** 342 μs; **d** 363 μs, (Hosseini et al. 1997)

of m = 6, seen in Fig. 5.50e. This indicates that the deformation Δr/R induced by individual mode numbers increases differently. The deformation induced by the larger m develops relatively slowly. This observation empirically agreed with Terao's observation (1973) that the detonation waves converged in a very stable manner. The detonation wave had many cellular structures distributed on its surface, this means that the mode number of the disturbances are numerous.

5.7.3 *Vertical Annular Co-axial Diaphragm-Less Shock Tube*

Figure 5.55 shows the second generation annular co-axial vertical shock tube constructed in late 1990s. The first generation vertical shock tube shown in Fig. 5.43 successfully eliminated mode number four, which was inherited in the horizontal structure. However, the vertical structure was so delicately designed that unexpected co-axial fringes were generated. Hence, the second generation vertical shock tube had a massive base and was supported rigidly. The high pressure driver chamber and the low pressure channel were sealed with a ring shaped rubber membrane which was a replacement of the polycarbonate ring shaped piston used in Fig. 5.43. The membrane was bulged loading with auxiliary high pressure helium from the other side. Upon the sudden release of the auxiliary helium, the rubber

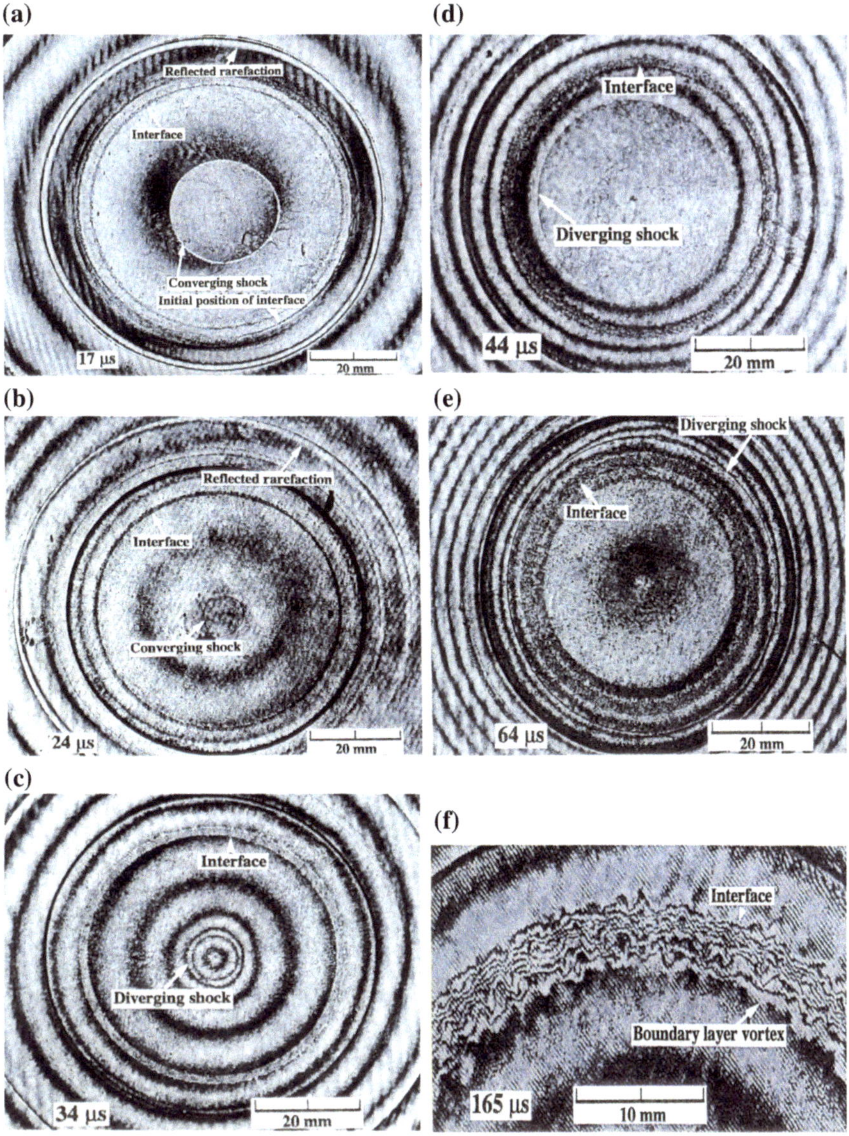

Fig. 5.57 Converging shock wave interaction with helium column for Ms = 1.21 in atmospheric air in helim at 1017.8 hPa: **a** 17 μs; **b** 24 μs; **c** 34 μs; **d** 44 μs; **e** 64 μs; **f** 165 μs. Enlarged view (Hosseini et al. 1997)

membrane shrank discharging the driver gas instantaneously into the channel. The annular co-axial shock wave was built up while propagating along the channel, turned 90° at the corner, and became a converging shock wave. The increase in number of fringes induced by the movement of the upper wall was minimized.

Nevertheless, this shock tube experienced slight vibrations during its operation at high pressures. Then the attainable shock wave Mach number, Ms was 2.5 in air. The optical arrangement was similar to the Twyman-Greene interferometry applied already in the previously described vertical shock tube shown in Fig. 5.43.

Figure 5.55 shows the evolution of converging shock wave using the vertical shock tube shown in Fig. 5.54 having mode number zero, m = 0, for Ms = 1.50 m = 0 in air. Figure 5.55a, b show perfectly cylindrical shock wave converges. Any signs of asymmetric wavelets were observed. At later stage in Fig. 5.55c, d, during the shock wave convergence and reflection a small dark spot was observed behind the diverging shock wave in Fig. 5.55c. This is a remnant of a density concentration occurred at 342 μs from trigger point. In Fig. 5.55d after 21 μs from Fig. 5.55c, the sign of the density concentration became faint. The diameter of the density concentration was very localized. Anyway the shock wave focused nearly perfectly at the center (Fig. 5.56).

5.7.3.1 Concentric Helium Column

Figure 5.57 shows the interaction of converging shock wave (Ms = 1.21) in atmospheric air with a 50 mm diameter helium column placed in a concentric position. The helium column was made by blowing a soap column with slightly pressurized helium at 1017.8 hPa.

When the converging shock wave impinged on the helium column, a shock wave was transmitted into the helium and an expansion wave was reflection from the helium interface as seen in Fig. 5.57a, b. The transmitted shock wave converged and became a diverging shock wave in Fig. 5.57c, d. The interface is gradually broadened with elapsed time. At the convergence early time, a remnant of density peak was still observable in Fig. 5.57e. Figure 5.57f shows a magnified view at later stage. The broadened interface did not show two-dimensional jagged surface but three-dimensionally deformed interface. This experiment motivates investigating the Richtmyer-Meshkov instability. The time attached to each picture indicates the elapsed time from the moment of the shock wave impingement on the helium column.

5.7.3.2 Eccentric Helium Column

For investigating the Richtmyer-Meshkov instability, a series of experiments were conducted using a vertical shock tube (Hosseini et al. 1997). Figure 5.58 shows the experimental arrangement of a converging shock wave interaction with a helium column positioned in an eccentric positon. The helium column was produced in the same way as seen in Fig. 5.57.

Figure 5.59 shows 50 mm diameter helium column impinged by a converging shock wave of Ms = 1.18 in atmospheric air. The acoustic impedance in air is 2.5 times as larger as that in pure helium. Although the helium column was

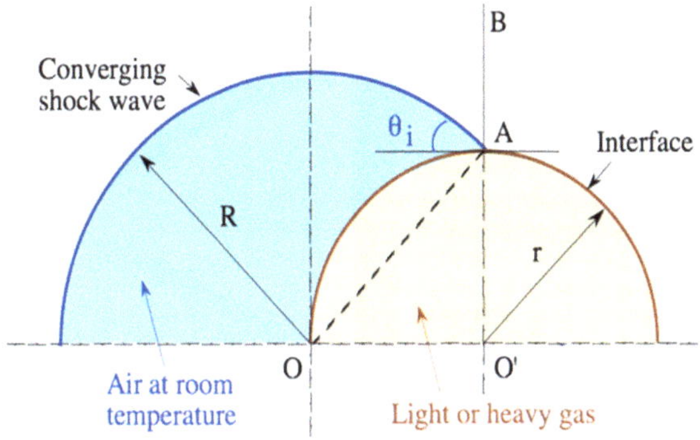

Fig. 5.58 Illustration of shock wave helium column interaction

contaminated by air presumably in at most 20% in volume, the weak shock waves in are reflected wave from the air/helium interface as an expansion wave and the transmitting wave in helium is as a shock wave. Hence the transmitted sock wave in helium is going to converge. Figure 5.59a shows a wave pattern at 30 μs after the impingement of converging shock wave with the helium column. Figure 5.59b shows its enlargement. Figure 5.59e shows a wave pattern at 75 μs after the shock wave impingement. The downstream side of the helium column moved to the center of convergence. At the same time, the helium interface contracts and its center slowly moves toward the center of the convergence. The transmitting shock wave in helium came out of the interface and going to converge. Hence at 177 μs, complex wave interactions occur simultaneously.

5.7.3.3 Focusing of a Transmitted Shock Wave Diffracting Over a Backward Facing Wall

In order to achieve a stable convergence, another compact vertical annular co-axial diaphrgam-less shock tube was constructed. Its photograph is presented in Fig. 5.60a, b. A rubber membrane was used for sealing the driver gas and the test gas. In this revised shock tube, the rubber membrane simply moved up and down between two curved grids as illustrated in Fig. 5.60a. Auxiliary high pressure helium bulged the rubber membrane making it attach tightly onto the upper grid. Then the test gas was sealed completely from the driver gas. When the auxiliary gas was quickly reduced, the membrane left from the upper grid and moved onto the lower grid. Meantime the high pressure driver gas rushed vertically into the low pressure channel. The space between two grids was wide enough to minimize pressure losses that usually occur at the diaphragm section of conventional shock tube. Figure 5.60b shows a photograph of the compact vertical shock tube. The exit was coated with fluoresce paint in order to conduct diffuse holographic observation. The height of the low pressure channel was about 1 m.

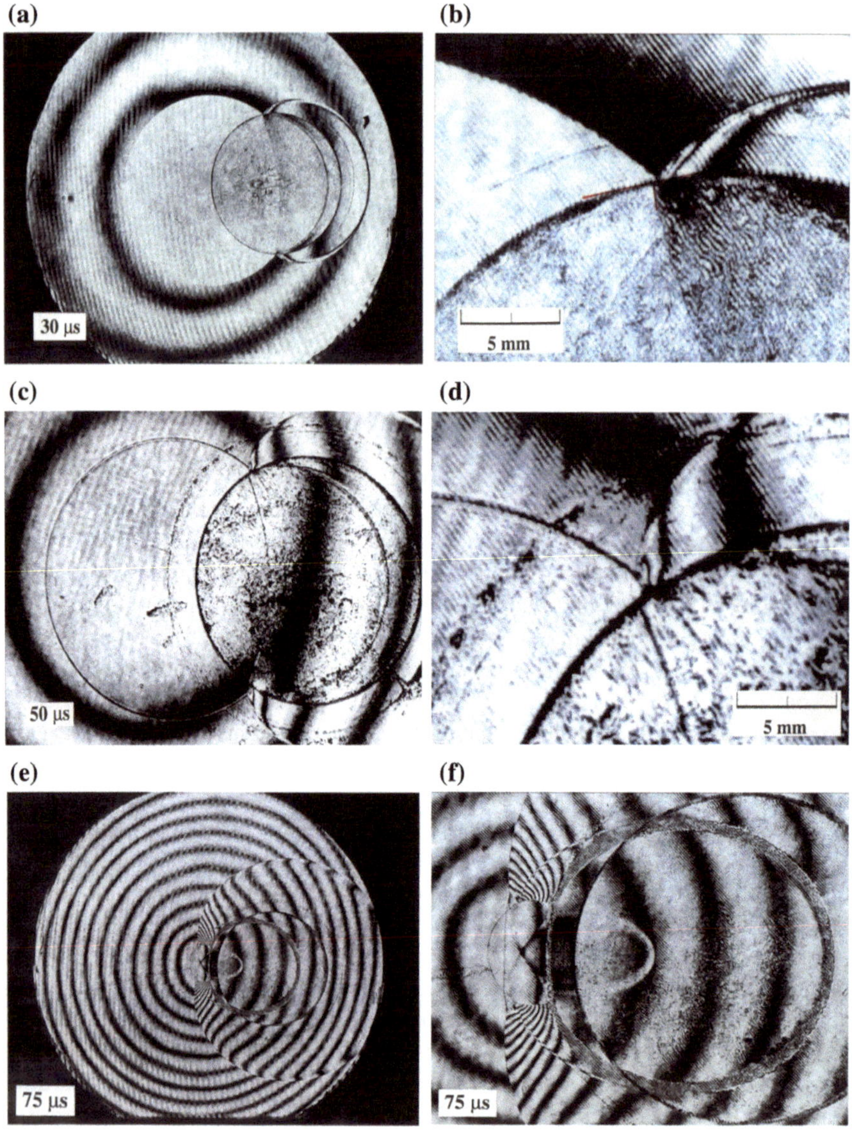

Fig. 5.59 The evolution of the converging shock wave interaction with 50 mm diameter helium soap bubble, Ms = 1.18 in atmospheric air: **a** 30 μs after impingement; **b** enlargement of (**a**); **c** 50 μs; **d** enlargement of (**c**); **e** 75 μs; **f** enlargement of (**e**); **g** 177 μs; **h** 345 μs

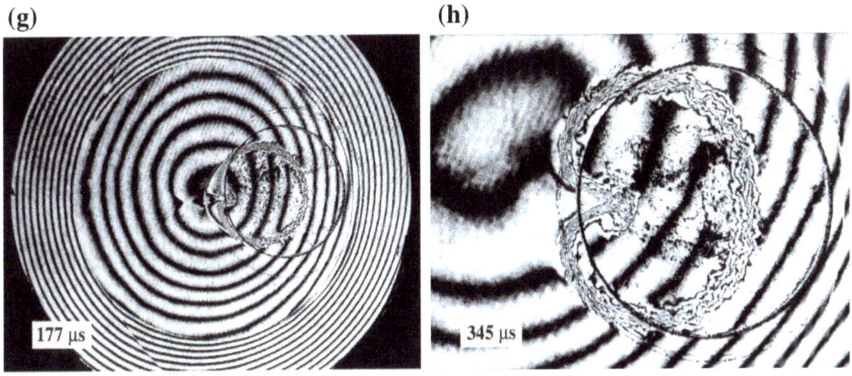

Fig. 5.59 (continued)

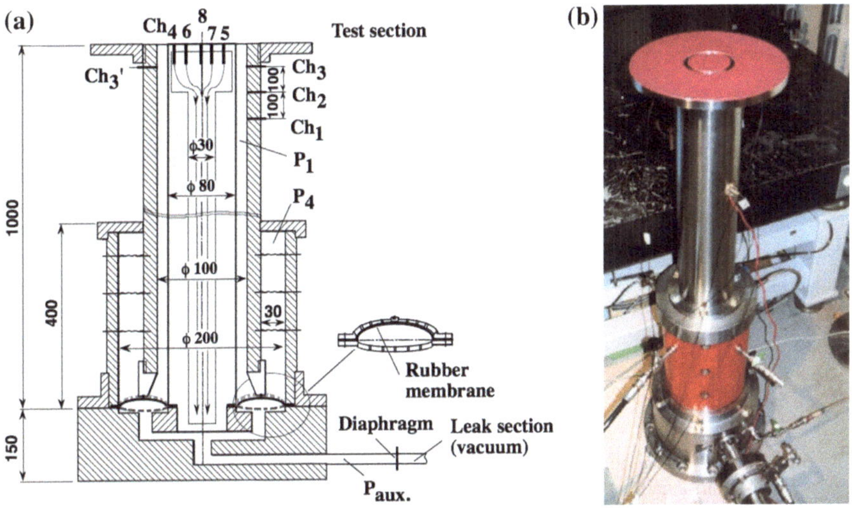

Fig. 5.60 Compact vertical annular shock tube and test section: **a** 80 mm × 100 mm vertical shock tube; **b** Photo of the vertical shock tube (Hosseini et al. 1999)

Figure 5.61 shows characteristic performance of the compact vertical shock tube. The ordinate denotes dimension-less high pressure. The abscissa denotes Ms measured at the end of the low pressure channel. A red broken line denotes numerical results obtained from the simple shock tube theory Gaydon and Hurle (1963). Black filled circles denote measured Ms. In a relatively limited range of Ms, an excellent agreement between the results is evident in Fig. 5.61 (Hosseini et al. 1999).

Figure 5.62 shows the propagation of the shock wave released from a 10 mm wide ring shaped opening into a 100 mm diameter cylindrical tube. Figure 5.62a, b shows the experimental arrangement and the aspheric lens shaped test section.

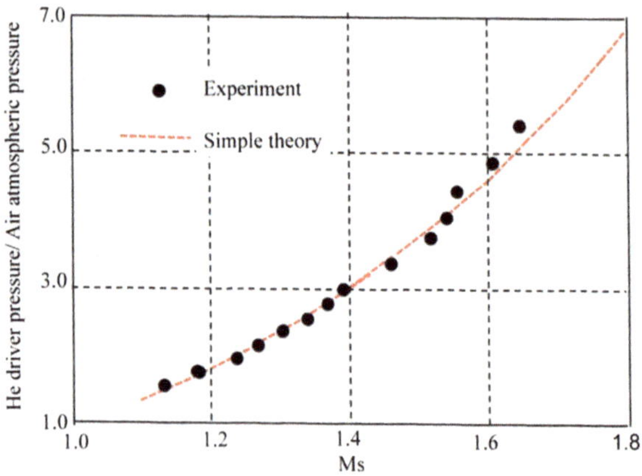

Fig. 5.61 Performance of the vertical compact shock tube shown in Fig. 5.59a (Hosseini et al. 1999)

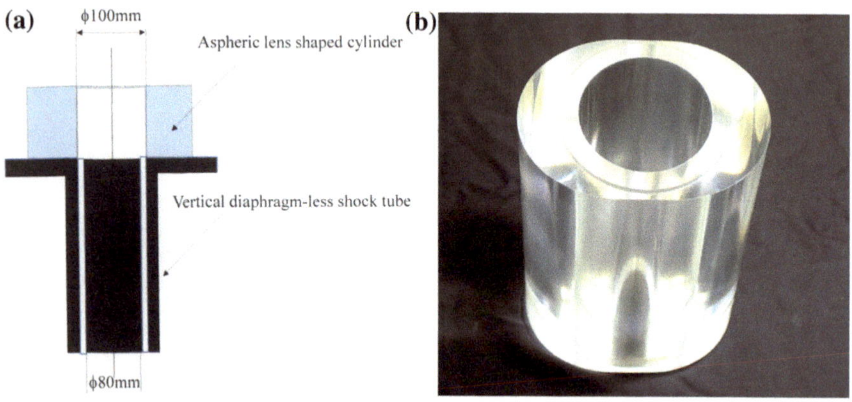

Fig. 5.62 Shock wave focusing released from a 10 mm ring shaped opening shown in Fig. 5.59b: **a** experimental arrangement; **b** 100 mm diameter aspheric lens (Hosseini et al. 1999)

In Chap. 3, the two-dimensional shock wave diffraction over a backward facing step was discussed. Figure 5.63 shows the evolution of an axially symmetric shock wave diffracting at the backward facing step for Ms = 1.50 in atmospheric air. In Fig. 5.63a, the images of the transmitted shock waves were superimposed on each other resulted in complex fringe distributions. Nevertheless, the shape of a diffracting shock wave and the resulting formation of corner vortices are well resolved. The circular diffracting shock waves approached toward the center as seen in Fig. 5.62c, d. When the circular diffracting shock waves merged at the center, the angle of their merger is shallow and hence the resulting reflected shock wave pattern is a RR. With elapsed time, the angle at which the reflected shock wave merged gradually increased and the transition to a MR occurred. In Fig. 5.63f the reflection pattern is a MR.

In Chap. 3, the shock wave diffractions from openings of various shapes were visualized using diffuse holographic interferometry. As seen in Fig. 5.60b, the shock wave diffracted from a 10 mm wide circular opening composed of 100 mm outer diameter and 80 mm inner diameter would form at first a toroidal shock wave. With elapsed time, the shock wave would be a diffracting shock wave from a circular cross sectional tube at discussed in Chap. 3. However, the inner part of the toroidal shock wave would converge toward the center of the test section.

The quantitative visualization of three-dimensional shock waves is not simple. Then it was decided to apply diffuse holographic interferometry already applied to shock wave diffraction experiments already shown in Fig. 3.19. The shock tube and its flange were coated with the pink color fluorescent paint as shown in Fig. 5.60b. Then diffuse holographic observations were conducted for Ms = 1.50 in atmospheric air by oblique illumination over the coated area with the *OB*. The reflection of the *OB* illuminated a holographic film. The test section was obliquely illuminated twice with the diffused *OB* at appropriate time interval, then double exposure diffuse holographic interferometry was completed (Figs. 5.61, 5.62).

Figure 5.63 shows reconstructed imaged showing the motion of transmitted shock waves. Figure 5.63a shows the shock waves at 30 μs from the trigger point. This indicated the time instant at which the second exposure was conducted. The diverging shock wave marked on the photo denotes the diffracting shock wave propagating outward. The imploding SW on the photo denotes the diffracting shock wave at the inner corner and is converging toward the center of the test section, Fig. 5.63b was taken at 50 μs after the time instant when Fig. 5.63a was taken. The initial diverging shock wave propagated further outward and a secondary shock wave followed. The converging shock wave was just going to converge at the center. Another secondary shock wave appeared behind the converging shock wave. These secondary shock waves were created as transmitted shock wave was reflected from the inner and outer wall surfaces. Figure 5.63c was taken at 16 μs after the time instant when Fig. 5.63b was taken. The diverging shock wave further propagated outward. The converging shock wave and the secondary shock wave imploded. Remnants of vortices were observed at the center. Figure 5.63d shows state long time afterward. The Magnified photo show, although very blurred, the accumulation of fringes (Fig. 5.64).

5.8 Explosion Induced Shock Wave Focusing from a Truncated Ellipsoidal Reflector

An ellipsoidal cavity has two focal points. Constructing a half truncated ellipsoidal cavity, having a geometry of the inner diameter of 135 mm and outer diameter of 190 mm and the aspect ratio is $2^{1/2}$. A spherical shock wave was generated by the explosion of a 10 mg AgN_3 pellet at the first focal point inside the truncated ellipsoidal cavity. Figure 5.65 shows the experimental setup. The resulting shock

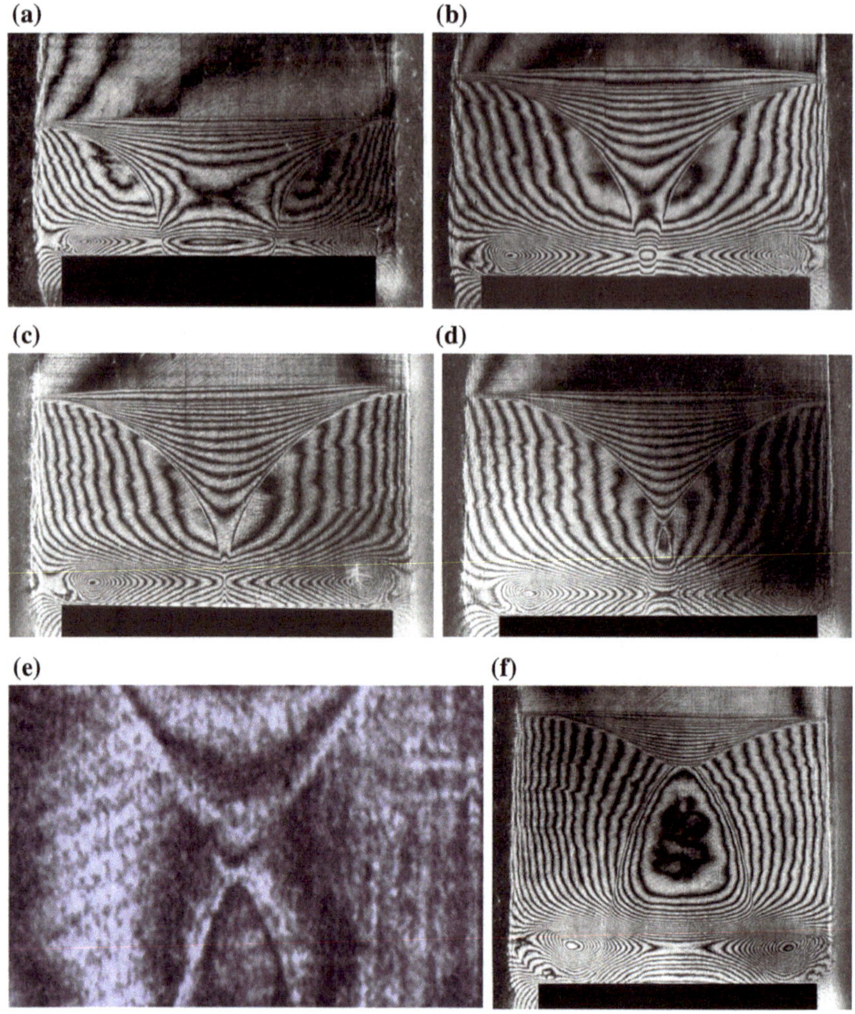

Fig. 5.63 Diffraction and convergence of a toroidal shock wave for Ms = 1.50 in atmospheric air: **a** 63 μs; **b** 92 μs; **c** 96 μs; **d** 104 μs; **e** enlargement of (**d**); **f** 148 μs

wave is reflected from the truncated ellipsoidal cavity and focuses at another focal point of outside the reflector. If it is an underwater shock wave, the resulting shock wave propagates at sonic speed and then the shock wave would focus at the second focal point. However, in air the shock wave propagates at high speed as Ms > 1, then the shock wave will not focus sharply at the second focal point.

Figure 5.66 shows the evolution of a reflected shock wave merging at an area outside the truncated ellipsoidal cavity. The shock wave propagates from right to

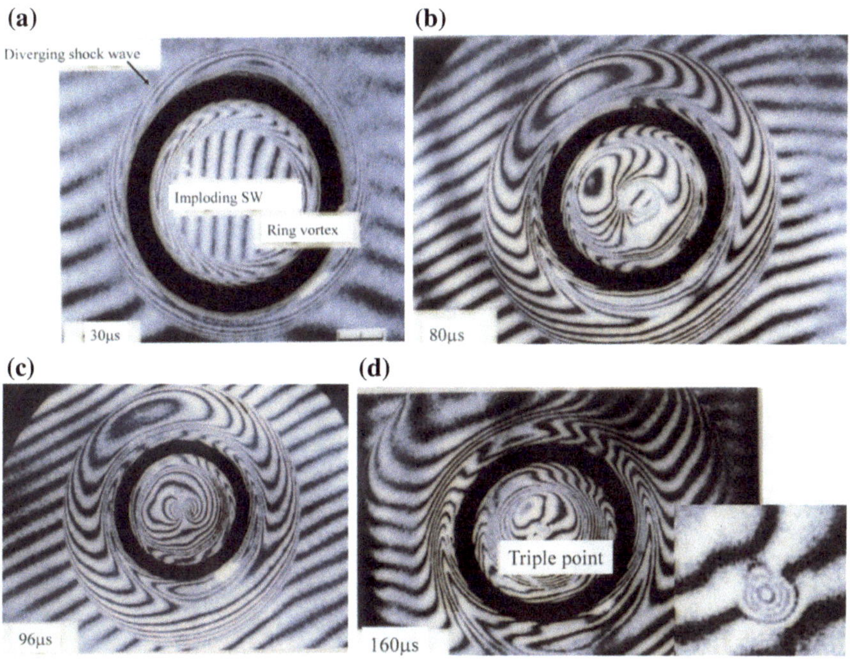

Fig. 5.64 Evolution of hemi-spherical shock wave released from a 10 mm ring shaped opening for Ms = 1.20 in atmospheric air: **a** 30 μs; **b** 80 μs; **c** 96 μs; **d** 160 μs

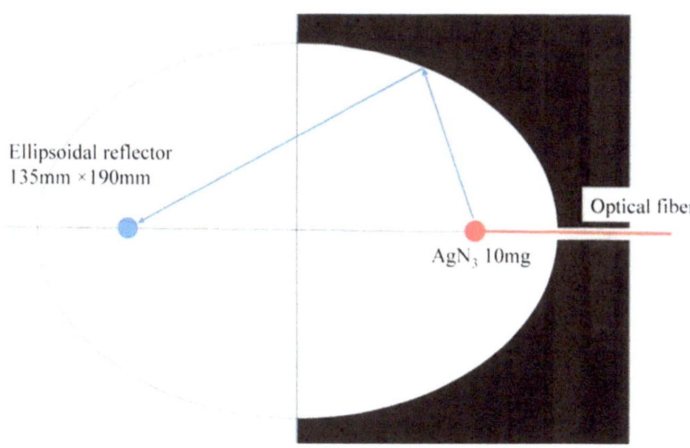

Fig. 5.65 Illustration of a spherical shock wave focusing from a truncated ellipsoidal cavity

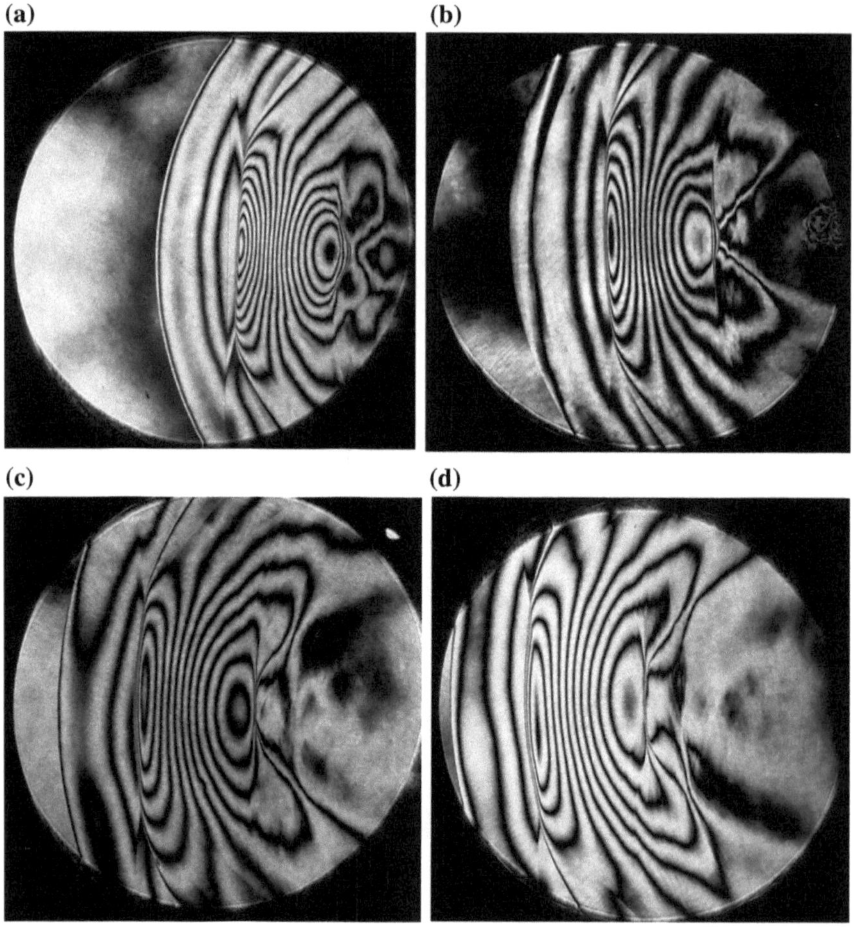

Fig. 5.66 Focusing of a spherical shock wave in air from ellipsoidal cavity, 10 mg AgN_3 in air at 294.0 K: **a** #88072508, 350 μs from trigger point; **b** #88072509, 360 μs; **c** #88072506, 450 μs; **d** #88072505, 500 μs

left. In Fig. 5.66a, the direct wave propagates much faster than the reflected wave. The reflection of the shock waves in air from the curved ellipsoidal cavity are very different from the shock wave reflection from the same shape of ellipsoidal cavity. Spherical incident shock waves with moderate strength in air will never reflect from ellipsoidal walls like spherical underwater shock waves. In Fig. 5.66a–d, the attenuation of the direct wave is clearly observed and very diversified patterns of fringes in the vicinity of the second focal point area.

References

Apazidis, N., Kjellander, M., & Tillmark, N. (2011). High energy concentration by symmetric shock focusing. In: K.Kontis, (Ed.), *Proceeding of 28th International Symposium on Shock Waves* (vol. 1, pp. 99–110). Glasgow.

Courant, R., & Friedrichs, K. O. (1948). *Supersonic flows and shock waves*. NY: Wieley Inter-Science.

Fujiwara, T., Sugiyama, T., Mizoguchi, K., & Taki, S. (1979). Stability of converging cylindrical detonation. *Journal of the Japan Society for Aeronautical and Space Sciences, 21,* 8–14.

Gaydon, A. G., & Hurle, I. R. (1963). *The shock tube high-temperature chemical physics*. London: Chapmam and Hall Ltd.

Guderley, G. (1942). Starke kugelige und zylindrische Verdichtungs Stoesse inder Naehe des kugelmittelpunktes bzw. der Zylinderachse. *Luftfahrtforschung, 19,* 302–312.

Honda, M., Takayama, K., & Onodera, O. (1977). Shock wave propagation over 90° bends, Rept Inst High Speed Mech. *Tohoku University 35,* 74–81.

Hoshizawa, Y. (1987). *Study of converging shock wave* (Master thesis). Graduate School of Engineering, Faculty of Engineering Tohoku University.

Hosseini, S. H. R., Takayama, K., & Onodera, O. (1999). Formation and focusing of toroidal shock waves in a vertical co-axial annular diaphragm-less shock tube. In: G. J. Ball, R. Hillier, & G. T. Robertz (Eds.), *Proceeding 22nd ISSW, Shock Waves* (vol. 2, pp. 1065–1070). London.

Hosseini, S. H. R., Onodera, O., Falcovitz, J., & Takayama, K. (1997). Converging cylindrical shock wave in an annular strut free diaphragm-lee shock tube. In: A. F. P. Houwing & A. Paul (Eds.), *Proceeding 21st ISSW, Shock Waves* (vol. 2, pp. 1511–1516). The Great Kepple Island.

Inoue, O., Imuta, G., Milton, B. E., & Takayama, K. (1995). Computational study of shock wave focusing in a log-spiral duct. *Shock Waves, 5,* 183–188.

Knystautus, R., & Lee, J. H. (1971). Experiments on the stability of converging and cylindrical detonation. *Combustion and Flame., 16,* 61–73.

Milton, B. E. (1989). Focusing of shock waves in two-dimensional and axi-symmetric ducts. In: K. Takayama (Ed.), *Proceeding International Workshop on Shock Wave Focusing* (pp. 155–192). Sendai.

Milton, B. E., Archer, D., & Fussey, D. E. (1975). Plane shock amplification using focusing profiles. In: G. Kamimoto (Ed.), *Proceeding 10th International Shock Tube Symposium, Modern Developments in Shock Tube Research* (pp. 348–355). Kyoto.

Neemeh, B. A., & Less, D. T. (1990). Stability analysis of initially weak converging cylindrical shock waves, In: Y. W. Kim (Ed.), *Proceedings of the 17th International Symposium on Shock Wave, Current Topics in Shock Waves* (pp. 957–962). Bethlehem.

Perry, R. W., & Kantrowitz, A. (1951). Production and stability of converging shock waves. *Journal of Applied Physics, 22,* 878–886.

Sun, M. (2005). Numerical and experimental study of shock wave interaction with bodies. (Ph. D. thesis). Graduate School of Engineering, Faculty of. Engineering Tohoku University.

Sun, M., Saito, T., Takayama, K., & Tanno, H. (2005). Unsteady drag on a sphere by shock wave loading. *Shock Waves, 14,* 3–9.

Takayama, K. (1983). Application of holographic interferometry to shock wave research. In *Internation Symposium of Industrial Application of Holographic Interferometry, Proceeding SPIE 298* (pp. 174–181).

Takayama, K. (1990). Proceedings of the international workshop on shock wave focusing, Inst High Speed Mechanics, Sendai.

Terao, K., & Sato, T. (1973). A study of a radially divergent-convergent detonation wave. In D. Bershader, & W. Griffice (Eds.), *Proceedings of the International Symposium on Shock Tube, Recent Developments in Shock Tube Research* (pp. 646–651).

Watanabe, M. (1993). Numerical and experimental study of converging shock wave (Ph.D. thesis). Graduate School of Engineering, Faculty of Engineering Tohoku University.

Whitham, G. B. (1959). A new approach to problems of shock dynamics. *Part II Three dimensional problems. JFM, 5,* 359–378.

Wu, J. H. T., Neemeh, R. A., Ostrowski, P. P., & Elabdin, M. N. (1978). Production of converging cylindrical shock waves by finite element conical contractions, In: Ahlborn, B., Hertzberg, A., Russell, D. (Eds.), *Proceeding of the 11th International Symposium on Shock Tubes and Waves, Shock Tube and Shock Wave Research* (pp. 107–114). Seattle.

Yamanaka, T. (1972). *An investigation of secondary injection of thrust vector control (in Japanese). NAL TR-286T*. Chofu, Japan.

Chapter 6
Shock Wave Mitigation

6.1 Introduction

Shock wave mitigation in air is one of the important research topics of the shock-wave research. Strong or moderate shock waves can be attenuated in a relatively straightforward manner, whereas weak shock waves take a longer process to be attenuated to sound waves. In this chapter experimental results of shock wave mitigations are presented.

6.2 Suppression of Automobile Engine Exhaust Gas Noise

The suppression of automobile engine exhaust noises (Sekine et al. 1989) and the suppression of train tunnel sonic booms (Sasoh et al. 1998) were interesting applications of shock wave research.

Matsumura visualized pressure waves built up along an exhaust gas pipe line connected to a 500 cc Subaru automobile engine operated at 2000 rpm. Gases discharged from the cylinder having high pressure and temperatures coalesced during their propagation, into weak shock waves and then entered into the manifold. Recorded overpressures in exhaust gases are about 20 kPa and the specific heats ratio of exhaust gas mixtures of air and combustion product gases and debris is 1.35; therefore, the shock wave Mach number, Ms, appearing in the exhaust pipe flow would be 1.1. This value is no longer considered as an acoustic wave but a weak shock wave. Car silencers are designed to suppress these waves were traditionally based on the acoustic theory. However, the acoustic theory would not be an appropriate approach for attenuating shock waves in the silencers.

In experiments aimed at suppressing exhaust gas pressure the 28 mm diameter exhaust pipe line of a half litter Subaru engine cylinder was connected to a 30 mm × 30 mm shock tube and a 40 mm × 80 mm shock section. The engine was operated at various rotation speeds. Sequential visualizations were conducted.

K. Takayama, *Visualization of Shock Wave Phenomena*,
https://doi.org/10.1007/978-3-030-19451-2_6

At that time, the engine rotation was synchronized with a Q-switch ruby laser irradiation. The motion of shock waves in the test section was sequentially visualized.

Figure 6.1 show holographic observation of a shock wave entering in the 40 mm × 80 mm shock tube test section. Twin vortices were generated at the corner, see Fig. 6.1a. In Fig. 6.1b–d, the transmitted shock wave propagated inside the test section and reflected from the upper and lower walls. Splashes composed of water vapors and debris contained in the exhaust gases covered the observation window glasses. The sequential images shown in Fig. 6.1 roughly simulate the condition in a real automobile engine manifold. Shapes of wall surface, wall conditions and partitions tested in the laboratory were helpful in acquisition of design data for practical silencers and mufflers (Matsumura 1995).

Compression waves in the exhaust gases coalesced into shock waves and were collected in the manifold. Then shock wave propagations in a manifold were visualized using two types of 2-D manifolds which modelled two half litter cylinders as shown in Fig. 6.2. The 2-D manifolds were sandwiched with two thick acrylic plates. In this analogue experiment, the engine produced shock waves having a strength of about Ms = 1.1 were produced by ignitions of 10 mg AgN_3 pellets which were confined in a space connected to the entry port and were sealed with a 9 μm thick Mylar diaphragm. For simulating an engine rotation speed of 2200 rpm by detonating explosives at frequency of about 13 ms interval.

Based on experimental data, the silence and mufflers of Subaru Legacy in 1993 version was designed. This was the first successful application of shock tube experiments for a practical automobile design. Today, prototype silencers and mufflers were designed numerically using a fine simulation scheme. The schemes were once quantitatively validated by comparing its predictions with the interferometric images.

6.3 Train Tunnel Sonic Boom

Japan is a mountainous country and then Japanese high-speed train network system has many short and long train tunnels. When a high speed train enters into a long tunnel, compression waves are built up ahead of the train. As the blockage ratio of the train cross section to the tunnel cross section is about 25%, the trains serve as a high speed piston having its blockage ratio of 25%; thereby producing trains of compression waves propagating at a sonic speed inside the tunnels. When the compression waves were released from the tunnel exit, train tunnel sonic blooms will startle people living in the vicinity of such long train tunnels. Then the train tunnel booms prevent complying the national demand to increase the train speed. For overcoming this problem, a collaborated research with the East Japan Railway Company was initiated aimed at mitigating train tunnel booms.

Figure 6.3a shows a 1:250 scale tunnel sonic boom simulator installed in the Shock Wave Research Center of the Institute of Fluid Science, Tohoku University.

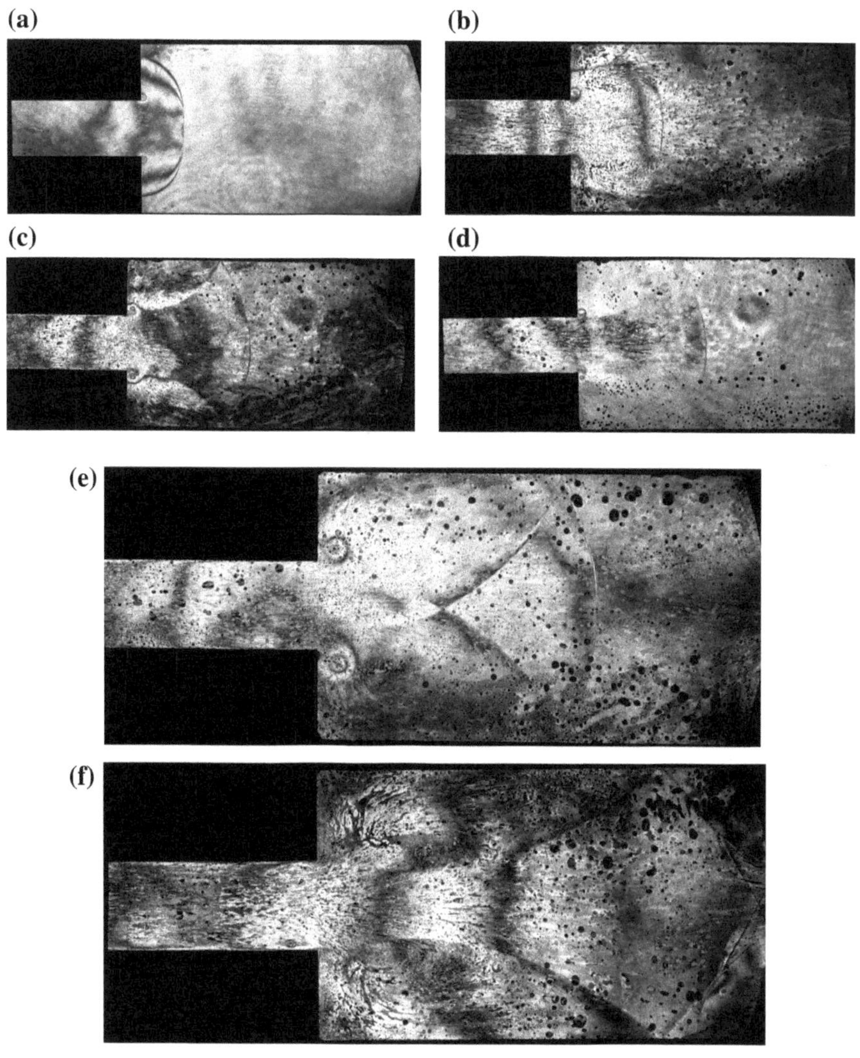

Fig. 6.1 Shock waves generated in the exhaust gas pipe line of a Subaru engine: **a** #87042808, engine rotation 6800 rpm; **b** #87042809, 6800 rpm; **c** #87042806, 5900 rpm; **d** #87042807, 5899 rpm; **e** #87042810, 6500 rpm; **f** #87042812, 7500 rpm (Sekine et al. 1989)

The simulator consisted of, in principle, a ϕ 40 mm and it is about 20 m long steel tube tilted by 8°. It has a piston launcher, and its recovery system at the tube's end. The piston was a flat head cylinder composed of polycarbonate. It had 20 mm diameter and 200 mm long. Its blockage ratio to the simulator's cross section was 26% which was nearly identical with the real train/tunnel blockage ratio.

The piston was supported by a sabot at its front end and the rear end was accelerated by high pressure nitrogen as shown in Fig. 6.3b. Upon its impact against a tapered entry, the frontal sabot split into four pieces and the piston pass

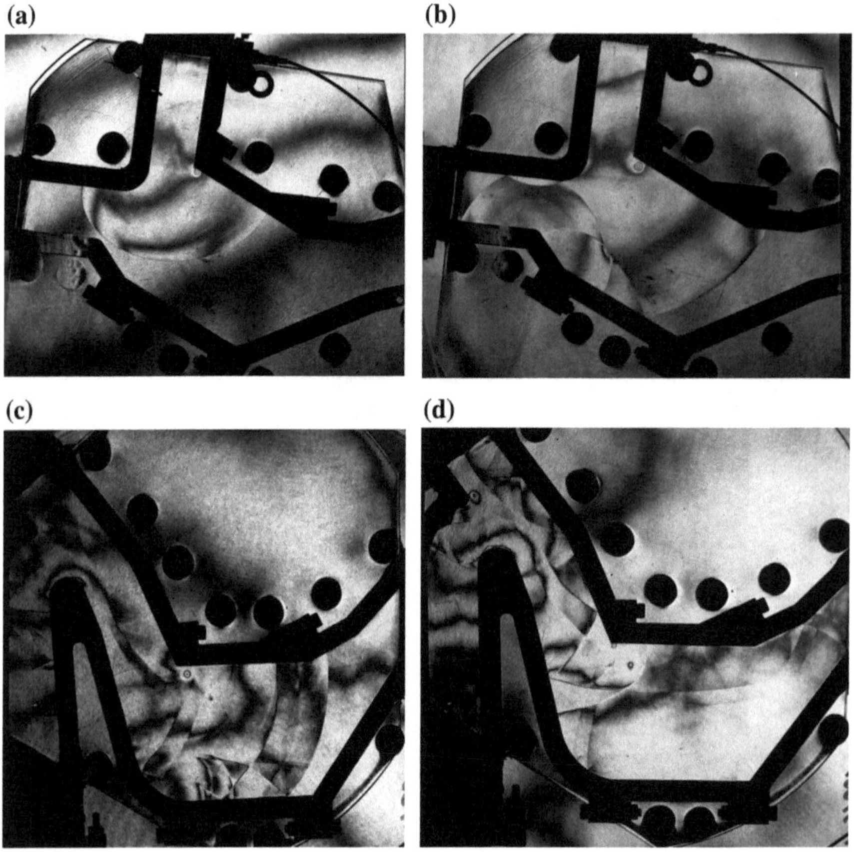

Fig. 6.2 Shock wave mitigation in two-dimensional manifolds for Ms = 1.1 generated by the ignition of a 10 mg AgN_3 pellet: **a** #94110903; **b** #94110905; **c** #94111114; **d** #94111110 (Sekine et al. 1989)

thorough the hole. A rear sabot was, with propagation, plugged into the tapered entry hole and stopped the high pressure nitrogen to flow into the tunnel simulator. By adjusting the pressure in the driver chamber, it was possible to precisely control a piston speed from 50 to 120 m/s, equivalent to the train speed of from 189 to about 430 km/h. The piston moved along an open section of about 1 m long and hence compression waves built up in front of the piston was attenuated prior to its entry to the test section.

Figure 6.4a, b show the pressure history formed in front of the moving piston. The pressures were measured by distributed pressure transducers along the test section at the piston entry speed of 60 and 100 m/s, respectively. In Fig. 6.4a, the amplitudes of overpressures gradually decrease with the piston propagation.

For examining the effect of wall perforation on mitigation of train tunnel sonic booms, the wall surface was covered with a 5 mm thick perforated aluminum wall, which is, in short, a roughed wall test section consisting of porous aluminum plate.

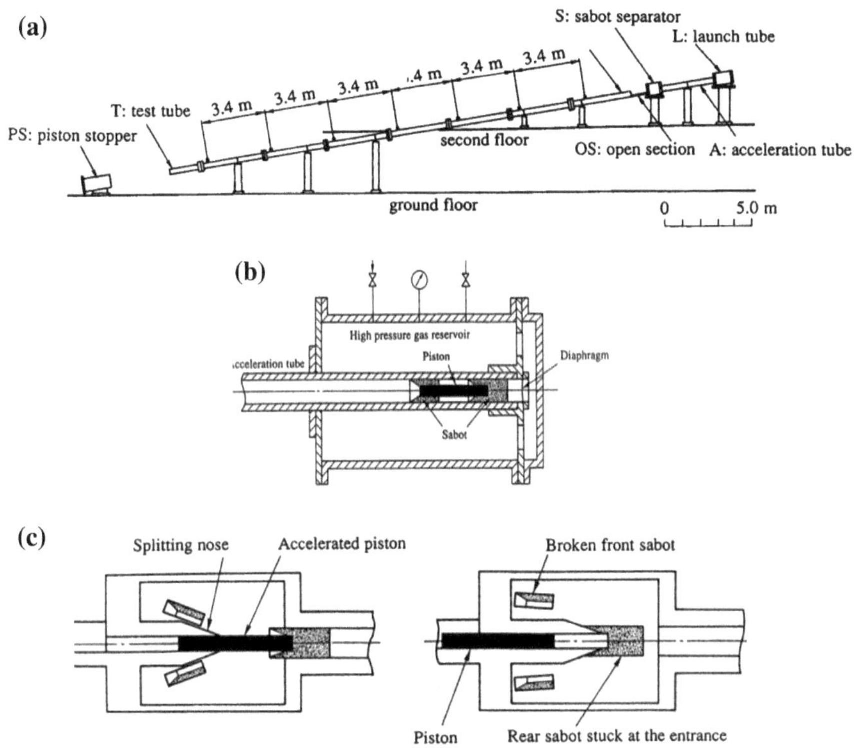

Fig. 6.3 Tunnel simulator: **a** tunnel simulator; **b** launching setup; **c** piston release and recovery

Pressure transducers were distributed at every 0.2 m intervals along the perforated wall. Figure 6.5a shows a perforated wall. The wall in the absence of the wall perforation is defined as b = 0. The wall 16% of surface area is covered by two 10 mm wide and 5 mm seep porous aluminum plates is defined as b = 1.0. Figure 6.5b show the effect of wall perforations on the shock wave mitigation for B = 0, smooth wall and b = 0.16 and b = 1.0.

Assuming that the Mach number of the sonic boom is Ms = 1 + ε, ε ≪ 1, the dimensionless overpressure of the boom Δp, can be related to ε, $\Delta p \sim (\gamma + 1)\varepsilon/4\gamma$, where γ is the specific heats ratio and γ = 1.4 in air. In Fig. 6.5b, for instance, at x = 23.8 m, Δp = 0.05 and hence ε = 0.02. In Fig. 6.5b, the high speed entry of the flat head piston created a weak shock wave of Ms = 1.02 at x = 10.2 m. A negative pressure profile is observed at a distance of x = 3.4 m indicates the existence of expansion wave caused at the rear end of the piston causing recovery to the ambient pressure during its propagation. In Fig. 6.5b, the experimental results are summarized. The ordinate denotes the maximum pressure Δp_{max} in kPa and the abscissa denotes the entry speed of a piston u_p in m/s. Filled circles denote smooth wall, in other word, no coverage of roughened wall or b = 0, open circles denote 16% coverage or b = 0.16, and filled squares denote 100% coverage with perforation

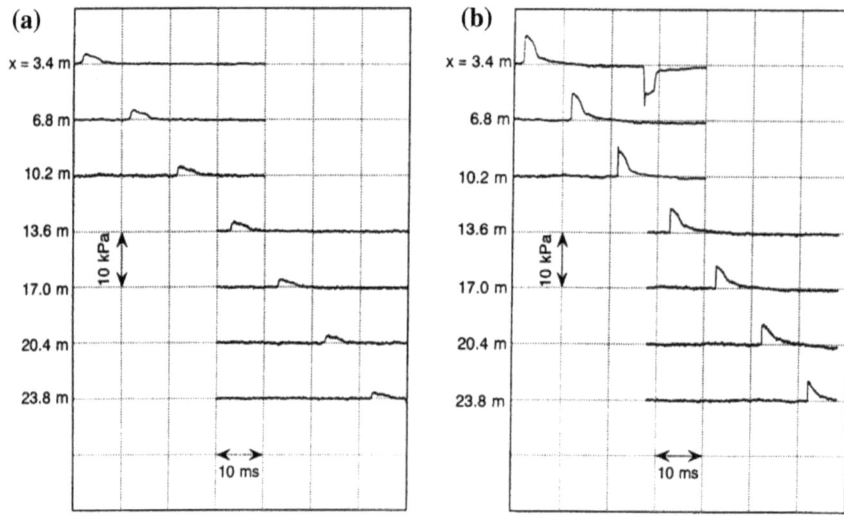

Fig. 6.4 Formation of shock wave driven by a flat head piston: **a** piston speed 60 m/s or 216 km/h; **b** piston speed of 100 m/s or 360 km/h

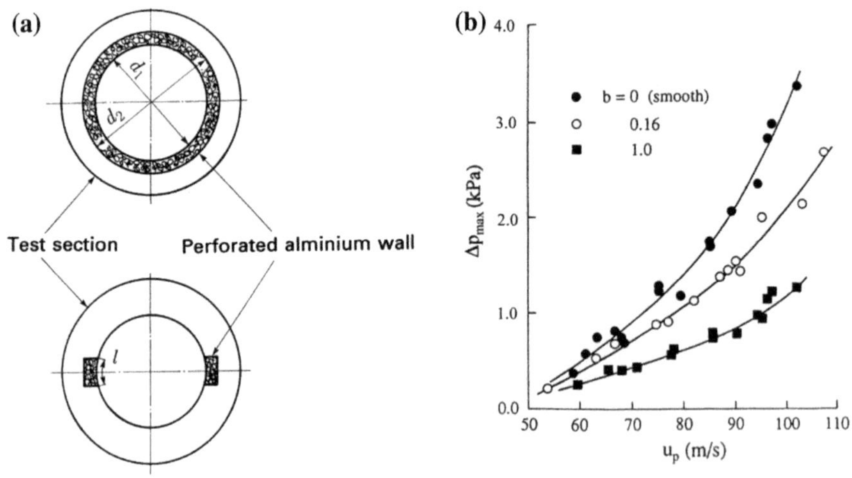

Fig. 6.5 The effect of wall perforation on the mitigation created by the piston motion: **a** the arrangement of wall roughness; **b** the effect of the wall perforation on the shock wave attenuation

b = 1.0. The wall perforation effectively mitigates the train tunnel sonic booms and this trend is enhanced with increasing the entry speed. At u_p = 70 m/s, the peak pressure along a partially roughened test section of b = 0.16 is about 15% of that along the smooth test section or b = 0, whereas the peak pressure along a totally roughened test section or b = 1.0 is reduced nearly 50% of that measured along the test section with a smooth wall.

Figure 6.6a, b show pressure histories along the roughened test section of b = 1.0 measured by distributed pressure transducers at the entry piston speeds of 98 and 65 m/s, respectively. The ordinate denotes the pressure variation on 5 kPa scale measured at the pressure transducers distributed at every 0.2 m from the entrance. The abscissa denotes the elapsed time on 10 ms scale. The pressure variation at x′ = 0.67 m shows the initial pressure variation along a smooth test section of b = 0. The initial steep pressure rise was smeared out with propagation because expansion wavelets induced from the roughened surface caught up the wave front gradually reducing its steepness. The negative pressure observed at the tail of the piston was gradually recovered as the expansion wave generated at the entry of the test section caught up the negative pressure. A similar trend is observed even at the slower entry speed of 65 m/s as seen in Fig. 6.6b. It is then concluded that roughened walls effectively attenuate the booms.

A 30 mm × 40 mm shock tube test section was connected to the 40 mm diameter smooth test section and the resulting weak shock waves running along the rectangular test section were visualized. Figure 6.7a, b show weak shock waves generated at the entry speeds of 75 and 60 m/s, respectively. The shock waves propagated from right to left. The peak over-pressures induced by these piston speeds were about 1.3 and 0.5 kPa and hence the corresponding shock Mach numbers were 1.06 and 1.02, respectively. Although a double path holographic interferometry was applied, the contrasts of these shock fronts seen in Fig. 6.7a, b are very faint because the path length of the *OB* path length was only 60 mm.

To examine the effect of wall roughness on train tunnel sonic boom's mitigation in real train's passing through a tunnel, aluminum panels each containing a 500 mm wide, 1000 mm long, and 50 mm thick aluminum foam, Alporous^{TM}, were arranged

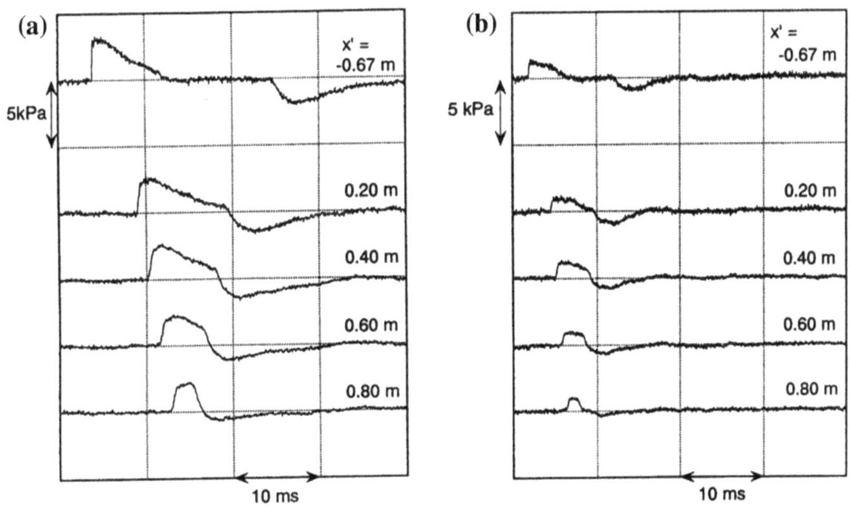

Fig. 6.6 Formation and attenuation of a shock wave driven by a flat head piston along a perforated test section: **a** piston speed 98 m/s or 353 km/h; **b** piston speed of 65 m/s or 234 km/h

(a)

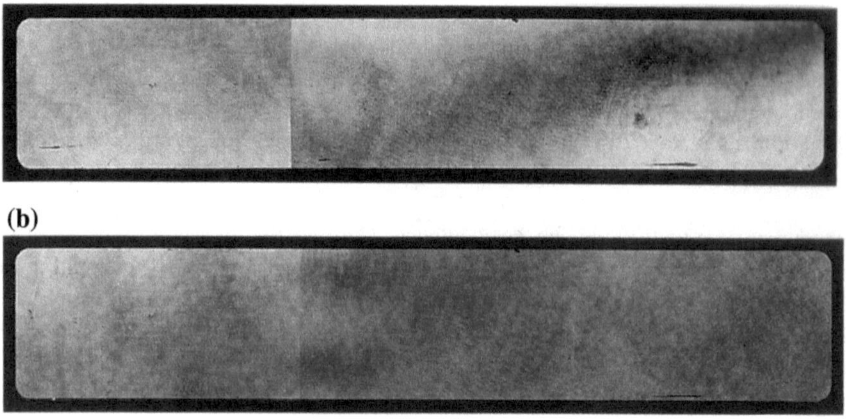

(b)

Fig. 6.7 Weak shock waves observed in a 30 mm × 40 mm test section: **a** #93051607 piston speed of 75 m/s; **b** piston speed of 60 m/s

along the western wall starting at the 50 m distance from the north entrance of the Ichinoseki Tunnel as seen in Fig. 6.8. The booms created by the north bound trains were measured using a 100 mm diameter PVDF pressure transducer. The pressure transducer was composed of piezo film, polyvinylidene difluoride, designed and manufactured in house and calibrated in the 60 mm × 150 mm diaphragm-less shock tube by the comparison of their output signals with those obtained by Kistler pressure transducer Model 603B (Sasoh et al. 1998).

Figure 6.9a, b show the time variation of the overpressures of real train tunnel sonic booms in the absence of the porous aluminum panels and at the train speed of 245 km/s or 68 m/s. Figure 6.8b shows the time variation of the overpressure of real train tunnel sonic booms in the presence of porous aluminum panels at the train speed of 240 km/h or 64 m/s are compared. The ordinate denotes over-pressures in kPa and the abscissa denotes the elapsed time in ms. Figure 6.9a shows a steep pressure rise which implies that the boom is not mitigated. The pressure transducer had about 40 mm thick disk shape so that the high pressure behind the reflected shock wave was diffracted at the edge of the disk shaped transducer. Then the expansion wave got focused at the center and a significant pressure decrease was recorded.

Figure 6.9b shows the presence of arrayed porous aluminum plates. The peak pressure was significantly smeared out. The measured peak pressure seen in Fig. 6.9a is higher than the signal seen in Fig. 6.9b. In conclusion, the introduction of porous walls, even if it is done partially, indeed mitigated the tunnel sonic boom. The installation of aluminum foam panels on the tunnel walls effectively mitigated the booms but the price of aluminum foam panels is not inexpensive and hence it will not necessarily be economical using it in all the long High Speed Train Tunnels.

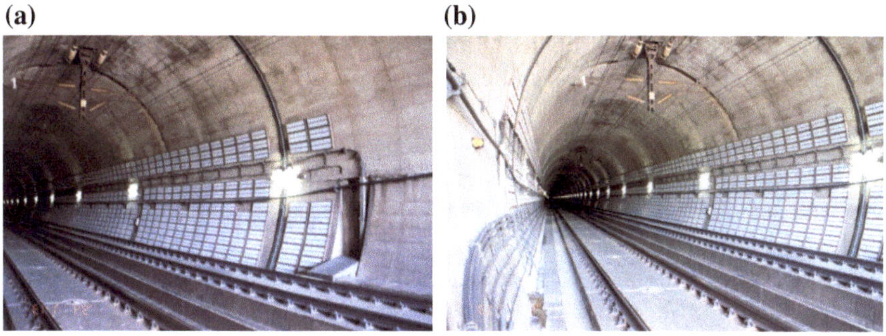

Fig. 6.8 Framed aluminum foam panels along the western side of the Ichinoseki tunnel viewed from the north entry

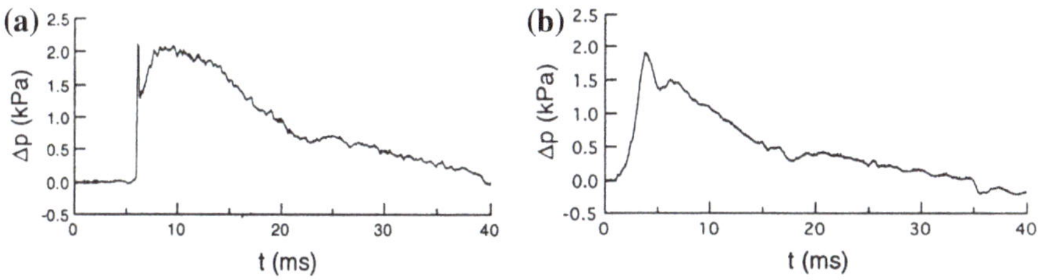

Fig. 6.9 Pressure history of booms monitored at the Ichinoseki Tunnel: **a** absence of aluminum foam wall at train speed of 245 km/h; **b** presence of aluminum foam wall at train speed of 240 km/h (Sasoh et al. 1998)

6.4 Shock Wave Mitigation Along Perforated Walls

The motivation for investigating shock wave interaction with perforated walls is due to the fact that this is the simplest ways for mitigating shock wave propagating inside ducts.

6.4.1 Shock Wave Mitigation Along Distributed I-Beams

Commercial aluminum I-beams of 30 mm × 100 mm and 5 mm in thickness were arranged at various combinations of heights (h) and intervals (d) inside the test section of the 60 mm × 150 mm diaphragm-less shock tube (Matsu'oka 1997). Figure 6.10a, b show the interactions of weak shock waves for Ms = 1.05 and 1.12 with 100 mm wide I-beams, 100 mm high and set at 60 mm separation. In such a long separation distance, the shock wave interacted almost independently with a single I-beam segment. The interaction with a neighboring segment is negligibly small. Figure 6.10c, d show the interaction of shock waves of Ms = 1.13 with arrayed 100 mm wide I-beams separated at 2 mm interval. This wall shape is similar to a wall having distributed 2 mm wide slits. In comparing such a narrow separation with Fig. 6.10b, a significant deviation of the transmitted shock wave patterns was observed.

Figure 6.11 show parametric visualizations of transmitted shock waves over arrayed I-beams of various combinations of h and d. Figure 6.11a–c show shock wave interaction with arrayed I-beams of h = 8 mm and d = 4 mm for Ms = 1.02. Figure 6.11d shows the interaction for Ms = 1.12.

Figure 6.12 show shock wave mitigation along arrayed I beams of h = 8 mm and d = 7 mm for Ms = 1.12 in atmospheric air at 292 K. At each interaction with the I-beam segment, the wavelets are accumulated behind the transmitted shock wave enhancing the mitigation.

6.4.2 40 mm Wide Opening with Roughened Surface

In Fig. 6.13, shock wave propagations along a 40 mm wide channel with distributed shallow slits are sequentially visualized for Ms = 1.22 in atmospheric air at 294 K. The test model was installed in the 60 mm × 150 mm diaphragm-less shock tube. At first the IS was diffracted at the entrance corners and then the resulting transmitted shock wave interacted with each slit showing complex wave interaction patterns and promoted mitigation. The test piece was installed in the 60 mm × 150 mm diaphragm-less shock tube. Figure 6.13c shows an enlargement of Fig. 6.13b: an earlier stage of shock wave propagation along the distributed slits. Expansion waves were generated from the slit openings and compression waves are generated from the bottom of the slits. These wavelets interacted with the transmitted shock wave and attenuated the transmitted shock wave. In Fig. 6.13e, f, the transmitted shock wave is seen from the exit and was diffracting at the exit corners. The evolution of the resulting pair vortices indicates that the attenuation of the shock wave is attenuated upon it a release from the exit.

6.4.3 25 mm Wide Opening with Smooth Surface

In Fig. 6.14, the evolution of shock wave along a 25 mm wide duct with smooth wall was visualized for Ms = 1.20 in atmospheric air at 296.8 K. Comparing to the case of shock wave interaction with a wall having distributed slits, as shown in Fig. 6.13, the transmitted shock wave was not strongly disturbed while progressing along the smooth wall. The fringes were uniformly distributed in this case. The shear flows in the boundary layer are released from the exit forming a series of vortices as seen in Fig. 6.14f, g.

(a)

(b)

(c)

(d)

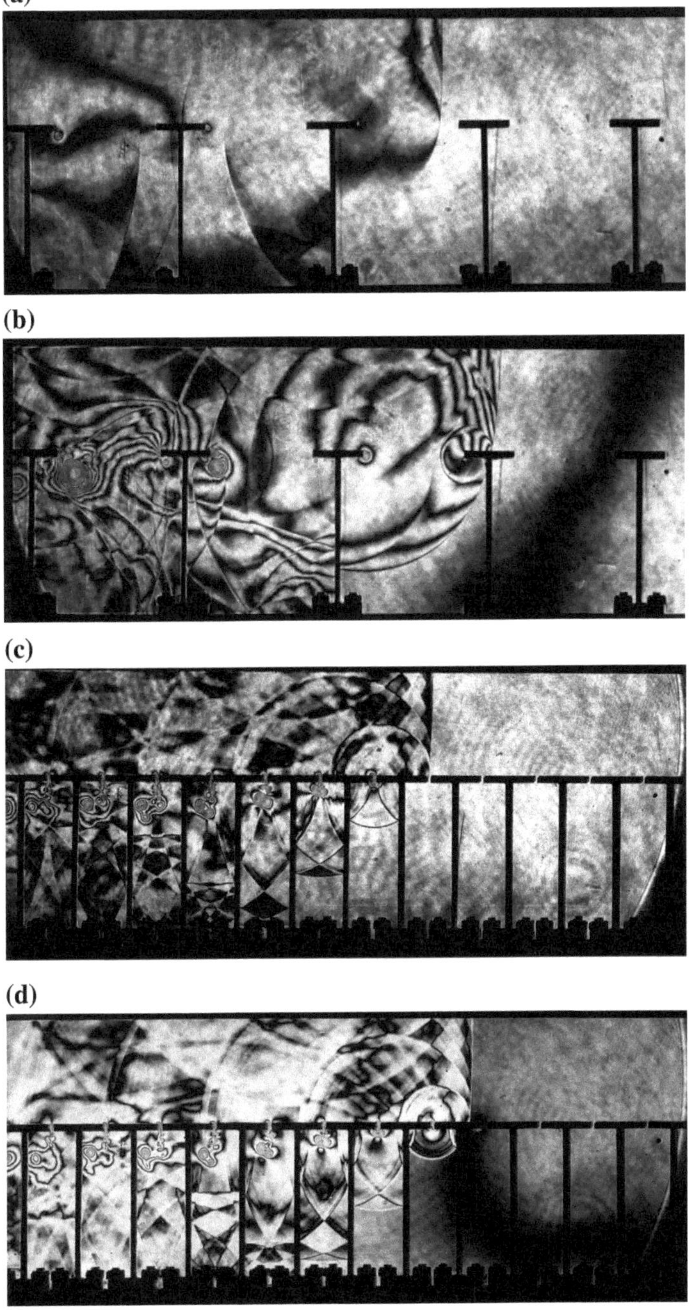

Fig. 6.10 Shock wave mitigation along a perforated wall of h = 100.0 mm in atmospheric air at 292 K: **a** #96050210, Ms = 1.048, d = 60 mm; **b** #96050204, Ms = 1.123, d = 60 mm; **c** #96043002, Ms = 1.13, d = 2 mm; **d** #96042542, Ms = 1.13, d = 2 mm (Matsuoka 1997)

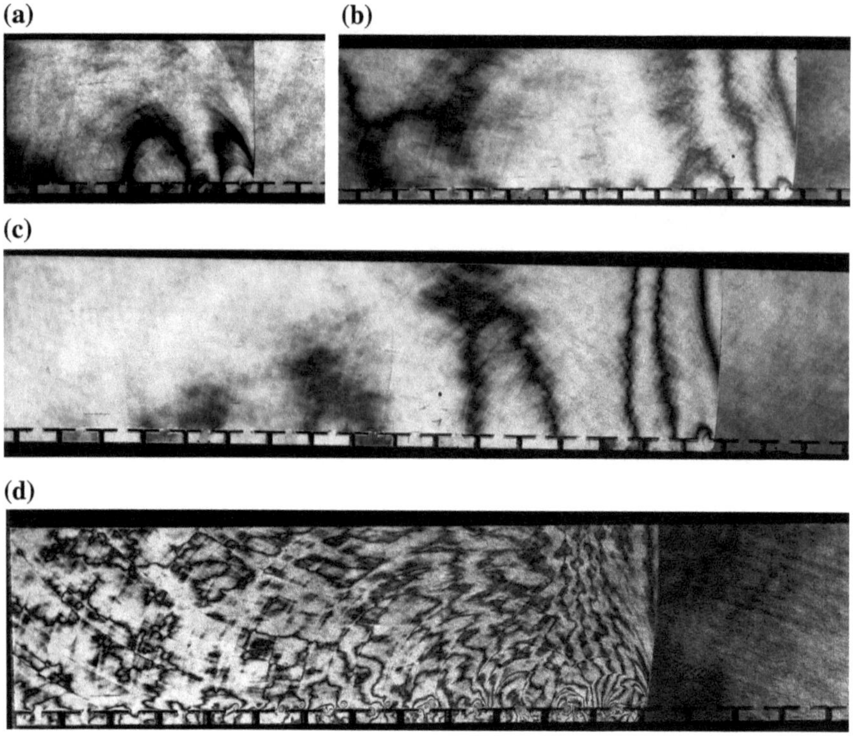

Fig. 6.11 Shock wave mitigation along arrayed I beams of h = 8 mm and d = 4 mm for Ms = 1.02 in atmospheric air at 292 K: **a** #96061708; **b** #96061705; **c** #96061707; and **d** #96061910 for Ms = 1.12

6.4.4 10 mm Wide Opening with Smooth Surface

Figure 6.15 show the evolution of shock wave propagation along a 10 mm wide opening for Ms = 2.7 in air at 100 hPa, 285.8 K. The transmitted shock wave created by diffraction at the entrance corner generated wavelets as seen in Fig. 6.15a, b. The fringe observed parallel to the wall exhibited shows the boundary layer developing along the smooth wall. It develops independently of the width of the opening. In the present experiments with smooth wall model it is the boundary layer that governs the shock wav attenuation. Hence the smaller the opening width is, the more effectively the shock wave would be attenuated.

Shock wave propagation along a 6.0 mm wide opening was sequentially observed. This arrangement was used to produce a narrow shock wave which was needed to interact with a 6 mm high circular soap column. For producing undisturbed shock tube flows propagating in a narrow shock tube, a so-called cookie cutter was used. The cookie cutter was installed in the 60 mm × 150 mm diaphragm-less shock tube.

(a)

(b)

(c)

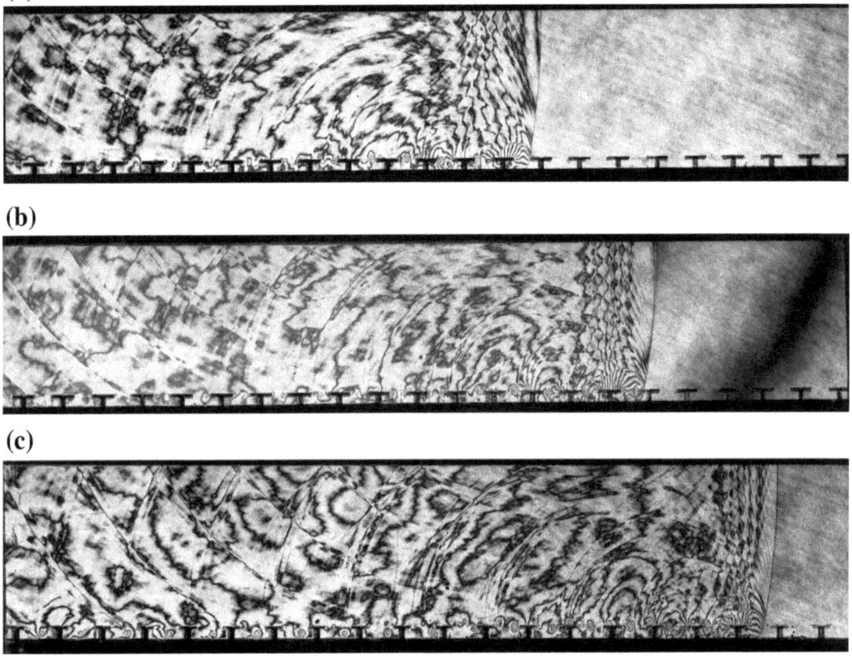

Fig. 6.12 Shock wave mitigation along arrayed truncated I-beams of h = 8 mm and d = 7 mm for Ms = 1.12 in atmospheric air at 292 K: **a** #96061906; **b** #96061201; **c** #96061205

Figure 6.16 show the evolution of shock wave propagating along a 6 mm wide opening for Ms = 1.4 in air at 100 hPa, 285.8 K. In Fig. 6.16a, b, as the leading edge has a sharp edge, the shock wave diffraction at the entrance is hardly seen. Hence, the transmitted shock wave is undisturbed and the flow behind it is uniform. This test section would be an effective cookie cutter, if a 6 mm wide test section is positioned at an appropriate distance from the leading edge.

6.5 Shock Wave Mitigation Passing Through a Small Hole

6.5.1 A Single Hole

Figure 6.17 show the evolution of the shock wave propagating through a 5 mm diameter hole drilled on a 30 mm thick wall for Ms = 1.27 in atmospheric air at 292 K. A 30 mm thick stainless steel wall was installed in the 60 mm × 150 mm diaphragm-less shock tube. The incident shock wave, IS, reflected from the steel plate but part of the IS propagated through the thin hole and a transmitted shock wave came from the hole.

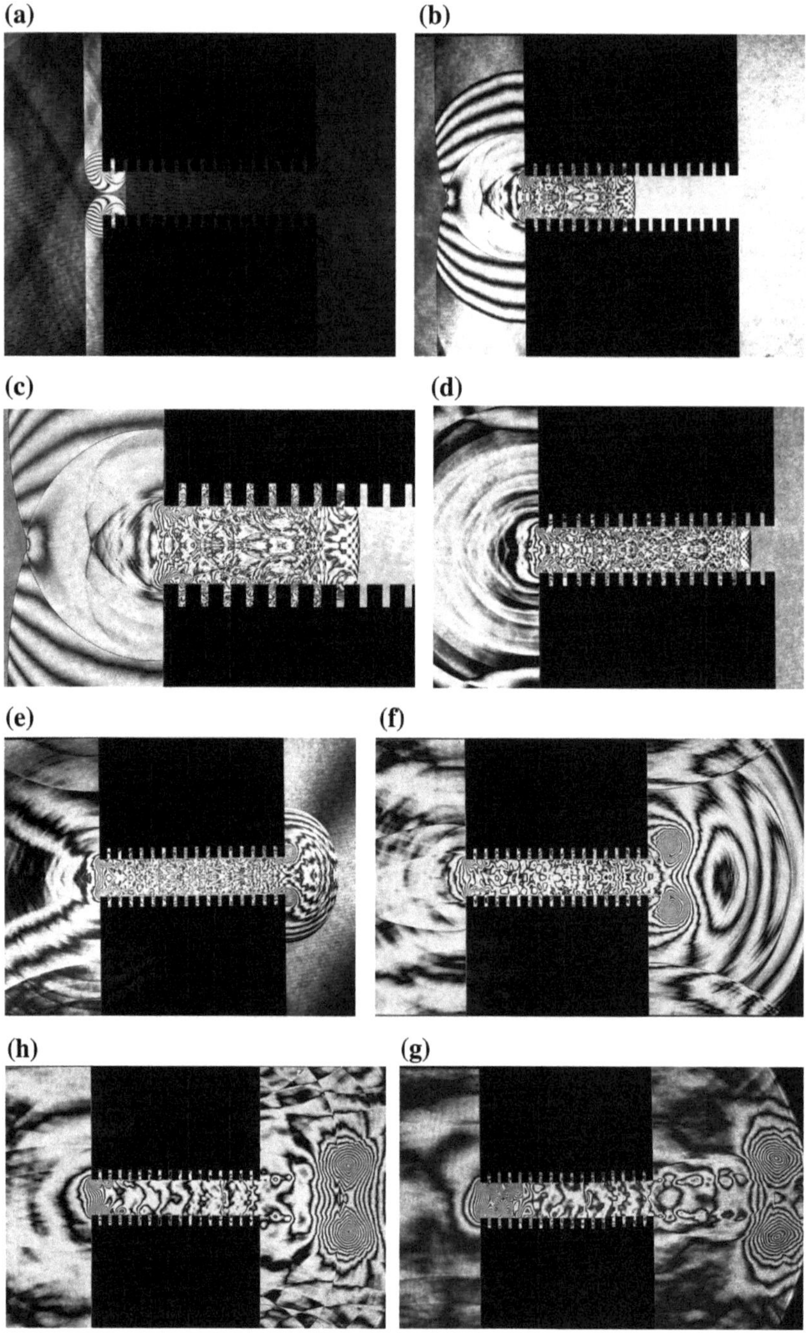

Fig. 6.13 Shock wave propagation along distributed slits for Ms = 1.22 in atmospheric air at 294 K: **a** #9304270; **b** #93042702, 500 μs from trigger point, Ms = 1.223; **c** enlargement of (**b**; **d** #93042703, 600 μs, Ms = 1.223; **e** #93042704, 700 μs, Ms = 1.218; **f** #93042705, 900 μs, Ms = 1.215; **g** #93042706, 1300 μs, Ms = 1.212; **h** #93042707, 1500 μs, Ms = 1.215

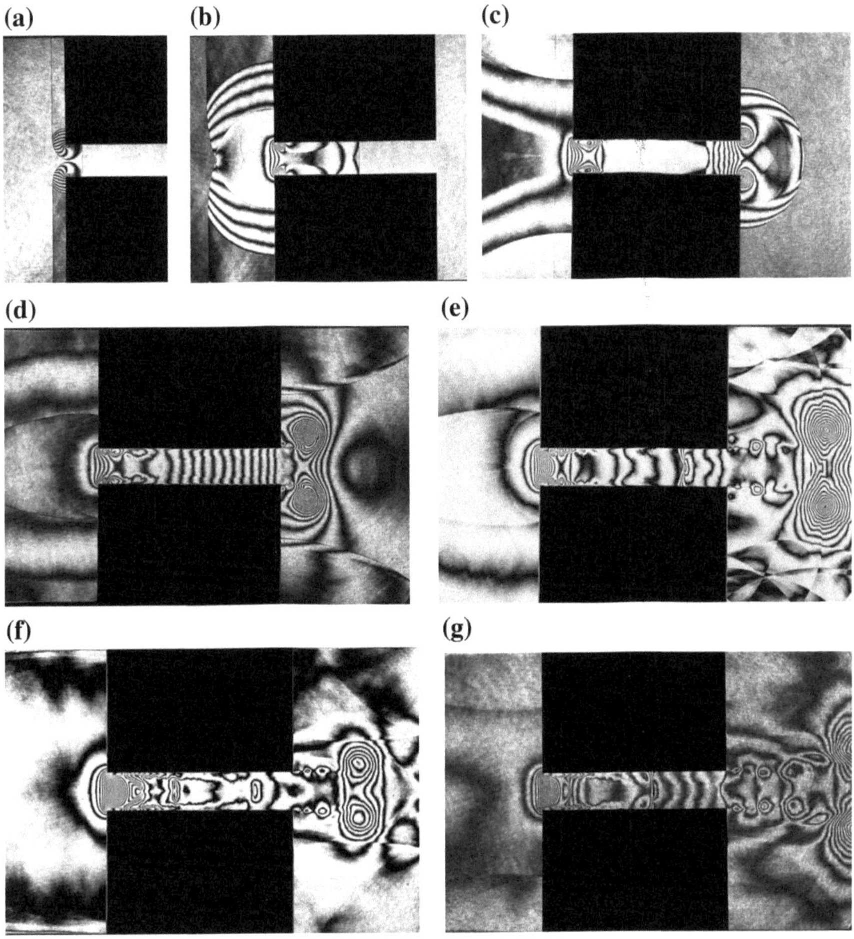

Fig. 6.14 The evolution of shock wave mitigation along a slotted wall with smooth surface for Ms = 1.20 in atmospheric air at 296.8 K: **a** #93042618, 100 μs from trigger point, Ms = 1.208; **b** #93042617, 300 μs, Ms = 1.212; **c** #93042615, 500 μs, Ms = 1.208; **d** #93042614, 900 μs, Ms = 1.208; **e** #93042613, 1300 μs, Ms = 1.208; **f** #93042612, 1300 μs, Ms = 1.267; **g** #93042611, 1500 μs, Ms = 1.219

6.5.2 Two Oblique Holes

Figure 6.18 show the evolution of shock wave mitigation through two 5 mm diameter obliquely drilled holes for Ms = 2.26 in air at 250 hPa, 290 K. The two holes were separated by 10 mm interval were drilled obliquely 30° on s 30 mm thick stainless steel plate. The plate was installed in the in the 60 mm × 150 mm diaphragm-less shock tube. This was a part of basic experiment related to the study of the suppression of automobile exhaust gas noise. The sequence of the merger of the two transmitted shock waves into a single shock wave was visualized.

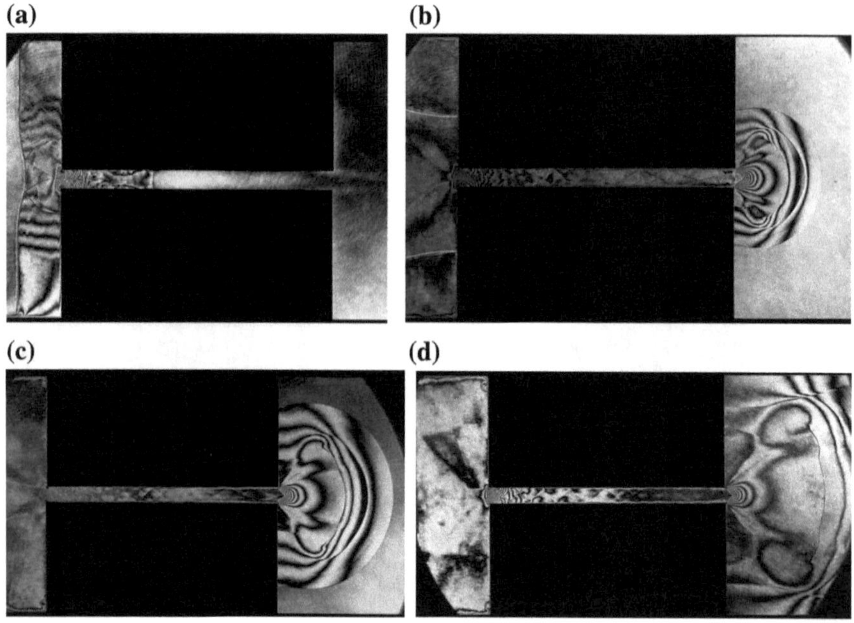

Fig. 6.15 Evolution of shock wave propagation along a 10 mm wide opening for Ms = 2.7 in air at 100 hPa, 285.8 K: **a** #85041502, 70 μs from trigger point, Ms = 2.215; **b** #85041504, 90 μs, Ms = 2.702; **c** #85041507, 130 μs, Ms = 2.711; **d** #85041508, 200 μs, Ms = 2.741

Figure 6.18a shows two spherical shock waves released from the oblique holes. Figure 6.18d–f show the process of two spherical shock waves gradually merging into a single shock wave.

Two 5 mm diameter holes were drilled obliquely at 7.5 mm interval vertical to the center line of a 30 mm thick stainless steel plate. The two holes merged at the center of the plate. Figure 6.19a shows the obliquely drilled holes and the early stage of the shock wave transmission through the oblique holes. The shock waves entering the oblique holes merged at the exit. Figure 6.19g explains two 5 mm diameter holes obliquely drilled at 7.5 mm interval at the frontal surface merged into a hole at the rear surface.

6.6 Sintered Stainless Steel Walls

1 mm thick sintered stainless steel plates were placed on the upper and lower walls of the 100 mm × 200 mm test section as seen in Fig. 6.20. This test section was accommodated in the 150 mm × 250 mm test section connected to the 60 mm 150 mm conventional shock tube. The sequential observation of shock wave attenuation along two 1 mm thick sintered stainless steel plates was conducted for Ms = 1.06 in atmospheric air at 290.8 K.

(a) (b) (c) (d) (e) (f)

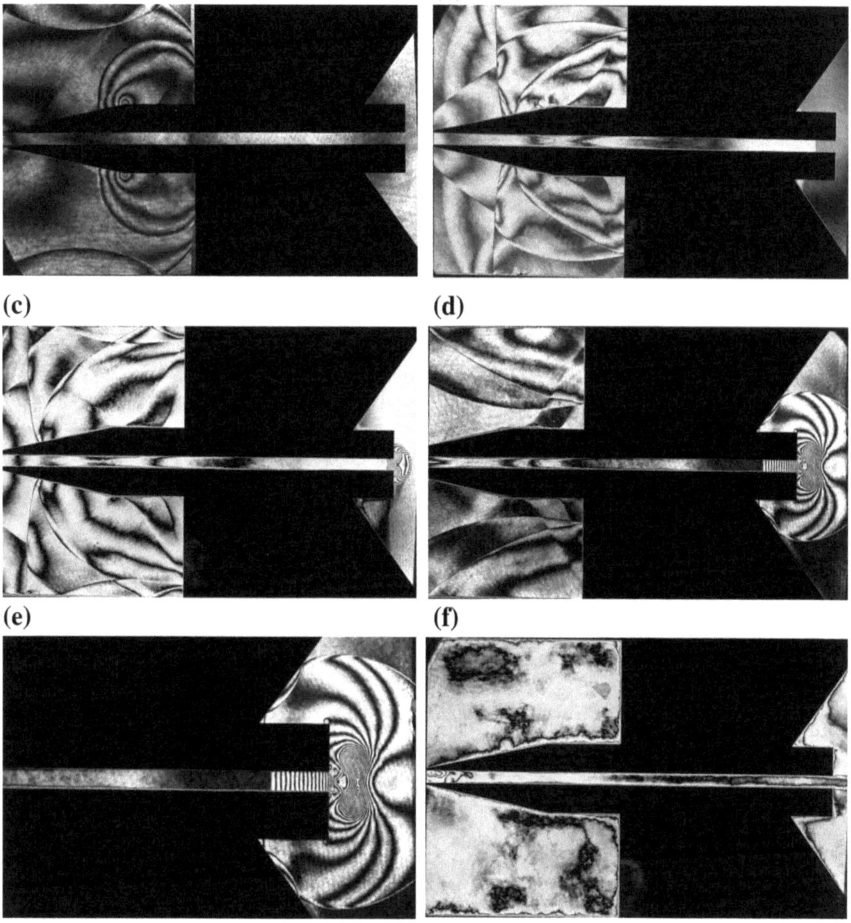

Fig. 6.16 The evolution of shock wave propagating along a 6 mm wide opening for Ms = 1.4 in air at 100 hPa, 285.8 K: **a** #86120204, 330 μs, Ms = 1.402; **b** #86120308, 440 μs, Ms = 1.420; **c** #86120301, 460 μs, Ms = 1.420; **d** #86120305, 480 μs, Ms = 1.420; **e** enlargement of (**d**); **f** #86120303, 500 μs, Ms = 1.408

The sintered stainless steel plates had many minute perforations evenly distributed on its surface but the surface was so smooth that it was an ideal material for attenuating weak shock waves like those seen induced in the automobile exhaust gas pipe line systems. The sintered stainless steel plates were strong enough to withstand against relatively high pressure loading. Although its physical properties were wonderful but his material was not easy to be used in applications like the one considered. In the series of experiments, the space behind the sintered stainless steel was open and then the transmitted shock waves were just reflected from the test section wall. For efficiently attenuating the transmitted shock wave, appropriate materials should be inserted behind the sintered stainless plates.

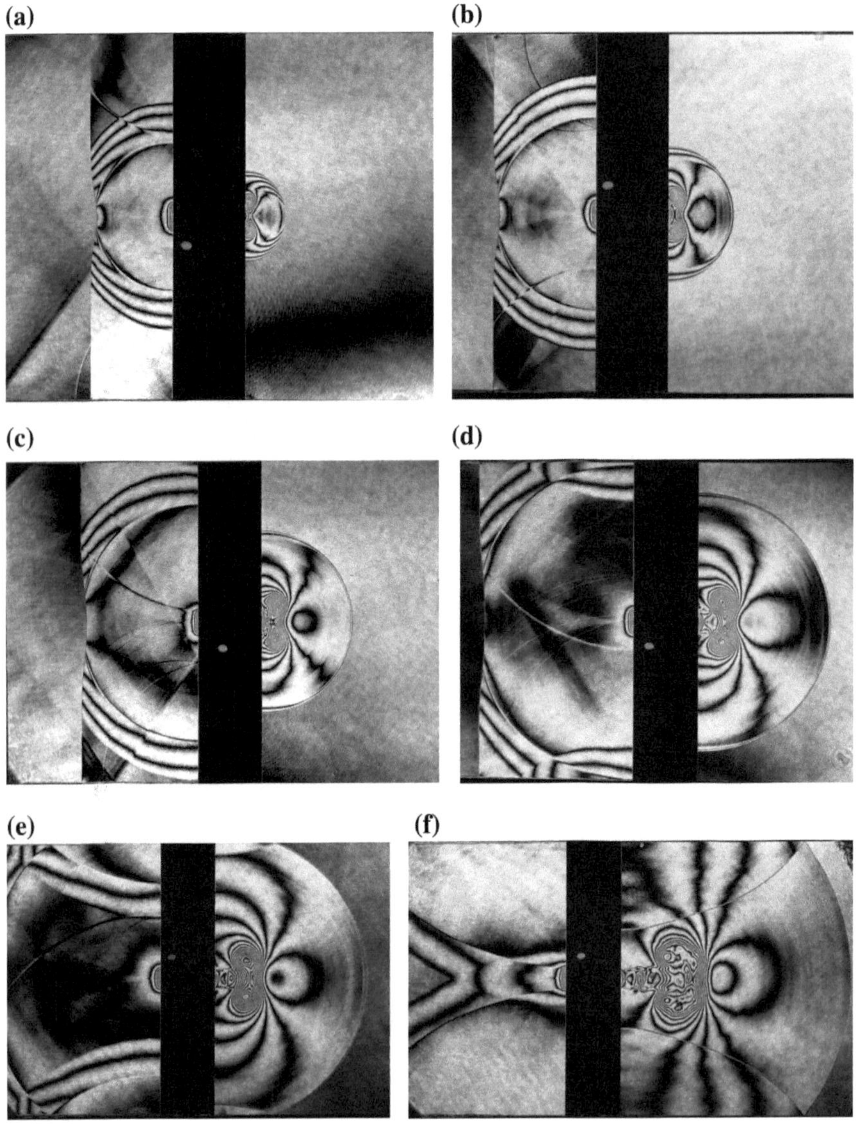

Fig. 6.17 The evolution of the shock wave propagating through a 5 mm diameter hole on a 30 mm thick wall for Ms = 1.27 in atmospheric air at 292 K: **a** #91111428, 330 μs; **b** #91111423, 360 μs; **c** #91111424, 400 μs; **d** #91111425, 450 μs; **e** #91111426, 510 μs; **f** #91111427, 620 μs

Figure 6.21 show the shock wave mitigation along the sintered stainless plates backed up with solid walls for moderate shock wave of Ms = 1.38 in air at 600 hPa, 290.8 K.

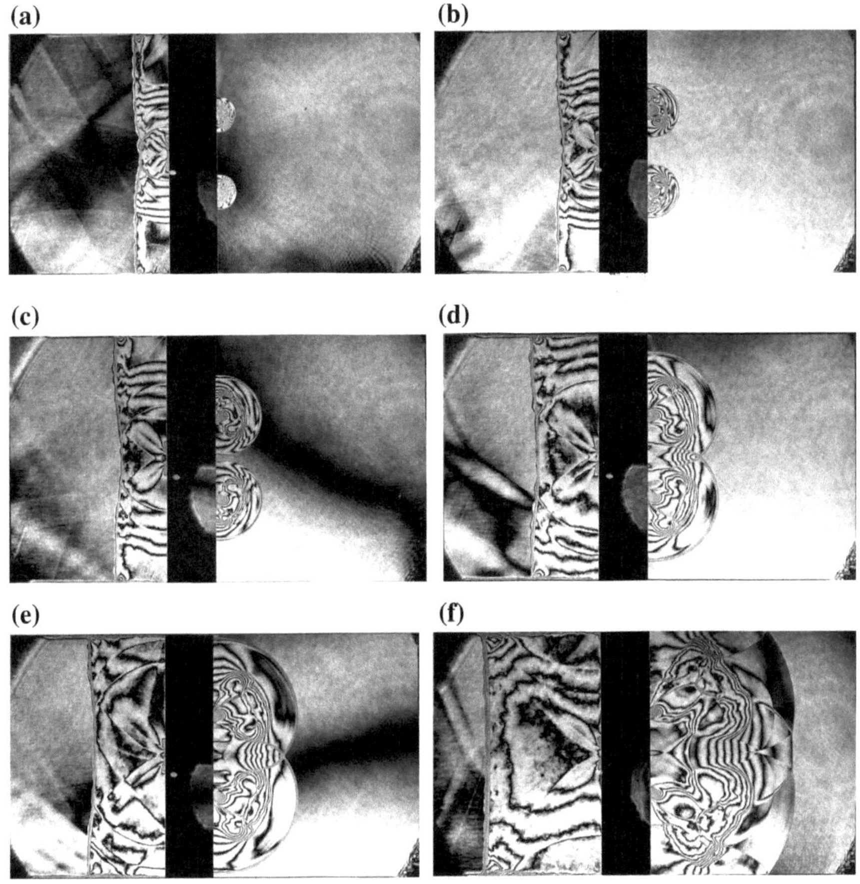

Fig. 6.18 Shock wave mitigation through two holes for Ms = 2.26 in air at 250 hPa, 290 K, two 5 mm diameter holes separated by 10 mm diverging 30°: **a** #91111819, 270 μs from trigger point, Ms = 2.185; **b** 91111820, 280 μs; **c** #91111821, 300 μs; **d** #91111822, 330 μs; **e** #91111823, 370 μs; **f** #91111824, 420 μs

6.6.1 Sintered Stainless Wall Backed up with Alporous^{TR}

In this sonic boom experiments, aluminum sponge, the trade name Alporous^{TR}, was frequently used. Figure 6.22 show a sequence of shock wave attenuation along the sintered stainless steel plates backed up with Alporous^{TR}.

6.6.2 Perforated Walls

Figure 6.23 show a sequential observation of shock wave propagation along a perforated wall. The perforated wall shown in Fig. 2.23 had 1.0 mm diameter holes

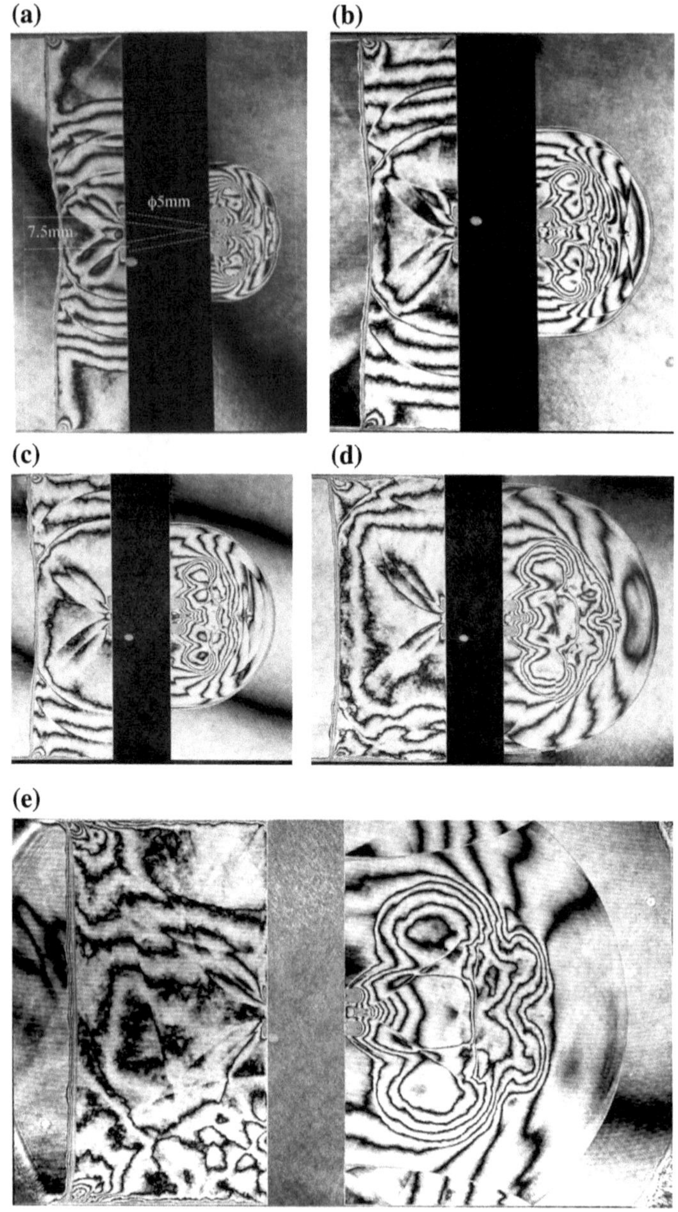

Fig. 6.19 The evolution of the shock wave attenuation through two 5 mm diameter holes separated in 7.5 mm interval for Ms = 2.20 in air at 250 hPa, 290.5 K: **a** #91111514, 280 μs, Ms = 2.188; **b** #91111515, 300 μs, Ms = 2.235; **c** #91111516, 330 μs, Ms = 2.188; **d** #91111517, 370 μs, Ms = 2.239; **e** #91111518, 420 μs, Ms = 2.239

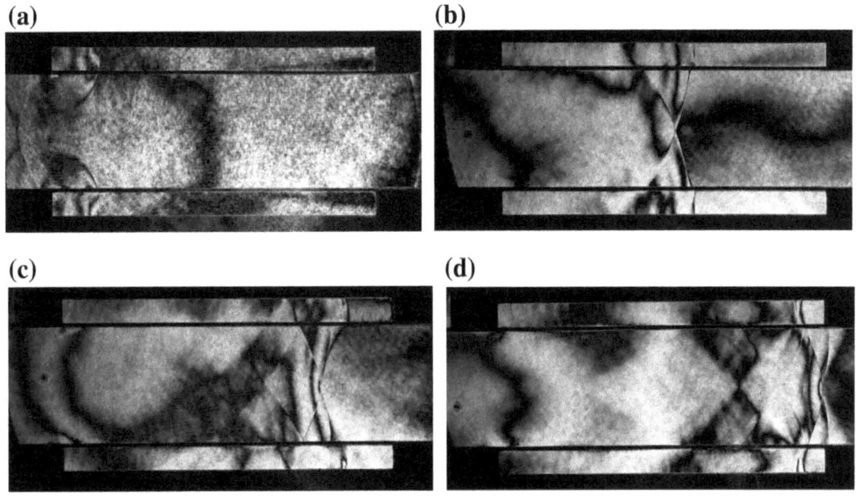

Fig. 6.20 Shock wave mitigation along sintered stainless plates backed up with solid walls for Ms = 1.06 in atmospheric air at 290.8 K: **a** #92011702, 800 μs, Ms = 1.069; **b** #92011703, 950 μs, Ms = 1.067; **c** #92011705, 1250 μs, Ms = 1.067; **d** #92011706, 1400 μs, Ms = 1.080

were uniformly distributed on a 2.0 mm thick brass plates; the perforation ratio was 0.5. The opening width was 30 mm and the supporting side wall was 40 mm wide.

Figure 6.24 show the evolution of shock wave attenuation along the perforated wall having perforation ratio of 0.2 for Ms = 2.50 in air at 333 hPa, 290 K. The inclination angle θ of the transmitted shock wave propagating along the upper and lower walls indicates approximately $\sin\theta = a/u_s$ where a is the sound speed in air and u_s is the speed of the transmitted shock wave. Therefore, measuring the θ, the transmitted shock wave attenuation is readily estimated.

6.6.3 Wall with Slotted Surfaces

The transition of the reflected shock waves from the slotted wedge was discussed in the Chap. 2 and the slot geometry of the slotted was described in Fig. 2.44. Figure 6.25a shows the slotted wall: the wall was 100 mm long and the slots were 7 mm deep: it was tested in the 60 mm × 150 mm conventional shock tube.

The shock wave propagation along a tested slotted wall placed at the bottom of the test section of the shock tube were sequentially visualized for Ms = 1.40 in atmospheric air at 293.7 K. Figure 6.25 show sequential observation of shock wave propagating along the slotted wall. The shock wave interacted with individual slots creating the wavelets that successively promoted mitigation. Figure 6.25 explains these generation of wavelets (Onodera and Takayama 1990).

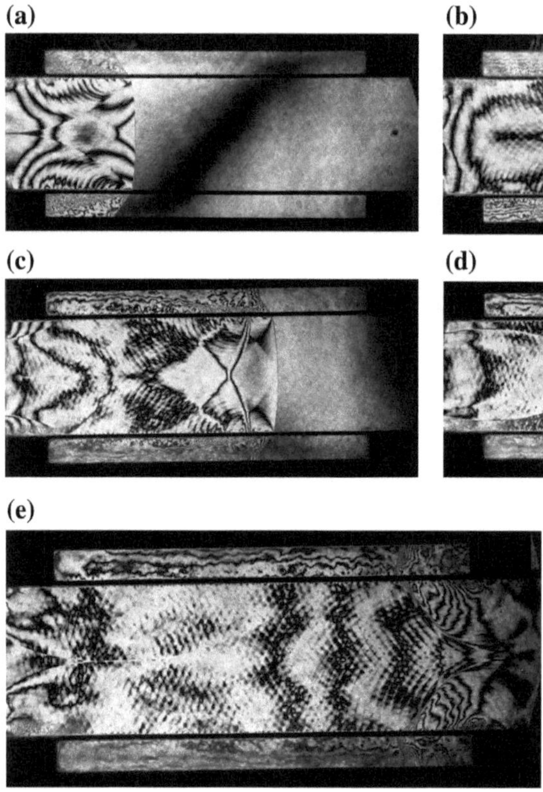

Fig. 6.21 Shock wave mitigation along the sintered stainless plates backed up with solid walls for Ms = 1.38 in air at 600 hPa, 290.8 K: **a** #92011711, 600 μs from trigger point, Ms = 1.392; **b** #92011712, 900 μs, Ms = 1.380; **c** #92011713, 800 μs, Ms = 1.390; **d** #92011714, 940 μs, Ms = 1.393; **e** #92011715, 1060 μs, Ms = 1.384

6.6.4 Distributed Slit Wall

A 150 mm × 230 mm test piece having distributed slits was placed on the upper and lower walls were inserted into the 150 mm × 250 mm test section of the 60 mm × 150 mm conventional shock tube. Figure 6.26 show shock wave propagation along slit walls for Ms = 1.40 in atmospheric air at 294.3 K. Irregularly shaped grey patterns observed in Fig. 6.26a were compression waves leaked through a narrow gap between the test piece and the shock tube side walls. Such compression waves merged with the transmitting shock wave and were not observable in Fig. 6.26b.

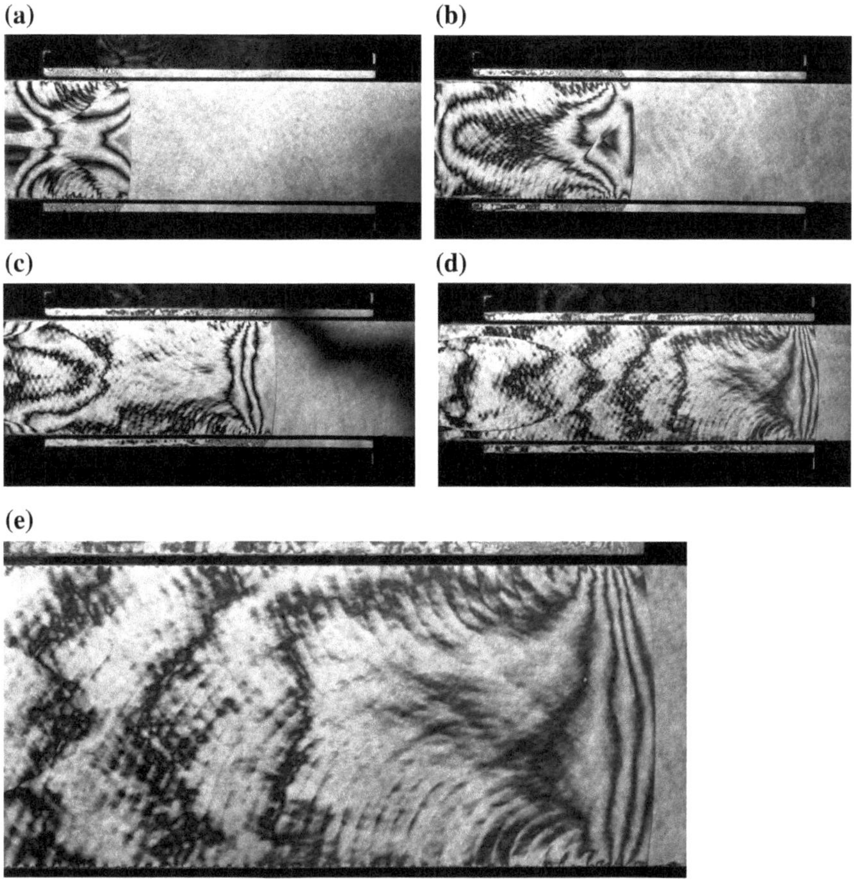

Fig. 6.22 Evolution of shock wave propagating along sintered stainless steel plates backed up with Alporous^{TR} wall for Ms = 1.40 in atmospheric air at 290.8 K: **a** #92011808, 600 μs from the trigger point, Ms = 1.399; **b** #92011809, 700 μs, Ms = 1.396; **c** #92011810, 800 μs, Ms = 1.396; **d** #92011811, 940 μs, Ms = 1.401; **e** enlargement of (**d**)

6.7 Alporous^{TR} Walls

Figure 6.27 show the evolution of the shock wave propagation along aluminum sponge Alporous^{TR} walls for Ms = 1.15 in atmospheric air at 282.0 K. The upper and lower walls and the both sides of the 150 mm × 250 mm test section were covered with a 10 mm thick Alporous^{TR}. It proved to be a useful material as liners for suppressing shock waves generated along engine exhaust gas pipe lines. Covering the test section with liners of Alporous^{TR} as shown in Fig. 6.27a does not eliminate the twin vortices at the test section entrance.

The shock wave was not reflected from the Alporous^{TR} liners but was transmitted from the bottom of the liners. It was reconfirmed to be a good material for

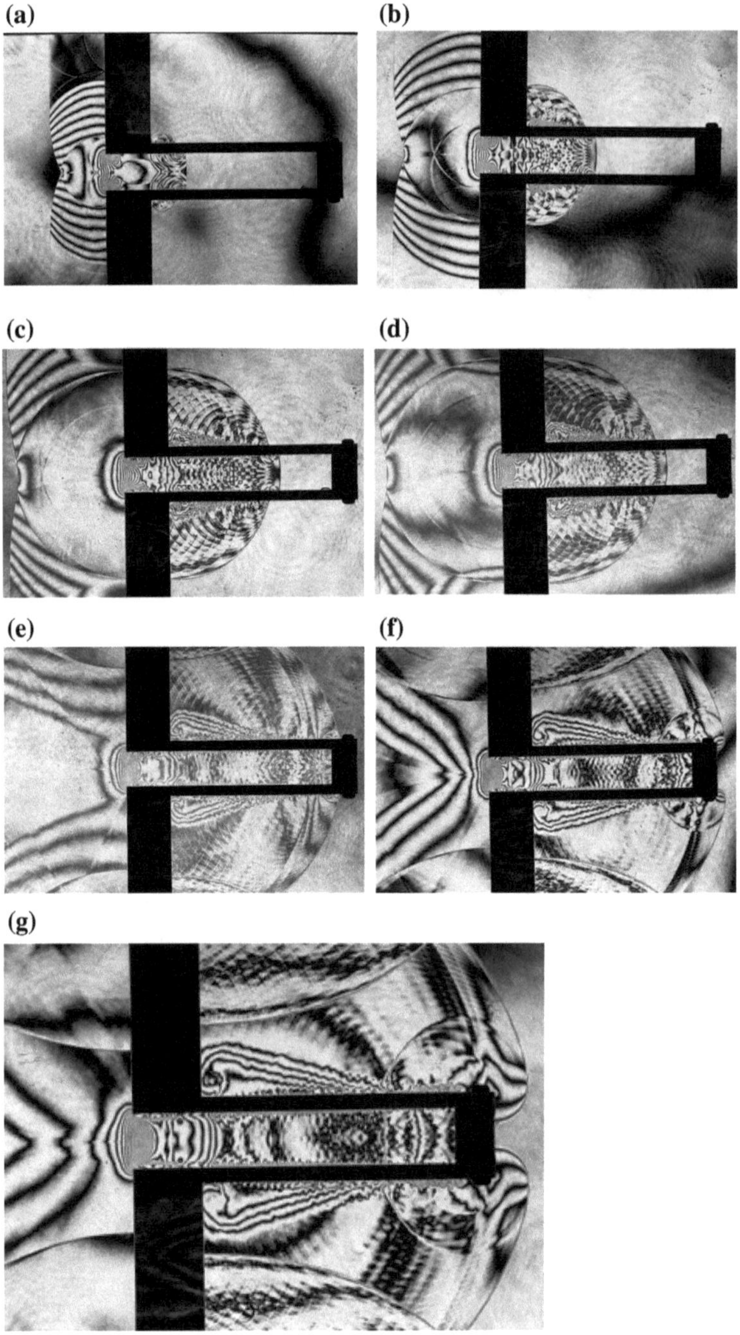

Fig. 6.23 Shock wave propagation along the perforated wall having perforation ratio 0.5 for Ms = 1.30 in air at 900 hPa, 292.7 K: **a** #87030415, Ms = 1.302; **b** #87030409, Ms = 1.260; **c** #87030303, Ms = 1.219; **d** #87030401, Ms = 1.217; **e** #87030405, Ms = 1.259; **f** #87030407, Ms = 1.257; **g** #87030413, 700 μs, Ms = 1.254

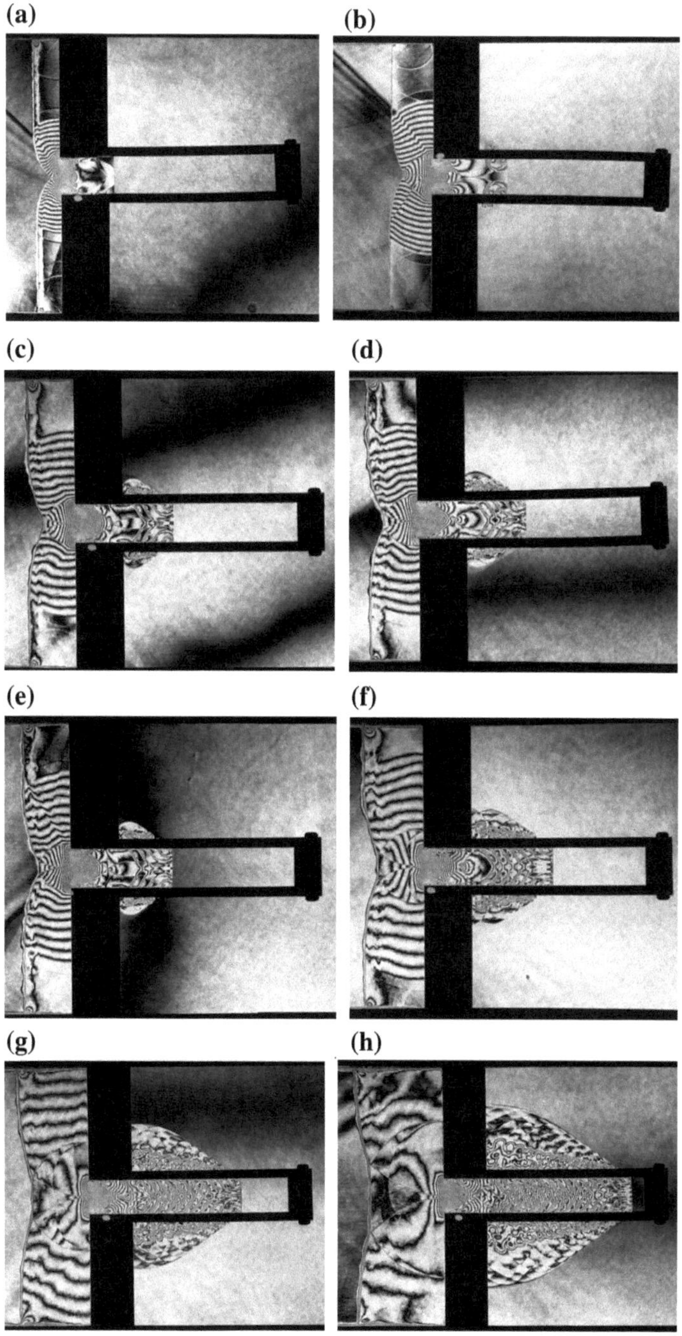

Fig. 6.24 The evolution of the shock wave propagating along a perforated wall having perforation ratio 0.2 for Ms = 2.50 in air at 333 hPa, 290 K: **a** #91111408, Ms = 2.560; **b** #91111416, Ms = 2.429; **c** #91111402, Ms = 2.560; **d** #91111409, Ms = 2.429; **e** #91111403, Ms = 2.560; **f** #91111404, Ms = 2.307; **g** #91111405, Ms = 2.209; **h** #91111406, Ms = 2.560

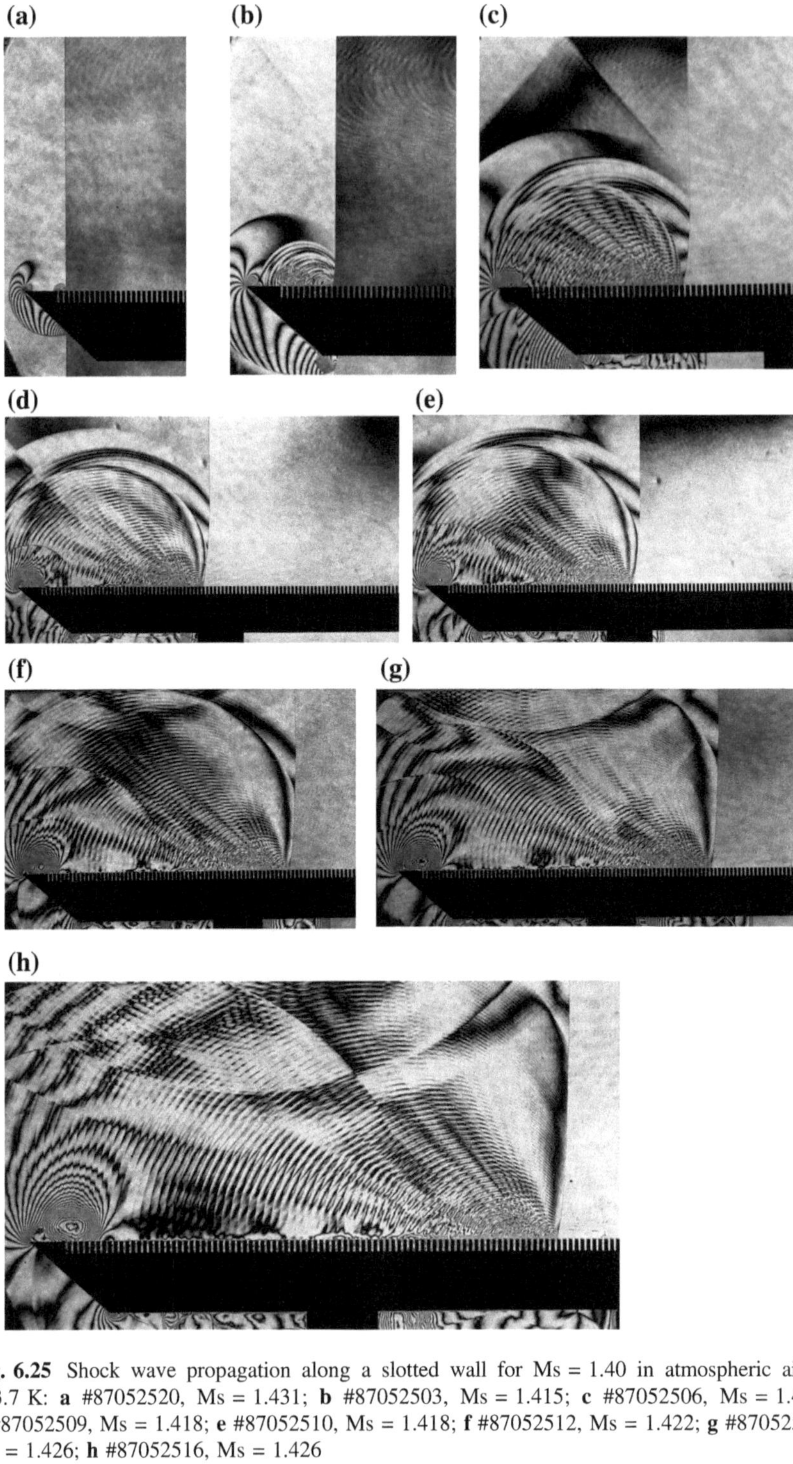

Fig. 6.25 Shock wave propagation along a slotted wall for Ms = 1.40 in atmospheric air at 293.7 K: **a** #87052520, Ms = 1.431; **b** #87052503, Ms = 1.415; **c** #87052506, Ms = 1.431; **d** #87052509, Ms = 1.418; **e** #87052510, Ms = 1.418; **f** #87052512, Ms = 1.422; **g** #87052514, Ms = 1.426; **h** #87052516, Ms = 1.426

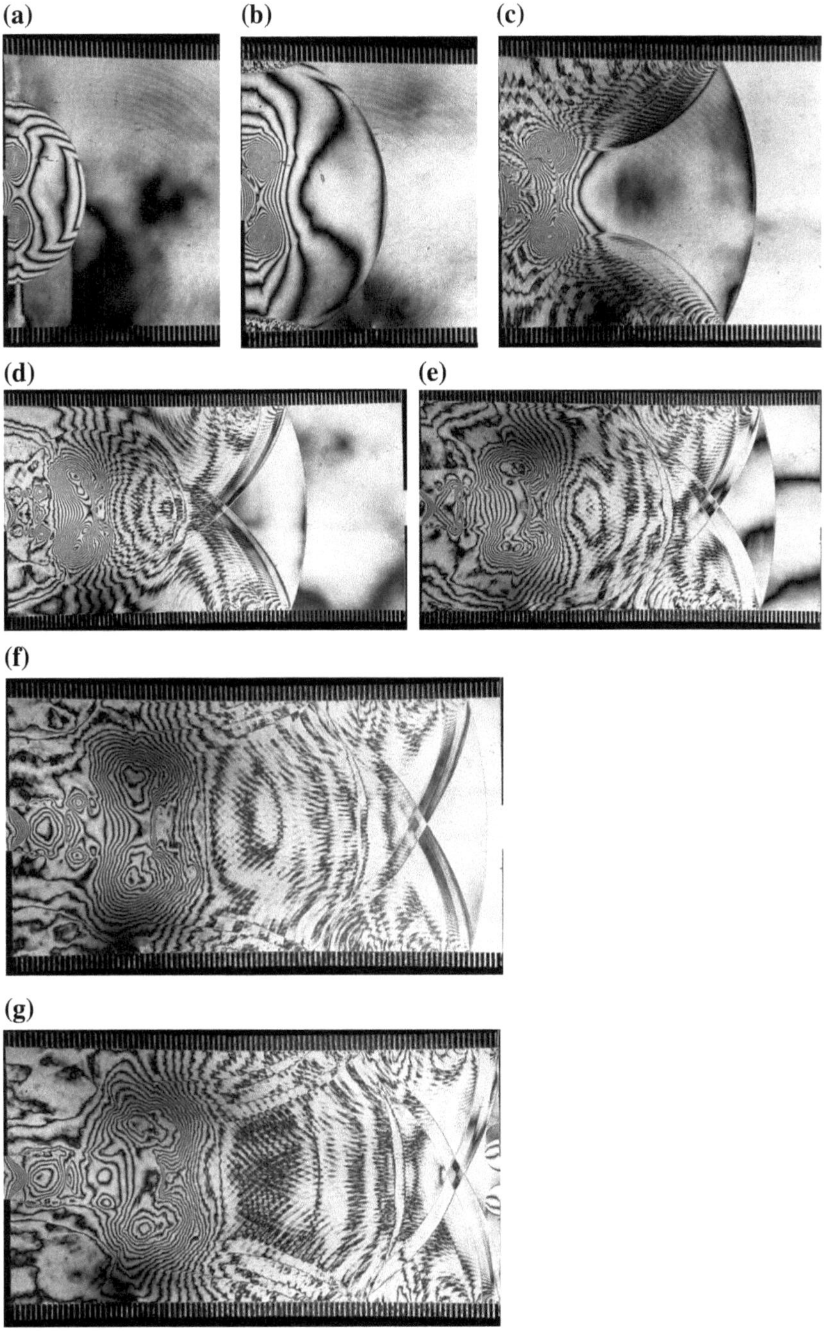

Fig. 6.26 Shock wave propagation along slit wall for Ms = 1.40 in atmospheric air at 294.3 K: **a** #90092811; **b** #90092810; **c** #90092809; **d** #90092807; **e** #90092807; **f** #90092808

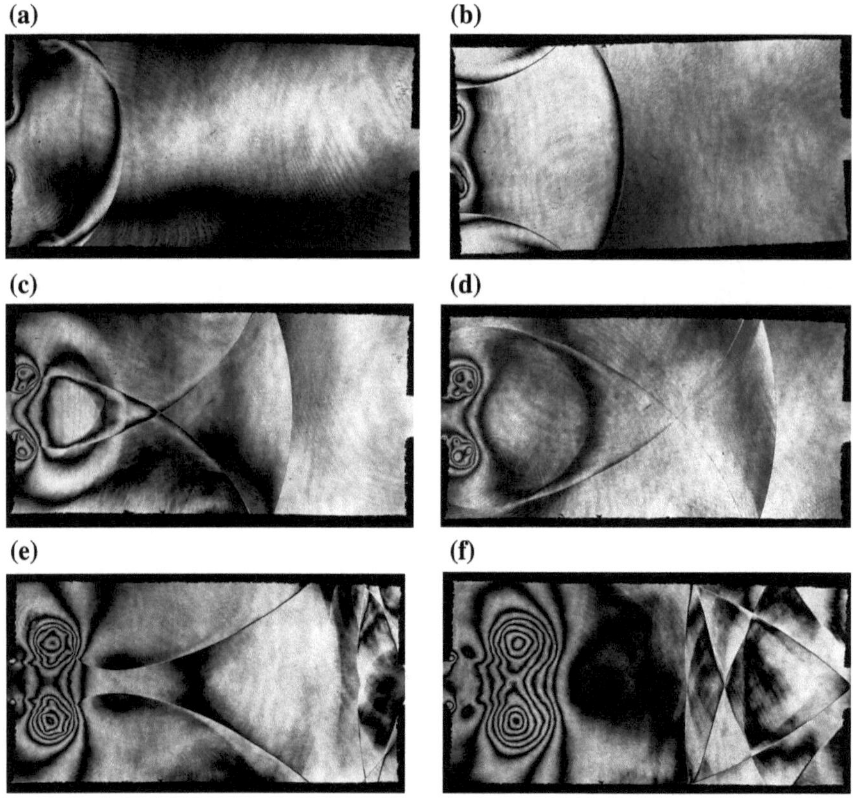

Fig. 6.27 The evolution of the shock wave propagation along aluminum sponge (Alporous[TR]) walls for Ms = 1.15 in atmospheric air at 282.0 K: **a** #88011907, 350 μs, Ms = 1.145; **b** #88011908, 430 μs, Ms = 1.153; **c** #88011910, 590 μs, Ms = 1.153; **d** #88011911, 670 μs, Ms = 1.147; **e** #88011913, 900 μs, Ms = 1.151; **f** #88011914, 1100 μs, Ms = 1.157

mitigating shock waves. Reflected shock waves became gradually faint during their propagation.

In the case of very weak shock waves shown in Fig. 6.28, the transmitted shock wave became faint. In Fig. 2.28f, no distinct reflections are observed. It should be emphasized that the Alporous[TR] is a good material for liners in silencers.

6.8 Shock Wave Mitigation While Passing Through Multiple Partitions

When entering into layers of partitions, the shock wave repeatedly reflected back and forth between the obstacles. Such multiple diffractions and reflections lead to promote the mitigation of shock waves. The transmitted shock waves also

(a) (b) (c) (d) (e) (f)

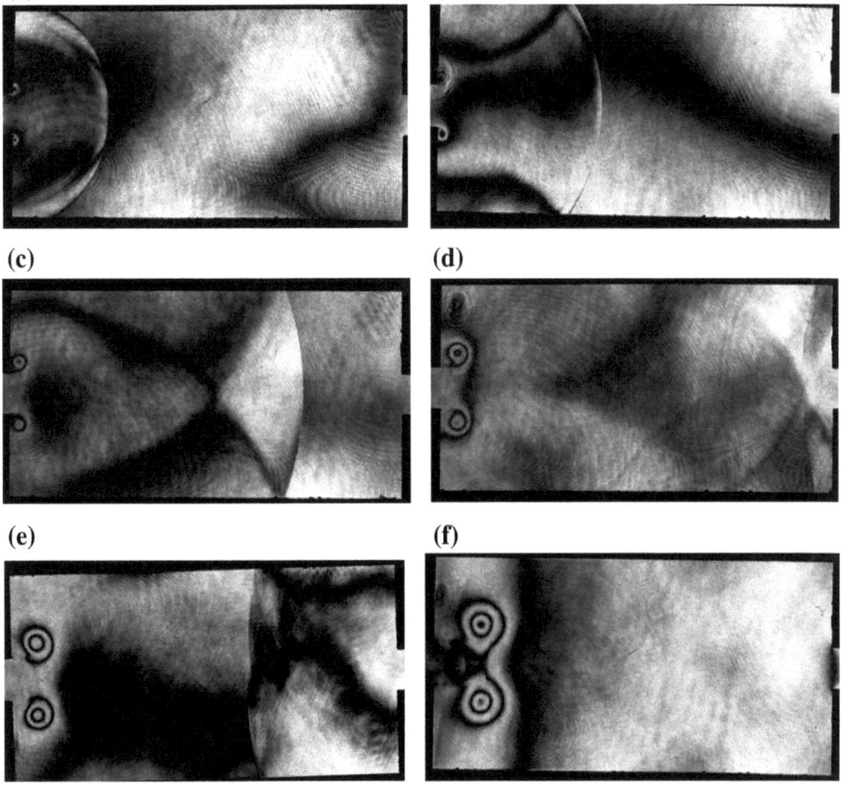

Fig. 6.28 The revolution of shock wave propagation along the Alporous^{TR} walls for Ms = 1.08 in atmospheric air at 282.0: **a** #88012005, 400 μs from trigger point, Ms = 1.082; **b** #88012007, 700 μs, Ms = 1.084; **c** #88012007, Ms = 1.084; **d** #88012008, 900 μs, Ms = 1.086; **e** #88012009, 1100 μs, Ms = 1.086; **f** #88012010, 1500 μs, Ms = 1.081

interacted with sidewalls. The entire wave interaction results in a complex flow field containing triple points accompanied with slip lines or vortices, which effectively promote the attenuation of the transmitted shock wave. The shock wave mitigation amount depends on the specific geometry of the partitions. Dependence of the shock wave mitigation on the partition geometry is discussed here. A comparison between numerical simulations and interferograms describing shock wave passing through different partitions would be an interesting bench mark test.

6.8.1 Mitigation Through Partitions

Figure 6.29a–j show one of the simplest partition: a 150 mm × 250 mm test section was split into 50 mm wide partitions. The transmitted shock wave created

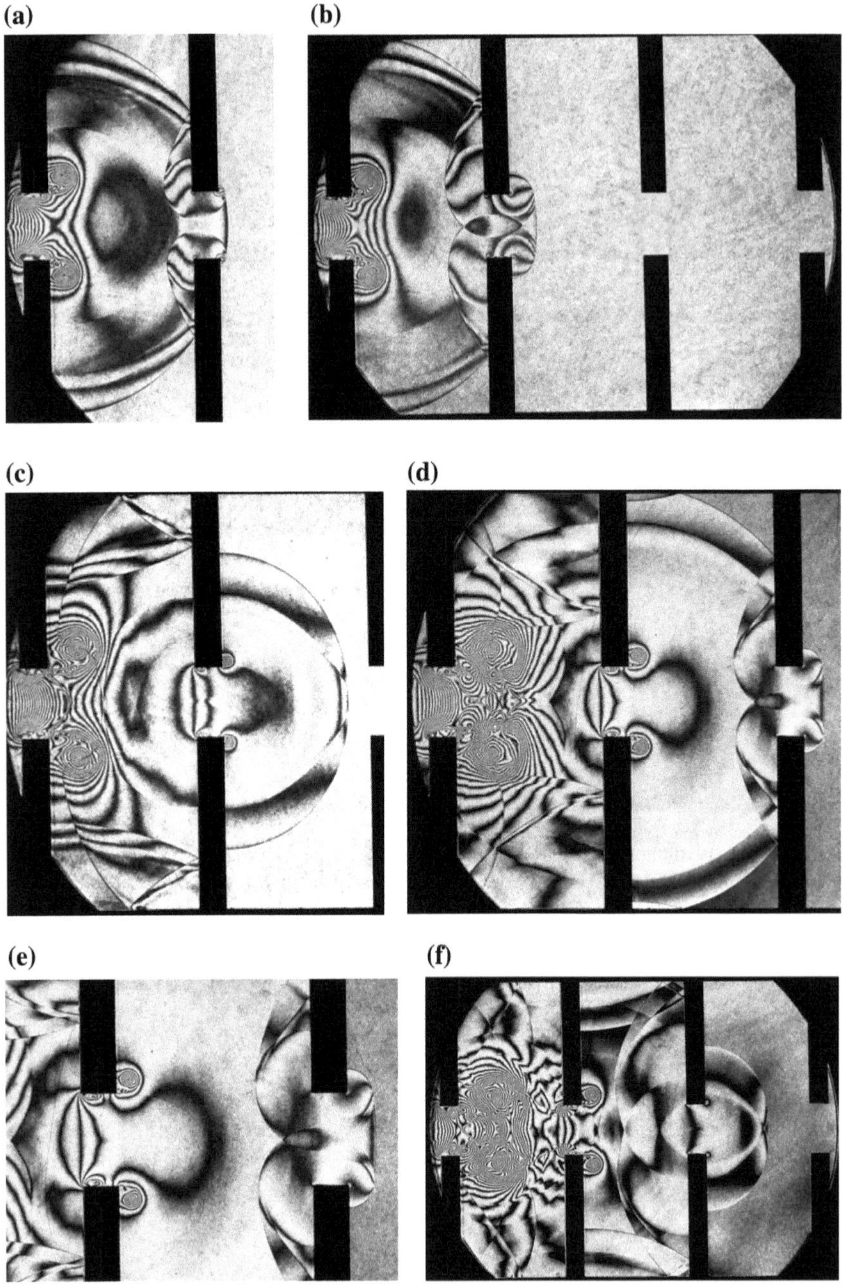

Fig. 6.29 The shock wave propagation through a partition for Ms = 1.20 in air: **a** #97122505, Ms = 1.201; **b** #97122510, Ms = 1.194; **c** #97122503, Ms = 1.198; **d** #97122201, Ms = 1.193; **e** enlargement of (**d**); **f** #98010602, Ms = 1.196; **g** enlargement of (**f**); **h** #98010604, Ms = 1.203; **i** #98010605, Ms = 1.197; **j** enlargement of (**i**)

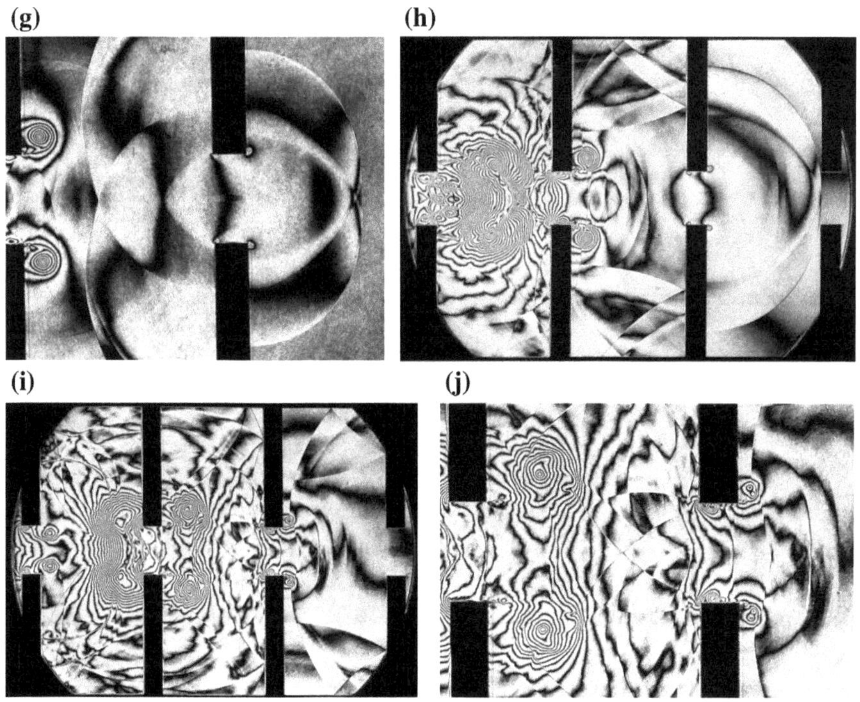

Fig. 6.29 (continued)

twin vortices at the entrance. The vortices interacted with the reflected shock waves to result in complex wave interactions. This leads to the attenuation of the transmitted shock wave.

6.8.2 Cookie Cutter Entry and Exit

The mufflers and silencers of automobile exhaust pipe lines were designed based mainly on acoustic theories. However, noises caused along the exhaust pipe lines were not necessarily acoustic noises but weak shock waves. Therefore, the acoustic theories are hardly effective. A partition having a staggered entrance and exist was installed in the 150 mm × 250 mm test section and sequentially visualized was conducted. In Fig. 6.30a–f, attenuation experienced by the transmitted shock wave passing through cookie cutter.

Figure 6.30f is an enlargement of Fig. 6.30e. The effect of cookie cutter on the shock wave attenuation was not promising because the sharp entrance shape would not effectively disturb the transmitted shock wave.

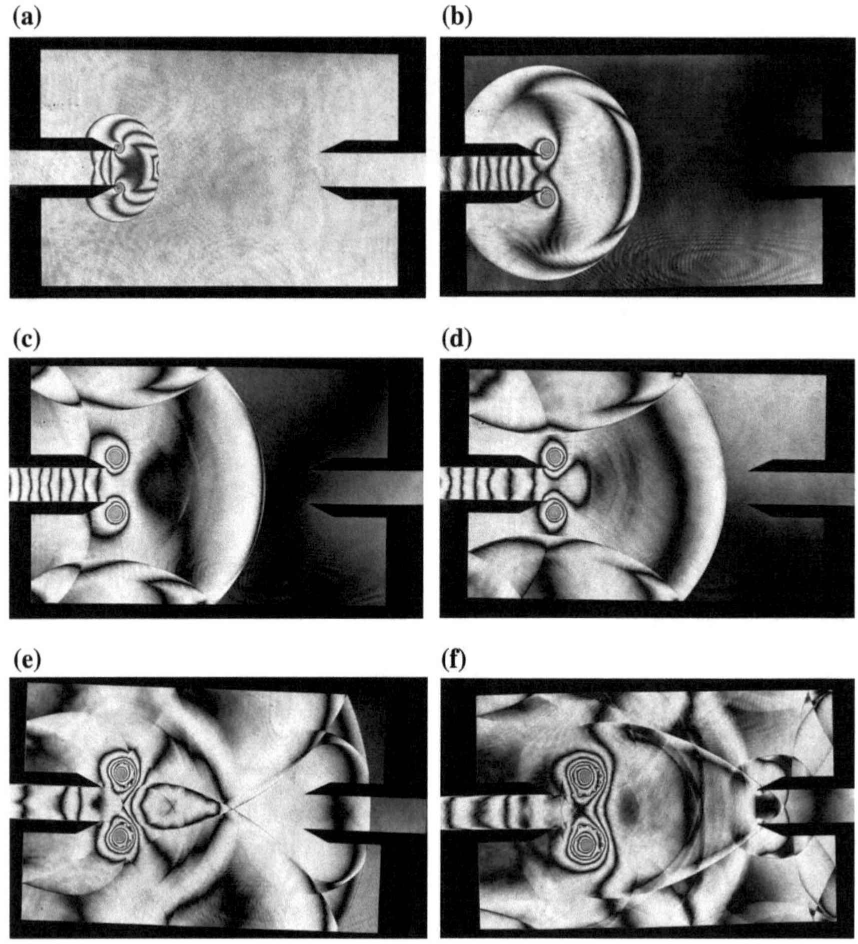

Fig. 6.30 Shock wave entry into a partition having a cookie cutter entrance and exit for Ms = 1.15 in atmospheric air at 296.7 K: **a** #87093002, Ms = 1.154; **b** #87093003, Ms = 1.150; **c** #87093004, Ms = 1.153; **d** #87093005, Ms = 1.142; **e** #87093007, 750 μs, Ms = 1.150; **f** #87093008, Ms = 1.151

6.8.3 Staggered Cookie Cutter Entry and Exit

The cookie cutter entry and exit were arranged in staggered manner. Experiments were conducted for Ms = 1.15 in atmospheric air at 296.7 K. The cookie cutter entry and exit did not contribute to significantly attenuate the transmitted shock wave but the staggered arrangement promoted attenuation. Figure 6.31a–h show shock wave attenuation in propagating in a full length partition having a staggered entry and exit for Ms = 1.15 in atmospheric air.

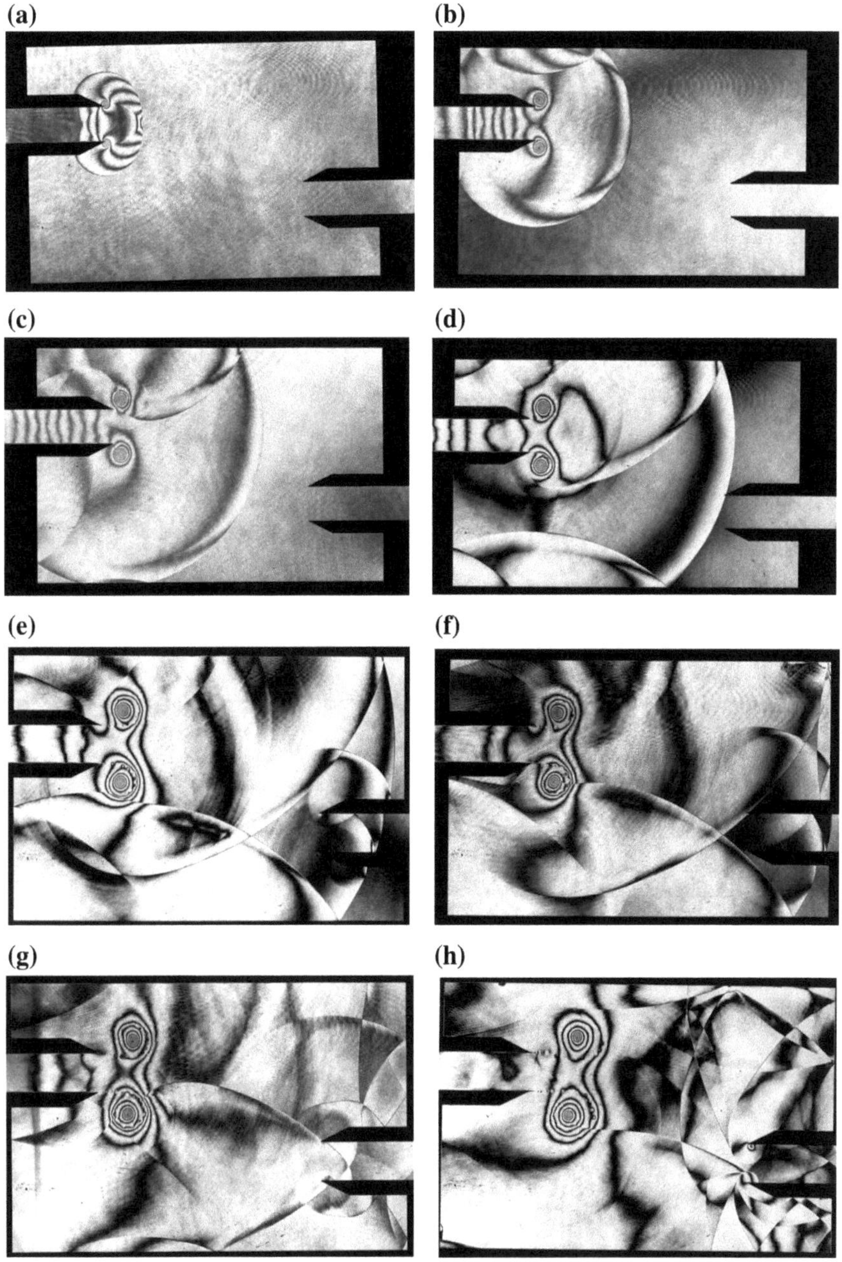

Fig. 6.31 Shock wave entry into a partition with cookie cutter entrance and exit in staggered position for Ms = 1.15 in atmospheric air at 296.7 K: **a** #87093011, Ms = 1.147; **b** #87093012, Ms = 1.152; **c** #87093013, Ms = 1.149; **d** #87093014, Ms = 1.147; **e** #87093015, Ms = 1.155; **f** #87093016, Ms = 1.140; **g** 87093017, Ms = 1.148; **h** #87093019, Ms = 1.15

6.8.4 Short Partition of Staggered Entry and Exit

See Fig. 6.32a–f show shock wave attenuation in propagating in a half length partition having a staggered entry and exit for Ms = 1.13 in atmospheric air. Shock waves in this arrangement attenuated more effectively than that in the full length partition.

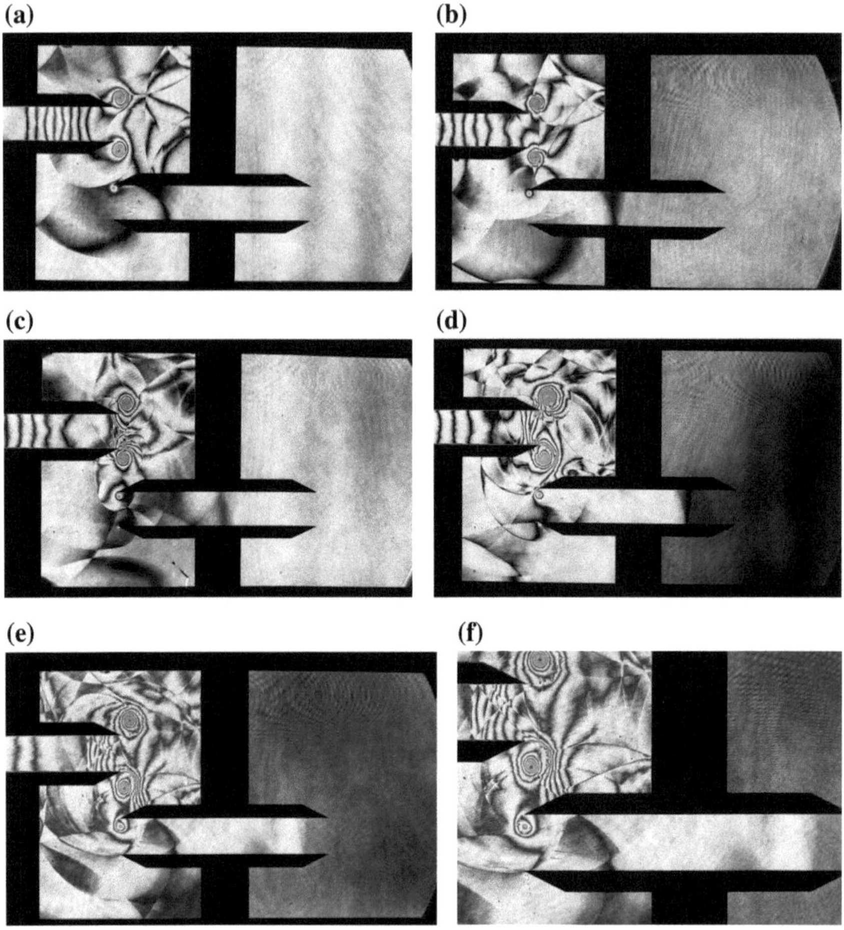

Fig. 6.32 Shock wave entry into a short partition composed of cookie cutter entrance and exit in staggered position for Ms = 1.13 in atmospheric air at 296.7 K: **a** #87100101, 500 μs from trigger point; **b** #87100102, 550 μs; **c** #87100103, 600 μs; **d** #87100104, 650 μs, Ms = 1.14; **e** #87100105, 770 μs, Ms = 1.14; **f** enlargement of (**e**)

6.8.5 *Short Partition with Straight Entry and Oblique Exit*

Figure 6.33 show the shock wave interaction with a short partition having the blunt shape entrance and exit arranged in staggered position for Ms = 1.38 in

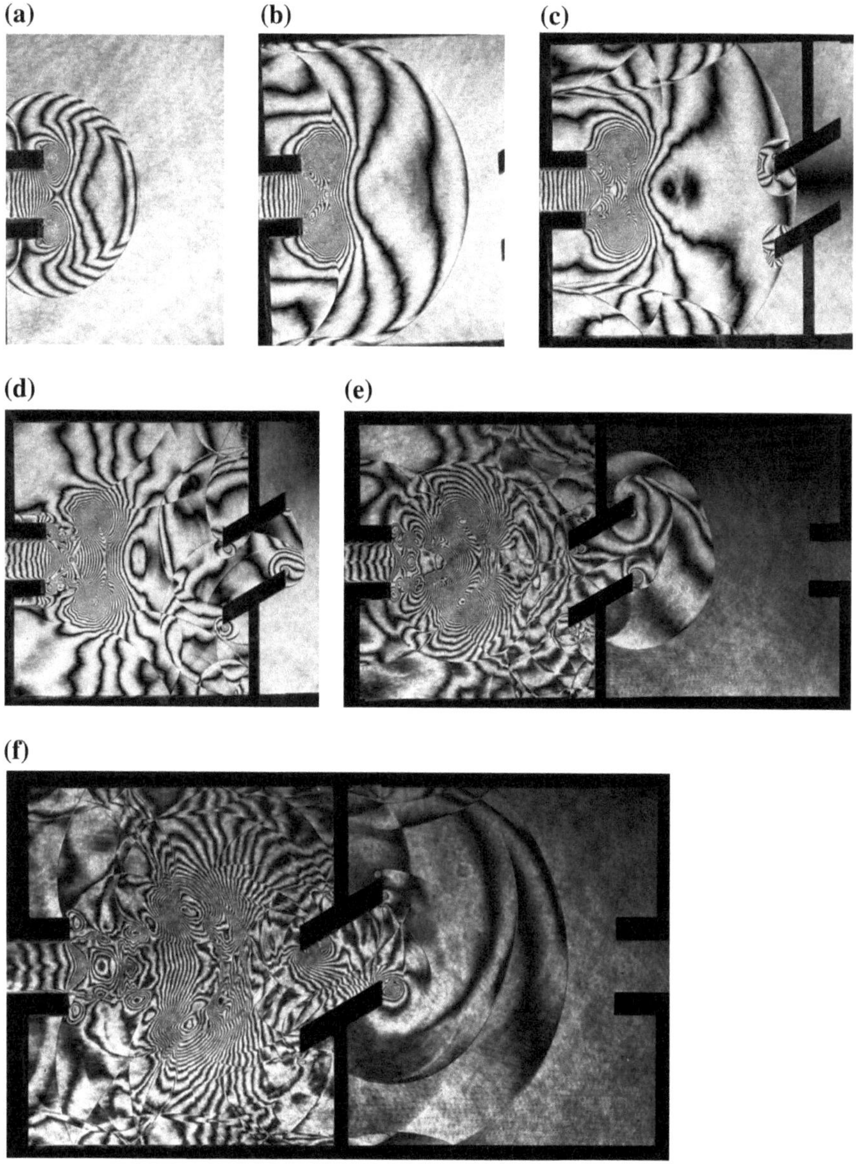

Fig. 6.33 Shock wave entry into a short partition with the entrance and exit arranged in staggered position for Ms = 1.38 in atmospheric air at 294.2 K: **a** #92012010; **b** #92012011; **c** #92012012; **d** #92012013; **e** #92012014; **f** #92012015

atmospheric air at 294.2 K. The test pieces were installed in the 60 mm × 150 mm diaphragm-less shock tube. Blunt edges promoted diffraction and reflection of the transmitted shock waves and effectively attenuated the transmitted shock wave. It is clear from Fig. 6.33 shows that the transmitted shock wave passing the tilted exit is attenuated.

6.8.6 Upright Baffle Plates

In a synchrotron radiation facility, intense synchrotron radiation is mitted by irradiations of high-energy beams on a thin beryllium plate. Then, the beryllium plate separates an ultra high-vacuum ring from the ambient air. In case the beryllium plate accidentally ruptured, the atmospheric air rushed into the ring. Then rushed air would break not only the high-vacuum but also contaminate the surface of the structures inside the ring (Ohtomo 1998).

Taking the pressure ratio of air in the ambient condition and the ultra-high-vacuum condition into account, this arrangement is analogous to a shock tube having very high pressure ratio. Therefore, the beryllium plate rupture creates a high speed air flow which is equivalent to a contact surface in the shock tube flows and drive a strong shock wave. The gate valve is equipped in a beam line, a duct connecting the ring and the beryllium plate, and is intended to immediately stop the shock wave and the contact surface in case the beryllium plate ruptured; it is important to retard the shock wave and contact surface arriving at the gate some time before the gate valve was closed. Usually, skimmers are arranged in a row along the beam line. The skimmers work as partition plates.

In the recorded interferograms, obliquely entering shock waves attenuated faster than those entering upright shock waves much more effectively than the upright entry. Therefore, even if it is a two-dimensional experiment, the effectiveness of oblique baffle plates would be verified testing it in the 180 mm × 1100 mm test section of the 100 mm × 180 mm diaphragm-less shock tube. Figure 6.34a, b show the test section of the refurbished UTIAS hypervelocity shock tube and the combination of four pieces of 250 mm × 250 mm plane mirrors, which formed 250 mm × 1000 mm mirror (Ohtomo 1998).

Figure 6.35 show the evolution of the shock wave propagating along an upright arrayed baffle plates for Ms = 3.0. The shock wave propagates from right to left and the tip of the upright baffles were sharpened. The field of view height vi is 180 mm and the opening width is 20 mm. Pressure transducers (Kistler 603B) were placed marked from a to h. Figure 6.35a shows that the diffracting shock wave is going to interact successively with the attenuating baffle plates. Leaving from the last baffle plate, the transmitted shock wave attenuated significantly as seen in the pressure history measured at port f in Fig. 6.35f. In Fig. 6.35h, the ordinate denotes dimension less pressure normalized by the ambient pressure. The abscissa denotes the elapsed time in μs. Blue line shows the IS pressure at t = 0. Red and green lines

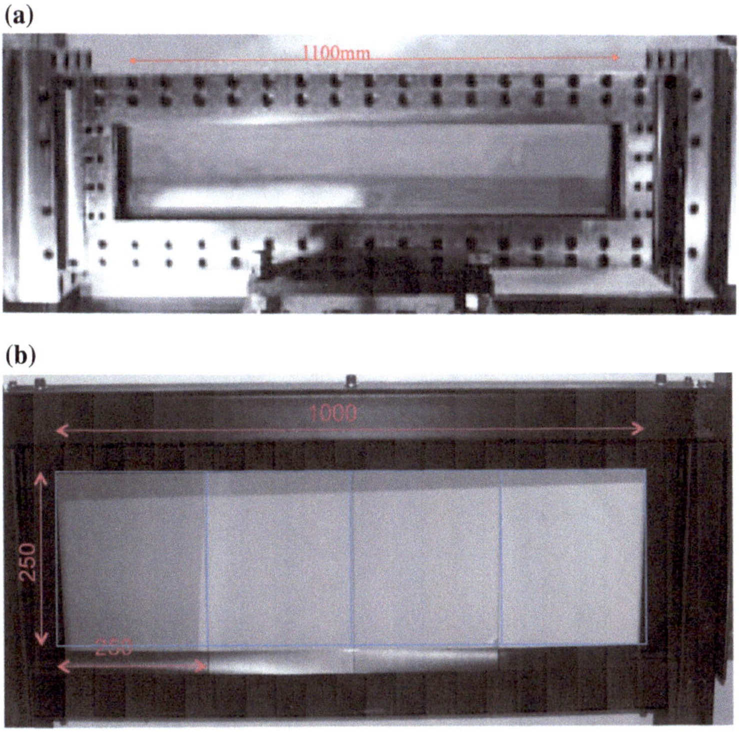

Fig. 6.34 The 1100 mm × 180 mm test section connected to the refurbishes 100 mm × 180 mm shock tube: **a** 180 mm × 1100 mm test section; **b** combination of four pieces of 250 mm × 250 mm plane mirrors (Ohtomo 1998)

denote appropriate numerical simulation (Voinovich et al. 1998) and measured data, respectively. They agree well with each other.

6.8.7 Staggered Baffle Plates

Figure 6.36 show the sequence of shock wave propagation along oblique baffle plates arranged staggered arrangement for Ms = 3.0 in air. The transmitted shock wave and the following contact region propagate along a zig-zag course. Figure 6.36a shows the staggered baffle plate arrangement and distribution of pressure measure by a transduces Kistler 603B.

Figure 6.36g shows the recorded pressure history measured at the port d by Kistler 603B. The ordinate denotes dimension-less pressure normalized by ambient pressure. The abscissa denotes the elapsed time in μs. Blue line denotes the incident shock pressure measure at t = 0. Red and green lines denote the numerical

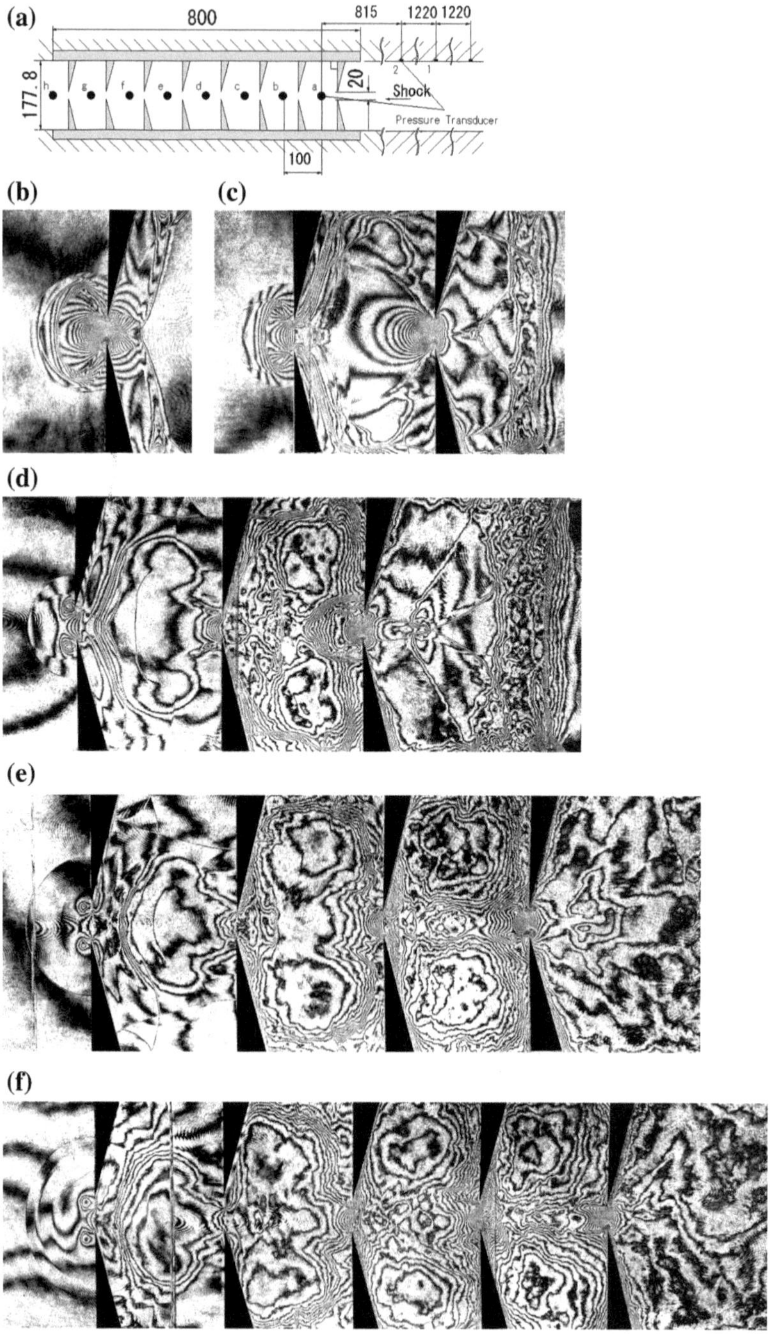

Fig. 6.35 The evolution of the shock wave attenuation along upright baffle plates for Ms = 3.0 in air: **a** baffle plate arrangement; **b** #02111103; **c** #02111104; **d** #02111102; **e** #02111101; **f** #0211010; **g** #02111106; **h** pressure history at the port (**f**) (Ohtomo 1998)

(g)

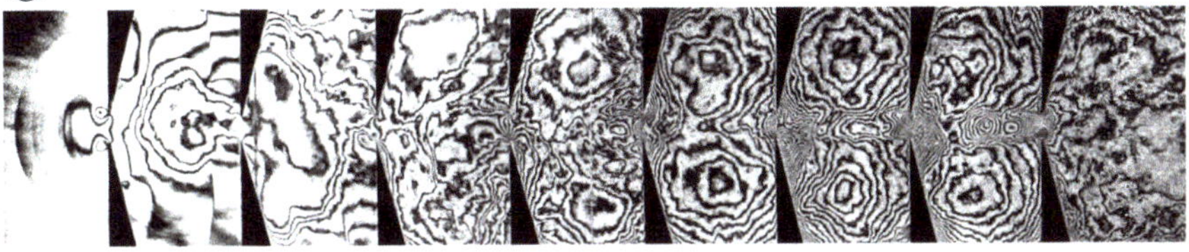

(h)

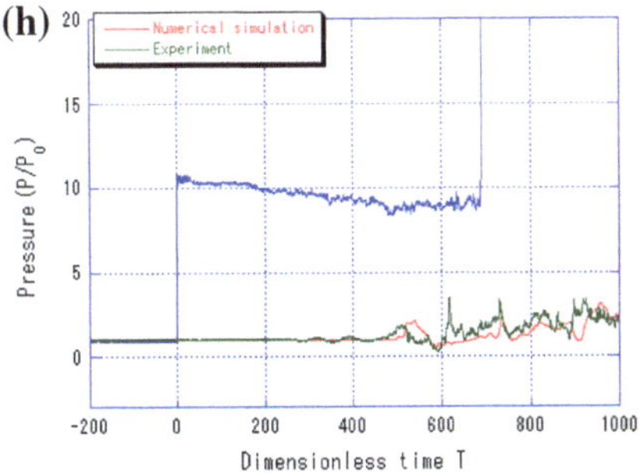

Fig. 6.35 (continued)

simulation and the measured pressure distribution, respectively. The numerical result agrees well with the measured ones.

Comparison of the present pressure history with the one shown in Fig. 6.35h, readily indicates that the oblique baffle plate arrangement effectively attenuated the transmitted shock waves. In conclusion, crossing obliquely arranges skimmers, gas particles would take a zig-zag course and then the contact surface would be slowed down resulting in effectively attenuating the passing shock wave.

6.8.8 Numerical Comparison Between Upright and Staggered Oblique Baffle Plates

The effect of the baffle plate alignments on the transmitted shock wave was investigated solving the Euler equations for incident shock Mach number Ms = 1.5 in ideal diatomic gas (Voinovich et al. 1998). Obtained results revealed that a staggered oblique array of baffle plates attenuated the transmitted shock wave and slowed down the contact region motion significantly better than the similar upright baffle plates. Figure 6.37a–d show numerically constructed interferograms of shock wave propagation along an array of the upright baffle plates as well as numerical results of an array of the staggered oblique It should be noticed that baffle plates' edge had a blunt end.

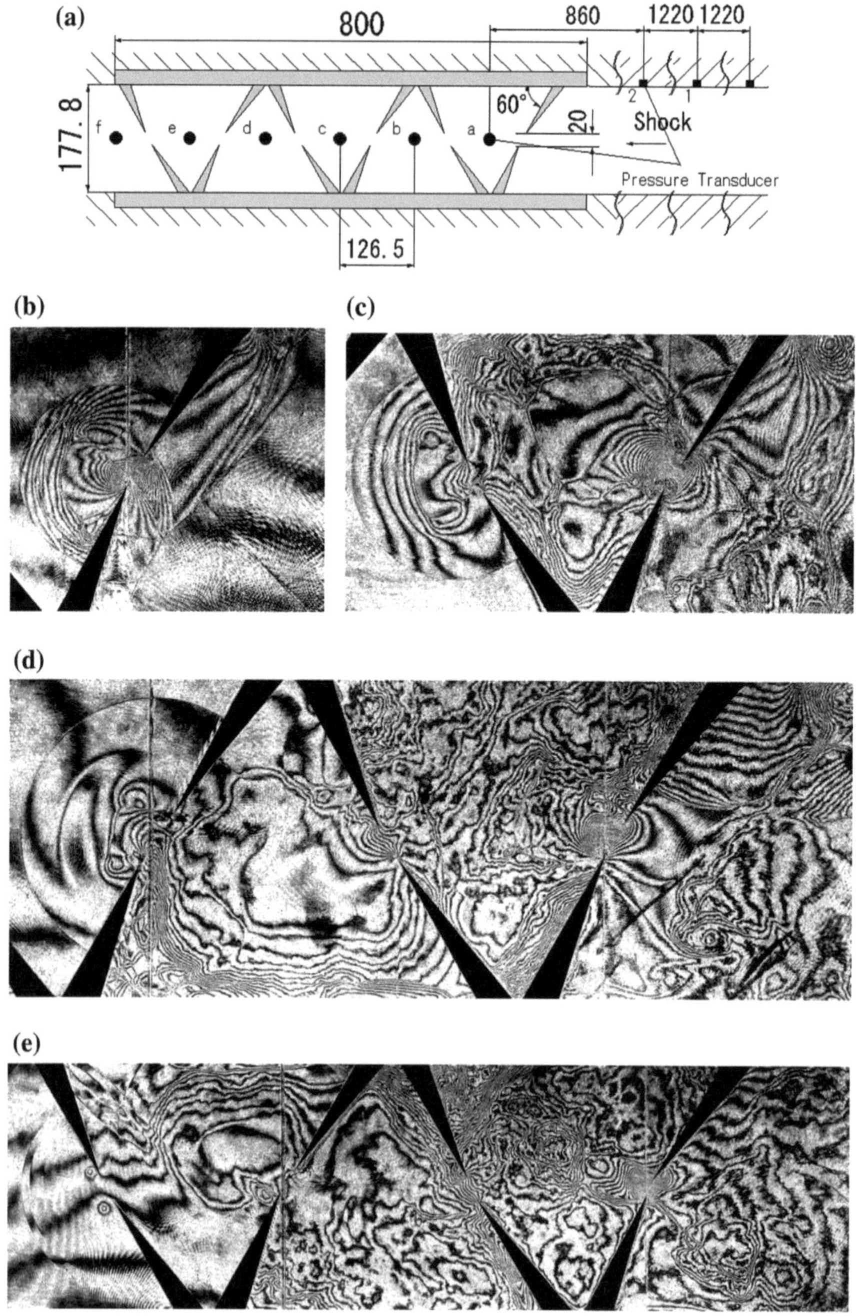

Fig. 6.36 The shock wave attenuation along staggered baffle plates for Ms = 3.0 in air: **a** baffle plate arrangement; **b** #02111108; **c** #02111107; **d** #02111109; **e** #02111105; **f** #02111111; **g** pressure history measured at the port (**d**) (Ohtomo 1998)

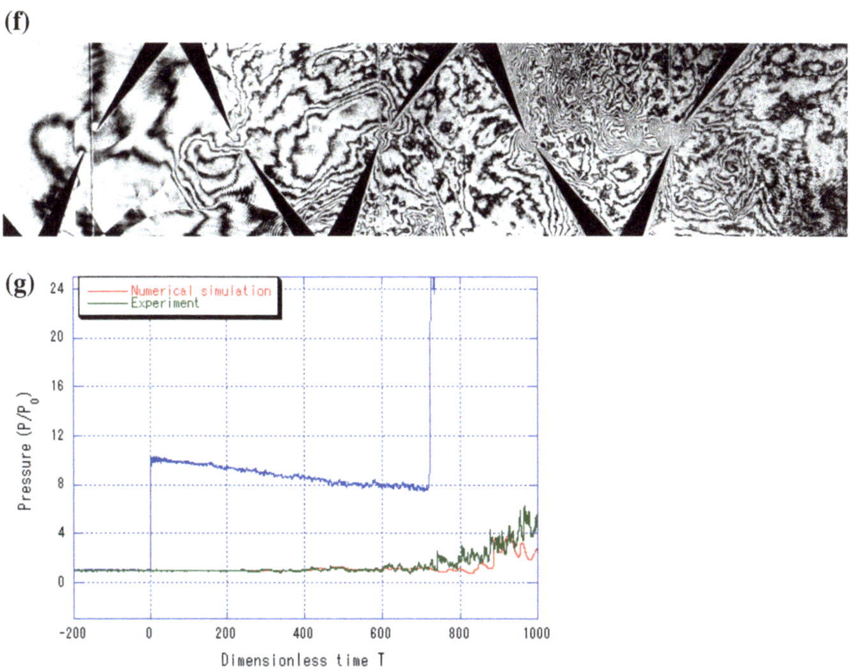

Fig. 6.36 (continued)

6.9 Shock Wave Propagation Along a Double Elbow

In order to experimentally investigate the shock wave propagation through relatively complex duct's geometries, the flow through a double elbow would be an appropriate topic as shock wave propagations along an elbow was already studied (Takayama 1993).

6.9.1 Double Elbows Having Smooth Surface

Figure 6.38 show the evolution of shock wave propagation through a double elbow duct having smooth surfaces for Ms = 1.21 in atmospheric air at 294 K. This is, in principle, a combination of two elbows. The incident shock wave was diffracted at the first entrance corners and repeated the diffraction at the second and the third corners. The transmitted shock wave was reflected from the first outside corner and repeated reflection at the second inside corner. The reflected shock waves interacted with the diffracted shock wave. The evolution of the shock wave propagation, eventually, became complex enough to make experimentalists amuse and to use numerical simulation to reproduce the prevailing flows. The transmitted shock wave attenuated when it exits the duct as seen in Fig. 6.38e.

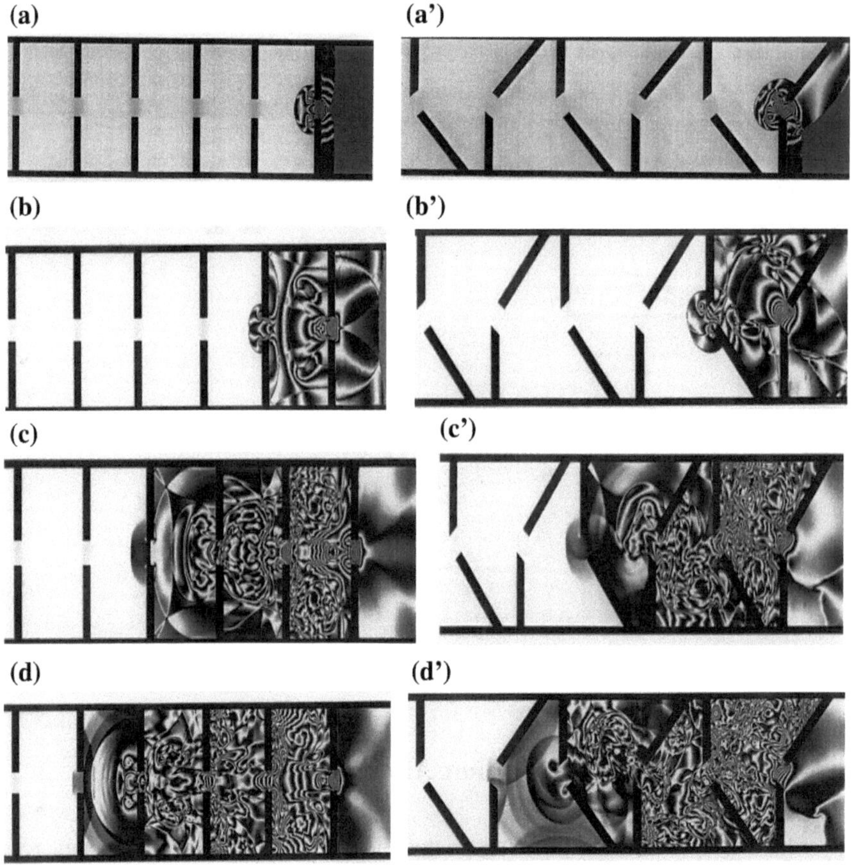

Fig. 6.37 Numerical comparison of shock wave mitigation between upright and staggered baffle plates for Ms = 1.5 in air (Voinovich et al. 1998)

6.9.2 Double Elbows Having Roughened Surface

As seen in Fig. 6.13, the shock wave was effectively attenuated through roughened surface. Figure 6.39 show the evolution of shock wave propagating along the double elbow having the same surface roughness as seen in Fig. 6.13. The double elbow having roughened surface attenuated the transmitted shock waves much more efficiently than that having smooth surface.

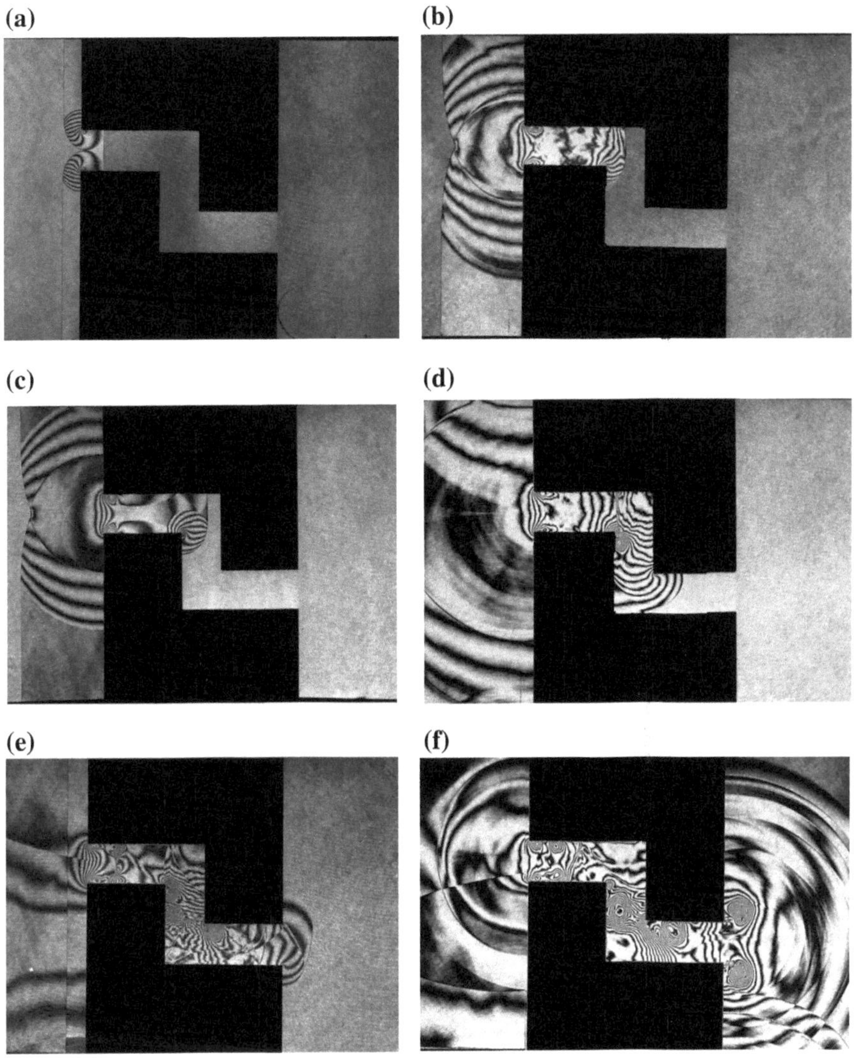

Fig. 6.38 The shock wave propagation through a double elbow having a smooth surface for Ms = 1.21 in atmospheric air at 294 K: **a** #93042001, 400 μs, Ms = 1.221; **b** #93042110, 500 μs from trigger point, Ms = 1.214; **c** #93042002, 500 μs, Ms = 1.220; **d** #93042111, 600 μs, Ms = 1.221; **e** #93042004, 700 μs, Ms = 1.219; **f** #93042006, 900 μs, Ms = 1.225

6.10 Arrayed Cylinders and Spheres

It is known that the train tunnel sonic booms released from tunnels having a gravel track are less-louder than those released from tunnels having a concrete floor or the so-called slab track. This difference is physically based on the fact that the gravel

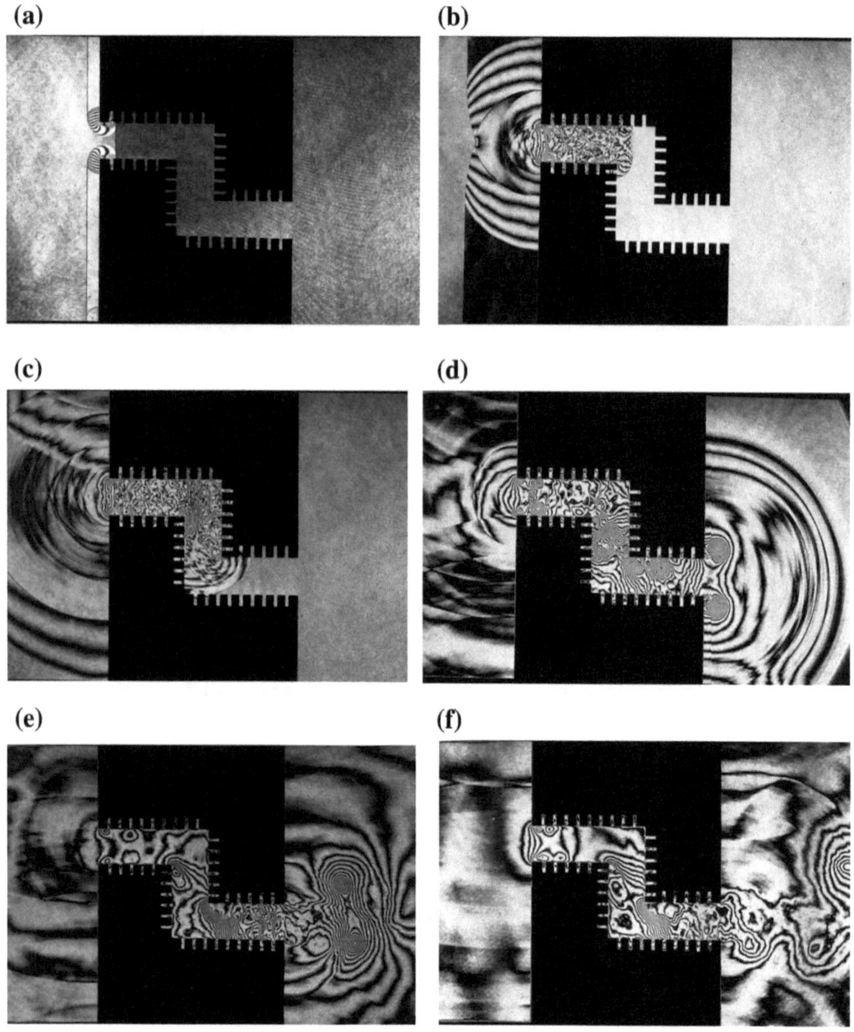

Fig. 6.39 Shock wave mitigation over a double elbow with perforations at Ms = 1.22 in air at 1013 hPa, 294 K: **a** #93042108, 400 μs from trigger point Ms = 1.219; **b** #93042107, 500 μs, Ms = 1.211; **c** #93042106, 600 μs, Ms = 1.215; **d** #93042103, 900 μs, Ms = 1.217; **e** #93042102, 1300 μs, Ms = 1.217; **f** #93042101, 1800 μs, Ms = 1.218

track is, in short, a layer of gravels and readily attenuates booms much more efficiently than the slab track.

During the operation of the Japan's first high-speed train network, no one claimed the train tunnel sonic booms. However, the high-speed train network was expanded to the southern Japan and the train speed was increased, people started to claim the train tunnel sonic booms. All the first generation train tunnels had gravel tracks, whereas all the upgraded train tunnels constructed in the second generation had slab tracks. The sonic booms were caused by the difference of the track's support. This became a motivation of the present series of experiments.

In an analogue experiment, the sequence of shock wave propagation along the arrayed cylinders or spheres was planned. At first the investigation was aimed at the visualization of shock wave propagating through three-dimensionally packed gravels. However, it was not a simple to reproduce such a situation by arranging spheres in three-dimensional space. Alternatively, an easy case was examined. Three 22 mm diameter spheres were connected in a line by truncated their edge and formed into 56 mm and installed the 60 mm × 150 mm diaphragm-less shock tube. The arrangement was shown in Fig. 6.40. Images of the shock wave propagation over these spheres were compared with that over 30 mm diameter cylinders (Abe 2002).

Figure 6.41 show the evolution of shock wave propagation along 30 mm diameter arrayed cylinders for Ms = 1.10 in atmospheric air at 293.5 K. As the reproducibility of experiments were good, then sequential observation using shorter delay time enabled to arrange the resulting interferograms in an animated display.

Figure 6.42 show the evolution of shock wave propagation along 30 mm diameter arrayed cylinders for Ms = 1.30 in air at 560 hPa, 289.5 K. A shock wave diffraction over a single cylinder was already discussed in Fig. 4.2. The diffraction and reflection occurred repeated as seen in Fig. 6.42b and its enlargement in Fig. 6.42c. The degree of the interaction among the transmitted shock waves became very complex. Then as already discussed, at every step when the transmitted shock waves passed the line of the cylinders, the fringe distribution behind the leading transmitted shock wave became complicated as seen in Fig. 8.35f.

Figure 6.43 show the evolution of shock wave propagation along 25 mm diameter arrayed spheres for Ms = 1.10 in atmospheric air at 295.5 K. The total projected area of the arrayed spheres to the shock tube cross section, the so-called

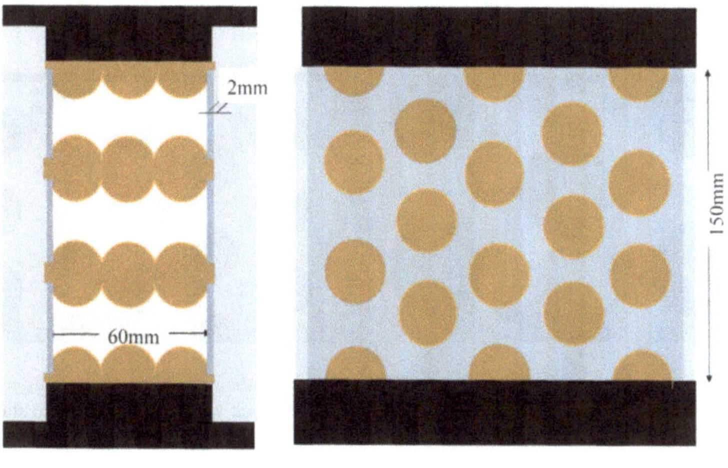

Fig. 6.40 Arrayed spheres installed in the 60 mm × 150 mm diaphragm-less shock tube

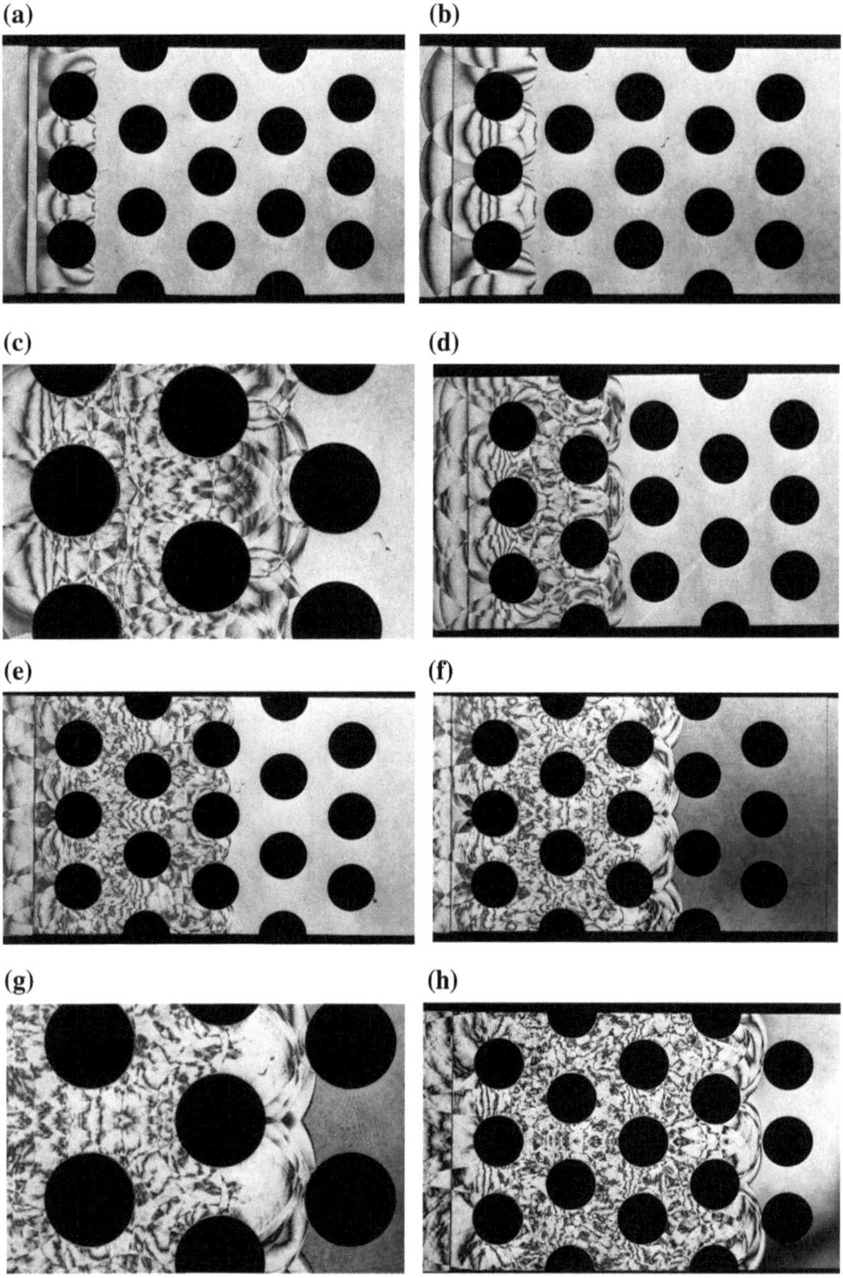

Fig. 6.41 Shock wave attenuation along an array of 30 mm diameter cylinders for Ms = 1.10 in atmospheric air at 293.5 K: **a** #98010910, 87 μs from the shock wave arrival at frontal stagnation point, Ms = 1.110; **b** #98010913, 116 μs, Ms = 1.104; **c** #98010908, 263.7 μs, Ms = 1.115; **d** #98010915, 245.5 μs, Ms = 1.104; **e** #98010909, 304.5 μs, Ms = 1.110; **f** #98010912, 381.8 μs, Ms = 1.103; **g** #98010912, Ms = 1.103; **h** #98010923, 497.1 μs, Ms = 1.111

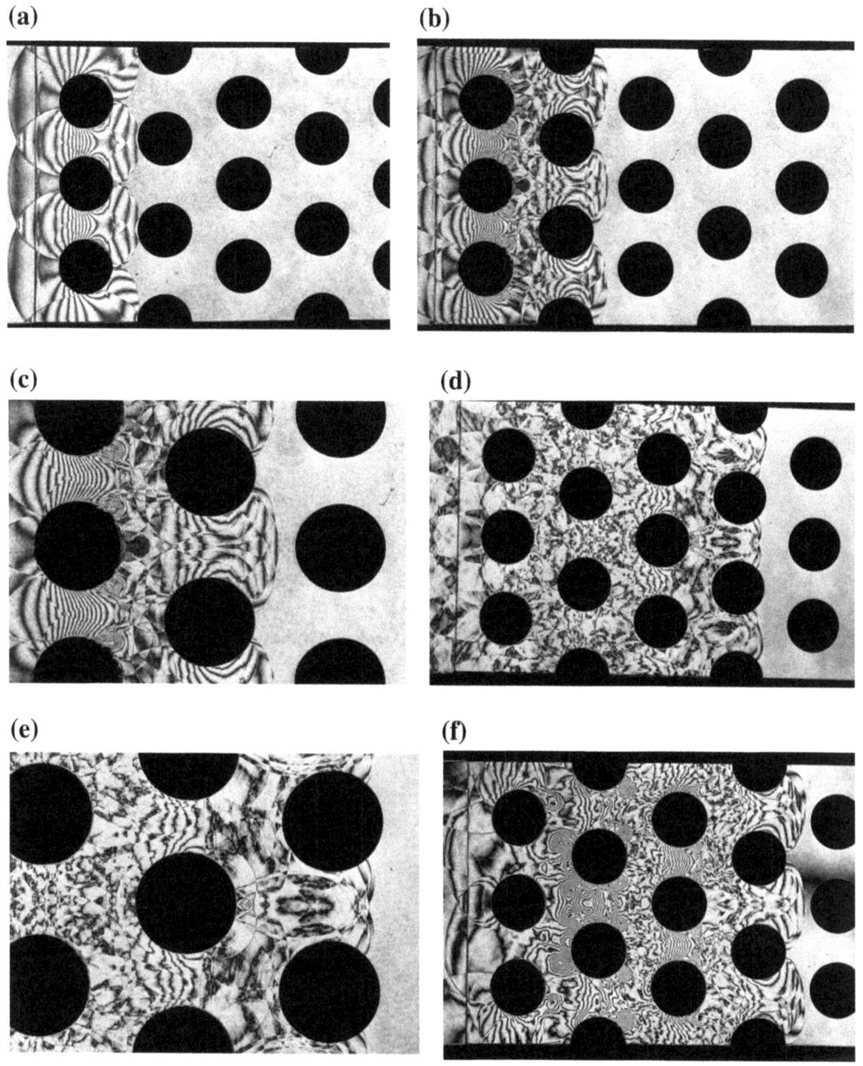

Fig. 6.42 Shock wave attenuation along an array of 30 mm diameter cylinders for Ms = 1.30 in air at 560 hPa, 289.5 K: **a** #98011205, 100 μs, Ms = 1.301; **b** #98011201, 184.3 μs, Ms = 1.298; **c** enlargement of (**b**); **d** #98011204, 346.4 μs, Ms = 1.294; **e** enlargement of (**d**); **f** #98011215, 364.5 μs, Ms = 1.306

the blockage ratio is 54% as seen in Fig. 6.43, whereas it is 63% for cylinders as seen in Fig. 6.42. In the interaction of a planar shock wave with a single sphere, the feature of the transmitted shock waves is different from the interaction of a planar shock wave with a single cylinder. Although the array of spheres are slightly similar to the case of arrayed cylinders, the fringe distributions observed in Fig. 6.43f look much more complicated than those observed in Fig. 6.42f.

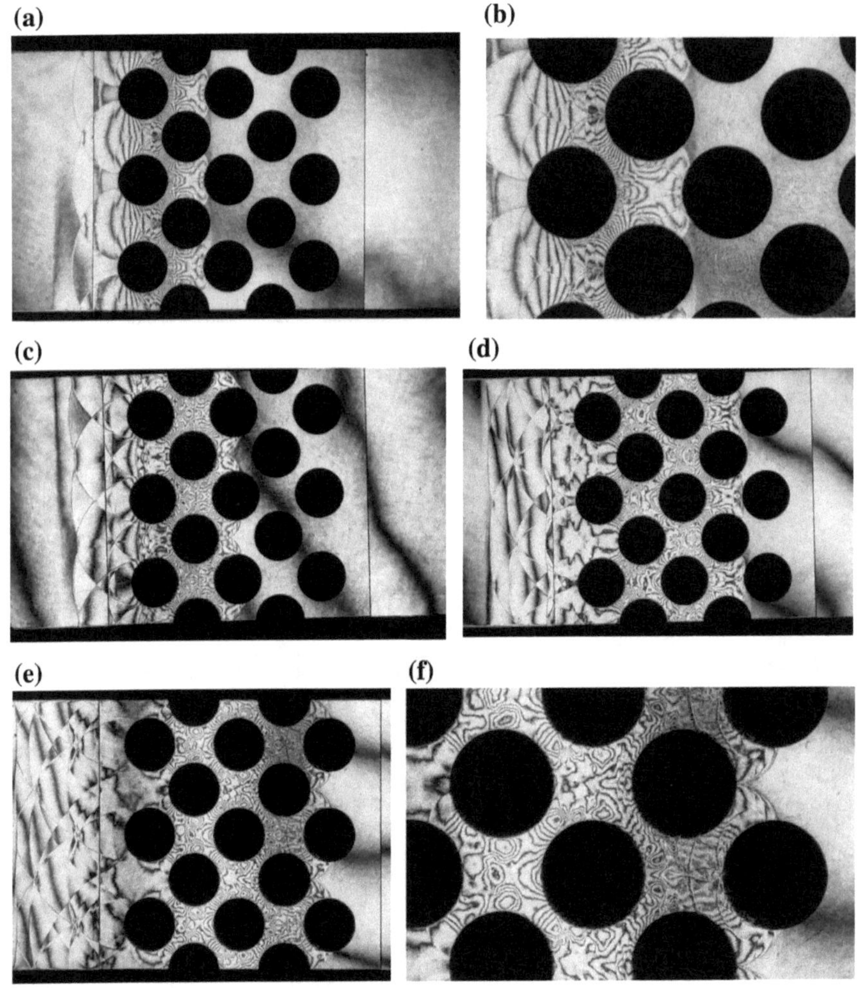

Fig. 6.43 Shock wave attenuation along an array of 22 mm diameter spheres for Ms = 1.10 in atmospheric air at 295.5 K: **a** #99102916, 575 μs from trigger point, Ms = 1.099; **b** enlargement of (**a**); **c** #99102917, 195 μs, Ms = 1.115; **d** #99102918, 272.5 μs, Ms = 1.099; **e** #99102920, 312.0 μs, Ms = 1.115; **f** enlargement of (**e**)

Figure 6.44 show an animated display of arrayed five 22 mm diameter spheres for Ms = 1.10. Sequential interferograms are edited for making an animated display. The blockage ratio of this particular arrangement is 75%. In this arrayed sphere five spheres were installed in a line, whereas four spheres were arranged in Fig. 6.43, therefore, the blockage ratio was 54%. The fringe distributions behind the leading shock wave became simpler with elapsing time as observed in Fig. 6.44r.

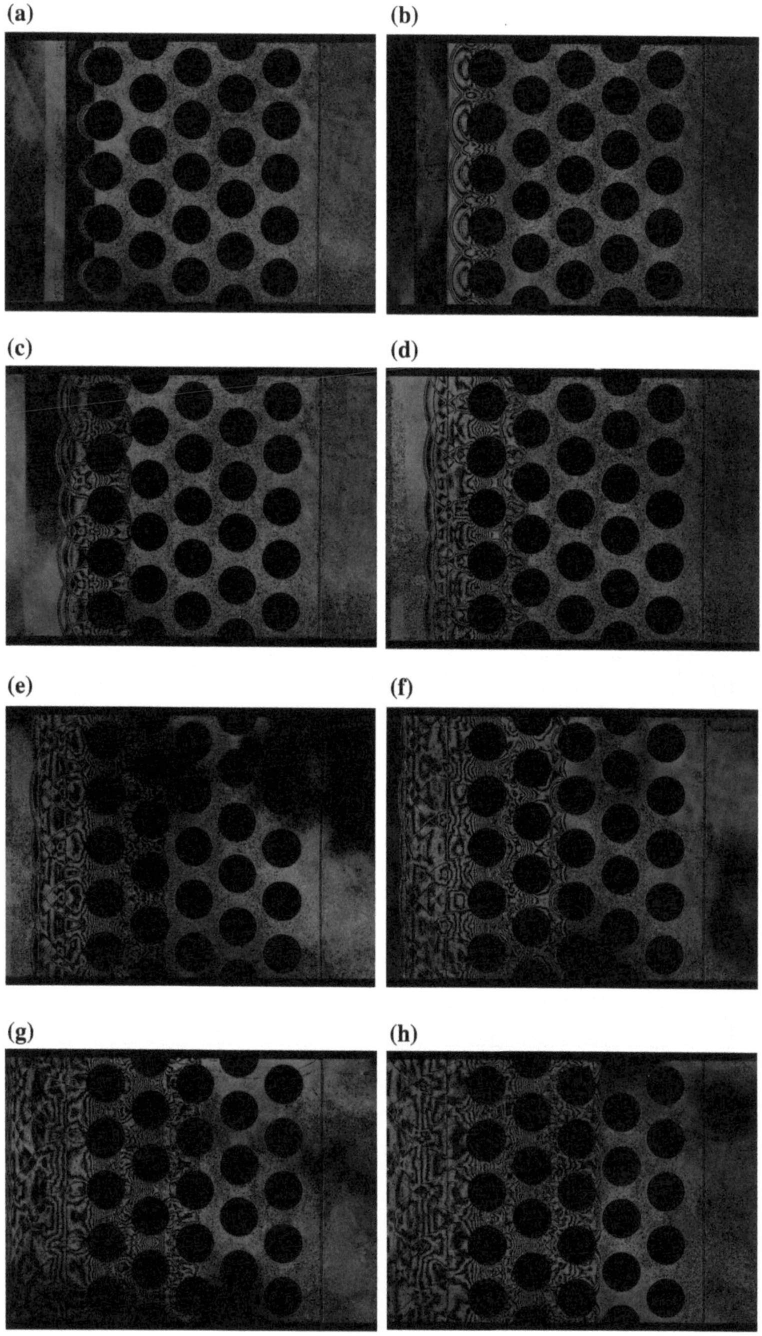

Fig. 6.44 Animated display of shock wave propagation along arrayed spheres, #99102901 for Ms = 1.10, 22 mm diameter spheres (Abe 2002)

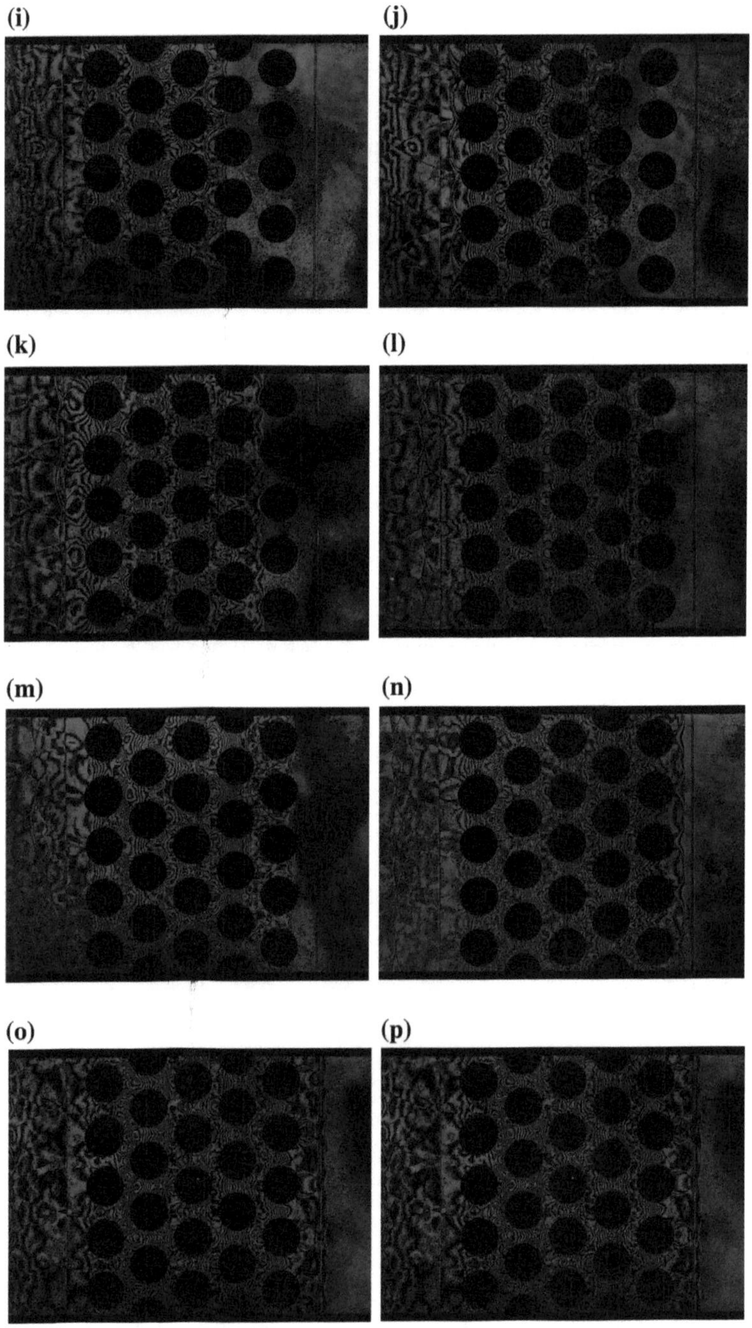

Fig. 6.44 (continued)

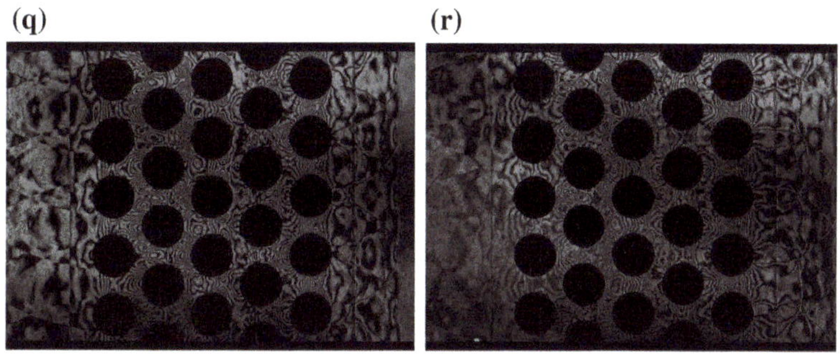

Fig. 6.44 (continued)

6.11 Shock Waves Released from Trailing Edges

When a shock wave passes along a slender body whose upper surface and lower surface have different surface roughness, the shock waves on the upper surface and lower surface would have different strength. If these shock waves are released from the trailing edge as shown in Fig. 6.38, a vortex will be formed some time after an elapsed time.

6.11.1 Vortex Formation

A 500 mm long splitting plate was inserted along the center line of the 60 mm 150 mm diaphragm-less shock tube. The trailing edge was positioned at the entrance of the test section as seen in Fig. 6.45. The upper part of the splitting plate

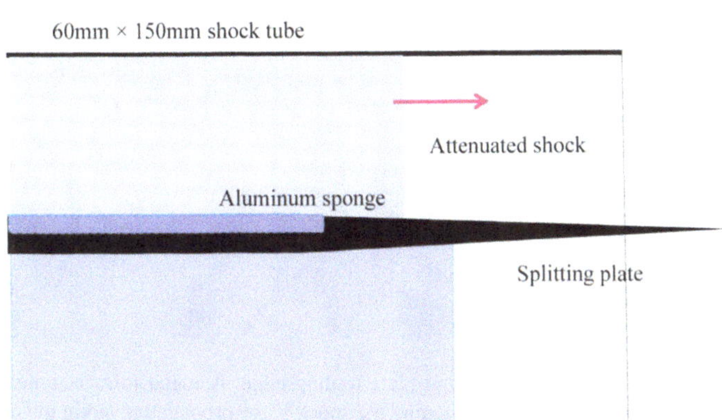

Fig. 6.45 A vortex released from a trailing edge of the 60 mm × 150 mm diaphragm-less shock tube

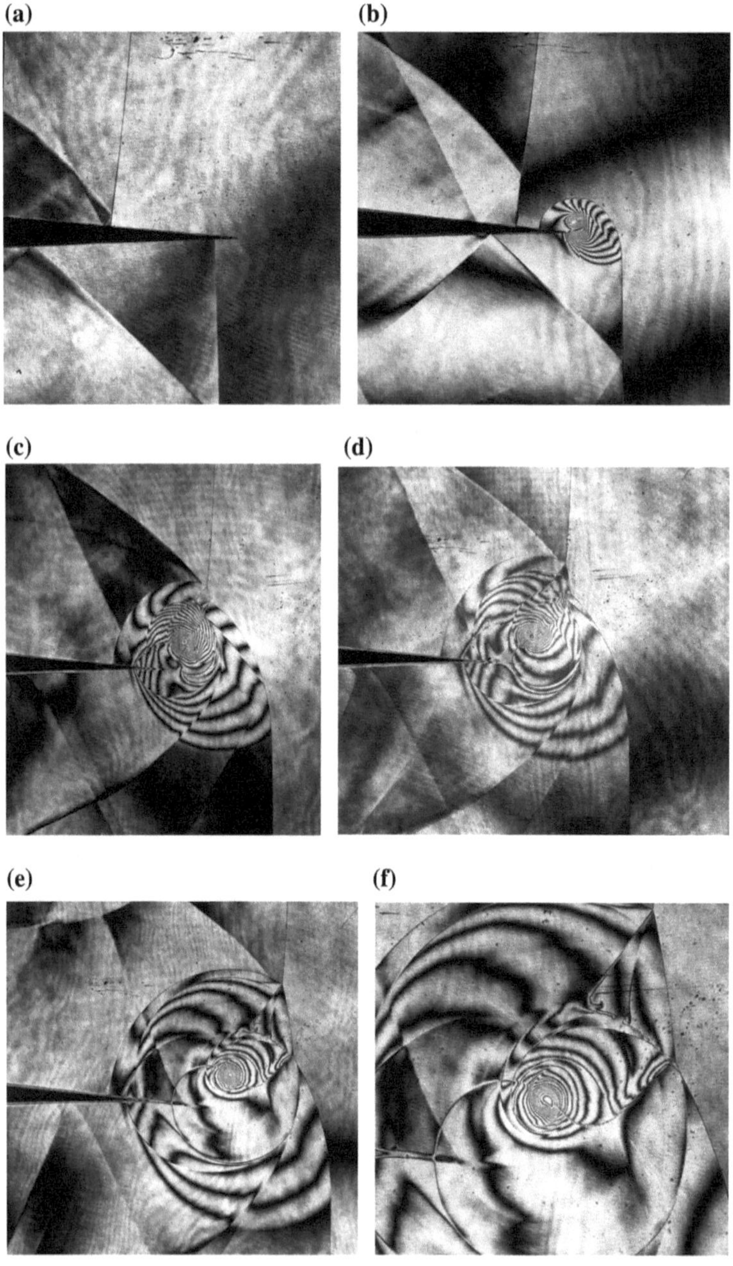

Fig. 6.46 Vortex formation at a splitting plate trailing edge. A metal form was installed on the upper side of the splitting plate for attenuating the shock wave propagating along the upper side for Ms = 2.2 in air at 300 hPa, 288.5 K: **a** #88021003, 635 μs from trigger point, Ms = 2.176: **b** #88021002, 670 μs, Ms = 2.183; **c** #88020909, 715 μs, Ms = 2.152; **d** #88020908, 730 μs, Ms = 2.159; **e** #88020907, 745 μs, Ms = 2.193; **f** #88020907, 745 μs, Ms = 2.193; **g** #88020904, 0.8 ms, Ms = 2.263; **h** #88020904, 0.8 ms, Ms = 2.263; **i** #88020903, 840 μs, Ms = 2.177; **j** #88020902, 1.9 ms, Ms = 2.193

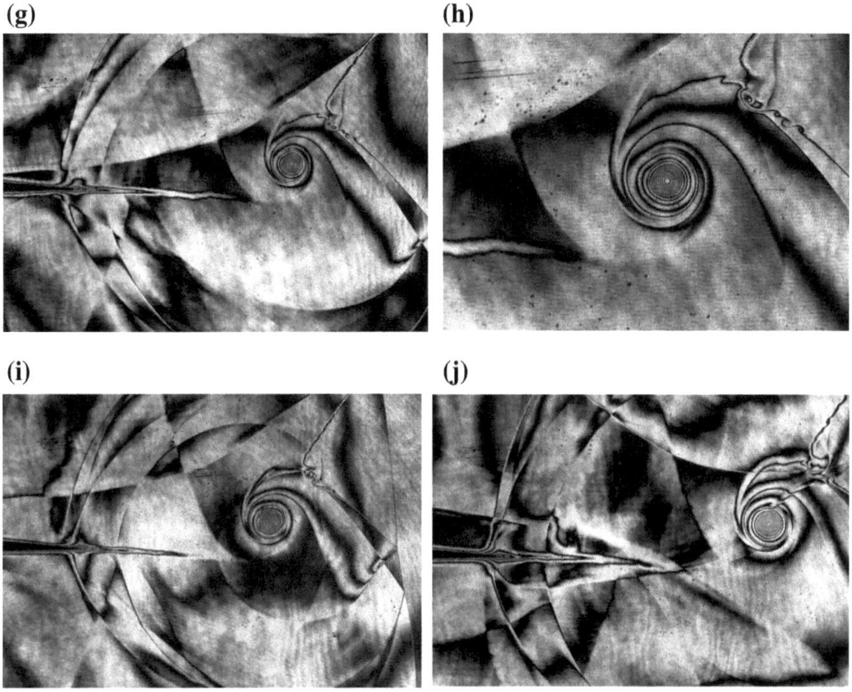

Fig. 6.46 (continued)

had a 200 mm long aluminum sponge so that the shock wave propagating along the upper surface would be attenuated. However, the shock wave propagating along the lower surface would remain at the same speed. The splitting plate worked to generate a shear flow behind the transmitted shock waves. Then some time later, a vortex would appear at the trailing edge. Figure 6.45 shows the experimental arrangement.

Figure 6.46 show the evolution of the vortex formation at the trailing edge for incident shock Ms $\approx$ 2.2 in air at 300 hPa, 288.5 K. The particle speed behind the transmitted shock wave is supersonic and that along the lower surface is about 10% faster than that along the upper surface. At first, a complex diffraction occurs at the trailing edge as seen in Fig. 6.46b. The diffracting shock wave initiates a distinct vortex. The vortex develops at elapsed time is seen in Fig. 6.46b–d. The vortex gradually depart from the trailing edge as seen in Fig. 6.46e–j.

6.11.2 Vortex Formation from a Two-Dimensional Separator

Replacing the sharp trailing edge as seen in Fig. 6.46 with a 50 mm wide two-dimensional backward facing splitting plate resulted in a different the vortex formation. Figure 6.47 show the evolution of diffraction at backward facing corner for Ms = 1.24 in atmospheric air at 292.8 K. At Ms = 1.24, even if the flow at the corner is accelerated, the corner flow remains subsonic and the boundary layer developed along the splitting plate is separated from the corner as seen in Fig. 6.47c–g.

6.11.3 Asymmetric Two-Dimensional Splitting Plate

Figure 6.48 show the evolution of asymmetric interaction of transmitted shock waves released from 50 mm wide splitting plate for Ms = 1.50 in atmospheric air at 294.2 K. At this value of Ms, the flow at the corner is accelerated and eventually becomes supersonic. Distributing the roughness along the lower surface, the shock wave propagating over it is attenuated. Therefore, the resulting twin vortex becomes asymmetrical. The fringe pattern was generated by the combined effect of the local supersonic flow and the vortices.

6.12 Reflection of Transmitted Shock Waves

Installing a splitting plate symmetrically in the 60 mm × 150 mm diaphragm-less shock tube, the evolution of the shock wave diffraction from the splitting plate for Ms = 1.335 in air at 271 hPa, 290.0 K. The plate was set carefully so that the shock wave would diffract precisely symmetrically at the corner of the splitting plate. This experiment was conducted for achieving the intersection of two shock wave having exactly the same strength without having any visible disturbances. As seen in Fig. 6.49, the intersection of the two shock waves forms at first a regular RR and with the increase in the intersecting angle the transition to a MR occur.

6.12.1 Head-on Collision of Two Spherical Shock Waves

Two 25 mm diameter shock tubes made of aluminum alloy, were placed to each other at 80 mm stand-off distance and were operated at the identical initial condition. The shock tubes were operated in a synchronized way by rupturing their diaphragms simultaneously. This is achieved by igniting micro-charges.

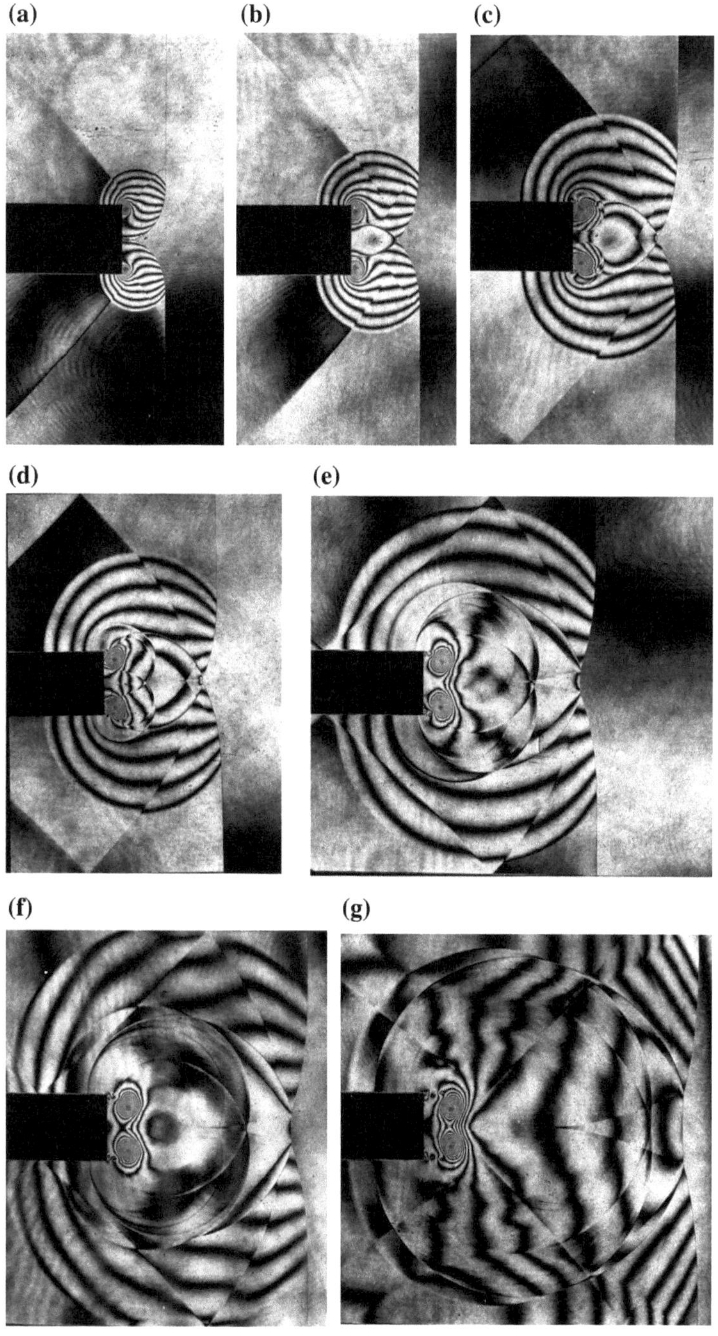

Fig. 6.47 The evolution of diffraction of the transmitted shock wave at backward facing corner for Ms = 1.24 in atmospheric air at 292.8 K: **a** #87111702, Ms = 1.249; **b** #87111701, Ms = 1.244; **c** #87111703, Ms = 1.244; **d** #87111704, Ms = 1.236; **e** #87111705, Ms = 1.250; **f** #87111706, Ms = 1.243; **g** #87111707, Ms = 1.252

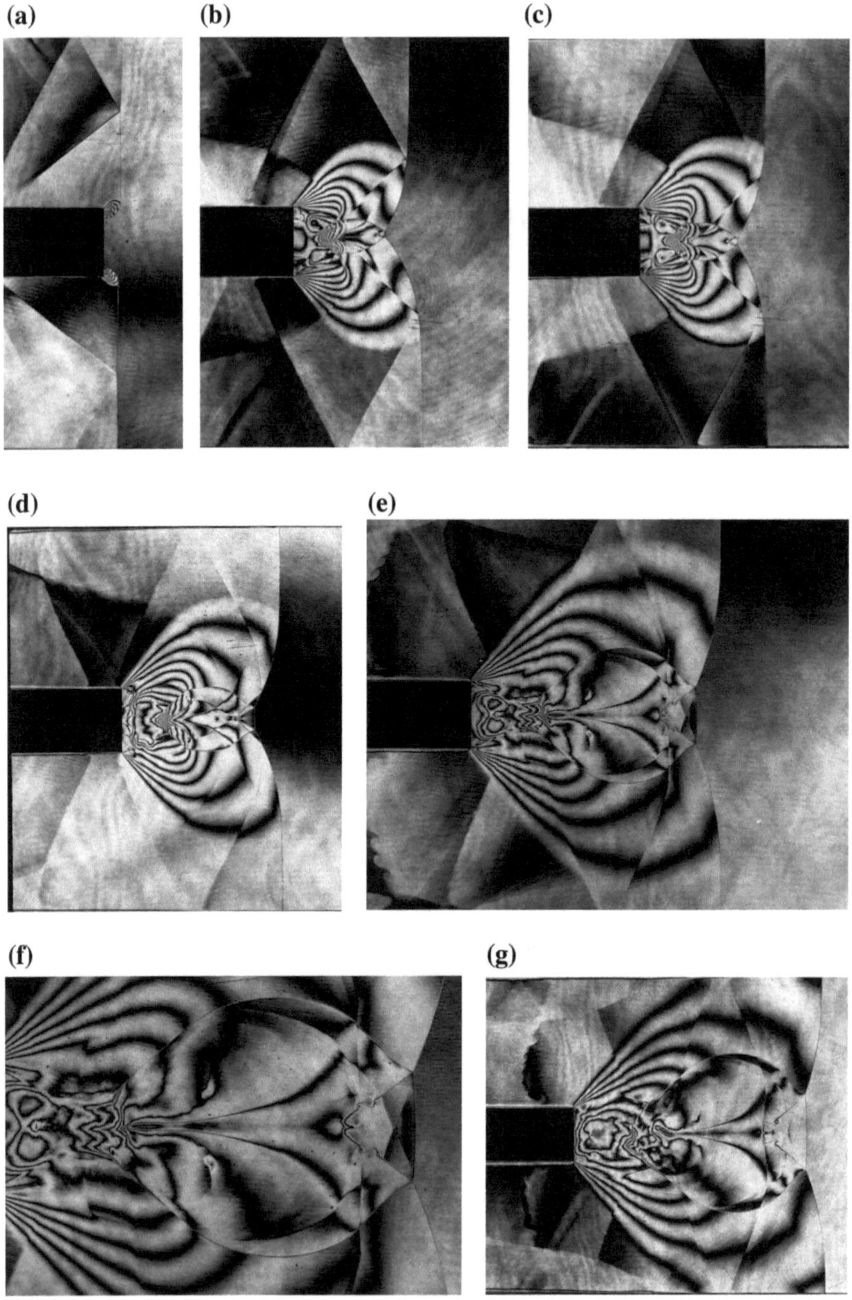

Fig. 6.48 The evolution of asymmetric interaction of transmitted shock waves from a 50 mm wide splitting plate for Ms = 1.50 in atmospheric air at 294.2 K: **a** #87111713, 390 μs from trigger point Ms = 1.496; **b** #87111714, 400 μs, Ms = 1.471; **c** #87111715, 450 μs, Ms = 1.494; **d** #87111716, 500 μs, Ms = 1.498; **e** #87111717, 520 μs, Ms = 1.501; **f** enlargement of (**e**); **g** #87111718, 540 μs, Ms = 1.501

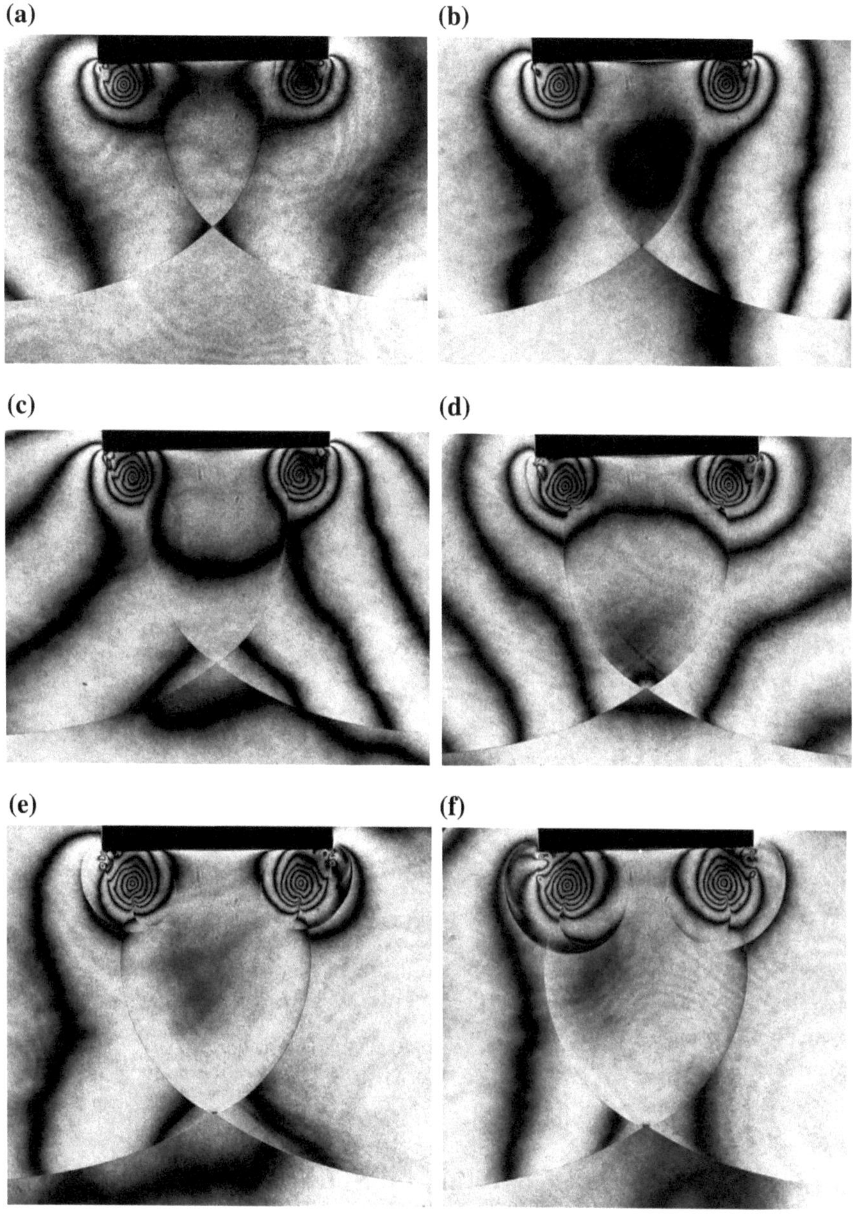

Fig. 6.49 The interaction of transmitting shock waves for Ms = 1.335 in air at 271 hPa, 290.0 K: **a** #97030301, 550 μs from trigger point Ms = 1.337; **b** #97030302, 575 μs, Ms = 1.33; **c** #97030303, 600 μs, Ms = 1.334; **d** #97030304, 625 μs, Ms = 1.336; **e** #97030305, 650 μs, Ms = 1.335; **f** #97030306, 675 μs, Ms = 1.337; **g** #97031006, 595 μs, Ms = 1.335; **h** #97031008, 600 μs, Ms = 1.339; **i** #97031009, 605 μs, Ms = 1.336; **j** #97031010, 610 μs, Ms = 1.335

(g) (h) (i) (j)

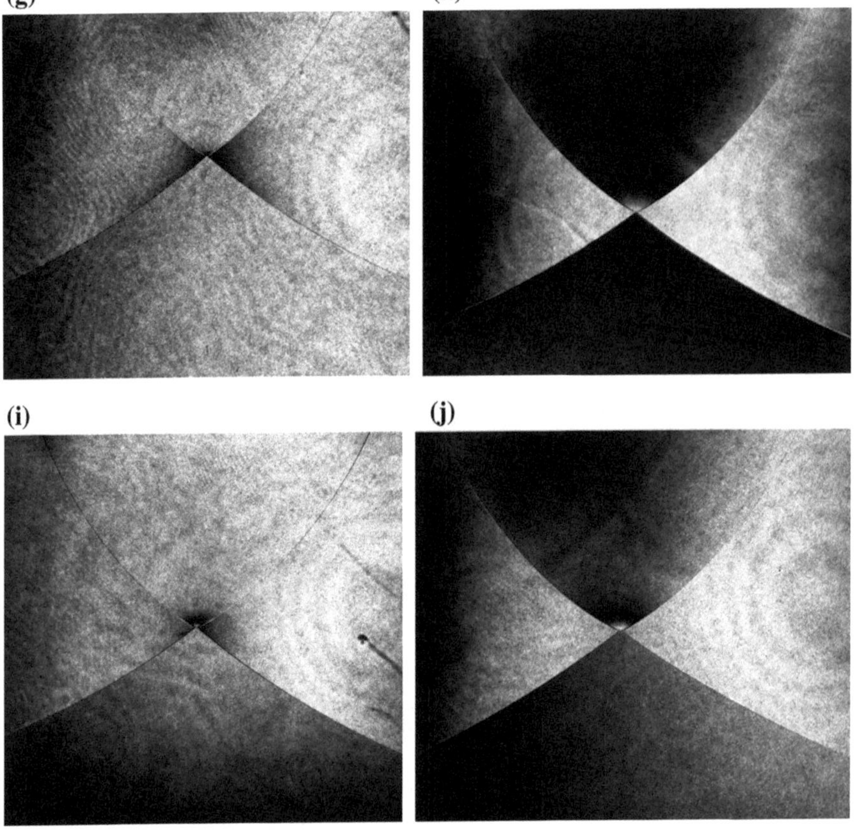

Fig. 6.49 (continued)

Figure 6.50 show steps in the head-on collision of two spherical shock waves for Ms = 1.22 in atmospheric air at 298 K. The two shock tubes produced shock waves of nearly identical strength. Figure 6.50 show the head-on collision of the two spherical shock waves. The reflection pattern so far observed was a RR.

6.13 Effects of Wall Condition on Shock Wave Mitigations

Although it is a laboratory scale experiment from a very narrow-sighted point of view, an effect of wall conditions on the shock wave mitigation was investigated (Abe 2002). Figure 6.51 shows a grassland model placed in the test section of the 60 mm × 150 mm diaphragm-less shock tube. Sequential visualizations were

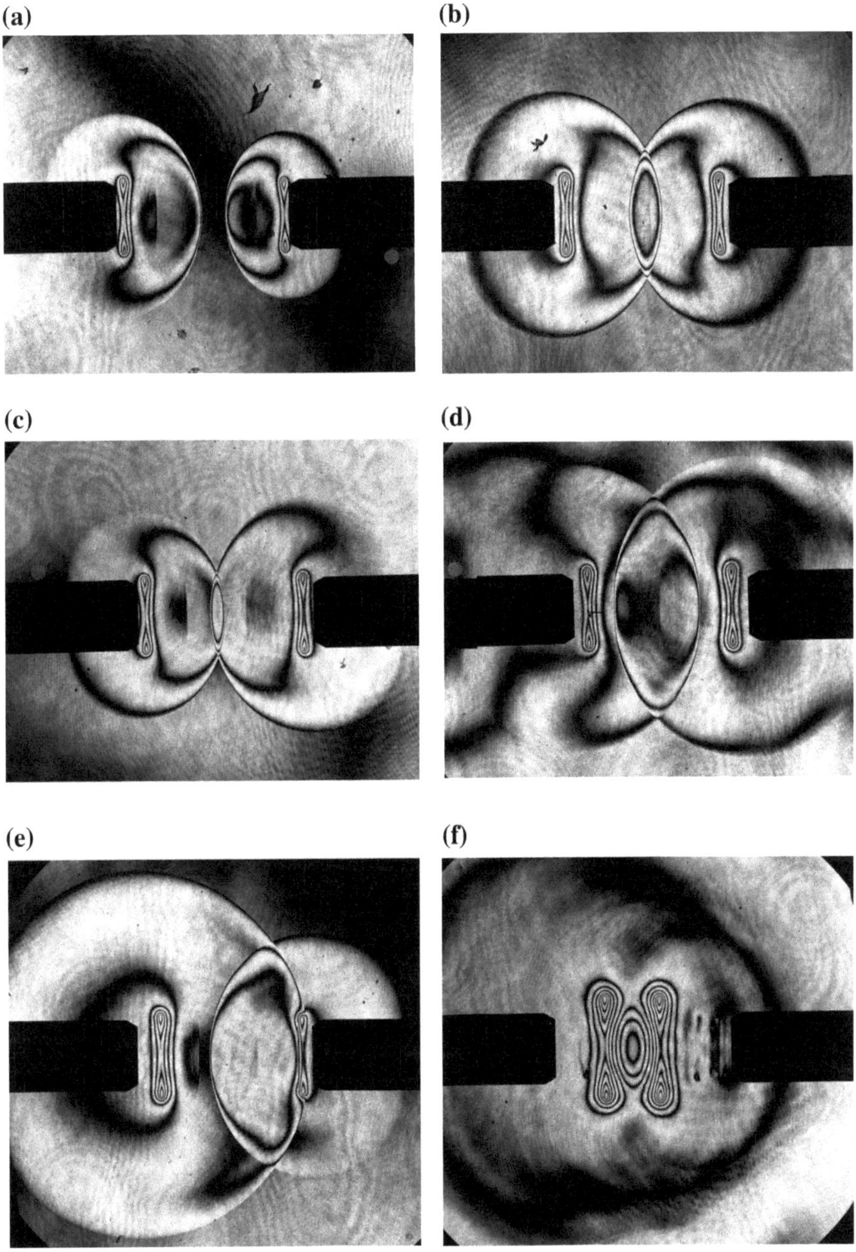

Fig. 6.50 Reflection of two spherical shock waves released from 25 mm diameter shock tubes separated 80 mm for Ms = 1.22 in atmospheric air at 298 K: **a** #87100911, 3.45 ms delay time from trigger point Ms = 1.233; **b** #87100912, 3.47 ms, Ms = 1.233; **c** #87100907, 3.50 ms, Ms = 1.232; **d** #87100909, 3.52 ms, Ms = 1.245; **e** #87100908, 3.53 ms, Ms = 1.221; **f** #87100905, Ms = 1.233; **g** enlargement of (**b**)

(g)

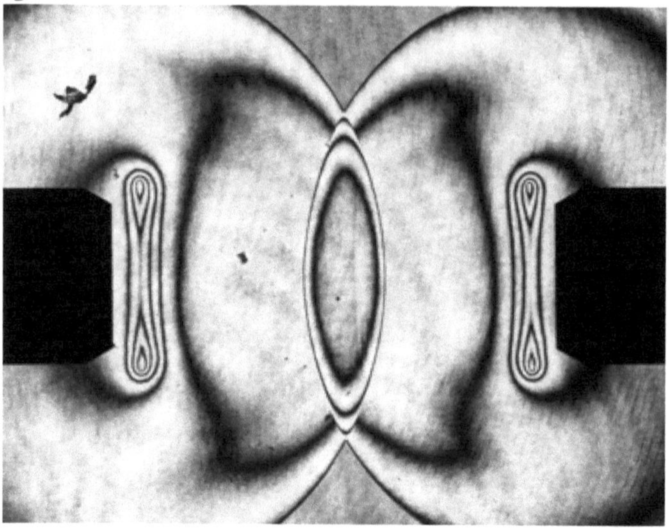

Fig. 6.50 (continued)

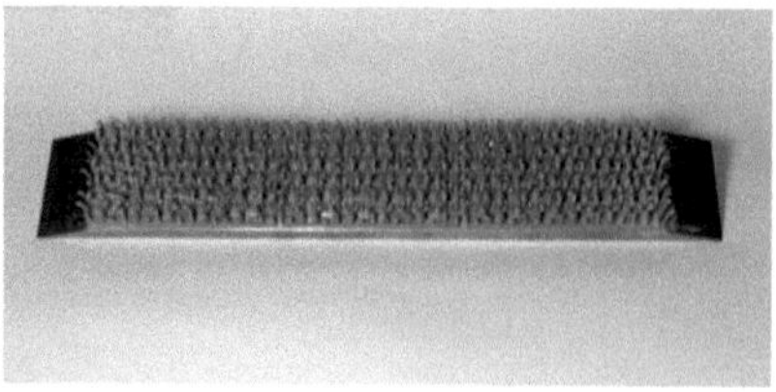

Grassland 60mm × 300mm,10mm height

(a)

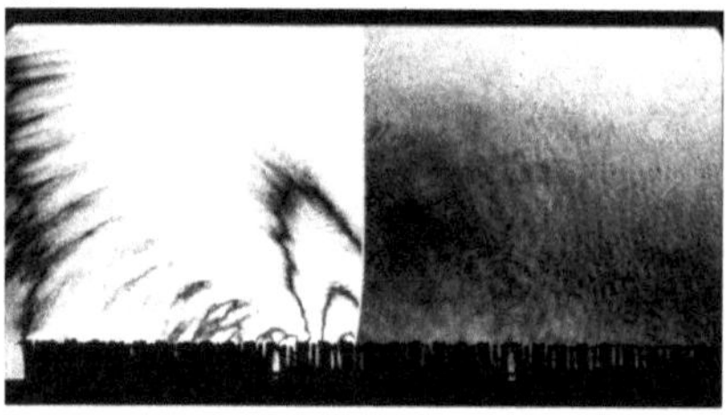

(b)

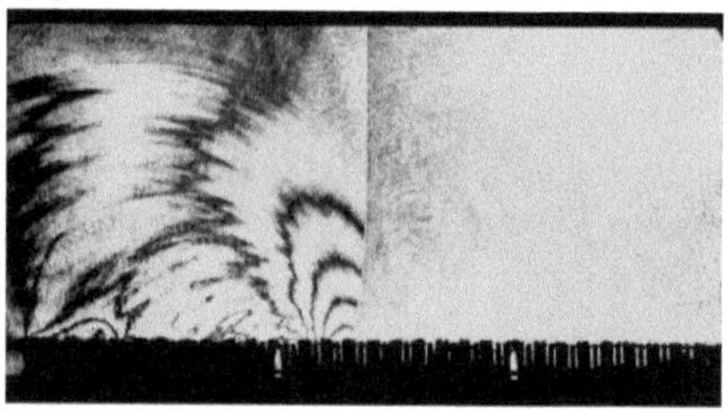

(c)

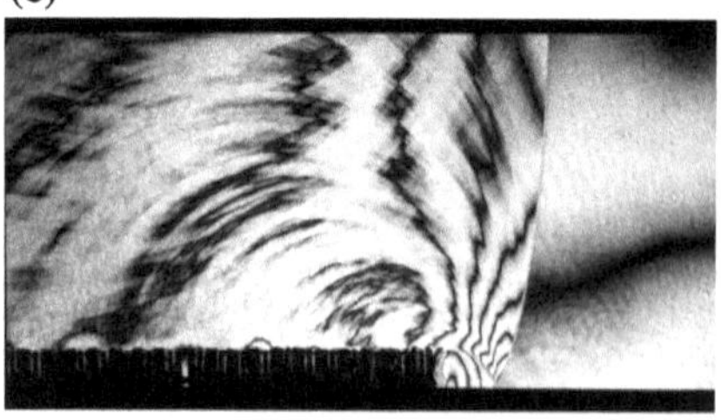

(d)

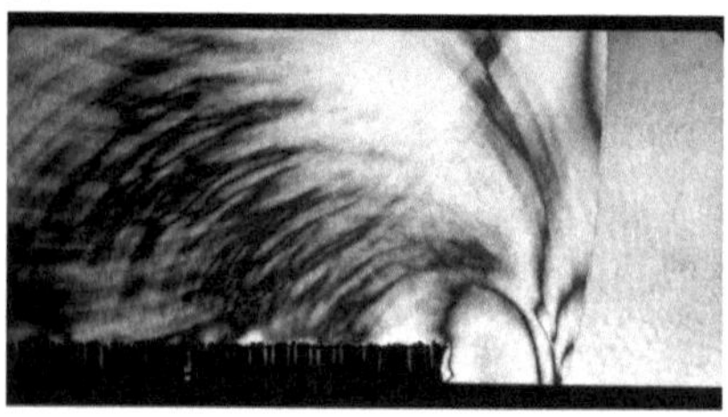

Fig. 6.51 Simulation of grassland 60 mm × 300 mm, 10 mm in height for Ms = 1.50 at atmospheric air (Abe 2002)

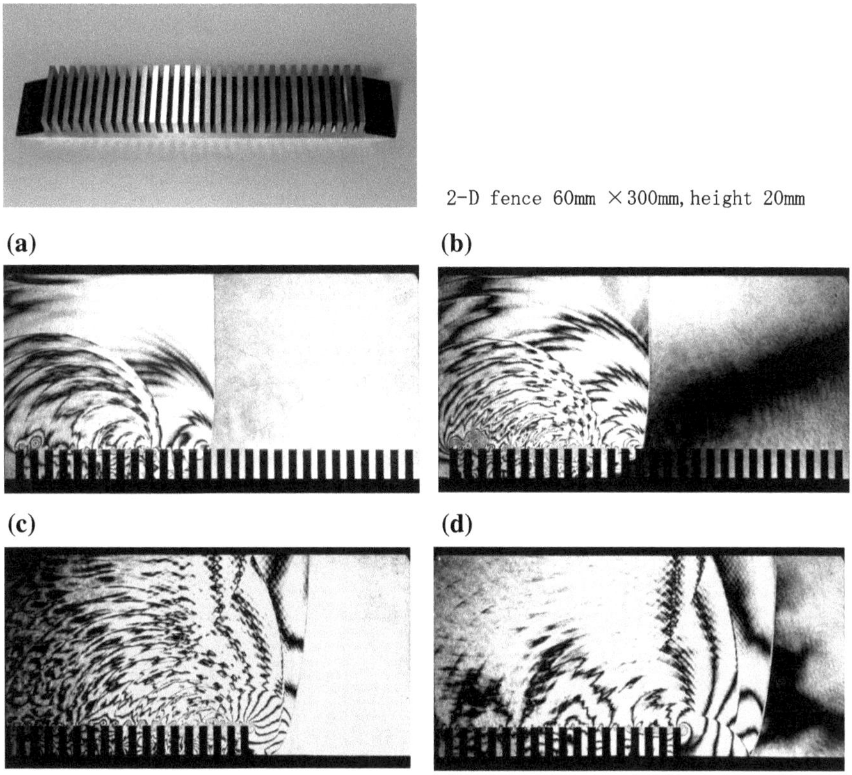

Fig. 6.52 Simulation of 2-D fence 60 mm × 300 mm, height 20 mm, width 5 mm, pitch 5 mm (Abe 2002)

conducted for Ms = 1.50 in atmospheric air. The grassland model consisted of an artificial turf made of plastic which was expected to move freely at the shock wave loading. Although the turf did moved as seen in Fig. 6.51, the plastic turf move less flexibly than initially expected. Hence, the effet of the grass land on the shock wave mitigation was not significant. If a very soft and flexible plastic turf may be used, the shock wave would be attenuated significantly (Fig. 6.52).

Figure 6.53 shows a 60 mm × 300 mm base made of brass, on which variously shaped fences or objects were installed. Shock waves passing over these obstacles attenuate. The degree of attenuation was more or less similar to the shock wave propagation along compactly distributed I-beams or attenuation along Alporous wall.

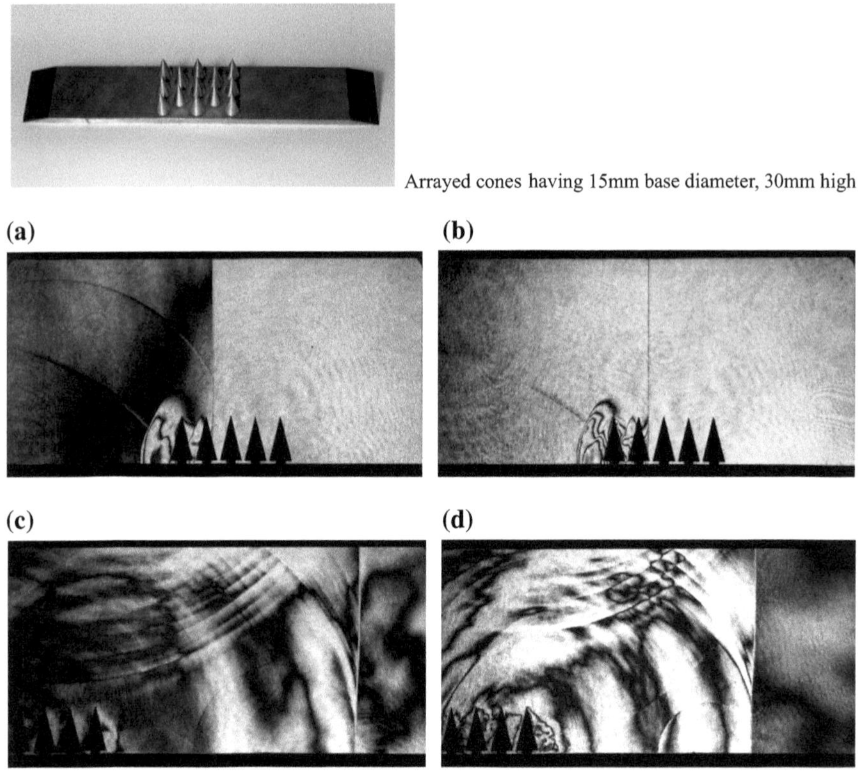

Fig. 6.53 Arrayed cones having base diameter ϕ15 mm and 30 mm in height for Ms = 1.5 in atmospheric air (Abe 2002)

A shock wave propagation over 25 mm base diameter and 30 mm high cones packed in a 60 mm × 300 mm stand made of brass. The shock Mach number was 1.50 in atmospheric air. Three-dimensional array of cones effectively attenuated the transmitted shock wave by the same physical background that the arrayed spheres mitigated the transmitted shock waves train much more efficiently than the arrayed cylinder did.

Figure 6.54 show sequential observation of arrayed cylinders of 15 mm diameter and 30 mm high and tested for Ms = 1.50 in atmospheric air. The shock wave is attenuated much more effectively than the transmitted shock wave propagation over the two-dimensional slotted wall. The interaction of shock waves with three-dimensionally arrayed cylinders and cones are not investigated well. The efforts of numerical simulation were just started.

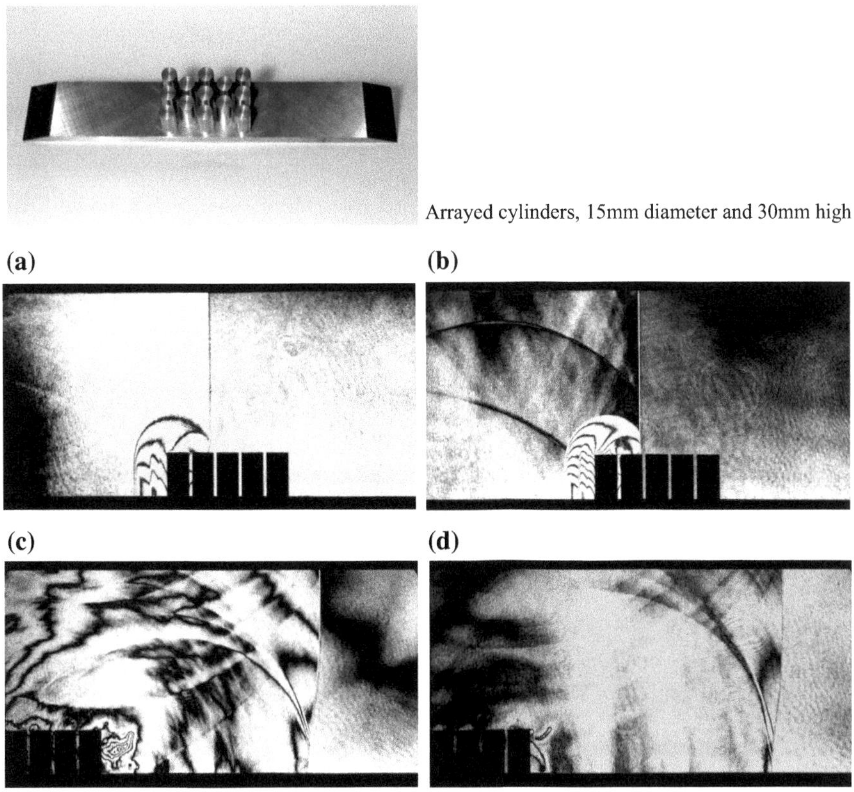

Fig. 6.54 Arrayed cylinders having ϕ15 mm in diameter and 30 mm in height. Tested for Ms = 1.5 in atmospheric air (Abe 2002)

During a field test of high-explosives, blast waves propagated over a forest. Then the resulting blast wave created a big bang noise. The reflection of the blast wave from the forest was not a startling noise but a noise similar to a sound hundreds of people murmured at the same time. The shock wave reflected from a moving boundary drastically mitigated and became a train of compression waves. This procedure is one of difficult taskes. The experimental result presented in Fig. 6.55 was intended to simulate the shock wave reflection from moving boundaries: 15 mm diameter and 37 mm high rubber tubes arrayed as shown in Fig. 6.55. The experiments were conducted for Ms = 1.5 in atmospheric air.

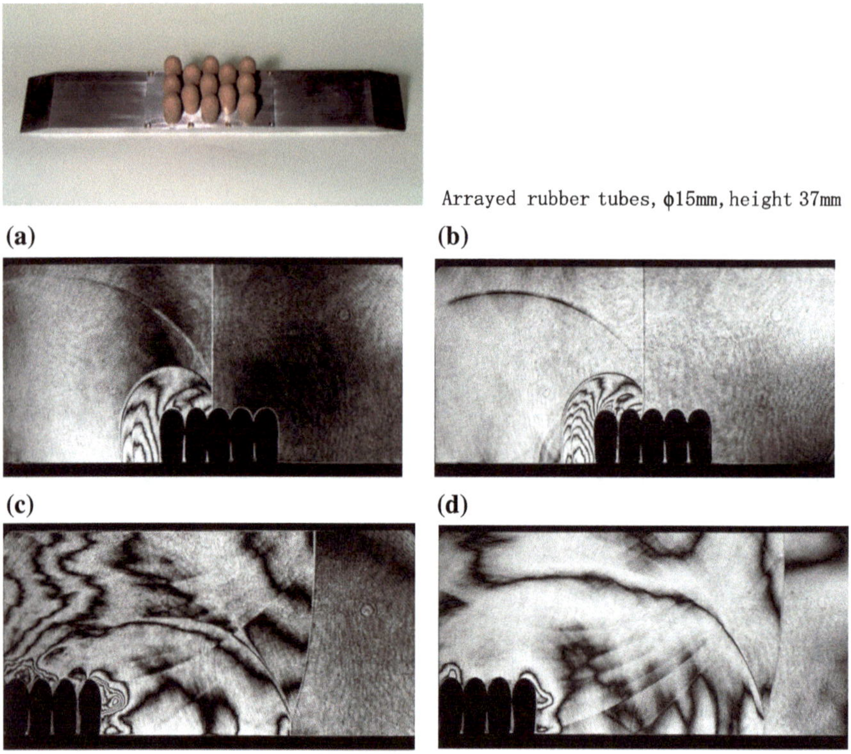

Fig. 6.55 Arrayed 15 mm diameter and 37 mm high rubber tubes for Ms = 1.5 in atmospheric air (Abe 2002)

References

Abe, A. (2002). *Experimental and analytical studies of shock wave attenuation over bodies of complex configurations*. (Ph.D. thesis), Graduate School of Engineering, Faculty of Engineering, Tohoku University.

Matsuoka, K. (1997). *Study of mitigation of high speed train tunnel sonic boom*. (Master thesis). Graduate School of Engineering, Faculty of Engineering, Tohoku University.

Matsumura, S. (1995). *Study of automobile exhaust gas induced shock waves and noises*. (Ph.D. thesis), Graduate School of Engineering, Faculty of Engineering, Tohoku University.

Ohtomo, T. (1998). *Study of shock wave attenuation along ducts of complex geometry*. (Master thesis), Graduate School of Engineering, Faculty of Engineering, Tohoku University.

Onodera, H., & Takayama, K. (1990). Shock wave propagation over slitted wedge. *Institute of Fluid Science, Tohoku University, 1,* 45–46.

Sasoh, A., Matsuoka, K., Nakashio, K., Timfeev, E. V., Takayama, K., Voinovich, P. A., et al. (1998). Attenuation of weak shock waves along partially perforated walls. *Shock Waves, 8,* 149–159.

Sekine, N., Matsumura, S., Aoki, K., & Takayama, K. (1989). Generation and propagation of shock waves in the exhaust pipe of a four cycle automobile engine. In Y. W. Kim (Ed.), *Current Topics in Shock Waves, Proceedings of 17th International Symposium on Shock Tubes and Shock Waves*, Bethlehem (pp. 671–676).

Takayama, K. (1993). Optical flow visualization of shock wave phenomena In R. Brun & L. Z. Dumitrescu (Eds.), *Proceedings of 19th ISSW* (Vol. 4, pp. 7–17). Marseille.

Voinovich, P. A., Timofeev, E. V., Saito, T., & Takayama, K. (1998). Supercomputer simulation of 3-D unsteady gas flows using a locally adoptive unstructured grid techniques. In *Proceedings of 16th International Conference on Numerical Method in Fluid Dynamics* (pp. 51–52).

Chapter 7
Shock Wave Interaction with Gaseous Interface

7.1 Introduction

Shock wave interaction with a gaseous interface is one of basic topics of the shock wave research (Abd-el-Fattah et al. 1978). Figure 7.1 shows a triangular shaped container accommodating a foreign gas interface. The foreign gas was tightly sealed in the container with a 30 μm thick Mylar membrane. The container was installed in the 60 mm × 150 mm conventional shock tube. A foreign gas was circulated through a supply system shown in Fig. 7.1 at the pressure slightly higher than the test pressure and circulated continuously for several minutes. Eventually the value of the foreign gas pressure was adjusted with the test gas pressure. Hence the level of impurity of the foreign gas was minimized by elongating the circulation period of time. The averaged level of the impurity was less than a few %.

The effect of the Mylar membrane on shock wave reflection is shown in Fig. 7.2. The shock wave propagates over a 45° air/air interface for Ms = 1.20. The IS passed through the interface without causing any disturbances. But the Mylar membrane was bent toward inward but the reflected shock wave pattern from the interface was a RR.

7.1.1 Air/He Interface

The reflection from the air/helium interface is called as the slow/fast interaction and that from the air/CO_2 interface is called as the fast/slow interaction. Figure 7.3 show the evolution of shock wave reflection from the air/helium interface for Ms = 1.20. The slow/fast interaction is analogous to the shock wave in air reflected from the water wedge.

Waves in the helium layer propagate at the sound speed in helium and hence are observed as the faint change of contrast. The speed of the shock wave propagating

K. Takayama, *Visualization of Shock Wave Phenomena*,
https://doi.org/10.1007/978-3-030-19451-2_7

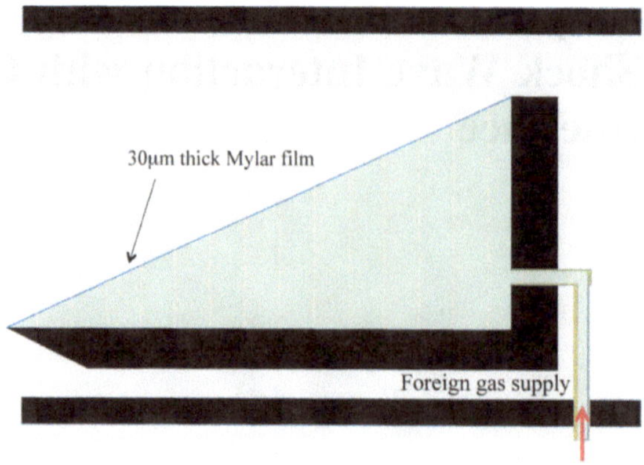

Fig. 7.1 The test section of a foreign gas interface installed in the 60 mm × 150 mm shock tube

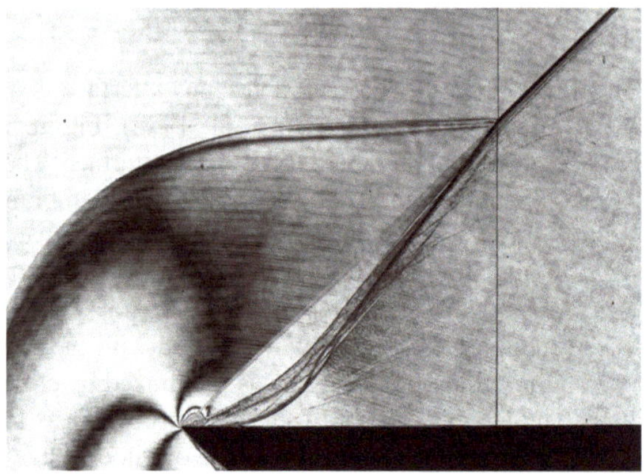

Fig. 7.2 Shock wave propagation over air/air interface at angle of 45° #93070101 for Ms = 1.198 in atmospheric air at 297.3 K

along the interface whose inclination angle is θ_w is defined as $u_{inter} = u_s/\cos\theta_w$, where u_s is the speed of the incident shock wave IS. Therefore, if the interface angle satisfies the condition of $\theta_w < \cos^{-1}(u_s/a_{helium})$, where the a_{helium} is the sound speed in helium, the compression waves induced by the IS propagate in helium head the IS.

In Fig. 7.3a, if the interface was a solid wedge, the reflection pattern should be a vNMR and the resulting triple point trajectory angle should be the glancing incidence angle θ_{glance}, of 25.5°. In the reflection from the present slow/fast interface, as seen in Fig. 7.3a it is larger than this value. In addition to this, the MS so far observed in Fig. 7.3a is not perpendicular to the interface but slightly tilted backward. It is clearly observed that the reflected shock wave from the interface behaved differently from that of a vNMR over solid wedges. As the disturbances propagated in helium at a_{helium} faster than the movement of the foot of the IS, it lifted up the interface slightly upward resulting in oblique fringes which are terminated at the MS. The presence of leading fringes uniquely appeared over this slow/fast interface but was not observed over the water wedges. The underwater shock wave

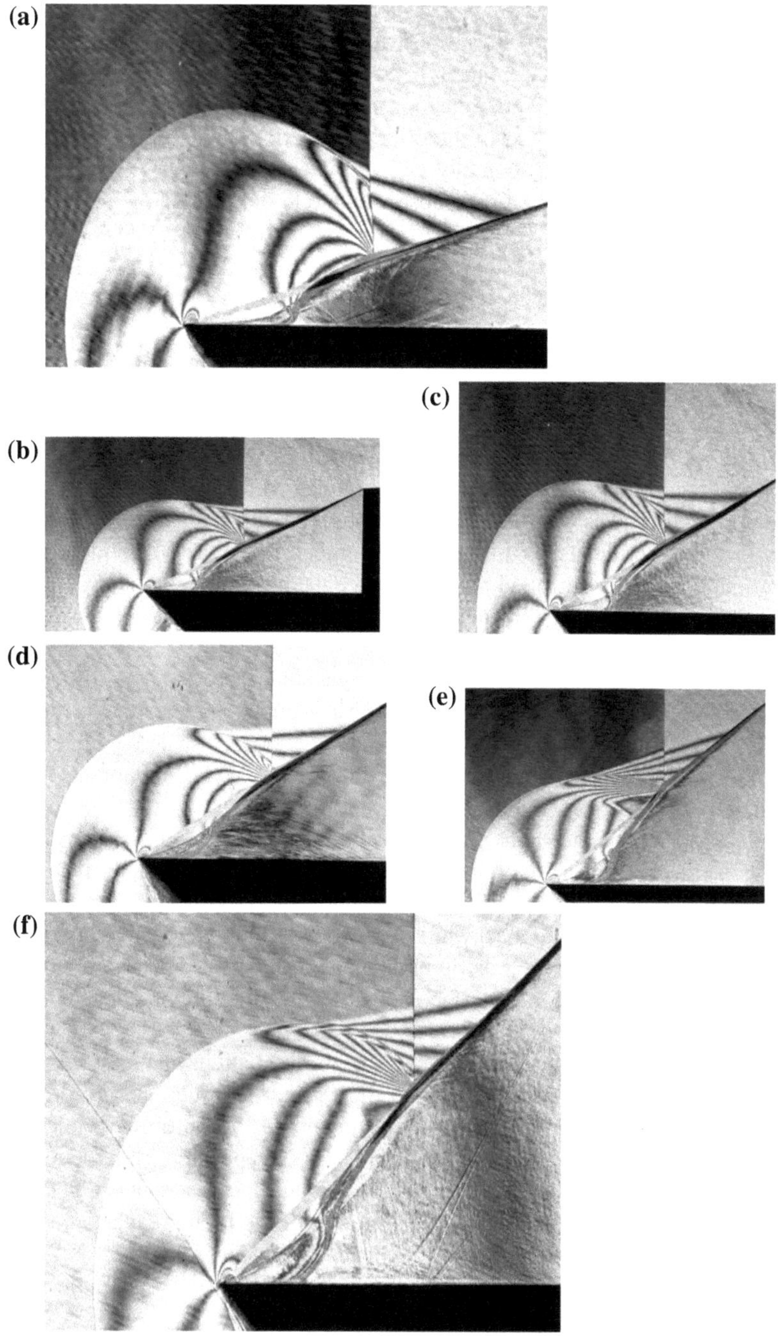

Fig. 7.3 Shock wave interaction with air/He interface, exchange time 10 min, Ms = 1.20 in atmospheric air at 297 K: **a** #93053108, Ms = 1.199, $\theta_w = 15°$; **b** #93052807, Ms = 1.203, $\theta_w = 25°$; **c** #93052801, Ms = 1.200, $\theta_w = 30°$; **d** #93053102, Ms = 1.200, $\theta_w = 35°$; **e** #93053103, Ms = 1.203, $\theta_w = 40°$; **f** #93052603, Ms = 1.207, $\theta_w = 44.6°$

propagation over a brass wall was a shock wave interaction with the slow/first interface, the stress waves were released from the brass wall to water.

A triple point, TP, formed on the reflected shock wave seen in Fig. 7.3a does not accompany a slip line SL. This reflection pattern is similar to a vNMR. When the interface angle θ_w approached to the critical transition angle, θ_{crit}, the TP approached to the interface and eventually MS terminated as seen in Fig. 7.3c, d. The precursory oblique fringes in air are observed and terminated on the reflected shock wave. When the $\theta_w \geqq \theta_{crit}$ is satisfied, the pattern of the reflected shock wave becomes a MR.

With increasing θ_w, when u_{inter} exceeds the a_{helium}, the compression waves created by the IS in helium coalesced into a weak oblique shock wave. This situation was closely related to the level of compression wave induced in the helium layer. Practically the situation was linked with the degree of deformation of the Mylar membrane. During the series of experiments conducted in May 1993 as presented in Fig. 7.3, the membrane deformed slowly in a controlled manner generating compression waves in the helium layer. Therefore, the head of the train of compression waves was not distinctly observed in Fig. 3.3. However, in the series of the experiments conducted in November 1990, the Mylar deformed and even ruptured in a uncontrolled manner. Hence the head of the train of compression waves became clearly visible in Fig. 7.4, whereas it was hardly observable in Fig. 7.3.

During the experiments of the foreign gas interface, the experiments were conducted in a facility having a circular test section connected to a horizontal cookie cutter but the test section was relatively loosely sealed with the Mylar membrane. Hence, at the impingement of the IS, the Mylar membrane deformed very largely.

Figure 7.4a shows the shock wave reflection from the air/helium interface at the interface angle of 25° for Ms = 1.40. The reflected shock wave pattern was a SMR and the TP accompanied the SL. However, the SL was not perpendicular to the interface but slightly tilted backward. When the IS impinged the interface, the resulting compression waves was running ahead of the IS. The compression waves precursory to the IS lifted the interface upward generating the precursory fringes in air.

Figure 7.4a explains the wave propagating in the helium layer. In Fig. 7.3, the membrane sustained the pressure behind the IS and deformed slightly inward. In Fig. 7.4, after the first deformation of the membrane, it deformed violently and drove an oblique shock wave. An oblique shock wave appeared from the foot of the MS to the leading edge of the interface. The oblique shock wave was reflected from the bottom wall. With increasing the θ_w, the foot of the IS approaches gradually to the precursory shock wave as shown in Fig. 7.4e. In Fig. 7.4f, the foot of the IS merged with the precursory wave in helium. The θ_{crit} of the air/helium interface would depend on the helium impurity. When $\cos\theta_w = u_s/a_{helium}$ is achieved, as seen in Fig. 7.4f, the transition to a RR occurs.

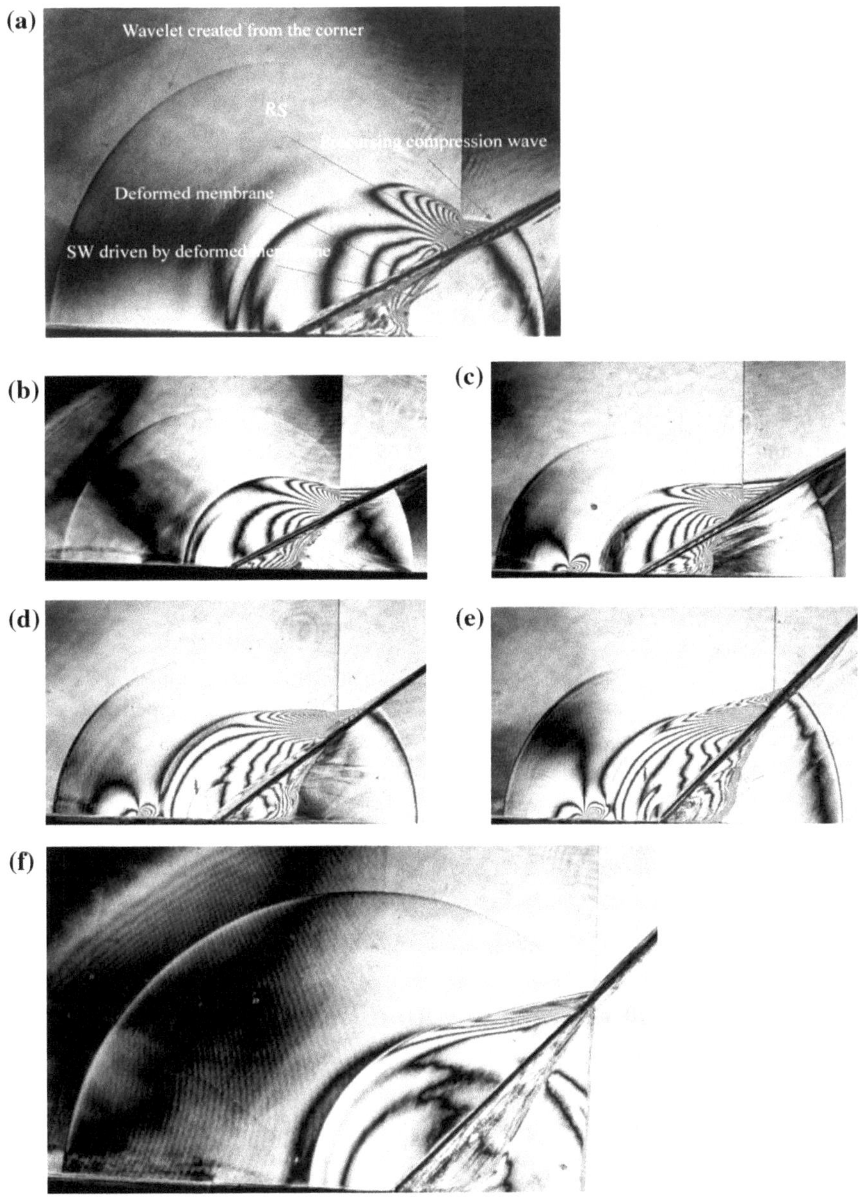

Fig. 7.4 The interaction of shock wave with air/helium interface, 10 min duration of time for circulating helium, for Ms = 1.40 in atmospheric air at 290.9 K: **a** #90111505, Ms = 1.405, $\theta_w = 25°$; **b** #90110901, Ms = 1.418, $\theta_w = 30°$; **c** #90111402, Ms = 1.391, $\theta_w = 30°$; **d** #90110903, Ms = 1.401, $\theta_w = 37°$; **e** #90110904, Ms = 1.399, $\theta_w = 45°$; **f** #90111503, 140 us, Ms = 1.405, $\theta_w = 47°$

7.1.2 Air/CO_2 Interface

Figure 7.5 show the evolution of shock waves interacting with air/CO_2 interface. The interface angles vary from 15° to 55° for Ms = 1.20. This is a fast/slow interaction: the sound speed in CO_2 is 280 m/s at 293 K and that in air is 345 m/s at 290 K. The oblique shock wave is generated in the slow gas. At a shallower interface angle as seen in Fig. 7.5a, b, the TP does not accompany a SL. The reflected shock wave pattern is vNMR but the MS slightly leaned forward. However, in Fig. 7.5d at θ_w = 30°, the transition already took place and the pattern of the reflected shock wave was a RR. In Fig. 7.5e at θ_w = 35°, the pattern of the reflected shock wave was a SuRR.

An oblique shock wave was formed in CO_2 and reflected from the bottom wall. The reflection patterns was always a SMR as shown in Fig. 7.5a–d. However, with increasing θ_w, its triple point gradually smeared out and the SL tends to vanish. The reflection pattern transits to a vNMR as seen in Fig. 7.5e–h.

7.1.3 Air/SF_6 Interface

Figure 7.6 shows the interaction from an air/SF_6 interface. This is a fast/slow interaction: the sound speed in SF_6 is 138 m/s at 293 K. A ring shaped test section was installed in the 60 mm × 150 mm conventional shock tube. The interface tightly sealed with a 30 μm Mylar membrane was rotating at any angle ranging from 0° to 90°. Figure 7.7 show the evolution of shock wave propagating along the interface varying the inclination angle for Ms = 1.40 at 286.7 K. In Fig. 7.7b–d, the initial reflected shock pattern was a SMR and at the critical transition angle of 34°, the transition to a RR occurs. In Fig. 7.7e, f, a SuRR were observed.

7.2 Shock Wave Interaction with a Helium Column

Figure 7.8 shows an experimental arrangement of the shock wave interaction with a soap column containing helium. In order to accommodate a vertical helium soap column, the 60 mm × 150 mm diaphragm-less shock tube was turned 90° sideway converting the 150 mm × 60 mm shock tube. The height of the soap bubble column was 60 mm and its diameter was 50 mm; it was stretched between the two walls of the test section by thin brass rings 4 mm in thick and 4 mm in width glued on the upper and bottom walls. Then, the soap bubble column was held vertically for a few minutes being blown by 1015.2 hPa high pressure helium (Nagoya 1995).

To visualize shock waves in the helium soap columns, the collimated object beam *OB* passed the test section from its top to bottom and then reflected from a plane mirror placed at the bottom of the test section. The reflected *OB* passed the

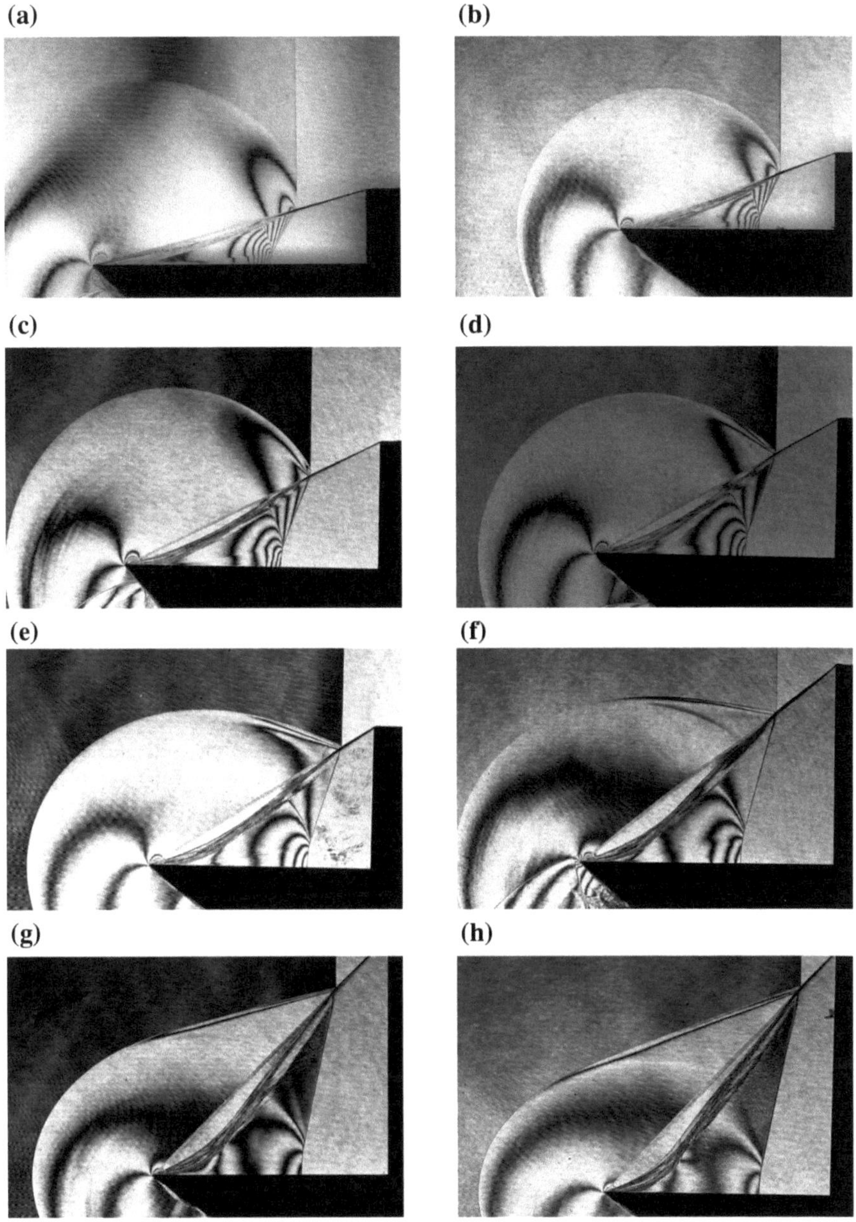

Fig. 7.5 Shock wave interaction with air/CO_2 interface for Ms = 1.20 in atmospheric air at 296.6 K, CO_2 exchange time of 10 min: **a** #9306102, Ms = 1.198, $\theta_w = 15°$; **b** #9306101, Ms = 1.201, $\theta_w = 20°$; **c** #9306103, Ms = 1.197, $\theta_w = 25°$; **d** #9306104, Ms = 1.200, $\theta_w = 30°$; **e** #9306105, Ms = 1.200, $\theta_w = 35°$; **f** #9306106, Ms = 1.198, $\theta_w = 40°$; **g** #9306108, Ms = 1.199, $\theta_w = 50°$; **h** #9306109, Ms = 1.199, $\theta_w = 55°$

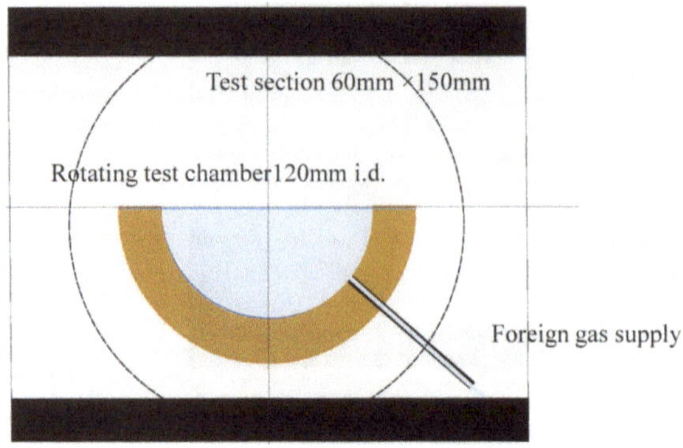

Fig. 7.6 Test section of air/SF_6 interface

test section again, then reflected from a half mirror located in the middle position of the collimated light path and illuminated the holographic film. Hence the *OB* passed the test section twice. With such a double path arrangement, even the 60 mm test section length was doubled. This optical arrangement is illustrated subsequently in Fig. 7.30b.

Figure 7.9 show sequential interferograms of shock/helium column interaction with helium filled soap bubble for Ms = 1.20 in atmospheric air at 295.7 K. A weak transmitted shock wave and a reflected shock wave propagated inside the bubble. The observed sequence of the interaction is very different from the interaction with a rigid cylinder. In Fig. 7.9c, the column started to contract. The transmitted shock wave in helium propagated faster than the IS in air and was released into air ahead of the IS. At the same time, the transmitted shock wave was reflected from the concave interface and focused: see in Fig. 7.9d–i. The patterns of focusing were similar to those observed during the focusing from concave rigid walls. Such effects were caused by the reflection from thin brass rings attached on the upper and lower walls. At the same time, the frontal surface of the column was deformed in Fig. 7.9e and the deformation was promoted.

The IS was diffracted along the rear part of the interface. At the same time, the transmitted shock wave came out from the interface as seen in Fig. 7.9d. The diffracting shock wave and transmitted shock wave interacted at the rear side of the interface in Fig. 7.9g. But such interaction patterns never occurred over a rigid cylinder.

At the same time, the deformed helium column moved to downstream at almost equal to the particle velocity behind the IS. The frontal side of the helium column was gradually flattened as shown in Fig. 7.9m. The deformation was accelerated and eventually became concave as shown in Fig. 7.9m, n. At this stage on, the interval of the double exposure was set to be 600 μs. In Fig. 7.9n, the first exposure was conducted at the moment when the IS arrived at the helium column and the second exposure was conducted when the transmitted shock wave reached at the right hand side of the observed field of view. Hence, the fringes in this interferogram correspond to the difference in the phase angles occurred during these time

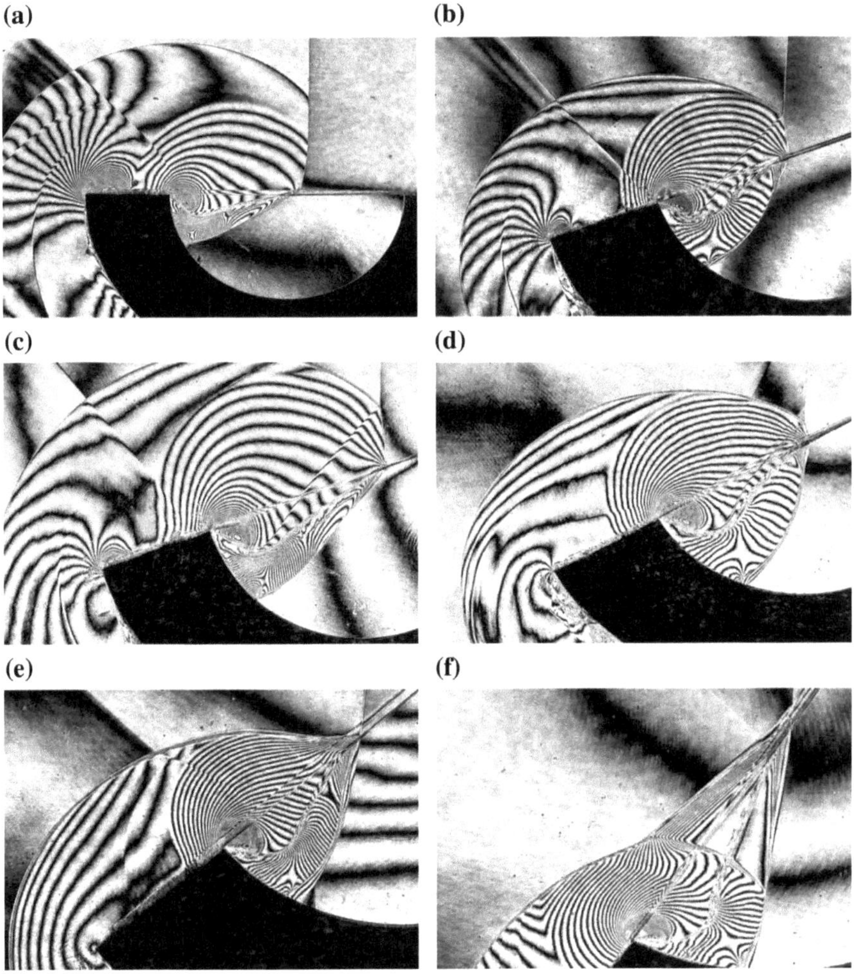

Fig. 7.7 Shock wave interaction with air/SF_6 interface, for Ms = 1.40 in atmospheric air at 286.7 K, Exchange time duration 10 min: **a** #90012308, Ms = 1.722, $\theta_w = 0°$; **b** #90012403, Ms = 1.443, $\theta_w = 20°$; **c** #90012401, Ms = 1.497, $\theta_w = 20°$; **d** #90012406, Ms = 1.380, $\theta_w = 30°$; **e** 90012404, Ms = 1.460, $\theta_w = 40°$; **f** #90012405, Ms = 1.480, $\theta_w = 60°$

intervals. Figure 7.9s shows a single exposure interferogram of the deformation of the helium column. Fringes describing the interfaces were jaggedly shaped, because the fringes were formed by the observation of three-dimensional phenomena with two-dimensional collimated *OB*.

Figure 7.10 show sequential deformations of the helium column at a later stage. With elapsed time, the helium column was deformed like the cross section of mushroom. The helium column moved at almost equal to the particle velocity behind the shock wave. Jaggedly shaped fringes never represented the

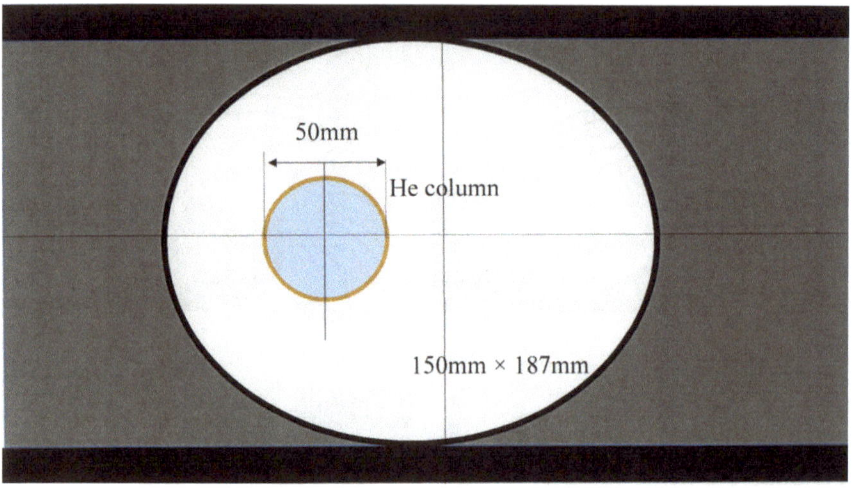

Fig. 7.8 A soap bubble column of 50 mm in diameter and 60 mm in height was installed horizontally in the 150 mm × 60 mm shock tube. The field of view had an elliptic shape of 150 mm × 187 mm (Nagoya 1995)

two-dimensionally deformation but the integration of the three-dimensional density distribution along the z-direction. The interfacial instability was described by the analytical model of the so-called Richtmyer-Meshkov Instability (RMI) (Nagoya 1995). Analytical description of the RMI was based on the theoretical background of the three-dimensional interaction between the density gradient and the pressure gradient. The RMI was induced in the interaction between the pressure jump across the shock waves and the density jump across the gaseous interfaces. As a result of RMI, three-dimensional disturbances developed and eventually the vortices were formed.

Fig. 7.9 Sequential observation of shock wave interaction with a helium column for Ms = 1.20 in ▶ atmospheric air at 295.7 K, He bubble 1015.2 hPa: **a** #93120801, 360 μs; **b** #93120701, 375 μs elapsed time from shock wave arrival at the frontal stagnation of the column, Ms = 1.200; **c** #93120811, 415 μs, Ms = 1.215; **d** #93120604, 430 μs, Ms = 1.197; **e** #93120602, 440 μs, Ms = 1.196; **f** #93120606, 410 μs, Ms = 1.196; **g** #93120806, 445 μs, Ms = 1.212; **h** #93120509, 450 μs, Ms = 1.209; **i** #93120607, 460 μs, Ms = 1.195; **j** #93120608, 470 μs, Ms = 1.204; **k** #93120609, 480 μs, Ms = 1.208; **l** #93120802, 500 μs, Ms = 1.206; **m** #93120803, 550 μs, Ms = 1.198; **n** #93120902, 650 μs, Ms = 1.204; **o** #93120903, 700 μs, Ms = 1.206; **p** #93120904, 750 μs, Ms = 1.211; **q** #93120905, 800 μs, Ms = 1.212; **r** #93120908, 950 μs, Ms = 1.208; **s** #93121004, 1.1 ms, Ms = 1.212 single exposure; **t** #93121106, 1.05 ms, Ms = 1.208

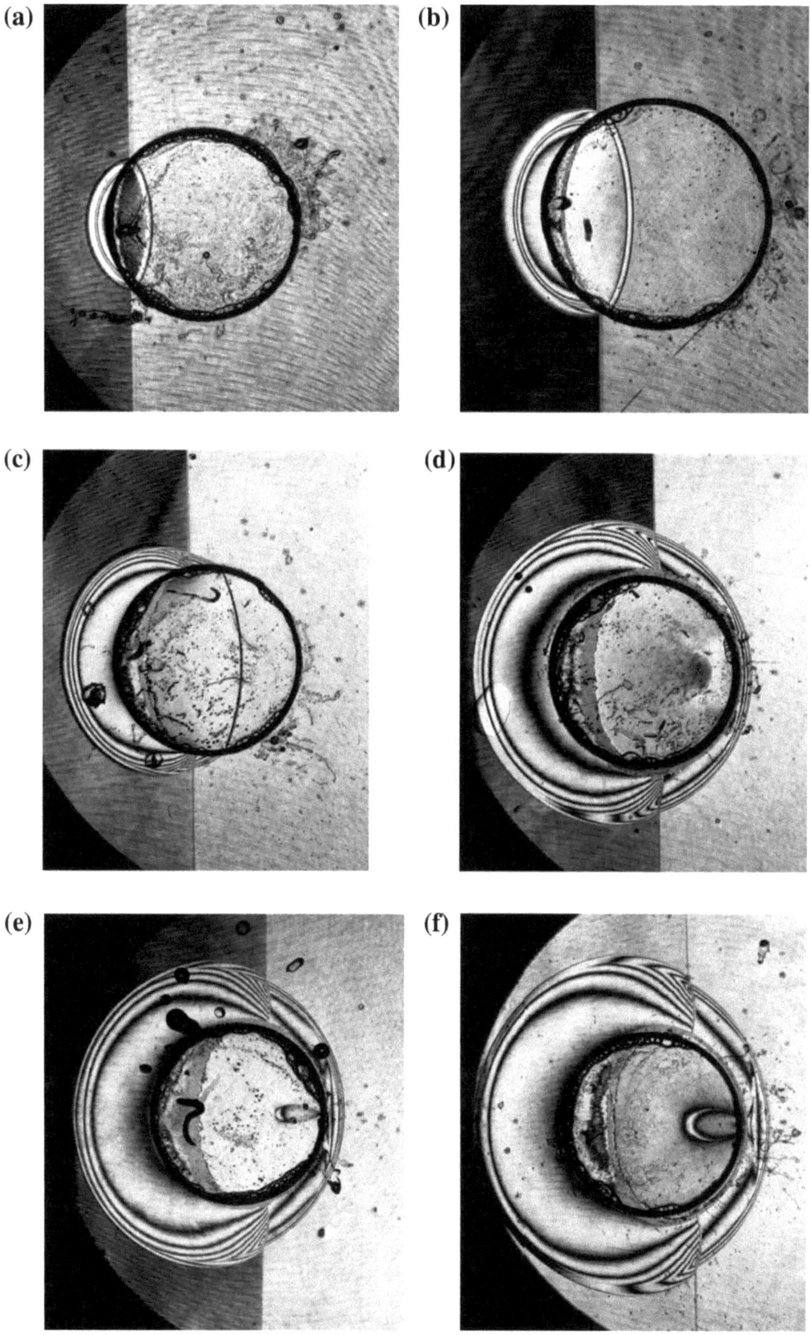
(a)
(b)
(c)
(d)
(e)
(f)

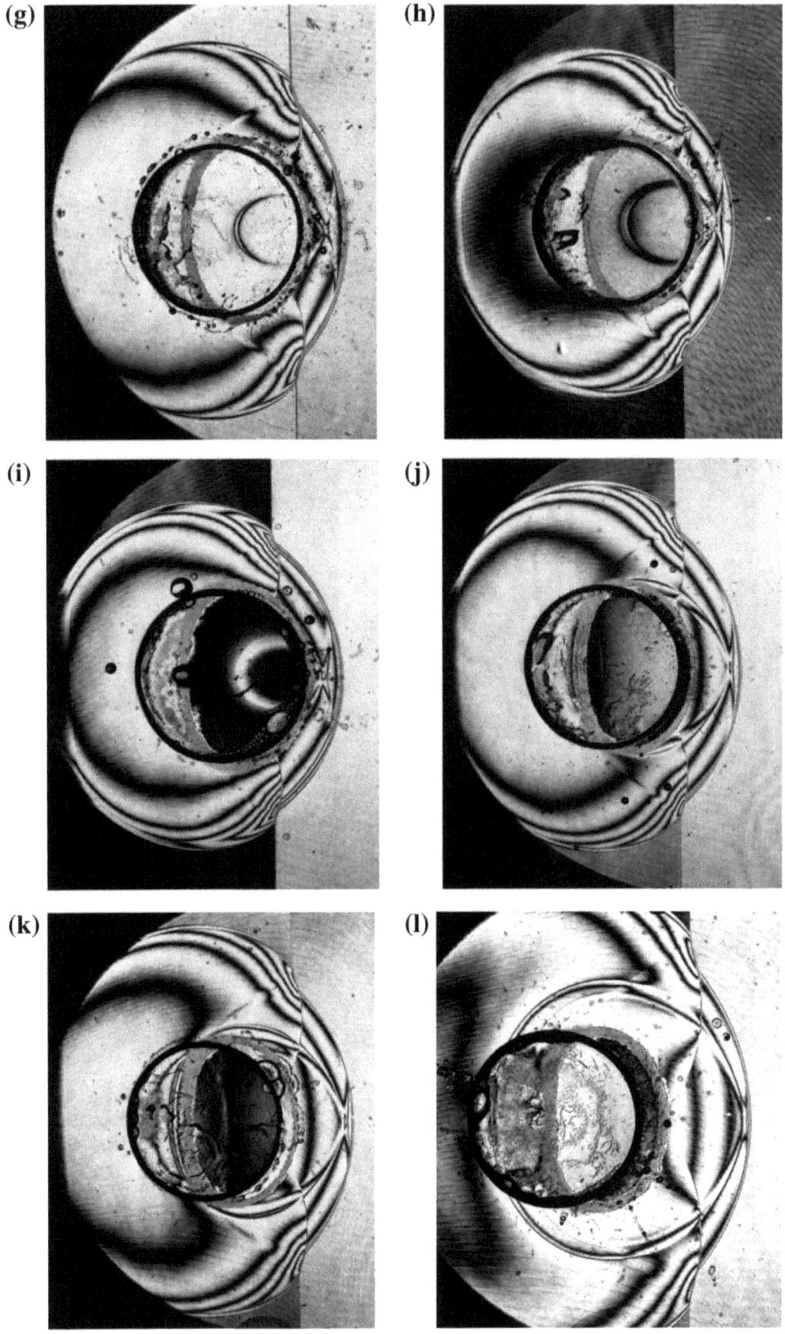

Fig. 7.9 (continued)

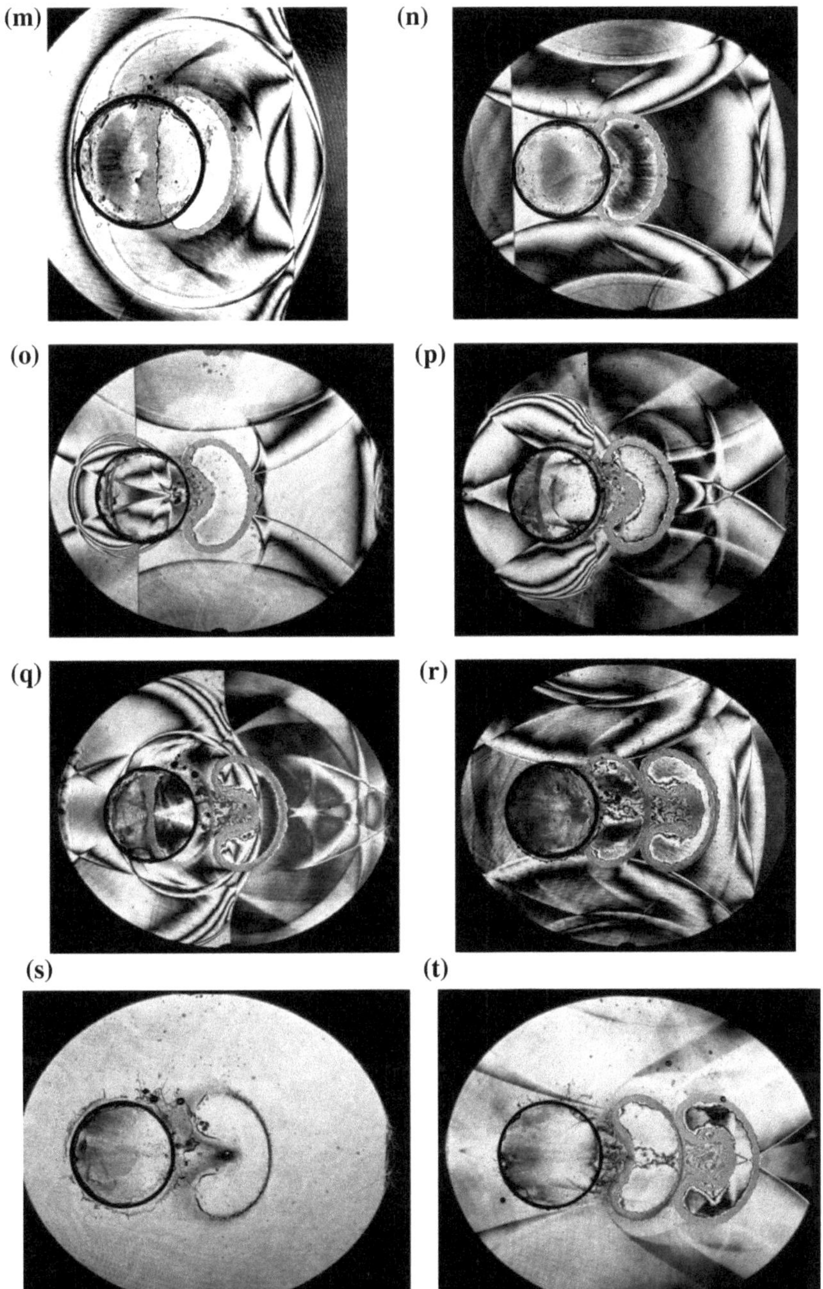

Fig. 7.9 (continued)

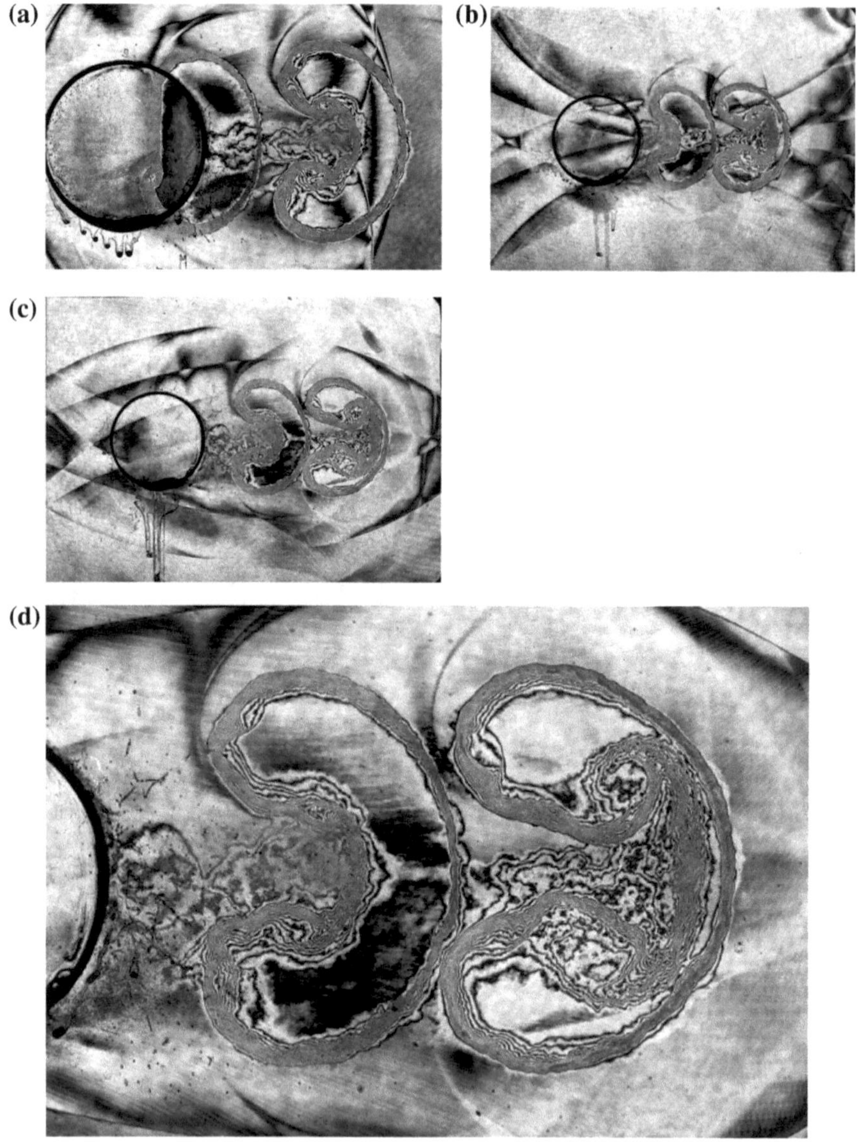

Fig. 7.10 Later stage of shock wave interaction with a helium column for Ms = 1.21 in atmospheric air at 295.0 K He bubble 1015.2 hPa, the interval of double exposures was 600 μs: **a** #93061106, 1.0 ms from arrival of IS at the helium column; **b** #93061108, 1.2 ms; **c** #93061109, 1.3 ms; **d** enlargement of (**c**)

Figure 7.11 summarizes the positions of helium columns observed in Figs. 7.9 and 7.10. The ordinate denotes the elapsed time in μs and the abscissa denotes the position of deforming helium columns in mm. The distance between the front side

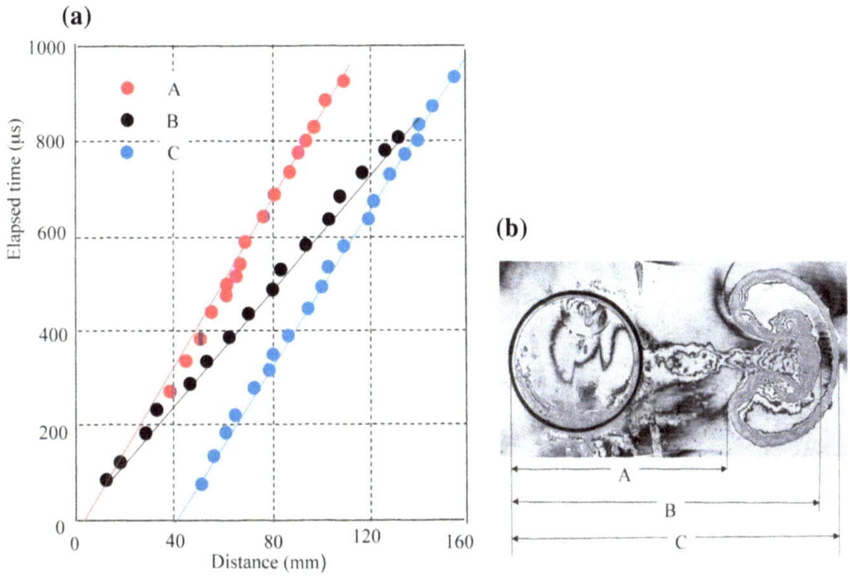

Fig. 7.11 The motion of helium columns, the summary of Figs. 7.9 and 7.10 (Nagoya 1995)

and the rear side of the helium columns seen in the pictures was almost unchanged and moved at the speed close to the particle velocity behind the IS, whereas the front side of the helium column is accelerated and eventually a mushroom shape of deformation was built up.

7.2.1 *Side View of a Helium Column*

The impingement of shock waves on the helium column was visualized from the top as seen in Fig. 7.10. Jaggedly observed fringes never meant the two-dimensional interfacial instability but a simple sequence of the integration of three-dimensional density fluctuations conducted along the interface. Therefore, the observation from the side of helium columns would be physically meaningful. Figure 7.12 shows a schematic observation from the side. The helium soap column is supported by thin brass rings from the upper and bottom walls and the IS is impinged from the left. Figure 7.13a shows the helium column prior to the shock wave arrival. Figure 7.13b shows a side view corresponding to Fig. 7.9o. Small protrusions and dents are distributed along the rear side of the column. These are vortices originated as a result of RMI. Straight and parallel fringes on the left correspond to the undisturbed helium column taken at the first exposure. Straight and parallel fringes on the right correspond to the transmitted shock wave visualized at the second exposure. It is emphasized again that the interfacial instability is

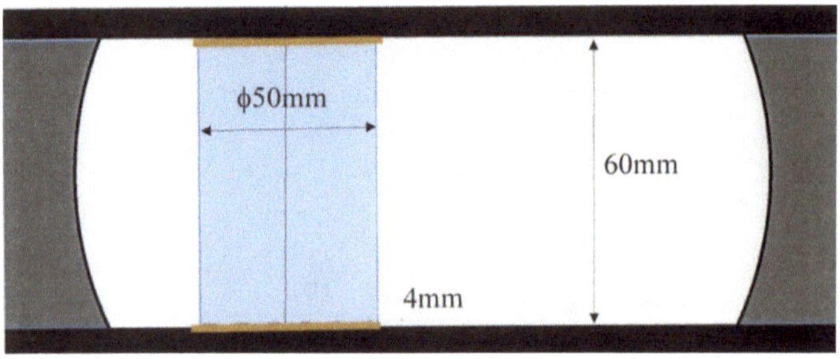

Fig. 7.12 Experimental arrangement for observing side view of the helium column

not two-dimensional but the three-dimensional which generated vortices. Figure 7.13c corresponds to an earlier stage of the interface deformation corresponding to Fig. 7.9l. Figure 7.13d was taken at 1.0 ms after the arrival of the shock wave at the leading edge of the helium column (Nagoya 1995).

7.3 Shock Wave Propagation Over Liquid Surface

7.3.1 Air/Silicone-Oil Interfaces

A 4 m long and 30 mm × 40 mm conventional shock tube made of extruded brass was connected to a 30 mm × 240 mm circular stainless steel test section and already used in the water wedge experiments. Figure 7.14 shows a direct shadowgram of air/silicone oil experiments. Silicone oil (Toshiba Silicone 10cSt) was filled in the hemi-circular space in the bottom and the shock wave propagates in air at supersonic speed in terms of the sound speed in silicone oil (Takayama et al. 1982). The IS was diffracted at the entrance and the transmitted shock wave was reflected from the silicone oil surface. The pattern of the reflected shock wave in air was DMR. The secondary triple point and the resulting Mach stem are clearly visible. When this photograph was taken, these wave patterns were mysterious. The shock wave propagated for Ms = 3.225 at the speed of 1.112 km/s. This is locally supersonic in silicone oil. Its Mach number is M = 1.13, relative to the sound speed in silicon oil. Then the shock wave in the silicone oil looks straight.

Figure 7.15 shows a sketch of the 40 mm deep and 150 mm long hem-circular test piece inserted in the lower part of the 30 mm × 240 mm circular test section. Silicone-oil (Toshiba Silicone 10cSt) was filled in the cavity.

Experiments were conducted for Ms ranging from 2.9 to 3.3 relative to shock wave speed ranging from 900 to 1100 m/s. Figure 7.16 shows shock waves propagating over the air/silicone-oil interface at variable Ms. When the incident shock wave speed, u_s, is lower than the sound speed in silicone oil, $a_{silicon}$,

(a)
(b)
(c)

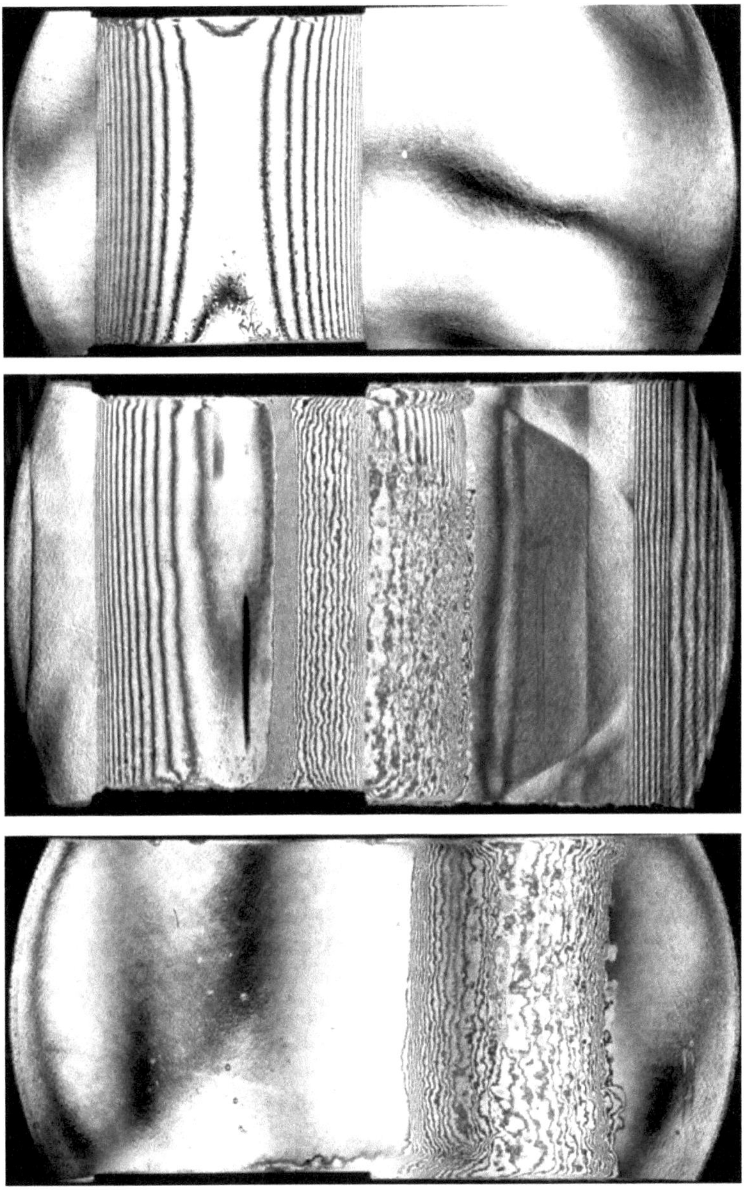

Fig. 7.13 Observation from side for Ms = 1.20 in air at 1013 hPa, 295 K, helium pressure at 1015.2 hPa: **a** #93112302, before shock loading; **b** #93111813, 500 μs after arrival of the shock wave at the column, Ms = 1.20 at 1015.2 hPa; **c** #93112211, 900 μs, Ms = 1.19; **d** #93112310, 1.0 ms, Ms = 1.21

(d)

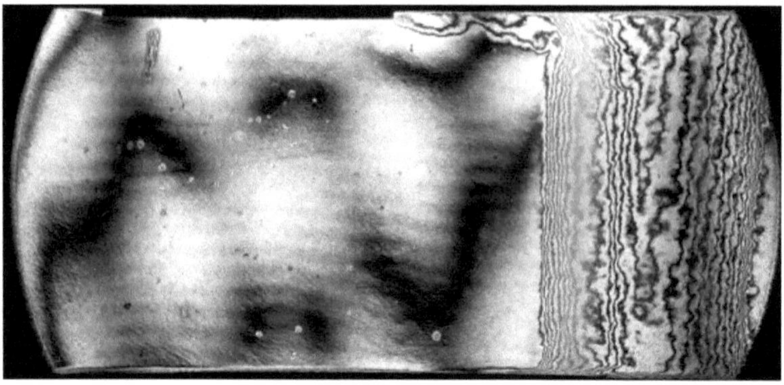

Fig. 7.13 (continued)

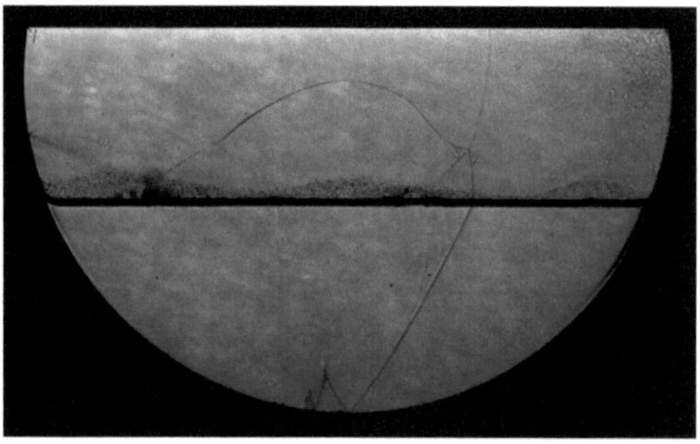

Fig. 7.14 Shock wave propagation along an air/silicone oil interface: Direct shadowgraph, #75050510, Ms = 3.225 in air at 150 hPa

$u_s < a_{silicon}$, compression wavelets are generated in the silicone oil and propagate at $a_{silicon}$. At $u_s \approx a_{silicon}$, the compression wave coalesced into a weak shock wave.

Figure 7.16 shows sequentially transmission of a shock wave at sonic speed $u_s = a_{silicon}$ in silicone oil. The shock wave in air and the transmitted wave in silicone wave are normal to the interface. For $u_s > a_{silicon}$, an oblique shock wave appears with an inclination angle θ, where $\sin\theta = a_{silicon}/u_s$, whereas $u_s < a_{silicon}$ the sonic wave propagates in silicone oil ahead of the shock wave in air.

As $a_{silicon}$, is 985 m/s at 290 K, the transmitted wave pattern inside silicone-oil varied depending their speeds. Figure 7.17 summarize the results. In Fig. 7.17a, b, c, the shock speed is subsonic $u_s < a_{silicone}$. The wave in silicone oil is visible in front of the shock wave in air. In Fig. 7.17d the shock wave in air propagates at almost the sonic speed in silicone oil and then it appears to be normal to the silicone oil surface $u_s \approx a_{silicon}$. In Fig. 7.17e–g, the shock wave in air propagates at supersonic speed $u_s > a_{silicone}$. Then shock wave in silicone oil appear to be oblique to the silicone oil surface.

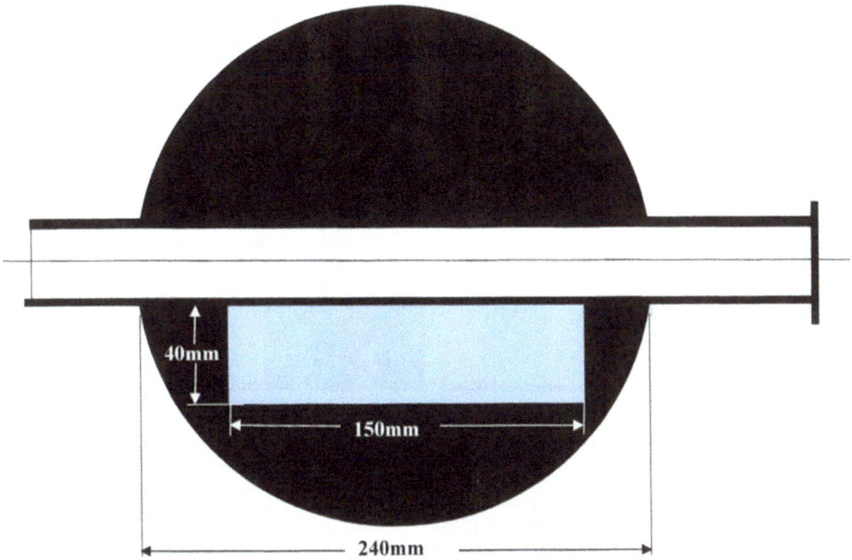

Fig. 7.15 Sketch of the test section: shock propagation along silicone oil surface (Takayama et al. 1982)

7.4 Shock Waves Induced by the Injection of High-Speed Jets

High speed water jets in air are often accompanied by high frequency noises when the jet speeds exceed the sound speed in air. To conduct analogue experiments of diesel fuel injections of a 7 mm bore compact gun was constructed in which the energy sources were convertible from gas gun to powder gun. In the gas gun mode operation, high pressure gas was used and in the powder gun mode operation, smokeless powder was used. The compact gun has a 7 mm bore and 10 mm long and can launch a 1.0 g weight high density poly-ethylene piston at muzzle speeds ranging from 0.1 to 1 km/s. The double exposure holographic interferometry was intensively used for visualization (Shi 1995).

Figure 7.18 show high speed liquid jets and shock waves induced in front of the jets. In Fig. 7.18a, b kerosene and water jets were ejected through a 0.5 mm diameter nozzle attached to a storage chamber of the gas gun, respectively. Grey shadows show the liquid jet boundary forming very irregular shape, whereas bow shock waves were formed smoothly at the head of liquid jets, respectively. In Fig. 7.18c–e, diesel fuel ejections through with a 0.5 mm diameter 90° two-hole nozzles were sequentially observed. Bow shock waves were formed ahead of impulsively ejected diesel fuel jets.

Figure 7.18 show bow shock waves formed in front of the jets and oblique shock waves were intermittently generated along jagged shaped jet boundary. The

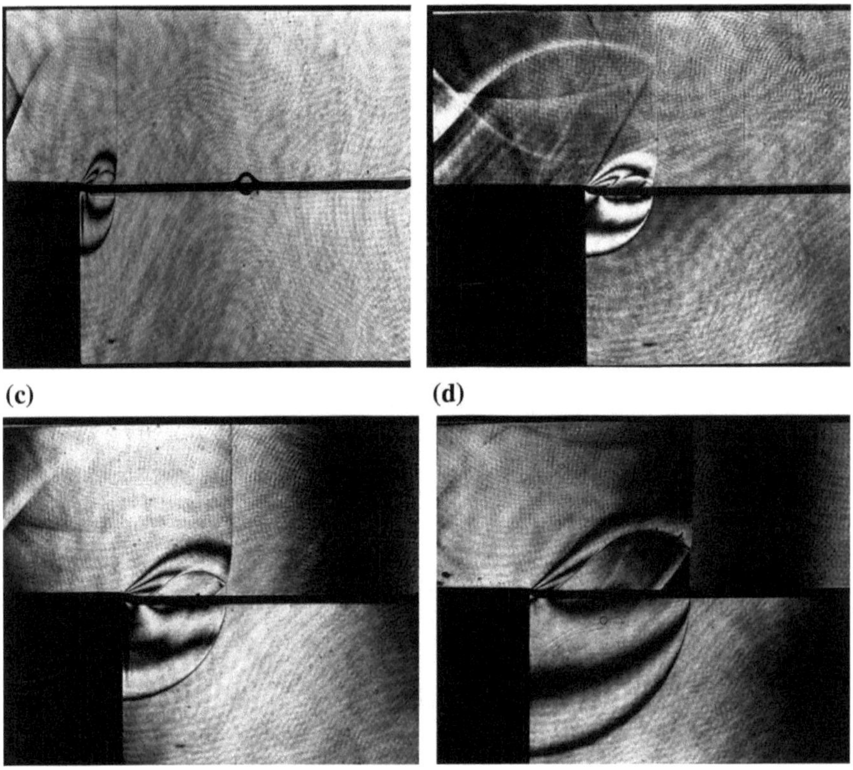

Fig. 7.16 Sequential observations of shock waves at sonic speed, $a_{silicon} = u_s$, Ms = 2.80 ± 0.05 in air at 125 hPa, 295 K: **a** #81052601, 8 μs from the entrance corner; **b** #81052202, 16 μs from the entrance corner; **c** #81052202, 22 μs from the entrance corner; **d** #81052605, 40 μs

irregularly shaped boundary was induced by not only turbulent mixing over the jet boundary and also by inherited intermittent stress wave propagation inside the titanium container that occurred when the projectile impacted the gun. The titanium container was not a rigid body but was deformed by the propagation of longitudinal and transversal waves upon the piston impingement. A 15 mm diameter powder gun was constructed and launched a 15 mm diameter and 25 mm long high-density polyethylene projectile weighing 4 g at a muzzle speed of 1.8 km/s and eventually achieved approximately 4 GPa in diesel fuel and water stored in a few cm^3 titanium container positioned at the end of the launch tube. In discharging the test liquid through a 0.5 mm diameter nozzle, the jet speed of approximately 3.0 km/s was readily obtained. In Fig. 7.18f–h, bow shock waves were formed in front of the jets and oblique shock waves were intermittently generated along jaggedly shaped jet boundary. Even though the edges of the jets looked very irregular, the shock waves looked smooth because the waves are formed from accumulation of disturbances

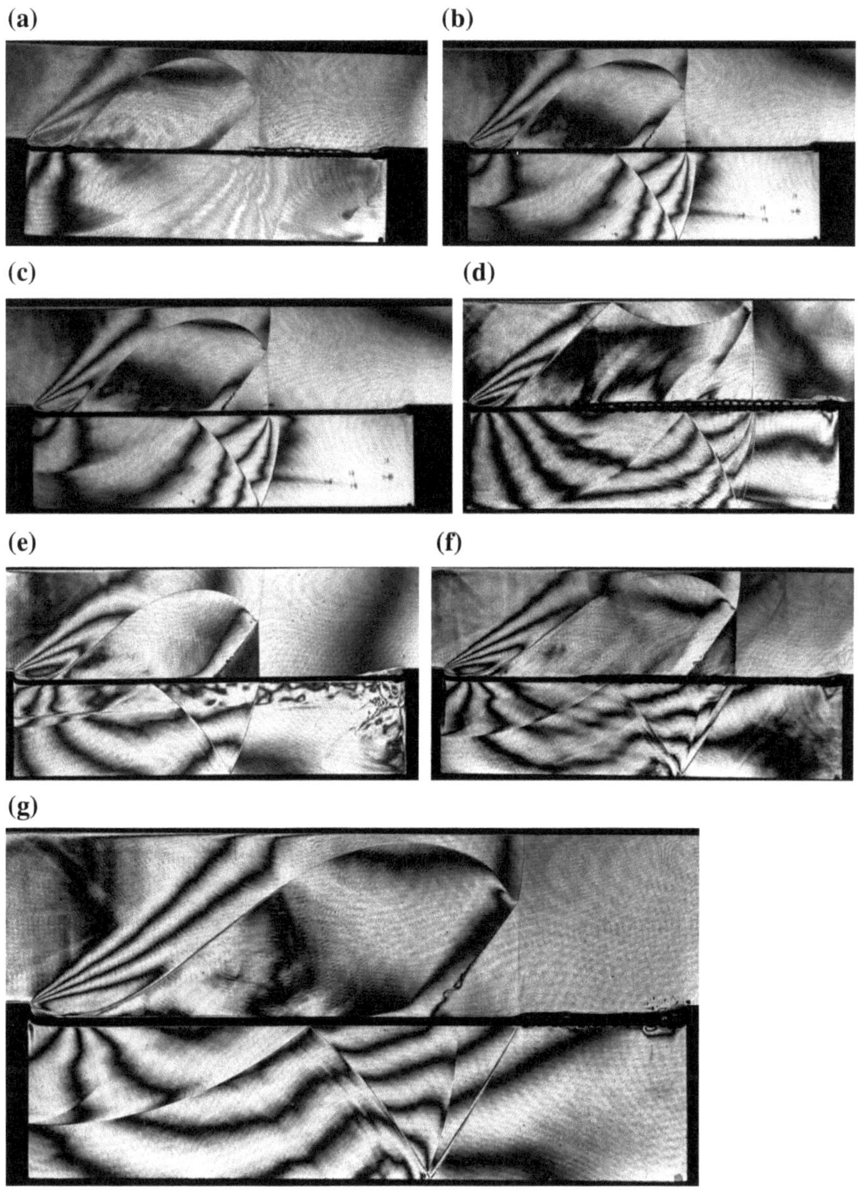

Fig. 7.17 Shock wave propagation over silicone oil: **a** #81020602, Ms = 3222, $u_s > a_{silicon}$; **b** #81020613, Ms = 3.016, u_s 0< $a_{silicon}$; **c** #81012901, Ms = 2.946, $u_s < a_{silicon}$; **d** #81012202, Ms = 3.061, $u_s \approx a_{silicone}$; **e** #81020605, Ms = 3.229, $u_s > a_{silicon}$; **f** #81012301, Ms = 3.350, $u_s > a_{silicon}$; **g** #81012302, Ms = 3.367, $u_s > a_{silicone}$

induced by the irregularities. The irregularly shaped boundary was induced due to not only turbulent mixing over the jet boundary but also inherited intermittent stress wave propagation inside the titanium container: These were created by the projectile impact. In Fig. 7.18h, the leading edge of the jet was not necessarily form a smooth blunt shape but became a jagged shape. Hence the shape of shock wave generated in front of the jet also took irregular shapes. The titanium container was not a rigid body but was readily deformed by the propagation of longitudinal and transversal waves induced by the piston impingement.

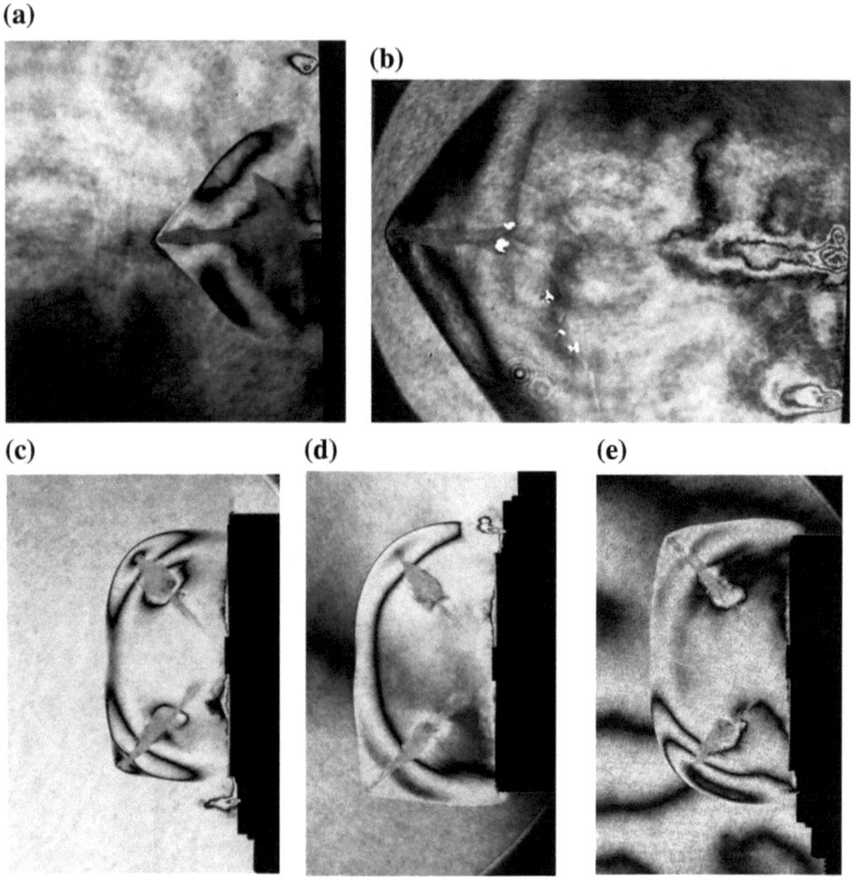

Fig. 7.18 Shock wave induced by high speed jet: **a** #92040603, injection pressure 12.3 atm, jet speed 480 m/s, fuel kerosene, nozzle diameter 0.5 mm; **b** #92040802, water jet, injection pressure 14.4 atm, nozzle diameter 0.5 mm; **c** #93032010, two-hole $\theta = 90°$ nozzle diameter 0.5 mm, diesel fuel, injection pressure 13.5 atm; **d** #93032011, two-hole $\theta = 90°$ nozzle diameter 0.5 mm, diesel fuel, injection pressure 13.5 atm; **e** #93032201, two-hole $\theta = 90°$ nozzle diameter 0.5 mm, diesel fuel, injection pressure 12.6 atm; **f** #94060303, nozzle diameter 2.5 mm, water jet driven by burning smokeless powder weighing 2 g; **g** #94060304, nozzle diameter 2.5 mm, water jet driven by smokeless powder 2 g; **h** #9406080, nozzle diameter 0.6 mm, light oil jet driven by smokeless powder 4 g

(f)

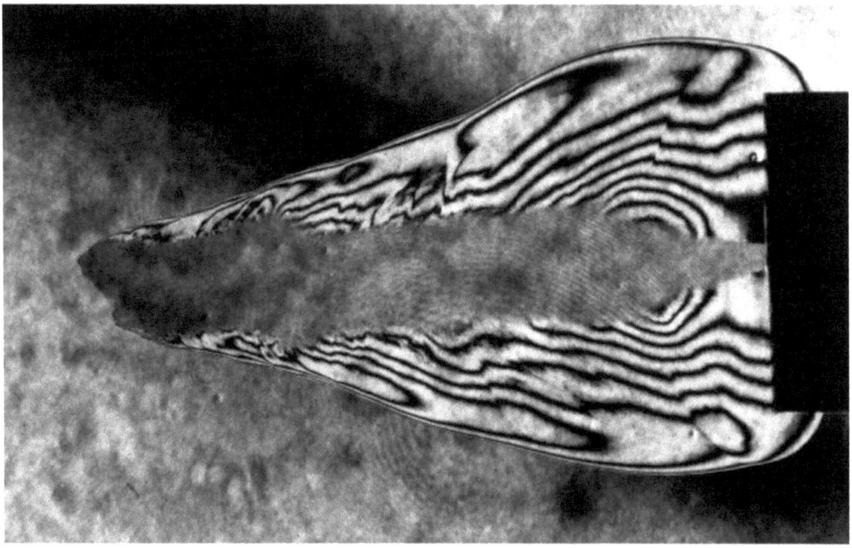

(g)

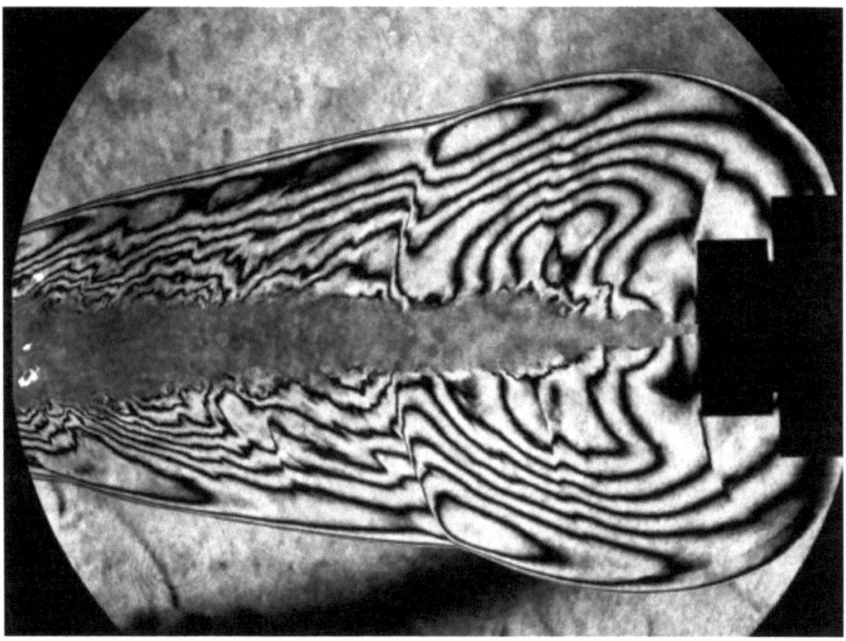

Fig. 7.18 (continued)

(h)

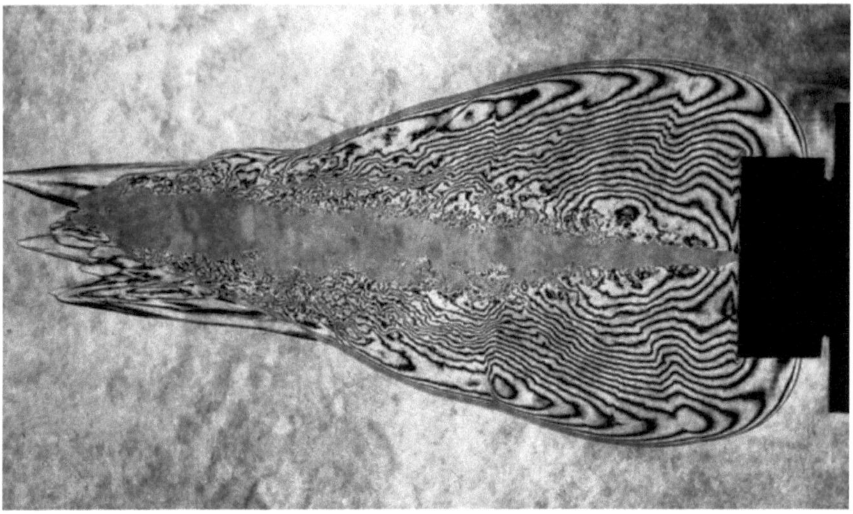

Fig. 7.18 (continued)

7.4.1 High-Speed Liquid Jet Induced by a Two-Stage Gas Gun

Pianthong (2002) worked on high-speed jet formation using a compact powder gun (Shi 1995) and measured the effect of stress wave propagations in the powder gun on the jet formation. Matthujack (2000) constructed a vertical two-stage gas gun and revealed that the irregular jet shapes were caused by the stress wave propagation in the nozzle. Figure 7.19a shows the two-stage gas gun supported vertically with two pillars. The gun was movable smoothly up and down suspended with linear guides which enabled to conduct fine measurements.

The gun consisted of a co-axially arranged 230 mm diameter and 1.5 m long high pressure chamber and a 50 mm diameter and 2.0 m long pump tube. A 50 mm diameter and 75 mm long high-density polyethylene piston weighing 130 g was accelerated driven by compressed helium filled in the pump tube. Then the resulting high pressure helium projected a 15 mm diameter, 20 mm long and weighing 4.2 g polyethylene projectile downward into an acceleration tube. Figure 7.19b shows the test section. The acceleration tube had 3 mm diameter holes distributed in a line which eliminate a detached shock wave created ahead of the projectile and at the same time worked to split the projectile from the sabot. The projectile hit a nozzle block in which the liquid under study was filled. The jets were ejected from the nozzle int the test section and visualized. Figure 7.19b shows the arrangement for pressure measurement using a fiber optic probe hydrophone.

Figure 7.20a shows a nozzle made of high strength carbon steel in which the test liquid is filled. Upon the impingement, the liquid was spontaneously pressurized

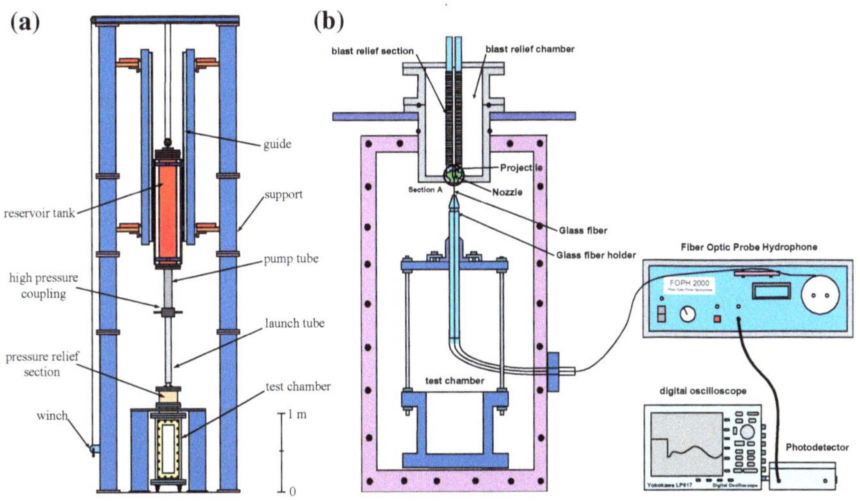

Fig. 7.19 Experimental setup: **a** gas gun accelerating projectile; **b** installation of the nozzle and pressure transducer (Matthujak 2007)

and ejected. The entire launch tube and the acceleration tube were accommodated in a 305 mm diameter and 850 mm long cylinder and installed in a rectangular observation chamber as shown in Fig. 7.19b. An optical fiber pressure transducer (FOPH2000 RP Acoustic Co. Ltd) was inserted into the exit of the nozzle facing to the liquid jet. The present pressure transducers are not using the piezo effect but detect optically the change in phase angle at the spot in water. The principle is the same as double exposure holographic interferometry. Irradiating the test water with a coherent laser beam through a 0.7 mm diameter optical fiber, the time variation of this laser beam is continuously monitored. This laser light beam is equivalent to the *OB* of the holographic interferometry. The time variation of the *OB* is simultaneously compared with undisturbed source laser beam which is equivalent to the *RB*. Then from the comparison of phase angle variation between the two beams, the density variation is obtained continuously. Trusting the Tait equation (Tait 1888), the density variations are converted readily to the pressure variations. If the equation of state and the relationship of refractive index and density is known for any liquids, this optical pressure transducer can be applied to measure pressure variations in any liquids.

The optical fiber has 0.7 mm diameter and response frequency of 10 MHz and hence this pressure transducer is wonderful for underwater shock wave study at the pressure range below 25 GPa. Figure 7.20b shows the result of the measured stagnation pressure of the jet. The ordinate denotes the stagnation pressure in MPa and the abscissa denotes the elapsed time in µs. Peak pressures A and B correspond to longitudinal and transversal stress waves released from the nozzle into the water. The values of peak pressures A and B are high. However, as these peak pressures were maintained for a very short time, then their values of the impulses are just

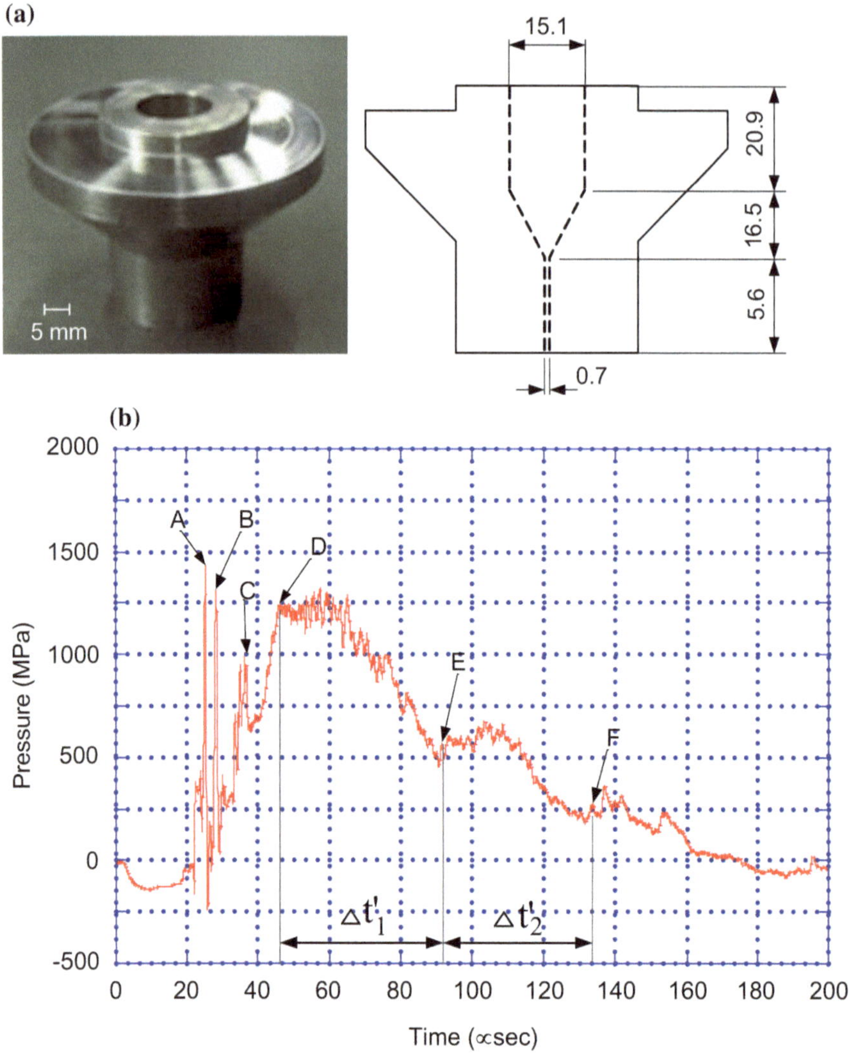

Fig. 7.20 Pressure measurement: **a** nozzle structure in (mm); **b** the results of pressure measurements (Matthujak 2007)

modest. These intermittent peak pressures D, E, and F correspond to the first, second, and third impulses.

Figure 7.21 summarize jet formation of water, Kerosene, Diesel fuel, and gasoline. The visualizations were conducted using direct shadowgraph and the images were recorded by a high speed video camera Shimadzu SH100 at the framing rate of 10^6 frame/s. Jets are released from the 0.5 mm diameter nozzle and are the fastest immediately after the ejections. Hence, on the first frames, the detached shock waves appeared in front of the leading jets. The oblique shock waves appeared from the nodes of the jets and their inclination angles are, so far as estimated from the pictures, range 7°–10°. The jet speed can reach about 2.0–3.0 km/s. With propagation, the inclination angles of the jets increase, which meant the attenuation the jet speeds. As seen in Fig. 7.20b, the peak stagnation pressures in the liquid container decrease intermittently. The intermittent acceleration induced nodes along the jet structure and shock waves are also generated at individual nodes. This sequence promotes the deformation of the jet and also positively contributes to the atomization of fuel jets. The sequence of formation and deformation of high speed jets differ significantly depending on liquid type. This implies the dependence of physical properties determining directly the process of the jet formation. For the jet speed at 2.0 km/s, for example, the stagnation temperature in the shock layer would exceed well over 3000 K. If such a condition is maintained longer duration of time, it would contribute to the atomization of fuel and might have induced auto-ignition. However, it has not happened so far. As seen in Fig. 7.21, the elapsed time of jet formation is at longest 600 μs. This is much shorter than the induction time for auto-ignition.

7.5 Shock Wave Interaction with Droplets

Holographic interferometric visualization is one of a useful methods of shock tube experiments. In the 1980 a collaboration started with Dr. T. Yoshida of Tohoku University Faculty of Engineering and visualized droplet shattering upon shock wave loading (Yoshida and Takayama 1985). The collaboration progressed when Dr. A. Wierzba of the Institute of Aviation Warsaw joint the project (Wierzba and Takayama 1987).

7.5.1 Shattering of Droplets Falling in a Line by Shock Wave Loading

When the liquid droplets were exposed suddenly to high speed flows, the droplets shattered. This is one of the fundamental research topics of shock wave research. Droplet breakup is categorized by a dimension-less similarity parameter, the so-called Weber number, We, which is defined as $We = \rho u^2 d/\sigma$ where u, d, ρ, and σ are particle velocity, droplet diameter, liquid density, and surface tension, respectively. The We number is, in other words, a ratio of dynamic energy to the surface tension.

Figure 7.22 shows schematic illustrations of droplet shattering. For a small We < 14, a vibrational type breakup occurs. At a larger We, droplets bulge into a bag shape and eventually collapse. Breakup patterns accompanying fragmentation

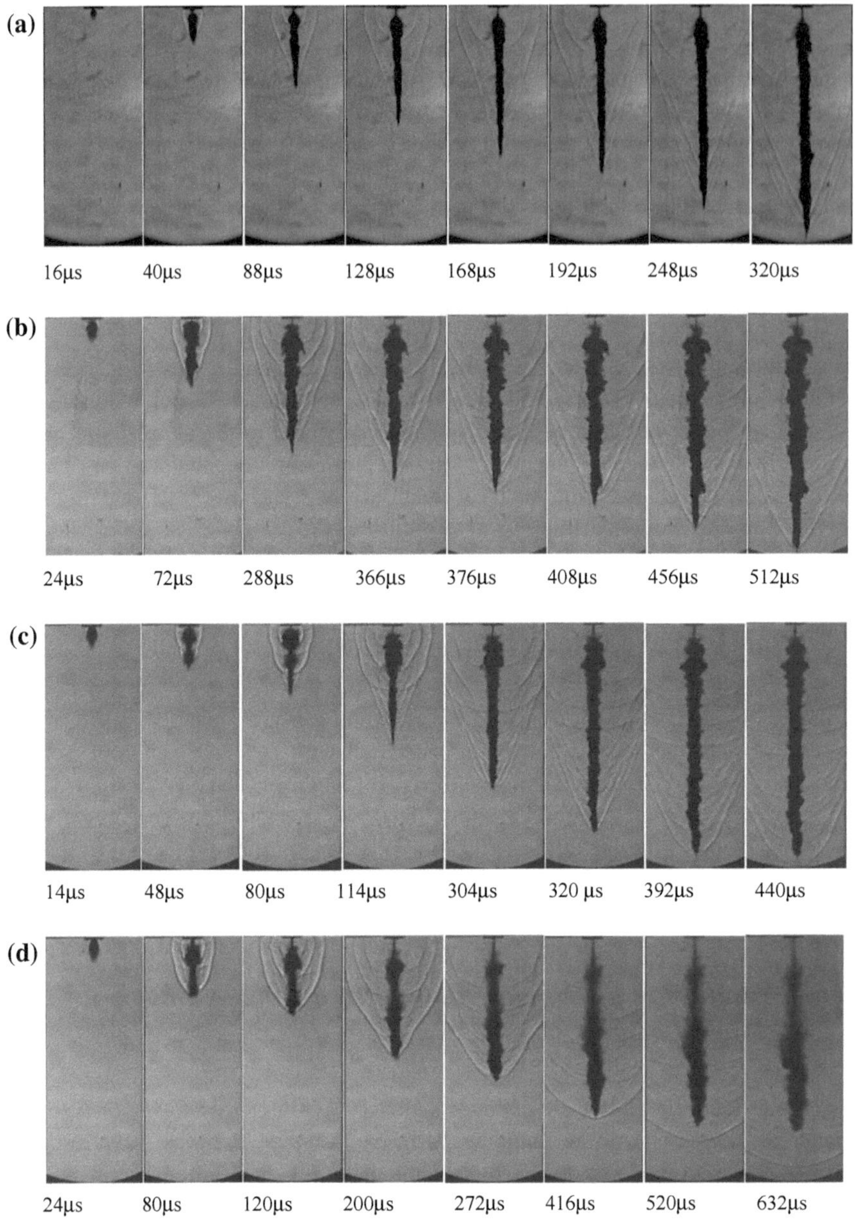

Fig. 7.21 Evolution of high speed liquid jets: **a** water $V_{piston} = 317$ m/s; **b** kerosene, $V_{piston} = 305$ m/s; **c** diesel oil, $V_{piston} = 295$ m/s; **d** gasoline, $V_{piston} = 295$ m/s

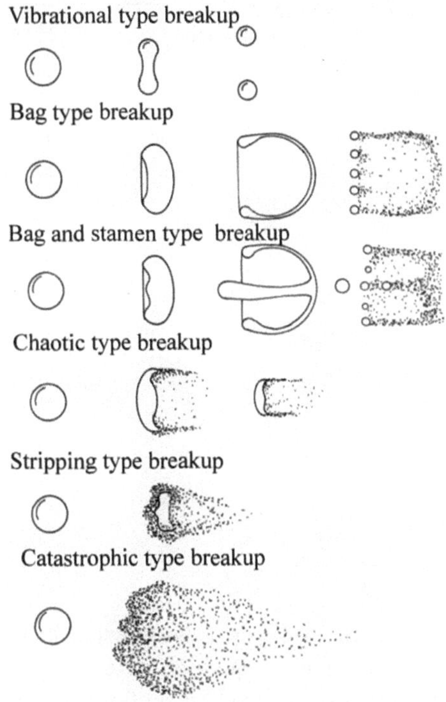

Fig. 7.22 Different patterns of droplet shattering (Wierzba and Takayama 1987)

the so-called stripping type breakup occurs at about We ~ 2000. For further larger We, a droplet is instantaneously fragmented into mist and hence is named as a catastrophic type breakup. A comprehensive reference survey was reported by Wierzba and Takayama (1987) and Gelfant et al. (2008). Droplet shattering is applied to science, technology, and industry, for example, chemical processes in two-phase flow systems, rain erosions in supersonic flights, spray combustions and many others applications.

Experiments were conducted in the 60 mm × 150 mm conventional shock tube and the breakup process was observed using double exposure or single exposure holographic interferometry. Test liquids were water and ethyl alcohol which were intermittently dripped from a small hole on the upper wall of the shock tube using an ultrasonic oscillator. Yoshida and Takayama (1985) designed the oscillator. Test liquids were filled in a capillary tube connected to a 0.7 mm diameter nozzle and oscillated at the frequency of 100–200 Hz. The droplet diameter and their interval were adjusted tuning the oscillator's frequency. Figure 7.23 shows an actuator for introducing droplets in a well controlled fashion into the test section. Selecting nozzle shapes and operational conditions, gas filled liquid bubbles of about 4 mm diameter can be introduced to the test section.

The droplets shattering is also conducted in high-speed wind tunnel flows. It was often argued that the droplet shattering in a high-speed wind tunnel flow is

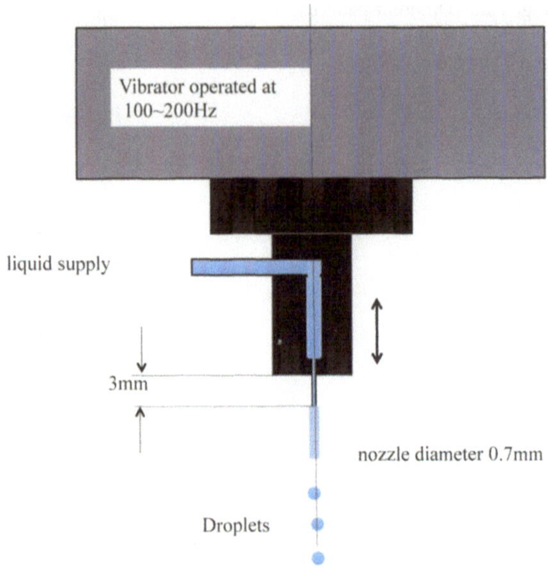

Fig. 7.23 Actuator for introducing droplets into the test section (Yoshida and Takayama 1985)

conducted in a steady flow, while that in shock tube flows is conducted in an unsteady flow. A question arises that there may be a difference in the shattering processes in the wind tunnel flows and the shock tube flows. However, the droplet shattering in the wind tunnels is not a steady process as the introduction of the droplets into the wind tunnel flows is unsteady. The unsteadiness governs the rest of the shattering procedure. In the shock tube flows, the introduction of the droplets exhibited unsteadiness and it remains until the early wave interaction around the droplets disappears and then the flow became steady for a while. At the later stage, the droplet shattering takes the same procedure appearing in the wind tunnel flows.

Figure 7.24 show a sequential observation of shattering ethyl alcohol droplets of 0.76 mm diameter and that are introduced into the shock tube with separation of about 6–8 mm. The droplets are exposed to interval exposed to a shock wave of Ms = 1.40 and consequently the flow behind the shock wave in atmospheric air at 289.6 K. The liquid jet was oscillated at frequency 144 Hz and became droplets in a line falling into the shock tube. At the earlier stage of the shock wave impingement, the procedure is similar to the shock wave/solid sphere interaction. In Fig. 7.24f, the boundary layer separation promotes the droplet deformation and wakes development. Figure 7.24k shows that the droplet volume increases monotonously with elapsing time. Although wave motions inside the droplet are not observed, transmitted waves and their reflections inside the droplet would promote the deformation process. Reinecke and Waldmann (1975) tried to visualize the droplet deformation by introducing radiography. Holographic interferometry also gave better resolution of images than conventional visualization methods used for estimating the wave motion inside droplets. Advance in numerical simulations someday would reproduce the wave motion inside deforming droplets even in a row.

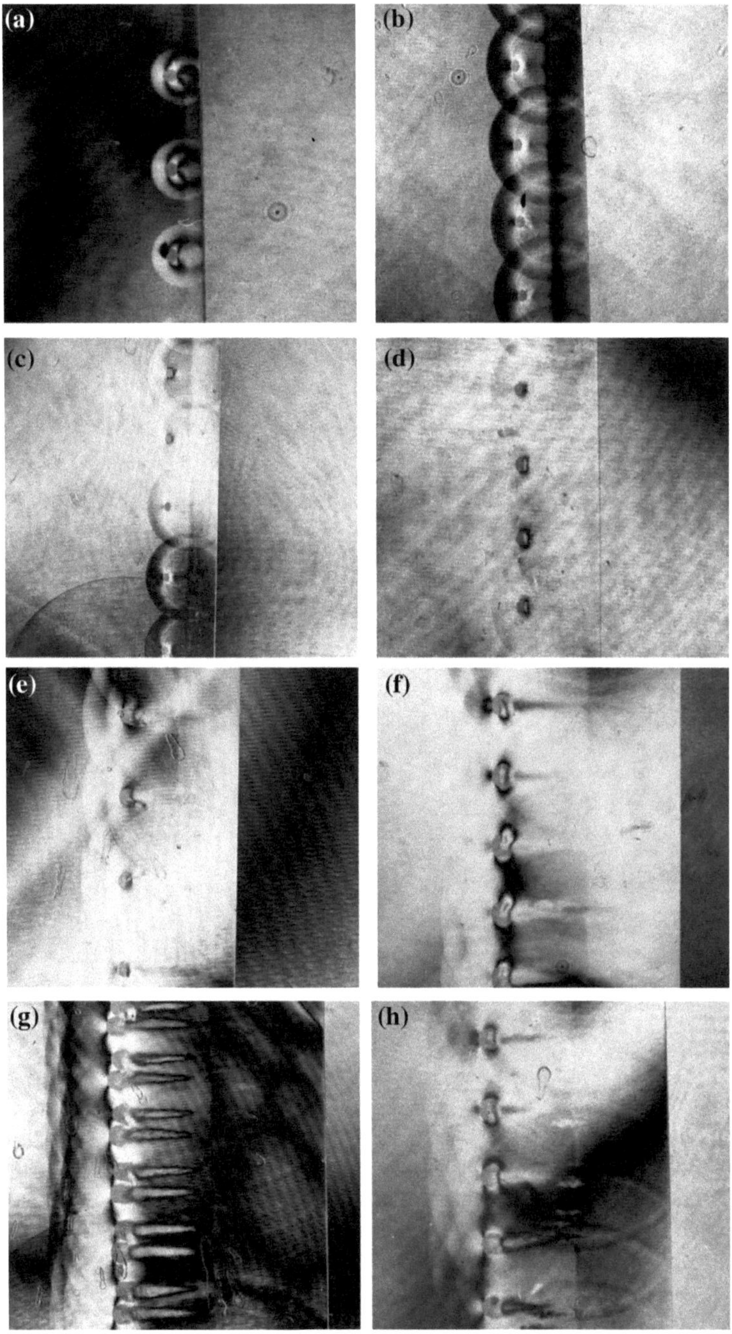

Fig. 7.24 Shattering of ethyl alcohol droplets of d = 0.76 mm for Ms = 1.601 in atmospheric air at 289.6 K, at frequency f = 144 Hz: **a** #83110707, 450 μs from trigger point; **b** #83111709, 485 μs; **c** #83111705, 490 μs; **d** #83110706, 470 μs; **e** #83111706, 500 μs; **f** #83110704, 490 μs; **g** #83111710, 485 μs; **h** #83110703, 510 μs; **i** #83111708, 520 μs; **j** #83110702, 530 μs; **k** #83110701, 550 μs

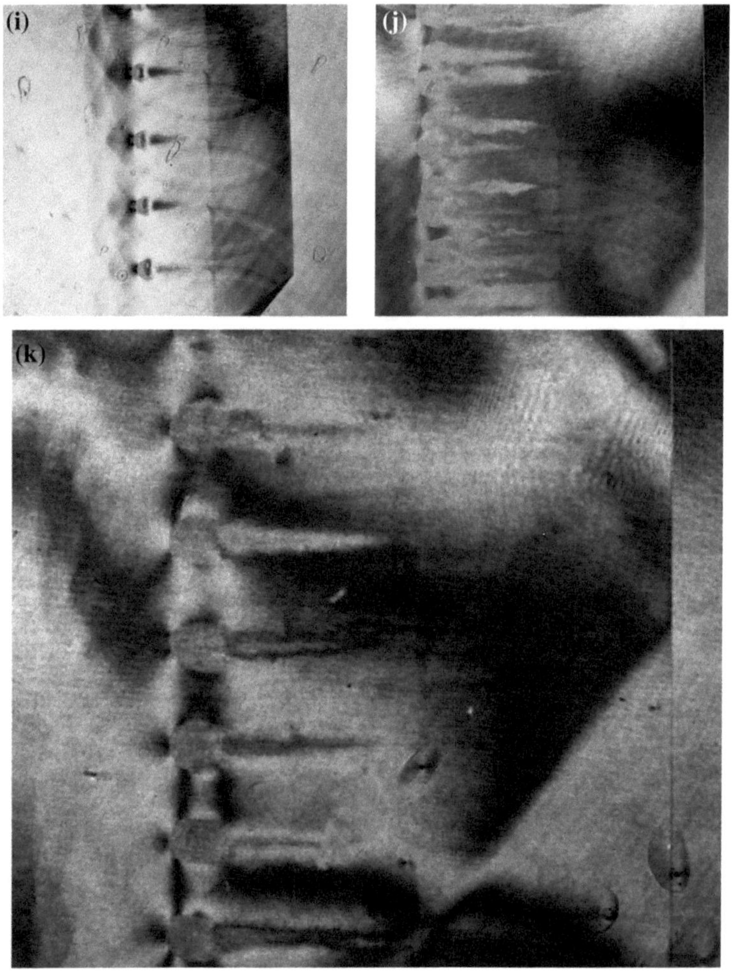

Fig. 7.24 (continued)

Comparing the background contrast with water droplets seen in Fig. 7.25, a slight deviation of grey contrast is visible between ethyl alcohol droplets and water droplets. Probably the contrast difference indicates change in the rate of evaporation which would decisively affect the refractive index of the gas mixture in vicinity of the droplets. Figure 7.25 show shattering of 1.0 mm diameter water droplets, introduced at space difference of about 5–7 mm and exposed to Ms = 1.58 shock wave in atmospheric air at frequency of 150 Hz. Figure 7.25d shows a single exposure interferogram, which shows a sign of boundary layer separation from the droplet's equator.

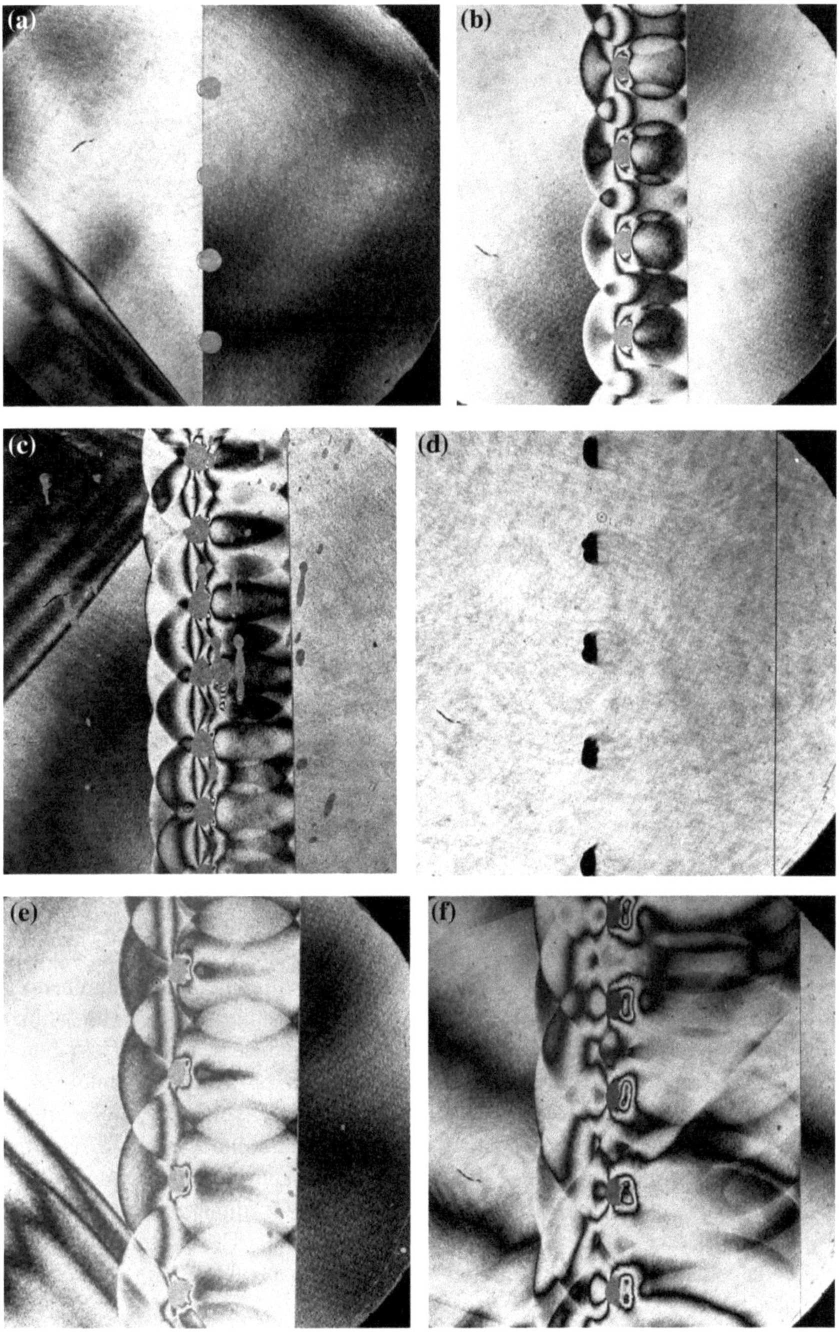

Fig. 7.25 Shattering of water droplets of d = 1.0 mm for Ms = 1.58 in air at 1013 hPa, 288 K, at f = 150 Hz: **a** #83121602; **b** #83121504; **c** #83121405; **d** #83121503, single exposure; **e** #83121505; **f** #83121603; **g** #83121903, single exposure; **h** #83121509

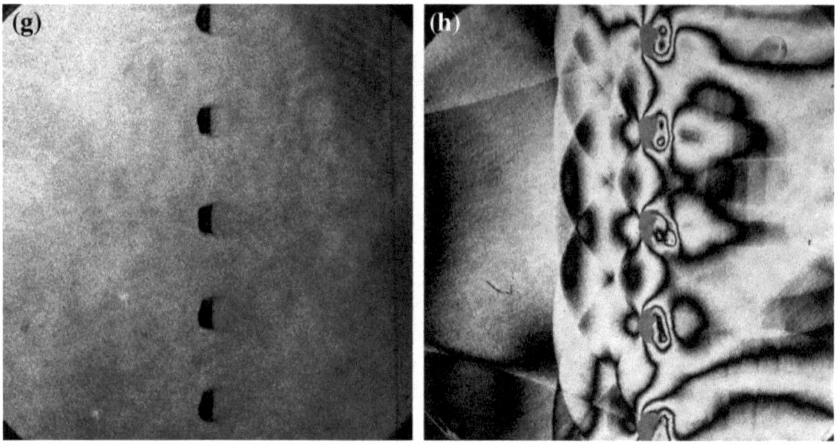

Fig. 7.25 (continued)

As large diameter droplets can not be spherical, Hence, 4 mm diameter air containing droplets were produced by blowing water at slightly pressurized air from a co-axial nozzle and oscillating at frequency of 144 Hz (Yoshida and Takayama 1985). Figure 7.26 show sequential single exposure interferograms of 4 mm diameter air containing water droplets interacting with a shock wave of Ms = 1.68. The transmitted shock wave induced the boundary layer separation over the droplet surface. At the water droplet's rear side, a wake was formed and accelerated by a recirculation vortex. The shadow of such a disintegrating droplet reminded of the shape of a tadpole. The frontal side of the droplet was suppressed and gradually blown off toward downstream. It took about 1 ms for a droplet to be almost fragmented into mist. The particle velocity of the shock wave of Ms = 1.68 is about 310 m/s. The speed of mist particles ranges, presumably, 50–100 m/s.

In Fig. 7.27, the time variations of shattering droplets of different diameters d = 1.03 mm and 4.0 mm and shock wave of different Mach numbers for Ms = 1.3–1.5 are summarized (Wierzba and Takayama 1987). The ordinate denotes droplet diameter normalized by their initial diameters. The abscissa denotes dimension-less time $t^* = tu(\rho/\rho_l)^{1/2}/\mu$, where t, u, ρ, ρ_l, and μ are time, particle velocity, air density, liquid density, and viscosity, respectively. The circles are data collected by the present holographic interferometry. A solid line summarizes their distribution. A dotted line and a dot-and-point line summarize the previous data collected by means of conventional visualization. Upon impingement, droplets are bulged almost to maximal and then started contracting. The present holographic results suggest distinctly for shorter time than that offered in Reinecke and Waldmann (1975).

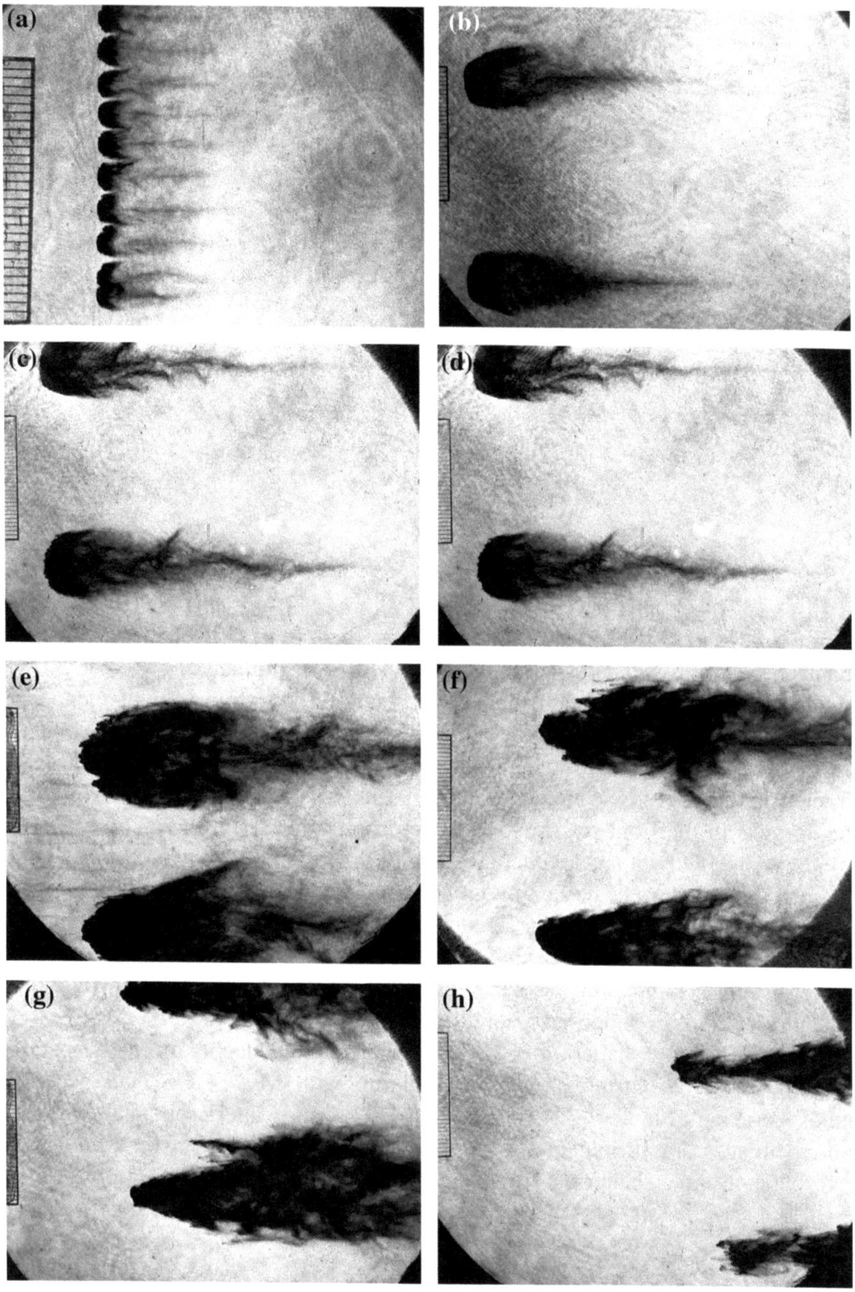

Fig. 7.26 Shattering of water droplets of d = 4.0 mm for Ms = 1.68 in atmospheric air at 289.6 K, at f = 144 Hz, single exposure: **a** #86102809, 200 μs from trigger point, Ms = 1.680; **b** #86102511, 300 μs, Ms = 1.644; **c** #86102504, 400 μs, Ms = 1.660; **d** #86102503, 500 μs, Ms = 1.670; **e** #86102420, 600 μs, Ms = 1.656; **f** #86102507, 700 μs, Ms = 1.673; **g** #86102418, 800 μs, Ms = 1.671; **h** #86102505, 1150 μs, Ms = 1.644

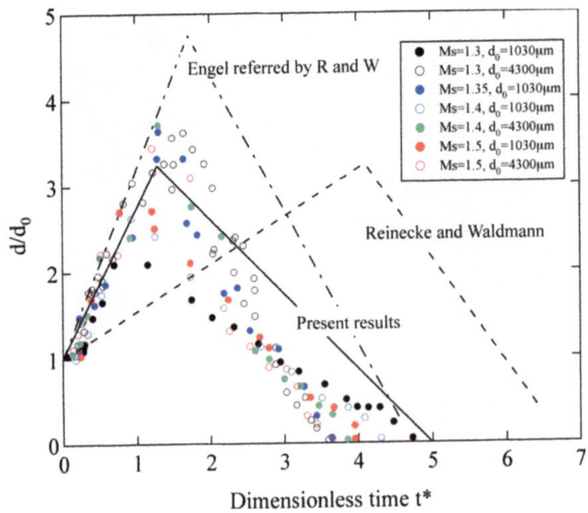

Fig. 7.27 Time variations in of droplet's diameter, summary of experiments (Wierzba and Takayama 1987)

7.5.2 *Shattering of Tandem and Triple Row Droplets*

Figure 7.28 show sequential visualization in the process of shattering 1.0 mm diameter water droplets, placed in 10 mm tandem position and dropped at 240 Hz frequency for Ms = 1.40 in atmospheric air. Figure 7.28b shows a double exposure interfrogram but all other photos are single exposure interferograms. Upon the shock wave loading, the droplets in the second row were exposed to the wake of droplets in the front row which have a lower relative speed. As seen in Fig. 7.28c–f, the droplets in the front row took over the droplets placed in the second row and eventually merged with each other as seen in Fig. 7.28e. Finally, the droplets in the first and second rows coalesced into a cloud as seen in Fig. 7.28f. In conventional shadow or schlieren photos, the shattering droplets appear as evenly illuminated grey expanding clouds, whereas in single exposure interferograms, the structures of shattering droplets are well resolved. The deviation of results shown in Fig. 7.26 was attributed to the inherited resolution of interferograms.

Figure 7.29 show shattering of 0.7 mm diameter ethyl alcohol droplets separated by 4 mm placed in triple rows and colliding with Ms = 1.25 shock wave in atmospheric air. Figure 7.29a, c are double exposure interferograms, in which blurred images are attributable to ethyl alcohol evaporation over the droplets. In Fig. 7.29e–g, three shattering droplets coalesced into one shattering droplet.

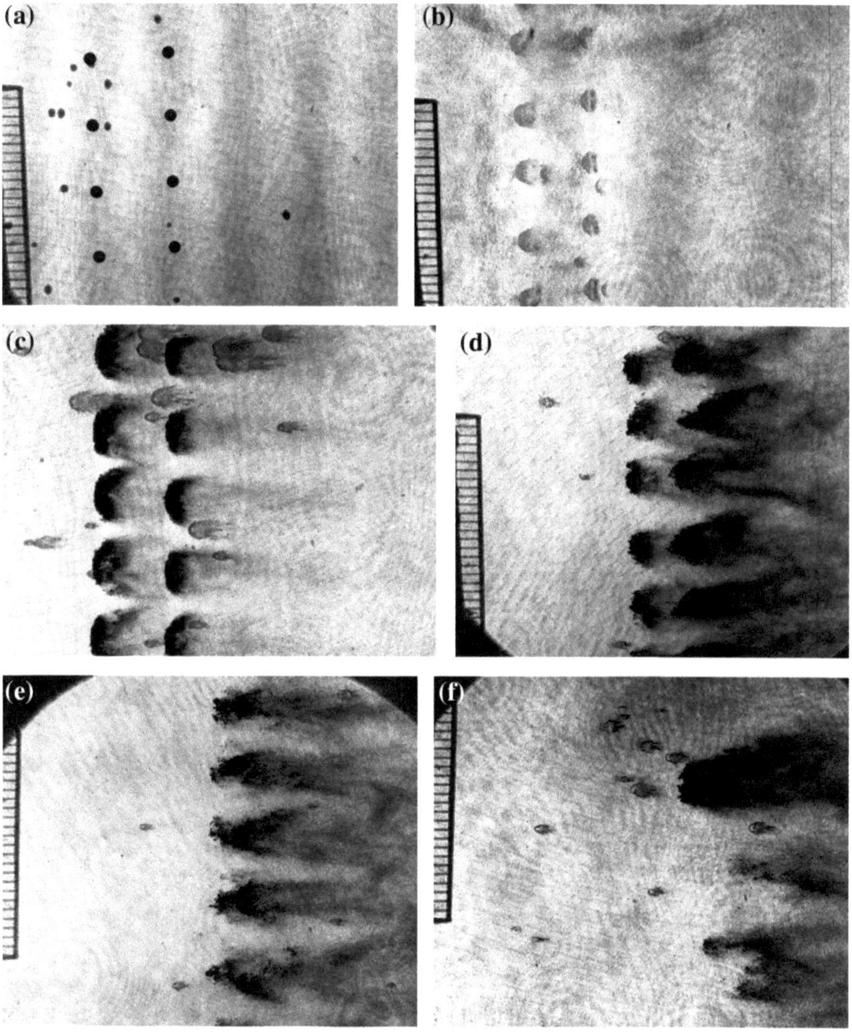

Fig. 7.28 Shattering of 1.0 mm diameter tandem water droplets separated by 10 mm impinged by a Ms = 1.40 shock wave in atmospheric air, at 298.5 K at frequency 240 Hz, single exposure: **a** #87100201, no flow picture: **b** #87100206, 280 μs from trigger point, Ms = 1.389; **c** #87100202, 400 μs from trigger point, Ms = 1389; **d** #87100211, 600 μs, Ms = 1406; **e** #87100215, 700 μs, Ms = 1401; **f** #87100212, 0.8 ms, Ms = 1.400

7.6 Shock Wave Interaction with a Water Column

In earlier stage of the stripping type breakup, waves transmitted into liquid droplets and shock waves diffracting over the droplet surface promoted their deformations. In particular, the unsteady drag force promoted deformation and then accelerated

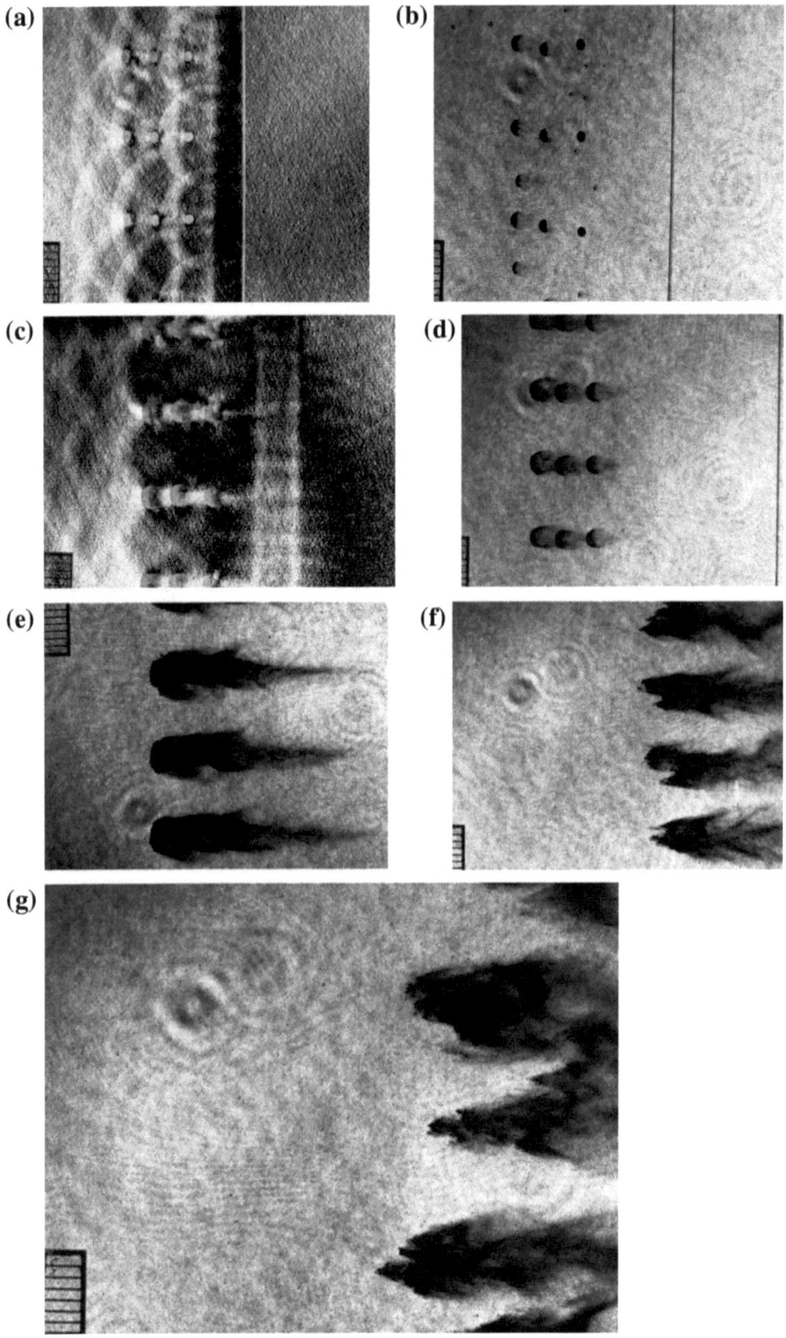

◂**Fig. 7.29** Shattering of 0.7 mm diameter triple row ethyl alcohol droplets separated by 4 mm for Ms = 1.250 in atmospheric air at 298.5 K. The droplets were dripped by three nozzles 4 mm interval at frequency 150 Hz: **a** #87040101, 250 μs from trigger point; **b** #87033116, 260 μs; **c** #87040102, 300 μs; **d** #87040103, 300 μs; **e** #87040104, 400 μs; **f** #87040113, 750 μs; **g** #87040108, 800 μs

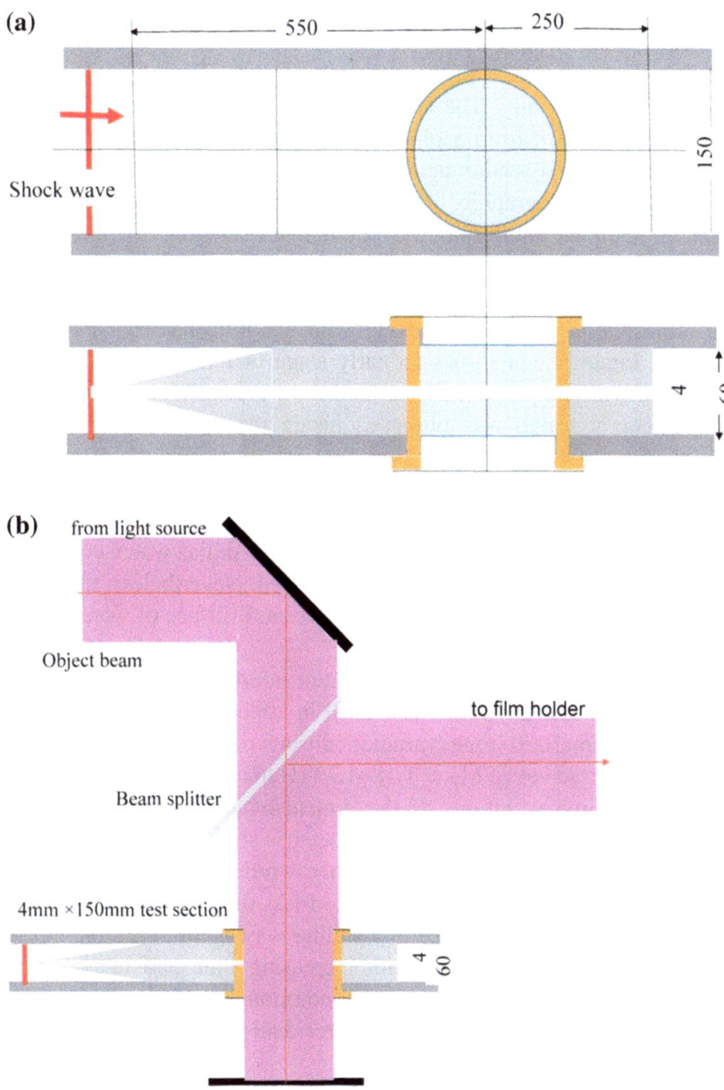

Fig. 7.30 Water column experiment: **a** 4 mm × 150 mm test section; **b** optical arrangement (Yamada 1992; Shitamori 1990)

the droplet shattering. In a case of very small droplets and volatile liquids, evaporation would contribute significantly to the droplet fragmentation. Therefore, in order to understand the droplet deformation sequence, an analogue experiment was conducted. Figure 7.30a shows a 4 mm × 150 mm test section consisting of a cookie cutter and observation windows inserted in the 150 mm × 60 mm conventional shock tube; Hamamura (1995), Igra (2000).

Supplying water with a syringe needle at the center of the test section a water column having 4 mm in height and 6 mm in diameter was installed. Turning the test section sideway, the water column was visualized from its top. The shock wave hits the water column from its side.

Figure 7.30b shows the used optical arrangement. The observation windows were positioned horizontally. The collimated *OB* illuminated the test section vertically and reflected back from a plane mirror placed at the other side of the test section. It passed the test section again and reflected from a half mirror redirecting the *OB* toward the holographic film. With this optical arrangement, Fig. 6.30b formed the double path interferometry. The height of the tested water column was doubled, to 8 mm (Shitamori 1990).

Figure 7.31 show sequential observations of shock wave (Ms = 1.17), propagating in atmospheric air and interacting with a 6 mm diameter and 4 mm high water column. Figure 7.31a shows an early stage of interaction. In Fig. 7.31b–d, the shock wave was just propagating around the water column and reflected. With this arrangement, a fringe was observed inside the water column. In a single exposure interferogram, the boundary layer separation around the droplet's equator is observed; as seen in Fig. 7.31f, g. The observed separation zones remind observation of dust free regions seen in the Sect. 4.1.4 **Cylinder in dusty gas**. In double exposure intereferogram, the shadow of the initial water column and the motion of the shock wave in water are superimposed. Nevertheless, the reflection of the transmitted shock wave from the interface frontal surface of water/air bubble is visible in Fig. 7.31i.

Figure 7.32 show sequential single exposure interferograms of a water column interacting with a shock wave for Ms = 1.40 in atmospheric air.

Figure 7.33 summarized time variation during deformation of water column colliding with shock waves of Ms = 1.18–1.73 (Hamamura 1995). The ordinates in Fig. 7.33a denote time variations of the dimensionless longitudinal diameter, in Fig. 7.33b time variation of transversal diameter is shown, and in Fig. 7.33c, time variation of dimension-less mass, or in other words, the residual area of water column is presented. The abscissa is dimension-less time as defined in Fig. 7.27. In Fig. 7.33a the data points are plotted in a similar way as the solid line in Fig. 7.27. The longitudinal diameter increase with increasing time and reach its maximal value at $t^* \sim 25$. Thereafter it decreases monotonously and eventually vanishes at $t^* \sim 35$. The transversal diameter and the residual mass decrease monotonously and vanish at $t^* \sim 35$.

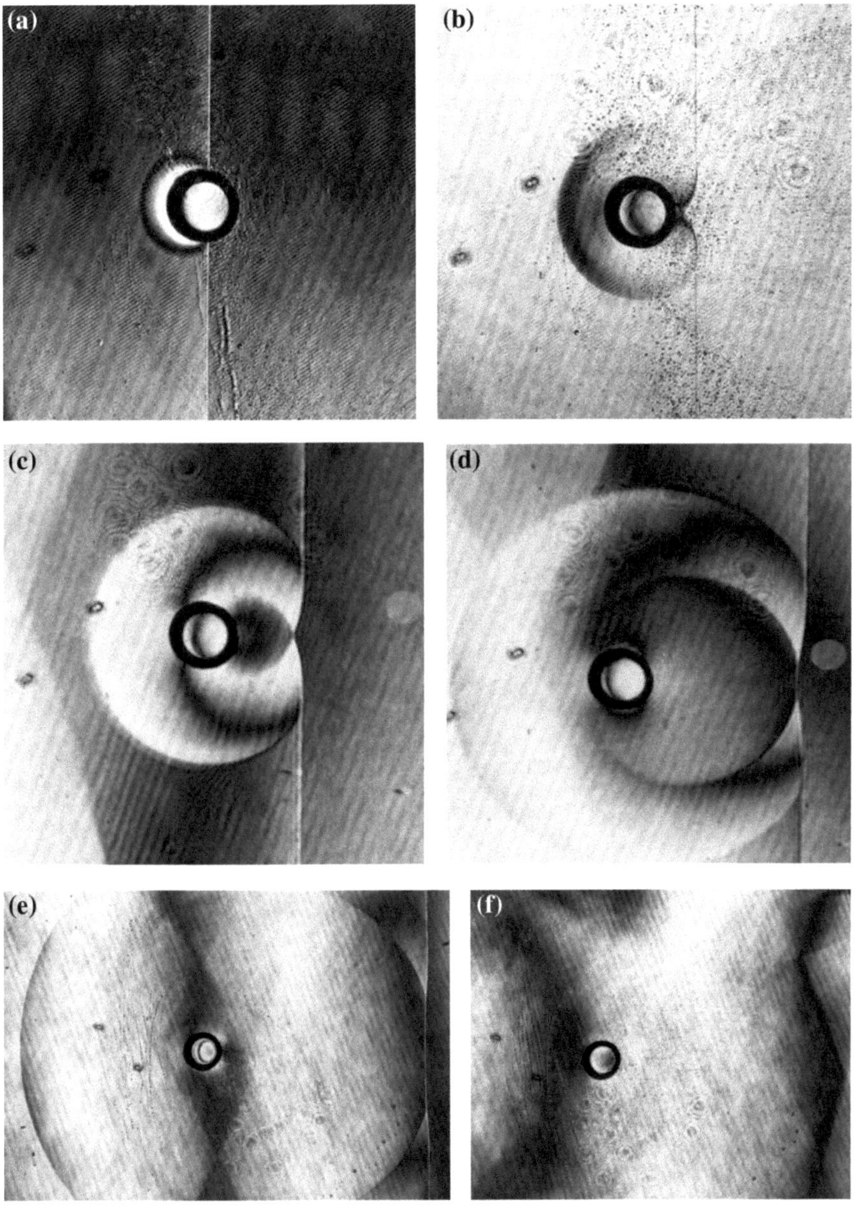

Fig. 7.31 Evolution of 6 mm diameter water column exposed to a shock wave for Ms = 1.17 in atmospheric air at 296 K: **a** #89062807, 290 μs from trigger point, Ms = 1.170; **b** #89062807, 290 μs, Ms = 1.170; **c** #89062701, 315 μs, Ms = 1.173; **d** #89062703, 340 μs, Ms = 1.171; **e** #89062704, 390 μs, Ms = 1.17; **f** #89062705, 490 μs, Ms = 1.172; **g** #89062108, 590 μs, Ms = 1.169, single exposure; **h** #89062114, 630 μs, Ms = 1.169, single exposure; **i** #89062707, 690 μs, Ms = 1.169; **j** #89062708, 790 μs, Ms = 1.174; **k** 89062118, 790 μs, Ms = 1.169, single exposure; **l** #89062709, 890 μs, Ms = 1.172; **m** #89062120, 865 μs, Ms = 1.174, single exposure; **n** #89062801, 990 μs, Ms = 1.169; **o** #89062802, 990 μs, Ms = 1.173; **p** #89062124, 1040 μs, Ms = 1.171, single exposure; **q** #89062120, 865 μs, Ms = 1.174

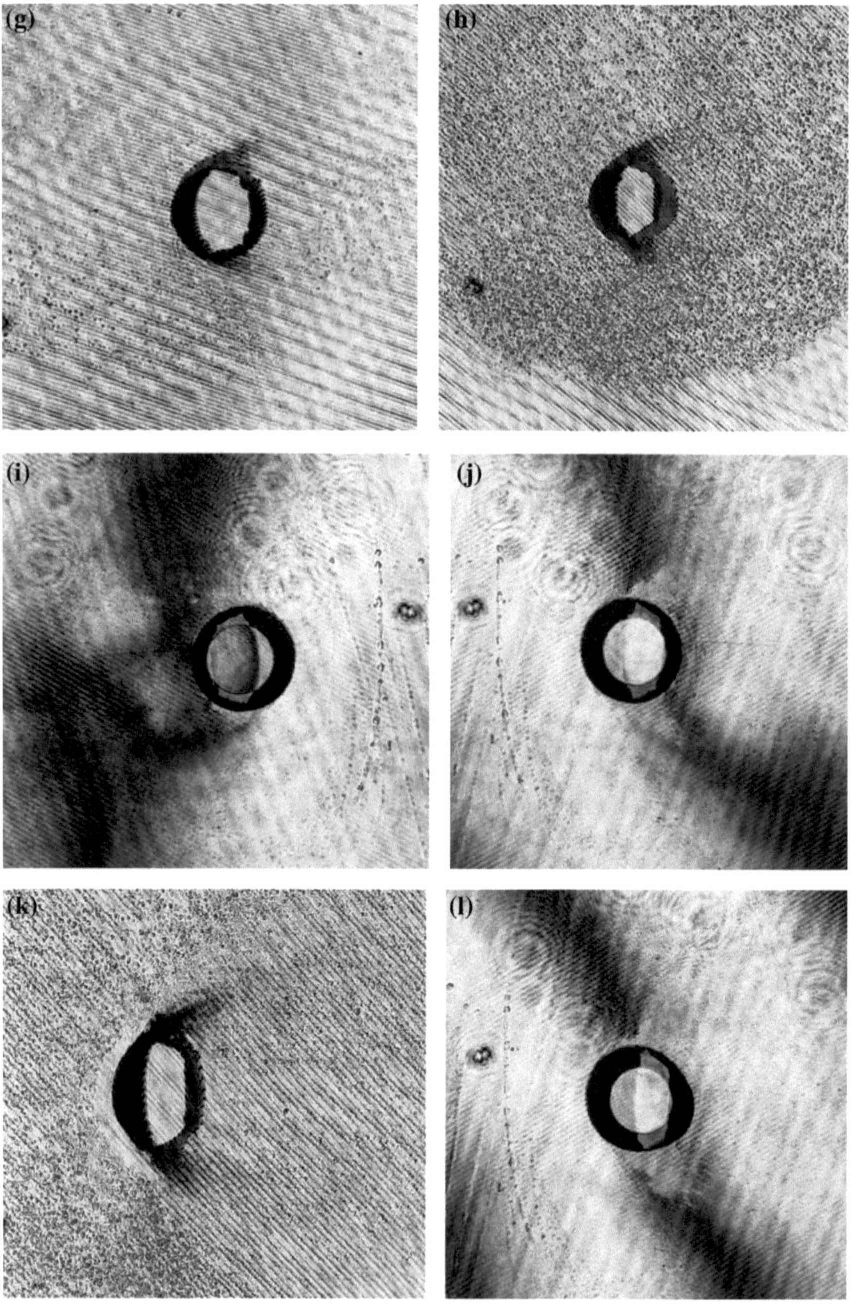

Fig. 7.31 (continued)

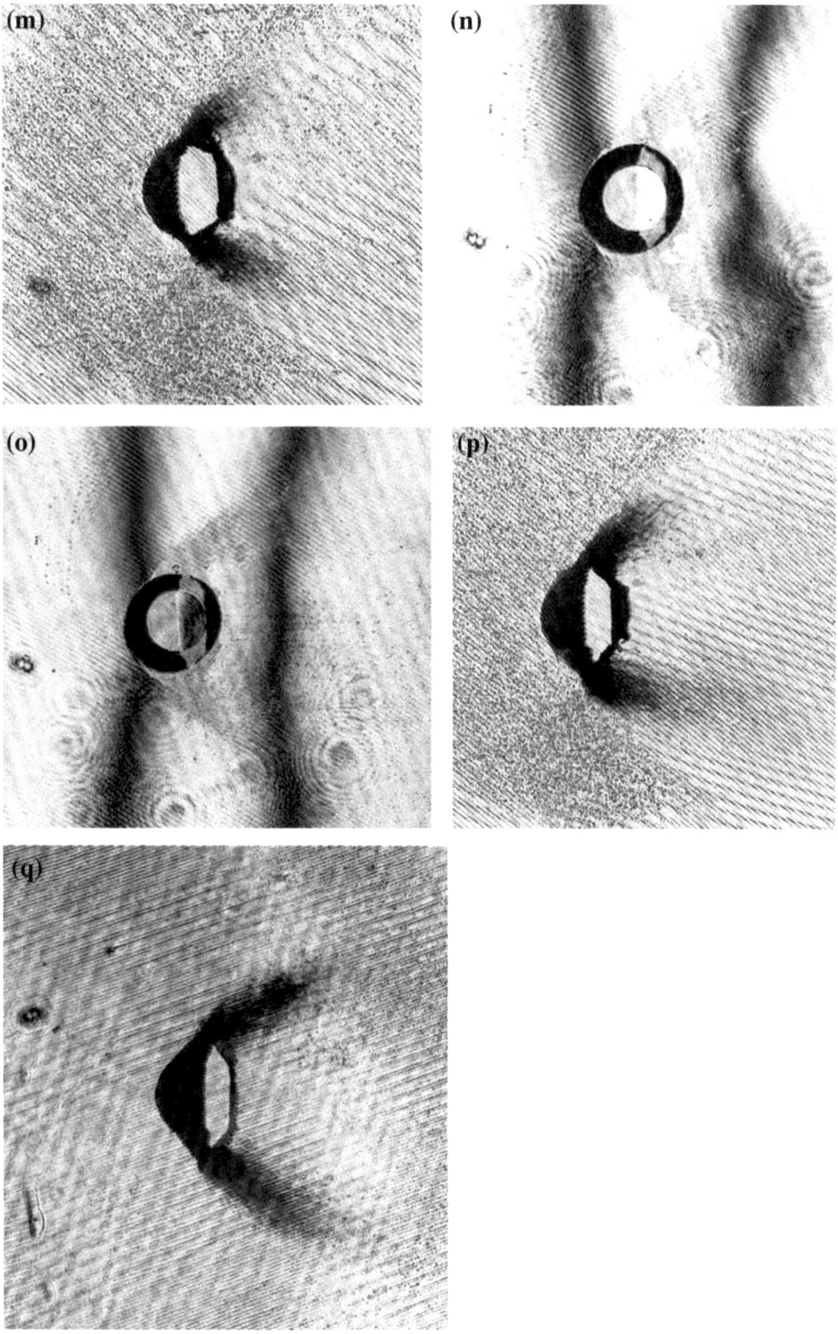

Fig. 7.31 (continued)

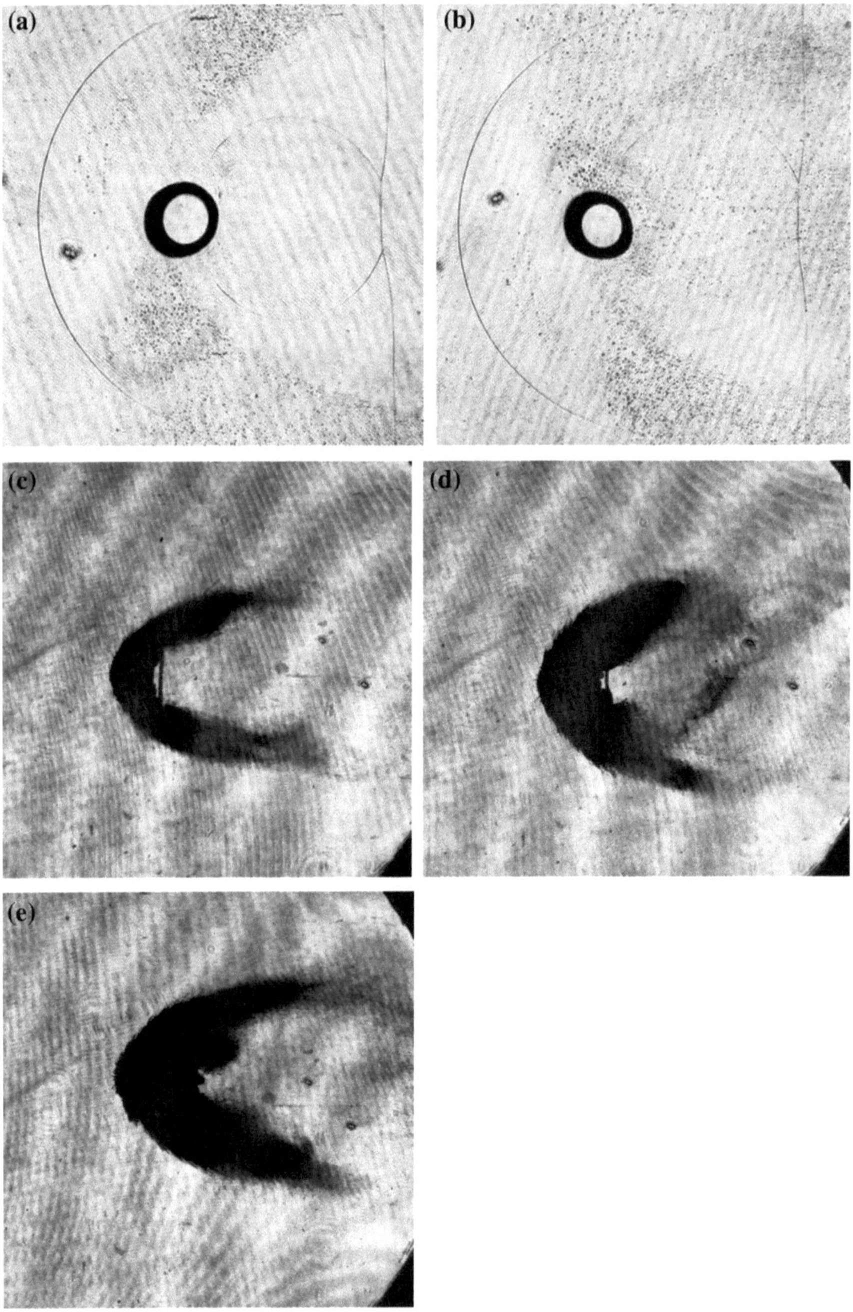

Fig. 7.32 Single exposure interferograms of shock wave interaction with a 6 mm diameter water column for Ms = 1.4 in atmospheric air: **a** #89062814, 170 μs from trigger point, Ms = 1.457; **b** #89062812, 190 μs, Ms = 1.428; **c** #90071210, 780 μs, Ms = 1.440; **d** #90071206, 860 μs, Ms = 1.446; **e** #90071202, 920 μs, Ms = 1.420

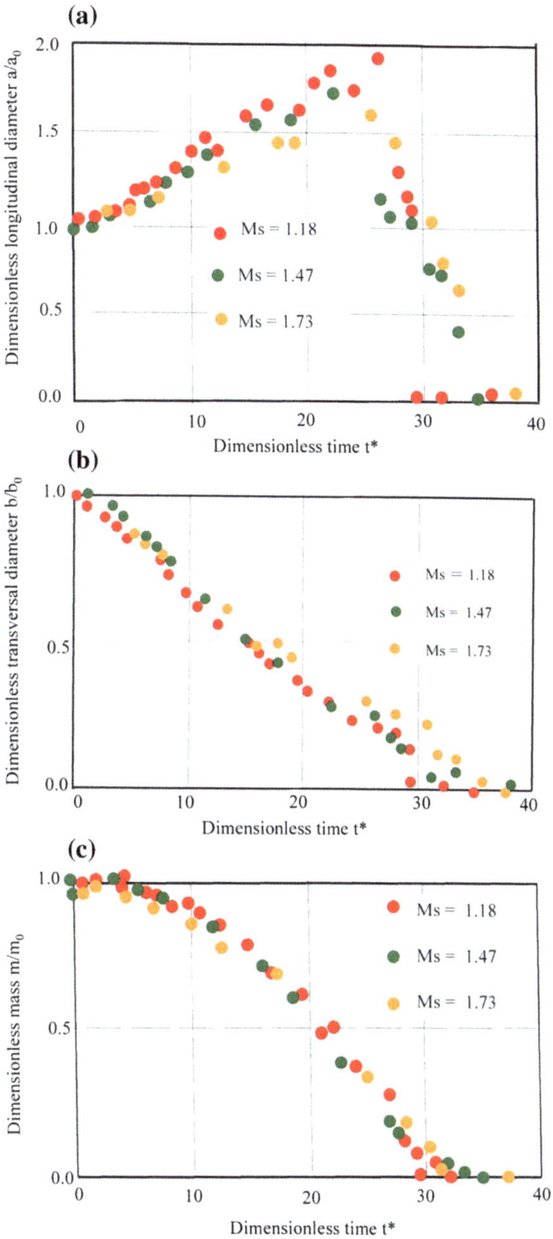

Fig. 7.33 Time variation of water column deformation: **a** longitudinal residual diameter a/a_0; **b** transversal residual diameter b/b_0; **c** residual mass m/m_0 (Hamamura 1995)

7.6.1 Shock Wave Interaction with Tandem Water Columns

In order to simulate 1.0 mm diameter tandem water droplets separated by 10 mm as seen in Fig. 7.28, an analogue experiment was conducted by replacing the droplet with two water column. Figure 7.34a shows the arrangement and a so-called no flow picture. Two 6 mm diameter water columns were placed at 20 mm separation distance and shock wave of Ms = 1.45 will be loaded in atmospheric air. Figure 7.34b shows the earlier interaction. The first water column was deformed and shattered at first. Then the perturbed shock wave interacted with the second water so that the shock wave transferred the memory of the interaction with the first water column to the second water column and eventually the second water started to shatter. It was noticed that the fringes were visible inside the column. Probably the wave motion generated inside water columns might contribute to the boundary layer separation and hence to the early stage of their shattering. The air flew from left to right, at later time the drag force working on the first water column was larger than that on the second water column. Then in Fig. 7.28, with the elapsing time, two water droplets in tandem merged. However, the positions of the two water columns were fixed and could not move. The degree of shattering was distinctly different, which readily demonstrated the difference in the drag force as clearly see in Fig. 7.34i–l.

Figure 7.35 summarizes results of visualization presented in Fig. 7.34 for Ms = 1.45. The ordinate designates dimension-less residual mass and the abscissa designates dimension-less time t*. Red filled circles denote the second column, downstream column. Green filled circles denote the first column, upstream column. Orange color filled circles denote a single column exposed to shock wave of Ms = 1.45. The first column shatters very similarly to a single column. However, the second column collapses differently from the first column. This experiment was performed at the stand-off distance L = 20 mm. The stand-off distance would be a parameter which may control the shattering of the second water column. For very large L, the two water column would shatter independently. If the stand-off distance approaches the two water column may merge and collapse simultaneously.

Fig. 7.34 Shattering of 6 mm diameter tandem water columns, L = 20 mm, Ms = 1.45 in atmospheric air at 300 K, single exposure: **a** #90071907, 150 μs from trigger point, Ms = 1.450, single exposure; **b** #90071907, 210 μs, Ms = 1.417; **c** #90071914, 240 μs, Ms = 1.450, single exposure; **d** #90072603, 270 μs, Ms = 1.443; **e** #90072604, 310 μs, Ms = 1.437; **f** #90072605, 380 μs, Ms = 1.447; **g** #90072310, 450 μs, Ms = 1.445, single exposure; **h** #90072314, 590 μs, Ms = 1.445, single exposure; **i** #90072316, 660 μs, Ms = 1.445, single exposure; **j** #90072402, 870 μs, Ms = 1.445, single exposure; **k** #90072410, 1120 μs, Ms = 1.442, single exposure; **l** #90072412, 1150 μs, Ms = 1.442, single exposure ►

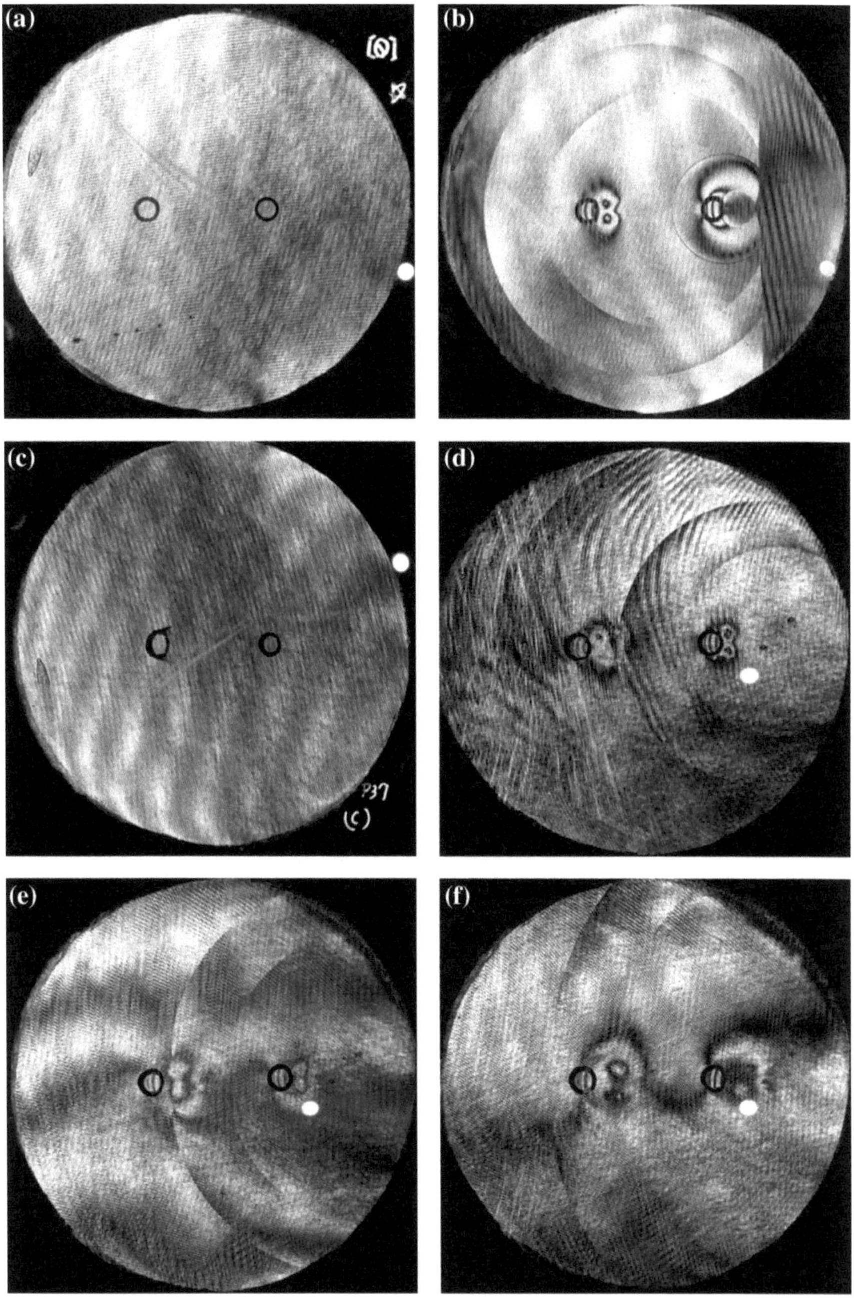
(a)
(b)
(c)
(d)
(e)
(f)

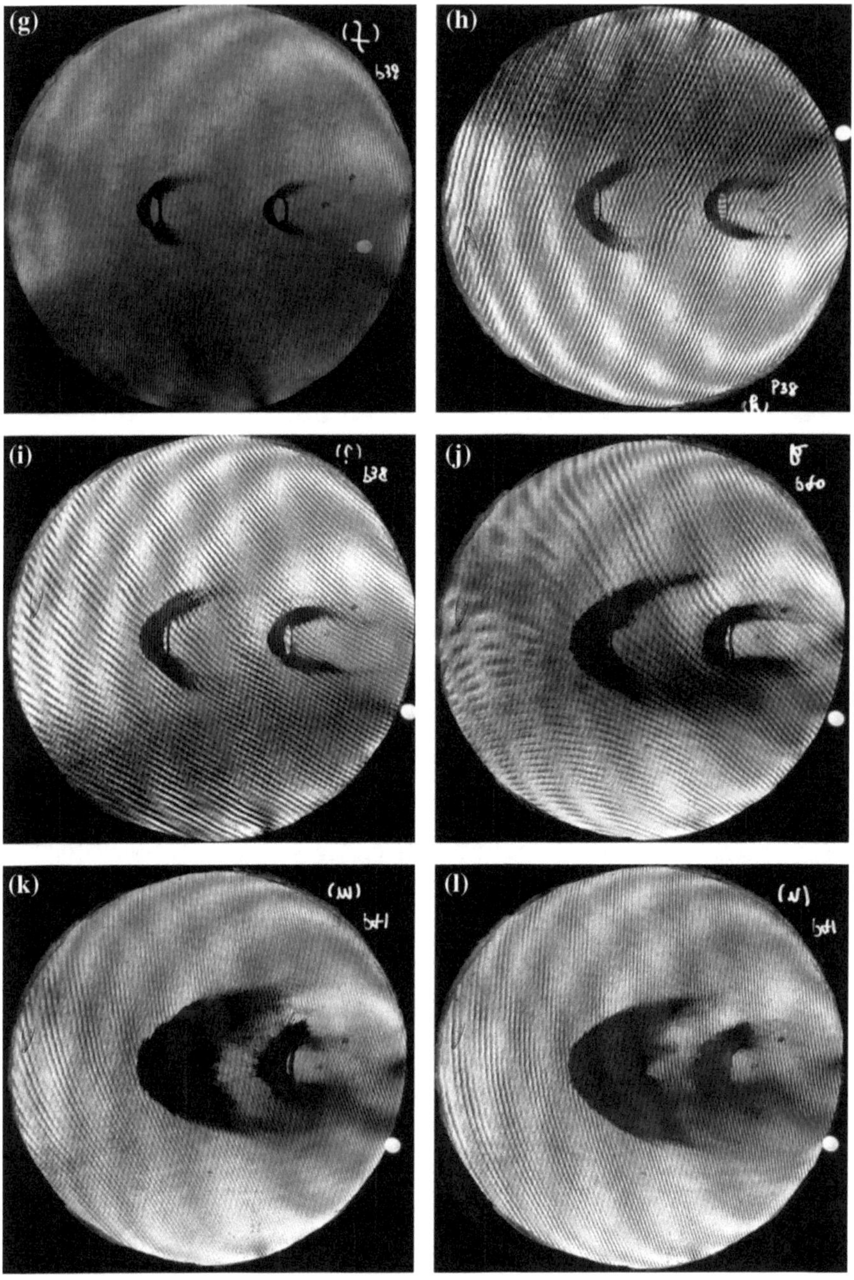

Fig. 7.34 (continued)

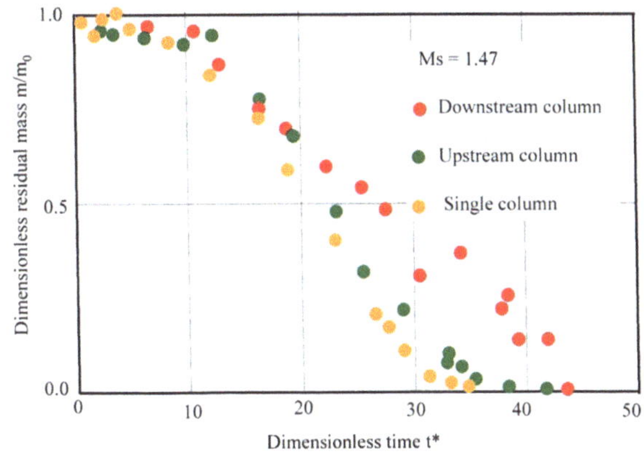

Fig. 7.35 Time variation of residual mass m/m_0 of water columns in tandem (Hamamura 1995)

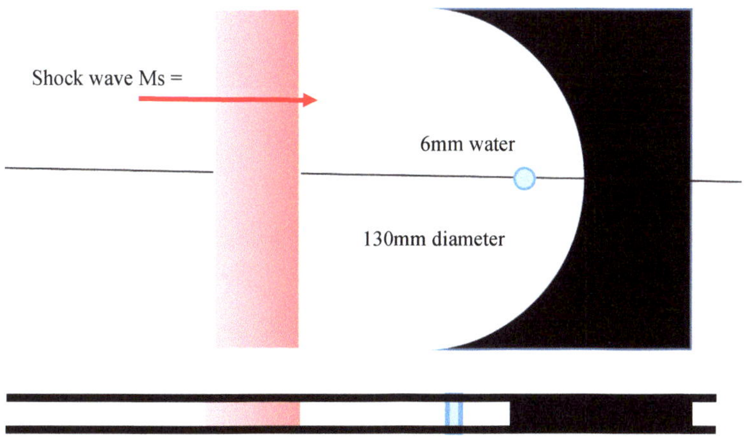

Fig. 7.36 Shock wave focusing on a 6 mm diameter and 4 mm high water column (Hamamura 1995)

7.6.2 *Interaction of Reflected Shock Wave with a Water Column Placed at Focal Point*

Figure 7.36 shows a 130 mm diameter and 4 mm wide circular reflector installed in a 60 mm × 150 mm conventional shock tube. This experiment is aimed at observation of the shattering of water column on which the shock wave of Ms = 1.45 is focused. The experimental condition is similar to the condition presented in Fig. 7.34.

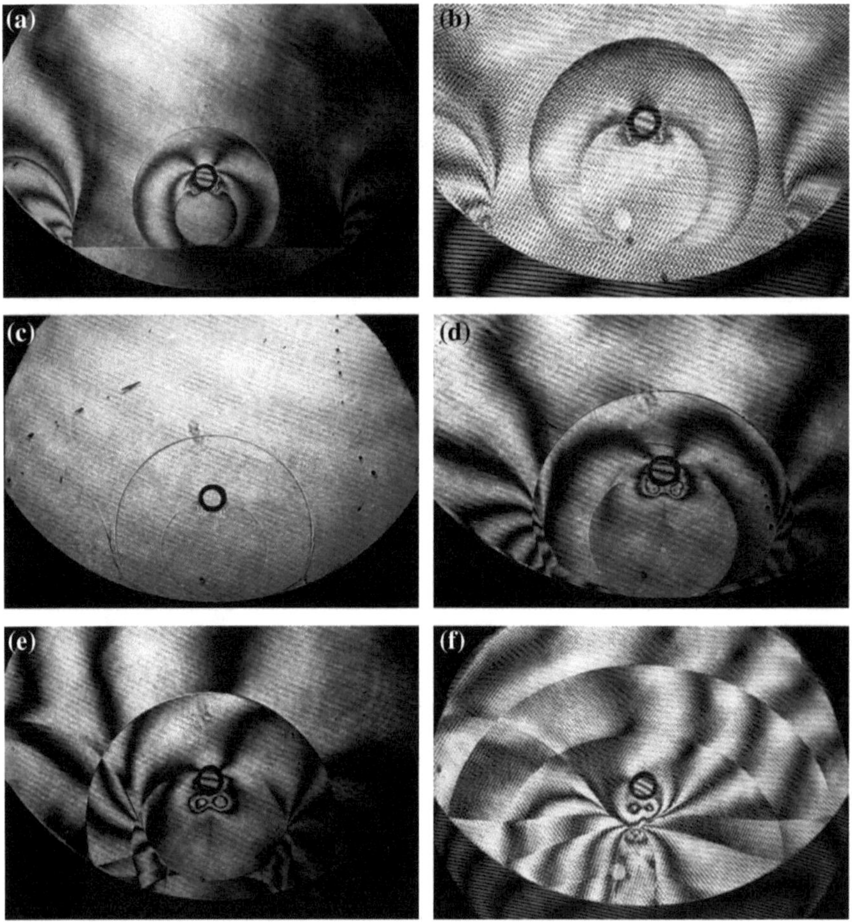

Fig. 7.37 Disintegration of 4 mm high and 6 mm diameter water column exposed to shock wave focusing from a 130 mm diameter concave reflector for MS = 1.45 in air at 1013 hPa, 300 K: **a** #89101703, 160 μs from trigger point, Ms = 1.450; **b** #90070915, 200 μs from trigger point, Ms = 1.447; **c** #89102502, 200 μs Ms = 1.458, single exposure; **d** #89101704, 180 μs, Ms = 1.450; **e** #89101705, 200 μs, Ms = 1.450; **f** #90070914, 270 υs, Ms = 1.447; **g** #89102506, 280 μs, Ms = 1.458, single exposure; **h** #89102512, 420 μs, Ms = 1.458; **i** #89102516, 620 μs, Ms = 1.458, single exposure; **j** #89102518, 720 μs, Ms = 1.458, single exposure; **k** #90070919, 920 μs, Ms = 1.447, single exposure; **l** #90071810, 1720 μs, Ms = 1.438, single exposure (Hamamura 1995)

Figure 7.37 show sequential observations of deformation of the water column positioned at a focal point of a 130 mm diameter circular. At first, the incident shock wave interacted with the water column as seen in Fig. 7.36a, b, similarly to

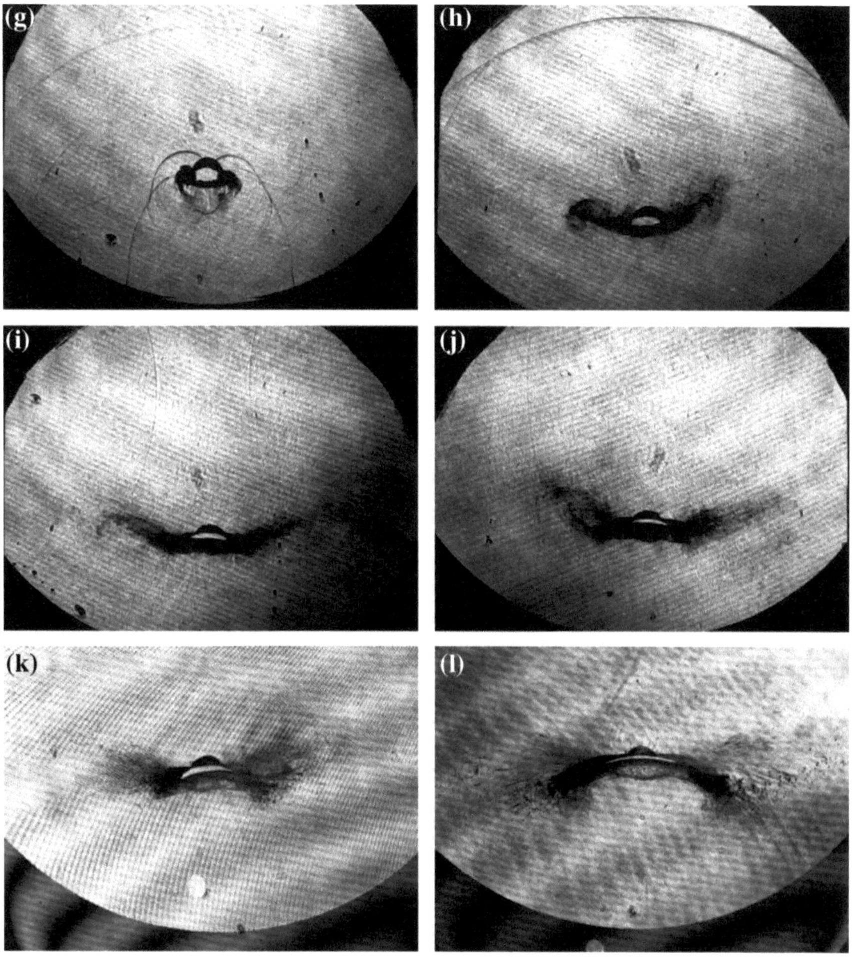

Fig. 7.37 (continued)

the shock wave/solid cylinder interaction. Soon the waves converging towards the focus area and deform the column. The water column is exposed to high pressure at the focal region and then the column is deformed without casing any flow separation. The water column is squeezed from outside and deformed.

References

Abd-el-Fattah, A. M., & Henderson, L. F. (1978). Shock wave at a fast-slow gas interface. *Journal of Fluid Mechanics, 86,* 15–32.
Gelfant, B. E., Silinikov, M. V., & Takayama, K. (2008). *Liquid droplet shattering*. Sanct Petersburg Technical University Press.
Hamamura, M. (1995). *Study of shock wave interaction with liquid column* (Master thesis). Graduate School of Engineering, Faculty of Engineering, Tohoku University.
Igra, D. (2000). *Experimental and numerical studies of shock wave interaction with gas-liquid interfaces* (Ph.D. thesis). Graduate School of Engineering, Faculty of Engineering, Tohoku University.
Matthujak, A. (2007). *Experimental Study of impact-generated high-speed liquid jets* (Ph.D. thesis). Graduate School of Engineering, Faculty of Engineering Tohoku University.
Nagoya, H. (1995). *Experimental study of Richtmyer-Meshkov instability* (Master thesis). Graduate School of Engineering, Faculty of Engineering, Tohoku University.
Pianthong, K. (2002). *Supersonic liquid diesel fuel jets generation, shock wave characteristics, auto-ignition feasibility* (Ph.D. thesis). School of Mechanical and Manufacturing Engineering, The University of New South Wales.
Reinecke, W. G., & Waldmann, G. D. (1975). *Shock layer shattering of cloud drops in reentry flight* (pp. 75–152). AIAA Paper.
Sachs, R. G. (1944). *The dependence of blast on ambient pressure and temperature*. BRL Report, No. 466.
Shi, H. H. (1995). *Study of hypersonic liquid jets* (Ph.D. thesis). Graduate School of Engineering, Faculty of Engineering, Tohoku University.
Shitamori, K. (1990). *Study of propagation and focusing of underwater shock focusing* (Master thesis). Graduate School of Tohoku University Faculty of Engineering, Tohoku University.
Tait, P. G. (1888). Report on physical properties of flesh and of sea water. Phys Chem Challenger Expedition, IV, 1–78.
Takayama, K., Onodera, O., & Esashi, H. (1982). Behavior of shock waves propagating along liquid surface. *Memoirs Institute High Speed Mechanics, Tohoku University, 419*, 61–78.
Yamada, K. (1992). *Study of shock wave interaction with gas bubbles in various liquids* (Doctoral Thesis). Graduate School of Tohoku University Faculty of Engineering, Tohoku University.
Yoshida, T., & Takayama, K. (1985). Interaction of liquid droplet and liquid bubbles with planar shock waves. In *International Symposium on Physical and Numerical Flow Visualization*, ASME.
Wierzba, A., & Takayama, K. (1987). Experimental investigation on liquid droplet breakup in a gas stream. *Report Institute of High Speed Mechanics, Tohoku University, 53*, 1–99.

Chapter 8
Explosion in Gases

In 1980, the shock wave laboratory received a license that permitted to used a small amount of explosives in scientific experiments. The National Chemical Laboratory at Tsukuba, Japan trained staffs of the laboratory to safely use lead azide PbN_6 pellets weighing from 4 to 10 mg. The density of the pellets was 2 mg/mm^3 and hence the size of the used pellets varied from about 1.3 to 1.7 mm^3. Various methods of igniting the pellets were examined. Eventually, the pellets were safely and reliably ignited in air and in water by a direct irradiation on them with a Q-switched laser beam. This method is much more reliable than the ignition by a spark discharge. Soon for conducting shock wave experiments in air and in water, the micro explosives are routinely ignited by directly irradiating a Q-switched laser beam on it or by transmitting a Q-switched Nd:YAG laser beam on it attached on the edge of a 0.6 mm diameter optical fiber (Esashi 1983). In the mid-1980, the Chugoku Kayaku Co. Ltd kindly provided us with silver azide pellets AgN_3 weighing 10 mg (Nagayasu 2002). At this time, shock waves created by micro-explosions in gases and liquids were visualized by double exposure holographic interferometry.

8.1 Micro-explosion in Air

The energy, density, and detonation speed of PbN_6 are 1.5 J/mg, about 2 g/mm^3 and 2.98 km/s, respectively. Figure 8.1 shows a 10 mg silver azide, AgN_3 pellet offered by the Chugoku Kayaku Co. Ltd. Its energy, density, and detonation speed are 1.8 J/mg, 3.8 g/mm^3, and 5.05 km/s, respectively. A silver azide pellet was also glued on the tip of a 0.6 mm diameter optical fiber and ignited by transmitting a Q-switch Nd:YAG laser beam of wave length 1064 nm, pulse width of 9.1 ns, and 14 mJ/pulse through it. This method was applied to ignite AgN_3 pellets of any size. For example, a size of pellet of 0.1 mm^3 weighing about 4 μg was ignited.

K. Takayama, *Visualization of Shock Wave Phenomena*,
https://doi.org/10.1007/978-3-030-19451-2_8

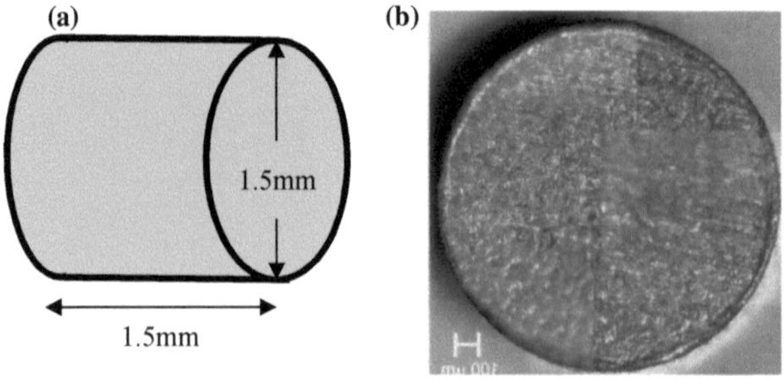

Fig. 8.1 AgN_3 pellet: **a** dimension of a 10 mg in weight; **b** frontal surface

Figure 8.2 shows a sequential observation of explosion of a 10 mg AgN_3 pellet glued on the edge of a 1.5 mm diameter optical fiber including its plastic covering. The laser induced detonation was recorded by a high-speed camera ImaCon D-200 at frame interval of 50 ns and exposure time of 10 ns. A luminous front propagates to the right. The bright spots indicated high temperature zones that sustain the detonation. Dark irregular shapes are the cloud of detonation product gases. It took 300 ns the detonation front to propagate a distance of about 1.5 mm. Then the detonation speed was about 5 km/s. Shock waves were not observed during this short period of time, as the shock waves was coupled with the cloud of detonation product gases.

Mizukaki (2001) irradiated the surface of a AgN_3 pellet with a diverging Q-switched Nd:YAG laser beam and detected the distance at which the explosion no longer occurred. Eventually the minimal energy was estimated from the longest distance at which the detonation prevails. This value was 3.6 mJ/cm^2. To ignite consistently 1.5 mm diameter AgN_3 pellet, the minimum energy to be deposited was 63 µJ. AgN_3 and PbN_6 pellets are detonated in a precisely controlled fashion in gases and liquids by directly or indirectly depositing such a minute energy on these pellets. Over-pressures behind the shock waves so far generated range those of the detonation pressures of these explosives to a few atmospheres depending on the

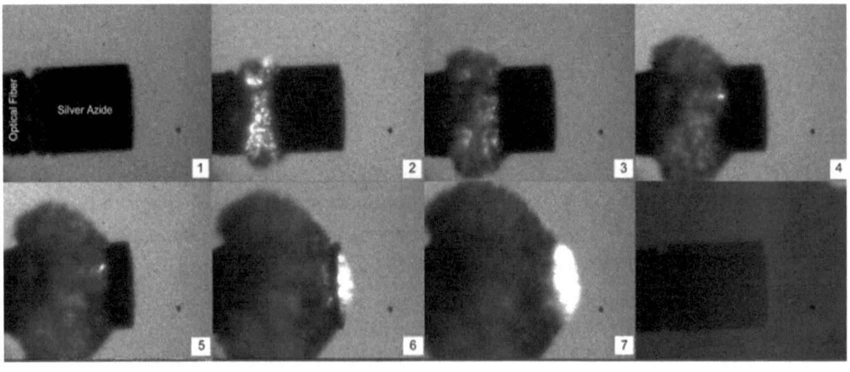

Fig. 8.2 The initiation of detonation of AgN_3 pellet. A shock wave is coupled with the cloud of detonation product gas cloud (Courtesy of Dr. Hamate)

amount of explosives. It is the most unique feature of the laser ignition system that it hardly generates any electric noises while creating shock waves. Figure 8.3 shows a correlation between the dimension-less over pressures in AgN_3 explosion and the scaled distance in m/kg$^{1/3}$ (Sachs 1944) where m is the distance from the explosive in m and kg is the weight of AgN_3 in kg. The measured over-pressures created by the detonation of 5 and 10 mg of AgN_3 are compared with appropriate TNT equivalence ratio. Empirically this is a ratio was estimated by comparison of the scaled explosion of 10 mg AgN_3 with the scaled TNT explosion. The empirical TNT equivalent ratio was in the range of 0.4–0.5.

Lead azide pellts were, at first, ignited by spark discharge of a commercial condenser bank of 3 kV and 0.3 μF. Electrical noises disturbed pressure measurements. Figure 8.4 show direct shadowgraphs of exploding 4 mg PbN_6 pellet using spark discharge. In Fig. 8.4a, the explosion produced a cloud of detonation product gas expanding at 2.9 km/s. A jaggedly shaped detonation product gas cloud accelerated the neighboring air and drove an irregularly shaped shock wave. However, at this stage, the detonation product gas cloud and the shock wave were closely couple. The same trend was already observable in Fig. 8.2.

The irregularly shaped cloud promoted the interfacial instability, which is so-called the Rayleigh-Taylor instability occurring when a high density medium accelerates a low density medium across their interface. However, the expanding detonation product gas cloud quickly attenuates and hence the diverging shock wave becomes spherical after running a few mm from the center of the explosion. The stand-off distance between the shock wave and the contact surface increased. The temperature and the pressure of the detonation product gas cloud were initially

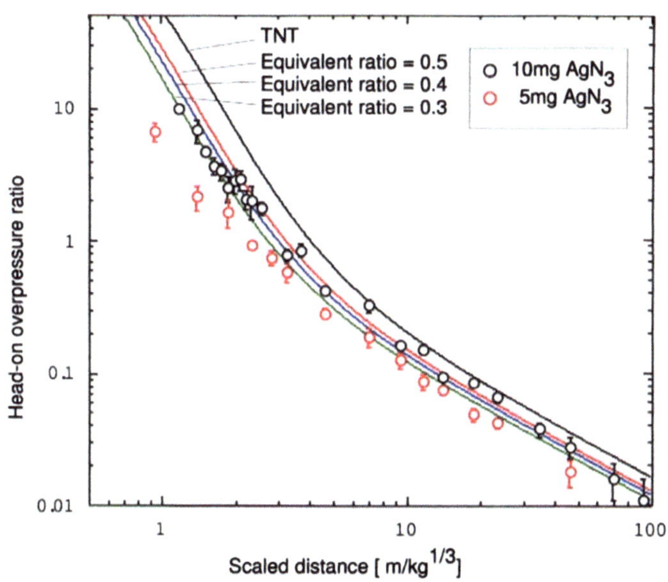

Fig. 8.3 Scaling effect of micro-explosions (Mizukaki 2001)

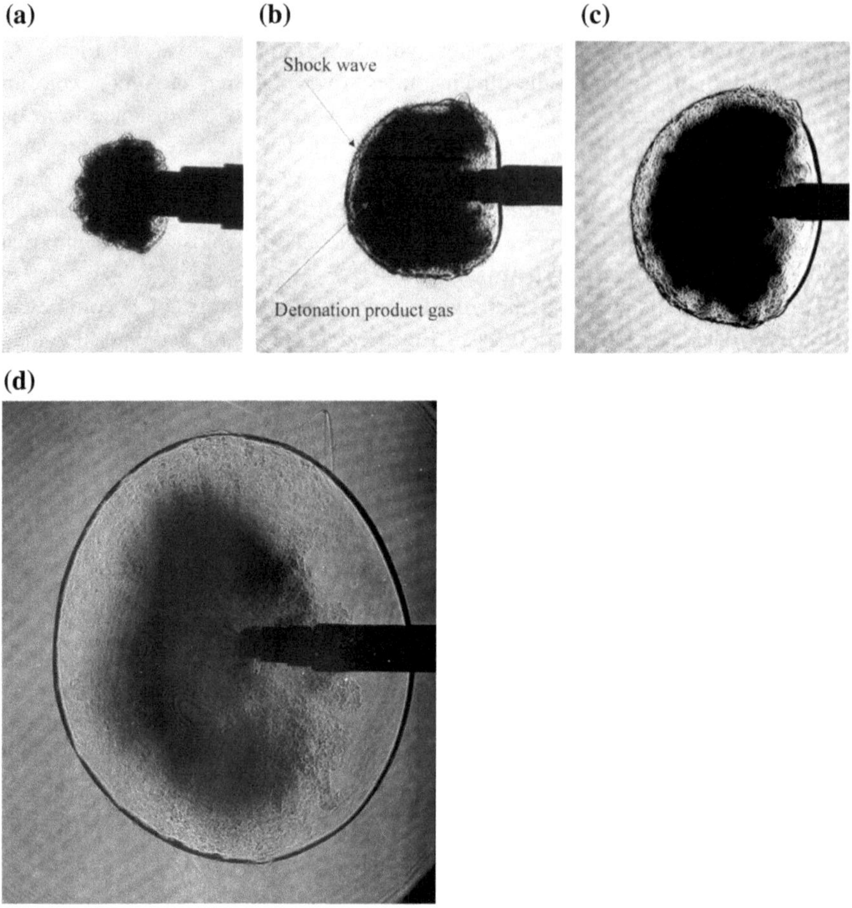

Fig. 8.4 Micro-explosion in air by detonating 4 mg PbN_6 pellet. Ignition is conducted by a spark discharge of a condenser bank of 3 kV and 0.3 μF, at 290.0 K: **a** #82101906; **b** #82101905, 34 μs from spark discharge; **c** #82101904, 44 μs; **d** #82101903, 64 μs

high but quickly decreased. Then a jaggedly shaped expansion wave propagated toward the center of explosion and then reflected from the center of the explosion forming eventually a secondary shock wave. Figure 8.4d shows the appearance of the secondary shock wave.

8.2 Reflection of Two Spherical Shock Waves

When irradiating the surface of a PbN_6 or AgN_3 pellet with a Q-switch laser beams, the micro-explosives were spontaneously ignited. Mizukaki (2001) formulized the trajectories of 10 mg AgN_3 pellets.

Then the trajectory of the resulting shock wave created by the explosion of 10 mg AgN_3 pellets in air is given by

$$T = aX^b, \tag{8.1}$$

where $T = t/t_0$, $X = x/x_0$, $a = 0.173$, $b = 1.51$.

$$\text{At } x = x_0, dX/dT = a_0 \tag{8.2}$$

where a_0 is the sound speed in air at $x_0 = 118$ mm, $t_0 = ax_0^b$.

The laser irradiation is a simple and safe method for igniting the micro-explosives. The critical transition angle θ_{crit} of reflected shock waves from wedges, cones and other two-dimensional body surfaces was investigated in detail, however, the transition of spherical shock wave reflections is one of the fundamental research topics. Figure 8.5 show the reflection of two identical spherical shock waves generated by 4 mg PbN_6 pellets at various stand-off distances L.

Two 4 mg PbN_6 pellets were simultaneously ignited by laser irradiation at a stand-off distance of L = 50 mm and the resulting reflection of the spherical shock waves was visualized. However, this arrangement differs from a reflection from a rigid wall at a stand-off distance of L = 25 mm.

In Fig. 8.5a, the intersection and collision of two spherical shock waves are shown. This arrangement is equivalent to the shock intersection with a wall set at a 45° angle and the reflection pattern is a RR. However, with decrease in the intersection angle, in other words, increasing the stand-off distance L, the reflection pattern will become a SMR and then the distinct triple points would appear. The reflection of two spherical shock waves of nearly identical strength forms a SMR but its earlier stage looks vNMR. However at later time as seen in Fig. 8.5e a trace of SL is observed. It should be noticed that the time attached to individual interferograms indicated the time instant of the second exposure which was synchronized with the arrival of shock wave at the targeting location.

8.3 Spherical Shock Wave Reflection from a Sphere

The reflection of spherical shock waves from a solid sphere is visualized. A 4 mg PbN_6 pellet was glued on a thin cotton thread using a cellulose-acetone solution and was suspended at a distance 50 mm above a 100 mm diameter brass sphere. Figure 8.6 show the evolution of a spherical shock wave reflected from the brass sphere. The resulting reflection pattern was initially a RR as seen in Fig. 8.6a and transited to a SMR as seen in Fig. 8.6b. At a later time as seen in Fig. 8.6c, d, conical shock waves penetrated the spherical shock wave. The conical shock waves were driven by small debris particles attached on the surface of the PbN_6 pellet and shattered at the explosion. The detonation product gas was the so-called a fire ball in a large scale explosion. At the early stage of the explosion, it had an irregular shape so that the resulting shock wave was slightly deviated from the spherical shape.

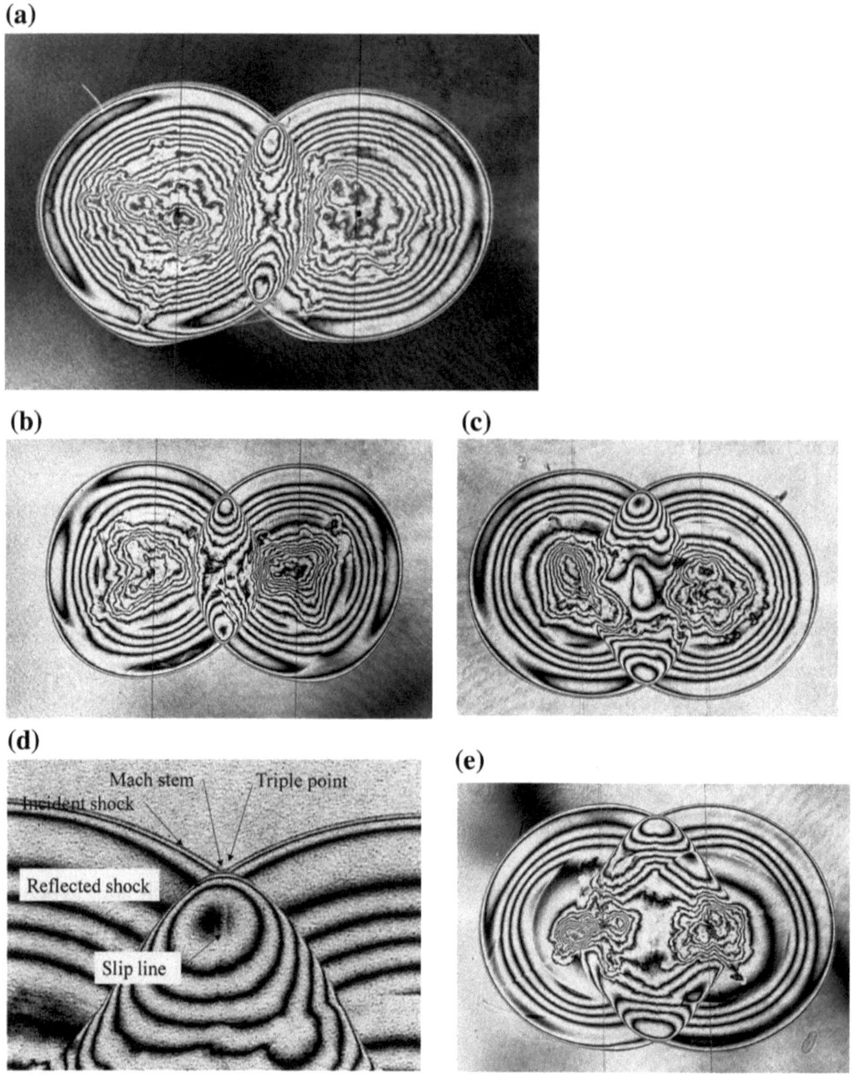

Fig. 8.5 Reflection of two spherical shock waves generated by simultaneous explosion of two 4 mg PbN_6 pellets placed at interval L = 50 mm: **a** #83032411, 50 μs from trigger point; **b** #83032410, 55 μs; **c** #83032407, 70 μs; **d** enlargement of (**e**); **e** #83032403, 100 μs

8.4 Spherical Shock Wave Interaction with a Soap Bubble

Figure 8.7 show the experimental arrangement used for visualizing the interaction of spherical shock waves with a soap bubble filled with slightly pressurized helium or SF_6. The air bubbles were blown by the mixture of water vapor and air. The

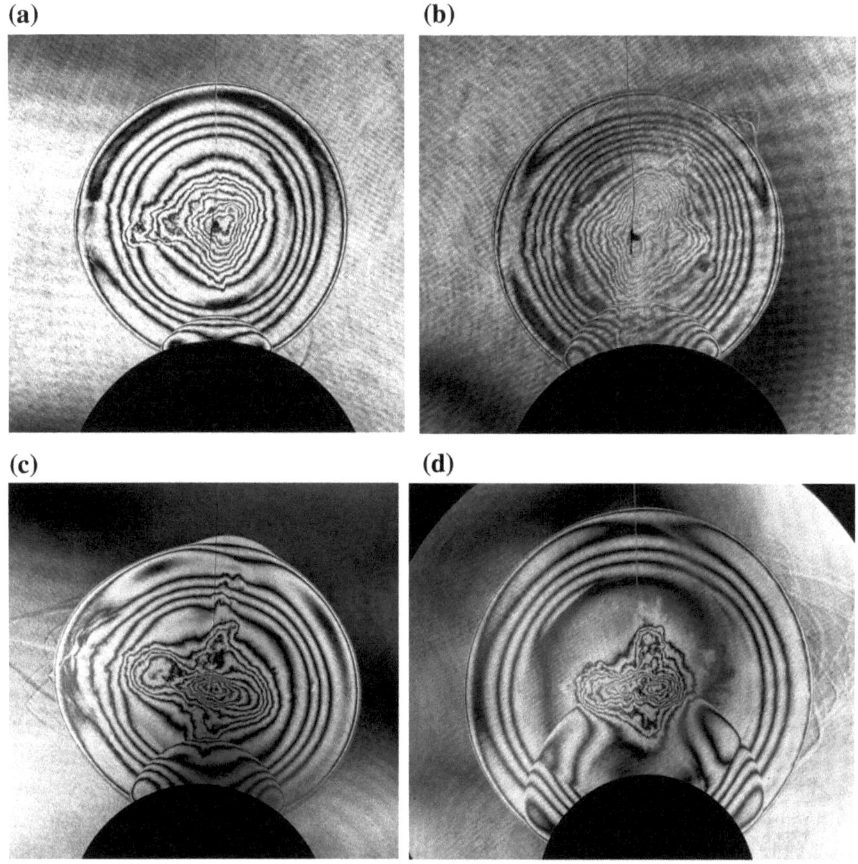

Fig. 8.6 The evolution of the spherical shock wave reflection from a 100 mm diameter brass sphere. The shock waves were generated by exploding 4 mg PbN_6 pellet placed 50 mm above the brass sphere: **a** #83032317, 70 μs from trigger point; **b** #83032315, 80 μs; **c** #83032312, 90 s; **d** #83032302, 160 μs

helium and SF_6 bubbles were blown by pressurized helium and SF_6, respectively. These gases were contaminated with air and hence the degree of the impurity of these gas bubbles was at most 5%.

Spherical shock waves were generated by exploding 10 mg AgN_3 pellets attached at the tip of a 0.6 mm core diameter optical fiber. The explosives were exploded by the transmission of a Q-switched Nd:YAG laser beam having 20 mJ pulse energy and 7 ns pulse width through the optical fiber. The distance between the center of the soap bubbles and the center of the explosive ranged from 85 to 90 mm. Helium bubbles tend to lift up slightly upward and hence the optimal diameter was 60 mm. SF_6 soap bubbles tend to sink slightly downward. As SF_6 was slowly absorbed by the soap film solution and supplied at slightly higher pressure at

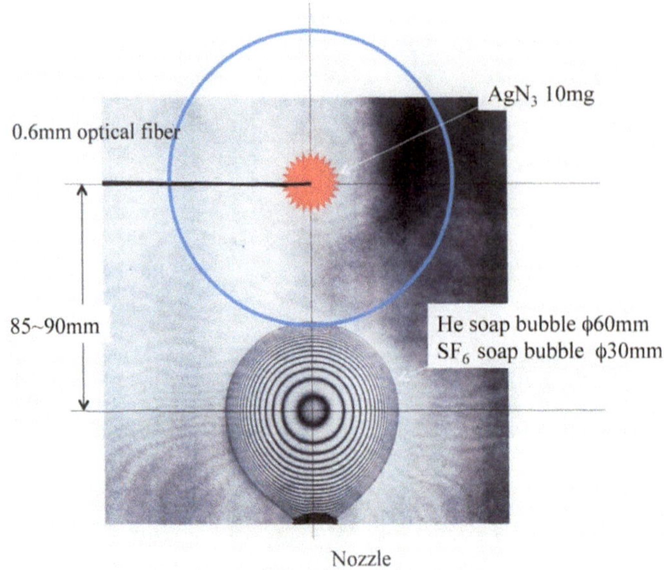

Fig. 8.7 Spherical shock wave interaction with a soap bubble

1015.2 hPa. Hence the maximal diameter was only 30 mm so that the visualizations were conducted as soon as the bubbles reached the desired shape and size.

At first, air bubbles in ambient air were tested. Figure 8.8 show sequential interaction of air soap bubbles with spherical shock waves. Spherical shock wave propagate along the air bubble and the transmitting shock waves propagate inside the air bubble. This was an easy experiment as any extra disturbances were created inside the bubbles.

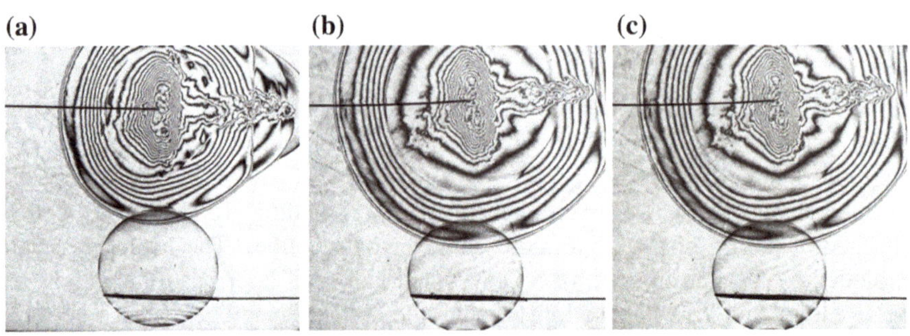

Fig. 8.8 Spherical shock wave interaction with air soap bubbles in air. The shock wave was created by the explosion of a 10 mg AgN_3 pellet: **a** #93022305; **b** #93022303; **c** #93022301

8.4.1 Helium Soap Bubble

Figure 8.9 show the sequential observation of spherical shock waves impinging horizontally on 60 mm diameter helium soap bubble. The visualization was conducted at every 100 μs interval. In Fig. 8.9a, talen at the early stage, an expansion wave was reflected from the bubble surface. The frontal side of the bubble surface started to contract. Small circular protrusions distributed over the bubble surface and grew with elapsed time were vortices appeared as a result of the Richtmyer-Meshkov Instability. In Fig. 8.9d, the first exposure was conducted when the spherical shock wave just reached the bubble surface and the second exposure was conducted after 500 μs from the first exposure.

Figure 8.10 show sequential deformations of the soap bubble. The visualization was carried out at relatively short time interval. Figure 8.10a shows the soap bubble before the shock wave exposure. The co-axial fringe distributions on the soap bubble surface shows a perfectly spherical shape. Figure 8.10b shows the generation of the

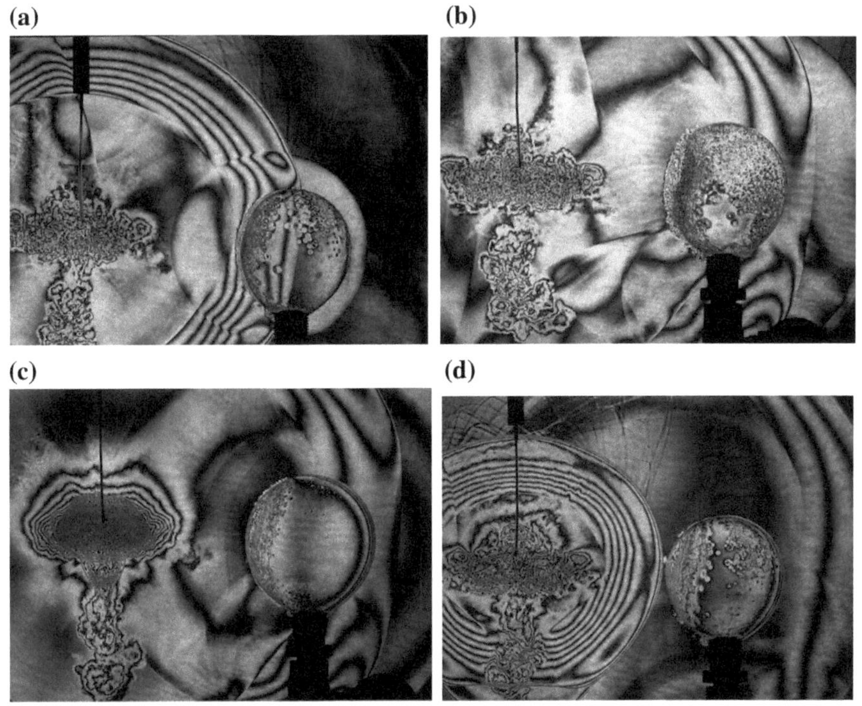

Fig. 8.9 Interaction of a helium bubble with a spherical shock wave generated horizontally by explosion of 10 mg AgN_3 pellet: **a** #93020920, 200 μs from trigger point; **b** #93020921, 300 μs; **c** #93020916, 500 μs; **d** #93020917, 500 μs

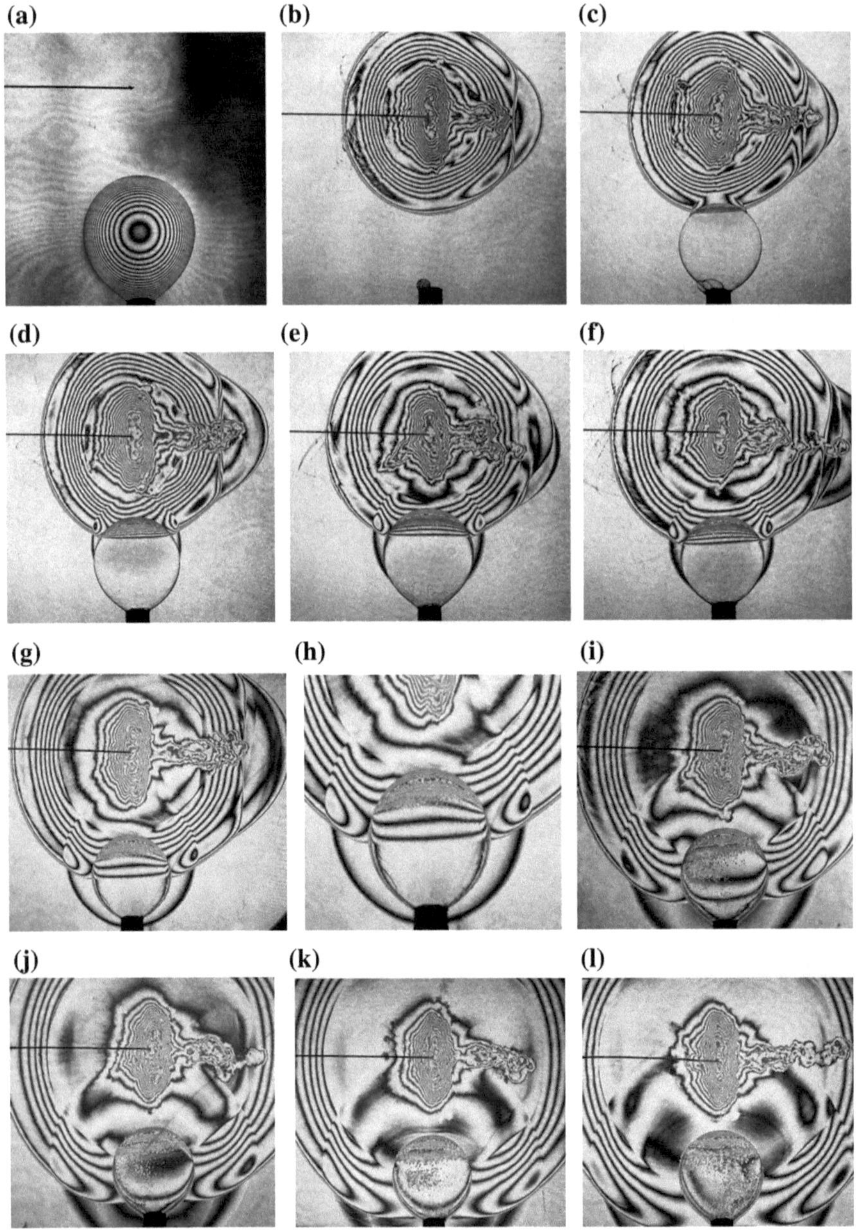

Fig. 8.10 Interaction of a helium bubble impinged by a spherical shock wave generated by the explosion of a 10 mg AgN_3 pellet: **a** #93020910; **b** #93020506; **c** #93020508, 0 μs; **d** #93020503, 10 μs after the shock wave impinement; **e** #93020504, 20 μs; **f** #93020505, 30 μs; **g** #93020804 60 μs; **h** enlargement of (**g**); **i** #93020806, 100 μs; **j** #93020807, 120 μs; **k** #93020808, 140 μs; **l** #93020809, 160 μs; **m** #93020810, 190 μs; **n** #93020811, 220 μs; **o** #93020814, 290 μs; **p** Enlargement of (**l**); **q** enlargement of (**o**)

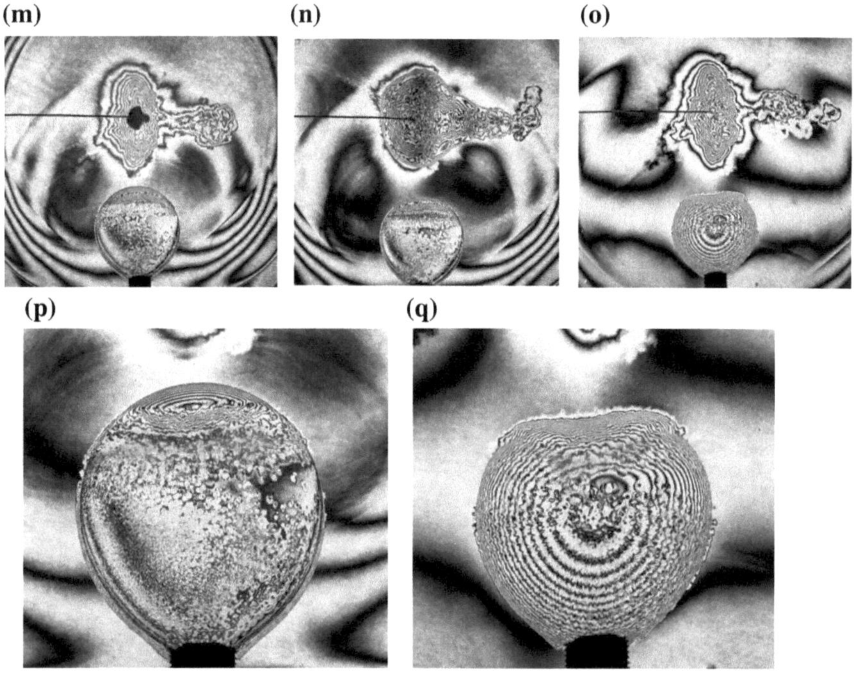

Fig. 8.10 (continued)

spherical shock wave. Figure 8.10c shows the shock wave impingement on the bubble surface. A convexly shaped precursory shock wave was observed inside the helium bubble. Along the bubble surface on which the shock wave impinged toward the helium therefore the reflected wave from the deformed bubble surface is a curved expansion wave. In Fig. 8.10e, f, the curved and broadened waves that are visible outside the bubble ahead of the intersection point of the foot of the spherical shock wave are compression waves. The presence of this precursory compression waves are already observed along the air/helium in Fig. 7.3. With elapsing time as seen in Fig. 8.10g–h, the precursory waves propagated toward the nozzle. A reflected expansion wave from the helium soap bubble interacted with the region behind the initial shock wave and even with the detonation product gas and attenuated. Figure 8.10h is a magnified image of Fig. 8.10g. Figure 8.10o, p are the enlargements of Fig. 8.10l, q, respectively and show the shapes of the soap bubble after 160 and 290 μs from the moment when the spherical shock wave impinged on the soap bubble. The circular rings that are observed on the soap bubble surface are vortices induced by the so-called Richtmyer-Meshkov Instability. Two magnified images show the deformation of the helium soap bubble and the growth of vortices during a time interval of 130 μs.

8.4.2 SF_6 Soap Bubble

Figure 8.11 show sequential observation of interactions of the soap bubble filled with SF_6 impinged with the spherical shock wave generated by the explosion of a 10 mg AgN_3 pellet This is a slow/fast interaction as discussed in Chap. 7. Shock waves are reflected and the transmitted shock wave in the SF_6 soap bubble retarded. Figure 8.11a–d show concavely shaped transmitted shock waves propagating inside the soap bubble. Figure 8.11f shows the resulting focusing of the concave transmitted shock wave.

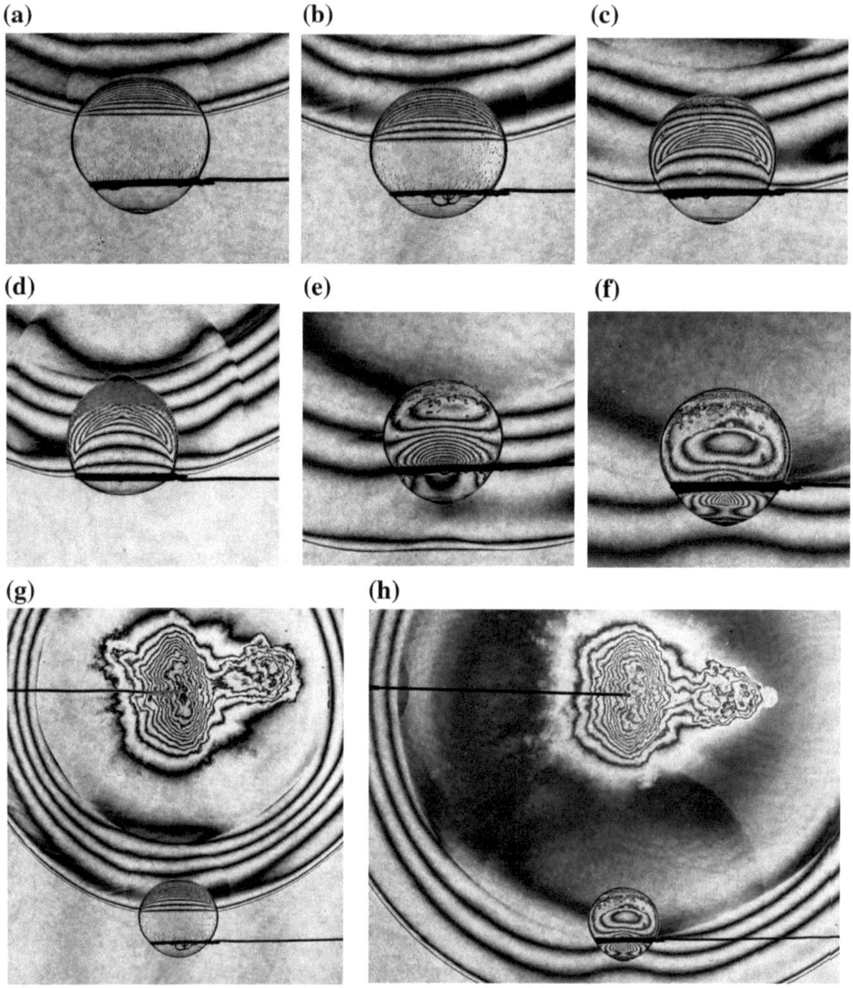

Fig. 8.11 The interaction of a SF_6 bubble impinged vertically with a spherical shock wave generated by explosion of a 10 mg AgN_3 pellet: **a** #93022409; **b** 93022407; **c** #93022404; **d** #93022402; **e** #93022406; **f** #93022405; **g** entire view of (**b**); **h** enlargement of (**f**)

8.5 Spherical Shock Wave Created by Explosion Inside an Aspheric Sphere

Shock wave propagations in a circular cross sectional duct are successfully visualized using the aspheric lens shaped test section. A 300 mm diameter aspheric test section was made of acryl and its schematic diagram is shown in Fig. 8.12a: a collimated *OB* passes the spherical test section parallel and came out parallel. Figure 8.12b shows its photo, Hosseini and Takayama (2005).

8.5.1 Explosion in an Aspheric Chamber

A spherical shock wave is generated by exploding a small AgN_3 pellet placed at the center of the aspheric chamber. If a perfectly spherical shock wave is reflected from the spherical test section as shown in Fig 8.12, the imploding shock wave would be experimentally visualized. Before manufacturing the aspheric chamber shown in Fig. 8.12b, a 150 mm diameter compact aspheric chamber was made. This was a pilot facility for training the research group. Through preparatory experiments, it was found that a small aspheric chamber had drawbacks which was so-called size effects. Then a 300 mm diameter aspheric chamber was manufactured as shown in Fig. 8.12b.

A cylindrical AgN_3 pellet was reformed to a conical shape having a 90° apex angle and a 1.5 mm bottom diameter. Then a pellet having a double conical shape was formed and was ignited form the bottom. Almost perfectly spherical shock waves were obtained in much shorter distance than the detonation of the previous cylindrically shaped AgN_3 pellets (Hosseini and Takayama 2005).

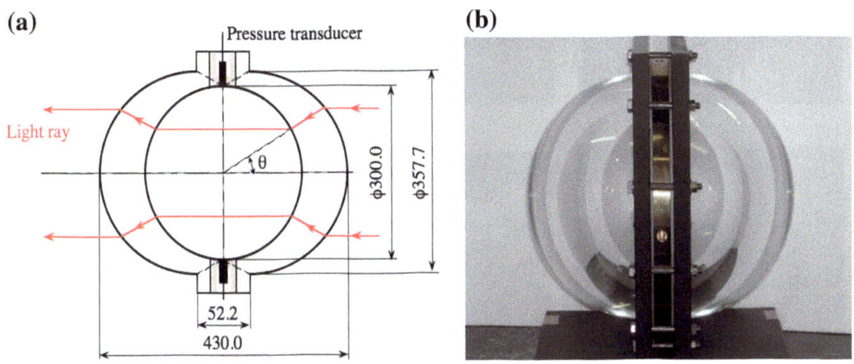

Fig. 8.12 Aspheric test chamber having a spherical test section: **a** 300 mm diameter test section; **b** photograph

The observation was conducted by direct shadowgraph and resulting images were recorded with a high-speed digital camera Shimadzu SH100 at 10^6 frames/s and exposure time of 250 ns. Figure 8.13 show the entire sequence of a spherical shock wave formation. The detonation product gas initially expanded at 5.8 km/s and emitted luminosity. At first, the fire ball and the shock wave were coupled. Then, the fireball attenuated and decoupled with the shock wave as seen in Fig. 8.13g. The shock wave became spherical and a secondary shock was seen in Fig. 8.13l. A perfectly spherical shock wave was obtained.

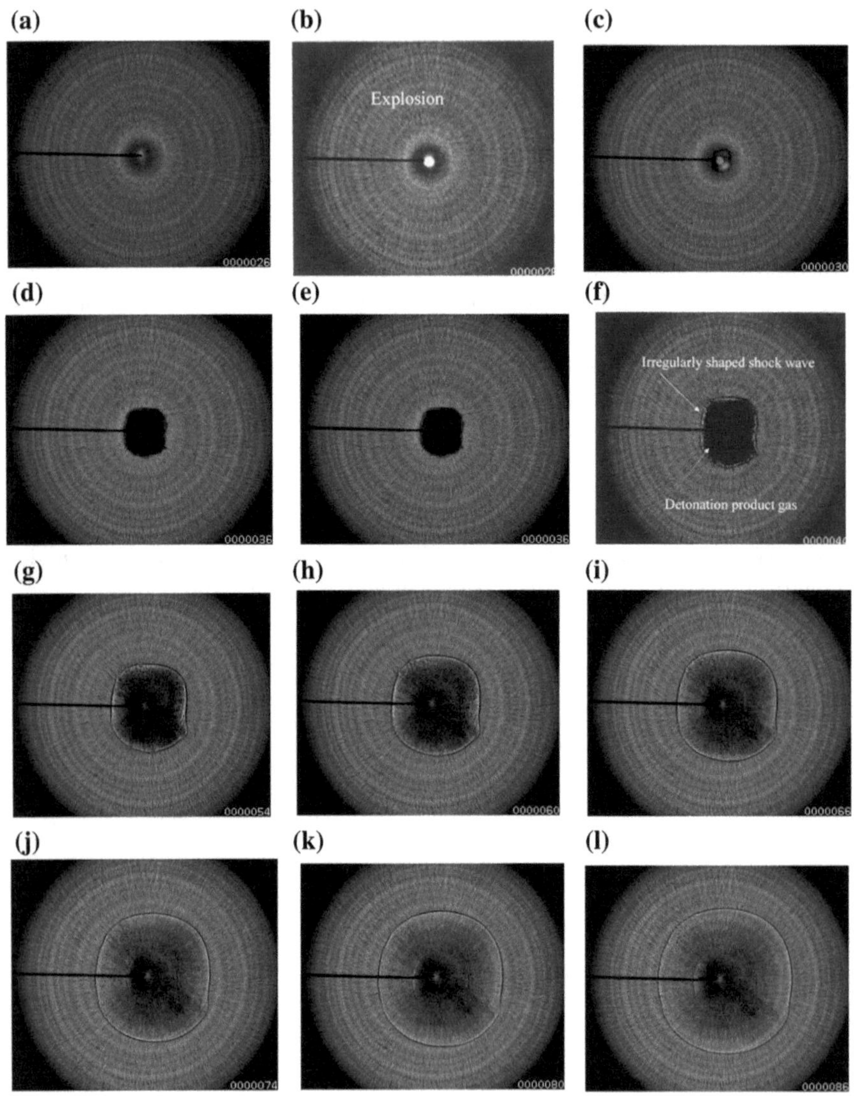

Fig. 8.13 Explosion in an aspheric chamber by detonating two conically shaped AgN_3 pellets at the center (Hosseini and Takayama 2005)

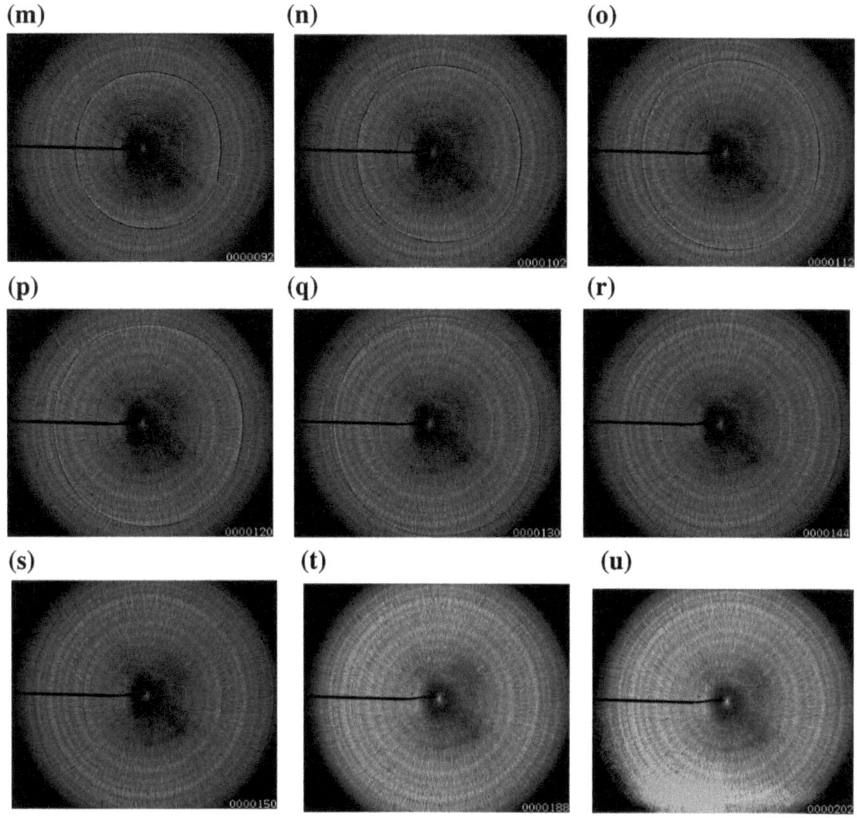

Fig. 8.13 (continued)

8.5.2 Implosion of a Reflected Spherical Shock Wave

Figure 8.13 show the formation of perfectly spherical shock wave and generation of a secondary shock wave. The spherical shock wave was well shaped and was co-axial to the spherical test section. The observation started at an appropriate time later when the diverging shock wave started converging.

Figure 8.14 show the evolution of the spherical shock wave during its imploding toward the center of the spherical chamber. The observation was conducted by direct shadowgraph and recorded with the Shimadzu SH100 with the frame rate of 10^6 frame/s. When two conically shaped AgN_3 pellets were ignited by irradiating laser beam at their base surfaces (Hosseini and Takayama 2005), an almost perfectly spherical shock wave was obtained as seen in Fig. 8.13. The spherical shock wave was reflected nearly perfectly from the spherical surface. Then a nicely shaped spherical imploding shock wave imploded toward the center of its convergence. Figure 8.14 shows its sequential observation using shadow graph. At first the imploding shock wave propagated slowly toward center. While approaching very close to the center, it

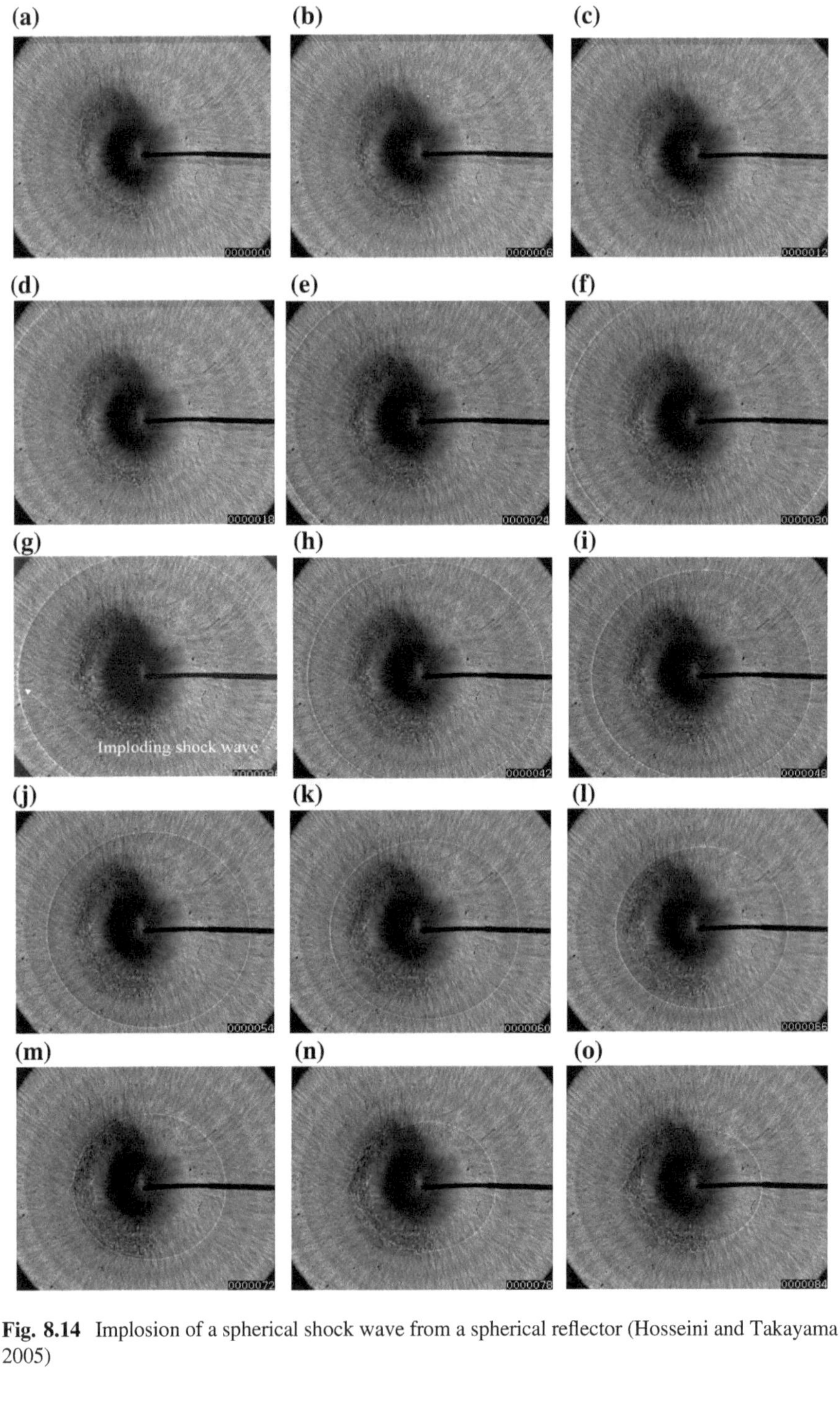

Fig. 8.14 Implosion of a spherical shock wave from a spherical reflector (Hosseini and Takayama 2005)

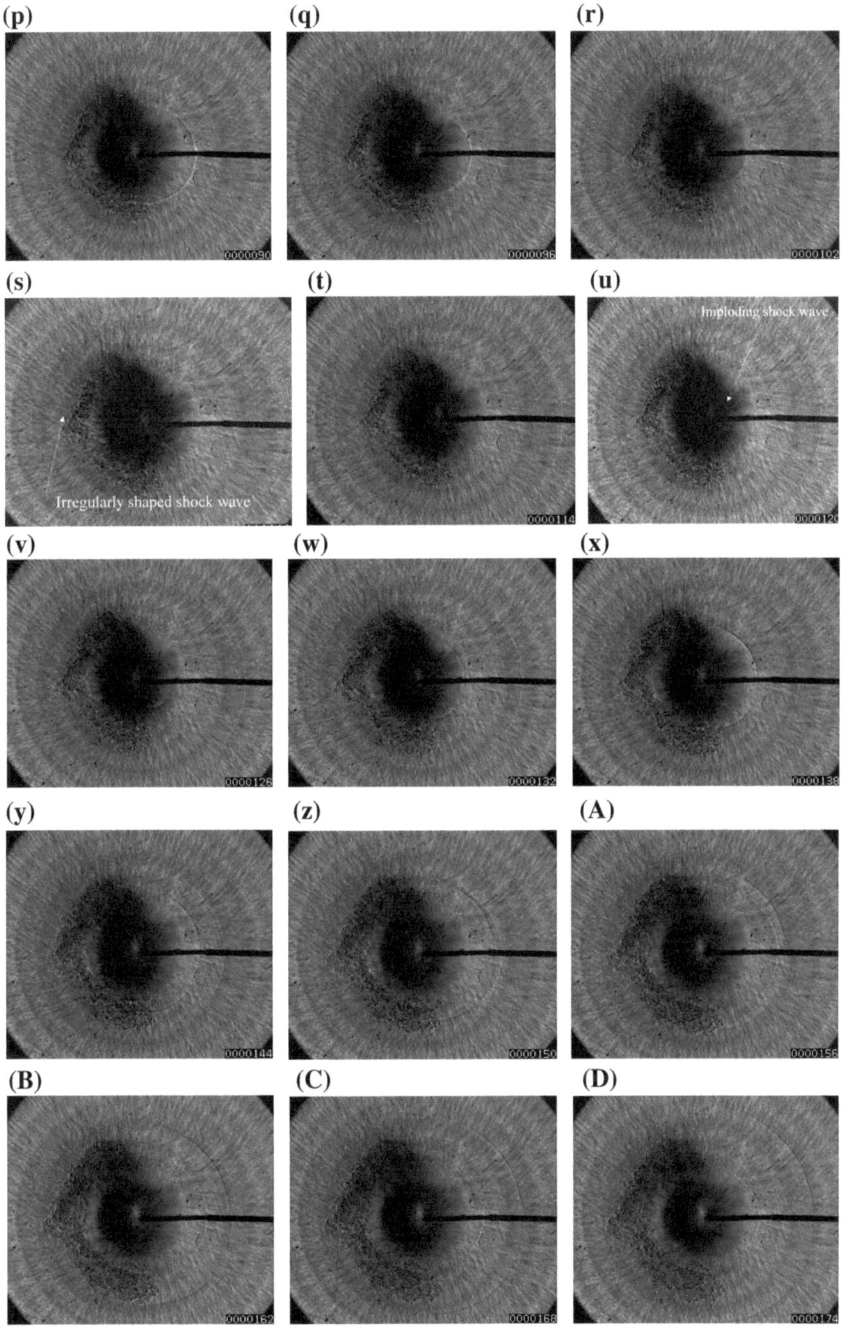

Fig. 8.14 (continued)

propagated exponentially at high speed. However, the imploding shock wave interacted with the so-called fire-ball of explosion product gases and its shape quickly became irregular. Figure 8.14 is a unique sequential observation of an imploding shock wave. Figure 8.14u shows at the moment of implosion. However, the spherical shock wave did not converge at the center of the chamber. Immediately after the convergence, it turned into a diverging shock wave.

8.6 Explosion Induced Detonation Waves

When a deflagration wave accelerating its velocity in a combustible gas mixture drives a shock wave, the shock wave transits to a detonation wave. The transition of a deflagration driven shock wave to detonation wave is called DDT. The investigation for determining the DDT criterion is one of the fundamental research topics in the detonation research. Hence, detonation waves are coupled with cellular structures. This indicates that the cell structures over 2-D detonation waves accompanied were arrayed simple Mach reflections. When cells disappeared at a condition, for instance, when a 2-D detonation wave accompanying cells passes from a two-dimensional duct to a diverging duct, the cells gradually disappear and the 2-D detonation wave is decoupled with the deflagration wave. In short, the 2-D detonation wave was quenched.

In a 2-D detonation tube, the surface area of a 2-D detonation wave is unchanged. Then the cell number is fixed and the cell size is also fixed. On the contrary, even though a 3-D detonation wave is slightly distorted from a perfectly spherical shape, it has a spherical shape and increases its averaged radius monotonously with its propagation. If cells cover entirely the 3-D detonation wave surface, with in creasing its radius and holding the constant cell size, the cell number should increase. These two options are hypothetical, spherical detonation waves neither could have cell structures of unlimited large size, nor could it have numerously large cell number. Therefore, the 3-D detonation wave will quench.

However, it is the most unlikely to imagine neither spherical detonation waves having unlimitedly large cell size nor have infinitely large cell numbers. Therefore, a three-dimensional detonation wave will quench sometime after its initiation.

In order to generate the 3-D detonation waves in a stoichiometric oxyhydrogen mixture, a 290 mm diameter and 300 mm wide cylindrical detonation chamber made of stainless steel was constructed. Figure 8.15 show a test chamber. The detonation waves were observed by double exposure holographic interferometry. Micro-explosives were ignited at the center of the test chamber. Debris particles were shattered at each explosion and impinged on 25 mm thick acrylic windows. Then, 5 mm thick acrylic plates protected the thick windows from the impact of high-speed debris particles (Komatsu 1999). AgN_3 pellets were glued on a 25 μm diameter cupper line and positioned at the center of the chamber. For conducting tests at higher initial pressure, a 100 mm diameter and 150 mm wide stainless steel chamber was placed inside the detonation chamber. The AgN_3 pellets weighing from a few μg to 20 mg were placed at the center of the chamber and were ignited

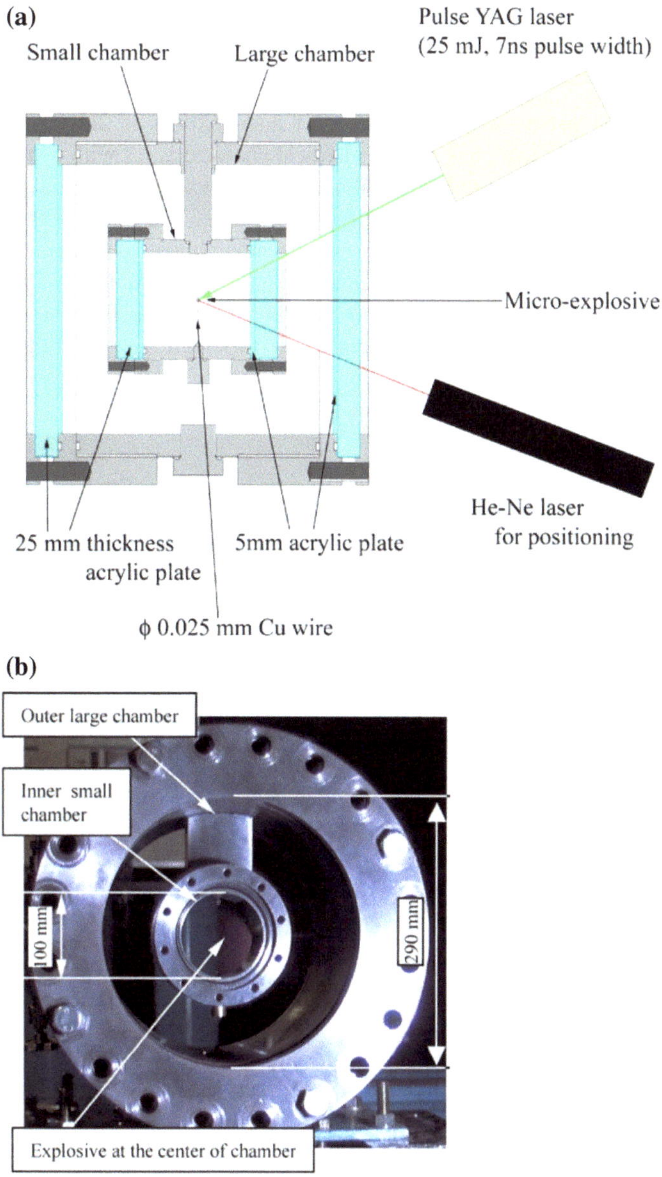

Fig. 8.15 Shock wave induced detonation wave: **a** schematic diagram of experimental setup; **b** experimental photo

by the irradiation of a Q-switched Nd:YAG laser beam of 25 mJ and 7 ns pulse width. To accurately irradiate the target explosives with the Q-switched laser beam, a thin He–Ne laser beam illuminated the target. Figure 8.15b shows a photo of a 100 mm diameter chamber inserted inside the 290 mm diameter main test chamber.

8.6.1 Explosion in Inert Gases of in $2H_2/N_2$ at 400 hP/200 hPa

A 10 mg AgN_3 pellet was ignited in inert gas mixture of $2H_2/N_2$ at 400 hPa/ 200 hPa. This gas mixture has almost identical characteristics with a stoichiometric oxyhydrogen $2H_2/O_2$ at 400 hPa/200 hPa. Figure 8.16 show sequential observations of the shock wave generation in this inert gas mixture. The shape of the detonation product gas was initially distorted but soon became almost spherical shape.

The surface of a 10 mg AgN_3 pellet was coated with minute glass particles and the explosive was ignited in the inert gas $2H_2/N_2$ at 400 hPa/200 hPa. Figure 8.17a shows a spherical shock wave generated in the inert gas mixture. Upon the

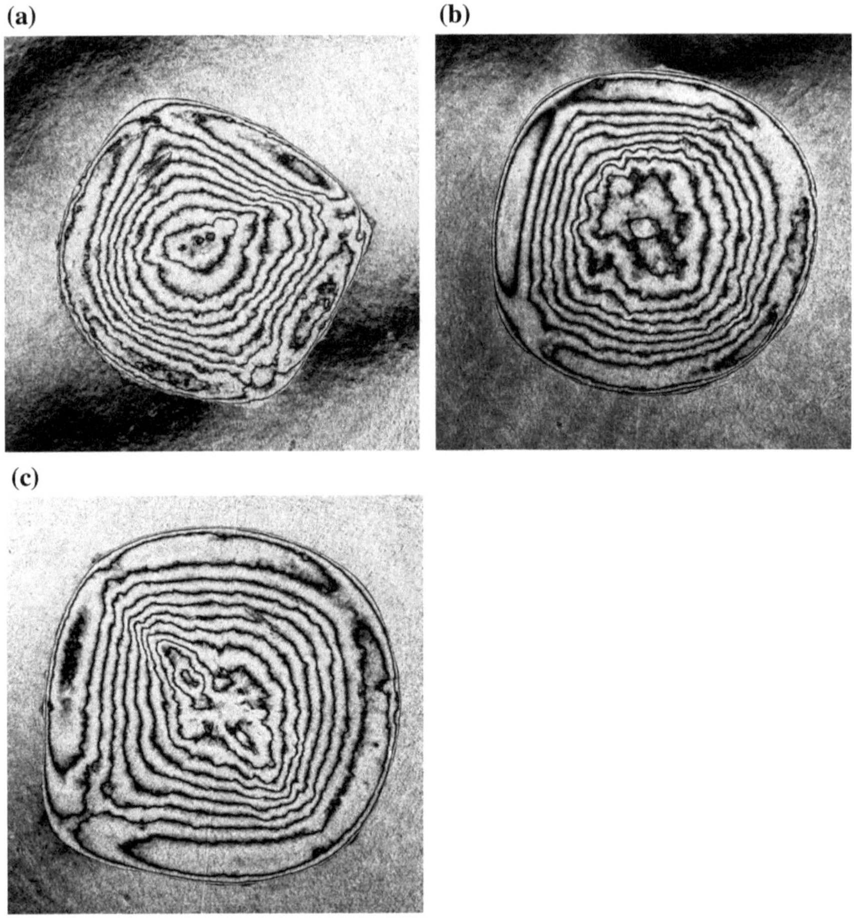

Fig. 8.16 Explosion of a 10 mg AgN_3 pellet in inert gases of $2H_2/N_2$ at 400 hPa/200 hPa. The shock wave was generated by the explosion: **a** #98020716, 15 μs from the ignition time; **b** #98020717, 20 μs; **c** #98020718, 25 μs

(a) (b)

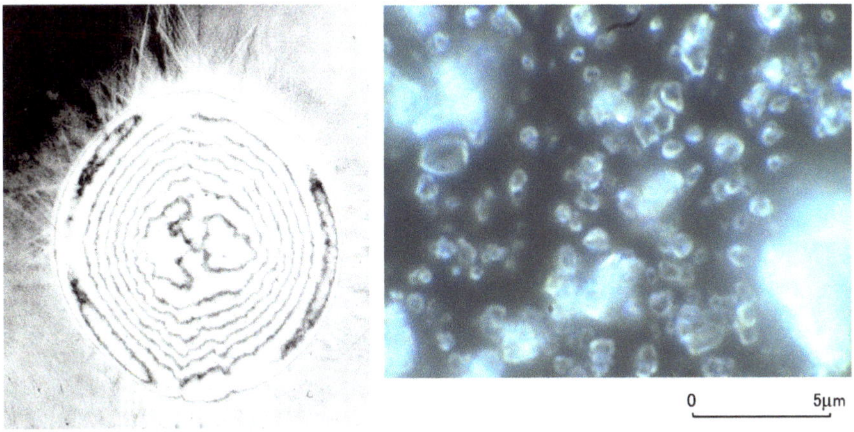

0 5μm

Fig. 8.17 Micro-explosion in inert gases. Micro glass particles were attached on a AgN_3 pellet: **a** shock wave in $2H_2/N_2$ at 400 hPa/200 hPa; **b** microscopic photo of glass particles

explosion, the glass particles shattered at supersonic speed. The high-speed shattering of the minute glass particles accompanied wakes and generated conical shock waves. Figure 8.17a shows the spherical shock wave overtaken by supersonic fragments. Figure 8.17b shows microscopic view of glass particles. The wakes induced behind the supersonic glass particles would produce a very well mixed detonable gas. It is a known fact that the detonation is promoted in well mixed detonable gases. Then the so-called DDT, the transition from the deflagration driven shock wave to detonation wave is fulfilled. Therefore, the supersonic flight of glass particles would decisively contribute to the initiation of the three-dimensional detonation waves.

8.6.2 *Stoichiometric $2H_2/O_2$ at 200 hPa/100 hPa*

Figure 8.18 show the spherical shock waves generated in $2H_2/O_2$ at 200 hPa/100 hPa by explosion of a 10 mg AgN_3. Unburnt debris particles were shattered at supersonic speed and created a conical shock waves which crossed the spherical shock wave. From the intersection of conical shock waves with the spherical shock wave, a detonation wave is initiated.

8.6.3 *Stoichiometric $2H_2/O_2$ at 400 hPa/200 hPa*

Figure 8.19 show the evolution of a spherical shock wave propagating in $2H_2/O_2$ at 400 hPa/200 hPa, respectively. The shock wave was produced by the ignition of a

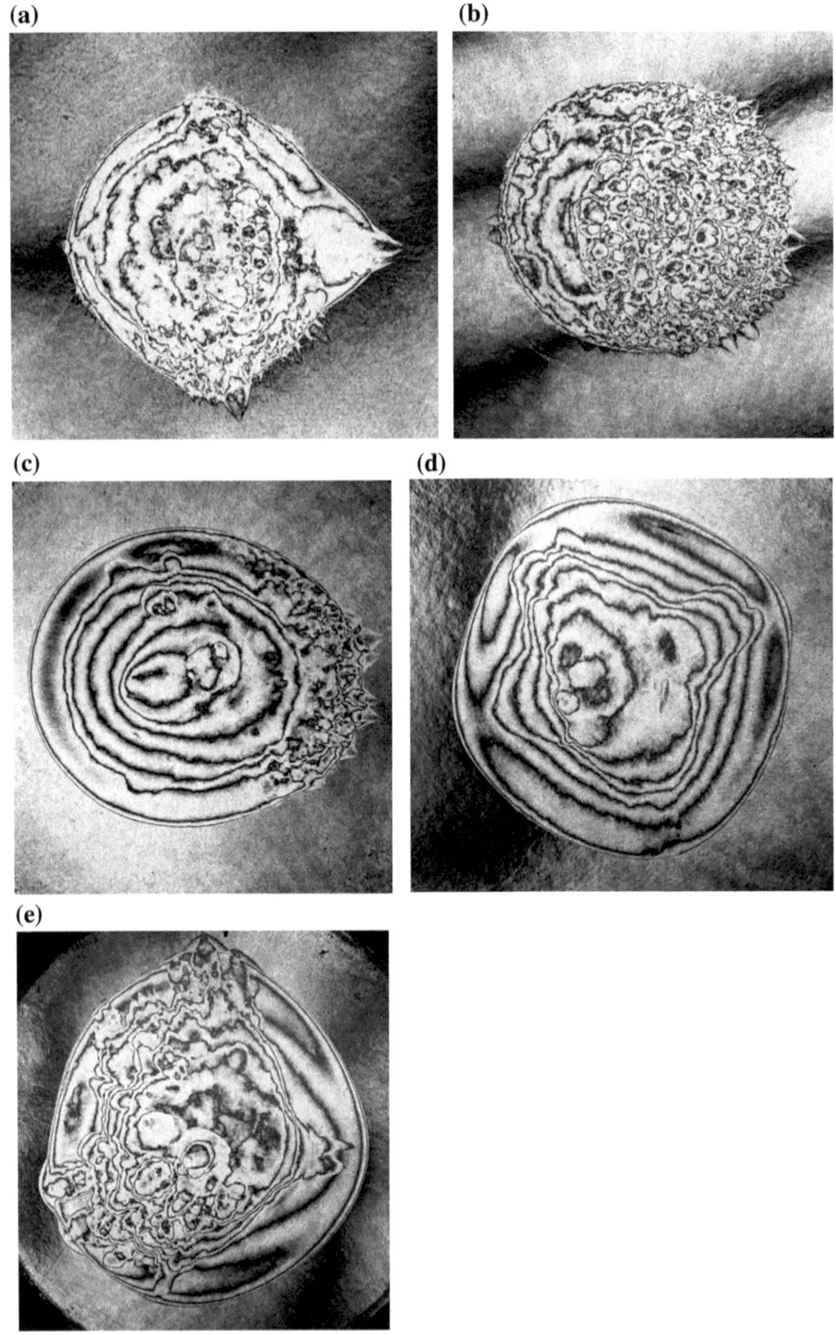

Fig. 8.18 Explosion induced detonation wave in stoichiometric $2H_2/O_2$ at 400 hPa/200 hPa. Shock waves are generated by ignition of a 10 mg AgN_3 pellet: **a** #98020505, 5 μs after ignition time; **b** #98020602, 15 μs; **c** #98020603, 25 μs; **d** #98020604, 35 μs; **e** #98020605, 45 μs

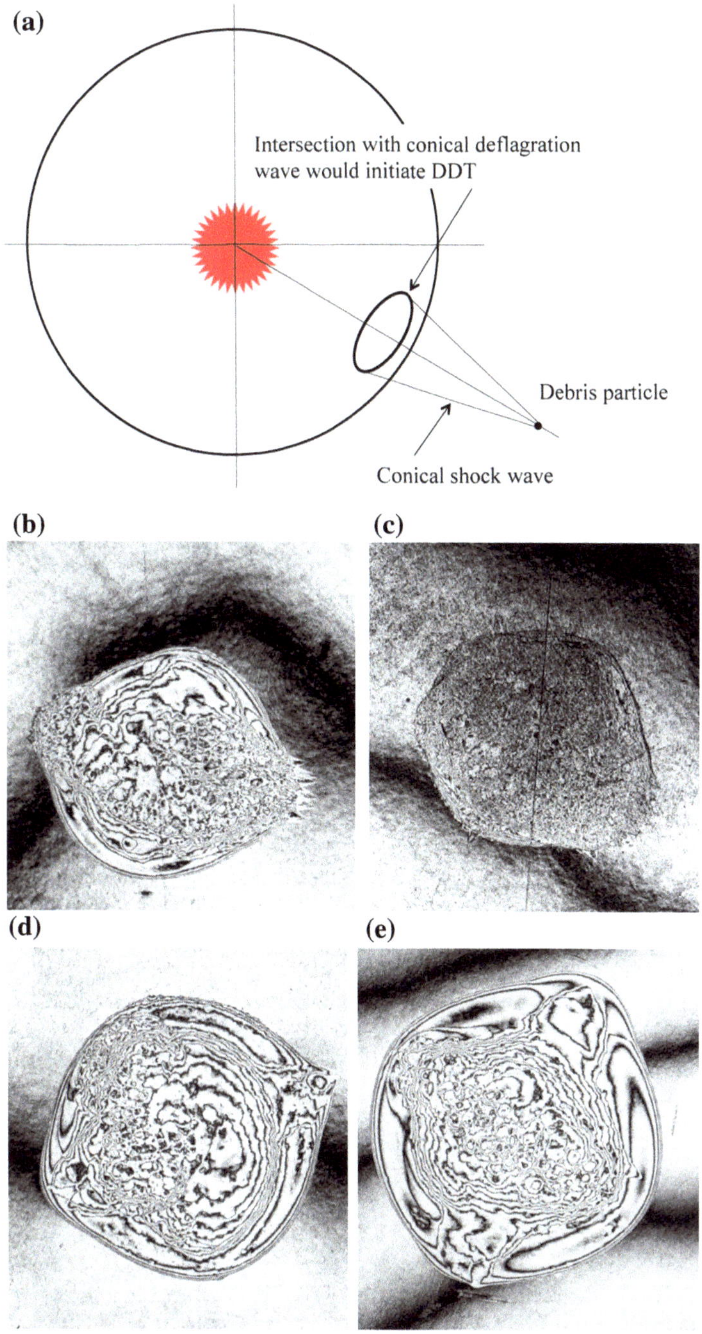

Fig. 8.19 The explosion induced detonation wave in stoichiometric oxyhydrogen $2H_2/O_2$ at 400 hPa/200 hPa, ignited by a 10 mg AgN_3 pellet: **a** an illustration of a conical shock wave penetrating the deflagration driven shock wave; **b** #98020606, 5 μs after ignition; **c** #98020607, 10 μs; **d** #98020608, 15 μs; **e** #98020609, 20 μs; **f** #98020610, 25 μs; **g** #98020611, 35 μs; **h** Enlargement of (**g**); **i** #98020612, 40 μs; **j** #98020613, 45 μs; **k** #98020701, 50 μs; **l** enlargement of (**j**); **m** #98020703, 60 μs

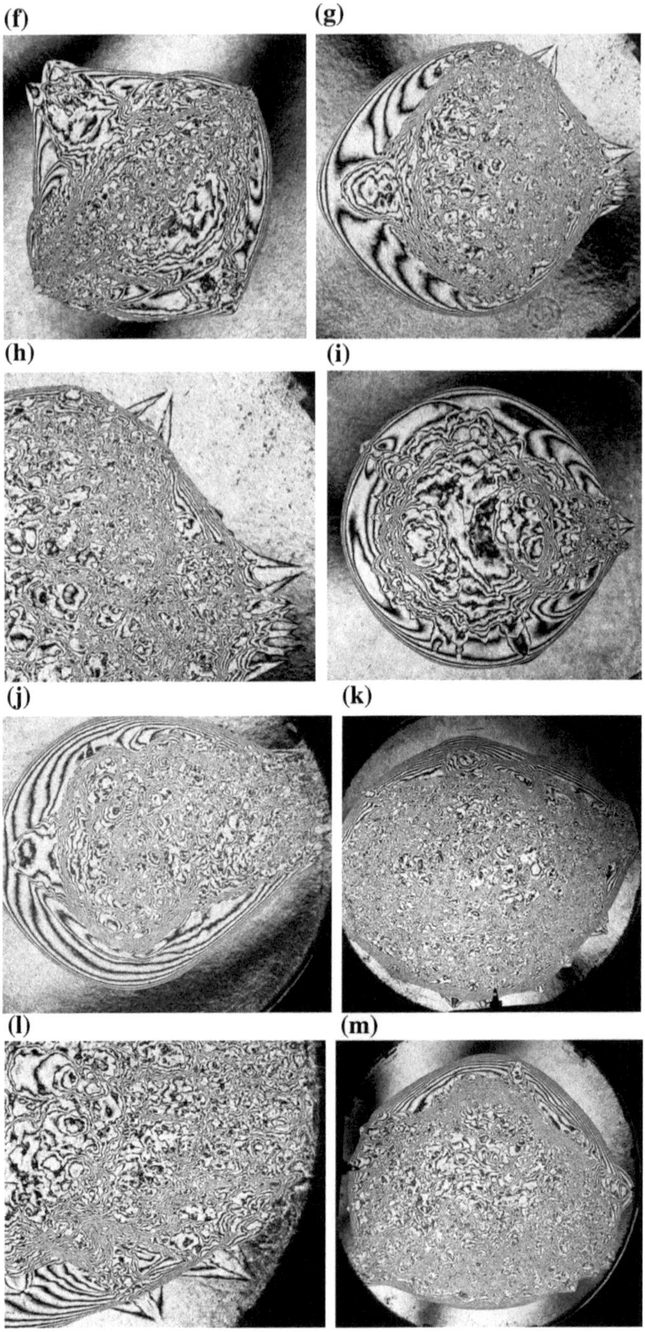

Fig. 8.19 (continued)

10 mg AgN_3 pellet. Conical shock waves overtook the precursory shock wave. Their half apex angle α are related to the speed of the supersonic debris u and the sound speed a in the oxyhydrogen mixture. The relationship between α and the speed ratio a/u is given by $\sin\alpha = a/u$. Measured values of α ranges from 10° to 20°. As the sound speed a in stoichiometric oxyhydrogen is about 850 m/s, the debris speed u ranges from 1.7 to 4.8 km/s. Figure 8.19a shows schematically a flight of a debris particle at supersonic speed. Then the resulting conical shock wave penetrated the deflagration driven shock wave. A wake produced behind the debris particle promote to effectively mix the stoichiometric oxyhydrogen and accelerate the transition of a deflagration wave driven shock wave to the detonation wave. The so-call DDT is completed. Therefore, a detonation is spontaneously initiated from the spot at which the conical shock wave penetrated the diverging shock wave.

8.6.4 Stoichiometric in $2H_2/O_2$ at 667 hPa/333 hPa

Figure 8.20 show the evolution of shock waves generated in $2H_2/O_2$ at 667 hPa/ 333 hPa ignited by the explosion of a 10 mg AgN_3. The propagation speed of the detonation wave in stoichiometric oxyhydrogen mixture is the Chapman-Jouget detonation velocity 2818 m/s. At the early stage, the shock wave propagated at about 5000 m/s but was attenuated exponentially to the C-J velocity. Therefore, the debris particles were projected at very high-speed but gradually attenuated. The half apex angle θ of a conical shock wave is defined as $\sin\theta = a/u_s$ where a, and u_s are the sound speed in $2H_2/O_2$ and the free flight speed of the debris, respectively. The value of θ varies depending on the distance from the center of explosion. Komatsu (1999) measured the values of θ and found that it varied from 10° to 22° at a later stage. This implies that the debris particles were ejected initially at 4.8 km/s and at the later stage attenuated to 2.2 km/s. Nevertheless, the wakes mixed $2H_2/O_2$ so well that the deflagration waves turned into detonation waves through the wake. The shaped of the detonation waves was not necessarily spherical shapes. Define the projected area of the detonation waves A and its periphery length L, the averaged radius R is defined as $2R = A/L$.

Figure 8.21 shows the evolution of a detonation wave in $2H_2/O_2$ at 1337 hPa/ 663 hPa. The initial pressure was 2000 hPa so that the test was conducted in a 100 mm diameter compact chamber. The sequential images were recorded by high-speed video camera Shimadzu SH100 at the framing rate of 10^5 frame/s and at the exposure time of 2.5 μs. The field of view of this test chamber was only 100 mm diameter so that the recording time was only up to 160 μs. The light source was a conventional flash lamp, whose brightness illuminating the field of view fluctuated through the observation period of time. In Fig. 8.21, images recorded on the frames from the 4th to 14th were clearly observed but other images were over exposed. At the early stage, a spherical shock wave turned into a detonation wave. The spherical detonation wave started to deform with elapsing time.

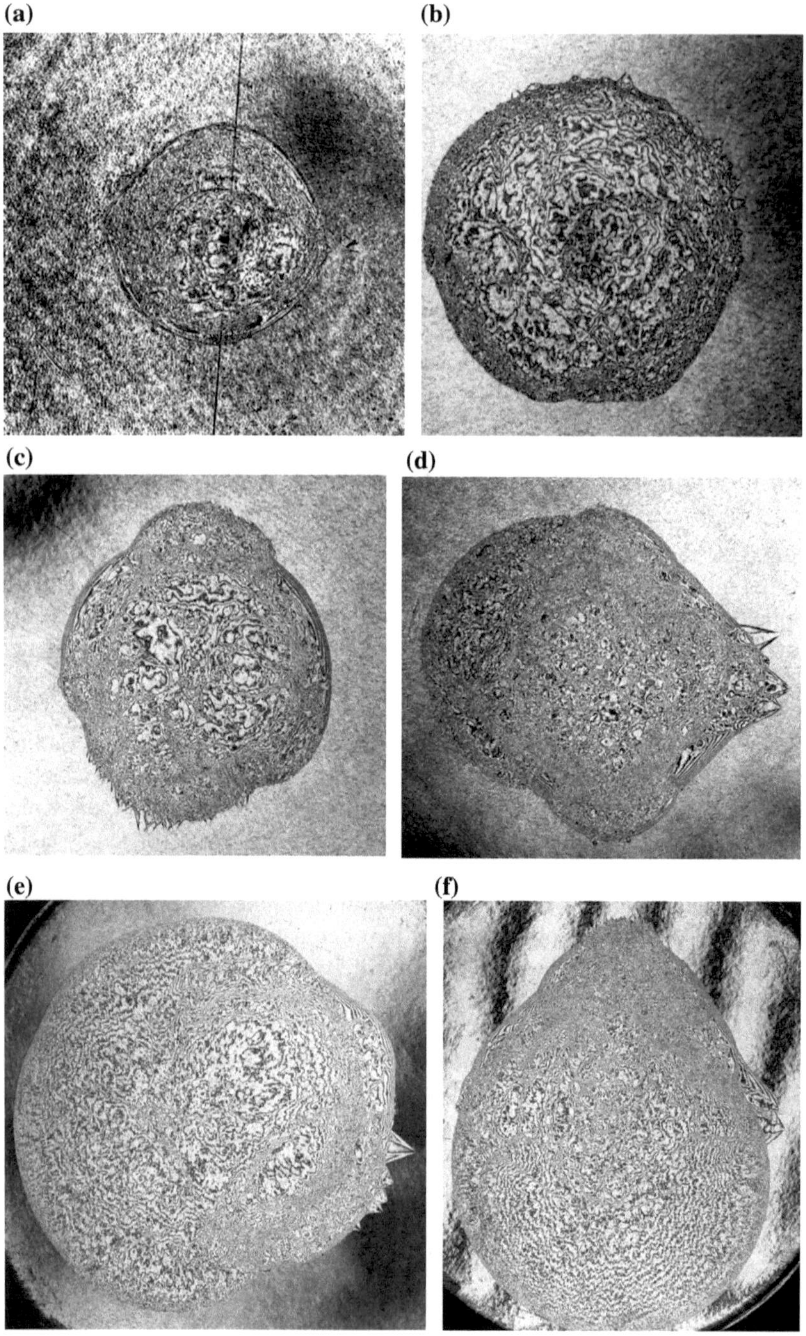

Fig. 8.20 The explosion induced detonation wave in stoichiometric $2H_2/O_2$ in 666 hPa/333 hPa. The shock wave was generated by the ignition of a 10 mg AgN_3 pellet: **a** #98020705, 1 μs after ignition; **b** #98020704, 5 μs; **c** #98020708, 10 μs; **d** #98020709, 15 μs; **e** #98020711, 25 μs; **f** #98020712, 30 μs; **g** #98020713, 35 μs; **h** #98020715, 45 μs

(g) **(h)**

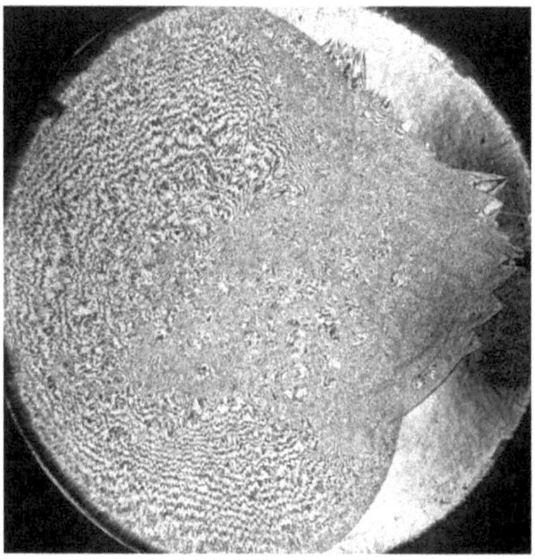

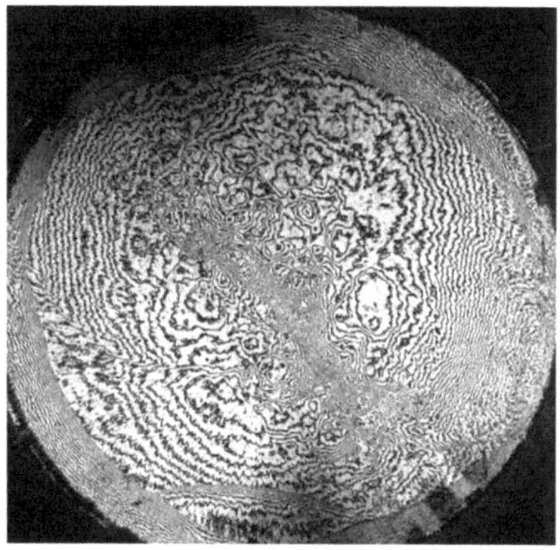

Fig. 8.20 (continued)

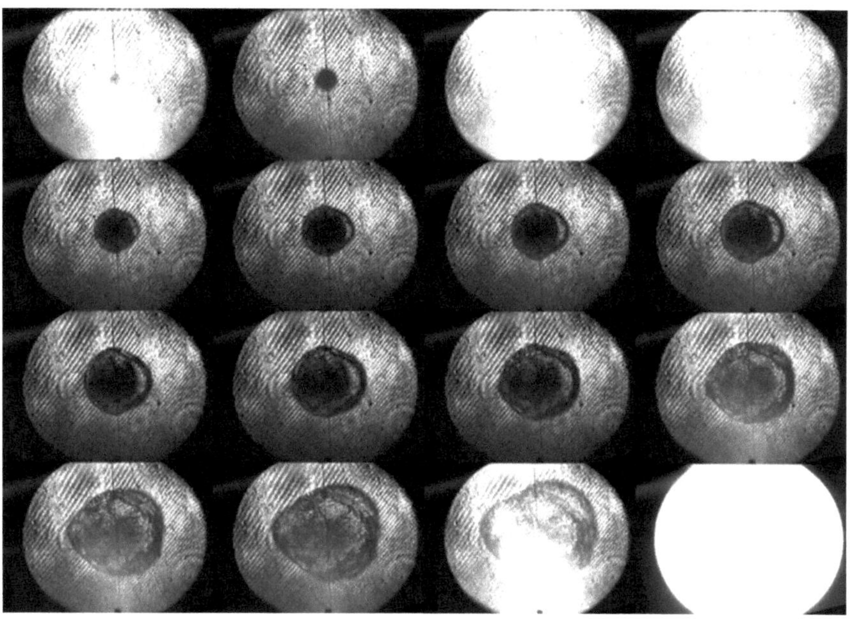

Fig. 8.21 Sequential observations of #00081105, the shock wave was generated by ignition of a AgN_3 7 mg, in stoichiometric $2H_2/O_2$ at 1337 hPa/663 hPa at framing rate of 10^5 frame/s and the exposure time of 2.5 μs

(a) (b)
(c) (d)
(e) (f)

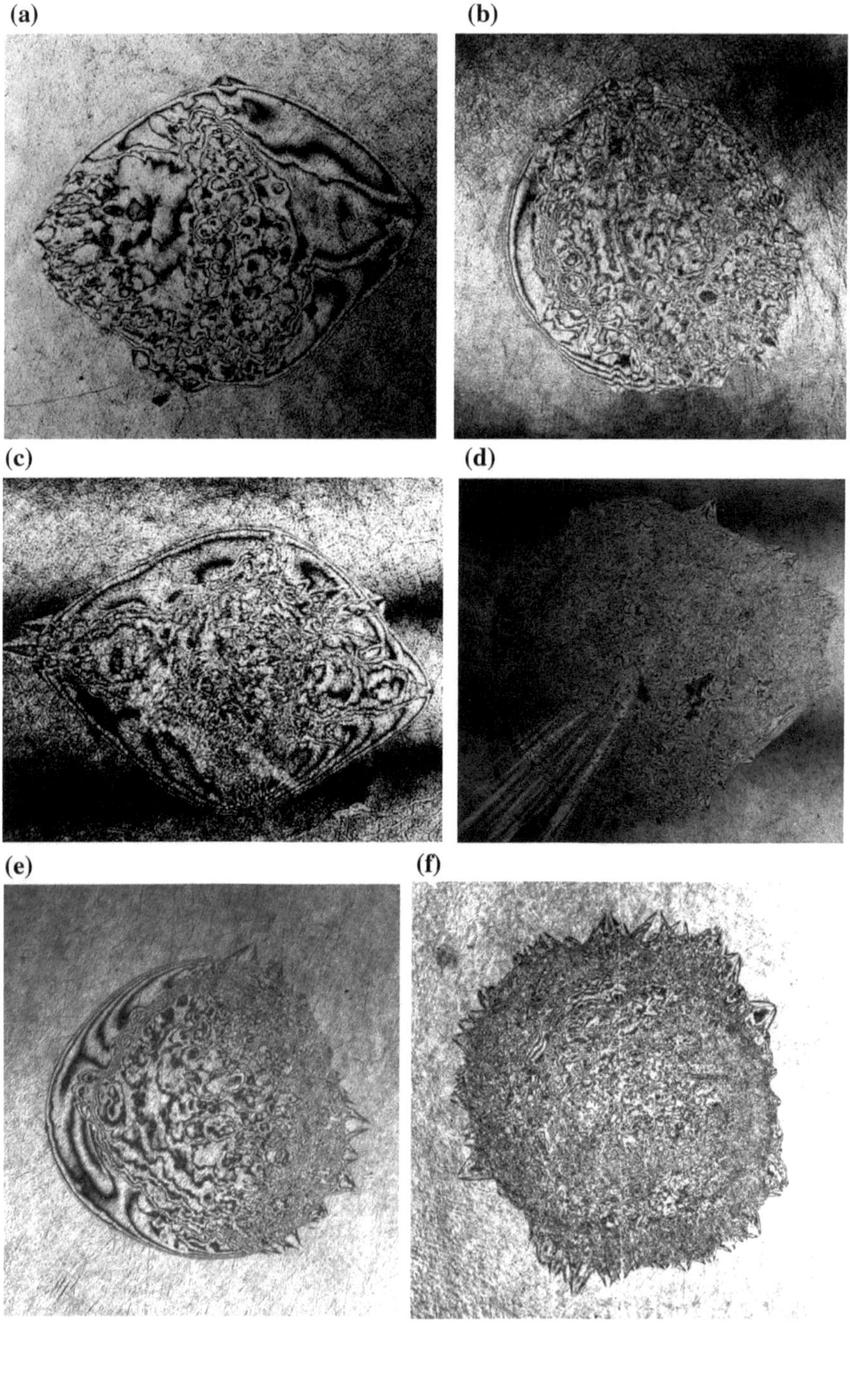

◂**Fig. 8.22** Detonation waves generated by explosion of a SiO_2 particle coated AgN_3 pellet: **a** 5 mg AgN_3, in $2H_2/O_2$ at 200 hPa/100 hPa; **b** 5 mg AgN_3, in $2H_2/O_2$ at 334 hPa/166 hPa; **c** 5 mg AgN_3, in $2H_2/O_2$ at 467 hPa/233 hPa; **d** 5 mg AgN_3, in $2H_2/O_2$ at 600 hPa/300 hPa; **e** 10 mg AgN_3, in $2H_2/O_2$ at 134 hPa/66 hPa; **f** 10 mg AgN_3, in $2H_2/O_2$ at 267 hPa/133 hPa; **g** 10 mg AgN_3, in $2H_2/O_2$ at 200 hPa/100 hPa

(g)

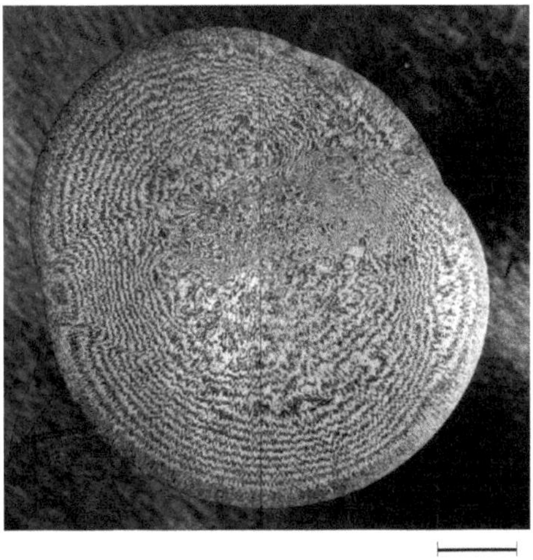

Fig. 8.22 (continued)

8.6.5 *Effects of SiO_2 Particle Coating in $2H_2/O_2$ at 30 μs After Ignition*

In Fig. 8.22, AgN_3 pellets of various weights coated with SiO_2 particles were ignited in $2H_2/O_2$ at various initial pressures. The observation was conducted at 30 μs after ignition were examined.

8.6.6 *Effects of SiO_2 Particle Coating in $2H_2/O_2$ at 50 μs After Ignition*

In Fig. 8.23, AgN_3 pellets of various weights coated with SiO_2 particles were ignited in $2H_2/O_2$ at various initial pressures. The observation was conducted at 50 μs after ignition were examined. The most of spherical shock wave surface was covered with cellular structures.

Figure 8.24 shows sequential observations of the initiation and development of a detonation wave interacted by explosion of a 10 mg AgN_3 pellet in $2H_2/O_2$ at

934 hPa/466 hPa. The images were recorded at 10^5 frames/s and exposure time of 2.5 μs. At first, the spherical detonation wave propagates at the C-J velocity but with elapsed time the spherical detonation wave gradually deformed.

8.6.7 Summary of Experiments

The averaged radius R of the detonation waves at any known time instant is defined as 2R = A/L where A and L are the projection area and of the periphery length of the detonation waves observed in individual interferograms. The velocity of the detonation was normalized by the C-J of the $2H_2/O_2$ mixture, 2818 m/s. Visualizations were conducted for the wide range of the initial pressure and the weight of explosives.

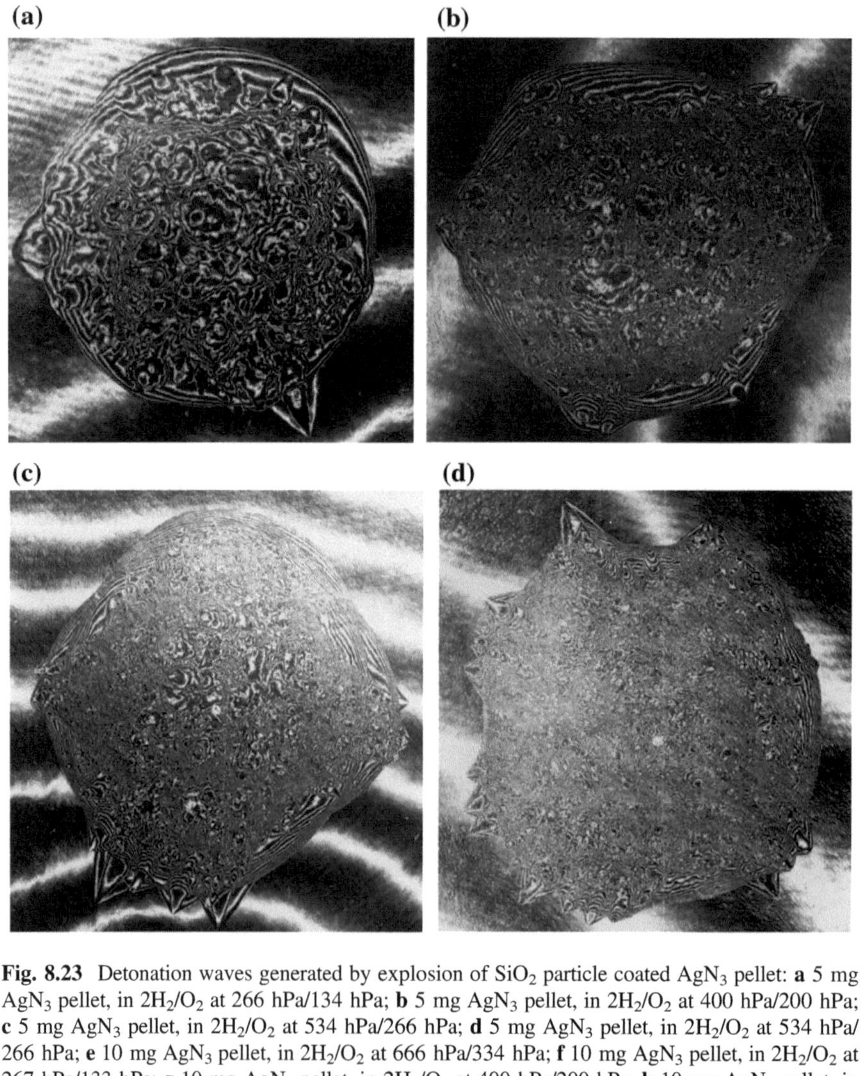

Fig. 8.23 Detonation waves generated by explosion of SiO_2 particle coated AgN_3 pellet: **a** 5 mg AgN_3 pellet, in $2H_2/O_2$ at 266 hPa/134 hPa; **b** 5 mg AgN_3 pellet, in $2H_2/O_2$ at 400 hPa/200 hPa; **c** 5 mg AgN_3 pellet, in $2H_2/O_2$ at 534 hPa/266 hPa; **d** 5 mg AgN_3 pellet, in $2H_2/O_2$ at 534 hPa/266 hPa; **e** 10 mg AgN_3 pellet, in $2H_2/O_2$ at 666 hPa/334 hPa; **f** 10 mg AgN_3 pellet, in $2H_2/O_2$ at 267 hPa/133 hPa; **g** 10 mg AgN_3 pellet, in $2H_2/O_2$ at 400 hPa/200 hPa; **h** 10 mg AgN_3 pellet, in $2H_2/O_2$ at 534 hPa/266 hPa; **i** 10 mg AgN_3 pellet, in $2H_2/O_2$ at 600 hPa/300 hPa

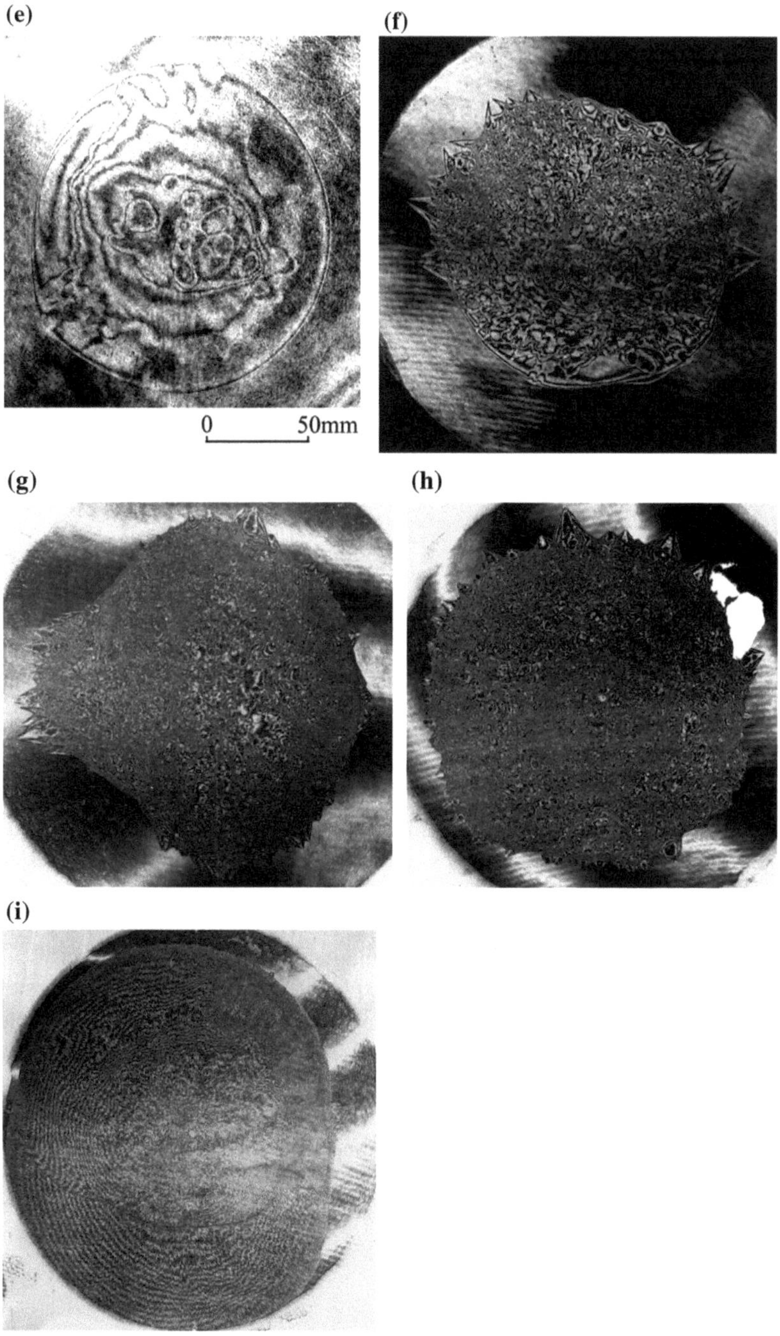

Fig. 8.23 (continued)

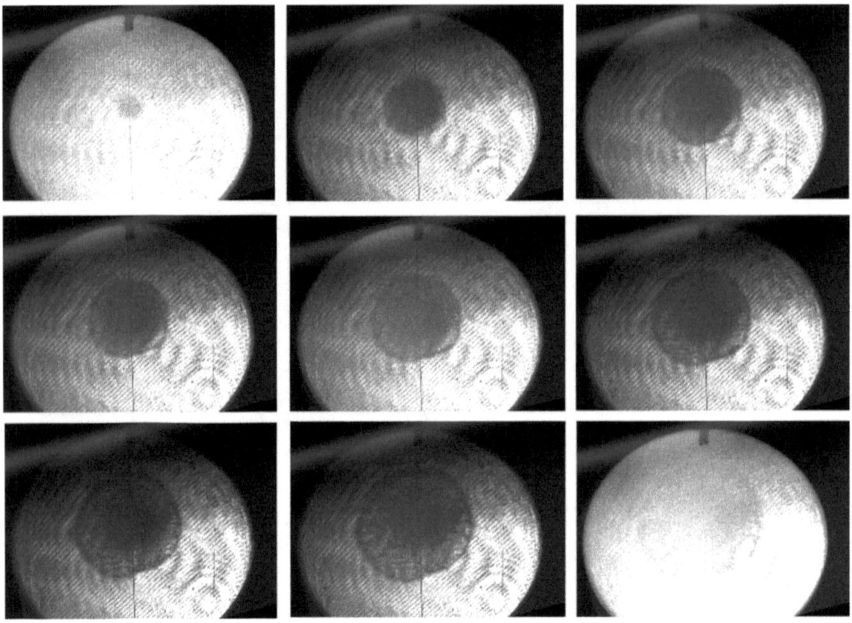

Fig. 8.24 Sequential observation of #00081704, 6 mg AgN_3 pellet in $2H_2/O_2$ at 934 hPa/466 hPa at framing rate of 10^5 frame/s, and exposure time of 2.5 μs

Figure 8.25a, b summarize the variation in normalized velocities against the initial pressure and the weight of the AgN_3 pellet in absence of SO_2 particles coating on the surface of the micro-explosives and in their presence, respectively. The ordinate denotes the normalized velocity, the x-axis and y-axis denotes the initial pressure in kPa and the weight of AgN_3 pellets in mg. The color scales indicate the level of the normalized velocity. Red color corresponds to velocity range exceeding the C-J velocity; light blue color corresponds to about half value of the C-J velocity.

Increasing the initial pressures and the explosives' weight resulted in normalized velocities exceeding the C-J velocity. In absence of attached SiO_2 particle, as shown in Fig. 8.25a, an intermediate and low speed regions are reached as shown in green and light blue color that are widely distributed. In the presence of attached SiO_2 particle, as shown in Fig. 8.25b, the value of peak velocity increases and the zones representing the moderate velocity range shown in yellow and light green colors are drastically widened. In general, the intermediately higher velocity range, the yellow and orange colored zones in Fig. 8.25b are widened.

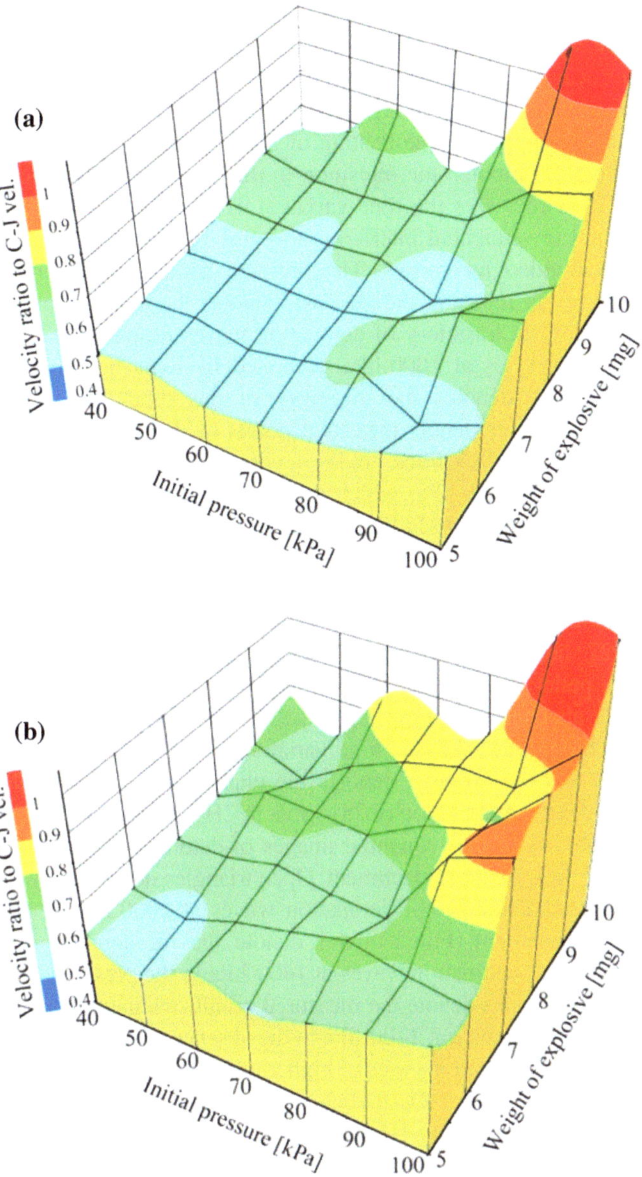

Fig. 8.25 Summary of experiments; **a** absence of SiO_2 particles on the AgN_3 surface; **b** presence of SiO_2 particles coating on the AgN_3 surface

In Fig. 8.25b, in the presence of SiO_2 particles attachment, the penetration of conical shock waves promoted the transition from three-dimensional deflagration driven shock waves to three-dimensional detonation waves even in a lower initial pressure and smaller amount of AgN_3 pellets. Intermediate and high speed regions are widened. In conclusion, the coating of SiO_2 particles on the surface of AgN_3 pellets effectively promoted the transition to the three-dimensional detonation.

Figure 8.26 summarizes the time variation of the trajectories of the detonation waves observed in sequential interferograms and also the trajectories observed in the high speed video recordings. The ordinate denotes the averaged radius of detonation waves in mm and the abscissa denotes the elapsed time in μs. Thick solid line denotes the trajectory of detonation waves propagating at the C-J detonation speed in $2H_2/O_2$ at 1000 hPa generated by explosion of a 10 mg AgN_3 pellet. Dashed line denotes the trajectory of the shock waves in $2H_2/N_2$ at 1000 hPa. Filled circles denote measured points of detonation wave generated in $2H_2/N_2$ at 1000 hPa. Black filled circles denote the measured points of detonation waves in $2H_2/O_2$ at 1000 hPa, which propagated at first at the C-J velocity but started to attenuate with elapsed time. Open circles denote the shock speed in $2H_2/N_2$ at 600 hPa and filled triangles denote the shock speed in $2H_2/N_2$ at 300 hPa. At low initial pressures, the average velocity remained at the deflagration velocity and the DDT was not achieved.

Figure 8.27 show the propagation of detonation waves generated by ignition of a 10 mg AgN_3 pellet at 1400 and 2000 hPa in tow different cases: the presence of SiO_2 particles attached on 10 mg AgN_3 pellets and the absence of SiO_2 particles attachment. The ordinate denotes time variations of averaged radius of detonation waves in mm and the abscissa denotes the elapsed time in μs. Thick solid line denotes the C-J detonation velocity 2818 m/s in $2H_2/O_2$. Filled triangles denote the measured points out of the sequential images of detonation waves at 2000 hPa in the presence of SiO_2 particle attachment. Open triangles denote the measured points out of the sequential images of detonation waves at 2000 hPa in the absence of SiO_2 particle attachment. Filled circles denote the measured points out of the sequential images of detonation waves at 1400 hPa in the presence of SiO_2 particle attachment. Open circles denote the measured points evaluated from the sequential images of detonation waves at 1400 hPa in the absence of SiO_2 particle attachment.

This trend was enhanced in cases of higher initial pressure or when the explosive pellets were covered with SiO_2 particles. Then, the detonation waves gradually departed from the C-J velocity. This trend is enhanced with decrease in the initial pressure and in the absence of particle attachment to the exploding pellet. When the detonation wave speeds departed from the C-J velocity, the cellular structures seen on the detonation wave surface decrease or disappears. However, the size of the present test chamber was so small that it was hard to conclude the quench of three-dimensional detonation wave.

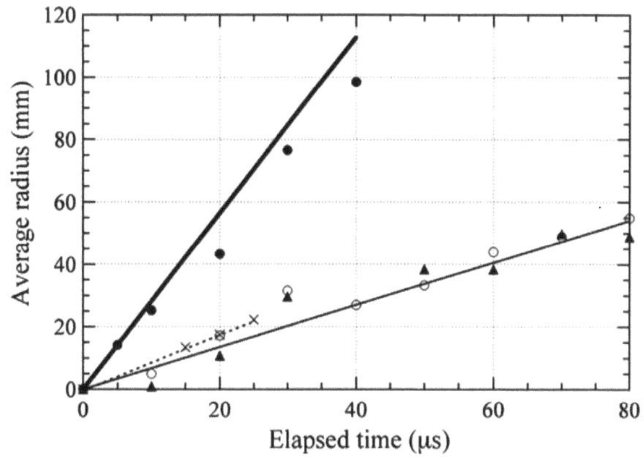

Fig. 8.26 Time variations in the averaged radius of the detonation wave

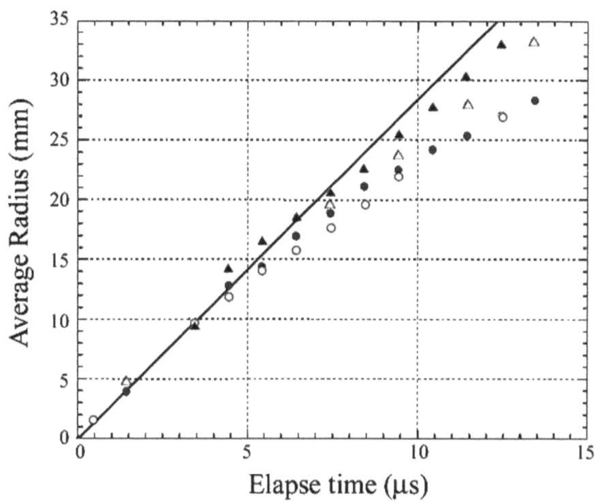

Fig. 8.27 Time variations in the averaged radius of a detonation wave at higher initial pressures

8.7 Shock Waves Generation by Laser Beam Focusing

A collimated Q-switched laser beam can be focused sharply at a spot in air. The instantaneous deposition of a high energy at a very small spot will ionize the gas molecules explosively generate a plasma cloud. The pulse laser beam focusing is equivalent to a micro-explosion and drives a spherical shock wave in air.

In focusing a collimated a Nd:Glass laser beam of energy 4 J and 18 ns pulse width (Model 511-D, BM Industry) in air, shock waves were generated. Figure 8.28 show sequential observations of resulting shock wave formation in air. Figure 8.28f shows a single exposure interferogram. An irregularly shaped dark shadow at the center shows a remnant of the plasma cloud, which is the equivalence to a fire ball in micro-explosions. This is not clearly visualized in double exposure interferometry. The plasma cloud disappears more quickly than the fire ball created

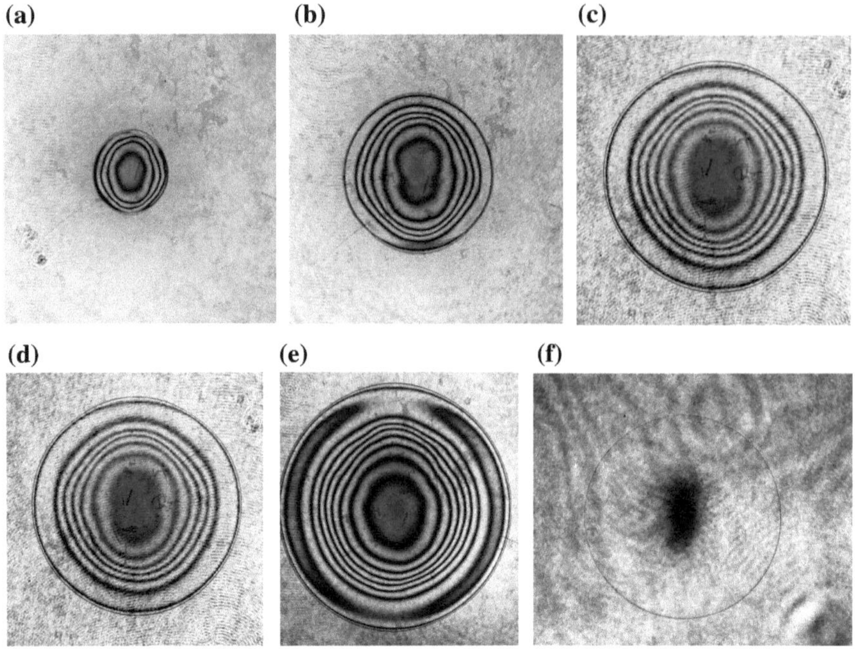

Fig. 8.28 The evolution of spherical shock wave formation by Nd:Glass laser beam focusing: **a** #95020215 initiation; **b** #95020216, 3 μs after initiation; **c** #95020217, 5μs; **d** #95020217, 8 μs; **e** #95020304, 10 μs later; **f** #95061501, single exposure

by the micro explosion and hence the laser induced shock wave would attenuate more quickly than the shock wave created by micro-explosions. The time attached to each intereferogram indicated the elapsed time after the laser beam irradiation.

Splitting the source laser beam into two parts and after propagating via identical light path length, the individual laser beams were focused at a spot on both sides on the surface of a 20 mm thick acrylic plate. Moosad et al. (1995) visualized the initiation and propagation of individual shock waves in the acrylic plate. Figure 8.29 show the shock wave evolution and propagation in the acrylic plate. Hemi-spherical shock waves were generated simultaneously in the acrylic plate. Hemi-spherical stress waves of almost identical strength propagated and interacted with each other inside the acrylic plate.

Figure 8.30 show laser induced spherical shock waves reflected from a steel plate.

Figure 8.31 show laser induced spherical shock wave interaction with a steel plate having 2 mm saw tooth surface roughness. The Nd:Glass laser beam of 4 J energy was focused at 5 mm stand-off distance from the steel plate. In Fig. 8.31b, the reflected pattern is a RR, whereas in Fig. 8.31c, the reflection pattern is a SMR. Figure 8.31c–f demonstrate three-dimensional wave interactions with two-dimensional saw tooth.

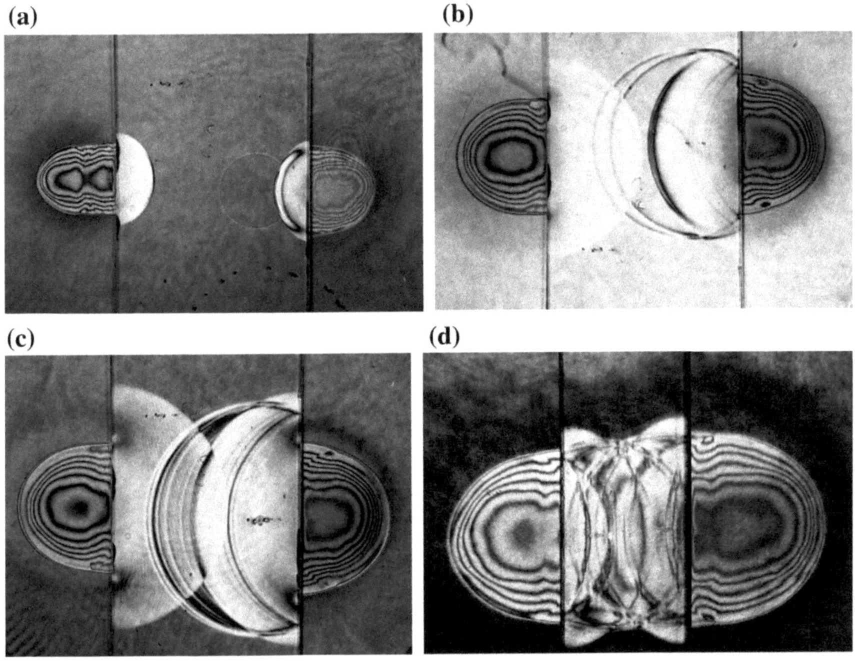

Fig. 8.29 Deposition of a Nd:Glass laser beam of 4 J energy on a 20 mm thick PMMA plate: **a** #95051705, 624 μs from trigger time. The input energy is 1.96 J and the reflected energy is 1.006 J; **b** #95051706, 625 μs, input energy 2.26 J reflected energy 1.005 J; **c** #95051707, 626 μs, input energy 2.21 J. reflected energy 0.996 J; **d** #95053007, 627 μs (Moosad et al. 1995)

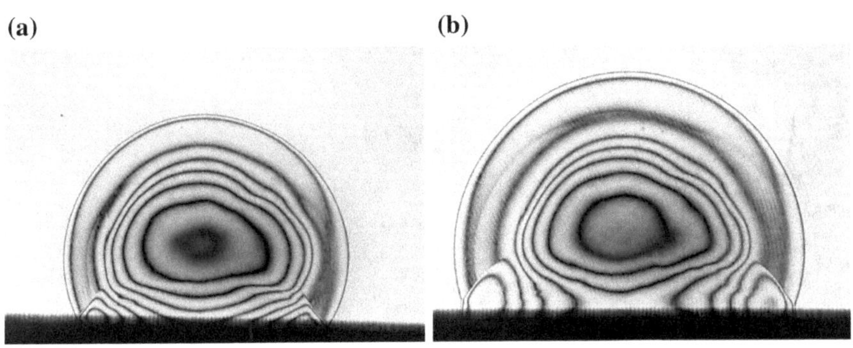

Fig. 8.30 Reflection of laser induced spherical shock wave from a steel plate. The center of focusing was 5 mm above the steel wall in air at 1013 hPa, 288 K: **a** #95022318, 615 μs from trigger point; **b** #95022320, 619 μs

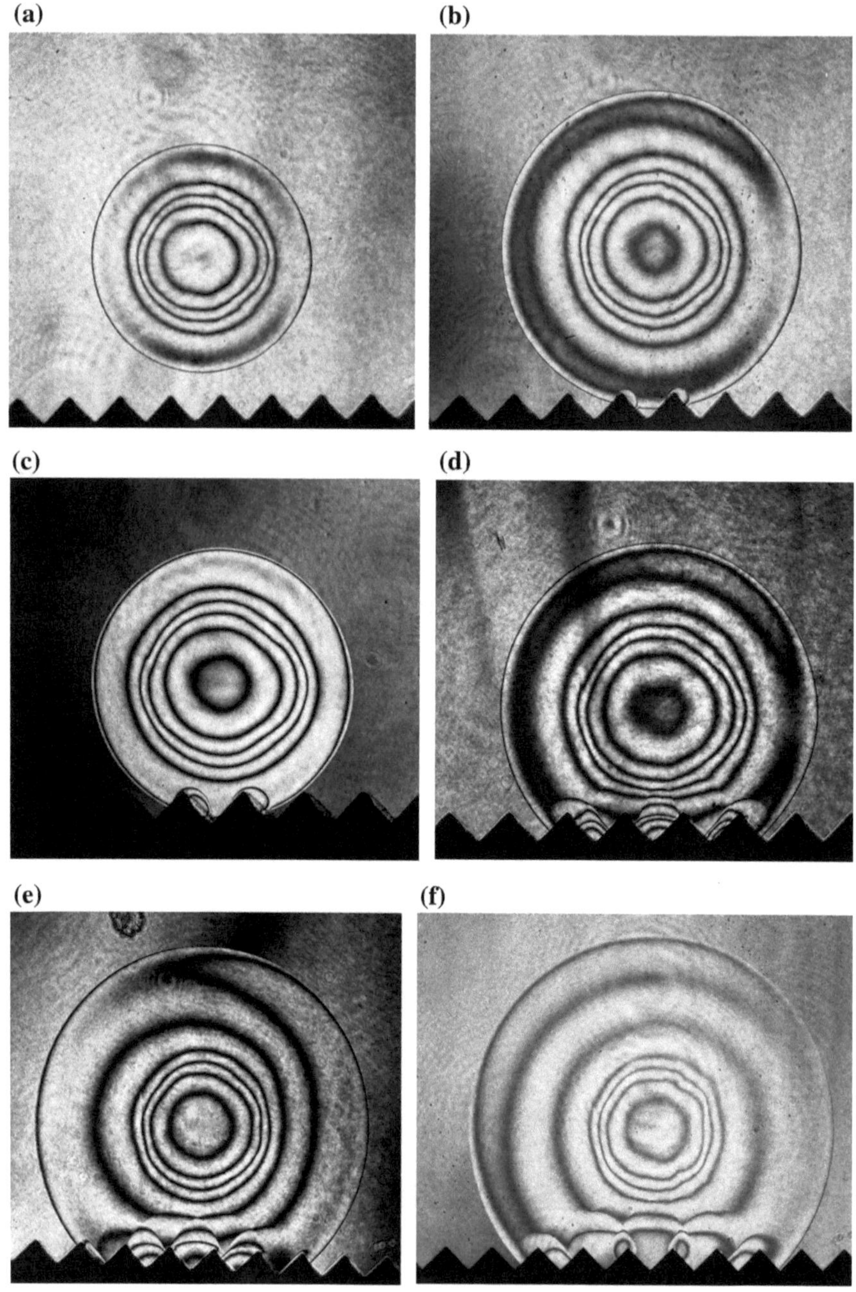

Fig. 8.31 Spherical shock wave interaction with a steel plate having 2 mm saw tooth roughness: **a** #99061407; **b** #99061404; **c** #99061402; **d** #99061405; **e** #99061411; **f** #99061703

References

Esashi, H. (1983). *Shock wave propagation in liquids* (Master thesis). Graduate School of Engineering, Faculty of Engineering, Tohoku University.

Hosseini, S. H. R., & Takayama, K. (2005). Implosion of a spherical shock wave reflected from a spherical wall. *Journal of Fluid Mechanics, 530,* 223–239.

Komatsu, M. (1999). *Experimental study of generation and propagation of spherical detonation waves* (Ph.D. thesis). Graduate School of Engineering, Faculty of Engineering, Tohoku University.

Mizukaki, N. (2001). *Study of quantitative visualization of shock wave phenomena* (Ph.D. thesis). Graduate School of Engineering, Faculty of Engineering Tohoku University.

Moosad, K. P. B., Jiang, Z. L., Onodera, O., & Takayama, K. (1995). Micro shock waves generated by focusing pulsed laser beams. In *Proceeding national symposium of shock waves* (pp. 217–220), Yokohama.

Nagayasu, N. (2002). *Study of shock waves generated by micro explosion and their applications* (Ph.D. thesis). Graduate School of Engineering, Faculty of Engineering, Tohoku University.

Sachs, R. G. (1944). *The dependence of blast on ambient pressure and temperature*. BRL Report, No. 466.

Chapter 9
Underwater Shock Waves

9.1 Introduction

Liquids are less compressible than gases. The compressibility is the measure of volume change when compressed by pressure. A pressure increment Δp is proportional to a relative volume change $\Delta V/V_0$,

$$\Delta p = -E\,\Delta V/V_0 \tag{9.1}$$

where V is specific volume, $V = 1/\rho$, V_0 is an initial volume, and E is the modulus of elasticity which has the dimension of pressure.

The sound speed is defined as

$$a^2 = (\partial p/\partial \rho)_s \text{ in isentropic condition} \tag{9.2}$$

The sound speed is related to the modulus of elasticity and it is defined as,

$$a = (E/\rho)^{1/2}. \tag{9.3}$$

Therefore, a pressure increase of 100 kPa in water at ambient condition reduces the volume of water only by 0.005 of the original volume, whereas when the same pressure applied to air the occupied volume is reduced to one half of its original volume. Water is less compressible than air and sometimes considered to be incompressible.

In an incompressible medium, as the density does not change and then $\Delta\rho = 0$. Hence, the sound speed is infinitely large so that the sound speed in water is much higher than that in air. In the incompressible flows, the information is transmitted

K. Takayama, *Visualization of Shock Wave Phenomena*,
https://doi.org/10.1007/978-3-030-19451-2_9

instantaneously everywhere in the flow field. Therefore, the flow Mach number is always zero. On the contrary, in compressible flows, a pressure perturbation Δp induces a particle velocity Δu as given by,

$$\Delta p = a\rho\, \Delta u, \tag{9.4}$$

where $a\rho$ is defined as the acoustic impedance. Equation (9.4) is derived from the conservation equations and the definition of sound speed.

The acoustic impedance is a coefficient which relates the pressure perturbation and the velocity perturbation. Recalling the Ohm's law in electricity and replacing the pressure perturbation and the velocity perturbation with the electrical voltage and the electrical current, respectively, the acoustic impedance is equivalent to the resistance in the flow.

The acoustic impedance is a measure of the compressibility. The acoustic impedance in water is approximately 3500 times larger than that of air. A given pressure increase in water induces particle velocity of 1/3500 of those induced in air. Glass and Heuckroth (1968) ruptured a glass spheres containing several ten atmospheres of gases in water and generated a low energy underwater shock wave. Today underwater shock waves are generated in a small scale underwater explosion but the pressure exceeded a few GPa.

In gases, the pressure decreases down to vacuum, whereas in liquids the pressure cannot decrease below the liquid's vapor pressure. When the pressures approach to the vapor pressure, the liquids start to evaporate. Water evaporates at 373 K at atmospheric pressure. The Bernoulli equation in incompressible flows is written as

$$p + \rho u^2/2 = P \tag{9.5}$$

where ρ, the density which is constant abd. When the flow velocity u increases, p decreases close to the water vapor pressure. Then water starts to boil even at room temperature. This phenomenon is named "Cavitation". Cavitation appears when water is suddenly exposed to a large tensile force. Upon the exposure to a high tensile stress, water shifts in the phase diagram from liquid phase to gas phase, in short, starts to evaporate inducing a cloud of water vapor. Therefore, the cavitation phenomenon plays an important role not only in the underwater shock wave research but also in the bubble dynamics.

Underwater shock wave research and bubble dynamic research are linked in their applications not only for engineering applications but also in biology and medical applications.

9.2 Underwater Micro-explosion

9.2.1 *Micro-explosives*

In 1981, the collaboration with Dr. Kuwahara of Tohoku University Hospital started. It was invited us to collaborate for the development of Extra corporeal shock wave lithotripsy, in short ESWL, using micro-explosives produced in house. A prototype lithotripter was designed (Takayama 1983). Meantime, the Chugoku Kayaku Co. Ltd. kindly supplied 10 mg AgN_3 pellets supporting the ESWL research (Nagayasu 2002).

Figure 9.1a shows an infinite fringe interferogram of a spherical shock wave generated by exploding a 4 mg of PbN_6 pellet taken after 34 μs from the ignition. Figure 9.1b shows results deduced from the fringe analysis. The pressure profile behind the shock wave is shown in Fig. 9.1a together with appropriate numerical simulation based on the Random Choice Method. Figure 9.1c shows a scaled distance diagram of the overpressures created by micro-explosions. The ordinate denotes overpressures normalized by the initial pressure and the abscissa denotes the scaled distance m/kg$^{1/3}$. Black filled circles denote shock overpressures obtained by the explosion of 4 mg PbN_6 pellets, red filled circles denote the pressures obtained by the exploding of AgN_3 pellets from 10 to 300 μg measured at stand-off distance of 10 mm, blue filled circles for AgN_3 pellets from 3 to 100 μg measured at stand-off distance 5 mm. It is noticed that the scaling law agrees well even in the case of small explosives.

A PbN_6 pellet was glued on a thin cotton thread and suspended in water as seen in Fig. 9.1a. Irradiation of a Q-switched ruby laser beam on to the explosive ignited it. A PbN_6 pellet has energy about 1.5 J/mg and hence a 4 mg PbN_6 has total energy of about 6 J, however, one third of its energy was consumed in an underwater shock wave formation. The ignition laser beam had energy of less than 10 mJ but the laser energy would not contribute to the shock wave formation. Later the laser ignition method was improved: A explosive pellet was glued on the edge of a 0.6 mm diameter quartz optical fiber. The transmission of a Q-switched Nd:YAG laser beam of energy of 10 mJ and 7 ns pulse width through it ignited the small explosive.

Figure 9.1a shows a co-axial fringe distribution taken at magnification of 0.5 time. Fringes are formed due to the integration of the density profile along the *OB* path but clear enough to resolve the interval between neighboring fringes. Assuming that the flow field is spherical, the density distribution behind the shock wave was determined by counting the fringes and their intervals. The measured densities and the numerical simulations were compared. When the density jump across the shock wave is determined, then the pressure profile is calculated using the Tait equation. The Tait equation is an empirical equation of state of water valid below 2.5 GPa (Tait 1888).

$$(p+B)/(p_0+B) = (\rho/\rho_0)^n, \tag{9.6}$$

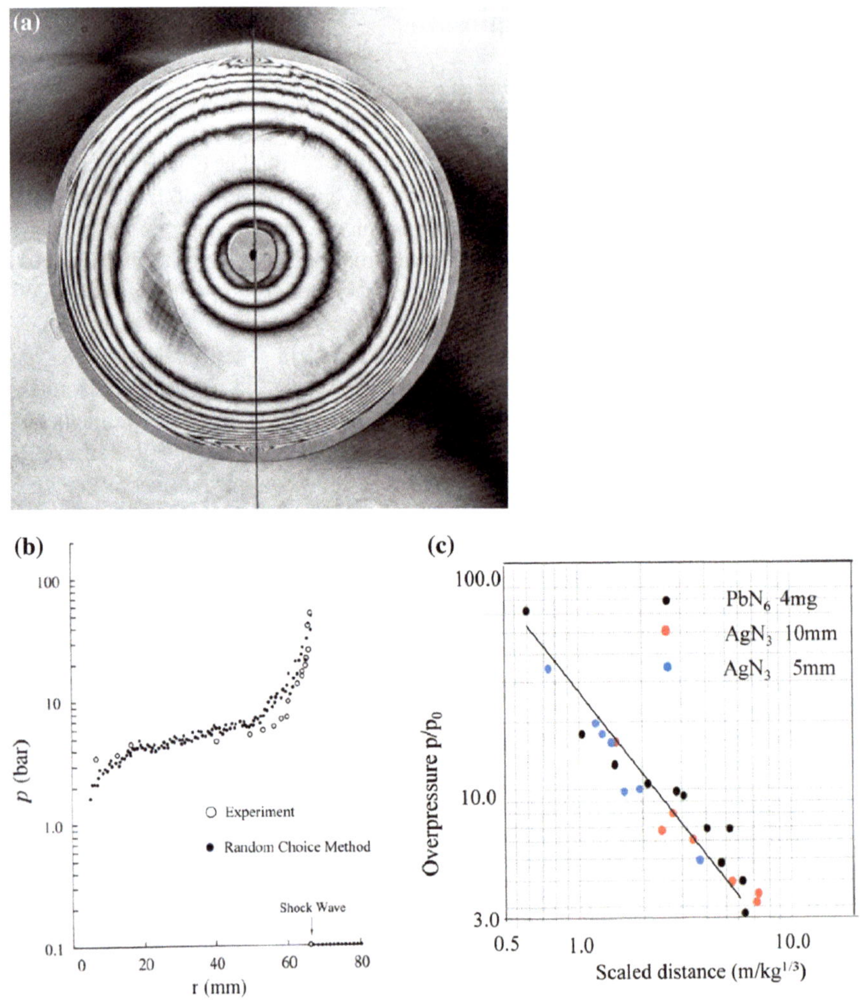

Fig. 9.1 A spherical underwater shock wave: **a** #83011712. Infinite fringe interferogram of 4 mg PbN_6 at 34 μs after ignition; **b** comparison with Random Choice Method; **c** dimension-less pressure versus scaled distance (Sachs 1944)

where p_0 and ρ_0 are ambient pressure and density, B = 300 MPa and n = 7.145. Once the peak density is given, the pressure profiles of the entire flow fields are determined by Eq. (9.6).

But it should be noticed that the *OB* path length was zero at the shock front therefore the density across the shock front could not be resolved. Then the pressure jump across the shock wave was taken from a pressure transducer reading.

9.2.2 Underwater Shock Waves

Figure 9.1c shows the scaling law of a shock wave overpressure. The ordinate shows an overpressure and the abscissa shows the scaled distance in m/kg$^{1/3}$ (Sachs 1944). Red and blue filled circles denote overpressures generated by explosions of AgN_3 pellets weighing from 3 to 300 μg at stand-off distance of 5 and 10 mm, respectively. Black filled circles denote overpressures of explosions of 4 mg PbN_6 pellets at stand-off distances of 10–80 mm. Normalized overpressures were expressed in the scaled distance regardless of types of explosives. The experimental results were compared with an appropriates simulation based on the Random Choice Method, RCM (Esashi 1983). The explosion of 4 mg PbN_6 pellet was assumed to be equivalent to a 2.0 mm diameter air bubble having the pressure and temperature at 1 GPa and 2800 K, respectively. The air bubble is placed at the center of a 100 mm radius water chamber at ambient condition of 100 kPa and 293 K. Pressure and the density variations are shown in Fig. 9.2a. The ordinate in Fig. 9.2a denotes the time and the abscissae denote the dimensionless distances normalized by the bubble radius. As soon as the bubble is released into water, the diverging underwater shock wave propagates outward and an expansion wave converges toward the center. The high pressure air becomes an expanding bubble. At the same time, the converging expansion is reflected at the center of the bubble and becomes a secondary shock wave.

The diverging shock wave is reflected from the outer wall at the 100 mm distance from the center. When the reflected shock wave encounters with the diverging bubble surface, the bubble starts contracting. Then the pressure outside the expanding bubble starts spontaneously contract. The sudden bubble contraction creates cavitation bubbles. The RCM (Esashi 1983), however, can not numerically reproduce the cavitation.

Figure 9.2b shows a streak recording of a shock wave generated by explosion of a 5 mg AgN_3 pellet. The ordinate denotes elapsing time in μs and the abscissa denotes a shock wave radius in mm. The scale of the streak recording was 10 μs/mm. A 5 mg AgN_3 pellet was glued at the edge of a 0.6 mm core diameter optical fiber and ignited by transmitting a Q-switched Nd:YAG laser beam through it. The detonation gas products appear as a dark shadow expanding vertically. Dark oblique lines running from the center are trajectories of the resulting spherical shock wave. The detonation speed of the AgN_3 pellet is about 2.89 km/s and hence its Mach number is $Ms = 1.93$. However, it attenuated exponentially with elapsed time to $Ms = 1 + \varepsilon$, and takes relatively long time to fully recover to a sound speed. Oblique lines appearing at later time are trajectories of reflected shock waves from the side wall.

Figure 9.3a shows a sequential observation of explosion of a 10 mg AgN_3 pellet using ImaCon 790 at framing rate of 10^5 frame/s. The detonation product gas expanded initially very fast as seen in Fig. 9.2b, but quickly attenuated. When it ceased to expand, the debris particles, probably non-reacted particles, were shattered from the surface of the explosive and penetrated the detonation product gas.

(a)

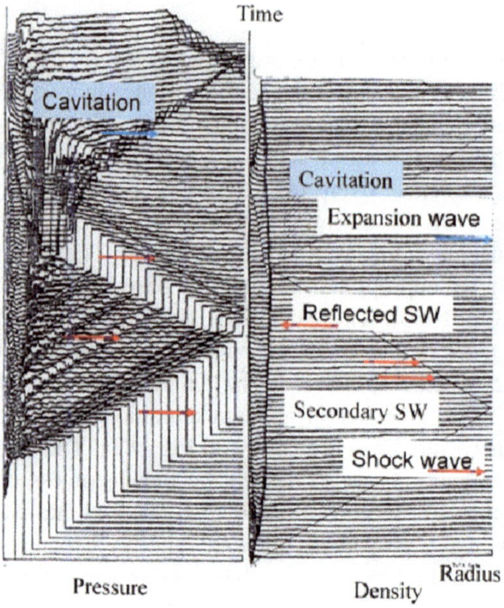

(b)

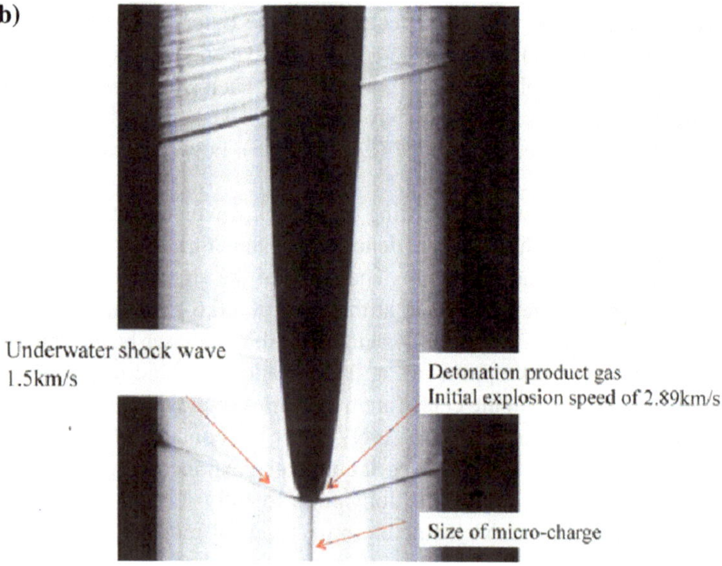

Fig. 9.2 Micro-explosion: **a** RCM numerical simulation (Esashi 1983): **b** streak recording, #90070614, of a shock wave generated by explosion of a 5 mg AgN_3 pellet at streak recording speed of 10 μs/mm

(a) (b)

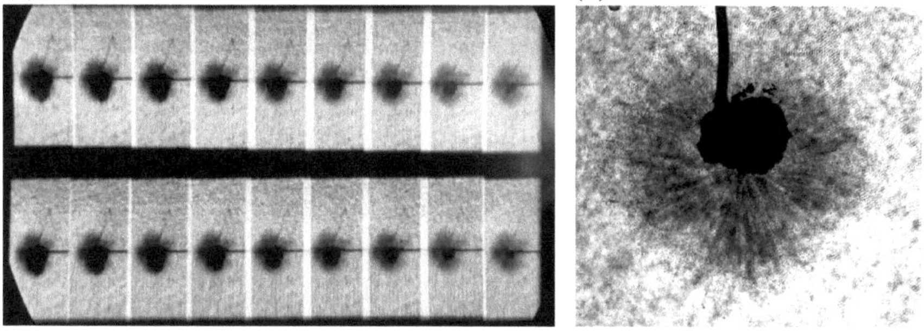

Fig. 9.3 Observation at a later stage of exploding a 10 mg AgN_3 pellet: **a** #90070908, 10^5 frame/s; **b** #92011611 taken at 4.2 ms after ignition at Tw = 290.1 K

At this time, it contracted and became slightly transparent. Figure 9.3b is a single exposure interferogram taken at 4.2 ms after the ignition. The optical fiber holding a AgN_3 pellet was significantly deformed.

9.2.3 Reflection of Underwater Shock Waves

Glass and Heuckroth (1968) intended to observe the presence of a MR in underwater shock wave reflection by rupturing simultaneously two pressurized glass spheres but were not successful to observed distinct triple points. Coleburn and Roslund (1970) visualized the head-on collision of two underwater shock waves by simultaneous detonations of two 63.3 mm diameter mixture of 50% TNT and 50% PETN weighing 225 g and eventually measured the critical transition angle of 37°, although the images they visualized were not necessarily clear.

As such amounts of high explosive are totally unsuitable to use in university laboratory, underwater shock wave reflections from metal walls were visualized using holographic interferometry, Fig. 9.4a, b shows underwater shock wave reflections from 10° and 150° brass cones. Spherical shock waves were created by explosion of 4 and 10 mg PbN_6 pellets at stand-off distance of 4 and 5 mm at over-pressures of over several ten MPa. In gases, the shock wave reflection pattern from these cone angles show either a MR or a RR. However, in underwater shock wave reflections from brass cones, longitudinal and transversal waves propagate much faster than the underwater shock wave. For example, in brass, the longitudinal wave propagates at 4.7 km/s and the transversal waves propagates at 2.1 km/s. The transversal waves transmitted into water created dense fringes at the foot of underwater shock waves. The fringes were too dense to identify patterns of reflected shock waves. Therefore, it is impossible to determine the critical transition angles by observing underwater shock wave reflections from metal cones.

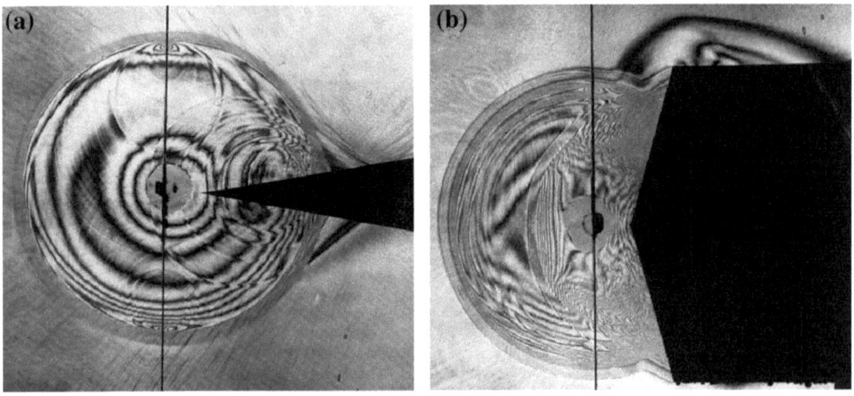

Fig. 9.4 Shock wave reflection from brass cones at 285.8 K: **a** #82122101, a shock wave was generated by explosion of a 4 mg PbN_6 pellet at stand-off distance of 4 mm; **b** #82122105, a shock wave was generated by explosion of a 10 mg PbN_6 pellet at stand-off distance of 5 mm

Eventually, it was decided to observe the reflection of two underwater shock waves having almost identical strengths obtained by exploding 5–6 mg PbN_6 pellets suspended from 4 to 10 mm interval in water. Figure 9.5 shows reflections of two underwater shock waves of identical strengths. Figure 9.5a shows a single exposure interferogram of two 6 mg PbN_6 pellets ignited at 6 mm interval. The resulting reflection pattern was a RR. Figure 9.5b shows the reflection conducted in the same condition as Fig. 9.5a. In the range of the intersection angle as seen in Fig. 9.5d–f, a MR would have appeared but the reflected pattern so far observed was vNMR. Condensed matter, including water, is less compressible than gases so that the particle velocities are always low and the slip lines are hardly generated. Hence the reflected patterns are always vNMR. In detonating two 5 mg AgN_3 pellets separated at 10 mm interval created the observed reflection patterns were vNMR as seen in Fig. 9.5f. Longitudinal waves were also observed along the optical fiber. Figure 9.5g shows enlargement of Fig. 9.5f.

9.2.4 *Reflection of Conical Shock Waves Generated by MDF Explosions*

Mild detonating fuse (MDF, Ensigh-Bichford Co. Ltd.) is a 2.0 mm aluminum sheath filled explosives HNS (Hexa-Nitro-Stilbene, detonation velocity of 6.8 km/s) (Nagayasu 2002). In order to determine θ_{crit} of reflecting conical shock wave, two 80 mm long MDF pieces were submerged in a test chamber intersecting at variable intersecting an angle α. Figure 9.6a shows the experimental arrangement. 10 mg AgN_3 pellets were glued at the edge of 80 mm long individual MDF, and ignited simultaneously with laser beams. Figure 9.6b shows conical shock wave

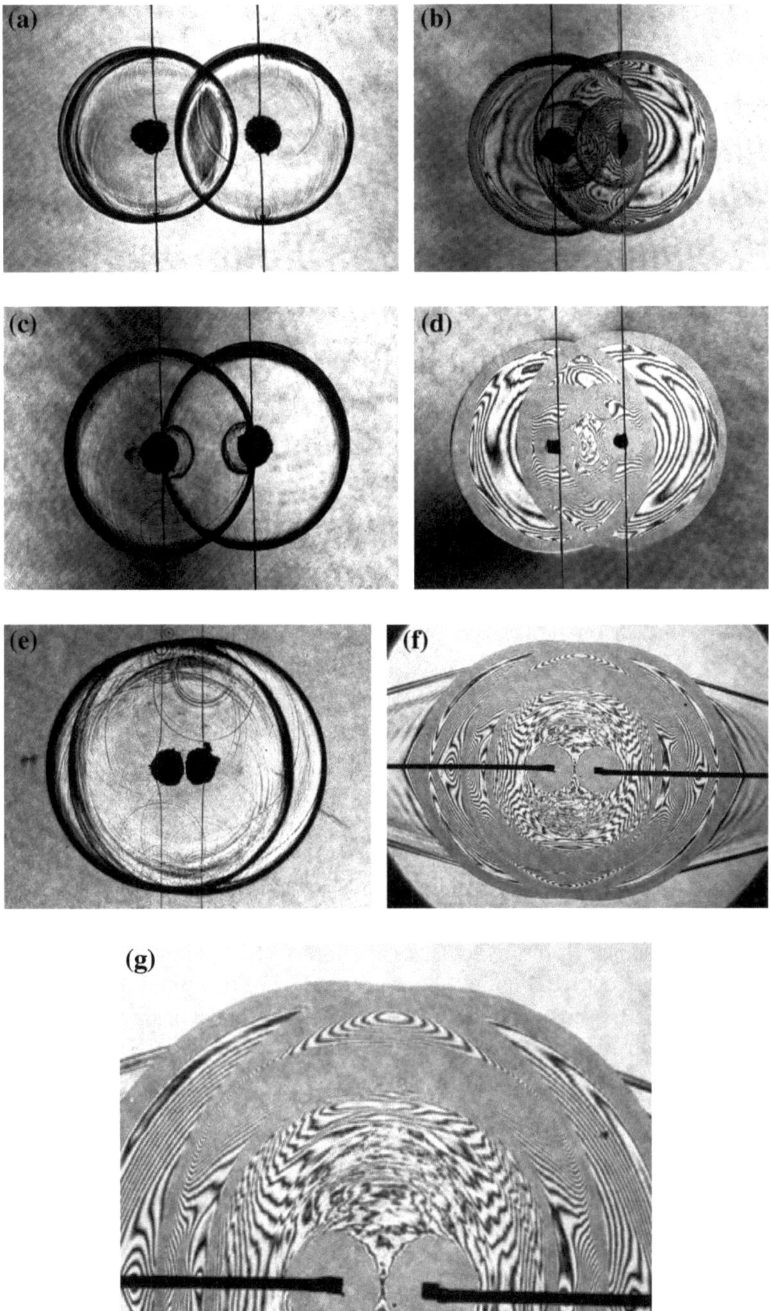

Fig. 9.5 Interaction of two spherical shock waves: **a** #83060208, 3.5 μs from ignition, PbN_6 6–6 mg, at interval L = 6 mm, single exposure, RR; **b** #83060206, PbN_6 6–6 mg, L = 6 mm, RR; **c** #83060302, L = 8 mm, single exposure, RR; **d** #83060102, PbN_6 6–6 mg, L = 10 mm, vNMR; **e** #83120504, PbN_6 5.7–5.9 mg, L = 8 mm, vNMR; **f** #93011205, AgN_3 5–5 mg, L = 10 mm, vNMR; **g** enlargement of (**f**)

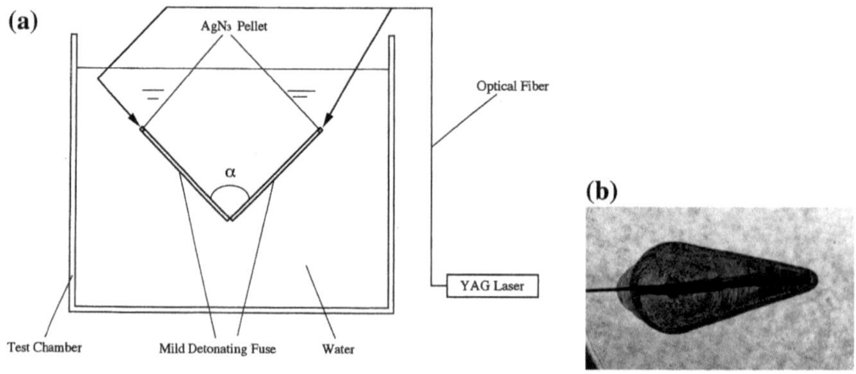

Fig. 9.6 Reflection of conical shock waves: **a** experimental setup; **b** explosion of MDF (Nagayasu 2002)

propagating along the MDF at propagation speed of 6.8 km/s. The shock wave so far generated along the MDF propagated at the constant speed at half apex angle of $\theta = 14.5°$.

Defining the detonation speed of MDF, U_{MDF}, and the conical shock speed U_{cone}, the relationship $\sin\theta = U_{cone}/U_{MDF}$ is valid. Then $U_{cone} = 1.7$ km/s. The sound speed of water being 1.5 km/s, the Mach number of the conical shock wave Ms is 1.14. During the experiments, to suppress bang noises occurred at each shot, the wall of the test chamber was covered with thick sponge sheets.

In Fig. 9.7a–d, the pattern of reflection pattern is a RR. However, in Fig. 9.7e–g, a vNMR appear at the points of the intersection of the incident shock waves, IS and Mach stem MS. Then the position of the triple can be plotted.

Figure 9.8 summarizes the locations of triple points. The ordinate denotes the normalized triple point height h/L, where h is the height of the estimated tripe point and L is the distance from the center of the intersection. The abscissa denotes the initial angle $\alpha/2$ in degree. Figure 9.8b is a summary of measured data. The ordinate denotes h/L and the abscissa shows the inclination angle in degree. The vNMR occurs at the transition angle of bout 30° (Nagayasu 2002).

9.3 Shock Wave Over a Liquid Surface

9.3.1 PDMS/Water Interfaces

Micro-explosives were frequently exploded in PDMS. PDMS is called silicone which has chemical structure, Polydimethylsiloxane, PDMS: $(CH_3)_3SiO–[(CH_3)_2SiO]_n–SiO(CH_3)_3$. To determine the isothermal compression test of PDMS. Figure 9.9a shows the experimental setup. Figure 9.9b shows an isothermal

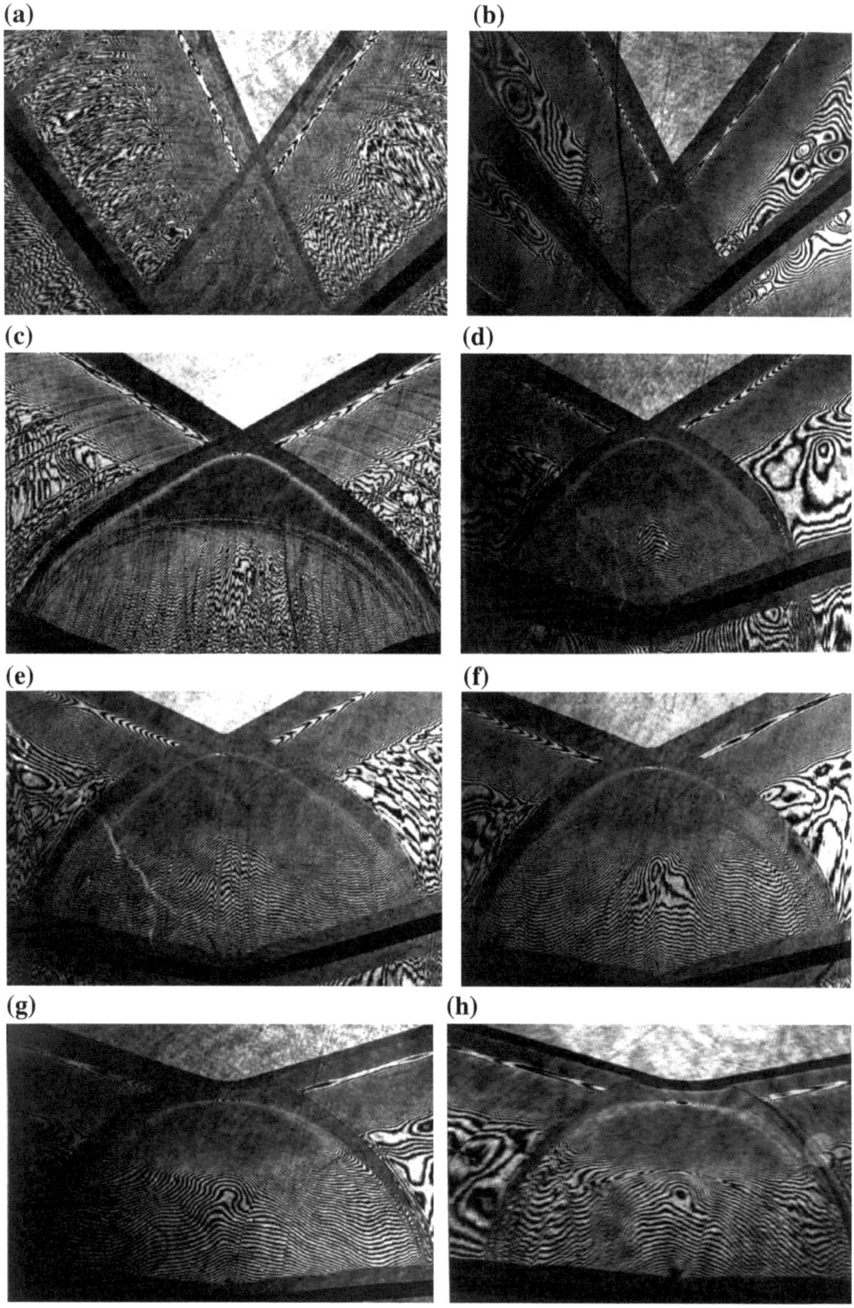

Fig. 9.7 Reflection of MDF generated conical shock waves: **a** #96111307, $\alpha = 58°$; **b** #96111906, $\alpha = 66°$; **c** #96111503, $\alpha = 95°$; **d** $\alpha = 114°$; **e** #96111904, $\alpha = 122°$; **f** #96111803, $\alpha = 125°$; **g** #96111902, $\alpha = 134°$; **h** #96111901, $\alpha = 143°$

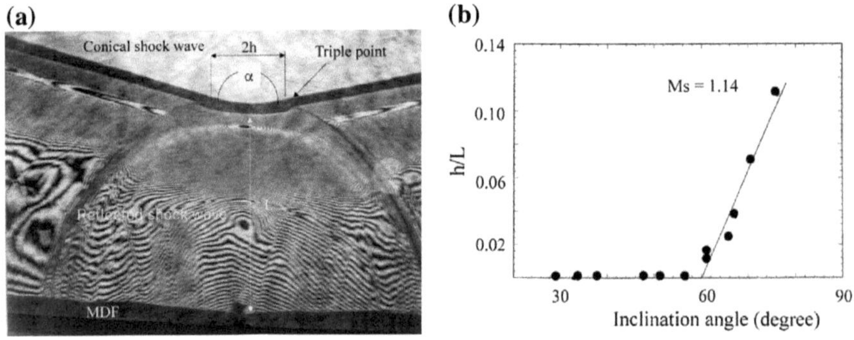

Fig. 9.8 Triple point height h against inclination angle; **a** intersection of two MDF; **b** summary

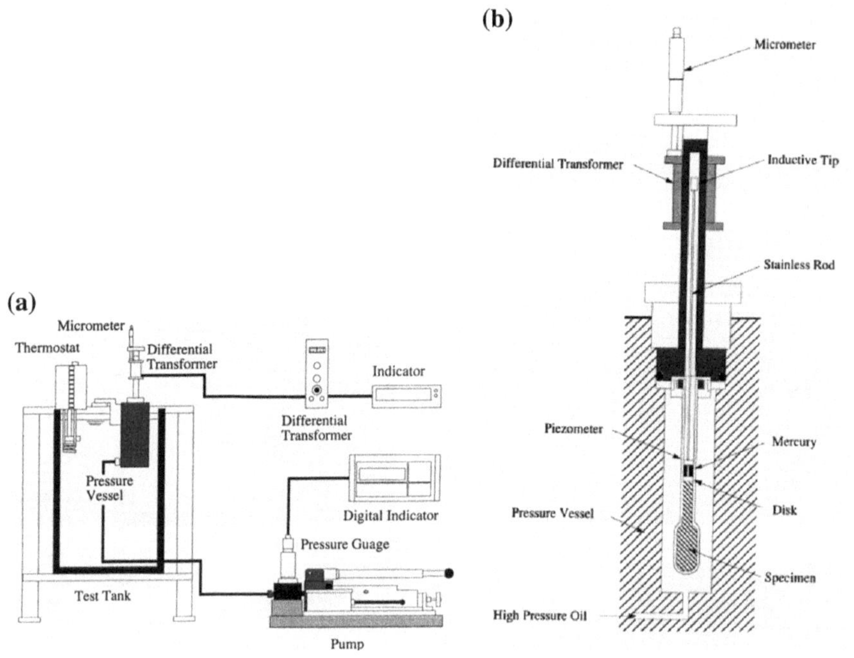

Fig. 9.9 Isothermal compression tester: **a** experimental setup; **b** compression chamber (Hayakawa 1987)

compression system. Filling PDMS in a small glass tester of known volume and compressing the tester with a hydraulic compressor starting from the ambient pressure and up to 250 MPa, the volume change of the tester was measured.

From the measured data, the equation of state of silicon oil can be determined assuming a formulation of the Tait type equation, that is, $(p + B)/(p_0 + B) = (\rho/\rho_0)^n$. The values of the exponent n and B in MPa for PDMS in isothermal conditions are: At 298 K, PDMS 1cSt, density of 0.818 g/cm^3, sound speed of 901.3 m/s; PDMS 10cSt (Shin-Etsu Chemical Industry Co., KF96-10, 10cSt), density of 0.932 g/cm^3, surface tension of 20.1×10^{-3} N/m, sound speed $a_{silicon}$ = 980 m/s at 298 K and parameters of Tait type equation of state, n = 9.36 and B = 95.4 MPa refractive index of 1.399 and kinematic viscosity ν = 1.0 10^{-5} m^2/s; PDMS 100cSt, density of 0.962 g/cm^3, surface tension of 20.9×10^{-3} N/m, and sound speed of 998.7 m/s, n = 10.24, and B = 93.6 MPa; PDMS 1000cSt, surface tension of 21.1×10^{-3} N/m, sound speed of 1000.4 m/s, whereas just for reference those of water at 298 K are viscosity of 0.89cSt, density of 0.997 g/cm^3, surface tension of 20.1×10^{-3} N/m, sound speed of 1496 m/s, n = 7.145 and B = 2.963 MPa (Hayakawa 1987; Yamada 1992).

Prior to the experiments, when micro-explosives were detonated in PDMS, the color in PDMS became yellowish probably due to the thermal decomposition of PDMS. In Fig. 9.10, shock waves in water propagate at 1500 m/s and shock waves in PDMS propagate at 980 m/s. Then the ratio m of acoustic impedances between water and 10cSt PDMS is 1.64. The energy transmitted from PDMS to water $I_{transmission}$ and the energy reflected $I_{reflection}$ is expressed in terms of m,

$$I_{transmission}/I_0 = 4m/(m+1)^2, \quad I_{reflection}/I_0 = (m-1)^2/(m+1)^2 \tag{9.7}$$

where I_0 is the total energy. In the planar wave interaction, approximately 94% of the total energy was transmitted from PDMS to water interface, and about 6% was

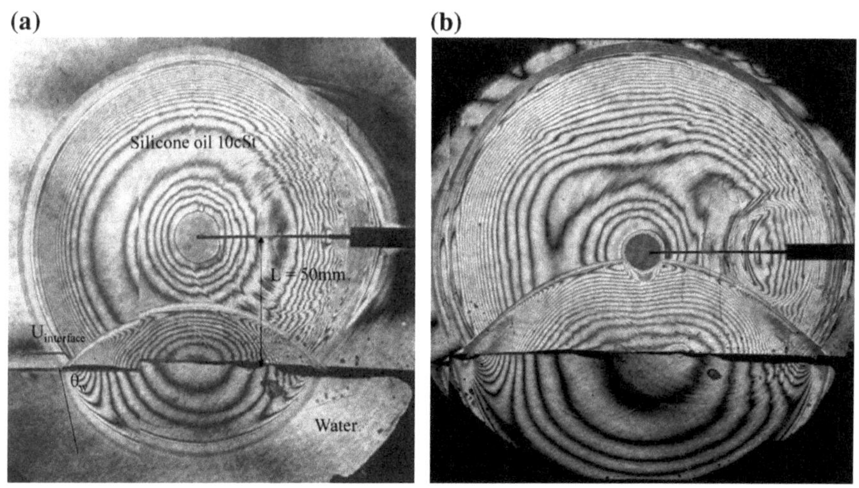

Fig. 9.10 Micro-explosion of 10 mg AgN_3 pellet at L = 50 mm above the interface at 1013 hPa, 296.8 K: **a** #90101603, 76.5 μs from the ignition; **b** #90101605, 92.0 μs from the ignition

reflected from water. Hence the water/PDMS interface would move just a little toward water. Then a shock wave was released into water. Although this estimate is valid only in the case of one dimensional acoustic waves, the predicted value would be a useful estimate of the energy transfer across the water/PDMS interface.

When detonating a 10 mg AgN_3 in PDMS at the distance L = 50 mm above the interface. The spherical shock wave arrived at the interface at $L/a_{silicone}$ = 51 μs and a spherical shock wave was transmitted. Denoting the intersection angle of the shock wave in PDMS with the interface by θ and defining $U_{interface}$ as the propagation speed of the foot of the shock wave along the interface, as seen in Fig. 9.10a, then $U_{interface}$ can be written as following:

$$U_{interface} = a_{silicone}/\sin\theta. \tag{9.8}$$

Equation (9.8) indicates that at the moment when the shock wave in 10cSt PDMS touches the plane interface, θ = 0 and hence $U_{interface}$ is higher than $a_{silicon}$. At $L/a_{silicon}$, the spherical shock wave in 10cSt PDMS induced a transmitted shock wave into the water, that propagated at an earlier stage along the interface at the same speed of $U_{interface}$. Denote θ_w as the intersection angle of the shock wave in water and a_{water} the sound speed of water, then

$$U_{interface} = a_{water}/\sin\theta_w. \tag{9.9}$$

As seen in Fig. 9.10a, at earlier stage θ_w is always larger than θ, hence the relation $a_{water}/\sin\theta_w = a_{silicone}/\sin\theta$ is valid. However, when this condition is violated, the shock wave in water propagates ahead of the shock wave in PDMS and a precursory oblique wave appears ahead of the shock wave in PDMS.

9.3.1.1 Explosion in PDMS Above the PDMS/Water Interface

Figure 9.11 shows the interaction of explosively generated spherical shock wave (by a 10 mg AgN_3 pellet), initiated 10 mm above the PDMS/water interface with PDMS over the water interface.

Figure 9.12 shows the evolution of the interaction of the spherical shock wave create by explosion of a 10 mg AgN_3 pellet at stand-off distance of L = 30 mm above a PDMS/water interface in PDMS over water interface.

9.3.1.2 Explosion in Water Below PDMS/Water Interface

Figure 9.13 shows the explosion in water at the stand-off distance of 10 mm below the PDMS/water interface. The pressure in water behind the reflected shock from the interface wouldn't decrease below the water vapor pressure, therefore, bubbles

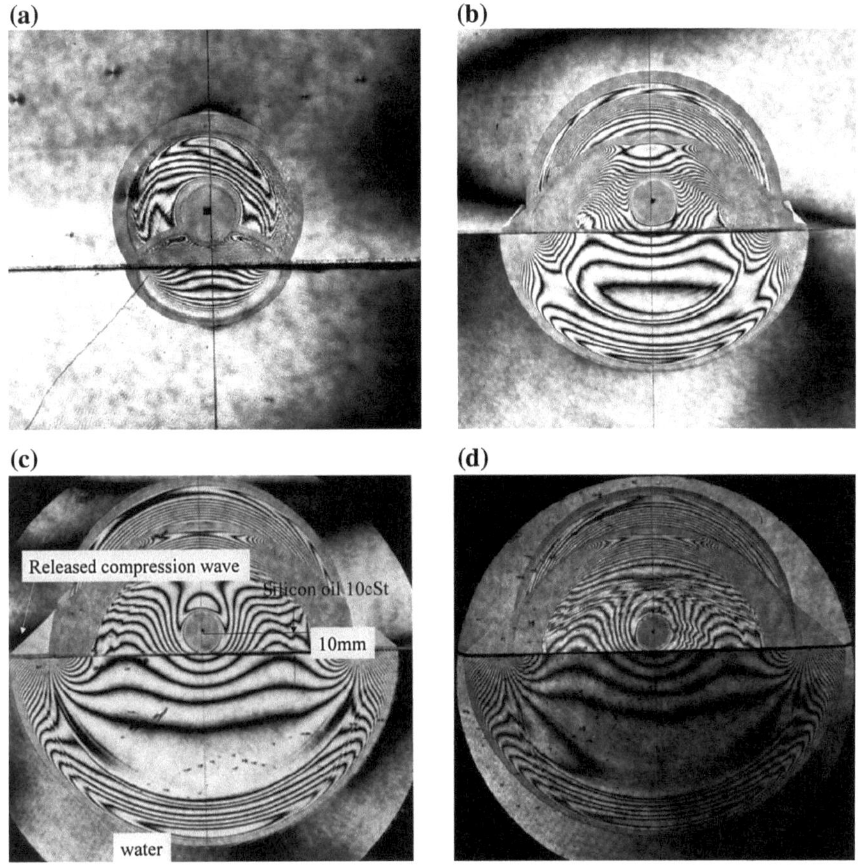

Fig. 9.11 Shock wave generated by explosion of a 10 mg AgN_3 pellet ignited at L = 10 mm above a PDMS/water interface at 1013 hPa, 297 K: **a** #92123002, 16 μs from ignition; **b** #92123102, 42.5 μs; **c** #92122901, 63 μs; **d** #92122902, 86 μs

are hardly initiated. A shock wave in PDMS is always followed by a precursory oblique wave. A secondary shock wave was generated behind the direct shock wave in water.

9.3.1.3 Explosion at PDMS/Water Interface

Figure 9.14a–d shows sequential single exposure interferograms. Shock waves were generated by exploding 8.7–9.0 mg AgN_3 pellets in 1cSt PDMS at L = 3 mm above the PDMS/water interface. Figure 9.14e–i also shows sequential single

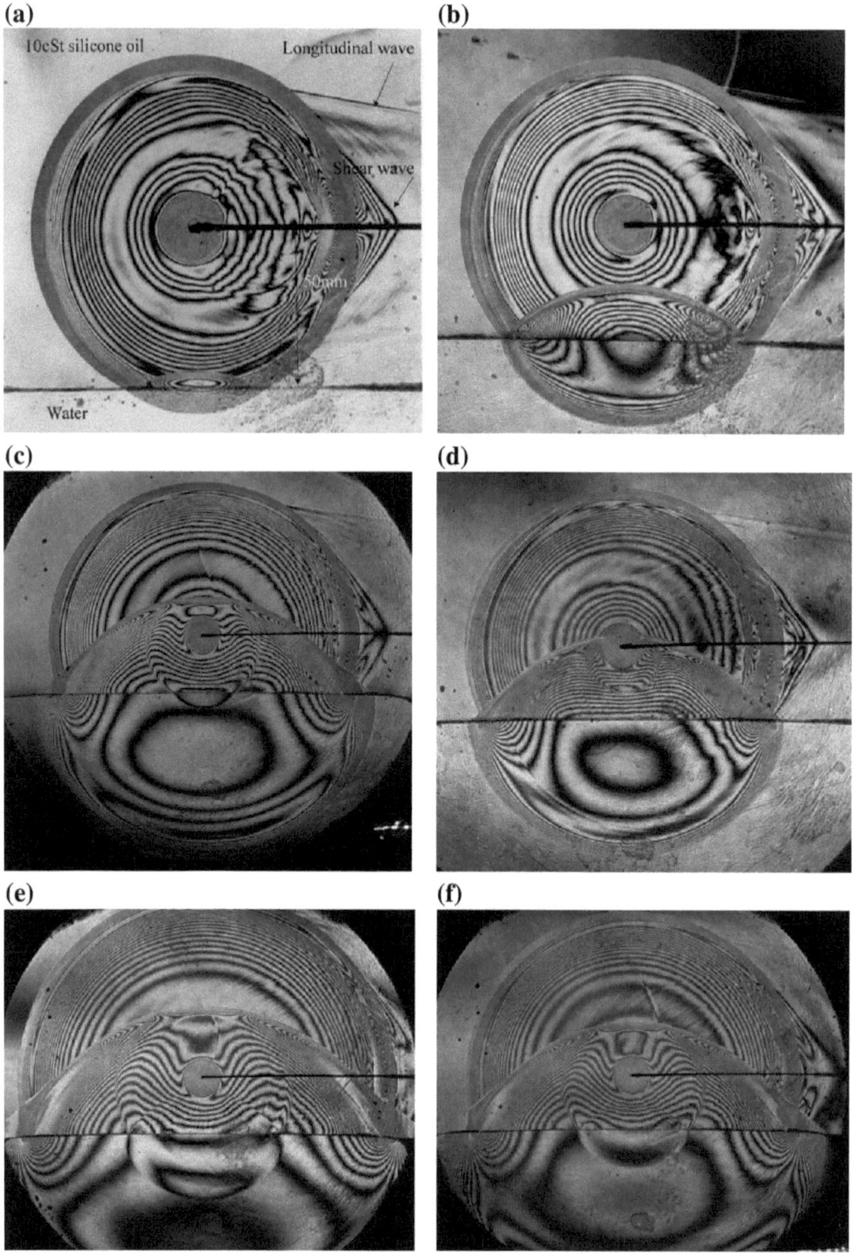

Fig. 9.12 Shock wave generated by explosion of a 10 mg AgN_3 pellet ignited at L = 30 mm above PDMS/water interface, Tw = 296.6 K: **a** #90101802, 38.2 μs from ignition; **b** #90101804, 42.0 μs; **c** #90101809, 47 μs; **d** #90101807, 61.2 μs; **e** #90101811, 90.5 μs; **f** #90101810, 96.7 μs

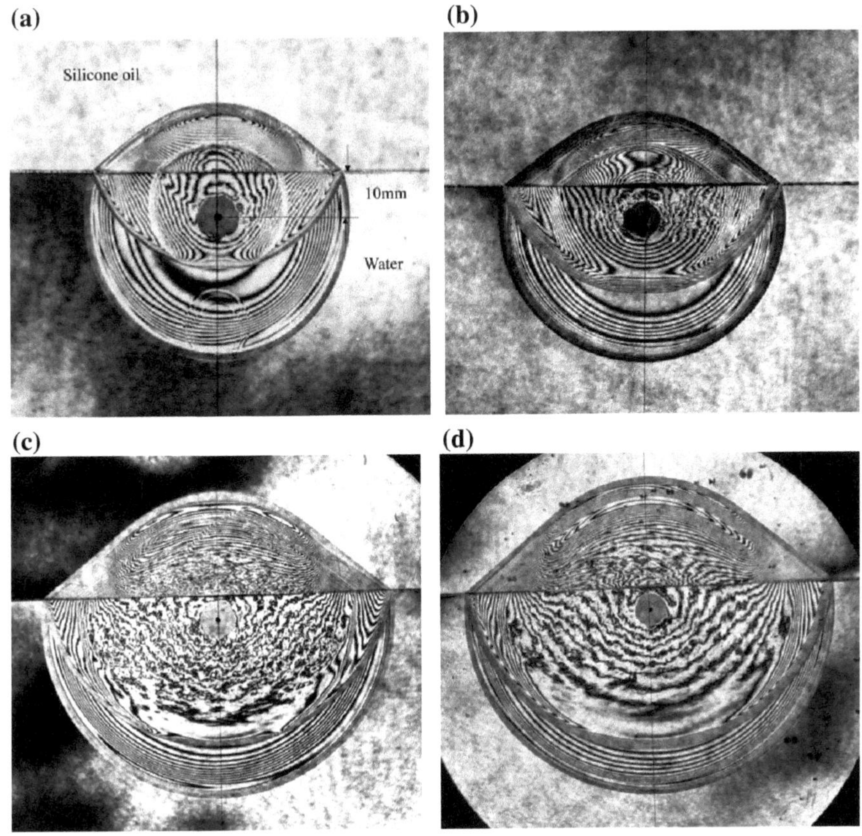

Fig. 9.13 Micro-explosion of 10 mg AgN_3 pellet at L = 10 mm below PDMS/water interface: **a** #92121507, 20 μs from ignition, see the generation of a secondary shock wave; **b** #92120806, 28 μs; **c** #92120904, 48 μs from ignition; **d** #92121504, 63 μs

exposure interferograms taken in the same condition as in Fig. 9.14a–d but L—10 mm. The difference in the magnitude of the stand-off distance, L does-not significantly affect the waves interaction.

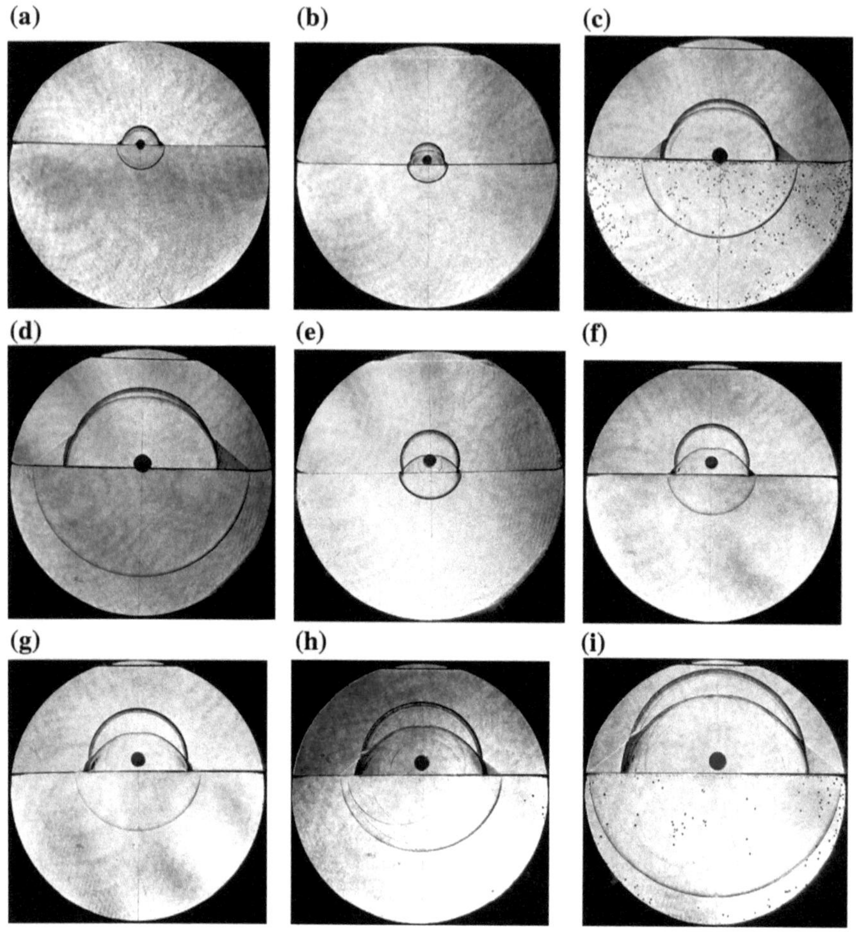

Fig. 9.14 Single exposure interferograms showing the interaction of 1cSt PDMS/water interface with shock waves generated by explosion of 8.7–9.0 mg AgN_3 pellets; **a–d** L = 3, **e–i**: L = 10 mm at 292.0 K: **a** #93102501, AgN_3 pellets, L = 3 mm; **b** #93102502, AgN_3 pellets, L = 3 mm; **c** #93102503, 8.7 mg AgN_3 pellets, L = 3 mm, L = 3 mm; **d** #93102504, L = 3 mm, 9.0 mg AgN_3pellet; **e** #93102506, L = 10 mm, 9.0 mg AgN_3 pellet; **f** #93102602, L = 10 mm, 9.0 mg AgN_3pellet; **g** #93102601, L = 10 mm, 8.9 mg AgN_3 8.9 mg; **h** #93102507, L = 10 mm, 8.1 mg AgN_3; **i** #93102804, L = 10 mm, 8.1 mg AgN_3

9.4 Underwater Shock Wave Focusing

An ellipsoid has two focal points. Hence a linear wave, for example, a sound wave or a light emitted at one of its focal points is reflected from the ellipsoidal wall and converges at its second focal point. A micro explosion placed at the first focal point creates a spherical shock wave, which focuses almost at the second focal point. The shock wave focuses at a point slightly deviated from the second focal point because the underwater shock wave interacts with the detonation product gas and cavitation phenomena perturb the shock wave propagation. Nevertheless, the deviation from proper focusing between linear waves and micro-explosion driven shock waves is negligibly small.

9.4.1 Two-Dimensional Elliptical Reflectors

Focusing of an underwater shock wave inside an elliptic cavity is discussed subsequently. Figure 9.15 shows the elliptic reflector used for visualization. The elliptical reflector has a minor radius of 45 mm and a major radius of 63.5 mm, ratio of the radii is square root of 2. The reflector is made of a 10 mm thick brass plate and sandwiched by two 15 mm thick acrylic plates.

The elliptical reflector shown in Fig. 9.15 is a two-dimensional test section. It was filled with water and a 4 mg PbN_6 pellet was glued on the edge of a 0.6 mm core diameter optical fiber at one of the focal points of elliptic reflector and detonated. The resulting shock wave became a cylindrical shock wave after repeated reflections between the two acrylic plates. The cylindrical shock wave propagated and eventually focused at another focal point (Shitamori 1990).

The time passed from ignition and until reaching complete the focusing took about 83 μs. Figure 9.16 shows sequential observations of the focusing process,

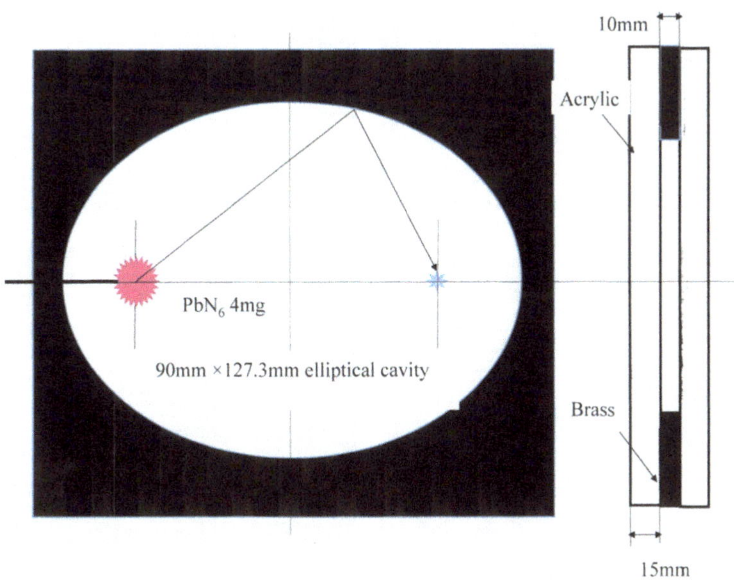

Fig. 9.15 Elliptical reflector, minor diameter 90 mm × major diameter 127.3 mm

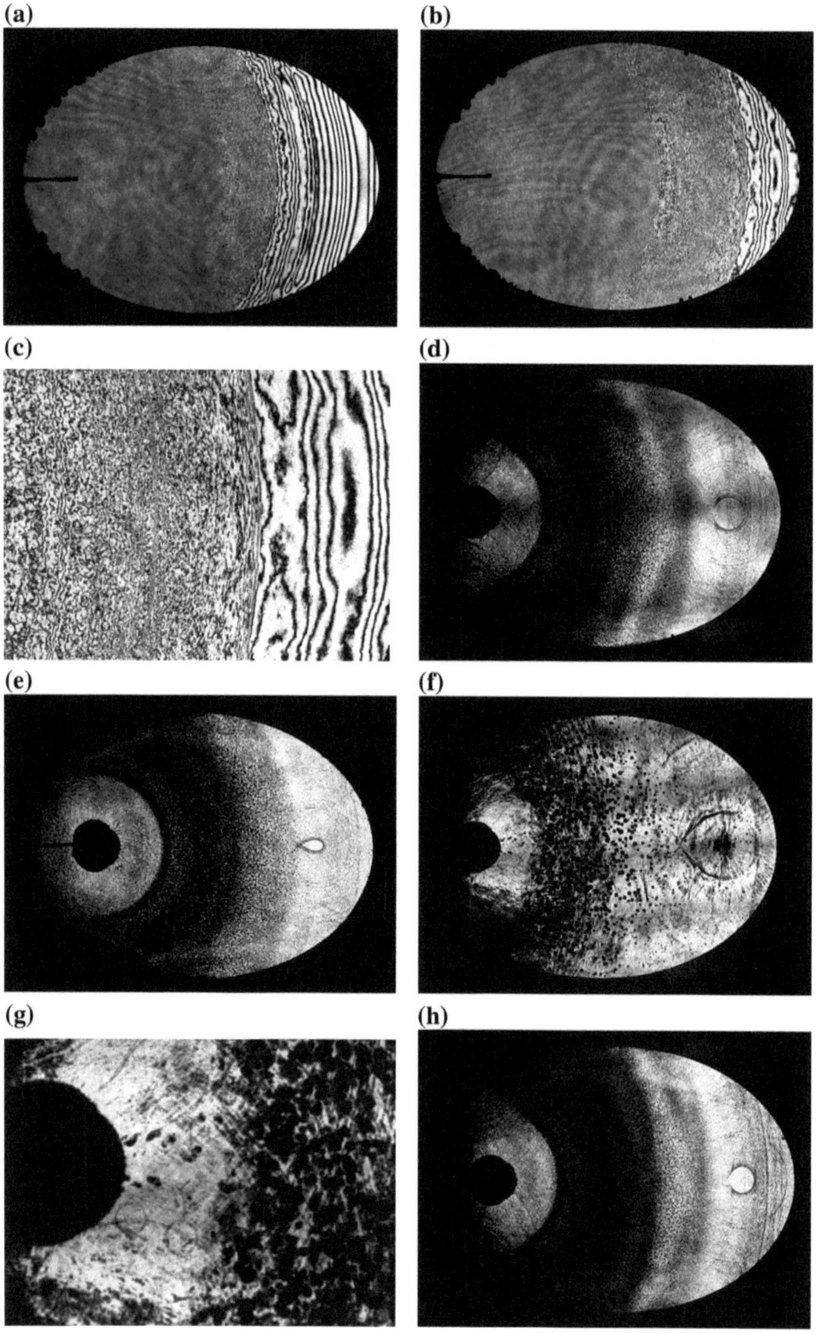

Fig. 9.16 Sequential observation of shock wave focusing in an elliptic cavity: **a** #87101502, 48 μs from explosion; **b** #87101503, 60 μs; **c** enlargement of (**b**); **d** #87101506, 65 μs; **e** 87101903, 73 μs; **f** #87101501, 83 μs; **g** enlargement of (**f**); **h** 87101605, 85 μs; **i** #87101606, 86 μs; **j** #87101710. 98 μs

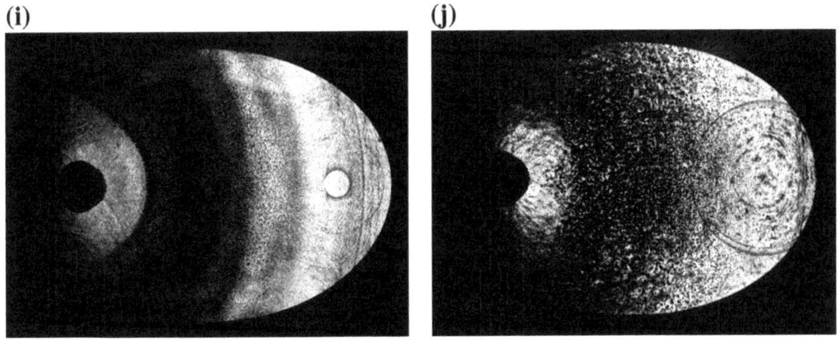

Fig. 9.16 (continued)

in particular the later stage of focusing. The overpressure measured at 5 mm from the center of point explosion was over 50 MPa hence such a sudden high-pressure loading deformed outward the acrylic plate spontaneously probably for less than a few μm. Such sudden deformations generated a strong tensile force in water and induced cavitation. Figure 9.16a, b shows double exposure interferograms. Fringes show stress waves in the acrylic plates and a grey region indicates cavitation bubble cloud. The other images in Fig. 9.16 are single exposure interferogram that are equivalent to shadowgraph. Dark regions show cavitation clouds. Figure 9.16c shows enlargement of Fig. 9.16b. Fine structure of bubbles are visible in these images. Figure 9.16g shows the enlargement of f. Small rings correspond to secondary shock waves created when the cavitation bubbles collapsed. The evolution of focusing of the reflected shock waves was clearly observed.

9.4.2 Shallow Spherical Reflectors

9.4.2.1 Concentric Circular Reflector

A 4 mg PbN_6 pellet was exploded at the center of a 50 mm diameter shallow concentric reflector of 70 mm in diameter. The resulting spherical shock wave was reflected from the reflector. The sequential interferograms (Esashi 1983) and their comparison with unreconstructed images, equivalence to direct shadowgraphs, are shown in Fig. 9.17.

The energy recovery efficiency measured at the focal point is defined as the ratio of energy carried in the reflected wave to that contained in the direct wave, which is the solid angle of the reflector viewed from the point of the explosion normalized by 4π. Then the efficiency of a 50 mm diameter concentric reflector which is composed of 70 mm diameter sphere as shown in Fig. 9.17a, is about 14%. In other words, about 14% of energy was recovered at the focal point.

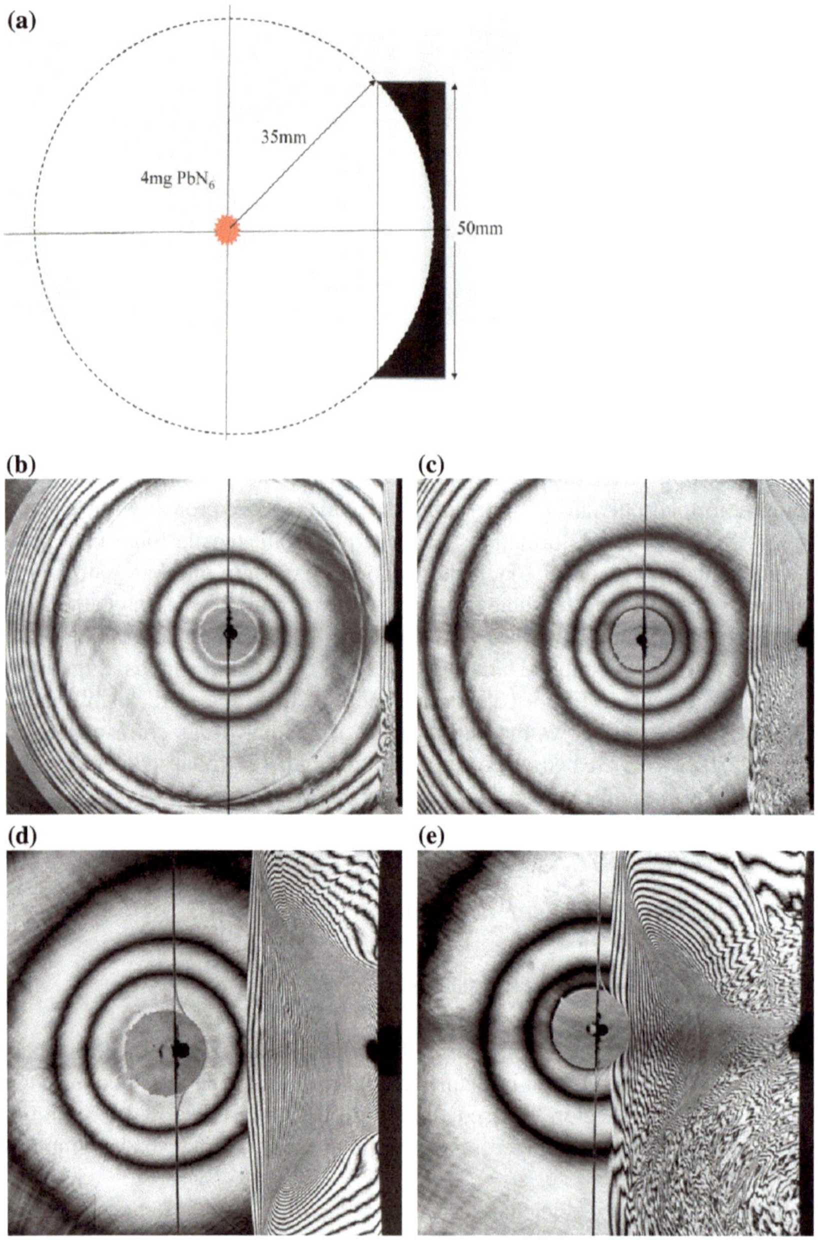

Fig. 9.17 Shock wave focusing from a 50 mm reflector PbN_6 4 mg at 287 K: **a** experimental setup; **b** #82121707, elapsed time 2.5 μs from the exit; **c** #82121706, 4.8 μs; **d** #82121715, 11 μs; **e** #82121710, 15 μs; **f** #82121713, 21 μs; **g** unreconstructed image of (**f**); **h** #82122007, 27 μs; **i** unreconstructed image of (**g**); **j** #82122010, 29 μs; **k** unreconstructed image of (**j**); **l** #82122011, 33 μs; **m** unreconstructed image of (**l**) (Esashi 1983)

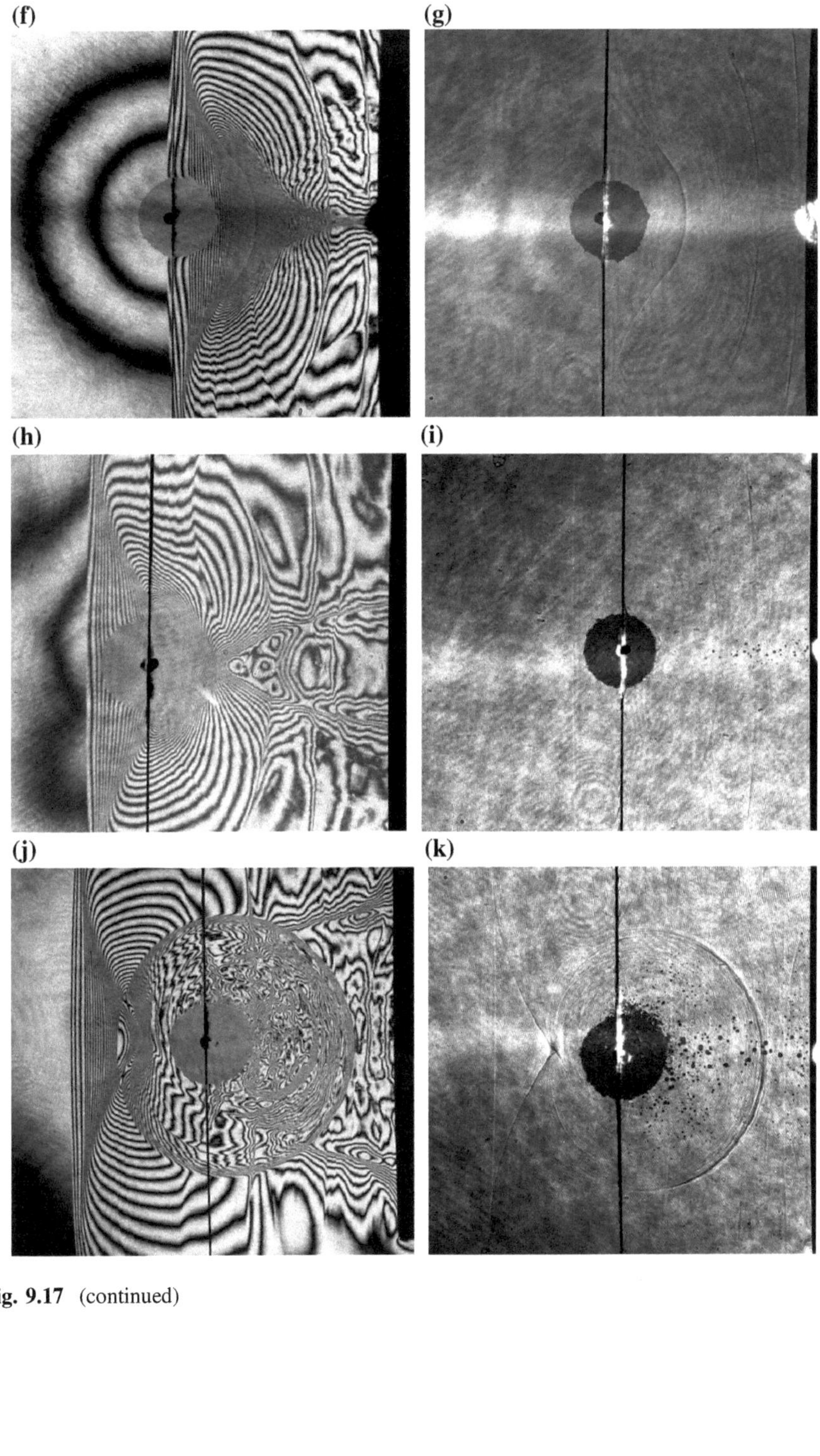

Fig. 9.17 (continued)

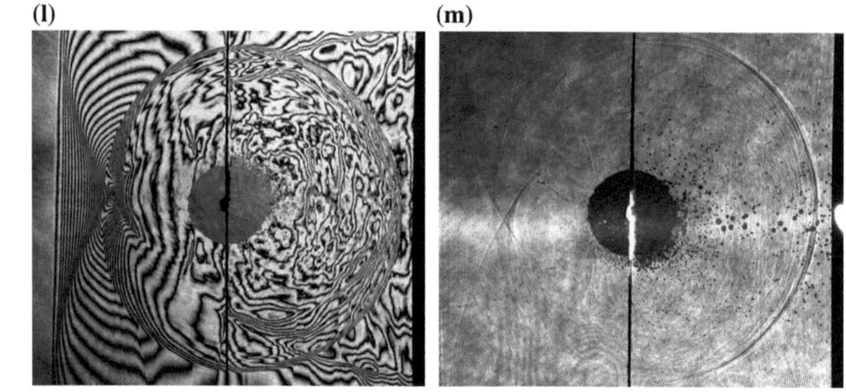

Fig. 9.17 (continued)

Figure 9.17g, i, k are unreconstructed interferograms which are equivalent to direct shadowgraphs. Double exposure interferograms describe quantitatively the density variation but sometimes the inception of cavitation bubbles are difficult to identify. The shock wave reflected from a concentric circular reflector moved toward the position at which the micro-explosive was ignited. Therefore, the shape of the reflected shock wave was concentric with the detonation product gas bubble which was expanding spherically. When the reflected shock wave approached to the spherical detonation product gas sphere, the spherical shock wave interacted with a water/gas interface. Then bubble clouds were generated along the center axis as seen in Fig. 9.17g, i, k. The detonation product gas bubble having a slightly irregular spherical shape was slowly expanding. At the same time, the reflected shock wave impinged, at first, on the bubble and converged toward the center of the bubble. Figure 9.17h, i shows the moment of convergence. In Fig. 9.18j, k, m, the frontal surface of the detonation product gas bubble bulged and bubble clouds were generated.

9.4.2.2 Shallow Eccentric Circular Reflectors

Figure 9.18a shows a reflection of shock wave generated by exploding 4 mg PbN_6 pellet at 50 mm away from a 35 mm radius circular shallow reflector. If a spherical sound wave is emitted at this off-centered point, the sound wave will focus at the conjugate focal point of this reflector.

Figure 9.18 shows the evolution of the shock wave reflected from the shallow reflector at 50 mm away from the 35 mm radius reflector. As seen in Fig. 9.18b–g, the reflected shock wave is converging at the position closer to the reflector (Esashi 1983).

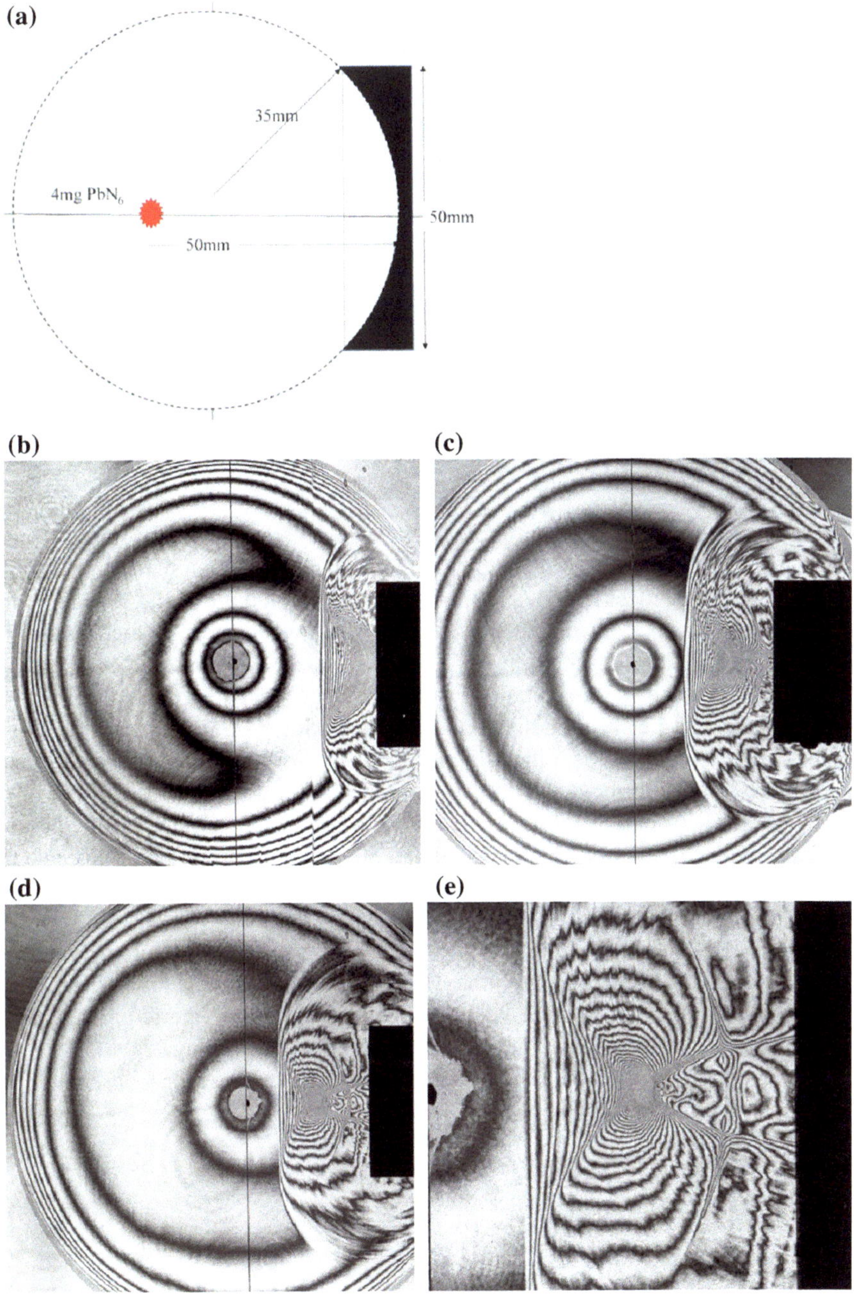

Fig. 9.18 Focusing of shock wave generated by exploding a 4 mg PbN_6 pellet at 50 mm from the reflector at 287 K: **a** experimental setup; **b** #83011906; **c** #83011907; **d** #83011905; **e** enlargement of (**d**); **f** #83011908; **g** enlargement of (**f**)

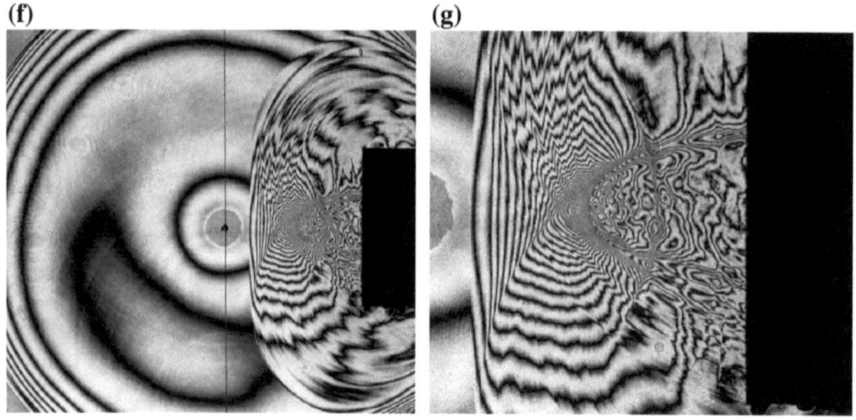

Fig. 9.18 (continued)

Figure 9.19a shows a reflection of shock wave generated by exploding a 10 mg PbN_6 pellet at the off-centered position at the 30 mm away from the 35 mm radius circular shallow reflector and 25 mm aside position. Figure 9.20b–e shows the evolution of the shock wave reflection from the shallow reflector and focusing at the off-centered position (Esashi 1983).

Figure 9.20a shows a reflection of shock wave generated by exploding a PbN_6 10 mg pellet at the 40 mm distance from the 35 mm radius circular shallow reflector and 25 mm aside position. Figure 9.20b–e shows sequential observation of focusing occurring at the conjugated point of the explosion center (Esashi 1983).

Figure 9.21a shows a reflection of shock wave generated at the off-centered position by exploding a 9 mg PbN_6 pellet at 50 mm away from the 35 mm radius circular shallow reflector and 25 mm aside position. Figure 9.21b–e shows focusing occurring at the conjugated point of the explosion center (Esashi 1983).

Figure 9.22a shows the reflection of shock wave generated by exploding a 9 mg PbN_6 pellet at 100 mm away from the 35 mm radius circular shallow reflector and 25 mm aside position. Figure 9.22b–g shows sequential interferograms. The reflected shock wave merges at a neighborhood to the conjugated focal point. However, the fringes coalesce not necessarily as sharply as what observable in Fig. 9.22 (Esashi 1983).

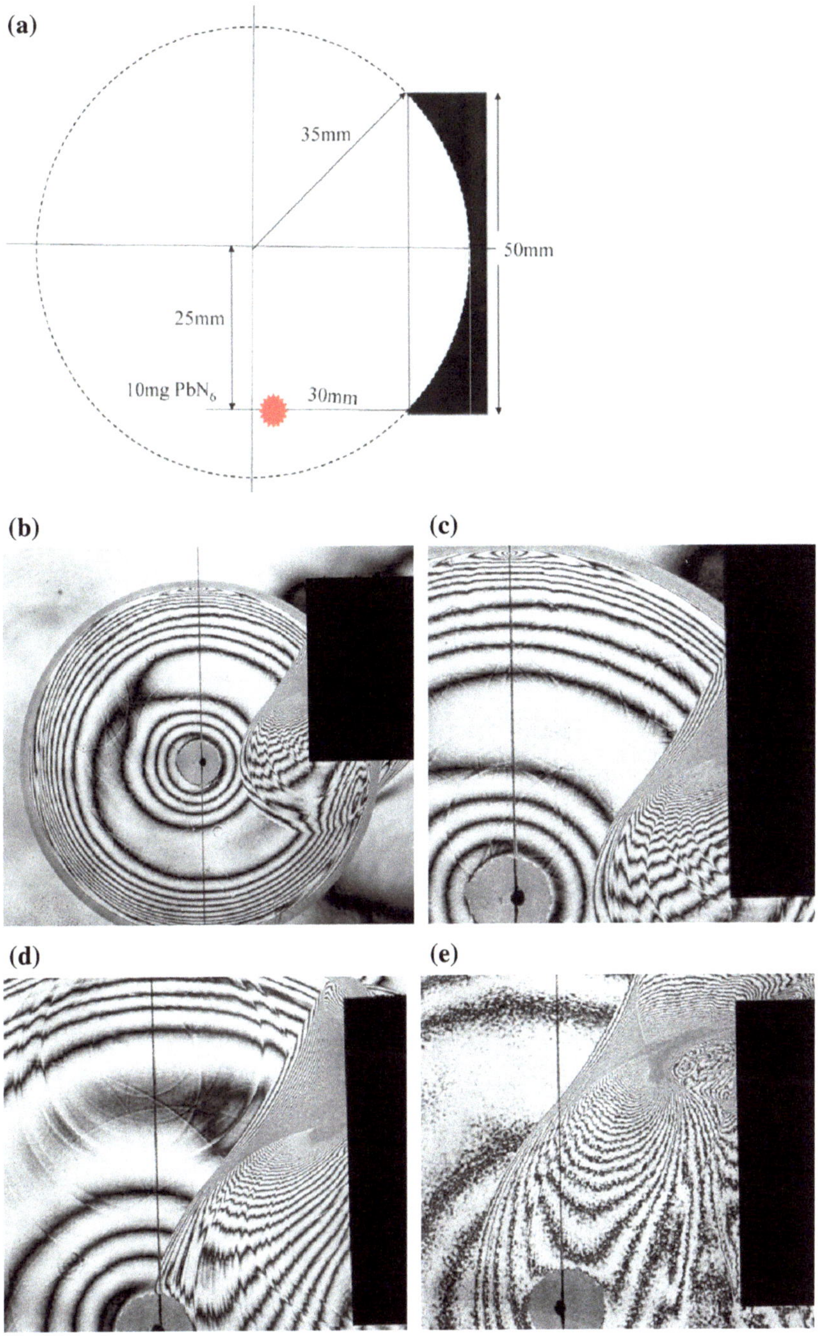

Fig. 9.19 Focusing of shock wave generated by exploding 10 mg PbN_6 pellet at 30 mm away from the 35 mm radius reflector and 25 mm aside position at 280.3 K: **a** experimental setup; **b** #84030909; **c** enlargement of (**b**); **d** #84030911; **e** #84030916

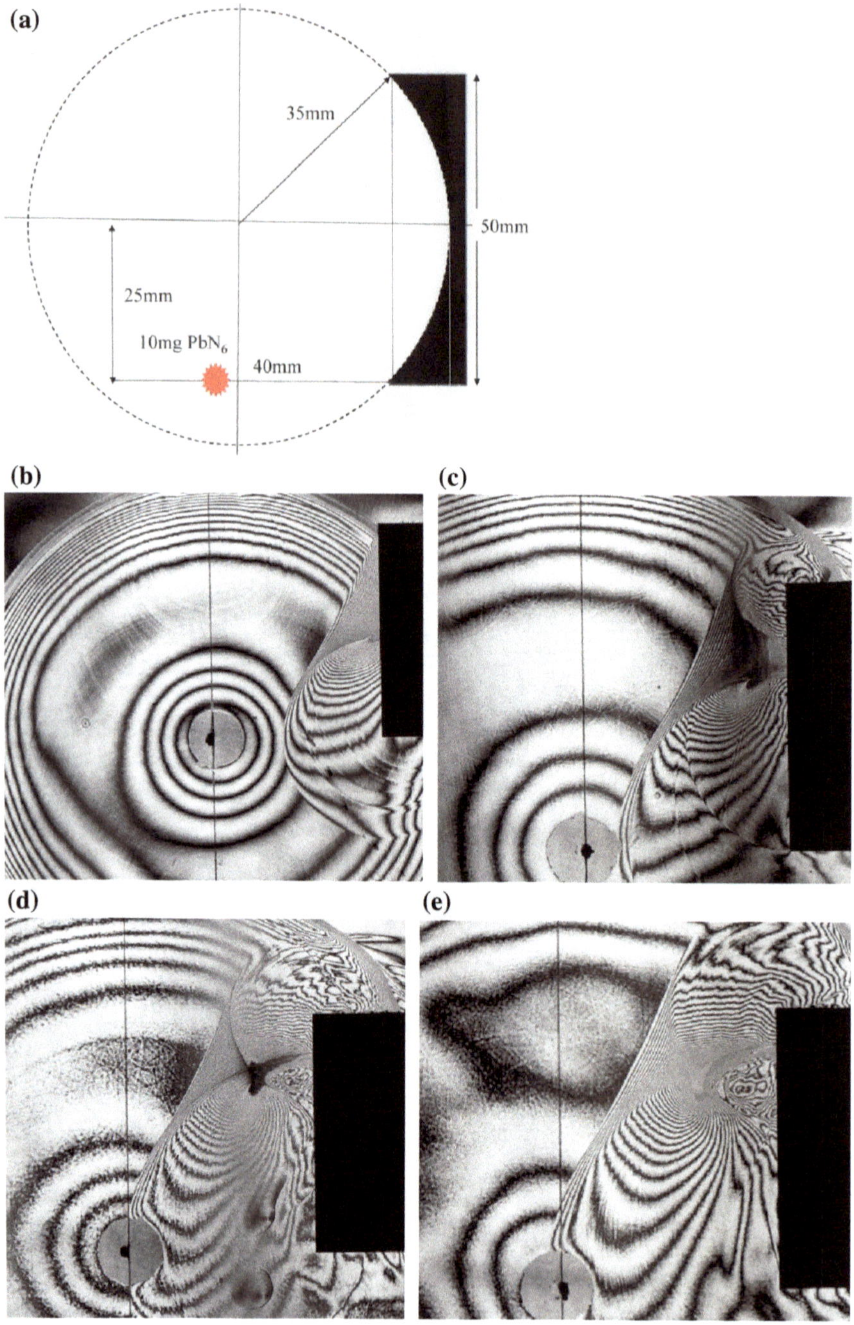

Fig. 9.20 Focusing of shock wave generated by exploding a 10 mg PbN_6 pellet at 40 mm from the reflector and 25 mm aside position: **a** experimental setup; **b** #84030902; **c** #84030907; **d** #84030908; **e** #84030905

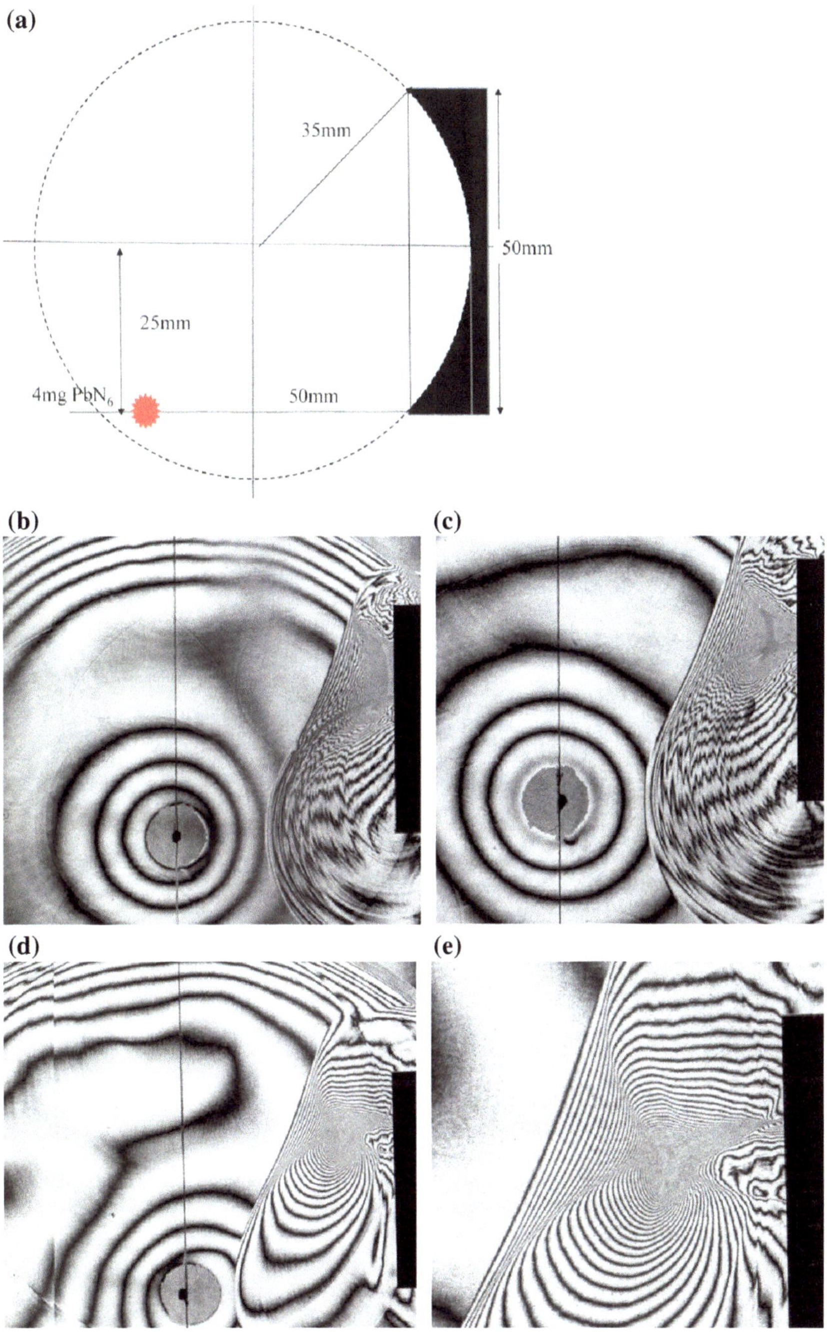

Fig. 9.21 Focusing of shock wave generated by exploding a 9 mg PbN_6 pellet at 50 mm away from reflector and 25 mm aside position, at 285 K: **a** experimental setup; **b** #8403081; **c** #84030812; **d** #84030901; **e** enlargement of (**d**)

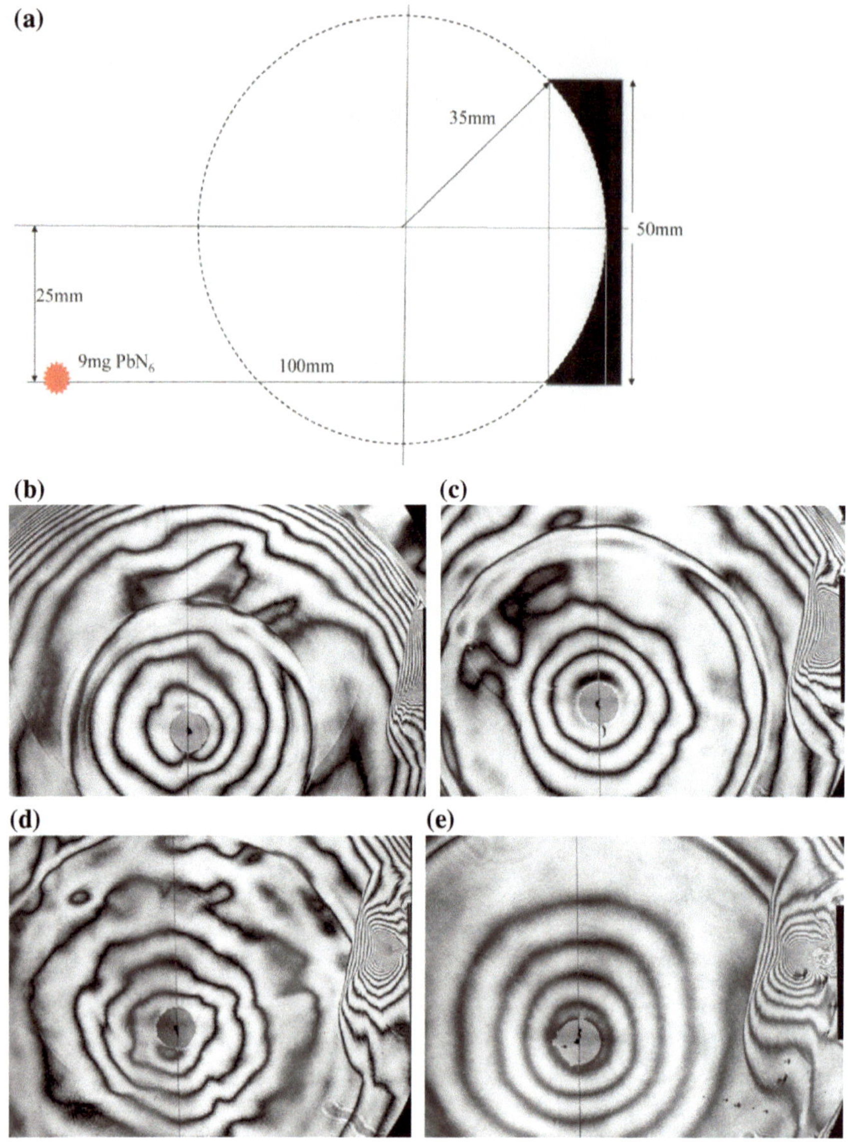

Fig. 9.22 Focusing of shock wave generated by exploding a 9 mg PbN_6 pellet at 100 mm away from the reflector and 25 mm aside position: **a** experimental setup; **b** #84031209; **c** #84031210; **d** #84031211; **e** #84031215; **f** enlargement of (**e**)

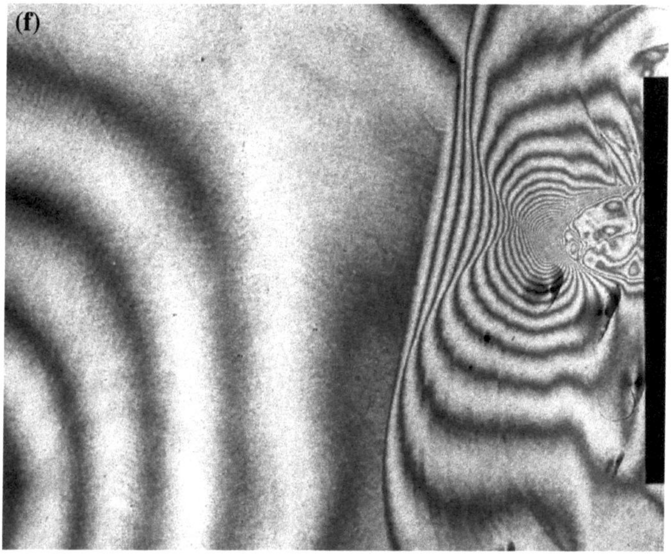

Fig. 9.22 (continued)

9.4.3 Shallow Circular Reflectors from 100 mm from the Explosion Point on the Center Line

Figure 9.23a shows the reflection of a shock wave generated by exploding a 4 mg PbN_6 pellet installed at 100 mm from the reflector. Figure 9.23b–k shows the evolution of the focusing of the reflected shock wave generated by explosion of PbN_6 4 mg from the 50 mm diameter reflector.

9.4.4 Shallow Ellipsoidal Reflector

Figure 9.24a shows a truncated 90 mm × 128 mm ellipsoidal reflector. A spherical shock wave generated by exploding a 4 mg PbN_6 pellet placed at the first focal point outside the truncated reflector is focused at the second focal point close to the reflector. Figure 9.24b–m shows the evolution of the shock wave focusing. In Fig. 9.24, fringe distributions show the whole sequence of propagation, reflection and focusing of the spherical shock wave (Esashi 1983).

As seen in Fig. 9.24, when a spherical shock wave is generated at the first focal point of the truncated 90 mm × 128 mm ellipsoidal reflector, the part of the diverging shock wave is recovered by focusing from the 50 mm diameter reflector. Only a 1.2% of the energy carried by the shock wave energy is recovered. If the shock wave would have been generated at a focal point inside a truncated

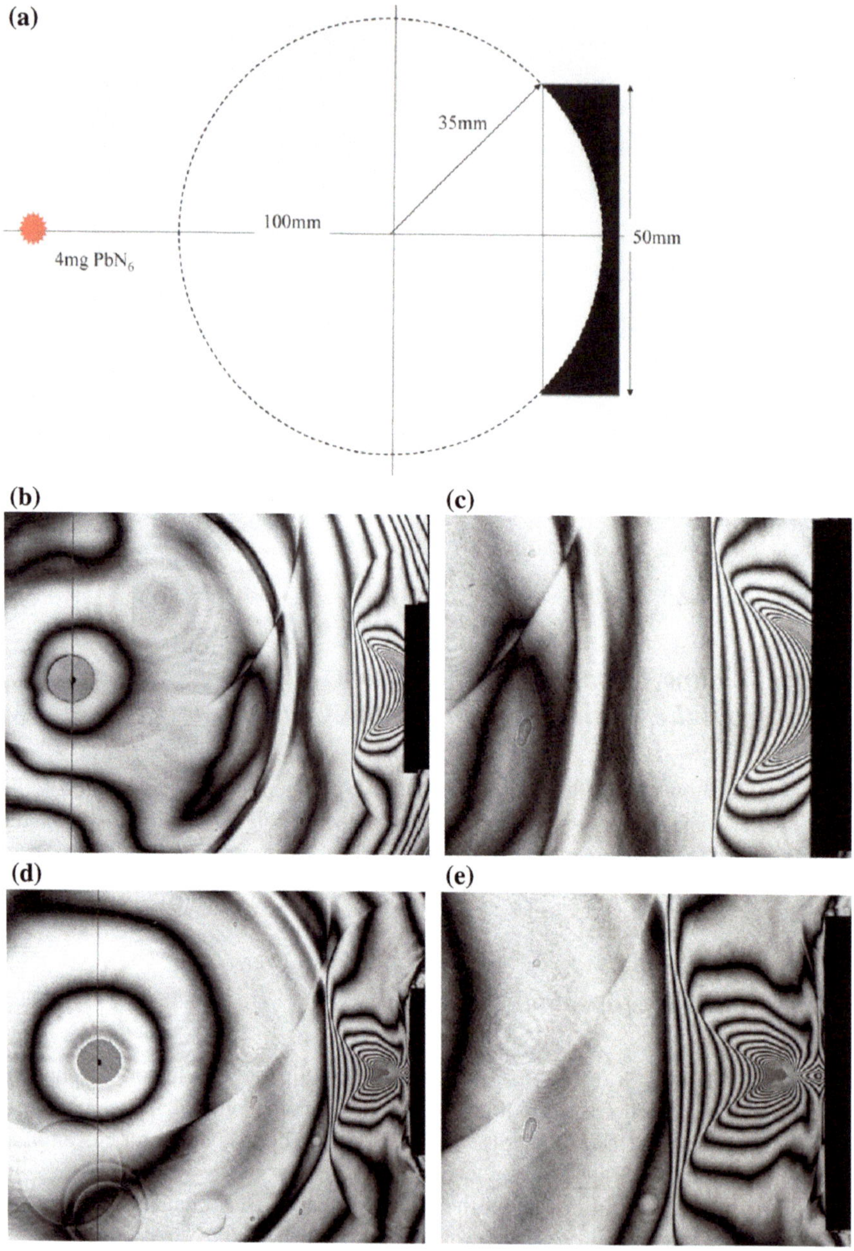

Fig. 9.23 Shallow circular reflectors from 100 mm away from point explosion on the center line, 4 mg PbN_6: **a** experimental setup; **b** #83012122; **c** enlargement of (**b**); **d** #83012103; **e** enlargement of (**d**); **f** #83012104; **g** enlargement of (**f**); **h** #83012105; **i** enlargement of (**h**); **j** #83012108

(f) (g) (h) (i) (j)

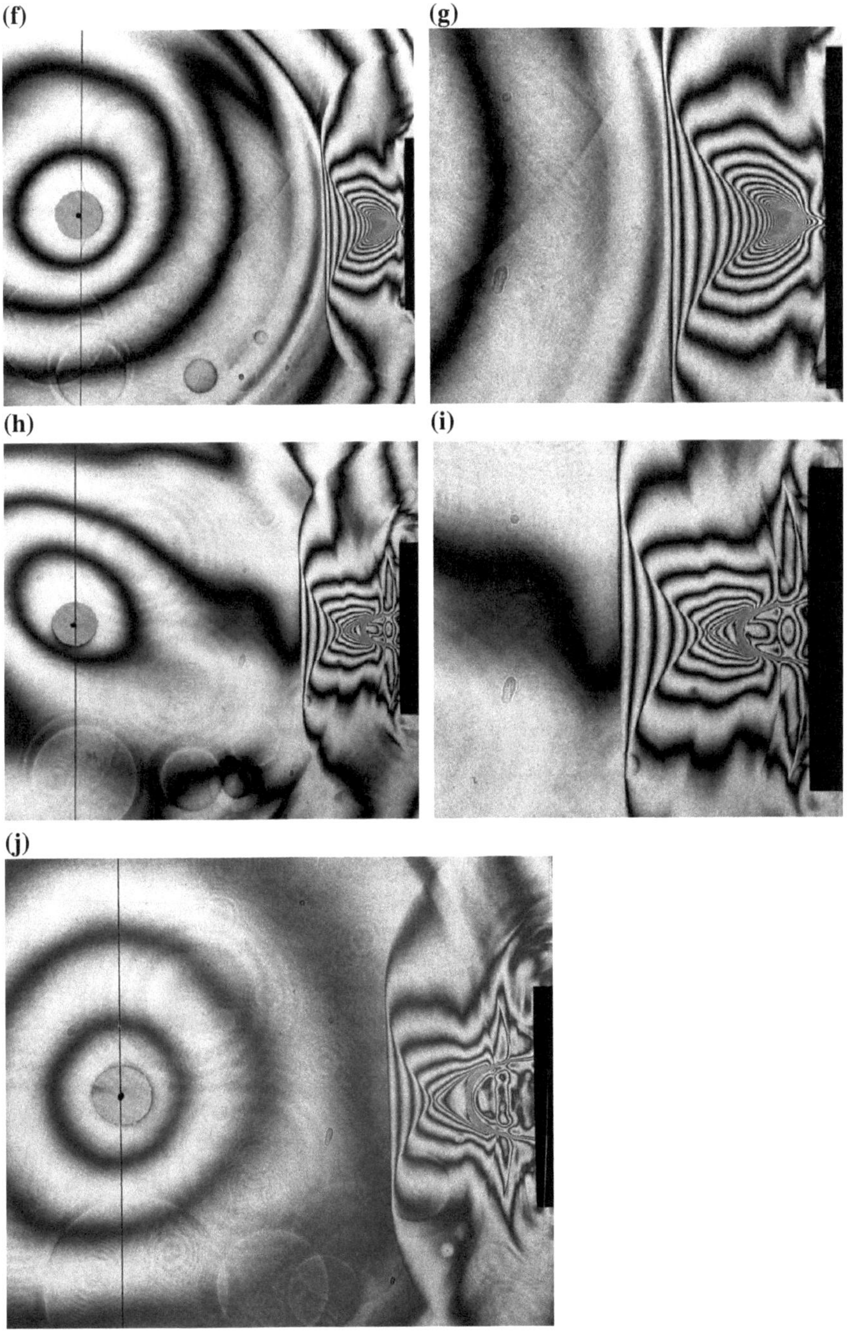

Fig. 9.23 (continued)

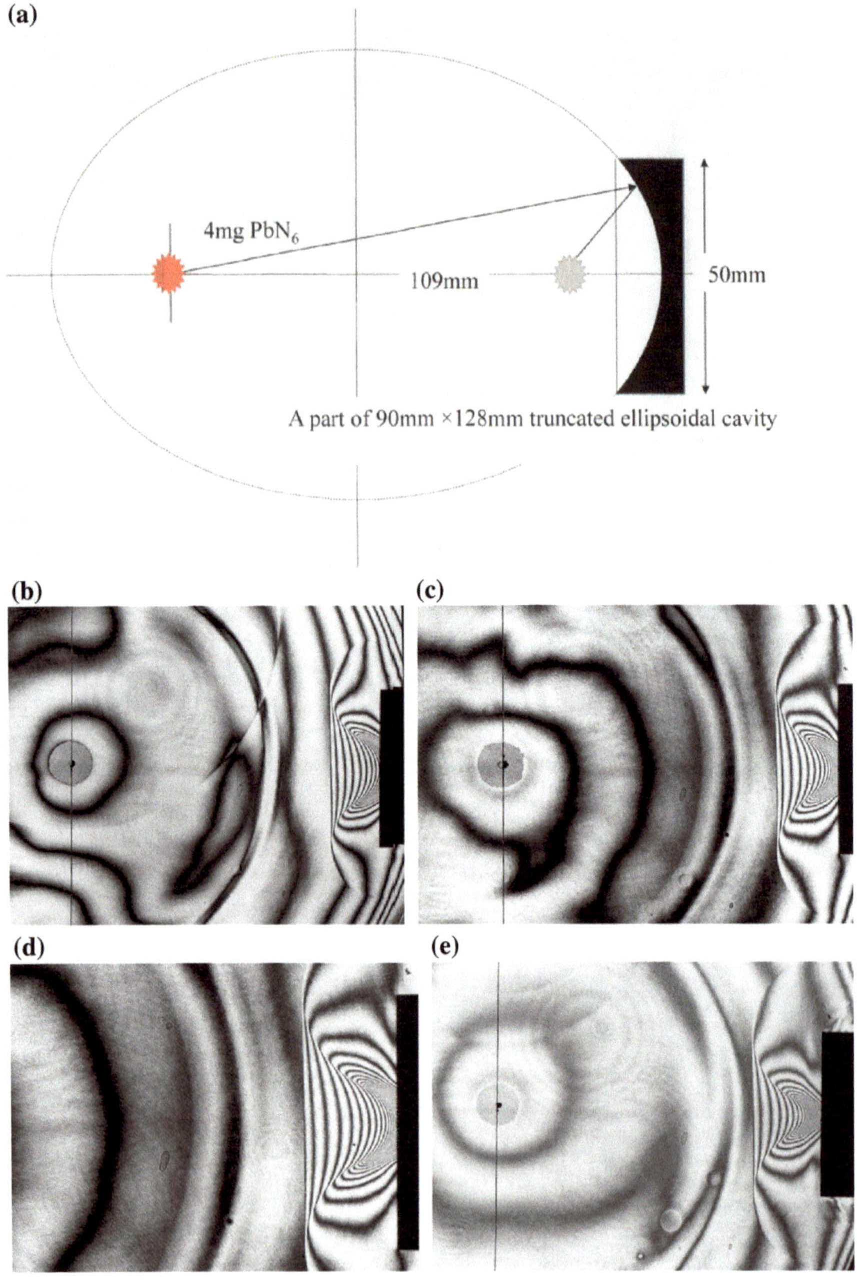

Fig. 9.24 Shock wave generation at the first focal point of a truncated 90 mm × 128 mm ellipsoidal reflector by exploding a 4 mg PbN_6 pellet and its focusing at the second focal point: **a** experimental setup; **b** #83012122; **c** #83012121; **d** enlargement of (**c**); **e** #83012120; **f** #83012103; **g** #83012104; **h** enlargement of (**g**); **i** #83012105; **j** #83012128; **k** enlargement of (**j**); **l** #83012123; **m** enlargement of (**l**)

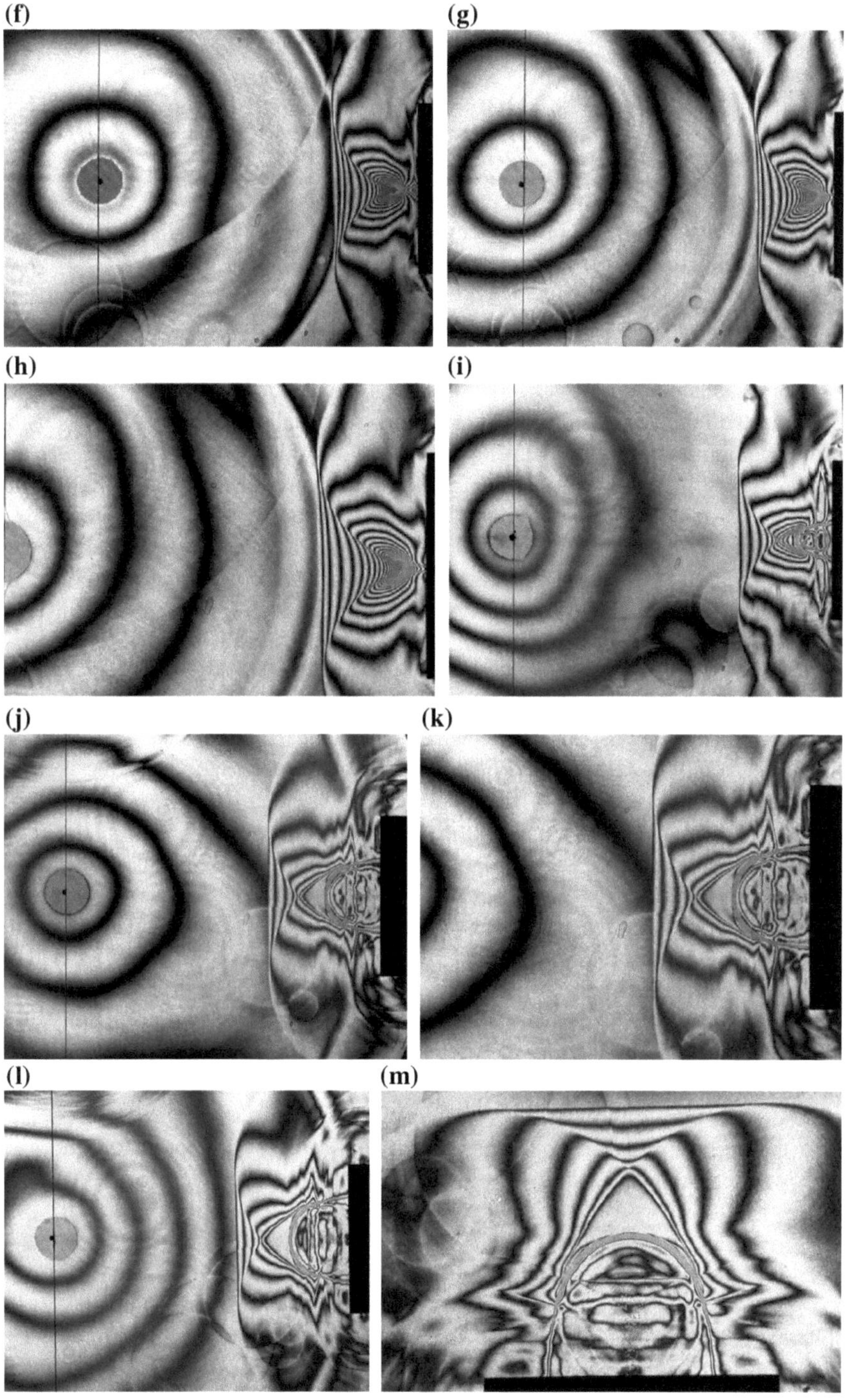

Fig. 9.24 (continued)

ellipsoidal cavity of the same geometry, the rate of energy recovered at the second focal point would have been much higher than this value (Esashi 1983).

9.4.5 Deep Ellipsoidal Reflectors

9.4.5.1 Diffuse Holographic Observation of Truncated Acrylic Reflector

When the shock wave is generated at a focal point inside a truncated 90 mm 128 mm ellipsoidal reflector, the reflected shock wave is focused at the second focal point outside the reflector. To visualize the wave motion inside the reflector, the truncated reflector was made of acryl and was visualized the wave motion using diffuse holographic interferometry, in which diffuse *OB* passed the transparent test section. The diffuse *OB* passed through the transparent reflector and carried holographic information onto the holographic film. Combining the optical arrangement with that of shadowgraph, the change in phase angles between the double exposures can be recorded as image holograms. A spherical shock wave was generated by exploding PbN_6 pellet attached on the edge of optical fiber and was inserted through a thin hole at the end wall of the reflector as shown in Fig. 9.25a.

A direct shock wave was released from the opening. The rest of the shock wave was transmitted into the acrylic reflector. For the given ratio of the acoustic impedance between the acryl and water, assuming, for the sake of simplicity one-dimensional waves, the ratio of the transmitted wave energy I_t to the incident wave energy I_i and the ratio of the reflected shock energy I_r to I_i. $I_t/I_i = 4m_1m_2/(m_1 + m_2)^2$, $I_r/I_i = (m_2 - m_1)^2/(m_1 + m_2)^2$, where m_1 and m_2 are acoustic impedance of acryl and water, respectively. However, this relation is valid only to one linear waves.

For reference, as stress waves in acryl propagate at 2.9 km/s and its density being 1.18, then the ratio m_{acryle}/m_{water} is 2.28 and hence $I_r/I_i = 0.152$ and $I_t/I_i = 0.848$. A fraction of energy is reflected into the acryl. In Fig. 9.25, a transmitted shock wave in acryl is visible as a discontinuous grey line. Compression stress waves in acryl are released into the water as grey lines. The dark irregularly shaped circle at the focal point inside the reflector is a detonation product gas. The direct wave passes outside through the opening, whereas the rest of waves is reflected from the curved wall and converges toward the second focal point. Vague grey shadows along the center axis visible in Fig. 9.25g–j are bubble clouds.

9.4.5.2 Focusing from a Truncated Ellipsoidal Cavity Made of Brass

Spherical shock waves created at a focal point inside a truncated 90 mm × 128 mm ellipsoidal reflector are focused at the second focal point outside the reflector and fringe concentration at the second focal point was observed. Figure 9.26a–l

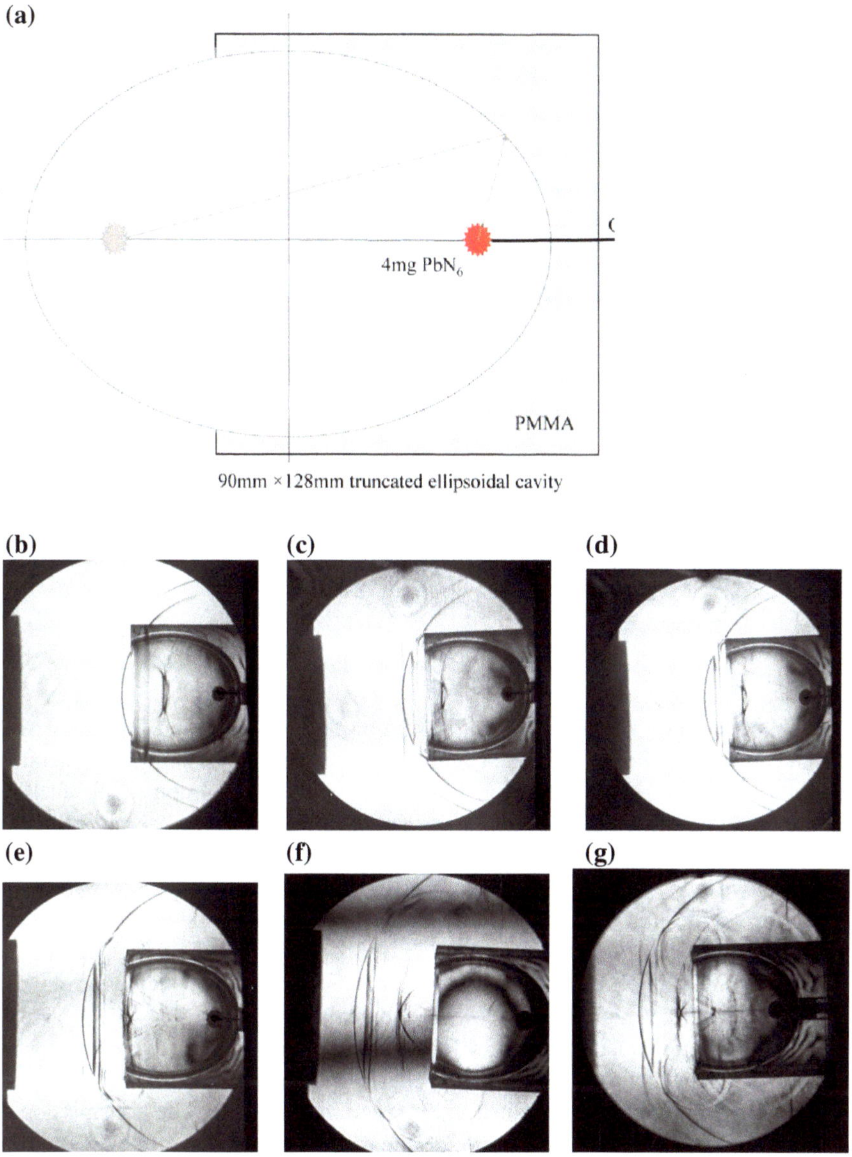

Fig. 9.25 Shock wave focusing from a deep reflector made of acryl, PbN_6 pellets weighing from 11.7 to 12.8 mg: **a** experimental setup; **b** #83070406, 55 μs from trigger point; **c** #83070405, 60 μs: **d** #83070404, 65 μs; **e** #83070503, 70 μs; **f** #83070403, 75 μs; **g** 83071317, 80 μs; **h** 83071204, 86 μs; **i** 83071314, 90 μs; **j** 83071302, 96 μs

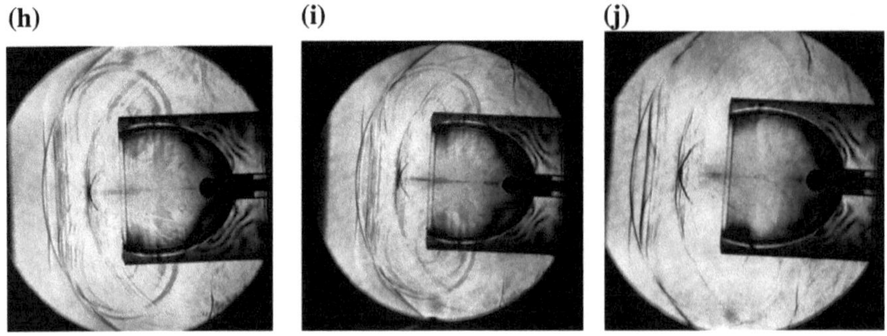

Fig. 9.25 (continued)

show sequential double exposure interferograms of shock wave focusing from a 90mm ×128mm truncated ellipsoidal reflector made of brass. When a 3.7mg to 5.6mg PbN6 pellet was ignited at the focal point inside the reflector, the direct shock wave came out first. Then the resulting reflected shock wave gradually focused at the focal point outside the reflector.

9.4.5.3 Shock Wave Focusing in a Closed Ellipsoidal Reflector

A stainless steel ellipsoidal reflector 500 mm × 700 mm and 100 mm wall thickness was manufactured (Takayama 1990). Figure 9.27a shows a schematic diagram of the ellipsoidal reflector. Figure 9.27b is a photograph of the facility. Figure 9.27c shows a PETN pellet supported by a thin optical fiber and placed at one of the focal points. The total weight of explosives including a PETN pellet and an igniter was well over 110 mg so that the reflector wall was very thick which was expected to withstand the high pressure and also to minimize the wall deformation. However, many pin holes were found densely populated on the upper part of the reflector. These pit holes were created due to cavitation erosion which were induced by disappearance of high pressure in very short time.

The PETN pellets had a cylindrical shape and had a 2 mm diameter hole as shown in Fig. 9.27c. A 10 mg AgN_3 pellet glued on the edge of 0.6 mm core diameter optical fiber was an igniter and inserted into the hole. The transmission of a Q-switch Nd:YAG laser beam through the optical fiber ignited the AgN_3 pellet and simultaneously exploded the PETN. The resulting spherical shock wave was so far generated inside the ellipsoidal cavity and focused at another focal point. The pressures at the focal point was measured with PVDF (poly vinyl denefluoride) pressure transducers manufactured in house. The pressure transducers were calibrated by comparison with Kistler pressure transducers Model 601H. Measured pressure history along the center axis recorded by the home-made transducers were compared with those obtained from the Kistler pressure gauge.

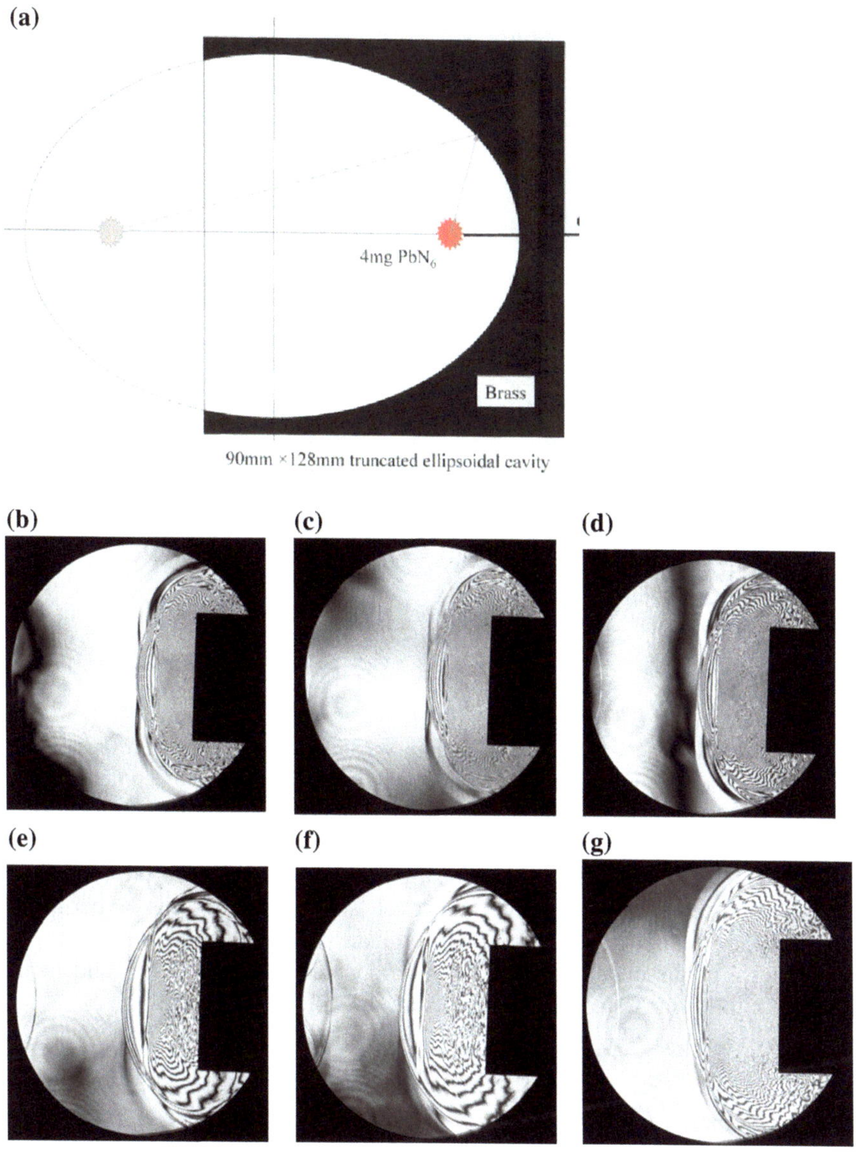

Fig. 9.26 Shock wave focusing from a truncated 90 mm × 128 mm ellipsoidal reflector made of brass. Shock waves were generated by exploding PbN_6 pellet weighing from 3.7 to 5.6 mg: **a** #83090710, 65 μs; **b** #83090709, 70 μs; **c** #83090708, 75 μs; **d** #83090707, 80 μs; **e** #83090706, 86 μs; **f** #83090705, 88 μs; **g** #83090704, 90 μs; **h** #83090703, 92 μs; **i** #83090702, 95 μs; **j** #83090701, 100 μs; **k** #83090607, 105 μs

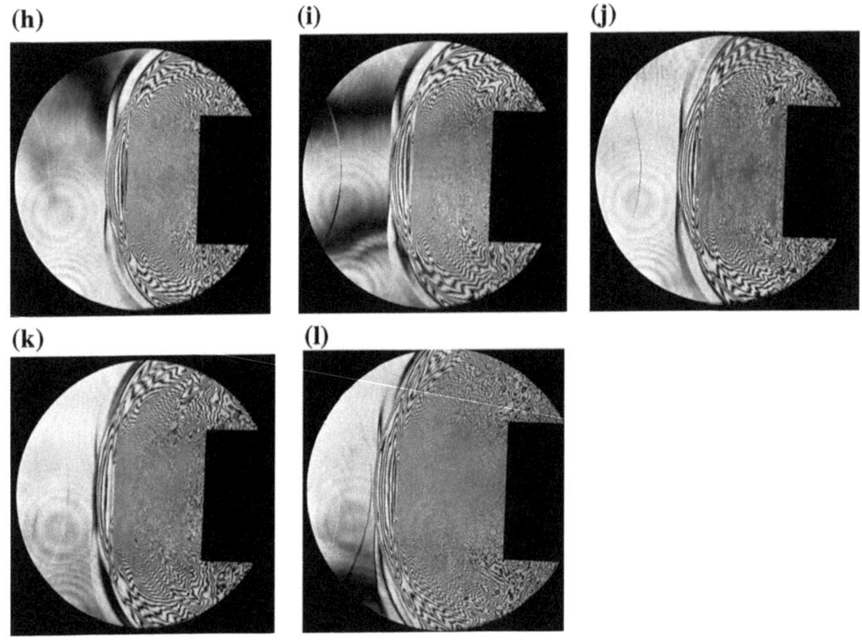

Fig. 9.26 (continued)

Three PETN pellets were tested. Figure 9.28a shows the peak pressure distribution in the neighborhood of the second focal point. The ordinate denotes the peak pressure in MPa and the abscissa denotes the distance from the second focal point in mm. A solid line denotes the result of the TVD numerical simulation (Shitamori 1990). Peak pressures at the second focal point and the measured pressure distributions agree well between the numerical simulation and the experiments. Figure 9.28b summarized the results. The ordinate denotes the peak pressures at the second focal points in MPa and the abscissa denotes weight of PETN and 10 mg AgN_3 in mg. So far tested the peak pressure of about 800 MPa.

9.4.6 Focusing of Underwater Shock Wave Generated by a Pulse Laser Beam

Figure 9.29a shows a schematic diagram of irradiation of a diverging Q-switched laser beam of 1 J intensity of 25 ns pulse width on a 60 mm diameter truncated 90 mm × 128 mm ellipsoidal dish shaped reflector. When the diverging laser beam irradiated the surface of the reflector, a hemi-spherical weak shock wave was created from the concave reflector surface. The shock wave created from the curved surface and propagated toward the focal point as shown in Fig. 9.28b. With

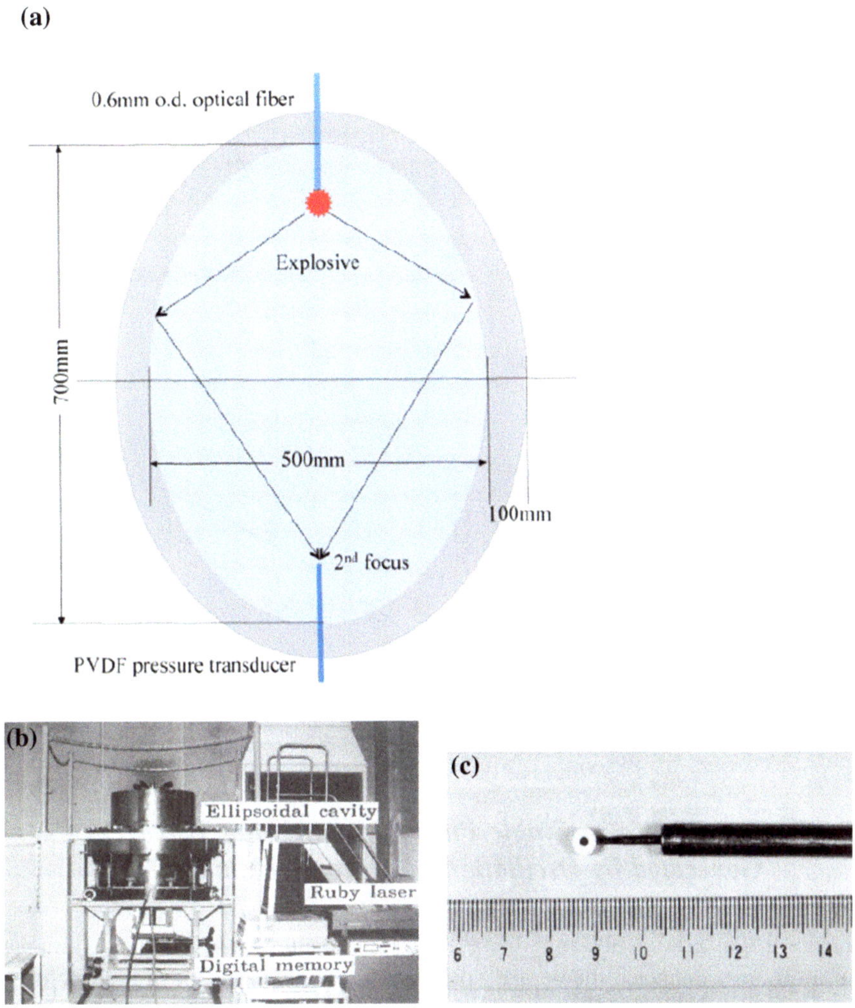

Fig. 9.27 Underwater shock wave focusing from a complete ellipsoidal reflector: **a** principle of underwater shock wave focusing; **b** facility; **c** explosive, PETN triggered by 10 mg AgN_3 igniter

elapsing time, the shock wave created from the curved wall followed the hemi-spherical shock wave and diffracted at the edge of truncated reflector. Figure 9.29b–k shows the evolution of the spherical shock wave propagation and focusing toward the focal point of the reflector. The laser induced shock wave was so weak that the cavitation bubbles were not created during the focusing.

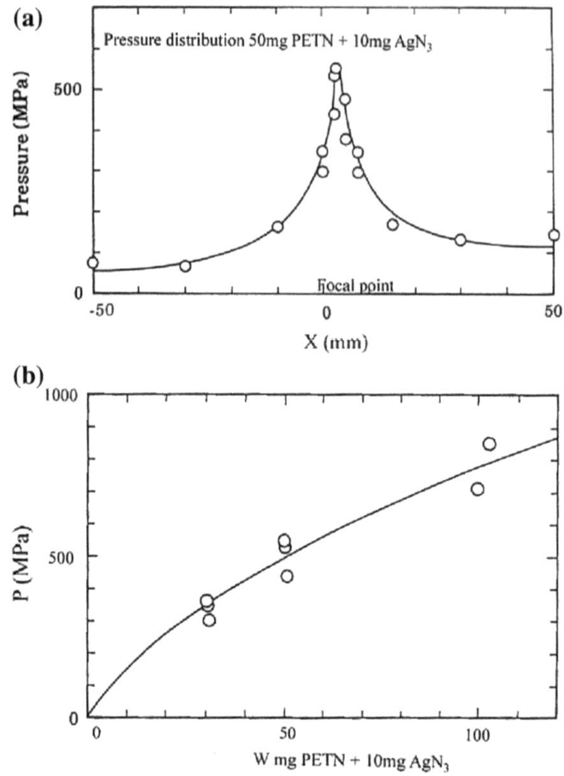

Fig. 9.28 Shock wave focusing in ellipsoidal cavity: **a** peak overpressure along the axis; **b** peak overpressure against mass of PETN pellet and 10 mg AgN_3

9.4.7 Focusing of a Single Pulse Strong Sound Wave Generated by Oscillation of a Piezo-ceramics

Figure 9.30a shows a 300 mm diameter barium titanium dish. The dish was a part of a 600 mm diameter sphere and composed of 24 segments, each of which was driven at maximum amplitude of 30 μm operated in precisely synchronized fashion in water. With properly tuning the electric impedance of the power supply circuit, the dish was oscillated only once and the second and third oscillations were damped very quickly. Hence, the electric energy was converted to drive the dish segments with higher efficiency (Okazaki 1989).

The impulsive motion of 24 segments of the ceramic dishes generated concavely shaped intense sound waves converging toward the center of the spherical dish as illustrated in Fig. 9.30a. The magnitude of acceleration was in the order of magnitude of $10^3 g$ so that a train of compression waves was generated. Figure 9.30b shows the train of compression waves converting to a shock wave in approaching close to the center. Then, the compression wave eventually turned into a shock wave that is intense enough to disintegrate kidney stones as observed in Fig. 9.30e. At the same time expansion waves generated from the edge of the ring shaped dish

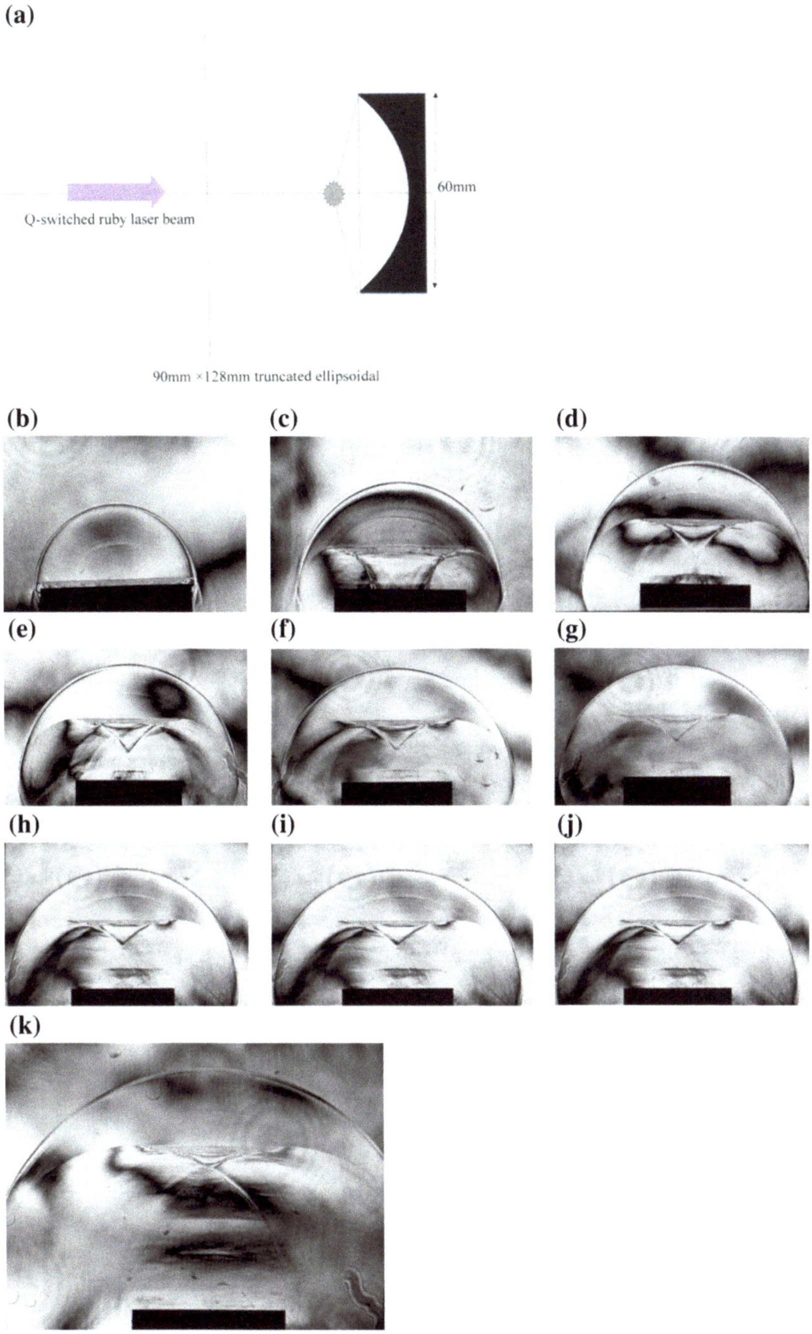

Fig. 9.29 Focusing of shock wave created by irradiation of a Q-switched laser beam of 1 J onto a truncated ellipsoidal reflector: **a** experimental arrangement: **b** #84030712, 15 μs elapsed time from trigger point; **c** #84030711, 25 μs; **d** #84030710, 35 μs; **e** #84030719, 36 μs; **f** #84030716, 37 μs; **g** #84030715, 37 μs; **h** #84030717, 40 μs; **i** #84030714, 41 μs; **j** #84030713, 43 μs; **k** #84030708, 55 μs

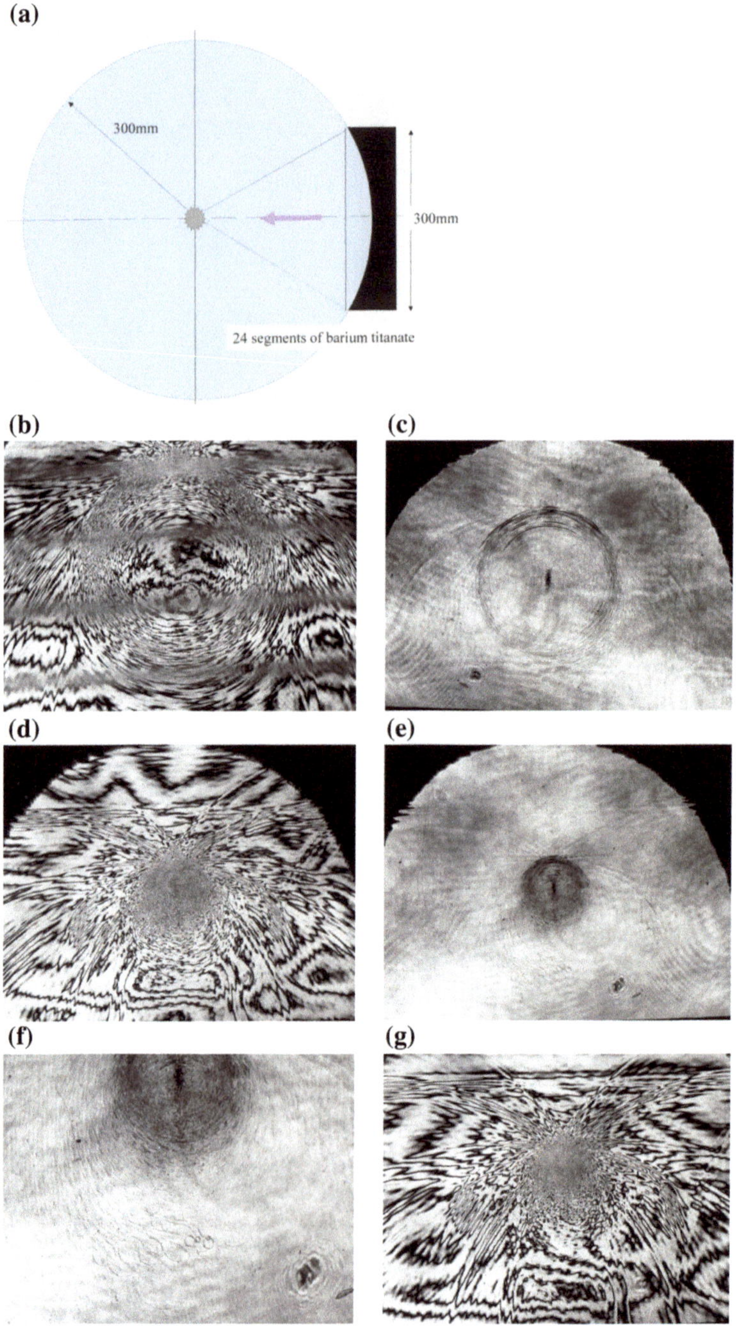

Fig. 9.30 Focusing of ultrasound waves: **a** experimental setup; **b** #88012205, 331 μs from trigger point; **c** #88012211, 331 μs; **d** #88012204, 351 μs; **e** #88012210, 351 μs; **f** enlargement of (**g**); #88012201, 358 μs; **h** enlargement; **i** #88012216, 358 μs; **j** #88012215, 358 μs; **k** #88012214, 358 μs; **l** enlargement of (**k**); **m** #88012206, 358 μs; **n** #88012202, 365 μs; **o** #88012203, 385 μs; **p** enlargement of (**m**); **q** enlargement of (**o**)

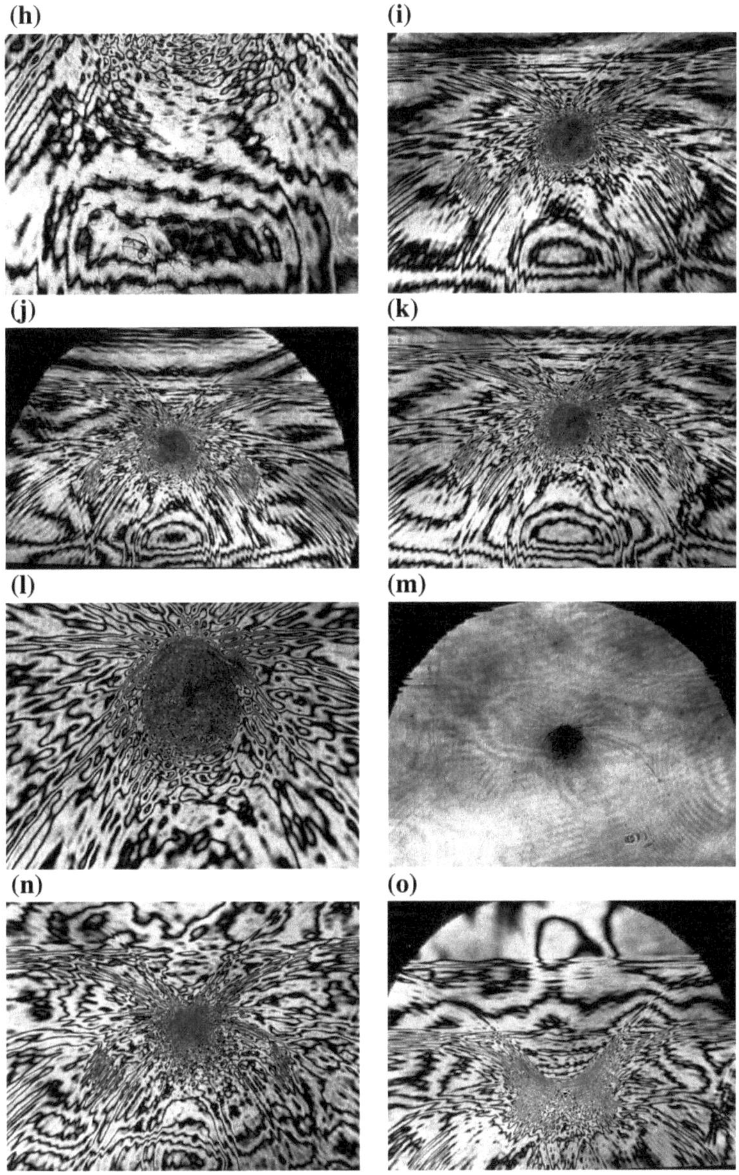

Fig. 9.30 (continued)

decreased the high pressures. The interferometric image clearly visualizes a high pressure spike appearing at the focal point, which would effectively crack kidney stone.

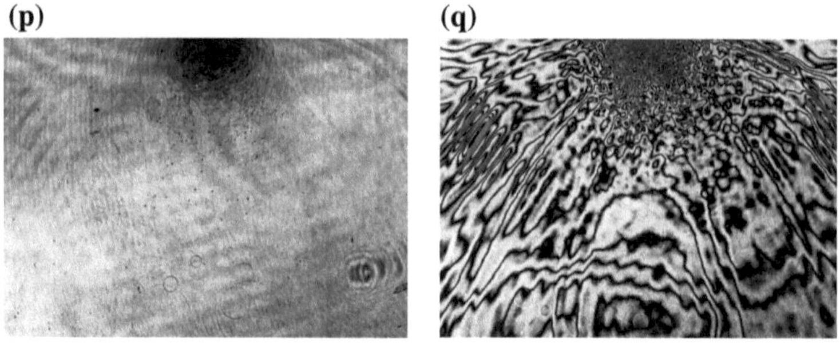

Fig. 9.30 (continued)

A train of compression waves driven by 24 segments eventually formed dense fringe accumulation at the center of the dish. Figure 9.30 shows sequential observations of a single oscillation of 2.7 kV input energy at the trigger time from about 330 to 385 μs. Figure 9.30c, e shows single exposure interfrograms. Fringe concentrations show localized pressure enhancements at focal area. The peak pressure so far measured were well over 100 MPa. When the expansion waves caught up the high pressures, a very sudden disappearance of high pressures induced a high tensile force, which spontaneously exceeded well over the so-called spalling pressure in water. Then, cavitation bubbles were generated and the bubbles immediately collapse generating secondary bubbles as seen in Fig. 9.30e, g. Figure 9.30f, h shows their enlargement. Figure 9.30p, q are enlargements of Fig. 9.30m, o, respectively. Figure 9.30b, c are double exposure interferograms and single exposure interferograms taken at the same delay time of 331 μs. Figure 9.30d, e are pairs taken at 351 μs. Figure 9.30k, l are pairs taken at 358 μs. In single exposure interferograms, the cavitation bubbles and secondary shock waves were observed, whereas in double exposure interferograms the density variations were readily estimated.

9.5 Underwater Shock Wave/Bubble Interaction

The interaction of underwater shock waves with bubbles is a fundamental research topic not only in the shock wave dynamics but also in the bubble dynamics (Shima 1997). The shock wave/bubble interaction is one of the mechanisms closely related to tissue damages occurring during the extracorporeal shock wave lithotripsy, ESWL.

9.5.1 A Single Spherical Air Bubble

In the hydrodynamic research, ot is known that cavitation phenomena occur not only in high-speed water flows but also are initiated by a sudden pressure decrease.

The so-called cavitation erosion was one of the important topics in the research of hydraulic machineries (Shima 1997). Tissue damages occurring during the ESWL treatments are generated by shock wave/bubble interactions (Chaussey et al. 1982; Kuwahara et al. 1986). Then an investigation of observing the shock wave interaction with a single bubble became a basic research topic.

A bubble placed in a quiescent water starts to oscillate exposed to a pressure fluctuation. Then such a symmetric bubble motion was summarized by Shima (1997). However, a bubble impinged by an underwater shock wave, unlike a symmetric bubble motion. Induced its complex deformation and eventually a micro-jet is generated.

If defining u the speed of the micro jet, a the sound speed in water, and ρ the water density, the resulting stagnation pressure p_{st} can be written as $p_{st} = a\rho u$. In particular, the diameter of the bubbles produced during the ESWL treatments is sub-millimeter diameter. If the jet speed is assumed to be 100 m/s, then this jet would create a stagnation pressure over 150 MPa, which would readily penetrate any tissues.

A slurry explosive comprises of thin shell glass balloons containing air mixed evenly in dissolved ammonium nitrate grains in kerosene. The glass balloons had diameters ranging a few 100 μm diameter containing air. When a detonation wave propagated in the slurry explosive and collapsed the thin shell glass balloons just like what observed during shock wave/air bubble interaction, the air in glass balloons had minimal volume, then the gas temperature enhanced high at a singular spot inside the bubble. The successively created high temperature maintained the detonation. The hot spots are not necessarily produced due to adiabatic compressions of contracting air bubble but created by the shock wave focusing from a concavely shaped air/liquid boundary. The shock wave/bubble interaction sustained the propagation of the detonation in slurry explosives. Shima (1997) assumed hot spot generations during due to adiabatic compression. However, it is the shock compression that enhanced the temperature much more efficiently than the adiabatic compression.

In order to sequentially observe a shock wave/bubble interaction, an analogue experiment was conducted. A 1.7 mm diameter air bubble interacted with a shock wave generated at 20 mm stand-off distance by exploding a 4 mg PbN_6 pellet attached on the edge of a 0.6 mm diameter optical fiber. The transmitted a Q-switched Nd:YAG laser beam thorough the optical fiber exploded the pellet. The experiment was conducted in a 500 mm × 500 mm × 500 mm stainless steel chamber. The air bubble was released from its bottom. The volume of the air bubble was determined by measuring from the length of air filled in a 0.3 mm diameter capillary tube prior to releasing it in the test chamber. A When the air bubble arrived at the 20 mm stand-off distance from the position of the explosive. The overpressure of the shock wave generated at the 20 mm stand-off distance was approximately 25 MPa.

The mechanism of the jet formation inside a contracting bubble is related to the generation of the so-called Neumann jet. When a cylindrical explosive was ignited from its flat surface, a plane transmitted detonation wave propagated through the explosive, for example, as seen in Fig. 8.2. If another surface had a reversely conical cavity, the transmitted detonation wave was diffracted from the concave

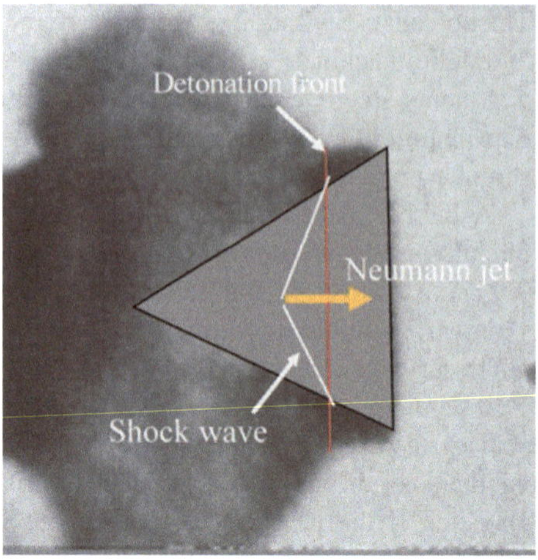

Fig. 9.31 The explanation of the Neumann effect

cavity wall and focused at the center axis creating a jet moving to the direction of the detonation wave propagation. At the same time, the detonation product gas at the top of the conical cavity was ejected to the same direction. Figure 9.31 shows the enlargement of exploding a 10 mg AgN_3 pellet shown in Fig. 8.2. A detonation wave propagated inside a cylindrical explosive shown in a red line and was diffracted at the concave wall. It focused at the center axis and the jet so far combined there was shown in a thick yellow arrow. The jet formation is called the Neumann effect which has an identical physical background as micro-jet formation inside the contracting air bubble. The Neumann effect is one of the representations of shock wave interactions at a fast/slow interface and its physical background is analogous to the jet formation occurring at the bubble collapse.

In Fig. 9.32, the evolution of shock wave interaction with a single bubble are shown. Visualizations were conducted by double exposure and single exposure holographic interferograms. In the double exposure interferograms, the first exposure recorded undisturbed bubble, the second exposure recorded the bubble deformed by the shock wave loading, and these two images were superimposed on one film. The double exposures were conducted at about 1 ms interval hence the bubble looked slightly blurred, whereas in a single exposure interferogram, the bubble shape was clearly observed.

When a shock wave hit a bubble, it started to contract and simultaneously the expansion wave was reflected from its surface as seen in Fig. 9.32b. The time allocated to individual double exposure interferograms indicates elapsed time from the explosion. The bubble continued to contract. When it reached a minimal volume, the pressure at its frontal stagnation point became maximal. The densely concentrated fringes at the frontal stagnation point show the pressure enhancement. At the next moment, when the high pressures were released into water, trains of compression waves propagated and coalesced into a secondary shock wave as shown in Fig. 9.32c–g.

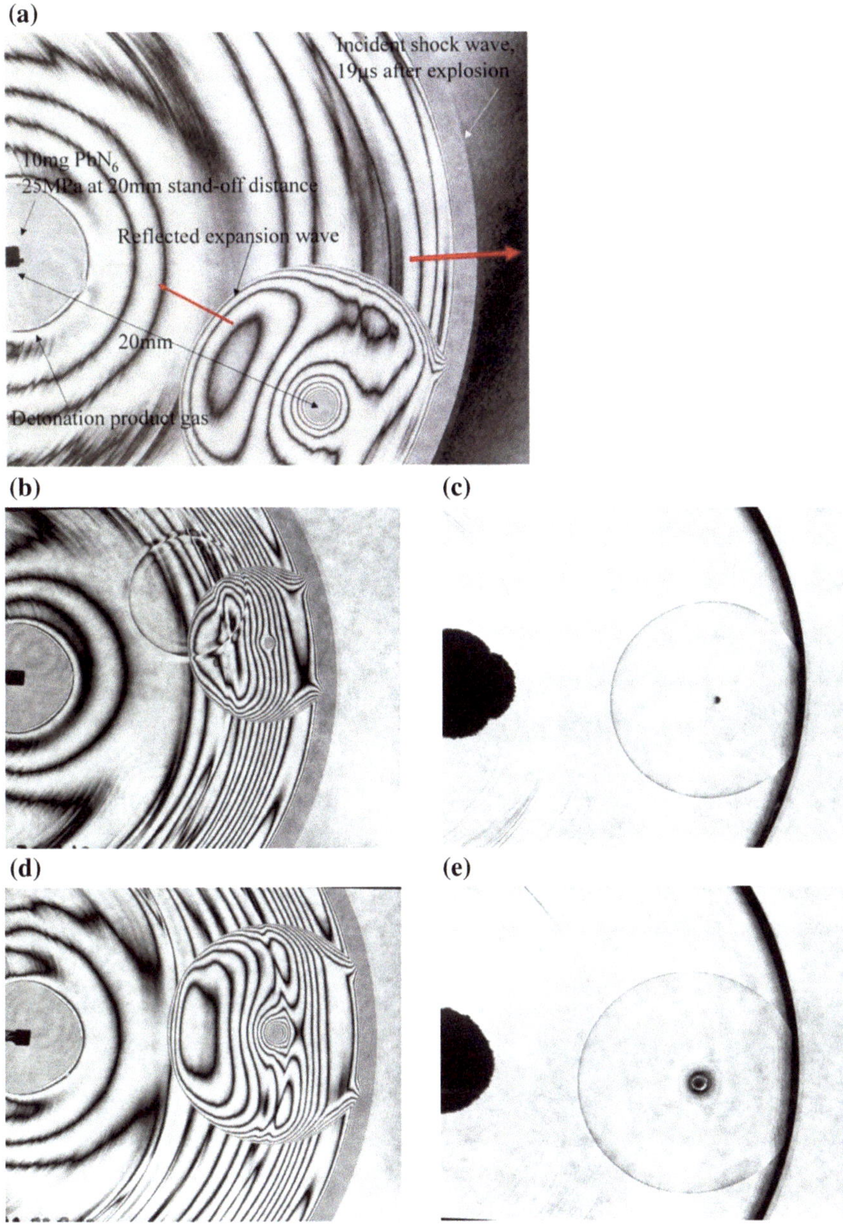

Fig. 9.32 Interaction of a shock wave with a single bubble of 1.7 mm diameter positioned at the 20 mm stand-off distance from the position of a 10 mg PbN_6 pellet at 296 K: **a** #86082608, 19.0 μs after shock wave impingement. The bubble rising speed was v = 0.187 m/s; **b** #86082507, 18.4 μs, v = 0.165 m/s; **c** #86082610, 23 μs, single exposure; **d** #86082506, 20.1 μs, v = 0.184 m/s; **e** #86082609, 25 μs, single exposure; **f** #86082505, 22.1 μs, v = 0.164 m/s; **g** #86082615, 25 μs, single exposure; **h** #86082504, 23.2 μs, v = 0.224 m/s; **i** #86082613, 26 μs, single exposure; **j** #86082606, 23.5 μs, v = 0.187 m/s (Ikeda et al. 1999)

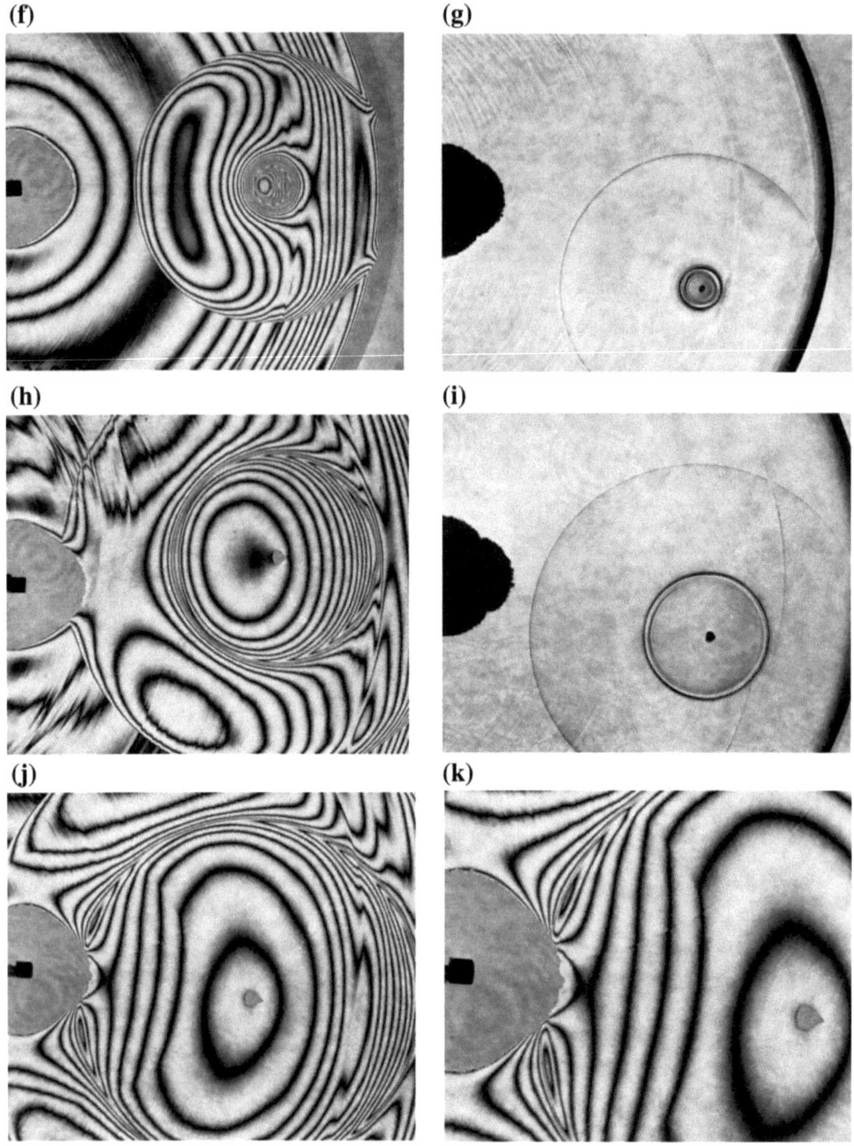

Fig. 9.32 (continued)

Following to the analogous mechanism to the Neumann effect as already explained in Fig. 9.31, a jet was formed inside the contracting bubble. Its formation depends on parameters such as the bubble size which is expressed in terms of the ratio of the deviated bubble radius to the perfectly spherical one and the ratio of sound speeds in the gas bubble to the liquid. Figure 9.32j, k show a later stage of the secondary shock wave impingement on the detonation product gas bubble.

The bubble surface showed an irregular shape because the interfacial instability occurred on the shock impinged bubble surface.

As the density gradient across the expansion wave is opposite to that across the shock wave in the double exposure interferograms, the direction of fringe shift across the expansion wave is different from that across the shock wave. Assuming that the fringe distributions in Fig. 9.32 are axial symmetry, the density distribution behind shock waves and expansion waves are readily determined by evaluating fringe distributions. Assuming isentropic flows, the pressure distribution is readily obtained from the density distribution (Abe 1989). Figure 9.33a–d are the pressure profiles along the distance from the center of explosion evaluated from the images presented in Fig. 9.32b, d, f, h, each of which was taken at 18.4, 20.1, 22.1, and 23.2 μs from the ignition, respectively. The ordinate denotes the pressure in bar and the abscissa denotes the distance normalized by the stand-off distance of 20 mm. Figure 9.33a shows the presence of a reflected expansion wave. Open circles denote experimental results estimated from fringe distributions. Figure 9.33b shows the generation of the peak pressure analogous to the pressure profile occurring at a point explosion. In Fig. 9.33c, d, a train of compression waves coalesced forming a secondary shock wave. The propagation of and the attenuation of the secondary shock wave are clearly observed in Fig. 9.33b–d.

When a bubble placed in a quiescent liquid responds to a pressure fluctuation caused by wavelets passing over the liquid surface and starts to symmetrically oscillate. Eventually, when it reached to a minimal volume, due to an adiabatic compression, the temperature inside the bubble reached maximal. Then the bubble emitted luminescence. This is the so-called sono-luminescence.

Another extreme case is that when a shock wave hit the bubble, the bubble responded sharply to the shock wave and started deforming asymmetrically. The degree of its deformation and the direction of the motion were governed by the shock wave strength. Regarding the wave motion inside the deforming bubble, a little is known. A numerical simulation would be possible approach to resolve asymmetrical wave motion inside contracting bubbles.

In the adiabatic compression, the dimension-less temperature T/T_0 normalized by the ambient temperature T_0 and dimension-less pressure p/p_0 normalized by the ambient pressure p_0 are related to,

$$T/T_0 = (p/p_0)^{2/7} \quad \text{for}\ \gamma = 1.4, \tag{9.10}$$

whereas in the shock compression, the temperature increases proportional to the pressure ratio as following,

$$T/T_0 \propto p/6p_0 \quad \text{for}\ p/p_0 \gg 1. \tag{9.11}$$

For example, in order to achieve $T/T_0 = 17$ by the adiabatic compression, p/p_0 can be 19,000, whereas in the shock compression, p/p_0 is only about 100, which can be obtained behind a shock wave of Ms = 8.3 in ideal diatomic gases.

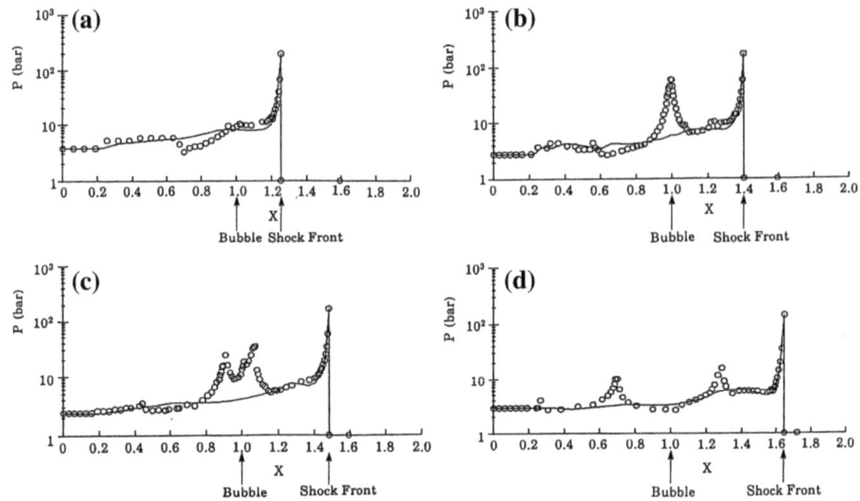

Fig. 9.33 Pressure distributions estimated from Fig. 9.32b, d, f, h: **a** 18.4 μs from the ignition; **b** 20.1 μs; **c** 22.1 μs; **d** 23.2 μs (Abe 1989)

In Fig. 9.32, the optical arrangement is based on, in principle, a direct shadowgraph. Hence holographic films are facing to the test section. Then although the luminous emission from the hot spot could be readily recorded, it was not recorded at all because the present holo-films were insensitive to the wavelength of the luminosity emission.

9.5.2 A Single Non-spherical Air Bubble in Water

Figure 9.34 shows sequential shadowgraphs of the shock wave interacting with a disk shaped air bubble having its aspect ratio of 1.5 mm × 2.5 mm. The images are recorded by ImaCon D-200. 14 images are recording for time interval ranging from 9.9 to 22.9 μs at 1 μs interval and the exposure time of 10 ns. The pressure is measured on the end wall by an optical fiber pressure transducer at the stand-off distance of 2 mm from the center of the air bubble. The shock wave is generated by exploding a 10 mg AgN_3 pellet at the 20 mm stand-off distance from the center of the bubble. The overpressure measured at 20 mm stand-off distance is about 25 MPa. The bubble starts to contract upon the shock wave loading. Although the reflected expansion wave is not distinctly visible but the secondary shock wave is observed from the moment when the bubble reached a minimal volume as seen in Fig. 9.34j, k. A sign of luminous emission was observed in Fig. 9.34i. The emission became brighter in Fig. 9.34j and maintained for about 1 μs. Figure 9.34j shows that the brightest point is slightly shifted toward the upstream, which implies that the luminous emission is caused due to wave motion occurring inside the

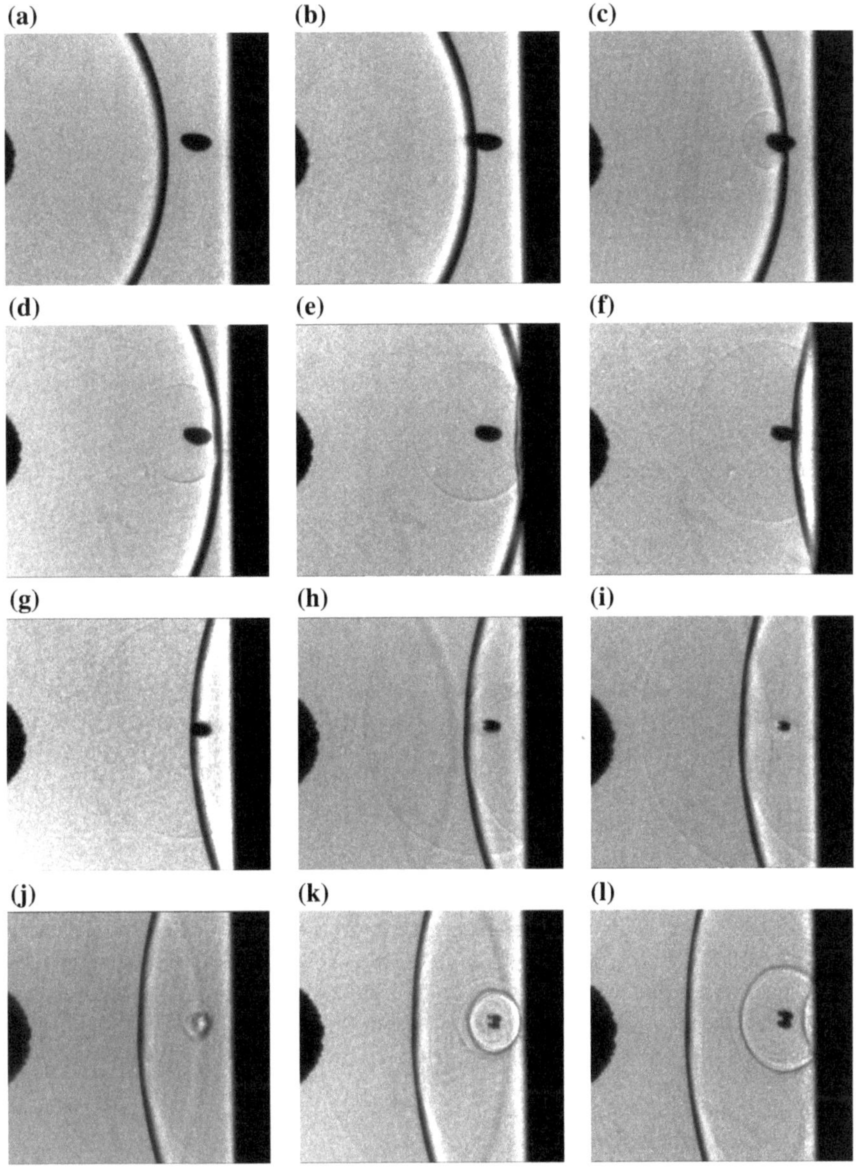

Fig. 9.34 Sequential observation of interaction of a non-spherical air bubble with an underwater shock wave generated by exploding a 10 mg AgN_3 pellet at stand-off distance of 20 mm: **a** 9.9 μs; **b** 10.9 μs; **c** 11.9 μs; **d** 12.9 μs; **e** 13.9 μs; **f** 14.9 μs; **g** 15.9 μs; **h** 16.9 μs; **i** 17.9 μs; **j** 18.9 μs; **k** 19.9 μs; **l** 20.9 μs; **m** 21.9 μs; **n** 22.9 μs; **o** time variation of pressure measured on the end wall (Courtesy of Dr. Ohtani)

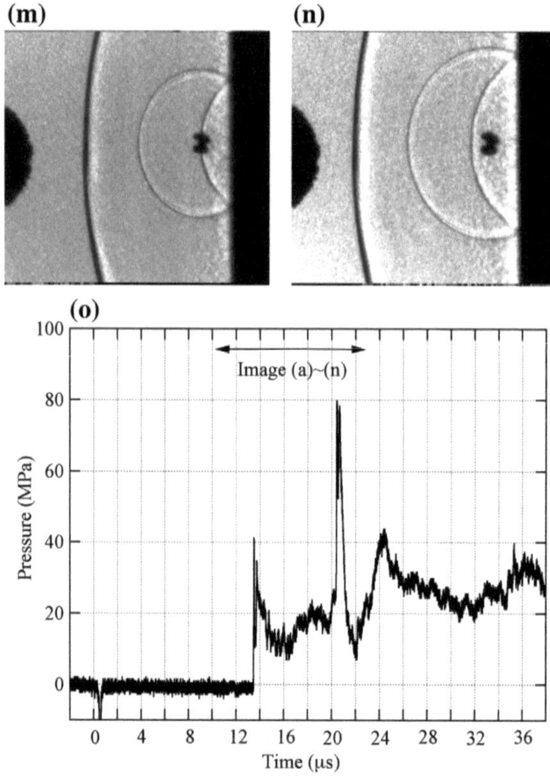

Fig. 9.34 (continued)

bubble. Figure 9.34o shows the pressure variation as a summary of the sequential visualization. The ordinate denotes pressure in MPa and the abscissa denotes elapsed time of shadow pictures in μs. At 13.5 μs shown in Fig. 9.34e, the incident shock wave arrived at the wall. At about 20.5 μs, the secondary shock wave arrived at the wall and its peak pressure reached about 80 MPa.

9.5.3 Luminous Emission

Figure 9.35 shows sequential observations of shock wave/bubble interaction recorded by a high speed video camera Shimadzu SH100 at the frame rate of 10^6 frame/s and exposure time of 125 ns. A 2.0 mm diameter spherical air bubble is impinged by a shock wave generated by exploding a 10 mg AgN_3 pellet at the stand-off distance of 20 mm. The black shadow on the right is an explosion product gas bubble, the so-called fire ball in the case of explosions in air. In Fig. 9.35b, a reflected expansion wave is shown as a faint grey circle. In Fig. 9.35d, faintly

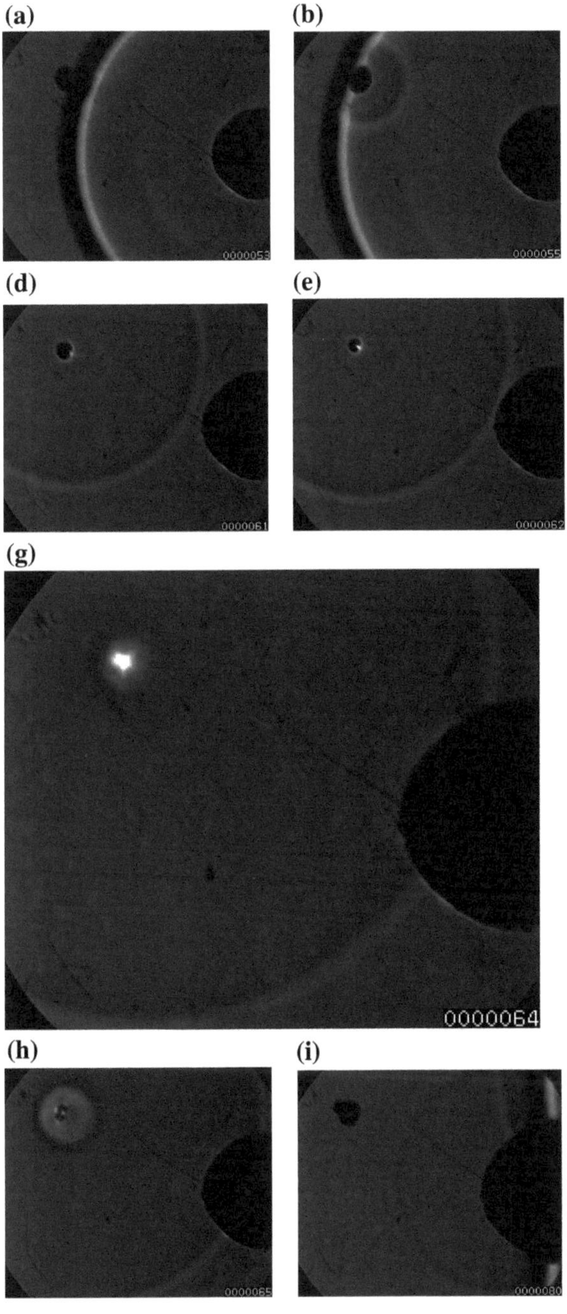

Fig. 9.35 Interaction of a 2.0 mm dia. Air bubble with a shock wave generated by exploding a 10 mg pellet at the stand-off distance 20 mm: **a** frame #53; **b** frame #55; **c** frame #58; **d** frame #61; **e** frame #62; **f** frame #63; **g** the brightest emission, frame #64; **h** frame #65; **i** frame #80 (Takayama et al. 2015)

shining point is observed at a frontal stagnation point. With elapsing time, the bubble contracts and its volume becomes minimal as seen in Fig. 9.35f. Then, the bubble started expanding and a secondary shock wave appeared in Fig. 9.35g, h. In Fig. 9.35g the luminosity is the brightest and the initiation of the secondary shock wave is faintly observed. The hot spot is located at the frontal side of the expanding bubble, which indicates that the hot spot is created by the wave interaction inside the bubble. An appropriate numerical simulation will reveal that the wave interaction decisively contributes to the hot spot formation.

Figure 9.36 summarizes the temporal variation of the brightness intensity of luminosity along the center axis. The ordinate denotes the brightness intensity of luminosity in arbitrary unit from the frame #60 to #67 for elapsed time from 60 to 67 μs. The abscissa denotes the distance between the center of the bubble and the center of the explosive in mm. The brightness of luminosity was maximal at the moment when the bubble volume was minimal at 65 μs and the secondary shock

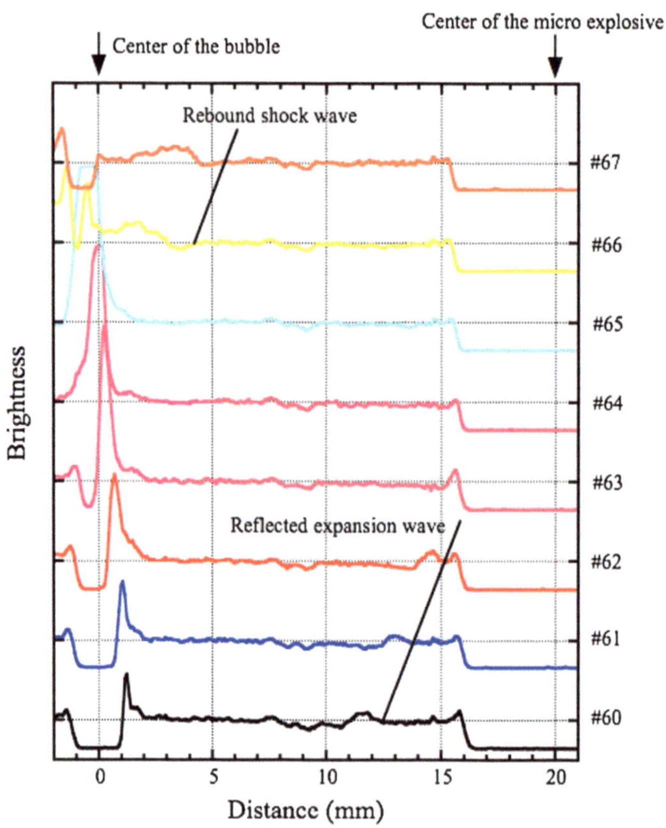

Fig. 9.36 Time variation of brightness of a luminous point in shock/bubble interaction. Shock wave was generated by exploding a 10 mg AgN_3 pellet at stand-off distance 20 mm (Takayama et al. 2015)

wave is initiated. At elapsed time of 67 μs, the luminous point moved toward the rear part of the bubble.

9.5.4 Shock Wave/Air Bubble Interaction in Silicone Oil

Figure 9.37 shows sequential observation of shock wave/bubble interaction in 1cSt, 10cSt, 100cSt and 1000cSt PDMS. The experiments were conducted in a 300 mm diameter and 300 mm wide test chamber by filling silicone oil of 1cSt, 10cSt, 100cSt and 1000cSt at 290 K in it.

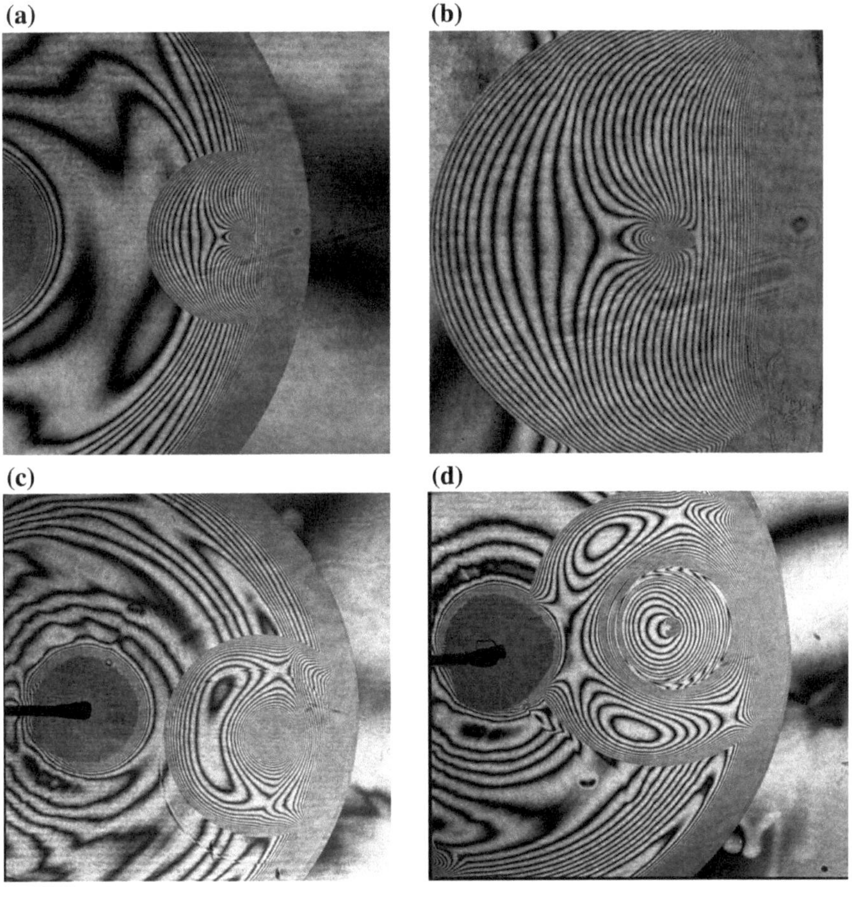

Fig. 9.37 Shock wave/air bubble interaction in 1cSt PDMS. Bubble diameter is 2.04 mm, at 297.9 K. A shock wave is generated by exploding a 10 mg AgN_3 pellet at stand-off distance 20 mm: **a** #91031103; **b** enlargement of (**a**); **c** #91031115; **d** #91031109

9.5.4.1 1cSt PDMS

Figure 9.37 shows sequential observations of a 2.04 mm air bubble in 1cSt PDMS interacting with shock waves generated by exploding a 10 mg AgN_3 pellet at the stand-off distance of 20 mm. The evolution of bubble motion is similar to the case of a 1.7 mm air bubble interacting with an underwater shock wave as seen in Fig. 9.32. Figure 9.37b is an enlargement of Fig. 9.37a, in which the bubble has a slightly ellipsoidal shape, whereas in Fig. 9.32, the bubbles were truly spherical. The slightly ellipsoidal bubble motion slightly differs from that of truly spherical one. In Fig. 9.37d, the final shape of the collapsing bubble had a convex shape (Hayakawa 1987).

9.5.4.2 10cSt PDMS

Figure 9.38 shows a sequential observation of shock wave/air bubble interaction of a 1.5 mm diameter air bubble in 10cSt PDMS. Shock waves were generated by exploding a 10 mg AgN_3 pellet at the stand-off distance of 20 mm (Hayakawa 1987). In Fig. 9.30f, at long time later the bubble reached almost a minimal volume again. The secondary shock wave interacted with the detonation product gas bubble. The expansion wave was reflected and the jaggedly shaped bubble surface were observed due to interfacial instability.

9.5.4.3 100cSt PDMS

Figure 9.39 shows the evolution of shock wave interaction with a 1.5 mm diameter spherical air bubble in 100cSt PDMS. The shock wave is generated by the explosion of a 10 mg AgN_3 pellet at stand-off distance 20 mm. The interaction of the bubble with the shock wave is very similar to that in 10cSt PDMS as seen in Fig. 9.38 (Hayakawa 1987). In Fig. 9.39a, the bubble reached a minimal volume and a very dense fringe accumulation was observed.

9.5.4.4 1000cSt PDMS

Figure 9.40 shows the evolution of shock wave interaction with a 1.5 mm diameter spherical air bubble in a highly viscous 1000cSt PDMS. The shock wave is generated by exploding a 10 mg AgN_3 pellet at stand-off distance of 20 mm. The bubble deformation is slightly retarded than that of less viscous PDMS (Hayakawa 1987).

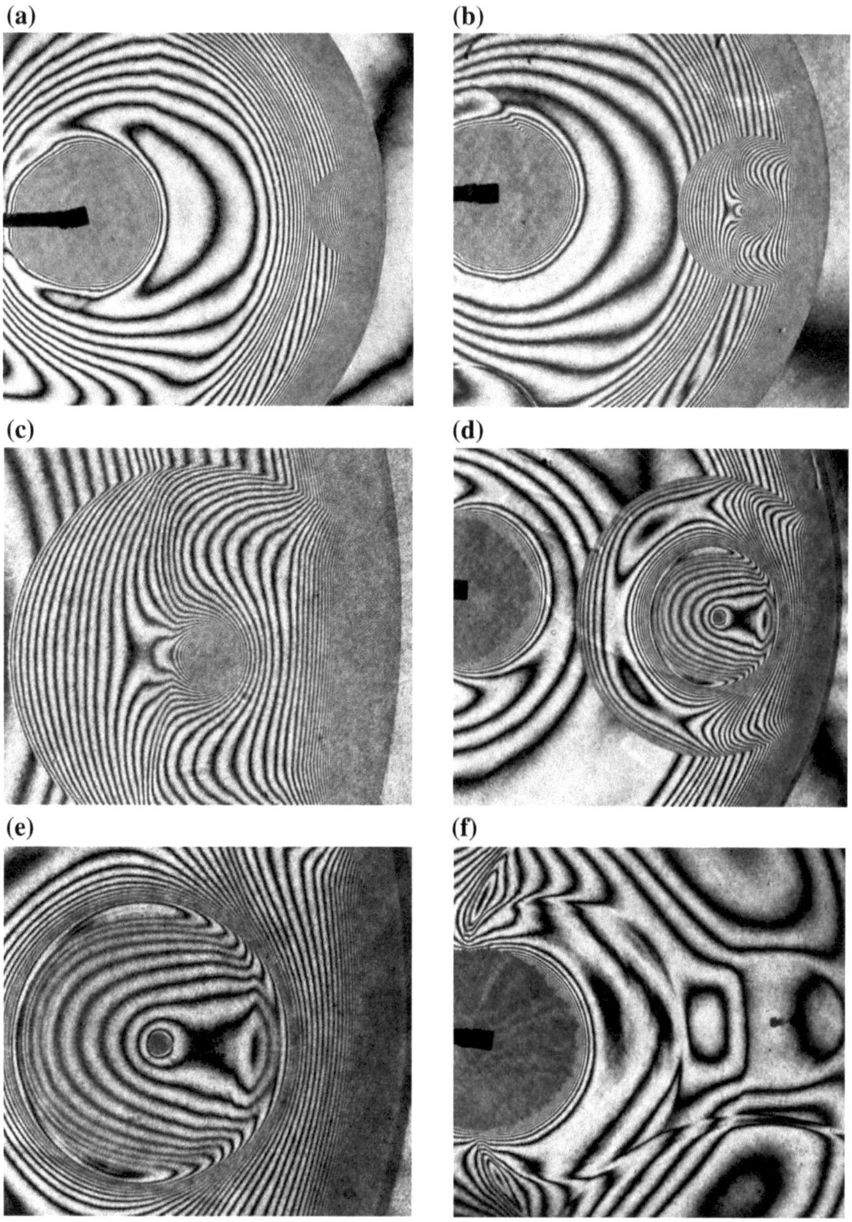

Fig. 9.38 Shock wave/air bubble interaction in 10cSt PDMS. Bubble diameter is 1.5 mm, at 298 K. A shock wave is generated by exploding a 10 mg AgN_3 pellet at stand-off distance 20 mm: **a** #98012102; **b** #98012003; **c** enlargement of (**b**); **d** #98012004; **e** enlargement of (**d**); **f** #98012006

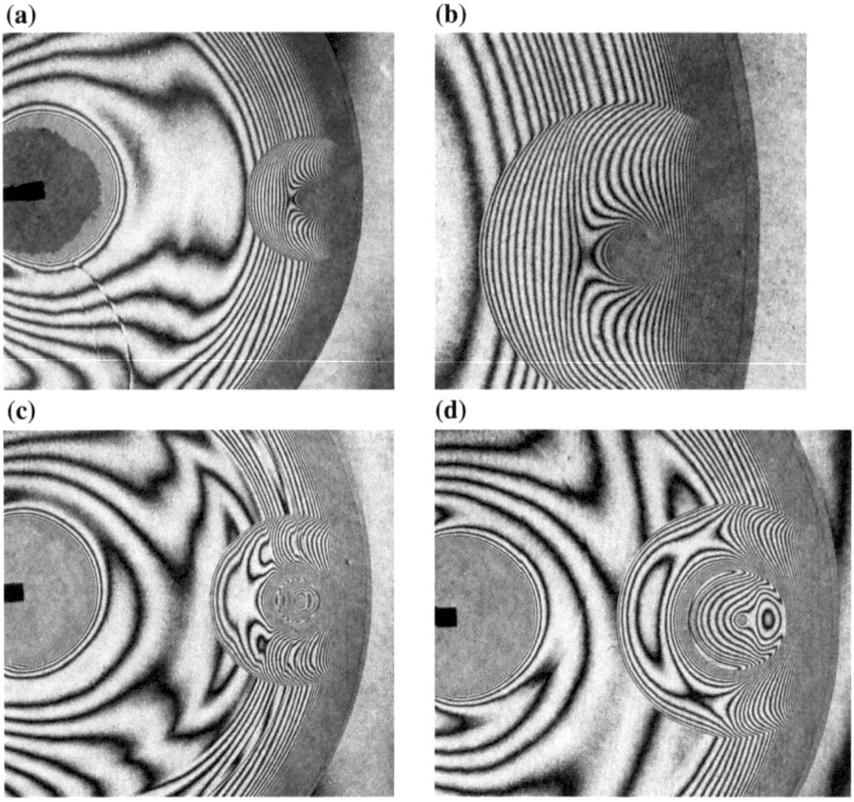

Fig. 9.39 Shock wave/air bubble interaction in 100cSt PDMS. Bubble diameter is 1.5 mm, at 298 K. A shock wave is generated by exploding a 10 mg AgN_3 pellet at stand-off distance 20 mm: **a** #98012104; **b** enlargement of (**a**); **c** #98012105; **d** #98012103 (Takayama et al. 2015)

9.5.5 Golden Syrup

Figure 9.41 shows the evolution of the shock wave interaction with a 2.0 mm diameter air bubble in 70 wt% golden syrup. The golden syrup contains 30% of water in weight and 70% of golden syrup in weight. The viscosity of this 70 wt% golden syrup is close to that of 1000cSt PDMS. The shock wave was generated by exploding a 10 mg AgN_3 pellet at the stand-off distance of 20 mm. The direct shadowgraph images were recorded with a high-speed video camera Shimadzu SH100 at framing rate of 10^6 frame/s and exposure time of 125 ns.

In Fig. 9.41, the shock wave propagates from right to left. A dark circular shadow on the right is the fire ball of a 10 mg AgN_3 pellet. A faintly grey circle surrounding the air bubble is a reflected expansion wave and the detonation product gas bubble is seen on the right hand side. In Fig. 9.41c, the air bubble contracts and the reflected expansion wave develops. In Fig. 9.41d, the air bubble has a minimal volume emitting luminosity. In Fig. 9.41e, the luminosity is emitted at the point

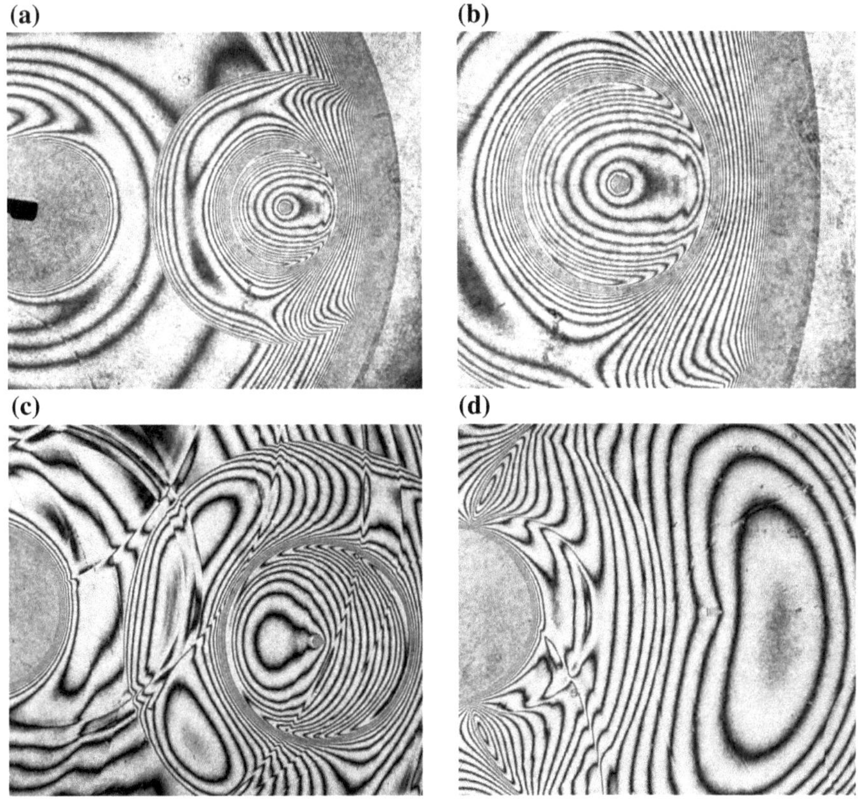

Fig. 9.40 Shock wave/air bubble interaction in 1000cSt PDMS. Bubble diameter is 1.5 mm, at 298 K. A shock wave is generated by exploding a 10 mg AgN_3 pellet at stand-off distance 20 mm: **a** #98032508; **b** enlargement of (**a**); **c** #98032509; **d** #98032510

located bubble's frontal side. This indicates that the hot spot is formed by a wave focusing inside the bubble (Takayama et al. 2015).

Figure 9.42 summarizes the time variation of luminosity along the distance between the bubble and the center of the explosive presented in Fig. 9.41. The ordinate denotes brightness of luminosity in arbitrary unit measured from the frame #52 to #61. The abscissa denotes elapsed time from 52 to 61 μs. At 57 μs corresponding to Fig. 9.41d, the luminosity is the brightest and the secondary shock wave is just generated at this time instant.

9.5.6 Helium Bubble in Silicon Oil

The wave motion inside the bubble impinged by a shock wave is governed by the ratio of the sound speed in a gas bubble to that in a liquid. Hence, the deformation

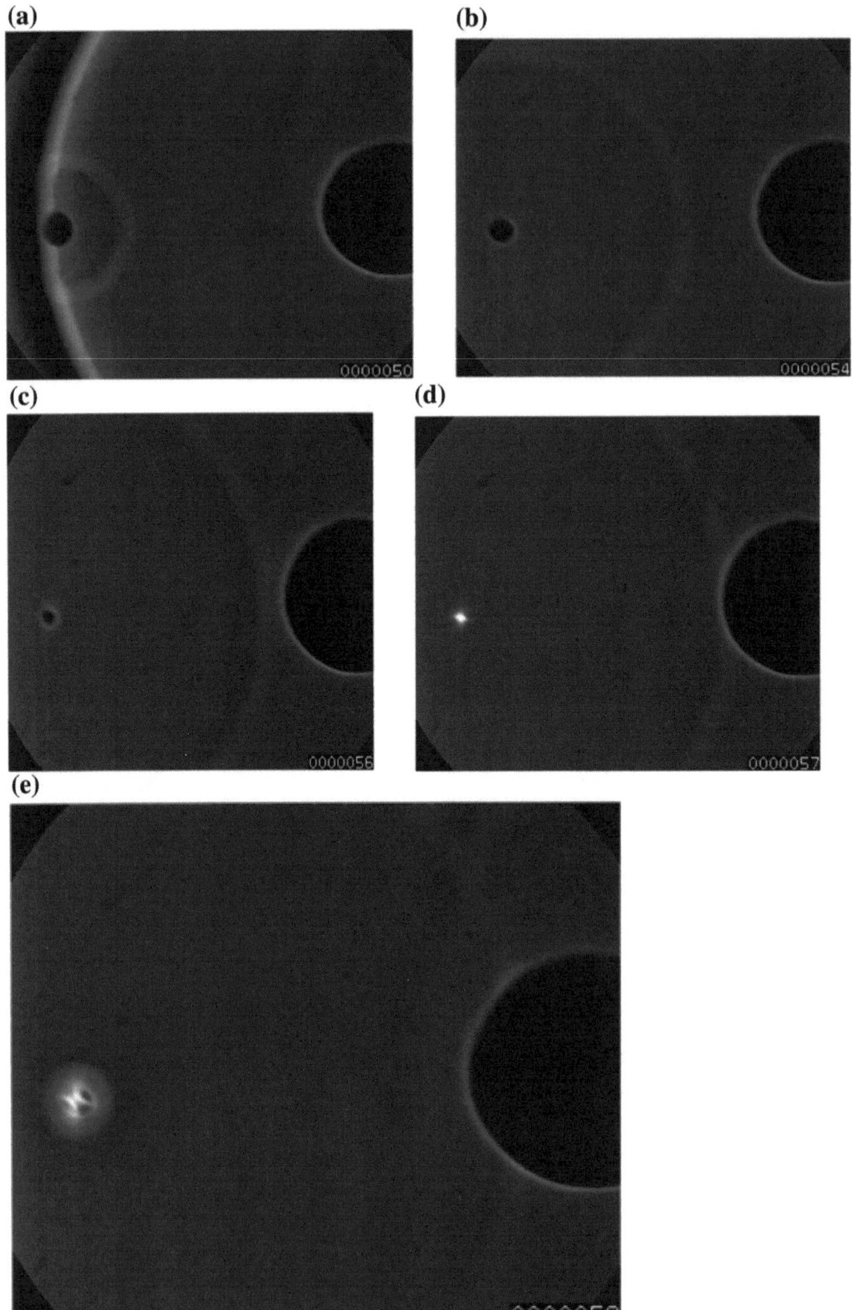

Fig. 9.41 Shock wave bubble interaction in a golden syrup of 70 wt%: **a** #53; **b** #55; **c** #56; **d** #57; **e** #58; **f** #59; **g** #60 (Hirano 2001)

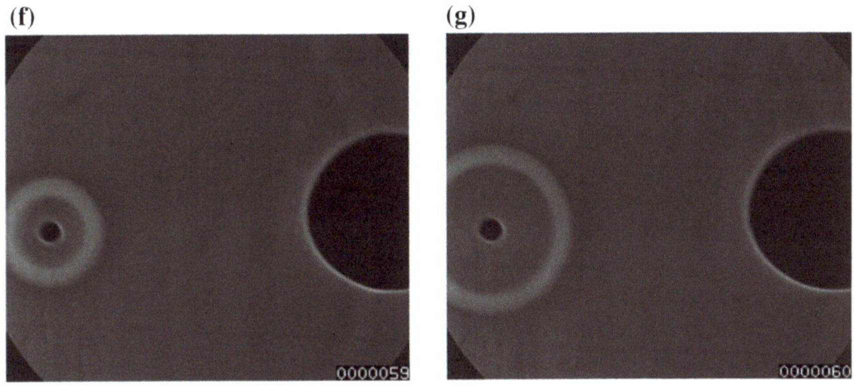

Fig. 9.41 (continued)

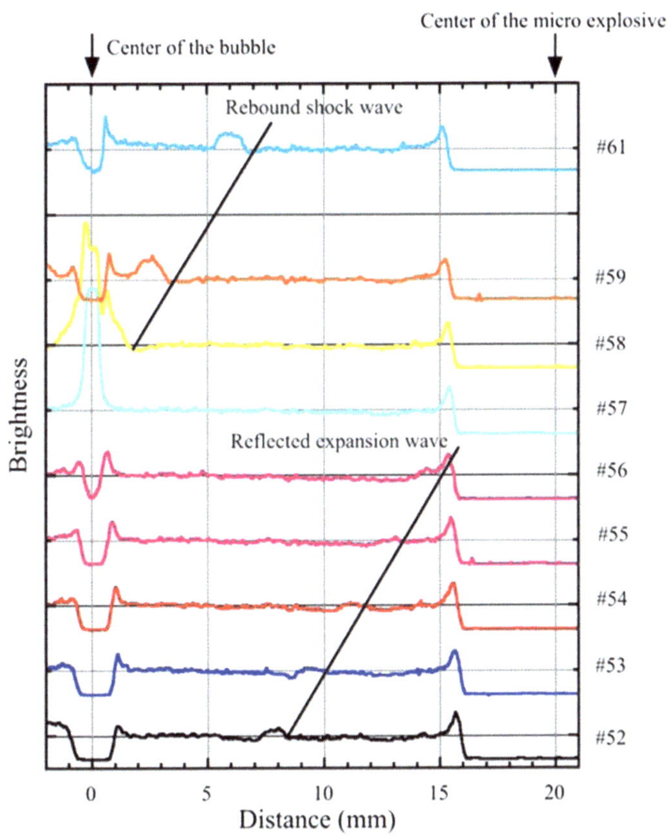

Fig. 9.42 Time variation of brightness of a luminous point in shock/bubble interaction in golden syrup. The summary of Fig. 9.41 (Hirano 2001)

of the gas bubble would be affected to this ratio. The interaction of an underwater shock wave with an air bubble is a typical fast/slow interaction. The ratio of the sound speed in air to that in water is about 0.23. Therefore, in the case of a slow/fast interaction, upon the impingement of a shock wave in a liquid having a slower sound speed, a gas bubble having a faster a sound speed would deform differently from that in the fast/slow interaction.

Yamada (1992) visualized the interaction of a helium bubble with a shock wave in a 1.0cSt PDMS using Ima Con 790 (John Hadland Ltd.) at the framing rate of 10^5 frame/s. The sound speed in helium is 970 m/s, whereas that in 1cSt PDMS is 901.3 m/s. However, the helium filled in the bubble is contaminated with air 50% in volume, then the ratio of the sound speed in the 50% helium-air mixture bubble to that in 1.0cSt PDMS is about 0.6.

The ratio of the acoustic impedance in air bubble to that in water is about 0.007, whereas that in a 50% helium-air mixture to that in 1cSt PDMS is 0.012. Figure 9.43 shows direct shadowgraphs of sequential interactions of a bubble containing a 50% helium-air mixture with a shock wave in 1.0cSt PDMS. The shock wave is generated by exploding a 10 mg AgN_3 pellet. Three sequential pictures were taken at the delay time of 596, 600, and 602 μs, respectively. Inter-frame time of these framing pictures was 2.5 μs. In Fig. 9.43a, the shock wave interacted with a 5% helium-air mixture bubble at the stand-off distance of 10 mm. The reflected expansion wave was observed. The frontal side moved into the bubble and the fain shadow of a reflected shock wave is observed in the seventh frame. The frontal side of the bubble is slightly fattened in the eighth frame. This time instant at which the eighth frame was taken was almost identical with the time instant at which the second frame in Fig. 9.43b was taken.

The bubble had a minimal volume as seen in the second frame in Fig. 9.43b. The bubble explosively expanded in the third frame, which indicates that the shock wave propagating in side the bubble focused at a point, when the bubble volume became minimal at a time instant between the second and third frames. The temperature at the focal point reached high enough to emit luminosity. At the same time, pressures at the focal point became high and spontaneously coalesced into a secondary shock wave. The point at which the secondary shock wave was originated and the point at which the luminosity emitted were almost identical points. This indicates that the reflected shock wave from the curved interface focuses at a point enhancing not only high pressures but also a high temperature enough to spontaneously emit luminosity. From the fourth to the fifth frames, the high pressures explosively blow the bubble as if the high pressures shatter its main structure. Although the main structure of the bubble is deformed and move away from the focal point, a part of the bubble stays at the focal point at which the particle speed is negligibly low. As the main structure of the bubble quickly moves away from the focal point, Therefore, this part looks like a counter jet ejected from the main structure of the bubble. This is not a counter jet but a finger. The slow/fast interaction never agrees with the Neumann effect that is valid only for a fast/slow interface. In conclusion, the sound speed ratio between gas and liquid can decisively govern the wave motion inside the gas bubble (Yamada 1992).

(a)

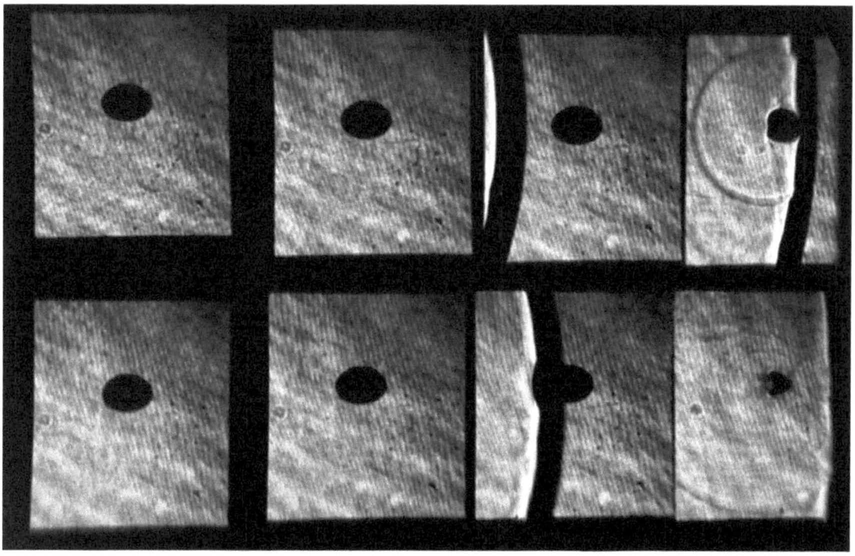

(b)

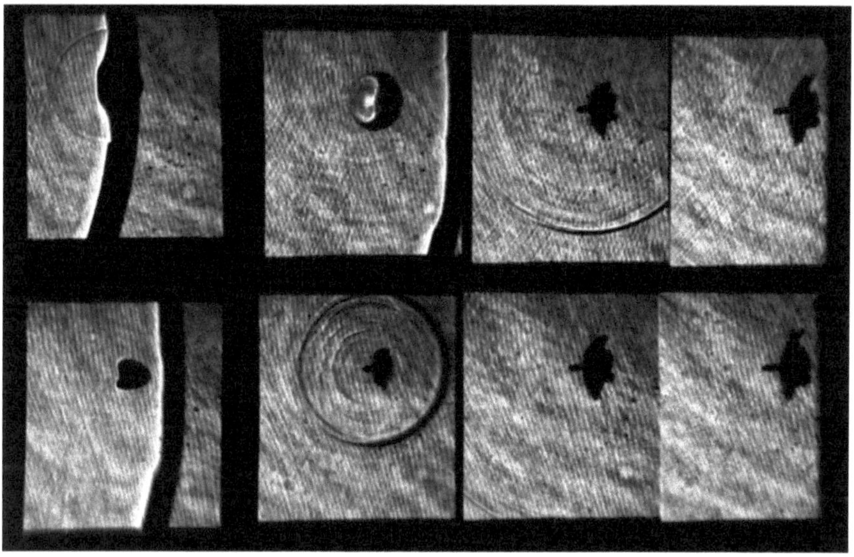

Fig. 9.43 Shock wave interaction with a helium/air bubble in 1cSt PDMA at stand-off distance 10 mm. The shock wave was generated by exploding a 10 mg AgN_3 pelelt. Images were recorded by ImaCon 790 at framing rate of 10^5 frame/s: **a** #92073005, 596 μs after trigger; **b** #92073003, 600 μs after trigger; **c** #92073001, 602 μs after trigger (Yamada 1992)

(c)

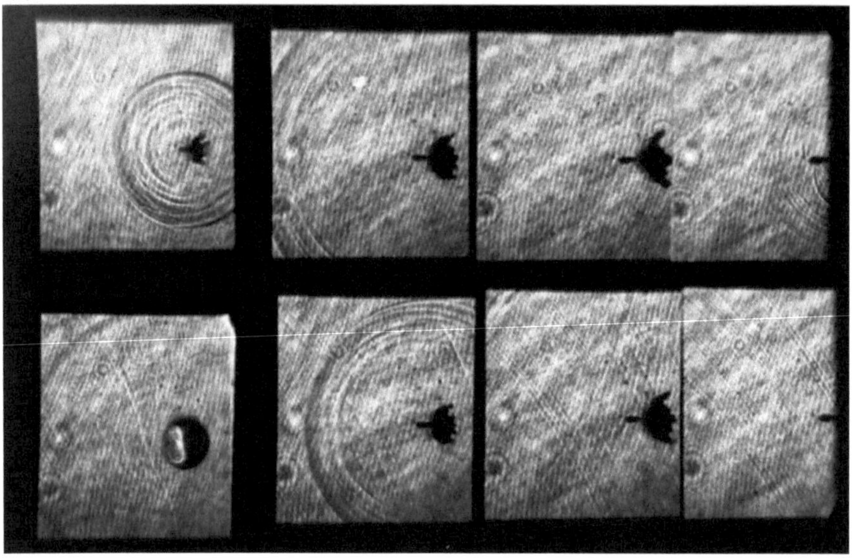

Fig. 9.43 (continued)

In order to demonstrate the slow/fast interaction as seen in Fig. 9.43 in a finer resolution, a sequential observation was carried out (Ohtani and Takayama 2010). Figure 9.44 shows a sequential interaction of a 0.87 mm × 1.22 mm helium bubble with a shock wave generated by exploding a 10 mg AgN_3 pellet in 1cSt PDMS at the stand-off distance of 20 mm. The visualization was conducted in shadowgraph and resulting images were recorded by Ima Con D200 for time interval ranging from 19.6 to 22.0 μs at 250 ns frame interval and at exposure time of 20 ns. Figure 9.45 shows sequential images ranging from 22 to 32 μs at the time interval of 1 μs and exposure time of 20 ns. In Fig. 9.44, the bubble had a minimal volume at 20.5 μs and emitted luminosity and at the same time generated the secondary shock wave. As seen in Fig. 9.43, when a high pressure generated at a focal point shattered the main structure of the bubble, a part of the bubble remained at the focal point forming a stretched finger which resembled a counter jet. The concept of the Neumann effect no longer applicable to interpret the formation of a stretched finger. Wave motions inside the helium-air bubble is only rudely speculated. It is expected that an appropriate numerical simulation would reproduce the formation of a stretched finger.

Figure 9.45 shows the interaction of helium-air bubble with a shock wave generated by exploding a 10 mg AgN_3 pellet in 1cSt PDMS at 1.0 μs frame interval. The jet formation to the stretched finger from the bubble was observed.

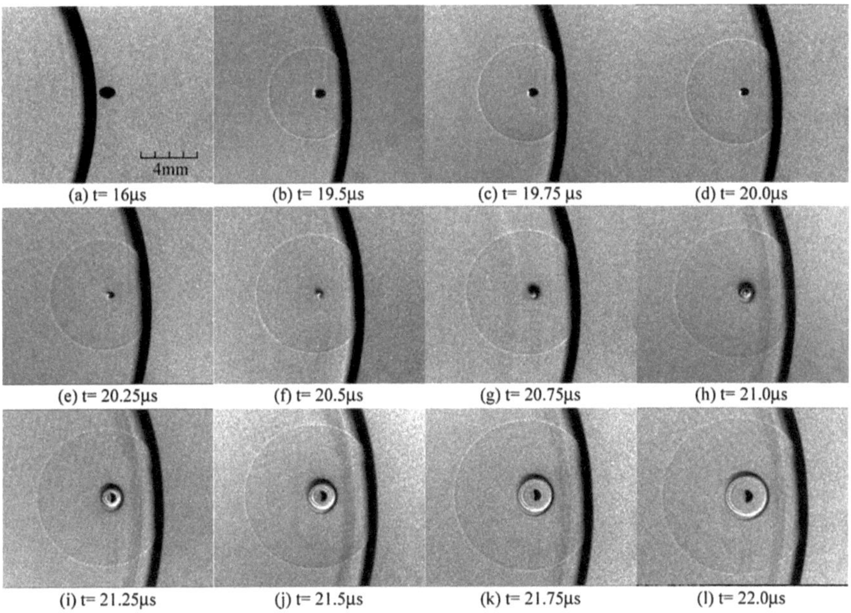

Fig. 9.44 Helium-air bubble interaction with a shock wave generated by exploding a 10 mg AgN_3 pellet in 1cSt PDMS at 500 ns frame interval (Ohtani and Takayama 2010)

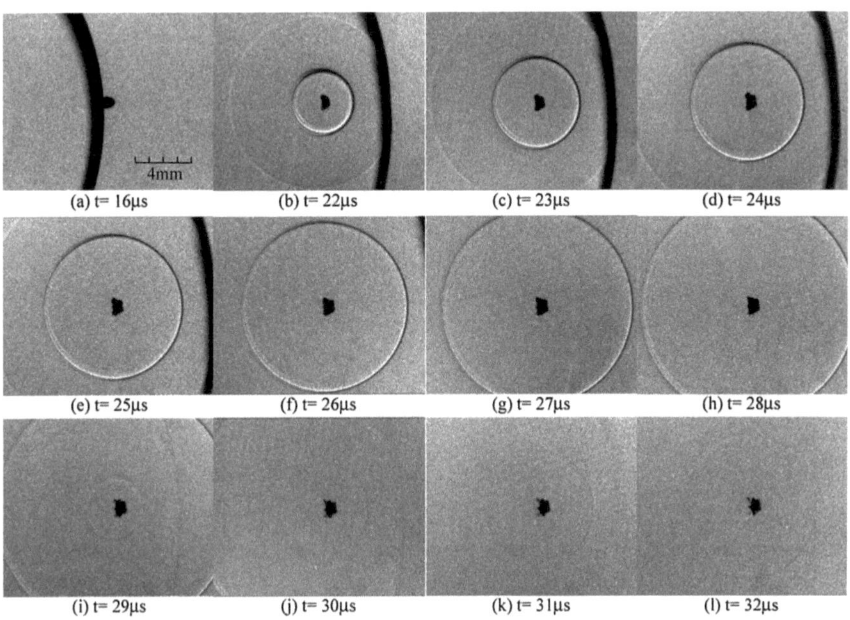

Fig. 9.45 Helium-air bubble interaction with a shock wave generated by exploding a 10 mg AgN_3 pellet in 1cSt PDMS at 1.0 μs frame interval (Ohtani and Takayama 2010)

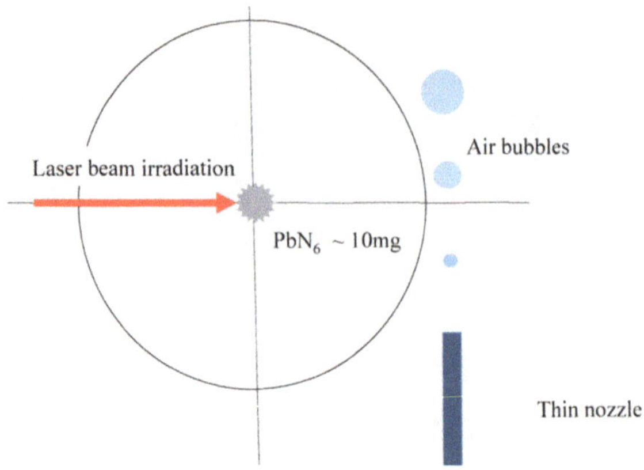

Fig. 9.46 Experimental arrangement of a shock wave interaction with bubble clouds at the stand-off distance 50 mm

9.5.7 Shock Wave Interaction with Bubble Cloud

Bubbles were randomly released into water from a capillary tube placed at the bottom of a test chamber and interacted with shock waves generated by exploding a 10 mg PbN_6 pellet. Figure 9.46 illustrates an experimental arrangement. A shock wave was generated at 50 mm stand-off distance from the center of rising bubbles.

Figure 9.47 shows a shock wave interaction with bubble clouds at the stand-off distance 50 mm: single exposure; double exposure interferograms; and their enlargements. Reflected expansion waves were observed only in the double exposure interferograms. The deformation of individual bubbles and the resulting wave interaction dissipated energy and eventually attenuated the shock wave. It is widely known that in maritime engineering and civil engineering, underwater blast waves are bubble effectively attenuated in passing a bubble curtain. Depending on the bubble size and their spatial distributions, an optimal combination of these parameters exists for mitigating underwater blast waves. This analogue experiment is aimed at searching for a similarity parameter between small scale laboratory experiments and field tests. Figure 9.47b, d, f, h are enlargements of Fig. 9.47a, c, e, g, respectively. It is observed that the bubbles are collapsed by the exposures of shock waves generated by neighboring bubble collapses. Smaller diameter bubbles respond quickly but larger diameter ones take time before their collapse.

9.5.8 Shock Wave Propagation in Bubbly Water

Japan imports crude oil from Middle East countries. On the way to the Middle East, the oil tankers carry hundred thousand tons of ballast water which contain an abundance of micro marine creatures. Their release into foreign sea water will cause a serious environmental pollution. The International Maritime Organization (retrieved from http://www/imo/home.asp) agreed to explore a technology for

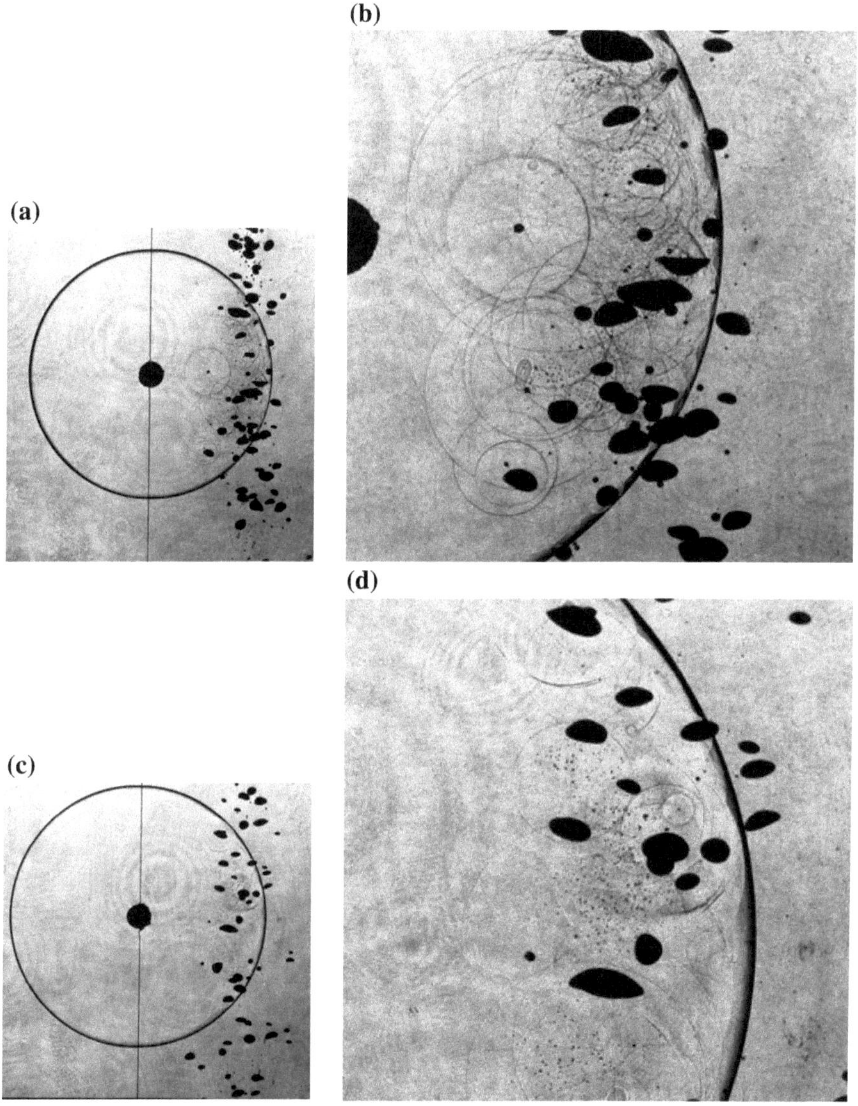

Fig. 9.47 Underwater shock wave interaction with bubble clouds at stand-off distance of 50 mm. Shock waves are generated by exploding a 10 mg PbN_6 pellet: **a** #83013117, 27 μs from trigger point, single exposure; **b** enlargement of (**a**); **c** #83013115, 30 μs shock wave/bubble interaction, single exposure; **d** enlargement of (**c**); **e** #83013113, 32 μs; **f** enlargement of (**e**); **g** #83013103, 40 μs; **h** enlargement of (**g**)

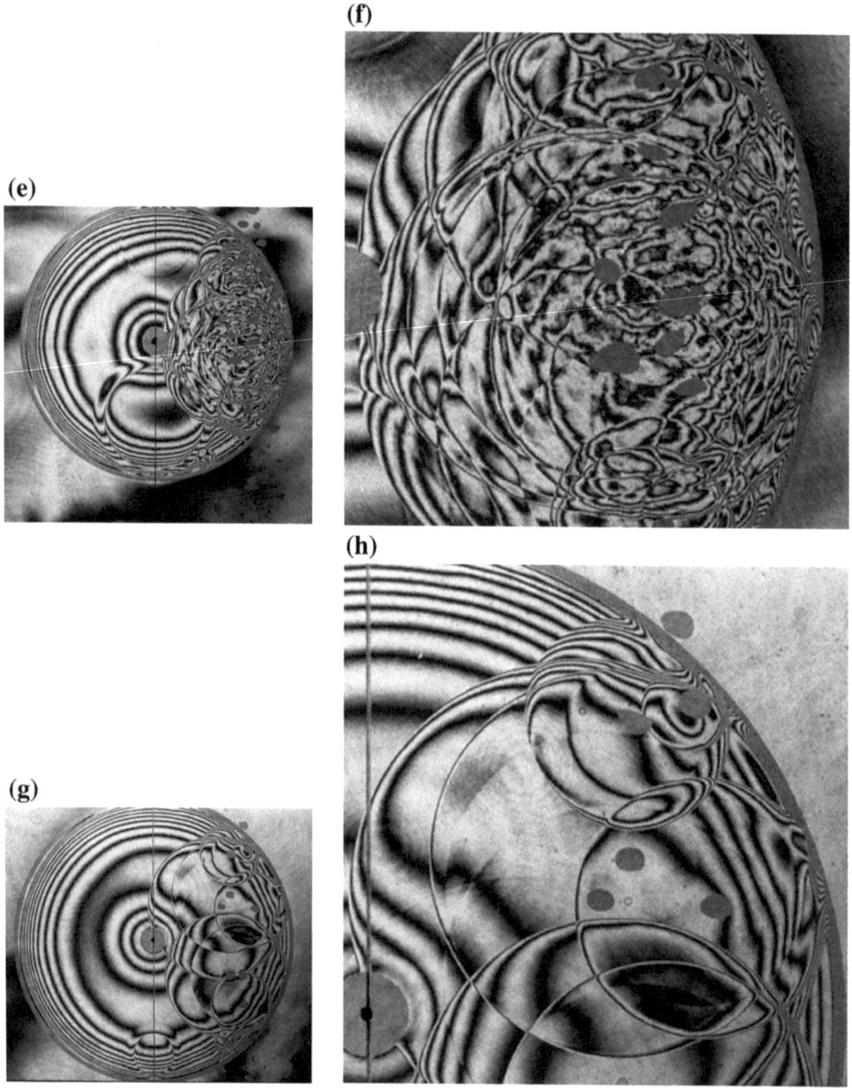

Fig. 9.47 (continued)

efficiently inactivating marine micro-creatures when releasing ballast water in foreign water. Professor Abe in Kobe University proposed a novel method of treating ballast water for inactivation of marine micro-creatures. An analogue experiment was conducted as illustrated in Fig. 9.48a. Micro-air bubbles were generated by swirling water at reasonably high-speed. Resulting micro-air bubbles were stored in a salty water chamber and then circulated into the test section. Figure 9.48b is a histogram of micro-bubbles. The ordinate denotes the production

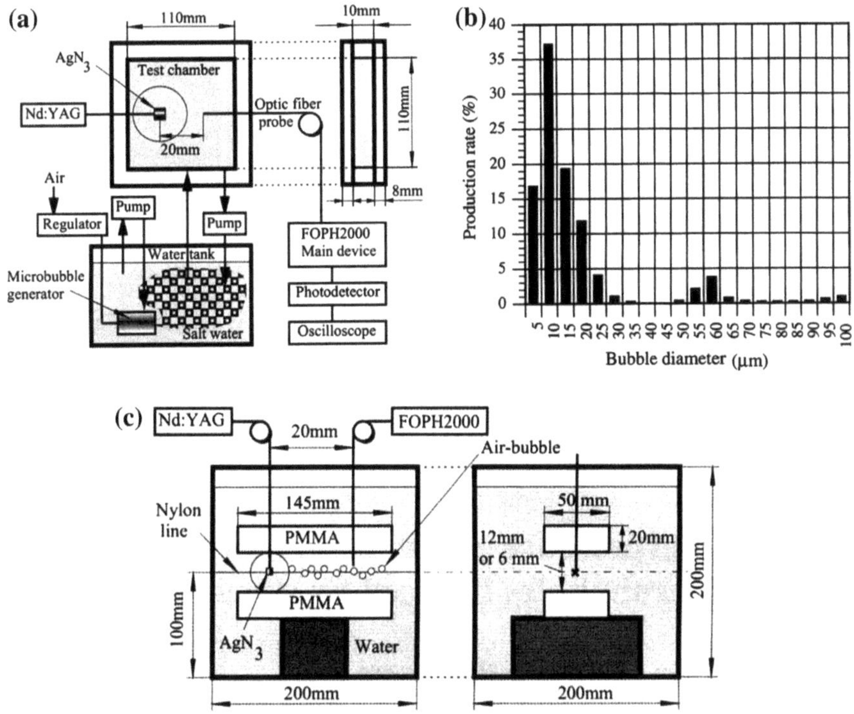

Fig. 9.48 Experimental setup: **a** experimental setup; **b** histogram of micro-bubbles; **c** a test section of 50 mm in width and 6 or 12 mm in height (Abe et al. 2010)

rate, which is a percentage of the bubbles having a specified bubble diameter. The abscissa denotes the bubble diameter in μm. The bubble diameters were distributed from 5 to 25 μm. 35% of the bubbles had a peak diameter of about 10 μm. Underwater shock waves were generated by exploding a 10 mg AgN_3 pellet in a 10 mm wide test section as shown in Fig. 9.48c. Over-pressures were measured with an optical fiber pressure transducer Model FOPH 2000. Shadowgraphs were taken and the images were recorded by Ima Con D200 at the framing rate of 5×10^6 frame/s. Figure 9.48c shows the test section having two 20 mm thick acrylic plates having dimension of 50 mm × 145 mm and the interval was selected either 6 mm or 12 mm.

Figure 9.49a shows a sequential observation of the shock wave propagation in a bubbly water. The shock wave was generated by exploding a 10 mg AgN_3 pellet at the stand-off distance of 20 mm as shown in Fig. 9.48a. The shock wave propagated from right to left. A spherical shock wave generated inside a 10 mm wide channel. The shock wave was spherical until it touched the side walls. The spherical shock wave repeatedly reflected and eventually became a cylindrical shock wave. In Fig. 9.49a, the two coaxial rings were projections of repeatedly reflected spherical shock waves from the side wall and were gradually flattened with elapsing time.

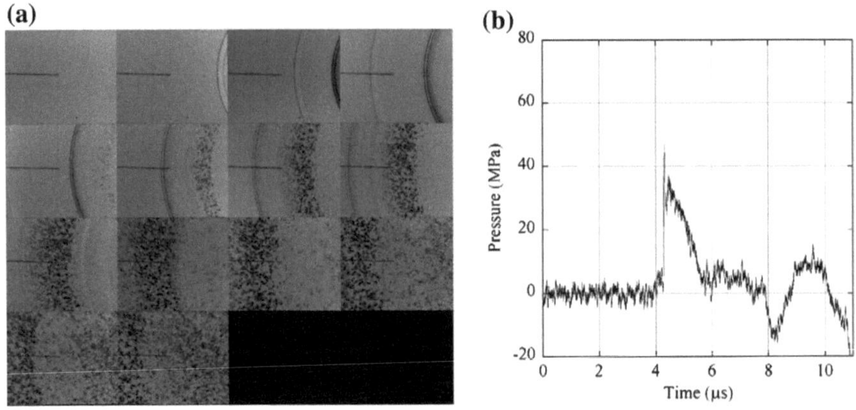

Fig. 9.49 The sequential observation of shock wave propagation in a bubbly water shown in Fig. 9.48a: **a** #090227001 shock wave interaction with bubbly water framing rete of 5×10^6 frame/s; **b** pressure profile at early stage (Abe et al. 2010)

Hence, their interval became narrower and eventually merged into a distinct ring shape, which indicated the completion of a two-dimensional cylindrical shock wave. The cylindrical shock wave then interacted with micro-bubbles. Although not observed by shadowgraph, the bubbles quickly collapsed and generated the secondary shock waves. In the third frame, dark double rings indicated the secondary shock waves. Even if they were thick, the side walls were bulged upon the spherical shock wave loading. The degree of the deformation of the side walls were negligibly minute but generated cavitation bubbles. The cavitation bubbles responded to the cylindrical shock wave and instantaneously collapsed. Figure 9.49b shows the time variation of pressures at earlier stage measured by the pressure transducer shown in Fig. 9.49a. The ordinate denotes the pressure in MPa and the abscissa denotes elapsed time in μs. The peak value of the pressure jump at the shock wave was about 40 MPa. The envelop of reflected expansion wave arrived at the pressure transducer at 4 μs after the shock wave and its peak value of expansion wave reached about 15 MPa.

Figure 9.50 shows sequential observations of shock wave/bubble interaction in the test section having 6 mm interval as shown in Fig. 9.48c. A spherical shock wave created by exploding a 10 mg AgN_3 pellet immediately became a two-dimensional shock wave. An arrow marked CS indicate a transient shock wave from a spherical to a two-dimensional cylindrical shock wave. Hence the wave front is slightly broadened. The frame interval is 1.5 μs and the first frame was taken at 6 μs after the ignition. A circular ring shown by an arrow marked B1 was a spherical secondary shock wave generated at a collapsed bubble attached on a thin nylon line placed along the center line. A longitudinal stress wave is transmitted in an acrylic wall, propagating at 2.9 km/s. Then, it was released into water as a precursory wave as shown by an arrow marked OS. Between the first and second frames, the reflection of the secondary shock wave reached the pressure transducer.

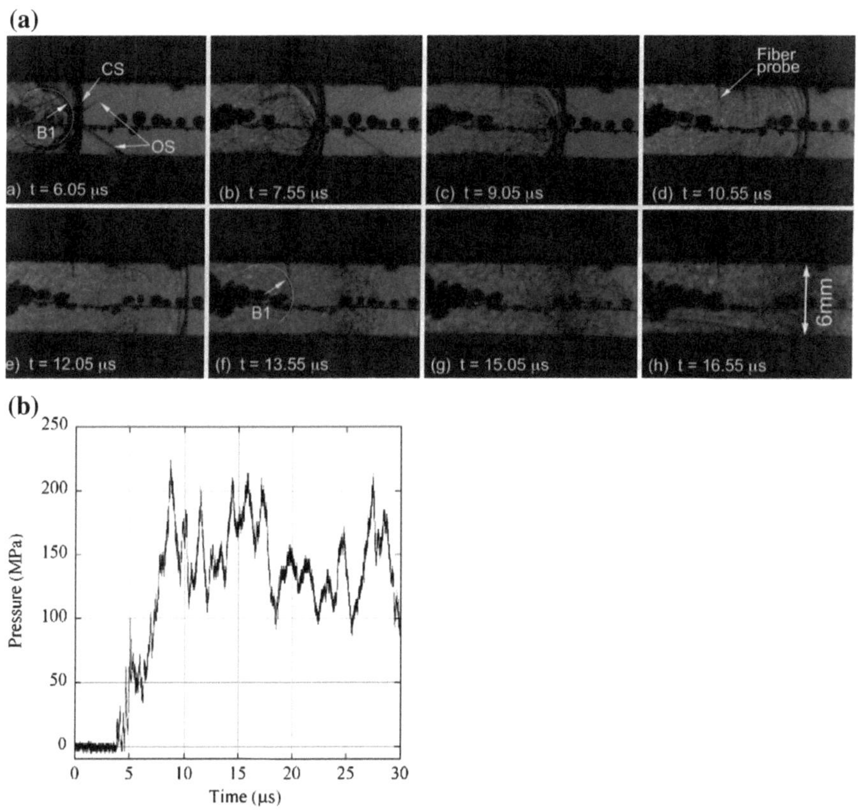

Fig. 9.50 Shock wave/bubble interaction in a two-dimensional duct having a 6 mm interval: **a** sequential observation with Ima Con D200. The shock wave propagated from left to right; **b** pressure variation (Abe et al. 2010)

The optical fiber pressure transducer was shown in the fourth frame. The shock wave successively impinged on air bubbles attached on the nylon line and created the secondary shock waves. In the meantime, its impingement on air bubbles created reflected expansion waves but these are not observed by shadowgraph. Blurred background noises show the reflected expansion waves and secondary shock waves generated by collapsing micro-bubbles. In the sixth frame it was shown by an arrow marked B1. But it was not as intense as that observed in the first frame shown by an arrow marked B1.

Figure 9.50b shows a result of pressure measured. The position of the pressure transducer is seen in the fourth frame in Fig. 9.50a. The ordinate denotes pressure in MPa and the abscissa denotes elapsed time in μs. The impingement of the direct shock wave enhanced pressure up to 60 MPa. However, the impingement of the reflected secondary shock waves from the wall eventually enhance over 200 MPa. If such a high-pressure is repeatedly exposed to micro-marine creatures, they would be gradually eradiated. In this analogue experiment, shock waves are generated by micro-explosions. Then the shock/bubble interaction resulted in the secondary shock wave. Its reflection from the wall of the confined space effectively produced high peak pressure well over 220 MPa. In conclusion, it would be possible to produce higher pressures exceeding 300 MPa, if the diameter and the number density of bubbles are optimized and the shock waves are repeatedly generated by frequent spark discharges in a relatively thin tube. When such an optimization can be achieved, the shock wave assisted eradication of micro-creatures would be practical.

9.5.9 Shock Wave Interaction with a Bubble on an Acrylic Plate

9.5.9.1 Flat Acrylic Plate

Figure 9.51 shows double exposure interferograms of sequential observations of shock wave interaction with a 5 mm diameter air bubble placed on a 7 mm thick acrylic plate. A shock wave was generated by exploding a 10 mg AgN_3 pellet at the stand-off distance of 50 mm (Yamada 1992). Upon a shock wave impingement on the bubble, it started to contract and its frontal side was flattened. The bubble shape was recorded twice firstly when it was set and secondly when synchronized with the event. The two images were superimposed. In Fig. 9.51c–l, the process of the bubble contractions is observed. Flattened bubbles show that the procedure of water jet formation and its impingement on the acrylic wall. The jet impingement on the acryl wall resulted in compression stress wave in Fig. 9.51g, h. Figure 9.51 demonstrates that the holographic interferometry gives quantitatively density distributions not only in water but also in acryl. Photo-elasticity can also produce fringes in photo-elastic materials but the fringes observed in a two-dimensional acrylic plate are iso-density contour lines.

Figure 9.52 shows a shock wave interaction with a 1.7 mm diameter air bubble placed on a 6 mm thick flat acrylic plate. Shock waves were generated by exploding a 10 mg AgN_3 pellet at stand-off distance of 15 mm. Figure 9.52a shows a single exposure interferogram. The expansion wave was reflected from the air bubble and propagated in water. The shock wave impinged on the acrylic plate and induced the stress wave which propagated faster than the transmitted shock wave in water. The reflected shock waves in water were accompanied by expansion wavelets which originated from small bubbles attached on the observation windows from the beginning. Figure 9.52d shows an enlargement of Fig. 9.52c. Figure 9.52f shows a double exposure interferogram: A transmitted shock wave in acryl and other waves were clearly observed but these were not visualized in single exposure interferograms. Figure 9.52g–h shows a later stage. The dark flattened shadow in the center indicates high pressures deforming the bubble surface toward the inside. This is a

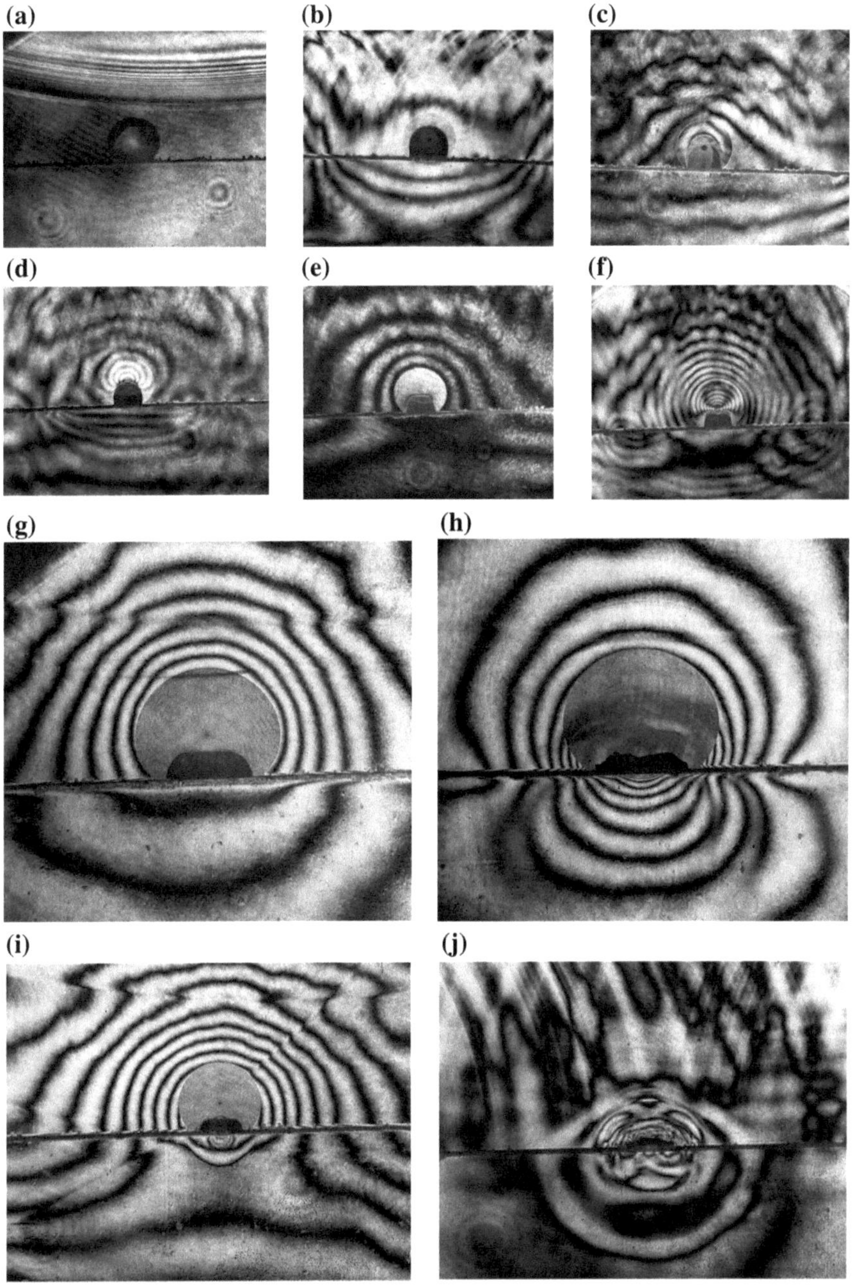

Fig. 9.51 Shock wave interaction with a 5 mm diameter air bubble placed on an acrylic plate and a 10 mg AgN_3 was detonated at stand-off distance of 50 mm: **a** #85012419; **b** #85012406; **c** #85012410; **d** #85012415; **e** #85012504; **f** #85012812; **g** #85012813; **h** #85012807; **i** #85012413; **j** #85012412; **k** #85012505; **l** #85012812

(k) (l)

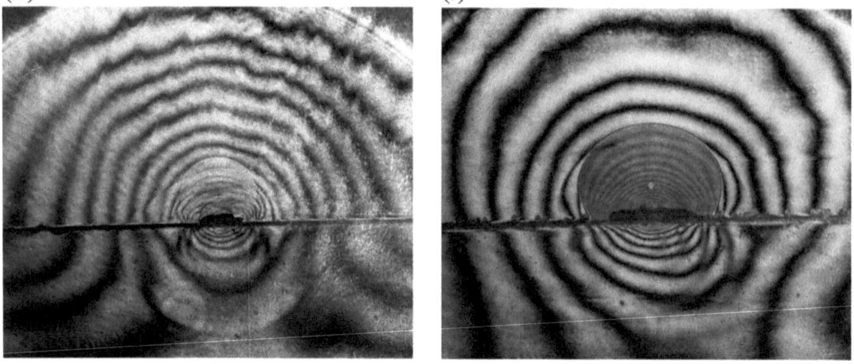

Fig. 9.51 (continued)

stage of jet formation. When the jet impinged the acrylic wall, a stress wave was transmitted in the acrylic plate. Reflected expansion waves and secondary shock waves propagated in water and also in acryl. Waves in acryl moved faster than that in water and was released in water forming an oblique wave. The similar shock wave/interface interaction was already observed when shock waves generated by exploding the explosive pellets interacted with the silicone oil/water interface.

9.5.9.2 V Shaped Groove

Figure 9.53 shows sequential shock wave/bubble interactions with a 1.7 mm diameter air bubble placed on a V-shape groove formed on a 6 mm thick acrylic plate. V shape angles are 45°, 100°, and 135°. A shock wave is generated by exploding a 10 mg AgN_3 pellet at stand-off distance of 15 mm. The bottom of the V shaped groove is a singular point. Hence, the response of the bubble would be unique.

9.5.9.3 Circular Groove

Figure 9.54a shows just before the shock impingement on a 1.7 mm diameter air bubble placed on the bottom of a 1.7 mm diameter two-dimensional groove on a 5 mm thick acrylic plate. Figure 9.54b shows the collapsing air bubble impinged by a shock wave generated by exploding a 10 mg AgN_3 pellet at the stand-off distance of 15 mm.

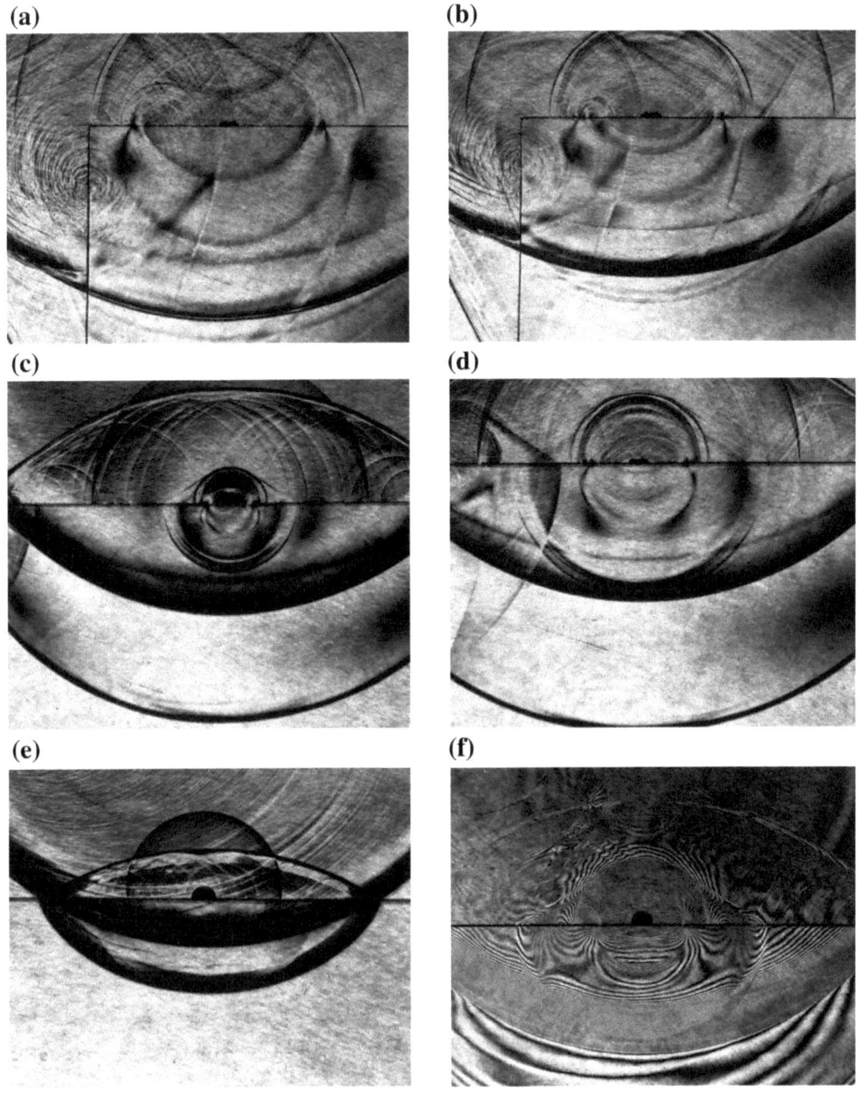

Fig. 9.52 Shock wave interaction with a 1.7 mm diameter air bubble placed on a 6 mm thick flat acrylic plate. Shock waves were generated by the explosion of a 10 mg AgN_3 pellet at stand-off distance of 15 mm, 295.0 K: **a** #86081104, 12 μs after shock wave impingement, single exposure; **b** #86081105, 11 μs, single exposure; **c** #86081106, 8 μs, stand-off distance of 30 mm, single exposure; **d** #86081108, 8 μs, stand-off distance of 30 mm, single exposure; **e** #86081107, 7 μs, single exposure; **f** #86081111, 8 μs; **g** #86081115, 8 μs; **h** enlargement of (**g**)

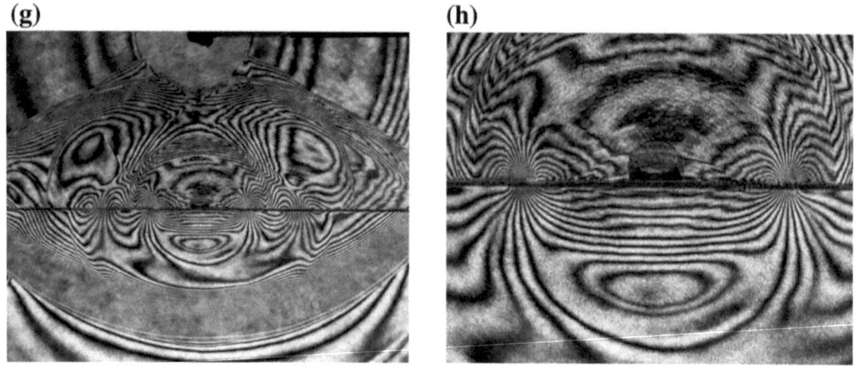

Fig. 9.52 (continued)

9.5.10 Two-Dimensional Bubble

The wave motion inside a spherical bubble cannot be visualized by a conventional visualization method. Then an analogue experiment was conducted to observe the deformation of a two-dimensional bubble impinged by a cylindrical shock wave. Figure 9.55 shows the motion of a 2-D air bubble impinged by a cylindrical shock wave (Yamada 1992). Figure 9.55a schematically shows a 2.2 mm diameter air bubble sandwiched between two acrylic plates separated 3 mm apart. The 2-D air bubble was positioned at 10 mm from the edge of the acrylic plates. A shock wave generated by exploding a 10 mg AgN_3 pellet at the stand off distance of 30 mm.

In Fig. 9.55b, a spherical shock wave impinged on two parallel acrylic plates separated 3 mm apart. The central part of the spherical shock wave propagated along the acrylic plates separated 3 mm interval and was diffracted and repeatedly reflected with its propagation. Eventually a cylindrical shock wave propagated inside the 2-D space. The longitudinal wave propagating in thin acrylic plates did not bother the 2-D bubble.

A ring shape seen in Fig. 9.55b is the edge of a 2-D bubble recorded at the first exposure. When the shock wave impinged, the bubble contracted and its frontal stagnation point was flattened. The deformation grew and eventually became a water jet penetrating the bubble.

In the two-dimensional observation, the fringe number N is given by

$$N = L\Delta n/\lambda, \tag{9.12}$$

where Δn is the variation of refractive index in the test section, L is the length of *OB* path which is 3 mm, and λ is the wavelength of ruby laser. The relationship between refractive index n and density ρ is defined as (Yamada 1992),

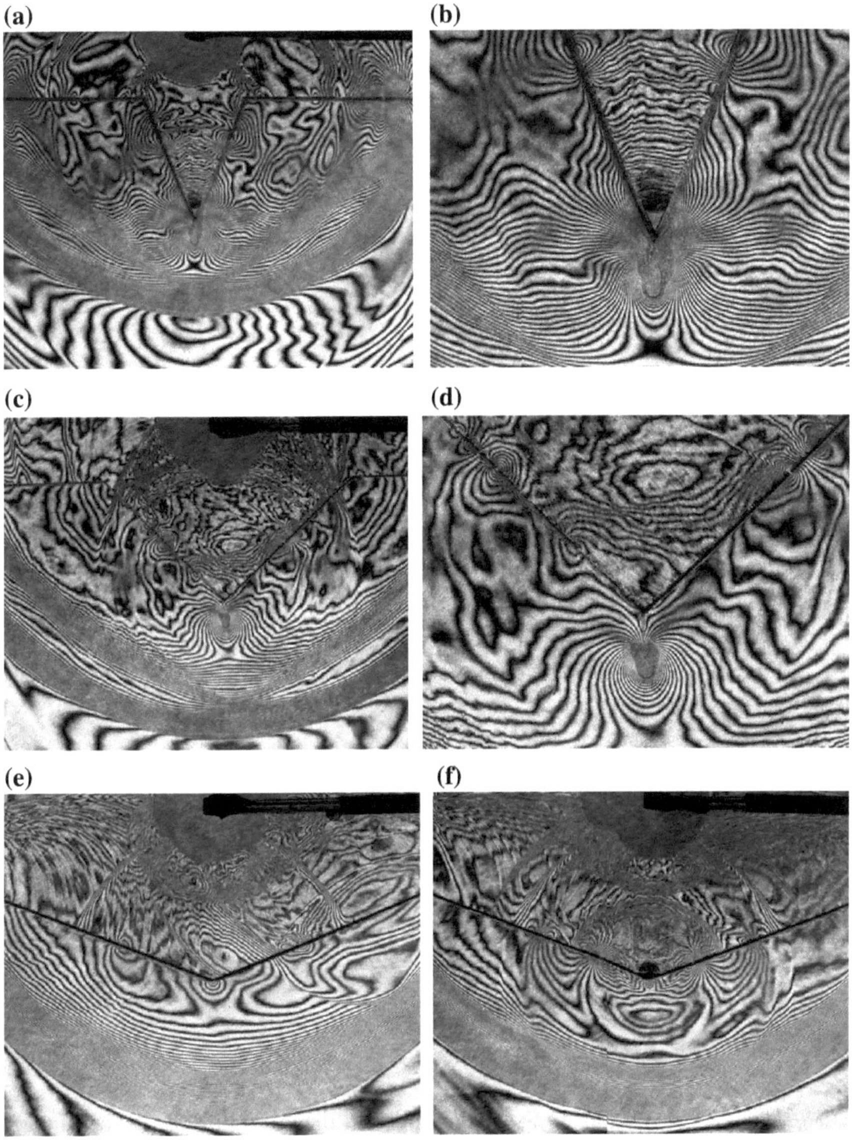

Fig. 9.53 The sequential observation of shock wave interaction with a 1.7 mm diameter air bubble placed at the bottom of V shaped grooves formed on a 6 mm thick acrylic plate. A shock wave is generated by explosion of a 10 mg AgN_3 pellet at stand-off distance of 15 mm, 295.0 K: **a** #86081204, 7 μs from trigger point, 45° groove; **b** enlargement of (**a**); **c** #86081201, 8 μs from trigger, 100° groove; **d** enlargement of (**c**); **e**#86081203, 8 μs from trigger time, 135° groove; **f** #86081202, 8 μs, 135° groove; **g** #86081205, 8 μs, 135° groove, single exposure; **h** #86081206, 8 μs, 135° groove; **i** enlargement of (**h**)

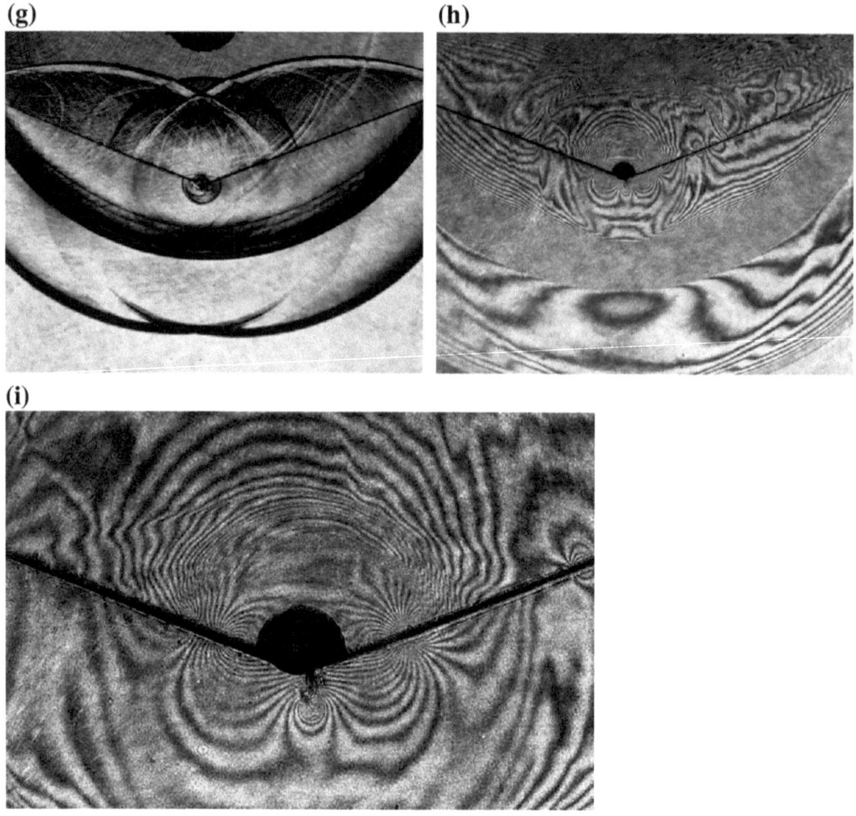

Fig. 9.53 (continued)

$$(n^2-1)/(n^2+2) \propto \rho. \tag{9.13}$$

Hence, one fringe shift in Fig. 9.55 is given by

$$\rho/\rho_0 = 1.266 \times 10^{-3}/\text{fringe}, \tag{9.14}$$

where ρ_0 is the water density at the room condition. Assuming the Tait equation, the density increment per a fringe can be rewritten in the form of pressure ratio p/p_0 per a fringe.

$$p/p_0 = 27.7/\text{fringe}, \tag{9.15}$$

where p_0 is ambient pressure. Hence, the fringe concentration seen in Fig. 9.55b corresponds to the total pressure increment of 13.5 ± 1.4 MPa. Unlike spherical bubbles, two-dimensional water jets do not penetrate the bubble. This is attributable to the fact that the stagnation pressure was too low to deform the 2-D bubble. A 2-D

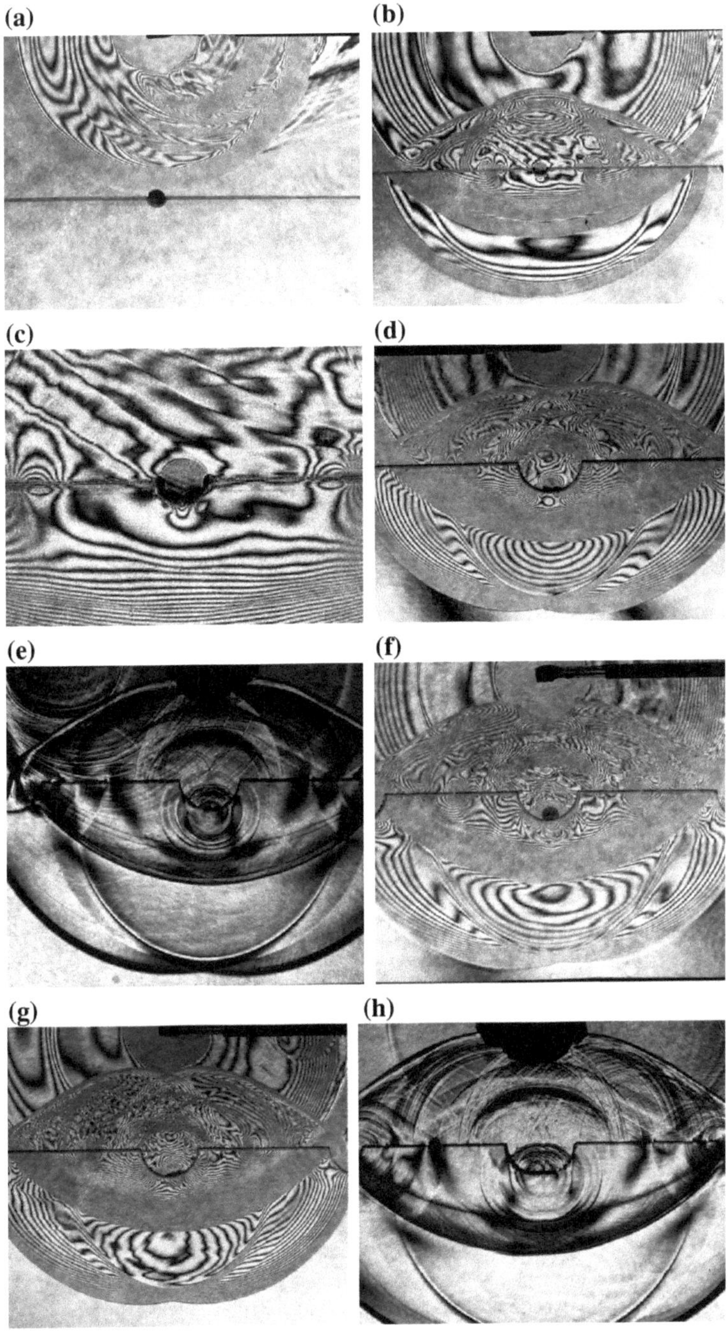

Fig. 9.54 The shock wave interaction with a 1.7 mm diameter air bubble placed on the bottom of a circular groove formed on a 5 mm thick acrylic plate. A shock wave was generated by explosion of a 10 mg AgN_3 pellet: **a** #86081802, 5 μs from trigger point; **b** #86081803, 7 μs; **c** enlargement of (**b**); **d** #86081213, 5 μs **e**; #86081210, 5 μs, single exposure; **f** #86081212, 5 μs; **g** #86081209, 6 μs; **h** #86081208, 6 μs, single exposure

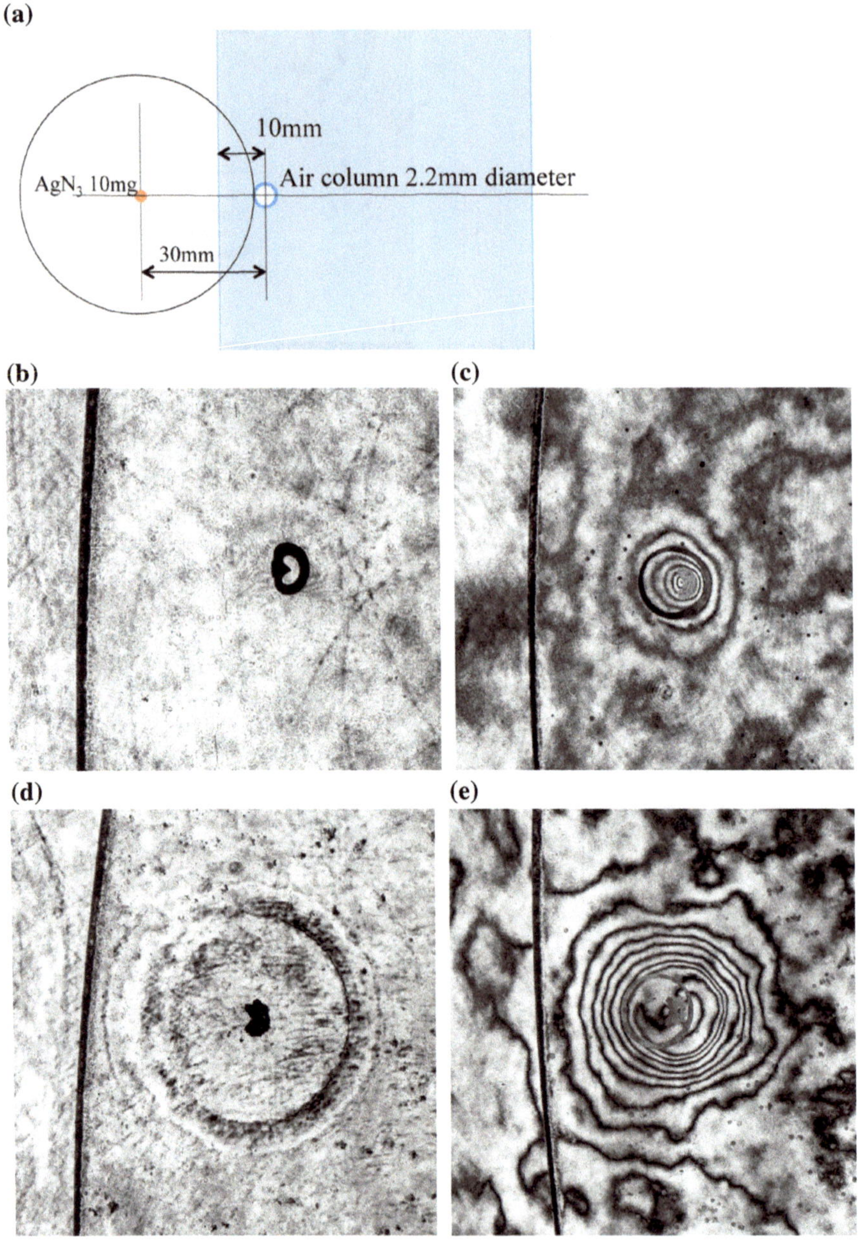

Fig. 9.55 2-D bubble interaction with a shock wave generated by exploding a 10 mg AgN_3 pellet at 1013 hPa, 288.5 K: **a** experimental arrangement; **b** #87031606, single exposure; **c** #87031630; **d** #87031604, single exposure; **e** #87031613

bubble expands more irregularly than a spherical bubble does (Takayama 1987). It should be noticed that although the 2-D bubble has 2.2 mm diameter and 3 mm high, their response to the shock wave exposure is different from an ideal 2-D air column. The edges of the 2-D bubble are stuck at the acrylic plate's surface and would not move as freely as a spherical bubble.

9.5.11 Sympathetic Explosion

If two explosives are in water and one of them was exploded, the resulting shock wave may ignite the neighboring explosive. This is the so-called sympathetic explosion (Nagayasu 2002). Depending on the stand-off distance between the two explosives, the ignition of the neighboring explosive hardly occurred. However, if air bubbles are attached on its surface, the explosive would be ignited spontaneously as soon as the shock wave of the neighboring explosive reached on its surface. Figure 9.56a demonstrates sympathetic explosions. Two PbN_6 pellets of 5.5 and 5.6 mg in weight were glued on a cotton threads at the stand-off distance of 4 mm. Air bubbles are attached on the lower one. The upper one was ignited first and about 2.6 μs later, the lower one detonated. The grey shadows are explosion product gas bubbles taken at the second exposure. The hot spot created inside the collapsing the air bubble triggered the lower explosive. The difference of the radii of the two spherical shock waves indicates the delay time of the ignition. Figure 9.56b shows a single exposure interferogram of sympathetic explosion of two PbN_6 pellets of about 6 mg in weight vertically arranged at stand-off distance of 8 mm. The upper one triggered the lower one. The shape of the detonation product gas bubble of the upper explosive became irregular initiated by the shock wave of the sympathetic explosion. Figure 9.56c shows a single exposure interferogram: two PbN_6 pellets of 6 mg each are positioned horizontally at the stand-off distance of 8 mm. The left one exploded first and the right one exploded. In Fig. 9.56d, two 5.4 mg PbN_6 pellets separated horizontally 8 mm apart. The right one exploded first and the left one exploded. Spherically shaped wavelets are remnants of secondary shock waves created by the collapse of small bubbles attached on the cotton thread. Figure 9.56e, f shows sympathetic detonation of explosives arrayed in a line on a cotton thread.

9.6 Ultrasonic Oscillatory Test

A preliminary study started in 1980 aiming at the development of Japan's geothermal power generation. An ultrasonic oscillatory test was adopted to test appropriate materials that would resist against corrosive steam and vapor mixtures extracted from underground heat sources. Metal pieces were oscillated in ultrasonic frequency in liquids for wide ranges of temperatures and pressures which would simulated such harsh environments (Sanada et al. 1983). Generations and collapses

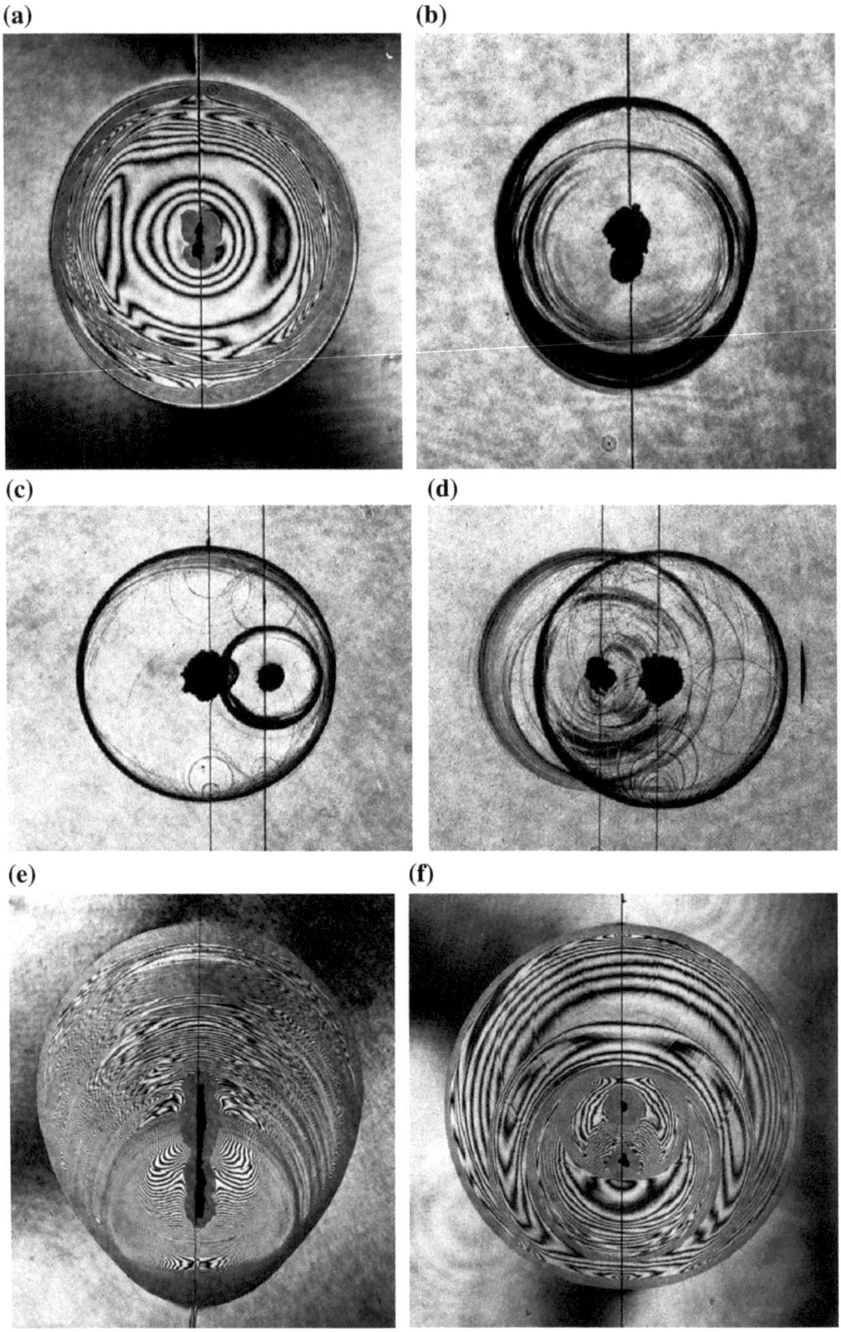
(a)
(b)
(c)
(d)
(e)
(f)

◂**Fig. 9.56** Sympathetic explosion caused by bubble collapse at 1013 hPa, 287.4 K: **a** #83112802, 16 μs from ignition, 5.5–5.6 mg PbN_6 pellets were glued vertically on a cotton thread at the stand-off distance of 4 mm; **b** #83112803, 9 μs from ignition, 5.8–5.9 mg PbN_6 pellets, vertically at the stand-off distance of 4 mm; **c** #83113004, 9 μs from ignition, 6.0–6.0 mg PbN_6 pellets, horizontally at stand-off distance of 8 mm; **d** #83120512, 7 μs from ignition, 5.4–5.4 mg PbN_6 pellets, horizontally at stand-off distance of 8 mm; **e** #83113007, 8 pieces of 5 mg PbN_6 pellets in a row; **f** #84031308, 5–5 mg PbN_6 pellets, vertically at stand-off distance of 8 mm, 25 μs

of bubbles on the metal specimens at a high-frequency are supposed to reproduce corrosive environments. Then ultrasonic oscillatory tests are extensions of shock/bubble interactions. Single exposure interferometry and high speed streak photography (ImaCon 790, John-Hadland) were intensively applied to the observation of bubble motions and subsequent generations of shock waves. For conducting the streak recording, an Argon-ion laser of 500 mW and a mechanical shutter having a 1 ms long opening time interval and a combination of 100 μs opening time and 100 μs closing time were used.

The experiment was conducted in a 200 mm diameter and 150 mm wide stainless steel chamber and the test liquid was ion exchanged water. The test chamber was designed to withstand against pressures ranging from 0.1 to 0.5 MPa and temperatures ranging from 272 to 373 K. Observation windows were 150 mm diameter and 20 mm thick optical glasses. An ultrasonic oscillator had a horn shaped oscillatory section which was driven by a 500 W motor and installed vertically into the test chamber. At the tip of the oscillatory horn, a 16 mm diameter test metal piece was attached and was immersed in test water by 3 mm in depth. The horn was driven at frequency of 17.7 kHz and a half amplitude of 17.5 μm. The phase angle of the horn's oscillation cycle was synchronized with the irradiation of a Q-switched ruby laser beam. The high-speed framing and streak observations were used to supplement holographic observations. The oscillator displacement x is defined by,

$$x = a \sin 2\pi\omega t, \tag{9.16}$$

where a is amplitude 17.5 μm and ω is frequency 17.7 kHz. One cycle of oscillation takes approximately 56 μs. Although the amplitude is very small, its acceleration eventually induces a force of the order of magnitude of 10^4 *g* where *g* is the gravitational acceleration. This force exceeds even the spalling strength in water attached to the test piece surface. Therefore, cavitation bubbles are created instantaneously. Sequential holographic images are presented in Fig. 9.57 and a streak photograph is shown in Fig. 9.58.

Figure 9.57a shows sequential single exposure interferograms synchronized with the phase angle of every π/4. The upward motion of the image corresponding to the oscillator's recess motion induced a very high tensile force on the oscillator's surface and hence cavitation bubble clouds are created. Dark small spots seen in Fig. 9.57 shows the temporal distribution of bubbles underneath the oscillator. The oscillator's forward motion compressed water and resulted in compression waves, which collapsed bubbles generating secondary shock waves on the oscillator's surface. Shock waves interacted then with the neighboring bubbles and collapsed

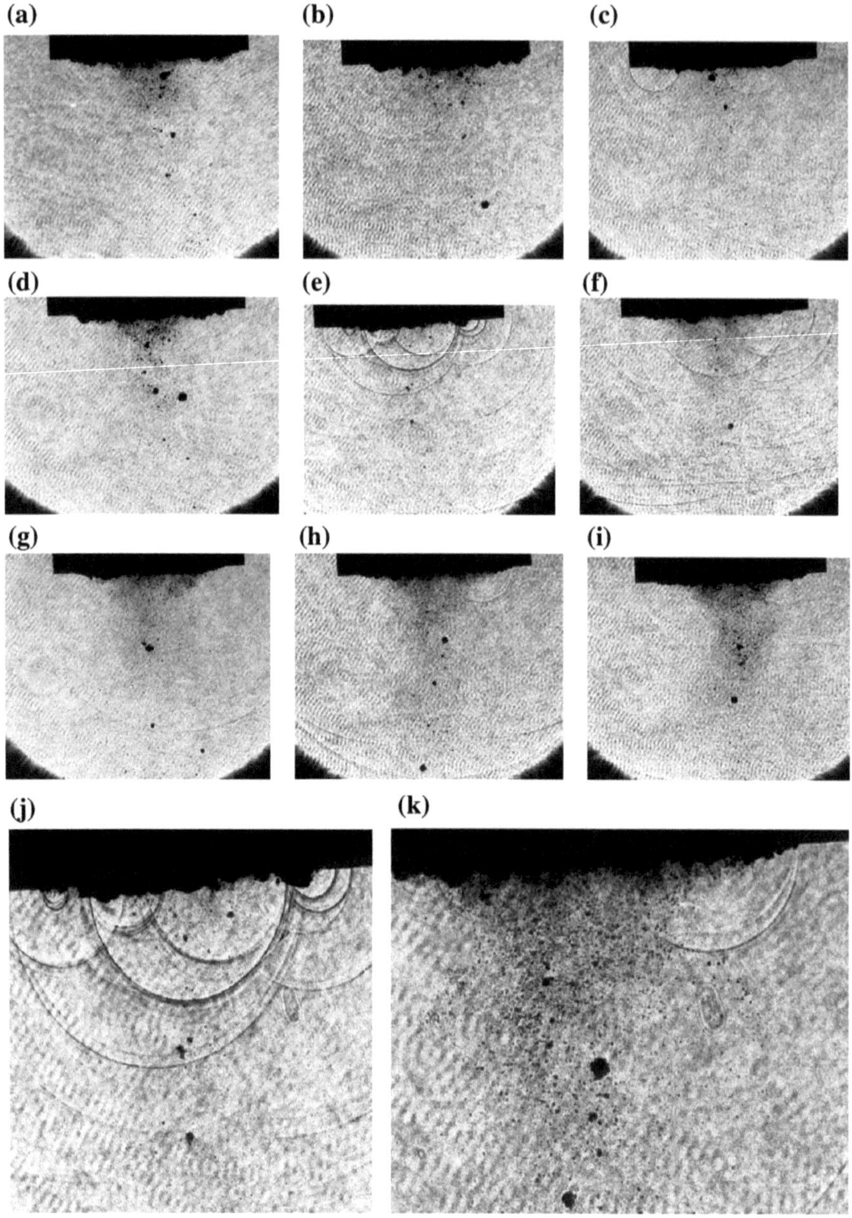

Fig. 9.57 The ultrasonic oscillation at 1013 hPa, 290 K: **a** #82112602, phase angle 0; **a** #82112603, π/4; **c** #82112604, π/2; **d** #82112605, 3π/4; **e** #82112606, π; **f** #82112607, 5π/4; **g** #82112608, 3π/2; **h** #82112609, 7π/4; **i** #82112610, 2π; **j** enlargement of (**e**); **k** enlargement of (**h**)

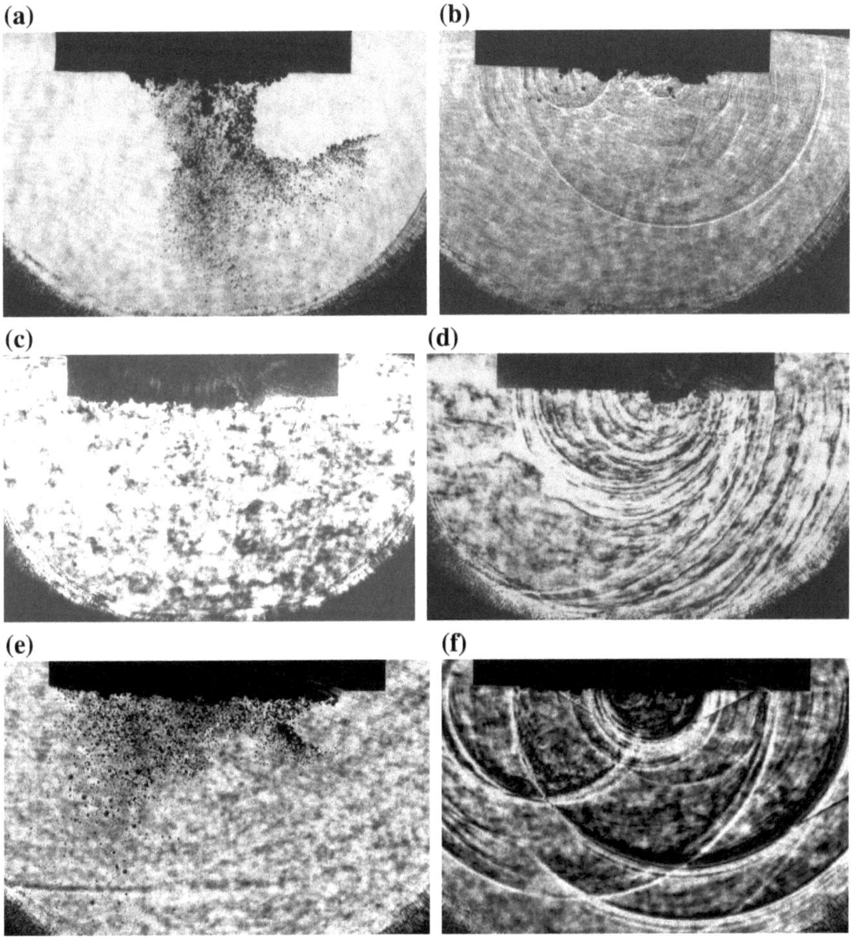

Fig. 9.58 The sequential observation of ultrasonic oscillation at 0.2 MPa, 283 K: **a** #81121765, $2\pi/3$; **b** #81121766, π; **c** #81121767, $4\pi/3$; **d** #811217658, $5\pi/3$; **e** #81121769, 2π; **f** #81121770, $\pi/3$

them. Such a chain reaction continues until the oscillator's forward motion ceases. The successive oscillatory motion creates a convective flow which slowly circulates in the entire test chamber. The circulation flow is directed at first downward and later moved upward close to the oscillator. Figure 9.57j, k shows enlargement of Fig. 9.57e, h. In Fig. 9.57a, the initial phase angle allocated was $\pi/4$ and the phase angle was increased every $\pi/4$ until in Fig. 9.57i the phase angle reached 2π. In Fig. 9.57e, the number of shock wave so far observed were maximal at the phase angle π.

Figure 9.58 shows sequential single exposure interferograms. It is observed that the generation of bubble clouds were not synchronized with the oscillator's cycle, this implies that the hysteresis exists between the oscillatory motion and the number of bubbles observed in the field of view. Figure 9.58a–f shows sequential observation at the test pressure of 0.2 MPa and the phase angle attached to the individual images increased from 2π/3 at every π/3. The phase angle 0 should imply the initiation of the oscillator nut. In Fig. 9.58a at the phase angle 2π/3, a bubble cloud is ejected downward and then due to the buoyancy, individual bubbles slowly moved upward. Eventually, by chance, the bubble cloud resembled a bonsai pine tree, which is indicating the presence of a convective flow. Bubbles migrate following the convective motion and periodically were exposed to pressure fluctuations. However, bubbles migrating in this convective flow look hardly collapsed and hence secondary shock waves are hardly observed. In Fig. 9.57b, the bubble clouds disappear and shock waves appear. The most of shock waves have their center attached to the oscillator's surface. A secondary shock wave so far created successively promoted the neighboring bubbles to oscillate. Eventually they collapsed and produced the secondary shock waves. Then, the bubbles attached to oscillator's surface collapses similarly to a chain reaction. The streak recording shown in Fig. 9.60a clearly explain the observation shown in Fig. 9.57.

Figure 9.59 shows the sequential observation in an elevated test condition at 0.3 PMa and 323 K. As the water temperature is higher than the room temperature, the number of bubbles generated at this condition are much more than those induced at room temperature. Then the number of secondary shock waves are increasing. Figure 9.59d shows a shock wave whose center was uniquely away from the oscillator' surface. The bubble is presumably attached to the window glass. The bubble formation and the secondary shock wave generation were strongly affected by the test condition.

Figure 9.60a is a streak photograph and shows cavitation bubbles induced by ultrasonic oscillation and trajectories of resulting secondary shock waves. Grey oblique lines indicate trajectories of secondary shock waves and a vague grey shadows indicate the initiation and extinction of cavitation bubbles appearing for about 40 μs. Secondary shock waves and also cavitation bubbles appear repeatedly at every 56 μs. It is estimated that by counting the number of secondary shock wave trajectories, about one third to one half of cavitation bubbles transited to the secondary shock wave. The ordinate denotes the diameter of the oscillator 16 mm. The abscissa denotes elapsed time. The bubble cloud is densely populated at the center of the oscillator and is maintained for about 40 μs. When the bubble cloud disappears, at several μs of pause secondary shock waves repeatedly appear and soon terminate. As seen in single exposure interferograms secondary shock waves have their center attached to the oscillator' surface. In the streak photograph, a 0.15 mm wide slit was placed on the oscillator surface. Hence the center of the most of bubbles are located inside the slit width. Therefore, the trajectory inside the slit width of 1.5 mm is not reliable.

Tracing the trajectory of a shock wave in Fig. 9.60a, the time variation of a bubble radius is determined as shown in Fig. 9.60b. The ordinate denotes a radius of a shock wave in mm and the abscissa denotes elapsed time in μs. In interpolating the trajectory, the speed of the shock wave can be obtained as shown in Fig. 9.60c. The shock wave attenuates very quickly to a sound wave already at 4 mm from the

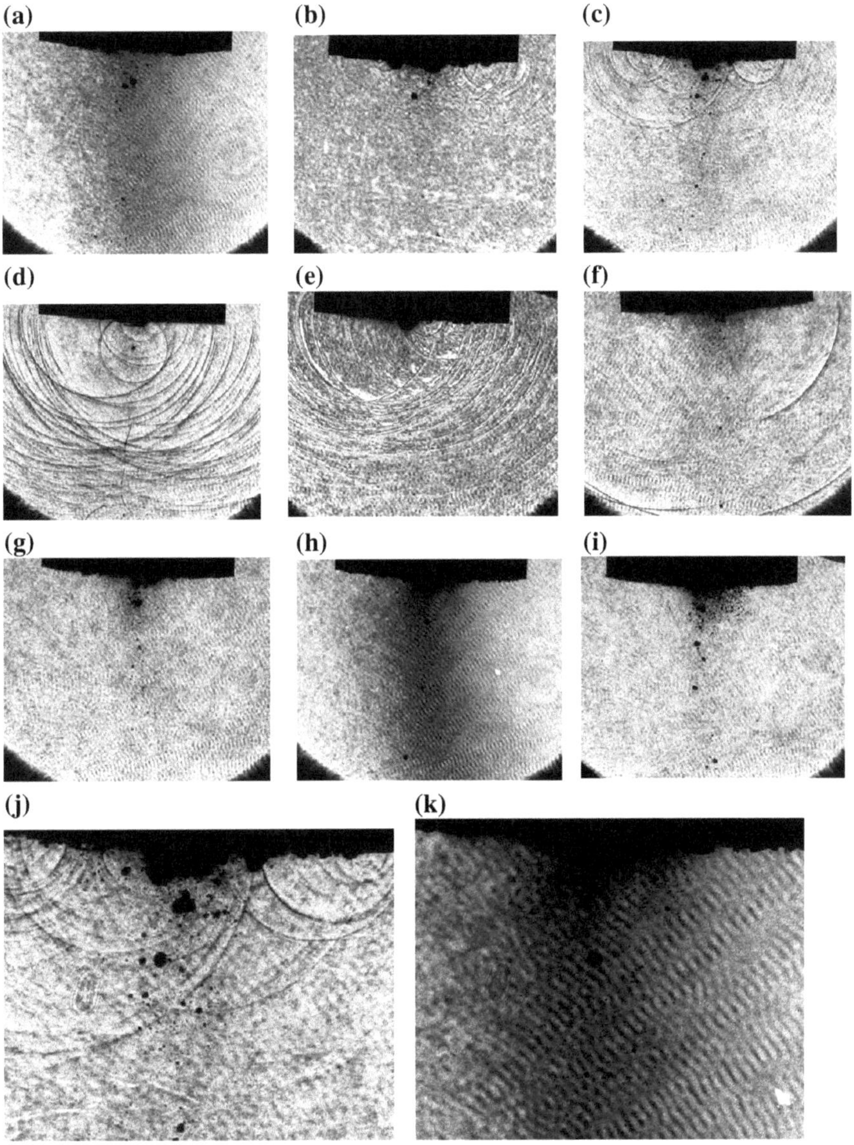

Fig. 9.59 Ultrasonic oscillation at 0.3 MPa, 323 K: **a** #82112532, phase angle 0; **b** #82112533, $\pi/4$; **c** #82112534, $\pi/2$; **d** #82112535, $3\pi/4$; **e** #82112536, π; **f** #82112537, $5\pi/4$; **g** #82112538, $3\pi/2$; **h** #82112539, $7\pi/4$; **i** #82112540, 2π; **j** enlargement of (**e**); **k** enlargement of (**h**)

center. At 4 mm from the center, the estimated shock Mach number Ms is 1.1. The pressure behind a shock wave of Ms = 1.1 is approximately 800 MPa. This overpressure is high enough to spontaneously erode the surface of hydraulic machine elements made of high-strength carbon steel. In the previous sequential

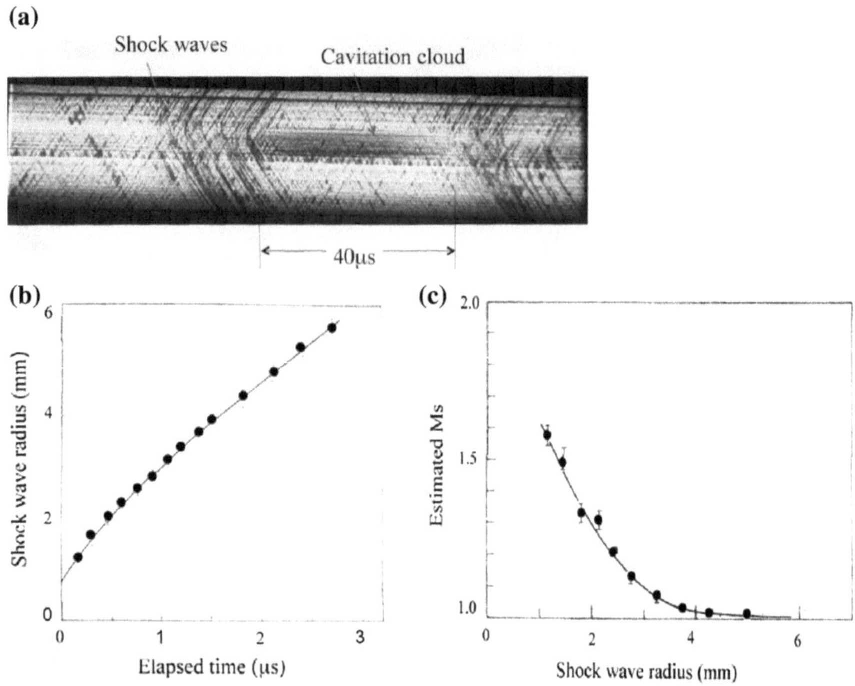

Fig. 9.60 Trajectory of secondary shock wave and its estimated speed: **a** streak photograph of bubble formation and cavitation bubble collapse, #82120820 at 1013 kPa, 293 K; **b** trajectory of a bubble; **c** speed of shock wave generated at collapsing bubble (Sanada et al. 1983)

observations, the oscillator's recess motion generated bubble clouds which were distributed in the vicinity of the surface. The shapes of secondary shock waves were hemi-spherical which indicated that most of collapsing bubbles were located on the oscillator's surface. The presence of circular shaped secondary shock waves was rather exceptional because all the bubbles so far generated did not necessarily collapse, but migrated downward in the convective flows and was eventually dissolved in water.

In order to examine the structure of bubble clouds, a shock wave was loaded over the oscillator. Figure 9.61 shows single exposure interferogram of shock wave propagation over the oscillator. The visualization was conducted to detect the distribution of migrating bubbles over the oscillator, a 10 mg PbN_6 pellet was exploded at about 13 mm away from the oscillator surface. The shock wave over-pressure at the stand-off distance of 20 mm was about 25 MPa so that all the water vapor hit by such a strong shock wave collapsed at once and created the secondary shock waves. Hemi-circular rings were spherical shock waves originated from bubbles which were distributed on the oscillator surface. Circular rings were spherical shock wave originated from bubbles distributed away from the oscillator

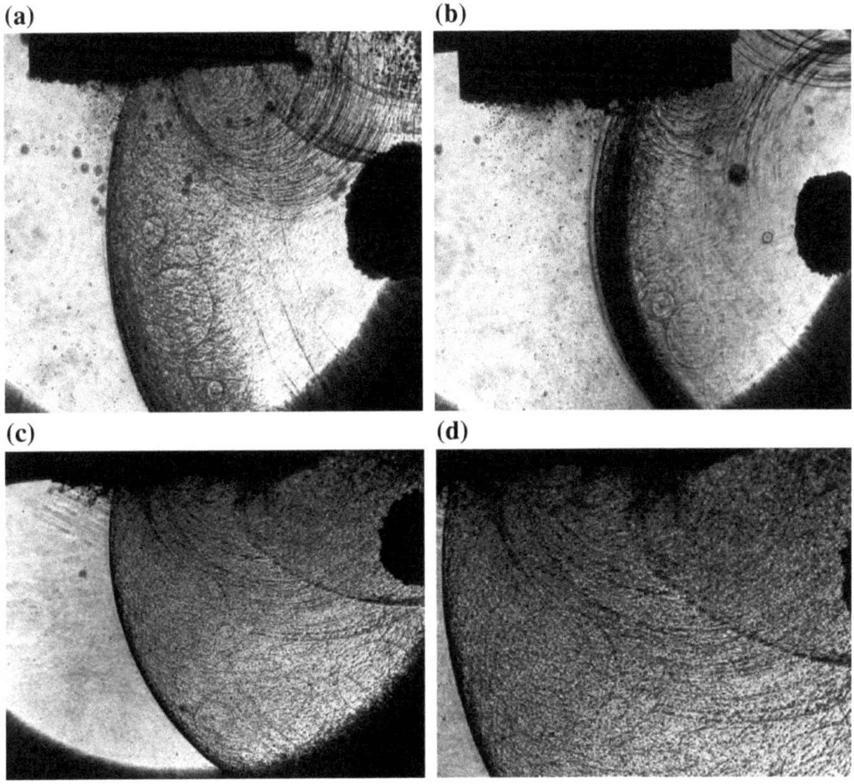

Fig. 9.61 Interaction of spherical shock waves with bubble cloud: **a** #82112973, 77 μs from trigger point; **b** #82112975, 91 μs; **c** #82113035, 31 μs; **d** enlargement of (**c**)

surface. It is readily observed that many bubbles are carried away from the oscillator surface and migrated in the convective flow. These bubbles never contributed to the erosion of the oscillator.

9.7 Interaction with Arrayed Acrylic Cylinders

Figure 9.62 shows stress wave propagations along 10 mm diameter and 10 mm long acrylic cylinders packed in a 5 × 6 arrangement and submerged in a 10 mm wide water chamber. A 9 mg AgN_3 pellet was exploded on the top of the arrayed cylinders. Underwater shock wave propagated at 1.5 km/s, whereas the stress wave propagated at 2.9 km/s in the cylinders. The stress wave transmitted into the cylinder through the contact point of the acrylic cylinders. However, cylinders touched tightly at their vertical contact points but did loosely at their side contact points. Therefore, the stress wave propagated selectively downward through the

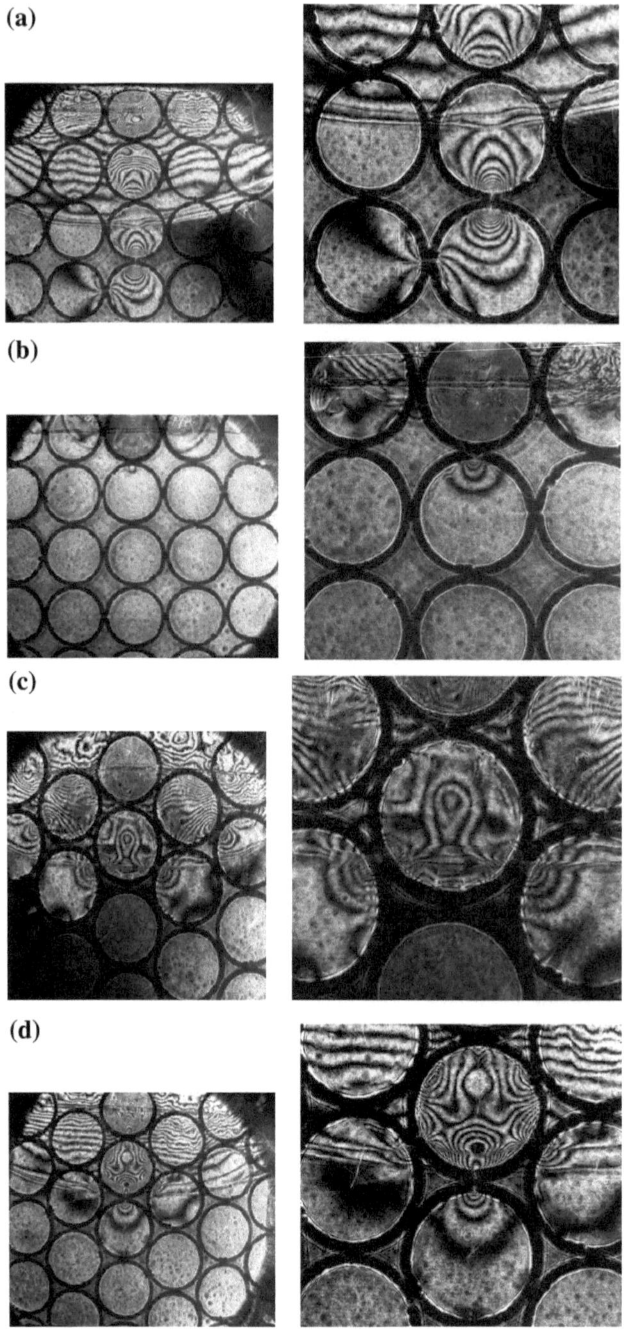

Fig. 9.62 Stress wave propagation in a 5 × 6 arrayed acrylic cylinders, 9 mg AgN_3: **a** #95100515, enlargement; **b** #95100517, enlargement; **c** #95100520, enlargement; **d** #95100522, enlargement

vertical contact points. Fringes appeared inside the cylinder propagating at much faster speed than that in water. Unlike stress waves visualized by photo-elasticity, the fringes appeared in the acrylic cylinders show the difference in phase angles which occurred during the double exposures. If the relationship between the refractive index and the density in acryl are known, the values of the stress would be determined directly by counting the fringe distributions.

9.8 Super-Cavitation

9.8.1 High-Speed Entry of a Slender Body into Water

Can a slender body move at a supersonic speed in water? The slender body was launched by a ballistic range into water at a supersonic speed. The slender body was launched at 1600 m/s into water (Saeki 1993). Figure 9.63a shows the slender body and a sabot. Figure 9.63b shows a test chamber installed in a ballistic range. The chamber had a 300 mm wide and 300 mm diameter cylindrical shape having 20 mm thick observation glass windows. The slender body was 10 mm diameter and 40 mm long and was made of titanium alloy. The slender body was launched into the test chamber from a 100 mm diameter entry port which was separated by a 30 μm thick Mylar membrane from the test chamber filled with water and accommodated in the ballistic range. As the entry speed was 1600 m/s, a bow shock appeared ahead of the slender body. The window glass of the test chamber was illuminated by a flood lamp and the images were recorded by a high-speed video camera Shimadzu SH100 at 10^6 frame/s. In Fig. 9.64 selective images were displayed at the frame interval of 16 μs.

Figure 9.64 shows sequential observation of the slender body projected horizontally into the test chamber. However, its head was slightly shifted upward so that a centrifugal force shifted its course and deformed it. Figure 9.64a shows an

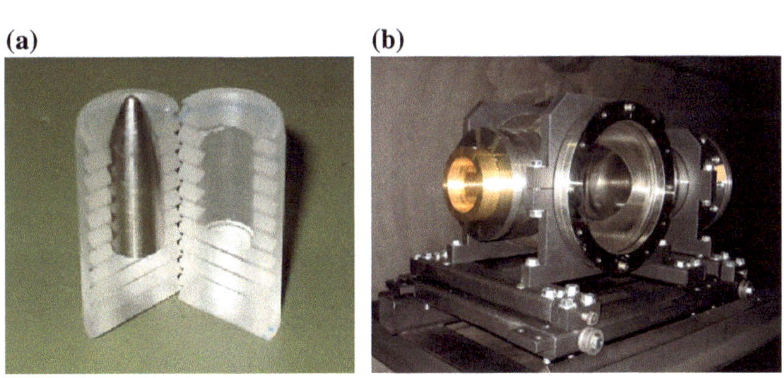

Fig. 9.63 Supersonic entry of a slender body projected into water: **a** a slender body made of titanium alloy; **b** test chamber

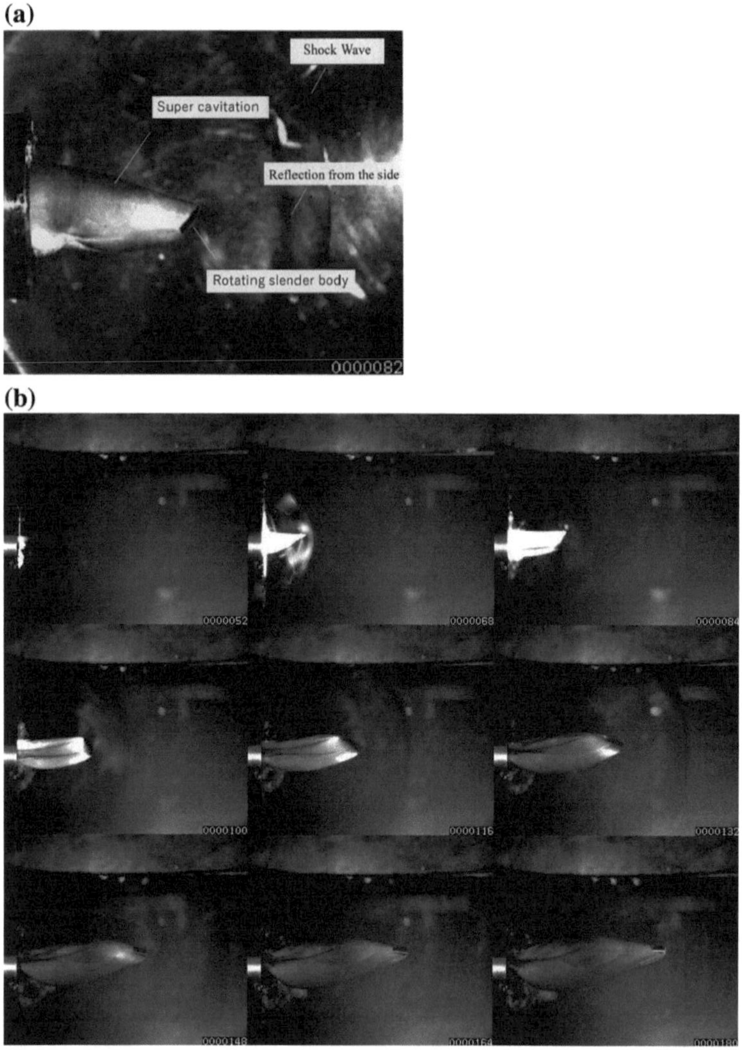

Fig. 9.64 High speed entry of a titanium slender body, sequential observation at frame interval of 16 μs: **a** enlarged view; **b** sequential observation (Saeki 1993)

enlarged image at later time. Immediately after the entry into water, a bow shock wave appeared in front of it. The bow shock wave attenuated quickly to subsonic speed relative to the sound speed in water. As soon as the slender body entered into water, its entire body was covered with water vapor bubble. It started to spin and immediately had an upright position seen in Fig. 9.64.

Inside the super-cavitation bubble, water vapor filled at vapor pressure which kept the balance with the pressure outside the cavitation bubble. Faintly grey

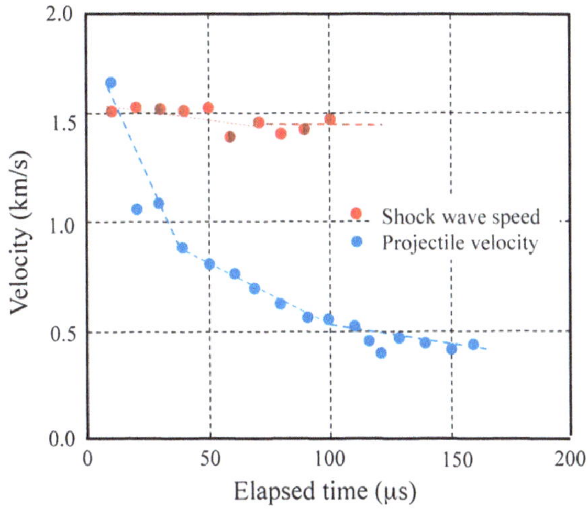

Fig. 9.65 Summary of Fig. 9.65, time variation of slender body velocity and shock wave speed

patterns are observable over its surface. The grey patterns are distributed not randomly but slightly coherently over the surface. The separation distance looked slightly elongated with elapsed time. Sequential observation of the grey patterns indicates that the space inside the super-cavitation bubble is not uniformly motion-less. In watching the time variation of the distributed grey patterns, it is speculated that high-speed flows are induced inside the super-cavitation. The quick attenuation and spinning motion would generate disturbances which would decisively produce high-speed flows and shock waves inside the super-cavitation bubble.

Figure 9.64b shows sequential images of the high-speed entry of the slender body into water. It entered into water at a supersonic speed and immediately decelerated to a subsonic speed. When the slender body hit the Mylar membrane placed at the entrance port, its impact made the slender body's head slightly upward. Its slightest upward inclination was amplified with propagation. The slender body immediately started to spin, which generated an enormous centrifugal force and bent the slender body as seen in Fig. 9.64.

Figure 9.65 summarizes images presented in Fig. 9.64 and shows the time variation of the speed of the slender body. The ordinate denotes the slender body velocity in km/s. The abscissa denotes elapsed time in μs. Filled red circles denote shock wave speed and filled blue circles denotes the speed of the slender body. The shock wave attenuated quickly from low supersonic speed to the sonic speed, whereas the slender body attenuated drastically from 1.6 km/s to approximately 0.5 km/s. The typical Reynolds number ranges of this flow region is from about 10^4 to 10^6. The drag coefficient C_D of this entry is so far estimated from the trajectory of the attenuating the slender body $C_D = 0.13$. The value of C_D for 1:4 cylinder in this Reynolds number range is $C_D = 0.87$ (Saeki 1993). It is concluded that the slender body moving in water at high-speed is surrounded by super-cavitation bubbles and the resulting value of C_D is significantly reduced to 0.13.

9.8.2 High-Speed Entry of a Sphere into Water

Figure 9.66a shows a 10 mm diameter tilted powder gun projecting a four split sabot made of polycarbonate and a 7.9 mm diameter stainless steel bearing ball contained in the sabot at any angles from 0° to 90°. Kikuchi (2011) projected the sabot and a 7.9 mm diameter ball vertically into water surface at the speed ranging from 300 to 1600 m/s. Figure 9.66b shows a direct shadowgraph of the entry of a 7.9 mm diameter sphere at 1143 m/s. Although the entry speed was subsonic in terms of the sound speed in water, at the moment of the projectile's water entry, the stagnation pressure momentarily reached at the air/water interface reached about 1.7 GPa which was almost one half of the detonation pressure of a AgN_3 pellet. Then, the high pressures propagate at the sonic speed in water. The projectile in water attenuated propagating at subsonic speed. In short, the projectile's water entry produced in front of the sphere a train of compression waves propagating at the sonic speed. This observation never contradicts the gas-dynamic interpretation, if the distance between the train of compression waves and the moving sphere is defined as a shock stand-off distance, the shock stand-off distance becomes infinitely long, when the sphere moved slowly.

Figure 9.67 shows sequential images of a 7.9 mm sphere's entry into water at the speed of 1143 m/s. The images were selected out of the recorded pictures taken by Shimadzu HS100 at 250,000 frame/s and exposure time of 500 ns. A detached shock was visible ahead of the sphere and the entry speed is equivalent to Ms = 0.763 in terms of the sound speed in water. If the train of compression waves

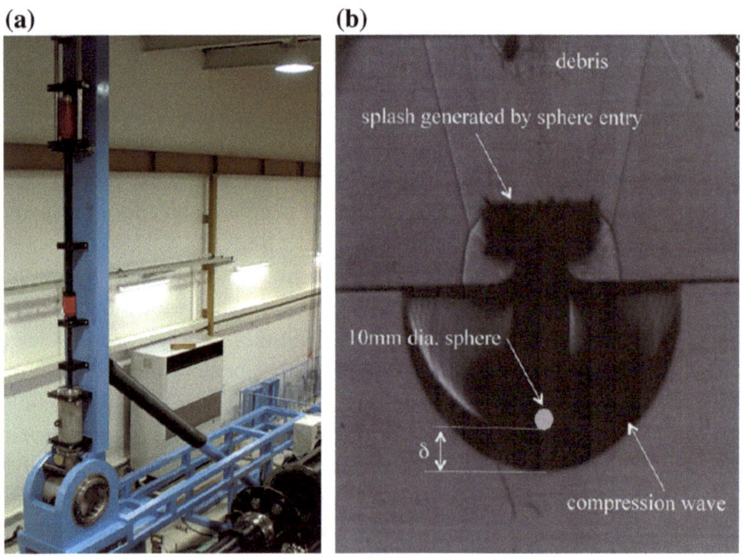

Fig. 9.66 Vertical entry of a sphere into water at entry speed of 1143 m/s: **a** a vertical two-stage light gas gun; **b** entry of 7.9 mm diameter sphere vertically into water at 1143 m/s

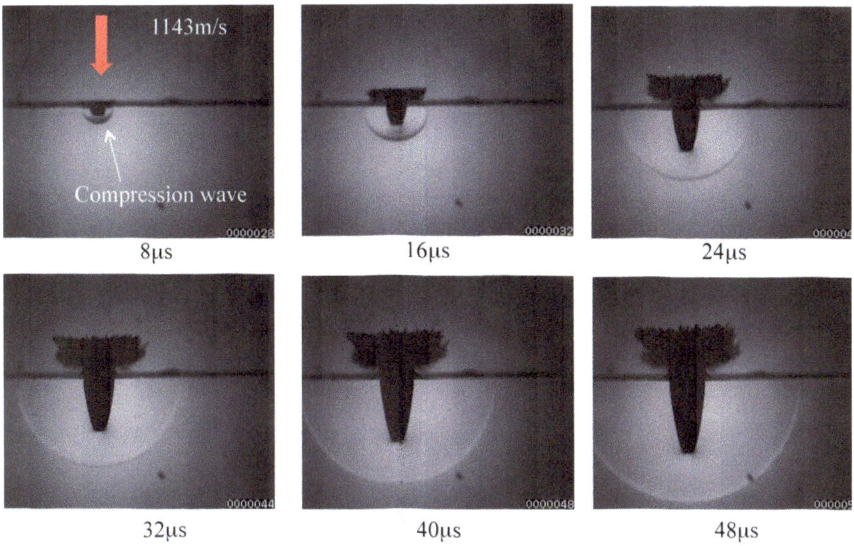

Fig. 9.67 Sequential observation of the entry of a 7.9 mm diameter sphere into water at 1143 m/s

is defined to be a bow shock propagating at the sonic speed. The dimension-less shock stand-off distance δ/d seen in Fig. 9.66 is about 2.0 in Fig. 9.67a. In Fig. 9.67, the sphere is decelerating and then the δ/d increases. In the 5th frame of Fig. 9.67, the δ/d is about 3.0. The similar trend was already discussed regarding the shock stand-off distance over a 40 mm diameter sphere launched in the ballistic range. The sphere dragged super-cavitation bubble whose surface looked a simple smooth interface. Although the visualization was conducted by a conventional direct shadow graph, if the test field was illuminated by a flood lamp in the similar manner as seen in Fig. 9.64, the surface irregularity of the super-cavitation would be observed.

Figure 9.68 shows the sequential observations of the supersonic entry of a 7.9 mm diameter sphere at 1539 m/s. Splashes from the water surface moved at the speed of the reflected shock wave from the water surface. This entry speed is equivalent to Mach number Ms = 4.46 in air and to Ms = 1.02 in water. While propagating in water, the sphere's speed varies from supersonic to subsonic. However, the dimension-less shock stand-off distance δ/d did not show any discontinuous change but increased monotonously.

Figure 9.69 summarizes results presented in Fig. 9.68. The ordinate denotes the sphere's penetration depth from water surface in mm. The abscissa denotes elapsed time in μs. Green filled circles denote positions of the sphere and red filled circles denotes the location of the shock wave. The sphere attenuates quickly with propagation and hence the dimension-less shock stand-off distance δ/d increases monotonously with elapsing time.

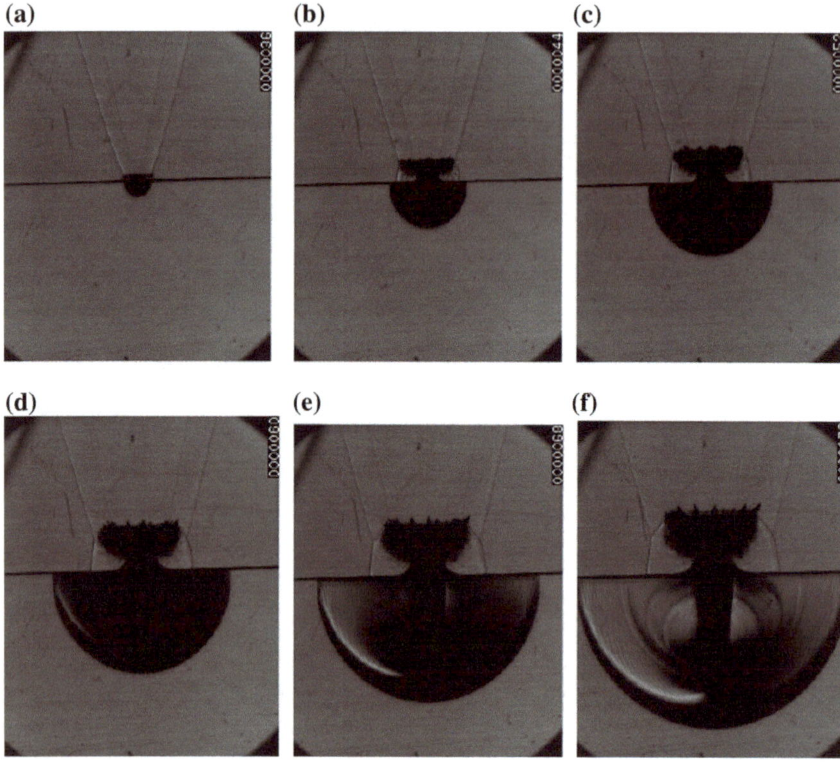

Fig. 9.68 Sequential observation of the entry of a 7.9 mm diameter sphere into water at 1539 m/s: **a** 8 μs; **b** 16 μs; **c** 24 μs; **d** 32 μs; **e** 40 μs; **f** 48 μs

Fig. 9.69 Trajectory of sphere and shock wave at entry speed of 1539 m/s

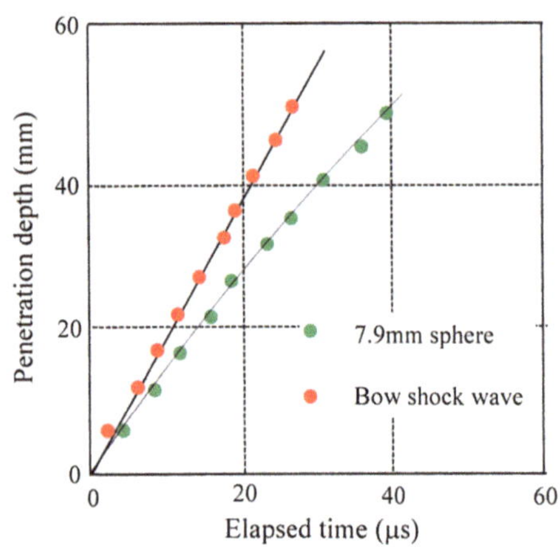

References

Abe, A. (1989). *Study of diffraction of shock wave released from the open end of a shock tube* (Ph.D. thesis). Graduate School of Engineering, Faculty of Engineering Tohoku University.

Abe, A., Ohtani, K., Takayama, K., Nishio, S., Mimura, H., & Takeda, M. (2010). Pressure generation from micro-bubble collapse at shock wave loading. *Journal of Fluid Science and Technology, 5,* 235–246.

Chaussey, Ch., Schmiedt, E., Jocham, D., Walter,V., Brendel, W., Forsmann, B., & Hepp, W. (1982). *Extracorporeal shock wave lithotripsy. New aspects in the treatment of kidney stone disease*. Kerger.

Coleburn, N. L., & Roslund, L. A. (1970). Interaction of spherical shock waves in water. In *Proceedings of 15th International Symposium on Detonation, Pasadena* (pp. 581–588).

Esashi, H. (1983). *Shock wave propagation in liquids* (Master thesis). Graduate School of Engineering, Faculty of Engineering, Tohoku University.

Glass, I. I., & Heuckroth, L. E. (1968). Low-energy spherical underwater explosions. *Physics of Fluids, 11,* 2095–2107.

Hayakawa, S. (1987). *Study of shock/bubble interaction in highly viscous liquids* (Master thesis). Graduate School of Engineering, Faculty of Engineering Tohoku University.

Hirano, T. (2001) *Development of revascularization of cerebral thrombosis using laser induced liquid jets* (M.D. thesis). Graduate School of Medicine, Tohoku University.

Ikeda, K., Matsuda, M., Tomita, K., & Takayama, K. (1999). Application of extracorporeal shock wave on bone. Basic and clinical study. In G. J. Ball, R. Hillier, & G. T. Robertz (Eds.), *Proceedings of 22nd ISSW, Shock Waves, London* (Vol. 1, pp. 623–626).

International Maritime Organization. *Science of ship and the sea.* Retrieved from http://www.imo/org/home.asp.

Kikuchi, T. (2011). *A shock dynamic study of high-speed impact onto condensed matter* (Ph.D. thesis). Graduate School of Engineering, Faculty of Engineering Tohoku University.

Kuwahara, M., Kambe, K., Kurosu, S., Orikasa, S., & Takayama, K. (1986). Extracorporeal stone disintegration using chemical explosive pellets as an energy source of underwater shock waves. *Journal of Urology., 135,* 814–817.

Nagayasu, N. (2002). *Study of shock waves generated by micro explosion and their applications* (Ph.D. thesis). Graduate School of Engineering, Faculty of Engineering, Tohoku University.

Ohtani, K., & Takayama, K. (2010) Shock wave interaction phenomena with a single helium gas bubble in liquid. In *7th International Conference on Flow Dynamics, Sendai* (pp. 120–121).

Okazaki, K. (1989). Fundamental study in extracorporeal shock wave lithotripsy using piezoceramics. *Japanese Journal of Applied Physics, 28*, 143–145.

Sachs, R. G. (1944). *The dependence of blast on ambient pressure and temperature* (BRL Report, No. 466).

Saeki, T. (1993). *Super-cavitation flows of a high speed projectile launched by two-stage light gas gun* (Master thesis). Graduate School of Engineering, Faculty of Engineering, Tohoku University.

Sanada, N., Ikeuchi, J., Takayama, K., & Onodera, O. (1983). Generation and propagation of cavitation induced shock waves in an ultrasonic vibration testing. In: D. Archer & B. E. Milton (Eds.), *Proceedings of 14th International Symposium on Shock Tubes and Waves, Sydney* (pp. 404–412).

Shima, A. (1997). Studies on bubble dynamics. *Shock Waves, 7,* 33–42.

Shitamori, K. (1990). *Study of propagation and focusing of underwater shock focusing* (Master thesis). Graduate School of Tohoku University Faculty of Engineering, Tohoku University.

Tait, P. G. (1888). Report on physical properties of flesh and of sea water. *Physics and Chemistry Challenger Expedition, IV*, 1–78.

Takayama, K. (1983). Application of holographic interferometry to shock wave research. In *International Symposium of Industrial Application of Holographic Interferometry, Proceedings of SPIE* (Vol. 298, pp. 174–181).

Takayama, K. (1987). Holographic interferometric study of shock wave propagation in two phase media, In H. Groenig (Ed.), *Proceedings of 16th International Symposium on Shock Tubes and Waves, Aachen* (pp. 51–62).
Takayama, K. (1990). High pressure generation by shock wave focusing in a confined ellipsoidal cavity. In K. Takayama (Ed.), *Proceedings of International Workshop on Shock Wave Focusing* (pp. 217–226). Institute of High Speed Mechanics, Sendai.
Takayama, K., Yamamoto, H., & Abe, A. (2015). Underwater-shock/bubble interaction and its application to biology and medicine. In R. Bonazza, D. Ranjan (Eds.), *Proceedings of 29th ISSW, Shock Waves, Madison* (Vol. 2, 861–868).
Yamada, K. (1992). *Study of shock wave interaction with gas bubbles in various liquids* (Doctoral thesis). Graduate School of Tohoku University Faculty of Engineering, Tohoku University.
Yutkin, L. A. (1950). Apparat YRAT-1 Medeport USSR Moscow.

Chapter 10
Applications of Underwater Shock Wave Research to Medicine

10.1 Extracorporeal Shock Wave Lithotripsy (ESWL)

In 1981, Professor M. Kuwahara of the Department of Urology, School of Medicine, Tohoku University invited us to develop a prototype lithotripter using micro explosions. Then the collaboration started applying results of the basic experiments to design a lithotripsy for clinical use.

A shock wave application to medicine was initiated by Yutkin (1950). He, for the first time, used electrical discharges to non-invasively remove urinary tract stones by insertion of a thin electrode through a urethra into a urinary bladder. Touching the electrode on urinary tract stones, he cracked the stones by the exposure of spark generated high-pressures. Later, fragments of the stones were removed using forceps.

Urologists in the world followed his novel technique. Various devices were developed and successfully applied to clinical treatments. A history of development of lithotripters are summarized in Loske (2007). Chaussey et al. (1986) commented that a basic idea of a lithotripsy was initiated by Heustler. His shock wave research was supervised by Professor Schardin who was a director of Ernst Mach Institute and was one of the students of Ernst Mach (Krehl 2009). The German tradition of the shock wave research was extended to this medical application.

An underwater shock wave generated by micro-explosion has a very high over-pressure but its Mach number is very close unity and behaves like a sound wave. The underwater shock wave generated at a focal point inside a truncated ellipsoidal reflector is focused at another focal point outside the reflector. As human bodies have slightly inhomogeneous structures but their acoustic impedances are more or less similar to that in water. Therefore, underwater shock waves generated at a focal point outside human bodies could focus on a kidney stone inside a human body. The pressures focused was high enough to disintegrate the kidney stone. Heustler developed his lithotripter based on this principle and the Dornier Systems succeeded his ideal and constructed the prototype lithotripter.

K. Takayama, *Visualization of Shock Wave Phenomena*,
https://doi.org/10.1007/978-3-030-19451-2_10

Chaussey et al. (1986) named this lithotripsy as extracorporeal shock wave lithotripsy (ESWL). 4–5% of people in the world are estimated to suffer urinary tract stone disease and then it would be wonderful to cure millions of people by applying with this non-invasive lithotripsy. Chaussey et al. (1982) used electric discharges for producing underwater shock waves. Kambe et al. (1986) generated underwater shock waves by detonating small PbN_6 pellets in a controlled way and proved that this method was suited for medical applications.

10.2 Truncated Ellipsoidal Reflector

Figure 10.1 shows visualizations of shock wave focusing and its numerical pressure distribution. A shock wave was generated at a focal point inside a truncated ellipsoidal reflector of minor radius of 45 mm × major radius of 63.6 mm made of acryl. The direct wave was released from the focal point and diverging, whereas the reflected shock wave propagates toward the second focal point outside the reflector. In the second picture, the reflected shock wave focused at the second focal point outside the reflector. A very sharp accumulation of pressures was observed at the second focal point (Obara 2001).

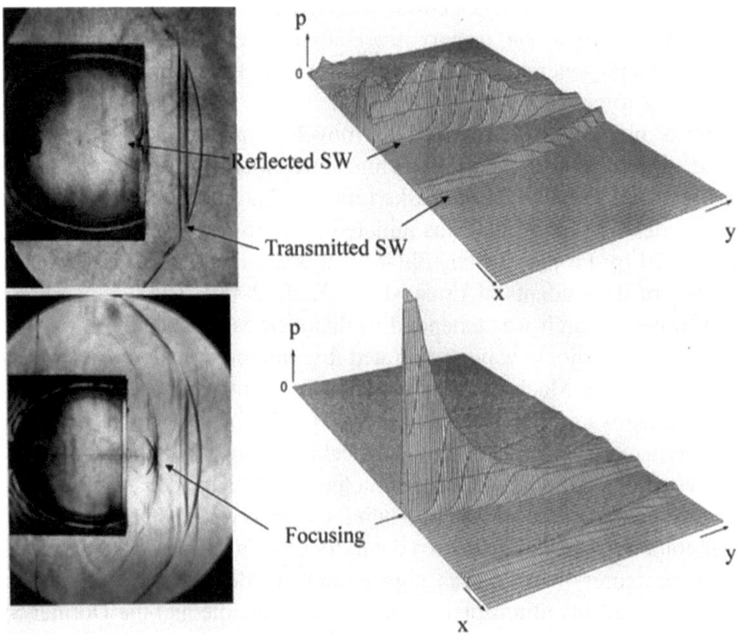

Fig. 10.1 Shock wave focusing using a truncated ellipsoidal reflector made of acryl and numerical simulation pressure distribution in x, y-plane (Obara 2001)

Prototype reflectors were larger than this test model but their character was almost the same as that of the test model. The prototype reflectors generated a maximal pressure at about 100 MPa but the test model generated 50 MPa which decreased to ambient pressure in a few μs. The half width of pressure profiles had approximately from 2.0 to 4.0 mm. High pressures at a focal point inside human bodies damage, at the same time, tissues. Therefore, the value of peak pressures and the pressure profile should be optimized hopefully to minimize the tissue damage. Kidney stones were disintegrated not by a single shock wave focusing. The shock waves were repeated focused and stones were gradually disintegrated. On the stone exposed to shock wave focusing, at first compression stress waves propagated inside the stone and the reflected stress wave from its rear surface became a tensile stress wave. Then the transmitted compression stress wave and the tensile stress wave combined and effectively crashed the stone. The stone was, therefore, disintegrated at its frontal side and at its rear side as well. Hence to maximize the value of the compression stress wave and the tensile stress wave, Chaussey et al. (1986) pointed out that the half width of the peak focusing pressure profile should have the length of about one half to one third of the stone diameter. The shape of the prototype reflector was optimized following his empirical observation.

In Fig. 10.2, eight half-truncated ellipsoidal reflectors were examined in which the major radii were varied while the minor radius was fixed to be 45 mm. The parameter e denotes the focal length from the exit, and the value ranges from 30 to 78 mm. f is the f-number of the reflector, a ratio of the focal distance to the diameter of the opening 90 mm. The value ranges from 0.33 to 0.87. A 10 mg PbN_6 pellet placed at the first focal point inside reflectors was detonated. The peak pressures were measured along the axis. Figure 10.3 summarize the results.

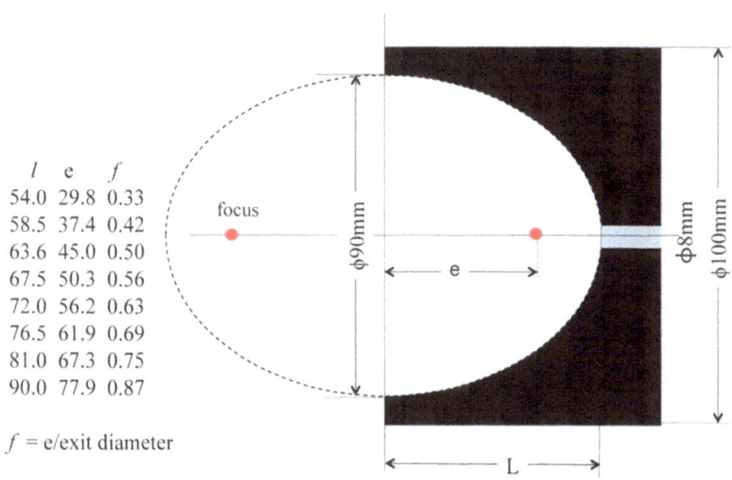

Fig. 10.2 Geometry of present half truncated reflector: minor radius = 45 mm. Major radius *l* variable from 54 mm to 90 mm (Obara 2001)

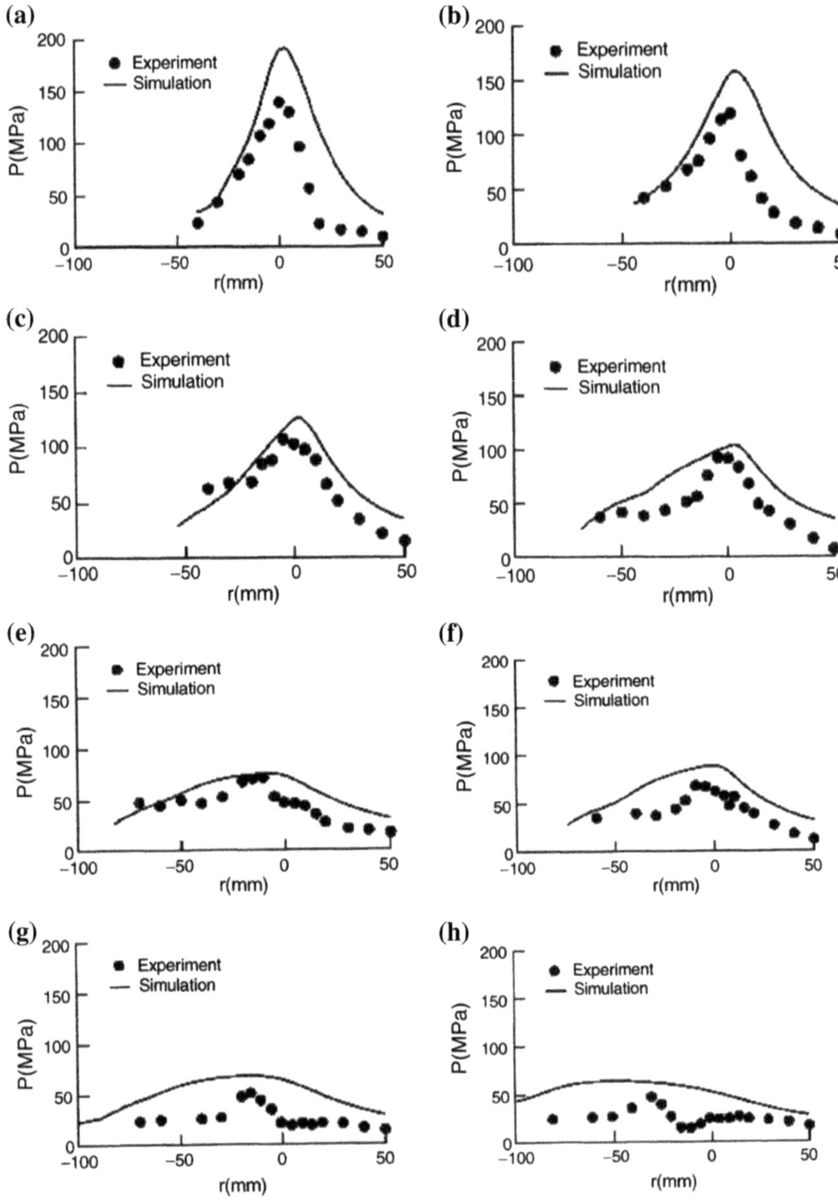

Fig. 10.3 Pressure profiles of reflectors presented in Fig. 10.2: **a** f = 0.33, L = 54.0 mm; **b** f = 0.42, L = 58.5 mm; **c** f = 0.50, L = 63.6 mm; **d** f = 0.56, L = 67.5 mm; **e** f = 0.63, L = 72.0 mm; **f** f = 0.69, L = 76.5 mm; **g** f = 0.75, L = 81.0 mm; **h** f = 0.87, L = 90.0 mm (Chaussey et al. 1986)

Figure 10.3 show peak pressure distributions along the major axis shown in Fig. 10.2 and the results of numerical simulations. The maximal peak pressure in Fig. 10.2a is approximately 160 MPa. This value is too high to safely apply to clinical treatments. Favorable pressures applicable to the ESWL in vivo is about 20–30 MPa, since the kidney stone's maximal yielding stress is at most 80 kg/cm^2 (Chaussey 1983). The peak pressure attenuates inside human bodies is about at the factor of 1/4 to 1/2. Hence, by taking these effects into account, the peak pressure focused at the second focal point should range in vivo from 60 to 100 MPa. The reflector shown in Fig. 10.3b, generates the peak pressure of approximately 140 MPa and the corresponding numerical result agrees well with the measured value. In the case of reflectors with larger f-numbers, the peak pressure so far generated is low and their focal point are slightly deviated from the exact second focal point. Expansion waves created at the reflectors' opening also converged at the second focal point. In the reflectors having large f-number, the procedure of focusing interacted with converging expansion waves and then the maximal peak pressure appeared in the position ahead of the second focal point. The slight deviation of the position at which the peak pressure appeared in the measurement and numerical simulation was attributable to the presence of detonation product gas, the so-called fire ball. In the experiments, the reflected shock wave interacted with the fire ball but this effect is neglected in the numerical simulations.

10.2.1 Prototype Ellipsoidal Reflector

Reflectors having larger f-number and longer stand-off distances may have been useful shape for clinical treatments, however, as seen in Fig. 10.3e–h, their peak pressures are about 50 MPa and their focal points are deviated toward the reflectors' openings. In conclusion, the reflectors are not suited for clinical use. The reflectors with a small f-number as seen in Fig. 10.3a, b, generate high peak pressure which will readily damage tissues. Hence, reflectors having shapes seen in Fig. 10.2c, d are appropriate.

Taking physical features of average Japanese adults into consideration, kidneys are positioned at about 100 mm from body surfaces. Hence, the distance from the reflector's opening to kidney ranges from at least 130–140 mm and the reflector's f-number is about 0.5.

Prototype reflectors of different configurations were made of brass and their pressure profiles were measured. Lastly, the #2 prototype reflector of f = 0.5 with the minor diameter of approximately 180 mm is manufactured. A 10 mg PbN_6 pellet was attached on the edge of 0.6 mm core diameter optical fiber and positioned at the first focal point inside the #2 reflector. Transmissions of a Q-switched Nd:YAG laser beam through the optical fiber ignited the pellet. Pressures were measured along the major axis with a pressure transducer, Kistler model 601H and PVDF (poly vinyl denefluoride) pressure transducers which were made of a 5 mm diameter sheet of polyvinyldene difluoride and manufactured in house.

The pressure distribution is shown in Fig. 10.4. The ordinate denotes the peak pressure in MPa and the abscissa denotes the axial distance in mm. The origin is the second focal point. The shock wave propagates to the left. The maximal peak pressure is about 85 MPa, which is high enough to safely disintegrate kidney stones. The maximal peak pressures appear at about 5 mm toward the focal point. The similar trend was also observed in model reflectors as seen in Fig. 10.2. The deviation of the position at which the peak pressure appeared was attributable to the interaction of the reflected shock wave with the fire ball.

In Fig. 10.5, double exposure and single exposure interferograms show the disintegration of a 10 mm diameter sintered alumina sphere placed at the secondary focal point of #2 reflector. The shock wave propagated from right to left. Upon the shock wave focusing on the alumina sphere, the pressure on the reflector side was so enhanced that fringes appeared densely on the stone model in Fig. 10.5a, b. In these double exposure interferograms, the shadow of the alumina sphere before and after the shock wave focusing were superimposed so that the shape of deformed sphere looked disturbed. The grey shadows visible downstream of the sphere were a bubble cloud. Figure 10.5c is a single exposure interferogram taken at 20 μs after the time instant when Fig. 10.5b was taken. On the frontal side of the sphere, a bubble cloud appeared and the alumina sphere was slightly deformed. Figure 10.4

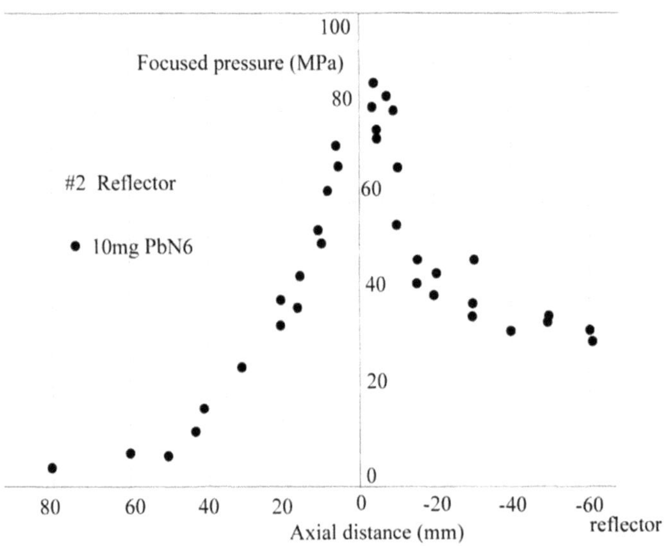

Fig. 10.4 Measured pressure profile of #2 reflector (Obara 2001)

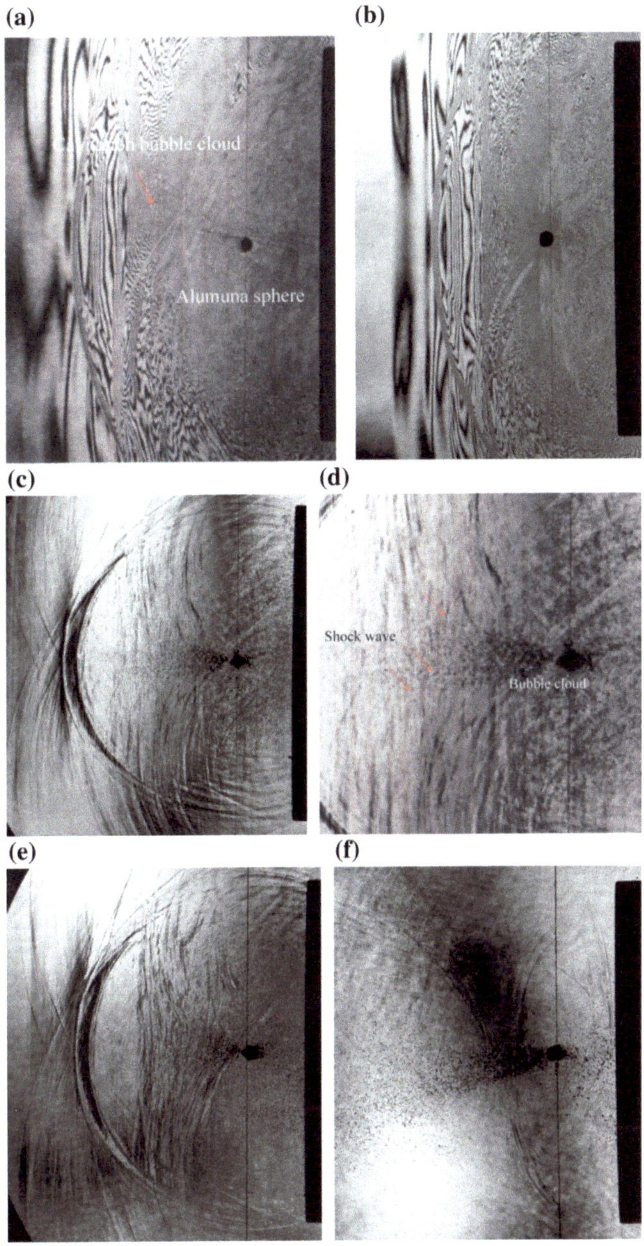

Fig. 10.5 Sintered 10 mm outer diameter alumina, focusing by ellipsoidal reflector, at 1013 hPa, 286.7 K: **a** #83122201, 90 μs, Shock waves are generated by exploding 9.1 mg PbN_6 pellet; **b** #83122202, 80 μs from trigger point, 9.1 mg PbN_6 pellet; **c** #83122203, 100 μs 7.9 mg PbN_6, single exposure; **d** Enlargement of (**c**); **e** #83122204, 140 μs, 7.7 mg PbN_6, single exposure; **f** #83122205, 249 μs, 7.5 mg, PbN_6, single exposure; **g** Enlargement of (**f**); **h** #83122206, 400 μs, 7.1 mg, PbN_6; **i** #83122210, 850 μs, 5.8 mg, PbN_6, single exposure; **j** #83122303, 1200 μs, 9.3 mg, PbN_6, single exposure; **k** #83122306, 2500 μs, 8.7 mg, PbN_6, single exposure

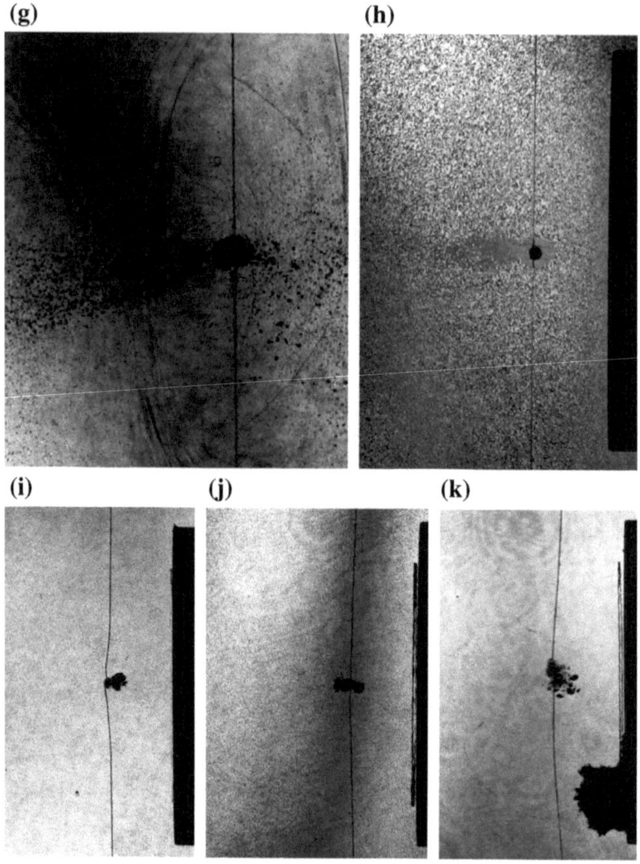

Fig. 10.5 (continued)

showed the higher peak pressure in the frontal side but decreasing pressure on the rear side. Hence this pressure gradient promoted the generation of cavitation bubble cloud which were shattered toward the downstream. Figure 10.5d is an enlarged image of Fig. 10.5c. Small rings observed at the tail of the cavitation cloud, which were secondary shock waves generated at the collapsing cavitation bubbles. Bubbles were spatially and temporally randomly collapsed responding to pressure fluctuations.

The distribution of cavitation bubbles is readily identified by observation of single exposure interferograms but the double exposure interferograms recorded all the changes in phase angles during the double exposure including subtle and major density variations, whereas the single exposure one could record only major density changes. Figure 10.5c–f are single exposure pictures and their enlargements.

In Fig. 10.5d, g ring shaped patterns are secondary shock waves occurring at bubbles collapse. It should be noticed that all the bubbles are not collapsed but only a selective number of bubbles collapsed.

Figure 10.5i–k show later stages. The alumina sphere was disintegrated. The fire ball contained lead vapor which was heavier than water. Therefore, long time after explosion, the fire ball came out along the reflector's lower wall as seen in Fig. 10.5k.

10.2.2 Preparatory Tests

Prior to the application of the prototype reflector to in vitro experiments, the shock wave mitigation through tissues was investigated. The shock wave attenuation through a slice of a swine meat of 15 mm × 100 mm, and 10 mm in thickness suspended at the exit of truncated reflector was observed as shown in Fig. 10.6.

Figure 10.7 show shock wave interactions with a ceramic sphere, a lib cartilage, and a rubber balloon containing swine liver. Shock waves propagated faster in the ceramic sphere and slightly faster over rubber balloons.

Figures 10.8 and 10.9 show shock wave interaction with extracted kidney stones. Depending on the type of kidney stones, interaction patterns differ from each other.

Figure 10.10 show sequential observations of shock wave focusing using the #2 reflector and the effect of sponge layer placed in front of the reflector opening. In Fig. 10.10a–c, the concentration of fringes at the second focal point are shown. Although the sponge sheet covered lower half of the opening, still the focusing was achieved. Figure 10.10d–g shows the procedure of focusing.

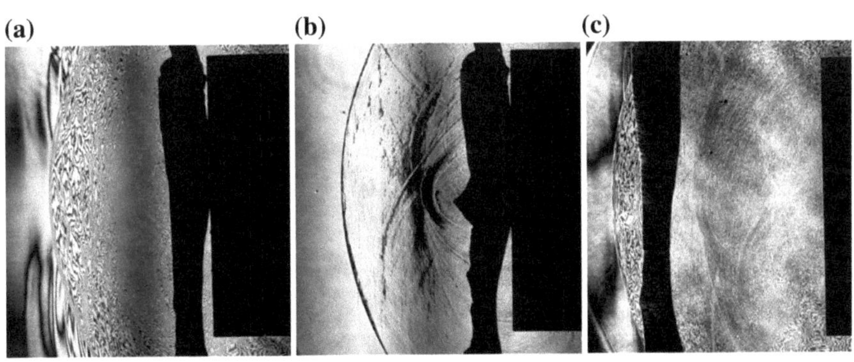

Fig. 10.6 Effects of a swine meat of 15 mm × 100 mm, thickness 10 mm on shock wave focusing: **a** #84030506, 31 μs from trigger point, 11.0 mg PbN_6; **b** #84030505, 31 μs, 11.8 mg PbN_6 single exposure; **c** #84030510, 39 μs, 9.0 mg PbN_6

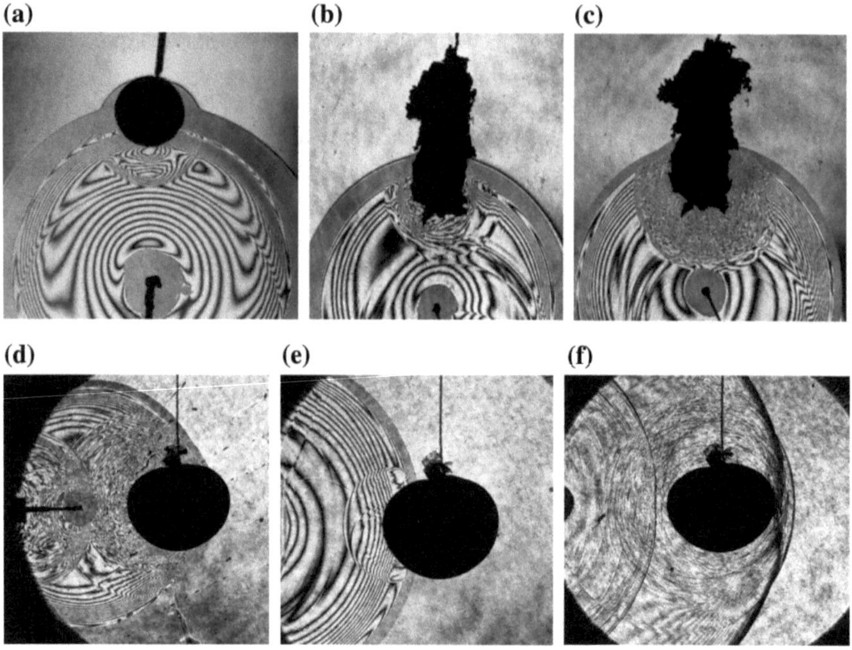

Fig. 10.7 Shock wave interaction with ceramic sphere and cartilage: **a** #84052105, 10 mm diameter ceramic sphere, AgN_3 11 mg; **b** #84051607, rib cartilage, 10 mg AgN_3; **c** #84051608, rib cartilage, 11 mg AgN_3; **d** #91121826, 387 μs, Swine liver confined in a rubber balloon, 10 mg AgN_3, stand-off distance L = 30 mm, 287.2 K; **e** #91121812, 383 μs, Swine liver confined in a rubber balloon, 10 mg AgN_3, L = 50 mm Tw = 301.2 K; **f** #91121815, 368 μs

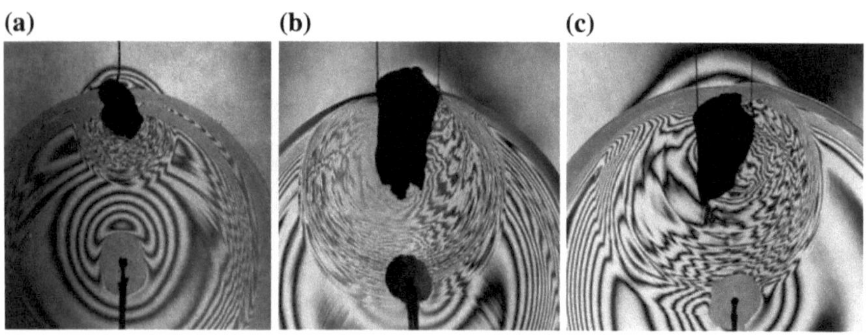

Fig. 10.8 Shock wave interaction with extracted kidney stones: **a** 84051703, dry weight 0.32 g 13 μs from trigger point, 8 mg PbN_6, at 288.7 K; **b** #84052201, 21 μs, 8.3 mg AgN_3, 290.3 K; **c** #84052202, same as #84052201 25 μs, 9.9 mg AgN_3, 290.3 K

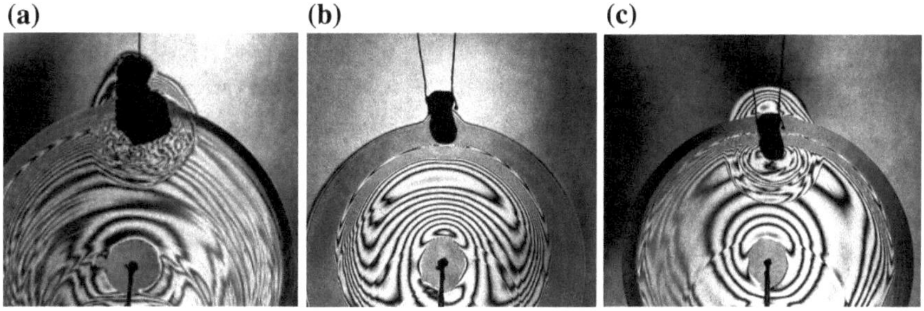

Fig. 10.9 Shock wave interaction with extracted kidney stone; **a** #84051801, dry weight 0.92 g, 12 μs from trigger point, 8 mg AgN_3, 288.4 K; **b** #84052109, 11 μs, 10.2 mg AgN_3, 291.2 K; **c** #84052109, the same as #84052109, 14 μs

10.2.3 In Vitro Experiments

Figure 10.11 shows sequential observation of the disintegration of an extracted kidney stone in vitro taken by 6000 frame/s high speed movie. The shock wave was generated by the explosion of a 10 mg PbN_6 pellet. The kidney stone of 10 mm in diameter and 20 mm in length was placed at the second focal point at 130 mm away from the opening. The inter-frame time was 130 μs. This frame interval was too long to resolve the shock wave. It is, hence, impossible to visualize both a shock wave and kidney stone deformation at the frame rate of 6000 frame/s. Nevertheless, the inception of cavitation bubbles was observed. When the shock wave passed the kidney stone, it instantaneously contracted, although not clearly identified, bulged and then was fragmented. When the stone bulged, the cracks were created on its frontal side.

During clinically experiments, patients did not claim any serious pains. The feeling of pains are totally different depending on gender, age and other factors. Even the identical shock wave over-pressures was loaded, patients claimed pain differently. Some people endured pains caused by over-pressure loading but other people could not. Very little is known about the correlation between feeling of pain and the peak overpressure.

Shock waves created by the explosion of a 10 mg AgN_3 pellet were focused on a kidney stone of the size 10 mm × 20 mm and visualized with a high speed movie with framing rate at 10,000 frame/s. The illumination was a flood light. The sequential images were displayed in Fig. 10.12 (Obara 2001). Unlike direct shadow pictures, the deformation of the kidney stone was still resolved. The shock wave propagated from the left hand side.

In No. 2–7 frames, a few μs after shock wave focusing, cavitation bubbles were generated ahead of the kidney stone. Exposed to the bubble cloud, the stone bulged. In No. 8–15 frames, the bubbles disappeared but still attached on the front side. Upon the exposure of high pressures, cracks were created on the frontal surface. When the bubble detached from the stone, it bulged and then started to contract.

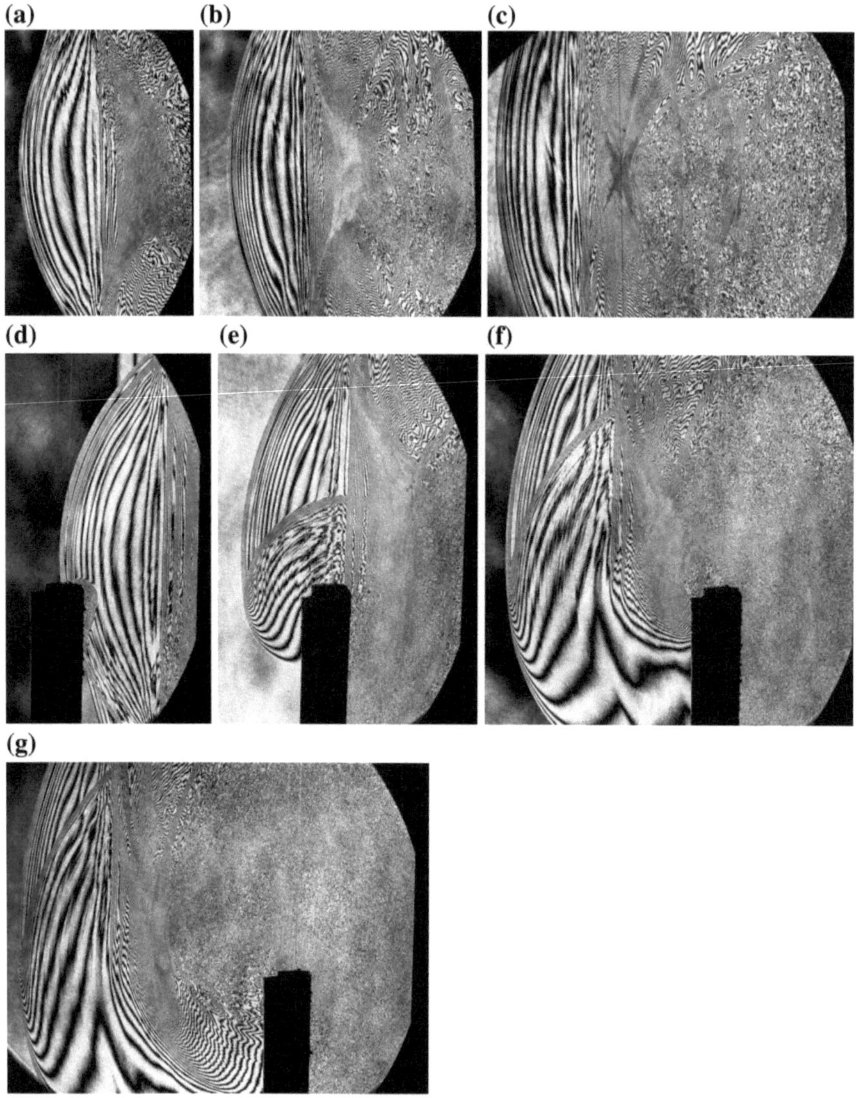

Fig. 10.10 Shock wave focusing from #2 reflector and blockage of focusing by inserting a sponge block at 50 mm from the exit, 10 mg AgN_3 at 288.3 K: **a** #86110108, 80 μs from trigger; **b** #86103003, 130 μs; **c** #86103008, 170 μs; **d** #86110108, 80 μs; **e** #86110106, 100 μs; **f** #86110107, 140 μs; **g** #86110109, 170 μs

The stone was cracked and at the moment when the stone started to expand, the stone started to shatter. The shattering speed is, so far monitored, about 3 m/s, which is too slow to damage tissues.

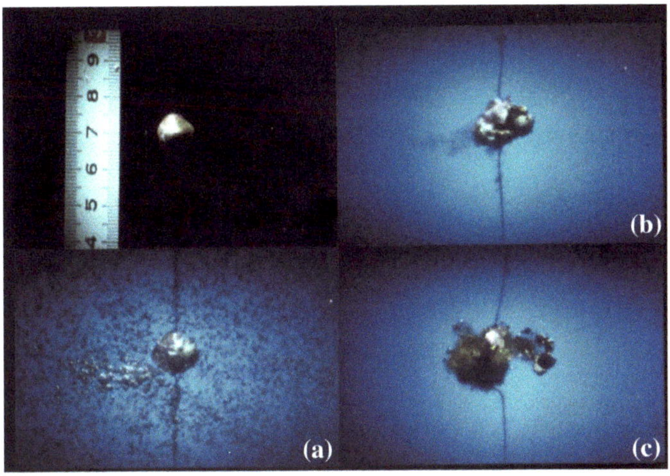

Fig. 10.11 High speed cinematography of disintegration of kidney stone, 6000 frame/s, #83122001

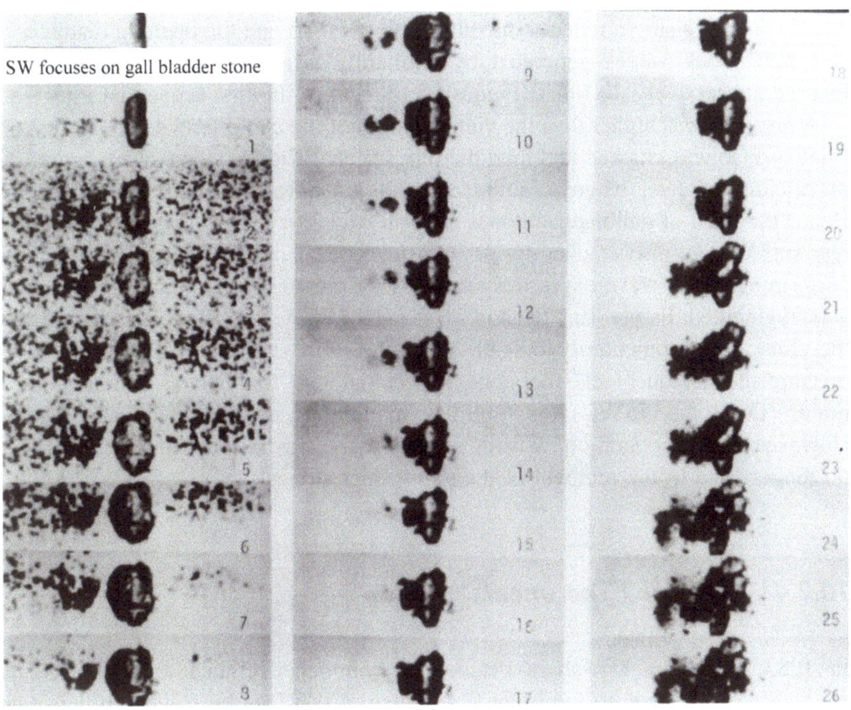

Fig. 10.12 Disintegration of a kidney stone observed at 10,000 frame/s

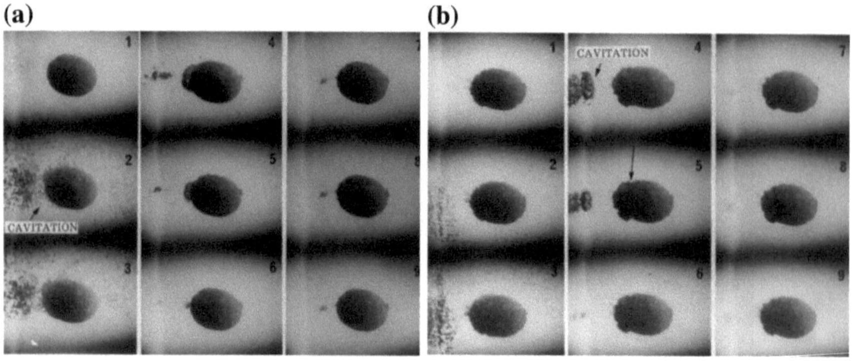

Fig. 10.13 Disintegration of a cholesterol stone 8000 frame/s: **a** first shock exposure; **b** 40th shock wave exposure (Abe et al. 1990)

Figure 10.13a, b show the disintegration of a gallbladder stone. The recording speed was 8000 frame/s and exposure time was approximately 30 μs. An extracted pure cholesterol stone of 20 mm × 25 mm was imbedded in a gelatin layer and placed at the second focal point of a reflector. The gelatin layer was to mimic a gall bladder tissue. Using a reflector having f-number 0.75 and the opening diameter of 180 mm, shock waves generated by exploding a 10 mg AgN_3 pellet and are focused successively 40 times. Figure 10.13a shows the first sequential images.

A pressure was higher than the yield strength of the cholesterol stone. At first the front surface was cracked and then the rear surface. Kidney stones were composed of calcium oxalate, whereas gallbladder stones were composed of cholesterol. Hence the most of gallbladder stones are brittle against a tensile stress, so that their rear surfaces are also cracked by the transmission of reflected tensile stress. After performing shock wave exposures successively for 40 times, the cholesterol stone was fragmented. Figure 10.13b shows a sequence of the 40th shot. In the 5th frame, the stone was momentarily bulged. After the 40th shock wave exposure when opening the gelatin block, the stone was completely fragmented into small sand grains. Once the ESWL was routinely applied to remove gallbladder stones. However, a novel operation by using an endoscope was introduced, the ESWL was no longer used to the removal of the gallbladder stones.

10.2.4 Clinical Experiments

In 1983, Professor Kuwahara decided to commence clinical experiments and applied its permission to the Ethic Committee of the Tohoku University Hospital. The application was granted. Figure 10.14 shows a result of the first clinical experiment (Kuwahara et al. 1986).

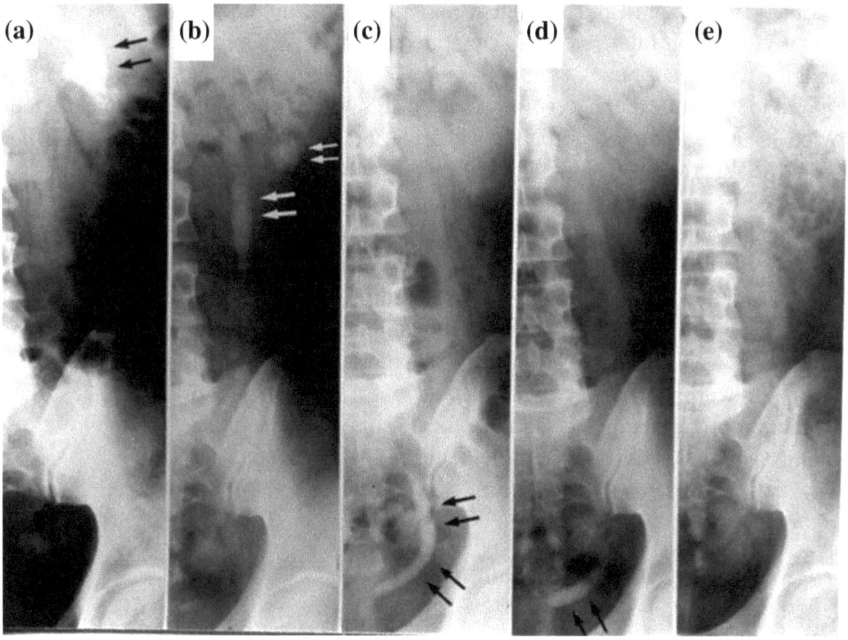

Fig. 10.14 The first clinical application of ESWL using micro-explosions (Kuwahara et al. 1986)

Arrows in Fig. 10.14a indicate a kidney stone before applying the ESWL treatment. After repeatedly focusing shock waves 230 times, the kidney stone was fragmented into sizes of sand grains. Then fragmented kidney stones already discharged out along a ureter. In Fig. 10.14b, fragmented stones formed a street of sand grain. The fragments moved into a bladder in Fig. 10.14c, d. Later, the sand grains were eventually discharged from the body as seen in Fig. 10.14e. Luckily, collaborators with Yachiyoda Kogyo Co. Ltd and Chugoku Kayaku Co Ltd were maintained and eventually a lithotripter using micro-explosives was applied to governmental authorization. In 1987, the Ministry of Health Japan granted the application and the lithotripter was officially used for clinical treatments. This system became one of the unique applications of underwater explosion to the human welfare.

10.2.5 Extracorporeal Shock Wave Induced Bone Formation

Success of ESWL treatments encouraged orthopedic surgeons to apply it to the orthopedic surgery. The treatments of delayed and nonunion of fracture bones are their clinical topics. Ikeda et al. (1999) noticed the ESWL treatment developed in Tohoku University and initiated a collaboration program with us. Dr. Ikeda who is an orthopedic surgeon engaged in the Kanazawa University Hospital, instructed us the state-in-art of clinical treatments of nonunion fracture bones Ikeda et al. (1999). He expected that if high-pressures generated by shock wave focusing was loaded on nonunion fractured bones, high pressures would stimulate bone formation. To respond to his request, the existing device of ESWL was modified and constructed a device so-named extracorporeal shock wave induced bone formation (ESWLIB).

The ESWL treatment aimed at selectively cracking the urinary tract stones but minimized tissue damages. On the contrary, the ESWLIB damaged tissues creating breeding in the controlled manner in order to unite fracture bones. Then, in the prototype ESWLIB, a strong shock wave is focused by exploding a 30 mg AgN_3 pellet placed at a focal point of a truncated ellipsoidal reflector used for the ESWL. The prototype ESWLIB was applied to animal experiments. Successful animal experiments were expanded to the clinical experiments. The results were successful. Today applications of underwater shock wave focusing to orthopaedic surgery became a routine therapeutic method for not only bone formation therapy but also healing pains in the elbows and the knees.

10.3 Tissue Damages Associated with ESWL

In ESWL treatments, the pressure behind the reflected shock wave increases in propagating toward the focal point. The pressure becomes a maximum when the focusing was completed on a kidney stone surface. In Fig. 10.4, the peak pressure gradually increases toward the focal point and exponentially enhanced in approaching very close to the focal point and became maximal, whereas the entry pressures are so low that the skin surface and the tissue at the entry regions are hardly damaged.

Figure 10.15a is a microscopic observation of damaged cross section of an artery of dog's kidney. In Fig. 10.5d, f, we saw a cavitation bubble cloud and secondary shock waves when bubbles were collapsed. In a bubble cloud, few bubbles survived for nearly one tenth of second (Kuwahara et al. 1989). The life time of bubbles is independent of their source and belongs to matter of probability. Bubbles interact with wavelets or subsequent shock waves and collapse. When collapsing, bubbles eventually create micro-jet. It is a micro-jet that pierces the tissue. A part of the dog's kidney artery looks as if ruptured by a needle in Fig. 10.15. Figure 10.15b shows the traces of shock wave converging into dog's kidney.

10.3.1 Shock Wave Interaction with a Bubble on Gelatin Surface

Tissue damages were caused by either by bubble collapse in the vicinity of the tissue surface or by the high pressure deposition on the tissue. Hence, to reduce the tissue damages, the number of shock wave exposures should be minimized while keeping the efficiency of disintegration of the stones high.

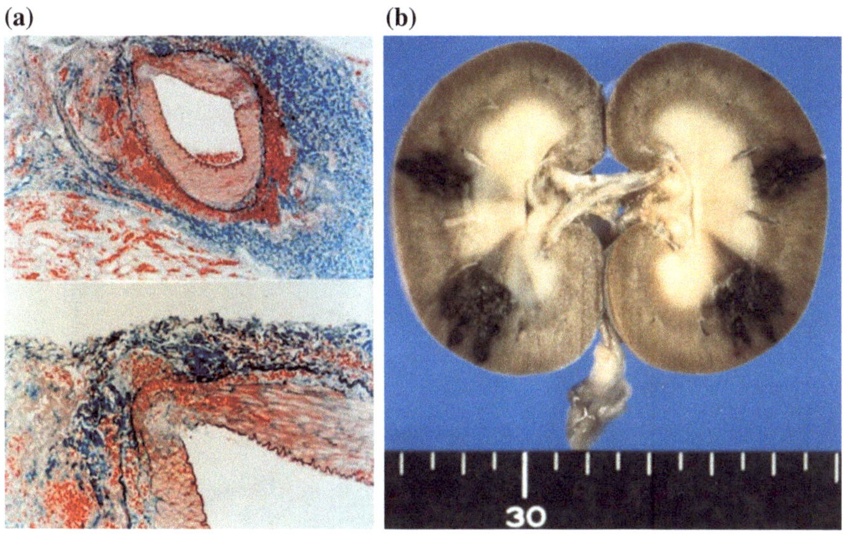

Fig. 10.15 Tissue damage: **a** damage on an artery of dog's kidney; **b** damages on dog's kidney tissues (courtesy of Professor Kuwahara)

To experimentally simulate tissue damage, 8 mm diameter air bubble was placed on a gelatin block and loaded a shock wave by exploding a 10 mg AgN_3 at stand-off distance of L = 30 mm. Figure 10.16 are sequential interferograms. The bubble was contracted forming a jet which penetrated the gelatin. At the same time, the shock wave propagating in the bubble was reflected from the gelatin surface and converged to reverse direction, which formed a multiple jet. It should be noticed that the collapse and jet formation in the bubble is governed by the bubble size and shock wave overpressure. The penetration depth was about 2.5 times as deep as the diameter of bubbles (Shitamori 1990; Obara 2001).

Figure 10.17 show sequential double exposure interferograms of a 1.5 mm diameter air bubble interacting with a shock wave generated by explosion of a 10 mg AgN_3 at L = 50 mm. Unlike Fig. 10.16, multiple reflections of the shock wave in the air bubble were not observable but the jet penetration into gelatin. The tissue damages seen in Fig. 10.15 were created by the collapse of small water vapor bubbles, presumably in the diameter of sub-millimeter. The jet formation presented here were just demonstrating the dynamic of jet penetrations.

Figures 10.17 and 10.18 are double exposure interferograms. A grey shadow was an air bubble placed on a gelatin surface and observed at the first exposure and shadows of deforming bubbles were observed at the second exposure. Upon the shock wave impingement, the bubble started to contract. At the same time the transmitted shock wave was propagating inside the contracting bubble. A jet formed due to the bubble contraction penetrated the gelatin layer. Figure 10.18e shows a remnant of jet penetration. The penetration depth would be affected by the

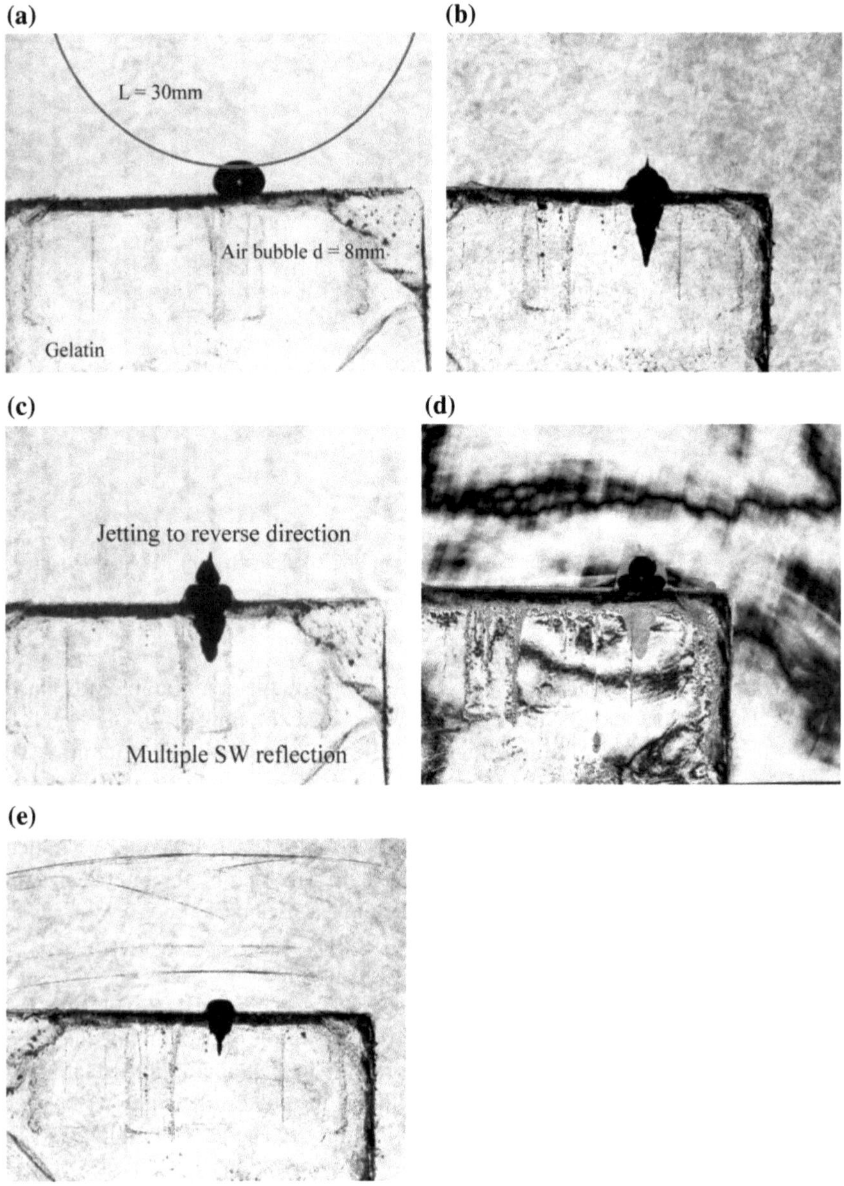

Fig. 10.16 Interaction of a shock wave with 8 mm diameter air bubble on gelatin surface at 296.0 K, 10 mg AgN_3, L = 30 mm: **a** #86090907, 50 μs from trigger point; **b** #86090908, 100 μs; **c** #86090912, 110 μs; **d** #86090913, 130 μs; **e** #86090914, 120 μs

magnitude of stagnation pressures on the gelatin surface. At the moment, no analytical model exists for predicting the stagnation pressure even though assuming a spherical bubble (Shitamori 1990; Obara 2001).

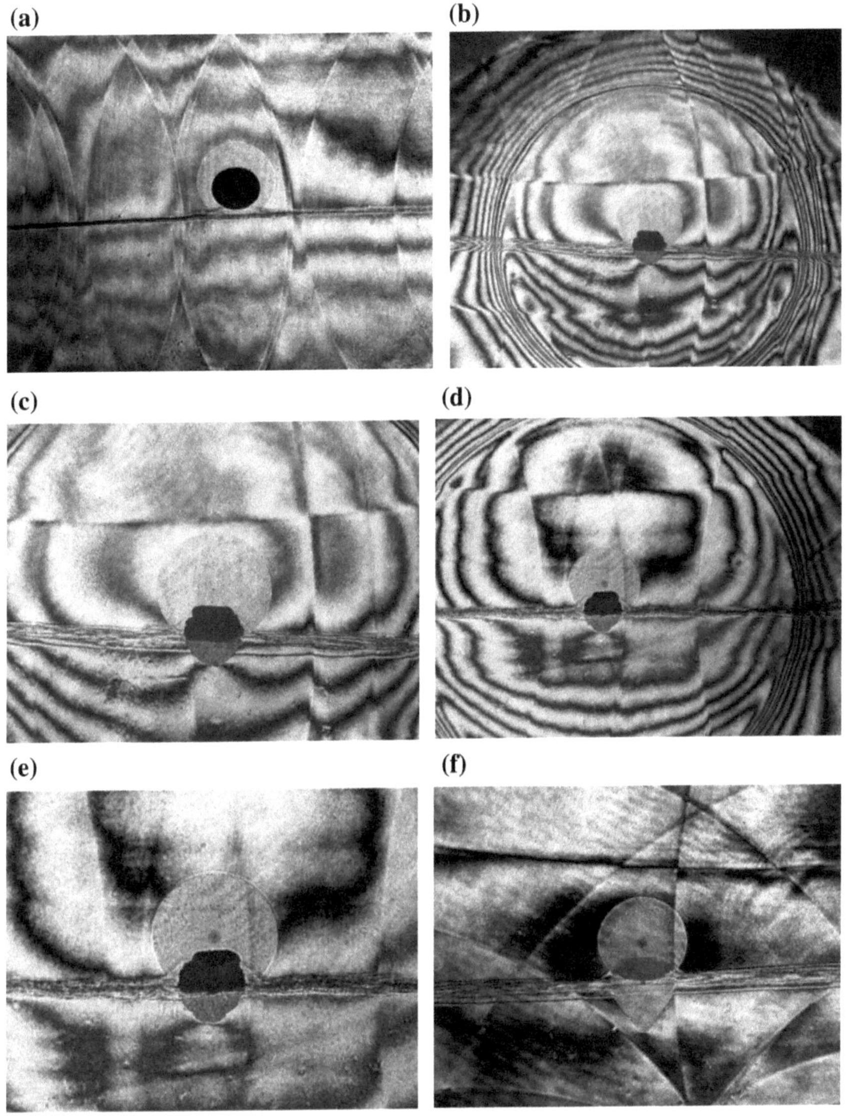

Fig. 10.17 Double exposure interferograms of shock wave interaction with a 1.5 mm diameter air bubble on gelatin surface, a 10 mg AgN_3 at the stand-off distance L = 50 mm at 286.4 K: **a** #85012913, 18 μs; **b** #85012919 23 μs; **c** Enlargement of (**b**); **d** #85012920, 24 μs; **e** Enlargement of (**d**); **f** #85012918, 23 μs

Figure 10.19 show sequential observations of shock wave/bubble interaction recorded by Ima Con High Speed Camera Model 790. Figure 10.19a, b show a 5 mm diameter air bubbles placed on 10 wt% gelatin surface, 5 wt% gelatin surface impinged by a shock wave generated by exploding a 10 mg AgN_3 pellet at the

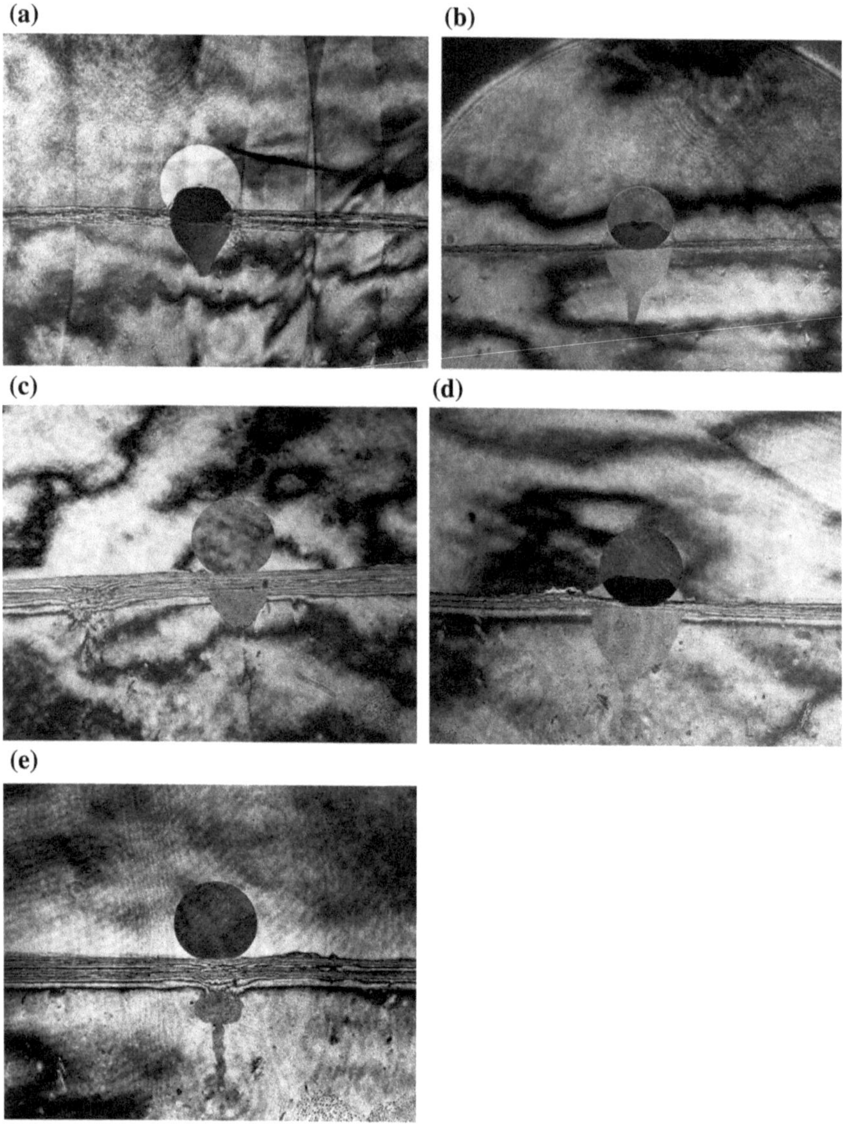

Fig. 10.18 Double exposure interferogram of shock wave interaction with a 3.0 mm diameter air bubble on gelatin surface. Shock wave was generated by exploding a 10 mg AgN_3 pellet at the stand-off distance L = 50 mm at 286.4 K: **a** #85013003 32 μs from trigger point; **b** #85013005 44 μs; **c** #85013105, 125 μs; **d** #85013103 65 μs; **e** #85013104, 95 μs

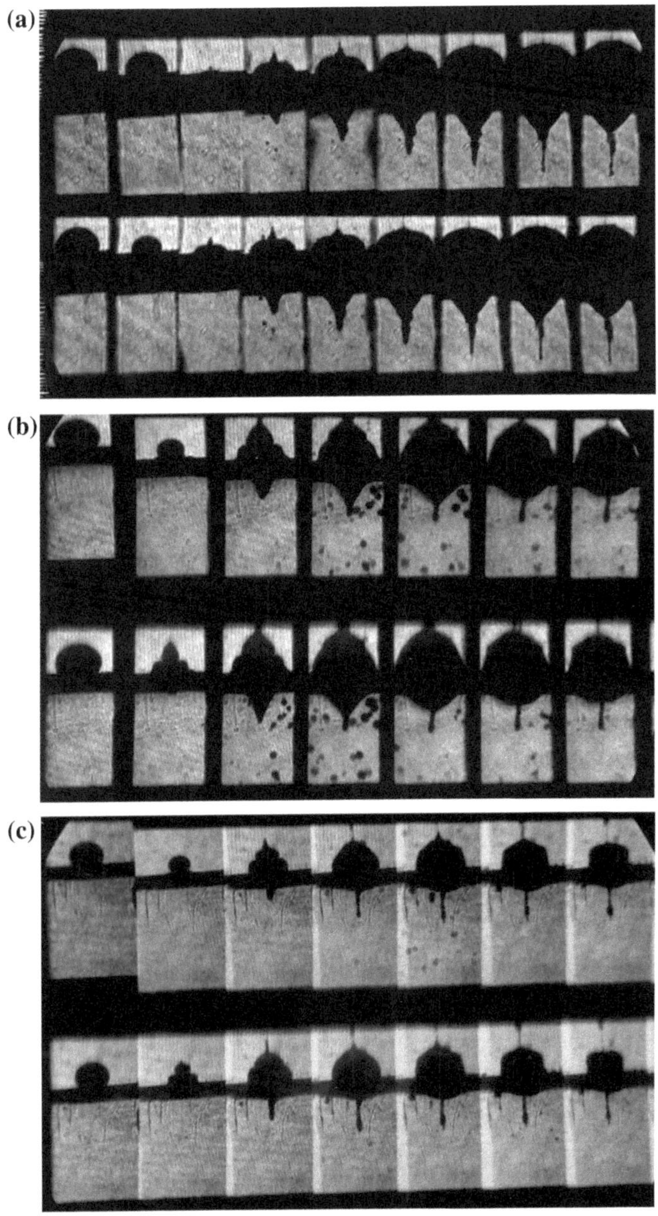

Fig. 10.19 High speed imaging, shock wave interaction with an air bubble placed on a gelatin surface, a 10 mg AgN_3 pellet was exploded at the stand-off distance of L = 50 mm, at atmospheric air, 286.4 K: **a** #90072201, 10 mg AgN_3, 5 mm air bubble at L = 50 mm, on 10 wt% gelatin surface, at 100,000 frame/s, exposure time of 1.25 μs; **b** #91051922, 10 mg AgN_3, 5 mm diameter air bubble at L = 50 mm, on 10 wt% gelatin surface, at 25,000 frame/s, exposure time of 5 μs; **c** #91052014, 3 mg AgN_3, 3 mm diameter air bubble at L = 50 mm, on 5 wt% gelatin surface, at 25,000 frame/s, exposure time of 5 μs

stand-off distance of L = 50 mm. The jet formation toward the reverse direction was observable (Shitamori 1990; Obara 2001). Figure 10.19c shows a 3 mm diameter air bubbles placed on 5 wt% gelatin surface or 10 wt% gelatin surface impinged by a shock wave generated by exploding a 3.0 mg AgN_3 pellet at the stand-off distance of L = 50 mm. The structure of gelatin affected very slightly to the jet penetration.

10.3.2 Domain and Boundary of Tissue Damage in ESWL

Figure 10.20 summarizes the relationship between overpressures and number of the shock exposures. The domain and boundary of tissue damage is shown. The shock waves were generated by focusing of strong sound waves created by a piezo-dish as discussed in Sect. 9.4.5. The ordinate denotes number of shock focusing and the abscissa denotes the over-pressure in MPa. The evaluation parameter of damages is the level of haematoma from a negative bleeding to serious bleeding. Red filled circles denote breeding areas larger than 30 mm^2. Filled orange color circles denote breeding areas from 0 mm^2 to smaller than 30 mm^2. Filled blue color circles denote negative breeding. There are distinct domains and boundaries among damage levels. It is desirable to disintegrate the stones with a minimum level of bleeding by exposing with moderately high peak pressure and at limited number of the shock exposure. Figure 10.20 indicates, however, that the level of haematoma will increase at higher peak pressures and smaller number of shock exposures. At lower peak pressures, the number of shock exposures will increase and at the same time the level of haematoma becomes high. Therefore, the optimum number of shock

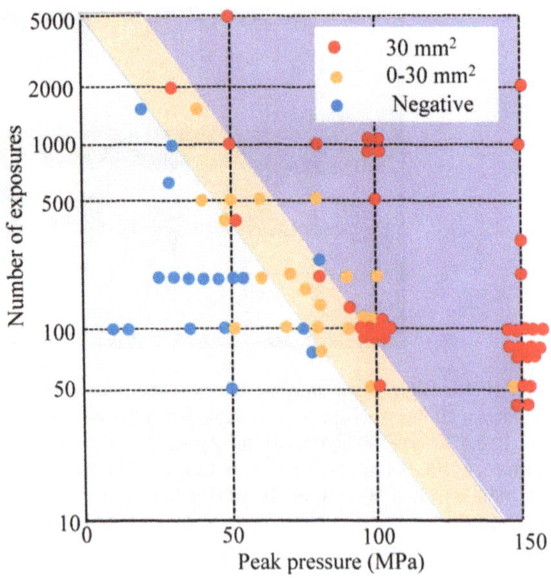

Fig. 10.20 Correlation between number of shock wave exposures and the peak pressures in terms damage level in electromagnetic ultrasound focusing (Okazaki 1989)

exposure will exist which will minimize the level of bleeding. However, the optimized number of shock exposures and the optimized shock strength would depend empirically on the type of stones and their dimension.

10.3.3 Shock Wave Induced Injury on Nerve Cells

Neuro-brain surgeons concerned nerve cell damages caused by shock wave loadings. However, the physical properties of human organs quoted from the literatures which were those collected from extracted tissues differ from those collected from living tissues in the presence of blood circulations (Kato 2004).

Therefore, it was decided to experimentally determine the threshold pressure at which nerve cells are damaged by shock wave loading. A 10 mg AgN_3 pellet was divided into small pieces by using a pair of bamboo tweezers and measured the amount of AgN_3 fragments from 2.5 to 300 μg with a digital precision balance. Figure 10.21 shows a crystal of AgN_3 weighing 2.5 μg. Fragments of AgN_3 were glued, with acetone-cellulose solution, on the edge of 0.6 mm core diameter optical fiber and were ignited by the transmission of a Q-switch Nd:YAG laser beam of total energy of 7 mJ and 20 ns pulse width (Nagayasu 2002).

In order to determine the threshold value of shock wave over-pressure at which nerve cells are damaged, a shock wave was focused on rats' brains accurately at a specified spot using a miniature ellipsoidal reflector. The miniature reflector is a 20 mm × 28.3 mm half truncated ellipsoidal reflector made of brass. Figure 10.22a shows its schematic diagram. Figure 10.22b shows a photograph viewed from its opening.

A shock wave generated by explosion of small AgN_3 pellet was focused at the first focal point of the truncated reflector by transmitting a Q-switched Nd:YAG laser beam through the optical fiber. Figure 10.23a, b show time variation of

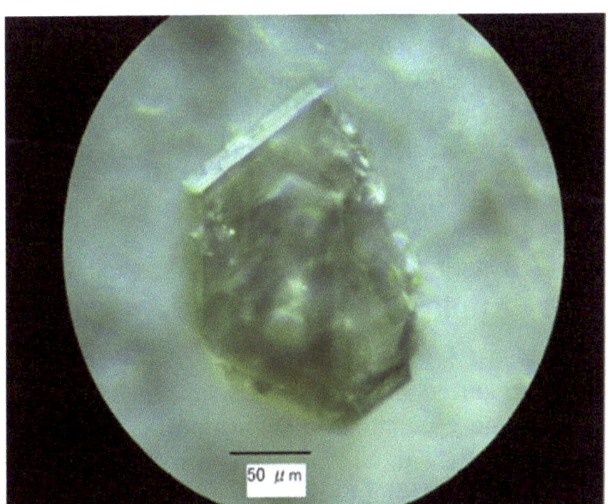

Fig. 10.21 A AgN_3 crystal 2.5 μg in weight

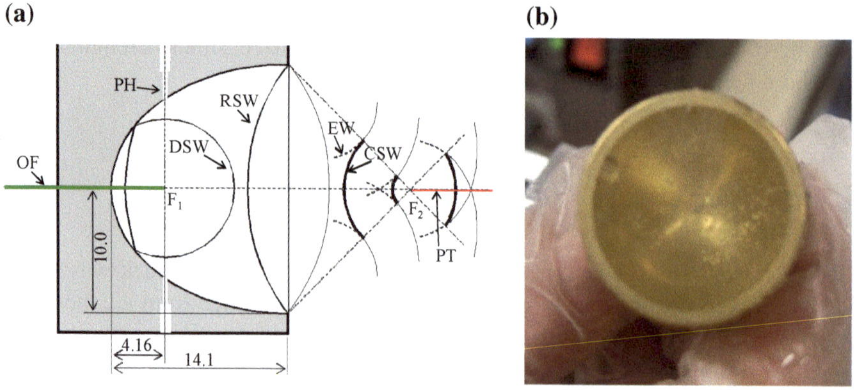

Fig. 10.22 A 20 mm × 28.4 mm truncated ellipsoidal reflector: **a** cross section of the reflector; **b** a view from opening (Kato 2004)

pressures for 2.5 and 15 μg AgN_3 pellets and sequential direct shadowgraphs of the focusing of 2.5 and 15.0 μg AgN_3 pellets. It was noticed that even such small explosives were detonated and their overpressures obeyed the scaling law. The pressures were measured by using optical fiber pressure trasducer.

Eight-week-old male rats were anesthetized and supported on a movable stage. Their right parietal bones were precisely exposed to the focal point. A bone defect of 3–5 mm diameter was made at the right convexity. The bone defect touched a water bath through which the shock wave was focused. A shock wave was generated exploding a 100 μg AgN_3 pellet and focused precisely on a rat's brain. The error of the positioning was less than ±0.1 mm. It was examined the level of overpressure at which the nerve cells were damaged. With this arrangement, high pressures were loaded on rats' nerve cells. The results collected here contributed well to the study of the blast wave injury (Kato 2004). Figure 10.24 shows an experimental arrangement for investigating the threshold pressure of nerve cell injury upon focusing shock wave onto rat's brain.

Figure 10.25 show test results. Shock wave focusing created apoptosis on rat's brain (Kato 2004). Depending on the focused pressure, apoptosis and necrosis occurred. This was an early result of shock/nerve cell interaction. Apoptosis occurs already at a shock wave exposure of 1 MPa.

10.4 Laser Induced Shock Waves for Medical Applications

It is recognized that the ESWL treatment always accompanies tissue damages. There is no exceptional combination of parameters that selectively disintegrates kidney stones without causing any tissue damages. Then it was decided to apply shock wave focusing to damage soft tissues in a controlled manner. Shock waves

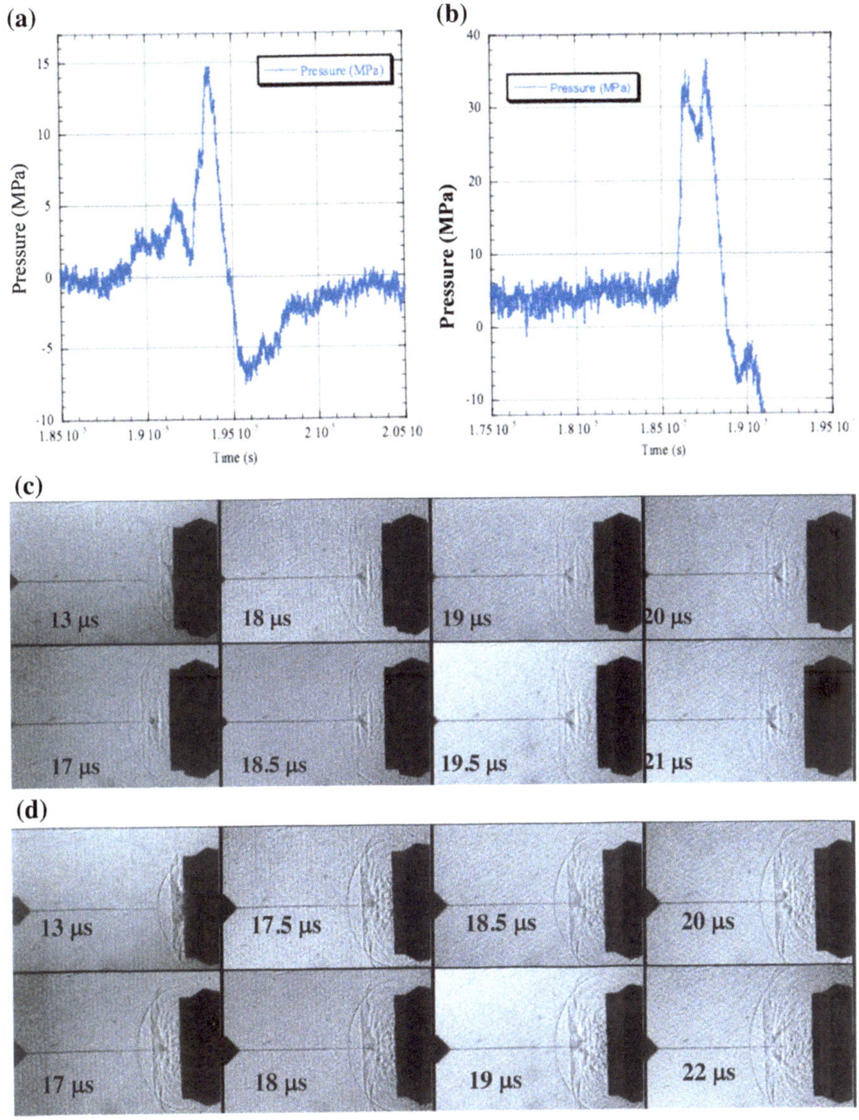

Fig. 10.23 Pressure profile and sequential observation of focusing of shock waves generated by the explosion of 2.5 and 15 μg AgN_3 pellets: **a** pressure profile of 2.5 μg AgN_3; **b** pressure profile of 15 μg AgN_3; **c** shadow pictures of 2.5 μg AgN_3; **d** shadow pictures of 15 μg AgN_3

created by focusing the pulse Ho:YAG laser beams are applied to the revascularization of cerebral thrombosis. This project started under collaboration with the Department of the Neuro-Brain Surgery in the School of Medicine of Tohoku University.

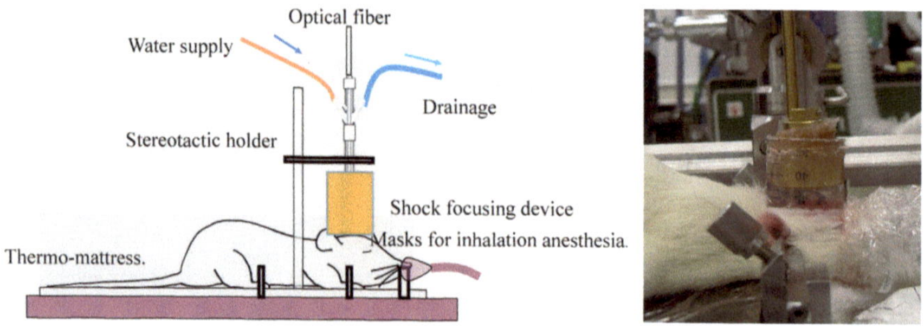

Fig. 10.24 The experimental arrangement of nerve cell injury focusing a shock wave on rat's brain

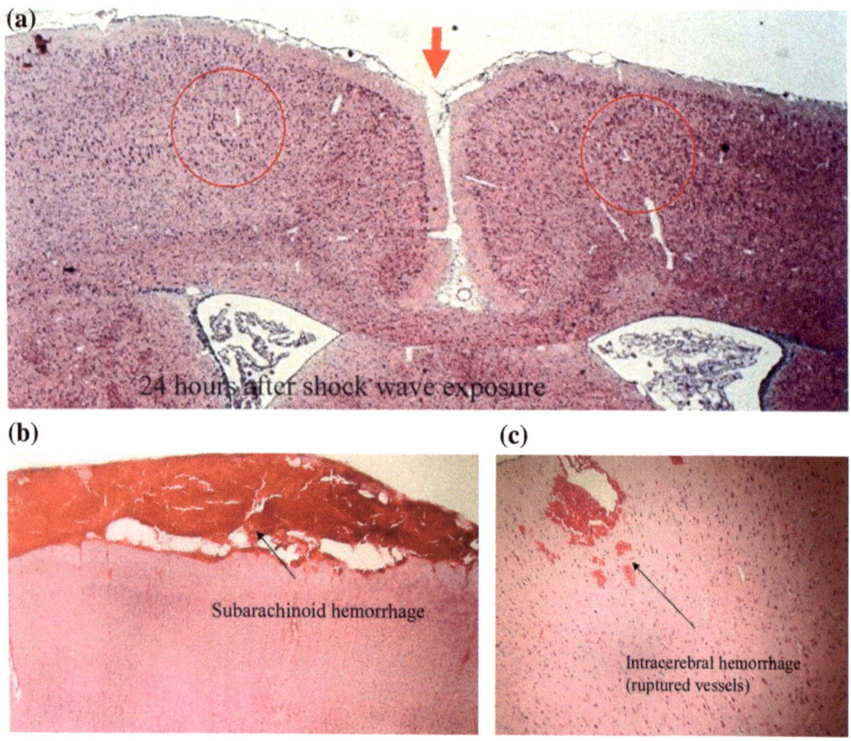

Fig. 10.25 Localized shock wave focusing created apoptosis (Kato 2004): **a** over view; **b**, **c** magnified sections

Laser beams irradiated in water were spontaneously absorbed in water molecules and elevated their energy level to the level of ionization. A water vapor bubble was spontaneously created, which was equivalent to a fire ball of exploding explosives in water spontaneously drove an underwater shock wave.

At first a pulsed Nd:YAG laser of 100 mJ, 130 ns pulse width, and wave length of λ = 1090 nm was focused. Figure 10.26a shows the dependence of absorption coefficient of the laser energy on wave length. The ordinate denotes absorption coefficients of light in water in 1/cm and the abscissa denotes wave lengths, λ in nm. Laser beams having wavelength of visible lights have low absorption coefficients as shown in color spectra which means that the deposition of a Q-switch ruby

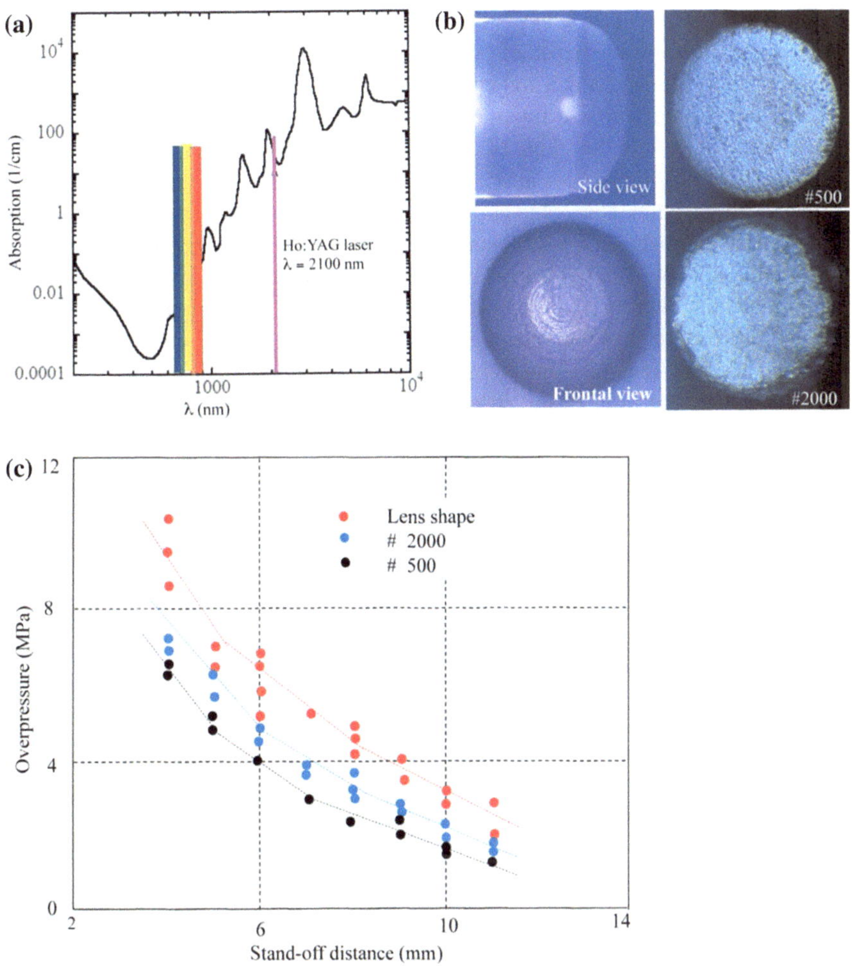

Fig. 10.26 Laser induced shock waves: **a** absorption of laser energy in water; **b** surface finish of ϕ0.6 mm core diameter optical fiber. Edge shape, lens shape, fine and coarse surface finish of #2000, and #500; **c** Shock wave overpressures transmitting 1.0 J Ho: YAG laser beam in MPa versus stand-off distance from the edge in mm

laser having λ = 694.3 nm is less effective for generating shock waves. A Ho:YAG laser having λ = 2100 nm is shown in pink color has the absorption coefficient over 1000 times higher than that of Nd:YAG laser having λ = 1064 nm. The focusing of Ho:YAG laser beam can effectively generate strong underwater shock waves. A Q-switched Ho:YAG laser beam of 1.0 J energy was transmitted through a 0.6 mm diameter quartz optical fiber.

The efficiency of laser beam focusing depends on the surface finish of optical fiber edge. Then the optical fiber edge was polished in a convex lens shape to effectively focus the laser beam and at a point outside the optical fiber.

Figure 10.26b shows shapes of optical fiber edge: lens shape; and flat fine finish edge of a #2000 finish, which indicates the surface roughness equivalence of 2000 grains per inch and a coarse finish #500 which indicates the surface roughness of 500 grains per inch. In Fig. 10.26c, the ordinate denotes the variation of over-pressures in MPa and the abscissa denotes the stand-off distance from the edge of optical fiber in mm. Red, blue and black filled circles denote lens shape, fine and coarse surface finishes of #2000 and #500, respectively. The lens shaped edge generated about 10 MPa over-pressures at the distance about 4 mm from the edge.

Optimal and durable shapes exist but the shapes shown in Fig. 10.26c were deviated far from the optimal one. Lens shaped edges, however, survived in average only for 100 laser transmissions. When inserting an optical fiber into a thin catheter and transmitting the laser beam, shock waves were generated in the vicinity of the edge and the creation of the vapor bubble induced micro water jets impacting the edge surface.

10.4.1 Revascularization of Cerebral Thrombosis

In order to visualize the generation of the bubble and formation of shock waves in a catheter, visualization was conducted using a 5 mm diameter and 60 mm long aspheric lens shaped cylinder made of acryl. Figure 10.27a shows an aspheric lens shaped thin tube. Figure 10.27b shows irradiations of a Q-switched laser beam having 91 mJ/pulse and 200 ns pulse width inside the aspheric lens shaped test section. A solid line along the axis show a lens shaped optical fiber. In Fig. 10.27b, the first 6 frames show sequential observation at the time interval of 1 μs. The arrows indicate the shock waves. In the second 6 frames, the first image was taken at 2 μs and the other frames started from 350 μs at the time interval of 50 μs, the increase in the bubble diameter was clearly observed. The shock wave propagation and flows in a 5 mm diameter tube are clearly observed (Ohki 1999).

Q-switched Ho:YAG laser beams repeatedly transmittomg through a 0.6 mm optical fiber can created shock waves at its edge. The shock wave can just penetrate the surface of artificial thrombus filled in a thin tube. Figure 10.28 show sequential penetration of the thrombus with successive shock wave impingements. Indeed it worked but not very efficiently. The optical fiber had a lens shaped edge so that the

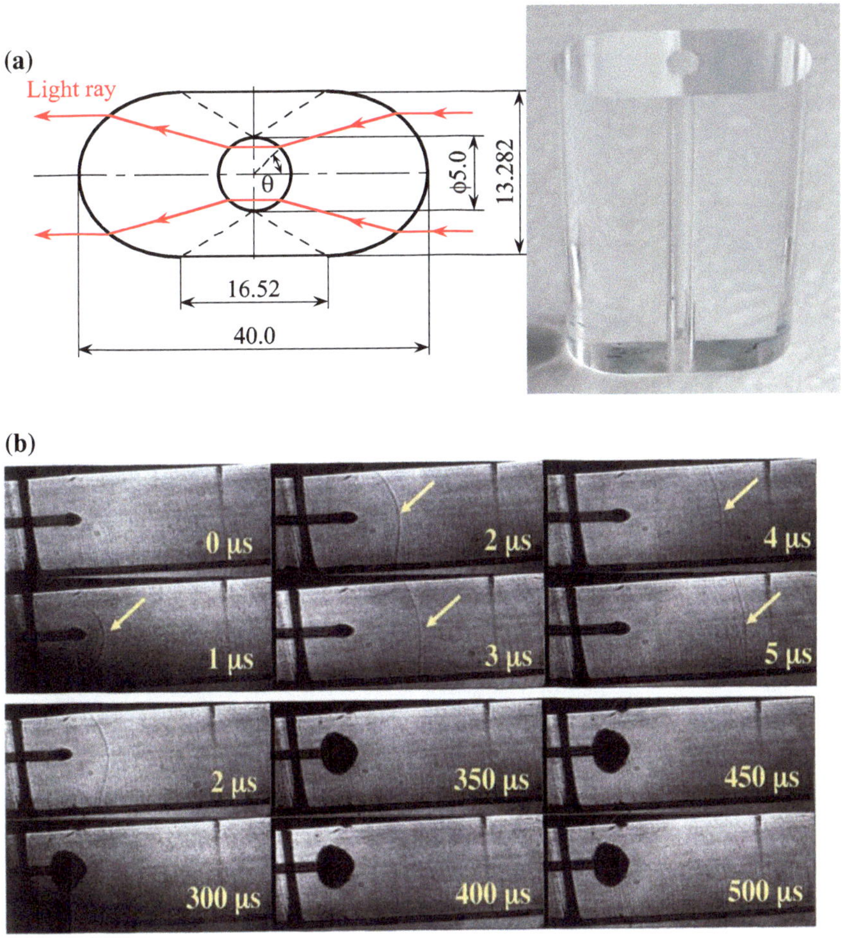

Fig. 10.27 Laser induced shock waves in a 5 mm diameter aspheric lens shaped tube, Q-switched Ho:YAG laser beams having 91 mJ/pulse and 200 ns pulse duration were transmitted through a 0.6 mm core diameter optical fiber: **a** aspheric lens shaped 5 mm diameter tube; **b** formation of a shock wave 0–5 μs, formation of bubble from 2 to 500 μs

resulting shock wave over-pressures were high enough to readily penetrate the thrombus. Figure 10.28 shows the optical fiber edge penetrating the artificial thrombus by successive irradiation of Q-switched Ho:YAG laser beams. The idea indeed worked but the penetration proceeded too slow. Encouraged by the result seen in Fig. 10.28, an extension tube was attached at the edge of the catheter. The extension tube transmitted a water jet to a position away from the edge of the optical fiber.

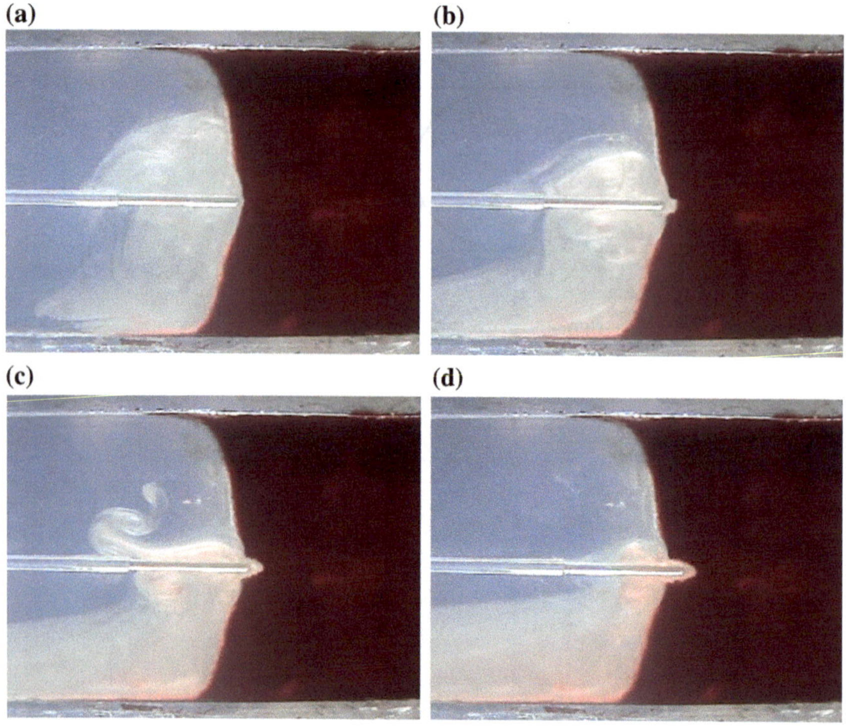

Fig. 10.28 Penetration of an artificial thrombus by Ho:YAG laser beam irradiation (Hirano et al. 2002)

10.4.2 Catheter of Dissecting Soft Tissue

10.4.2.1 Laser Induced Dissector

After conducting in vitro tests, a relatively long extension tube was connected to the open end of a catheter so as to reach the middle cerebral artery. Figure 10.29 shows the prototype catheter.

Pulsed Ho:YAG laser beams having pulse width of 350 ns and frequency of 3 Hz, and maximum energy of 1.3 W were transmitted through a 0.6 mm core diameter optical fiber. The optical fiber was inserted in a 4 Fr and 20 mm long stainless steel catheter and a 2.7 Fr and about 300 mm long flexible tube was connected at the edge of the catheter. Then water jets were ejected at the volume rate of about 13 mm^3/s at each laser irradiation. This volume rate was little readily to aspirate. Then to assure its effectiveness, an in vivo experiment was conducted: a swine artery was choked with thrombi and a prototype catheter was applied to penetrate the thrombi.

Figure 10.30 shows the sequence of X-ray monitors. Yellow circles show the artery under study. At first the artery was choked with thrombi and the blockage of

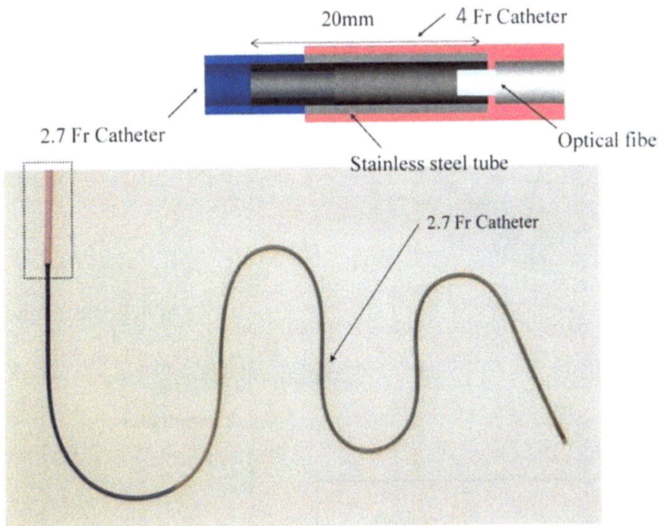

Fig. 10.29 Ho:YAG laser induced dissection catheter

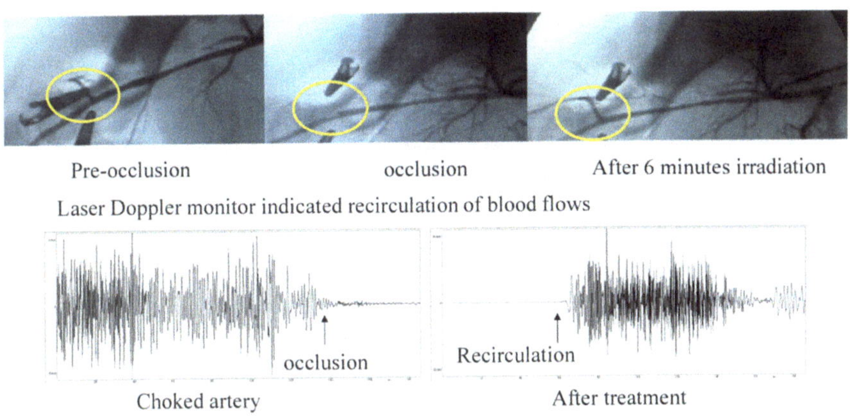

Fig. 10.30 In vivo experiment of revascularization of thrombosis of a swine artery

the blood circulation was confirmed. After applying laser irradiation for 6 min at 3 Hz, the blood circulation restarted. It was convinced that the catheter succeeded the revascularization of thrombosis. The prototype catheter consumed extremely small amount of water only a few cc of water for one sequence.

Throughout in vivo experiments, the final goal of revascularization of cerebral thrombosis was found to be a benign subject to achieve (Hirano et al. 2002). Hence, the goal of the project was shifted to develop a device for dissecting soft tissues: the removal of brain tumors. Figure 10.31a shows an illustration of a prototype catheter

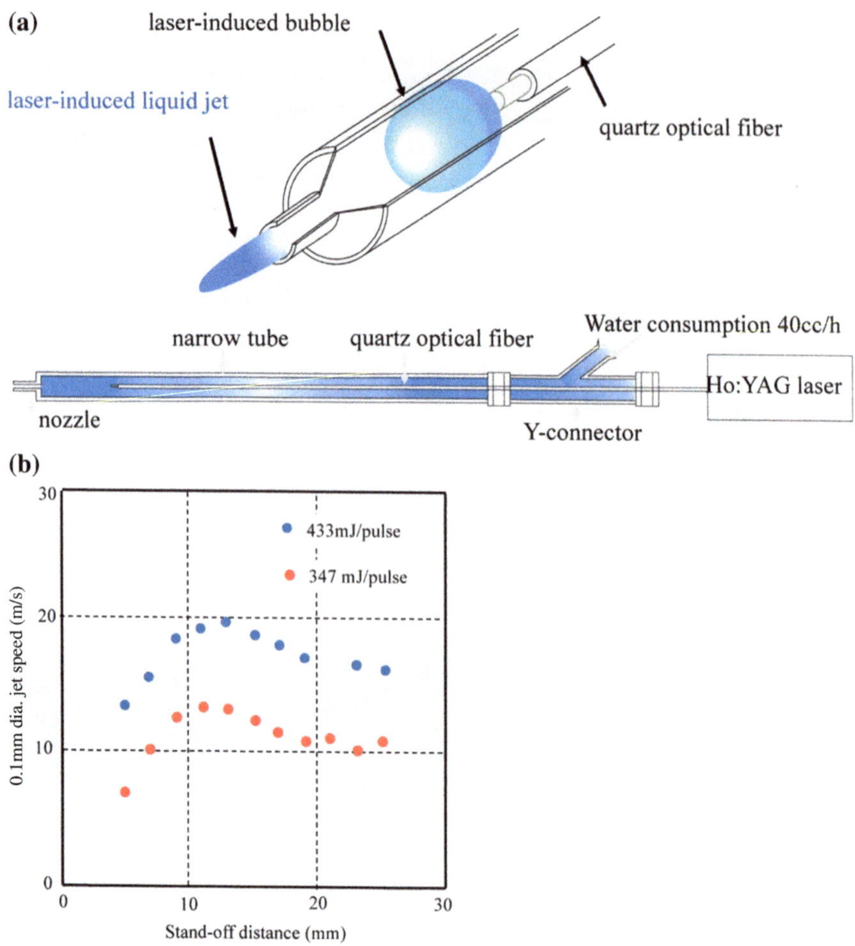

Fig. 10.31 A prototype of dissection catheter: **a** schematic structure of a catheter; **b** jet velocity versus stand-off distance (Hirano et al. 2002)

equipped with a 0.1 mm diameter nozzle. A pulsed Ho:YAG laser of wave length of 2100 nm, pulse width of 350 ns, and 3 Hz was transmitted through a 0.6 mm quartz optical fiber and inserted into a narrow tube from its end. Water was ejected from a 0.1 mm diameter nozzle intermittently at volume rate of 13 mm^3/s at every laser irradiation. Water was continuously supplied from the end of a Y-connector. Hence, the jet speed varies depending on the amount of the laser energy. Figure 10.31b shows the relationship between the jet speed and the laser energy. The ordinate denotes jet speed in m/s and the abscissa denotes the distance from the nozzle exit. Blue and red filled circles denote the laser energy of 433 and 347 mJ/pulse, respectively. The jet speed increases with increasing laser energy. However, the jet speed is maximal at the stand-off distance of about 18 mm.

Figure 10.32a show a sequential observation of a 0.1 mm diameter jet penetrating into a gelatin block visualized every 256 μs interval. The laser energy was 433 mJ/pulse at frequency of 3 Hz. The penetration speed was about 0.5 mm/shot. It is observed that the jet slowly penetrates the gelatin block. Figure 10.32b shows a result of in vivo experiments. A piece of extracted pig liver was dissected using a prototype dissection catheter. The specimen was successfully dissected but the thin blood vessels were preserved. It is concluded that 0.1 mm diameter jets are able to successfully dissect soft tissues and to preserve over 0.2 mm diameter blood vessels. In clinical tests of removing brain tumors, the brain tumor was removed with slightest bleeding as blood vessels of diameter over 0.2 mm were preserved. The catheter had an aspiration tube. The catheter weighed less than 100 g so that it was easy to hold in hand (Nakagawa 1998).

Figure 10.33a shows a clinical application of the dissection catheter to the removal a temporal insular glioblastoma (brain tumor). As the catheter had an aspiration, jet water and the remnant of bleeding were not observed in the field of view. The jets preserved blood vessels of diameter less than 0.2 mm so that the field of view was not disturbed by bleeding not a flood of jet water. The water jet speed can reach about 10–15 m/s and hence its stagnation pressures are high to readily rupture or pierce thin blood vessels. However, it will occur, if 0.1 mm diameter water jet may impinge sharply perpendicularly on the surface of blood vessels exceeding 0.2 mm diameter. During the surgical operation, the catheter was held by hand so that the water jet would impinge obliquely on thin blood vessels. Therefore, blood vessels over 0.2 mm diameter are preserved.

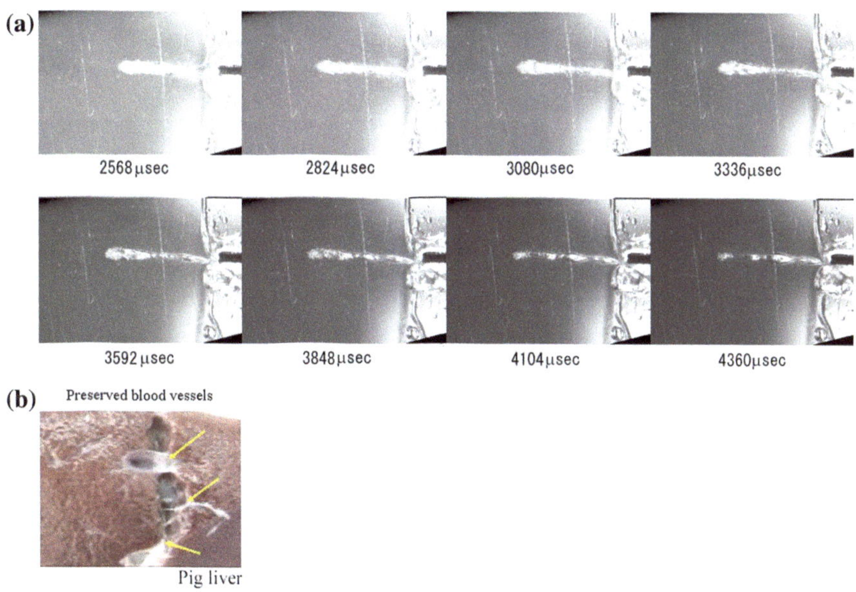

Fig. 10.32 Examples of dissecting catheter: **a** penetration into a gelatin block; **b** dissection of pig liver

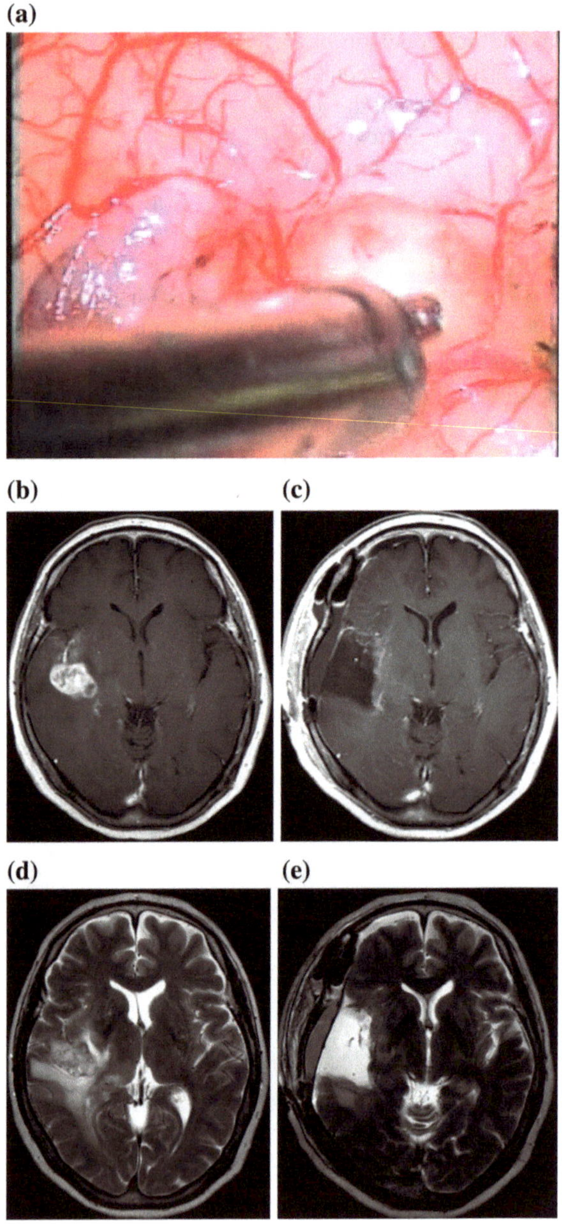

Fig. 10.33 The results of clinical treatment of a jet dissection catheter applied to temporal insular glioblastoma: **a** application to clinical test; **b** X-ray image before treatment; **c** X-ray image after treatment; **d** CT image, before treatment; **e** CT image, after treatment (Nakagawa et al. 2008)

Figure 10.33b, c show X ray images before and after the operation. Figure 10.33d, e show CT images before and after the operation. The tumor was removed but neighboring vessels were not damaged. The results were successful and hence it is expected to have the approval of the Ministry of Health, Japan (Nakagawa 2008).

10.4.2.2 Piezo Actuator

The amount of water ejected at a single movement of the actuator moving robot arms was at most a 0.1 mm^3. This volume water is equivalence of the amount of water required to dissect soft tissues needed for laser induced soft tissue dissectors. Then a robot arm technique was converted to intermittently drive micro water jets applied to the dissection of soft tissue. Figure 10.34a, b show an actuator driven jet generator and its structure. The actuators are inexpensive and commercially available so that a prototype catheter can be compact and has a light weight.

Figure 10.35 shows streak photographs of actuator driven jets at 800 Hz. The jets were generated using nozzles diameter 0.1, 0.15, and 0.2 mm. The ordinate denotes the distance from the jet opening in mm and the abscissa denotes elapsed time in ms. Hence, the inclination angles of the streak images indicate the jet speed. In the case of 0.15 mm nozzle diameter, the jet is ejected in a regulated manner, and

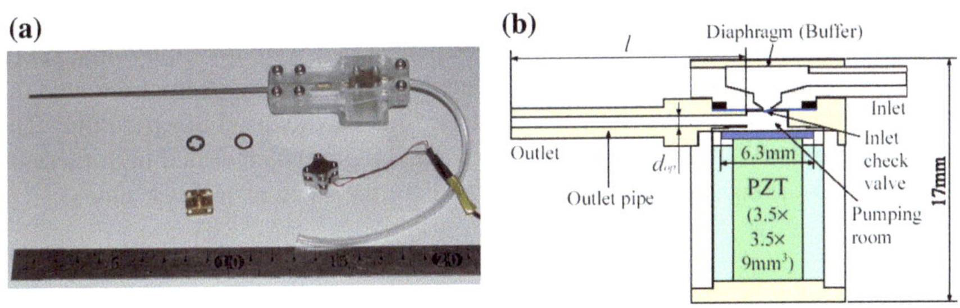

Fig. 10.34 Soft tissue dissection device: **a** prototype device; **b** Illustration of structure

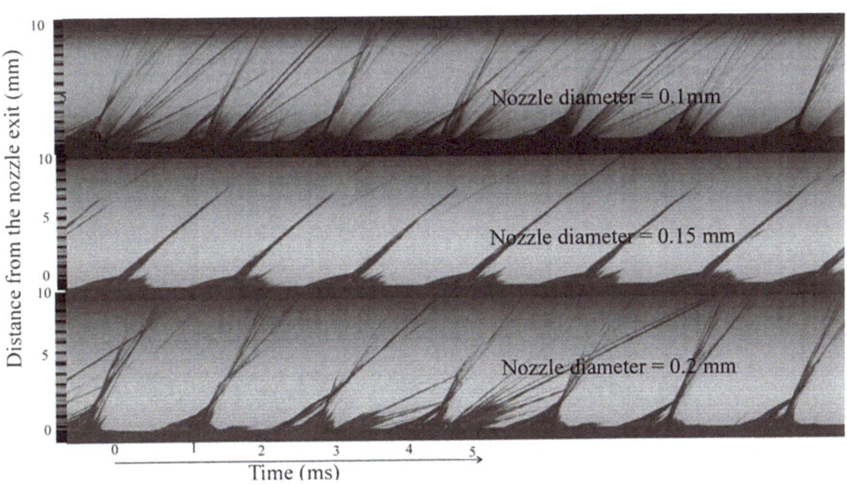

Fig. 10.35 Streak recording of jet formation at variable nozzle diameters

its ejection speed is about 45 m/s, whereas in the case of 0.10 and 0.2 mm diameter nozzles, the shadow of the jet fluctuate randomly which corresponds to the fluctuation of the jet speeds. Hence the nozzle diameter 0.15 mm was found to achieve the optimum combination of the parameters.

10.4.3 Laser Assisted Drug Delivery

Application of a compact gun to a drug delivery method was reported, for the first time, in Nature (1987), in which minute drug particles attached on a metal plate were ejected at high speed, The metal plate is impinged by a high-speed projectile from behind. Particles moving at high-speed penetrated into tissues and created a drug delivery effect. Therefore, this system was named as a particle gun. Figure 10.36 shows mechanical drug delivery methods reported in open literatures. These methods are more or less related to high-speed flows or applications of shock waves.

Gold particles of 1 μs in diameter coated with DNA were driven by high speed flows and impacted into plant seeds. This method proposed by DuPont initiated the recombination of DNA in plant cells and today intensively used in agriculture. The high speed flow was created by rupturing a diaphragm which sealed high pressure helium and working air. The principle is exactly the same as a shock tube.

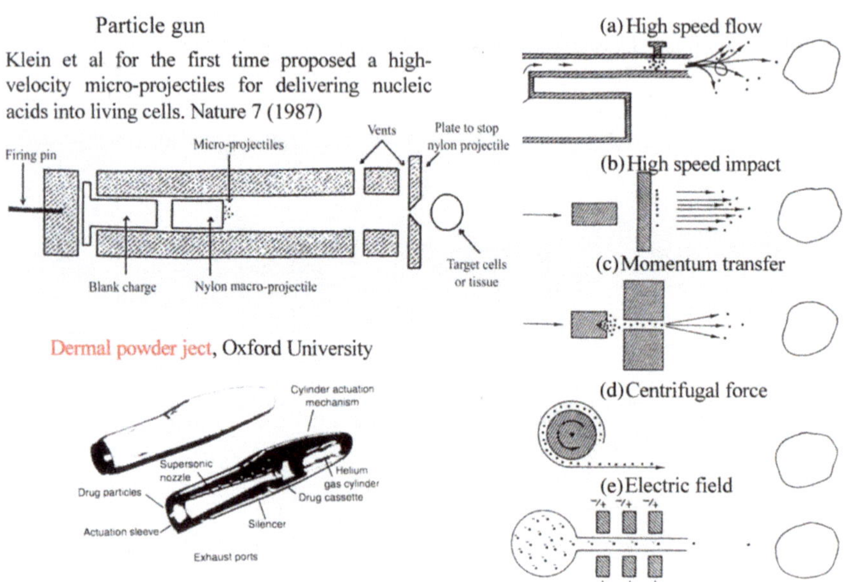

Fig. 10.36 The methods of previous drug delivery

The dermal powderject was invented by Professor Bellhouse et al. (1997) of Oxford University and was already used in clinical treatments. Various drug delivery methods are proposed as shown in Fig. 10.36. As reported in this Chapter, a laser induced drug delivery system is presented.

Figure 10.37 shows, as an analogue experiment of a laser induced drug delivery, shattering of polyethylene beads of 4.8 mm diameter attached vertically in a line on a steel plate. The steel plate was impacted with a nylon piston from behind. The steel plate's sudden deformation ejected the beads. The images were recorded by direct shadowgraph at the framing rate of 10^6 frame/s by Shimazu digital video camera SH100. In Fig. 10.37a, when a 50 mm diameter and 50 mm long nylon cylinder impacted the target steel plate at 340 m/s, an impact flash was emitted at the reverse side. The beads vertically arrayed were just about to be projected. The nylon cylinder squeezed into the target and quickly bulged the thin steel plate. Then the beads were ejected in air. Bow shock waves were formed in front of the beads as seen in Fig. 10.37c–f.

Figure 10.38 schematically explain the laser induced drug delivery system and shows a 100–150 μm thick aluminum foil on which 1 μm diameter gold particles were attached and overlaid a 10 μm thick BK7 glass plate. When a Q-switched high power laser beam having a 2 mm diameter flat head shape irradiated the aluminum foil through the overlaid glass plate, the aluminum plate absorbed the laser energy and a plasma cloud was instantaneously generated on its surface between the foil and the BK7 glass plate. The laser energy deposition was very similar to the explosion of a 2 mm diameter explosive on the surface of the aluminum plate and then the equivalence of the explosion product gas was the aluminum plasma confined between the glass plate and aluminum foil. The plasma cloud explosively

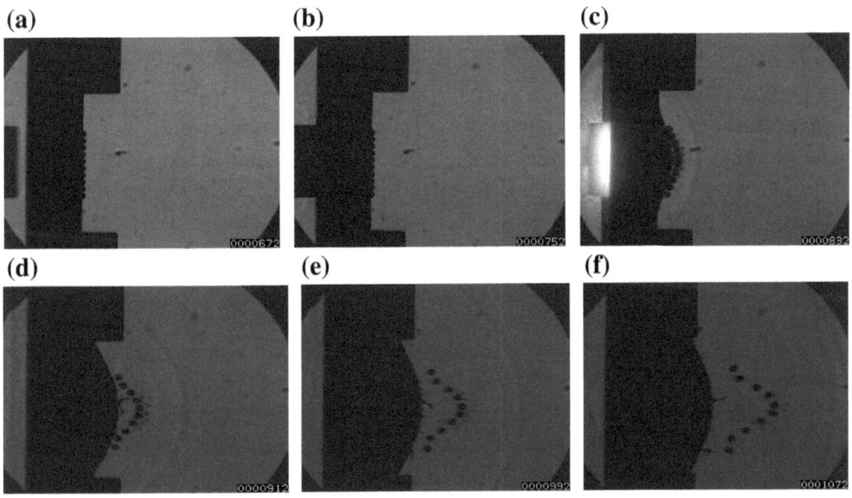

Fig. 10.37 Propulsion of 4.8 mm polyethylene beads attached ln a line on a steel plate hit by a nylon cylinder at 340 m/s

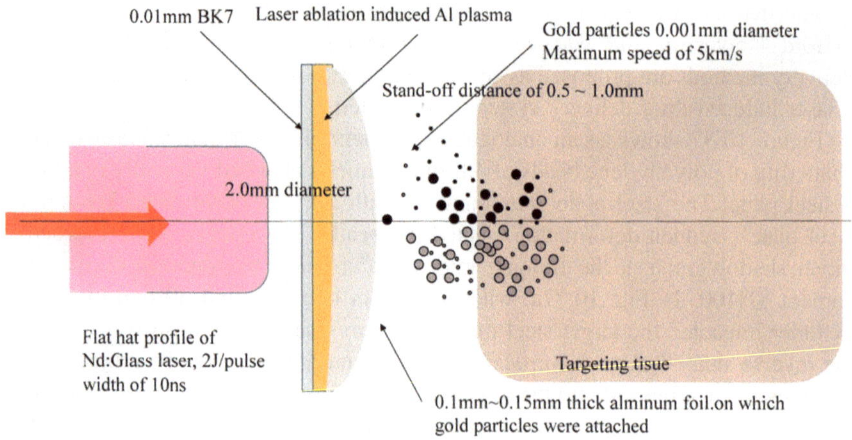

Fig. 10.38 Schematic illustration of pulse laser induced drug delivery

bulged the Al foil and ejected 1 μm diameter gold particles attached on the surface at maximum speed of 5 km/s (Menezes et al. 2008). The gold particles were ejected to targeting tissues positioned at about 1–1.5 mm stand-off distance and penetrated into the targeting tissue. The penetration depth ranged from 0.1 to 0.15 mm stand-off distance.

When dry drug particles are attached on the surface of the aluminum foil as a replacement of gold particles, the abovementioned method becomes a laser ablation assisted particle drug delivery system. The dry drug articles would fly at hypersonic speed when ejected from the foil and soon attenuated to a low supersonic speed. The particles would then penetrate targeting tissues deeply. Just from gas dynamic curiosity, a flow having Ms = 10 and the Reynolds number Re = 100, and Knutsen number Kn = 0.1 is defined as a Hypersonic Stokes flow. However, the distance the dry particles fly would range from 0.5 to 1.5 mm. Although the flow under study is a Hypersonic Stokes flow, effects of aerodynamic heating would not affect the penetration of dry drug delivery aa the rate of heat transfer is too slow.

Figure 10.39a shows an experimental result of shattering of 1 μm tungsten particles in air. The ordinate denotes the speed of particle clouds in m/s and abscissa denotes the flight distance in mm. The open circles denote results measured from the high-speed images. The particles were ejected using a similar system as shown in Fig. 10.38. The motion of the cloud of tungsten particles visualized by direct shadowgraph and images were recorded by high speed video camera Shimadzu SH100.

The particles moved at hypersonic speeds for a few mm distance and soon attenuated. Figure 10.39b shows the penetration of 1 μm diameter tungsten particles into a pig liver placed at 1 mm stand-off distance from the foil on which tungsten particles were attached. The particles were shattered but the central part of the distributed particle penetrated deeply into the pig liver. The particles penetrated

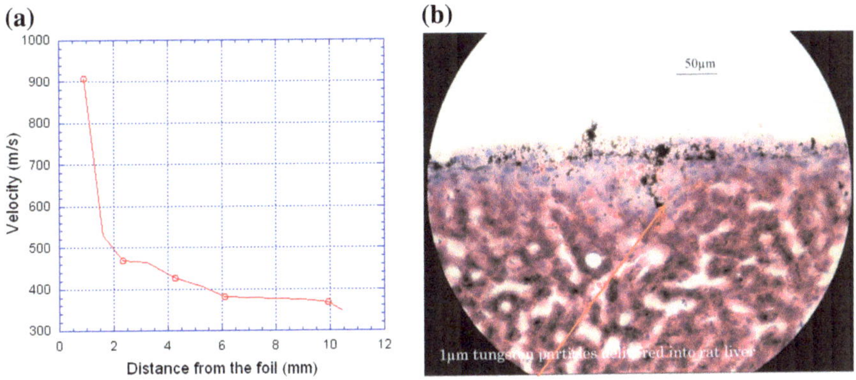

Fig. 10.39 Injection of 1 μm diameter tungsten particles: **a** the particle cloud velocity m/s versus the stand-off distance; **b** penetration test (Menezes et al. 2008)

about 100 μm deep. When this system is applied to the DNA recombination, the use of metal particles is not desirable. It would be a wonderful improvement in the drug delivery methods, if liquid droplets of diameter ranging from 10 to 20 μm can be shattered at reasonably high speed and can penetrate into soft tissue at 50 μm in depth.

Figure 10.40a shows a microscopic observation. It is clearly observed that a 1 μm diameter tungsten particle penetrated a cell line. If the tungsten particle was coated with DNA, there will be chances to have the DNA recombines. In using a shock wave induced drug delivery system, it became a routine methodology to recombine plant cell DNA by means of dry drug delivery system. Figure 10.40b shows the injection of 1 μm gold particles coated with a plasmid DNA into onion cells. In Fig. 10.40c, a gene expression in onion cells was observed. The colored spots indicate the transformed cells in the onion cells. The observation indicates that the present laser ablation induced drug delivery method worked (Nakada et al. 2008).

10.4.4 Shock Wave Ablation Catheter

To cure arrhythmia, the high radio frequency ablation catheter, the so-called ablation catheter is used clinically. In this treatment, high radio frequency eradicated an arrhythmia source in the heart and then the nerve cell at the arrhythmia source was completely destroyed and then the symptom disappeared. However, the radio frequency ablation induced high temperature on the spot at which the ablations were repeatedly applied subsequently inducing thrombi. This would create embolisms in a high probability.

In order to overcome such side effects and eradicate arrhythmia, a shock wave ablation catheter was proposed. In this proposed method, high pressures were

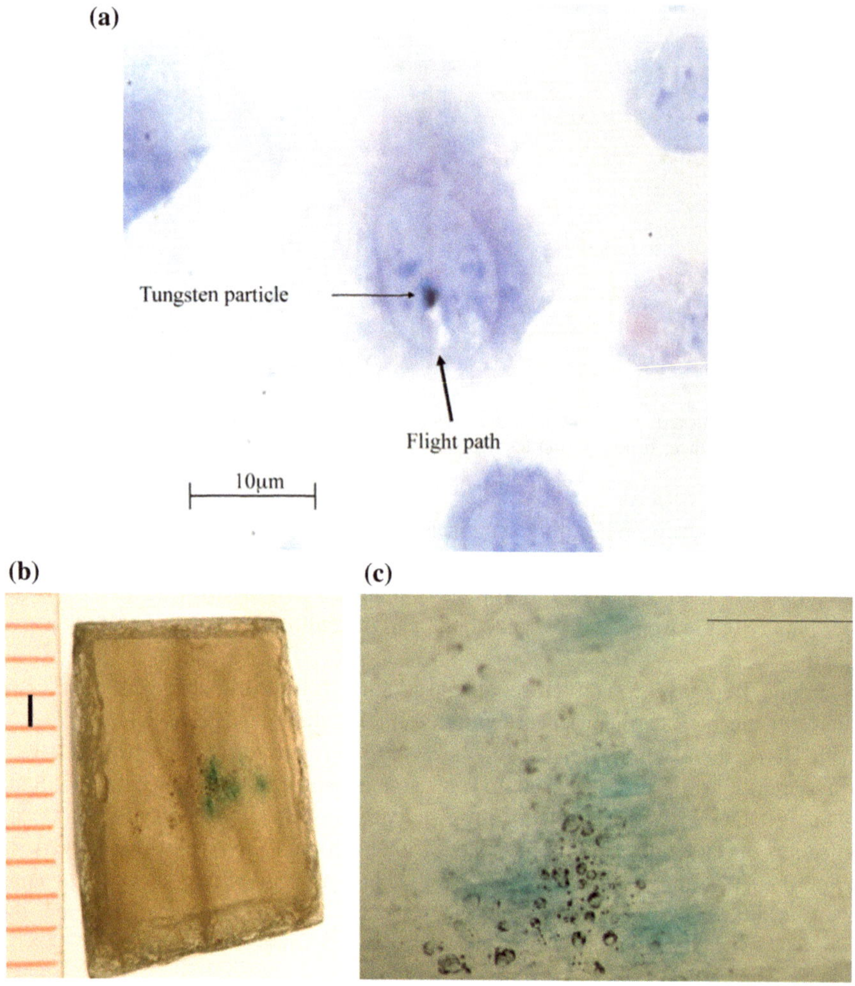

Fig. 10.40 Laser ablation induced drug delivery: **a** 1 μm diameter tungsten particles injected into cell line; **b** gene expression appeared in onion cells; **c** enlargement of (**b**) (Nakada et al. 2008)

applied very locally to the arrhythmia source by the shock wave focusing already applied to ESWL treatments. However, the proposed catheter was inserted into patients' artery to the inside heart so that the diameter of the truncated ellipsoidal reflector was less than 4 mm opening diameter and the source of shock wave generation was a Q-switched Ho:YAG laser beam transmitted via 0.4 mm diameter optical fiber (Yamamoto et al. 2015). Figure 10.41 shows a truncated ablation catheter which is going to be inserted through an artery to a spot inside the heart at which the arrhythmia source is located. The truncate ellipsoidal cavity had its inner diameter of 4.0 mm and outer diameter of 5.6 mm, whose *f*-number was about 0.5.

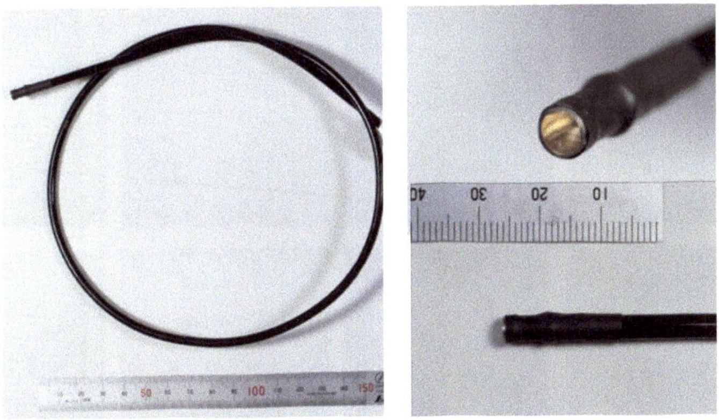

Fig. 10.41 Prototype of a shock wave ablation catheter (Yamamoto et al. 2015)

Empirically the energy transmission of the reflector in this *f*-number is maximal. The laser beams were transmitted repeatedly in such a confined space for 100 times during one sequence and then the water temperature in the space was elevated. Then saline water is continuously circulated. However, the degree of the elevated temperature was negligibly low.

Then, the catheter is covered with a thin film which is not only disconnecting the cooling saline water circulating inside the reflector but also its leakage into the blood inside the heart. While treatments, the cavity touched the internal membrane of the heart but the pressure at the exit of the cavity was low so that the internal membrane was not damaged. During the shock wave focusing, the peak pressure was gradually increased and exponentially enhanced in the vicinity of the focal point so that the arrhythmia source area was selectively damaged.

Pulsed Q-switched Ho:YAG laser beams of 35 mJ/pulse were repeatedly transmitted at 3 Hz through a 0.4 mm diameter optical fiber. Laser beams were focused through a lens shaped edge and created repeatedly shock waves. Figure 10.42a shows a sequential observation of a shock wave focusing. The ellipsoidal cavity was positioned at the bottom. The shock wave was going to focus at about 2 mm away from the opening of the truncated ellipsoidal cavity and focusing pressure was measures by an optical pressure transducer placed at the upper part in Fig. 10.42a. Figure 10.42b shows the resulting focused pressure. The ordinate denotes pressure in MPa. The abscissa denotes time in μs. The origin was the time instant when the pressure was maximal. Peak focus pressure was about 50 MPa and its half width is about 100 ns, which is strong enough to damage the arrhythmia source. As seen in Fig. 10.42a, the pressure gradually increases toward the arrhythmia source and exponentially enhanced at about 2 mm depth. Hence the tissue at the surface is not damaged seriously.

Figure 10.43 shows the result of a shock wave focusing on rat's heart from three different directions. For each direction, shock waves were focused at peak pressures

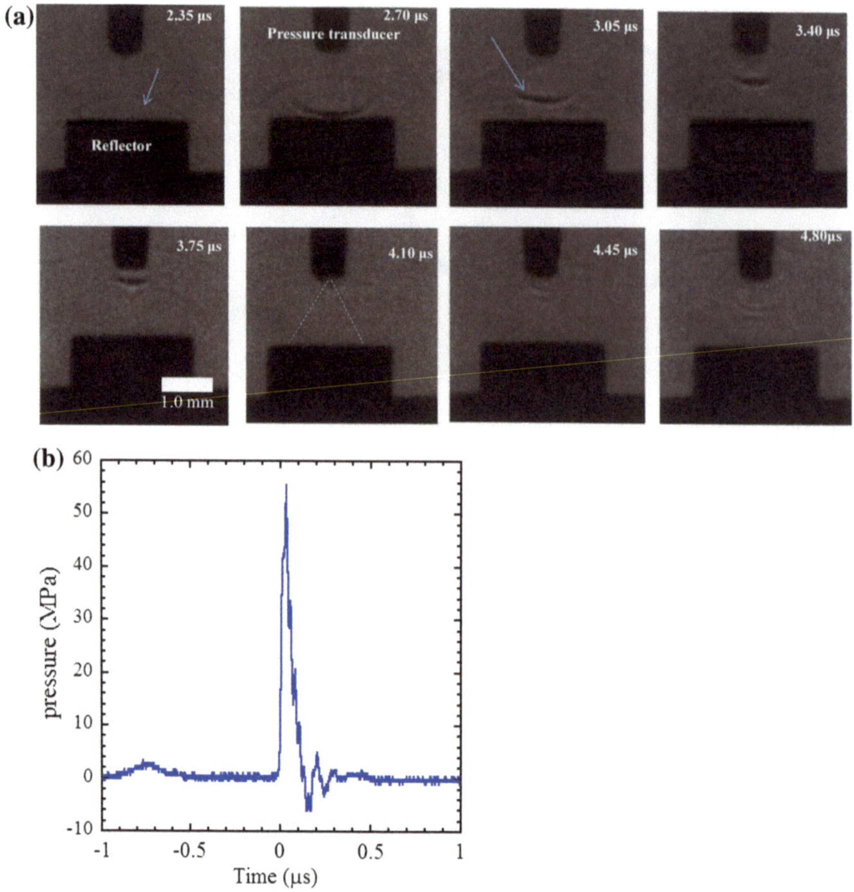

Fig. 10.42 Ho:YAG laser focusing: **a** focusing of a Q-switched Ho:YAG laser; **b** pressure history (Yamamoto et al. 2015)

ranging from 25 to 35 MPa at 3 Hz for 5 s. The internal membranes were not damaged but a bleeding was observed at the focal point in 2 mm depth.

10.5 Applications of Numerical Simulation to Clinical Purposes

Those who have unstable aneurysms that may rupture anytime would need immediate treatments. On the other hand, stable aneurysms do not need to have urgent treatments. However, it is not easy to determine the urgent treatment simply by observation of images of the aneurysms. Then neuro-brain surgeons requested numerical analysts whether or not the blood circulations in a three-dimensional

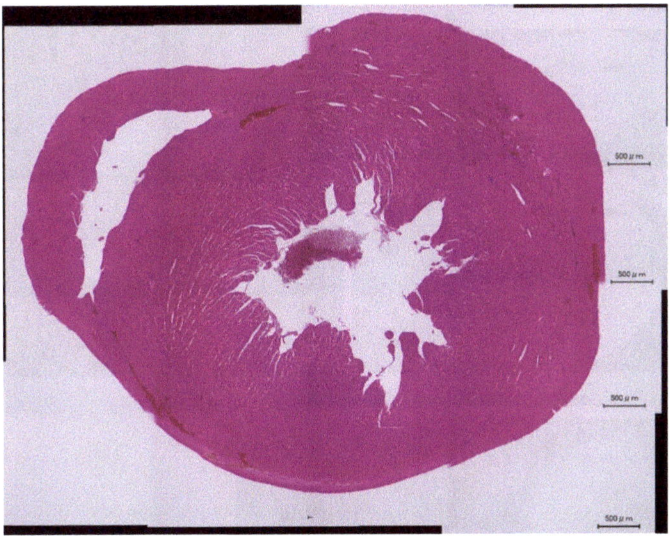

Fig. 10.43 Shock wave focusing on rat's heart. Overpressure at the focal point ranges from 25 to 35 MP (Yamamoto et al. 2015)

replica of patients' aneurysms are possible. In 2000 a project started constructing three-dimensional meshes out of the X-ray images or CT scan images of blood vessels in brains viewed from various directions. At first a software was developed for constructing three-dimensionally distributed blood vessels. Figure 10.44a shows three-dimensional meshes of the blood vessel having an aneurysm in the central part given. At first, the wall boundary of the blood vessel was a solid boundary. At that time, reliable physical properties of aneurysms and living brain blood vessels were not known and hence the simulation was just to examine a blood circulation through complexly shaped three-dimensional pipe line. Elastic deformation was ignored but the Navier-Stokes Equations were solved at non-slip boundary condition. Blood pressures were given according to empirical data. Simulations were conducted, corresponding to various clinical conditions, such as blood pressures and blood mass flow. Figure 10.44b shows one of the primary results. A three-dimensional velocity distribution in blood vessel and an aneurysm. A region with red color indicates blood flow is faster than a region with blue color. Hence in the red color region, the wall shear force is larger than the blue color region. The blood vessel expressed with red color has a higher risk to rupture. Therefore, the numerical simulation would help decision making. However, the result presented in Fig. 10.44 is a preliminary study and many parts should be refined in the near future.

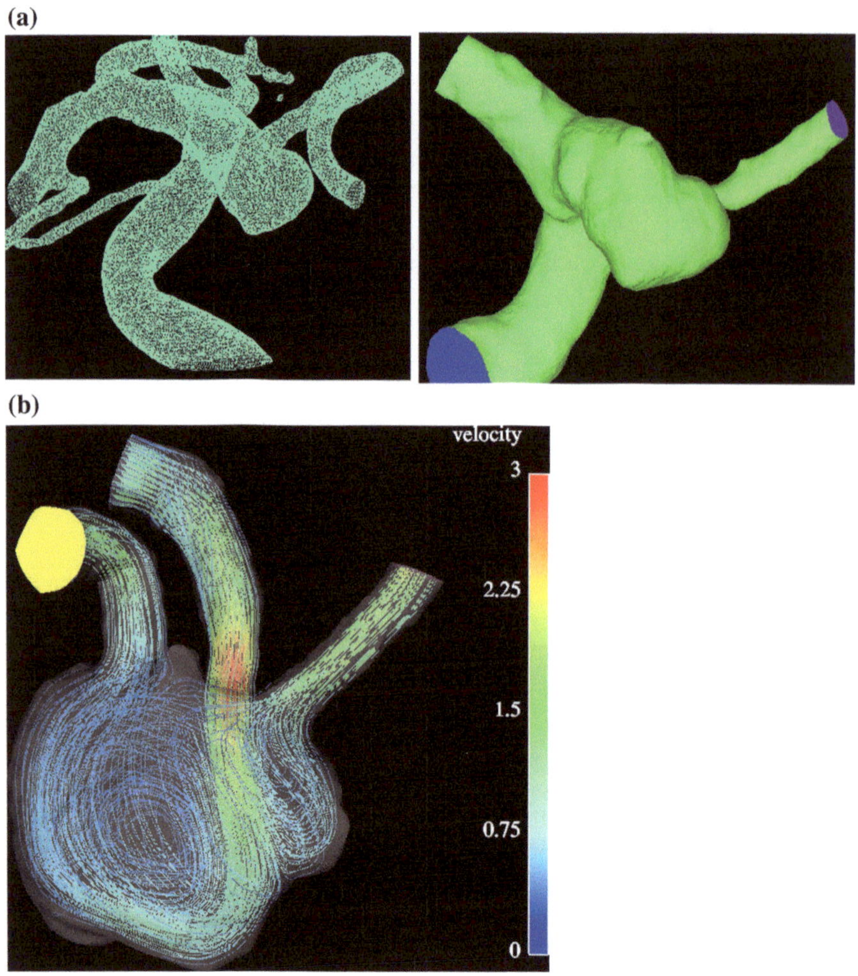

Fig. 10.44 Computational blood flow: **a** computational replica of aneurysm; **b** velocity distribution inside an aneurysm (Hassan et al. 2004)

When the three-dimensional meshes are constructed, we simulated assuming the laminar boundary layer and then imposing appropriate input and outlet flow conditions at the ends of individual blood vessels.

References

Abe, Y., Ise, H., Kitayama, O., Usui, R., Suzuki, N., Matsuno, M., et al. (1990). Disintegration of gallbladder stones by ESWL. *Gallstone, 4,* 451–459.

Bellhouse, H. J., Quikan, N. J., & Ainsworth, R. W. (1997). Needle-less delivery of drugs, in dry powder form, using shock waves and supersonic gas flow. In A. F. P. Houwing, & A. Paul, (Eds.) Proc. 21st ISSW, (Vol. 1, pp. 51–56). Australia: The Great Keppel Island.

Chaussey, C. H., Schmiedt, E., Jocham, D., Walter, V., Brendel, W., Forsmann, B., et al. (1982). *Extracorporeal shock wave lithotripsy. New aspects in the treatment of kidney stone disease.* Muenchen: Karger.

Chaussey, C. H., Schmidt, J. E., Joachim, D., Ferbes, G., Brundel, W., Forsmann, B., et al. (1986). *Extracorporeal shock wave lithotripsy.* Muenchen: Karger.

Hassan, M., Ezura, M., Timfeev, E. V., Tominaga, T., Saito, T., Takahashi, A., et al. (2004). Computational simulation of therapeutic parent artery occlusion to treat giant vertebrobasilor aneurysm. *AINR American Journal of Neuroradiology, 25,* 63–68.

Hirano, T. (2001). *Development of revascularization of cerebral thrombosis using laser induced liquid jets* (MD thesis). Graduate School of Medicine, Tohoku University.

Hirano, T., Uenohara, H., Nakagawa, A., Sato, S., Takahashi, A., Takayama, K., & Yoshimoto, T. (2002). A novel drug delivery system with Ho:YAG laser induced liquid jet. In *Proceedings of the International Federation for Medical and Biological Engineering. 2nd European Conference* (pp. 1006–1007).

Ikeda, K., Matsuda, M., Tomita, K., & Takayama, K. (1999). Application of extracorporeal shock wave on bone. Basic and clinical study. In G. J. Ball, R. Hillier & G. T. Robertz (Eds.), *Shock Waves. Proceedings of 22nd ISSW, London* (Vol. 1, pp. 623–626).

Kambe, K., Kuwahara, M., Kurosu, S., Orikasa, S., & Takayama, K. (1986). Underwater shock wave focusing, an application to extracorporeal lithotripsy. In D. Bershader & R. Hanson (Eds.), *Shock Waves and Shock Tubes, Proceedings of the 15th International Symposium on Shock Waves and Shock Tubes, Berkeley* (pp. 641–647).

Kato, K. (2004). *Study of mechanism and damage threshold of brain nerve cells by shock wave loading* (MD thesis). Graduate School of Medicine, Tohoku University.

Krehl, P. O. K. (2009). *History of shock waves, explosions and impact.* Berlin: Springer.

Kuwahara, M., Kambe, K., Kurosu, S., Orikasa, S., & Takayama, K. (1986). Extracorporeal stone disintegration using chemical explosive pellets as an energy source of underwater shock waves. *The Journal of Urology, 133,* 814–817.

Kuwahara, M., Ioritani, M., Kambe, K., Shirau, S., Taguchi, K., Sitoh, S., et al. (1989). Hyperechoic region induced by focused shock waves in vivo in vitro possibility of acoustic cavitation. *Journal of Lithotripsy and Stone Disease, 1,* 282–287.

Loske, A. M. (2007). *Shock wave physics for urologists.* Universidad National Autonoma de Mexico.

Menezes, V., Takayama, K., Gojani, A., & Hosseini, S. H. R. (2008). Shock wave driven micro-particles for pharmaceutical applications. *Shock Waves, 18,* 393–400.

Nakada. M., Menezes, V., Kanno, A., Hosseini, S. H. R., & Takayama, K. (2008). Shock wave based biolistic device for DNA and drug delivery. *Japanese Journal of Applied Physics, 47,* 1522–1526.

Nagayasu, N. (2002). *Study of shock waves generated by micro explosion and their applications* (Ph.D. thesis). Graduate School of Engineering, Faculty of Engineering, Tohoku University.

Nakagawa, A. (1998). *Basic study of shock wave assisted therapeutic devises in the field of neuro brain surgery* (MD thesis). Graduate School of Medicine, Tohoku University.

Nakagawa, A., Kumabe, T., Kanamori, M., Saito, R., Hirano, T., Takayama, K., et al. (2008). Clinical application of pulsed laser-induced liquid jet: Preliminary report in glioma surgery. *Neurological Surgery, 36,* 1005–1010.

Obara, T. (2001). *A study of applications of underwater shock waves to medicine* (Ph.D. thesis). Graduate School of Engineering, Faculty of Engineering Tohoku University.

Ohki, T. (1999). *Study of medical applications of pulsed Ho:YAG laser induced underwater* (Master thesis). Graduate School of Engineering, Faculty of Engineering Tohoku University.
Okazaki, K. (1989). Fundamental study in extracorporeal shock wave lithotripsy using piezoceramics. *Japanese Journal of Applied Physics, 28,* 143–145.
Shitamori, K. (1990). *Study of propagation and focusing of underwater shock focusing* (Master thesis). Graduate School of Tohoku University Faculty of Engineering, Tohoku University.
Yamamoto, H., Hasebe, Y., Kondo, M., Fukuda, K., Takayama, K., & Shimokawa, H. (2015). Development of a novel shock wave catheter ablation system. In R. Bonazza & D. Ranjan (Eds.), *Shock Waves, Proceedings of the 29th ISSW, Madison* (Vol. 2, pp. 855–860).
Yutkin, L. A. (1950). *Apparat YRAT-1 Medeport USSR Moscow.*

Chapter 11
Miscellaneous Topics

11.1 Hypersonic Flows

In the advent of the space exploitation, the shock tube supported the atmospheric re-entry of space vehicles. Hence the space technology owed its success immensely to the shock wave research. Initially the design of the heat shield of re-entry vehicles became a main topic in the shock wave research but later it was shifted to hypersonic propulsions of SCRAM jet engines.

In the 1990s, a free piston shock tunnel was constructed in the Shock Wave Research Center of the Institute of Fluid Science in house. Figure 11.1 shows a sketch of the free piston shock tunnel. Hypersonic flow experiments were carried out under the collaboration with the National Aerospace Laboratory, Kakuda Branch. This shock tunnel was a pilot facility aiming to construct the high enthalpy shock tunnel (HEIEST) in the NAL Kakuda Branch in the near future. This free piston shock tunnel achieved the stagnation enthalpy of 4.8 MJ/kg producing the nozzle flow speed u_∞ of 2750 m/s, the stagnation temperature T_∞ of 387 K, the stagnation pressure p_∞ of 2.26 kPa, the stagnation density of ρ_∞ 2×10^{-2} kg/m^3, and resulting flow Mach number M_∞ of 6.99. A relatively uniform hypersonic nozzle flow was sustained for approximately 300 μs. Koremoto (2000) investigated the starting process of the nozzle flows. Hashimoto (2003) visualized flows over double wedges and double cones installed in the nozzle flow by shadowgraph and recorded in a high-speed video camera Shimadzu SH100.

11.1.1 Flows Over Double Wedges and Double Cones

Figure 11.2a, b show a double wedge model and a double cone model installed in the test section, respectively. The first wedge angle and the first half cone angle were 25° and the second wedge angles were varied to be θ = 40, 50, and 68° and

K. Takayama, *Visualization of Shock Wave Phenomena*,
https://doi.org/10.1007/978-3-030-19451-2_11

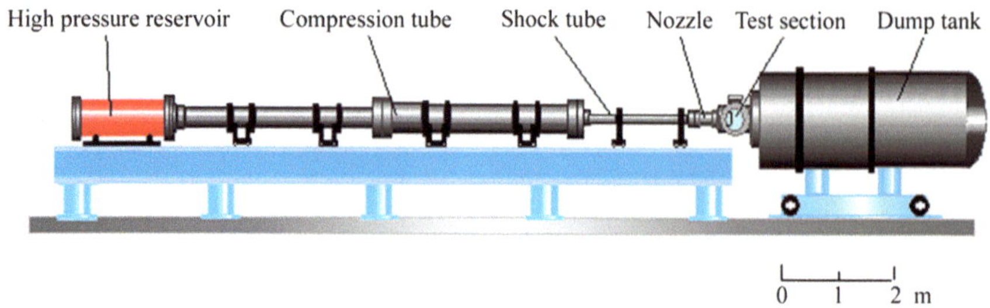

Fig. 11.1 The free piston shock tunnel of the SWRC, Tohoku University

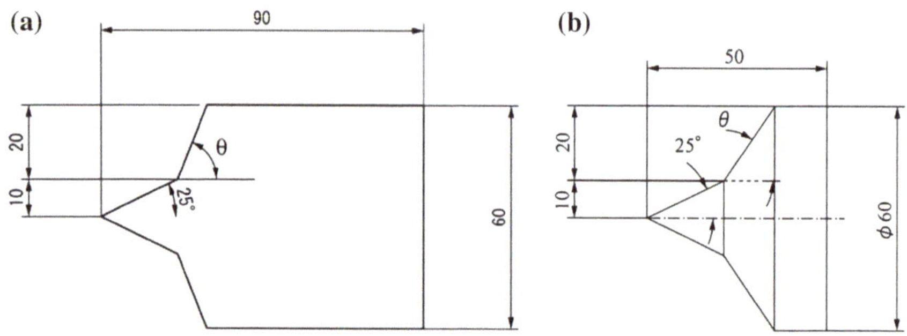

Fig. 11.2 Test models: **a** double wedge θ = 40, 50, 68°; **b** double cone θ = 40, 65°

second cone angles were varied to be θ = 40 and 65° as shown in Fig. 11.2. The double wedge model was 90 mm long and 60 mm diameter. The double cone model was 50 mm long and 60 mm diameter. Both of them were made of brass.

Figure 11.3a, c show images of direct shadowgraphs of the hypersonic flow over a double wedge of 25/50° and a double wedge of 25/68° double wedges, respectively, recorded by a high speed digital camera Shimadzu SH100 at the framing rate of 10^6 frame/s. A shock wave was attached at the leading edge of the first wedge. A detached shock appeared at the corner of the first and the second wedge. The boundary layer developed along the first wedge surface and separated ahead of the corner of the first and second wedges. A separation zone appeared at the corner, forming a recirculation region. However, it is not easy to identify flow features in recirculation regions over the double wedges and double cones by observing the framing pictures. To precisely interpret flow features, it was decided to make the flaming pictures rearrange in a streak display.

0.5 mm wide slices of the individual framing pictures were made along the double wedge surface or the double cone surface seen in Figs. 11.3a, c and 11.4a, c. The slices of sequential images were arranged in a sequential order, which eventually formed a streak picture. Figures 11.3b, d and 11.4b, d show resulting streak

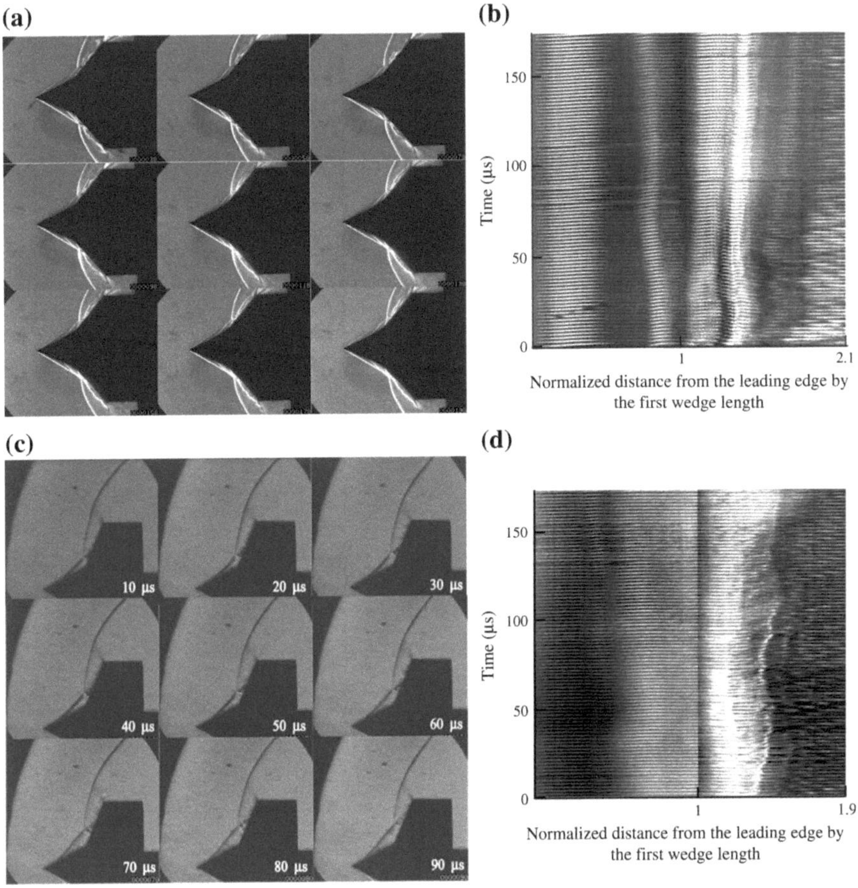

Fig. 11.3 High-speed video images of a hypersonic flow over double wedges: **a** direct shadow images over a 25/50° wedge; **b** streak display of (**a**); **c** direct shadow images over a 25/68° wedge; **d** streak display of (**c**) (Hashimoto 2003)

displays corresponding to Figs. 11.3a, c and 11.4a, c. The ordinate denotes elapsed time in μs and the abscissa denotes the distance normalized by the distance from the leading edge to the corner.

Figure 11.4a–d show hypersonic flows over double cones and direct shadow images and their streak displays are presented. The streak display indicates that the recirculation region fluctuates at higher frequency over the 25/65° double cone than over the 25/50° double cone. A similar trend is observed over the double wedges. This explains the usefulness of the streak display out of the high speed framing images.

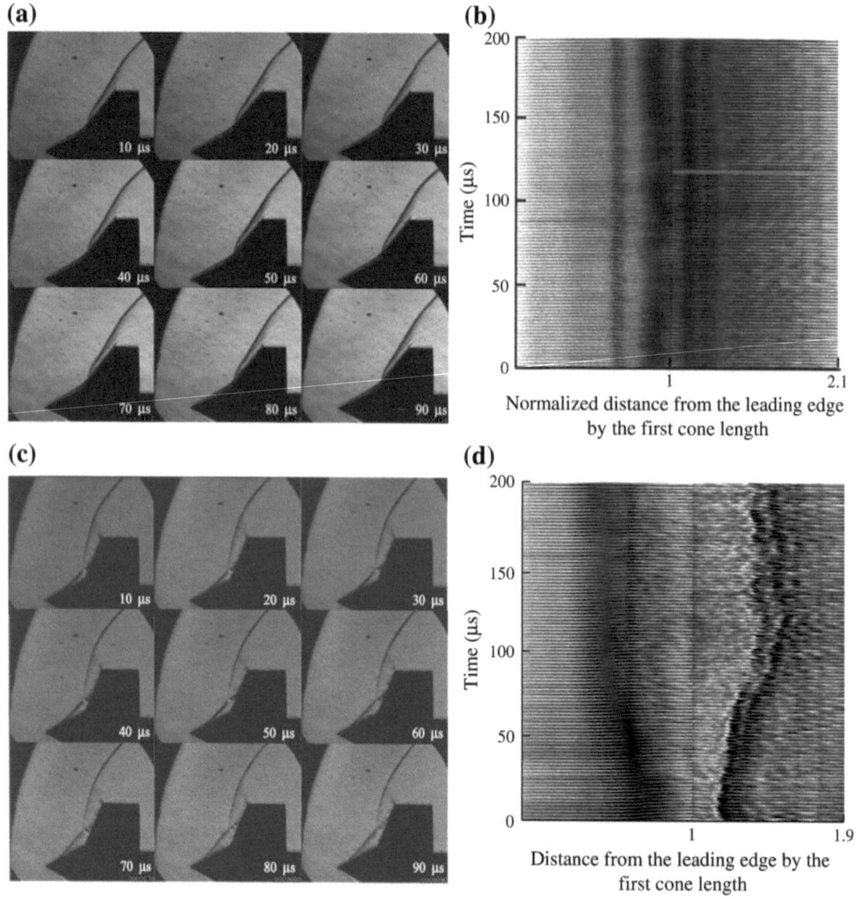

Fig. 11.4 High-speed video images of a hypersonic flow over double cones of 25/50° at 2.3 kPa, 390 K flow velocity of 2759 m/s: **a** direct shadow images over a 25/50° cone; **b** streak display of (**a**); **c** direct shadow images over a 25/65° cone; **d** streak display of (**c**) (Hashimoto 2003)

11.2 Ballistic Ranges

In the late 1980, a single stage powder gun was constructed and flights of projectiles were visualized. Projectiles' motions were measured with a VISAR (velocity interferometry from surface of any reflectors) and compared by an appropriate numerically simulation. The numerical code for simulating a two-stage light gas gun was based on the Random Choice Method and developed by Professor Gottlieb of the UTIAS. Figure 11.5 shows a two-stage light gas gun installed in the Shock Wave Research Center (Matsumura et al. 1990).

Smoke-less powder weighing 150 g was filled in a propellant chamber. A 60 mm diameter poly-carbonate piston weighing from 0.5 to 2.0 kg was placed

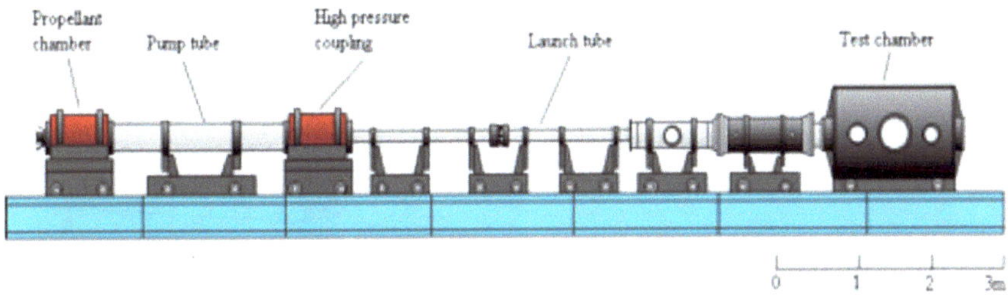

Fig. 11.5 Two-stage light gas gun (Matsumura et al. 1990)

in a 60 mm diameter and 3 m long pump tube. Helium at initial pressure of 0.5 MPa was filled in the pump tube. The heavy piston was accelerated by high pressures generated by controlled combustion of the smoke-less powder eventually compressed helium filled in high-pressure coupling up to 0.5 GPa elevating the temperature up to a few thousand K. The inertia mass of the high pressure coupling absorbed the high stress wave impulsively generated upon the impact of the heavy piston and maintained the high pressure and high temperature helium for a very short time. Then, a 14 mm diameter nylon projectile was accelerated along the acceleration tube and readily reached to maximum speed of 5 km/s.

11.2.1 Bow Shock in Front of Free Flight Blunt Bodies in Air

Figure 11.6 show bow shock waves appearing ahead of a free flight of blunt cylinder. Figure 11.6d–f are reconstructed three-dimensional holograms. Figure 11.6d was taken immediately after the projectile entering into the test chamber. The hologram was taken at the moment when the blunt cylinder just took over the precursory shock wave. The inclination angle θ of individual bow shock wave indicated the flight speed, that is $\sin\theta = a/u$, where a and u are the sound speed in air and the free flight speed, respectively.

In hot hypersonic experiments, a reliable data acquisition is one of the important tasks. Professor Park of NASA Ames advised that the shock stand-off distance over spheres demonstrated real gas effects, because the shock stand-off distances in an intermediate hypersonic flow region is sharply affected by the real gas effects. Shock stand-off distances in front of free flight spheres intermediate hypersonic flow region were visualized in the two-stage light gas gun (Nonaka 2000) and also in the free piston shock tunnel (Hashimoto 2003). Figure 11.7 summarizes previous experiments conducted in the Shock Wave Research Center. ρR is a hypersonic similarity parameter. The ordinate denotes dimension-less shock stand-off distance. The abscissa denotes the free flight speed in km/s. Red, olive color, light blue, and dark blue filled circles denote results of ballistic range experiments corresponding

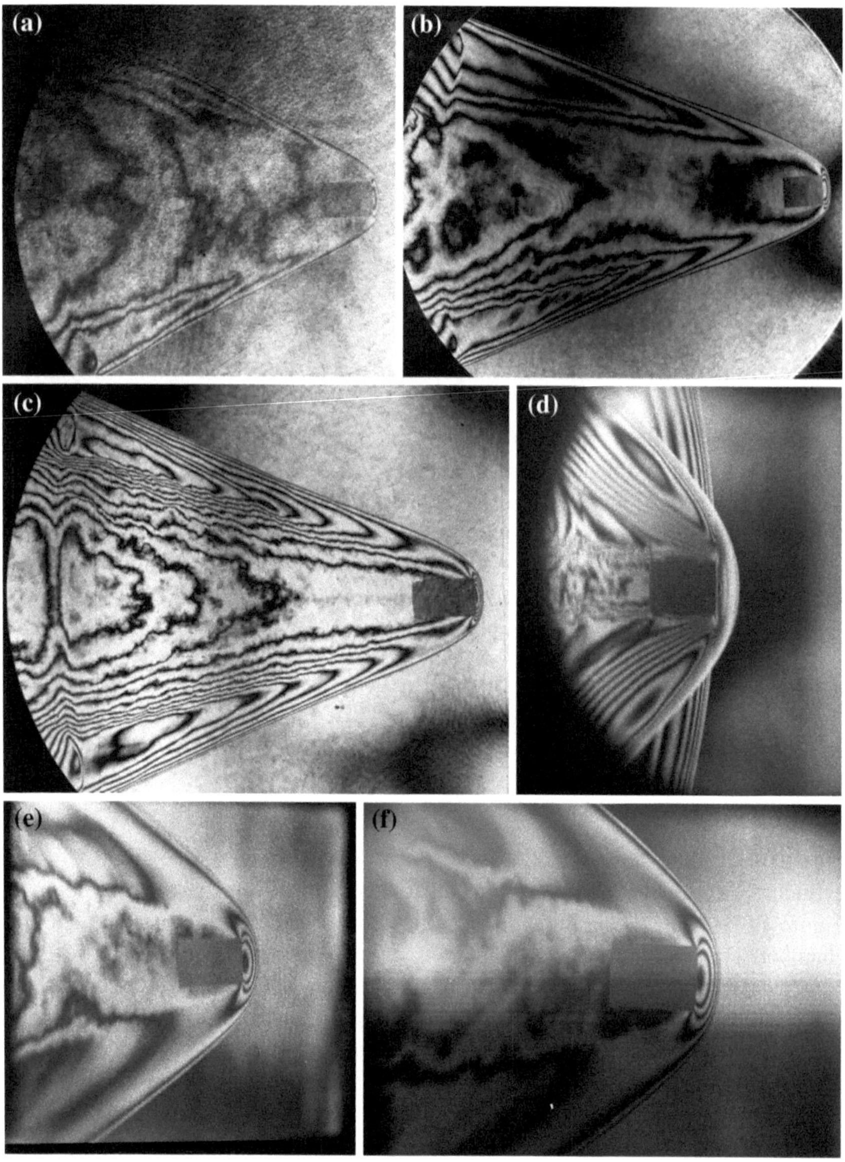

Fig. 11.6 Free flight of 10 mm diameter polycarbonate projectiles launched from a two-stage light gas gun: **a** #92101304, Ms = 2.56; **b** #92101601, Ms = 2.50; **c** #93022101, Ms = 2.77; **d** #98102901, Ms = 1.41, at 1013 hPa, 3-D hologram; **e** #98102902, Ms = 1.56, 3-D hologram; **f** #98102903, Ms = 1.49, 3-D hologram (Matsumura et al. 1990)

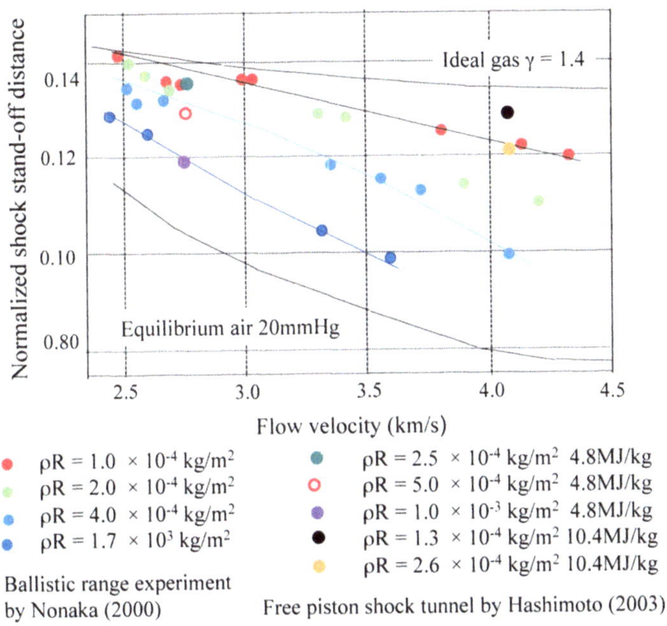

Fig. 11.7 Shock stand-off distance at intermediate hypersonic flow ranges collected by the ballistic range

to $\rho R = 1.7 \times 10^{-3}$ kg/m^2, $\rho R = 1.0 \times 10^{-4}$ kg/m^2, $\rho R = 2.0 \times 10^{-3}$ kg/m^2, $\rho R = 4.0 \times 10^{-4}$ kg/m^2 (Nonaka 2000), respectively, and black, red, and light blue open circles, a black filled circle, and a yellow filled circle denote result of free piston shock tunnel experiments corresponding to 2.5×10^{-4} kg/m^2, at 4.8 MJ/kg, 5.0×10^{-4} kg/m^2, at 4.8 MJ/kg, 1.0×10^{-3} kg/m^2, at 4.8 MJ/kg, 1.3×10^{-4} kg/m^2, at 10.4 MJ/kg, and 2.6×10^{-4} kg/m^2, at 10.4 MJ/kg (Hashimoto 2003), respectively. Experimental results corresponding to individual similarity parameters consistently lie on a line. The lines departed, depending on the values of similarity parameter, from the line of ideal gas having $\gamma = 1.4$ to the line estimated in the equilibrium air at 20 mmHg.

In 1998, a ballistic range was installed. Figure 11.8 shows an illustration of the SWRC ballistic range, which consists of a two-stage light gas gun and a 1.8 m diameter and 12 m long observation chamber. The two-stage light gas gun consists of a powder chamber, a high-pressure coupling, a 50 mm diameter and 3 m long pump tube, a launch tube, and a test chamber (Numata 2009). The launch tube is 3.4 m long and is convertible to either 15 or 50 mm diameter tube. The smokeless powder is installed in the powder chamber and is ignited following the US army standard. The high pressure coupling has a pre-stressed structure so that it has very compact if compare with the previous two-stage light gas gun. The observation chamber has a large space and hence various test sections can be accommodated.

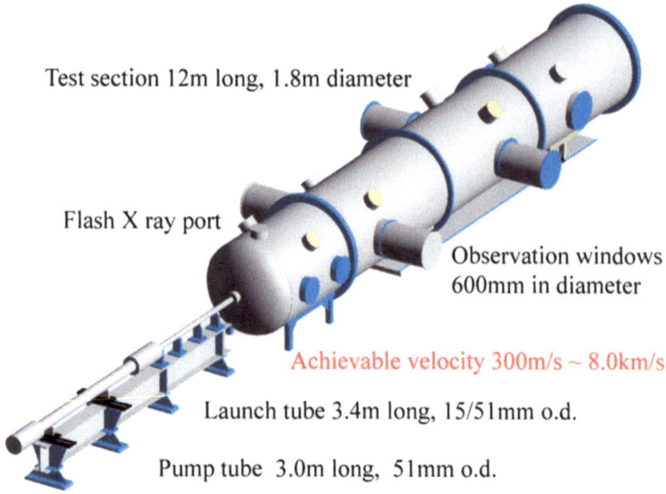

Fig. 11.8 SWRC ballistic range

Some test sections were filled not only with foreign gases but also with water. The observation chamber has two sets of pairs of 600 mm diameter optical windows and four sets of flash X-ray sources and detectors. Optical flow visualizations are conducted mostly by double exposure holographic interferometry and shadowgraph. For sequentially recording images, high speed camera ImaCon D-200 and Shimadzu digital high-speed camera SH100 were used. 15 mm diameter and 50 mm diameter projectiles are launched at speed ranging from the sonic speed to 8.0 km/s.

11.2.2 Free Flight in Combustible Mixtures

In order to visualize detonation waves induced by a high speed projectile, a compact test chamber filled with oxyhydrogen mixture was installed in the observation chamber. A 40 mm diameter nylon sphere was launched at 2.2 km/s into stoichiometric oxygen and hydrogen mixture $2H_2/O_2$ at 333 hPa/167 hPa. The experimental arrangement for launching a 40 mm diameter sphere into the test section was already explained in Fig. 4.23.

Figure 11.9a shows a detonation wave over a 40 mm sphere in fuel rich mixture ($2H_2$: 360 hPa, O2: 140 hPa). Figure 11.9b shows a detonation wave in stoichiometric mixture ($2H_{2:}$ 333 hPa, $O_{2:}$167 hPa) at 2.2 km/s. In Fig. 11.9a, b, these interferograms comprised three interferograms taken at slightly different delay time. Figure 11.9c, d were enlargements of the shock layers corresponding to the images shown in Fig. 11.9a, b, respectively. It should be noticed that a slight deviation of shock stand-off distance was observed which were caused due to the slight

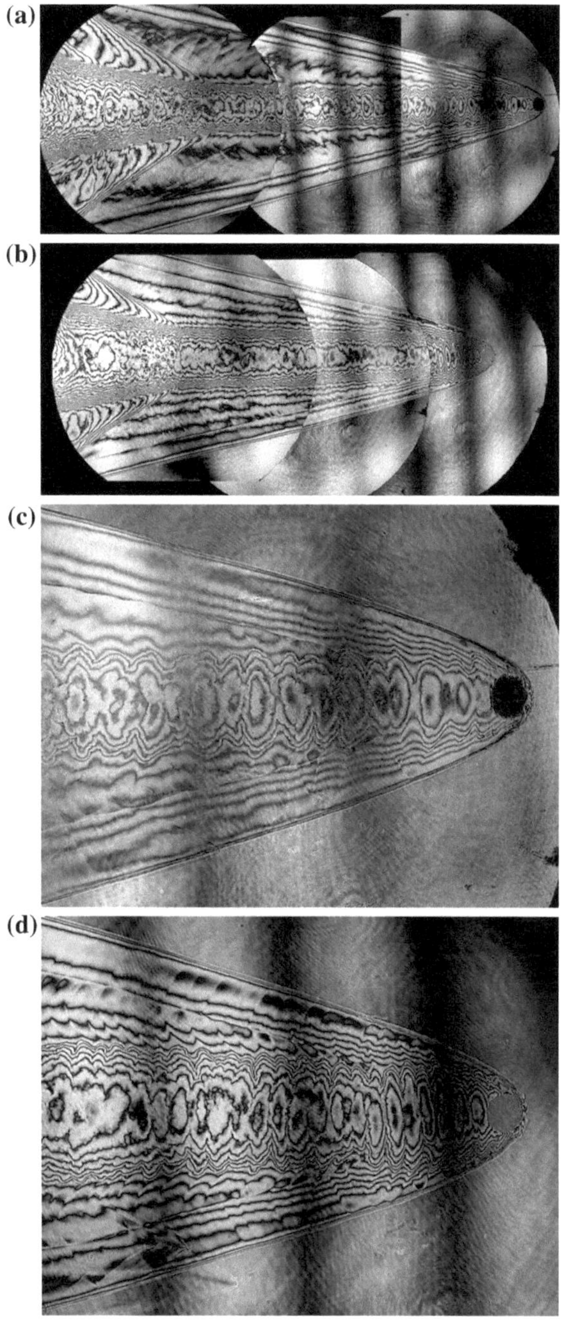

Fig. 11.9 Free flight in a combustible mixture: **a**, **c** projectile speed of 2.2 km/s in $2H_2$ = 360 hPa/O_2 = 140 hPa, fuel rich mixture; **b**, **d** projectile speed of 2.2 km/s in $2H_2$ = 333 hPa/O_2 = 167 hPa, stoichiometric mixture

difference in the constituents in combustible gas mixtures. The wakes have a coherent structure but its cycles are slightly different depending on the constituent of the combustible gas mixtures.

11.2.3 Space Debris Bumper Shields

Basic experiments of space debris bumper shields were conducted installing a test section in the observation chamber. A 15 mm launch tube was used projecting a 10 mm diameter stainless steel bearing ball. Figure 11.10 show sequential observation of shadow pictures recorded with a high speed video camera Shimadzu SP100 at framing rate of 10^6 frame/s and the exposure time of 125 ns. The pictures were displayed at every 24 μs interval. The target plate was aluminum alloy. In Fig. 11.10c, an impact flash was emitted and sustained for about 75 μs. The hypervelocity impact created stress waves in the targeting aluminum alloy. The stress created in the aluminum alloy well exceeded its yielding stress and hence the deformation in the aluminum alloy looked very much like that of liquid motion. Patterns of splashes observable on the target surface closely resembled in the splashes observed at high speed impact into a water surface.

Figure 11.11 show sequential observations of the penetration of a 10 mm diameter bearing ball at 2 km/s against a 10 mm thick composition of aluminum and Kevler. Images were recorded in the same way as described in Fig. 11.10. The target plate is the so-called stuffed bumper shield routinely used for the protection of the space structures from the debris attacks. The composite plate tested here are so designed as to stop the penetration of fragmented debris particles against the main shield.

11.2.4 Space Debris Bumper Shield at Cryogenic Temperature

The environmental temperature at which the satellites circulated varies widely from the cryogenic temperature about 100 K in the shadow of the earth to over 350 K at the sunny side. The bumper shield experiments are usually conducted at room temperature. Then the effect of the environmental temperature on high speed impact was investigated. At the cryogenic temperature, how can the high speed impact be affected? (Numata 2009).

Figure 11.12a shows a cryogenic test chamber and the installation of a frontal bumper made of aluminum alloy positioned at 100 mm separation distance from the main wall. Circulating liquid nitrogen, the temperature of the entire chamber and the test section were reached down to cryogenic temperature of 120 K was achieved. Figure 11.12a shows a frontal wall and main wall cooled by tightly contacting the supporting metal pieces as seen in Fig. 11.12b.

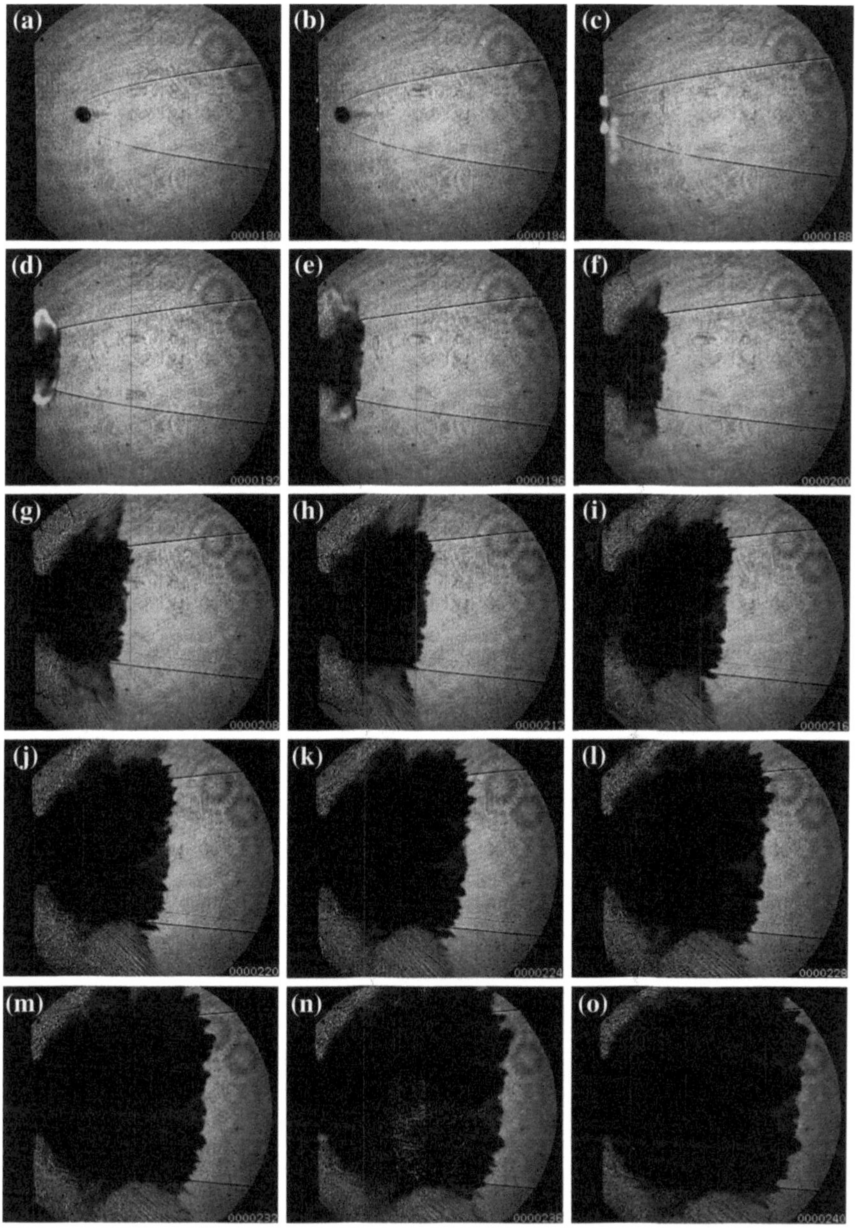

Fig. 11.10 Sequential observation of hypersonic impact of 10 mm diameter sphere against an aluminum bumper shield at entry speed of 3 km/s. Notice splash of impact fragments (Numata 2009)

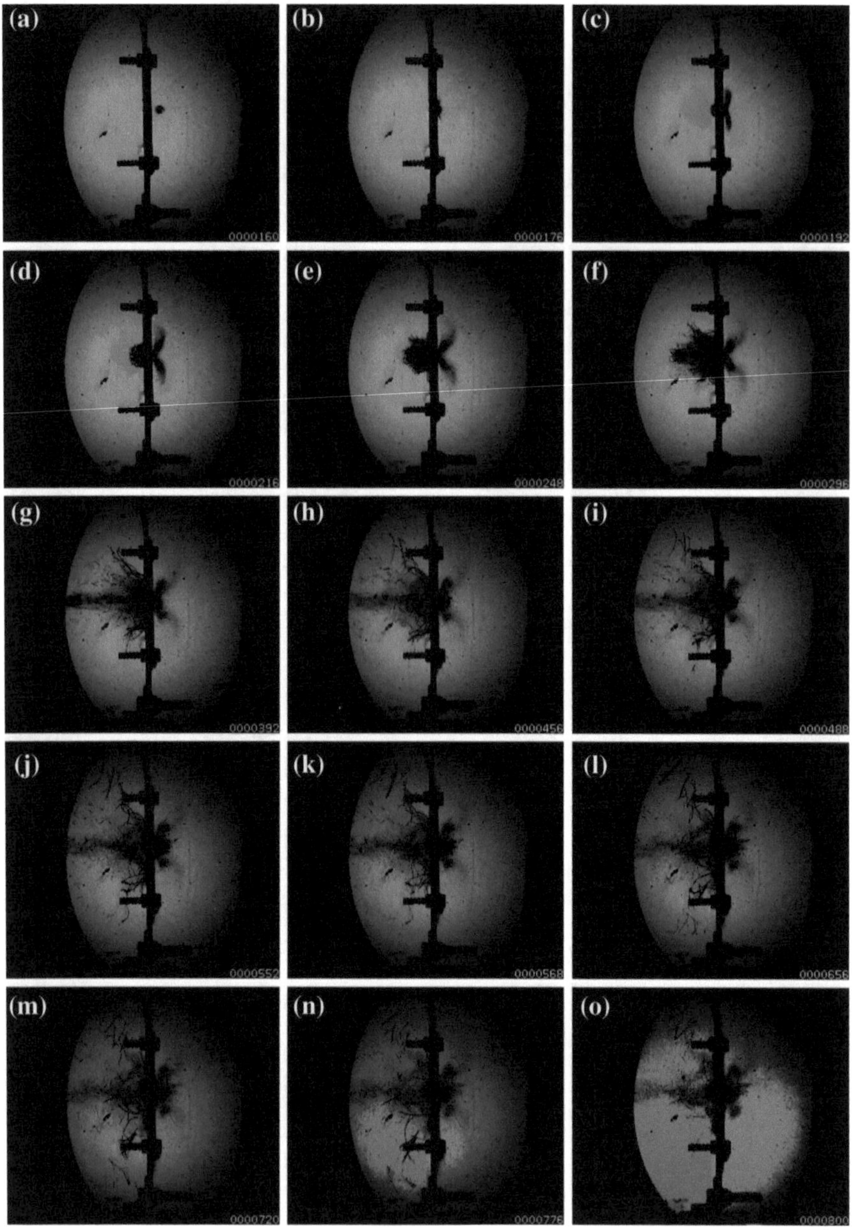

Fig. 11.11 Penetration of an aluminum/Kevler composite plate upon the impact of a projectile flying at 2 km/s (Numata 2009)

(a)

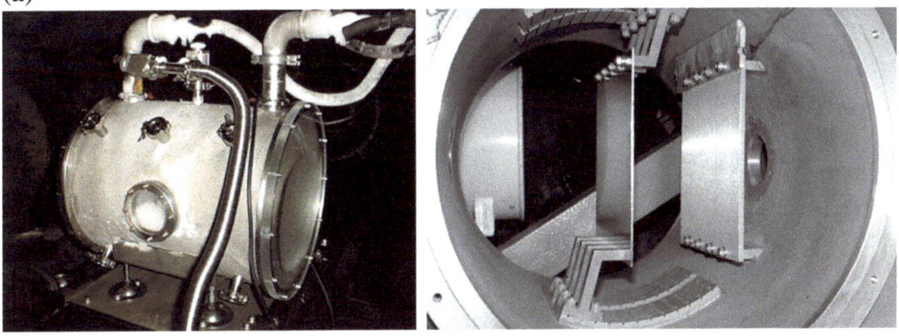

(b)

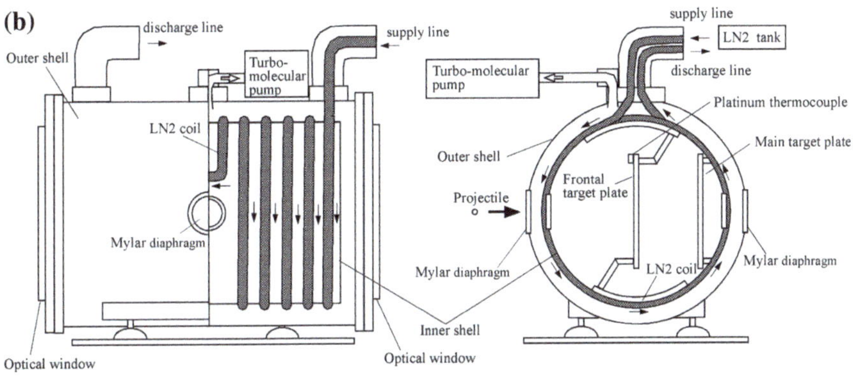

Fig. 11.12 Cryogenic test chamber: **a** test chamber; **b** internal structure of the test chamber (Numata 2009)

Figure 11.13a–c show images selected from high speed impact tests at projectile speeds of 2.7, 3.3, and 3.7 km/s against the wall at about 293 K, respectively. Figure 11.13d–f show images selected from high speed impact tests at projectile speeds at the impact speed of 2.8, 3.4, 3.7 km/s against the wall at about 118 K, respectively. The distributions of debris clouds did not differ significantly depending on the wall temperatures. However, the distribution of the debris clouds seemed to be more elongated at the impact speed 2.8 km/s, than that of 3.7 km/s.

It is concluded that the present analogue experiment revealed that the wall temperature did not affect significantly the debris cloud structure but it was strongly affected by the impact speed of the projectile.

Figure 11.14 show the result of an analogue experiment: the evolution of the debris cloud when a projectile impacted at 3.7 km/s against a bumper shield at 120 K. The images were recorded by Shimadzu SH 100 at the framing rate of 10^6 frame/s. At first, high-speed impacts of small fragments on the cryogenic wall created points of impact flashes. Then the main part of the debris cloud impacted following the generation of the intense impact flash. The main structure of the debris cloud was reflected. However, the main debris clouds were reflected from the main wall made of composite materials but never penetrated the main wall at this speed range.

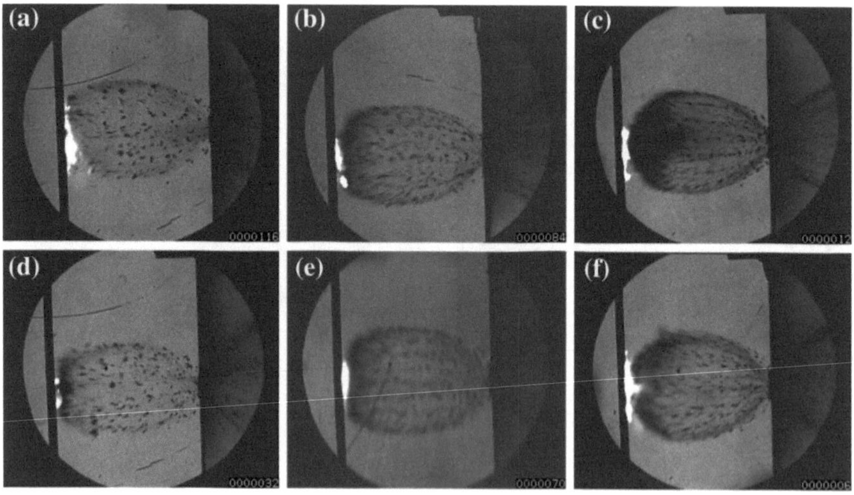

Fig. 11.13 Impacts on the main wall, comparison of wall temperatures: **a** projectile speed of 2.72 km/s and wall temperature at 295.2 K; **b** 3.33 km/s, 292.2 K; **c** 3.71 km/s, 294.2 K; **d** 2.78 km/s, 116.7 K; **e** 3.39 km/s, 118.8 K; **f** 3.70 km/s, 119.9 K (Numata 2009)

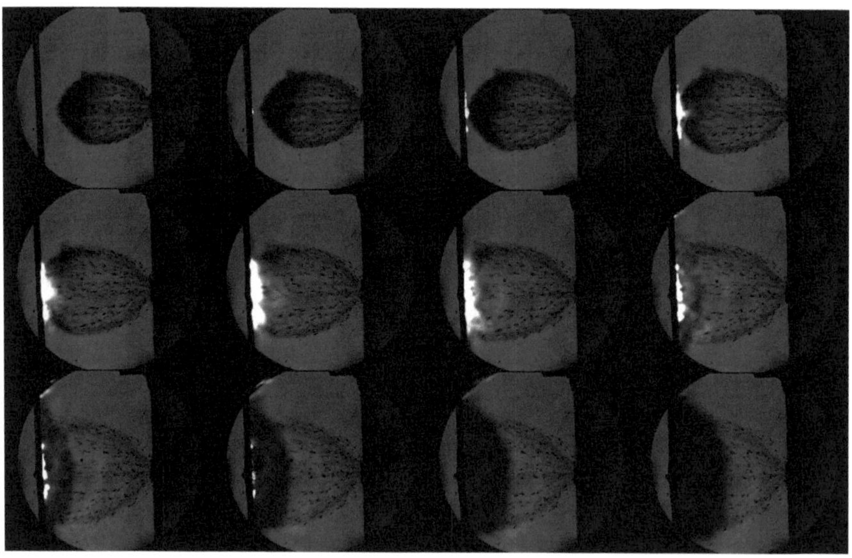

Fig. 11.14 The analogue experiment of space debris bumper shield at impact speed of 3.7 km/s against an aluminum plate at 120 K. Images were recorded by a high speed video camera Shimadzu SH100 (Numata 2009)

11.3 Shock Waves in Glass Plates

Shock waves were loaded onto glass plates by exploding AgN_3 pellets and resulting stress wave propagation in the glass plates were visualized using single and double exposure interferometry (Aratani 1998).

11.3.1 Shock Propagation in Tempered Glass Plates

A 10 mg AgN_3 pellet was placed at one of the focal points of an 8 mm thick and 106 mm × 150 mm elliptic tempered glass plate. The resulting high-pressures spontaneously generated a stress wave propagating inside the glass plate. Figure 11.15 show the evolution of stress wave propagation. Fringes represent density changes in the glass plates. At first, a hemi-spherical compression stress wave was generated at the center of the explosion and was reflected from the glass/ air interface forming a spherical tensile stress wave. The spherical tensile stress wave was reflected from the other side of the glass plate forming a cylindrical tensile stress wave. Eventually the reflection of the spherical reflected tensile stress wave from another side of the glass plate formed a cylindrical compression stress wave. Trains of cylindrical tensile and compression stress waves propagating at the longitudinal speed were generated repeatedly as seen in Fig. 11.15b. The stress waves are reflected from its elliptic edge and eventually converged at the second focal point as seen in Fig. 11.15i. In Fig. 11.15b–e, transversal stress waves propagating behind the longitudinal stress wave and then cracked the focal area. It is noticed that cracks so far formed at first focal point have fine and course structures (Aratani 1998).

Figure 11.16 show an evolution of a compression stress wave propagating inside a glass plate. A 20 mg AgN_3 pellet was ignited placed at an edge of an 8 mm thick and 250 mm diameter tempered glass plate and spontaneously created circular compression stress waves as seen in Fig. 11.16a and the reflected tensile stress wave is about to focus in Fig. 11.16b.

A 10 mg AgN_3 pellet was attached on a side of a 10 mm thick and 150 mm 150 mm rectangular tempered glass plate and detonated forming trains of cylindrical tensile and compression stress waves as seen in Fig. 11.17. Figure 11.17 show sequentially the evolution of trains of these cylindrical stress waves propagation inside the rectangular glass plate and their reflection from the vertical side edges as seen in Fig. 11.17f–h. It should be noticed that the stress waves attenuate consistently with propagation, which shows a gradual decrease in the contrast of the intensity of the shadow as seen in Fig. 11.17i, j. The interactions of stress wavelets became very complex. The elapsed time is readily estimated from the distance the initial shock wave traveled on the upper side of the test piece.

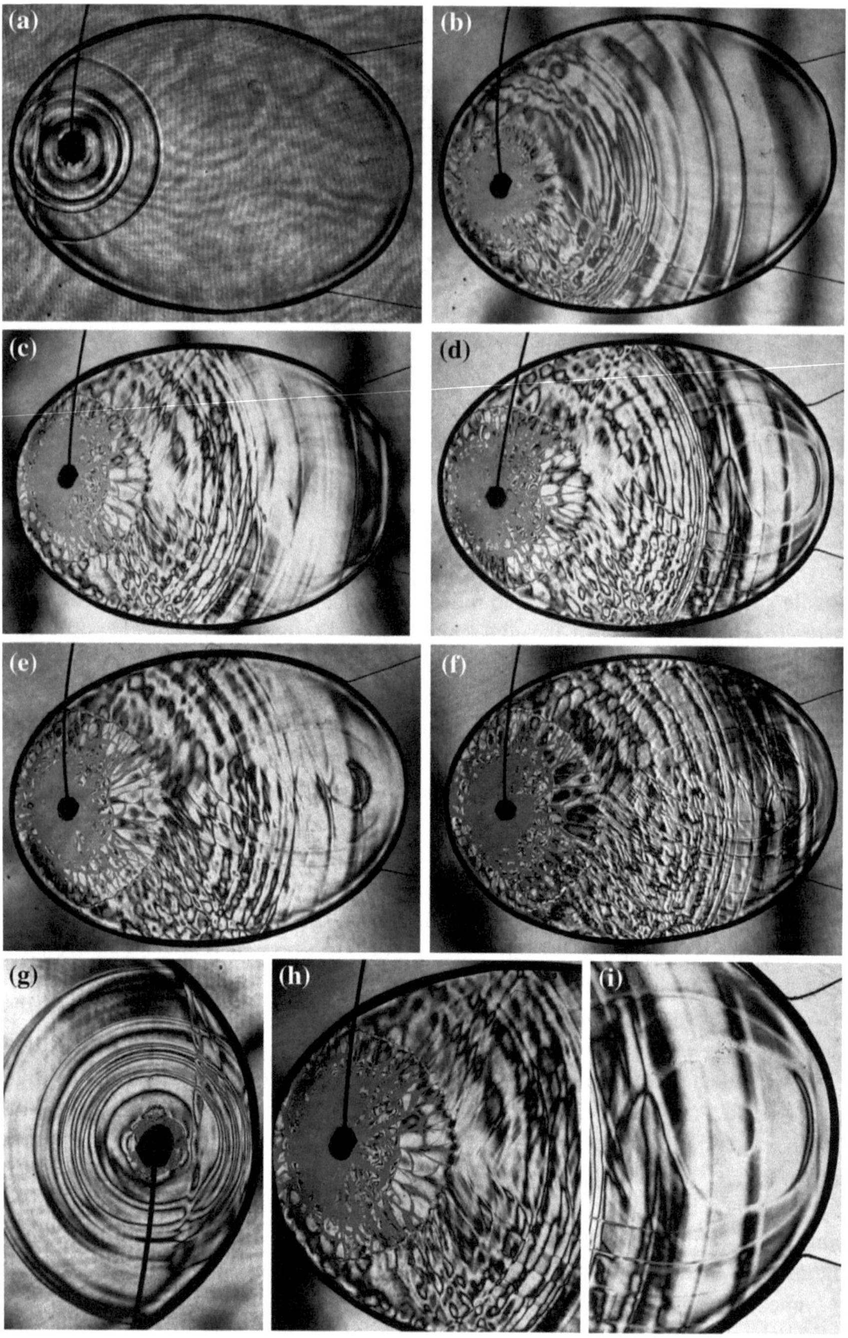

Fig. 11.15 Propagation of a stress wave created by a explosion of a 10 mg AgN_3 pellet on a 106 mm × 150 mm elliptic tempered 8.0 mm thick tempered glass plate: **a** #89012701, 125 μs from trigger point; **b** #89013006 137 μs; **c** #89013008, 143 μs; **d** #89013013, 145 μs; **e** #89013011, 146 μs; **f** #89013010, 147 μs; **g** #89013001, 127 μs; **h** #89013008 143 μs; **i** #89013013 145 μs (Aratani 1998)

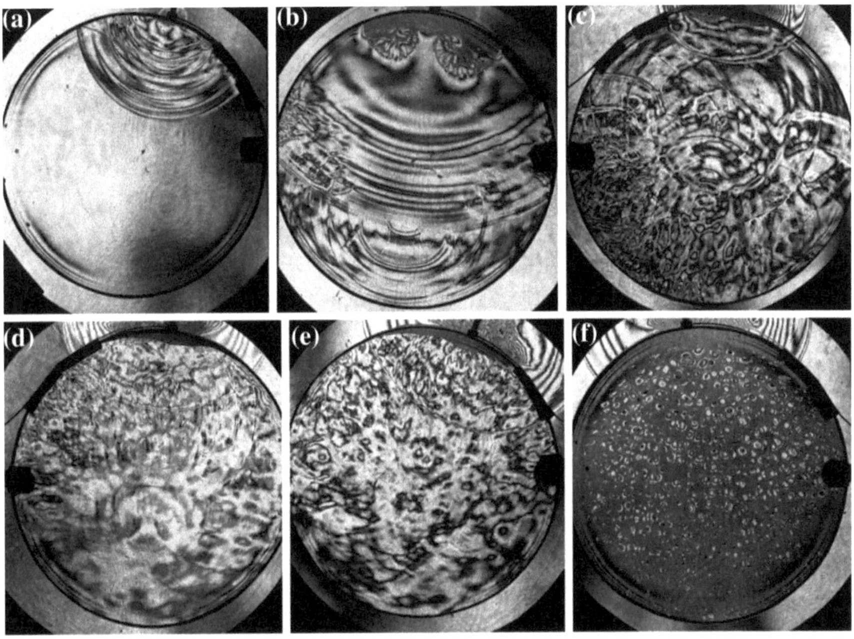

Fig. 11.16 Shock wave propagation over an 8 mm thick tempered glass plate. Shock wave is generated by exploding a 20 mg AgN_3 pellet at 291.0 K: **a** #87011404, 7 μs after ignition; **b** #87011603, 12 μs; **c** #87011407, 7 μs; **d** #87011604, 11 μs; **e** #87011406, 17 μs; **f** #87011606, 10 μs (Aratani 1998)

11.3.2 Laser Induced Shock Wave Propagation in Acrylic Blocks

A collimated Q-switched ruby laser beam was focused at 10 mm distance from the edge of a 50 mm × 90 mm × 150 mm acrylic plate as seen in Fig. 11.18a. When a Q-switched ruby laser having energy of 1 J/pulse and pulse width of 25 ns was focused in a point of about 0.1 mm in diameter, such an intense energy deposition in such a confined space instantaneously generates high temperatures and high pressures enough to vaporize the acryl and to create micro cracks which drive shock waves or compression waves even in acryl. However, the laser beam was not necessarily focused sharply at a point but the zone of focusing was slightly elongated to the direction of its irradiation. Hence the resulting cavity in the acryl was not a point but stretched cavity accompanying cracks over it. Trains of compression wavelets were generated from the cavities as seen in Fig. 11.18a. Figure 11.18c shows the double exposure interferogram of a cylindrical compression stress wave. Figure 11.18d shows a single exposure interferogram of the cylindrical shock wave. The laser beam was focused from a vertical direction and hence a vertically stretched cavity was observed.

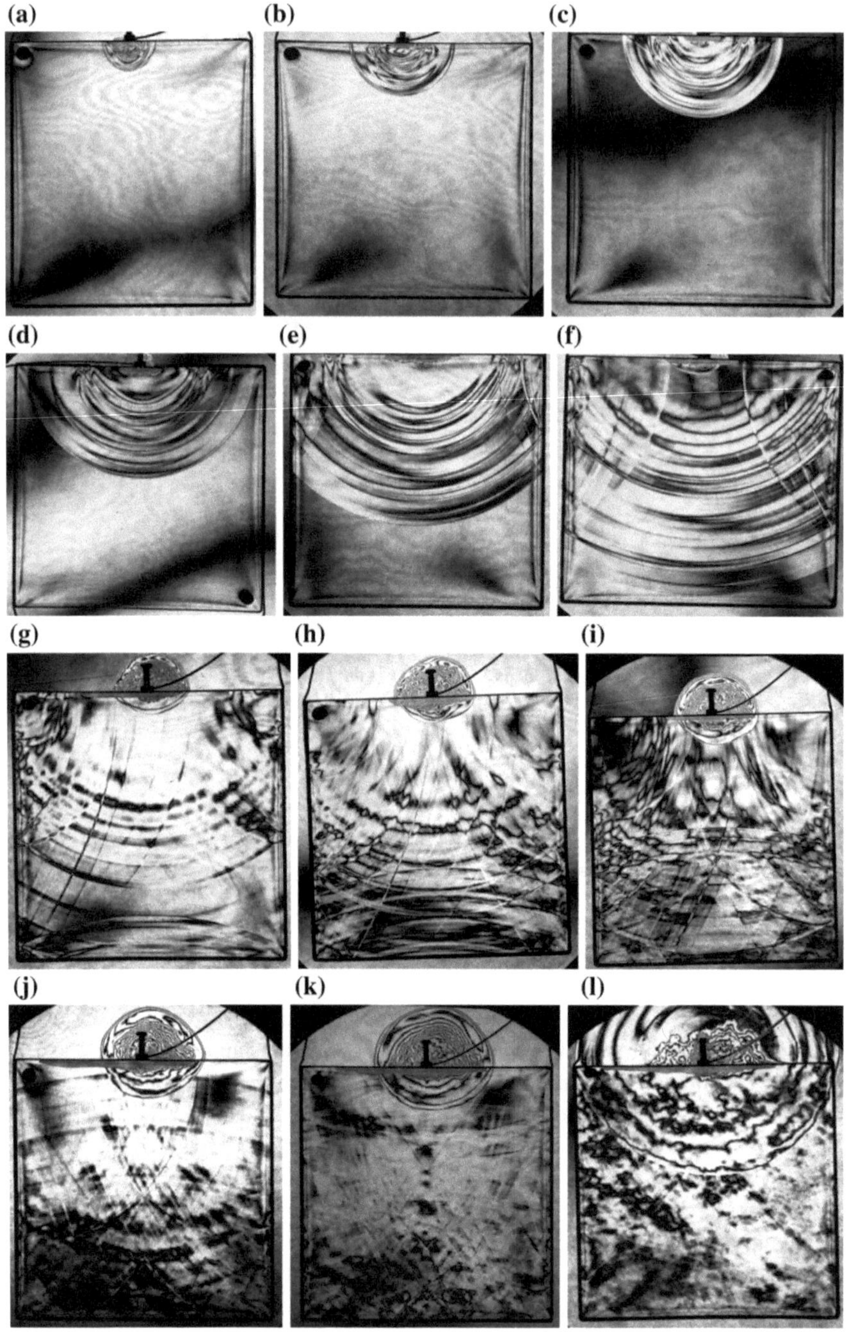

Fig. 11.17 Propagation of stress waves in a 150 mm × 150 mm and 10 mm in thickness rectangular #90082701; **b**#90082804; **c** #90082801; **d** #90082802; **e** #90082901; **f** #90082903; **g** #90082904; **h** #90083001; **i** #90083002; **j** #90083003; **k** #90083101; **l** #90083103 (Aratani 1998)

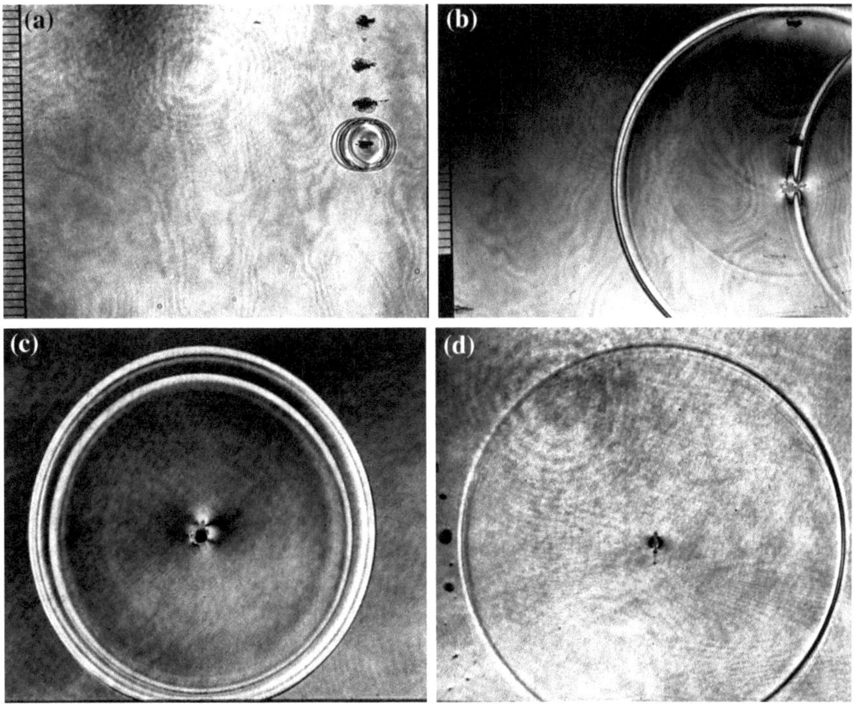

Fig. 11.18 Laser focusing induced stress waves. 1 J Q-switched ruby laser beam was focused in a 50 mm × 90 mm × 150 mm acrylic plate: **a** #85120206, laser bean was focused at 10 mm from the edge; **b** #85120210 at 8 μs from laser focusing; **c** #85120614; **d** #85111904

11.3.3 Shock Wave Propagation in Foam

Figure 11.19 shows a diaphragm-less shock tube consisting of a 230 mm diameter high pressure chamber, a co-axially arranged 60 mm × 150 mm low pressure channel, and about 1700 mm long test section. The structure of this diaphragm-less shock tube was already described by Yang (1995). The test section was specially modified to accommodate a 60 mm × 150 mm and 1200 mm long polyurethane foam, which was tightly inserted into the 1700 mm long test section as shown in Fig. 11.19 (Kitagawa et al. 2006).

Figure 11.20a shows a 60 mm × 150 mm polyurethane foam inserted in the 60 mm × 150 mm shock tube. In order to visualize the interaction of the foam with the reflected shock wave of Ms = 1.50 in the vicinity of the end wall of the 60 mm × 150 mm shock tube, a diffuse holographic interferometry was adopted. The single exposure collimated *OB* uniformly illuminated the foam's surface on which yellow fluorescent paint was sprayed. Then, the reflected *OB* from the deformed foam carried the holographic information of the deformed foam and was recorded on a holographic film. To ensure the deformation of the foam, 2 mm

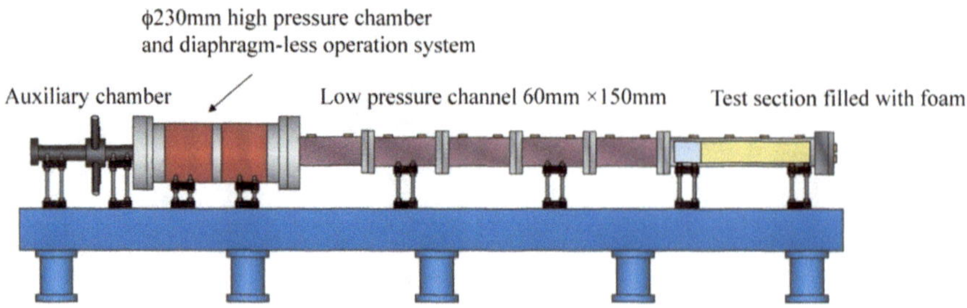

Fig. 11.19 A 60 mm × 150 mm diaphragm-less shock tube modified to perform shock wave propagation in polyurethane foam

diameter black dots were printed in a space of 5 mm × 5 mm on the foam surface as seen in Fig. 11.20a. Figure 11.20b shows the deformation of the foam interacted with the reflected shock wave. In the analytical model, the wave motion in the foam and its deformation should be one-dimensional. The predicted pressures and temperatures in the foam and in air in the vicinity of the end wall can reach much higher values than those predicted in a pure gas. Indeed in the shock tube having a polyurethane foam inserted at its end wall, the foam temperature at the end wall became so high that the foam indeed melted. This finding agreed well with the prediction.

In Fig. 11.20b the foam deformed two-dimensionally. The deformation pattern resembles very much like the bifurcation of a reflected shock wave occurring in the vicinity of the shock tube end wall in pure gas. In the case of a foam shock tube, the pressure behind the reflected shock wave became so high that the criteria proposed by Mark (1956) would be satisfied and hence an analogous pattern to the bifurcation would appear. If a two-dimensional analytical model of a foam shock tube may exist, it would be good to reproduce the pattern as shown in Fig. 11.20b.

11.3.4 Shock Waves in Sand Layers

It is appropriate to apply an established method of visualizing shock waves in gases to shock wave propagations in sand layers. Sand grains under study are standard Eglin sands. Figure 11.21 shows a microscopic view of the Eglin sand. Their diameter ranges from 330 to 500 μm (Yamamoto et al. 2015).

11.3.4.1 Point Explosion in a Sand Layer

At first a 10 mg AgN_3 pellet was exploded at the center a cylindrical chamber having 100 mm in diameter and 155 mm in height filled with Eglin sands. Eglin sand grains were poured into the test section and the bulk density was about

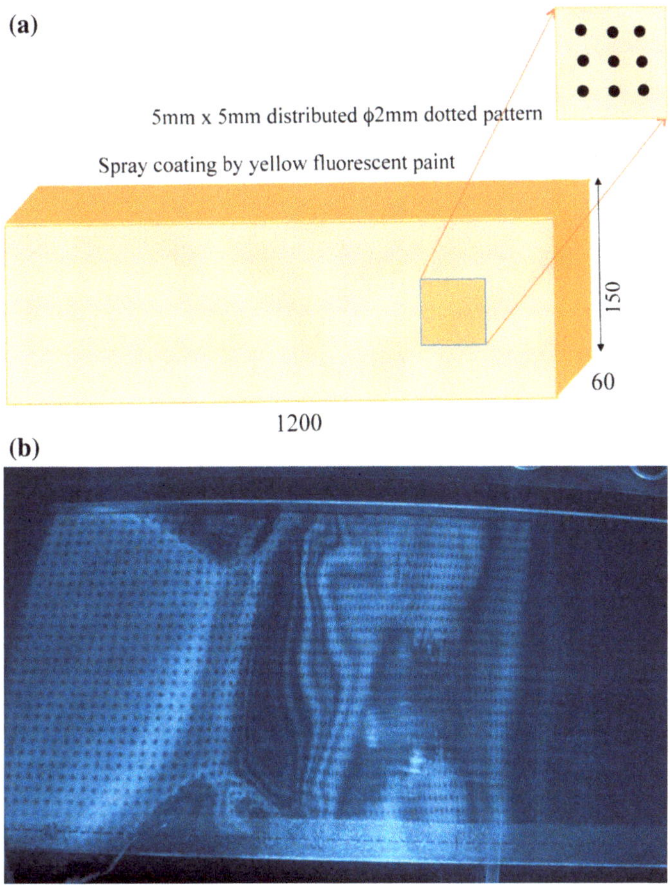

Fig. 11.20 Deformation of polyurethane foam upon shock wave loading: **a** foam model; **b** diffuse holographic observation (Kitagawa et al. 2006)

Fig. 11.21 Microscopic view of Eglin sand grains (Yamamoto et al. 2015)

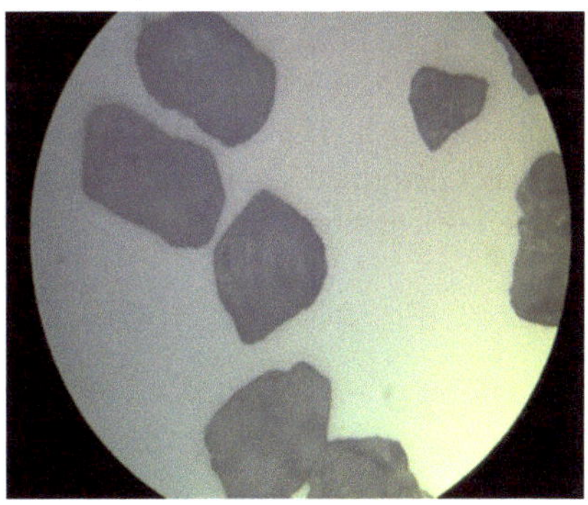

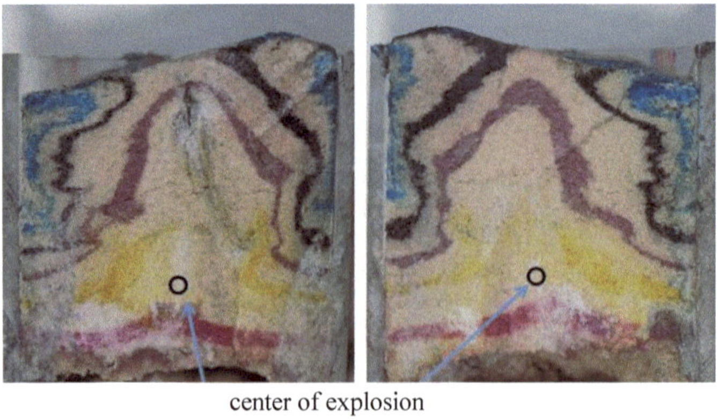

Fig. 11.22 The cross sections of a point explosion in a sand layer filled with Eglin sands (Yamamoto et al. 2015)

1.55 g/cm^3. In order to quantitatively identify the deformations of the sands layer generated by a shock wave loading, thin layers of colored sand grains of the blue upper layer, black, brown, yellow, red, and the last brown layer are placed horizontally and in a nearly equal interval in the sand layer. The explosion at the center of the chamber spontaneously moved the sand layer, which eventually deformed the colored sand layers. As soon as the explosion was over and the movements of sand grains were ceased, the entire test piece was submerged into a liquid named Permeate™ (D&D Corp. Japan), which is volatile liquid and has a so low surface tension that it eventually permeated into the spaces among sand grains. When the liquid evaporated and the sand grains were tightly solidified, the frozen specimen was divided into two pieces.

Figure 11.22 shows the cross sections of the half-cut test piece. The sand grains were moved due to the exposure of high pressures at approximately several hundred MPa in the vicinity of the explosion center. The deformation of the color layers indicated the memory of motions of sand grains, in short, shattered by the spherical shock wave at first and by the reflected shock wave from the acrylic wall. The blue and black colored layers initially in the first and second layers from the upper surface were shattered upward and squeezed toward the side. The brown layer initially in the third layer was deformed upward but remained its closed shape. The sand grains located close to center of explosion changes were fragmented and became whitish. The color layers located below the position of the explosion center were not deformed significantly because the shock wave reflection from the bottom of the test chamber.

11.3.4.2 Penetration of High Speed Sphere into Sand Layers

Encouraged by the results of point explosions in a sand layer, the impingement of a high speed sphere vertically into a sand layer was investigated. A compact vertical powder gun was constructed as illustrated in Fig. 11.23. The basic part of this powder gun was once used to create the high-speed fuel jets experiment. Since this is a compact gun, preparatory experiments were carried out to control the delay time of the ignition smoke-less powder. Initially, a method of the laser ignition of a 10 mg AgN_3 pellet was adopted to ignite black powder and then to initiate the simultaneous combustion of the smoke-less powder. However, it was eventually revealed that the uncertainty existing in simultaneous combustion of the smokeless powder was uncontrollable. Then this compact gun followed the traditional method of impacting the primer with a mechanical impact. A primer containing smokeless powder (HS-7) weighing 3 g and a propellant comprising of black powder weighing 1.2 g started the ignition of smokeless powder (H50-BMG) weighing 5.0 g. The high-pressure so far created in a powder chamber drove a 9.5 mm diameter stainless steel bearing ball which was contained in a 14 mm diameter four-split sabot made of polycarbonate. The combination of the sabot and the bearing ball was accelerated through a vertical launch tube and propagated through a perforated tube. Compression waves and a resulting shock wave were expected to attenuate through the perforation. The sabot remover was placed at the end of the

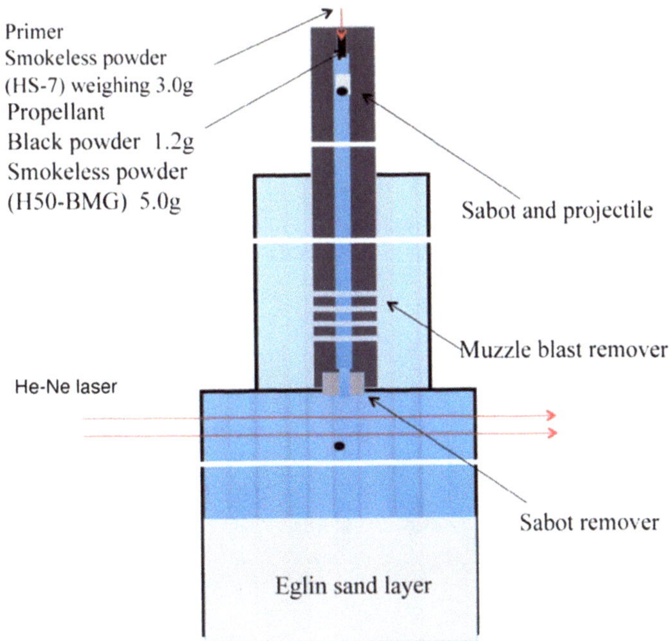

Fig. 11.23 Experimental setup of launching a 9.5 mm diameter sphere into a sand layer

perforation tube. Upon impacting on the sabot remover, only the bearing ball could pass through the sabot remover but the main part of the sabot squeezed into the edge of the sabot remover. The bearing ball speed was measured in the position located immediately after passing through the sabot remover or just before impacting the sand layer by measuring the time interval of the bearing ball intersecting the two He–Ne laser beams.

Figure 11.24 shows a rectangular cross sectional test section. This test section was designed to monitor the deformation of the sand layer occuring the penetration of 9.5 mm diameter bearing ball by distributing thin layers of colored sand grains horizontally embedded in the test section as illustrated. The test had a 33 mm 100 mm cross section and 250 mm in depth and comprised of a 25 mm thick frontal acrylic plate for the visualization, a 25 mm thick stain-less steel rear wall, and 50 mm wide and 50 mm thick brass side walls. A 9.5 mm diameter bearing ball made of stainless steel was projected vertically at 1.01 km/s as shown in Fig. 11.24 and its entry speed was measured in front the test section using the time of flight method. The test section was obliquely illuminated with a flood lamp and the variation of images monitored on the acrylic plate was recorded with a high speed digital camera Shimadzu SH100 with the framing rate of 10^6 frame/s and the exposure time of 125 ns. When the projectile propagated through the sand layer, a luminosity was emitted in the vicinity of the projectile. The luminosity was created by the disintegration quartz grains exposed to a shock wave.

0.2 mm diameter optical fibers transmitting a He–Ne laser beam were horizontally suspended and were distributed at given intervals along the center axis.

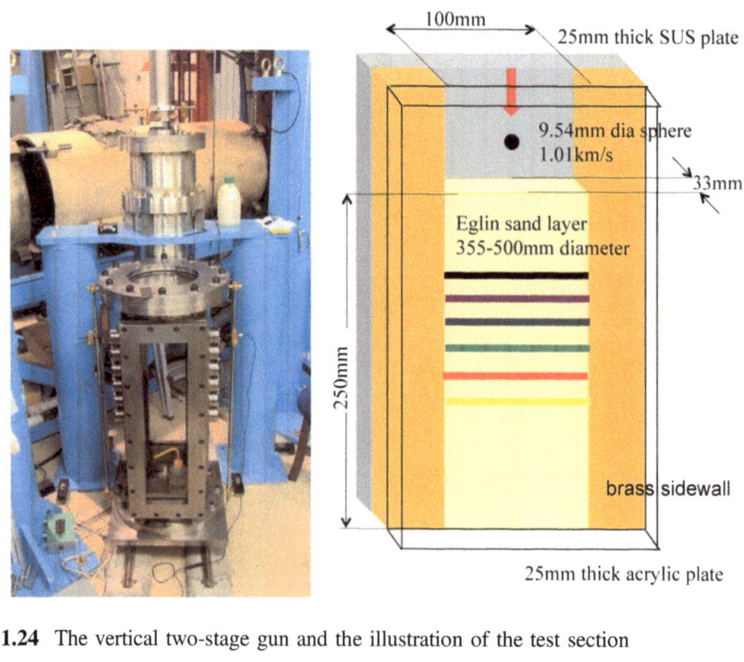

Fig. 11.24 The vertical two-stage gun and the illustration of the test section

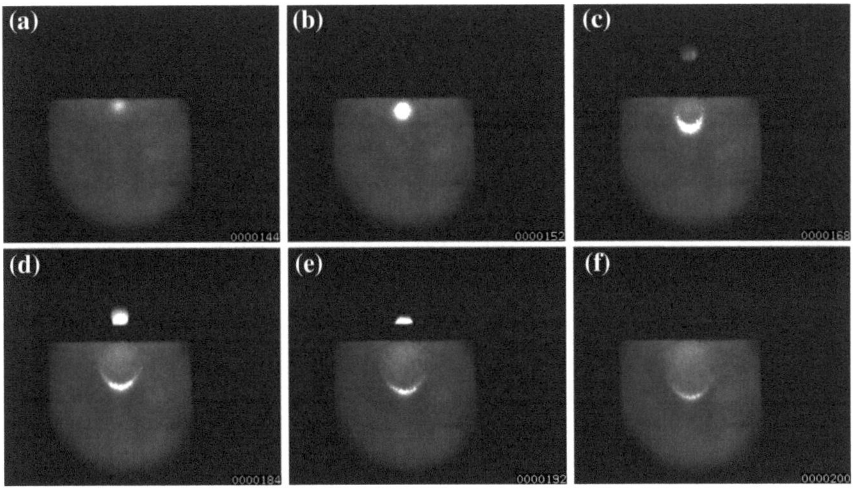

Fig. 11.25 The luminous emission at a projectile impact into sand layer: **a** 8 μs from entry; **b** 16 μs from entry; **c** 32 μs from entry; **d** 40 μs from entry; **e** 48 μs from entry; **f** 56 μs from entry (Yamamoto et al. 2015)

The on-off signals created at the moment when the projectile ruptured the optical fibers would provide the attenuation of the projectile through the sand layer. The measured projectile motion was compared with the result of visualization.

Figure 11.25 show the projectile movement observed at every 8 μs up to elapsed time of 56 μs from the impingement of the projectile on the sand layer. Figure 11.26 show the sequential observation up to 168 μs. Figure 11.25a shows the state at 8 μs from the impingement. A quartz crystal generates electric charges when exposed to high pressures or deformation. It is known that a quartz crystal emits a luminosity of wave length of 654 nm when cracked by an external force. This is the inherited character of a quartz crystal and the so-called piezo effect.

When the projectile entered into the sand layer, the surface of an impacted sand layer was shining faintly as seen in Fig. 11.25a. It should be noticed that as seen in Fig. 11.24, the sphere penetrated inside a 33 mm wide sand layer and the luminosity occurred about 10 mm away from the inner wall. When the sphere fully penetrated the sand layer, the surface of the entire sphere was exposed to significantly high shear stress so that the sand grains would emit intense luminosity. Then a circular luminous front was observed as seen in Fig. 11.25b.

The brightness of the luminosity is related to the degree of fragmentation of the sand grains. The finest size of corn starch powders was observed among the fractured sand grains attached to the sphere surface. The fragmented sand grains inside the shock layer which were located a few mm away from the sphere surface had sizes of sugar grains. In Fig. 11.25c, the shape of luminous front resembles a bow shock wave observed at 10 mm away from the side wall. With elapsing time, the radius of the detached bow shock wave gradually was enlarged and gradually

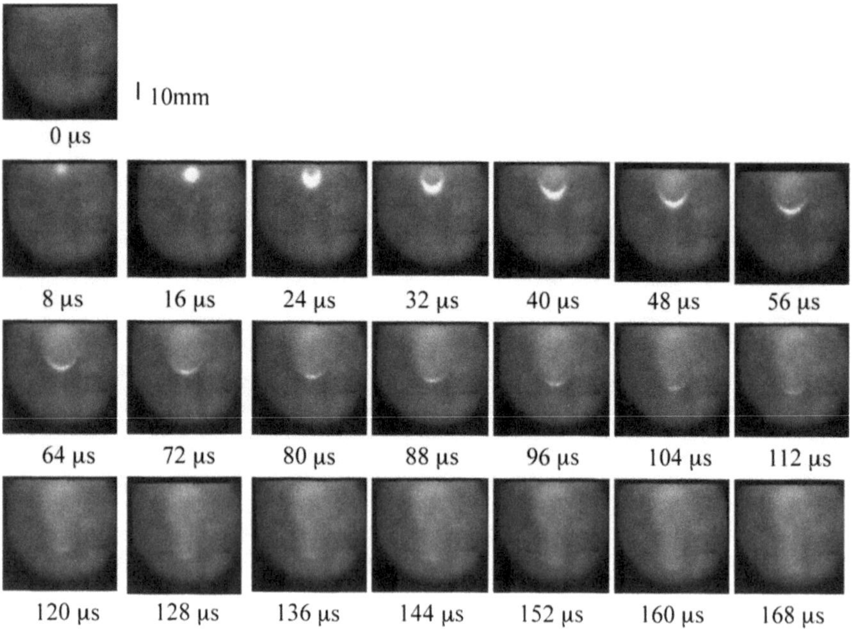

Fig. 11.26 Sequential observation of projectile penetration #13031102 (Yamamoto et al. 2015)

broadened. At the same time the luminous front became fainter. A faint front appearing above the sphere observed in Fig. 11.25c was caused by a remnant of the deformed sabot. The flood lamp illuminated the field of view not necessarily uniformly and then the sabot fragment looked faint as the edge of the flood lamp illuminated, whereas as seen in Fig. 11.25d the remnant of the sabot looked brighter as illuminated with central part of the flood lamp. It should be noticed that at 10 mm above the sand surface, the field of view was blocked. The bulk density of the sand layer is 1.55 g/cm^3. This value is significantly lower than that in air/sand mixtures tested in sand compaction experiments. In evaluating the time variation of luminous fronts, the sound speed in the sand layer was found about 180 m/s. This value is much slower than that in air.

Figure 11.26 show the sequential observation up to later time. The entry speed was 1.01 km/s which is equivalent to Ms = 5.5 based on the estimated sound speed in the present sand layer of 180 m/s. Therefore, the bow shock is detached closely when impacted the sand layer. If the shock wave is detached in a similar manner as it appears in air, the ratio of a shock stand-off distance δ to the sphere of diameter D, $\delta/D = 0.1$ (Liepmann and Roshko 1960). Therefore, the shining bow shock waves are located very close to the sphere as seen in Fig. 11.26b at 16 μs. The sphere decelerated and the corresponding Ms quickly decreases to sonic speed. Then the shock stand-off distance is elongated quickly as already discussed in penetration of a supersonic sphere into water as shown in Fig. 9.69 in Chap. 9.

Figure 11.26c, d were taken at 24–32 μs from the projectile impact on the sand layer. The sphere is further decelerated and the radii of curvature of the detached shock wave increased. Eventually at 48 μs as seen in Fig. 11.26f, the detached shock wave would be repeatedly reflected from the sidewalls and eventually turned into a cylindrical shock wave. The central part of the luminous front is still faintly shining. Even though the speed of the detached shock wave approached gradually to the sonic speed, the luminosity started fading. At 168 μs, the sphere stopped completely. Measuring the distribution of the brightness along the center line, the trajectory of the sphere would be estimated. If analyzing the time variation of the trajectory of luminous front, the attenuation of the sphere and the transition of the bow shock wave to the sonic wave would be readily determined.

Figure 11.27 summarizes the variation of the brightness along the center line of the individual pictures shown in Fig. 11.26. The ordinate denotes the degree of brightness in the individual frames in arbitrary unit. The abscissa denotes the distance in pixels. The origin corresponds to the sand layer surface. The number on the

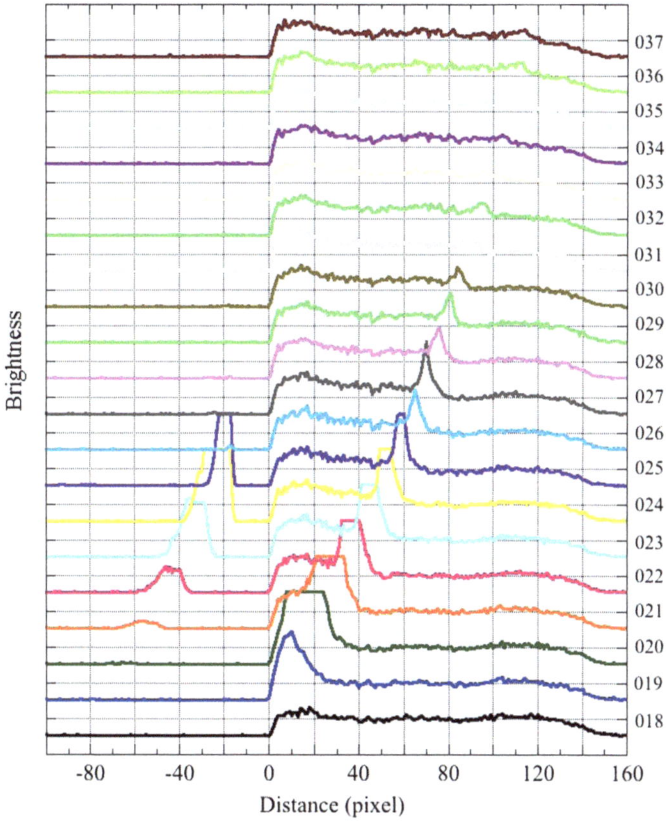

Fig. 11.27 The time variation of the luminous front profile, summary of Fig. 11.26

right shows frame number. The ordinate denotes the elapsed time every 8 μs interval. The intensity profile of the frame #18 corresponds to Fig. 11.26a at 0 μs. The intensity profile of the frame No. 037 corresponds to elapsed time of 168 μs. The luminosity sharply increases from #20 to #22 and the maximal luminosity appears at the frame #20 or 16 μs. and the peak brightness gradually decreases. The luminous front started widening and stretching from #26 to #37, indicating the increase of the shock stand-off distance of the bow shock. Tracing the trajectory of luminous front, the trajectory of bow shock wave can be estimated. Then the resulting speed of luminous front is about 150–200 m/s. This value is almost in the range of the sound speed in the sand layer. It is noticed that from the #36 to #37 at 160–168 μs, the projectile moved very slowly but the luminous front was still resolved.

In conclusion, a method routinely used in the shock wave research was successfully applied to the diagnostics of sand dynamics. The observation of luminous emissions appearing in shock layer is found a useful method. Improving the present experimental arrangement, the reflection of bow shock wave from the solid boundary may have been observable. The critical transition of reflected shock waves in sand layers will have been measured from these compact experiments.

Immediately after the impact experiment, test pieces were carefully immersed deeply in a chamber filled with the Permeate™ (D&D Corp. Japan). The specimens were divided into two segments. Then, structures of specimens were found well preserved the projectile motion. Figure 11.28a, b show the left and the right sides of specimen of a 100 mm diameter × 120 mm long cylindrical sand layer impacted at speed of 1.6 km/s. The surface of the sand layer was inclined by 30°, then the sphere impacted obliquely and propagated toward the shallower side. Initially horizontally distributed color layers were deformed indicating the sphere's oblique movement by observing the pattern of dragging the color layers.

White sand grains were observed along the sphere's trajectory. The white sand powders adhered to the sphere's surface were caused due to the presence of a very high shear force and high temperature at the sphere surface. During sand grains' fragmentations, the quart grains emitted luminosity forming a shining circular area. The white powders filling the space between stretched blue color layers and crashed sand grains would definitely emit the intense luminosity during the sphere's high-speed penetration. The sphere was probably spinning and then ground the sand grains into powders. Due to the spinning motion, the sphere bounced from the bottom of the sand layer and stopped. It should be noticed that during the early experiments the sabot was successfully plugged at the entrance of the sabot remover and only the sphere impinged the sand layer. Then the thermal decomposition never happened to occur. This is the reason why fragmented sand grains remained fresh white powder as seen in Fig. 11.28a, b.

Figure 11.28c shows the penetration onto a 30° tilted 100 mm diameter × 120 mm long cylindrical sand layer surface at the penetration speed of 1.01 k/s. Only the first black color layer was significantly deformed, unlike observation in Fig. 11.28a, b, the other color layers deformed symmetrically. At the penetration depth of 80 mm, the sphere stopped but slightly reflected obliquely

Fig. 11.28 Penetration of sphere into sand layers: **a**, **b** impact against oblique surface at 1.93 km/s: **c** #11030201, 1.02 km/s; **d** #13032901, 1.01 km/s; **e** enlargement of (**d**) (Yamamoto et al. 2015)

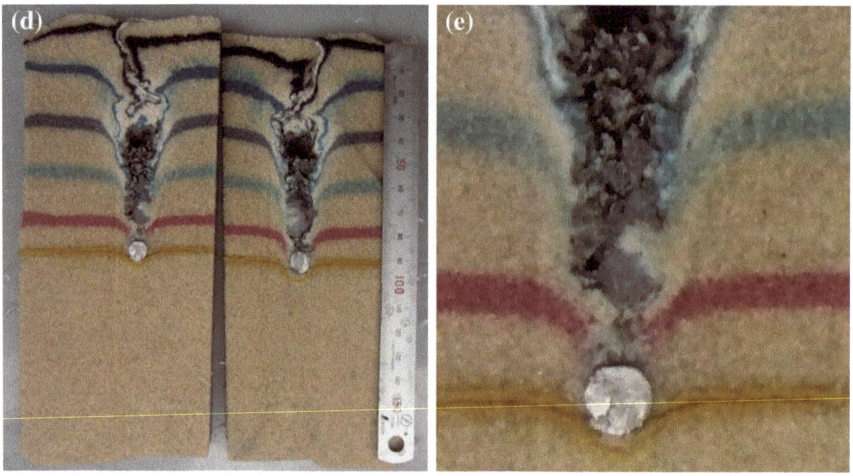

Fig. 11.28 (continued)

from the bottom. The sphere moved at high-speed and penetrated colored layers symmetrically toward the bottom forming a cavity. The internal cavity wall is covered with white sand grains. The white powders were created by the cracked sand grains due to the contact with the high-speed sphere or the exposure of the high pressures behind the bow shock wave. The luminous emission was created due to the fracture of the sand grains. The shape of the cavity resembled the shape of a wake flow but the wall was covered with carbon soot. The carbon soot was created by the thermal decomposition of polycarbonate sabot which followed the sphere as observed in Fig. 11.27b–d. The sphere's trajectory was straight so that the deformation of the color layers resembled the velocity profile in boundary layers.

A strong earthquake attacked the laboratory on March 11, 2013 and damaged the facility for conducting the sand impact experiment. Immediately after this incident, the recovered started. The image shown in Fig. 11.28d was the results of the experiment collected after the incidence and the author's final experiment taken on March 29th, 2013 just before his retirement on March 31st, 2013.

At the moment of the projectile impact on the sand layer, a splash of sand grains covered the entire open space and then the impact crater was covered by a falling fresh sand grains. The pattern of color layer's deformation became asymmetrically even though the experimental arrangement was symmetrical. The asymmetrical sedimentation would be attributable to the shape of the test section of 33 mm 100 mm. Figure 11.28d shows the result of the impact of a 9.5 mm diameter sphere at the impact speed of 1.01 km/s. The whole piece of the sabot impacted following the sphere and thermally decomposed inside the cavity created by the sphere. Hence the cavity was totally blackened by carbon soot. The deviation of the deformed color layers from their original horizontal positions would explain the history of deceleration of the sphere speed.

11.4 Shock Waves in Volcanic Eruptions

Volcanic eruptions occur, when the energies deposited in magma are released suddenly and the magma fragmentations take place. Then volcanic eruptions are more or less linked with shock wave phenomena (Glass 1975). Then volcanic eruptions would be reviewed from the point of view of the shock wave research. Analogue experiments of shock wave generations during explosive eruptions and the resulting shock wave propagation in air are readily investigated in shock tubes. However, shock tube experiments are limited only in the laboratory scale. Appropriate numerical simulations would be only a possible approach to reproduced volcanic eruptions. Volcanic eruptions are reviewed from the shock wave research (Takayama and Saito 2004).

11.4.1 In Situ Observation of Eruption

A project started to measure shock wave over-pressures in situ when Mt. Aso, Nakadake erupts. A pressure transducer by using a PVDF (poly vinyl denefluoride) piezo film was manufactured in house and was calibrated by comparing its output signal with that obtained by Kistler pressure transducer model 603B. The pressure transducer was installed at the edge of a stainless steel bar and was positioned in a bunker located on the top of the cliff overlooking a caldera of Mt. Aso Nakadake. Output pressure signals were then converted to optical signals, transmitted to the Aso Volcano Museum located 4 km away from the volcanic mouth, and stored in a digital memory placed in the Museum. The data stored in the digital memory were updated via the telephone line from the Institute of Fluid Science in Sendai. Figure 11.29 is an illustration of the data acquisition and their transmission from the volcano to the Museum. With this arrangement, the time variation of the pressure released from the volcanic conduit would be monitored in Sendai.

Figure 11.30 show the installation of a pressure transducer in 1995 inside the bunker, which was constructed at the top of the cliff overlooking the caldera of Mt. Aso, Nakadake. It was anticipated at that time that the eruption would occur shortly but Mt. Aso was quiet until the project was cancelled in 2013. When a progress report of this project was presented in a conference, a volcanologist commented that it would be a nice effort to study the eruption by installing such an equipment on the Mt. Aso. Cynically, when this project was cancelled, Mt. Aso erupted violently.

11.4.2 Numerical Simulation

At the same time, a super-computation was conducted to simulate numerically over-pressures and to compare the numerical result with the measured ones if Mt.

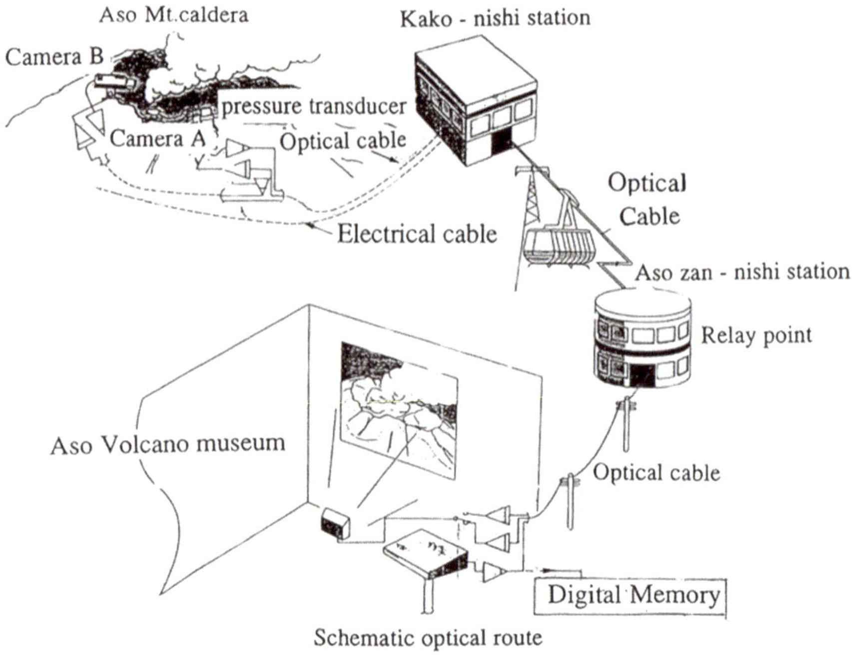

Fig. 11.29 The installation of a pressure transducer at the volcanic mouth of Mt. Aso. The signal is transmitted to the Aso Volcano Museum located 4 km away from the Mt. Aso (Takayama and Saito 2004)

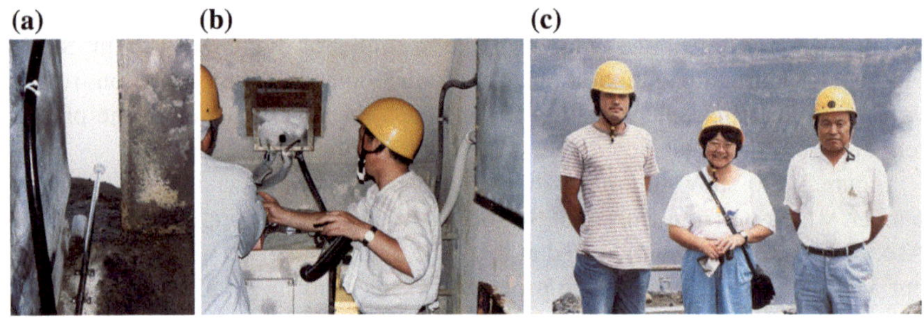

Fig. 11.30 The in situ pressure measurement of Mt. Aso in 1995; **a** A pressure transducer looking down a caldera of Mt. Aso; **b**, **c** installation of a pressure transducer in a bunker

Aso, Nakadake may erupt any time. Figure 11.31a shows a three-dimensional computational mesh describing Mt. Aso, Nakadake. The mesh was constructed by referring the digital map published by the former Geographical Survey Institute, presently the Geospatial Information Authority of Japan. In Fig. 11.31a, the position of the volcanic mouth was shown by a red arrow and the position of the pressure transducer was installed was shown by a blue arrow. Figure 11.31b–d

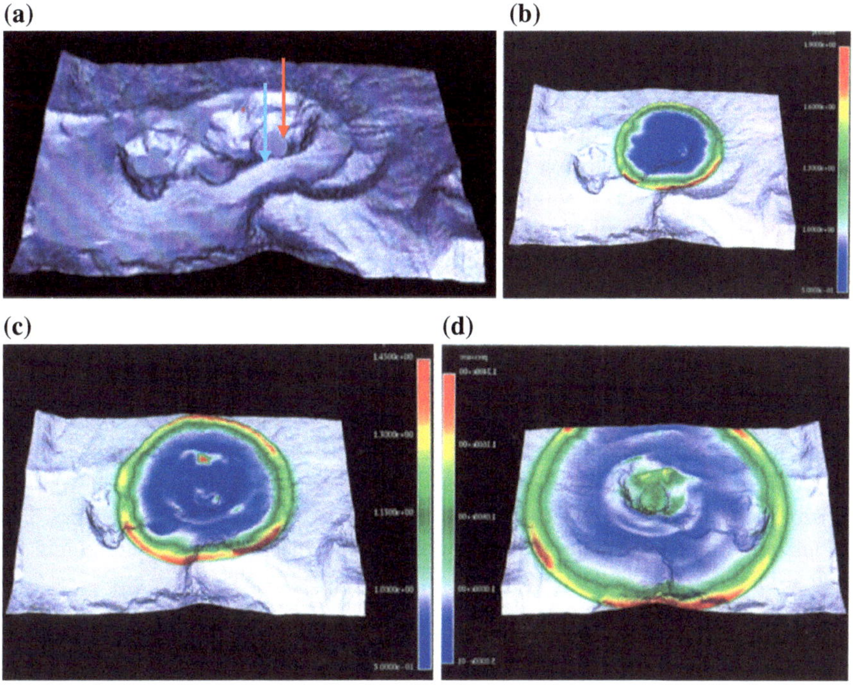

Fig. 11.31 The numerical pressure distributions behind a blast wave expected to occur at Mt. Aso eruption: **a** three-dimensional meshes from digital topological map; **b** 0.62 s; **c** 0.95 s; **d** 1.73 s

show temporal pressure distributions simulated by the late Professor Voinovich of the Ioffe Institute, Academy of Science Russia (Voinovich et al. 1999). The simulations were conducted. Different initial conditions were assumed in the volcanic conduit which was a vertical shaft filled with high temperatures and high pressures mixtures of air and foreign gases. Numerical scheme was the three-dimensional Euler solver on unstructured meshes which was flexible enough to accurately express the complex ground geometry. The initial conditions were tuned to match with the pressure history measured at the position as shown with a blue arrow in Fig. 11.31a. In Fig. 11.31b–d, the shock wave propagated along complex geometry and reflected locally from three-dimensional boundaries. These complex geometries significantly contributed to shock wave attenuation. In Fig. 11.31d the shock wave already passed the point at which the pressure transducer was installed.

11.4.3 Water Vapor Explosion

A water vapor explosion is one of the mechanisms which trigger volcanic eruptions. An analogue experiment was carried out for understanding the procedure of interactions between a melted metal droplet and gas bubbles. Observations were conducted to confirm whether or not molted tin droplets falling into water would shatter. It often happened in steel industry that if a large volume of melted metal dropped into water, the contact between the melted metal explosively vaporize water. This is the so-called water vapor explosion and is closely linked with a magma water vapor explosion which would trigger the eruption.

Figure 11.32a–c show single exposure interferograms of melted tin droplets falling in water. The droplet was exposed to underwater shock wave generated by the explosion of a 10 mg AgN_3 pellet. Figure 11.32c shows the generation of reflected shock waves and wavelets from bubble surfaces but successive explosion was not observed (Kitamura 1995). There is a so-called a critical mass of the melted tin below which the successive disintegration never happens to occur. The experiment was initiated firstly to determined the critical mass. At first, the critical mass is believed to be based on the parameters; mass of melted tine; over-pressure of shock waves; and the bubble size. It was expected that through these parametric study, the critical mass of melted tin could be estimated. However, it was not possible to estimate the critical mass simply by assuming these parameters. The visualization presented here is a preliminary parametric study for determining the critical mass.

11.4.4 Magma Fragmentation

An analogue experiment of the shattering of fragmented debris particles was conducted in the 60 mm × 150 mm diaphragm-less vertical shock tube. Figure 11.33a shows the vertical shock tube designed for conducting an analogue experiment in which the dynamics of magma fragmentation is visualized. This shock tube was designed to conduct dusty gas shock tube experiment having a relatively large cross section. A high pressure chamber had a diaphragm-less structure, in which a quickly moving piston served as a diaphragm opening. The high pressure chamber was placed horizontally on the first floor. The vertical low pressure channel had a 60 mm 150 mm cross section and 6000 mm in length and connected to acrylic observation windows having 150 mm × 1700 mm field of view. Polystyrene beads having density 20 kg/m^3 and diameter of 4.0 ± 0.2 mm, Styrocell M551X, product of Shell International Petroleum Company were filled in the test section. In order to trace the motion of beads, some of the beads were dyed various colors. The experiment was conducted for Ms = 1.36 in atmospheric air and at room temperature. Figure 11.33b show the sequential observation of shock wave loading on a polyethylene bead layer. The test section was illuminate with a flood light and the observation was conducted with high-speed video camera Shimadzu SH100 at the framing rate of 10^5 frame/s. Upon the shock wave loading, the polyethylene bead layers were compressed. Soon the beads on the surface were lifted up into air driven by the shock wave reflected from the bottom wall. The bead motion would be analogous to the magma fragmentation in the volcanic conduits (Kitagawa et al. 2006).

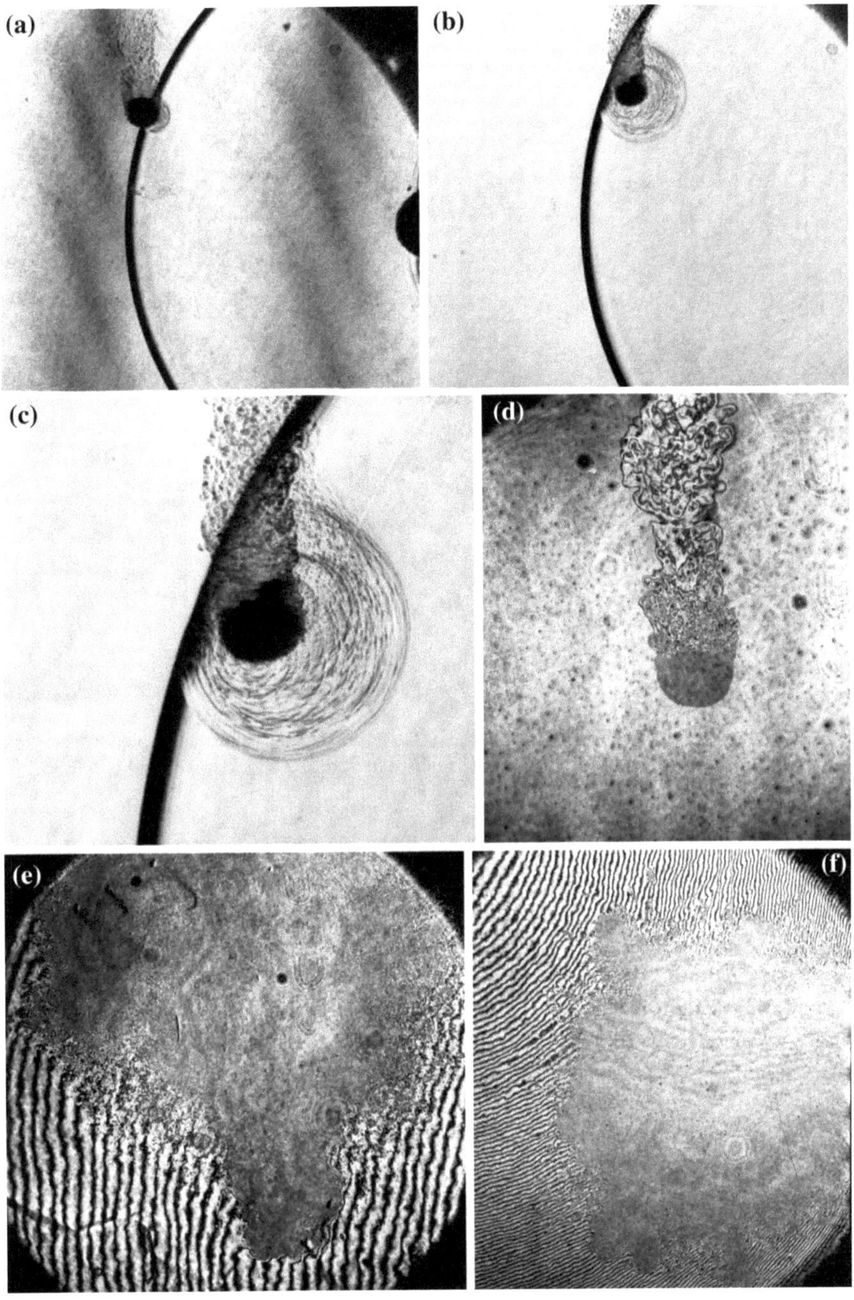

Fig. 11.32 Analogue experiments of water vapor explosion of a melted tin droplet falling into water. A Shock wave was generated by exploding a 10 mg AgN_3 pellet: **a** #94010409; **b** #94010501; **c** enlargement of (**b**); **d** #93101105; e #93093001; **f** #93093005 (Kitamura 1995)

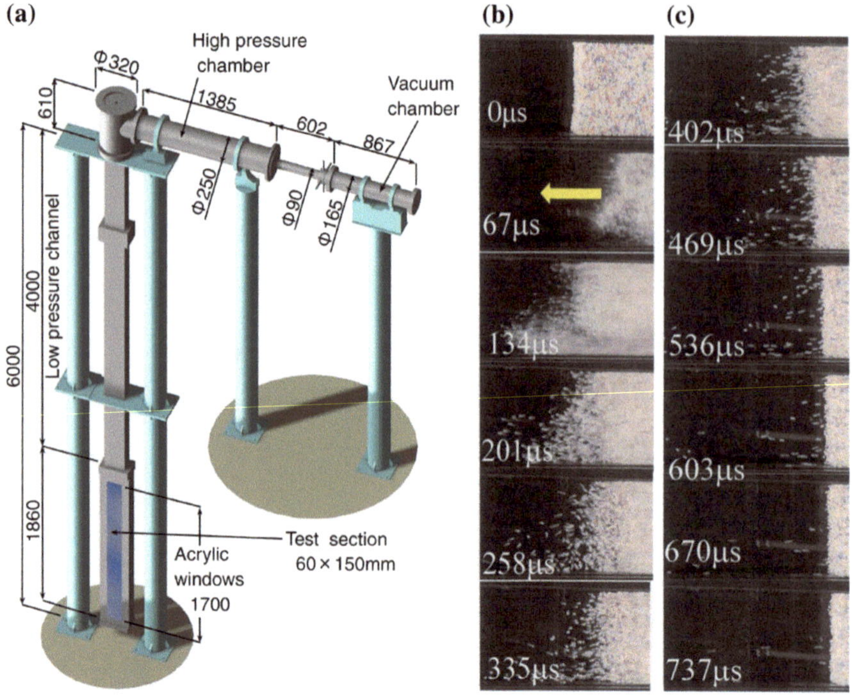

Fig. 11.33 A 60 mm × 150 mm diaphragm-less vertical shock tube: **a** vertical shock tube; **b**, **c** sequential observation of polyethylene beams exposed to a shock wave for Ms = 1.36

In Fig. 11.34a, three pressure transducers, Kistler Model 603B, were distributed along the vertical test section and a transducer was placed at the end wall. In Fig. 11.34b, the pressure histories from these transducers are presented for Ms = 1.36. The pressure history from the first #1 transducer shows sharp pressure rises corresponding to the incident and reflected shock waves but the pressure behind the incident shock wave is followed by an expansion wave due to the short high pressure chamber. However, the arrival of the expansion wave promoted the shattering of the beads. The signal out of the second #2 transducer shows the pressure enhancement only at the earlier stage and then a continuous decrease due to the shattering of the polyethylene beads toward the upstream. These pressure histories agree well with the results of observation seen in Fig. 11.34b.

It is concluded that the wave motion occurring in a high pressure chamber is similar to the wave motion and also the movement of debris in a volcanic conduit. However, during the eruptions, the reflected expansion wave from the opening of the volcanic mouth not only promotes the process of degassing in magma but also accelerating the fragmentation of a lave layer located at the bottom of the volcanic conduit. An analogue experiment of the fragmentation of a lava layer was conducted in a vertical shock tube (Yamamoto et al. 2008).

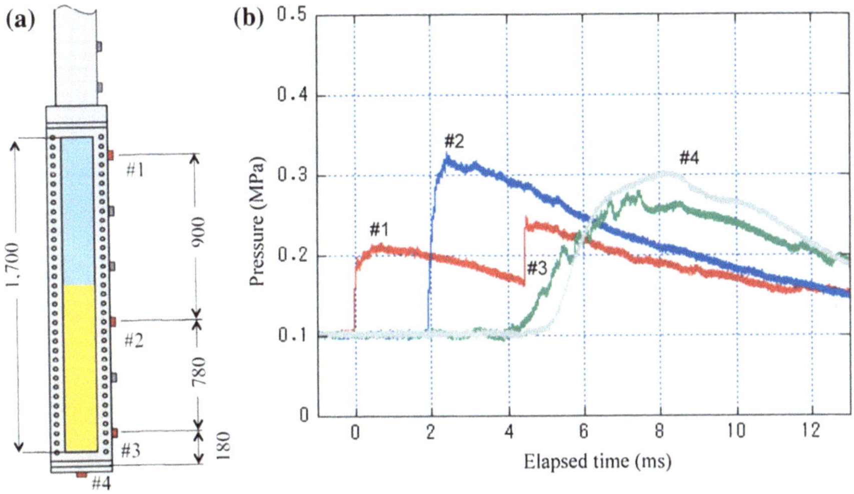

Fig. 11.34 Vertical shock tube test section: **a** pressure transducer distribution; **b** pressure recording

11.5 Shock Wave Interaction with Letters SWRC

The shock wave interaction with letters SWRC which stand for Shock Wave Research Center was sequentially visualized by double exposure holographic interferometry for Ms = 1.20 in atmospheric air at 294.0 K. Capital letters of SWRC of 60 mm in thickness made of carbon steel were arrayed in the test section of the 60 mm × 150 mm diaphragm-less shock tube and the result of sequential interferograms were edited in an animated display (Abe 1989) in Fig. 11.35. Interaction with curved boundaries, reflections and focusing from concave walls, and diffraction at corners of shock waves created very interesting wave patterns. At the same time, it would be nice if these patterns are reproduced numerically.

11.6 Shock Waves Generated in Daily Life

There are many phenomena in our daily life that create shock waves and their characters are very analogous to the shock wave phenomena. Although many of them miss mathematical formulations, the phenomena indeed have analogous characters to shock wave and then are called shock wave-like phenomena. The propagation of rumors, the transmission of information, panic motion in human cloud, and traffic flows are typical shock wave like phenomena.

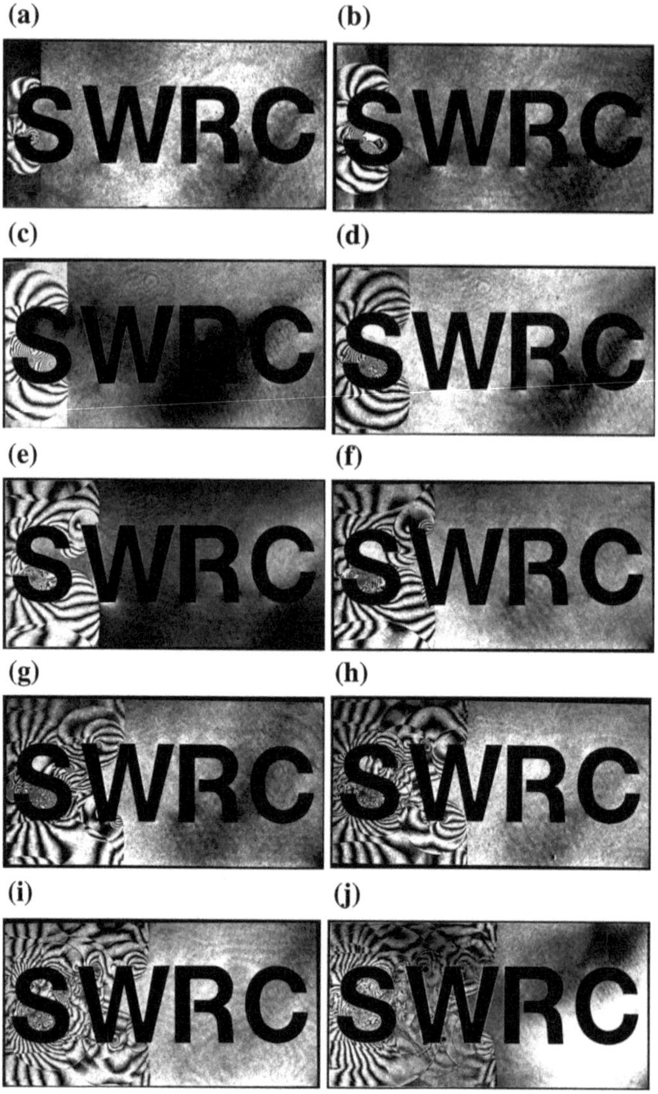

Fig. 11.35 Shock wave propagation along letters SWRC at Ms = 1.20 in atmospheric air at 294.0 K: **a** #97112103, 240 μs from trigger point, Ms = 1.195; **b** #97112112, 270 μs, Ms = 1.208; **c** #97112115, 300 μs, Ms = 1.197; **d** #97112118 320 μs, Ms = 1.200; **e** #97112130, 360 μs, Ms = 1.205; **f** #97112505, 390 μs, Ms = 1.201; **g** #97112510, 420 μs, Ms = 1.200; **h** #97112513, 460 μs, Ms = 1.202; **i** #97112518, 490 μs, Ms = 1.201; **j** #97112522, 530 μs, Ms = 1.209; **k** #97112527, 580 μs, Ms = 1.204; **l** #97112533, 600 μs, Ms = 1.201; **m** #97112615, 700 μs, Ms = 1.198; **n** #97112618, 730 μs, Ms = 1.20; **o** #97112705, 770 μs, Ms = 1.205; **p** #97112710, 820 μs, Ms = 1.204; **q** #97112713, 850 μs, Ms = 1.202; **r** #97112719, 900 μs, Ms = 1.204; **s** #97112722, 930 μs, Ms = 1.204; **t** #97112726, 970 μs, Ms = 1.198; **u** #97112734, 1030 μs, Ms = 1.198; **v** #97112736, 1050 μs, Ms = 1.206

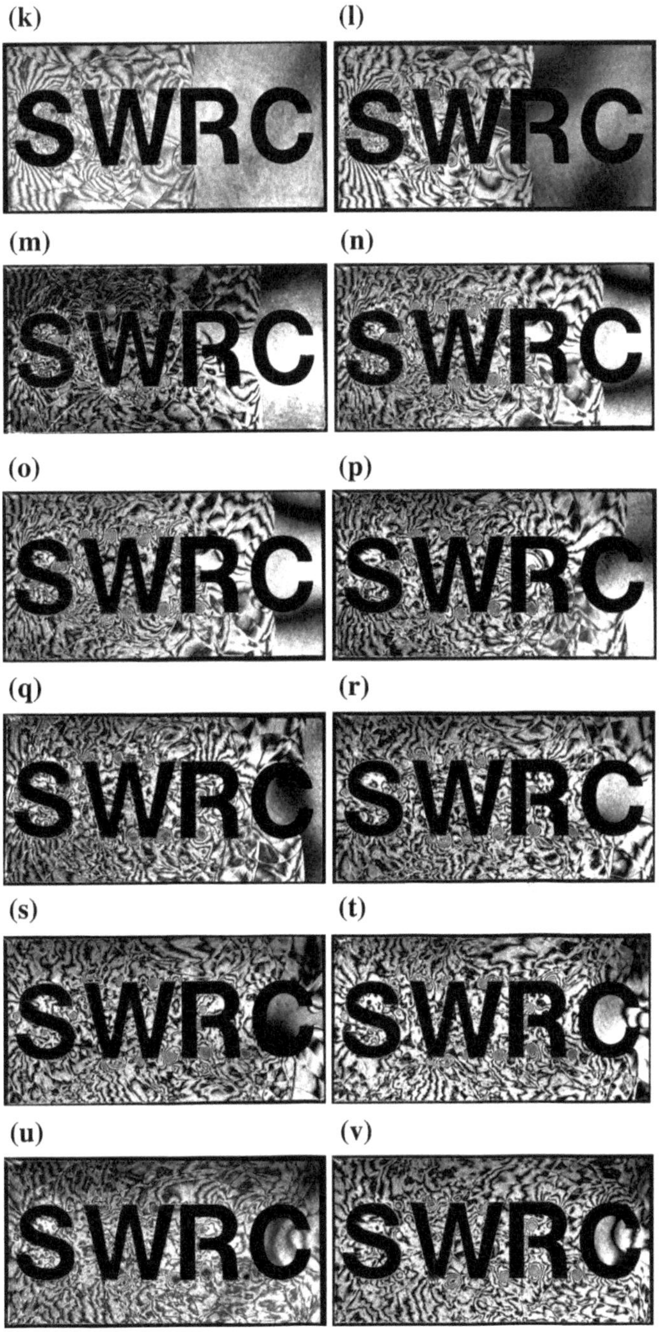

Fig. 11.35 (continued)

11.6.1 Swing of a Whip

Glass commented in his book (Glass 1975) that the swing of a whip generated a cracking noise which was similar to a shock wave. The cracking noise was visualized by using a pair of 1000 mm diameter schlieren mirrors and was recorded with high-speed digital camera Shimadzu SH100. The light source was a conventional flash lamp. In swinging a whip, its tuft eventually moved at supersonic speed driving a shock wave. Figure 11.36 sequentially show the shock wave formation during the swing of a whip. The tuft moved at approximately 450 m/s and the resulting shock Mach number Ms was about 1.30 in atmospheric air. If a planar shock wave is assumed, the over-pressure of the shock wave of Ms = 1.36 would be 180 kPa, which is destructive enough to damage targets.

11.6.2 Blow of Trobone

Some composers of symphonies, for example, Mahler favored trombone blown in *fff*. Considering the structure of the trombone, it can generate a startlingly intense

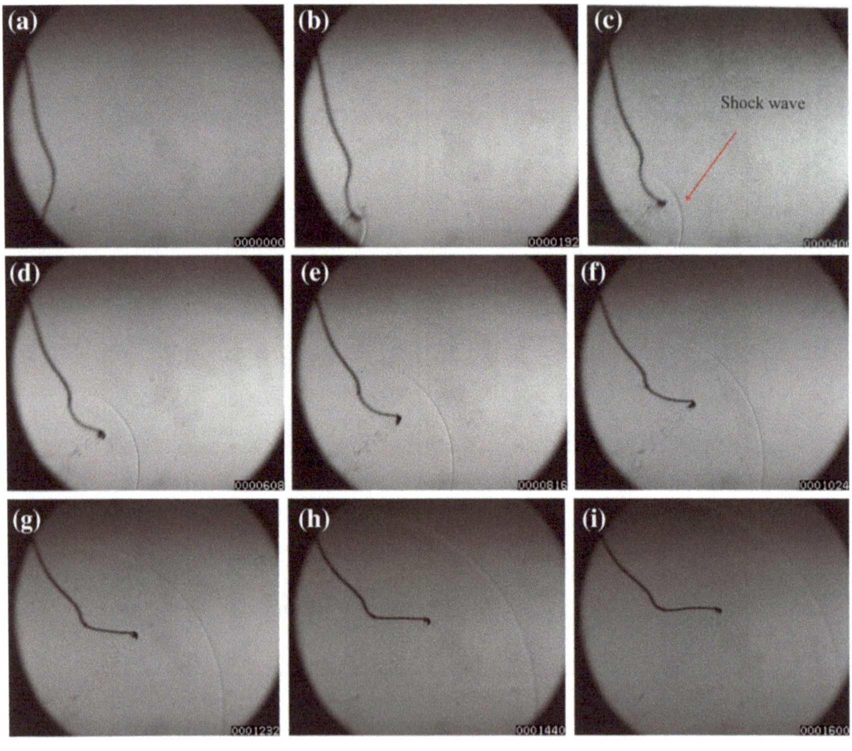

Fig. 11.36 A shock wave created at the tuft of a whip recorded by Shimadzu SP100 at 10^6 frames/s

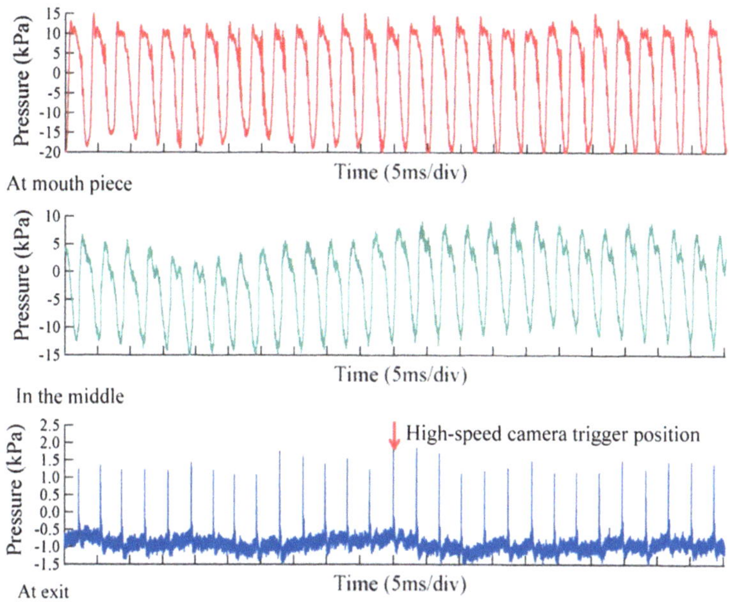

Fig. 11.37 Comparison of pressure profiles at trombone blow measured at three position: at mouth piece, in the middle position and at the exit

sound, in other words, a compression wave having a profile very close to that of a weak shock wave. It is described in a book of Physics of Musical Instruments (Fletcher et al. 1998) that players may have their physical risk when they try to produce strong sounds in blowing a trombone. Nevertheless, a trombone blow can generate weak shock waves. The sound pressures generated by a trombone blow was measured by pressure transducers, Kistler 603B placed at the mouth piece, the center of the tube length; and at the exit. Figure 11.37 shows variations of pressure measured at these positions. The ordinates denote pressures in MPa and the abscissae denote elapsed time in 5 ms in division. Measured peak pressures reach to about 10 kPa, which means strong breaths are blown into the mouth piece. The pressures so far blown into the tube attenuate quickly but the wave heads are steepened when waves reached peak pressures of about 5 kPa in the middle position. At the exit, the pressure profile is so steepened forming a weak shock wave having the peak pressure of 1.5 kPa. This over-pressure corresponds to a shock wave of Ms = 1.009. The trombone player in this experiment was a member of a university student orchestra.

The resulting weak shock waves were visualized by interferogram and direct shadowgraph recorded by the high speed video camera Shimadzu SH100 at the framing rate of 10^6 frame/s. Figure 11.38 show double exposure interferograms. Slightly dark shadow seen in Fig. 11.38a shows a weak shock wave of Ms = 1.009. Figure 11.38a show a stand holding a pressure transducer facing to the exit of the trombone and a shock wave. Figure 11.38b was taken at 300 μs after the

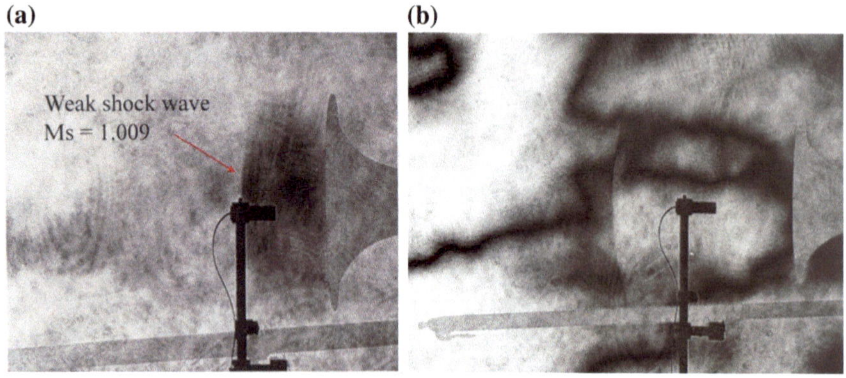

Fig. 11.38 Double exposure holographic observation of compression waves released from a trombone: **a** #10091612; **b** #10091609

observation in Fig. 11.38a. A weak shock wave is shown as a discontinuous change in the grey contrast. Such a weak shock wave was observed in a grey shadow released from the exit of the train tunnel simulator.

Figure 11.39 show sequential direct shadowgraphs of the transmission of a weak shock wave released from a trombone.

11.6.3 Visualization of Flows Around an Arrow of Japanese Archery

The Japanese archery was one of the traditional marshal arts inherited from old time. Today it is a popular sport. The length of arrows is close to 900 mm so that the visualization of the whole length of the arrows in free flight requires 1000 mm diameter schlieren mirrors. The speed of arrows in free flight is about 50 m/s but is too slow to exhibit any effects of compressibility. In order to create an effect of compressibility, it was decided to decrease its surface temperature down to nitrogen temperature.

Professor Sato of Tohoku University, a master of Japanese archery, was invited to shoot arrows. He shot an arrows so as to fly it across 1000 mm diameter schlieren mirrors. Figure 11.40a shows his bowing in front of the schlieren mirror. Figure 11.40b shows an experimental arrangement of measuring the arrow speed by a time of flight method.

In order to create the density gradient over an arrow, it was immersed in liquid nitrogen contained in a Dewar for 5 min. As soon as the arrow's surface temperature became close to the liquid nitrogen temperature, it was taken out of the Dewar. Professor Sato shot it within 20 s as seen in Fig. 11.40a. Figure 11.41 shows the temperature variation on the arrow surfaces monitored by using

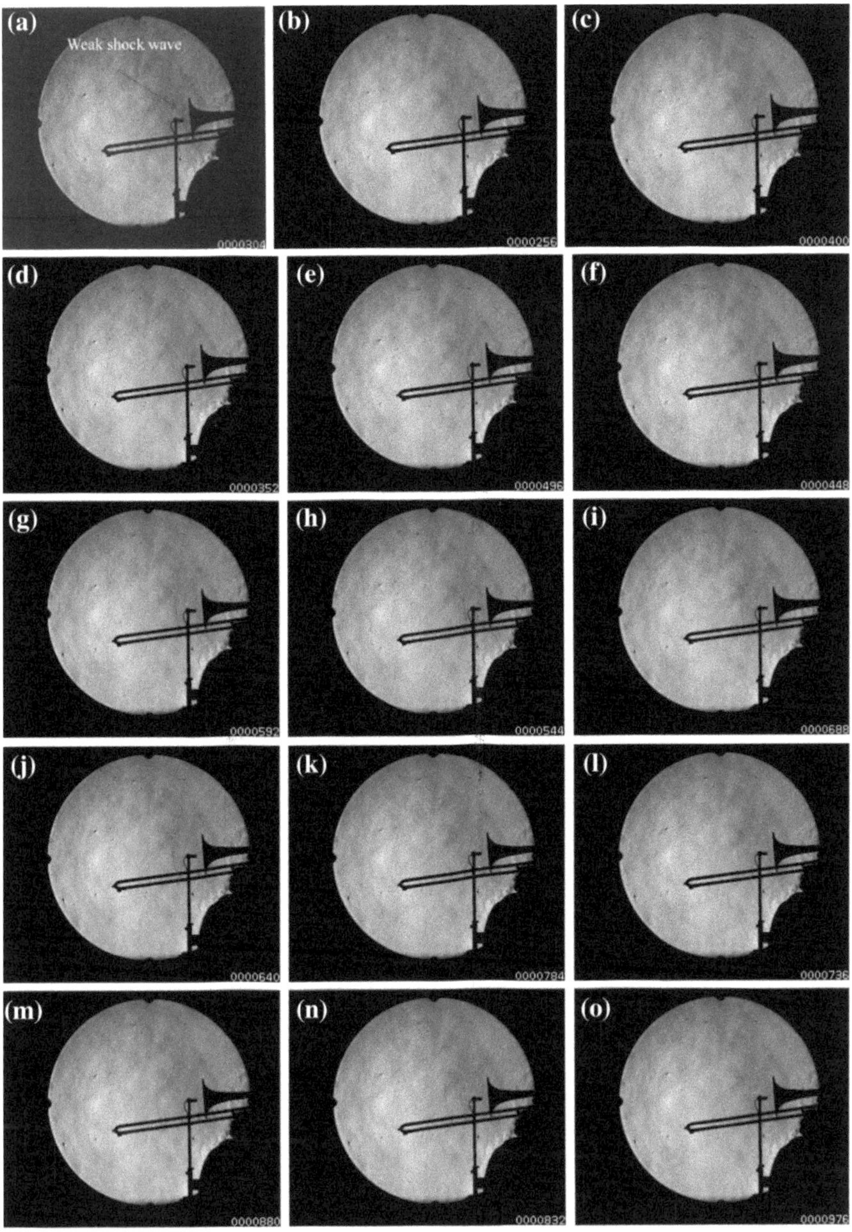

Fig. 11.39 The sequential observation of weak shock waves released from at a trombone

thermocouples at the arrow head, position A and at 200 mm from the arrow head, position B. The ordinate denotes the surface temperature in K. The ordinate denotes elapsed time in s.

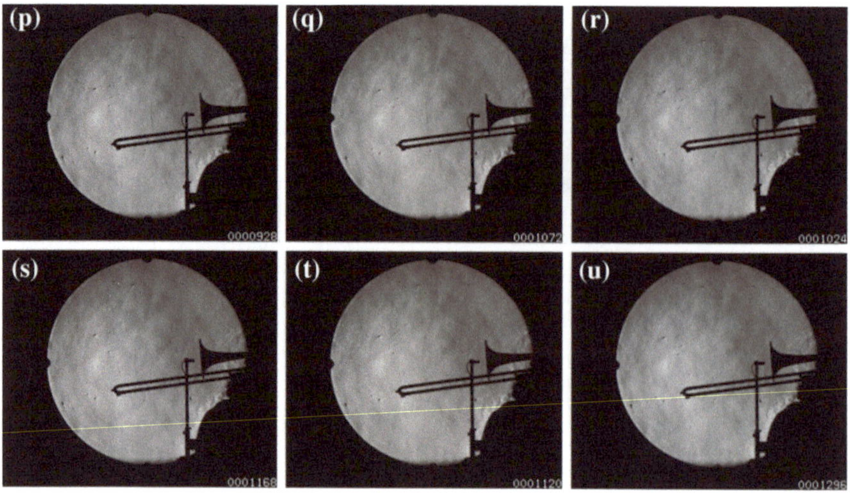

Fig. 11.39 (continued)

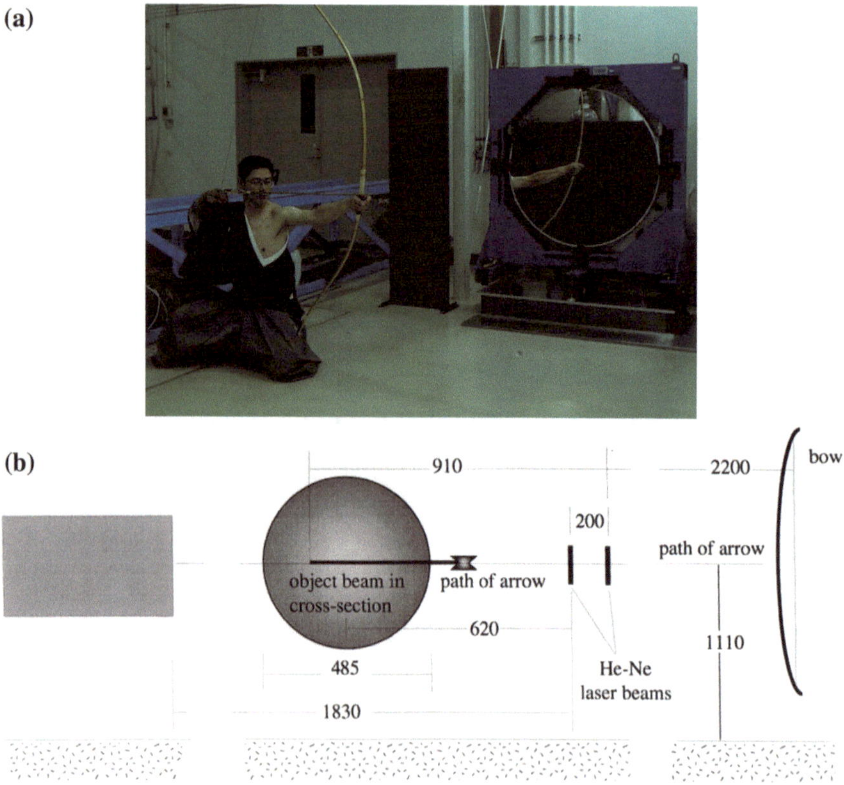

Fig. 11.40 The visualization of an arrow of Japanese archery: **a** professor Sato's shooting an arrow in front of a 1000 mm diameter schlieren mirror; **b** arrangement of measuring the arrow speed (Sato and Takayama 1999)

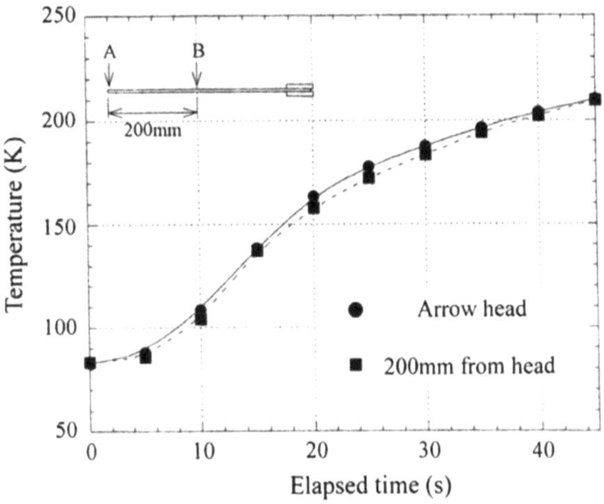

Fig. 11.41 Time variation of surface temperature on the arrow

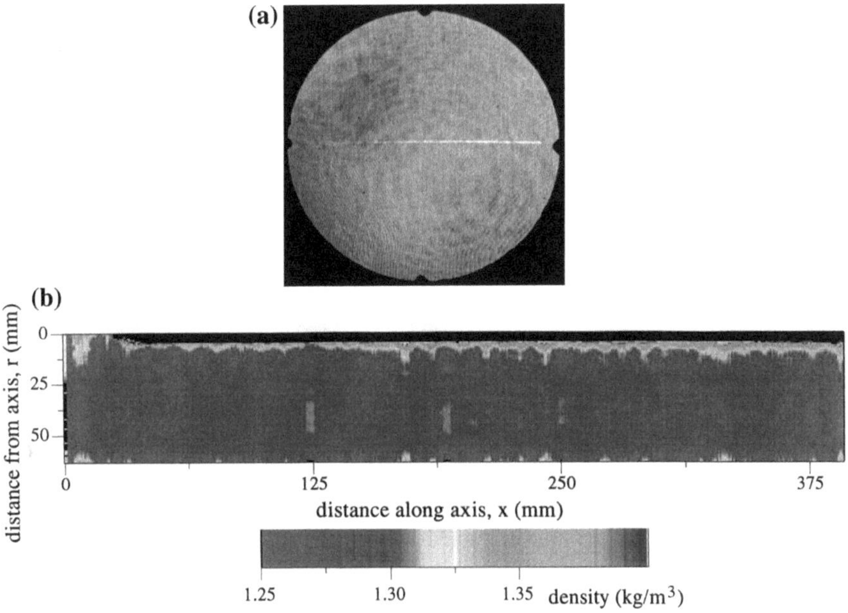

Fig. 11.42 Temperature gradient along an arrow: **a** #02070507; **b** density profile

When taking it out of the Dewar, the temperature increased almost identically at the positions A and B. When the surface temperature recovered to be about 160 K after 20 s. The temperature gradient across the thermal boundary layer developing along the arrow in free flight would be a good maker to demonstrate the variation of the density gradient across the boundary layer in the infinite fringe interferometry. Figure 11.42 shows the flight of an arrow at the speed of 50 m/s. It should be noticed the entire length of the arrow was visualized for the first time. Applying the Fourier fringe analysis to the fringe distribution seen in Fig. 11.42, the density and temperature distributions over the arrow were determined (Houwing et al. 2005). Assuming a laminar boundary layer and the Prantdl number Pr = 1, the velocity profile is readily obtained from the temperature profile. Figure 11.40b shows the density profile. The present research was a preliminary work demonstrating the capability of holographic interferometry to incompressible flow study.

11.7 Mass Extinction

In the history of the earth, every 25.6 million years on average, giant asteroids impacted on the earth. The frequent impacts, in a geological time scale, were enough to discontinuously change chronological periods of the earth and decisively affected the course of the evolution of creatures on the earth (Hosseini et al. 2016). A devastating mass extinction of species took place in about 65 million years ago. Geological survey revealed that the asteroid with an estimated diameter over 10 km obliquely entered into the earth atmosphere and impacted on the Yucatan Peninsula creating a 170 km diameter crater (Rampino 1999).

Underwater shock waves and seismic waves contributed to mass extinction. Marine creatures having lungs and balloons were killed spontaneously by the exposure of underwater shock waves but ocean bottom dwellers such as foraminifera survived. Then it was hypothesized that underwater shock waves only locally contributed to the mass extinction.

Ocean bottom dwellers, ostracodas, about 50 μg in weight and 0.4 mm × 0.4 mm in size were sampled from the silt layer of the Matsushima Bay near Sendai. The sampled ostracodas were kept in the original sea water in the refrigerator for a while. These were micro-creatures representing ocean bottom dwellers.

Figure 11.43a shows a 500 mm × 500 mm × 500 mm stainless steel test chamber equipped with a needle hydrophone and a support detonating a 100 mg PETN pellet. Containers made of polyamide accommodating ostracodas were placed on the bottom of the test chamber. Figure 11.44b shows the polyamide container of 30 mm in diameter and 10 mm in thickness having 5 pits of 3 mm in diameter and 1.5 mm in depth. Five ostracodas were placed in each pit. The container was covered with a thin latex rubber membrane disconnecting the pit from the test environment. The experimental procedure is described in Hosseini et al. (2016).

Figure 11.44a shows time variation of over-pressures generated by explosion of 100 mg PETN pellets. The over-pressures were measured at the stand-off distance 60 and 120 mm using a 0.5 mm diameter PVDF needle hydrophone (Mueller Ingeniertechnik). The loading pressures were controlled by varying the stand-off distance between the center of the explosion and the ostracoda containers. Figure 11.44b shows the results. The ordinate denotes number of ostracodas survived

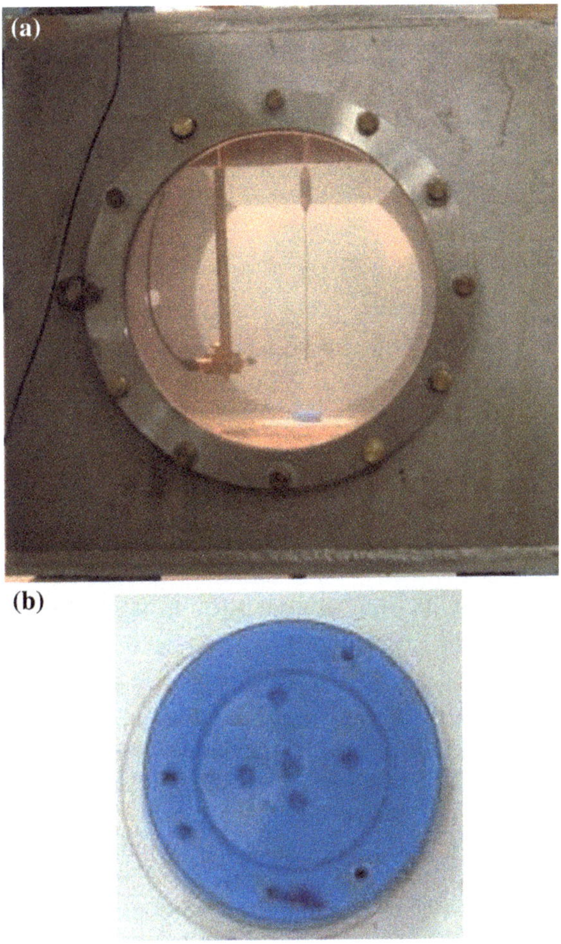

Fig. 11.43 Experimental arrangement: **a** test chamber; **b** ostracoda placed inside each test samples

or dead. The abscissa denotes the over-pressures in MPa. Red color denotes survivors in a week after the shock exposure, blue denotes survivors only for a first week and dead afterward, black denotes spontaneous dead. 13 samples at 0 MPa are alive as reference without shock wave exposures. The number of survivors decreases with increase in the over-pressures. At the over-pressure ranging from 12 to 15 MPa, the number of survivors decreases and at over 17 MP almost all ostracodas dead. The results indicated that upon the direct exposure of underwater shock waves having over-pressures exceeding 17 MPa, the ostracodas were killed.

The ocean bottom layers are composed of sediments of marine snow and layers of silt in which osracodas and foraminiferas live. The silt layer would effectively mitigate shock waves while their propagation. In the silt layer. Hence the ocean bottom dwellers living inside the shallow silt layer would survive against much stronger underwater shock wave exposure. In conclusion, the present analogue experiment verified the survival of the ocean bottom dwellers against the asteroid impact.

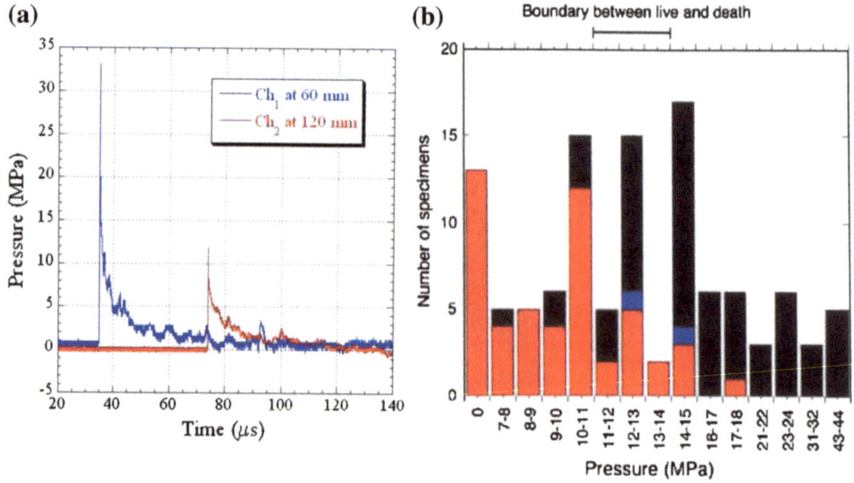

Fig. 11.44 Results: **a** pressure history at stand-off distances of 60 and 120 mm; **b** mortality

11.8 A Water Wave: Shock Wave-like Phenomenon

The motion of a sea wave approaching toward a shore has its inherent speed of transmitting the wave motion that is proportional to a square root of the water depth. This is equivalent to a sound speed in media and defined in Courant and Friedrichs (1948). When a wave is propagating at a constant speed toward the shore the depth of which becomes gradually shallower, the equivalent sound speed becomes gradually slower. Hence, the ratio of the propagation speed of water wave to the equivalent sound speed exceeds unity. Physically, in applying gas-dynamic analogy to the motion of a water wave, the water wave changes its character from a subsonic flow to a supersonic flow when the water wave speed became supersonic. In a supersonic water wave, its shape is gradually steepened. The wave motion and its shape were equivalent to the formation of a shock wave phenomenon in gas-dynamic. The appearance of a dispersed but discontinuous wave front moving over shallow shores is an analogous phenomenon to a shock wave in gas dynamics: the so-called shock wave-like phenomena (Courant and Friedrichs 1948; Glass 1975; Takayama et al. 1994). There are many shock wave-like phenomena in our daily life. However, the most of the shock wave-like phenomena have no mathematical models. On the contrary in the case of water waves, the mass continuity across water waves and the motion of water waves were mathematically expressed in the form of conservation equations as following:

$$\partial E/\partial t + \partial F/\partial x + \partial G/\partial y = 0$$

where t, u, v are time and velocity components in the x, y direction: $\phi = gH$; g and H are the gravitational constant, and the depth of a water wave, respectively:

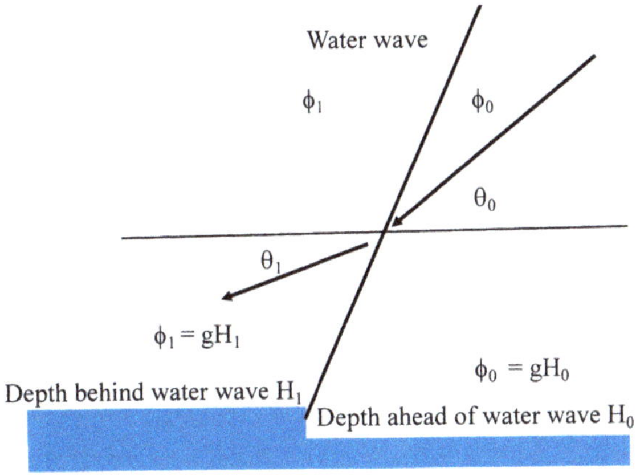

Fig. 11.45 Conditions across an oblique water wave

$E = (\phi, \phi u, \phi v)$, $F = (\phi u, \phi u^2 + \phi^2/2, \phi uv)$, $G = (\phi v, \phi uv, \phi v^2 + \phi^2/2)$. These partial differential equations are valid in the wave motions over shallow water and are equivalent to a special case of the gas-dynamics conservation equations having the specific heats ratio of $\gamma = 2$. Due to this analogy, shallow water tables were once used as replacements of supersonic wind tunnels in a working gas having the specific heats ratio of $\gamma = 2$. However, it should be noticed that the energy conservation is not taken into consideration in this formulation.

Figure 11.45 schematically describes an oblique water wave propagating from the right to the left at its inclination angle of θ_0 and deflected by θ_1. Across the oblique water wave, the water depth changes from H_0 to H_1.

The relationship of the continuity and the equation of motion across an oblique water wave are analogous to the Rankin-Hugoniot equations in gas-dynamics: Equations of continuity are given by

$$\phi_0 u_0 \sin\theta_0 = \phi_1 u_1 \sin(\theta_0 - \theta_1) \tag{11.1}$$

and

$$u_1 \cos\theta_0 = u_1 \cos(\theta_0 - \theta_1); \tag{11.2}$$

the equation of motion is given by

$$\phi_0 u_0^2 \sin^2\theta_0 + \phi_0^2/2 = \phi_1 u_1^2 \sin^2(\theta_0 - \theta_1) + \phi_1^2/2. \tag{11.3}$$

Then the jump of a water wave, H_1/H_0, is readily determined by solving abovementioned equations:

$$\begin{aligned} H_1/H_0 &= \{(1 + 8\mathrm{Fr}\,\sin^2\theta_0)^{1/2} - 1\}/2 \\ &= \tan\theta_0/\tan(\theta_0 - \theta_1), \end{aligned} \tag{11.4}$$

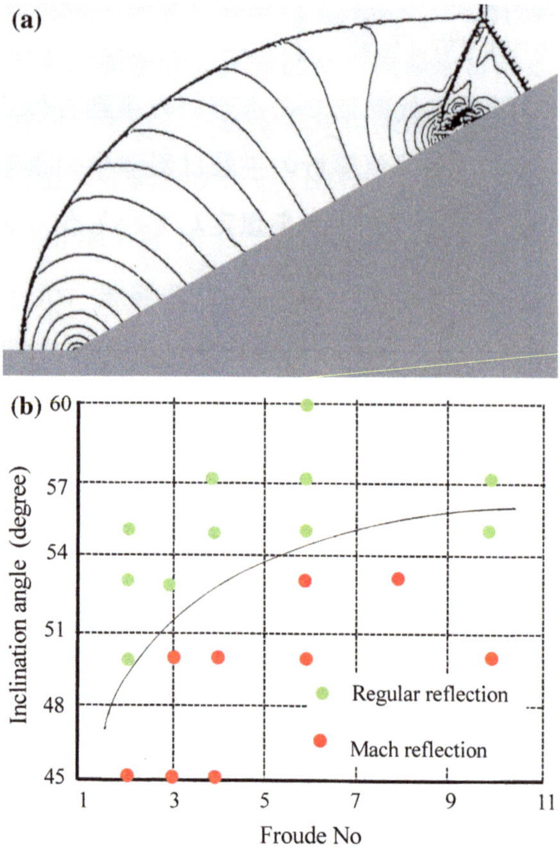

Fig. 11.46 Numerical simulation of water wave reflection: **a** typical Mach reflection pattern over wedge of 30° for Fr = 9.0; **b** the domain and boundary of reflected shock waves

where Fr is called as the Froude number defined as $Fr = u_0^2/\phi_0$. The Fr governs the dynamics of the water wave and is equivalent to Mach number in gas-dynamic shock waves. The Fr becomes larger for higher water wave speed or the shallower the wave depth.

Figure 11.46a shows contour lines of wave heights reflected over a 60° inclined water wave for Fr = 9. Each line shows a line of equal height and is obtained by solving the above mentioned conservation equations. This display was adjusted to match with the gas dynamic shock wave reflection from a wedge of wedge angle 30°. The obtained reflection pattern is similar to a Mach reflection. Wave heights become highest at the triple point and along the Mach stem.

The conservation equations were solved numerically for selective combinations of the Fr number and the inclination angles. The obtained reflected wave patterns were either a Mach reflection or a regular reflection in gas-dynamic shock waves. Figure 11.46b summarizes computational results. The results were displayed to match with the gas-dynamic shock wave reflections from wedges. The ordinate denotes the wedge angle in degree, that is, $90° - \theta_0$. The abscissa denotes the Fr

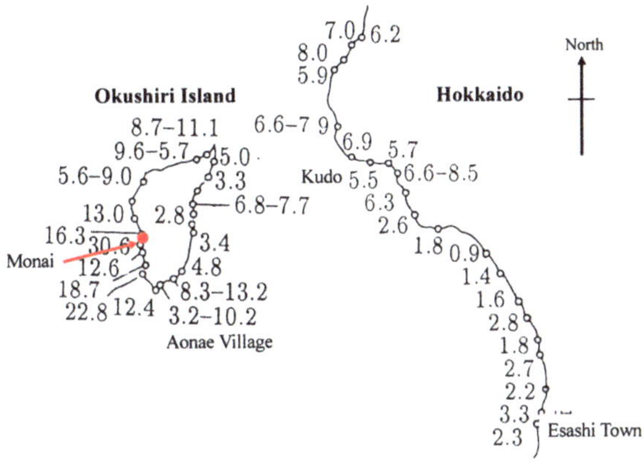

Fig. 11.47 The height of a Tsunami when it attacked the Okushiri Island on July 12th, 1993. Notice that the Monai beach uniquely recorded a wave height over 30 m

which is equivalence to Ms. Red filled circles denote Mach reflection, MR. Green filled circles denote regular reflection, RR. From the observation of these reflection patterns, a detachment criterion was empirically estimated. Mach reflection patterns were clearly identified as seen in Fig. 11.46a but regular reflection patterns were so smeared that they are not necessarily clearly identified. Reflected water waves merge at a localized point and result in focusing of water wave, in which the height of the water wave becomes locally very high. The focusing of the water wave was observed in nature when the tsunami attacked the shore.

At 7:30 p.m., on July 12th, 1993, an earthquake in the Richter scale 7.5 occurred at the western offshore near the Okushiri Island in Hokkaido, Japan. Immediately after the earthquake, a destructive tsunami attacked the west coast of the Okushiri Island. Figure 11.47 shows the Okushiri Island and its opposite shore on the western coast of Hokkaido. Small circles allocated along the coast shown in Fig. 11.47 were the location of the spots at which the wave heights were monitored. Along the eastern coast of the Okushiri Island, the higher second peak values of the wave heights indicated reflected waves from the western coast of Hokkaido. It should be noticed that the wave height at the Monai shore shown in a red filled circle on the coast of the Okushiri Island was singularly over 30 m. Taking the shape of the bottom of the coast in the neighborhood of the Monai shore into consideration, the appearance of such a wave height would be generated due to focusing of the water wave. The focusing of the tsunami would be numerically simulated, if the topographies of the bottom of the Monai coast are known.

References

Abe, A. (1989). *Study of diffraction of shock wave released from the open end of a shock tube* (Ph.D. thesis). Graduate School of Engineering, Faculty of Engineering Tohoku University.

Aratani, S. (1998). *Study of effects of shock waves on thin tempered glass manufacturing* (Ph.D. thesis). Graduate School of Engineering, Faculty of Engineering Tohoku University.

Courant, R., & Friedrichs, K. O. (1948). *Supersonic flows and shock waves*. New York: Wiley Inter-Science.
Fletcher, N. H., & Rossing, T. D. (1998). *The physics of musical instruments*. New York: Springer.
Glass, I. I. (1975). *Shock wave and man*. Canada: Toronto University Press.
Hashimoto, T. (2003). *Analytical and experimental study of hypersonic nozzle flows in free piston shock tunnel* (Doctoral thesis). Graduate School of engineering, Faculty of Engineering, Tohoku University.
Hosseini, S. H. R., Kaiho, K., & Takayama, K. (2016). Response of ocean bottom dwellers exposed to underwater shock waves. *Shock Waves, 26*, 69–73.
Houwing, A. F. P., Takayama, K., Jiang, Z., Sun, M., Yada, K., & Mitobe, H. (2005). Interferometric measurement of density in nonstationary shock wave reflection flow and comparison with CFD. *Shock Waves, 14*, 11–19.
Kitagawa, K., Takayama, K., & Yasuhara, M. (2006). Attenuation of shock waves propagating in polyurethane foams. *Shock Waves, 15*, 437–445.
Kitamura, T. (1995). *A study of water vapor explosion* (Master thesis). Graduate School of Engineering, Faculty of Engineering, Tohoku University.
Koremoto, K. (2000). *Experimental and analytical study of optimization of performances of a high enthalpy free piston shock tunnel* (Doctoral thesis). Graduate School of Engineering, Faculty of Engineering, Tohoku University.
Liepmann, H. W., & Roshko, A. (1960). *Element of gas-dynamics*. New York: Wiley.
Mark, H. (1956). *The interaction of a reflected shock wave with the boundary layer in a shock tube*. NACA TM 1418.
Matsumura, T., Inoue, O, Gottlieb, J. J., & Takayama, K. (1990). A numerical study of the performance of a two-stage light gas gun. *Report Institute of Fluid Science, Tohoku University, 1*, 121–133.
Nonaka, S. (2000). *Experimental and numerical study of hypersonic flows in ballistic range* (Ph.D. thesis). Graduate School of Engineering, Faculty of Engineering Tohoku University.
Numata, D. (2009). *Experimental study of hypervelocity impact phenomena at low temperature in a ballistic range* (Ph.D. thesis) Graduate School of Engineering, Faculty of Engineering Tohoku University.
Rampino, M. R. (1999). Evidence of periodic cosmic showers and mass extinction on earth. In Y. Miura (Ed.), *International Symposium on PIEC, Yamaguchi*.
Sato, A., & Takayama, K. (1999). Measurement of flight of an arrow, measurement and control. *Japan SoC, Automatic Control, 4*(4), 255–263.
Takayama, K., & Saito, T. (2004). Shock wave/geophysical and medical applications. *Annual Review Fluid Machine, 36*, 345–370.
Takayama, K., Miura, Y., Olim, M., Saito, T., & Toro, E. F. (1994). Mach reflection of water waves and the Okushiri Tsunami. In *Proceedings of the 1993 National Shock Wave Symposium* (pp. 487–490).
Voinovich, P., Timofeev, E. V., Saito, T., Takayama, K., Hyodo, Y., & Galyukv, A. O. (1999). An adoptive shock capturing method in real 3-D applications. In *Proceedings of the 2nd International Symposium on Shock Waves* (Vol. 1, pp. 641–646).
Yamamoto, H., Takayama, K., & Cooper, W. (2015). Evolution of luminous front at impact of a 1 km/s projectile into sand layers. In R. Bonazza & D. Ranjan (Eds.), *Shock Waves, Proceedings of the 29th ISSW, Madison* (Vol. 1, pp. 763–767).
Yamamoto, H., Takayama, K., & Kedrinskii, V. (2008). An analogue experiment of magma fragmentation behavior of rapidly decompressed starch syrup. *Shock Waves, 17*, 371–385.
Yang, J.-M. (1995). *Experimental and analytical study of behavior of weak shock waves* (Doctoral thesis). Graduate School of Tohoku University, Faculty of Engineering.

Chapter 12
Concluding Remarks

This book summarized results of the shock wave research the author was involved since 1973. He visualized shock wave phenomena at first by conventional direct shadowgraphs and became interested in double exposure holographic interferometry. The double exposure holographic interferometry was found to be appropriate for a quantitative visualization of underwater shock wave phenomena, in particular, created by the explosion of small explosives weighing from 3 μg to 20 mg. The small explosives were ignited by irradiating the small explosives with a Q-switch laser beam. In the meantime, the underwater shock wave focusing was applied to the non-invasive disintegration of kidney stones. The basic shock dynamic experiments were conducted mostly by visualizations. It was lucky that the success of these preliminary studies initiated the interdisciplinary collaborations between the shock wave research and the medicine and biology.

Glass stated (1975) "A shock tube is a test tube of modern aeronautics." The author trusted his word and constructed horizontal and vertical shock tubes having geometries ranging from miniature sizes to large diameters. He developed diaphragm-less shock tubes with various geometries, which improved the characteristics of existing shock tubes. The shock tubes became no longer a simple tool of gas-dynamic experiments but became a useful tool for producing reliable data not only in high-speed gas dynamics but also in the field of high temperature chemical kinetics.

Shock wave phenomena are typical representations of gas dynamic non-linearity and are always associated with flow unsteadiness. In the study of the transition of reflected shock waves over wedges from the Mach reflection to the regular reflection and vice versa, the analytical models are historically based on the shock dynamics in steady flows. In ideal shock tube flows, the shape of shock waves propagating over the wedge is assumed always to be self-similar. However, due to the flow unsteadiness and the presence of wall boundary layers in real shock tube flows, analytical predictions and experimental findings never agree with each other.

K. Takayama, *Visualization of Shock Wave Phenomena*,
https://doi.org/10.1007/978-3-030-19451-2_12

Then experimental results obtained in compact shock tubes are not necessarily identical with those obtain in large shock tubes, even if the experimental conducts are exactly identical.

Shock wave phenomena appear in nature in the wide range of dimensions. Shock waves appear even from a sub-millimeter scale during shock/bubble interactions in water or during the focusing of ultra short pulse laser beams in water. Gigantic bow shocks are observed in the astronomical scale during super-nova explosions Takayama (1995). Lighthill (1953) delivered a lecture at the Gas Dynamic of Cosmic Clouds Symposium held in 1953 in Cambridge pointing out "The cosmic turbulence consists of not only usual vortex motions but also of three-dimensional statistical assemblage of N-waves." Gas dynamic non-linearity commonly plays a role in the motion of these shock wave phenomena.

In our daily life, many phenomena exist which have analogous features to the shock waves, although the most of them miss mathematical formulations Glass (1975). The author wanted to describe a gas-dynamic non-linearity appearing in the shock wave dynamics and to demonstrate its colorful applications of shock wave research to interdisciplinary fields. This book aimed to emphasize a pleasure of visualizing shock wave phenomena appearing in various gas-dynamic topics. Hence in this context, this book would serve as a continuation of the book "Shock Wave and Man" written by Professor Glass (1975).

References

Glass, I. I. (1975). *Shock Wave and Man*. Toronto University Press.

Lighthill, M. J. (1953). On the energy scattered from the interaction of turbulence with sound or shock wave. *Proceedings of the Cambridge Philosophical Society, 49,* 531–551.

Takayama, K. (1995). Shock Wave Hand Book (in Japanese) Springer Verlag, Tokyo.

Bibliography

Abe, A., Ohtani, K., Takayama, K., Nishio, S., Mimura, H., & Takeda, M. (2010). Pressure generation from micro-bubble collapse at shock wave loading. *Journal of Fluid Science and Technology, 5,* 235–246.

Bellhouse, H. J., Quikan, N. J., & Ainsworth, R. W. (1997). Needle-less delivery of drugs, in dry powder form, using shock waves and supersonic gas flow. In A. P. F. Houwing & A. Paul (Eds.), *Proceeding of 21st ISSW* (Vol. 1, pp. 51–56). The Greate Keppel Island.

Ben-Dor, G., Takayama, K., & Needham, C. E. (1987). The thermal nature of the triple point of a Mach reflection. *Physics of Fluids, 30,* 1287–1293.

Bryson, A. E., & Gross, R. W. F. (1961). Diffraction of strong shocks by cone, cylinder, and spheres. *Journal of Fluid Mechanics 10*, 1–16.

Chaussey, Ch., Schmiedt, E., Jocham, D., Walter, V., Brendel, W., Forsmann, B., & Hepp, W. (1982). *Extracorporeal shock wave lithotripsy*. Kerger: New aspects in the treatment of kidney stone disease.

Hornung, H. G., & Kychakoff, G. (1978). Transition from regular to Mach reflection of shock waves in relaxing gases. In B. Ahlborn, A. Hertzberg, & D. Russell (Eds.), *Proceedings of the 11th International Symposium on Shock Tubes and Waves, Shock Tube and Shock Wave Research* (pp. 296–302). Seattle.

Krehl, P., & van der Geest, M. (1991). The discovery of the Mach reflection effect and its demonstration in an auditorium. *Shock Waves, 1,* 3–15.

Matsuoka, K. (1997). *Study of mitigation of high speed train tunnel sonic boom* (Master thesis). Graduate School of Engineering, Faculty of Engineering, Tohoku University.

Meguro, T., Takayama, K., & Onodera, O. (1997). Three-dimensional shock wave reflection over a corner of two intersecting wedges. *Shock Waves, 7,* 107–121.

Suguyama, H., Takayama, K., Shirota, R., & Doi, H. (1986). An experimental study on shock waves propagating through a dusty gas in a horizontal channel. In D. Bershader & R. Hanson (Eds.), *Proceedings of the 15th International Symposium on Shock Waves and Shock Tubes, Shock Waves and Shock Tubes* (pp. 667–673), Berkeley.

Sun, M. (2005). *Numerical and experimental study of shock wave interaction with bodies* (Ph.D. thesis). Graduate School of Engineering, Faculty of Engineering Tohoku University.

Takayama, K., & Sekiguchi, H. (1977). An experiment on shock diffraction by cones. *Reports of the Institute of High Speed Mechanics, Tohoku University, 36,* 53–74.

Wu, J. H. T., Neemeh, R. A., Ostrowski, P. P., & Elabdin, M. N. (1978). Production of converging cylindrical shock waves by finite element conical contractions. In B. Ahlborn, A. Hertzberg, & D. Russell (Eds.), *Proceedings of 11th International Symposium on Shock Tubes and Waves, Shock Tube and Shock Wave Research* (pp. 107–114). Seattle.

Yamamoto, H., Takayama, K., & Kedrinskii, V. (2008). An analogue experiment of magma fragmentation behavior of rapidly decompressed starch syrup. *Shock Waves, 17,* 371–385.